Climate-Resilient Agriculture, Vol 2

Mirza Hasanuzzaman
Editor

Climate-Resilient Agriculture, Vol 2

Agro-Biotechnological Advancement for Crop Production

Springer

Editor
Mirza Hasanuzzaman
Department of Agronomy
Sher-e-Bangla Agricultural University
Dhaka, Bangladesh

ISBN 978-3-031-37427-2 ISBN 978-3-031-37428-9 (eBook)
https://doi.org/10.1007/978-3-031-37428-9

This Springer imprint is published by the registered company Springer Nature Switzerland AG
The registered company address is: Gewerbestrasse 11, 6330 Cham, Switzerland

Paper in this product is recyclable.

Preface

Climate change has been identified as one of the most significant threats to biological systems. Recent climate change patterns indicate that the earth is witnessing unprecedented levels of warming. Climate change has a profound impact on agriculture, altering food production and food security. Numerous environmental stresses have been accelerated due to rapid changes in the climatic patterns that affect crop growth and productivity, as well as increase the chances of crop failure. The extreme climate conditions that are imposed on plant species result in significant physiological, biochemical, morphological, and molecular changes, which eventually hinder plant growth and yield attributes. For example, wheat yields are likely to drop by 4–6% for every degree of global temperature increase, while maize productivity is projected to decline by the end of the century in areas producing 56% of the world's maize. Moreover, increased soil erosion and nutrient loss are common phenomena caused by changes in temperature and precipitation, which have an adverse effect on soil health. As a result, it becomes harder to cultivate crops, and the productivity of global agricultural land is declining.

Climate change is projected to worsen the issue of future food security by putting more pressure on agriculture. Climate change is expected to have a considerable influence on agriculture and food security. With a growing global population and in the face of persistent and worsening climate change, it has become crucial to investigate sustainable adaptation mechanisms that can address the adverse effects of stressful environments on plants. Therefore, it is urgent to find strategies to reduce the impacts of climate change through mitigation and adaptation for improving the resilience of agriculture. Adoption of climate-resilient agriculture might be the most sustainable way to address this climate change-related agricultural loss. The goal of climate-resilient agriculture is to increase long-term agricultural yields and productivity by sustainably exploiting the natural resources already accessible through crop and livestock production systems. It reduces and/or eliminates greenhouse gas emissions while also responding to climate change and fostering resilience in the agricultural sector.

Changes in agronomic practices, the introduction of improved crop varieties, and alterations to plant physiology and biology are all examples of adaptation strategies.

The introduction of genomics and other 'omics technologies and the modulation of transcription factors in the crop breeding strategies may be some effective way to increase the environmental stress tolerance of many crops, and targeting these within traditional and new breeding technologies could be an effective strategy to produce better crops. The adaptation of agronomic management methods to climate change encompasses a wide range of activities such as soil nutrient management, tillage intensity, crop choice and rotation, water management, and agricultural diversification. Understanding plants' physiological, cellular, and molecular mechanisms to climate change, which include significant alterations in the transcriptome, proteome, and metabolome of plants, has advanced significantly. With the rapid technological advancements, likewise in other sectors, it is now possible to overcome the detrimental effects of climate change in agriculture by the adaptation of tolerant crop varieties, sustainable agronomic practices, and improved crop physiology and biology. However, the success rate of crop breeding strategies is still slow because the issues in public perception and policy remain as limitations to the effective use of the tools like genome editing. Therefore, there still remains huge scope for crop improvement for the future adaptation to climate change for this ever-growing global population.

This is the second volume (Agro-Biotechnological Advancement for Crop Production) of the two-volume book *Climate-Resilient Agriculture*. It contains 43 comprehensive chapters on adaptive strategies of plants under adverse environments and technological advancement toward climate-resilient agriculture.

I would like to give special thanks to the contributors for their outstanding and timely work in producing such fine chapters. We are highly thankful to Kenneth Teng (Senior Book Editor) and Shanthini Kamaraj (Project Coordinator) Springer Nature, New York, and all other editorial staff for precious help in formatting and incorporating editorial changes in the manuscripts. We acknowledge my research students, Md. Rakib Hossain Raihan and Ayesha Siddika, Department of Agronomy, Sher-e-Bangla Agricultural University, Bangladesh, for their generous help in formatting the manuscripts. Special thanks to Dr. Rajib Roychaudhury, University of Haifa, Israel, for his valuable help during the initial proposal writing. The editors and contributing authors hope that this book will include a practical update on our knowledge of climate-resilient agriculture.

Dhaka, Bangladesh Mirza Hasanuzzaman

Contents

1 **Climate Change and Global Crop Production: An Inclusive Insight** 1
Saswati Ghosh, Sukamal Sarkar, Sourav Garai, Anirban Roy, Saikat Saha, Saikat Dey, Subhas Chandra Santra, Debojyoti Moulick, Mousumi Mondal, and Koushik Brahmachari

2 **Uptake and Use Efficiency of Major Plant Nutrients for Climate-Resilient Agriculture** 35
Dong Qin, Rumi Tominaga, and Hirofumi Saneoka

3 **Improving Land-Use Efficiency for Climate-Resilient Agriculture** . . 51
Waqas Ahmad, Bisma Muneer, Dua Zahra Khan, Niaz Ahmad Wahocho, and Muzamil Farooque Jamali

4 **Climate-Resilient Fertilizer Management for Crop Production** 61
Ajay Saroha, Amit Kotiyal, and Aditi Thakur

5 **Modern Agronomic Measurement for Climate-Resilient Agriculture** 81
Abdul Jabbar, Hina Ibrahim, Athar Mahmood, Muhammad Mansoor Javaid, Allah Wasaya, Muhammad Ather Nadeem, Mujahid Ismael, Muhammad Awais, and Shahid Raza Khan

6 **Crop Management for Sustainable Wheat Production** 107
Rehan Jameel, Maria Naqve, Muhammad Anjum Zia, Athar Mahmood, Muhammad Mansoor Javaid, and Muhammad Ather Nadeem

7 **Climate Resilient Weed Management for Crop Production** 125
Dibakar Roy, Sourav Ghosh, Debarati Datta, Dasari Sreekanth, Deepak Pawar, Pijush Kanti Mukherjee, Dibakar Ghosh, Subhas Chandra Santra, and Debojyoti Moulick

8 Climate-Resilient Technology for Maize Production 157
Muaz Ameen, Asma Zafar, Muhammad Mansoor Javaid, Muhammad Anjum Zia, Athar Mahmood, Maria Naqve, and Safura Bibi

9 Climate Resilience Technologies for Wheat Production 189
Zahoor Ahmad, Ejaz Ahmad Waraich, Celaleddin Barutçular, Aiman Hina, Asim Abbasi, Muhammad Mohiuddin, Inzamam Ul Haq, Muhammad Ashar Ayub, and Sundas Sana

10 Improving Plant Nutrient Use Efficiency for Climate-Resilient Agriculture 209
Paramita Deb, Anwesha Mandal, Harendra, Suborna Roy Choudhury, Anupam Das, Swati Hazra, Arkabanee Mukherjee, Dibakar Ghosh, Shuvasish Choudhury, Subhas Chandra Santra, and Debojyoti Moulick

11 Biochar for Plant Stress Tolerance for Climate-Resilient Agriculture 245
Kazim Ali, Hafiz Muhammad Mubasher, Umair Rasheed, Amjid Hussain, Ahmad Sher, and Abdul Qayyum

12 Chitosan for Plant Growth and Stress Tolerance 259
Muhammad Saad Ullah, Athar Mahmood, Muhammad Mansoor Javaid, Maria Naqve, Safura Bibi, Zain Ul Abidin, Ikram ul Haq, and Shahid Raza Khan

13 Exogenous Application of Biostimulants and Commercial Utilization 281
Bushra Sarwar, Ahmad Sher, Muhammad Ijaz, Muhammad Irfan, and Sami Ul-Allah

14 Cross Talk of Biostimulants with Other Signaling Molecules Under Abiotic Stress 295
Shruti Rohatgi, Riya Jain, Shivangi Mathur, Deeksha Singh, and Rajiv Ranjan

15 Abiotic Stress Sensitivity and Adaptation in Field Crops 319
Sukamal Sarkar, Saikat Saha, Saswati Ghosh, Sujay Kumar Paul, Saikat Dey, Debojyoti Moulick, Subhas Chandra Santra, and Koushik Brahmachari

16 Biostimulants for Plant Abiotic Stress Resistance and Climate-Resilient Agriculture 363
Anwesha Chatterjee and Harshata Pal

17 Soil Salinity and Sustainable Agriculture 391
Muhammad Nauman, Safura Bibi, Athar Mahmood, Muhammad Mansoor Javaid, Muhammad Azeem, and Muhammad Ather Nadeem

18 Mechanism and Approaches to Enhance Salt Stress Tolerance in Crop Plants 407
Saima Asghar, Safura Bibi, Asif Mukhtiar, Muhammad Anjum Zia, Maria Naqve, Athar Mahmood, Muhammad Mansoor Javaid, and Muhammad Azeem

19 Mechanisms and Approaches of Enhancing Drought Stress Tolerance in Crop Plants 427
Nono Carine Temegne, Esaïe Tsoata, Godswill Ntsomboh-Ntsefong, Atabong Paul Agendia, Francis Emmanuel Ngome, and Emmanuel Youmbi

20 Conferring Plant Tolerance to Drought and Salinity by the Application of Biochar 451
Ayesha Batool, Maria Naqve, Muhammad Anjum Zia, Athar Mahmood, Sahar Mumtaz, and Muhammad Mansoor Javaid

21 Accumulation and Toxicity of Arsenic in Rice and Its Practical Mitigation 463
Asanka Tennakoon, G. A. H. Galahitigama, P. W. M. Tharindi, T. Suthajini, N. P. M. Abeysinghe, Salinda Sandamal, Karthika Rajendran, Adel A. Hadifa, Ayman El Sabagh, and Disna Ratnasekera

22 Mechanism and Approaches to Enhancing Heat Stress Tolerance in Crop Plants 499
Maira Tanveer, Athar Mahmood, Bushra Sarfraz, Muhammad Anjum Zia, Muhammad Mansoor Javaid, Safura Bibi, Maria Naqve, Muhammad Ather Nadeem, Muhammad Azeem, and Abdul Jabbar

23 Mechanisms and Responses to Enhancing Pollutants Stress Tolerance in Crop Plants 521
Arun Kumar Maurya

24 Phytohormones as Stress Mitigator in Plants 547
Zain Ul Abidin, Athar Mahmood, Safura Bibi, Muhammad Mansoor Javaid, Muhammad Anjum Zia, Muhammad Saad Ullah, Muhammad Azeem, Muhammad Ather Nadeem, and Bilal Ahmad Khan

25 Role of Plant Extracts and Biostimulant in Mitigating Plant Drought and Salinity Stress . . . 571
Shaimaa I. M. Elsayed, Ali S. Sabra, and Elsayed A. Omer

26 Secondary Metabolism and Its Role in Enhancing Drought Stress Tolerance . . . 603
Rasha Fouad, Hend Fouad, Ahmed H. El-Desoky, and Elsayed A. Omer

27 Seed Priming for Abiotic Stress Tolerance . . . 641
Kazim Ali, Hafiz Muhammad Mubasher, Ahmad Sher, Abdul Sattar, and Abdul Manaf

28 Advances in Biotechnological Tools and Their Impact on Global Climate Change and Food Security . . . 667
Zafar Iqbal, Asad Azeem, Sami Ul-Allah, Ahmad Sher, Muhammad Qadir Ahmad, Bilal Haider, and Muhammad Asghar

29 Biotechnological Attributes of Bio-stimulants for Relieving Abiotic Stress . . . 677
Proma Ghosh and Harshata Pal

30 Biotechnological Techniques for Sustainable Waste Management . . . 689
Mujahid Farid, Sheharyaar Farid, Muhammad Arslan Aslam, Khadija Azhar, Khadija Tariq, Zainab Akbar Ali, Zunera Ijaz, Zoya Sarfraz, Muqaddas Khalid, and Mohid Ahmad

31 Role of Biotechnology in Management of Solid Waste . . . 713
Mujahid Farid, Wajiha Sarfraz, Zaki ul Zaman Asam, Mohsin Abbas, Fatima Abdul Sattar, Khadija Asif, Shabnam Abbas, Malaika Iqbal, Zahra Afgun, and Aiza Afzal

32 Bioremediation of Sites Contaminated with Heavy Metals, Techniques, and Their Application . . . 733
Mujahid Farid, Muhammad Zubair, Numra, Muhammad Bilal Shakoor, Mubeen Fatima, Sheharyaar Farid, Zaiba Saher, Sehar Bashir, and Nafeesa Liaqat

33 MicroRNAs (miRNAs): Crosstalk with Regulatory Networks of Abiotic Stress Tolerance in Plants . . . 751
Deepu Pandita

34 Orchestration of Omics Technologies for Crop Improvement . . . 771
Deepu Pandita

35 Transgenic Approaches for Stress Tolerance in Crops . . . 793
Deepu Pandita

36 Translationally Controlled Tumor Protein and Its Relationship with Responses of Plants to Abiotic Stresses 819
Deyvid Novaes Marques, Nicolle Louise Ferreira Barros, and Cláudia Regina Batista de Souza

37 Plant Tissue Culture and Crop Improvement 841
Amna Aslam, Ayesha Bibi, Safura Bibi, Athar Mahmood, Muhammad Mansoor Javaid, Hafeez Ur Rehman, and Muhammad Ather Nadeem

38 Nanotechnology for Climate-Resilient Agriculture 863
Somali Dhal and Harshata Pal

39 Mitigation of Plant Abiotic Stress by Plant Growth-Promoting Bacteria, Hormones, and Plant Extracts 881
Muhammad Saqlain, Safura Bibi, Athar Mahmood, Muhammad Anjum Zia, Muhammad Mansoor Javaid, and Javaria Nargis

40 Bioremediation and Phytoremediation Aspects of Crop Improvement 903
Nadeem Iqbal, Malik Tahir Hayat, Muhammad Dawood, Waqar-Un Nisa, Shamim Akhtar, Nazneen Bangash, Abdul Nazeer, and Muhammad Ahsan Asghar

41 Ecofriendly Management of Insect Pests for Sustainable Agriculture 931
Asim Abbasi, Aqsa Asif, Zahoor Ahmad, Inzamam Ul Haq, Asad Aslam, Ramish Saleem, Aliza Saleha, and Zeenat Zafar

42 Ecofriendly Management of Disease for Sustainable Agriculture ... 959
Dipika Mal, Gunuguntla Veera Narayana, and Malige Bhavani

43 Use of Advance Composting Techniques and Areas of Improvement in Pakistan 975
Mujahid Farid, Nouman Munawar, Gulfam Anjum, Muhammad Abubakar, Umar Shahzad, Muhammd Sheraz Asif, Hina Batool, Sajal Mir, Sara Javed, and Hafiz Khuzama Ishaq

Correction to: Advances in Biotechnological Tools and Their Impact on Global Climate Change and Food Security C1
Zafar Iqbal, Asad Azeem, Sami Ul-Allah, Ahmad Sher, Muhammad Qadir Ahmad, Bilal Haider, and Muhammad Asghar

Index 993

Contributors

Mohsin Abbas Department of Environmental Sciences, University of Gujrat, Hafiz Hayat Campus, Gujrat, Pakistan

Shabnam Abbas Department of Environmental Sciences, University of Gujrat, Hafiz Hayat Campus, Gujrat, Pakistan

Asim Abbasi Department of Environmental Sciences, Kohsar University Murree, Murree, Pakistan

N. P. M. Abeysinghe Department of Export Agriculture, Faculty of Agricultural Sciences, Sabaragamuwa University of Sri Lanka, Balangoda, Sri Lanka

Zain Ul Abidin Department of Botany, Faculty of Sciences, University of Agriculture, Faisalabad, Pakistan

Muhammad Abubakar Department of Environmental Sciences, University of Gujrat, Hafiz Hayat Campus, Gujrat, Pakistan

Zahra Afgun Department of Environmental Sciences, University of Gujrat, Hafiz Hayat Campus, Gujrat, Pakistan

Aiza Afzal Department of Environmental Sciences, University of Gujrat, Hafiz Hayat Campus, Gujrat, Pakistan

Atabong Paul Agendia Department of Plant Biology, University of Yaoundé I, Yaoundé, Cameroon

Mohid Ahmad Department of Environmental Sciences, University of Gujrat, Hafiz Hayat Campus, Gujrat, Pakistan

Muhammad Qadir Ahmad Department of Plant Breeding and Genetics, Bahauddin Zakariya University, Multan, Pakistan

Waqas Ahmad College of Agriculture, University of Layyah, Layyah, Pakistan

Zahoor Ahmad Department of Botany, Constituent Punjab College, University of Central Punjab, Bahawalpur, Pakistan

Shamim Akhtar Department of Botany, University of Gujrat, Gujrat, Pakistan

Kazim Ali College of Agriculture, University of Layyah, Layyah, Pakistan

Zainab Akbar Ali Department of Environmental Sciences, University of Gujrat, Hafiz Hayat Campus, Gujrat, Pakistan

Muaz Ameen Department of Botany, Facility of Sciences, University of Agriculture, Faisalabad, Pakistan

Gulfam Anjum Department of Environmental Sciences, University of Gujrat, Hafiz Hayat Campus, Gujrat, Pakistan

Zaki ul Zaman Asam Department of Environmental Sciences, University of Gujrat, Hafiz Hayat Campus, Gujrat, Pakistan

Muhammad Asghar College of Agriculture, University of Layyah, Layyah, Pakistan

Muhammad Ahsan Asghar Department of Biological Resources, Agricultural Institute, Centre for Agricultural Research, ELKH, Martonvásár, Hungary

Saima Asghar Department of Botany, Faculty of Sciences, University of Agriculture, Faisalabad, Pakistan

Aqsa Asif Department of Zoology, Cholistan University of Veterinary and Animal Sciences, Bahawalpur, Pakistan

Khadija Asif Department of Environmental Sciences, University of Gujrat, Hafiz Hayat Campus, Gujrat, Pakistan

Muhammd Sheraz Asif Department of Environmental Sciences, University of Gujrat, Hafiz Hayat Campus, Gujrat, Pakistan

Amna Aslam Department of Botany, Faculty of Sciences, University of Agriculture, Faisalabad, Pakistan

Asad Aslam Key Laboratory for Sustainable Forest Ecosystem Management-Ministry of Education, College of Forestry, Northeast Forestry University, Harbin, Heilongjiang, People's Republic of China

Muhammad Arslan Aslam Department of Energy, Environment and Climate Change, Asian Institute of Technology (ATI), Pathum Thani, Thailand

Faculty of Science and Technology, University of the Basque Country, Basque, Spain

Muhammad Awais College of Agriculture, Bahauddin Zakariya University, Layyah, Pakistan

Muhammad Ashar Ayub Institute of Agro-Industry and Environment, The Islamia University of Bahawalpur, Bahawalpur, Punjab, Pakistan

Asad Azeem Department of Plant Breeding and Genetics, University of Agriculture, Faisalabad, Pakistan

Muhammad Azeem Biosaline Research Laboratories, Department of Botany, University of Karachi, Karachi, Pakistan

Khadija Azhar Department of Environmental Sciences, University of Gujrat, Hafiz Hayat Campus, Gujrat, Pakistan

Nazneen Bangash Department of Biosciences, COMSATS University Islamabad, Islamabad, Pakistan

Nicolle Louise Ferreira Barros Departamento de Genética, Universidade Federal do Rio Grande do Sul (UFRGS), Porto Alegre, RS, Brazil

Celaleddin Barutçular Department of Field Crops, Faculty of Agriculture, Cukurova University, Adana, Turkey

Sehar Bashir Department of Environmental Sciences, University of Gujrat, Hafiz Hayat Campus, Gujrat, Pakistan

Ayesha Batool Department of Botany, Faculty of Sciences, University of Agriculture, Faisalabad, Pakistan

Hina Batool Department of Environmental Sciences, University of Gujrat, Hafiz Hayat Campus, Gujrat, Pakistan

Malige Bhavani Lovely professional University, Domain of Horticulture, School of Agriculture, Phagwara, Punjab, India

Ayesha Bibi Department of Botany, Faculty of Sciences, University of Agriculture, Faisalabad, Pakistan

Safura Bibi Department of Botany, Facility of Sciences, University of Agriculture, Faisalabad, Pakistan

Department of Botany, Faculty of Sciences, University of Agriculture, Faisalabad, Pakistan

Koushik Brahmachari Department of Agronomy, Faculty of Agriculture, Bidhan Chandra Krishi Viswavidyalaya, Mohanpur, Nadia, West Bengal, India

Department of Agronomy, Faculty of Agriculture, Bidhan Chandra Krishi Viswa Vidyalaya, Mohanpur, Nadia, West Bengal, India

Anwesha Chatterjee Amity Institute of Biotechnology, Amity University Kolkata, Kolkata, West Bengal, India

Shuvasish Choudhury Plant Stress Biology and Metabolomics Laboratory, Department of Life Science and Bioinformatics, Assam University Silchar, Silchar, Assam, India

Suborna Roy Choudhury Department of Agronomy, Bihar Agricultural University, Bhagalpur, Bihar, India

Anupam Das Department of Soil Science and Agricultural Chemistry, Bihar Agricultural University, Bhagalpur, Bihar, India

Debarati Datta ICAR-Central Research Institute for Jute and Allied Fibres, Barrackpore, India

Muhammad Dawood Department of Environmental Sciences, Bahauddin Zakariya University, Multan, Pakistan

Cláudia Regina Batista de Souza Instituto de Ciências Biológicas, Universidade Federal do Pará (UFPA), Belém, PA, Brazil

Paramita Deb GD Goenka University, Gurgaon, Haryana, India

Saikat Dey School of Agriculture and Rural Development, Ramakrishna Mission Vivekananda Educational and Research Institute, Narendrapur, West Bengal, India

Somali Dhal Department of Biotechnology and Medical Engineering, National Institute of Technology (NIT) Rourkela, Rourkela, Odisha, India

Ayman El Sabagh Department of Field Crops, Faculty of Agriculture, Siirt University, Siirt, Turkey

Department of Agronomy, Faculty of Agriculture, Kafrelsheikh University, Kafr Elsheikh, Egypt

Ahmed H. El-Desoky Department of Pharmacognosy, National Research Centre, Giza, Egypt

Shaimaa I. M. Elsayed Horticultural Crops Technology Department, National Research Centre, Dokki, Cairo, Egypt

Mujahid Farid Department of Environmental Sciences, University of Gujrat, Hafiz Hayat Campus, Gujrat, Pakistan

Sheharyaar Farid School of Natural Sciences and Ryan Institute, University of Galway, Galway, Ireland

Mubeen Fatima Department of Environmental Sciences, University of Gujrat, Hafiz Hayat Campus, Gujrat, Pakistan

Hend Fouad Department of Medicinal and Aromatic Plants Research, National Research Centre, Giza, Egypt

Rasha Fouad Department of Medicinal and Aromatic Plants Research, National Research Centre, Giza, Egypt

G. A. H. Galahitigama Department of Export Agriculture, Faculty of Agricultural Sciences, Sabaragamuwa University of Sri Lanka, Balangoda, Sri Lanka

Sourav Garai School of Agriculture and Rural Development, Ramakrishna Mission Vivekananda Educational and Research Institute, Narendrapur, West Bengal, India

Dibakar Ghosh ICAR-Indian Institute of Water Management, Bhubaneswar, India

Indian Institute of Tropical Meteorology, Pune, Maharashtra, India

Proma Ghosh Amity Institute of Biotechnology, Amity University Kolkata, Kolkata, West Bengal, India

Saswati Ghosh Department of Environmental Science, Kanchrapara College, University of Kalyani, Kanchrapara, West Bengal, India

Sourav Ghosh ICAR-Central Research Institute for Jute and Allied Fibres, Barrackpore, India

Adel A. Hadifa Department of Rice Research, Field Crops Research Institute, Agricultural Research Center (ARC), Giza, Egypt

Bilal Haider College of Agriculture, University of Layyah, Layyah, Pakistan

Ikram ul Haq Department of Plant Breeding and Genetics, College of Agriculture, University of Sargodha, Sargodha, Pakistan

Inzamam Ul Haq College of Plant Protection, Gansu Agricultural University, Lanzhou, China

Harendra GD Goenka University, Gurgaon, Haryana, India

Malik Tahir Hayat Department of Environmental Sciences, COMSATS University Islamabad, Abbottabad Campus, Islamabad, Pakistan

Swati Hazra School of Agricultural Sciences, Sharda University, Greater Noida, Uttar Pradesh, India

Aiman Hina Ministry of Agriculture (MOA), National Centre for Soybean Improvement, State Key Laboratory for Crop Genetics and Germplasm Enhancement, Nanjing Agricultural University, Nanjing, China

Amjid Hussain College of Agriculture, University of Layyah, Layyah, Pakistan

Hina Ibrahim Department of Botany, University of Agriculture Faisalabad, Faisalabad, Pakistan

Muhammad Ijaz College of Agriculture, University of Layyah, Layyah, Pakistan

Department of Agronomy, Bahauddin Zakariya University, Multan, Pakistan

Zunera Ijaz Department of Environmental Sciences, University of Gujrat, Hafiz Hayat Campus, Gujrat, Pakistan

Malaika Iqbal Department of Environmental Sciences, University of Gujrat, Hafiz Hayat Campus, Gujrat, Pakistan

Nadeem Iqbal Department of Environmental Sciences, COMSATS University Islamabad, Abbottabad Campus, Islamabad, Pakistan

Doctoral School of Environmental Sciences, University of Szeged, Szeged, Hungary

Zafar Iqbal College of Agriculture, University of Layyah, Layyah, Pakistan

Muhammad Irfan College of Agriculture, University of Layyah, Layyah, Pakistan

Hafiz Khuzama Ishaq Dipartimento di Scienze e Tecnologie Ambientali, Biologiche e Farmaceutiche, Università degli Studi della Campania Luigi Vanvitelli, Caserta, Italy

Mujahid Ismael Department of Botany, University of Agriculture Faisalabad, Faisalabad, Pakistan

Abdul Jabbar Department of Botany, University of Agriculture Faisalabad, Faisalabad, Pakistan

Department of Botany, Facility of Sciences, University of Agriculture, Faisalabad, Pakistan

Riya Jain Plant Biotechnology Lab, Department of Botany, Faculty of Science, Dayalbagh Educational Institute (Deemed to be University), Dayalbagh, Agra, India

Muzamil Farooque Jamali Department of Horticulture, Sindh Agriculture University Tandojam, Tandojam, Pakistan

Rehan Jameel Department of Botany, Faculty of Sciences, University of Agriculture, Faisalabad, Pakistan

Muhammad Mansoor Javaid Department of Agronomy, College of Agriculture, University of Sargodha, Sargodha, Pakistan

Department of Agronomy, Faculty of Agriculture, University of Agriculture, Faisalabad, Pakistan

Sara Javed Department of Environmental Sciences, University of Gujrat, Hafiz Hayat Campus, Gujrat, Pakistan

Muqaddas Khalid Department of Environmental Sciences, University of Gujrat, Hafiz Hayat Campus, Gujrat, Pakistan

Bilal Ahmad Khan Department of Agronomy, College of Agriculture, University of Sargodha, Sargodha, Pakistan

Department of Soil and Crop Sciences, Texas A&M University, College Station, TX, USA

Dua Zahra Khan College of Agriculture, University of Layyah, Layyah, Pakistan

Shahid Raza Khan Department of Agronomy, College of Agriculture, University of Sargodha, Sargodha, Pakistan

Amit Kotiyal Department of Horticulture, Lovely Professional University, Phagwara, Punjab, India

Nafeesa Liaqat Department of Environmental Sciences, University of Gujrat, Hafiz Hayat Campus, Gujrat, Pakistan

Athar Mahmood Department of Agronomy, Faculty of Agriculture, University of Agriculture, Faisalabad, Pakistan

Dipika Mal Lovely professional University, Domain of Horticulture, School of Agriculture, Phagwara, Punjab, India

Abdul Manaf Department of Agronomy, PMAS–Arid Agriculture University, Rawalpindi, Pakistan

Anwesha Mandal GD Goenka University, Gurgaon, Haryana, India

Deyvid Novaes Marques Departamento de Genética, Universidade de São Paulo (USP), Escola Superior de Agricultura "Luiz de Queiroz" (ESALQ), Piracicaba, SP, Brazil

Shivangi Mathur Plant Biotechnology Lab, Department of Botany, Faculty of Science, Dayalbagh Educational Institute (Deemed to be University), Dayalbagh, Agra, India

Arun Kumar Maurya Department of Botany, Multanimal Modi College, Ghaziabad, Uttar Pradesh, India

Sajal Mir Department of Environmental Sciences, University of Gujrat, Hafiz Hayat Campus, Gujrat, Pakistan

Muhammad Mohiuddin Department of Environmental Sciences, Kohsar University Murree, Murree, Pakistan

Mousumi Mondal School of Agriculture and Allied Sciences, Neotia University, Sarisha, India

Debojyoti Moulick Department of Environmental Science, University of Kalyani, Nadia, West Bengal, India

Plant Stress Biology and Metabolomics Laboratory, Department of Life Science and Bioinformatics, Assam University Silchar, Silchar, Assam, India

Hafiz Muhammad Mubasher College of Agriculture, University of Layyah, Layyah, Pakistan

Arkabanee Mukherjee Indian Institute of Tropical Meteorology, Pune, Maharashtra, India

Pijush Kanti Mukherjee ICAR-Directorate of Weed Research, Jabalpur, India

Asif Mukhtiar Department of Botany, Faculty of Sciences, University of Agriculture, Faisalabad, Pakistan

Sahar Mumtaz Department of Botany, Division of Science and Technology, University of Education, Lahore, Pakistan

Nouman Munawar Department of Environmental Sciences, University of Gujrat, Hafiz Hayat Campus, Gujrat, Pakistan

Bisma Muneer College of Agriculture, University of Layyah, Layyah, Pakistan

Muhammad Ather Nadeem Department of Agronomy, College of Agriculture, University of Sargodha, Sargodha, Pakistan

Maria Naqve Department of Botany, Faculty of Sciences, University of Agriculture, Faisalabad, Pakistan

Department of Botany, Facility of Sciences, University of Agriculture, Faisalabad, Pakistan

Gunuguntla Veera Narayana Lovely professional University, Domain of Horticulture, School of Agriculture, Phagwara, Punjab, India

Javaria Nargis Department of Botany, University of Sargodha, Sargodha, Pakistan

Muhammad Nauman Department of Botany, Faculty of Sciences, University of Agriculture, Faisalabad, Pakistan

Abdul Nazeer Department of Environmental Sciences, COMSATS University Islamabad, Abbottabad Campus, Islamabad, Pakistan

Francis Emmanuel Ngome Institute of Agricultural Research for Development (IRAD), Yaoundé, Cameroon

Waqar-Un Nisa Center for Interdisciplinary Research in Basic Sciences, International Islamic University, Islamabad, Pakistan

Godswill Ntsomboh-Ntsefong Department of Plant Biology, University of Yaoundé I, Yaoundé, Cameroon

Numra Department of Environmental Sciences, University of Gujrat, Hafiz Hayat Campus, Gujrat, Pakistan

Elsayed A. Omer Medicinal and Aromatic Plants Researches Department, National Research Centre, Dokki, Cairo, Egypt

Department of Medicinal and Aromatic Plants Research, National Research Centre, Giza, Egypt

Harshata Pal Amity Institute of Biotechnology, Amity University Kolkata, Kolkata, West Bengal, India

Deepu Pandita Government Department of School Education, Jammu, Jammu and Kashmir, India

Sujay Kumar Paul Department of Agriculture, MITS Institute of Professional Studies, Bhujabala, Rayagada, Odisha, India

Deepak Pawar ICAR-Directorate of Weed Research, Jabalpur, India

Abdul Qayyum Department of Agronomy, The University of Haripur, Haripur, Pakistan

Dong Qin Graduate School of Integrated Sciences for Life, Hiroshima University, Higashi-hiroshima, Japan

Karthika Rajendran VIT School of Agricultural Innovations and Advanced Learning (VAIAL), Vellore Institute of Technology (VIT), Vellore, Tamil Nadu, India

Rajiv Ranjan Plant Biotechnology Lab, Department of Botany, Faculty of Science, Dayalbagh Educational Institute (Deemed to be University), Dayalbagh, Agra, India

Umair Rasheed College of Agriculture, University of Layyah, Layyah, Pakistan

Disna Ratnasekera Department of Agricultural Biology, Faculty of Agriculture, University of Ruhuna, Matara, Sri Lanka

Hafeez Ur Rehman Department of Botany, Faculty of Sciences, University of Agriculture, Faisalabad, Pakistan

Shruti Rohatgi Plant Biotechnology Lab, Department of Botany, Faculty of Science, Dayalbagh Educational Institute (Deemed to be University), Dayalbagh, Agra, India

Anirban Roy School of Agriculture and Rural Development, Ramakrishna Mission Vivekananda Educational and Research Institute, Narendrapur, West Bengal, India

Dibakar Roy ICAR-Directorate of Weed Research, Jabalpur, India

Muhammad Saad Ullah Department of Botany, Faculty of Sciences, University of Agriculture, Faisalabad, Pakistan

Ali S. Sabra Medicinal and Aromatic Plants Researches Department, National Research Centre, Dokki, Cairo, Egypt

Saikat Saha Hooghly Krishi Vigyan Kendra, Bidhan Chandra Krishi Viswavidyalaya, Indian Council of Agricultural Research, Chinsurah, West Bengal, India

Zaiba Saher Department of Environmental Sciences, University of Gujrat, Hafiz Hayat Campus, Gujrat, Pakistan

Ramish Saleem Department of Zoology, Constituent Punjab College of University of Central Punjab, Bahawalpur, Pakistan

Aliza Saleha Department of Zoology, The Islamia University of Bahawalpur, Bahawalpur, Pakistan

Sundas Sana Department of Botany, Sub Campus Rahim Yar Khan, The Islamia University of Bahawalpur, Bahawalpur, Pakistan

Salinda Sandamal State Key Laboratory of Systematic and Evolutionary Botany, Institute of Botany, Chinese Academy of Sciences, Beijing, China

University of Chinese Academy of Sciences, Beijing, China

Hirofumi Saneoka Graduate School of Integrated Sciences for Life, Hiroshima University, Higashi-hiroshima, Japan

Subhas Chandra Santra Department of Environmental Science, University of Kalyani, Nadia, West Bengal, India

Muhammad Saqlain Department of Botany, Faculty of Sciences, University of Agriculture, Faisalabad, Pakistan

Bushra Sarfraz Department of Botany, Facility of Sciences, University of Agriculture, Faisalabad, Pakistan

Wajiha Sarfraz Department of Botany, Government College Women University, Sialkot, Pakistan

Zoya Sarfraz Department of Environmental Sciences, University of Gujrat, Hafiz Hayat Campus, Gujrat, Pakistan

Sukamal Sarkar School of Agriculture and Rural Development, Ramakrishna Mission Vivekananda Educational and Research Institute, Narendrapur, West Bengal, India

Ajay Saroha Department of Horticulture, Lovely Professional University, Phagwara, Punjab, India

Bushra Sarwar College of Agriculture, University of Layyah, Layyah, Pakistan

Abdul Sattar College of Agriculture, University of Layyah, Layyah, Pakistan

Department of Agronomy, Bahauddin Zakariya University, Multan, Pakistan

Fatima Abdul Sattar Department of Environmental Sciences, University of Gujrat, Hafiz Hayat Campus, Gujrat, Pakistan

Umar Shahzad Department of Environmental Sciences, University of Gujrat, Hafiz Hayat Campus, Gujrat, Pakistan

Muhammad Bilal Shakoor College of Earth and Environmental Sciences, The University of the Punjab, Lahore, Pakistan

Ahmad Sher College of Agriculture, University of Layyah, Layyah, Pakistan

Department of Agronomy, Bahauddin Zakariya University, Multan, Pakistan

Deeksha Singh Plant Biotechnology Lab, Department of Botany, Faculty of Science, Dayalbagh Educational Institute (Deemed to be University), Dayalbagh, Agra, India

Dasari Sreekanth ICAR-Directorate of Weed Research, Jabalpur, India

T. Suthajini Department of Biosystems Technology, Faculty of Technology, Eastern University, Chenkalady, Sri Lanka

Maira Tanveer Department of Botany, Facility of Sciences, University of Agriculture, Faisalabad, Pakistan

Khadija Tariq Department of Environmental Sciences, University of Gujrat, Hafiz Hayat Campus, Gujrat, Pakistan

Nono Carine Temegne Department of Plant Biology, University of Yaoundé I, Yaoundé, Cameroon

Asanka Tennakoon Department of Biosystems Technology, Faculty of Technology, Eastern University, Chenkalady, Sri Lanka

Aditi Thakur Department of Horticulture, Lovely Professional University, Phagwara, Punjab, India

P. W. M. Tharindi Department of Export Agriculture, Faculty of Agricultural Sciences, Sabaragamuwa University of Sri Lanka, Balangoda, Sri Lanka

Rumi Tominaga Graduate School of Integrated Sciences for Life, Hiroshima University, Higashi-hiroshima, Japan

Esaïe Tsoata Department of Plant Biology, University of Yaoundé I, Yaoundé, Cameroon

Zain Ul Abidin Department of Botany, Faculty of Sciences, University of Agriculture, Faisalabad, Pakistan

Sami Ul-Allah College of Agriculture, University of Layyah, Layyah, Pakistan

Department of Plant Breeding and Genetics, Bahauddin Zakariya University, Multan, Pakistan

Niaz Ahmad Wahocho Department of Horticulture, Sindh Agriculture University Tandojam, Tandojam, Pakistan

Ejaz Ahmad Waraich Department of Agronomy, University of Agriculture Faisalabad, Faisalabad, Pakistan

Allah Wasaya College of Agriculture, Bahauddin Zakariya University, Layyah, Pakistan

Emmanuel Youmbi Department of Plant Biology, University of Yaoundé I, Yaoundé, Cameroon

Asma Zafar Department of Botany, Facility of Sciences, University of Agriculture, Faisalabad, Pakistan

Zeenat Zafar Department of Zoology, The Islamia University of Bahawalpur, Bahawalpur, Pakistan

Muhammad Anjum Zia Department of Biochemistry, Faculty of Sciences, University of Agriculture, Faisalabad, Pakistan

Muhammad Zubair Department of Chemistry, University of Gujrat, Hafiz Hayat Campus, Gujrat, Pakistan

About the Editor

Mirza Hasanuzzaman is a Professor of Agronomy at Sher-e-Bangla Agricultural University, Dhaka, Bangladesh. He completed his Bachelor of Science in Agriculture (Hons.) from Sher-e-Bangla Agricultural University, where he achieved First Class, received a Gold Medal for being first in his class, and earned a Sher-e-Bangla Agricultural University Award. He also completed a Master of Science in Agronomy from the same university, where he was once again at top of his class with a CGPA 4.0. In 2012, he received his Ph.D. with a dissertation on 'Plant Stress Physiology and Antioxidant Metabolism' from the United Graduate School of Agricultural Sciences, Ehime University, Japan, with a Japanese Government (MEXT) Scholarship. Later, he completed his postdoctoral research in the Center of Molecular Biosciences (COMB), University of the Ryukyus, Okinawa, Japan, with a 'Japan Society for the Promotion of Science (JSPS)' postdoctoral fellowship. Subsequently, he became an Adjunct Senior Researcher at the University of Tasmania with an Australian Government's Endeavour Research Fellowship. Dr. Mirza Hasanuzzaman is one of the Highly Cited Researchers recognized by Clarivate Analytics and the only Bangladeshi researcher with this recognition. Prof. Hasanuzzaman published over 250 articles in the high-profile international peer-reviewed journals. He edited 33 books and written 45 book chapters on important aspects of plant physiology, plant stress tolerance, and crop production. These books have been published by world-renowned publishers such as Springer, Elsevier,

CRC Press, and Wiley. According to Scopus, Prof. Hasanuzzaman's publications have already received over 19000 citations with an *h*-index of 73. Prof. Hasanuzzaman's current research foci include the crop responses to climate change and physiological and molecular mechanisms of environmental stress tolerance. He completed several research projects funded by the World Bank, FAO, the University Grants Commission of Bangladesh, the Ministry of Science and Technology (Bangladesh), and others. Prof. Hasanuzzaman supervised the dissertations of 41 Masters and 3 Ph.D. students. Dr. Hasanuzzaman is an editor and reviewer of more than 100 international journals and was a recipient of the 'Publons Global Peer Review Award 2017, 2018 and 2019' which is managed by Web of Science. He has presented 45 papers, abstracts, and posters at international conferences in many countries. Prof. Hasanuzzaman is a member of 50 professional societies and is the acting Treasurer Secretary of the Bangladesh JSPS Alumni Association and Publication Secretary of Bangladesh Society of Agronomy (BSA) and Weed Science Society of Bangladesh. He is a Fellow of Bangladesh Academy of Sciences, The Linnean Society of London, Royal Society of Biology, and International Society of Environmental Botanists. He received the World Academy of Science (TWAS) Young Scientist Award 2014, University Grants Commission (UGC) Gold Medal 2018, Global Network of Bangladeshi Biotechnologists (GNOBB) Award 2021, Bangladesh Academy of Sciences (BAS) Gold Medal Award-2022 (Senior Group), and Society for Plant Research Young Scientist Award (Agriculture) 2023.

Chapter 1
Climate Change and Global Crop Production: An Inclusive Insight

Saswati Ghosh, Sukamal Sarkar, Sourav Garai, Anirban Roy, Saikat Saha, Saikat Dey, Subhas Chandra Santra, Debojyoti Moulick, Mousumi Mondal, and Koushik Brahmachari

Abstract Climate change predictions may benefit from an awareness of how current changes have impacted food availability. Many agricultural activities throughout the globe have seen fast climate change over the last few decades, and greenhouse gas (GHG) levels in the atmosphere have also surged. Because of the near inevitability that climate and CO_2 trends will continue in the future, several concerns about food security remain unanswered, one of which is whether or not global agriculture's overall productivity will be influenced. It was also observed that climate change might have a significant impact on global food security in the next decades. Global anthropogenic trends, such as expanding populations, the diversion of grains to biofuels, increased protein consumption, and weather extremes, are imposing increasing pressure on agricultural productivity. In order to lower global CO_2 levels and store atmospheric CO_2, sustainable farming systems and management practises should be adopted. For resource-strapped farmers vis-à-vis policymakers, in this chapter, we documented a comprehensive list of

---

S. Ghosh
Department of Environmental Science, Kanchrapara College, University of Kalyani, Kanchrapara, West Bengal, India

S. Sarkar · S. Garai · A. Roy · S. Dey
School of Agriculture and Rural Development, Ramakrishna Mission Vivekananda Educational and Research Institute, Narendrapur, West Bengal, India

S. Saha
Hooghly Krishi Vigyan Kendra, Bidhan Chandra Krishi Viswavidyalaya, Indian Council of Agricultural Research, Chinsurah, West Bengal, India

S. C. Santra · D. Moulick (✉)
Department of Environmental Science, University of Kalyani, Nadia, West Bengal, India

M. Mondal
School of Agriculture and Allied Sciences, Neotia University, Sarisha, India

K. Brahmachari
Depertment of Agronomy, Faculty of Agriculture, Bidhan Chandra Krishi Viswavidyalaya, Mohanpur, Nadia, West Bengal, India

M. Hasanuzzaman (ed.), *Climate-Resilient Agriculture, Vol 2*,
https://doi.org/10.1007/978-3-031-37428-9_1

possible implications of climate change on key worldwide field crop yield, mitigation techniques, and long-term prospects.

Keywords Ghgs · Climate change · Yield · Food security · Rice

1 Introduction

The climate change has been perceived as one of the biggest threats to socioeconomic development, the impact being more pronounced in the developing nations (Manivannan et al. 2017; Kennicutt et al. 2019). While the developed nations have trained their system to implement climate smart agricultural (CSA) practices, the resource-challenged farmers of the developing countries may not be able to implement many of the climate-resilient technologies at the cost of immediate profit. The environmental safety and sustainability are put as the last option by the resource-poor farmers. Thus, the small and marginal land holders continue to remain vulnerable to climate change and related disasters. The developing countries cannot afford large investment for implementation of CSA technologies for their priority towards infrastructural development as well as obligation for providing livelihood support to the poorer sections of population through subsidies and incentives (Bai and Tao 2017). There remains ample scope for evaluating a wide range of climate-resilient policies and initiatives considering the diversity of agroclimatic situations, perceivable risks associated across the ecological regions, and nevertheless the vulnerability of the farming community at both short and long term. The fast-swelling population growth along with over-exploitation and inefficient management of the resources have reduced the per capita availability of food grain, arable land, as well as water and have also led to massive degradation and pollution of the land and water resources at a deplorable level. Finally, we must realise that, at the dawn of the new century, there is compelling evidence that a change in agriculture is required immediately in order to address the worldwide issues of food demand. This transformation must increase production per unit area per unit time while maintaining the soil-water-crop-animal-human continuum and efficient resource use through the careful application of adaptable agro-techniques.

The world's most climate change susceptible area is Southeast Asia. Agriculture is undoubtedly the backbone of this region. Thus, in brief, the fact is that the development of agriculture is the hard core of economic growth. This area has a relatively high population density, and a sizeable section of the population depends on agriculture for a living (Kazemi et al. 2018; Hossain et al. 2021). According to the IPCC study from 2007, Southeast Asia's temperature has risen by 0.1 to 0.3 °C every decade while the region's yearly rainfall has reduced by 1–3 mm/year. Southeast Asia's average surface air temperature would rise by 0.75–0.87 °C by 2039, 1.32–2.01 °C by 2069, and 1.96–3.77 °C by 2100 (IPCC 2020). By 2050, precipitation in Southeast Asia would rise from 1% to 2.25%, predicts the IPCC. On the other side, it was predicted that there would be fewer wet days, which reflects the intensification of rainfall events and the diminished efficiency of rainfall. Natural water resources are abundant in Southeast Asia and are particularly important for agricultural development. Under the predicted climate scenario, rising temperatures will cause evaporation and transpiration to occur at a faster pace in the future. Water will be scarce for irrigation purposes as a result of the irregular rainfall pattern. Saltwater intrusion brought on by sea level rise, land and water resources become more salinised. Thus, the quantity and quality of water supplies are negatively impacted by climate change. The IPCC predicts that low soil moisture and high temperature stress will cause agricultural yields in tropical nations to drop (Sarkar et al. 2020). Moreover, from agro-ecosystem point of view, crops that are usually exposed to a wide range of stressor (biotic/abiotic or both) in various magnitudes will pose additional obstacles in achieving steady yield (Moulick et al. 2023; Hossain et al. 2023). The present chapter was framed to summarise the ill impact of climate change on the global crop production and the possible adaption mechanisms and mitigation techniques in both breeding and agronomic strategies for the resource-strapped farmers vis-à-vis policymakers.

2 Climate Change and Global Crop Production

2.1 Global Climate Change Trends

The weather in a particular region over a longer period of time is considered the climate of the region. Climate change, therefore, is the long-term alteration of average weather conditions, including the variability of their properties, that exists for an extended period (IPCC 2007). Alternately, abnormal variation in climatic phenomena and its subsequent effect on human livelihood, farming activities, and the ocean and polar environment are also referred to as climate change. Over time, the Earth's climate has changed. The alarming report was first made by Hubert Horace Lamb (1913–1997) regarding the aberrant climatic changes and their upcoming effect on human civilisation (Kelly 1997). Charles David Keeling (1928–2005), an eminent scientist, measured the changes in CO_2 concentration accurately in the atmosphere at the mountain station on Mauna Loa, Hawaii, which was an iconic document indicating future climate change (Heimann 2005). In the last few decades, the mammoth population bursting, industrialisation, and anthropogenic activities of humans have led to a rapid increment of average surface temperature, sea level rise, and melting of glaciers. Higher CO_2 absorption makes the ocean more acidic and creates problem for living animals like corals, which face difficulties in making their shells and skeletons (Hoegh-Guldberg 1999). Svante Arrhenius and Arvid Högbom demonstrated the influence of greenhouse gases on global temperature at the end of the nineteenth century (Heimann 2005). In a Nolte et al. (2008) study, reported that climate change would result in higher future ozone levels over polluted areas. The changes in the Indian climate were noted by Lockyer (1910). This study raised awareness in scientific communities and created a basis for sociopolitical acceptance regarding the threat of climate change, resulting in the meeting of international potentates in the United Nations Framework Convention on Climate Change in 1992 (Kelly 1997). Interestingly, over the past several decades, a tremendous increment in the number of publications on global climate change has been observed in authoritative scientific journals such as Nature and Science (Lobell et al. 2008), reflecting its deep impact on scientists and policymakers. Considering the threat of climatic anomalies to society, the world's powers committed to reducing anthropogenic greenhouse gas emissions by at least 5% from 1990 levels in the period of 2008 to 2012 (Bohringer 2003). James Hansen, a scientist for NASA, warned about the blatant impact of global warming almost 20 years ago and suggested that CO_2 should be reduced from its current 385.0 mg L^{-1} to at most 350.0 mg L^{-1} to save civilisation on this planet (Hansen et al. 2008). At present, the National Centers for Environmental Information reported that each month of 2021 was warmer than the average temperature, even the temperature in the coolest month, "February," was 1.15 F more than average (NOAA 2021). The study also revealed that since 1880, the total land and ocean temperature has risen at an average pace of 0.08 °C each decade; however, since 1981, the average increase in temperature has been 0.18 °C, which is more than double that rate (Fig. 1.1).

In the future, the trend of global warming will depend on human anthropogenic activities such as fossil fuel burning, deforestation, intensive farming, and others that emit carbon dioxide and other greenhouse gases. At the moment, the aforementioned activities emit nearly 40 billion metric tonnes of carbon dioxide into our

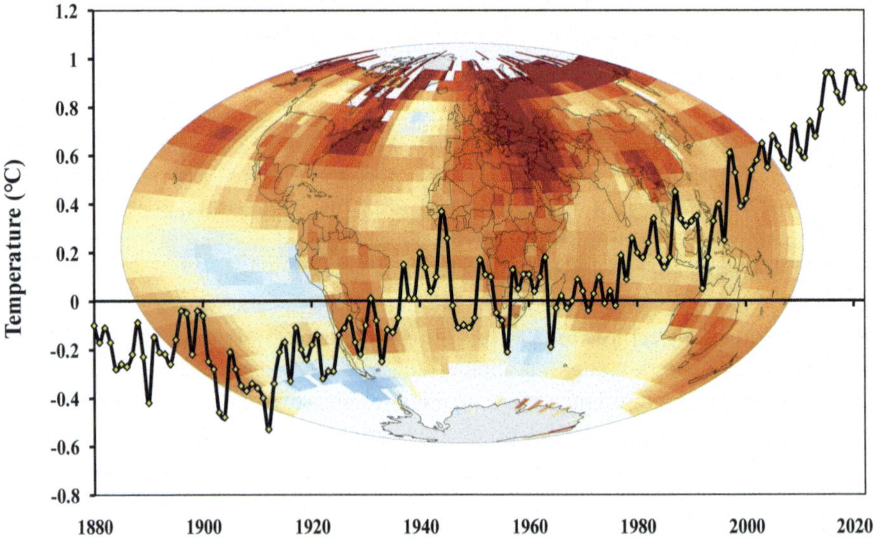

Fig. 1.1 Yearly surface temperature compared to the 20th-century average from 1880–2020. Negative values indicate cooler-than-average years; positive values show warmer-than-average years. (Data obtained from National Oceanic and Atmospheric Administration, USA, website: www.noaa.gov)

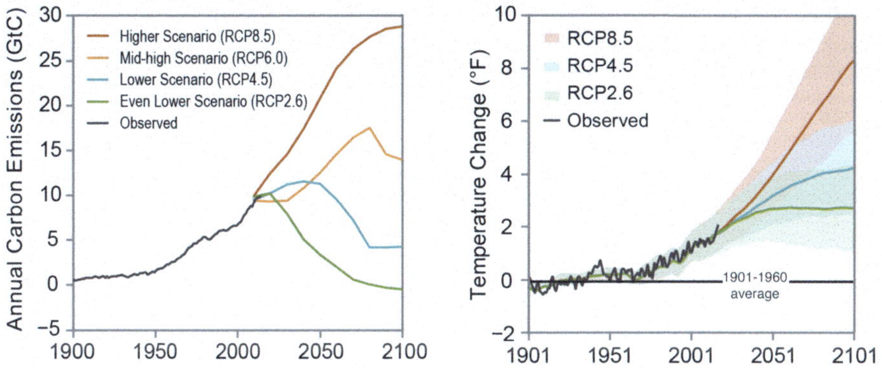

Fig. 1.2 Observed and projected annual C emission (**a**) and temperature change (**b**) during 1900–2100 time periods based on the central estimate (lines) and a range (shaded areas, two standard deviations) as simulated by the full suite of CMIP5 global climate models. (Data obtained from National Oceanic and Atmospheric Administration, USA, website: www.noaa.gov)

atmosphere each year (NOAA 2021), as emission exceeds carbon sequestration. The U.S. Climate Science Special Report predicted that if the emissions continue at the same pace as they have since 2000, the earth's temperature will be 5 °F as compared to the 1901–1960 average (USGCRP 2017). In contrast, if the pace of emissions is slowed down and begins to decline significantly by 2050, then the projected temperature will be at a minimum of 2.5 °F higher than in the early twentieth century (Fig. 1.2).

2.2 *Trend of Global Crop Production*

Agriculture and global food production have changed tremendously over last five decades (Fig. 1.3) with the advancement of technologies, rapid industrialisation along with green revolution in 1960s (FAO 2000). Moreover, expansion of irrigated land, arable land, and land under permanent crops contribute almost 70% yield increment as compared to pre-green revolution period (FAO 2000).

Food and Agriculture Organisation showed that almost 50% production increase (Fig. 1.4) happened in between 2000 and 2018, reflecting 2.9 billion tonnes more than in 2000 (FAO 2020). Among the crops, cereal accounted one-third of total production, followed by sugar crops, vegetables, oilseeds, fruits, and tuber crops. The introduction of improved varieties, more fertiliser uses in assured irrigation system, and pesticide application were the major factors for above-mentioned yield enhancement though horizontal productive land intensification was less in between the period (FAO 2020).

The similar report revealed that among the individual groups, only four crops viz. sugarcane (21%), maize (13%), rice (9%), and wheat (8%) accounted half of world total production in 2018 (Fig. 1.5). Only 4% production came from potato

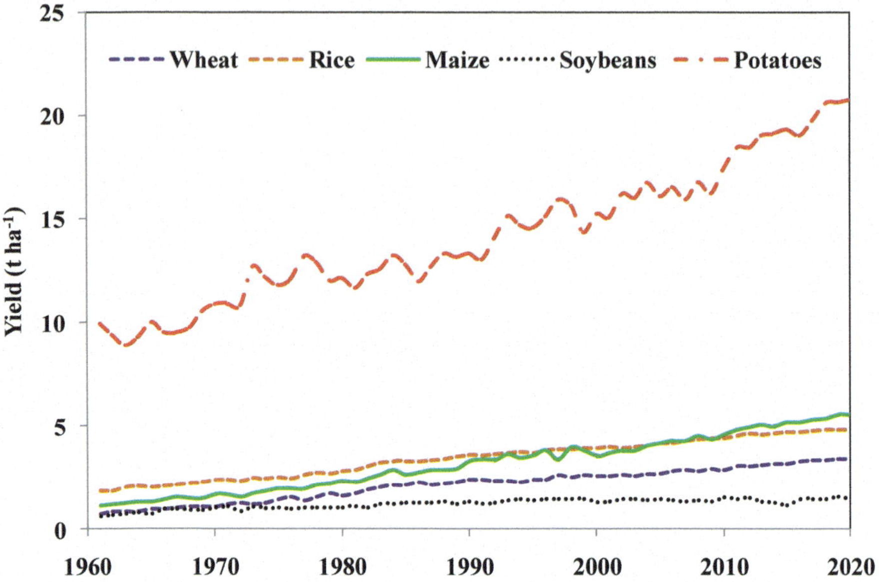

Fig. 1.3 Trend of major crop production in Asian countries. (Data obtained from Food and Agriculture Organisation, website: www.fao.org)

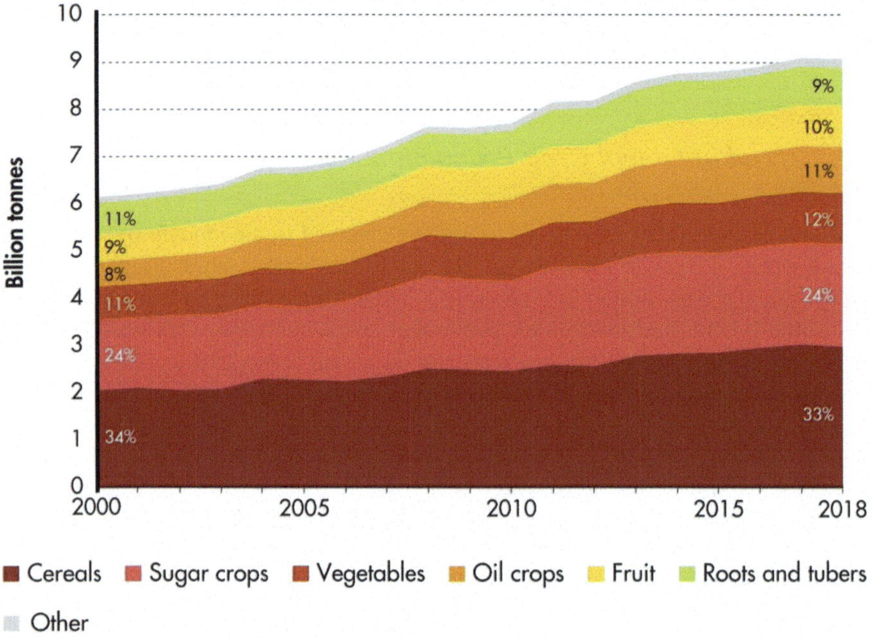

Fig. 1.4 Global crop production (Commodity groups) trend from 2000 to 2018. (Data obtained from Food and Agriculture Organisation, website: www.fao.org)

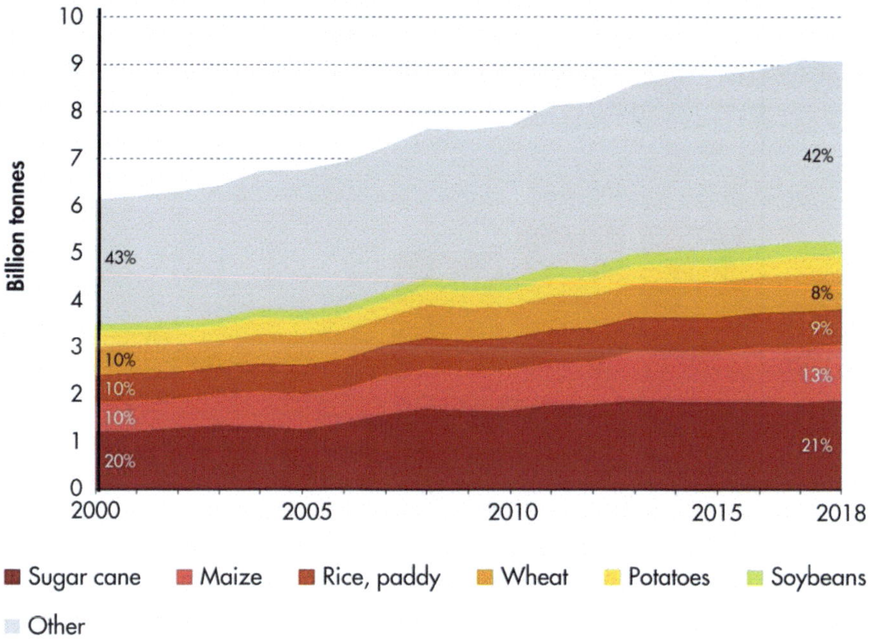

Fig. 1.5 Global crop production trend (Main commodities) from 2000 to 2018. (Data obtained from Food and Agriculture Organisation, website: www.fao.org)

and soybeans. Production share by major produces for the last two decades is presented in Fig. 1.6 where maximum sugarcane production was accounted by Brazil followed by India and rice; wheat and potato production were accounted by China followed by India.

In future, the demand for food and agricultural products has been expected to rise 50% as compared to present as the much higher current volume of food demand and the much larger world population need to be fed (Alexandratos and Bruinsma 2012).

2.3 *Impact of Climate Change to the Global Crop Production*

Global crop production has been noticeably hampered in recent times and might be carried out for the next few decades as a consequence of climatic abnormalities such as irregularities in rainfall, increasing CO_2 concentration, and temperature (Abbass et al. 2022). Integration of these factors affects normal growth duration, physiological responses of crops, brings pest outbreaks and unpredictable phenomena, and squeezes the available resources for agriculture, which increases the price of raw products for agro-industries (Panda et al. 2003; Fand et al. 2012; Biswas et al. 2022). Rising temperatures significantly hampered the wheat yield from 1962 to 2002, as reported by Lobell and Field (2007). Furthermore, Gourdji et al. (2013)

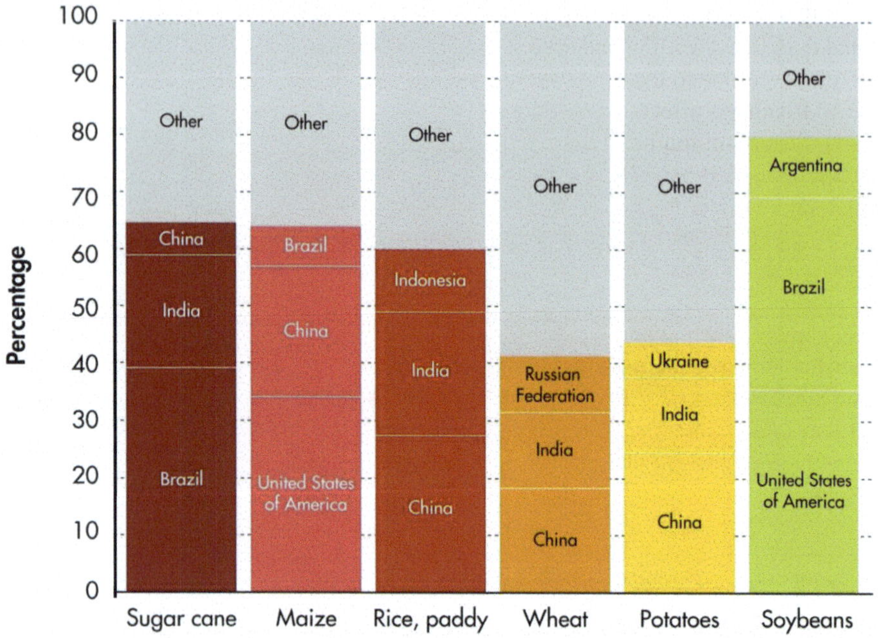

Fig. 1.6 Global crop production trend (Main producers) from 2000 to 2018. (Data obtained from Food and Agriculture Organisation, website: www.fao.org)

showed the decreasing wheat production trend in South Asian countries under extreme temperature events during 1980–2011. The aforesaid documentation is also supported by Ortiz et al. (2021). High temperatures hasten plant growth arrangements (Blum et al. 2001), reduce photosynthetic processes (Salvucci and Crafts-Brandner 2004), and have a significant impact on reproductive operations (Farooq et al. 2011).

Temperature rising during night affects the rice yield because it has been revealed that almost 5% yield loss happens with the increment of 1 °C temperature (Tao et al. 2006). In the Philippines, a dramatic yield reduction was recorded in Indian rice cultivars due to night temperature enhancement during the last 25 years (Peng et al. 2004; Banerjee et al. 2022; Sarkar et al. 2022). Interestingly, it has been well predicted that global temperature rise will significantly reduce rice yields in the near future (Lobell and Gourdji 2012). An average temperature increase of 1–4 °C in the middle of summer reduces the length of crop phenophases, which affects final yield of rice (Hatfield and Prueger 2015). Humid and subtropical regions may have to pay the price of upcoming heat waves as expected by world climate models (Battisti and Naylor 2009). Global warming has a negative impact on two important yield-attributing traits, such as the number of grains per m^2 and average grain weight and grain quality. Exposure of rice in high temperatures shows disparities in flowering patterns, and seed set lessens and lowers grain weight (Qasim et al. 2020). Direct exposure to solar radiation during the daytime reduces the anthesis period and quickens the peak flowering period (Tao et al. 2006). High temperatures are inversely proportional to pollen shedding and seed setting in rice plants. Lengthier seed setting time was observed due to the high daytime temperature in rice (Matsui et al. 2001).

Cold damage affects more than 15 million ha of rice seedbed (Sharma et al. 2021). Seven million ha of potential rice-growing land in Asia and Southeast Asia remain unplanted because of cold temperatures (Sharma et al. 2021). Additionally, cultivars belonging to the indica subspecies are more susceptible to damage from low temperatures (Mondal et al. 2021). In addition, recent climate change has increased the frequency of cold waves between November and mid-February, when the seedbed for winter rice is being prepared. Traditional, long-standing nursery management, low temperature stress, and foggy weather all appear to be significant contributors to the main field for delayed transplanting. These decrease the vegetative period with poor tiller development in the main field and lead to fewer GDD and less solar radiation (Biswas et al. 2019). Due to delayed transplanting, winter rice may be negatively affected by terminal heat stress and cyclonic storms, which typically occur from March onwards.

Another major global food source, maize production, is also in trouble due to temperature enhancement. The lesser number of grains formed in maize may be related to low photosynthetic accumulation and reproductive phenomena. Maize exposed to heat (30 °C) appeared to have fewer anthesis-silking intermissions during the flowering phase (Edreira et al. 2011). In another study, it was observed that high temperatures above 35 °C are the principal reason for non-viability of pollen, less pollination potential, spikelet dropping, and abnormal kernel number that are directly linked with the active reproductive phase (Vega et al. 2001). Global warming-induced weather extremes include betel vine leaf shedding and discoloration (Chaudhary et al. 2011), lemon leaf squeezing (Pautasso et al. 2012), and pineapple root rot (Vedwan and Rhoades 2001), caused by chilling injury and extreme fog.

3 Adaptation to Crop Overall Extreme Climate Stresses

With an increase in the Earth's temperature, the climate change scenario became worse day by day. The climate change endures several alterations and slowly became abiotic stress (Raza et al. 2019). These environmental changes in turn have a detrimental impact on crop species and imposing numerous threats in agricultural production system globally has been presented schematically in Fig. 1.7 (Zhu 2016). It is estimated in various research works that under field conditions, heat and drought are considered as the most important stress elements to influence plant growth and productivity (Bellard et al. 2012). Several plant species require an optimum temperature for their vegetative growth and blooming. Plant physiological processes are heavily influenced with a fluctuation in temperature (Pereira 2016). Besides, cold stress often results in sterility of seeds and drought stress has negative impact on plant morphology. These climatic conditions produce enormous responses related to physiological, biochemical, morphological, as well as molecular modification in plant species and

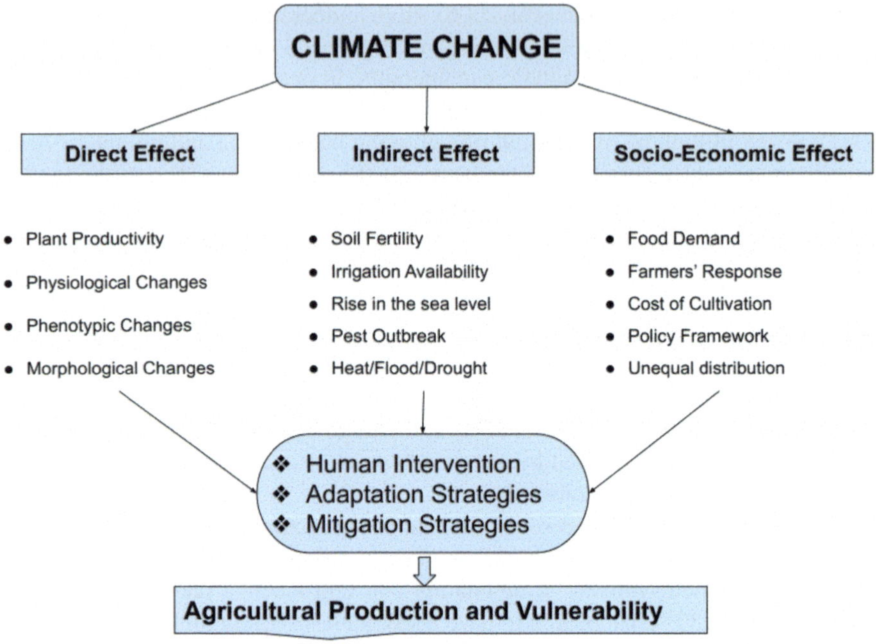

Fig. 1.7 Impact of Climate Change on Agricultural Production

ultimately hampers the plant development and yield attributes (Gull et al. 2019). Moreover, the global warming and climate change scenario have both the negative and positive impact on crop production. Different adaptation strategies were discussed in this chapter to understand the crop overall extreme climate stresses.

3.1 Heat Stress

There are various types of abiotic stresses in which plants are exposed to nature; heat stress has a self-regulating mode of action in plant physiological mechanism as well as in the metabolism of plant cells (Lafitte et al. 2007). It has been found that the mode of action produces through heat stress is very often compounded by other stress factors (i.e., drought stress and salt stress) and ameliorates the combined effects of abiotic stress on plants (Barnabás et al. 2008). The vulnerability to high temperatures in plants varies with various developmental stages of plant growth and heat stress affecting a certain level of plant growth in vegetative and reproductive stages of plant growth. These observed impacts largely depend on different crop species and genotypes, with inter and intra specific variations (Ladha et al. 2003; Cramer et al. 2011).

Numerous physiological injuries have been reported under high temperature in plants (i.e., inhibition of plant root and shoot growth, scorching of leaves and other plant parts, fruit damage, and abscission of leaves) very often leading to a sharp decrease in productivity (Bita and Gerats 2013; Lamaoui et al. 2018). In many cases, the architecture of the plant changes due to temperature fluctuations, plant growth reduces, and total dry weight of the plant affected due to the lower shoot net assimilation rates (Vollenweider and Günthardt-Goerg 2005). Higher plants generally exhibit characteristics set of cellular as well as metabolic responses due to exposure to excess heat, at least 5 °C above their optimum growing temperature (Bellard et al. 2012; Sulmon et al. 2015). These excess heat phenomena change in the cellular structure, metabolic as well as physiological functions of the plants and help the plant to produce different phytohormones, i.e., abscisic acid (ABA) and antioxidants to withstand excess heat (Zhu 2016).

The structural change in photosynthetic membrane occurring due to high temperature is the swelling of grana and ion-leakage is accompanied by the event reduced by the energy allocation in photosystem. Thus, it is obviously an essential function to maintain the cellular membrane function under heat stress condition (Wahid and Shabbir 2005; Allakhverdiev et al. 2008; He et al. 2018). Various important phytohormones including ABA, ethylene, and salicylic acid create a negative impact on plant growth under temperature stress condition (Talanova et al. 2003; Larkindale and Huang 2004). While auxin, cytokinin, and gibberellic acid decrease the impact of heat stress on plant through premature plant senescence (Larkindale and Huang 2004).

Plants have developed various stress tolerance mechanism strategies under heat stress conditions and are classified under long-term phenological and morphological adaptation processes. These processes include change in leaf orientation, transpirational change, or alteration in cellular lipid composition. For an economical benefit of a crop variety under heat stress, increased photosynthetic rate coupled with heat avoidance and membrane thermostability is an important strategy to stand with (Nagarajan et al. 2010; Scafaro et al. 2019).

Various research articles have portrayed that a high carbohydrate availability help in improving an important physiological trait to withstand against heat stress. Sucrose is the end product in photosynthesis, and it translocates into the organs through phloem, and it also helps in stress signalling into the plant (Hossain et al. 2015; Khan et al. 2015). A study on heat-tolerant tomato genotype suggested that the activity of sucrose increased the resistance of heat tolerance in young tomato fruits manifold due to increasing cell wall strength and sugar signalling in plants. Similarly, other pigments (i.e., carotenes, xanthophylls, and some other terpinoids) show a greater tolerance against heat stress (Velikova et al. 2011).

3.2 Cold Stress

Cold stress is considered as one of the the major abiotic stress which significantly reduces the crop yield and affects the quality and shelf-life of the produce. Cold stress mainly impacts on the cellular function of the plants (Kumar Yadav 2010;

Solanke and Sharma 2008). Therefore, plants are categorised into three categories on cold stress; first, chilling delicate plants; second, plants are susceptible within a range of temperature above 0 °C and below 15 °C; third, the crop plants are capable of withstanding in low temperature but suffering when ice formation begins in tissues (Sanghera et al. 2011; Hussain et al. 2018; Hassan et al. 2021a, b). Lastly, frost resistance plants, which can tolerate very low temperatures when exposed to field condition (Sun et al. 2021; Bu et al. 2021). Cold stress very often damages the cellular structure and causes injury to plants by changing its membrane structure and rupturing the protoplasm in the plant cell. Moreover, reduced plant growth to death of plants takes place due to cellular damage and transformation in plant metabolic activity (Abe et al. 2003; Agarwal et al. 2005; Abdel Kader et al. 2011; Hajihashemi et al. 2018).

Various research articles have portrayed that traditional breeding methods have a limited success of improvement against cold tolerance of important crop species. Conventional breeding methods includes the inter-generic or inter-specific hybridisation and in vitro fertilisation processes to improve the cold tolerance of numerous plant species, but there was not so much success (Zarka et al. 2003; Puijalon et al. 2011; Jutsz and Gnida 2015). Due to low genetic variability of yield traits under cold stress, the scope of conventional breeding methods is very low. Thus, there is a huge scope to look for an alternative approach to mechanise the cold stress tolerance of crop plants (Zhen and Ungerer 2008; Wang et al. 2018a, b). The new transgenic crop plants with cold stress tolerance mechanism can be developed through plant biotechnological, metabolomics-based approaches (Choudhury et al. 2021). The plant-specific recombinant DNA technology and situation-specific gene transfer protocols can effectively address the problem of cold tolerance mechanism (Aguilar et al. 2001; Ding et al. 2019b). The genes responsible for cold stress in plants can be easily identified and isolated through biotechnological interventions and development of transgenic lines for cold stress can be made within a short period of time (Cantrel et al. 2011; Bao et al. 2014; Bremer et al. 2017). Different research articles suggested that the cold tolerance and accumulation in plants is a complex process and there is narrow scope to develop the cold tolerance in important plant species. Therefore, there must be use of appropriate methods and approaches to reach the objective of sustainable food production. The genotype designed will be much better than the existing crop species (Ding et al. 2018, 2019a).

3.3 *Raising CO_2 Concentration*

The changing global atmosphere in recent times increases the CO_2 concentration at a double rate. This increasing rate hampers the plant metabolism and ultimately affects the plant growth and development. Many research articles have portrayed the stress-reducing effect of elevated CO_2 in case of plant abiotic stress management but the underlying mechanism for drought tolerance remains elusive (AbdElgawad et al. 2016; Zinta et al. 2016). It is also revealed from different research articles that the addition of extra C from elevated CO_2 will help in inducing stomatal closing. This in turn improves the water use efficiency coupled with protecting the plant against drought stress and help in increasing plant metabolic activity (Avramova et al. 2017; Wada et al. 2020; Shabbaj et al. 2022). However, increased C concentration resulting in an improved supply of plant leaf defence molecules. These molecules in turn help in improving the protection against oxidative damage in high CO_2 concentration in the environment (Pandey et al. 2015).

Many of the scientific reports indeed supported the theory regarding high concentration of elevated CO_2 showing increased antioxidant levels and/or improve plant protection against oxidative damage but on the other side, a number of scientific reports have suggested that there is no role of elevated CO_2 on antioxidants (Ainsworth et al. 2008; Albert et al. 2011). Thus, increased antioxidant defence of elevated CO_2 for stress mitigation cannot be attributed universally (Erice et al. 2007; Dieleman et al. 2012; Das and Roychoudhury 2014). Moreover, photorespiration may be a key alternative process for oxidative stress management under elevated CO_2 (Farfan-Vignolo and Asard 2012; Feng et al. 2014). Elevated CO_2 in general reduces the reactive oxygen species (ROS) formation or ROS and promotes carboxylation over oxygenation (Gonzàlez-Meler and Siedow 1999; Geissler et al. 2010). It has been observed that ROS beyond permissible limits (irrespective of stressors) considered as toxic for the cellular environment often resulted in poor germination and weak seedling establishment (Choudhury et al. 2022a, b, c; Saha et al. 2019; Sahoo et al. 2019; Chowardhara et al. 2019a, b; Mazumder et al. 2022a, b; Moulick et al. 2016, 2017, 2018c); inhibition in photosynthesis, delay in flowering followed by yield reduction have already been reported (Moulick et al. 2018a, b). Therefore, elevated CO_2 reduces stress impact through either increasing the defence against oxidative damage or decreasing the challenge of oxygenation.

3.4 *Rainfall Abnormalities*

Plants are very often exposed to rainfall abnormalities during a crop spell, which in another term called as drought condition. Among the other abiotic stress agents, drought is also considered as one of the major stress elements which affects the plant productivity (Brodersen et al. 2019; Seleiman et al. 2021). Plant fresh biomass mainly comprised 80–90% of water and plays a vital role in many physiological processes and influences the plant growth and development (Daryanto et al. 2017; O'Connell 2017). In today's world, quality water is a valuable resource to maintain the pace of agricultural production and global food security (Gray and Brady 2016). But, rainfall abnormalities, uneven distribution of rainfall, and indiscriminate depletion of water resources signalling a threat towards maintaining future food security and agricultural production scenario across the globe. Many researchers have represented water as a critical threat for global food security and an important element for past famines (Osakabe et al. 2014; Kaur and Asthir 2017; Zandalinas et al. 2020).

The global climate change is escalated due to the increasing air temperature coupled with raising CO_2 level, ultimately affecting the rainfall and its distribution across the globe (Wu et al. 2007; Imran et al. 2021). Although, the deficient amount of water mainly creates the drought stress phenomena but there are some other factors, i.e., loss of soil water through evaporation, dry wind with high light intensity exacerbates the drought event manifold (Deng et al. 2005; Cramer et al. 2011). Therefore, drought will be considered as an important stress element in those areas where crop production is solely dependent on rainfall. Therefore, over a certain period, high chances of water stress are observed due to an uneven rainfall distribution in some recent past years (Zhang et al. 2022; dos Santos et al. 2022). Climate change due to industrialisation, urbanisation, and many other associated factors prominently affect the rainfall patterns and these in turn scarce the water availability to the plants. The change in seasonal boundary shortening the monsoon spell in south-east Asian countries often make a disastrous situation for plant growth (Demidchik 2015; Menezes-Silva et al. 2017; Bryant et al. 2021). This ultimately hampers the crop production and threatens the world food security issues. Therefore, the adjustment of planting time of crop plants and selection of drought-tolerant varieties can help us to cope up with the drought stress situation. Further, in dryland areas, specific crops (i.e., Millets) suitable for those areas must be selected (Close 1997; Chandra Babu et al. 1999; Verslues et al. 2006; Choudhury et al. 2022c).

3.5 *Raising Sea Level*

Salt is essential for making a tasty food but excessive deposition of salt in soil can make the field unproductive for growing crops. Due to a rapid climate change, the low-lying coastal areas are exposed to deluge with saltwater contamination (Chen and Mueller 2018). Though, this salt can be dispersed by rainfall, but climate change scenario is also increasing the extreme weather events, i.e., heat waves, thunderstorm, uneven rainfall, and shortage of rainfall. This in turn, leads to more use of ground water for irrigation purpose and human use (Shrivastava and Kumar 2015). Thus, the more depletion of water table, set aside more salt into the soil profile. Climate change and other devastating phenomena already make farming a razor-thin venture for coastal region farmers worldwide. With this the salt contamination, staggered plant growth is making farming an unprofitable source of income for the farmers (Hasegawa et al. 2018). Many of the researchers have estimated that already 20 percent of cultivable land is under the salt contamination and gradually becomes unproductive. Besides, with the raising atmospheric temperature, the oceanic temperature is also increasing, and the glaciers and ice sheets are also melting (Chen et al. 2017). This salt water requires extra space and lying over the low land coastal region (Chen et al. 2017, 2019). Therefore, under this devastating situation, selection of salt-tolerant varieties must be an important approach to cope up with the situation. Change in planting time according to the monsoon spell will be an admirable approach.

4 Breeding Adaptation to the Climate Change

Climate change is the major concern for twenty-first century because of unpredicted changes in rainfall pattern, winter duration, and frequency of rainfall causing crops vulnerable to changing environment. Developing new variety involves continuous selection from competitive genotypes and a series of evaluation for identification of a stable variety. Climate change created a focused challenge for selection of suitable stable genotypes. Climate changes have increased the ranking change under cross-over interactions differently for favourable and unfavourable environments which invites more genetic improvements introgressing new set of germplasm (Xiong et al. 2021). Plant breeding creates new variations by conventional or advanced techniques to make genotype responsive to climate change. Adaptive changes and climate-responsive plastic genotype identification are the new goals exploiting evolutionary changes (Williams et al. 2008). Evolutionary significance of outcrossing in cross-pollinated crops and to some extent in self-pollinated crop enhances adaptation under climate change (Lawrie et al. 2006). Genotype responsive to higher CO_2 at reproductive stage creates advantage to the crop though combined with high temperature affects grain yield (Chaturvedi et al. 2017).

4.1 Breeding Strategies Under Climate Change

Classical plant breeding is a continuous pipeline for developing new types considering primarily high yield and biotic and abiotic stress tolerance capacity as an added advantage of respective genotype. Genetic variations are the primary component that can enhance climate change-responsive crop variety development. Crop wild relative (CWR) is the genetic reservoir of allelic diversity for crop improvement; studies on CWR can reveal how plant adapts under climate change (Cronin et al. 2007). Reducing the gap of target environment and the areas of crop evaluation can be done through screening in artificial conditions. Prediction of future environment and selection of trait specific to the target environment is advantageous (Dreccer et al. 2013). Drought-tolerant genotype is tolerant which is specific to a particular stage of crop growth but developing drought tolerance in more than one stage is beneficial as timing of drought stress occupies a significant part of genotype and environment out of total variation. Reproducible environments like CO_2 tunnel, temperature tunnel, growth chamber, and moisture content gradient field can be used for predicting genotype performance. Focused identification of germplasm sets based on varying climate, and soil and production data based on multiple locations can enhance allelic richness for further variability creation. Exposure of combination of stress like heat, drought, heavy element, and water stress, biotic stress like fungus and insect to assess the genotype response how stress combination is affecting has been recommended for climate resilience (Rivero et al. 2022). Exploitation of higher photosynthesis capacity, high light use efficiency, and higher CO_2 assimilation can enhance productive potential in crops under climate change.

4.2 Physiological and Biochemical Traits for Climate Change

Coleoptile length to harness moisture at deep furrow, enhanced photosynthesis rate, and water-soluble carbohydrate can boost production under future climate change in areas with drought prediction along with warmer winter. ICC 4958, a major chickpea variety, was used for development of some introgression line which performed better under drought stress condition; these lines showed early vigour, early flowering along with high transpiration efficiency suitable under variable climate (Barmukh et al. 2022). Precise phenotyping of stress involving spectral reflectance for simultaneous measurement of nutrient status, water deficit, and canopy cover can enhance early selection (Fischer et al. 1998; Penuelas et al.1998). Considering positive and negative interaction between traits improving nutrient and water use efficiency, plant rhizosphere and microbiome interactions, photo assimilate translocation, stomatal density, etc., can enhance plant suitability under climate change (Rivero et al. 2022). Developing different crop variety with specific adaptation like low light in rice has the potentiality to grow under climate change during kharif (Ganguly et al. 2020). Genotype-specific alleles contributing higher enzyme activity can enhance nutrient acquisition in several crops like rice (Zhang et al. 2011), lentil (Ganguly et al. 2021), etc.

4.3 Climate-Responsive Breeding

Genetic loci responsive to external changes helps plants to adapt immediately. Flowering locus (FT) gene is a major regulator of plants' response to climate change which determines variation in flowering time pathway and growth (Wigge 2011). Natural variation of FT gene can enhance local adaptation during seasonal environmental changes in Arabidopsis acting as a sensitive quantitative trait variation source in plants (Li et al. 2010). Cross species transferability of genetic markers and exploiting wide genetic variations in fruit crop can enhance adaptation under climate change (Sarkar et al. 2021). Easier involvement of crop wild relative (CWR) from transboundary sources along with exploiting in-situ germplasm for addressing climate change has been instrumental for new genotype development (Galluzzi et al. 2020). Careful phenotyping at early crop growth stage integrating multi-trait improvement using adaptive breeding in more than one location can enhance breeding adaptation under climate change. Selection using genomic selection strategy from wide population showed quick gain (Laverdière et al. 2022). Intensity of natural stress environment is less influenced externally which demands controlled environment phenotyping including image-derived trait acquisition, automated digital trait assessment, etc. (Langstroff et al. 2022). Artificial intelligence (AI)-based plant breeding technologies have the capacity to accelerate the development of climate-resilient crop with better adaptation (Roy et al. 2021).

4.4 Biotechnological Interventions

Additive genetic analysis along with classical improvement can enhance genetic gain. Identification of QTLs related to yield, disease resistance, and drought can be exploited for improvement of quantitative traits through introgression breeding. Identification and multiple QTL can be exploited for multiple stress tolerance ability. Multiple stress tolerance can accompany climate-resilient plant breeding. Association mapping and model-based prediction and whole genome genotyping can enhance exploitation of a wide range of variations (Jaccoud et al. 2001). Crop genomics has resulted quicker molecular breeding for new advanced genotypes. Genotyping and sequencing resulting in admixture population can tell the story of adaptation in new environment as described in Switch grass (Lovell et al. 2021). Selection of climate-adaptive traits for micro geographic location using location-specific mapping population is required for crop adaptation study (Frachon et al. 2018). Rapid breeding is the key to increase the genetic gain and quick product delivery to farmer's field. Some of the techniques for managing population advancement like pedigree breeding with no off-season nursery, pedigree breeding with off-season nursery, haploid breeding, and single seed descent with controlled environment can be applied (Atlin et al. 2017). Rapid generation advancement in rice and different crops along with speed breeding standardised in several crops like lentil, wheat, and pea can enhance their yield potential (Ghosh et al. 2018; Roy et al. 2021; Bhattacharyya et al. 2022). Common gene based on pan genome study can enhance evolutionary study for biotic and abiotic stress tolerance loci which remained conserved for generation after generation (Bayer et al. 2020). Exploiting phenotypic plasticity under climatic variations in high-altitude plant species *Primula sikkimensis* through transcriptome sequencing generated for identification of new loci contributing to climate change adaptation (Gurung et al. 2019). Disease susceptibility makes crop more challenging under climate change. Sustainable approach to deliver tolerant crop invites genome editing with precise genome manipulation as developed in rice with OsSWEET14. Quick selection with advanced genomics tool and prediction-based phenotyping and introgression of wide genomic variations can enhance quick variety development. In recent years, researchers are also exploring the possibilities of next-generation sequencing techniques, application of metabolomics, potentials of metabolites, and to incorporate the beneficial aspect in developing better climate resilience in a wide range of crops (Hossain et al. 2021a, b; Choudhury et al. 2021).

5 Agronomic Mitigation Strategies for Sustainable Crop Production

5.1 Agronomic Approaches for Improving the Effect of Heat Stress

Several agronomic management practices have the potential to mitigate the detrimental effects of extreme stresses. Examples of agronomic management practices include soil fertility management, use of suitable cultivars, interference with sowing time, sowing method and depth of sowing, scheduling of water management practices, synchronisation of fertiliser use, use of several resource conservation technologies (RCTs), etc. Alteration in planting time, use of photo-thermo-insensitive varieties, crop diversification, and irrigation water management are the major sustainable approaches to minimise environmental aberration and impart a harmless ecosystem to the crop ecology (Ali and Erenstein 2017). Under extreme heat stress, inadequate application of minerals and fertilisers has detrimental effects on diminished plant vigour and yield, which creates difficulties for global food production. While plants are subjected to extreme heat stress, several essential nutrients like K, Ca (macronutrient) and B, Se, and Mn (micronutrients) have significant roles in improving stomatal resistance by regulating osmotic balance in the cell and inducing heat resistance potential in the plant (Waraich et al. 2012). Exogenous application of potassic fertiliser has significant contributions to the plant under heat stress, such as improving various metabolic and physiologic activities, maintaining cell membrane integrity, translocating photosynthetic products, and increasing antioxidant concentration and enzyme activities (Verma et al. 2020). Selection and implementation of climate-resilient crops and heat resistance cultivars for different regions to adapt and mitigate heat stress is the most sustainable approach for the agro-ecosystem (Sarkar et al. 2021). Introduction of photo-thermo-insensitive varieties is well suited for various agro-ecological zones without compromising crop production and yield.

Time of planting is one of the sustainable approaches to mitigate heat stress by synchronising the most sensitive growth stages. Sowing time helps to escape the heat stress, thereby increasing the crop yield. Several researchers determined that late sowing of wheat drastically reduces the crop yield due to high temperature stress during late developmental stages, and that early sowing of wheat maximises the crop yield (Bassu et al. 2009; Bannayan et al. 2013). In coastal saline zone, sowing time plays a pivotal role in mitigating the heat stress of the winter and summer crops. Due to a prolonged and asymmetrical pattern of rainfall in West Bengal, the transplanting of rice during the rainy season is sometimes delayed by 3–4 weeks beyond the usual procedure (Ghosh et al. 2022a, b). Farmers are left with no choice but to await the planting of lentils by a month in order to include this pulse in the cropping sequence. Delayed sowing is sensitive to environmental conditions like extremely high temperatures, which have a detrimental impact on the development and production of crop. The crop flowers and matures under high temperature and terminal drought conditions. Lentil is cultivated in wide parts of India using residual moisture (Ghosh et al. 2022a, b). According to Tawaha and Turk (2001), the shorter growing period accessible to the late-sown crop can also be ascribed to the drop in grain production as a result of the delayed sowing due to the reduced maturation time of the crop. Significant reduction in seed yield under water deficit was reported earlier by Panahyan et al. (2009). Tawaha and Turk (2002) mentioned that the delay of sowing time affected grain yield, plant height, primary branches per plant, and the number of pods per plant. The cause of plant height reduction might be due to rapid increase in temperature as well as reduction of soil moisture. Hence, the sowing time of crops must be suitably adjusted with the occurrence of hot summers to escape the high temperature stress. Resource conservation technologies have played a significant role in mitigating the high temperature stress on crop production. Mulching is one of the most important soil moisture conservation technologies. It acts as a nutrient management, weed management, and water management tool for soil by reducing the evaporation loss from soil and allowing it to remain cool under hot summer conditions (Samui et al. 2020; Mondal et al. 2020). Heat stress is normally interrelated with water stress, so efficient management of water should be important for crop growth and yield (Hassan et al. 2021a, b). Thus, scheduling irrigation based on available water resources and applying water when it is most needed by the crop have a greater impact under heat stress (Zhang et al. 2002; Tahar et al. 2011).

5.2 *Agronomic Approaches to Mitigate Cold Stress*

Many researchers suggested that optimum application of potassic fertiliser plays an important role to increase the adaptability against cold stress in *Alfalfa* and it is because potassium has the capacity to regulate the stomatal openings, turgidity of cell interior, translocation of uptake water and minerals, and to increase the photosynthetic rate, carbohydrate assimilation, and protein synthesis (Lissbrant et al. 2009; Oosterhuis et al. 2014; Kaiser 2018; Cakmak 2005; Jungers et al. 2019). Under cold stress, optimum soil moisture needs to be maintained to escape the chilling injury. Excess moisture or water logging condition in the soil decrease the soil water temperature during cool season leading to improper root growth and water uptake pattern. Thus, adequate drainage is essential to prevent the crop from chilling stress. Mulching also plays a significant role in controlling the soil temperature of the winter crops by trapping.

5.3 *Agronomic Approaches to Mitigate the Negative Impact of Raising CO_2 Concentration*

Globally, increasing atmospheric CO_2 concentration over a long period of geological time possesses a major environmental constraint and is responsible for raising temperatures and causing climate change. It has been determined that increasing CO_2

levels in the atmosphere is one of the major causes of global climate change and alterations of atmospheric temperature (Shakun et al. 2012). Atmospheric carbon fixation is emerging as the most effective method of mitigating excess CO_2 concentrations via plants and other sustainable management practises (Lal 2004). Several agronomic management practises are being followed to reduce the rising CO_2 concentration through soil carbon sequestration, whereas several anthropogenic activities are the major causes of diminishing soil organic carbon (Sombrero and de Benito 2010). It has been estimated that up to 30–50% of soil organic carbon is lost through conventional tillage (Schlesinger 1985). Intensive tillage practises keep the soil bare, leading to the oxidisation of organic carbon and the secretion of carbon dioxide into the environment (Al-Kaisi and Yin 2005). Conservation tillage is a soil disturbing process that is limited to the plant root zone only, keeps the crop residues as such with or without incorporation, and maintains the soil microbial habitat by manipulating and hampering the soil less. Several reports state that up to 25 Gt of carbon can be isolated from soil worldwide through conventional tillage in the next 50 years. Hence, it helps to reduce the secretion of carbon dioxide while isolating atmospheric carbon into the soil. Rice fields are the primary source of greenhouse gas emissions under waterlogging conditions, including CO_2, CH_4, and N_2O, which are ultimately responsible for global climate change. It is reported that, along with CH_4, up to 20% of total greenhouse gases have the potential to cause global warming (IPCC 2007). Direct-seeded rice and the system of rice intensification are becoming the most significant approaches to reduce methanogenesis by eliminating the puddled and waterlogged conditions in the rice field. It is reported that judicious application of chemical fertiliser along with residual crop biomass can fix about 21.3–32.5% of carbon into the soil (Windeatt et al. 2014). Similarly, it lends itself well to integrated nutrient management, which aids in increasing crop yield while decreasing CO_2 and other greenhouse gas emissions. Afforestation is found to be a significant way to mitigate climate change by increasing ground cover with forest plants. Forest trees can diminish the emission of atmospheric carbon by sequestering carbon into the plant biomass. Agroforestry is the combining of agricultural cultivation systems in which forest trees and agricultural crops are grown together. Agroforestry is one of the most effective and sustainable approaches for restoring atmospheric carbon (Biswas et al. 2022).

5.4 *Agronomic Approaches to Mitigate the Rainfall Abnormalities*

Rainfall is one of the major factors for agricultural crop production being shifted towards the uncertainty due to the impact of climate change and global warming. Agricultural crop production is largely dependent on the amount, intensity, and distribution of rainfall. Hence, to mitigate the adverse effects of aberrant weather condition and abnormalities of rainfall, appropriate contingency planning is required (Fig. 1.8). Thus, implementation of contingency crop strategies is required to manage the uneven distribution of rainfall in a given area by using suitable crop and varieties, adjusting cropping system and various agronomic applications (Addy 2009).

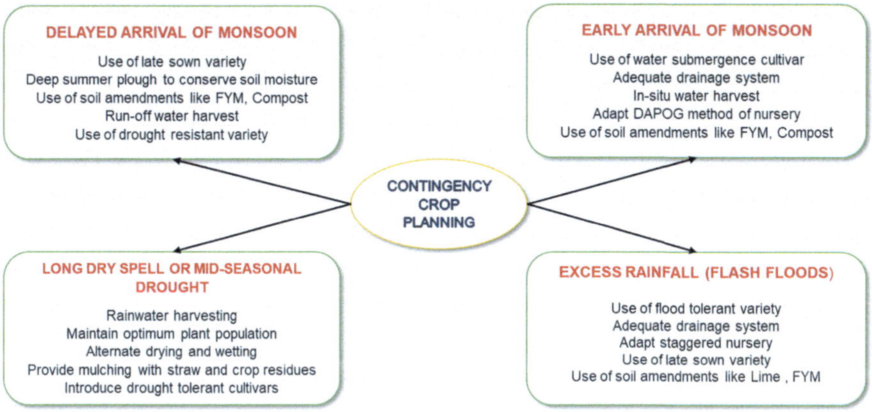

Fig. 1.8 Schematic representation of the contingency Crop Planning

Climate change and global warming has created a detrimental threat throughout the world, resulting in melting of glacier and rising the sea level. Rising sea level affects coastline areas by entering into the crop field, not only affecting the crop yields, making the soil unproductive, destroying the habitat, and affecting the trade market (Darwin and Tol 2001). Southeast Asia, East Asia, South Asia, and the Southeast US are more liable to be affected by rising sea level (Dasgupta et al. 2009). Soil salinity increased due to inundation of land, making the soil unproductive; therefore, several agricultural management practices should be adapted to mitigate the effect of sea water. Several agronomic management practices can be adapted to escape from the effect of sea water by reducing the soil salinity level including leaching of saline water from the plant root zones, draining the excess flood water, using soil amendments, and other relevant management practices. Use of salt-tolerant varieties is the most significant way to mitigate the soil salinity without affecting the crop yields. Coastline areas mostly experienced the accumulation of extreme flood water during the rainy season, while in summer and winter season, soil salinity level increased due to higher rate of evaporation from the soil. Therefore, salinity level from soil can be diminished through the avoiding of summer fallow. Soil reclamation by using soil amendments like farm yard manure (FYM), compost, organic manure, and growing salt accumulating plant (phytoremediation) is becoming most easy, cost-effective management practices throughout the world.

6 Organisation Initiatives Towards Climate Change and Global Crop Production

Globally, a number of regional and international organisations are working intensively to mitigate the negative impact of the climate change on multidisciplinary aspects. While some organisations are only beginning their path, several of the organisations have already achieved carbon neutrality. Others, who have made progress in their sustainability approach, are now taking it a step further by pledging to achieve complete carbon neutrality. Organisations working primarily on the theme of climate change and crop production aspects are summarised in Table 1.1.

Table 1.1 Major international organisations working primarily on theme of climate change and crop production aspects

Sl. no.	Name of the organisation	Working arena	Website
1.	Food and Agriculture Organisation (FAO), United Nation	To promote better management of trade-offs and synergies between adaptation and mitigation in the agriculture sectors. By providing technical advice, information, and resources for better decision-making and the implementation of adaptive measures, FAO assists its member nations in these endeavours.	www.fao.org
2.	Rainforest Action Network (RAN), United States	To make large firms accountable for preserving human rights, maintaining healthy forests, and safeguarding biodiversity. Their climate and energy team works to end the most polluting kinds of energy and pushes the financial sector to stop funding the production of fossil fuels.	www.ran.org
3.	Project Drawdown	Policymakers, academic institutions, businesses, and activists may all access Project Drawdown as an open-source, peer-reviewed resource for climate and food production solutions.	https://drawdown.org/
4.	Indigenous Environmental Network (IEN)	To talk about the environmental threats to people, towns, lands, and rivers. The organisation favours a fair transition, carbon pricing, and global protection of holy places and natural resources.	https://www.ienearth.org/
5.	CGIAR and its umbrella organisations	The CGIAR is a global cooperation that brings together multinational organisations conducting food security research. The goals of CGIAR research are to decrease rural poverty, boost food security, enhance human health and nutrition, and manage natural resources sustainably.	https://www.cgiar.org
6.	Asian Development Bank (ADB)	To provide financial and technical support to the member countries of Asia for achieving climate and food sustainability.	https://www.adb.org/
7.	Western Africa Development Bank (BOAD)	To provide financial and technical support to the member West African countries for achieving climate and food sustainability.	https://www.boad.org/en/
8.	Australian Centre for International Agricultural Research (ACIAR)	To help and inspire Australian agricultural experts to utilise their expertise to identify and resolve agricultural issues in developing nations.	www.aciar.gov.au
9.	United States Agency for International Development (USAID)	To give sustainable development assistance and foreign aid to civilians in developing nations.	www.usaid.gov

7 Conclusion

If we carefully evaluate the current research trend in mitigating fluctuating climate-induced adverse consequences, few dominant trends can be seen: (i) assessment of the impact of single stressor on certain crops, (ii) evaluation of a particular remedial measures on limited number of crops. However, still we are lacking the knowledge of (a) the consequences of multiple stressors on a particular crop/cropping system, (b) trailing behind in answering how a particular stress management is working on multiple environments. Another important concern is differences among the '*pro and anti GMO lobby*' and the respective policymaking bodies. This not only creates a huge vacuum in terms of knowledge gap but also creates lots of confusion. Moreover, we are focusing our strategies for a short time span (say 2030); instead, we should focus on longer time span and orient our efforts accordingly. Climate change is not a short time phenomenon, so our efforts should not be focused on short-time remedial options.

References

Abbass K, Qasim MZ, Song H et al (2022) A review of the global climate change impacts, adaptation, and sustainable mitigation measures. Environ Sci Pollut Res Int 29:42539–42559

Abdel Kader DZ, Saleh AAH, Elmeleigy SA, Dosoky NS (2011) Chilling-induced oxidative stress and polyamines regulatory role in two wheat varieties. J Taibah Univ Sci 5:14–24. https://doi.org/10.1016/S1658-3655(12)60034-X

AbdElgawad H, Zinta G, Beemster GTS, Janssens IA, Asard H (2016) Future climate CO_2 levels mitigate stress impact on plants: increased defense or decreased challenge? Front Plant Sci 7:556. https://doi.org/10.3389/fpls.2016.00556

Abe H, Urao T, Ito T et al (2003) Arabidopsis AtMYC2 (bHLH) and AtMYB2 (MYB) function as transcriptional activators in abscisic acid signaling. Plant Cell 15:63–78. https://doi.org/10.1105/TPC.006130

Addy K (2009) Aberrant -weather-situations-and.html. Aberrant weather situations and contingency crop planning. From http://addykraipur.blogspot.com

Agarwal S, Sairam RK, Srivastava GC, Meena RC (2005) Changes in antioxidant enzymes activity and oxidative stress by abscisic acid and salicylic acid in wheat genotypes. Biol Plant 49:541–550. https://doi.org/10.1007/S10535-005-0048-Z

Aguilar PS, Hernandez-Arriaga AM, Cybulski LE, Erazo AC, de Mendoza D (2001) Molecular basis of thermosensing: a two-component signal transduction thermometer in *Bacillus subtilis*. EMBO J 20:1681–1691. https://doi.org/10.1093/EMBOJ/20.7.1681

Ainsworth EA, Beier C, Calfapietra C et al (2008) Next generation of elevated [CO_2] experiments with crops: a critical investment for feeding the future world. Plant Cell Environ 31:1317–1324. https://doi.org/10.1111/J.1365-3040.2008.01841.X

Albert KR, Mikkelsen TN, Michelsen A, Ro-Poulsen H, van der Linden L (2011) Interactive effects of drought, elevated CO_2 and warming on photosynthetic capacity and photosystem performance in temperate heath plants. J Plant Physiol 168:1550–1561. https://doi.org/10.1016/J.JPLPH.2011.02.011

Alexandratos N, Bruinsma J (2012) World agriculture towards, 2012 revision, vol 2030/2050. FAO, Rome, ESA, Working Paper No. 12-03

Ali A, Erenstein O (2017) Assessing farmer use of climate change adaptation practices and impacts on food security and poverty in Pakistan. Clim Risk Manag 16:183–194. https://doi.org/10.1016/j.crm.2016.12.001

Al-Kaisi MM, Yin X (2005) Tillage and crop residue effects on soil carbon and carbon dioxide emission in corn-soybean rotations. J Environ Qual 34:437–445. https://doi.org/10.2134/jeq2005.0437

Allakhverdiev SI, Kreslavski VD, Klimov VV et al (2008) Heat stress: an overview of molecular responses in photosynthesis. Photosynth Res 98:541–550. https://doi.org/10.1007/S11120-008-9331-0

Atlin GN, Cairns JE, Das B (2017) Rapid breeding and varietal replacement are critical to adaptation of cropping systems in the developing world to climate change. Glob Food Sec 12:31–37. https://doi.org/10.1016/j.gfs.2017.01.008

Avramova V, Abdelgawad H, Vasileva I et al (2017) High antioxidant activity facilitates maintenance of cell division in leaves of drought tolerant maize hybrids. Front Plant Sci 8:84. https://doi.org/10.3389/fpls.2017.00084

Babu RC, Pathan MS, Blum A, Nguyen HT (1999) Comparison of measurement methods of osmotic adjustment in rice cultivars. Crop Sci 39:150–158. https://doi.org/10.2135/cropsci1999.0011183X003900010024x

Bai H, Tao F (2017) Sustainable intensification options to improve yield potential and eco-efficiency for rice-wheat rotation system in China. Field Crop Res 211:89–105. https://doi.org/10.1016/j.fcr.2017.06.010

Banerjee H, Sarkar S, Dutta SK et al (2022) Nitrogen management trade-offs in hybrid rice for agronomy, carbon, and energy efficiency. Nutr Cycl Agroecosyst 123:5–18. https://doi.org/10.1007/s10705-022-10199-0

Bannayan M, Eyshi Rezaei E, Hoogenboom G (2013) Determining optimum planting dates for rainfed wheat using the precipitation uncertainty model and adjusted crop evapotranspiration. Agric Water Manag 126:56–63. https://doi.org/10.1016/j.agwat.2013.05.001

Bao F, Huang X, Zhu C et al (2014) Arabidopsis HSP90 protein modulates RPP4-mediated temperature-dependent cell death and defense responses. New Phytol 202:1320–1334. https://doi.org/10.1111/NPH.12760

Barmukh R, Roorkiwal M, Dixit GP et al (2022) Characterization of "QTL-hotspot" introgression lines reveals physiological mechanisms and candidate genes associated with drought adaptation in chickpea. J Exp Bot 73:7255–7272. https://doi.org/10.1093/jxb/erac348

Barnabás B, Jäger K, Fehér A (2008) The effect of drought and heat stress on reproductive processes in cereals. Plant Cell Environ 31:11–38. https://doi.org/10.1111/J.1365-3040.2007.01727.X

Bassu S, Asseng A, Motzo R, Giunta F (2009) Optimizing sowing date of durum wheat in a variable Mediterranean environment. Field Crops Res 111:109–118

Battisti DS, Naylor RL (2009) Historical warnings of future food insecurity with unprecedented seasonal heat. Science 323:240–244

Bayer PE, Golicz AA, Scheben A, Batley J, Edwards D (2020) Plant pan-genomes are the new reference. Nat Plants 6:914–920. https://doi.org/10.1038/s41477-020-0733-0

Bellard C, Bertelsmeier C, Leadley P, Thuiller W, Courchamp F (2012) Impacts of climate change on the future of biodiversity. Ecol Lett 15:365–377. https://doi.org/10.1111/J.1461-0248.2011.01736.X

Bhattacharyya P, Chakraborty K, Molla KA et al (2022) Climate resilient technologies for rice based production systems in Eastern India. ICAR-National Rice Research Institute, Cuttack, Odisha, p 408

Biswas B, Patra SR, Timsina J et al (2019) Climate change-resilient rice production technology: a high yielding, water efficient and remunerative option for South Asian farmers. Global J Agric Allied Sci 1:20–29

Biswas B, Chakraborty D, Timsina J et al (2022) Agroforestry offers multiple ecosystem services in degraded lateritic soils. J Clean Prod 365:132768. https://doi.org/10.1016/j.jclepro.2022.132768

Bita CE, Gerats T (2013) Plant tolerance to high temperature in a changing environment: scientific fundamentals and production of heat stress-tolerant crops. Front. Plant Sci 4:273. https://doi.org/10.3389/FPLS.2013.00273

Blum A, Klueva N, Nguyen HT (2001) Wheat cellular thermotolerance is related to yield under heat stress. Euphytica 117:117–123

Bohringer C (2003) The Kyoto protocol: a review and perspectives. Oxf Rev Econ Policy 19:451–466

Bremer A, Wolff M, Thalhammer A, Hincha DK (2017) Folding of intrinsically disordered plant LEA proteins is driven by glycerol-induced crowding and the presence of membranes. FEBS J 284:919–936. https://doi.org/10.1111/FEBS.14023

Brodersen CR, Roddy AB, Wason JW, McElrone AJ (2019) Functional status of xylem through time. Annu Rev Plant Biol 70:407–433. https://doi.org/10.1146/ANNUREV-ARPLANT-050718-100455

Bryant C, Fuenzalida TI, Brothers N et al (2021) Shifting access to pools of shoot water sustains gas exchange and increases stem hydraulic safety during seasonal atmospheric drought. Plant Cell Environ 44:2898–2911. https://doi.org/10.1111/PCE.14080

Bu X, Wang X, Yan J et al (2021) Genome-wide characterization of B-box gene family and its roles in responses to light quality and cold stress in tomato. Front Plant Sci 12:698525. https://doi.org/10.3389/fpls.2021.698525

Cakmak I (2005) The role of potassium in alleviating detrimental effects of abiotic stresses in plants. J Plant Nutr Soil Sci 168:521–530. https://doi.org/10.1002/jpln.200420485

Cantrel C, Vazquez T, Puyaubert J et al (2011) Nitric oxide participates in cold-responsive phosphosphingolipid formation and gene expression in *Arabidopsis thaliana*. New Phytol 189:415–427. https://doi.org/10.1111/J.1469-8137.2010.03500.X

Chaturvedi AK, Bahuguna RN, Shah D, Pal M, Jagadish SVK (2017) High temperature stress during flowering and grain filling offsets beneficial impact of elevated CO_2 on assimilate partitioning and sink strength in rice. Sci Rep 7:8227

Chaudhary P, Rai S, Wangdi S et al (2011) Consistency of local perceptions of climate change in the Kangchenjunga Himalaya landscape. Curr Sci 101:504–513

Chen J, Mueller V (2018) Coastal climate change, soil salinity and human migration in Bangladesh. Nat Clim Chan 8:981–985. https://doi.org/10.1038/S41558-018-0313-8

Chen JJ, Mueller V, Jia Y, Tseng SKH (2017) Validating migration responses to flooding using satellite and vital registration data. Am Econ Rev 107:441–445. https://doi.org/10.1257/AER.P20171052

Chen J, Kosec K, Mueller V (2019) Moving to despair? Migration and well-being in Pakistan. World Dev 113:186–203. https://doi.org/10.1016/J.WORLDDEV.2018.09.007

Choudhury S, Sharma P, Moulick D, Mazumder MK (2021) Unrevealing metabolomics for abiotic stress adaptation and tolerance in plants. J Crop Sci Biotechnol 24:479–493. https://doi.org/10.1007/s12892-021-00102-8

Choudhury S, Mazumder MK, Moulick D et al (2022a) A computational study of the role of secondary metabolites for mitigation of acid soil stress in cereals using dehydroascorbate and monodehydroascorbate reductases. Antioxidants 11:458. https://doi.org/10.3390/antiox11030458

Choudhury S, Moulick D, Mazumder MK et al (2022b) An in vitro and in silico perspective study of seed priming with zinc on the phytotoxicity and accumulation pattern of arsenic in rice seedlings. Antioxidants 11:1500. https://doi.org/10.3390/antiox11081500

Choudhury S, Moulick D, Ghosh D et al (2022c) Drought-induced oxidative stress in pearl millet (*Cenchrus americanus* L.) at seedling stage: survival mechanisms through alteration of morphophysiological and antioxidants activity. Life 12:1171. https://doi.org/10.3390/life12081171

Chowardhara B, Borgohain P, Saha B et al (2019a) Assessment of phytotoxicity of zinc on Indian mustard (*Brassica juncea*) varieties during germination and early seedling growth. Ann Plant Soil Res 21:239–244

Chowardhara B, Borgohain P, Saha B, et al (2019b) Phytotoxicity of cd and Zn on three popular Indian mustard varieties during germination and early seedling growth. Biocatal Agric Biotechnol 21:101349. https://doi.org/10.1016/j.bcab.2019.101349

Close TJ (1997) Dehydrins: a commonalty in the response of plants to dehydration and low temperature. Physiol Plant 100:291–296. https://doi.org/10.1111/J.1399-3054.1997.TB04785.X

Cramer GR, Urano K, Delrot S, Pezzotti M, Shinozaki K (2011) Effects of abiotic stress on plants: a systems biology perspective. BMC Plant Biol 11:163. https://doi.org/10.1186/1471-2229-11-163

Cronin JK, Bundock PC, Henry RJ, Nevo E (2007) Adaptive climatic molecular evolution in wild barley at the Isa defense locus. Proc Natl Acad Sci U S A 104:2773–2778. https://doi.org/10.1073/pnas.0611226104

Darwin RF, Tol RSJ (2001) Estimates of the economic effects of sea level rise. Environ Resour Econ 19:113–129. https://doi.org/10.1023/A:1011136417375

Daryanto S, Wang L, Jacinthe PA (2017) Global synthesis of drought effects on cereal, legume, tuber and root crops production: a review. Agric Water Manag 179:18–33. https://doi.org/10.1016/J.AGWAT.2016.04.022

Das K, Roychoudhury A (2014) Reactive oxygen species (ROS) and response of antioxidants as ROS-scavengers during environmental stress in plants. Front Environ Sci 2. https://doi.org/10.3389/FENVS.2014.00053

Dasgupta S, Laplante B, Meisner C, Wheeler D, Yan J (2009) The impact of sea level rise on developing countries: a comparative analysis. Clim Chan 93:379–388. https://doi.org/10.1007/s10584-008-9499-5

Demidchik V (2015) Mechanisms of oxidative stress in plants: from classical chemistry to cell biology. Environ Exp Bot 109:212–228. https://doi.org/10.1016/J.ENVEXPBOT.2014.06.021

Deng XP, Shan L, Inanaga S, Inoue M (2005) Water-saving approaches for improving wheat production. J Sci Food Agric 85:1379–1388. https://doi.org/10.1002/JSFA.2101

Dieleman WIJ, Vicca S, Dijkstra FA et al (2012) Simple additive effects are rare: a quantitative review of plant biomass and soil process responses to combined manipulations of CO_2 and temperature. Glob Chang Biol 18:2681–2693. https://doi.org/10.1111/J.1365-2486.2012.02745.X

Ding Y, Jia Y, Shi Y et al (2018) OST1-mediated BTF3L phosphorylation positively regulates CBFs during plant cold response. EMBO J 37(8):e98228. https://doi.org/10.15252/EMBJ.201798228

Ding Y, Lv J, Shi Y et al (2019a) EGR2 phosphatase regulates OST1 kinase activity and freezing tolerance in *Arabidopsis*. EMBO J 38(1):e99819. https://doi.org/10.15252/EMBJ.201899819

Ding Y, Shi Y, Yang S (2019b) Advances and challenges in uncovering cold tolerance regulatory mechanisms in plants. New Phytol 222:1690–1704. https://doi.org/10.1111/NPH.15696

Dreccer MF, Bonnett D, Lafarge T (2013) Plant breeding under a changing climate. In: Christou P, Savin R, Costa-Pierce BA, Misztal I, Whitelaw CBA (eds) Sustainable food production. Springer, New York. https://doi.org/10.1007/978-1-4614-5797-8_307

Edreira JIR, Budakli Carpici E, Sammarro D, Otegui ME (2011) Heat stress effects around flowering on kernel set of temperate and tropical maize hybrids. Field Crops Res 123:62–73

Erice G, Irigoyen JJ, Sánchez-Díaz M, Avice J, Ourry A (2007) Effect of drought, elevated CO_2 and temperature on accumulation of N and vegetative storage proteins (VSP) in taproot of nodulated alfalfa before and after cutting. Plant Sci 172:903–912. https://doi.org/10.1016/J.PLANTSCI.2006.12.013

Fand BB, Kamble AL, Kumar M (2012) Will climate change pose serious threat to crop pest management: a critical review. Int J Sci Res Publ 2:1–14

FAO (2000) The state of food and agriculture. Food and Agriculture Organization of the United Nations (FAO), Rome. From http://www.fao.org/docrep/x4400e/x4400e00.htm

FAO (2020) World food and agriculture – statistical. Yearbook, Rome. https://doi.org/10.4060/cb1329en

Farfan-Vignolo ER, Asard H (2012) Effect of elevated CO_2 and temperature on the oxidative stress response to drought in *Lolium perenne* L. and *Medicago sativa* L. Plant Physiol Biochem 59:55–62. https://doi.org/10.1016/J.PLAPHY.2012.06.014

Farooq M, Bramley H, Palta JA, Siddique KHM (2011) Heat stress in wheat during reproductive and grain-filling phases. Crit Rev Plant Sci 30:491–507

Feng GQ, Li Y, Cheng ZM (2014) Plant molecular and genomic responses to stresses in projected future CO_2 environment. CRC Crit Rev Plant Sci 33:238–249. https://doi.org/10.1080/07352689.2014.870421

Fischer RA, Rees D, Sayre KD et al (1998) Wheat yield progress associated with higher stomatal conductance and photosynthetic rate, and cooler canopies. Crop Sci 38:1467–1475

Frachon L, Bartoli C, Carrère S et al (2018) A genomic map of climate adaptation in *Arabidopsis thaliana* at a micro-geographic scale. Front Plant Sci 9:967. https://doi.org/10.3389/fpls.2018.00967

Galluzzi G, Seyoum A, Halewood M, López Noriega I, Welch EW (2020) The role of genetic resources in breeding for climate change: the case of public breeding programmes in eighteen developing countries. Plan Theory 9:1129. https://doi.org/10.3390/plants9091129

Ganguly S, Saha S, Vangaru S et al (2020) Identification and analysis of low light tolerant rice genotypes in field conditions and their SSR-based diversity in various abiotic stress tolerant lines. J Genet 99:24

Ganguly S, Roy A, Murmu SK et al (2021) Variation in P-acquisition ability and acid phosphatase activity at the early vegetative stage of lentil and their validation on P-deficiency field. Acta Physiol Plant 43:109. https://doi.org/10.1007/s11738-021-03280-8

Geissler N, Hussin S, Koyro HW (2010) Elevated atmospheric CO_2 concentration enhances salinity tolerance in *Aster tripolium* L. Planta 231:583–594. https://doi.org/10.1007/S00425-009-1064-6

Ghosh S, Watson A, Gonzalez-Navarro OE et al (2018) Speed breeding in growth chambers and glasshouses for crop breeding and model plant research. Nat Protoc 13:2944–2963. https://doi.org/10.1038/s41596-018-0072-z

Ghosh D, Brahmachari K, Sarkar S et al (2022a) Impact of nutrient management in rice-maize-greengram cropping system and integrated weed management treatments on summer greengram productivity. Indian J Weed Sci 54:25–30. https://doi.org/10.5958/0974-8164.2022.00004.1

Ghosh D, Brahmachari K, Skalický M et al (2022b) The combination of organic and inorganic fertilizers influence the weed growth, productivity and soil fertility of monsoon rice. PLoS One 17:e0262586. https://doi.org/10.1371/journal.pone.0262586

Gonzàlez-Meler MA, Siedow JN (1999) Direct inhibition of mitochondrial respiratory enzymes by elevated CO_2: does it matter at the tissue or whole-plant level? Tree Physiol 19:253–259. https://doi.org/10.1093/TREEPHYS/19.4-5.253

Gourdji SM, Sibley AM, Lobell DB (2013) Global crop exposure to critical high temperatures in the reproductive period: historical trends and future projections. Environ Res Lett 8:024041

Gray SB, Brady SM (2016) Plant developmental responses to climate change. Dev Biol 419:64–77. https://doi.org/10.1016/J.YDBIO.2016.07.023

Gull A, Lone AA, Ul N, Wani I (2019) Biotic and abiotic stresses in plants. Abiotic Biot Stress Plants. https://doi.org/10.5772/INTECHOPEN.85832

Gurung PD, Upadhyay AK, Bhardwaj PK, Sowdhamini R, Ramakrishnan U (2019) Transcriptome analysis reveals plasticity in gene regulation due to environmental cues in *Primula sikkimensis*, a high altitude plant species. BMC Genomics 20:989

Hajihashemi S, Noedoost F, Geuns JMC, Djalovic I, Siddique KHM (2018) Effect of cold stress on photosynthetic traits, carbohydrates, morphology, and anatomy in nine cultivars of *Stevia rebaudiana*. Front Plant Sci 9:1430. https://doi.org/10.3389/FPLS.2018.01430/FULL

Hansen J, Sato M, Kharecha P et al (2008) Target atmospheric CO_2: where should humanity aim? Open Atoms Sci J 2:217–231

Hasegawa T, Fujimori S, Havlík P et al (2018) Risk of increased food insecurity under stringent global climate change mitigation policy. Nat Clim Chan 8:699–703. https://doi.org/10.1038/S41558-018-0230-X

Hassan MA, Xiang C, Farooq M et al (2021a) Cold stress in wheat: plant acclimation responses and management strategies. Front Plant Sci 12:676884. https://doi.org/10.3389/FPLS.2021.676884

Hassan MU, Chattha MU, Khan I et al (2021b) Heat stress in cultivated plants: nature, impact, mechanisms, and mitigation strategies—a review. Plant Biosyst An Int J Dealing Aspects Plant Biol 155:211–234. https://doi.org/10.1080/11263504.2020.1727987
Hatfield JL, Prueger JH (2015) Temperature extremes: effect on plant growth and development. Weather Clim Extremes 10:4–10
He M, He CQ, Ding NZ (2018) Abiotic stresses: general defenses of land plants and chances for engineering multistress tolerance. Front Plant Sci 9:1771. https://doi.org/10.3389/FPLS.2018.01771
Heimann M (2005) Obituary: Charles David Keeling 1928–2005. Nature 437:331
Hoegh-Guldberg O (1999) Climate change, coral bleaching and the future of the world's coral reefs. Mar Freshw Res 50:839–866
Hossain MA, Bhattacharjee S, Armin SM et al (2015) Hydrogen peroxide priming modulates abiotic oxidative stress tolerance: insights from ROS detoxification and scavenging. Front Plant Sci 6:420. https://doi.org/10.3389/fpls.2015.00420
Hossain A, Bhatt R, Sarkar S et al (2021) Cost-effective and eco-friendly agricultural technologies in rice-wheat cropping systems for food and environmental security. In: Sustainable intensification for agroecosystem services and management. Springer, Singapore, pp 69–96
Hossain A, Ahmed S, Ahmad Z et al (2021a) Next-generation genetic engineering tools for abiotic stress tolerance in plants. In: Frontiers in plant-soil. Interaction. Academic Press, London, pp 153–197. https://doi.org/10.1016/B978-0-323-90943-3.00008-0
Hossain A, Pramanick B, Bhutia KL et al (2021b) Emerging roles of osmoprotectant glycine betaine against salt-induced oxidative stress in plants: a major outlook of maize (*Zea mays* L.). In: Frontiers in plant-soil. Interaction. Academic, London, pp 567–587. https://doi.org/10.1016/B978-0-323-90943-3.00015-8
Hossain A, Maitra S, Sarker S, Mahmud AA, Ahmad Z, Emon RM, Vemuri H, Malek MA, Alam MA, Rahman MA, Alam MJ, Jahan N, Bhadra P, Moulick D, Saha S, Skalicky M, Brestic M (2023) Aluminium stress in plants: consequences and mitigation mechanisms. In: Tripathi DK, Pandey S, Singh VP, Chauhan DK, Sharma S (eds) Beneficial chemical elements of plants: recent developments and future prospects, 1st edn. Wiley, UK pp 1–46
Hussain HA, Hussain S, Khaliq A, et al (2018) Chilling and drought stresses in crop plants: implications, cross talk, and potential management opportunities. Front Plant Sci 9:393. https://doi.org/10.3389/FPLS.2018.00393
Imran QM, Falak N, Hussain A, Mun B, Yun B (2021) Abiotic stress in plants; stress perception to molecular response and role of biotechnological tools in stress resistance. Agronomy 11:1579. https://doi.org/10.3390/AGRONOMY11081579
IPCC (2007) 2007: Synthesis report. In: Core Writing Team, Pachauri RK, Reisinger A (eds) Contribution of working groups I, II and III to the fourth assessment report of the Intergovernmental Panel on Climate Change. Climate change. IPCC, Geneva, 104 pp
IPCC (2020) Global climate data and information. In: Wrold Bank. From https://climateknowledgeportal.worldbank.org/country/india/climate-data-projections. Accessed 21 Jan 2020
Jaccoud D, Peng K, Feinstein D, Kilian A (2001) Diversity arrays: a solid state technology for sequence information independent genotyping. Nucleic Acids Res 29:E25
Jungers JM, Kaiser DE, Lamb JFS et al (2019) Potassium fertilization affects alfalfa forage yield, nutritive value, root traits, and persistence. Agron J 111:2843–2852. https://doi.org/10.2134/agronj2019.01.0011
Jutsz AM, Gnida A (2015) Mechanisms of stress avoidance and tolerance by plants used in phytoremediation of heavy metals. Arch Environ Prot 41:104–114. https://doi.org/10.1515/AEP-2015-0045
Kaiser DE (2018) Potassium for crop production. UMN extension. Online. UMN Extension. From http://extension.umn.edu/phosphorus-and-potassium/potassium-crop-production
Kaur G, Asthir B (2017) Molecular responses to drought stress in plants. Biol Plant 61:201–209. https://doi.org/10.1007/S10535-016-0700-9

Kazemi S, Eshghizadeh HR, Zahedi M (2018) Responses of four rice varieties to elevated CO_2 and different salinity levels. Rice Sci 25:142–151. https://doi.org/10.1016/j.rsci.2018.04.002
Kelly M (1997) Hubert Horace Lamb (1913–97): climatologist who pioneered the study of climate chang. Nature 388:836
Kennicutt MC, Bromwich D, Liggett D et al (2019) Sustained Antarctic research: a 21st century imperative. One Earth 1:95–113. https://doi.org/10.1016/j.oneear.2019.08.014
Khan MIR, Fatma M, Per TS, Anjum NA, Khan NA (2015) Salicylic acid-induced abiotic stress tolerance and underlying mechanisms in plants. Front Plant Sci 6:462. https://doi.org/10.3389/fpls.2015.00462
Ladha JK, Dawe D, Pathak H et al (2003) How extensive are yield declines in long-term rice-wheat experiments in Asia? Field Crops Res 81:159–180. https://doi.org/10.1016/S0378-4290(02)00219-8
Lafitte HR, Yongsheng G, Yan S, Li ZK (2007) Whole plant responses, key processes, and adaptation to drought stress: the case of rice. J Exp Bot 58:169–175. https://doi.org/10.1093/jxb/erl101
Lal R (2004) Soil carbon sequestration impacts on global climate change and food security. Science 304:1623–1627. https://doi.org/10.1126/science.1097396
Lamaoui M, Jemo M, Datla R, Bekkaoui F (2018) Heat and drought stresses in crops and approaches for their mitigation. Front Chem 6:26. https://doi.org/10.3389/FCHEM.2018.00026
Langstroff A, Heuermann MC, Stahl A, Junker A (2022) Opportunities and limits of controlled-environment plant phenotyping for climate response traits. Theor Appl Genet 135:1–16. https://doi.org/10.1007/s00122-021-03892-1
Larkindale J, Huang B (2004) Thermotolerance and antioxidant systems in *Agrostis stolonifera*: involvement of salicylic acid, abscisic acid, calcium, hydrogen peroxide, and ethylene. J Plant Physiol 161:405–413. https://doi.org/10.1078/0176-1617-01239
Laverdière JP, Lenz P, Nadeau S et al (2022) Breeding for adaptation to climate change: genomic selection for drought response in a white spruce multi-site polycross test. Evol Appl 15:383–402. https://doi.org/10.1111/eva.13348
Lawrie RG, Matus-Cádiz MA, Hucl P (2006) Estimating out-crossing rates in spring wheat cultivars using the contact method. Crop Sci 46:247–249
Li Y, Huang Y, Bergelson J, Nordborg M, Borevitz JO (2010) Association mapping of local climate-sensitive quantitative trait loci in *Arabidopsis thaliana*. Proc Natl Acad Sci U S A 107:21199–21204. https://doi.org/10.1073/pnas.1007431107
Lissbrant S, Berg WK, Volenec J, et al (2009) Phosphorus and potassium fertilization of alfalfa. Purdue University Extension Bulletin. AY-331-W
Lobell DB, Field CB (2007) Global scale climate–crop yield relationships and the impacts of recent warming. Environ Res Lett 2:014002
Lobell DB, Gourdji SM (2012) The influence of climate change on global crop productivity. Plant Physiol 160:1686–1697
Lobell DB, Burke MB, Tebaldi C et al (2008) Prioritizing climate change adaptation needs for food security in 2030. Science 319:607–610
Lockyer WJS (1910) Does the Indian climate change? Nature 84:178
Lovell JT, MacQueen AH, Mamidi S et al (2021) Genomic mechanisms of climate adaptation in polyploid bioenergy switchgrass. Nature 590:438–444. https://doi.org/10.1038/s41586-020-03127-1
Manivannan S, Thilagam VK, Khola OPS (2017) Soil and water conservation in India: strategies and research challenges. J Soil Water Conserv 16:312. https://doi.org/10.5958/2455-7145.2017.00046.7
Matsui T, Omasa K, Horie T (2001) The difference in sterility due to high temperatures during the flowering period among japonica rice varieties. Plant Prod Sci 4:90–93
Mazumder MK, Moulick D, Choudhury S (2022a) Iron (Fe^{3+})-mediated redox responses and amelioration of oxidative stress in cadmium (Cd^{2+}) stressed mung bean seedlings: a biochemical and computational analysis. J Plant Biochem Biotechnol 31:49–60. https://doi.org/10.1007/s13562-021-00654-4

Mazumder MK, Sharma P, Moulick D, Tata SK, Choudhury S (2022b) Salicylic acid ameliorates zinc and chromium-induced stress responses in wheat seedlings: a biochemical and computational analysis. Cereal Res Commun 50:407–418. https://doi.org/10.1007/s42976-021-00201-w

Menezes-Silva PE, Sanglard LMVP, Ávila RT et al (2017) Photosynthetic and metabolic acclimation to repeated drought events play key roles in drought tolerance in coffee. J Exp Bot 68:4309–4322. https://doi.org/10.1093/JXB/ERX211

Mondal M, Skalicky M, Garai S et al (2020) Supplementing nitrogen in combination with rhizobium inoculation and soil mulch in peanut (*Arachis hypogaea* L.) production system: part I. Effects on productivity, soil moisture, and nutrient dynamics. Agronomy 10:1582. https://doi.org/10.3390/agronomy10101582

Mondal M, Biswas B, Garai S et al (2021) Raising climate-resilient embolden rice (*Oryza sativa* L.) seedlings during the cool season through various types of nursery bed management. Sustainability 13:12910

Moulick D, Ghosh D, Chandra Santra SC (2016) Evaluation of effectiveness of seed priming with selenium in rice during germination under arsenic stress. Plant Physiol Biochem 109:571–578. https://doi.org/10.1016/j.plaphy.2016.11.004

Moulick D, Santra SC, Ghosh D (2017) Seed priming with Se alleviate as induced phytotoxicity during germination and seedling growth by restricting as translocation in rice (*Oryza sativa* L. c.v. IET-4094). Ecotoxicol Environ Saf 145:449–456. https://doi.org/10.1016/j.ecoenv.2017.07.060

Moulick D, Santra SC, Ghosh D (2018a) Effect of selenium induced seed priming on arsenic accumulation in rice plant and subsequent transmission in human food chain. Ecotoxicol Environ Saf 152:67–77. https://doi.org/10.1016/j.ecoenv.2018.01.037

Moulick D, Santra SC, Ghosh D (2018b) Rice seed priming with se: a novel approach to mitigate as induced adverse consequences on growth, yield and as load in brown rice. J Hazard Mater 355:187–196. https://doi.org/10.1016/j.jhazmat.2018.05.017

Moulick D, Santra SC, Ghosh D (2018c) Seed priming with se mitigates as-induced phytotoxicity in rice seedlings by enhancing essential micronutrient uptake and translocation and reducing as translocation. Environ Sci Pollut Res Int 25:26978–26991. https://doi.org/10.1007/s11356-018-2711-x

Moulick D, Ghosh D, Mandal J, Bhowmick S, Mondal D, Choudhury S, Santra SC, Vithanage M, Biswas JK (2023) A cumulative assessment of plant growth stages and selenium supplementation on arsenic and micronutrients accumulation in rice grains. J Cleaner Product 135764. https://doi.org/10.1016/j.jclepro.2022.135764

Nagarajan S, Jagadish SVK, Prasad ASH et al (2010) Local climate affects growth, yield and grain quality of aromatic and non-aromatic rice in northwestern India. Agric Ecosyst Environ 138:274–281. https://doi.org/10.1016/J.AGEE.2010.05.012

NOAA National Centers for Environmental Information, State of the Climate. Global Climate Report for 2021, online January 2021. Accessed 20 Oct 2022. From https://www.ncdc.noaa.gov/sotc/global/202113

Nolte CG, Gilliland AB, Hogrefe C, Mickley LJ (2008) Linking global to regional models to assess future climate impacts on surface ozone levels in the United States. J Geophys Res 113:D14307, https://doi.org/10.1029/2007JD008497.

O'Connell E (2017) Towards adaptation of water resource systems to climatic and socio-economic change. Water Resour Manag 31:2965–2984. https://doi.org/10.1007/S11269-017-1734-2

Oosterhuis DM, Loka DA, Kawakami EM, Pettigrew WT (2014) The physiology of potassium in crop production. Adv Agron Elsevier 126:203–233. https://doi.org/10.1016/B978-0-12-800132-5.00003-1

Ortiz AMD, Outhwaite CL, Dalin C, Newbold T (2021) A review of the interactions between biodiversity, agriculture, climate change, and international trade: research and policy priorities. One Earth 4:88–101

Osakabe Y, Osakabe K, Shinozaki K, Tran LSP (2014) Response of plants to water stress. Front Plant Sci 5:86. https://doi.org/10.3389/FPLS.2014.00086

Panahyan-e-Kivi M, Ebadi A, Tobeh A, Jamaati-e-Somarin SH (2009) Evaluation of yield and yield components of lentil genotypes under drought stress. Res J Environ Sci 3(4):456–460
Panda RK, Behera SK, Kashyap PS (2003) Effective management of irrigation water for wheat under stressed conditions. Agric Water Manag 63:37–56
Pandey R, Zinta G, AbdElgawad H et al (2015) Physiological and molecular alterations in plants exposed to high [CO_2] under phosphorus stress. Biotechnol Adv 33:303–316. https://doi.org/10.1016/J.BIOTECHADV.2015.03.011
Pautasso M, Döring TF, Garbelotto M, Pellis L, Jeger MJ (2012) Impacts of climate change on plant diseases—opinions and trends. Eur J Plant Pathol 133:295–313
Peng S, Huang J, Sheehy JE et al (2004) Rice yields decline with higher night temperature from global warming. Proc Natl Acad Sci U S A 101:9971–9975
Pereira A (2016) Plant abiotic stress challenges from the changing environment. Front Plant Sci 7:1123. https://doi.org/10.3389/FPLS.2016.01123
Puijalon S, Bouma TJ, Douady CJ et al (2011) Plant resistance to mechanical stress: evidence of an avoidance-tolerance trade-off. New Phytol 191:1141–1149. https://doi.org/10.1111/J.1469-8137.2011.03763.X
Qasim MZ, Hammad HM, Abbas F et al (2020) The potential applications of picotechnology in biomedical and environmental sciences. Environ Sci Pollut Res Int 27:133–142
Raza A, Razzaq A, Mehmood SS et al (2019) Impact of climate change on crops adaptation and strategies to tackle its outcome: a review. Plan Theory 8. https://doi.org/10.3390/PLANTS8020034
Rivero RM, Mittler R, Blumwald E, Zandalinas SI (2022) Developing climate-resilient crops: improving plant tolerance to stress combination. Plant J 109:373–389. https://doi.org/10.1111/tpj.15483
Roy A, Purkaystha S, Bhattacharyya S (2021) Advancement in molecular and fast breeding programs for climate-resilient agriculture practices. In: Husen A (ed) Harsh environment and plant resilience, Springer, Singapore. https://doi.org/10.1007/978-3-030-65912-7_4
Saha B, Chowardhara B, Kar S et al (2019) Advances in heavy metal-induced stress alleviation with respect to exogenous amendments in crop plants. In: Priming and pretreatment of seeds and seedlings. Springer, Singapore, pp 313–332. https://doi.org/10.1007/978-981-13-8625-1_15
Sahoo S, Borgohain P, Saha B et al (2019) Seed priming and seedling pretreatment induced tolerance to drought and salt stress: recent advances. Priming and pretreatment of seeds and seedlings. Springer, Singapore, pp 253–263. https://doi.org/10.1007/978-981-13-8625-1_12
Salvucci ME, Crafts-Brandner SJ (2004) Inhibition of photosynthesis by heat stress: the activation state of RuBisCO as a limiting factor in photosynthesis. Physiol Plant 120:179–186
Samui I, Skalicky M, Sarkar S et al (2020) Yield response, nutritional quality and water productivity of tomato (*Solanum lycopersicum* L.) are influenced by drip irrigation and straw mulch in the coastal saline ecosystem of Ganges Delta, India. Sustainability 12:6779. https://doi.org/10.3390/su12176779
Sanghera GS, Wani SH, Hussain W, Singh NB (2011) Engineering cold stress tolerance in crop plants. Curr Genomics 12:30–43. https://doi.org/10.2174/138920211794520178
dos Santos TB, Ribas AF, de Souza SGH, Budzinski IGF, Domingues DS (2022) Physiological responses to drought, salinity, and heat stress in plants: a review. Stress 2:113–135. https://doi.org/10.3390/stresses2010009
Sarkar S, Ghosh A, Brahmachari K (2020) Application of APSIM model for assessing the complexities of rice-based cropping systems of South-Asia. In: Maitra S, Pramanick B (eds) Advanced agriculture, 1st edn. New Delhi Publishers, New Delhi, pp 212–233
Sarkar T, Roy A, Choudhary SM, Sarkar SK (2021) Impact of climate change and adaptation strategies for fruit crops. In: Islam van Amstel MNA (ed) Climate change impacts, India, mitigation and adaptation in developing countries. Springer, Berlin. https://doi.org/10.1007/978-3-030-67865-4_4

Sarkar S, Gaydon DS, Brahmachari K et al (2022) Testing APSIM in a complex saline coastal cropping environment. Environ Model Softw 147:105239. https://doi.org/10.1016/j.envsoft.2021.105239

Scafaro AP, Bautsoens N, den Boer B, Van Rie J, Gallé A (2019) A conserved sequence from heat-adapted species improves RuBisCO Activase thermostability in wheat. Plant Physiol 181:43–54. https://doi.org/10.1104/PP.19.00425

Schlesinger WH (1985) Changes in soil carbon storage and associated properties with disturbance and recovery. In: Trabalha JR, Reichle DE (eds) The changing carbon cycle: a global analysis. Springer, New York, pp 194–220

Seleiman MF, Al-Suhaibani N, Ali N et al (2021) Drought stress impacts on plants and different approaches to alleviate its adverse effects. Plan Theory 10:1–25. https://doi.org/10.3390/PLANTS10020259

Shabbaj II, Madany MMY, Balkhyour MA, Tammar A, AbdElgawad H (2022) CO_2 enrichment differentially upregulated sugar, proline, and polyamine metabolism in young and old leaves of wheat and sorghum to mitigate indium oxide nanoparticles toxicity. Front Plant Sci 13. https://doi.org/10.3389/fpls.2022.843771

Shakun JD, Clark PU, He F et al (2012) Global warming preceded by increasing carbon dioxide concentrations during the last deglaciation. Nature 484:49–54. https://doi.org/10.1038/nature10915

Sharma N, Reinke R, Sacks EJ (2021) Comparison of methods to evaluate rice (*Oryza sativa*) germplasm for tolerance to low temperature at the seedling stage. Agronomy 11:385

Shrivastava P, Kumar R (2015) Soil salinity: a serious environmental issue and plant growth promoting bacteria as one of the tools for its alleviation. Saudi J Biol Sci 22:123–131. https://doi.org/10.1016/J.SJBS.2014.12.001

Solanke AU, Sharma AK (2008) Signal transduction during cold stress in plants. Physiol Mol Biol Plants 14:69–79

Sombrero A, de Benito A (2010) Carbon accumulation in soil. Ten-year study of conservation tillage and crop rotation in a semi-arid area of Castile-Leon. Spain Soil Till Res 107:64–70. https://doi.org/10.1016/j.still.2010.02.009

Sulmon C, Van Baaren J, Cabello-Hurtado F et al (2015) Abiotic stressors and stress responses: what commonalities appear between species across biological organization levels? Environ Pollut 202:66–77. https://doi.org/10.1016/J.ENVPOL.2015.03.013

Sun S, Fang J, Lin M et al (2021) Comparative metabolomic and transcriptomic studies reveal key metabolism pathways contributing to freezing tolerance under cold stress in kiwifruit. Front Plant Sci 12:628969. https://doi.org/10.3389/FPLS.2021.628969

Tahar A, Ali B, Hwary EI, Yagoub S (2011) Effect of skipping irrigation on growth, yield, yield components and water use efficiency of wheat in semi-arid region of Sudan. Agric Biol J N Am 2:1003–1009

Talanova VV, Akimova TV, Titov AF (2003) Effect of whole plant and local heating on the ABA content in cucumber seedling leaves and roots and on their heat tolerance. Russ J Plant Physiol 50:90–94. https://doi.org/10.1023/A:1021996703940

Tao F, Yokozawa M, Xu Y, Hayashi Y, Zhang Z (2006) Climate changes and trends in phenology and yields of field crops in China, 1981–2000. Agric Forest Meteorol 138:82–92

Turk MA, Tawaha AM (2001) Wheat response to 2, 4-D application at two growth stages under semi-arid conditions. Acta Agronomica Hungarica 49(4):387–391

USGCRP (2017) Climate science special report: fourth national climate assessment [Wuebbles DJ, Fahey DW, Hibbard KA, Dokken DJ, Stewart BC, Maycock TK (eds)], Volume I. U. S. Global Change Research Program, Washington, DC, 470 pp, https://doi.org/10.7930/J0J964J6

Vedwan N, Rhoades RE (2001) Climate change in the Western Himalayas of India: a study of local perception and response. Clim Res 19:109–117

Vega CRC, Andrade FH, Sadras VO, Uhart SA, Valentinuz OR (2001) Seed number as a function of growth. A comparative study in soybean, sunflower, and maize. Crop Sci 41:748–754

Velikova V, Várkonyi Z, Szabó M et al (2011) Increased thermostability of thylakoid membranes in isoprene-emitting leaves probed with three biophysical techniques. Plant Physiol 157:905–916. https://doi.org/10.1104/PP.111.182519

Verma S, Kumar N, Verma A et al (2020) Novel approaches to mitigate heat stress impacts on crop growth and development. Plant Physiol Rep 25:627–644. https://doi.org/10.1007/s40502-020-00550-4

Verslues PE, Agarwal M, Katiyar-Agarwal S, Zhu J, Zhu JK (2006) Methods and concepts in quantifying resistance to drought, salt and freezing, abiotic stresses that affect plant water status. Plant J 45:523–539. https://doi.org/10.1111/j.1365-313X.2005.02593.x

Vollenweider P, Günthardt-Goerg MS (2005) Diagnosis of abiotic and biotic stress factors using the visible symptoms in foliage. Environ Pollut 137:455–465. https://doi.org/10.1016/J.ENVPOL.2005.01.032

Wada S, Miyake C, Makino A, Suzuki Y (2020) Photorespiration coupled with CO_2 assimilation protects photosystem I from photoinhibition under moderate poly(ethylene glycol)-induced osmotic stress in rice. Front Plant Sci 11:1121. https://doi.org/10.3389/FPLS.2020.01121

Wahid A, Shabbir A (2005) Induction of heat stress tolerance in barley seedlings by pre-sowing seed treatment with glycinebetaine. Plant Growth Regul 46:133–141. https://doi.org/10.1007/S10725-005-8379-5

Wang L, Ma R, Yin Y, Jiao Z (2018a) Role of carbon ion beams irradiation in mitigating cold stress in *Arabidopsis thaliana*. Ecotoxicol Environ Saf 162:341–347. https://doi.org/10.1016/J.ECOENV.2018.07.013

Wang L-J, Zhang P, Wang R-N, Wang P, Huang S (2018b) Effects of variety and chemical regulators on cold tolerance during maize germination. J Integr Agric 17:2662–2669. https://doi.org/10.1016/S2095-3119(17)61880-X

Waraich EA, Ahmad R, Halim A, Aziz T (2012) Alleviation of temperature stress by nutrient management in crop plants: a review. J Soil Sci Plant Nutr 12:221–244. https://doi.org/10.4067/S0718-95162012000200003

Wigge PA (2011) FT, a mobile developmental signal in plants. Curr Biol 21:R374–R378. https://doi.org/10.1016/j.cub.2011.03.038

Williams SE, Shoo LP, Isaac JL, Hoffmann AA, Langham G (2008) Towards an integrated framework for assessing the vulnerability of species to climate change. PLoS Biol 6:2621–2626. https://doi.org/10.1371/journal.pbio.0060325

Windeatt JH, Ross AB, Williams PT et al (2014) Characteristics of biochars from crop residues: potential for carbon sequestration and soil amendment. J Environ Manag 146:189–197. https://doi.org/10.1016/j.jenvman.2014.08.003

Wu G, Zhang C, Chu LY, Shao HB (2007) Responses of higher plants to abiotic stresses and agricultural sustainable development. J Plant Interact 2:135–147. https://doi.org/10.1080/17429140701586357

Xiong W, Reynolds MP, Crossa J et al (2021) Increased ranking change in wheat breeding under climate change. Nat Plants 7:1207–1212. https://doi.org/10.1038/s41477-021-00988-w

Yadav SK (2010) Cold stress tolerance mechanisms in plants. A review. Agron Sustain Dev 30:515–527. https://doi.org/10.1051/agro/2009050

Zandalinas SI, Fritschi FB, Mittler R (2020) Signal transduction networks during stress combination. J Exp Bot 71:1734–1741. https://doi.org/10.1093/JXB/ERZ486

Zarka DG, Vogel JT, Cook D, Thomashow MF (2003) Cold induction of Arabidopsis CBF genes involves multiple ICE (inducer of CBF expression) promoter elements and a cold-regulatory circuit that is desensitized by low temperature. Plant Physiol 133:910–918. https://doi.org/10.1104/PP.103.027169

Zhang X, Pie D, Li Z, Li J, Wang Y (2002) Management of supplemental irrigation of winter wheat for maximum profit. Deficit Irrig Prac 22:57–66

Zhang Q, Wang C, Tian J, Li K, Shou H (2011) Identification of rice purple acid phosphatases related to phosphate starvation signalling. Plant Biol (Stuttg) 13:7–15. https://doi.org/10.1111/j.1438-8677.2010.00346.x

some, such as cobalt and silicon, being considered beneficial elements (Ma 2005). The mineral nutrients supplied by the soil play significant roles in the entire plant life cycle. Plants absorb minerals via roots and transfer them to the other plant parts to maintain their metabolism. Thus, mineral nutrients constitute a major limit for plant growth, and certain environments, such as the tropics, have extremely limited content of nutrients, impacting crop productivity (Dotaniya and Meena 2015). When a plant is challenged with one or multiple mineral deficiencies, appropriate strategies are triggered to increase its utilization efficiency. Among these strategies, first, a plant regulates the expression of genes related to nutrient-specific transporters in its roots and modulates its root architecture, resulting in enhanced root growth and increased branching for expanding the contact area with the soil (Gojon et al. 2009; Giehl and von Wirén 2014). To absorb increased amounts of mineral nutrients from the soil, certain plants also develop a relationship with microbes in the rhizosphere by secreting exudates through their roots (Dotaniya and Meena 2015). In most situations, a plant faces complex environmental stress that impacts mineral use efficiency. Climate change has increased the frequency of extreme weather events and extended dry periods in many cultivated lands. As a result, a long drought imposes dry stress, which is also associated with salinity stress on the crop (Hu and Schmidhalter 2005). Drought and salinity decrease the nutrient uptake ability of the plant: (1) transpiration rates, repair capacity of transporters, and membrane permeability are decreased under drought stress (Alam 1999); (2) lower soil water content decreases the diffusion rate and root-effective absorbing area of minerals (Pinkerton and Simpson 1986; Alam 1999); (3) higher Na^+ concentration causes specific toxicities and disturbance in ionic balances, ultimately impacting plant growth (Hu and Schmidhalter 2005).

Among soil properties, the two most important are the composition of organic compounds and pH, both of which affect nutrient use efficiency and thus are directly related to plant yields (Sanchez et al. 1997; Herencia et al. 2008). The opposite relationship between the solubility of minerals and organic compounds is important for forming different soil properties (Tittonell et al. 2005; Herencia et al. 2008). The causes behind the variability in soil properties are complex, including location, climate, fertility management, and several other factors (Tittonell et al. 2005). Maintaining abundant levels of organic compounds in soil contributes to agricultural sustainability (Whitbread et al. 2003). Organic compounds form stable complexes with metals through their functional groups, concentrating various metals in the soil (Herencia et al. 2008). To achieve increased crop production and agricultural lands of higher quality products, higher input of chemical fertilizers rather than organic fertilizers is preferred, leading to decreased soil quality and disturbances in the balance of soil organic matter (Liu et al. 2009; Savci 2012). Specifically, P is considered a nonrenewable resource and is expected to be exhausted in the next 50–100 years (Cordell et al. 2009). Thus, improving mineral use efficiency is critical for modern agriculture. The application of organic amendments improved the availability of minerals, as suggested in a previous study (Madrid 1999). Thus, using soil amendments, such as biochar and compost, to improve soil quality can improve nutrient use efficiency. Both the nutrient function and metabolism have been described in several previous studies and reviews (Amtmann and Armengaud 2009; Hänsch and Mendel 2009). This study reviews and discusses (1) the means to improve nutrient use efficiency, especially for N, P, and K, which strongly affect crop productivity and quality, and (2) the effect of climate change on agriculture.

2 Nutrient Uptake Strategies of Plants

Mineral nutrients, including N, K, P, calcium (Ca), magnesium (Mg), sulfur (S), boron (B), chloride (Cl), iron (Fe), copper (Cu), manganese (Mn), molybdenum (Mo), zinc (Zn), and nickel (Ni) are essential for plant growth. In addition, some minerals, such as sodium (Na), cobalt (Co), aluminum (Al), and silicon (Si), have specific functions in some plants and are thus considered to be beneficial elements (Pilon-Smits et al. 2009). Furthermore, agricultural production requires abundant mineral nutrients to maintain productivity and address the increasing food demand (Meena and Meena 2017). Therefore, enhancing nutrient use efficiency is a potential approach for achieving adequate crop productivity for the growing population worldwide (Meena and Meena 2017).

2.1 Nitrogen (N)

In plants, nitrogen is mainly found in the photosynthetic machinery, where it plays a role in carbon fixation, whereas, during maturation, nitrogen is translocated to the grain (Hawkesford 2012). Improving nitrogen use efficiency (NUE) leads to less environmental degradation and increased crop productivity with precisely the same N input to croplands. Therefore, NUE has become an indicator of achieving sustainable development goals (Zhang et al. 2015). The Haber–Bosh process, which produces mineral N fertilizers, is highly energy-consuming (Erisman et al. 2008). However, N has become an affordable fertilizer for most farmers, with the abundant application of N guaranteeing an improved yield (Kant 2018). There are several forms of N, and nitrate (NO_3^-) is preferentially absorbed by most cereals (Kant et al. 2011). However, not all of the N fertilizer applied to the field is utilized by plants, resulting in 50–75% being lost in the soil via leaching (Raun and Johnson 1999; Hodge et al. 2000; Asghari and Cavagnaro 2011). Thus, improving NUE is a critical issue in terms of improving crop production and saving energy. However, there are several factors involving NUE (Fig. 2.1).

Four families of transporters/channels contribute to nitrate transport: nitrate transporter 1/peptide (NPF), nitrate transporter 2 (NRT2), channel homolog, and chloride channel families (Léran et al. 2014; Noguero and Lacombe 2016; O'Brien et al. 2016). The NPF and NRT2 transporter families, which perform their function in the roots, are mainly involved in the transport of nitrate in plants (Kant 2018). Compared with the NPF family consisting of approximately 50–139 genes, the

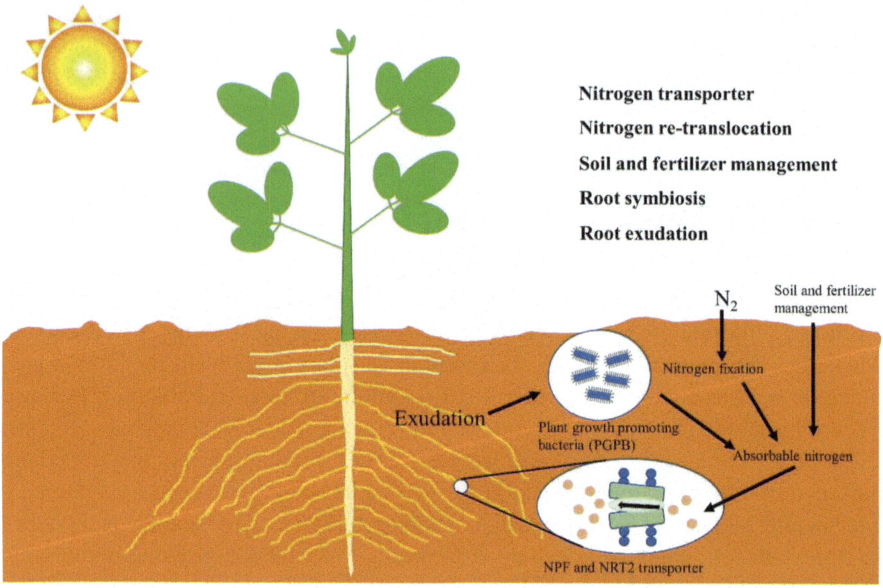

Fig. 2.1 Schematic representation of factors that influence N use efficiency (NUE)

NRT2 transporter family has only eight members with a high affinity for nitrate in plants (Léran et al. 2014; Von Wittgenstein et al. 2014). Besides their role in nitrate transport, transporters NPF6.3 and NRT2.1 are considerably important for nitrate signaling and sensing (Bouguyon et al. 2012). These two transporters influence each other and regulate the root system architecture by enhancing lateral root development (Little et al. 2005; Remans et al. 2006; Krouk et al. 2010). Thus, understanding the mechanisms by which the plant responds to nitrate is critical for increasing NUE. Unfortunately, commercially high NUE lines based on the modulation of the expression of genes involved in N uptake, transport, and assimilation are unavailable for breeding to improve the yield (McAllister et al. 2012; Kant 2018). Although changing individual genes may not improve NUE, understanding all processes involved in N uptake, transport, assimilation, and translocation to sink tissues is needed to improve NUE (Kant 2018).

Some microorganisms form symbiotic systems with plants when the latter is in a state of nutrient imbalance (Tikhonovich and Provorov 2011). The most known model is the legume-rhizobia nodular symbiosis that can access extra N using biological N_2 fixation (Franche et al. 2009). The efficient utilization of N_2 fixation was unexpectedly identified in wheat and sugar cane plantations in Brazil. Notably, N_2 fixation provides up to 30% of the required N in wheat and up to 150 kg N/ha/year for sugar cane fields (Döbereiner 1997). Thus, breeding or using genetically modified plants with more efficient symbiotic systems with microorganisms might improve NUE, resulting in increased crop productivity. Several groups of bacteria known as plant growth-promoting bacteria (PGPB), such as rhizobium, contribute to the nutritional uptake of plants, facilitating plant growth (Kraiser et al. 2011). PGPBs improve nutrient use efficiency in various ways. Some PGPBs produce molecules, such as indole acetic acid, gibberellins, and cytokinin, that modulate the levels of phytohormones in the plant (Kraiser et al. 2011). For example, 1-aminocyclopropane-carboxylate produced by PGPB was shown to reduce the level of ethylene and unrestrict the inhibition of root and seedling growth (Glick et al. 1998). Mantelin et al. (2006) reported that the PGPR *Phyllobacterium* strain STM196 facilitated overcoming high external nitrate-induced inhibition in *Arabidopsis* lateral roots and increased the root volume, thus enhancing plant growth under different N concentrations. In a previous study, the application of bacteria belonging to the *Achromobacter* genus (PGPB) in the cultivation of *Brassica napus* also improved nitrate uptake by plants (Bertrand et al. 2000). Thus, understanding the relationship between bacteria and plants may contribute to the development of improved NUE approaches with reduced cost.

The fertilizer type used is an important factor affecting NUE. For instance, mineral fertilizers are substantially used as a means of N supply to crops in developed countries (Robertson and Vitousek 2009). Commercial mineral N fertilizers are mainly composed of ammonium nitrate, ammonium sulfate, urea, and anhydrous ammonia, which are soluble and easily absorbed and assimilated by crops (Hirel et al. 2011). However, these mineral N fertilizers also easily leach into the groundwater before the mineral N is converted to nitrate (Hirel et al. 2011). Thus, to improve NUE for crops, mineral N should be used at the proper time and place (Jarvis et al. 2011). Manure is the second most used fertilizer in croplands, except mineral N fertilizers. However, over half of the amount of manure is lost during transport and storage, and of the remaining amount, 25% is lost during crop cultivation (Hirel et al. 2011). Nonetheless, using manure as a nutrient supply for crops improves the physical and chemical properties and biological activity of soil (Watson et al. 2002). To improve NUE, various technologies have been implemented to minimize manure loss and improve manure quality, such as providing cover to manure during storage, reducing soiled surface areas in animal houses, incineration of solid manures, and reducing the pH of slurry or separating the slurry (Jarvis et al. 2011).

2.2 Phosphorus (P)

Phosphorus (P) is involved in many important physiological functions. As a major structural component of nucleic acids, such as DNA and RNA, and membrane lipids, P plays a role in all aspects of life (Amtmann and Armengaud 2009). Notably, P has low mobility and concentration in soil, substantially limiting plant growth and crop yield (Han et al. 2022). Although P is heavily used in modern agriculture to maintain the productivity of crops, only up to 30% of input P is utilized by plants (Cordell et al. 2009; Han et al. 2022). Moreover, the unabsorbed P is lost, ending up and impacting the aquatic ecosystems (Zak et al. 2018). Compared with other

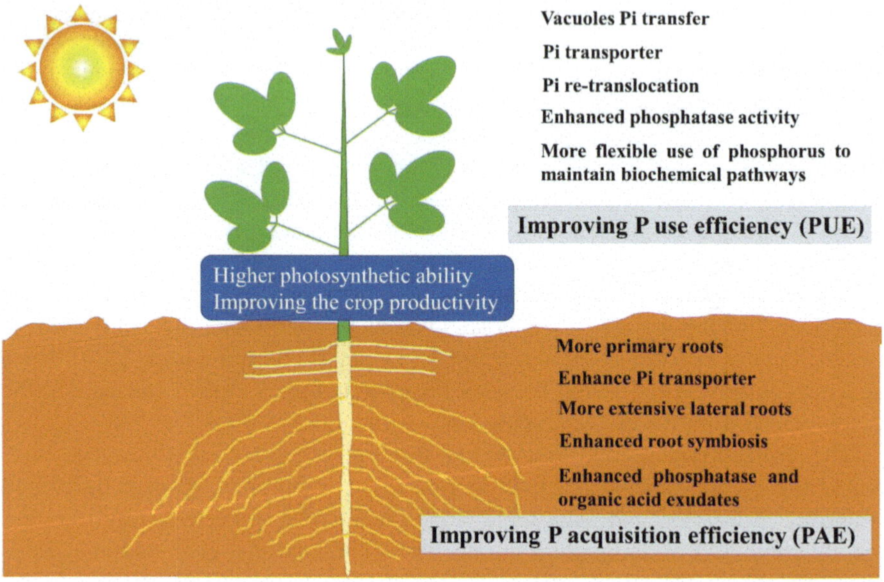

Fig. 2.2 Schematic representation of improvements in P utilization efficiency. Enhancing the P acquisition efficiency (PAE) and P use efficiency (PUE) is important for improving crop productivity

macronutrients, P is nonrenewable and could be depleted in the next 50–100 years (Cordell et al. 2009). Using industry data, Cordell et al. (2009) made a conservative analysis that showed that P fertilizers have a similar production pattern to oil: phosphorus production could reach its peak in 2033, after which it will be significantly reduced. Thus, increasing the efficient use of P in the future is urgently needed.

Improving P use efficiency (PUE; Fig. 2.2) can be accomplished in the following ways: (1) plants can enhance their P acquisition efficiency (PAE) by increasing the uptake of P from soil; and (2) plants can enhance internal PUE, which could improve plant growth and accelerate conversion into harvestable products (Veneklaas et al. 2012; Han et al. 2022). Various studies have focused on the mechanisms by which plants can improve PAE. These mechanisms include: (1) secreting more organic anions and phosphatase and decreasing the pH of the rhizosphere to increase the mobility of P from organic and inorganic P sources (Jain et al. 2007; George et al. 2011; Han et al. 2022); (2) modifying the root architecture, such as enhancing root growth and increasing the number of lateral roots to increase the contact area with the soil and facilitate the uptake of more P (Wang et al. 2010b; George et al. 2011; Jia et al. 2018; Dong et al. 2020); (3) modulating the P transport systems of root cells and enhancing the uptake ability under low P conditions (Bucher 2007; Jain et al. 2007; George et al. 2011); and (4) enhancing the interaction with microorganisms, such as mycorrhizal fungi to increase the ability of plants to explore the soil media and increase P mobility (Bucher 2007; George et al. 2011). Improving PAE results in increased crop yields and reduced P losses. However, increased accumulation of P in the plant biomass can exhaust the P source, resulting in almost all P being stored in grains in the form of phytate (Han et al. 2022; Dong et al. 2022). In particular, monogastric animals cannot utilize phytate, and a large amount of P is thus directly released to arable regions, causing eutrophication and P pollution (Dong et al. 2020; Han et al. 2022). Thus, the most sustainable way to improve P use efficiency and maintain the productive agricultural system is to enhance PUE (Veneklaas et al. 2012).

Briefly, PUE is defined as the increase in crop production per unit of input P source (Hammond et al. 2009). Improved utilization refers to more efficient use of the internal P of the plant, resulting in higher crop productivity compared with that of normal commercial crops. Plants have various strategies for using the internal P smartly under limited external P conditions. External P exists in plant cells as free inorganic orthophosphate (Pi) or organic P, with Pi being affected by the external Pi supply (Veneklaas et al. 2012). When a plant grows under P deficiency conditions, it translocates the P from older leaves to supply the photosynthesis in young leaves while maintaining the nutrient uptake by roots during the vegetative stage (Veneklaas et al. 2012; Han et al. 2022). As young leaves have a low abundance of ribosomal RNA (rRNA), use less P to generate rRNA, and have low activities of Calvin–Benson cycle enzymes and reduced content of photosynthetic machinery (delayed greening), they may contribute to P use efficiency (Sulpice et al. 2014). In addition, delayed greening conserves P, allowing young leaves to produce more cells, form stable structures, and invest in chloroplast development, thus protecting young leaves from photodamage and oxidation (Kuppusamy et al. 2020). These findings revealed that Pi is preferentially distributed to new leaves to improve PUE and contribute to carbon metabolism, especially in plants under P deficiency (Han et al. 2022). The vacuole is considered a major Pi pool and provides most of the Pi supply in the form of cellular Pi buffering for maintaining cellular metabolism (Veneklaas et al. 2012; Liu et al. 2015; Han et al. 2022). Several studies have identified the tonoplast Pi transporter, which is involved in the vacuolar accumulation of polyphosphate in some organisms, such as yeast, *Arabidopsis,* and rice (Liu et al. 2015; Wang et al. 2015). The yeast has four protein complexes, ScVTC1, 2, 3, and 4, responsible for vacuolar P accumulation, with ScPHO91 responding under low P conditions (Hürlimann et al. 2007; Secco et al. 2012). In *Arabidopsis*, mutations in *AtVPT1,* which are involved in vacuolar Pi transport, caused lower vacuolar Pi concentrations, and reduced vacuole/cytoplasmic Pi ratio (Liu et al. 2015; Liu et al. 2016b). In rice, a high-affinity Pi transporter *OsPht1;8* was shown to transport Pi from old to young leaves, and silencing of this gene led to an increased accumulation of P in older blades under P-deficient conditions (Li et al. 2015). The development of high PUE crops via transgenic approaches has been discussed in previous reviews (Wang et al. 2010a; Wang et al. 2010b; Veneklaas et al. 2012; Han et al. 2022). Various studies have revealed that adjusting the P transporter in tissues and modulating Pi mobilization can improve the PUE of crops. In soybean, *GmPT1* is involved in Pi transport; overexpression of *GmPT1* enhances the remobilization ability of Pi from older to young leaves under P deficiency conditions, improving the soybean yield (Song et al. 2014). The OsVPE1 and OsVPE2 are Pi efflux transporters in rice; overexpression of both transporters was reported to reduce the amount of vacuolar Pi and increase its release into the cytoplasm to maintain cellular Pi requirements under P deficiency conditions (Xu et al. 2019). To avoid the contamination of agricultural lands by genetically modified crops, developing new crop lines with high PUE is necessary. Wissuwa et al. (2015) reported that several genetic loci are involved in PUE and contribute to improved plant growth in field systems under P-limited conditions.

2.3 *Potassium (K)*

Although K is not metabolized, it is essential for various metabolic processes, such as transport, translation, and direct activation of enzymes (Amtmann and Armengaud 2009). After nitrogen, K is the most abundant essential nutrient in plants and plays a role in crop quality and yield (Wang et al. 2021). The K content in plants ranges from 0.8% to 8%, with the cytosolic K concentration usually maintained up to 200 mM (Maathuis 2009). Unlike N and P, which form complex molecular structures, K is only found in the form of soluble K^+; thus, K is more easily leached than N and P in soil (Sardans and Peñuelas 2015). Although abundant amounts of K are stored in the soil, K deficiency still occurs in large agricultural areas (Schroeder 1978; Römheld and Kirkby 2010). Usually, depending on plant availability, the K in soil is classified into four groups (Fig. 2.3): water-soluble, structural form,

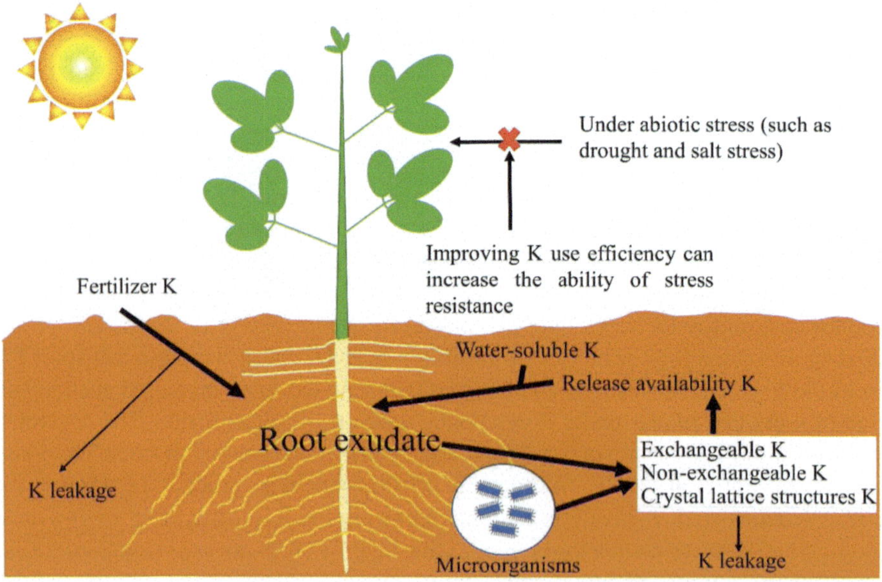

Fig. 2.3 Schematic representation of various K forms in soil and their flow. Improved K use efficiency could improve stress resistance ability

non-exchangeable, and exchangeable (Zörb et al. 2014). Most soil K minerals (90–98%) are found in the form of crystal lattice structures that plants cannot directly absorb, and thus the availability of K is greatly affected by the soil type, its physicochemical properties, and the microbial activities in the rhizosphere (Zörb et al. 2014). K use efficiency can be improved by the inoculation of K-solubilizing microorganisms to croplands that enhances K solubility via the secretion of organic acids, increasing its utilization by plants (Basak and Biswas 2009; Abou-el-Seoud and Abdel-Megeed 2012; Zörb et al. 2014). In addition, the plant secretes root exudates to utilize the non-exchangeable K source; for example, maize secretes oxalic acid depending on the K supply (Kraffczyk et al. 1984; Wang et al. 2011). Mineral forms of K also activate changes in root architecture, such as the enhancement of root growth and length (Wang et al. 2011). Thus, the ability of roots to uptake K markedly affects K use efficiency. Regarding their K acquisition ability, different plant species have varied K utilization efficiency (Guoping et al. 1999). In addition, valuable genetic loci involved in the efficient utilization of K have been identified in a previous study (Moriconi and Santa-María 2013) and could be used to improve the amounts of available K in the soil. In *Arabidopsis*, two transmembrane proteins, AtHAK5 and AKT1, are responsible for K^+ transport in the root under low K conditions (Pyo et al. 2010). Another study selected K-efficient genotypes in canola, leading to improved yields under low K conditions (Damon et al. 2007).

Mild K deficiency does not show immediate symptoms in plants because of the high mobility and redistribution between tissues (Römheld and Kirkby 2010). However, it should be clear that the yield production and crop quality depend on the efficiency of K utilization (Pettigrew 2008; Zörb et al. 2014). Previous studies have reported that the corn yield was a response to K fertility (Ebelhar and Varsa 2000), and the soybean plant weight and grain yield were increased by K application (Bharati et al. 1986). With the new exploration of K function, K affected most biotic and abiotic stresses (Zörb et al. 2014). Among the physiological and metabolic functions of K, one of the functions is the maintenance of osmotic pressure and cellular turgor (White 2013; Sardans and Peñuelas 2015). Although many compounds exist in plants to execute osmotic adjustment, such as amino acids, sugar, and alcohols (Hu and Schmidhalter 2005), energy for synthesis is more costly for these compounds than uptake of K directly (Zörb et al. 2014). With the increase in extreme weather worldwide, dryness will greatly limit crop production. Abundant K could increase the resistance ability of crops under water stress (Grzebisz et al. 2013). Because K is considered an osmoticum, the opening of stomata depends on the water influx of the guard cell vacuole, which is driven by osmotic pressure (Peiter 2011). Thus, the supply of K can highly affect the CO_2 assimilation of plants under high light stress (Kwak et al. 2001). Meanwhile, salt stress also impacts a third of the cropland area worldwide (Zörb et al. 2014). Na is toxic to plants because Na can displace K, leading to the inactivation of K-dependent enzymes and massive leakage of K from roots (Munns and Tester 2008; Shabala and Cuin 2008). High external K could reduce Na influx, and a high K/Na ratio is an important enzymatic function (Demidchik and Maathuis 2007; Shabala and Cuin 2008). K is involved in the detoxification function (Wang et al. 2013). Under drought, the K-deficient plants produce more reactive oxygen species (ROS) during CO_2 fixation; however, the reduction of ROS is related to adequate K supply (Sardans and Peñuelas 2015). Thus, improving K use efficiency can also improve plant resistance to various stresses (Fig. 2.3). Nevertheless, genetic and breeding techniques for improving the utilization of K are currently very limited, and further research is required on the mechanisms by which the utilization of K can be improved.

3 Interaction Between Nutrient Use Efficiency in Plants and Climate Change

Agricultural productivity is affected by many factors, including land and crop management and biotic and abiotic conditions. However, among these factors, humans cannot control climate conditions with the current technology. For example, a 1 °C increase in global temperature reduced wheat yield by between 4.1% and 6.4% worldwide (Liu et al. 2016a). The increasing global temperature has also impacted maize, rice, and soybean productivity (Zhao et al. 2017). In addition, climate change-induced global climatic anomalies, such as drought, floods, and heat and cold stresses, have highlighted the impact of geographical variations to crop productivity (Anderson et al. 2020). From the point of view of plant nutrition, these extreme abiotic stresses could impact crop nutrient uptake. We discuss ways of buffering these climate change-induced anomalies that decrease nutrient use efficiency.

Suitable water availability and temperature play a major role in plant growth and seed maturation (Bradford 1994; Teixeira et al. 2013). Drought is caused by long-term extreme high temperatures and water deficit, which impact yield productivity and food security (Hussain et al. 2019). Under drought stress, plants decrease their transpiration rate, stomatal conductance, and plant biomass accumulation (Yokota et al. 2006). Furthermore, water supply affects all metabolic processes and impacts the absorption of nutrient elements from roots, thus significantly decreasing the NUE (Alam 1999). In addition to the problem of insufficient water caused by drought and high temperatures, lack of oxygen and root diseases caused by waterlogging and flooding owing to prolonged heavy rains contribute to reduced crop yield (Anderson et al. 2020). Therefore, to sustain the growing demand for food during climate change, several adaptation strategies must be proposed, such as optimizing land and fertilizer management and selecting and cultivating tolerant crops (Fig. 2.4).

The challenges of land and fertilizer management for farmers include timing and uncertainties during the growing season of crops (Anderson et al. 2020). Although various studies have developed process-based crop models to evaluate the impact of climate change, these dramatic environmental changes have increased the uncertainty of such adaptation strategies (Angulo et al. 2013; Bassu et al. 2014). However, the development of gene editing technology enables us to acquire crops containing specific resistance. However, public acceptance of genetically modified foods is not yet high, and the policy restrictions in various countries limit the application of this

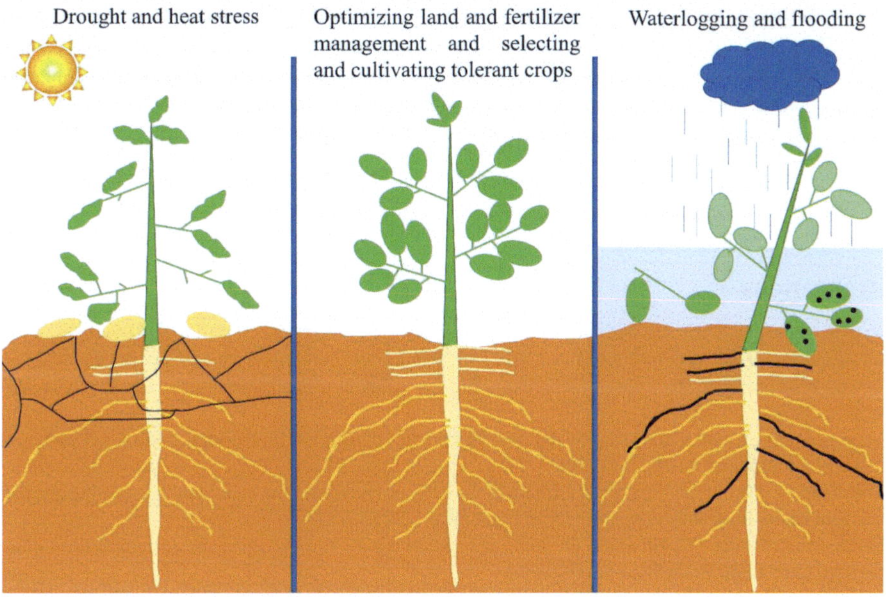

Fig. 2.4 Schematic representation of climate change impact on crop plants. Optimizing land and fertilizer management and selecting and cultivating tolerant crops could maintain crop production

technology (Scheben and Edwards 2018). To overcome the reduction in crop productivity owing to climate change, soil nutrient management, especially those of N, P, and K elements constitute important factors for maintaining crop productivity in developing countries. In addition, selecting and breeding new crop varieties with high yields under stress conditions are important for supporting the growing population (Anderson et al. 2020).

4 Conclusion and Future Perspectives

With the growing global population facing the crisis of depletion of nonrenewable resources, we must improve the efficiency of resource utilization. Improving the efficiency of using highly demanded elements in plants, such as potassium, P, and K will be necessary for overcoming the impact of resource scarcity on crop production in the future. The development of transgenic technology and breeding for the selection of crops that use mineral nutrients more efficiently will contribute to sustainable agriculture. Despite the irreversible global climate change, humans are making efforts to mitigate and adapt to the effects of future weather extremes. However, for crops, the impact of global climate change will be devastating. Various weather extremes will further reduce crop yields and severely impact global food security. Therefore, developing appropriate countermeasures to allow crops to withstand fewer impacts from extreme weather conditions is critical. Currently, we are trying to prevent the effects of extreme weather in advance by managing land and fertilizers more precisely. Genetic and breeding techniques for creating crops that can tolerate extreme weather and maintain their yields will significantly contribute to future food security. As humanity becomes more aware of global climate change, we will develop more technologies to address the coming climate change crisis.

References

Abou-El-Seoud II, Abdel-Megeed A (2012) Impact of rock materials and biofertilizations on P and K availability for maize (*Zea Maize*) under calcareous soil conditions. Saudi J Biol Sci 19:55–63

Alam SM (1999) Nutrient uptake by plants under stress conditions. In: Pessarakli M (ed) Handbook of plant and crop stress. CRC Press, New York, pp 285–313

Amtmann A, Armengaud P (2009) Effects of N, P, K and S on metabolism: new knowledge gained from multi-level analysis. Curr Opin Plant Biol 12:275–283

Anderson R, Bayer PE, Edwards D (2020) Climate change and the need for agricultural adaptation. Curr Opin Plant Biol 56:197–202

Angulo C, Rötter R, Trnka M, Pirttioja N, Gaiser T, Hlavinka P, Ewert F (2013) Characteristic 'fingerprints' of crop model responses to weather input data at different spatial resolutions. Eur J Agron 49:104–114

Asghari HR, Cavagnaro TR (2011) Arbuscular mycorrhizas enhance plant interception of leached nutrients. Funct Plant Biol 38:219–226

Basak B, Biswas D (2009) Influence of potassium solubilizing microorganism (*Bacillus mucilaginosus*) and waste mica on potassium uptake dynamics by Sudan grass (*Sorghum vulgare* Pers.) grown under two Alfisols. Plant Soil 317:235–255

Bassu S, Brisson N, Durand JL, Boote K, Lizaso J, Jones JW, Rosenzweig C, Ruane AC, Adam M, Baron C (2014) How do various maize crop models vary in their responses to climate change factors? Glob Change Biol 20:2301–2320

Bertrand H, Plassard C, Pinochet X, Touraine B, Normand P, Cleyet-Marel J (2000) Stimulation of the ionic transport system in Brassica napus by a plant growth-promoting rhizobacterium (*Achromobacter* sp.). Can J Microbiol 46:229–236

Bharati M, Whigham D, Voss R (1986) Soybean response to tillage and nitrogen, phosphorus, and potassium fertilization. Agron J 78:947–950

Bouguyon E, Gojon A, Nacry P (2012) Nitrate sensing and signaling in plants. Semin Cell Dev Biol 23:648–654

Bradford KJ (1994) Water stress and the water relations of seed development: a critical review. Crop Sci 34:1–11

Bucher M (2007) Functional biology of plant phosphate uptake at root and mycorrhiza interfaces. New Phytol 173:11–26

Cordell D, Drangert JO, White S (2009) The story of phosphorus: global food security and food for thought. Glob Change Biol 19:292–305

Damon P, Osborne L, Rengel Z (2007) Canola genotypes differ in potassium efficiency during vegetative growth. Euphytica 156:387–397

Demidchik V, Maathuis FJ (2007) Physiological roles of nonselective cation channels in plants: from salt stress to signalling and development. New Phytol 175:387–404

Döbereiner J (1997) Biological nitrogen fixation in the tropics: social and economic contributions. Soil Biol Biochem 29:771–774

Dong Q, Echigo K, Raboy V, Saneoka H (2020) Seedling growth, physiological characteristics, nitrogen fixation, and root and nodule phytase and phosphatase activity of a low-phytate soybean line. Plant Physiol Biochem 149:225–232

Veneklaas EJ, Lambers H, Bragg J, Finnegan PM, Lovelock CE, Plaxton WC, Price CA, Scheible WR, Shane MW, White PJ (2012) Opportunities for improving phosphorus-use efficiency in crop plants. New Phytol 195:306–320
Von Wittgenstein NJ, Le CH, Hawkins BJ, Ehlting J (2014) Evolutionary classification of ammonium, nitrate, and peptide transporters in land plants. BMC Evol Biol 14:1–17
Wang X, Shen J, Liao H (2010a) Acquisition or utilization, which is more critical for enhancing phosphorus efficiency in modern crops? Plant Sci 179:302–306
Wang X, Yan X, Liao H (2010b) Genetic improvement for phosphorus efficiency in soybean: a radical approach. Ann Bot 106:215–222
Wang HY, Shen QH, Zhou JM, Wang J, Du CW, Chen XQ (2011) Plants use alternative strategies to utilize nonexchangeable potassium in minerals. Plant Soil 343:209–220
Wang M, Zheng Q, Shen Q, Guo S (2013) The critical role of potassium in plant stress response. Int J Mol Sci 14:7370–7390
Wang C, Yue W, Ying Y, Wang S, Secco D, Liu Y, Whelan J, Tyerman SD, Shou H (2015) Rice SPX-major facility Superfamily3, a vacuolar phosphate efflux transporter, is involved in maintaining phosphate homeostasis in rice. Plant Physiol 169:2822–2831
Wang Y, Chen YF, Wu WH (2021) Potassium and phosphorus transport and signaling in plants. J Integr Plant Biol 63:34–52
Watson CA, Atkinson D, Gosling P, Jackson LR, Rayns F (2002) Managing soil fertility in organic farming systems. Soil Use Manag 18:239–247
Whitbread A, Blair G, Konboon Y, Lefroy R, Naklang K (2003) Managing crop residues, fertilizers and leaf litters to improve soil C, nutrient balances, and the grain yield of rice and wheat cropping systems in Thailand and Australia. Agric Ecosyst Environ 100:251–263
White PJ (2013) Improving potassium acquisition and utilisation by crop plants. J Plant Nutr Soil Sci 176:305–316
Wissuwa M, Kondo K, Fukuda T, Mori A, Rose MT, Pariasca-Tanaka J, Kretzschmar T, Haefele SM, Rose TJ (2015) Unmasking novel loci for internal phosphorus utilization efficiency in rice germplasm through genome-wide association analysis. PLoS One 10:e0124215
Xu L, Zhao H, Wan R, Liu Y, Xu Z, Tian W, Ruan W, Wang F, Deng M, Wang J (2019) Identification of vacuolar phosphate efflux transporters in land plants. Nat Plant 5:84–94
Yokota A, Takahara K, Akashi K (2006) Water stress. In: Rao K, Raghavendra A, Reddy K (eds) Physiology and molecular biology of stress tolerance. Springer, Dordrecht, pp 15–39
Zak D, Kronvang B, Carstensen MV, Hoffmann CC, Kjeldgaard A, Larsen SE, Audet J, Egemose S, Jorgensen CA, Feuerbach P (2018) Nitrogen and phosphorus removal from agricultural runoff in integrated buffer zones. Environ Sci Technol 52:6508–6517
Zhang X, Davidson EA, Mauzerall DL, Searchinger TD, Dumas P, Shen Y (2015) Managing nitrogen for sustainable development. Nature 528:51–59
Zhao C, Liu B, Piao S, Wang X, Lobell DB, Huang Y, Huang M, Yao Y, Bassu S, Ciais P (2017) Temperature increase reduces global yields of major crops in four independent estimates. PNAS 114:9326–9331
Zörb C, Senbayram M, Peiter E (2014) Potassium in agriculture–status and perspectives. J Plant Physiol 171:656–669

Moriconi JI, Santa-María GE (2013) A theoretical framework to study potassium utilization efficiency in response to withdrawal of potassium. J Exp Bot 64:4289–4299
Munns R, Tester M (2008) Mechanisms of salinity tolerance. Annu Rev Plant Biol 59:651
Noguero M, Lacombe B (2016) Transporters involved in root nitrate uptake and sensing by *Arabidopsis*. Front Plant Sci 7:1391
O'brien JA, Vega A, Bouguyon E, Krouk G, Gojon A, Coruzzi G, Gutiérrez RA (2016) Nitrate transport, sensing, and responses in plants. Mol Plant 9:837–856
Peiter E (2011) The plant vacuole: emitter and receiver of calcium signals. Cell Calcium 50:120–128
Pettigrew WT (2008) Potassium influences on yield and quality production for maize, wheat, soybean and cotton. Physiol Plant 133:670–681
Pilon-Smits EA, Quinn CF, Tapken W, Malagoli M, Schiavon M (2009) Physiological functions of beneficial elements. Curr Opin Plant Biol 12:267–274
Pinkerton A, Simpson J (1986) Interactions of surface drying and subsurface nutrients affecting plant growth on acidic soil profiles from an old pasture. Aust J Exp Agric 26:681–689
Pyo YJ, Gierth M, Schroeder JI, Cho MH (2010) High-affinity K+ transport in *Arabidopsis*: AtHAK5 and AKT1 are vital for seedling establishment and postgermination growth under low-potassium conditions. Plant Physiol 153:863–875
Raun WR, Johnson GV (1999) Improving nitrogen use efficiency for cereal production. Agron J 91:357–363
Remans T, Nacry P, Pervent M, Girin T, Tillard P, Lepetit M, Gojon A (2006) A central role for the nitrate transporter NRT2. 1 in the integrated morphological and physiological responses of the root system to nitrogen limitation in *Arabidopsis*. Plant Physiol 140:909–921
Robertson GP, Vitousek PM (2009) Nitrogen in agriculture: balancing the cost of an essential resource. Annu Rev Environ Resour 34:97–125
Römheld V, Kirkby EA (2010) Research on potassium in agriculture: needs and prospects. Plant Soil 335:155–180
Sanchez PA, Shepherd KD, Soule MJ, Place FM, Buresh RJ, Izac AMN, Uzo Mokwunye A, Kwesiga FR, Ndiritu CG, Woomer PL (1997) Soil fertility replenishment in Africa: an investment in natural resource capital. Replenish Soil Fertil Afr 51:1–46
Sardans J, Peñuelas J (2015) Potassium: a neglected nutrient in global change. Glob Ecol Biogeogr 24:261–275
Savci S (2012) Investigation of effect of chemical fertilizers on environment. APCBEE Procedia 1:287–292
Scheben A, Edwards D (2018) Bottlenecks for genome-edited crops on the road from lab to farm. Genome Biol 19:1–7
Schroeder D (1978) Structure and weathering of potassium containing minerals. IPI Res Top:5–25.
Secco D, Wang C, Shou H, Whelan J (2012) Phosphate homeostasis in the yeast *Saccharomyces cerevisiae*, the key role of the SPX domain-containing proteins. FEBS Lett 586:289–295
Shabala S, Cuin TA (2008) Potassium transport and plant salt tolerance. Physiol Plant 133:651–669
Solomon D, Lehmann J, Fraser JA, Leach M, Amanor K, Frausin V, Kristiansen SM, Millimouno D, Fairhead J (2016) Indigenous African soil enrichment as a climate-smart sustainable agriculture alternative. Front Ecol Environ 14:71–76
Song H, Yin Z, Chao M, Ning L, Zhang D, Yu D (2014) Functional properties and expression quantitative trait loci for phosphate transporter *GmPT1* in soybean. Plant Cell Environ 37:462–472
Sulpice R, Ishihara H, Schlereth A, Cawthray GR, Encke B, Giavalisco P, Ivakov A, Arrivault S, Jost R, Krohn N (2014) Low levels of ribosomal RNA partly account for the very high photosynthetic phosphorus-use efficiency of Proteaceae species. Plant Cell Environ 37:1276–1298
Teixeira EI, Fischer G, Van Velthuizen H, Walter C, Ewert F (2013) Global hot-spots of heat stress on agricultural crops due to climate change. Agric For Meteorol 170:206–215
Tikhonovich I, Provorov N (2011) Microbiology is the basis of sustainable agriculture: an opinion. Ann Appl Biol 159:155–168
Tittonell P, Vanlauwe B, Leffelaar P, Rowe EC, Giller KE (2005) Exploring diversity in soil fertility management of smallholder farms in western Kenya: I. Heterogeneity at region and farm scale. Agric Ecosyst Environ 110:149–165

Kant S (2018) Understanding nitrate uptake, signaling and remobilisation for improving plant nitrogen use efficiency. Semin Cell Dev Biol 74:89–96

Kant S, Bi YM, Rothstein SJ (2011) Understanding plant response to nitrogen limitation for the improvement of crop nitrogen use efficiency. J Exp Bot 62:1499–1509

Kraffczyk I, Trolldenier G, Beringer H (1984) Soluble root exudates of maize: influence of potassium supply and rhizosphere microorganisms. Soil Biol Biochem 16:315–322

Kraiser T, Gras DE, Gutiérrez AG, González B, Gutiérrez RA (2011) A holistic view of nitrogen acquisition in plants. J Exp Bot 62:1455–1466

Krouk G, Lacombe B, Bielach A, Perrine-Walker F, Malinska K, Mounier E, Hoyerova K, Tillard P, Leon S, Ljung K (2010) Nitrate-regulated auxin transport by NRT1. 1 defines a mechanism for nutrient sensing in plants. Dev Cell 18:927–937

Kuppusamy T, Hahne D, Ranathunge K, Lambers H, Finnegan PM (2020) Delayed greening in phosphorus-efficient *Hakea prostrata* (Proteaceae) is a photoprotective and nutrient-saving strategy. Funct Plant Biol 48:218–230

Kwak JM, Murata Y, Baizabal-Aguirre VM, Merrill J, Wang M, Kemper A, Hawke SD, Tallman G, Schroeder JI (2001) Dominant negative guard cell K^+ channel mutants reduce inward-rectifying K^+ currents and light-induced stomatal opening in Arabidopsis. Plant Physiol 127:473–485

Léran S, Varala K, Boyer JC, Chiurazzi M, Crawford N, Daniel-Vedele F, David L, Dickstein R, Fernandez E, Forde B (2014) A unified nomenclature of NITRATE TRANSPORTER 1/ PEPTIDE TRANSPORTER family members in plants. Trends Plant Sci 19:5–9

Li Y, Zhang J, Zhang X, Fan H, Gu M, Qu H, Xu G (2015) Phosphate transporter OsPht1; 8 in rice plays an important role in phosphorus redistribution from source to sink organs and allocation between embryo and endosperm of seeds. Plant Sci 230:23–32

Little DY, Rao H, Oliva S, Daniel-Vedele F, Krapp A, Malamy JE (2005) The putative high-affinity nitrate transporter NRT2. 1 represses lateral root initiation in response to nutritional cues. PNAS 102:13693–13698

Liu M, Hu F, Chen X, Huang Q, Jiao J, Zhang B, Li H (2009) Organic amendments with reduced chemical fertilizer promote soil microbial development and nutrient availability in a subtropical paddy field: the influence of quantity, type and application time of organic amendments. Appl Soil Ecol 42:166–175

Liu J, Yang L, Luan M, Wang Y, Zhang C, Zhang B, Shi J, Zhao FG, Lan W, Luan S (2015) A vacuolar phosphate transporter essential for phosphate homeostasis in *Arabidopsis*. PNAS 112:E6571–E6578

Liu B, Asseng S, Müller C, Ewert F, Elliott J, Lobell DB, Martre P, Ruane AC, Wallach D, Jones JW (2016a) Similar estimates of temperature impacts on global wheat yield by three independent methods. Nat Clim Change 6:1130–1136

Liu TY, Huang TK, Yang SY, Hong YT, Huang SM, Wang FN, Chiang SF, Tsai SY, Lu WC, Chiou TJ (2016b) Identification of plant vacuolar transporters mediating phosphate storage. Nat Commun 7:1–11

Ma JF (2005) Plant root responses to three abundant soil minerals: silicon, aluminum and iron. Crit Rev Plant Sci 24:267–281

Maathuis FJ (2009) Physiological functions of mineral macronutrients. Curr Opin Plant Biol 12:250–258

Madrid L (1999) Metal retention and mobility as influenced by some organic residues added to soils: a case study. In: Iskandar IK, Selim HM (eds) Fate and transport of heavy metals in the vadose zone. CRC Press, New York, pp 201–223

Mantelin S, Desbrosses G, Larcher M, Tranbarger TJ, Cleyet-Marel JC, Touraine B (2006) Nitrate-dependent control of root architecture and N nutrition are altered by a plant growth-promoting *Phyllobacterium* sp. Planta 223:591–603

Mcallister CH, Beatty PH, Good AG (2012) Engineering nitrogen use efficient crop plants: the current status. Plant Biotechnol J 10:1011–1025

Meena SK, Meena VS (2017) Importance of soil microbes in nutrient use efficiency and sustainable food production. In: Vijay SM, Pankaj KM, Jaideep KB, Arunava P (eds) Agriculturally important microbes for sustainable agriculture. Springer, Singapore, pp 3–23

Dong Q, Sho N, Rumi T, Akihiro U, Victor R, Hirofumi S (2022) Aberrant RNA splicing of the phytic acid synthesis gene *inositol-1, 3, 4 trisphosphate 5/6-kinase* in a low phytic acid soybean line. Soil Sci Plant Nutr 68(5–6):553–562
Dotaniya M, Meena V (2015) Rhizosphere effect on nutrient availability in soil and its uptake by plants: a review. Proc Natl Acad Sci India B Biol Sci 85:1–12
Ebelhar S, Varsa E (2000) Applications in sustainable production: tillage and potassium placement effects on potassium utilization by corn and soybean. Commun Soil Sci Plant Anal 31:2367–2377
Erisman JW, Sutton MA, Galloway J, Klimont Z, Winiwarter W (2008) How a century of ammonia synthesis changed the world. Nat Geosci 1:636–639
Franche C, Lindström K, Elmerich C (2009) Nitrogen-fixing bacteria associated with leguminous and non-leguminous plants. Plant Soil 321:35–59
George TS, Fransson AM, Hammond JP, White PJ (2011) Phosphorus nutrition: rhizosphere processes, plant response and adaptations. In: Else B, Astrid O, Emmanuel F (eds) Phosphorus in action. Springer, Berlin/Heidelberg, pp 245–271
Giehl RF, Von Wirén N (2014) Root nutrient foraging. Plant Physiol 166:509–517
Glick BR, Penrose DM, Li J (1998) A model for the lowering of plant ethylene concentrations by plant growth-promoting bacteria. J Theor Biol 190:63–68
Gojon A, Nacry P, Davidian JC (2009) Root uptake regulation: a central process for NPS homeostasis in plants. Curr Opin Plant Biol 12:328–338
Grzebisz W, Gransee A, Szczepaniak W, Diatta J (2013) The effects of potassium fertilization on water-use efficiency in crop plants. J Plant Nutr Soil Sci 176:355–374
Guoping Z, Jingxing C, Tirore EA (1999) Genotypic variation for potassium uptake and utilization efficiency in wheat. Nutr Cycl Agroecosyst 54:41–48
Hammond JP, Broadley MR, White PJ, King GJ, Bowen HC, Hayden R, Meacham MC, Mead A, Overs T, Spracklen WP (2009) Shoot yield drives phosphorus use efficiency in *Brassica oleracea* and correlates with root architecture traits. J Exp Bot 60:1953–1968
Han Y, White PJ, Cheng L (2022) Mechanisms for improving phosphorus utilization efficiency in plants. Ann Bot 129:247–258
Hänsch R, Mendel RR (2009) Physiological functions of mineral micronutrients (Cu, Zn, Mn, Fe, Ni, Mo, B, Cl). Curr Opin Plant Biol 12:259–266
Hawkesford MJ (2012) Improving nutrient use efficiency in crops. In: eLS. Wiley, Chichester
Herencia JF, Ruiz JC, Morillo E, Melero S, Villaverde J, Maqueda C (2008) The effect of organic and mineral fertilization on micronutrient availability in soil. Soil Sci 173:69–80
Hirel B, Tétu T, Lea PJ, Dubois F (2011) Improving nitrogen use efficiency in crops for sustainable agriculture. Sustainability 3:1452–1485
Hodge A, Robinson D, Fitter A (2000) Are microorganisms more effective than plants at competing for nitrogen? Trends Plant Sci 5:304–308
Hu Y, Schmidhalter U (2005) Drought and salinity: a comparison of their effects on mineral nutrition of plants. J Plant Nutr Soil Sci 168:541–549
Hürlimann HC, Stadler-Waibel M, Werner TP, Freimoser FM (2007) Pho91 is a vacuolar phosphate transporter that regulates phosphate and polyphosphate metabolism in *Saccharomyces cerevisiae*. Mol Biol Cell 18:4438–4445
Hussain HA, Men S, Hussain S, Chen Y, Ali S, Zhang S, Zhang K, Li Y, Xu Q, Liao C (2019) Interactive effects of drought and heat stresses on morpho-physiological attributes, yield, nutrient uptake and oxidative status in maize hybrids. Sci Rep 9:1–12
Jain A, Vasconcelos MJ, Raghothama K, Sahi SV (2007) Molecular mechanisms of plant adaptation to phosphate deficiency. Plant Breed Rev 29:359
Jarvis S, Hutchings N, Brentrup F, Olesen JE, Van Der Hoek K (2011) Nitrogen flows in farming systems across Europe. In: Mark AS, Clare MH, Jan WE, Gilles B, Albert B, Peringe G, Hans VG, Bruna G (eds) The European nitrogen assessment. Cambridge University Press, Cambridge, pp 211–228
Jia X, Liu P, Lynch JP (2018) Greater lateral root branching density in maize improves phosphorus acquisition from low phosphorus soil. J Exp Bot 69:4961–4970

Chapter 3
Improving Land-Use Efficiency for Climate-Resilient Agriculture

Waqas Ahmad, Bisma Muneer, Dua Zahra Khan, Niaz Ahmad Wahocho, and Muzamil Farooque Jamali

Abstract In addition to providing a setting for agricultural production, land is a prerequisite for better environmental management, including the hydrologic cycle's functions of transmitting and purifying water, recycling nutrients, ameliorating, filtering pollutants, and serving as a source and sink for greenhouse gases. The use of land to accommodate shifting human requirements (agricultural, forestry, conservation) while maintaining the future socioeconomic and ecological consequences of the land is known as sustainable land management. Sustainable agricultural development requires sustainable land management as a fundamental building block. In this chapter, the effects of increasing population, industrialization, and climate change on agricultural land use are discussed. Moreover, techniques to enhance the land-use efficiency under reducing agricultural land resources have been discussed. This chapter will increase the understanding of agricultural land-use systems and its efficient management for sustainable agriculture under climate change scenario.

Keywords Land-use system · Industrialization · Rural–urban interaction

1 Introduction

Land uses the modification of natural environment and its management through agricultural, transport, recreational, commercial, and residential activities. Some physical factors including topography, climate, soil type, human factors, and population density have adverse effect on land which play a role in the destruction of land and reduce the efficiency of land.

Land-use efficiency is the efficient use of land by using natural resources and getting a maximum yield through minimum production cost. By increasing

W. Ahmad (✉) · B. Muneer · D. Z. Khan
College of Agriculture, University of Layyah, Layyah, Pakistan

N. A. Wahocho · M. F. Jamali
Department of Horticulture, Sindh Agriculture University Tandojam, Tandojam, Pakistan

M. Hasanuzzaman (ed.), *Climate-Resilient Agriculture, Vol 2*,
https://doi.org/10.1007/978-3-031-37428-9_3

efficiency of agriculture land use, agriculture area could be improved. The efficiency of land use is considered according to the economic situation of country. Land-use efficiency is the increase of gross income by using natural expenditures of living.

Land-use efficiency is adversely effected day by day because of many problems The use of fertilizers is increased to control pesticides, but unlimited use of fertilizers has a negative effect on earth which is the major reason of destruction of agriculture lands. The artificial modification of products used for crop improvement is bad for crops and land. All these problems are reducing the effectiveness of agriculture production.

These problems of land-use efficiency are due to improper land monitoring system, technical equipment issues, and some financial issues. The problem of affectual use of land expedient in Russia, first of all connected of soil used in agricultural sector of the agronomy. The land and asset capacity have importunity reduced over the last decades as a result of the removal of big area of agricultural land from communication. In Russia, the share of agriculture land does not be greater in number, 10% of the Russian. A geographic area is belonging to under the jurisdiction of a governmental authority (Lysenko EA 2007). At the same time, the area of agricultural land in Russia is much higher than in the largest countries of European Union. This indicates huge reserve in agricultural productivity, which regrettably is not get used effectively enough. The concept introduced 100 years ago showing a relationship between plant productivity and water use for agricultural production.

Climate-resilient agriculture is the long-term crop productivity by using natural resources and improving livestock production system with increasing farmer's income under climatic changes. It is difficult to manage the land-use efficiency under the climatic changes because climatic changes result in destruction of land, reduce soil fertility, etc., which directly decrease land uses.

It is a field of research that studies the primary synthesis of complex life like structure from primordial lifeless origins and changes of product. The problems decrease the efficiency of agriculture land used in the export of product from (CIS) countries.

Negative consequences of anthropogenic impact on agricultural land are soil erosion, acidification, desertification, soil salinity, heavy metal pollution, oil and oil product, and pesticides (Alston E et al. 2018). During the climatic changes, the concentration of CO_2 increases which increases the agricultural productivity and also improves the efficiencies of resources (water, land, etc.).

The main objective of this chapter is to improve the soil structure, agriculture crops production, and land management under different climatic changes.

2 Land-Use Efficiency

To evaluate the land-use measures, it is necessary to provide a sustainable land management practice. Land-use efficiency (LUE) is applied framework that recognizes the conditions, integrated methods, drawback of structure and its classification, and indicator system although evaluation procedure.

Now, in these days, the rational land-use measuring is necessary both nationally and globally. To enhance the value and resource of land, it is essential to focus on land use rights and restrictions and responsibilities according to follow the land policies for sustainability. The land-use challenges are studied to increase and measuring the efficiency of land within the framework. All the management of land efficiency serves as governing and monitoring of land according to the objective which is followed by socioeconomic and environmental development. Therefore, land-use management practices are interrelating with different parties, e.g., institutional arrangements (play important role to ensure process).

According to European cohesion policies since 2007, and land policy guidelines in the republic of Latvia (Therefore it is necessary to introduced more understandable indicators in land management system (LMS). The measuring techniques of LUE are based on comparison of inputs and outputs, and also, they recognize multidimensional valuable resources and are used in evaluation, development, analyzing, and synthesis.

3 Land-Use Sustainability

Land is used by people for the development of ecosystem or receiving ecological activities. As land can create the landscapes of aesthetics with high economic and agricultural and environmental values, under many threats it can also lead to soil losses, i.e., soil degradation and soil erosion with unbalanced ecosystem. Land use is basically a society–nature interaction—the interaction between societies and environment with the ecosystem. Thus, the land problems lower the sustainability of the land.

Sustainability indicators are needed to monitor the current situation of area or a land and to check out the society–nature interaction (Haberl H et al. 2004). Sustainability development is the process to fulfill the present needs of society without competing the future needs (Turner and Romme 1994). The major objective of sustainability land use is to maintain a balance in the production of higher yields of crops while protecting the environment from natural disasters and well-being of economy (Aznar-Sanchez JA et al. 2019).

4 Key Issues of Land Use and Sustainability

Land-use sustainability is important for the economic sustainability development, but due to some to key issues, the land sustainability is disturbed; some irregular development activities have negative effect on land sustainability, i.e., land deterioration contamination of land and rural decline. Urban problems also damage the land sustainability, i.e., air pollution, waste management issues, and water pollution. In order to avoid these issues, land-use planning should be done to protect the land and economy.

In rural area, the agriculture is considered as the major activity of population, but in urban area, the living depends on the industrial activities. A huge number of urban areas depend on the rural resources, and increasing population is not getting very much from agriculture because of low rate of crop yield and productivity for the higher rate of population (Tacoli C 1998). For the developing world, urban–rural interaction is considered as a challenge for population.

In China, urbanization and farmland protection problems are getting attentions. Research has shown that urbanization has adverse impact on the farmland protection because urbanization has caused so much farmland losses of medium farmers. In China, the global urbanization level is above 50%, while the farmland level is just 7% with the great losses of farmland.

5 Climate Factor Effecting Land-Use Efficiency

The rapid urbanization (the process of making an area more urban) in China has a large effect on land use. The scarcity of land resource has become sustainability constraints that are associated with the protection of nature, biodiversity, and soil water which are ecologically sustainable potential. The urban development there are 12 form an idea of the amount % land use efficiency. Number of asses the USA socioeconomic development of land use and comprehend effective land strategies. This paper includes 184 cities; we analyze first the land-use performance of China and the USA from different perspective (Yu J et al. 2019).

Average urban land use efficiency of USA and China was not very high furthermore the average efficiency fluctuation in 2008 and 2015. The USA and central China have more repair power for efficiency improvement. Landscape ecology focuses on the spatial patients and process of ecological and human interactions primary production, respiration, energy, reproduction, and decomposition.

6 Land Management

Domestic studies of land management started rather late. Research achievement focuses on the macrolevel land management issues and strategies, optimization, government land management function, and so on.

On the basis of land-use planning, human-induced activities are destroying the socioeconomic and ecological disturbance of land systems globally, including different climatic regions. In these climatic regions, the land management planning monitors the threatening and economic conditions which are based on agribusiness, farmers, and other land users.

As optimizing the land pressure and other natural resources which is affected by human generated activities and also caused by climatic changes is increasing the tropical threats and rising serious concerns all the sustainability of management instruments and resources. Land-use planning in topical areas helps the land users in selecting and putting it into those areas which needed the best and also helps in safeguarding natural resources and ecosystem. Maintenance between sustainability of generations of farmers for past and future is necessary.

Land-use planning supports the public and private services to make the strategic decisions on how to manage the land. Land-use practices are also helping the environmental behavior. Land areas defined are based on combination of soil, landform, and climatic characteristics. Such process can also consider the link between different lands uses.

7 Land Resource Engineering and Land-Use Policy

On the basis of coordination among land policy and land engineering, the concept of definition of land engineering is the contents and demonstrates the identity and development of land engineering. On the other hand, land policy helps and guides the establishment of the land engineering, and it is an important mean to improve and able to permit the land policy. The land engineering is based on the content, which is conversion of non-agricultural land into agricultural land, conversion of low standard use land into a high standard land use, conversion of current land into human construction use, and conversion of polluted and damaged land into usable land (Wu et al. 2018).

Land policy is a basic principle, directions, and instruction for specific economic and social interests to attain the land use and management in a definite historical period. As resources of land are non-renewable, and the base of objectives of land policy adjustment to achieve the support of land resources. Land policy associated with the class of production relationship and productivity is the changes between adaptation and non-adaptation.

8 Factors Influencing the Use of Agricultural Land

The many factors that affect how land is used for particular reason.

8.1 Socioeconomic Factors

The socioeconomic element is defined by the access and utilization for land resource, the introduction of new technologies, the development of the economic sector and territory including infrastructure and environment land productivity and capacity as well as living environments and the population, investment, and credit facilities, and the increase in competitiveness and making proper and efficient usage of renewable resources of energy (Auzins et al. 2013). The land that is used for the agricultural purposes is a key element in acknowledgment facilities, as the efficient consumption of these credit facilities is able to enhance farms by the usage of new technologies, since they can enhance the speed of management of agricultural land, reduce time resources, and increase yields. Land productivity and capacity play a significant role when it comes to utilizing land for agriculture, as the higher the productivity of the land the more expected this can be easily handled and consumed. In addition, it is important to consider the area, the place in which the land situated, and the habitat area or population. Some pattern of the land use and the habitat area, for instance the greater the number of people the greater amount of land utilized in the area. The smaller the area, the more likely to be the land used for farming could not be utilized to its fullest extent in that zone. The possibility of using land can also affected by the environmental aspect.

8.2 Environmental Factors

Environmental aspects are defined by biodiversity and eco-integrity (ecosystem organization and process) and natural assets (Daba et al. 2015). An essential role of usage of land in agricultural purposes is the environmental element which includes plants genetic variety. For example, grazing is a great way to increase diversity in plants and the higher the likelihood that would provide an energy-efficient value. It is equally important to provide a particular area that is rich in natural capital. For instance, the availability of raw material to construct in the area as well as the quality of soil (impacts the kind of use of the land for agriculture and the productivity) as well as whether it has access to the water or else it needs to be given a boost which could raise the cost of using the land.

8.3 Institutional Factors

The land use is affected in a way by an institutional aspect which is described by the interaction between the norms of regulation and the organizations (Auzins et al. 2013). It is crucial to align the land-use system with laws and regulations to ensure that the manner of using land does not violate the rules set forth by the law. This can play essential to be aware the land that is not used for agriculture can be subjected to an enhance of tax rate, however, also land-usage efficiency, features and factors that affect it, as well being an efficient and sustainable way to manage land. The utilization of agricultural land is crucial for viable land managing which is designed to sustain or enhance the quality of production and delivery of services, safeguard natural sources, assure sustainable economics, avoid the degradation of water quality, and also decrease soil erosion, thereby reducing the risk of production (Auzins et al. 2013). To identify the elements that affect the usage of land for agriculture, it is essential to examine the variables which influence other elements in addition.

8.4 Management Practices for Increasing the Efficiency of Land Use

The efficiency of land use varies based on requirements for land use in particular, such as the extent to which the land can be utilized for the huge businesses of agriculture like cooperatives, associations, research, and educational institutions, or is utilized by larger or smaller private households. Enhancing the efficiency of land use is a hot topic across many countries. Hence, displays of land efficacy measurements and methods of measuring economic efficiency of land have been shaped. The primary requirement for efficient utilization of land for agriculture is an increase in fertility of soil. It is totally depending on enhancement of the agricultural systems within the farms and organization of land areas, the development for rotational cropping, investigation of the layout of the area sown and the development of soil treatment systems as well as the management and installation irrigation systems and fertilizer system pest and disease control and production of seed and environmental security measures (Cintina V and Pukite V 2018).

The system of agriculture must satisfy the society demands, the ecological demands of crops and agriculture as well as the natural climate circumstances, the rate of increase in production as well as economic structure, and the lowest hazard of ecological pollution that has to be protected. Affordable use of land for agriculture includes both internal and external requirements that must be fulfilled or improved in order for efficient usage land. A key element in the efficacy of land used for farming is the way in which parcels of land, which guarantees an appropriate proportion of land, the accessible resources of labor, financial resources, and an appropriate use of land used for agricultural purposes as well as the areas of high production, and the specialty of agricultural land. The use of agricultural land is not as effective in many areas because there is not a well-adjusted production variable. Because of the absence of jobs in the rural areas, the population does not oppose working for low wages. This can lead to lower yield, and ultimately, poor efficiency in the usage of land, since worker's absence inspiration to do their job. Efficiency in land use is also a factor that affects the quality of h-support, so it is imperative to establish an extensive set of economic and organizational measures in addition to ensuring that there is a coordinated approach at every stage of the government and local governments. Landowners and consumers are also expected to offer the state with support. A mechanical soil cultivator is an important component of the efficacy for using land used in agriculture because it increases fertility of soil. It is responsible for 30–50% of the expenses associated with production of crop. The basis of resource-saving technology is cost reduction by minimizing soil treatment.

9 Policy Suggestion

We have demonstrated that the creation of the carbon market together with the established markets for agricultural commodities with plausible global perspectives could result in the efficient distribution of emissions reduction and production of agricultural products. But, the provision of biodiversity and water resources services was not effective. We developed a policy that requires the new land use to be accounted for in relation to water used by purchasing entitlements. This is a policy that is in place in the Australia and other countries (Connor JD et al. 2008). Based on other research results, this policy did have some influence at price levels, but did not provide a significant factor in the usage of land. A functional water market that has an upper limit on the use of resources in which reforestation is competing with other users of water (e.g., industrial, urban, or irrigation of agriculture) to get water resources that are scarce could yield better results (Nordblom TL et al. 2010) but also do not examine this policy because it only operates within entirely contributed catchments within the area of study. In the same way, the absence of a constant, widespread market policy on biodiversity services severely hampered productivity and led to a number of negative results when high biodiversity area was transformed into monoculture type of cultivation. Other strategies of usage of land that improved the efficiency of production across multiple objectives could be conceivable, but they came with significant expenditures in the term of lost economic opportunities. The issue for policymakers is to address these expenses and improve efficiency across multiple goals. Another challenge is cutting down on the costs of transactions related to administration of policy and the targeting of many ecosystem facilities. Two options are to mass credit to provide ecosystem services and equally involve the expansion of the markets that provide ecosystem-related services (Robertson M et al. 2014). Credit bundling is the process of the collation and sale of all services, including in this instance the case of the new use of land, such as ecological plantings into the distinct marketplace including a way to pay to ecosystem service providers (Connor JD et al. 2008). The findings illustrate how the design of metric significantly affects the quality of the aids for society. Stacking is the process of selling credits to single markets to provide ecology services (Robertson M et al. 2014). It can be stacked or bundled with a careful planning of markets, and the accompanying institution-based arrangements (e.g., multiobjective benefit metric, pure rights to property, low transaction cost) are essential for enhancing productivity efficiency across many ecosystem amenities.

References

Alston E, Alston LJ, Mueller B, Nonnenmacher T (2018) Institutional and organizational analysis: concepts and applications. Cambridge University Press, Cambridge

Auzins A, Geipele I, Stamure I (2013) Measuring land-use efficiency in land management. Adv Mater Res 804:205–210

Aznar-Sanchez JA, Piquer-Rodriguez M, Velasco-Munoz JF, Manzano-Agugliaro F (2019) Worldwide research trends on sustainable land use in agriculture. Land Use Policy 87:104069

Cintina V, Pukite V (2018) Analysis of influencing factors of use of agricultural land. Res Rural Dev 1:181–187

Connor JD, Ward JR, Bryan B (2008) Exploring the cost effectiveness of land conservation auctions and payment policies. Aust J Agric Resour Econ 52:303–319

Daba kā kapitāls (Nature as capital) (2015) Eiropas komisijas mājaslapa (European Commission home page). Retrieved March 5, 2018, from: http://ec.europa.eu/environment/basics/natural-capital/index_lv.htm. (in Latvian)

Haberl H, Fischer-Kowalski M, Krausmann F, Weisz H, Winiwarter V (2004) Progress towards sustainability? What the conceptual framework of material and energy flow accounting (MEFA) can offer. Land Use Policy 21:199–213

Lysenko EA (2007) Plant sigma factors and their role in plastid transcription. Plant Cell Rep 26:845–859

Nordblom TL, Christy BP, Finlayson JD, Roberts AM, Kelly JA (2010) Least cost land use changes for targeted catchment salt load and water yield impacts in southeastern Australia. Agric Water Manag 97:811–823

Robertson M, BenDor TK, Lave R, Riggsbee A, Ruhl JB, Doyle M (2014) Stacking ecosystem services. Front Ecol Environ 12:186–193

Tacoli C (1998) Rural-urban interactions: a guide to the literature. Environ Urban 10:147–166

Turner MG, Romme WH (1994) Landscape dynamics in crown fire ecosystem. Landscape Ecology 9:59–77

Wu X, Yang H, Waugh DW, Orbe C, Tilmis S, Lamarque J-F (2018) Spataial and temporal variability of interhemispheric transport time. Atmos Chem Phys 18:7439–7452

Yu J, Zhou K, Yang S (2019) Land use efficiency and influencing factors of urban agglomerations in China. Land Use Policy 88:104143

Chapter 4
Climate-Resilient Fertilizer Management for Crop Production

Ajay Saroha, Amit Kotiyal, and Aditi Thakur

Abstract Fertilizer is an important input in agriculture production. As per the increasing demand of food for population, fertilizer is a boon. However, current agricultural producers are put under immense stress to end hunger as a result of expanding population and shrinking cultivable land. To fulfill the demand, different agricultural inputs, like fertilizers and other chemicals, are being exaggerated, causing environmental damage. The tremendous use of these inputs without consciousness about ecology sustainability which influences the soil health causes land degradation and leads to climate change. Continuous change in climate by such activities causes variation in the temperature, rainfall pattern, microclimate of crops, soil microbial activity, etc. So, for the future food demand of increasing population, we must target significant agricultural production limited by a sustainable environment. In this concern, new alternates like use of supportable resources with the fertilizer help in reducing fertilizer dose and maintain the soil health and environment such as green manuring crops, panchagavya, vermicompost, and FYM. Ultimately for the better future, farmers should adopt the strategies like integrated nutrient management (INM), integrated soil fertility management (ISFM), sustainable water management, ridge furrow mulching, conservation agriculture, and breeding strategies.

Keywords Soil · Biochar · Vermicompost · Soil temperature · Nutrient · Biofertilizer · Erosion

1 Introduction

The agriculture sector makes a significant contribution to GDP and is vital to the security of the world's food supply. Somewhere about a two-third majority of the labor force is employed there. Farms provide the raw materials for several manufacturing companies including the sugar, textile, jute, food, and milk processing

A. Saroha · A. Kotiyal (✉) · A. Thakur
Department of Horticulture, Lovely Professional University, Phagwara, Punjab, India

M. Hasanuzzaman (ed.), *Climate-Resilient Agriculture, Vol 2*,
https://doi.org/10.1007/978-3-031-37428-9_4

industries (Reddy 2015). The ecological, economic, and social impacts of climate change on agriculture are substantial in the world's ecology. Scientific studies indicate that human activities contribute to climate change drastically. According to experts, agriculture will have a harder time producing enough food in the future due to the stresses of climate change and increasing climatic unpredictability. The IPPC Report 2014 (Burkett et al. 2014) provides an authoritative reaffirmation of the fact that climate change and variability affect food and fiber production around the world because of increased temperature, CO_2, weeds, pests, pathogens, and disrupted precipitation and transpiration regimes (Altieri et al. 2015). Climate change threatens food production systems, threatening the lives and food security of billions of people worldwide. Climate change will disproportionately affect marginalized groups in emerging economies compared to wealthy and industrial countries. Because of its dependence on weather and the fact that farmers, on average, are older and less well off than their urban counterparts, agriculture is especially vulnerable to climate change (Srinivasa Rao et al. 2019).

Agriculture, especially crop production, is directly influenced by the condition of soil. The soil is a lively natural body with three-dimensional structures made up of minerals, organic compounds, and living forms in which plants develop. Soil is the fine material that accumulates on land after water, wind, or ice has eroded larger particles of rock or after in-situ weathering of rock components (Nortcliff et al. 2011). One of the most important materials at our disposal is soil. Not only it provides the plants with water and nutrients, but also it acts as a stable base for them to grow from. Numerous tiny organisms call soil their home. Each soil has unique physical, biological, and chemical characteristics that interact in a complex hierarchy to determine its overall quality and potential uses (McCauley et al. 2005). It is not possible to directly measure soil quality; therefore, scientists use indirect methods, called biomarkers, to infer soil and environmental changes. Different types of soil properties are as follows: (a) The first three characteristics of soil are its texture, structure, and porosity; they are all considered physical features, (b) chemical properties are exchange capacity, soil pH, and salt-affected soils, and (c) biological properties are soil fauna, soil flora, and soil microorganisms.

2 Nutrients

The crust of earth is composed of sedimentary rocks and igneous rocks (19:1), with the latter consisting of around 80% shales, 15% sandstone, and 5% limestone. Based on the amount used by plants, nutrients are divided into macronutrient and micronutrient elements. Plants need macroelements in more quantity (more than 100 ppm) for plant life such as oxygen, hydrogen, and carbon (the three basic nutrients), phosphorus (P), nitrogen (N), and potassium (K) and (the three primary nutrients), and sulfur (S), calcium (Ca), and magnesium (Mg) (the three secondary nutrients). Plants also need micronutrients, which are nutrients present at levels lower than 100 ppm. They are also called as trace elements. These elements are iron

(Fe), manganese (Mn), zinc (Zn), copper (Cu), boron (B), molybdenum (Mo), chlorine (C), and nickel (Ni). Nevertheless, when the concentration of these trace elements increases, it has a detrimental impact on soil physicochemical and biological qualities (Sudhakaran et al. 2018). While trace elements are found in soil naturally, it is often contaminated due to human activities such as burning fossil fuels or municipal waste, using polluted water for irrigation, using synthetic fertilizers or manures, or spraying pesticides that include heavy metals (Fytianos et al. 2001).

3 Major Soil Problem

Increase in the growth and poor development of plants is one of the indications of problematic soil. Soil compaction (thick soil that drains water slowly), topsoil loss, and erosion are the three main problems with soil. Fertilization of plants which are deteriorating and not thriving well can help in overcoming the symptoms of deficiency of soil nutrients.

3.1 Soil Compaction

Heavy machinery, automobiles, gardening tools, and people are the major causes of soil compaction. This reduces the size and number of soils pores and removes air from the soil. Compaction and a high proportion of clay can make slow draining worse. Another factor is hardpans, which are layers of compacted soil that may occur at any depth and are impervious to water, air, and nutrients. Inhibiting root growth and plant establishment, soil compaction causes soggy roots, root and crown rots, and higher acidity.

3.2 Topsoil Loss

Topsoil is fertile which is up to 4–8 inches. Most of the organic stuff, plant roots, as well as microbial and animal life, are present in topsoil. When storms occur, topsoil that is exposed or on slopes is more likely to be carried away.

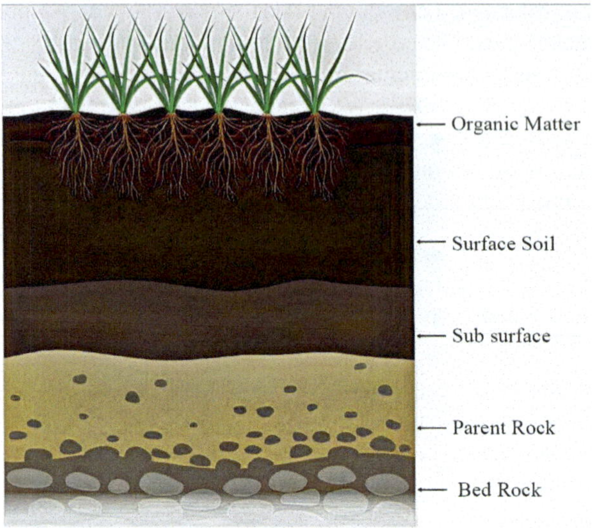

Fig. 4.1 Various layer of soil. (Source: FAO)

3.3 Erosion

Erosion is one of the major components which leads to the depletion of crop production. Slopes, areas of exposed soil, and areas near downspouts are particularly vulnerable to erosion because they are places where water may easily flow onto or over the land. Coastlines may erode due to storms and high tides.

4 Soil Horizons

The most recognizable visual characteristics of a soil profile might be its soil horizons. There are types of horizons, namely O, A, B, E, C, and R (Owens and Rutledge 2003). These horizons can be explained as follows: (i) horizon O—layers with a heavy organic content, and some have been submerged in water for extended periods of time; (ii) horizon A—mineral horizons that developed near the surface or beneath an O horizon and that show a significant loss of the original rock structure; (iii) horizon E—mineral horizons that are distinguished beside the disappearance of silicate clay, iron, aluminum, or several mishmashes of these types of elements and that keep up the accumulation of sand and silt particles; (iv) horizon B—soil structure can be found in horizons that are formed below an A, E, or O horizon; and in an illuvial concentration, iron, aluminum, humus, carbonates, gypsum, silica, or any combination of these elements, (v) horizon C—a horizon that does not contain hard bedrock—is not considerably affected by pedogenic processes and does not meet the criteria for an O, A, E, or B horizon. The C layers' makeup may be like or dissimilar from the substance assumed to have created the solum and (vi) R layers: Last layer of horizon is hard bedrock (Fig. 4.1).

5 Climate Change Affecting Soil Properties

Jenny proposed the most possible soil property–climatic factor correlations in 1941 (Carolina and Kimble 1990).

5.1 Soil Temperature

The IPPC 2007 study predicted that the global average air temperature will rise by 1.8–4.0 °C in response to mounting evidence of global climate change (GCC) (Reddy 2015). The shifting patterns of environmental features such as temperature rises and precipitation change patterns, excess UV radiation, and increased occurrence of severe weather phenomena like floods and droughts are increasing as major threats to tropical vegetable production. Soil temperatures will also be influenced by the sort of plant which grows on the surface, which may transform as a result of climate change (Karmakar et al. 2016).

5.2 Soil Chemical

Climate change creates significant changes. Most rapidly accelerated chemical or mineralogical changes occur because of changes in external conditions; these include salt and nutrient cation loss where leaching increases and salinization where net upward water movement occurs due to increased evapotranspiration, decreased rainfall, or irrigation water supply. In areas with less precipitation or more dryness, salinization and alkalinization would rise (Karmakar et al. 2016).

6 General Impact of Climate Change on Soil Health

"The ability of a given sort of soil to operate, within natural or managed ecosystem limits, to sustain plant and animal production, maintain or enhance water quality, and support human health and habitation," as defined by the Soil Science Society of America (SSSA). Unlike soil health, which is a dynamic condition in which human influence causes a change in soil quality, soil quality is a static (qualitative) assessment of soil capacity. Climate change can affect soil health via physical, chemical, and biological aspects (Shourie and Singh 2021). Climate change may accelerate soil erosion, thus limiting food supply. Increased rainfall will hasten the pace of soil loss, thus limiting farm production. Varieties of global forest changes may emerge from climate change's direct and indirect impacts (Raison and Khanna 2011).

6.1 Major Direct and Indirect Climate Change's Consequences

The consequences of climate change can be pointed as (i) increase in temperature and concentration of carbon dioxide (CO_2) in atmosphere, (ii) precipitation changes, as well as the frequency and the severity of extreme weather events including frosts, droughts, and storms (wind, rain, and ice), heat waves, and changes in the severity of pest and pathogen outbreaks, and (iii) alteration in forest growth and soil organic matter (OM) and nutrient input and changes in the impacts of pests and pathogens on soil health.

7 Impact of Fertilizer on Sustainability of Environment

Environmental sustainability is a concept that encompasses activities to conserve physical resources, the use of recyclable materials and renewable resources, a reduction in the use of harmful compounds during production, and environmental preservation. The green revolution focused on advanced high-yielding cultivars, extensive fertilizer use, and irrigation, resulting in a considerable rise in crop output. Prolonged soil nutrient exploitation through intense agriculture without enough external application might contribute to soil deterioration. Agrochemicals utilization in growing crops is one of the most efficient tools to boost soil nutrients (McArthur and McCord 2017). Chemical fertilizers overuse has an adverse impact on the environment and human and animal health (Wangunci and Zhao 2019). Unbalanced use of chemical nutrients is one example of a problematic input that poses a danger to the long-term viability of agricultural systems. Agricultural production is reduced when soils are deficient in macronutrients. So, chemical fertilizer use necessitates a substitution between agricultural yield and long-term environmental sustainability. Appropriate chemical nutrient application and organic manure usage are valid measures for determining agricultural system environmental sustainability (Bora 2022).

8 Total Supply of Ammonia, Phosphoric Acid, and Potash Globally 2016–2020

Globally, primary nutrients containing fertilizers are used rapidly. Fertilizers in the form of potash, phosphoric acid, and ammonia are provided by various nations to ensure enough K, P, and N intake, which in turn leads to increased crop yields and national incomes. The major fertilizers' demand and production are increasing in the world, and it shows the nitrogen fertilizer production increased above 150 tons

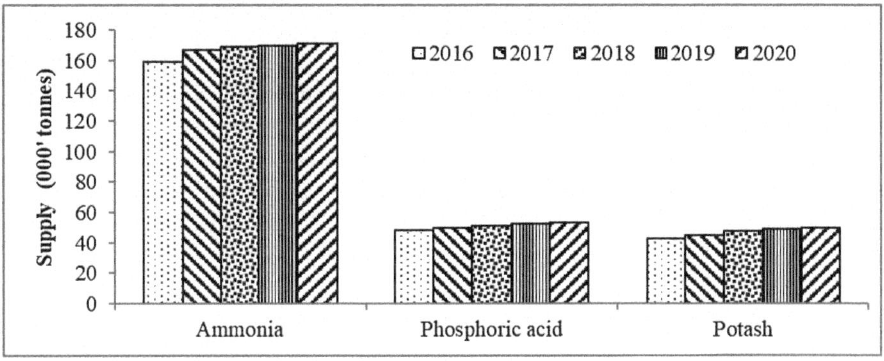

Fig. 4.2 Trends in supply of chemical fertilizers (FAO, 2020)

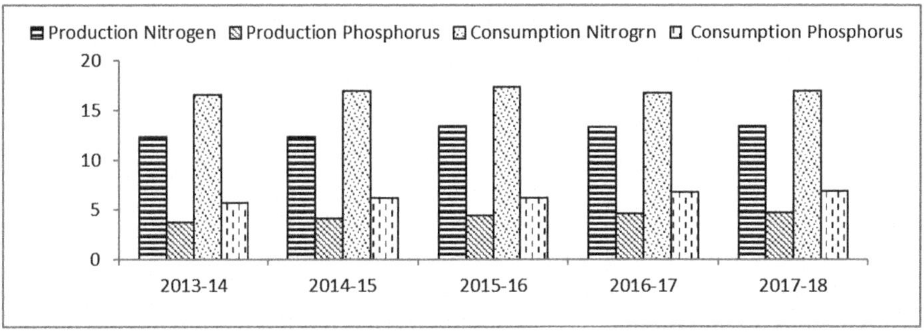

Fig. 4.3 Production and consumption of nitrogen and phosphorus in India (Source - Department of fertilizer Government of India)

in the year 2016–2020, whereas phosphorus production and potash production are increasing slower as compared to nitrogen (Fig. 4.2). It shows the dependency on three major fertilizers by the world farmers.

8.1 Production and Consumption in India

In India, alone the dependency on chemical fertilizers increased, which is known for its agriculture-based economy. It is drastic that the production is lesser as compared to the consumption of these chemicals by the agriculture sector (Fig. 4.3).

9 Alternates of Chemical Fertilizers

At present, the agriculture sector exhibits 30–35% of worldwide greenhouse gas (GHGs) emissions and is the greatest source of non-anthropogenic gas emission. Non-CO_2 GHGs which mostly because of methane (CH_4) emissions from cattle and rice farming, as well as nitrous oxide (N_2O) emissions from fertilized soils which also contributed to climate change (Lagomarsino et al. 2022). The usage of nitrogen fertilizers directly impacts the quantity of NH_4^+ or NO_3^- accessible in topsoil, which in turn affects the activities of nitrification and denitrification (BragadoCarmo et al. 2012). A sustainable technique for enhancing soil fertility and recycling nutrients is to use crop waste, animal leftovers, green manuring crops, compost, and biofertilizers such as *Azospirillum, Rhizobium,* and *Azotobacter*. Decomposition of agricultural debris in the absence of oxygen and sludge's is a potential technique for producing products that help in improving the efficiency of farming land by being rich in plant's macronutrients and micronutrients (Alburquerque et al. 2012). Anaerobic (without presence of oxygen) remains can be utilized as fertilizers and soil cleansers, with nutrients in digestates more readily accessible than those in slurry or fresh manure (Martin et al. 2014). Some of the alternates are as follows.

9.1 Biochar

Biochar amendment to soil has been reported as a strategy to boost carbon storage of land and minimize global N_2O emissions from soil. Biochar may enrich soils with both a little mineralizable proportion as well as a more recalcitrant and less mineralizable component of carbon. Furthermore, biochar has been proven to retain NO_3 inside its pores. As a result, adding biochar to soils may alter the conditions that govern nitrification, denitrification, and other N transformation and loss processes (Borchard et al. 2019).

9.2 Sewage and Sludge

Sewage sludge is a semi-solid by-product of industrial or municipal wastewater treatment. The application of modern remediation technologies like pyrolysis and carbonization of biowaste products has the potential to generate linked solutions for organic C and P concerns in food processing in the future. When compared to the input material, the pyrolysis materials generated have a higher (two to three times) overall C and other element content, viz., P, Zn, and Ca (Frišták et al. 2022). P levels in pyrogenic materials obtained from sewage sludge that is applied to the subsoil range from 1% to 20%. Pyrogenic carbonaceous materials produced from sewage sludge represent a potential alternative for conventional P inorganic fertilizers and organic C suppliers (Xiao et al. 2022).

9.3 Composting

Composting is an ecologically sound and beneficial process that may successfully minimize the emissions generated by livestock and poultry dung by converting it into manure that can increase the organic carbon matter of cultivable land (Zhao et al. 2022). Composting livestock and poultry waste is important for preserving agroecological stability and the sustainable development of agricultural waste. Composting has been shown to be an effective method for converting biodegradable waste into beneficial organic manure for soils, and it may be divided into many processes. Composting is a broad mechanism of biological breakdown of various complex organic materials to create heat, water, and simple water-soluble chemicals under high-temperature conditions (Bao et al. 2021).

9.4 Nanofertilizer

Nanofertilizers have shown considerable promise for long-term usage in soil fertility and agricultural productivity, with little or no environmental impact. Nanofertilizers have nanoscopic particles, a significant surface area-to-volume ratio, the ability to encapsulate nutrients, and better mobility, which may boost plant nutrient availability and agricultural productivity (Jakhar et al. 2022). Most of the cultivable land is losing fertility and degrading because of climate change and the utilization of intensive farming methods. Considering the significance of macronutrient fertilizers, crop fertilization nanotechnology research has mostly concentrated on the assessment of nanoparticles as micronutrient carriers, particularly Zn-, Cu-, Mn-, and Fe-based nanoparticles (BragadoCarmo et al. 2012). Nanoparticle-encapsulated fertilizers will increase the availability and uptake of nutrients by crop plants. Nanoparticles with a size of less than 100 nm can be used as fertilizer for effective nutrient management, which is more environmentally friendly and less polluting. The primary reasons that fertilizers are so popular are their superior size, surface area, and penetration ability when compared to the same material found in bulk form (Dharam Singh et al. 2017).

9.5 Green Manuring

Green manure is a prominent method for increasing the amount of organic matter and its components in soil. Many researchers have indicated that growing green manure may improve soil's physical, chemical, and biological qualities; modify the variety and community composition of soil microorganisms; ease fertilizer resource scarcity; and reduce fertilizer input costs (He et al. 2020). It is indeed a significant effective substitute of fertilizer for the present "Chemical Fertilizer Zero Increase" initiative and the organic matter increase plan, although both are critical to agriculture's long-term development (Hu et al. 2022).

9.6 Meat and Bone Meal (MBM)

Organic waste is one of the best approaches for the supply of nutrients to the plants such as meat and bone meal. Meat and bone meal are prepared from raw animal components and contains slaughterhouse by-product that is not safe and suitable for human food, including as skulls, hooves, blood, fat, feathers, bones, and giblets. Such by-product contributes to 30% of the overall weight (Andriamanohiarisoamanana et al. 2018). Because it is rich in macronutrients including N, P, and Ca, as well as micronutrients and OM, meat and bone meal have the potential to be used in agriculture (Załuszniewska and Nogalska 2022). On a dry matter (DM) basis, MBM incorporates four times more nitrogen, ten times more phosphorus and eight times more calcium than manure. But it lacks essential nutrients like K and Mg. MBM may be used in farming without causing damage to the environment since it recycles nutrients from organic waste (Jastrzębska et al. 2021).

9.7 Vermicompost

Soil health and crop yield benefit from the addition of vermicompost, an organic fertilizer-rich in nutrients, beneficial microbes, and plant growth hormones. Farmers are not able to use organic manures as much as they could because of their low nutrient content, high bulk, limited market availability, little awareness of their beneficial influence, and difficult handling (Yatoo et al. 2021). The addition of numerous organic substrates such as specific microbial inoculants, bone meal, eggshell and banana, and maintaining moisture, pH, and optimal temperature can increase the recycling performance and quality of vermicompost (Fig. 4.4).

Furthermore, vermicompost has been shown to have several indirect impacts on plant development, including insect and parasite worm control and disease suppression (Basco et al. 2017). The vermicomposting process yields nutrient-rich vermicompost suited for long-term agricultural output.

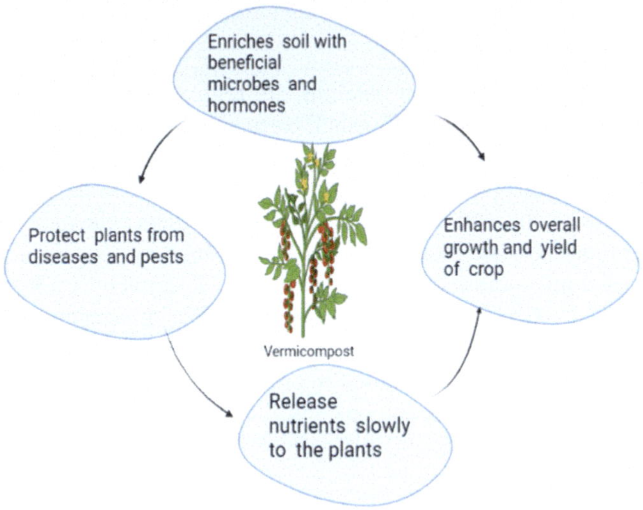

Fig. 4.4 Benefits of vermicompost. (Source: Department of Fertilizer, Government of India)

9.8 Biofertilizer

Organic farming is frequently promoted as the response to the issues with emerging intensive agriculture. The use of biofertilizers will generally reduce the global warming potential of agriculture. Because nitrogen fertilizers require a lot of energy to produce and because nitrogen in biofertilizers can reduce the need for mineral fertilizers, the use of biofertilizers will reduce the global warming potential of agriculture. All fertilizers should be absorbed by plants when they require them, with the least amount of fertilizer loss to the environment. Leaching, runoff, and, in the case of nitrogen, gaseous losses are all examples of losses to the environment. Biofertilizers or fertilizers made from organic waste could take the place of some mineral fertilizers, which would cut down on energy use and resource extraction (Michler et al. 2019).

9.9 Panchagavya

Panchagavya is an organic product that is made from five different cow bi-products, including cow dung, cow urine, cow milk, cow ghee, and cow curd. It may play a role in fostering growth and supplying immunity to the plant system, conferring resistance to pests and diseases. Panchagavya contains a variety of nutrients, including macronutrients like N, P, and K and micronutrients like various amino acids, vitamins, and growth regulators like auxins and gibberellins, as well as advantageous microorganisms like pseudomonas, azotobacter, and phosphor bacteria, among others, which are necessary for plant growth and development (Raghavendra et al. 2015).

9.10 Other Sources

Eggshells, an important source of calcium, are regularly discarded as garbage in homes, hotels, and other facilities. Eggshell particularly accounts for around 11% of the overall mass of the entire egg and comprises approximately 91% $CaCO_3$ (King'ori 2011). When this calcium-rich material is added to the soil, it not only provides calcium to the soil but also raises its pH levels (Gaonkar et al. 2007). Banana peel is another essential biowaste which is high in potassium followed by magnesium, calcium, sodium, and other minerals, respectively (Budhalakoti 2019; Aboul-Enein et al. 2016). Traditionally, the degrading of banana peels has been employed as a fertilizer in order to boost soil nutrients. Tea trash is another significant biowaste that is routinely discarded in open places following tea processing. It has been observed that tea waste can be controlled using vermicomposting, thereby improving the content of nitrogen (Bhuvaneswari et al. 2021).

10 Effects of Fertilizer on Human and Soil Health

Fertilizers are an efficient way for the agricultural sector to boost crop yields. The use of fertilizers is a common practice that could be hazardous to both human and environmental health. Some major constraints are as follows.

10.1 Ground Water Contamination

Large crops that receive repeated fertilizer treatments may suffer the effects on the surrounding soil ecosystem. Concentrated mixtures of N, P, and K make up fertilizer materials, which over time may put stress on soil ecosystems. Particularly N is a highly water-soluble substance that can move quickly through soil layers. Combining fertilizers and pesticides can exacerbate the stress on soil ecosystems, resulting in chemical overexposure that eventually contaminates public drinking water sources.

10.2 Nitrate Concentration

The commercial fertilizer manufacturing process generates very high amounts of nitrate compounds. When organic nitrogen materials break down in the soil, nitrates, which are naturally occurring ions, are created. The chemically charged molecules known as nitrate ions disrupt the soil environment's natural balance. Nitrates are an especially dangerous threat to fetal development, infant health, and the development of the brain and immune system.

10.3 Food Contamination

Crops and water both take up nitrates from the soil as fertilizers permeate soil ecosystems and seep into groundwater supplies. Certain plants, such as cauliflower, broccoli and spinach naturally absorb more nitrate elements than other plant kinds. Up to 70% of the nitrates found in a typical diet are found in the vegetables that people eat. Color-enhancing substances, which are used as meat preservatives, can also contaminate food. Nitrate materials are also present in commercial baby food products and the water used to thin infant formula.

10.4 Destruction of Soil Biodiversity

Applying fertilizer causes the role of bacteria that fix nitrogen to be diminished, while the role of everything that consumes nitrogen is amplified, which leads to the destruction of soil biodiversity. The decomposition of humus and organic matter is then accelerated by these feeders. Organic matter loss alters the physical makeup of soil. Soils are less effective at storing water and air because they have fewer pore spaces and fewer of their sponge-like characteristics. The complex ecosystem of biological exchanges disintegrates as there is less oxygen available for soil microbiology to grow in.

10.5 Negative Impact on Soil Structure

The long-term use of synthetic fertilizers could harm the soil's structure. Because it has an impact on how well water can permeate the soil and how resistant it is to erosion, soil aggregate stability is a key indicator of soil structural quality. The synthetic fertilizer's ammonium ions, which make soil particles separate rather than aggregate, may be to blame for the reduced soil aggregates. This could negate any gains from higher soil organic carbon content and increase fertilizer runoff, which pollutes waterways.

11 Mitigation

The climate change effect on fertilizer use efficiency (FUE) is a crucial concern nowadays. FUE can be concluded by observation of biotic and abiotic component of plant and environment relationship (Barłóg et al. 2022). Reduction of climate change effect on fertilizer use can be avail by different methods.

11.1 Breeding for Climate-Resilient Varieties

One of the best ways to reduce the effects of climate change and other environmental changes is to breed climate-resilient varieties. There is a possibility of reducing the growing seasons for some rabi season crops due to the later onset and/or shorter duration of winter. Breeding for improved NUE (nutrient use efficiency) will be influenced by requiring measurements and determining whether fertilizer treatments are necessary or not. Breeders must prioritize plants with high uptake efficiency (the rate at which plants absorb nutrients from the soil) and utilization or physiological efficiency, i.e., the rate at which plants convert those nutrients into grain (Lightfoot 2018). Nutrient uptake mechanisms include rhizosphere microbial population interactions, rhizosphere chemical and biological property changes that increase nutrient availability, soil volume expansion as a result of increased root growth and altered root architecture, and alterations to the expression of ion transporters in the roots.

11.2 Fertigation

Fertilizers increase the depth to which roots grow and aid plant nutrient absorption in soils with low fertility. Increase in crop yield and quality can only be achieved by optimizing the plant's utilization of the nutrients delivered to it, since only 20–40% of the fertilizer applied is digested by the crops, with the remaining proportion lost via multiple processes. Farmers must be made aware of new techniques and resources for feeding plants nutrition in order to increase yield and increase input use effectiveness (Ranjan and Sow 2021). The use of fertigation, a technique that enables the simultaneous application of water and fertilizers through the irrigation system, maximizes their responsible and efficient use and allows for the maximum efficiency of both without any losses. The benefits of fertigation are as follows: (a) improvement of the fertilizer application process, (b) use only the necessary amount of water and fertilizers, (c) increases the profitability of the use of inputs applied, (d) respect for the environment, and (e) better use of poor-quality soil and water.

11.3 GIS-Based Risk Mapping of Crop Pests

Better resource management through precision agriculture increases the possibility of rising yields and financial gains in agricultural production. When GIS-GPS-RS technologies are used in conjunction with precision farming practices and site-specific crop management, the goal is to maximize the use of soil resources. The use of Geographic Information Systems (GIS), an immersing technology for agriculturalists, assists farmers in overcoming fertilizer overuse applications and other agrochemical applications. (GIS helps in providing proper data at specific interval for the proper utilization of the chemical and identifying agroecological hot spots and potential pest risk areas (Yadav et al. 2010). GIS analytical capabilities can be used to assess variable parameters that may have an impact on agricultural production.

11.4 Integrated Nutrient Management (INM)

In order to maximize crop production, integrated nutrient management applies both chemical and organic fertilizers, including biofertilizers, green manure, organic manure, and other organic decomposable materials. The fundamental principles of INM are the optimization of benefits from all potential sources of plant nutrients in an integrated manner in order to maintain soil fertility and supply of plant nutrients to an optimal level for maintaining desired crop productivity. The term INM refers to the practice of employing microorganisms and the right proportions of organic and inorganic fertilizers to increase the availability and efficiency of nutrient while protecting the soil's natural nutrients and keeping yields high (Selim 2020). INM methods may include a wide variety of subheadings, such as the application of farmyard manure, natural and mineral fertilizers, soil amendments, crop residues, farm waste recycling, agroforestry, green manure, and compost (Selim 2018).

There are five main ideas used by INM which are as follows: (a) plant reserves of nutrients, (b) crops, organic manure, and municipal solid waste all provide nutrients useful to plants, (c) purchased plant nutrients from farm outside, (d) nutrient losses, such as those that occur when crops are harvested and when nitrogen and ammonia in the soil are released as gas or leached away (nitrate, sulfate, etc.), and (e) nutrient outputs from plants, such as crop absorption of nutrients.

11.5 Soil Nutrient Conservation Practices

Protecting soils and watersheds, enhancing water quality, rehabilitating plant, and animal habitats, and promoting healthier soil are all crucial to long-term environmental sustainability. Some soil nutrient conservation practices are no-tillage, terrace farming, crop rotation, contour farming, windbreaks, wetland restoration, and forest cover reestablishment. Unsustainable agricultural practices may harm soil quality, which in turn affects the global climate cycle. Overproduction of carbon dioxide, a greenhouse gas that contributes to global warming, may be emitted from poorly maintained soil. Carbon may be successfully sequestered via agriculture through soil restoration and conservation practices, making it more resistant to climate change.

11.6 Rescheduling of Crop Calendars

Climate change has an impact on rainfall patterns, which shortens the growing season for crops and ultimately encourages farmer to use more fertilizer to increase yield. Early insect infestations are now a much bigger problem as a result of this. Therefore, it is necessary to modify the crop calendars in accordance with the shifting crop environment. With the help of weather forecasting, farmers came to know about the types of crops they can sow for the upcoming season.

12 Future Thrust

A drastic change in climate due to intensive use of agricultural inputs results in intense weather events which ultimately impact agricultural productivity. Weather events like an increase in temperature which will likely have a significant impact on the yield of the crop, uneven rainfall which leads to flood-like conditions in one place and drought-like conditions in another increase crop failure, and chilling injury during the winters ultimately affect the crops yield, microbial population, soil organic matter, availability of water to crop plants, and depletion in soil fertility and soil erosion are simple and complex consequences of climate change. Such kinds of events are major obstacles to agriculture and food security. Change in climate increases nutrient mineralization which majorly affects fertilizer use effectiveness and speeds up evapotranspiration. Conservation agriculture (CA) is one example of the "climate-smart" agricultural technology pushed by governments and development organization to combat the decline in crop yields caused by global warming (Michler et al. 2019). CA is facing serious financial difficulties as a result of increased crop production demand, expensive energy-based input costs, and a downward trend in farm earnings. Agriculture methods are the single most influential element in global food production, which are generally responsible for the drastic climate change (Di Benedetto et al. 1988). Most of these methods involve increasing the usage of inorganic fertilizers and other potentially harmful chemicals in order to achieve the same level of crop yield, which is a major threat to the sustainability of the environment. These chemical fertilizers and pesticides have a residual effect that remains under the surface of the ground for a considerable amount of time. As a result of these inputs, the quality of soil degrades and comes up with biomagnification and eutrophication for aquatic habitat (Cassman 1999). However, current agricultural practices that have significantly increased the world's food supply have unintentionally harmed the ecology, spotlighting the practice of sustainable agriculture.

Therefore, for the prospects of aiming for higher agriculture production without hampered the climate and soil, various innovative steps should be followed such as the use of organic manure like compost, oilcake, vermicompost, green manuring crops, crop rotation, fertigation, panchagavya, and biofertilizers. Various strategies, such as soil and crop management methods, site-specific nutrient management, integrated pest management (IPM), integrated nutrient management (INM), organic mulches for conserving moisture, and breeding techniques can significantly contribute to enhancing crop production. Environmental protection could be achieved by reducing chemical use like pesticides and fertilizers as well as increasing the effectiveness of crop inputs. Sustainable farming has great potential for both humanity and the environment.

References

Aboul-Enein AM, Salama ZA, Gaafar AA, Aly HF, Bou-Elella FA, Ahmed HA (2016) Identification of phenolic compounds from banana peel (*Musa paradisiaca* L.) as antioxidant and antimicrobial agents. J Chem Pharm Res 8(4):46–55

Alburquerque JA, de la Fuente C, Campoy M, Carrasco L, Nájera I, Baixauli C, Caravaca F, Roldán A, Cegarra J, Bernal MP (2012) Agricultural use of digestate for horticultural crop production and improvement of soil properties. Eur J Agron 43:119–128. https://doi.org/10.1016/j.eja.2012.06.001

Altieri MA, Nicholls CI, Henao A, Lana MA (2015) Agroecology and the design of climate change-resilient farming systems. Agron Sustain Dev 35(3):869–890. https://doi.org/10.1007/s13593-015-0285-2

Andriamanohiarisoamanana FJ, Saikawa A, Kan T, Qi G, Pan Z, Yamashiro T, Iwasaki M, Ihara I, Nishida T, Umetsu K (2018) Semi-continuous anaerobic co-digestion of dairy manure, meat and bone meal and crude glycerol: process performance and digestate valorization. Renew Energ 128:18. https://doi.org/10.1016/j.renene.2018.05.056

Bao Y, Feng Y, Qiu C, Zhang J, Wang Y, Lin X (2021) Organic matter- and temperature-driven deterministic assembly processes govern bacterial community composition and functionality during manure composting. J Waste Manag 131(71):31–40. https://doi.org/10.1016/j.wasman.2021.05.033

Barłóg P, Grzebisz W, Łukowiak R (2022) Fertilizers and fertilization strategies mitigating soil factors constraining efficiency of nitrogen in plant production. Plan Theory 11:1855. https://doi.org/10.3390/plants11141855

Basco MJ, Bisen K, Keswani C, Singh HB (2017) Biological management of Fusarium wilt of tomato using biofortified vermicompost. Mycosphere 8(3):467–483. https://doi.org/10.5943/mycosphere/8/3/8

Bhuvaneswari A, Kalaivanan K, Durairaj S, Selladurai G (2021) Effect of bisphenol-A on the bioconversion of tea waste into vermicompost by *Eudrilus eugeniae* (Kinberg, 1867) at different intervals. Int J Biol Sci 8(1):5–12. https://doi.org/10.26438/ijsrbs/v8i1.512

Bora K (2022) Spatial patterns of fertilizer use and imbalances: evidence from rice cultivation in India. Environ Chall 7:e100452. https://doi.org/10.1016/j.envc.2022.100452

Borchard N, Schirrmann M, Cayuela ML, Kammann C, Wrage-Mönnig N, Estavillo JM, Fuertes-Mendizábal T, Sigua G, Spokas K, Ippolito JA, Novak J (2019) Biochar, soil and land-use interactions that reduce nitrate leaching and N_2O emissions: a meta-analysis. Sci Total Environ 651:2354–2364. https://doi.org/10.1016/j.scitotenv.2018.10.060

BragadoCarmo J, Filoso S, Zotelli LC, Neto ED, Pitombo LM, Duarte-Neto PJ, Vargas VP, Andradek CA, Gava GJC, Raffaellarossetto CH, Neto AE, Martinelli LA (2012) Infield greenhouse gas emissions from sugarcane soils in Brazil: effects from synthetic and organic fertilizer application and crop trash accumulation. Glob Change Biol Bioenergy 5(3):267–280. https://doi.org/10.1111/j.1757-1707.2012.01199.x

Budhalakoti N (2019) Formulation and standardisation of banana peel extracted insoluble dietary fibre based buns. Int J Appl Sci 32:1–9. https://doi.org/10.9734/CJAST/2019/45832

Burkett VR, Suarez AG, Bindi M, Conde C, Mukerji R, Prather MJ, St Clair AL, Yohe GW (2014) Point of departure. In: Field CB, Barros VR, Dokken DJ, Mach KJ, Mastrandrea MD, Bilir TE,

Chatterjee M, Kl E, Estrada YO, Genova RC, Girma B, Kissel ES, Levy AN, MacCracken S, Mastrandrea PR, White LL (eds) Climate Change 2014: impacts, adaptation, and vulnerability. Cambridge University Press, Cambridge, pp 169–194

Carolina N, Kimble JM (1990) Predicted impact of climatic warming on soil properties and use. ASA Special Pub 53:71–82. https://doi.org/10.2134/asaspecpub53.c6

Cassman KG (1999) Ecological intensification of cereal production systems: yield potential, soil quality, and precision agriculture. Proc Natl Acad Sci U S A 96(11):5952–5959. https://doi.org/10.1073/pnas.96.11.5952

Dharam Singh M, Chirag G, Prakash PO, Hari Mohan M, Singh R, Manish Kumar D (2017) Nano-fertilizers is a new way to increase nutrients use efficiency in crop production. Int J Agric Sci 9(7):3831–3833

Di Benedetto L, Maugeri G, Poli Marchese E (1988) Sezione and principal linee di ricerca: Dinamismo della vegetazione del piano mediterraneo della sicilia. Giornale Botanico Italiano 122:190–239. https://doi.org/10.1080/11263508809430602

FAO (2020) World fertilizer trends and outlook to 2020: summary report. Food and Agriculture Organization of United Nations, p 38. http://www.fao.org/3/i6895e/i6895e.pdf

Frišták V, Pipíška M, Koperová D, Jagerhofer R, Soja G, Bell SM (2022) Utilization of sewage sludge-derived pyrogenic material as a promising soil amendment. Agriculture 12(3):1–13. https://doi.org/10.3390/agriculture12030360

Fytianos K, Katsianis G, Triantafyllou P, Zachariadis G (2001) Accumulation of heavy metals in vegetables grown in an industrial area in relation to soil. Bull Environ Contam Toxicol 67(3):0423–0430. https://doi.org/10.1007/s00128-001-0141-8

Gaonkar M, Chakraborty AP, Professor A (2007) Application of eggshell as fertilizer and calcium supplement tablet. Int J Innov Res Technol Sci Eng 3297(3):3520–3525. https://doi.org/10.15680/IJIRSET.2016.0503183

He HB, Li WX, Zhang YW, Cheng JK, Jia XY, Li S, Yang HR, Chen BM, Xin GR (2020) Effects of Italian ryegrass residues as green manure on soil properties and bacterial communities under an Italian ryegrass (*Lolium multiflorum* L.)-rice (*Oryza sativa* L.) rotation. Soil Till Res 196:e104487. https://doi.org/10.1016/j.still.2019.104487

Hu A, Huang R, Liu G, Huang D, Huan H (2022) Effects of green manure combined with phosphate fertilizer on movement of soil organic carbon fractions in tropical sown pasture. Agron J 12(5):1101. https://doi.org/10.3390/agronomy12051101

Jakhar AM, Aziz I, Kaleri AR, Hasnain M, Haider G, Ma J, Abideen Z (2022) Nano-fertilizers: a sustainable technology for improving crop nutrition and food security. NanoImpact 27:e100411. https://doi.org/10.1016/j.impact.2022.100411

Jastrzębska M, Kostrzewska MK, Saeid A, Jastrzębski WP (2021) Do new-generation recycled phosphorus fertilizers increase the content of potentially toxic elements in soil and plants? Fortschr Mineral 11(9):999. https://doi.org/10.3390/min11090999

Karmakar R, Das I, Dutta D and Rakshit A (2016) Potential Effects of Climate Change on Soil Properties: A Review Science International 4(2):51–73. https://doi.org/10.17311/sciintl.2016.51.73

King'ori AM (2011) A review of the uses of poultry eggshells and shell membranes. Int J Poult Sci 10(11):908–912. https://doi.org/10.3923/ijps.2011.908.912

Lagomarsino A, Valagussa M, Scotti C, Borrelli L, Becagli C, Tosca A (2022) Mitigation of GHG emissions from soils fertilized with livestock chain residues. Agronomy 12(7):1593. https://doi.org/10.3390/agronomy12071593

Lightfoot DA (2018) Developing crop varieties with improved nutrient-use efficiency. In: Shrawat A, Zayed A, Lightfoot DA (eds) Engineering nitrogen utilization in crop plants. Springer, Cham, pp 1–14. https://doi.org/10.1007/978-3-319-92958-3

Martin SL, Clarke ML, Othman M, Ramsden SJ, West HM (2014) Biochar-mediated reductions in greenhouse gas emissions from soil amended with anaerobic digestates. Biomass Bioenergy 79:39–49. https://doi.org/10.1016/j.biombioe.2015.04.030

McArthur JW, McCord GC (2017) Fertilizing growth: agricultural inputs and their effects in economic development. J Dev Econ 127:133–152. https://doi.org/10.1016/j.jdeveco.2017.02.007

McCauley A, Jones C, Jacobsen J (2005) Basic soil properties. In: Soil and water management. http://landresources.montana.edu/SWM/PDF/Final_proof_SW1.pdf
Michler JD, Baylis K, Arends-Kuenning M, Mazvimavi K (2019) Conservation agriculture and climate resilience. J Environ Econ Manag 93:148–169. https://doi.org/10.1016/j.jeem.2018.11.008
Nortcliff S, Hulpke H, Bannick CG, Terytze K, Knoop G, Bredemeier M, Schulte-Bisping H (2011) Soil, 1. Definition, function, and utilization of soil. Ullmann's Encyc Ind Chem 33:399–422
Owens PR, Rutledge EM (2003) Soil morphology. In: Encyclopedia of soil science in the environment. Elsevier Publishing, Oxford, pp 511–520
Raghavendra KV, Gowthami R, Shashank R, Harish Kumar S (2015) Panchagavya in organic crop production. Pop Kheti 2(2):233–266
Raison RJ, Khanna PK (2011) Possible impacts of climate change on forest soil health. In: Singh BP, Al C, Chan KY (eds) Soil health climate change, Soil biology, vol 29. Springer, Berlin/Heidelberg, pp 257–285
Ranjan S, Sow S (2021) Fertigation: an efficient means for fertilizer application to enhance nutrient use efficiency. Food Sci Rep 2(5):23–27
Reddy PP (2015) Climate resilient agriculture for ensuring food security. Springer, New Delhi, pp 2–13
Selim M (2018) Potential role of cropping system and integrated nutrient management on nutrients uptake and utilization by maize grown in Calcareous soil. Egypt J Agron 40(3):297–312. https://doi.org/10.21608/agro.2018.6277.1134
Selim MM (2020) Introduction to the integrated nutrient management strategies. Int J Agron 2020:e821678. https://doi.org/10.1155/2020/2821678
Shourie A, Singh A (2021) Impact of climate change on soil fertility. In: Choudhary DK, Mishra A, Varma A (eds) Climate change and the microbiome, Soil biology, vol 63. Springer, Cham, pp 49–62. https://doi.org/10.1007/978-3-030-76863-8_4
Srinivasa Rao C, Kareemulla K, Krishnan P, Murthy GRK, Ramesh P, Ananthan PS, Joshi PK (2019) Agro-ecosystem based sustainability indicators for climate resilient agriculture in India: a conceptual framework. Eco Indic 105:621–633. https://doi.org/10.1016/j.ecolind.2018.06.038
Sudhakaran M, Ramamoorthy D, Savitha V, Balamurugan S (2018) Assessment of trace elements and its influence on physico-chemical and biological properties in coastal agroecosystem soil, Puducherry region. Geol Ecol Landsc 2(3):169–176. https://doi.org/10.1080/24749508.2018.1452475
Wangunci Y, Zhao G (2019) Life cycle assessment of potential pollutant-induced human capital loss caused by different agricultural production systems in Beijing. China J Clean Prod 240:e118141. https://doi.org/10.1016/j.jclepro.2019.118141
Xiao Y, Raheem A, Ding L, Chen WH, Chen X, Wang F, Lin SL (2022) Pretreatment, modification and applications of sewage sludge-derived biochar for resource recovery- a review. Chemosphere 287(1):e131969. https://doi.org/10.1016/j.chemosphere.2021.131969
Yadav DS, Chander S, Selvaraj K (2010) Agro-ecological zoning of brown planthopper [Nilaparvata lugens (Stal)] incidence on rice (*Oryza sativa* L.). J Sci Ind Res 69(11):818–822
Yatoo AM, Ali MN, Baba ZA, Hassan B (2021) Sustainable management of diseases and pests in crops by vermicompost and vermicompost tea. A review. Agron Sustain Dev 41:7. https://doi.org/10.1007/s13593-020-00657-w
Załuszniewska A, Nogalska A (2022) The effect of meat and bone meal (MBM) on phosphorus (P) content and uptake by crops, and soil available P balance in a six-year field experiment. Sustainability 14(5):2855. https://doi.org/10.3390/su14052855
Zhao X, Li J, Che Z, Xue L (2022) Succession of the bacterial communities and functional characteristics in sheep manure composting. Biology 11(8):1181. https://doi.org/10.3390/biology11081181

Chapter 5
Modern Agronomic Measurement for Climate-Resilient Agriculture

Abdul Jabbar, Hina Ibrahim, Athar Mahmood, Muhammad Mansoor Javaid, Allah Wasaya, Muhammad Ather Nadeem, Mujahid Ismael, Muhammad Awais, and Shahid Raza Khan

Abstract Agronomy covers nearly 30% of the earth and intricately tied to environmental quality. The environmental quality section is primarily concerned with understanding how agriculture affects our environment and improving agricultural management to increase air, soil, and water quality. Climate risk assessment is critical for effective adaption activity, and several techniques have emerged. The dynamics of the individual component have received little consideration in climate risk and vulnerability assessments. The article describes the major problem of climate which occurs due to modern agriculture techniques. Crop sensitivity increases with an increase in weather variability. Heat and drought are the most common stressors encountered in the field, and they have a substantial impact on plants. Over uses of pesticides may result in the degradation of biodiversity. Deforestation is considered as the second largest source of anthropogenic cause of CO_2 in the atmosphere. Amazon forests have supplied an important carbon sink service. The mature forest may currently be jeopardized due to a variety of activities; for decades, mature forests throughout Amazonia have made an important contribution to the mitigation of climate change. Efforts to mitigate climate change through Reduced Emissions from Deforestation and Degradation (REDD) rely on the mapping and monitoring of tropical forest carbon stocks and emissions across large geographical areas.

Keywords Climate risks · Modern agricultural system · Pesticides and insecticides · Water management · Organic agricultural system

---

A. Jabbar · H. Ibrahim · M. Ismael
Department of Botany, University of Agriculture Faisalabad, Faisalabad, Pakistan

A. Mahmood (✉)
Department of Agronomy, Faculty of Agriculture, University of Agriculture, Faisalabad, Pakistan
e-mail: athar.mahmood@uaf.edu.pk

M. M. Javaid · M. A. Nadeem · S. R. Khan
Department of Agronomy, College of Agriculture, University of Sargodha, Sargodha, Pakistan

A. Wasaya · M. Awais
College of Agriculture, Bahauddin Zakariya University, Layyah, Pakistan

M. Hasanuzzaman (ed.), *Climate-Resilient Agriculture, Vol 2*,
https://doi.org/10.1007/978-3-031-37428-9_5

1 Introduction

Agronomy is a subfield of agricultural science that examines the theories and methods of crop, soil, and water management. Taken from the Greek words "agros," which means "field," and "nomos," which means "to manage" (Maliwal et al. 2007). In the subject of agriculture known as agronomy, methods are used to make the crop more hospitable so that they produce more (Fadiji et al. 2022). Norman (1980) defined agronomy as the science of altering the crop environment, which is complicated with the dual objectives of enhancing agricultural output and understanding the mechanism at work (Daughtry et al. 1984).

1.1 History of Agronomy

The drive for bigger and better harvests in the typical American landscape led to the development of agronomy as a science in the twentieth century (Davidson et al. 2012). The field underwent rapid development as a field science with the establishment an American society of agronomy in 1907. In 1938, Sir R.G. Staplendon published a remark in Herbage Abstract (Insua et al. 2019).

1.2 Relation of Agronomy with Other Sciences

Agronomy comprises soil science, agriculture chemistry, crop anatomy, plant ecology, and metabolism. It is very crucial to have a thorough grasp of the physiological, synthetic, and biotic characteristics of agricultural land in order to influence the environment. In a similar line, it is must to understand the physiology of crops in order to their needs. Advanced economic analysis has benefited agricultural productivity. Agronomy is a significant area of study in agriculture science. While there are numerous traditions associated with soil science, it is often perceived as a research-based and technological field that strives to comprehend and influence processes at the molecular level. Particularly in the anglophone tradition, agronomy is often thought to be the application of plant and soil science to crop production. Due to the fact that the majority of agronomic research is applied and practical, occurring in experiment stations, laboratories, and on farmers' fields, agronomists seldom find themselves in the political or public spotlight. Agronomy is a phrase that has multiple meanings. It engages with husbandry on a regional and technical level, focusing on the most recent state of the art. It is a general phrase that, on a global scale, is used to describe the work of experts from a variety of fields, such as soil science, plant breeding, and crop protection. In this chapter, it refers to a critical evaluation of sugar beet production methods in light of environmental science and plant crop physiology (Wuest et al. 2021).

1.3 Understanding the Climate Risk

The added risk to farm business and output is climate change risk caused by the increase in weather variability and extremes in the coming years (Zougmoré et al. 2021). It will probably become increasingly crucial to agricultural productivity. In the next coming several centuries, the whole United State will warm by 2-to-5-degree Fahrenheit (Raza et al. 2019). That increase in temperature is significantly greater than rate of change that took place in the twentieth century. Winter and spring precipitation will likely be higher in the Northern and in the South, whereas summer so fall cloudburst will most likely be the same or less in most places (Jeyasri et al. 2021). Although, determining the impact of climate changes in throughout the year production and finances can be difficult (Grippa et al. 2019).

1.4 Understanding Crop Sensitivity to Climates

The sensitivity of crops increases with increased weather variability, forcing growth to adapt in a variety of ways, frequently by switching cultivate they cultivate or by adjusting the period of their rotations (Freitas et al. 2021). Heat waves and warm nights are already affecting fruit set in vegetable crops and lowering grain crop production in several regions of the country. Crop development is also being affected by warmer winters and more unpredictable spring weather, particularly in crops (Roudier et al. 2011). Crop production is quite weather-patterned. It is influenced by interannual climatic changes, shocks during particular phonological stages, long-term trends in average rainfall patterns temperature, and extreme weather events (Javaid et al. 2022a). Various forms of pressures impact different crop species and drastically affect at each morphological and physiological stage, whereas some crops are more resistant to those stress than others. Climate carbon dioxide quantity changes from 281 to 401 mol at the end of continuous erosion and unsustainable fossil fuel use (Erans et al. 2022; Javaid et al. 2022b). At the end of the twenty-first decade, the CO_2 concentration is estimated to rise by two or up to 800 moles (Weller et al. 2021). The poor environmental conditions primarily affect agricultural yield in underdeveloped nations, and intense temperatures and excessive CO_2 buildup have led scientists to discover novel solutions to less foreseeable problems (Pivello et al. 2021). The yield of fresh smart climate cultivars of crop is required to tackle these problems and ensure sustainability. Additional abiotic conditions that influence stress include UV-B, light intensities, flooding, gas emissions, and physical and chemical problem. An average temperature of the earth is predicted to rise by 2 °C in the twenty-first century (Ramirez-Villegas et al. 2013). The harsh environmental conditions seen all around of globe have an impact on all living things such as humans, fish, animals, and plants. Everyone is concerned with the threat to global climatic conditions because crop yields may be jeopardized by changes in many environmental elements, which could jeopardize food security. According to recent studies, industrialized nations are more susceptible to climate shifts (8–11%) than emerging nations (Zou et al. 2019).

1.5 Plant Yield and Climate Change

Weather variability had significant effect on plant physiology because of number of reasons. Climate uncertainty and environmental stress increase the likelihood of various (Wasaya et al. 2021). Challenges for plants are shown in Fig. 5.1; crop production is impacted by climate change through direct, indirect, and socioeconomic effects. Furthermore, as documented by the Food and Agriculture organizing (FAO) and as visualized at Fig. 5.1, weather-changing events (dry spell, swamping, increasing in temperature, etc.) are increasing tremendously. The current agricultural systems have limited variation and high input concentration, and unstable production brought on by environmental changes on crops has exacerbated climate influences. It is projected that the increase in intensity of rainfall, temperature fluctuations, salt, and insecticides pesticides attacks will reduce crop output, raising the chances for famine. In addition to temperature differences, rainfall has had an impact on crop adaptation (Fig. 5.1).

The climate dramatically changes when the earth's temperature rises because of non-living stress. Weather changing can be exceedingly harmful and put naturally occurring crop species in danger in many different ways (Alengebawy et al. 2021). In case of field, heat and drought are the most common stressors which have a large

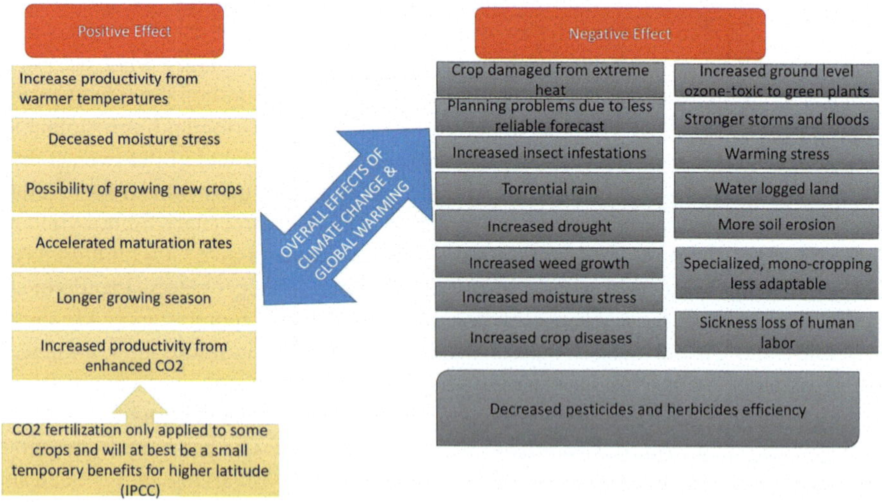

Fig. 5.1 Climate change has direct, indirect, and socioeconomic implications for agriculture production. Between 1990 and 2016, there was an enlarged number of extreme weather-related incidents. (Porfirio et al. 2018)

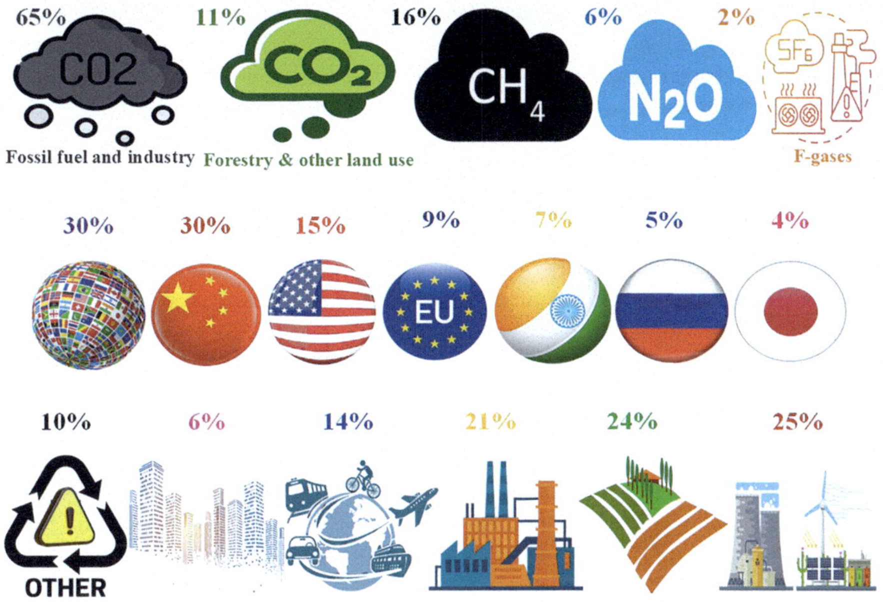

Fig. 5.2 Impact of weather variability and global warming on crops and people. (Raza et al. 2019)

impact on plants. According to reports, plants need a specific temperature to grow and blossom normally (Jeyasri et al. 2021; Mahmood et al. 2021). The production and yield of grains are negatively impacted by heat stress, sterility is triggered by cold stress, and the morpho-physiology of plants is negatively impacted by drought stress. These climatic issues significantly impede plant growth and yield, leading to extensive responses involving molecular, biochemical, physiological, and morphological changes. These impacts are elaborated upon in the context of "Global Warming and Weather Variability: Pleasant and Unpleasant Effects on Agriculture and Humans" (Fig. 5.2).

1.6 Excess Use of Pesticides and Insecticides

Pesticides are used to minimize insects and pests that attack and injure crops. From decades, various insecticides have been made to protect crops. Pesticides help the crops, but they have a detrimental effect on the ecosystem (Mahmood et al. 2016). They degrade biodiversity because of the overuse of pesticides. The survival of many organisms especially birds and aquatic life is threatened by toxic substances. Concerns about pesticides have been raised regarding environmental stability and sustainability (Rani et al. 2021). This chapter will address pesticides and their wide range of applications and related environmental issues. This chapter will also cover the long-time impact of pesticides on the climate and the pollution caused by pesticide misuse. As the chapter draws to a close, it addresses the strategies for ending pesticide use before examining how the world will change in the future once pesticides have been eradicated (Aqueel et al. 2015; Sarkar et al. 2021; Karimi et al. 2022). With the overall world, population will be increased by about 10 billion people at the time of 2050, rising food production for every nation's top priority. Evidence suggests that the world's population has been increasing by 97 million people every year; because of multiple treatments of the water, air, and soil, ecosystem has been contaminated by the continual application of persistent and non-biodegradable pesticides. Furthermore, pesticides have been taken up by plants in the upper tropic level of the food chain (Heong et al. 2015).

1.7 Pesticides Use (From Past to Present)

Pesticides have already been employed since the time of the Ancient Romans, who used salts, ashes, and bitters to control weeds and burned sulfur to kill pests (Mahmood et al. 2016). According to a Roman naturalist, arsenic is being used as a pesticide (Li et al. 2022). In the 1600s, arsenic and honey were used to control ants. Farming in the USA began employing synthetics such as calcium arsenate, sulfur, and nicotine sulfate for field-related purposes in the late 1800s, but their efforts were unsuccessful due to the crude steps. Arsenic, an alloyed form of copper, was introduced in the USA in 1867 to combat the Colorado potato beetle infestation (Kaur et al. 2019).

1.8 Pesticide Registration and Safety

Pesticide registration is a complex, time-consuming, and resource-intensive legal and administrative procedure that requires the expertise and capabilities of both the registration authority and pesticide producers. In order to ensure the safety of both the active and inert chemicals used in the manufacturing of pesticides, potential impacts connected with the usage of weed killers on human health and the environment are analyzed in this process.

1.9 Risks Associated with Pesticide Use

The hazards of using pesticides have outweighed any potential benefits. Pesticides severely harm non-target species, the biodiversity of flora and fauna, aquatic and terrestrial food webs, and ecosystems (Gonçalves et al. 2021). Within a few hours of application, 80–90% of the insecticides used can volatilize, according to

(Majewski and Capel et al. 2019). It frequently happens when utilizing a sprayer. The pesticides that are volatilized escape into air and may subsequently be harmful for non-target organisms. The use of pesticides that volatilize from treated plants, where even the droplets can adversely affect other plants, serves as a strong illustration of this phenomenon (Kaur et al. 2019). Pesticide exploitation has led to the extinction of a number of terrestrial and aquatic species of flora and fauna. Additionally, they put certain endangered species, including the bald eagle, peregrine falcon, and osprey, in danger of going extinct. Therefore, dangerous levels of these contaminants have been discovered in water bodies, soil, and air (Rathi et al. 2021; Shrestha et al. 2021).

1.10 Impact of Pesticides and Insecticides on Wildlife and the Environment

Pesticides have an impact on weather variability change during all stages of production, delivery, and use. Pesticide manufacture emits three major greenhouse gases CO_2, methane, and N_2O (Panchasara et al. 2021; Zhuang et al. 2022). Phosphate ore is mined, purified, and synthesized into glyphosate, the active ingredient in the well-known weed killer Roundup. Phosphate mining impacts the biosphere by eliminating wildlife habitats and polluting the air and water (Javaid et al. 2022b). The effects of even trace levels of pesticides in waterways on fish and other aquatic fauna worsen as water temperatures rise (Jensen et al. 2021). For instance, as the water's temperature rose from 45 to 63 °F, the toxicity of glyphosate in bluegill and rainbow trout doubled. According to a scientific investigation, pesticides can affect endangered salmon's swimming ability, growth, development behavior, and reproduction even at low doses (Jenkins 2021). Farm-found insects have enhanced winter survival and expanded the number of generations they can generate (Rathore and Nollet 2012). Farmers now use toxic pesticides like neonicotinoids slathered on seeds to regulate these insects as their population grows (Wilson and Fox 2021). Neonicotinoids are pesticides that kill insects by inhibiting nerve impulses, which resulted in a 40% decrease in insects throughout, excluding pollinators (Danion et al. 2018; Yuan et al. 2019).

1.11 Solution and Hope for the Future

We have presented a bleak picture of what would happen if pesticide use and climate change proceed in their current patterns thus far. Embracing change may seem unattainable, yet there is good news (White et al. 2019). By making little improvements to our ordinary routine, such as what we eat, and by supporting laws that encourage the production of pesticide-free foods, we can all work together to find solutions. Organic agricultural methods are vital since it is the main plan to fight weather variability. Many of these approaches have their roots in indigenous and black farming traditions. In my neighborhood, several small farmers are working with land trusts to maintain farmland through regenerative agriculture (White et al. 2019). Now till agriculture and cover crops are examples of organic techniques that lower down methane, nitrogen dioxide, and carbon dioxide emissions. Additionally, by restoring the health of the soil, they aid in the conversion of CO_2 into oxygen and strive to stop climate change (Tully and McAskill 2020). Healthy soil may store more carbon and increase water filtration, resulting in cleaner water since it is alive with microorganisms. It could be difficult to control pests and weeds without using harmful pesticides, but it is doable (Tully et al. 2021). Numerous science-based techniques will help you use fewer pesticides, such as flame weeding, mulching, mechanical and manual weed removal, and choosing the least hazardous insecticides. Beneficial insect populations can also be increased to eat pest insects (Mazzoncini et al. 2010; Hassan et al. 2018).

1.12 High Emission of Carbon Dioxide and Methane in the Environment

Crops may be affected directly by increase in atmospheric carbon dioxide levels or indirectly by climatic changes, according to the results of controlled experiments (Sarkar et al. 2021; Skendžić et al. 2021). Carbon dioxide enhances dry matter production, mostly through enhancing photosynthetic response, minimizing transpiration, and excessive water use efficiency (Amitrano et al. 2021; Mahmood et al. 2022). For a doubling of CO_2 concentrations, for instance, yields of C_3 crops (like wheat), for example, might increase by 10–50% (Hassan et al. 2019; Govere et al. 2022). Through a variety of methods, including crop effect study, marginal spatial analysis and agricultural systems analysis, it has been determined how sensitive crop yields are to variation in climate (Nobre et al. 2016). Such research suggests that an average temperature of 2 °C may outcome in a 3–17% decline in potential yields for the central mid-latitude grain regions (Phan et al. 2014) According to research of agricultural systems, agronomic, policy, and market feedback mechanisms could absorb or prevent a significant portion of the anticipated negative consequences of climate change. These findings are merely preliminary (Karki and Gurung 2012). The combined net impact of increased CO_2 and climate change on world agriculture is now difficult to predict. Agricultural emissions primarily methane and nitrous oxide contribute significantly to anthropomorphic climate change (Nisbet et al. 2021). Therefore, lowering these emissions could have a big impact on mitigating climate change. Figure 5.3 shows the total emissions of greenhouse gases by different source (Jiri et al. 2015).

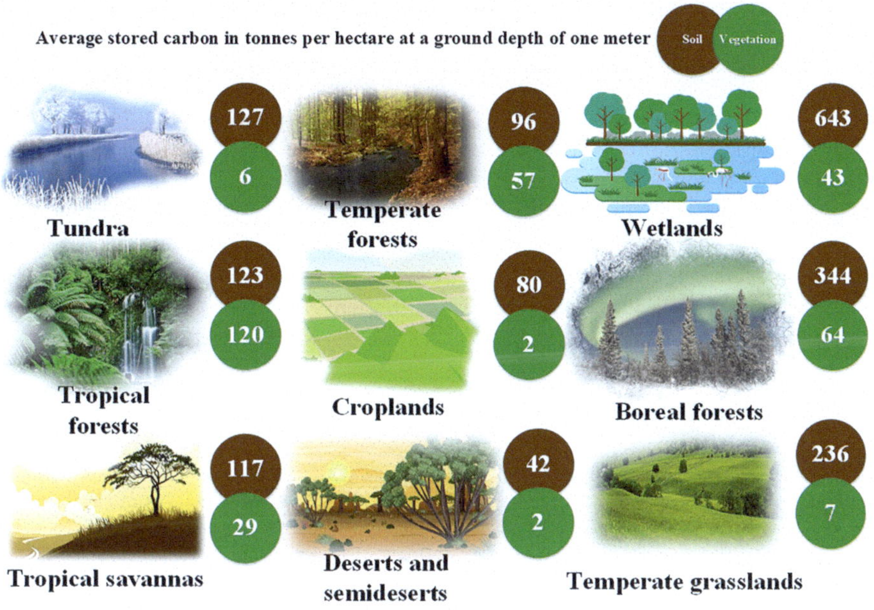

Fig. 5.3 Emission of different gases in purpose of energy sources. (Song et al. 2019)

1.13 Agricultural Greenhouse Gas Emission

Multiple climate toxicants such as CO_2, CH_4, and N_2O are the three main contributors to global warming that result from anthropogenic climate change (Javaid et al. 2022a). These three greenhouse gases have linked with agriculture and food-productive items, but direct agricultural emissions are exceptional in that CH_4 and N_2O predominate (Etminan et al. 2016). Reports described that these potential gases from 21% to 37% emitted into the atmosphere from the world food system (Clark et al. 2020). Methane and nitrous oxide, two non-CO_2 gases, make up a disproportionately substantial portion of agricultural emissions (Smith et al. 2021). In another way to comprehend how agronomy contributes to climate change and what may be accomplished by reducing agricultural emissions, we must therefore grasp how emissions of these gases affect temperature change (Hanif et al. 2022). Our common reporting metric of GWP100 Scientists developed GWP100 in 1990 as a way to balance greenhouse gases' ability to cause global warming (Myhre et al. 2014). It is necessary to give more attention to the applications and constraints across various metrics (Jackson et al. 2019). We urge the adoption of various alternative emission measurement methodologies, modeling of the pertinent impacts, and other future environmentally robust approaches (Le Quéré et al. 2020; Weller et al. 2021).

2 Principal Resource for Managing Climate Risk

The risk of climate unpredictability is always present in agricultural productivity. Producers are frequently at the whim of uncontrollable environmental forces, especially variations in rainfall from season to season and between years. Changes from the "normal" climate might provide the conditions for the occurrence of bugs and outspread of diseases, among other production concerns. Climate data and projections can be used to optimize resource utilization, lower production risk, and boost agricultural businesses' profitability. However, simply supplying potential customers with improved climate projections is insufficient (Fraisse et al. 2016).

2.1 Natural Resources

Before the food arrives on our plates, food is produced, preserved, processed, packaged, delivered, prepared, and saved. At every level, the food supply releases greenhouse gases into the atmosphere. Large volumes of the powerful greenhouse gases nitrous oxide and biogas are emitted specifically by agriculture. Belching is the method through which methane produced by enteric fermentation during the digestion of livestock is released into the environment. Additionally, it might escape from manure and organic waste storage areas of landfills. Nitrous oxide emissions from organic and mineral base nitrogen fertilizers are created indirectly. Pressure on finite water resources increases as the population increase and water requirements for agriculture, residential, industrial, and municipal use rise (Tompkins and Adger 2004). According to estimates, the agricultural sector alone uses over 85% of the total water consumed by humans, with certain emerging nations seeing this percentage reach over 90%. Water withdrawals rise as a result of rising community requirements for grain and livestock production as shown in (Fig. 5.4).

2.2 Human Resources

To better understand human resources management and its evolution, it is very pertinent to understand the social elements that are affecting agricultural industry (Dietz et al. 2020). Globalization and commercialization continuously define new requirements for agricultural labor to improve the productivity and knowledge in addition to land, labor, and capital which are crucial (Usman et al. 2021). Multination corporations are playing a bigger role in the world's economy and varied cultures having a bigger impact on their HR policies and planning a solution for the mitigation of environmental pollution. Current HR policies are compatible with existing acceptable behavior and patterns that transcend the workplace, a certain degree of matching HR practices and local culture increase performance (Moustaghfir et al.

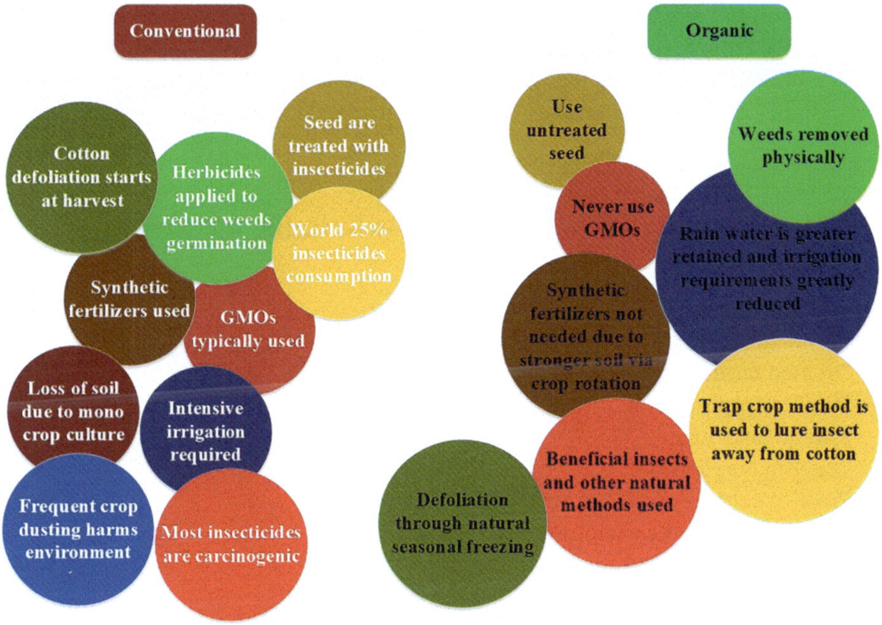

Fig. 5.4 World population and freshwater use. (Abdul-Razak and Kruse 2017)

2020). Staff employees must not only feel appreciated but also engaged with their education if the mission of managing natural resources in the face of climate change is to be completed (Amushila and Bussin 2021). Many government workers in public administration started out as hands-on protectors of nation's natural resources; this passion has to be fully realized because the danger of climate change might exacerbate organization issues. Current personnel may be well-equipped (or have been given professional positions) to solve climate change concerns within the framework of their current work procedures and management system (West et al. 2009).

2.3 *Financial Resources*

Measures to mitigate and adapt to climate change are funded locally, nationally, or internationally with funds from public, private, or alternate sources of finance (Khan and Munira 2021). The party to the convention, Kyoto protocol, and Paris Agreement is necessary for providing financial support to people who are less wealthy and more vulnerable. This acknowledges the wide variation in how much each country contributes to greenhouse gas emissions and how well-equipped they are to stop them and deal with their effects. However, the responsibilities of the National Development Bank (NDBs) in a developed and developing nation are seldom compared in the existing research in the role of NDBs in weather finance (Geddes et al. 2018). Public policies are essential for financing renewable energy. Renewable energy investment requires many policies since there are several marketplace system failures, bottlenecks, hazards, and actors involved (Kern et al. 2017).

3 The Following Techniques Will Be Suitable for Boosting the Climate

- Sustainable forests
- Organic farming
- Diverse crop rotation
- Rotational grazing
- Water management
- Practice such as conservation tillage
- Integrated system agriculture
- Rehabilitation of degraded pastures
- Amazon native tree species
- Use of organic fertilizers in sustainable agriculture
- Cultivating recovery capacity

3.1 Sustainable Forests

European forest institute and the Food and Agriculture Organization of the United Nations (FAO) have just released a new report that forests are vital resources for people's livelihoods and are essential to solving many global concerns, such as climate catastrophe and poverty (EFI) (Katila et al. 2019). Sustainable forestry, also known as sustainable forest management, was developed to fulfill demands for forest resources without jeopardizing their availability for future generations (Kumar et al. 2021). Conceptually, this entails using procedures that may create essential goods and services while also regenerating in accordance with nature's regular pattern without forests, and humanity could not exist (Gremmen 2022). The oxygen we breathe is produced by trees through natural processes, and they also use carbon dioxide from the atmosphere (Aitken and Simard 2015). Because trees can help to absorb a significant amount of fossil fuel emissions generated by human activity, sustainable forestry is the most important way that people will be reduced the effects on weather variability (Jhariya et al. 2019). To reduce carbon emissions, our forests are home to a significant portion of the terrestrial creatures on earth, which are vital to our natural ecosystems. However, despite all of these causes, human activity is gradually depleting our natural forest resources (Regnier et al. 2022). Due to this, it is an imperative that people learn and adopt sustainable forestry practices and laws as soon as possible. Once our forests are gone, we will not be able to support ourselves and our future generations in a sustainable way (Hahn and Knoke 2010). Acknowledging the world carbon cycle and consequently weather change requires quantifying. Forests play reasonable role toward the absorption, storage, and release of carbon (Regnier et al. 2022).

3.2 Organic Farming

Compared to conventional farming, organic farming offers a comprehensive approach that lessens its influence on the environment. Increasing the amount of land used for organic farming can improve farming systems' resilience, improve soil health, and maintain or enhance biodiversity (Knapp and van der Heijden 2018; Basavalingaiah et al. 2022). It can also help with climate change mitigation. The amount of land used for organic farming can improve farming systems' resilience, improve soil health, and maintain or enhance biodiversity (Lin et al. 2011; Clark 2020). It can also help with climate change mitigation (MacRae et al. 2010).

3.3 Organic Consumes Less Energy and Reduces GHG Emissions

Organic farming relies on creating closed nutrient cycles and reducing nitrogen losses rather than being reliant on external, fossil fuel-intensive fertilizer or pesticide inputs. This has the ability to cut agricultural GHG emissions by about 20% globally (Maraseni et al. 2021). Animals in organic systems have access to free-range areas and are permitted to graze as much as possible, and 65% of the food has to come from the farm (Knapp and van der Heijden 2018). Refraining from using synthetic fertilizer reduces N_2O emissions from the soil by 40% per hectare in organic farming (Burger et al. 2016). As a result of fewer animals and grassland-based systems, emissions are reduced and soil carbon reserves are improved (Guyomard et al. 2021).

3.4 Environmental Benefits of Organic

Organic farming is done through a process that works to conserve the environment while simultaneously enhancing it (Wittwer et al. 2021). Organic agriculture strengthens soil through techniques like composting, cover crops, and crop rotation rather than relying on chemical fertilizers and pesticides can drain the soil of essential nutrients and hasten, and this will hasten environmental degradation.

Additionally, organic farmers work to conserve and safeguard the natural environment using the knowledge that a variety of biological environments contributes in the world's ability to feed its population as well as the planet's (Singh 2021). Organic strategies aid in preventing hazardous and persistent chemical runoff from entering our water sources (Spahr et al. 2020). Additionally, organic farming contributes to the reduction of our carbon footprint and fight against weather variability by forbidding the use of petroleum-based fertilizers and by absorbing CO_2 from the atmosphere (Elbasiouny et al. 2020).

The data also show that different forms of energy were used in different ways. In organic systems, 63% of the total energy was used for direct energy consumption (labor, fuel, and equipment), compared to 27.5% for conventional systems (Khanali et al. 2021). The production and delivery of off-farm inputs like seed, soil fertilizer, and herbicide represented in diverse shades of green consumed the most indirect energy in traditional systems, making up 72.5% of the system's overall energy needs. The results are anticipated to differ for permanent crops, and it should be highlighted that these data represent annual agricultural output (Fess and Benedito et al. 2018).

3.5 Diverse Crop Rotation

George Washington Carver, an agricultural chemist, created crop rotation techniques to preserve soil nutrients and identified hundreds of new applications for crops like peanuts and sweet potatoes (Benitez-Alfonso 2022). Crop rotation is the technique of successively growing different crops on the same plot of land in an effort to improve soil health, increase its nutrient content, and decrease the number of weeds and insects that grow there (Aziz et al. 2022). The capacity to disrupt the disease process in topsoil is very advantageous in diversified crop rotation on the same farmland, and monoculture facilitates pathogen growth (Ripoche et al. 2021). Without alternation, these pathogens can multiply in the soil and outpace the spread making plant disease outbreaks worse. A host plant illness might be the monoculture farming technique (Dhingra 2021). Pest control involves halting the pest cycle, reducing weeds and disease, enhancing soil quality, and protecting the ecosystem. In several parts of Europe, crop rotations are a crucial component of agricultural management. A realistic carbon budget for an agricultural ecosystem with intercropping can only be obtained when at least a full rotation is taken into account; this is because cropping and fallow period are irregular and rarely coincide with the scheduled year and because of different crops and related agricultural practices in various carbon inputs into the soil.

3.6 Alternating Grazing

The bulk density of rotational grazing was lower than that of regular grazing (Baronti et al. 2022). Rotational grazing was identical to no grazing and had a higher SOC than continuous grazing. Rotational grazing's beneficial effects on SOC may offer prospects for climate change mitigation. The aridity class had little influence on grazing tactic comparison (Benitez-Alfonso 2022). We discovered that grazing treatment can considerably affect soil performance and health outcomes when compared between continuous and no grazing strategies. Furthermore, site-specific environmental variables play crucial moderating affects (Kremen and Miles 2012). These knowledge gaps would be greatly reduced, and our collective understanding of the effects of grazing on soil quality would be greatly improved, resulting in greater management and policy impacts. Consistent guidelines for soil health analysis and coordination across regional, national, and international efforts would help close these gaps in knowledge (Byrnes et al. 2018).

3.7 Core Idea

Grazing causes soil compaction when compared to areas with no grazing. Rotational grazing techniques may offer a chance to reduce climate change (Byrnes et al. 2018). Through continual grazing, rotation increases soil bulk density and organic matter content. Lowering grazing intensity raises soil organic carbon and bulk density. Environmental factors that are unique to a certain site are significant moderators. The thicknesses show the flow of heat and water exchange, and the dotted lines indicate the transport of water (blue) and heat (red). The highland pasture in position A is in good condition; the meadow in position B is light to moderately degraded and has a few tiny "Bare Land" patches, and the meadow in position C is severely and substantially degraded and has several big "Bare Land" patches. According to the University of Wisconsin, pastures are the greatest crop for lowering runoff, erosion, and phosphorus in comparison to other land uses. As the grassland vegetation gets thicker and the soil quality improves, the water quality also gets better (Gayer et al. 2021).

3.8 Water Management

According to a recent UN Water Policy Brief, water management should be a major component of the struggle to minimize the worst threat of environmental change and cut greenhouse gas emissions. In addition to other anomalies such as water scarcity, lower agricultural productivity, rising sea levels, melting ice caps, and loss of biodiversity, the increased global temperature is considered one of the primary adverse outcomes of environmental change (Srivastav et al. 2021). The rising temperature may cause the troposphere to produce more ozone at a faster pace Liu and Wang 2020). In particular, environmental issues have been thought of as only an increase in temperature, but as mentioned above, there are other aspects that have been noted, such as changes in water availability, excess, decreasing rising per capita water requirements, changes in land use pattern, impacts on food security, etc. Weather variability may have impacts in availability of water resources, energy production, temperature, agricultural productivity, yearly rainfall pattern, and public health among other things (Aggarwal et al. 2019). According to a number of studies, water shortage is becoming a global issue due to climate change indicating that creating resilience against water scarcity brought on by climate change requires more focus on potable water control techniques than any other strategy (Howard et al. 2010). Water stress region with severely low water supplies is made worse by climate change growing competition for water potentially provoking violence. Water stress or regions with severely low water supplies are made worse by climate change, growing competition for water, and potentially provoking violence. The world must become more water-smart (Ali et al. 2013; Naumann et al. 2018). We cannot afford to wait since everyone has a party to play. Currently, there is climate change. Water is a component of the answer, and we must act. The health and lives of children will be protected by adjusting to the water consequence of climate change. Reducing greenhouse gas emissions and switching to solar-powered water system would help to save the children future.

3.9 Practice Such as Conservation Tillage

Conservation tillage was first advocated to mitigate soil erosion, enhance soil organic compound content, and increase soil water storage. It is now recognized as a component of the answer to alleviate the impacts of global warming and to ensure sustainable farming (Harper et al. 2018). A definition from the Sustainable Technology Resource Centre is among the most regularly used (CTIC). The parts that follow will describe how conservation tillage works to improve soil health from the perspectives of water interplay, soil physiochemical qualities, and the atmosphere as well as by lowering greenhouse gases for lessening the consequences of climate variability. Climate science is characterized by increased temperature, uncertain rainfall, floods, droughts, windstorm, and other peak weather patterns. These phenomena have shown a negative impact on agriculture by disrupting soil structure, depleting soil nutrients, sinking biodiversity, spreading plant disease and pests as well as lowering crop yield (Geisen et al. 2019).

More than 20 years ago, it was understood that when there is no-tillage, the proportion of phosphorus and potassium that can be extracted from the soil rises, and distinct gradients of these nutrients appear in the soil's surface layers (Yadav et al. 2020). Some trends have arisen despite the fact that the outcomes of these tillage studies show a lot of diversity (Piazza et al. 2019). A review of long-term trials assessing the impact of tillage on crop growth and production takes into consideration variations in weather patterns, potential gradual changes in soil properties, and the learning curve often observed with the adoption of new tillage methods. The evaluation focuses exclusively on studies conducted in North America, Europe, and New Zealand over the past 10 years or more (Akhter et al. 2020). When utilized for bio-tillage, cover crops with deep, thick roots may significantly enhance the soil's shape and H_2O and air transmission by creating bio-pores. The roots of a bio-tillage cover crop should be deep and thick, growing quickly, decomposing leftover roots quickly, adapting well to soil and climate conditions (Zhang and Peng 2021). Tillage among the primary agronomic practices is believed to lower SOC stock and has a significant impact on soil carbon emission. According to estimates, converting all European farmland to zero tillage would eliminate all direct emissions of carbon from fossil fuels (Panchasara et al. 2021). Numerous investigations claim that tillage boosts the oxidation of organic carbon releasing a significant amount of CO_2 into the atmosphere over small duration of time (Baveye et al. 2020). The amount of CO_2 emitted right after tillage rose with plowing depth and was in very instance much higher than the amount from the no-tillage condition. Through the cleavage of organic compounds in soil (SOM), which also produce CO_2, intensive soil farming lowers the overall carbon concentration (Hamidi et al. 2021).

Alteration in dissolved organic carbon (DOC) and dissolved organic nitrogen (DON) pools, which serve as a platform and source of energy for heterotrophic microbes, has an impact on greenhouse gases (Chow et al. 2022).

3.10 Integrated System Agriculture

Agriculture may be combined with cattle, livestock, and fish in an integrated agricultural system to create year-round work opportunities, produce additional revenue, and utilize the excreta of chickens, for instance, by placing them on the top layer (van der Velden et al. 2022). Pigs are on the bottom layer, and agriculture and the growth of fodder crops use the pond's leftover groundwater. Rainwater harvesting agriculture (RHA) is an intergrade method for semi-arid water management on rain-fed land that was developed by experts in Gansu Province more than 10 years ago. This system is made up of three primary components; a system for collecting rainwater; an irrigation system that uses less water; and a system for producing crops with exceptional efficiency. (Munir et al. 2021). An integrated farming system should be considered to provide the basic needs of a household such as food (grains, pulses, oilseed, milk, vegetables, honey, meat, etc.), feed, fodder, and fiber (Choudhary et al. 2019). Undoubtedly, the bulk of farmers have been engaged in farming for a very long time, but their main focus has been on separate elements rather than an integrated approach. Numerous efforts have been made at the ICAR and State Agriculture Universities' level to increase the productivity of the various farming system parts, such as crops, dairy, livestock, poultry, goat maintaining, ducker, honey bees, agro-base, horticulture, and mushroom cultivation, but their integration into the farming system as whole has lagged (Choudhary et al. 2019).

3.11 Amazon Native Tree Species

Amazon's financial cooperation has relied heavily on the extraction of both renewable and nonrenewable resources and environmental assets in the past 50 years, which has resulted in substantial environmental factors throughout the area (Palahi et al. 2020; Sharifi et al. 2021). It is possible for climate change brought on by rising greenhouse gas emissions and another human force to raise air temperatures and alter precipitation patterns in complicated ways (Sun et al. 2022). Identifying present and likely future changes in climate factors, particularly those associated with the hydrological cycle, in these places has not been a simple understanding, the significance of knowing how climate influences the arrangement and operation of the rainy forest that we know it today. This challenge for the Amazonian Forest is brought on its part by the lack of theoretical analysis and in part by the precipitation's inherent fluctuation (Giardina et al. 2018; Powell et al. 2018). In comparison to the world's current annual human-induced carbon emissions, which amount to approximately 9–14 decades' worth (Makarov et al. 2020). If carbon emissions from tropical forests throughout the world and the Amazon, in particular, are not drastically reduced in the upcoming years, it will be very difficult to reduce global warming and prevent overall temperature. The average temperature from rising by no 2 °C (Smith et al. 2020). By setting the photosynthetic assimilation rate at 25% of the maximal photosynthetic assimilation rate, the CO_2 fixation impacts are taken into account in the computations (Huntingford et al. 2013).

4 Conclusion

Modern agriculture is an ever-evolving approach to agricultural innovation and farming practices that empowers farmers to enhance efficiency while reducing the natural resources needed to meet the global demands for food, fuel, and fiber. Farmer can enhance productivity whilst reducing environment effect by using modern farming practice. In this chapter, we explore the modern agriculture and its impact on environment. Here, we employ various techniques to mitigate the adverse effects of pesticides, insecticides, and high carbon dioxide emissions on the environment. These efforts involve utilizing human resources, including natural, social, and financial resources, to effectively manage climate risks. Many techniques that can help to boost up our climate are sustainable forest management. Without forests, humanity could not exist. Because trees can help to absorb a significant amount of fossil fuel emissions generated by human activities. Sustainable forestry is one of the most important ways that people can lessen the effect of climate change. Organic farming is also a way to improve the farming system and soil health and also enhance biodiversity. Practices such as conservation tillage, rotational grazing, and water management are also providing beneficial impacts on our soil condition as well as boost up the climate.

References

Abdul-Razak M, Kruse S (2017) The adaptive capacity of smallholder farmers to climate change in the Northern Region of Ghana. Clim Risk Manag 17:104–122

Aggarwal P, Vyas S, Thornton P, Campbell B, Kropff M (2019) Importance of considering technology growth in impact assessments of climate change on agriculture. Glob Food Secur 23:41–48

Aitken S, Simard S (2015) Restoring forests: how we can protect the water we drink and the air we breathe. Altern J 41(1):30–35

Akhter MJ, Jensen PK, Mathiassen SK, Melander B, Kudsk P (2020) Biology and management of Vulpia myuros—an emerging weed problem in no-till cropping systems in Europe. Plan Theory 9(6):715

Alengebawy A, Abdelkhalek ST, Qureshi SR, Wang M-Q (2021) Heavy metals and pesticides toxicity in agricultural soil and plants: ecological risks and human health implications. Toxics 9(3):42

Ali H, Iqbal N, Shahzad AN, Sarwar N, Ahmad S, Mehmood A (2013) Seed priming improves irrigation water use efficiency, yield, and yield components of late-sown wheat under limited water conditions. Turk J Agric For 37:534–544

Amushila J, Bussin MH (2021) The effect of talent management practices on employee retention at the Namibia University of Science and Technology: administration middle-level staff. SA J Hum Resour Manag 19:11

Amitrano C, Rouphael Y, Pannico A, De Pascale S, De Micco V (2021) Reducing the evaporative demand improves photosynthesis and water use efficiency of indoor cultivated lettuce. Agronomy 11(7):1396

Aqueel MA, AbM R, Balal RM, Shahid MA, Mustafa I, Javaid MM, Leather SR (2015) Tritrophic interactions between parasitoids and cereal aphids are mediated by nitrogen fertilizer. Insect Sci 22:813–820

Aziz A, Khan A, Asif M, Nadeem MA, Ashraf M, Raza A, Javaid MM, Abbas T, Akhtar N, Munir S (2022) Impact of organic mulches on weed dynamics and productivity of rainfed wheat (*Triticum aestivum* L.). Pak J Weed Sci Res 28(1):1–17

Baronti S, Ungaro F, Maienza A, Ugolini F, Lagomarsino A, Agnelli AE, Calzolari C, Pisseri F, Robbiati G, Vaccari FP (2022) Rotational pasture management to increase the sustainability of mountain livestock farms in the Alpine region. Reg Environ Chang 22(2):1–12

Basavalingaiah K, Paramesh V, Parajuli R, Girisha H, Shivaprasad M, Vidyashree G, Thoma G, Hanumanthappa M, Yogesh G, Misra SD (2022) Energy flow and life cycle impact assessment of coffee-pepper production systems: An evaluation of conventional, integrated and organic farms in India. Environ Impact Assess Rev 92:106687

Baveye PC, Schnee LS, Boivin P, Laba M, Radulovich R (2020) Soil organic matter research and climate change: merely re-storing carbon versus restoring soil functions. Front Environ Sci 8:579904

Benitez-Alfonso Y (2022) George Washington Carver: a plant scientist's perspective. Curr Biol 32(1):R9–R13

Bertram C, Hilaire J, Kriegler E, Beck T, Bresch D, Clarke L, Cui R, Edmonds J, Charles M, Zhao A (2021) NGFS climate scenario database: technical documentation V2.2. Institute for Applied Systems Analysis (IIASA), Laxenburg, Austria

Burger M, Horwath WR, Six J, Waterhouse H (2016) Evaluating mitigation options of nitrous oxide emissions in California cropping systems. California Air Resources Board, Research Division, Sacramento

Byrnes RC, Eastburn DJ, Tate KW, Roche LM (2018) A global meta-analysis of grazing impacts on soil health indicators. J Environ Qual 47(4):758–765. https://doi.org/10.2134/jeq2017.08.0313

Choudhary S, Kumar R, Gupta S (2019) Integrated farming system (IFS) is possible way out for double farmer's income. J Pharmacogn Phytochem 5:282–289

Chow ATS, Ulus Y, Huang G, Kline MA, Cheah WY (2022) Challenges in quantifying and characterizing dissolved organic carbon–sampling, isolation, storage, and analysis. J Environ Qual 51:837

Clark S (2020) Organic farming and climate change: the need for innovation. Sustainability 12(17):7012

Clark MA, Domingo NG, Colgan K, Thakrar SK, Tilman D, Lynch J, Azevedo IL, Hill JD (2020) Global food system emissions could preclude achieving the 1.5 and 2 C climate change targets. Science 370(6517):705–708

Danion M, Le Floch S, Cabon J, Louboutin L, Morin T (2018) Transchem project–part II: transgenerational effects of long-term exposure to pendimethalin at environmental concentrations on the early development and viral pathogen susceptibility of rainbow trout (Oncorhynchus mykiss). Aquat Toxicol 202:126–135

Daughtry C, Gallo K, Biehl L, Blad BL, Norman JM, Gardner BR, Kanemasu E, Asrar G (1984) Spectral estimates of agronomic characteristics of crops. LARS Technical Reports, Purdue University, Purdue e-Pubs, Indiana, USA, 63

Davidson EA, David MB, Galloway JN, Goodale CL, Haeuber R, Harrison JA, Howarth RW, Jaynes DB, Lowrance RR, Nolan BT (2012) Excess nitrogen in the US environment: trends, risks, and solutions. Issues Ecol (15)

Deng X, Yang Q, Zhang D, Dong S (2022) Application of conservation tillage in China: a method to improve climate resilience. Agronomy 12(7):1575

Dhingra IC (2021) Green economy: opportunities and challenges: an international perspective. Taylor & Francis Group, Oxfordshire, England, UK

Dieleman CM, Rogers BM, Potter S, Veraverbeke S, Johnstone JF, Laflamme J, Solvik K, Walker XJ, Mack MC, Turetsky MR (2020) Wildfire combustion and carbon stocks in the southern Canadian boreal forest: implications for a warming world. Glob Chang Biol 26(11):6062–6079

Dietz T, Shwom RL, Whitley CT (2020) Climate change and society. Annu Rev Sociol 46:135–158

Elbasiouny H, Elbanna BA, Al-Najoli E, Alsherief A, Negm S, El-Nour A, Nofal A, Sharabash S (2020) Agricultural waste management for climate change mitigation: some implications to Egypt. In: Waste management in MENA regions. Springer, Cham, pp 149–169

Erans M, Sanz-Pérez ES, Hanak DP, Clulow Z, Reiner DM, Mutch GA (2022) Direct air capture: process technology, techno-economic and socio-political challenges. Energy Environ Sci 15(4):1360–1405

Etminan M, Myhre G, Highwood E, Shine K (2016) Radiative forcing of carbon dioxide, methane, and nitrous oxide: a significant revision of the methane radiative forcing. Geophys Res Lett 43(24):12,614–612,623

Fadiji AE, Mthiyane DMN, Onwudiwe DC, Babalola OO (2022) Harnessing the known and unknown impact of nanotechnology on enhancing food security and reducing postharvest losses: constraints and future prospects. Agronomy 12(7):1657

Fess TL, Benedito VA (2018) Organic versus conventional cropping sustainability: a comparative system analysis. Sustainability 10(1):272

Fraisse C, Andreis JH, Borba T, Cerbaro V, Gelcer E, Pavan W, Pequeno D, Perondi D, Shen X, Staub C (2016) AgroClimate-tools for managing climate risk in agriculture. Agrometeoros 24(1):121–129

Freitas TR, Santos JA, Silva AP, Fraga H (2021) Influence of climate change on chestnut trees: a review. Plan Theory 10(7):1463

Gayer C, Berger J, Dieterich M, Gallé R, Reidl K, Witty R, Woodcock BA, Batáry P (2021) Flowering fields, organic farming and edge habitats promote diversity of plants and arthropods on arable land. J Appl Ecol 58(6):1155–1166

Geddes A, Schmidt TS, Steffen B (2018) The multiple roles of state investment banks in low-carbon energy finance: An analysis of Australia, the UK and Germany. Energy Policy 115:158–170

Geisen S, Wall DH, van der Putten WH (2019) Challenges and opportunities for soil biodiversity in the anthropocene. Curr Biol 29(19):R1036–R1044

Giardina F, Konings AG, Kennedy D, Alemohammad SH, Oliveira RS, Uriarte M, Gentine P (2018) Tall Amazonian forests are less sensitive to precipitation variability. Nat Geosci 11(6):405–409

Gonçalves AM, Rocha CP, Marques JC, Gonçalves FJ (2021) Enzymes as useful biomarkers to assess the response of freshwater communities to pesticide exposure–a review. Ecol Indic 122:107303

Govere S, Nyamangara J, Nyakatawa E (2022) Beneficial effect of climate change on wheat yield and water footprints in the Middle-Manyame sub-catchment, Zimbabwe. J Water Clim Change 13(5)

Gremmen B (2022) Regenerative agriculture as a biomimetic technology. Outlook Agric 51(1):39–45

Grippa P, Schmittmann J, Suntheim F (2019) Financial risk. Finance and development.

Guyomard H, Bouamra-Mechemache Z, Chatellier V, Delaby L, Détang-Dessendre C, Peyraud J-L, Requillart V (2021) Why and how to regulate animal production and consumption: the case of the European Union. Animal 15(1):100283

Hahn WA, Knoke T (2010) Sustainable development and sustainable forestry: analogies, differences, and the role of flexibility. Eur J For Res 129(5):787–801

Hamidi NH, Ahmed OH, Omar L, Ch'ng HY (2021, 1801) Soil nitrogen sorption using charcoal and wood ash. Agronomy 11(9)

Hanif S, Lateef M, Hussain K, Hyder S, Usman B, Zaman K, Asif M (2022) Controlling air pollution by lowering methane emissions, conserving natural resources, and slowing urbanization in a panel of selected Asian economies. PLoS One 17(8):e0271387

Harper JK, Roth GW, Garalejić B, Škrbić N (2018) Programs to promote adoption of conservation tillage: a Serbian case study. Land Use Policy 78:295–302

Hassan MU, Chattha MU, Mahmood A, Sahi ST (2018) Performance of sorghum cultivars for biomass quality and biomethane yield grown in semi-arid area of Pakistan. Environ Sci Pollut Res 25:12800–12807

Hassan MU, Chattha MU, Barbanti L, Chattha MB, Mahmood A, Khan I, Nawaz M (2019) Combined cultivar and harvest time to enhance biomass and methane yield in sorghum under warm dry conditions in Pakistan. Ind Crop Prod 132:84–91

Heong KL, Wong L, Delos Reyes JH (2015) Addressing planthopper threats to Asian rice farming and food security: fixing insecticide misuse. In: Rice planthoppers. Springer, Dordrecht, pp 65–76

Howard G, Charles K, Pond K, Brookshaw A, Hossain R, Bartram J (2010) Securing 2020 vision for 2030: climate change and ensuring resilience in water and sanitation services. J Water Clim Change 1(1):2–16

Huntingford C, Zelazowski P, Galbraith D, Mercado LM, Sitch S, Fisher R, Lomas M, Walker AP, Jones CD, Booth BB (2013) Simulated resilience of tropical rainforests to CO2-induced climate change. Nat Geosci 6(4):268–273

Insua JR, Utsumi SA, Basso B (2019) Assessing and modeling pasture growth under different nitrogen fertilizer and defoliation rates in Argentina and the United States. Agronomy 111(2):702–713

Jackson R, Friedlingstein P, Andrew R, Canadell J, Le Quéré C, Peters G (2019) Persistent fossil fuel growth threatens the Paris Agreement and planetary health. Environ Res Lett 14(12):121001

Javaid MM, Florentine SK, Ashraf M, Mahmood A, Sattar A, Wasaya A, Li FM (2022a) Photosynthetic activity and water use efficiency of *Salvia verbenaca* L. under elevated CO_2 and water-deficit conditions. J Agron Crop Sci 208:536–551

Javaid MM, Mahmood A, Alshaya DS, AlKahtani MD, Waheed H, Wasaya A, Khan SA, Naqve M, Haider I, Shahid MA, Nadeem MA (2022b) Influence of environmental factors on seed germination and seedling characteristics of perennial ryegrass (*Lolium perenne* L.). Sci Rep 12:1–11

Jenkins DM (2021) Chandra Jenkins Chief, CA Delta Section Regulatory Division US Army Corps of Engineers 1325 J Street

Jensen D, Keith L, Wilbert M (2021) Bright green lies: how the environmental movement lost its way and what we can do about it. Monkfish Book Publishing, New York

Jeyasri R, Muthuramalingam P, Satish L, Pandian SK, Chen J-T, Ahmar S, Wang X, Mora-Poblete F, Ramesh M (2021) An overview of abiotic stress in cereal crops: negative impacts, regulation, biotechnology and integrated omics. Plan Theory 10(7):1472

Jhariya MK, Yadav DK, Banerjee A, Raj A, Meena RS (2019) Sustainable forestry under changing climate. In: Sustainable agriculture, forest and environmental management. Springer, Singapore, pp 285–326

Jiri O, Mafongoya P, Mubaya C, Mafongoya O (2015) Seasonal climate prediction and adaptation using indigenous knowledge systems in agriculture systems in Southern Africa: a review. J Agric Sci 8(5):156

Karimi H, Mahdavi S, Asgari Lajayer B, Moghiseh E, Rajput VD, Minkina T, Astatkie T (2022) Insights on the bioremediation technologies for pesticide-contaminated soils. Environ Geochem Health 44(4):1329–1354

Karki R, Gurung A (2012) An overview of climate change and its impact on agriculture: a review from least developing country. Nepal Int J Ecosyst 2(2):19–24

Katila P, Colfer CJP, De Jong W, Galloway G, Pacheco P, Winkel G (2019) Sustainable development goals. Cambridge University Press, Cambridge

Kaur R, Mavi GK, Raghav S, Khan I (2019) Pesticides classification and its impact on environment. Int J Curr Microbiol Appl Sci 8(3):1889–1897

Kern F, Kivimaa P, Martiskainen M (2017) Policy packaging or policy patching? The development of complex energy efficiency policy mixes. Energy Res Soc Sci 23:11–25

Khan MR, Munira S (2021) Climate change adaptation as a global public good: implications for financing. Clim Chang 167(3):1–18

Khanali M, Akram A, Behzadi J, Mostashari-Rad F, Saber Z, Chau K-w, Nabavi-Pelesaraei A (2021) Multi-objective optimization of energy use and environmental emissions for walnut production using imperialist competitive algorithm. Appl Energy 284:116342

Knapp S, van der Heijden MG (2018) A global meta-analysis of yield stability in organic and conservation agriculture. Nat Commun 9(1):1–9

Kremen C, Miles A (2012) Ecosystem services in biologically diversified versus conventional farming systems: benefits, externalities, and trade-offs. Ecol Soc 17(4)

Kumar M, Phukon SN, Singh H (2021) The role of communities in sustainable land and forest management. In: Forest resources resilience and conflicts. Elsevier, Amsterdam, pp 305–318

Le Quéré C, Jackson RB, Jones MW, Smith AJ, Abernethy S, Andrew RM, De-Gol AJ, Willis DR, Shan Y, Canadell JG (2020) Temporary reduction in daily global CO_2 emissions during the COVID-19 forced confinement. Nat Clim Chang 10(7):647–653

Li A, Wang M, Kroeze C, Ma L, Strokal M (2022) Past and future pesticide losses to Chinese waters under socioeconomic development and climate change. J Environ Manag 317:115361

Lin BB, Chappell MJ, Vandermeer J, Smith G, Quintero E, Bezner-Kerr R, Griffith DM, Ketcham S, Latta SC, McMichael P (2011) Effects of industrial agriculture on climate change and the mitigation potential of small-scale agro-ecological farms. CAB Rev Perspect Agric Veterinary Sci Nutr Nat Res 6(020):1–18

Liu Y, Wang T (2020) Worsening urban ozone pollution in China from 2013 to 2017–part 1: the complex and varying roles of meteorology. Atmos Chem Phys 20(11):6305–6321

MacRae R, Lynch D, Martin R (2010) Improving energy efficiency and GHG mitigation potentials in Canadian organic farming systems. J Sustain Agric 34(5):549–580

Mahmood I, Imadi SR, Shazadi K, Gul A, Hakeem KR (2016) Effects of pesticides on environment. In: Plant, soil and microbes: volume 1: implications in crop science. Springer, Cham, pp 253–269

Mahmood A, Wang X, Shahzad AN, Fiaz S, Ali H, Naqve M, Javaid MM, Mumtaz S, Naseer M, Dong R (2021) Perspectives on bioenergy feedstock development in Pakistan: challenges and opportunities. Sustainability 13:8438

Mahmood A, Wang X, Ali A, Awan MI, Ali L, Fiaz S, Naqve M, Hassan MU, Javaid MM, Wang H, Shoujun Y (2022) Bio-methane production from maize with varying nitrogen levels and harvesting times under semi-arid conditions of Pakistan. Pol J Environ Stud 31:2213

Majewski MS, Capel PD (2019) Pesticides in the atmosphere: distribution, trends, and governing factors. CRC Press, Boca Raton

Makarov I, Chen H, Paltsev S (2020) Impacts of climate change policies worldwide on the Russian economy. Clim Pol 20(10):1242–1256

Maliwal P, Maliwal P, Mundra S (2007) Agronomy at a glance, vol 2. Agrotech Publishing Academy, Udaipur

Maraseni T, An-Vo D-A, Mushtaq S, Reardon-Smith K (2021) Carbon smart agriculture: An integrated regional approach offers significant potential to increase profit and resource use efficiency, and reduce emissions. J Clean Prod 282:124555

Mazzoncini M, Canali S, Giovannetti M, Castagnoli M, Tittarelli F, Antichi D, Nannelli R, Cristani C, Bàrberi P (2010) Comparison of organic and conventional stockless arable systems: a multidisciplinary approach to soil quality evaluation. Appl Soil Ecol 44(2):124–132

Moustaghfir K, El Fatihi S, Benouarrek M (2020) Human resource management practices, entrepreneurial orientation and firm performance: what is the link? Meas Bus Excell 24(2):267–283

Munir MS, Bajwa IS, Ashraf A, Anwar W, Rashid R (2021) Intelligent and smart irrigation system using edge computing and IoT complexity. Open Access 2021:6691571

Myhre GS, Drew Pongratz, Julia (2014) Anthropogenic and natural radiative forcing. In: Intergovernmental Panel on Climate C (ed) Climate Change 2013 – The Physical Science Basis: Working Group I Contribution to the Fifth Assessment Report of the Intergovernmental Panel on Climate Change. Cambridge University Press, Cambridge, pp 659–740. https://doi.org/10.1017/CBO9781107415324.018

Naumann G, Alfieri L, Wyser K, Mentaschi L, Betts R, Carrao H, Spinoni J, Vogt J, Feyen L (2018) Global changes in drought conditions under different levels of warming. Geophys Res Lett 45(7):3285–3296

Nisbet EG, Dlugokencky EJ, Fisher RE, France JL, Lowry D, Manning MR, Michel SE, Warwick NJ (2021) Atmospheric methane and nitrous oxide: challenges along the path to net zero. Philos Trans R Soc 379(2210):20200457

Nobre CA, Sampaio G, Borma LS, Castilla-Rubio JC, Silva JS, Cardoso M (2016) Land-use and climate change risks in the Amazon and the need of a novel sustainable development paradigm. Proc Nat Acad Sci 113(39):10759–10768

Norman J (1980) Interfacing leaf and canopy light interception models [Plant photosynthesis]. 1:20. CRC Press, United States

Palahi M, Pantsar M, Costanza R, Kubiszewski I, Potocnik J, Stuchtey M, Nasi R, Lovins H, Giovannini E, Fioramonti L (2020) Investing in nature to transform the post COVID-19 economy: a 10-point action plan to create a circular bioeconomy devoted to sustainable wellbeing. Solutions 2(11)

Panchasara H, Samrat NH, Islam N (2021) Greenhouse gas emissions trends and mitigation measures in Australian agriculture sector—a review. Agriculture 11(2):85

Phan T-HD, Brouwer R, Davidson M (2014) The economic costs of avoided deforestation in the developing world: a meta-analysis. J For Econ 20(1):1–16

Piazza G, Ercoli L, Nuti M, Pellegrino E (2019) Interaction between conservation tillage and nitrogen fertilization shapes prokaryotic and fungal diversity at different soil depths: evidence from a 23-year field experiment in the Mediterranean area. Front Microbiol 10:2047

Pivello VR, Vieira I, Christianini AV, Ribeiro DB, da Silva ML, Berlinck CN, Melo FP, Marengo JA, Tornquist CG, Tomas WM (2021) Understanding Brazil's catastrophic fires: causes, consequences and policy needed to prevent future tragedies. Perspect Ecol Conserv 19(3):233–255

Porfirio LL, Newth D, Finnigan JJ, Cai Y (2018) Economic shifts in agricultural production and trade due to climate change. Palgrave Commun 4(1):1–9

Powell TL, Koven CD, Johnson DJ, Faybishenko B, Fisher RA, Knox RG, McDowell NG, Condit R, Hubbell SP, Wright SJ (2018) Variation in hydroclimate sustains tropical forest biomass and promotes functional diversity. New Phytol 219(3):932–946

Ramirez-Villegas J, Challinor AJ, Thornton PK, Jarvis A (2013) Implications of regional improvement in global climate models for agricultural impact research. Environ Res Lett 8(2):024018

Rani L, Thapa K, Kanojia N, Sharma N, Singh S, Grewal AS, Srivastav AL, Kaushal J (2021) An extensive review on the consequences of chemical pesticides on human health and environment. J Clean Prod 283:124657
Rathi BS, Kumar PS, Vo D-VN (2021) Critical review on hazardous pollutants in water environment: occurrence, monitoring, fate, removal technologies and risk assessment. Sci Total Environ 797:149134
Rathore HS, Nollet LM (2012) Pesticides: evaluation of environmental pollution. CRC Press, Boca Raton
Raza A, Razzaq A, Mehmood SS, Zou X, Zhang X, Lv Y, Xu J (2019) Impact of climate change on crops adaptation and strategies to tackle its outcome: a review. Plan Theory 8(2):34
Regnier P, Resplandy L, Najjar RG, Ciais P (2022) The land-to-ocean loops of the global carbon cycle. Nature 603(7901):401–410
Ripoche A, Autfray P, Rabary B, Randriamanantsoa R, Blanchart E, Trap J, Sauvadet M, Becquer T, Letourmy P (2021) Increasing plant diversity promotes ecosystem functions in rainfed rice based short rotations in Malagasy highlands. Agric Eco Environ 320:107576
Roudier P, Sultan B, Quirion P, Berg A (2011) The impact of future climate change on West African crop yields: what does the recent literature say? Glob Environ Change 21(3):1073–1083
Sarkar S, Gil JDB, Keeley J, Jansen K (2021) The use of pesticides in developing countries and their impact on health and the right to food. European Union
Sharifi A, Ahmadi M, Ala A (2021) The impact of artificial intelligence and digital style on industry and energy post-COVID-19 pandemic. Environ Sci Pollut Res 28(34):46964–46984
Shrestha A, Anwar MP, Islam AM, Gurung T, Dhakal S, Tanveer A, Javaid MM, Nadeem M, Ikram NA (2021) Weed science as a new discipline and its status in some south Asian universities and colleges: examples from Bangladesh, Bhutan, Nepal and Pakistan. CAB Rev Perspect Agric Veterin Sci Nutr Nat Res 4:2021
Singh M (2021) Organic farming for sustainable agriculture. Indian J Organ Farm 1(1):1–8
Skendžić S, Zovko M, Živković IP, Lešić V, Lemić D (2021) The impact of climate change on agricultural insect pests. Insects 12(5):440
Smith MN, Taylor TC, van Haren J, Rosolem R, Restrepo-Coupe N, Adams J, Wu J, de Oliveira RC, da Silva R, de Araujo AC (2020) Empirical evidence for resilience of tropical forest photosynthesis in a warmer world. Nat Plants 6(10):1225–1230
Smith P, Reay D, Smith J (2021) Agricultural methane emissions and the potential for mitigation. Phil Trans R Soc A 379(2210):20200451
Song K, Qu S, Taiebat M, Liang S, Xu M (2019) Scale, distribution and variations of global greenhouse gas emissions driven by US households. Environ Int 133:105137
Spahr S, Teixidó M, Sedlak DL, Luthy RG (2020) Hydrophilic trace organic contaminants in urban stormwater: occurrence, toxicological relevance, and the need to enhance green stormwater infrastructure. Environ Sci Water Res Technol 6(1):15–44
Srivastav AL, Dhyani R, Ranjan M, Madhav S, Sillanpää M (2021) Climate-resilient strategies for sustainable management of water resources and agriculture. Environ Sci Pollut Res 28(31):41576–41595. https://doi.org/10.1007/s11356-021-14332-4
Sun Y, Zhang X, Ding Y, Chen D, Qin D, Zhai P (2022) Understanding human influence on climate change in China. Natl Sci Rev 9(3):nwab113
Tompkins EL, Adger WN (2004) Does adaptive management of natural resources enhance resilience to climate change? Ecol Soc 9(2)
Tully KL, McAskill C (2020) Promoting soil health in organically managed systems: a review. Organ Agric 10(3):339–358
Tully K, Mcaskill C, Shade J (2021) Promoting soil health in organically managed systems.
Usman M, Hameed G, Saboor A, Almas LK (2021) Research and Development spillover, irrigation water use and agricultural production in Pakistan. WSEAS Trans Environ Dev 17:840–858
van der Velden R, da Fonseca-Zang W, Zang J, Clyde-Smith D, Leandro WM, Parikh P, Borrion A, Campos LC (2022) Closed-loop organic waste management systems for family farmers in Brazil. Environ Technol 43(15):2252–2269

Wasaya A, Abbas T, Yasir TA, Sarwar N, Aziz A, Javaid MM, Akram S (2021) Mitigating drought stress in sunflower (*Helianthus annuus* L.) through exogenous application of β-aminobutyric acid. J Soil Sci Plant Nut 21:936–948

Weller S, Florentine S, Javaid MM, Welgama A, Chadha A, Chauhan BS, Turville C (2021) *Amaranthus retroflexus* L. (Redroot Pigweed): effects of elevated CO_2 and soil moisture on growth and biomass and the effect of radiant heat on seed germination. Agronomy 11:728

West JM, Julius SH, Kareiva P, Enquist C, Lawler JJ, Petersen B, Johnson AE, Shaw MR (2009) U.S. natural resources and climate change: concepts and approaches for management adaptation. Environ Manag 44(6):1001. https://doi.org/10.1007/s00267-009-9345-1

White K, Hardisty DJ, Habib R (2019) The elusive green consumer. Harv Bus Rev 11(1):124–133

Wilson RJ, Fox R (2021) Insect responses to global change offer signposts for biodiversity and conservation. Ecol Entomol 46(4):699–717

Wittwer RA, Bender SF, Hartman K, Hydbom S, Lima RA, Loaiza V, Nemecek T, Oehl F, Olsson PA, Petchey O (2021) Organic and conservation agriculture promote ecosystem multifunctionality. Sci Adv 7(34):eabg6995

Wuest SE, Peter R, Niklaus PA (2021) Ecological and evolutionary approaches to improving crop variety mixtures. Nat Eco Evolution 5(8):1068–1077

Yadav GS, Lal R, Meena RS (2020) Vehicular traffic effects on hydraulic properties of a Crosby silt loam under a long-term no-till farming in Central Ohio, USA. Soil Tillage Res 202:104654

Yuan ZQ, Fang C, Zhang R, Li FM, Javaid MM, Janssens IA (2019) Topographic influences on soil properties and aboveground biomass in lucerne-rich vegetation in a semi-arid environment. Geoderma 344:137–143

Zhang Z, Peng X (2021) Bio-tillage: a new perspective for sustainable agriculture. Soil Tillage Res 206:104844

Zhuang M, Caro D, Qin W, Wang C, Yang X, Liu R, Zhang L (2022) Spatial heterogeneity of greenhouse gas emissions from cereal crop production in China. Environ Chem Lett 1-6

Zou X, Zhang X, Lv Y, Xu J (2019) Impact of climate change on crops adaptation and strategies to tackle its outcome: a review. Plants (Basel) 8(2). https://doi.org/10.3390/plants8020034

Zougmoré RB, Läderach P, Campbell BM (2021) Transforming food systems in Africa under climate change pressure: role of climate-smart agriculture. Sustainability 13(8):4305

Chapter 6
Crop Management for Sustainable Wheat Production

Rehan Jameel, Maria Naqve, Muhammad Anjum Zia, Athar Mahmood, Muhammad Mansoor Javaid, and Muhammad Ather Nadeem

Abstract Climate change is a serious issue over all the world as it directly influences on agriculture sector. Abiotic and biotic stresses occurred in response to environmental changes. Crop yield is highly vulnerable to environmental changes. Wheat is the major staple food globally and severely affected by changing climate. Wheat is one of the most extensively grown crops in the world; however, its productivity lags behind that of other staple food crops. The main reason for this crop loss is the environmental changes which badly affect the wheat crop. 1 °C increase in temperature lowers 6% yield of wheat, and due to drought, 4.4% of yield loss is seen in previous years. Food security and ecosystem resilience are the most serious global issues. Climatic-smart agriculture is the only approach to reduce the detrimental impact of climate changes on wheat crop adaptability before they have a significant influence on global crop production. In this book chapter, we summarize the negative impacts of environment on wheat and climate-resilient technologies for alleviating negative impact of climate on wheat production.

Keywords Climate change · Stressors · Agriculture · Wheat · Food security

R. Jameel · M. Naqve
Department of Botany, Faculty of Sciences, University of Agriculture, Faisalabad, Pakistan

M. A. Zia
Department of Biochemistry, Faculty of Sciences, University of Agriculture, Faisalabad, Pakistan

A. Mahmood (✉)
Department of Agronomy, Faculty of Agriculture, University of Agriculture, Faisalabad, Pakistan
e-mail: athar.mahmood@uaf.edu.pk

M. M. Javaid (✉) · M. A. Nadeem
Department of Agronomy, College of Agriculture, University of Sargodha, Sargodha, Pakistan
e-mail: mansoor.javaid@uos.edu.pk

M. Hasanuzzaman (ed.), *Climate-Resilient Agriculture, Vol 2*,
https://doi.org/10.1007/978-3-031-37428-9_6

1 Introduction

Agriculture is the only enterprise that is directly influenced by environmental changes. Climate change is a global issue, but in developing countries of the world, it is the worse shape because of inadequate measurements to combat these changes and less information about diverse food production methods (Fahad and Wang 2018; Javaid et al. 2022a). Depending upon crop variety and location of investigation, 25% loss is projected in short period of time, although some researchers estimated up to 50% loss by the year 2080 (Challinor et al. 2014; Javaid et al. 2022b). Climatic variability accounts for 60% of crop yield reduction, posing a threat to sustainable agricultural schemes and food security globally. Environmental changes negatively influence income of small farmers, who rely entirely on crop farming for a living. A recent research anticipated that 20% worldwide will rise in hungry people by the year 2050 (Carty and Magrath 2013).

The condition is severe in Asia, where agriculture is particularly sensitive to climate change. Due to climatic changes, the region has faced a variety of issues, including heat waves, rising temperatures, droughts, and floods (Hossain et al. 2020). Crop cultivation in countries of South Asia, such as Pakistan, India, Bangladesh, and Nepal, is under serious threat of climate change (Hossain et al. 2019). The rise in average maximum temperature is expected to 1.4–1.8 °C (12%) by 2030, besides 2.1–2.6 °C (21%) by 2050, in South Asia (Tesfaye et al. 2017; Javaid et al. 2022c). According to one analysis, if suitable climate change adaptation measures are not adopted by South Asian countries, their GDP will fall about 1.8% by the year 2050 and 8.8% by the year 2100 (Fig. 6.1) (Ahmed and Suphachalasai 2014).

Pakistan is an agricultural-based country, which shares 18.9% of its GDP by agriculture. Agricultural industry also employs 42.3% of the total population of

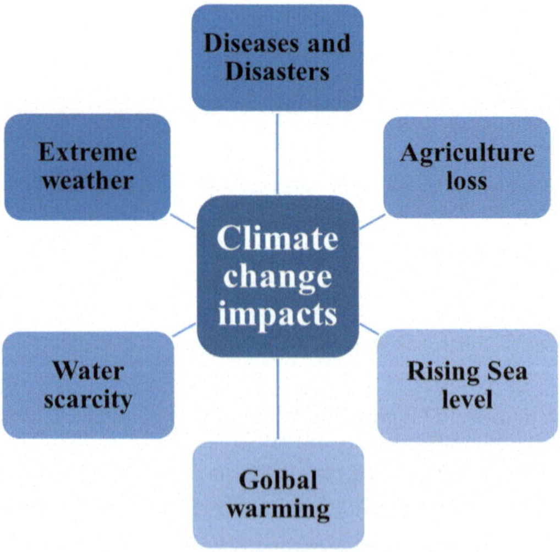

Fig. 6.1 Effects of environmental change on earth. (Ahmed and Suphachalasai 2014)

country and thus regarded as spine of Pakistan's economy. Agriculture comprises mostly the cattle, poultry, and fisheries sectors, as well as agricultural production, with important cereal and other crops. The wheat is a principal food crop of Pakistan and is grown all over the state to fulfill the staple food requirements. Agriculture of Pakistan is facing significant danger from climate change, as Pakistan consistently ranked between top 10 most vulnerable countries globally for more than 5 years (Kreft et al. 2016; Mahmood et al. 2021).

Wheat is the first ever domesticated crop since 7000 and 9000 BC, and its cultivation is expanded throughout the world (Farooq et al. 2021; Kavamura et al. 2021). Among the cereal crop, wheat is called king of cereals (Iqra et al. 2020). Wheat is the staple food of 40% population of the world. Total cultivated area for wheat is 218 million hectares globally with total 765 million tons production, and its trade in the world is more than all other crops combined. Wheat has 13% protein content which is comparatively higher among other major cereals (Giraldo et al. 2019) and supplies 40% energy calories globally (Hussain et al. 2021). As the wheat is major consumable crop of the world, so by the year 2026, it is necessary to increase its production up to 11%, with only increase in 1.8% of cultivated area (Kavamura et al. 2021). A major part of wheat production is utilized by humans, while the other is used as animal feed and industry processing (Pequeno et al. 2021).

Environmental conditions harm wheat production and many other crops (Javaid et al. 2018). According to the Intergovernmental Panel on Climate Change (IPCC), globally average temperature might be increased by up to 2 °C by 2050 (and 5 °C by the year 2100), while traditional rainfall patterns are altered (Reynolds et al. 2021). Changes in temperature are highly dangerous for food security. In Pakistan, there was a 4.4% decrease in wheat output in 2019 as compared to the previous year, and it is also noted the wheat's highly dubious prospects in 2019 because of severe water deficit. Despite drought stress, increase in every 1 °C of temperature is expected to reduce world wheat yield by 6% (Zhao et al. 2017).

If proper and early initiatives are not adopted to mitigate the detrimental effects of climate change, Pakistan may continue to endure yield reduction and food shortage and lower farm income (Mahmood et al. 2020). Most attempts to enhance food production in the last part of the twentieth century focused on enhancing and expanding modern agricultural inputs such as variety of new seeds, irrigation, and artificial fertilizers and pesticides (Si et al. 2017; Adhikari et al. 2018).

Climatic resilience is defined as an agricultural system's ability to maintain its structures and assure the delivery of its services, with environmental variations and extremes. This may be accomplished through agricultural practices, e.g., increasing resilience capacities, durability, adjustability, and transformability (Meuwissen et al. 2019). Some of the environmental resilient practices include crop rotation, reduced tillage strategies, crop rotation, cover crops, decreased usage of mineral resources, and organic agricultural production methods as shown in (Fig. 6.2; Behnke et al. 2018). Unfortunately, due to greater risks and economic concerns, as well as general opposition to modify in farm methods, these therapies have not acquired widespread acceptance among farmers (Javaid et al. 2016; Roesch-McNally et al. 2018).

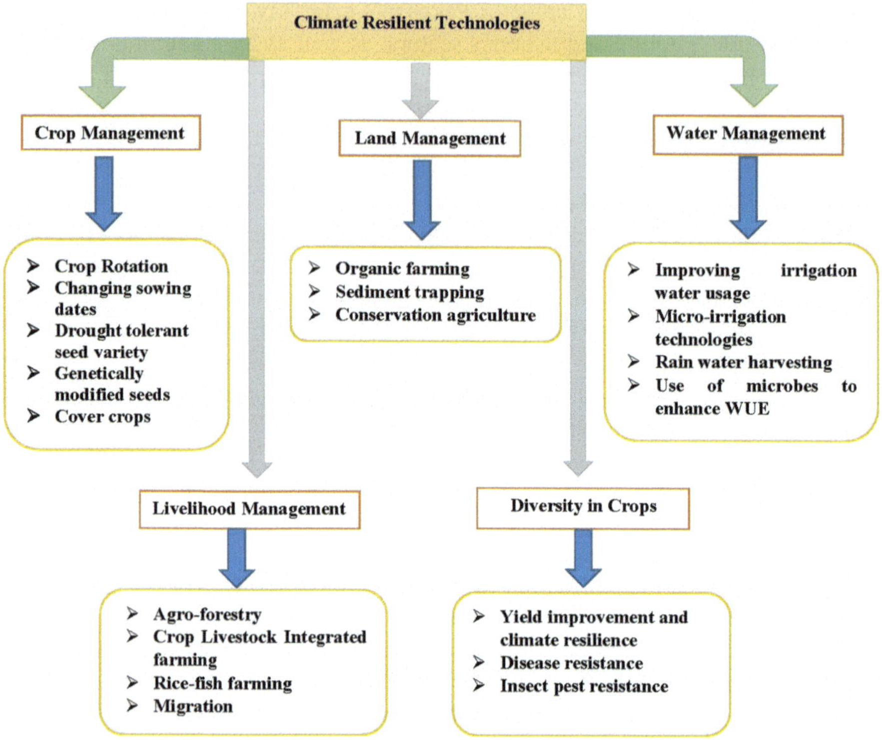

Fig. 6.2 Climate-resilient technologies to overcome impact of climate change on wheat. (Behnke et al. 2018)

2 Crop Management Approaches

2.1 Crop Rotation

Crop rotation, utilization of drought-resistant seed varieties, short-period crops, crop diversity, and altering planting dates are common climate-resilient sustainable farming methods used by farmers (Viswanathan et al. 2020). Crop rotation or cultivating a series of crops on the same land is commonly practiced and aids in the management of many agroecological issues such as deteriorating soil conditions and emissions of greenhouse gases (Dury et al. 2012). Changing cultivated crop combined with "no plowing" minimizes agricultural loss by 84% (Deuschle et al. 2019). Crop rotations with nitrogen-fixing crops like groundnuts, beans, and cowpeas improve soil potency and boost availability of nutrients to succeeding crops, resulting in higher crop production (Branca et al. 2013). Rotation of wheat with sunflower improved soil organic matter in rain-fed regions of Spain (Pedraza et al. 2015). Selection of proper crops for rotation is vital, which boosts soil nitrogen levels, increases plant biomass, and increases fertility of soil (Raphael et al. 2016).

Improving soil fertility is a long-term solution for increasing agricultural yield while increasing carbon storage and managing the carbon revolution (Bryan et al. 2013). The data from 3 years planting rotation of wheat after peas showed significant wheat yield improvements up to 1.3 t ha^{-1}. Wheat rotation with leguminous plants increased wheat production by 0.9–2.7 t ha^{-1} (Partal and Paraschivu 2020).

2.2 *Changing Sowing Time*

Changing sowing dates had positive effect on crop yield and is widely used as a viable method in rice and wheat-growing areas (Jalota et al. 2012). To deal with rising temperatures, farmers may shift sowing date to a cooler time of year (Gorst et al. 2018). Changing sowing dates (reducing or delaying) in flowering may enhance grain weight, yield of wheat, and protect from environmental variations (Garcia et al. 2011). In the case of rising temperatures, extending the planting time might improve wheat production from 4% to 6% (Mustafa et al. 2021).

2.3 *Drought-Tolerant Seed Varieties*

Approximately 80% of the total cultivated land is affected by drought globally. Drought is serious limiting factor for wheat production. In Australia, 46% wheat loss is due to drought by the year 2006 (Dai 2013). About 33% loss of wheat crop in Pakistan is due to drought. Wheat performance may be genetically enhanced by introducing improved alleles at existing loci by traditional crossing, helped by marker and other technology. In general, increase of 10–50 Kg/ha is seen in grains and legumes all over the world by genetically modified drought-resistant cultivars. By genetically modified wheat and barley cultivars, 2–5 tons/ha increase in grain yield has been observed (Mohammadi 2018).

2.4 *Genetically Modified Seeds*

Genetically modify (GM) seeds have several benefits in agronomy, and their application in food crops is fiercely debated (Saab 2016). Several organizations, policy makers, and scholars are suspicious to adopting GM food crops, despite the fact that no adverse effects have been recorded in the past which raised concerns about their usage (Viswanathan et al. 2020). Genetic diversity is examined as a critical tool for developing novel cultivars which have genetic differences and resemblances (Raza et al. 2018). Landraces are an important source of genetic research. For instance, wheat landrace stored in a data bank includes more genomic diversity and solid foundation for stress tolerance since it contains cultivars that are adaptable to a variety of climate stresses (Lopes et al. 2015). Wheat is an example of temperate crop which is especially vulnerable to the effects of climate change. The wheat crop has genetic ability to drought or heat stress tolerance, as well as a combination of both stresses, by modifying its physiological and biochemical properties (Viswanathan et al. 2020). Wheat breeders can adopt GM strategies to generate superior wheat varieties that can withstand drought and heat stress. Advances in phenotype and genomic selection provide new approaches for boosting genetic gains in stressful conditions (Langridge and Reynolds 2021).

2.5 Cover Crops

Cover crops are described as extra crops sown on the land after harvest or crops intercropped with the primary crop (Branca et al. 2013). Crop cover is significant for ecological balance because it prevents soil from erosion, increases water availability, controls weeds, and manages mineral levels of soil (Meyer et al. 2019). Several studies were conducted to demonstrate the advantages of crop covers (Srivastav et al. 2021). It was reported that cover cropping with mucuna benefited the maize yield. In Brazil, 198–246% increase is found in maize yield by using cover crops (Branca et al. 2013). Multicrops cultivated as crop cover use less water than a single crop cover (Meyer et al. 2020). Kaye and Quemada (2017) studied cover cropping, highlighting that it contributes to climate change management, enhances climate resilience, and reduces soil erosion following high rainfall in central Spain and Pennsylvania. In the northern Great Plains and humid agricultural systems of the USA, the effects of cover crops on soil health and production have been extensively reported (Mesbah et al. 2019). In semiarid Montana, pea enhanced winter wheat production by 5.2% (Miller et al. 2011). Lablab bean as cover crop with wheat increased wheat output up to 14% (Northupl and Rao 2015).

3 Water Management Techniques

3.1 Improving Irrigation Water Usage

Globally, declining water levels are viewed as a serious danger to crop yield reduction, particularly in rice cultivation. Irrigation is most widely used to boost yields while also reducing production variability. The selection of a suitable irrigation system is critical to ensure the sustainability of limited water supplies (Viswanathan et al. 2016). In India, irrigation water use for mitigating the rising temperature effects on wheat growth and yield. Irrigated crop productivity is typically greater than rain-fed crop. Irrigation has resulted in the greatest increase in yields in Madhya Pradesh, Maharashtra, and Bihar (Zaveri and Lobell 2019). Indian government policies have supported irrigation expansion for decades to enhance agricultural growth, reducing production risk, and decreasing rural poverty (Si et al. 2017).

3.2 Micro-irrigation Technologies

Drought susceptible states of India are progressively using technologies for water conservation known as micro-irrigation system (MIS), particularly sprayer and drop irrigation, as part of the National Micro Irrigation Mission (Tanveer et al. 2015; Viswanathan et al. 2016). Numerous studies have revealed that using various types of micro-irrigation systems (MISs) increases agricultural productivity and quality. The MIS success examples can be found all over the world. Extensive usage of drip irrigation aided Israeli farmers in maintaining agriculture on dry land (Tal 2016). Implementation of MIS along different states of India, as a result of state and federal policies, encouraged the use of sprayer and drop irrigation methods to combat environmental water crisis (Bahinipati and Viswanathan 2019). According to reports, greater use of MIS increased water savings by 39% and energy savings to 58% (Kumar and Palanisami 2019). According to a research in Morocco, drip irrigation produced 28% more wheat yield than surface irrigation (Ali et al. 2015; Suryavanshi et al. 2015).

3.3 Rain Water Harvesting

Rainwater harvesting is an ancient method used to conserve water in dry areas. Rainwater harvesting can help India and other underdeveloped countries to better response for climate change. Rainwater storage also minimizes the usage of underground water for agriculture (Srivastav et al. 2021). Researchers from the United States and Zambia indicated that rainwater collected for cultivation is also used for drinking purpose during times of great water scarcity (Viswanathan et al. 2020). Rainwater collection and management, together with food and water security, have become critical activities in the battle against climate change (Glendenning et al. 2012; Shahid et al. 2014). A study indicated that irrigation with rainwater up to 40–59 mm may improve wheat production from 2000 to 2250 Kg ha^{-1} (Saleem and Ali 2015).

3.4 Use of Microbes to Enhance Water Use Efficiency

Many different species of soil microorganisms help to protect crops from a variety of soilborne diseases and water management issues. Mycorrhizal associations can help improve water use efficiency and fertilizer management in the agroecosystem.

Microorganisms help in attaining many forms of adaptations to climate change, such as building resistance to harsh weather events, sustaining genetic variety, and boosting water use efficiency under several biotic and abiotic stresses. The development of osmoprotectants and the synthesis of proteins toward heat waves and droughts can help to minimize the amount of stress in crops. The significance of bacteria in the regulation of various forms of greenhouse gases will also be maintained, as well as climate change-resilient activities (Srivastav et al. 2021). Microbes have the ability to reduce the effect of salinity up to 50% in wheat and increase yield of wheat (EL Sabagh et al. 2021). There are two strategies for coping with climate change via microbial biodiversity, one for providing natural equilibrium and the other for developing adaptations in many habitats via various forms of biogeochemical cycles (Wallenstein and Hall 2012; Tanveer et al. 2013).

4 Strategies for Sustainable Land Management

4.1 Sediment Retention

Sustainable land management is defined by the United Nations as "the utilization of natural resources, such as land, water, livestock, and plants, to produce things to fulfil changing human requirements, while assuring long-term production capabilities of these resources and the preservation of their environmental services" (Sanz et al. 2017). The government of India established the "Soil Health Card program" in 2014 that provides agriculturalists with information on fertilizers use and crop planting depending on soil condition. Furthermore, natural fertilizer (compost), rewilding, and sedimentation retention were demonstrated for durable land management approaches. Nutrients retention in soil may increase wheat production significantly (Keesstra et al. 2018).

4.2 Organic Farming

Fertilizers are essential for mitigating the effects of climate change and the adaptation of plants. It offers significant energy to plants and maintains soil quality and enhances yield. So, the significance of fertilizers in sustaining the world cannot be overstated (Henderson et al. 2018). Compost positively improves crop yield without causing any damage to soil. It is reported that 100% (2–4 t ha^{-1}) increase was seen in maize yield in Kenya, groundnut yield by 0.3–0.9 t ha^{-1}, millet yield 0.3–1 t ha^{-1}, and Bolivia potato yield increased by 4–15 t ha^{-1}. In Latin America, the use of green manure and composting increased maize/wheat yields up to 198–250% (Viswanathan et al. 2020). In Ethiopia, utilization of compost as compared to chemically fertilized plots increased yield of wheat up to 20%, barley 9%, maize 7%, and finger millet

3% (Branca et al. 2013). Despite the lower production as compared to traditional farming, organic farming is regarded as a sustainable farming approach (Andersen et al. 2015). According to Indian experts, organic agriculture improved soil health and increased profit up to 22% despite loss of 9% in productivity due to the higher price paid for verified organic food (Viswanathan et al. 2020). Because of some issues organic farming provides less profits per unit product than conventional farming, it takes a bigger area to generate the same output (Muller et al. 2017).

4.3 Conservation Agriculture

Another ecofriendly approach of improving resilience to climate change is conservation agriculture, which delivers crop yields comparable to traditional farming due to reduced tillage, soil cover protection, and crop diversification (rotation) (Williams et al. 2018). Independent landowners and environmental organizations in Europe, notably the Great Britain, regard rewilding as a land management method. It often includes methods such as establishing naturalistic grazing and fire regimes on grasslands or in boreal forests, as well as modifying flood patterns in river systems. Rewilding further includes "passive management, natural recolonization, aided migration, and the restoration of missing species from a system" (Viswanathan et al. 2020). According to studies, rewilding in Australia aims to improve ecosystems and promote self-reliant habitats (Sandom et al. 2019). Sediment trapping is a successful approach for mitigating the harmful impacts of soil degradation and conserving the soil (Mekonnen et al. 2015).

5 Strategies for Managing Livelihoods

5.1 Agro-forestry

In some countries, integrated farming and agroforestry use as livelihood management services (Singh and Singh 2017). According to studies, agroforestry is used by 1.2 billion people in poor nations to sustain agricultural production and livelihoods. Agroforestry systems include multiple trees mixed with agricultural crops, nurseries, farmhouse plantings, soil conservation, and reclamation by trees or plants, shelterbelts, windbreaks, and hydro-forestry (Viswanathan et al. 2020). Such integrated practices are frequently used because they have the potential to increase agricultural yield and food security through climate adaptation and mitigation techniques (Mbow et al. 2014). Vietnam is an example of adoption of successful agroforestry because forest are less sensitive to low productivity due to adverse weather, providing revenue, food, fodder, and other environmental benefits (Nguyen et al. 2013).

5.2 *Crop Livestock Integrated Farming*

Integrated farming strategies are seen as a sustainable solution since they combine crop production with a variety of livelihood strategies, such as livestock management. It has been established that sustainable farming supports agricultural biodiversity and other food-based land use patterns that increase agroecosystems resilience to climate change (Singh and Singh 2017). Due to increased urban demand for meat and associated goods, animal farming or livestock management is predicted to develop greatly in the future. Due to changing climates, farmers are needed to make conscious decisions on specific livestock types or a hybrid of such sorts that better matches the local location (Viswanathan et al. 2020). As compared to cow dung and dairy slurry, poultry manure boosts wheat output (Aranguren et al. 2021). In maize–wheat production system, when 50 percent of mineral-N was replaced with solid manure or liquid manure, economic profits improved significantly to 17.2% and 19.1%, respectively, while endpoint environmental consequences fell by 24.6% and 37.9% (Li et al. 2020).

5.3 *Migration*

Analysis reveals that rural families throughout the world are experiencing various climate-resilient techniques linked to agricultural land and water management, as well as livelihood maintenance. The success and sustainability of these methods vary with area (Viswanathan et al. 2020). Technological interventions, such as laser land leveling and desalination, have also aided in the development of climate-resilient farming techniques. Simultaneously, successful implementation of most of these techniques needs strong laws as well as financial and institutional help from governments, private sectors producers of the technology, and related services (Viswanathan et al. 2020).

6 Diversity in Crops

6.1 *Yield Improvement and Climate Resilience*

Grain yield is a polygenic trait influenced by environment or other factors. Landraces have been shown to provide genes for grain production enhancement in irrigated areas as well as in high temperature and drought stress conditions (Lopes et al. 2015). Landrace studies from throughout the world have found possible avenues for increasing grain productivity and climate resistance (Mondal et al. 2016). Mexican landraces are capable of adapting temperature and water stress (Vikram et al. 2016). Synthetic hexaploid wheat, which is molded from wild tetraploid and diploid varieties, has modern traits. In diploid wheat, genomic areas can lead to a roughly 10%

increase in grain weight and grain production. In diploid wheat, genomic areas lead to 10% increase in grain weight and grain production (Börner et al. 2015). Such beneficial genetic variants can be transferred through synthetic wheat. According to research, synthetic wheat cultivars may draw more water from deep soil, which is a great adaptive characteristic under drought stress (Viswanathan et al. 2020). Other synthetic variants that are more resistant to water-logged soils, high temperatures, freezing, and wild variant of wheat are also rich in diversity (Mondal et al. 2016).

6.2 Disease Resistance

Diseases produced by fungus and fungal pathogens are a significant danger to wheat productivity. New pathogenicity is evolved in all diseases by migration, variation, selection, and recombination of pathogenic strains, although it is more common in species that cause powdery mildew and rust. Diseases can cause harvest loss by 70% in vulnerable types. Identifying and using multiple sources of long-lasting resistance is one strategy for mitigating disease risk. The three rusts are globally significant fungal diseases of wheat caused by obligatory parasites: stripe or yellow rust, leaf, or brown rust and stem or black rust (Viswanathan et al. 2020). Though most of the genes against rust, derived from hexaploid wheat, many are derived from wild varieties and other genera, e.g., Aegilops, Dasypyrum, Secale, and Thinopyrum (Mondal et al. 2016).

6.3 Insect Pest Resistance

Global agricultural losses due to insect pests were estimated to approximately 5.1% in the pre-green revolution era, but increased to 9.3% in the post-green revolution era in the 1990s (Dhaliwal et al. 2010). Insects and pests are versatile and dynamic. Changing temperature of environment can alter their morphology, activity, and distribution. For example, high temperature in winters may increase aphid generations along each wheat growth cycle (Hulle et al. 2010). Fluctuations in temperature may change the activity of aphid, while progress on resistance against disease greatly influenced to wheat yield protection and control of insects have mostly relied on pesticides usage. The introduction of resistant genes (single or combination with several genes) to provide broad-spectrum defense against numerous diseases and insect might have a major positive impact (Alford et al. 2014).

7 Climate-Resilient Agriculture (CRA)

Including FAO, several international and national organizations are continually involved in resolving local and global development challenges caused by environmental changes on agriculture by developing suitable strategies and institutional reforms (Viswanathan et al. 2020). Many countries of the world relied on agriculture and its related sectors for livelihood. Extreme weather conditions are also the hazards that agronomists must factor into their decisions against environmental challenges (Lybbert and McPeak 2012). Apparently, farmer faces difficulties to take any decision due to lack of information about crops, fertilizers, climatic variability, diseases, and pest attack. This highlights the importance of broadening institutional assistance, technical solutions, and overcoming techniques, technological solutions, and strategies to overcome barriers against climate-resilient agriculture promotion (Viswanathan et al. 2020; Imran et al. 2022).

Technological advancements can help in the usage of CRA methods besides compromising farmer's livelihood. Technologies under-develop for actual weather forecasting, and decision-making is crucial for controlling irrigation schedules, fertilizer usage, and crop selection (Sidhu et al. 2011). Application of various technologies, like laser leveling tensiometers and direct rice sowing, may improve farm-level efficiency and the utilization of water and other essential inputs. (Trærup 2012; Mahmood et al. 2022). Social innovations have been demonstrated to be effective in enabling climate change responses. Social innovations are important in tackling climate change challenges because they help to develop links through collaborative activities on local scale, thereby providing a new aspect to climate change action to institutional dynamics (Rodima-Taylor 2012). As compared to traditional practices, the soil conservation strategy increased wheat output about 57% (Mustafa et al. 2021).

Climate-smart villages (CSVs) are the concept that uses participatory techniques to integrate administrative and technological choices for coping with climate change in agriculture. This strategy seeks to find and improve CRA alternatives that perform effectively at local environment, so assisting policymakers, agricultural experts, and global and regional investors in developing future agriculture policies. CRA must evaluate the stability of its outcomes in order to generate positive and long-term results (Viswanathan et al. 2020).

8 Conclusion and Future Perspectives

Climate change is the most serious limiting factor for production of wheat. Wheat is highly sensitive to climatic variations. In response to these variations, a major yield loss occurs globally every year. Several biotic and abiotic stresses occurred in response to climate change. Drought is one of the major limiting factors as a result of climate change which affects the growth and yield of wheat. Increase in temperature cause early maturation of wheat which greatly reduced wheat grain size and weight. So here is the need to overcome the effect of these climatic changes to ensure global food security. Batter understanding of climate-resilient technologies will help to ameliorate the negative impact of environmental change on wheat crop. We can improve the production by educating farmers about soil resilience (improving soil health), adaptations in crop varieties (genetically modified varieties which can withstand by the extremities of climate), proper utilization of water, tillage practices (adding organic manure in soils), and adaptations of livestock system.

References

Adhikari P, Araya H, Aruna G, Balamatti A, Banerjee S, Baskaran P, Barah BC, Behera D, Berhe T, Boruah P, Dhar S (2018) System of crop intensification for more productive, resource-conserving, climate-resilient and sustainable agriculture: experience with diverse crops in varying agroecologies. Int J Agri Sustain 16:1–28

Ahmed M, Suphachalasai S (2014) Assessing the costs of climate change and adaptation in South Asia. Asian Development Bank, Mandaluyong City

Alford L, Andrade TO, Georges R, Burel F, Van Baaren J (2014) Could behaviour and not physiological thermal tolerance determine winter survival of aphids in cereal fields. PLoS One 9:114982

Ali HH, Tanveer A, Naeem M, Jamil M, Iqbal M, Javaid MM, Kashif MS (2015) Efficacy of pre-emergence herbicides in controlling *Rhynchosia capitata*, an emerging summer weed in Pakistan. Philipp Agric Sci 98:301–311

Andersen MM, Landes X, Xiang W, Anyshchenko A, Falhof J, Østerberg JT, Olsen LI, Edenbrandt AK, Vedel SE, Thorsen BJ, Sandøe P (2015) Feasibility of new breeding techniques for organic farming. Trends Plant Sci 20:426–434

Aranguren M, Castellon A, Besga G, Ojinaga M, Aizpurua A (2021) Influence of wheat crop on carbon and nitrogen mineralization dynamics after the application of livestock manures. Geoderma 402:115351

Bahinipati CS, Viswanathan PK (2019) Incentivizing resource efficient technologies in India: evidence from diffusion of micro-irrigation in the dark zone regions of Gujarat. Land Use Policy 86:253–260

Behnke GD, Zuber SM, Pittelkow CM, Nafziger ED, Villamil MB (2018) Long-term crop rotation and tillage effects on soil greenhouse gas emissions and crop production in Illinois, USA. Agric Ecosyst Environ 261:62–70

Börner A, Ogbonnaya FC, Röder MS, Rasheed A, Periyannan S, Lagudah ES (2015) Aegilops tauschii introgressions in wheat. In: Alien introgression in wheat. Springer, Cham, pp 245–271

Branca G, Lipper L, McCarthy N, Jolejole MC (2013) Food security, climate change and sustainable land management. A review. Agron Sustain Dev 33:635–650

Bryan E, Ringler C, Okoba B, Roncoli C, Silvestri S, Herrero M (2013) Adapting agriculture to climate change in Kenya: household strategies and determinants. J Environ Manag 114:26–35

Carty T, Magrath J (2013) Growing disruption: climate change, food and the fight against hunger. Oxfam International, Oxford

Challinor AJ, Watson J, Lobell DB, Howden SM, Smith DR, Chhetri N (2014) A meta-analysis of crop yield under climate change and adaptation. Nat Clim Chang 4:287–291

Dai A (2013) Increasing drought under global warming in observations and models. Nat Clim Chang 3:52–58

Deuschle D, Minella JP, Hörbe T d AN, Londero AL, Schneider FJ (2019) Erosion and hydrological response in no-tillage subjected to crop rotation intensification in southern Brazil. Geoderma 340:157–163

Dhaliwal GS, Jindal V, Dhawan AK (2010) Insect pest problems and crop losses: changing trends. Indian J Ecol 37:1–7

Dury J, Schaller N, Garcia F, Reynaud A, Bergez JE (2012) Models to support cropping plan and crop rotation decisions. A review. Agron Sustain Dev 32:567–580

EL Sabagh A, Islam MS, Skalicky M, Ali Raza M, Singh K, Anwar Hossain M, Hossain A, Mahboob W, Iqbal MA, Ratnasekera D, Singhal RK (2021) Salinity stress in wheat (*Triticum aestivum* L.) in the changing climate: adaptation and management strategies. Front Agron 3:661932

Fahad S, Wang J (2018) Farmers' risk perception, vulnerability and adaptation to climate change in rural Pakistan. Land Use Policy 79:301–309

Farooq N, Mehmood A, Tanveer A, Nadeem MA, Sarwar G, Abbas T, Javaid MM, Matloob A (2021) Exploiting fodder and nutritive value of alternanthera philoxeroides (alligator weed) at different times of harvest: a potential fodder weed. Pak J Bot 53:1101–1106

Garcia GA, Serrago RA, Appendino ML, Lombardo LA, Vanzetti LS, Helguera M, Miralles DJ (2011) Variability of duration of pre-anthesis phases as a strategy for increasing wheat grain yield. Field Crops Res 124:408–416

Giraldo P, Benavente E, Manzano-Agugliaro F, Gimenez E (2019) Worldwide research trends on wheat and barley: a bibliometric comparative analysis. Agronomy 9:352

Glendenning CJ, Van Ogtrop FF, Mishra AK, Vervoort RW (2012) Balancing watershed and local scale impacts of rain water harvesting in India. A review. Agric Water Manag 107:1–13

Gorst A, Dehlavi A, Groom B (2018) Crop productivity and adaptation to climate change in Pakistan. Environ Dev Econ 23:679–701

Henderson B, Cacho O, Thornton P, van Wijk M, Herrero M (2018) The economic potential of residue management and fertilizer use to address climate change impacts on mixed smallholder farmers in Burkina Faso. Agric Syst 167:195–205

Hossain MS, Arshad M, Qian L, Zhao M, Mehmood Y, Kächele H (2019) Economic impact of climate change on crop farming in Bangladesh: an application of Ricardian method. Ecol Econ 164:106354

Hossain MS, Arshad M, Qian L, Kächele H, Khan I, Islam MDI, Mahboob MG (2020) Climate change impacts on farmland value in Bangladesh. Ecol Indic 112:106181

Hulle M, Acier AC, Bankhead-Dronnet S, Harrington R (2010) Aphids in the face of global changes. C R Biol 333:497–503

Hussain D, Asrar M, Khalid B, Hafeez F, Saleem M, Akhter M, Ahmed M, Ali I, Hanif K (2021) Insect pests of economic importance attacking wheat crop (*Triticum aestivum* L.) in Punjab, Pakistan. Int J Trop Ins Sci 42:9–20

Imran K, Zafar H, Chattha MU, Mahmood A, Maqbool R, Athar F, Alahdal MA, Farhana B, Mahmood F, Hassan MU (2022) Seed priming with different agents mitigate alkalinity induced oxidative damage and improves maize growth. Notulae Botanicae Horti Agrobotanici Cluj-Napoca 50:12615

Iqra L, Rashid MS, Ali Q, Latif I, Malik A (2020) Evaluation of genetic variability for salt tolerance in wheat. Biol Clin Sci Res J 2020:1–8

Jalota SK, Kaur H, Ray SS, Tripathi R, Vashisht BB, Bal SK (2012) Mitigating future climate change effects by shifting planting dates of crops in rice–wheat cropping system. Reg Environ Chang 12:913–922

Javaid MM, Tanveer A, Ali H, Shahid M, Balal R, Aqeel M (2016) Wheat yield loss in a two species competition with Emex australis and Emex spinosa. Planta Daninha 34:35–46

Javaid MM, Florentine S, Ali HH, Weller S (2018) Effect of environmental factors on the germination and emergence of *Salvia verbenaca* L. cultivars (verbenaca and vernalis): An invasive species in semi-arid and arid rangeland regions. PLoS One 13:e0194319

Javaid MM, Florentine SK, Ashraf M, Mahmood A, Sattar A, Wasaya A, Li FM (2022a) Photosynthetic activity and water use efficiency of *Salvia verbenaca* L. under elevated CO2 and water-deficit conditions. J Agro Crop Sci 208:536–551

Javaid MM, Mahmood A, Alshaya DS, AlKahtani MD, Waheed H, Wasaya A, Khan SA, Naqve M, Haider I, Shahid MA (2022b) Influence of environmental factors on seed germination and seedling characteristics of perennial ryegrass (*Lolium perenne* L.). Sci Rep 12:1–11

Javaid MM, Wang X, Florentine SK, Ashraf M, Mahmood A, Li FM, Fiaz S (2022c) Effects on photosynthetic response and biomass productivity of Acacia longifolia ssp. longifolia under elevated CO2 and water-limited regimes. Front Plant Sci 13:817730

Kaye JP, Quemada M (2017) Using cover crops to mitigate and adapt to climate change. A review Agron Sustain Dev 37(1):4. https://doi.org/10.1007/s13593-016-0410-x.

Kavamura VN, Mendes R, Bargaz A, Mauchline TH (2021) Defining the wheat microbiome: towards microbiome-facilitated crop production. Comput Struct Biotechnol J 19:1200–1213

Keesstra S, Nunes J, Novara A, Finger D, Avelar D, Kalantari Z, Cerdà A (2018) The superior effect of nature based solutions in land management for enhancing ecosystem services. Sci Total Environ 610:997–1009

Kreft S, Eckstein D, Melchior I (2016) Global climate risk index 2017. Who suffers most from extreme weather events? Weather-related loss events in 2015 and 1996 to 2015

Kumar DS, Palanisami K (2019) Adoption of water management technologies-managing the water-energy nexus in agriculture. Econ Polit Wkly 54(14)

Langridge P, Reynolds M (2021) Breeding for drought and heat tolerance in wheat. Theor Appl Genet 134:1753–1769

Li S, Wu J, Wang X, Ma L (2020) Economic and environmental sustainability of maize-wheat rotation production when substituting mineral fertilizers with manure in the North China Plain. J Clean Prod 271:122683

Lopes MS, El-Basyoni I, Baenziger PS, Singh S, Royo C, Ozbek K, Aktas H, Ozer E, Ozdemir F, Manickavelu A, Ban T (2015) Exploiting genetic diversity from landraces in wheat breeding for adaptation to climate change. J Exp Bot 66:3477–3486

Lybbert TJ, McPeak J (2012) Risk and intertemporal substitution: livestock portfolios and off-take among Kenyan pastoralists. J Dev Econ 97:415–426

Mahmood N, Arshad M, Kaechele H, Shahzad MF, Ullah A, Mueller K (2020) Fatalism, climate resiliency training and farmers' adaptation responses: implications for sustainable rainfed-wheat production in Pakistan. Sustainability 12:1650

Mahmood A, Wang X, Shahzad AN, Fiaz S, Ali H, Naqve M, Javaid MM, Mumtaz S, Naseer M, Dong R (2021) Perspectives on bioenergy feedstock development in Pakistan: challenges and opportunities. Sustainability 13:8438

Mahmood A, Bibi S, Naqve M, Javaid MM, Zia MA, Jabbar A, Ud-Din W, Attia KA, Khan N, Al-Doss AA (2022) Physiological, biochemical and yield responses of Linseed (*Linum usitatissimum* L.) in α-tocopherol-mediated alleviation of salinity stress. Front Plant Sci 13:867172

Mbow C, Van Noordwijk M, Luedeling E, Neufeldt H, Minang PA, Kowero G (2014) Agroforestry solutions to address food security and climate change challenges in Africa. Curr Opin Environ Sustain 6:61–67

Mekonnen M, Keesstra SD, Stroosnijder L, Baartman JE, Maroulis J (2015) Soil conservation through sediment trapping: a review. Land Degrad Dev 26:544–556

Mesbah A, Nilahyane A, Ghimire B, Beck L, Ghimire R (2019) Efficacy of cover crops on weed suppression, wheat yield, and water conservation in winter wheat–sorghum–fallow. Crop Sci 59:1745–1752

Meuwissen MP, Feindt PH, Spiegel A, Termeer CJ, Mathijs E, De Mey Y, Finger R, Balmann A, Wauters E, Urquhart J (2019) A framework to assess the resilience of farming systems. Agric Syst 176:102656

Meyer N, Bergez JE, Constantin J, Justes E (2019) Cover crops reduce water drainage in temperate climates: a meta-analysis. Agron Sustain Dev 39:1–11

Meyer N, Bergez JE, Constantin J, Belleville P, Justes E (2020) Cover crops reduce drainage but not always soil water content due to interactions between rainfall distribution and management. Agric Water Manag 231:105998

Miller PR, Lighthiser EJ, Jones CA, Holmes JA, Rick TL, Wraith JM (2011) Pea green manure management affects organic winter wheat yield and quality in semiarid Montana. Can J Plant Sci 91:497–508

Mohammadi R (2018) Breeding for increased drought tolerance in wheat: a review. Crop Pasture Sci 69(3):223–241

Mondal S, Rutkoski JE, Velu G, Singh PK, Crespo-Herrera LA, Guzman C, Bhavani S, Lan C, He X, Singh RP (2016) Harnessing diversity in wheat to enhance grain yield, climate resilience, disease and insect pest resistance and nutrition through conventional and modern breeding approaches. Front Plant Sci 7:991

Muller A, Schader C, El-Hage Scialabba N, Brüggemann J, Isensee A, Erb KH, Smith P, Klocke P, Leiber F, Stolze M, Niggli U (2017) Strategies for feeding the world more sustainably with organic agriculture. Nat Commun 8:1–13

Mustafa G, Mahmood H, Iqbal A (2021) Environmentally friendly farming and yield of wheat crop: a case of developing country. J Clean Prod 314:127978

Neate PJ (2013) Climate-smart agriculture: success stories from farming communities around the world. CTA Success Stories.

Nguyen Q, Hoang MH, Oborn I, van Noordwijk M (2013) Multipurpose agroforestry as a climate change resiliency option for farmers: an example of local adaptation in Vietnam. Clim Chang 117:241–257

Northupl BK, Rao SC (2015) Green manures in continuous wheat systems affect grain yield and nitrogen content. Agron J 107:1666–1672

Partal E, Paraschivu M (2020) Results regarding the effect of crop rotation and fertilization on the yield and qualities at wheat and maize in South of Romania. Sci Pap-Ser A-Agron 2:184–189

Pedraza V, Perea F, Saavedra M, Fuentes M, Castilla A, Alcántara C (2015) Winter cover crops as sustainable alternative to soil management system of a traditional durum wheat-sunflower rotation in Southern Spain. Procedia Environ Sci 29:95–96

Pequeno DN, Hernandez-Ochoa IM, Reynolds M, Sonder K, MoleroMilan A, Robertson RD, Lopes MS, Xiong W, Kropff M, Asseng S (2021) Climate impact and adaptation to heat and drought stress of regional and global wheat production. Environ Res Lett 16:054070

Raphael JP, Calonego JC, Milori DMB, Rosolem CA (2016) Soil organic matter in crop rotations under no-till. Soil Tillage Res 155:45–53

Raza A, Shaukat H, Ali Q, Habib M (2018) Assessment of RAPD markers to analyse the genetic diversity among sunflower (*Helianthus annuus* L.) genotypes. Turkish J Agric Food Sci Technol 6:107–111

Reynolds MP, Lewis JM, Ammar K, Basnet BR, Crespo-Herrera L, Crossa J, Dhugga KS, Dreisigacker S, Juliana P, Karwat H, Kishii M (2021) Harnessing translational research in wheat for climate resilience. J Exp Bot 72:5134–5157

Rodima-Taylor D (2012) Social innovation and climate adaptation: local collective action in diversifying Tanzania. Appl Geogr 33:128–134

Roesch-McNally GE, Arbuckle JG, Tyndall JC (2018) Barriers to implementing climate resilient agricultural strategies: the case of crop diversification in the US Corn Belt. Glob Environ change 48:206–215

Saab A (2016) Climate-resilient crops and international climate change adaptation law. Leiden J Int Law 29(2):503–528. https://doi.org/10.1017/S0922156516000121

Saleem M, Ali S (2015) Aquifer modelling for feasibility and design of rainwater harvesting – a review. Int J Sci Res 4:504–508

Sandom CJ, Dempsey B, Bullock D, Ely A, Jepson P, Jimenez-Wisler S, Newton A, Pettorelli N, Senior RA (2019) Rewilding in the English uplands: policy and practice. J Appl Ecol 56:266–273

Sanz M, De Vente J, Chotte JL, Bernoux M, Kust G, Ruiz I, Almagro M, Alloza J, Vallejo R, Castillo V (2017) Sustainable land management contribution to successful land-based climate change adaptation and mitigation: a report of the science-policy interface. United Nations Convention to Combat Desertification (UNCCD), Bonn

Shahid, MA, Balal RM, Javaid M, Aquel MA, Abbas T, Mattso N, Islam S (2014) Mitigation of stress induced by salinity and nickel in pea (*Pisum sativum* L.) by exogenous application of synthetic proline and proline-enriched *Lolium perenne* (L.) leaf extract. In 2014 ASHS annual conference

Si W, Li P, Muhammad A, Mo F, Hui X, Asfa B, Javaid M, Muhammad A, Zhang J, Li F (2017) Micro-field rain-harvesting farming system in semiarid Pakistan: a critical topic to address. Pak J Bot 49:1217–1221

Sidhu RS, Vatta K, Lall U (2011) Climate change impact and management strategies for sustainable water-energy-agriculture outcomes in Punjab. Indian J Agric Econ 66:329–339

Singh R, Singh GS (2017) Traditional agriculture: a climate-smart approach for sustainable food production. Energy Ecol Environ 2:296–316

Srivastav AL, Dhyani R, Ranjan M, Madhav S, Sillanpää M (2021) Climate-resilient strategies for sustainable management of water resources and agriculture. Environ Sci Pollut Res 28:41576–41595

Suryavanshi P, Buttar GS, Brar AS (2015) Micro irrigation for sustainable agriculture: a brief review. Indian J Econ Dev 11:147–155

Tal A (2016) Rethinking the sustainability of Israel's irrigation practices in the drylands. Water Res 90:387–394

Tanveer A, Khaliq A, Javaid M, Chaudhry M, Awan I (2013) Implications of weeds of genus euphorbia for crop production: a review. Planta Daninha 31:723–731

Tanveer A, Javaid MM, Irfan M, Khaliq A, Yaseen M (2015) Yield losses in chickpea with varying densities of dragon spurge (*Euphorbia dracunculoides*). Weed Sci 63:522–552

Tesfaye K, Zaidi PH, Gbegbelegbe S, Boeber C, Rahut DB, Getaneh F, Seetharam K, Erenstein O, Stirling C (2017) Climate change impacts and potential benefits of heat-tolerant maize in South Asia. Theor Appl Climatol 130:959–970

Trærup SL (2012) Informal networks and resilience to climate change impacts: a collective approach to index insurance. Glob Environ Change 22:255–267

Vikram P, Franco J, Burgueno-Ferreira J, Li H, Deepmala S, Saint Pierre C, Ortiz C, Sneller C, Tattaris M, Guzman C, Sansaloni CP (2016) Unlocking the genetic diversity of creole wheats. Sci Rep 6:1–13

Viswanathan PK, Kuma MD, Narayanamoorthy A (eds) (2016) Micro irrigation systems in India: emergence, status and impacts. Springer, Singapore

Viswanathan PK, Kavya K, Bahinipati CS (2020) Global patterns of climate-resilient agriculture: a review of studies and imperatives for empirical research in India. Rev Dev Change 25:169–192

Wallenstein MD, Hall EK (2012) A trait-based framework for predicting when and where microbial adaptation to climate change will affect ecosystem functioning. Biogeochemistry 109:35–47

Williams A, Jordan NR, Smith RG, Hunter MC, Kammerer M, Kane DA, Koide RT, Davis AS (2018) A regionally-adapted implementation of conservation agriculture delivers rapid improvements to soil properties associated with crop yield stability. Sci Rep 8:1–8

Zaveri E, Lobell DB (2019) The role of irrigation in changing wheat yields and heat sensitivity in India. Nat Commun 10:1–7

Zhao C, Liu B, Piao S, Wang X, Lobell DB, Huang Y, Huang M, Yao Y, Bassu S, Ciais P, Durand JL (2017) Temperature increase reduces global yields of major crops in four independent estimates. Proc Natl Acad Sci 114:9326–9331

Chapter 7
Climate Resilient Weed Management for Crop Production

Dibakar Roy, Sourav Ghosh, Debarati Datta, Dasari Sreekanth, Deepak Pawar, Pijush Kanti Mukherjee, Dibakar Ghosh, Subhas Chandra Santra, and Debojyoti Moulick

Abstract Weeds are ubiquitous in all cropping situations and can inflict major losses in yield and production quality. In recent years, climate variability has caused major impacts on natural and human ecosystems. Agriculture is directly impacted by climate change amid rising atmospheric CO_2, elevated temperatures, and changes in rainfall patterns which have a substantial effect on growth and physiology of crops and weeds. Weeds unlike other pests share a similar trophic level with crop plants and pose competition for essential resources thereby causing substantial crop yield losses. Climate change influences crop–weed interactions by favoring C_4 weeds in the increased temperature scenario and poses serious yield penalties. Since majority of the competitive weeds like Cyperus, Amaranthus, etc., are C_4 plants, these would cause increased problem in future. Although CO_2 and temperature are the major contributing factors for climate change, shifts in rainfall patterns also pose weed management challenges and increased crop–weed competition. Reduced water availability, due to frequent droughts, would shift the competitive balance between crops and favor some xerophytic weed species. Climate change vis-à-vis continuous and indiscriminate usage of herbicides poses several adverse effects on biodiversity, environment, and human health. Integrated weed management (IWM) components including weed prevention, optimal fertilizer schedule, summer tillage, crop rotation, modified land preparation, altering crop geometry, seed rate and sowing time, stale seedbed technique, use of weed competitive cultivars, cover crops, residue, allelopathy, etc., which complement reduced and judicious herbicide application should be promoted to combat weed problems under changing climate.

D. Roy · D. Sreekanth · D. Pawar · P. K. Mukherjee
ICAR-Directorate of Weed Research, Jabalpur, India

S. Ghosh · D. Datta
ICAR-Central Research Institute for Jute and Allied Fibres, Barrackpore, India

D. Ghosh
ICAR-Indian Institute of Water Management, Bhubaneswar, India

S. C. Santra · D. Moulick (✉)
Department of Environmental Science, University of Kalyani, Nadia, West Bengal, India

M. Hasanuzzaman (ed.), *Climate-Resilient Agriculture, Vol 2*,
https://doi.org/10.1007/978-3-031-37428-9_7

Keywords Climate change · Weed · Crop · Elevated CO_2

1 Introduction

In the era of drastic climate change regimes, apart from unpredictable scarcity of water, onset of sudden flood, UV-induced damage soil management remains the most sort after arena of concern. Besides intense population growth, degradation of arable soil, deprivation of adequate nutrient pool remains a challenge for the researchers across the world. Abundance of heavy metals (HM)/metalloids in arable soil beyond the threshold limit is another important factor that perhaps exaggerated the challenges of soil management (Moulick et al. 2019a, 2020, 2021; Saha et al. 2019). As a result of HM contamination, severe consequences can be seen in wide range of field crops and thus hinders their sustainable production (Moulick et al. 2018a, b). These adverse effects imposed by HMs can be seen in almost every stage of crops life cycle and even well documented in the post-harvest phases also (Moulick et al. 2016a, 2018d, 2020, 2022). Moreover, exposure of crops in HM-contaminated agroecosystem caused (a) inhibition of germination and poor seedling establishment (Moulick et al. 2016b, 2017, 2018c, 2019b; Chowardhara et al. 2019a, b); (b) hampers the redox homeostasis (Choudhury et al. 2021a, b; Mazumder et al. 2021, 2022; Choudhury et al. 2022a, b). In reality, crops usually experience combination(s) of stress(es). As a result of exposure to stress the internal redox homeostasis gets derailed, often beyond any repair and ultimately resulted into death of the crop (or crops in cropping system) and drastic decline in yield.

At present, global climate change has brought the scientific fraternity in front of great challenges to secure yield stability, by avoiding/impart resilience to multiple stress(es) operating simultaneously. The temperatures are predicted to have risen by 0.1–0.3 °C/decade worldwide, according to long-term warming patterns since pre-industrial times (IPCC 2014). The Inter-Governmental Panel on Climate Change (IPCC) has projected the temperature increase to be between 1.1 °C and 6.4 °C by the end of the twenty-first century (IPCC 2007). On the other hand, the concentrations of CO_2 in the atmosphere increased at a record-breaking speed of 419.05 parts per million (ppm) in 2021 (IPCC 2021). This rise will continue in the near future and reports are stating that it could exceed 600 ppm (Schimel et al. 1996), while the IPCC also proposed, as a conservative prediction of 700 ppm by the end of the twenty-first century (IPCC 2007).

Weeds seem to be more genetically diverse and physiologically flexible than crops. As a result, weeds adapt rapidly and flourish in a variety of harsh environments. Climate change is anticipated to surge weed competition, resulting in larger output losses if weed control is not properly implemented (Miri et al. 2012; Valerio et al. 2013; Ghosh et al. 2021, 2022). Seasonal variations in temperature and CO_2 may modify the crop–weed–pest relations and livestock agricultural systems, in addition to directly affecting crop growth and productivity. Further various unpredictable hazardous natural events like draught, flood, heat stress, and cold waves become more severe and will negatively affect the productivity of agroecosystems. On the other hand, especially higher CO_2 levels will probably have a positive effect on the productivity of some selected C3 crops such as wheat, rice, barley, and soybean. Higher CO_2 levels are anticipated to increase agricultural yields by up to 13% by 2050 (Jaggard et al. 2010). The elevated CO_2 results in partial stomatal closure, which raises plant tissue temperature and has a negative effect on plant growth and production. Aside from these directly connected problems, agricultural output and quality may also be affected by unreliable rainfall patterns along with rise temperatures during the growing season (Hatfield et al. 2011; Mahajan et al. 2012; Kadam et al. 2014). The detrimental impact of concurrent temperature increases on most food crops offset the positive effects of higher CO_2 on growth and production (Prasad et al. 2001, 2005).

There is a global range expansion (introduction or migration into different regions) of weedy vegetation, along with changes in species life cycles ultimately leading to development of diverse climate smart weed population. Migration of weeds will further lead to differences in the qualitative and quantitative nature of weed populations in natural as well as in managed ecosystems too. In this chapter, we focused our discussion on effect of elevated CO_2, temperature and erratic rainfall on weed dynamics, weed shift and future abundance under various environmental conditions, behavior of herbicides on weed flora and their effectiveness under climate change and probable integrated weed management practices under context of climate change.

2 Climate Change and Weed Dynamics

Weed having greater physiological elasticity as well as intraspecific genetic multiplicity as compared with most cultivated crops provides them with a modest edge in catastrophic events such as cyclones, flooding, drought, and fires (from high temperature) and followed by faster reestablishment after these unpredictable natural calamities (Singh et al. 2011). The impact of climate change on single weed species as well as community depends on a complex set of parameters. Future climatic scenario may increase geographic expansion of aggressive weed species of tropical and subtropical regions into temperate or cooler areas (e.g., *Cassia, Amaranthus, Sesbania, Crotalaria, Rottboellia, Imperata, Panicum, and Striga*). Along with elevated CO_2 condition will help C3 weeds to dominate over C4 weeds which will again alter species diversity and density. Various facets of the climate change which had an influence on the weeds may be due to the alteration in life cycle, changes in crop–weed competition pattern and period, expansion in infestation area (introduction to new areas), and weed dynamics.

2.1 Effects of Elevated CO_2 on Weed Dynamics

The change in CO_2 level may be favorable for net photosynthesis of C3 crops, as most of the global cultivated crops are C3 whereas plants having C_4 photosynthetic pathway may be less responsive to this projected elevated CO_2 level. Thus, elevated level of CO_2 will have an essential role in crop–weed competition, degree of weed infestation, and crop yield loss under climate change scenario. Numerous studies give an idea about the nature of crop–weed competition between C3 crop and C4 weed, where a greater atmospheric CO_2 level at the vegetative growth of C3 crop (like rice and wheat) generally favors over the growth of C4 weeds. However, when both crop and weeds belongs to similar photosynthetic pathway, weed produced higher biomass than crop under elevated CO_2 condition. It was reported that in wheat (C3) crops, C3 weeds like *Phalaris minor* and *Avena ludoviciana* grew more rapidly than wheat crops grown in environments with increasing CO_2 levels. On the other hand, many economic important C4 crops like maize, sugarcane, and sorghum face significant competition from C3 weeds like *Chenopodium album* L under elevated CO_2 condition. (Ziska 2000). Further, Ziska et al. (2004) also reported that the biomass of the sorghum yield-reducing velvet leaf weed (*Abutilon theophrasti*) rose as a result of rising CO_2 levels. Growth and reproduction of *Parthenium hysterophorus* had positive correlation with CO_2 levels (Naidu and Paroha 2008).

In rice (C3), sole enhancement in CO_2 levels led to higher competitiveness against *Echinochloa glabrescens*, although the reverse was true in case of combined increase in CO_2 and temperature. Increasing atmospheric CO_2 and related changes in climate can directly affect weed physiology and alter crop weed interactions thereby reduces the efficiency of weed control measures. However, the dominance and appearance of weeds depends upon agro-climatic regions, associated crops and management practices adopted, physiological suitability, and genetic variations between weed and crop (Treharne 1989). For example, *P. minor* a C3 species is a dominating weed in wheat in western Indo-Gangetic Plains of India.

2.2 Effects of Elevated Temperature on Weed Dynamics

The distribution of weed species across different agro-ecological regions depends greatly on atmospheric temperature. Under raised up temperature, C4 weeds will offer higher competition to the C3 crops. Dominant weeds in rice crop are C4, and under elevated temperature, they become problematic in many areas. For example, *Ischaemum rugosum* is a common weed of rice in tropical region but now it is also an important weed of northern India. Under elevated temperature, parasitic weeds like *Striga* spp. and *Orobanche* could be a serious threat for global crop production (Mohamed et al. 2006, 2007; Phoenix and Press 2005). Under changing climatic condition, weeds required higher temperature for their growth and development like *Datura stramonium* L. would become more competitive (Cavero et al. 1999). By promoting growth and reproduction, elevated temperature also caused *Hieracium aurantiacum* L. to be more abundant under climate change (Brinkley and Bomford 2002).

2.3 *Effects of Erratic Rainfall and Drought Weed Dynamics*

The amount of rainfall plays a major role on weed emergence, dominance, and its competition with crops be it C_3 or C_4 type (Choudhury et al. 2022c). The alteration of rainfall period and amount could alter the occurrence of various weeds. For instance, due to increased rainfall, the weed species *Leptochloa chinensis* and *Marsilea quadrifolia* have taken over the space once occupied by *Echinochloa* spp. The prevalence of weeds and their effects on agricultural productivity could change due to irregular rainfall patterns, increased aridity, and high temperatures. Changes in distribution of rainfall favor hydromorphic weeds; however, C4 weed would be more benefited over C3 weeds under drought spells. Under water stress condition or dry spell during rainy season caused more crop–weed competition and resulted significant amount of crop yield loss due to inferior crop growth. When comparing well-watered soybeans to those under water stress, *Xanthium strumarium* L.'s impact on yield was more pronounced in the soybean crop (Mortensen and Coble 1989).

3 Weed Shift Under Climate Change Scenario

Climate change will alter distribution of plant species, as well as the overall effectiveness and production of ecosystem. In forests all over the world, increasing enormous number of woody vines as a result of CO_2 enrichment has been linked to higher tree mortality and impaired tree regeneration (Phillips et al. 2002). Preliminary findings revealed that many weeds were more tolerant of cold temperatures under elevated CO_2 (Boese et al. 1997), indicating that many weed species may expand toward polar side (McDonald et al. 2009; Ziska and Dukes 2011). Similarly, it was predicted that *P. hysterophorus* L, an invasive alien weed, may spread to colder region in response to climate change, particularly high CO_2 levels (Naidu 2013).

A surge in parasitic weeds would also represent a serious threat to the yield of the rice and sorghum crops in rainfed agriculture (Rodenburg et al. 2011). Due to temperature suitability, most of the harmful C3 and C4 weeds of the arable land are restricted to tropical and subtropical regions and not seen in cooler regions (Holm et al. 1997). Under climate change scenario, such species may adapt climatic conditions in situ or transfer their range to colder regions of higher latitude in near future. Several studies reported that such weeds often benefit in CO_2-enriched environment (Polley et al. 2002; Ziska and George 2004). Weeds like *Lonicera sempervirens*

L. and *Pueraria lobata* (Lour.) Merr. have their geographic ranges expanded in the past, and it is now a reality (Patterson 1995). Similarly, *Ischaemum rugosum* Salisb. was predominantly reported to be found in warmer regions of India, but it is now widespread in cooler regions of Northern India (Mahajan et al. 2012). The ongoing cultivation of host crops, such as corn, sorghum, rice, sunflower, legumes, and vegetables, combined with inadequate soil fertility, particularly in tropical areas, is causing an increase in parasitic weed concerns, such as *Striga* spp. and *Orobanche* spp. Likewise, climate change has induced altered weed distribution, such as the emergence of *Marsilea* spp. in India under the wet environments of rice. Severe drought forces the transition to direct-seed rice, encouraging recalcitrant grass weeds such as *Dactyloctenium aegyptium, Eleusine indica, Leptochloa chinensis*, and aerobic rice (*O. sativa*) (Matloob et al. 2015). Under forecast climate change, these weeds are expected to expand their geographic range, impacting the productivity of rain fed crops.

Due to their complex dispersion and higher adaption capacities, aggressive weeds have the potential to detect climatic change (Bergmann et al. 2010). Weeds strive to endure, once they have established themselves in order to maintain a new environment (Smith et al. 2012). As a result, spectrum of shifts is frequently supported by natural selection, which results in genetic and evolutionary adaptations to the new habitat (Richardson et al. 2013). Due to a lack of rainfall and protracted drought, arable crops and pastures will develop slowly, leaving barren land and allowing more robust drought-tolerant weeds to invade. Additionally, the impact of increased CO_2 on the geographic spread of weeds in controlled ecosystems needs to be taken into consideration (McDonald et al. 2009).

4 Herbicide Efficacy Under Climate Change Scenario

Herbicides are the synthesized chemicals which used to kill weeds and minimize agricultural output losses. The application and usage of herbicides has been gradually enhanced worldwide in traditionally low herbicide using developing countries where availability hand labor is low and expensive (Gianessi 2013). It is already reported that elevated CO_2 and temperature can change herbicide efficacy (Varanasi et al. 2016; Korres et al. 2016). For instance, elevated CO_2-induced weed growth could lower the time at seedling stage, which is the most critical and sensitive stage to herbicide action. Herbicide absorption also declined due to CO_2-induced stomatal conductance. However, elevated temperature condition triggers herbicide efficacy by increasing their absorption and followed by translocation of foliar herbicides, but faster translocation increases the rate of metabolism, which ultimately minimize the herbicidal action in the target plants (Johnson and Young 2002). It is reported that increased CO_2 and temperature can modify the leaf's structure and morphology by increasing leaf thickness or changing the cuticle wax's viscosity, which would reduce the absorption of herbicides (Ziska and Bunce 2006). Therefore, there is a need to focus on effect of climate change on weed flora shift and herbicide resistance to assure the global food security needs.

4.1 Effect of Elevated CO_2 on Herbicide Efficacy

Herbicide efficacy is influenced by the climate factors, and herbicides, predominantly foliage sprayed post-emergence (Kudsk and Kristensen 1992). A study by Archambault et al. (2001) noticed that enrichment of CO_2 lowered the effectiveness of the glyphosate. The impact on agriculture will be greater if these impacts are typical and widespread.

The number and conductance of stomata decreased in C3 plants, but leaf thickness increased, perhaps interfering with herbicide foliar absorption (Ainsworth and Long 2005), as well as significant upsurge in starch buildup on the leaf surface (Patterson 1995). Furthermore, if vegetative growth is accelerated which leads to enhanced photosynthesis in because of increasing CO_2, under such circumstances perennial weeds may create more trouble in coming future. Higher vegetative growth will cause dilution effect of the herbicide absorbed and therefore ultimately reduce the effectiveness of the commonly used foliage applied herbicide such as glyphosate; however, the specific mechanism causing enhanced resistance to glyphosate remains unknown (Manea et al. 2011). This might be linked with decrease in translocation rate as the root system matures. Furthermore, an upsurge in the root–shoot ratio is one of the responsible factors for herbicide effectiveness (Ziska et al. 2004). Therefore, future research should be focused on various molecular aspects of herbicidal defense mechanism operated in both C3 and C4 plants, resource distribution to the below ground plant parts, source–sink interactions, and mitochondrial respiration in the context of climate change scenarios.

4.2 Effect of Elevated Temperature on Herbicide Efficacy

The volatility of certain herbicides, such as trifluralin, increases as the temperature rises, making them less effective (Beestman and Deming 1974). It was temperatures (day/night 32/22 °C and 26/16 °C) which had less impact on *acifluorfen* (diphenyl ether group) phytotoxicity in *Xanthium strumarium* L. and *Ambrosia artemisiifolia* L. than relative humidity (Ritter and Coble 1981). However, the herbicides like (imidazolinone group), flumetsulam (sulfonanilide family), and thifensulfuron (sulfonylurea group) degradation significantly affected by elevated temperature in soil (Mcdowell et al. 1997). The herbicide glyphosate preoccupation is reliant on the temperature, as clear indication from *Desmodium tortuosum* (Sw.) DC a C3 weed (Sharma and Singh 2001). An increase in relative humidity/temperature boosted the efficacy of mesotrione on *X. strumarium* and *A. theophrasti* by threefold (Johnson and Young 2002). It was observed that when temperatures go beyond the range of

20–34 °C, it reduced the action of pyrithiobac (a pyrimidinylthiobenzoic acid group) on *Amaranthus palmeri* L. (Mahan et al. 2004). Anderson et al. (1993) reported that relative humidity had the greatest impact on glufosinate ammonium phytotoxicity, and this is due to alterations in cuticle hydration and droplet drying (Ramsey et al. 2005). *Raphanus raphanistrum* L., growing at low temperatures of 5/10 °C, was poorly managed with 1200 g ai ha^{-1} of glufosinate in controlled environmental chambers in Australia under varied night/day temperatures of 5/10, 15/20, and 20/25 °C in comparison; mortality (%) was attained for the same dosage at 15/20 and 20/25 °C (Kumaratilake and Preston 2005), suggesting that glufosinate is more efficient at higher temperatures (Table 7.1). These minimized herbicide efficacies under climate change ultimately developed herbicide-resistant weeds under climate change scenarios (Table 7.2).

5 Integrated Weed Management for Imparting Climate Resilience

Climate variability is likely to alter growth of crops and weeds differentially thereby having an impact on the weed control methods. Moreover, changes in numerous climate factors may modify the crop–weed interactions and tilt the balance either crop or weed. Opportunistic weed species show higher persistence under changing climate in terms of superior seed dispersal and better adaptation abilities, and thus, integrating various weed control practices is imperative for imparting climate resilience.

5.1 Weed Prevention

Preventive weed management measures are those measures where weeds are restricted from infesting an area. Conditions are created which reduce weed appearance within the crop field and limit the spreading and perseverance of weed propagules. Plant quarantine and weed legislation help to restrict exotic weeds from possible entry into a new country. Prevention may include preventing weeds not to germinate from soil and to form seeds, tubers, rhizomes, corms/cormlets, or other propagules for dissemination; avoiding seed rains getting accumulated from elsewhere; and facilitating exhaustion of seed bank. This approach is better if followed before sowing of crops.

Principles of preventive weed control include:

(i) *Minimizing seed production*: An annual weed can produce up to several thousands of seeds per plant which add to the soil seed bank and may cause problems for producers and future users. Seed production from weeds should be prevented during the fallow period and on bunds and channels because these

Table 7.1 Impact of temperature on herbicide efficacy

Climate parameter	Herbicide	weed	Impact	Reference
Temperatures (day/night 32/22 °C and 26/16 °C)	Acifluorfen (diphenylether group) phytotoxicity in	*Xanthium strumarium* L. and *Ambrosia artemisiifolia* L.	Reduced the efficacy	Ritter and Coble (1981)
An increase in relative humidity or temperature	Mesotrione	*X. strumarium* and *A. theophrasti*	Enhanced the efficacy by threefold	Johnson and Young (2002)
Temperatures beyond the range of 20–34 °C	Pyrithiobac (pyrimidinylthiobenzoic acid group)	*Amaranthus palmeri* L.	Reduced the efficacy	Mahan et al. (2004)
Low temperatures of 5/10 °C	Glufosinate ammonium	*Raphanus raphanistrum* L.	Reduced the efficacy	Kumaratilake and Preston (2005)
Elevated CO_2	Glyphosate	*Chenopodium album* L.	Increased tolerance to glyphosate	Ziska et al. (2004)
Elevated CO_2	Glyphosate	*Chloris gayana, Eragrostis curvula, Paspalum dilatatum*	Reduced the efficacy	Manea et al. (2011)
Elevated CO_2	Glyphosate	*Cirsium arvense*	Reduced the efficacy	Ziska et al. (2004)
Elevated temperature	Flumiclorac	*Amaranthus retroflexus* L.	Increased herbicide efficacy	Fausey and Renner (2001)
Elevated temperature	Mesotrione	*Digitaria sanguinalis*	Reduced the efficacy	Johnson and Young (2002)
Elevated temperature	Mesotrione	*Amaranthus palmeri*	Reduced the efficacy	Godar et al. (2015)
High humidity	Glufosinate ammonium	*Setaria faberi*	Increased herbicide efficacy	Anderson et al. (1993)
High humidity	Mesotrione	*Amaranthus Rudis* L.	Increased herbicide efficacy	Johnson and Young (2002)

can contribute remarkably to the soil seed bank densities. Growing grasses, legumes, or vegetables on bunds accompanied by appropriate weed management measures may reduce weed seed production besides additional farm income (Rao et al. 2017).

(ii) *Minimizing seed dispersal*: Weeds having same maturity with the crop result in contamination. Under weed-infested environments, farm-saved seeds, manures or compost, and irrigation water can also get contaminated with weed seeds.

Table 7.2 Herbicide-resistant weed cases under climate change scenario

Exposed condition	Weed	Herbicide	Reference
Increased temperatures and elevated CO_2	*Conyza canadensis Chenopodium album*	Glyphosate	Matzrafi et al. (2019)
	Echinochloa colona	Cyhalofop-butyl	Refatti et al. (2019)
	Phalaris minor	Sulfosulfuron	DWR (2014–15); DWR (2015–16)
	Phragmites australis (Gulf Coast type)	Glyphosate	Prince et al. (2018)
Elevated CO_2	*Chenopodium giganteum*	Glyphosate	Ziska et al. (1999)
	Chloris Gayana Eragrostis curvula Paspalum dilatatum	Glyphosate	Manea et al. (2011)
	Cirsium arvense	Glyphosate	Ziska et al. (2004)
	Elytrigia repens	Glyphosate	Ziska and Teasdale (2000)
	Sinapis arvensis	4-chloro-2-methylphenoxyacetic acid (MCPA)	Žaltauskaitė et al. (2023)
	Parthenium hysterophorus	Glyphosate	Cowie et al. (2020)
	Parthenium hysterophorus	Glyphosate	Bajwa et al. (2019)
	Phalaris minor	Sulfosulfuron	DWR (2014–15); DWR (2015–16)
Increased temperatures	*Amaranthus Rudis, Digitaria sanguinalis*	Mesotrione	Johnson and Young (2002)
	Amaranthus palmeri	Mesotrione	Godar et al. (2015)
	Phalaris minor	Sulfosulfuron	DWR (2014–15); DWR (2015–16)
	–	Trifluralin	Beestman and Deming (1974)
	Amaranthus palmeri	Pyrithiobac	Mahan et al. (2004)
	Kochia scoparia	Dicamba	Ou et al. (2018)
	Kochia scoparia	Glyphosate	Ou et al. (2018)
	Alternanthera philoxeroides	Penoxsulam	Willingham et al. (2008)

Using certified seeds and well-decomposed manures and compost can significantly reduce weed seed ingression into crop fields. Weed seed dispersal through irrigation can be minimized by analyzing irrigation water in canal-irrigated areas and controlling weed proliferation in irrigation channels. Farm equipment/machineries should be cleaned from attached soils and before moving from weed-infested area to other fields. Patches of new weed seedlings, tubers, rhizomes, etc., may be destroyed by digging deep and burning the weed along with its roots.

Safeguard measures consist of using pure and clean crop seed, certified seeds, maintaining clean farm machineries and animals, well-decomposed farmyard manure (FYM)/compost, weed control in nurseries, restricting weed proliferation in farm bunds and irrigation channels, and enacting plant/ weed quarantine law (Das 2008). Agronomic practices and the weed control measure adopted for raising crops have inherent weed prevention approach. Apart from the above-mentioned measures to check weed control, several emerging technologies/arenas could be tried to develop resilience against weed-induced damages. Among the comparatively unexplored arena in weed management (a) utilization of crops wild relatives (Hossain et al. 2022); (b) potential applications of metabolomics and/or metabolites (Choudhury et al. 2021b; Hossain et al. 2021b); (c) utilization of next-generation sequencing (Hossain et al. 2021a) could be a matter of interest.

5.2 *Cultural/Ecological Weed Management*

Cultural weed management can provide selective stimulation to crops toward better competitiveness like higher crop vigor, which, ultimately, results in smothering of weeds. It is well known that a healthy crop is the finest weed destroyer. Various recommended cultural weed control practices include competitive crop cultivars, modifying crop geometry and sowing time, scheduling fertilizers application and irrigation to provide competitive advantage to crop, intercropping, stale seedbed, residue mulching, and crop rotation (Behera et al. 2019; Das and Yaduraju 2012; Singh et al. 2016). Under elevated CO_2 conditions, stimulation of belowground growth of plants is higher than that of shoot growth. Therefore, enhanced root or rhizome growth in such species makes manual elimination a tough task. In weeds with asexual reproduction habit, higher CO_2 may stimulate additional plant propagation from the rhizosphere and will have adverse consequences on weed regulation (Ziska and Goins 2006). Extended growing seasons of weeds under climate variation scenario requires modification in cultural practices. For example, shifting from puddled transplanting to dry direct-seeded rice under with aerated soil conditions necessitates post-emergence weed management measures in order to sustain a steady crop yield.

5.2.1 Cultivar Selection

An ideal cultivar should have rapid seedling emergence, quick growth rate, higher plant height, tillering or branching habit, rapid leaf area expansions, dense canopy, rapid canopy closure, more developed root system, efficient nutrient use with tall height, and indeterminate growth habit. Crops having allelopathic potential should be opted for suppressing weeds (Sangeetha and Baskar 2015). Basmati rice cultivars have greater competitive capacity against weeds when compared with non-basmati varieties may attributed to their early vigor, faster canopy coverage, and tall stature (Singh et al. 2009). Tall-growing wheat genotype had a considerable suppressive effect on the development of *P. minor* and caused more reduction in density, height, and total dry matter of this weed than dwarf varieties (Walia and Singh 2005).

5.2.2 Modifying Crop Geometry and Sowing Time

Modifying row spacing and planting geometry in crops can alter crop–weed competition. Reducing the inter-row layout and increasing seeding density can limit weed proliferation through faster canopy closure and reducing light availability to weeds (Chauhan and Johnson 2011). Seeding rate >150 kg ha^{-1} (Bhullar and Walia 2004) and minimize the row spacing by 15 cm (Mahajan and Brar 2001) led to lesser dry matter accrual of *P. minor* followed by increased yield of wheat over recommended crop geometry. Compared to normal row spacing of 23 cm, there was a 25.0% decrease in weed dry weight when rice was sown using the paired-row sowing (15-30-15 cm) method (Mahajan and Chauhan 2011). Das (2008) found that early sowing between October 25 and November 10 was advantageous to wheat against *P. minor* since soil temperature was still high for germination of *P. minor*.

5.2.3 Fertilizer, Manures, and Irrigation Management

Most weed exhibited upgraded growth and modest ability with improved availability of nitrogen, leading to rivalry for natural resources. Placement of fertilizers and time of application significantly reduce weed dry biomass due to lesser availability of applied nutrient in inter-row spaces (Pandey et al. 2006). Seed cum fertilizer drills lead to placement of basal fertilizer below the seeds, thereby reducing fertilizer availability to surface-lying weeds. Increased grain yield and effective control of *P. minor* with cross sowing of wheat and placement of fertilizer below the seed were noticed compared to unidirectional sowing and broadcasting of fertilizers (Pandey and Kumar 2005). Some of the nutrient management strategies to reduce early competition from weeds include deep placement of fertilizer, basal nitrogen application as Neem cake/ Karanja cake instead of urea (Ghosh et al. 2020a, b).

Effective soil moisture management is a tool to manage weeds in crop field. The sub-surface drip irrigation results lesser weed density compared to surface flooding owing to minimal wetting of soil surface and lower amount of moisture available to weeds. Ideal time and irrigations practices reported to be essential in minimizing the weed density and weight (Das and Yaduraju 2007). Maintaining 2–5 cm water level in the paddy field minimized weed emergence and lowered weed pressure specially *Cyperus rotundus* (Chhokar et al. 2014). Good land leveling is critical to avoid high spots where weeds can become established.

5.2.4 Crop Rotation, Intercropping, and Crop Diversification

Crop rotation with plants having dissimilar agro-ecology or cultural conditions leads to interruption of normal life cycles of awkward weeds like *P. minor* and lower emergence (Chhokar et al. 2008). Growing berseem after wheat harvesting in rice–wheat cropping system lowered seed stock of *P. minor*, since the *P. minor* were also cut while harvesting of berseem which prevented seed setting and shedding (Singh et al. 1999).

Similarly, in potato cropping systems, emerged *P. minor* plants can be uprooted during soil manipulation (Singh et al. 1999). Intercropping in sugarcane with legume (mungbean/black gram) and legumes in sorghum, wheat, and maize have been reported beneficial toward weed management (Das, 2008). Intercropping is ideal for wide row-spaced crops with narrow row-spaced crops and of tall-growing crops with short-stature crops. Cowpea, green gram, black gram, soybean when intercropped with maize, sorghum, and pearl millet could manage weeds to a large extent. *A. ludoviciana, P. minor,* etc., are existing in wheat crop (Singh et al. 1999), and *Echinochloa* sp. existing in rice crop under rice–wheat cropping system were largely controlled when wheat was replaced with berseem, mustard, or winter maize for 3–4 years. Other promising crop rotations, which could reduce *P. minor* to a large extent in rice–wheat system, are as follows: rice–fallow–sugarcane–ratoon sugarcane–sunflower–rice–wheat–sugarcane (4 years), rice–potato–sunflower (1 year), rice–potato–onion (1 year), and rice–mustard–sugarcane (2 years) (Chhokar et al. 2008). Including low land rice in crop rotation can reduce problematic weeds like *Cyperus rotundus* owing to anoxic conditions.

5.2.5 Cover Crops

Cover crop can modify the weed's microclimate as it can block available light, change soil temperature, deprive the weeds of moisture and nutrients, and thus suppress weeds and reduce soil seed bank. Cover crops like cowpea, lucerne, lablab bean, rye, oat, clovers, and buckwheat had a smothering action on weeds (Price et al. 2006). Additionally, cover crops act as a weed-suppressive mulch after their termination or incorporation (green manuring). Allelopathic effects are also evident in certain cover crops like Buckwheat, oat, cereal rye, and sunflower (Jabran et al. 2015). Leguminous crops may be employed as mulch for weed suppression with 60–80% establishment, when incorporated can lower the nitrogen fertilizer demand (~75 kg N ha^{-1}) (Korres et al. 2010).

5.2.6 Stale Seed Bed

Stale seedbed is an important strategy for exhausting 2–3 initial flushes of weeds before crops are sown (20–25 days prior to normal sowing date), thus reducing weed pressure in a crop. Before sowing/transplanting, weed germination is stimulated through shallow tillage followed by irrigation and thereafter those are eliminated by using a non-selective herbicide (paraquat, glyphosate, or glufosinate) or plowing. This practice not only reduces weed pressure during the crop season but also exhausts the weed seed bank. Stale seedbed with a stimulating irrigation once along with glyphosate application reduced grassy weed, broadleaf weeds density, and sedges by 44%, 58%, and by 56%, respectively, with an overall 62% reduction in weed growth rate in comparison with recommended practices (Kumar et al. 2013).

5.2.7 Allelopathy

Allelopathy is the adverse impacts of biochemicals (direct/indirect) generated by one plants or microorganisms on the growth and development of the adjacent plants and microorganisms in the vicinity (Molisch 1937). Plants with allelopathic capability are regarded as an alternative, sustainable substitutes effective for weed control and are means to decline reliance on herbicides (Appiah et al. 2015). Allelopathic crops can be integrated in crop rotations, intercropping, or mulch crops for under integrated weed management modules. Allelopathic impact of rice stubble and straw was demonstrated by Tamak et al. (1994) against *P. minor* and other weeds. They found that 10% concentration of straw-stubble extract caused greater inhibition of seed germination 2 days after sowing (DAS), compared to 6 and 10 DAS, respectively.

5.2.8 Brown Manuring

Brown manuring involves co-culturing of legumes such as dhaincha (*Sesbania*) or sunnhemp (*Crotalaria*) with cereals and desiccating them before flowering (Maitra and Zaman 2017). In order to execute brown manuring in rice, *Sesbania* (25 kg/ha) is grown together as co-culture for initial 25–30 days and thereafter selectively controlled by herbicides 2,4-D (0.50 kg a.i/ha or bispyribac-Na @ 20–25 g a.i/ha) (Das et al. 2019). *Sesbania* plants performs twin role of dead and live mulch leading to weed minimizing and trim down herbicide usage in rice. This procedure may be practiced in maize for better profitability and weed management. Brown manuring lower broad-leaved weed densities by 76–83% and grass densities by 20–33%, respectively. Upon comparison with only a rice crop, it also cuts infestations of problematic *Cyperus rotundus* considerably weeds (Das et al. 2019).

5.2.9 Weed Seed Predation and Decay

Weed seed predation and decay are effective tools for depleting weed seed banks. In general, conservation tillage led to a greater fraction of weed seeds lying on the topsoil where they are more inclined to seed predation, while residue mulch offers a promising habitat for seed predators and decaying agents (Chauhan et al. 2010). Similarly, zero tillage with higher soil moisture and higher microbial diversity in surface favors microbial-mediated seed decay (Das et al. 2020).

5.3 Physical/Mechanical Weed Management

Physical methods of weed control, i.e., tillage/discing/harrowing, hand weeding/pulling, hoeing, digging, sickling, burning, flooding are being followed since ancient times. Mechanical weeding is machine-intensive and can be adopted using tractor-drawn equipment in large farms under conventional agriculture. Some tractor-operated weeders are standard/high residue rotary hoe, spike-tooth/spring tine harrow, flex-tyne weeder, finger weeder, rotating wire weeder, and pneumatic weeder (Bond et al. 2003). Rotational tillage is a better style for weed management, particularly under several cropping systems. Hand weeding using small hand tools/implements is drudgery-some and labor-dependent. It is efficient enough against shallow-rooted annual and biennial weeds but not against deep-rooted and perennial weeds. Digging-out underground perennating structures from deep soil layers can reduce perennial weeds considerably but is labor-intensive and less economical (Brainard et al. 2013). Similarly, there is scope for thermal weed control (Bauer et al. 2020), but selectivity achieved through a certain heat tolerance of the crop is difficult to actuate in fields which may pose jeopardy of crop damage and fire from dry plant residues. Although most conventional physical methods are less economical and labor-intensive, they offer enough potential for location-based integration as an element of the IWM.

5.3.1 Tillage and Crop Establishment

Summer plowing or off-season tillage is an efficient weed management procedure for disrupting the growth of perennial weed population by desiccating the weed propagule through exposure in hot sun. While inversion tillage disk and moldboard plow result in depth burial and dispersal of seed in the plow layer, non-inversion tillage like chiseling and reduced tillage leave a higher proportion of seed close to surface. Higher fraction of weed seed is left near the soil surface after non-inversion tillage results in higher weed germination and establishment in the initial years.

Intensive tillage may negatively impact soil conservation, promote soil erosion, reduce soil biological diversity, and deplete/oxidize soil organic carbon from topsoil (Das et al. 2020). Revolving tillage system with green manuring, cover crops, soil amendments, residue incorporation, and organic manures may potentially reduce the negative effect of tillage. Tillage also provides germination promoters for weeds through scarification, light flashes, temperature variation, ambient CO_2 concentrations, and/or greater nitrate concentrations for breaking dormancy (Benech-Arnold et al. 2000). Under conservation agriculture system, advocating no till, shift in weed community from annual broadleaf species to annual grasses, and perennials have been reported (Erenstein and Laxmi 2008; Nichols et al. 2015). Lower grassy weed density with *P.* minor and *A. ludoviciana* was reported under zero tillage in wheat in contrast to reduced and conventional tillage, while density of *Rumex denticulatus* was higher in zero tillage (Brar and Walia 2009; Chhokar et al. 2007).

Bed planting proved advantageous method over flat conventional sowing of wheat as bed planting reduced the number of *P. minor* by 12.5% over flat sowing (Walia et al. 2003). Raised bed raising of wheat provides an opportunity for mechanical weeding and also reduces the weed interference by burial the weed seeds deep during bed preparation. Also, the inter-row bed space is used to manage weeds by mechanically during the early vegetative growth of weeds. Reshaping the bed before planting of wheat can kill the first level of *P. minor* seedlings (Das et al. 2014b).

5.3.2 Soil Solarization

Soil solarization employs a hydrothermal process for raising soil temperatures by trapping short-wave solar radiation (Das and Yaduraju 2008). In this process, a transparent polyethylene film is placed over the field after proper leveling which increases temperature of soil to a level lethal to weed seeds, insect pests, pathogens, and nematodes. During summer (April to June), solarization can be done for a minimum of 2 weeks' period for optimum control of weeds, while prolonged treatment would result in elongated weed free periods. It may control weeds in the wet-season (*kharif*) and the effect may continue in the winter (*rabi*) crops unless soil disturbed.

5.3.3 Mulching and Crop Residue

Mulches control weeds through light exclusion, physical barrier to seeding emergence, and allelopathy (Das 2008). Mulch includes clean straw, hay or manure, saw dust, crop stubbles, black plastic, etc. Datta et al. (2017) observed that mulch @ 5 t/ha after the first weeding and mulch @ 5 t/ha fb sulfentrazone @ 0.3 kg/ha recorded a noteworthy reduction in weed density as well as weed dry weight and therefore higher weed control efficiency at harvesting. Wheat residue mulch of 5 t/ha declined the density of grass (73–76%), broadleaf (65–67%), and sedge (22–70%), associated with no residue control in zero till direct-seeded rice (Kumar et al. 2013). To compensate the limited availability of residues for mulch during the rice season, growing short-duration catch crops such as mungbean [*Vigna radiata* (L.) R. Wilczek] during the fallow period between wheat harvest and rice planting and retaining the entire mungbean residue as mulch in rice is an effective IWM practice in RW system (Baghel et al. 2020). Black polythene mulch recorded significantly less weed density and dry biomass over water hyacinth, paddy straw, and wheat straw mulch, respectively (Goswami and Saha 2006).

5.3.4 Use of Weeders

In the current scenario, non-availability of agricultural labours coupled with higher wage is causing mechanical weeders to become more and more popular. The machineries like mini-weeders, power tillers, mini-tractor-drawn rotavator have applicability for weed control in wider spaced field crops like sugarcane, cotton, and orchards. Cono-weeder is popular for managing the submerged or water loving weeds in the puddled transplanted rice field. In the cotton crop, the use of mini weeder and power tillers found use full to manage different types of weeds. CRIJAF Nail weeder and cycle weeder can effectively control weeds in jute besides reducing drudgery, increasing productivity, and conserving moisture. Weeders play an imperative role under organic farming systems. In case of cono-weeder, amalgamation of *Sesbania* and *Azolla* resulted in greater weed control during vegetative stages of rice crop particularly for SRI or System of Rice Intensification.

5.3.5 Thermal Weed Control

Thermal weed control methods comprise of (a) hot water treatment, steaming and flaming, infrared weeders or hot air and (b) indirect heating by using microwaves, electrocution, laser radiation, or UV light (Korres et al. 2019). Steaming has been advocated to kill weed seeds and soil-borne pests and pathogens. Steaming is more effective as compared to the stale seedbed technique and soil solarization for reducing the weed seed bank and can execute weed seeds suppressed up to 20 cm deep in soil (Barberi et al. 2009). Flaming is one of the most extensively used thermal weed control method, particularly in organic agriculture. Exposure to intense heat (with flame temperature of 800–1000 °C for 1 s) causes osmolysis and cell death resulting in reasonable weed control (Ascard 1997).

5.4 *Biological Weed Management*

Biological control fosters a prey–predator relationship between the weed and bioagent (insects, pathogens) and follows the natural law of homeostasis, the science of check, and balance (Das 2008). It conveys not to eradicate weeds completely but bring weeds density within the economic permissible level (Das 2008). Biological weed management is pocket friendly, self-sustaining, and eco-friendly in contrast to other popular of weed management methods. The control of *Eichhornia crassipes* in large aquatic bodies by using *Neochetina eichhorniae* and other bioagents

(Coetzee et al. 2011) and the control of *P. hysterophorus* (Parthenium) by using *Zygogramma bicolorata* (Mexican beetle) and other insects (Singh et al. 2017) are the successful cases of biological control of weeds, globally. Biological control is relatively cheap, least toxic to humans/animals and environment, and effective and adaptable for controlling perennial, parasitic, and invasive weeds. Growth, development, and reproductive fluctuations in the selected weedy target due to varying climate could alter the efficacy of the biocontrol agents.

Bioherbicides are inoculums of plant pathogens, mainly fungi, capable of in vitro culturing in artificial media and mass production and applied directly to target weeds to kill/ reduce their population/growth. The pathogen bioagent generally remains active only on target weed in selective crop; therefore, it needs to be sprayed on target weed in every season. Bioherbicides research gained attention in 1980s, when some potent pathogens were successfully utilized to make effective formulations for weed control.

Biological/bioherbicide approach has inherent limitations for higher implementation in crops. Most bioagents kill single weed; therefore, weed problem in a crop infested with a large variety of weeds remains hardly resolved. For all weeds, specific bioagents or bioherbicides are not available. Furthermore, it is a slow process of killing of weeds; early weed competition may cause sufficient damage to crops; environment and ecology greatly affect their stability and performance; the shelf life of bioherbicides is very less compared to herbicides; high production cost involved in their rearing/culturing, mass production, application; and bioherbicides need registration with Environmental Protection Agency (EPA), which is a long time-taking process. Despite all these, this method offers potential for application in crop/location-based management of single dominant weed (annual/perennial) such as *P. minor* in wheat in India; *P. hysterophorus* in terrestrial non-cropped situations; *Eichhornia crassipes*, *Salvinia molesta* in water bodies (Das et al. 2014a) across the world.

5.4.1 Bioagents

These are commercially formulated and sprayed like herbicides over crop and weeds in the field. The rust fungus *Puccinia canaliculata* (Schw.) Lagerh has the probable to control the yellow nutsedge (*C. esculentus*). Release of the pathogen, in the spring on the seedlings of yellow nutsedge lessened the plant population, hampers tuber formation and flowering (Phatak et al. 1987). Deleterious rhizobacteria (DRB) colonizing plant root surfaces decrease their root growth and proliferation. Biological control of downy brome weed in winter wheat by *Pseudomonas* spp. has been demonstrated under field conditions (Kennedy et al. 1991).

5.4.2 Biobased Herbicide Products

Essential oils having herbicidal activity and approved for organic farming include pine, clove, lemongrass, and manuka oil (Korres et al. 2019). These oils although having quick action are non-selective and volatilize in short time span. Dead cells and metabolites of *Streptomyces acidiscabies* were found to have selective pre- and post-emergence control of numerous weed species (Korres et al. 2019). "Sarmentine" is a contact herbicide with broad-spectrum (Huang et al. 2010). Brassica (Rapeseed and mustard) seed meal containing glucosinolates gain herbicidal properties upon enzymatic hydrolysis by myrosinase. They provide adequate control of several weed species including wild oat (*A. fatua*), *Lolium multiflorum*, *Lactuca*, and *Amaranthus retroflexus* (Handiseni et al. 2011).

5.5 *Chemical Weed Control*

Agriculture systems have wide necessity on herbicides. Application of herbicide is cost-effective compared to conventional weeding methods, requires less labor, tackles difficult-to-control weeds, and allows flexibility in weed management. Usually non-residual, non-selective herbicides like glyphosate, paraquat, or glufosinate AM are recommended under ZT conditions before crop sowing for eliminating existing weed flush. Subsequently, pre-emergence herbicide is essential for reducing flush of germinating annual weeds coming up with the crops. Moreover, selective post-emergence herbicides, if available, may be applied at an appropriate crop growth stage depending on the weed intensity.

Effectiveness of herbicide largely depends on the climatic condition. Drought and increased temperatures can reduce uptake of herbicides, volatilization, and structural disintegration thereby reducing its effectiveness. Exposure to elevated CO_2 leads to anatomical, morphological, and physiological changes in weeds that in turn affect overall effectiveness of herbicides due to changes in uptake rates and translocation of herbicides. Manea et al. (2011) reported increased glyphosate tolerance under elevated CO_2 among three of four C4 grass species. The reasons might be that increasing CO_2 caused leaf thickness and reduced stomatal number and conductance that restricts the uptake of foliar-applied herbicides. Moreover, greater dry matter accumulation under elevated CO_2 levels could cause dilution of applied herbicide and thereby lowering its efficacy. CO_2 enrichment led to higher starch deposits especially in C3 plants which cause interference with herbicide activity (Patterson et al. 1999).

5.6 *Precision Weed Management*

SSWM or site-specific weed management, i.e., advocating control methods only where weeds are found/detected insignificant quantities compare to those causing economic losses, offers economic and environmental benefits. For this, the instrument linked with technologies to detects weeds in a crop / cropping system, considering into account of previously identified factors takes action to successfully control them. Under usual patchy and scattered weed distribution in crop fields, site-specific, weed patch-specific, or spot spray of herbicide is greater economical and less degrading to environment than blanket application. This will reduce amount of herbicides and their intake into environment. Band application sustain herbicide treatment at a half of suggested rate along with mechanical weed control brought a satisfactory total weed reduction by 83–87% (Kneievic et al. 2003).

Recently, artificial intelligence (AI) and robotics research have geared up for weed management (Young et al. 2014). Weed control using advance robotics is a four-step process. This involves multiple phases of operations including guidance, documentation, precision weed elimination, and plotting of weeds (Young et al. 2014). The machines are equipped with sophisticated high-resolution computer vision systems (cameras and sensors), AI, global positioning system (GPS), software and robotic arms/ nozzles that enable guidance, detection and identification of weeds through a real-time machine vision system, precise weed control, and mapping (Gonzalez-de-Soto et al. 2016). Robotic machines are used to control weeds mechanically, chemically, or through flame. "Weed master" an advanced machine uses GPS to navigate precisely and a digital image recognition system. These systems aid differentiate between the intended crop and weed. It is powered by methanol fuel and uproots weeds autonomously. "Weed controller" is a four-wheel-drive machine with specialized weed pursuing robot integrated which recognizes weeds based on color photography. In contrast, *intelligent hoe* utilizes visual recognition system (with previously embedded details of weeds) to find rows of crops and steer itself among them (Korres et al. 2019).

Use of robotics may reduce herbicides use and their environmental impact and hence can improve sustainability, particularly in vegetable crops and organic agriculture (Korres et al. 2019). The practicability of a robotic weed control system is governed upon machine learning (vision analyses, robotic efficacy, adjustable rate application technology, decision support system) along with equipped with the capability to assess the strength of weed-sensing tools, directly or indirectly. Upon detection a weed, the robotic part selectively/precisely ejects micro-dose (in small quantity) of herbicides on targeted sites /weeds. Further, these automated machines can target weeds in patches rather than entire fields and reduce usage of herbicides. Some agricultural robots for weed control are as follows: WeedMaster®, WeedSeeker® (for pot spraying), Tertill, RIPPA, Hortibot, SwagBot, ASTERIX, AgBot II, Blue River LettuceBot2, Naïo Technologies, ecoRobotix. However, several barricades prevent their large-scale adoption, most important being the absence of automated weed detection and identification method in crop fields and other determining factors.

6 Probable IWM Modules for Climate Resilient Weed Management in Crops

Integrated weed management (IWM) approach aims at better, longer, and efficient weed management, while maintaining an ecological balance. IWM models should be specifically designed considering crop, weed flora, and regional features and should have high social and economic acceptability among farmers.

Generally, the following preventive and agronomic measures may be adopted as applicable to crops for providing selective advantage/ stimulus to crops to develop more competitive against weeds from the commencement of their emergence.

- Use certified and clean seeds, avoid undecomposed farm yard manure, keep irrigation channels free from weeds.
- Cleaning seeds by dipping in 2% brine solution to separate weed seeds by floating (in rice).
- Deep summer plowing in April–May to expose and desiccate vegetative propagules of perennial grassy and sedge weeds.
- Follow scientific crop rotation to break/disrupt established life cycle of particular weed or a group of weeds.
- Green manuring before crop season or brown manuring co-existing with crops for initial 25–30 DAS with *Sesbania* spp. or *Crotalaria* spp. (as applicable).
- Adoption of stale seed beds to eliminate initial weed flushes.
- Avoid basal application of heavy dose of nitrogen in cereals, which may stimulate weed growth and apply N in 3 or more split doses.

Rice

(a) Direct-seeded rice (DSR) [conventional till (CT)/zero till(ZT)]

(i) Green manuring with *Sesbania* spp. (~*Dhaincha*) during summer for conventional till DSR followed by pre-emergence (pendimethalin, butachlor, pretilachlor, pyrazosulfuron-ethyl) herbicides at 2–3 days after sowing (DAS)—residue mulching immediately after pre-emergence herbicides—post-emergence herbicides (bispyribac-Na, cyhalofop-butyl, penoxsulam, almix) or hand weeding at 25–30 DAS or finger weeder twice at 20-day interval.

(ii) Under conservation agriculture (ZTDSR): Previous crop residue retention + brown manuring with *Dhaincha/Crotalaria* + pre-emergence herbicides followed by post-emergence herbicides /hand pulling/weeding at 25–30 DAS.

(iii) Stale seed bed with glyphosate 1.0 kg/ha followed by pre-emergence herbicide at 2–3 DAS – hand weeding/finger weeder at 25–30 DAS.

(b) Puddled transplanted rice (PTR)

(i) Maintain 2–3 cm standing water throughout crop growing period.

(ii) Plant 25- to 30-day-old seedlings at closer spacing of 20 × 15 or 15 × 15 cm with 2–3 seedlings per hill.

(iii) Residue incorporation/green manuring + pre-emergence herbicides + post-emergence herbicides/hand weeding/rotavation/rotary weeding at 30–35 days after transplanting (DAT).

(iv) Pre-emergence herbicides at 2–3 DAT + hand weeding /rotavation/rotary weeding at 30–35 days after transplanting (DAT).

(v) Pre-emergence at 2–3 DAT /early post-emergence application at 10–15 DAT of herbicides followed by cono-weeder at 25–30 DAT.

(vi) Residue incorporation/green manuring + high-density planting + post-emergence herbicides – hand weeding/rotavation/rotary weeding at 40–45 DAT.

Wheat

Greater seed rate, narrow row spacing, fertilizers mainly nitrogen and its slightly higher concentration of doses and placement along the crop rows, zero tillage with and without previous crop residue, soil solarization in the earlier season—all have positive impact on the overall weed management in wheat. Crop rotation (rice–mustard/pea) can control resistant as well as susceptible *P. minor*.

(i) Zero tillage with closer row spacing/skip row spacing coupled residue retention and application of preemergence herbicide (like pendimethalin and pyroxasulfone) under moist soil conditions or otherwise where resistant Phalaris minor present, apply post-emergence broad-spectrum herbicides or herbicide or apply pre-mix or tank-mix herbicide mixture under heavy infestation in wheat.

(ii) Furrow-irrigated raised bed system with residue + post-emergence broad-spectrum herbicides or herbicide pre-mix or tank-mix herbicide mixture + need-based hand weeding (if at all required).

(iii) Under conventional till wheat, early sowing (last week of October or first week of November under Northern Indian conditions), closer row spacing/skip row sowing with residue retention—post-emergence herbicides.

(iv) For CT or ZT sown wheat crop apply post-emergence broad-spectrum herbicide or herbicides; or use pre-mix or tank-mix herbicide mixture followed by hand weeding to control three problematic thorny weeds like *Cirsium arvense*, *Carthamus oxycantha*, *Argemone mexicana* and other problematic non- throny weeds like *Convolvulus arvensis* (Field bind weed), *Cannabis sativa* (Bhang), *Asphodelus tenuifolius* (Wild onion) which might have escaped herbicidal action due to late germination or tolerated herbicide application to a greater extent.

Maize

(i) CT/ZT + early/timely sowing+ pre-emergence herbicide tank-mixture (atrazine+pendimethalin) alone is capable of complete broad-spectrum control of weeds + tembotrione for perennial *Cyperus rotundus* or hand weeding at 25–30 DAS.

(ii) CT/ZT + early/timely sowing + bed planting with residue retention followed by tank-mixture of atrazine+topramezone 500 + 25.2 g/ha or atrazine+tembotrione 500 + 120 g/ha 20–25 DAS.

(iii) Soil solarization during hot summer months + tank-mixture of atrazine+topramezone 500 + 25.2 g/ha or atrazine+tembotrione 500 + 120 g/ha 20–25 DAS or hand weeding (with little soil disturbance).

(iv) Residue incorporation/retention + timely sowing+ pre-emergence herbicides tank mixture (atrazine +pendimethalin) followed by hand weeding at 25–30 DAS (if at all required).

(v) Intercropping with soybean/green gram/black gram + pre-emergence pendimethalin – hand weeding.
(vi) Brown manuring with *Sesbania* under both CT and ZT + pre-emergence pendimethalin – hand weeding at 30–35 DAS.

Pulses

Following a good crop growing such as proper time, method, rate, and depth of sowing, crop establishment methods such as CT, ZT, residue mulching, fertilizer placement along the crop rows, line sowing, seed treatment for biological N fixation along with disease management is a mandatory requirement for growing a healthy crop, which in turn may lead to smothering weeds. Soil solarization during summer followed by pre-emergence herbicide (mainly pendimethalin) at 2–3 DAS + hand weeding (if required) at 25–30 DAS.

(i) Deep summer cultivation with irrigation and residue (wheat, maize) incorporation followed by pre-emergence herbicides + hand weeding at 25–30 DAS.
(ii) Pre-emergence herbicides + hand weeding at 25–30 DAS for problematic weeds such as *Cynodon dactylon* and *Cyperus rotundus*.
(iii) Previous crop residue incorporation + higher seed rate + pre-emergence herbicides followed by post-emergence herbicides or hand weeding at 25–30 DAS for problematic weeds such as *Cynodon dactylon* and *Cyperus rotundus*.
(iv) In soybean and pigeon pea, bed planting with residue + pendimethalin pre-emergence (2–3 DAS) followed by imazethapyr 100 g/ha or quizalofop-ethyl 50 g/ha as post-emergence (at 25–30 DAS) would lead to better weed control.
(v) In chickpea, previous crop residue retention + pendimethalin pre-emergence followed by quizalofop-ethyl @ 100 g/ha or clodinafop-propargyl @ 150 g/ha can prove better.
(vi) In blackgram, pendimethalin + imazethapyr (pre-mix) 0.9 kg/ha as pre-emergence or imazethapyr 100 g/ha as early post-emergence or imazethapyr + imazemox 50 g/ha as early post-emergence combined with 1 hand weeding at 30–35 DAS.

Oilseeds

(i) Following a good crop husbandry such as proper time, method, rate, and depth of sowing, crop establishment methods such as CT, ZT, residue mulching, fertilizer placement along the crop rows, line sowing, seed treatment for pest and disease management is a mandatory requirement for growing a healthy crop, which in turn may lead to smothering weeds. In most oilseed crops, apply pre-emergence herbicides (mainly, pendimethalin)—hand weeding at 30–35 DAS (for late-emerging weeds that escaped herbicide treatment and for problematic perennial weeds).
(ii) In groundnut, apply pre-emergence herbicide (pendimethalin, oxyfluorfen)—earthing-up—hand weeding if required.
(iii) In groundnut, apply imazethapyr 125 g/ha at 2–3 leaf stage of weeds followed by 1 hand weeding at 45 DAS.

(iv) In mustard and rapeseed crop, apply pre-emergence herbicide (pendimethalin, oxadiargyl) or early post-emergence herbicide (e.g., isoproturon)—hand weeding at 30–35 DAS.

Sugarcane

(i) Atrazine at 1.50 kg/ha or metribuzin 0.5 kg/ha at 2–3 days after planting (DAP) before emergence of weeds followed by one hoeing to destroy emerged weeds at 30–35 DAP.
(ii) Atrazine at 1.50 kg/ha at 2–3 days after planting (DAP) followed by halosulfuron-ethyl 90 g/ha at 30–35 DAP or one hoeing/manual weeding to destroy emerged weeds.
(iii) Apply atrazine at 1.50 kg/ha pre-emergence followed by one hoeing/manual weeding or chlorimuron ethyl + metsulfuron methyl 8 g/ha post-emergence at 2–4 leaf stage for controlling broad-leaved weeds.
(iv) Intercropping with black gram + pre-emergence pendimethalin followed by earthing-up and post-emergence herbicide.

Vegetables

(i) In brinjal, apply pre-emergence herbicides (e.g., pendimethalin) followed by hand weeding at 35–40 DAT.
(ii) In onion, apply pre-emergence pendimethalin 1.0 kg/ha or oxadiargyl 90 g/ha or oxyfluorfen 250 g/ha followed by 1 hand weeding at 30–35 DAT or quizalofop-ethyl 50 g/ha at 30–35 DAT.
(iii) In potato, apply metribuzin 0.4 kg/ha as pre-emergence followed by earthing-up at 25–30 DAP.

Although its influences on agricultural weeds have not been thoroughly investigated, climate change is a pressing global concern. Old school thought on carbon cycling in plants and nitrogen management in crops could partially address the effects of climate change, but weed issues could potentially be made worse by rising CO_2 concentrations, high temperatures, and, most crucially, water stress. These circumstances can call for the emerge of innovative agronomic techniques to tackle weed competitiveness under the background of climate change. The stimulatory or inhibitory effect of the climate variables on crops should typically hold true for weeds because crops and weeds are on the same trophic level. It has been developed that an upsurge in air temperature promotes both weed growth and pesticide effectiveness. Although C4 weeds predominate in agriculture, substantial crop–weed competition may soon arise from major weeds with C3 and C3-C4 intermediary routes. Importantly, due to species interactions, it is essential to research all potential combinations of crop–weed competitive interactions, including those involving C3 crops and C3 weeds, C4 crops and C4 weeds, C3 crops and C4 weeds, and C4 crops and C3 weeds. Under the scenario of climate change, a number of weeds will put more pressure on crop–weed competition. In forthcoming decades, more adaptable research studies with challenging research situations might provide helpful answers for controlling yield reduction.

References

Ainsworth EA, Long SP (2005) What have we learned from 15 years of free-air CO_2 enrichment (FACE)? A metaanalytic review of the responses of photosynthesis, canopy properties and plant production to rising CO_2. New Phytol 165: 351–372. https://doi.org/10.1111/j.1469-8137.2004.01224.x

Anderson DM, Swanton CJ, Hall JC, Mersey BG (1993) The influence of temperature and relative humidity on the efficacy of glufosinate-ammonium. Weed Res 33(2):139–147. https://doi.org/10.1111/j.1365-3180.1993.tb01927.x

Annual Report (2014-15) ICAR-Directorate of Weed Research, Jabalpur, 116 p

Annual Report (2015-16) ICAR-Directorate of Weed Research, Jabalpur, 137 p

Appiah KS, Amoatey CA, Fujii Y (2015) Allelopathic activities of selected Mucuna pruriens on the germination and initial growth of lettuce. Int J Basic Appl Sci 4(4):475–481

Archambault DJ, Li X, Robinson D, O'Donovan JT, Klein KK (2001) The effects of elevated CO_2 and temperature on herbicide efficacy and weed/crop competition. Rept. Prairie Adapt. Res. Collab 29

Ascard J (1997) Flame weeding: effects of fuel pressure and tandem burners. Weed Res 37:77–86

Baghel JK, Das TK, Pankaj MI, Nath CP, Bhattacharyya R, Ghosh S, Raj R (2020) Impacts of conservation agriculture and herbicides on weeds, nematodes, herbicide residue and productivity in direct-seeded rice. Soil Till Res 201:104634

Bajwa AA, Wang H, Chauhan BS, Adkins SW (2019) Effect of elevated carbon dioxide concentration on growth, productivity and glyphosate response of parthenium weed (Parthenium hysterophorus L.). Pest Manag Sci 75(11):2934–2941

Barberi P, Moonen AC, Peruzzim A, Fontanelli M, Raffaelli M (2009) Weed suppression by soil steaming in combination with activating compounds. Weed Res 49(1):55–66

Bauer MV, Marx C, Bauer FV, Flury DM, Ripken T, Streit B (2020) Thermal weed control technologies for conservation agriculture—a review. Weed Res 60(4):241–250

Beestman GB, Deming JM (1974) Dissipation of acetanilide herbicides from soils. Agronomy 66(2):308–311

Behera B, Das TK, Ghosh S, Parsad R, Rathi N (2019) Effects of brown manure species, seed rate and time of application of 2,4-D on weed control efficiency, productivity and profitability in maize. Indian J Weed Sci 51(4):393–397

Benech-Arnold RL, Sánchez RA, Forcella F, Kruk BC, Ghersa CM (2000) Environmental control of dormancy in weed seed banks in soil. Field Crops Res 67(2):105–122

Bergmann J, Pompe S, Ohlemüller R, Freiberg M, Klotz S, Kühn I (2010) The Iberian Peninsula as a potential source for the plant species pool in Germany under projected climate change. Plant Ecol 207(2):191–201

Bhullar MS, Walia US (2004) Effect of seed rate and row spacing on the efficacy of clodinafop for combating isoproturon resistant *Phalaris minor* Retz. in wheat. Plant Prot Q 19(4):143–146

Boese SR, Wolfe DW, Melkonian JJ (1997) Elevated CO_2 mitigates chilling-induced water stress and photosynthetic reduction during chilling. Plant Cell Environ 20(5):625–632

Bond W, Turner RJ, Grundy AC (2003) A review of non-chemical weed management. HDRA, the Organic Organisation, Ryton Organic Gardens, Coventry, UK, p 81

Brainard DC, Haramoto E, Williams MM, Mirsky S (2013) Towards a no-till no-spray future? Introduction to a symposium on nonchemical weed management for reduced-tillage cropping systems. Weed Technol 27(1):190–192

Brar AS, Walia US (2009) Weed dynamics and wheat (*Triticum aestivum* L.) productivity as influenced by planting techniques and weed control practices. Indian. J Weed Sci 41(1):161–166

Brinkley TR, Bomford M (2002) Agricultural sleeper weeds in Australia: what is the potential threat? Bureau of Rural Sciences, Canberra

Cavero J, Zaragoza C, Suso ML, Pard A (1999) Competition between maize and *Datura stramonium* in an irrigated field under semi-arid conditions. Weed Res 39:225–240

Chauhan BS, Johnson DE (2011) Growth response of direct-seeded rice to oxadiazon and bispyribac-sodium in aerobic and saturated soils. Weed Sci 59(1):119–122

Chauhan BS, Migo T, Westerman PR, Johnson DE (2010) Post dispersal predation of weed seeds in rice fields. Weed Res 50(6):553–560

Chhokar RS, Sharma RK, Jat GR, Pundir AK, Gathala MK (2007) Effect of tillage and herbicides on weeds and productivity of wheat under rice-wheat growing system. Crop Prot 26:1689–1696

Chhokar RS, Sharma RK, Singh RK, Gill SC (2008) Herbicide resistance in little seed canarygrass (*Phalaris minor*) and its management. In: Proceedings of the 14th Australian agronomy conference. Australian Society of Agronomy, Adelaide, p 106

Chhokar RS, Sharma RK, Gathala MK, Pundir AK (2014) Effects of crop establishment techniques on weeds and rice yield. Crop Prot 64:7–12

Choudhury S, Moulick D, Mazumder MK (2021a) Secondary metabolites protect against metal and metalloid stress in rice: an *in silico* investigation using dehydroascorbate reductase. Acta Physiol Plant 43(1):1–10

Choudhury S, Sharma P, Moulick D, Mazumder MK (2021b) Unrevealing metabolomics for abiotic stress adaptation and tolerance in plants. J Crop Sci Biotechnol 24(5):479–493

Choudhury S, Mazumder MK, Moulick D, Sharma P, Tata SK, Ghosh D, Ali MH, Siddiqui HM, Brestic M, Skalický M, Hossain A (2022a) A computational study of the role of secondary metabolites for mitigation of acid soil stress in cereals using dehydroascorbate and mono – dehydroascorbate reductases. Antioxidants 11(3):458

Choudhury S, Moulick D, Mazumder MK, Pattnaik BK, Ghosh D, Vemireddy LR, Aldhahrani A, Soliman MM, Gaber A, Hossain A (2022b) An *in vitro* and *in silico* perspective study of seed priming with zinc on the phytotoxicity and accumulation pattern of arsenic in rice seedlings. Antioxidants 1(8):1500

Choudhury S, Moulick D, Ghosh D, Soliman M, Alkhedaide A, Gaber A, Hossain A (2022c) Drought-induced oxidative stress in pearl millet (*Cenchrus americanus* L.) at seedling stage: survival mechanisms through alteration of morphophysiological and antioxidants activity. Life 12(8):1171

Chowardhara B, Borgohain P, Saha B, Awasthi JP, Moulick D, Panda SK (2019a) Assessment of phytotoxicity of zinc on Indian mustard (*Brassica juncea*) varieties during germination and early seedling growth. Annu Plant Soil Res 21(3):239–244

Chowardhara B, Borgohain P, Saha B, Awasthi JP, Moulick D, Panda SK (2019b) Phytotoxicity of Cd and Zn on three popular Indian mustard varieties during germination and early seedling growth. Biocatal Agric Biotechnol 21:101349

Cowie BW, Venter N, Witkowski ET, Byrne MJ (2020) Implications of elevated carbon dioxide on the susceptibility of the globally invasive weed, Parthenium hysterophorus, to glyphosate herbicide. Pest Manag Sci 76(7): 2324–2332

Coetzee JA, Hill MP, Byrne MJ, Bownes A (2011) A review of the biological control programmes on *Eichhornia crassipes* (C. mart.) solms (Pontederiaceae), *Salvinia molesta* DS Mitch. (Salviniaceae), *Pistia stratiotes* L. (Araceae), *Myriophyllum aquaticum* (vell.) verdc. (Haloragaceae) and *Azolla filiculoides* Lam. (Azollaceae) in South Africa. Afr Entomol 19(1):451–468

Das TK (2008) Weed science: basics and applications. Jain Brothers, New Delhi

Das TK, Yaduraju NT (2007) Effect of irrigation and nitrogen levels on grassy weed competition in wheat and comparative eco-physiology of *Phalaris minor* Retz. and *Avena sterilis* ssp. *Ludoviciana* Dur. in wheat. Indian. J Weed Sci 39(3& 4):178–184

Das TK, Yaduraju NT (2008) Effect of soil solarization and crop husbandry practices on weed species competition and dynamics in soybean-wheat cropping system. Indian J Weed Sci 40(1&2):1–5

Das TK, Yaduraju NT (2012) The effects of combining modified sowing methods with herbicide mixtures on weed interference in wheat crops. Int J Pest Manag 58(4):310–319

Das TK, Ahlawat IPS, Yaduraju NT (2014a) Little seed canary grass (*Phalaris minor*) resistance to clodinafop-propargyl in wheat fields in North-Western India: appraisal and management. Weed Biol Manag 14:11–20

Das TK, Bhattacharyya R, Sudhishri S, Sharma AR, Saharawat YS, Bandyopadhyay KK, Sepat S, Bana RS, Aggarwal P, Sharma RK, Bhatia A (2014b) Conservation agriculture in an irrigated cotton–wheat system of the western Indo-Gangetic Plains: crop and water productivity and economic profitability. Field Crops Res 158:24–33

Das TK, Ghosh S, Nath CP (2019) Brown manuring optimization in maize: impacts on weeds, crop productivity and profitability. J Agric Sci 157:599–610

Das TK, Nath CP, Das S, Biswas S, Bhattacharyya R, Sudhishri S, Raj R, Singh B, Kakralia SK, Rathi N, Sharma AR, Dwivedi BS, Biswas AK, Chaudhari SK (2020) Conservation agriculture in rice-mustard cropping system for five years: impacts on crop productivity, profitability, water-use efficiency, and soil properties. Field Crops Res 250:107781

Datta D, Saxena SC, Ghosh S (2017) Boosting productivity and profitability of soybean (*Glycine max*) in Uttarakhand by weed management through herbicides and mulch. Indian J Agron 62(4):538–540

Erenstein O, Laxmi V (2008) Zero tillage impacts in India's rice–wheat systems: a review. Soil Till Res 100:1–14

Fausey JC, Renner KA (2001) Environmental effects on CGA-248757 and flumiclorac efficacy/soybean tolerance. Weed Sci 49(5): 668–674

Godar AS, Varanasi VK, Nakka S, Prasad PV, Thompson CR, Mithila J (2015) Physiological and molecular mechanisms of differential sensitivity of Palmer amaranth (Amaranthus palmeri) to mesotrione at varying growth temperatures. PloS one 10(5): https://doi.org/10.1371/journal.pone.0126731

Ghosh D, Brahmachari K, Brestic M, Ondrisik P, Hossain A, Skalicky M, Sarkar S, Moulick D, Dinda NK, Das A, Pramanick B (2020a, 1906) Integrated weed and nutrient management improve yield, nutrient uptake and economics of maize in the rice-maize cropping system of Eastern India. Agronomy 10(12)

Ghosh D, Brahmachari K, Skalicky M, Hossain A, Sarkar S, Dinda NK, Das A, Pramanick B, Moulick D, Brestic M, Raza MA (2020b) Nutrients supplementation through organic manures influence the growth of weeds and maize productivity. Molecules 25(21):4924

Ghosh D, Brahmachari K, Das A, Hassan MM, Mukherjee PK, Sarkar S, Dinda NK, Pramanick B, Moulick D, Maitra S, Hossain A (2021) Assessment of energy budgeting and its indicator for sustainable nutrient and weed management in a rice-maize-green gram cropping system. Agronomy 11(1):166

Ghosh D, Brahmachari K, Skalický M, Roy D, Das A, Sarkar S, Moulick D, Brestič M, Hejnak V, Vachova P, Hassan MM, Hossain A (2022) The combination of organic and inorganic fertilizers influence the weed growth, productivity and soil fertility of monsoon rice. PLoS One 17(1):e0262586

Gianessi LP (2013) The increasing importance of herbicides in worldwide crop production. Pest Manag Sci 69(10):1099–1105

Gonzalez-de-Soto M, Emmi L, Perez-Ruiz M, Aguera J, Gonzalez-de-Santos P (2016) Autonomous systems for precise spraying–evaluation of a robotised patch sprayer. Biosyst Eng 146:165–182

Goswami SB, Saha S (2006) Effect of organic and inorganic mulches on soil-moisture conservation, weed suppression and yield of elephant-foot yam (*Amorphophallus paeoniifolius*). Indian J Agron 51(2):154–156

Handiseni M, Brown J, Zemetra R, Mazzola M (2011) Herbicidal activity of Brassicaceae seed meal on wild oat (*Avena fatua*), Italian ryegrass (*Lolium multiflorum*), redroot pigweed (*Amaranthus retroflexus*), and prickly lettuce (*Lactuca serriola*). Weed Technol 25(1):127–134

Hatfield JL, Boote KJ, Kimball BA, Ziska LH, Izaurralde RC, Ort D, Thomson AM, Wolfe D (2011) Climate impacts on agriculture: implications for crop production. Agronomy 103(2):351–370

Holm LG, Doll J, Holm E, Pancho J, Herverger J (1997) Worlds weeds: natural histories and distribution. Wiley, New York, p 1129

Hossain A, Ahmed S, Ahmad Z, Alam MJ, Moulick D, Saha B, Samanta S, Maitra S, Bhadra P, Bukhari MA, Aftab T (2021a) Next-generation genetic engineering tools for abiotic stress tolerance in plants. In: Frontiers in plant-soil interaction. Academic, pp 153–197. https://doi.org/10.1016/B978-0-323-90943-3.00008-0

Hossain A, Pramanick B, Bhutia KL, Ahmad Z, Moulick D, Maitra S, Ahmad A, Aftab T (2021b) Emerging roles of osmoprotectant glycine betaine against salt-induced oxidative stress in plants: a major outlook of maize (*Zea mays* L.). In: Frontiers in plant-soil interaction. Academic, pp 567–587. https://doi.org/10.1016/B978-0-323-90943-3.00015-8
Hossain A, Maitra S, Pramanick B, Bhutia KL, Ahmad Z, Moulick D, Syed MA, Shankar T, Adeel M, Hassan MM, Aftab T (2022) Wild relatives of plants as sources for the development of abiotic stress tolerance in plants. In: Plant perspectives to global climate changes. Academic, pp 471–518. https://doi.org/10.1016/B978-0-323-85665-2.00011-X
Huang H, Morgan CM, Asolkar RN, Koivunen ME, Marrone PG (2010) Phytotoxicity of sarmentine isolated from long pepper (*Piper longum*) fruit. J Agric Food Chem 58:9994–10000
IPCC (2007) Climatic change 2007: the physical science basis. Cambridge University Press, Cambridge and New York
IPCC (2014) Climate change 2014: synthesis report. Contribution of working groups I, II and III to the fifth assessment report of the Intergovernmental Panel on Climate Change. IPCC, Geneva
IPCC (2021) Summary for policymakers. In: Climate Change 2021: The physical science basis. Contribution of working group I to the sixth assessment report of the Intergovernmental Panel on Climate Change. WMO, IPCC Secretariat, Geneva
Jabran K, Mahajan G, Sardana V, Chauhan BS (2015) Allelopathy for weed control in agricultural systems. Crop Prot 72:57–65
Jaggard KW, Qi A, Ober ES (2010) Possible changes to arable crop yields by 2050. Philos Trans R Soc B Biol Sci 365(1554):2835–2851
Johnson BC, Young BG (2002) Influence of temperature and relative humidity on the foliar activity of mesotrione. Weed Sci 50(2):157–161
Kadam NN, Xiao G, Melgar RJ, Bahuguna RN, Quinones C, Tamilselvan A, Prasad PVV, Jagadish KS (2014) Agronomic and physiological responses to high temperature, drought, and elevated CO_2 interactions in cereals. Adv Agron 127:111–156
Kennedy AC, Young FL, Elliott LF, Douglas CL (1991) Rhizobacteria suppressive to the weed downy brome. Soil Sci Soc Am J 55(3):722–727
Knežević M, Đurkić M, Knežević I, Lončarić Z (2003) Effects of pre-and post-emergence weed control on weed population and maize yield in different tillage systems. Plant Soil Environ 49(5):223–229
Korres NE, Singh A, Nizami AS, Murphy JD (2010) Is grass biomethane a sustainable transport biofuel? Biofuels Bioprod Biorefin 4(3):310–325. https://doi.org/10.1002/bbb.228
Korres NE, Norsworthy JK, Tehranchian P, Gitsopoulos TK, Loka DA, Oosterhuis DM, Gealy DR, Moss SR, Burgos NR, Miller MR, Palhano M (2016) Cultivars to face climate change effects on crops and weeds: a review. Agron Sustain Dev 36:12
Korres NE, Burgos NR, Travlos I, Vurro M, Gitsopoulos TK, Varanasi VK, Duke SO, Kudsk P, Brabham C, Rouse CE, Salas-Perez R (2019) New directions for integrated weed management: modern technologies, tools and knowledge discovery. Adv Agron 155:243–319
Kudsk P, Kristensen JL (1992) Effect of environmental factors on herbicide performance. In: Proceedings of the first international weed control congress, Melbourne, pp 173–186
Kumar V, Singh S, Chhokar RS, Malik RK, Brainard DC, Ladha JK (2013) Weed management strategies to reduce herbicide use in zero-till rice–wheat cropping systems of the Indo-Gangetic Plains. Weed Technol 27(1):241–254
Kumaratilake AR, Preston C (2005) Low temperature reduces glufosinate activity and translocation in wild radish (*Raphanus raphanistrum*). Weed Sci 53(1):10–16
Mahajan G, Brar LS (2001) Studies on herbicide resistance in *Phalaris minor* under Punjab conditions. Indian J Weed Sci 33(1&2):1–4
Mahajan G, Chauhan BS (2011) Effects of planting pattern and cultivar on weed and crop growth in aerobic rice system. Weed Technol 25(4):521–525
Mahajan G, Singh S, Chauhan BS (2012) Impact of climate change on weeds in the rice-wheat cropping system. Curr Sci 102(9):1254–1255

Mahan JR, Dotray PA, Light GG (2004) Thermal dependence of enzyme function and inhibition; implications for, herbicide efficacy and tolerance. Physiol Plant 120(2):187–195

Maitra S, Zaman A (2017) Brown manuring, an effective technique for yield sustainability and weed management of cereal crops: a review. Int J Bioresour Sci 4(1):1–5

Manea A, Leishman MR, Downey PO (2011) Exotic C_4 grasses have increased tolerance to glyphosate under elevated carbon dioxide. Weed Sci 59(1):28–36

Matloob A, Khaliq A, Chauhan BS (2015) Weeds of rice in Asia: problems and opportunities. Adv Agron 130:291–336

Matzrafi M, Brunharo C, Tehranchian P, Hanson BD, Jasieniuk M (2019) Increased temperatures and elevated CO_2 levels reduce the sensitivity of Conyza canadensis and Chenopodium album to glyphosate. Sci Rep 9: 2228. https://doi.org/10.1038/s41598-019-38729-x

Mazumder MK, Sharma P, Moulick D, Tata S, Choudhury S (2021) Salicylic acid ameliorates zinc and chromium induced stress responses in wheat seedlings: a biochemical and computational analysis. Cereal Res Commun 50(3):407–418

Mazumder MK, Moulick D, Choudhury S (2022) Iron (Fe^{3+})-mediated redox responses and amelioration of oxidative stress in cadmium (Cd^{2+}) stressed mung bean seedlings: a biochemical and computational analysis. J Plant Biochem Biotechnol 31(1):49–60

McDonald A, Riha S, DiTommaso A, DeGaetano A (2009) Climate change and the geography of weed damage: analysis of U.S. maize systems suggests the potential for significant range transformations. Agric Ecosyst Environ 130(3&4):131–140

Mcdowell RW, Condron LM, Main BE, Dastgheib F (1997) Dissipation of imazapyr, flumetsulam and thifensulfuron in soil. Weed Res 37(6):381–389

Miri HR, Rastegar A, Bagheri AR (2012) The impact of elevated CO_2 on growth and competitiveness of C_3 and C_4 crops and weeds. Eur J Exp Biol 2(4):1144–1150

Mohamed KI, Papes M, Williams R, Benz BW, Peterson AT (2006) Global invasive potential of 10 parasitic witchweeds and related Orobanchaceae. Ambio 35(6):281–288

Mohamed KI, Bolin JF, Musselman LJ, Peterson AJ (2007) Genetic diversity of striga and implications for control and modeling future distributions. In: Ejeta G, Gressel J (eds) Integrating new technologies for striga control: towards ending the witch-hunt. World, Singapore, pp 71–84

Molisch H (1937) The influence of one plant on another (Der Einfuss Einar Pflanze Auf Die Andere-allelopathie). Fishcher, Jena

Mortensen DA, Coble HD (1989) The influence of soil water content on common cocklebur (*Xanthium strumarium*) interference in soybeans (Glycine max). Weed Sci 37(1):76–83

Moulick D, Ghosh D, Santra SC (2016a) An assessment of some physicochemical properties and cooking characteristics of milled rice and associated health risk in two rice varieties of arsenic contaminated areas of West Bengal, India. Int J Agric Food Sci 6(2):44–55

Moulick D, Ghosh D, Santra SC (2016b) Evaluation of effectiveness of seed priming with selenium in rice during germination under arsenic stress. Plant Physiol Biochem 109:571–578. https://doi.org/10.1016/j.plaphy.2016.11.004

Moulick D, Ghosh D, Santra SC (2017) Seed priming with Se alleviate as induced phytotoxicity during germination and seedling growth by restricting as translocation in rice (*Oryza sativa* L cv IET-4094). Ecotoxicol Environ Saf 145:449–456

Moulick D, Santra SC, Ghosh D (2018a) Effect of selenium induced seed priming on arsenic accumulation in rice plant and subsequent transmission in human food chain. Ecotoxicol Environ Saf 152:67–77. https://doi.org/10.1016/j.ecoenv.2018.01.037

Moulick D, Santra SC, Ghosh D (2018b) Rice seed priming with se: a novel approach to mitigate as induced adverse consequences on growth, yield and as load in brown rice. J Hazard Mater 355:187–196. https://doi.org/10.1016/j.jhazmat.2018.05.017

Moulick D, Santra SC, Ghosh D (2018c) Seed priming with se mitigates as-induced phytotoxicity in rice seedlings by enhancing essential micronutrient uptake and translocation and reducing as translocation. Environ Sci Pollut Res 25:26978–26991. https://doi.org/10.1007/s11356-018-2711-x

Moulick D, Santra SC, Ghosh D (2018d) Consequences of paddy cultivation in arsenic-contaminated paddy fields of lower Indo-Gangetic plain on arsenic accumulation pattern and selected grain quality traits: a preliminary assessment. In: Hasanuzzaman M, Nahar K, Fujita M (eds) Mechanisms of arsenic toxicity and tolerance in plants. Springer, Singapore. https://doi.org/10.1007/978-981-13-1292-2_3

Moulick D, Chowardhara B, Panda SK (2019a) Agro ecotoxicological aspect of arsenic (as) and cadmium (cd) on field crops and its mitigation: current status and future prospect. In: Plant-metal interactions. Springer, Cham, pp 217–246. https://doi.org/10.1007/978-3-030-20732-8_11

Moulick D, Santra SC, Ghosh D, Panda SK (2019b) An assessment of efficiency of zinc priming in rice (cv. MTU-7029) during germination and early seedling growth. In: Priming and pretreatment of seeds and seedlings. Springer, Singapore, pp 495–507. https://doi.org/10.1007/978-981-13-8625-1_24

Moulick D, Samanta S, Saha B, Mazumder MK, Dogra S, Panigrahi KC, Banerjee S, Ghosh D, Santra SC (2020) Salinity stress responses in three popular field crops belonging to Fabaceae family: current status and future prospect. In: The plant family Fabaceae. Springer, Singapore, pp 519–541. https://doi.org/10.1007/978-981-15-4752-2_20

Moulick D, Samanta S, Sarkar S, Mukherjee A, Pattnaik BK, Saha S, Awashy JP, Bhowmick S, Ghosh D, Samal AC, Mahanta S, Mazumder MK, Choudhury S, Bramhachari K, Biswas JK, Santra SC (2021) Arsenic contamination, impact and mitigation strategies in rice agro-environment: an inclusive insight. Sci Total Environ 800:149477. https://doi.org/10.1016/j.scitotenv.2021.149477

Moulick D, Ghosh D, Skalický M, Gharde Y, Mazumder MK, Choudhury S, Biswas JK, Santra SC, Brestic M, Vachova P, Hossain A (2022) Interrelationship among rice grain arsenic, micronutrients content and grain quality attributes: an investigation from genotype × environment perspective. Front Environ Sci 10:857629. https://doi.org/10.3389/fenvs.2022.857629

Naidu VSGR (2013) Invasive potential of C-3-C-4 intermediate *Parthenium hysterophorus* under elevated CO_2. Indian J Agric Sci 83(2):176–179

Naidu VSGR, Paroha S (2008) Growth and biomass partitioning in two weed species, *Parthenium hysterophorus* (C_3) and *Amaranthus viridis* (C_4) under elevated CO_2. Ecol Environ Conserv 14(4):9–12

Nichols V, Verhulst N, Cox R, Govaerts B (2015) Weed dynamics and conservation agriculture principles: a review. Field Crops Res 183:56–68

Ou J, Stahlman PW, Jugulam M (2018) Reduced absorption of glyphosate and decreased translocation of dicamba contribute to poor control of kochia (Kochia scoparia) at high temperature. Pest Manag Sci 74(5):1134–1142

Pandey IB, Kumar K (2005) Response of wheat (*Triticum aestivum*) to seeding methods and weed management. Indian J Agron 50(1):48–51

Pandey IB, Dwivedi DK, Prakash SC (2006) Impact of method and levels of fertilizer application and weed management on nutrient economy and yield of wheat (*Triticum aestivum*). Indian J Agron 51(3):193–198

Patterson DT (1995) Weeds in a changing climate. Weed Sci 43(4):685–701

Patterson DT, Westbrook JK, Lingren PD, Rogasik J (1999) Weeds, insects, and diseases. Clim Change 43(4):711–727

Phatak SC, Callaway MB, Vavrina CS (1987) Biological control and its integration in weed management systems for purple and yellow nutsedge (*Cyperus rotundus* and *C. esculentus*). Weed Technol 1(1):84–91

Phillips OL, Vásquez Martínez R, Arroyo L, Baker TR, Killeen T, Lewis SL, Malhi Y, Monteagudo Mendoza A, Neill D, Núñez Vargas P, Alexiades M (2002) Increasing dominance of large lianas in Amazonian forests. Nature 418(6899):770–774

Phoenix GK, Press MC (2005) Effects of climate change on parasitic plants: the root hemiparasitic Orobanchaceae. Folia Geobot 40:205–216

Polley HW, Johnson HB, Tischler CR (2002) Woody invasion of grasslands: evidence that CO_2 enrichment indirectly promotes establishment of *Prosopis glandulosa*. Plant Ecol 164(1):85–94

Prasad PVV, Craufurd PQ, Kakani VG, Wheeler TR, Boote KJ (2001) Influence of high temperature during pre-and post – anthesis stages of floral development on fruit-set and pollen germination in peanut. Aust J Plant Physiol 28(3):233–240

Prasad PVV, Allen LH Jr, Boote KJ (2005) Crop responses to elevated carbon dioxide and interaction with temperature: grain legumes. In: Tuba Z (ed) Ecological responses and adaptations of crops to rising atmospheric carbon dioxide. Haworth Press Inc., New York, pp 113–155

Price AJ, Reeves DW, Patterson MG (2006) Evaluation of weed control provided by three winter cover cereals in conservation-tillage soybean. Renew Agric Food Syst 21(3):159–164

Prince CM, MacDonald GE, Erickson JE (2018) Effects of elevated temperature and carbon dioxide concentrations on the response of two common reed (Phragmites australis) haplotypes to glyphosate. Invasive Plant Sci Manag 11(4):181–190

Ramsey RJL, Stephenson GR, Hall JC (2005) A review of the effects of humidity, humectants, and surfactant composition on the absorption and efficacy of highly water-soluble herbicides. Pestic Biochem Phys 82(2):162–175

Rao AN, Brainard DC, Kumar V, Ladha JK, Johnson DE (2017) Preventive weed management in direct-seeded rice: targeting the weed seedbank. Adv in Agron 144:45–142

Refatti JP, De Avila LA, Camargo ER, Ziska LH, Oliveira C, Salas-Perez R, Rouse CE, Roma-Burgos N (2019) High [CO_2] and temperature increase resistance to cyhalofop-butyl in multiple-resistant Echinochloa colona. Front Plant Sci 10:529. https://doi.org/10.3389/fpls.2019.00529

Richardson AD, Keenan TF, Migliavacca M, Ryu Y, Sonnentag O, Toomey M (2013) Climate change, phenology, and phenological control of vegetation feedbacks to the climate system. Agric For Meteorol 169:156–173

Ritter RL, Coble HD (1981) Influence of temperature and relative humidity on the activity of acifluorfen. Weed Sci 29(4):480–485

Rodenburg J, Meinke H, Johnson DE (2011) Challenges for weed management in African rice systems in a changing climate. J Agric Sci 149(4):427–435

Saha B, Chowardhara B, Kar S, Devi SS, Awasthi JP, Moulick D, Tanti B, Panda SK (2019) Advances in heavy metal-induced stress alleviation with respect to exogenous amendments in crop plants. In: Priming and pretreatment of seeds and seedlings. Springer, Singapore, pp 313–332. https://doi.org/10.1007/978-981-13-8625-1_15

Sangeetha C, Baskar P (2015) Allelopathy in weed management: a critical review. Afr J Agric Res 10:1004–1015

Schimel D, Ives D, Enting I, Heimann M, Joos F, Raynaud D, Wigley T (1996) CO_2 and the carbon cycle. In: Houghton JT, Pilho LGM, Callendar BA, Harris N, Kattenberg A, Maskell K (eds) Climate change 1995. IPCC. Cambridge University Press, Cambridge, pp 65–131

Sharma SD, Singh M (2001) Environmental factors affecting absorption and bio-efficacy of glyphosate in Florida beggar weed (*Desmodium tortuosum*). Crop Prot 20(6):511–516

Singh S, Kirkwood RC, Marshall G (1999) A review of the biology and control of *Phalaris minor* Retz. (littleseed canary grass) in cereals. Crop Prot 18:1–16

Singh S, Chhokar RS, Gopal R, Ladha JK, Gupta RK, Kumar V, Singh M (2009) Integrated weed management: a key to success for direct-seeded rice in the Indo-Gangetic plains. In: Integrated crop and resource management in the rice–wheat system of South Asia. International Rice Research Institute, Los Banos, pp 261–278

Singh RP, Singh RK, Singh MK (2011) Impact of climate and carbon dioxide change on weeds and their management–a review. Indian J Weed Sci 43(1&2):1–11

Singh V, Jat ML, Ganie ZA, Chauhan BS, Gupta RK (2016) Herbicide options for effective weed management in dry direct-seeded rice under scented rice-wheat rotation of western Indo-Gangetic Plains. Crop Prot 81:168–176

Singh SA, Gill PK, Dhaliwal HS, Kumar VI (2017) Life cycle and effectiveness of *Zygogramma bicolorata* Pallister (Chrysomelidae: Coleoptera) on *Parthenium hysterophorus* eradication. J Glob Agric Ecol 7(2):60–65

Smith C, Van Klinken RD, Seabrook L, McAlpine C (2012) Estimating the influence of land management change on weed invasion potential using expert knowledge. Divers Distrib 18(8):818–831

Tamak JC, Narwal SS, Singh L, Singh I (1994) Effect of aqueous extracts of rice stubbles and straw and stubbles on the germination and seedling growth of wheat, oat, berseem and lentil. Crop Res 8:180–185

Treharne K (1989) The implications of the 'greenhouse effect' for fertilizers and agrochemicals. In: Bennet RC (ed) The greenhouse effect and UK agriculture. University of Reading Press, Reading, pp 67–78

Valerio M, Lovelli S, Perniola M, Di Tommaso T, Ziska L (2013) The role of water availability on weed–crop interactions in processing tomato for southern Italy. Acta Agric Scand B Soil Plant Sci 63(1):62–68

Varanasi A, Prasad PV, Jugulam M (2016) Impact of climate change factors on weeds and herbicide efficacy. Adv Agron 135:107–146

Walia US, Singh M (2005) Competitive ability of wheat cultivars with associated weeds. J Plant Sci Res 21:32–39

Walia US, Brar LS, Jand S (2003) Integrated effect of planting methods and herbicides on *Phalaris minor* and wheat. Indian J Weed Sci 35(3&4):169–172

Willingham SD, Senseman SA, McCauley GN, Chandler JM (2008) Effect of temperature and propanil on penoxsulam efficacy, absorption, and translocation in alligatorweed (Alternanthera philoxeroides). Weed Sci 56(6):780–784

Young SL, Meyer GE, Woldt WE (2014) Future directions for automated weed management in precision agriculture. In: Young SL, Pierce FJ (eds) Automation: the future of weed control in cropping systems. Springer, Dordrecht, pp 249–259

Žaltauskaitė J, Dikšaitytė A, Miškelytė D, Kacienė G, Sujetovienė G, Januškaitienė I, Juknys R (2023) Effects of elevated CO_2 concentration and temperature on the mixed-culture grown wild mustard (Sinapis arvensis L.) response to auxin herbicide. Environ Sci Pollut Res 30(5): 13711–13725

Ziska LH (2000) The impact of elevated CO_2 on yield loss from a C_3 and C_4 weed in field-grown soybean. Glob Change Biol 6(8):899–905

Ziska LH, Teasdale JR (2000) Sustained growth and increased tolerance to glyphosate observed in a C3 perennial weed, quackgrass (Elytrigia repens), grown at elevated carbon dioxide. Funct Plant Biol 27(2): 159–166

Ziska LH, Bunce JA (2006) Plant responses to rising atmospheric carbon dioxide. In: James ILM, Michael DM (eds) Plant growth and climate change. Blackwell Publishing, Oxford, pp 17–47

Ziska LH, Dukes J (2011) Weed biology and climate change. Wiley, Ames

Ziska LH, George K (2004) Rising carbon dioxide and invasive, noxious plants: potential threats and consequences. World Resour Rev 16(4):427–447

Ziska LH, Goins EW (2006) Elevated atmospheric carbon dioxide and weed populations in glyphosate treated soybean. Crop Sci 46(3):1354–1359

Ziska LH, Faulkner S, Lydon J (2004) Changes in biomass and root: shoot ratio of field-grown Canada thistle (*Cirsium arvense*), a noxious, invasive weed, with elevated CO_2: implications for control with glyphosate. Weed Sci 52(4):584–588

Chapter 8
Climate-Resilient Technology for Maize Production

Muaz Ameen, Asma Zafar, Muhammad Mansoor Javaid, Muhammad Anjum Zia, Athar Mahmood, Maria Naqve, and Safura Bibi

Abstract Agriculture's vulnerability to changing climate needs to be the focus of scientific, economic, political, and social efforts to ensure viable existence on this planet. Climate change impacts all types of life, but the influence on plant life is much more noticeable because plants are the primary energy source and matter fixer on land. Producing climate resilience in plants, particularly in crops, is crucial for all living. Climate-resilient agriculture has been boosted by biotechnology. Wheat, rice, and maize contribute a minimum of 30% of the nutritional calories for 4.5 billion people in 94 emerging economies. Cereals are crucial for providing global dietary, fodder, and nutrient intake. Maize is an important element in livestock feed and bioenergy. Rising demand and supply gaps have intensified market instability and pushed up global maize pricing. Long-term corn production development might be adversely impacted by the change in climate, which is increasing current issues including drought and heat stress, insect, pests and diseases, and soil depletion. In tropical areas, regional climate uncertainty and variation, and the increase in biotic and abiotic stressors, further muddle the situation. Even though these difficulties are resolved, and production improved, millions of impoverished people will go starving. Climate change adaptation will need suitable technologies, legislation, and structural reforms, but this chapter focuses on genetic techniques and tactics that might improve the changes over time and the budget for breeding climatic resilience

M. Ameen · A. Zafar · M. Naqve · S. Bibi
Department of Botany, Faculty of Sciences, University of Agriculture, Faisalabad, Pakistan

M. M. Javaid (✉)
Department of Agronomy, College of Agriculture, University of Sargodha, Sargodha, Pakistan
e-mail: mansoor.javaid@uos.edu.pk

M. A. Zia
Department of Biochemistry, University of Agriculture, Faisalabad, Pakistan

A. Mahmood (✉)
Department of Agronomy, Faculty of Agriculture, University of Agriculture, Faisalabad, Pakistan
e-mail: athar.mahmood@uaf.edu.pk

M. Hasanuzzaman (ed.), *Climate-Resilient Agriculture, Vol 2*,
https://doi.org/10.1007/978-3-031-37428-9_8

maize. The chapter will explain how crop modification will help to fulfill increasing maize requirements by overcoming biotic and abiotic stressors. Combining traditional and molecular breeding may produce high-yielding, stress-tolerant, extensively adaptable maize cultivars. Enhanced germplasm alone will not boost productivity or facilitate climate-related adaption; crops and agricultural techniques must also be enhanced.

Keywords Climate tolerant · Abiotic stress · Cereal crops · Drought stress · Heat stress · Genetic modification

1 Introduction

Agriculture must feed 3.5 billion more humans in the coming 50 years. Rising rural and urban populations will need 70% more wheat, maize, and rice before 2050 (Searchinger et al. 2019; Grote et al. 2021). Maize is an important supplier of food supply and financial growth in Africa, Central America, and Asia (Islam and Karim 2019). Maize contributes at minimum of 30% of the dietary requirements for 4.5 billion individuals in 94 emerging economies. These comprise 900 million impoverished maize-eaters (Dowswell et al. 2019). A 67% of the world's maize output originates from lower and moderate nations; thus, it is vital to millions of impoverished farmers. The population of the planet is 7.7 billion and will be 9.3 billion by 2050 (Withers et al. 2018). The emerging world's maize consumption will double by 2050. Maize is a significant food crop; however, its usage as fodder has risen. Drastic economic expansion in Asia, the Mideast, and Central America has raised the demand for poultry and cattle supplies from wealthy customers. Maize is a crucial element in livestock feed; therefore, excess supply has pushed up costs and caused it less accessible for impoverished customers (Sanchez 2019).

Changes in climate affect human actions that either directly or indirectly influence the nature of the atmosphere, as well as natural climate unpredictability. Anthropogenic impacts, including fossil fuel emissions, forest destruction, unsustainable land use, agribusiness, and industrialization, modify the atmospheric composition, leading to climate change (Mercure et al. 2019; Pichler et al. 2021). Burning energy sources and other anthropogenic factors have increased atmospheric greenhouse gas emissions. According to worldwide climate effects estimates on agriculture, a 2 °C warming by 2050 would put 50 million people at threat of starvation because of agricultural consequences (Hasegawa et al. 2018; Yerlikaya et al. 2020). Climate change will affect South and Southeast Asia heavily. If present patterns continue till 2050, the productivity of main crops in South Asia would decline by 10% (rice), 12% (wheat), and 17% (maize) due to drought and heat stressors from changing climate (Aryal et al. 2020; Shahzad et al. 2021). The maize feeder industry is expanding in China, India, and Pakistan, where economic expansion puts dairy, poultry, and beef in a more affordable manner. Globally maize supply disruptions and rising input costs affect emerging nations. Maize prices have climbed 43%

since 2008, in addition to other products, notably wheat is expected to rise further due to rising demand and limited supply (Ben Hassen and El Bilali 2022). Furthermore, trailing domestic output would impose pressure on emerging nation economies, pushing up their maize importation from 7% of today's requirement to 24% in 2050, paying approximately USD 30 billion (Chapagain et al. 2020). Although if the yielding rise is accelerated, poor maize users will have less inexpensive food. Farm owners, government entities, researchers, and input providers are addressing maize requirements. Asia, Latin America, and sub-Saharan Africa significantly boosted maize output by 6% yearly from 2003 to 2008 (Leitner et al. 2020).

Area growth is not feasible and frequently land destruction occurs. At the present pace of area expansion, agricultural diversification and conservation of forests will suffer (Wilson et al. 2019). Upcoming food supply expansion must not compromise human safety, environment protection, or agricultural industry sustainable development. Plant modification is not a cure, but it may help satisfy the demand for more feed by generating high productive variants. Maize productivity nearly doubled in the last 40 years owing to better crop types, fertilizers, irrigation, and herbicides (Benbi 2018). Changing climate complicates efforts to boost agricultural output. Climate-resilient maize has been developed for greater tolerance to climate-related characteristics, production, resistance, and consumption demands. The International Maize and Wheat Improvement Center (CIMMYT) data from over 20,000 maize experiments in Africa, along with daily meteorological data, indicated that every degree's day spent over 30 °C decreased the ultimate yield by 1% under ideal rainfed and by 1.7% under water shortages conditions (Allen-Sader et al. 2019; Khaki et al. 2019). A rise of 1° in seasonal climate generates a 3.1–7.4% drop-in maize, wheat, soybean, and rice yield (Rolla et al. 2018).

Climate-resilient farming must also encourage consumer, regional entity, and financial organization cooperation to boost crop yield (Koopmans et al. 2018). Plant researchers may generate variants with adaptive features and climate tolerance by studying physiological processes, molecular pathways, and heredity of maize crop. Upcoming breeders require plant genome libraries, massive data process management, biotechnology treatments, and high-throughput molecular engineering to address the issue (Niazian and Niedbała 2020). It allows scientists to employ appropriate ideotypes to satisfy breeder needs and to find better alleles/haplotypes for next-generation breeding operations. It allows accurate and robust genomic alteration to generate pest-resistant crops to pathogens and environmental stressors (Tripathi et al. 2019a, b). In recent decades, better grain variants tailored to various agroecologists and addressing the varying stakeholder expectations have been developed and delivered, mainly via traditional breeding (Munaweera et al. 2022). With the global population expected to reach 9 billion by 2050, cereals output must expand by 37% yearly to fulfill food requirements. To nourish the world's rising rural and urban communities, wheat, maize, and rice production must rise by 70% by 2050 (Mottaleb et al. 2018).

Changing climate has prompted farmers to adapt from old agricultural practices to contemporary climate-adaptable agronomic systems such as diversification of crops, seeding heat- and drought-resistant types, caring for the ecosystem, and gathering freshwater for food and nutrition security (Jiri et al. 2017). The implementation of climate-adaptable agriculture practices must also incorporate potential losses, farming tools, enhanced administration, and organizations improving their effectiveness (Patel et al. 2020). Climate change presents equal challenges and opportunities for maize growers. Temperature and precipitation changes will increase dryness, exposing small-scale farmers to severe consequences inherent with rainfed cultivation (Yang et al. 2019). Ironically, resource-poor producers in emerging economies struggle to sustain output balance and help to crops viable amid unpredictable prices of commodities and growing energy expenses (Purushothaman and Patil 2019). Small-scale farmers depend on OPVs (open-pollinated varieties) or outmoded hybrids produced over 30 years ago, restricting their nutritional and food security (Donovan et al. 2022). Growing demands, continued hunger, and poverty depletion of natural resources and environmental degradation will force the globe to double maize-based agricultural system efficiency (Tanumihardjo et al. 2020). This necessity will only be satisfied by agrarian societies, both national and international academics, policymakers, the business sector, and numerous other development agencies working together to advance pro-poor agricultural societies (Fischer and Connor 2018).

2 Importance of Maize in World Food Security

Maize is cultivated in variety of elevations and latitudes compared to other food crops as it is growing in mild to warm conditions, on moist to semi-arid regions, and in numerous soil conditions. Maize covers 150 million acres globally with an annual production of 750 million metric tons (Shahzad et al. 2021). Maize contributed 27% in cereal zone, 34% in cereal yield, and 8% in all basic crop. Maize is a major food source and nourishment for millions in Asian, Africa, and Latin America (Table 8.1). Calories gained from maize varied from area to area; for example, 61% in Mesoamerica, 45% in ESA (Eastern and Southern Africa), 29% in the Andean region, 21% in WCA (West and Central Africa), and 4% in South Asia. Twenty-five percent of maize consumption in underdeveloped nations and 15% worldwide is used as foodstuff (Ngurumwa 2016; Tanumihardjo et al. 2020). In South Asia, 46% is utilized as foodstuff for impoverished families that cannot buy wheat and rice. Maize helps the impoverished in Latin America, Africa, and Asia, overcomes starvation, and increases agricultural production. Huge maize production makes acceptable to farmers in regions with limited space and heavy population explosion (Sanjana Reddy 2017; Giller et al. 2022).

Maize yield moved outside the USA and Canada from Mesoamerica in the second millennium. The sub-Saharan Africa (SSA) and Mexico and Central America have two aspects. Firstly, Africans located south of the Sahara favor white maize, except in West Africa (Shawa 2019; Elisante 2020). This tendency affects the interchange of genotypes and trading since 96% of global maize trading is yellow (Fitting 2021). In contrary to other key maize producing countries, SSA and

Table 8.1 Maize area, production, and yield across maize growing regions. Computed by authors based on United States Department of Agriculture (USDA)

	Area (Million hectares)	Yield (Metric tons per hectare)	Production (Million metric tons)	Change in Production
Region	2021/22	2021/22	2021/22	From last year (Percent)
World	206.77	5.89	1218.76	−3.21
United States	34.56	11.11	383.94	−5.00
China	43.32	6.29	272.55	−0.57
European Union	9.24	7.69	70.98	−15.47
South Africa	3.00	5.43	16.30	6.13
Ukraine	5.49	7.68	42.13	−28.79
Russia	2.90	5.25	15.23	−1.48
India	9.90	3.33	33.00	−4.55
Pakistan	1.50	6.33	9.50	−6.32
Indonesia	3.90	3.26	12.70	1.57
Mexico	7.10	3.88	27.55	0.18
Canada	1.39	10.05	13.98	0.11
Turkey	0.54	12.04	6.50	−4.62

Mesoamerica cultivate maize as a food product, but they trade it for income. SSA has 25 million hectares of maize, mostly in small-holder farms that yield 38 million metric tons for food. South Africa grows 2.8 million acres, largely for economic livestock feed (Ochami 2021).

In many nations, reliance on maize threatens sustainable food security, particularly when alternative dietary supplements are insufficient. Maize is weak in digestible proteins and its leucine hinders the body's ability to absorb niacin, which leads to protein shortage (Day et al. 2021). Breeders at CIMMYT spent years developing "Quality Protein Maize" (QPM) with greater amounts of tryptophan and lysine in the endosperm and improved amino acid imbalance (Maqbool et al. 2021). Randomized studies in Ethiopia, Tanzania, and Uganda demonstrated enhanced weight and height of children eating these types, especially in Southern Ethiopia, in which the people depend primarily on maize (Jada and van den Berg 2022). Maize is not just eaten. A 63% of worldwide maize consumption is for animal feed, whereas 56% is in emerging nations. A 70% of maize is utilized as fodder and 3% for human utilization in developed nations (Makkar 2018). Maize's expanding application in bioenergy drives industrial demand. Maize-based bioethanol expenditure is growing in 50 emerging nations with ethanol-blending objectives (Oliveira et al. 2017). Maize is a major raw material for the United States "corn ethanol" program. Food costs and children hunger will rise with massive maize-based production of bioethanol, notably in sub-Saharan Africa (SSA) and South Asia (SA) (Afzal et al. 2021).

Current market statistics show that the two forces of bioethanol development (rising fuel costs and energy security measures) may accelerate in the foreseeable future (Nazir et al. 2020). Maize is a significant feed ingredient for the livestock industry, particularly in Asia, where usage for broiler chickens and pig feeding has surged. Pakistan, Nepal, China, Indonesia, India, Philippines, Thailand, and Vietnam ate 20 million metric tons of pig and chicken in 1980 and 77 million in 2000 (Ashraf et al. 2018). Broiler feeds comprise 60–65% maize, 28–30% soybeans, and 2–3% oils. The seven primary Asian nations used 29 million metric tons of maize for feeding in 1980 and 109 million tons in 2000 (Bailey and Wellesley 2017). Maize as animal feed is expected to constitute a substantial part of expected growth, expanding about 6% annually. World trade has been a key approach for solving maize supply deficits as supply and use diversify. Behind wheat, maize is the most-traded crop (Wade et al. 2022).

3 Influence of Climatic Variation on Agricultural Productivity

Changes in climate will worsen agricultural production. Severe heat, massive flooding, repeated water shortages, and salt stress impact agricultural production and also reduce the capacity of ecological systems to adjust to climate change. Extreme weather is the primary driver of food availability in emerging regions, affecting farming and other agricultural-related equipment (Campbell et al. 2018; Mojid 2020). Smaller nations, like desert African and South Asian, are starting to suffer. These areas depend on subsistence farming and are not financially or technically prepared for extreme weather events (Khatun 2019). Because of droughts, several African nations face a food crisis. The IWMI (International Water Management Institute) predicts that wheat production will drop by 50% by 2050 in Southern Asia (Islam and Karim 2019). This is approximately 7% of global agricultural output; therefore, hunger might intensify. Globally, maize production has been declined by 3.1% because of temperature variation and 0.7% because of rainfall patterns from 1980 to 2008 which leads to a 3.8% reduction in total yield (Arunrat et al. 2022).

South Asia and SSA are extremely hungry nations. All of those are climate-vulnerable zones. Asian and African governments fail to take action; the circumstance is expected to worsen. Providing nutritious meals to a large population is difficult. World's food availability has failed due to unregulated population expansion, rising food costs, and a worldwide food crisis (Bloem and de Pee 2017). To face the demands, yields of main crops must rise 1.1–1.3% yearly. According to the European Commission's Directorate General for Agriculture, extreme water-deficit regions are expected to expand from 19% to 35% by 2070 (Abdelmajid et al. 2021).

According to the Peterson Institute, climate change will diminish agricultural production in developing countries by 10–25%. Elevated CO_2, high temperatures, high tropospheric ozone (O_3), frequent droughts, and strong precipitation may cause catastrophic floods (Abdelmajid et al. 2021). In sensitive places, changing climate might produce waterlogging, soil erosion, salination, and sodic soil. Changes in seasonal rainfall, intensities, and temperatures might affect the global distribution of agricultural pests and diseases. Deforestation, overgrazing, and chemical fertilizers add 25% to emissions of greenhouse gases (Cirino 2021). Environmental issues may be mitigated by reducing greenhouse gas emissions and boosting food-secure crops. By 2080, watering needs may climb by 45%, and available water losses may surge by 20%, although with infrastructure upgrades (Sehgal and Khan 2020). Genetic engineering may reduce environmental issues through sequestering carbon, energy-efficient agriculture, and chemical fertilizer elimination. Genetically modified (GM) crops may help with food self-reliance, nutrient supply, alleviating poverty, stability, starvation elimination, and global climate change (Behera et al. 2019).

4 Current and Future Research Necessities Regarding Climate-Improved Germplasm

Considering the time gap between enhanced germplasm creation and farmer accessibility, breeders' approaches must be strengthened to counteract expected yield reductions and achieve predicted yield improvements (Xu et al. 2017). Geographic information systems (GIS) will help to focus genetic efforts via forecasting sensitive locations, regulating genetic mobility, and predicting future crop yield conditions (Raffini et al. 2020). To boost adaptive capability, future susceptible areas must concentrate on maize breeding. Projected climate predictions will facilitate the understanding of present regional analogs of these climates to promote genetic transfer for breeding (Dempewolf et al. 2017). Mega-environments combine maize-yielding locations with identical environmental features and adaption behaviors should be focused to maximize the performance of using genetic assets (Oladosu et al. 2017). The highest temperature, soil pH, and rainfall define maize mega-environments, along with genetic effectiveness in multi-environment investigations (Katsenios et al. 2021).

GIS can help to identify places with comparable types of soil and rainfall patterns to promote and adopt sustainable agricultural approaches. By combining the traditional, biological, and genetic engineering, we may provide climate-adapted genomes (Harms et al. 2021). Promoting traditional breeding techniques, including phenotypes and double-haploid innovation, and expertise in these techniques, will accelerate climate-adapted genotype creation. Developing molecular techniques, like genome-wide association analysis, aim to accelerate the development of cultivars that may be tolerant to numerous biotic and abiotic factors (Uffelmann et al. 2021). Major advancements in genotype and phenotype will help to investigate the immense unutilized genetic diversity in maize gene banks, particularly landraces and progenies (Volk et al. 2021). A better knowledge of these biological resources might lead to their more efficient use in elaboration new stress-resilient genes and

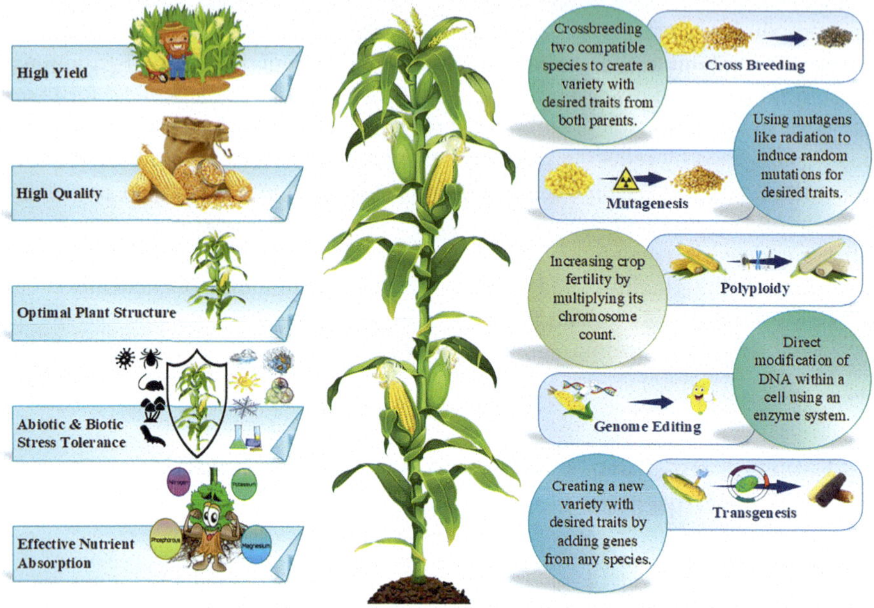

Fig. 8.1 Popular plant modification strategies used to improve top crop varieties. (Gao 2021)

diversifying the germplasm of top crops (Prasanna et al. 2021; Singh et al. 2021). Transgenic techniques (Fig. 8.1; Gao 2021) might potentially boost genetic variety for major abiotic stressors in hybrid crops, notably for characteristics with minimal genetic differences in genetic pool of maize (Khan et al. 2021).

South African producers need drought- and heat-tolerant maize. Small area of East and South Asia (ESA) producers may need drought and waterlogging resistance varieties (Prasanna et al. 2021). Isolated breeding programs utilizing just one stressor may lose traits or genetic diversity for other stressors. Considering that both drought and heat stresses are dissimilar, crop development agencies must identify characteristics and factors linked with combined resilience to both stressors (Wu et al. 2018). To create climate-adapted cultivars, controlled stress testing facilities and defined methodologies for particular combinations of pressures are required. Agricultural phenotyping for both drought and heat stressors will need to mitigate stress severity and duration (Abdelrahman et al. 2018). Climate variability may also intensify severe heat outbreaks. Climate-adapted germplasm breeding techniques should account for local temperature variance. Warmer temperatures and lower rainfall may minimize production in Africa by 10–20% till 2050 (Mulungu and Ng'ombe 2019).

To increase output, highland germplasm with better temperature resistance is needed. This may be done by enhancing highland germplasm's heat resistance and enabling farmers' accessibility to tropical and subtropical genotypes (Girma et al. 2020). Under ideal circumstances, days to maturation affect crop productivity. In rainfed locations, small-holder farmers choose early-maturing maize (Krell et al. 2021). Early maturing maize helps plants to complete blooming, the most drought-sensitive phase, before the start of dryness and escape severe dehydration (Mbanyele et al. 2021) whereas genetic techniques have increased early maize yields during low rainfall situations, still it has yield disadvantage under normal rainfall conditions. Rising temperatures may further reduce the production. In Zimbabwe's drought-prone plains, producers without the supply of water choose extra-early cultivars that bloom in 60–65 days (Kinfe and Tesfaye 2018). Ideally, a daily rise of 2.9 °C during the growth period might cut the days to the maturation of an extremely early variant from 65 to 58, while an early maturity maize variety might mature in around 64 days, as opposed to 72 days within favorable conditions (Singh 2020). Considering this drop-in growth time, breeders projects addressing drought-prone locations, like the lowlands of Zimbabwe, may need to concentrate their efforts to early maturing germplasm instead of extremely early maturing types (Khan and Akhtar 2015).

5 Materials, Methods, and Techniques to Build Climate-Resistant Maize

Crops should be resistant to environmental and biological stresses to adjust to a changing climate. Crop-specific and crop-shared genomic approaches and methods must accommodate for climate variations (Cortés and López-Hernández 2021). Locate and transfer desirable genomes, characteristics, and haplotypes to elite genetic variants are desirable. Further, special techniques that combine molecular markers with elite germplasm development are needed to accelerate the production and distribution of improved genotypes (Fig. 8.2; Gao 2021). Today's technology and techniques are listed below.

5.1 Genetic Modification (GM)

Plant biotechnology utilizes biological entities or intracellular elements in agriculture. Convectional breeding, genome editing, and molecular breeding are now being utilized (Dilawari et al. 2021). Genetic manipulation improves gene diversity for developing food plant types with better features. Water stress resilience, salt resilience, extreme heat resilience, water-efficient utilization, early vigor, effective nitrogen usage, waterlogging, cold resilience, resilience to pests, and insect management trigger the seedling growth or blooming of crop. Genetic manipulation may reduce greenhouse gas (GHG) emissions, biofuel consumption, fertilizer usage, carbon sequestration, and biotic or abiotic stressors (Kumari et al. 2022). In the past couple of decades, the development of "Golden Rice" achievement via

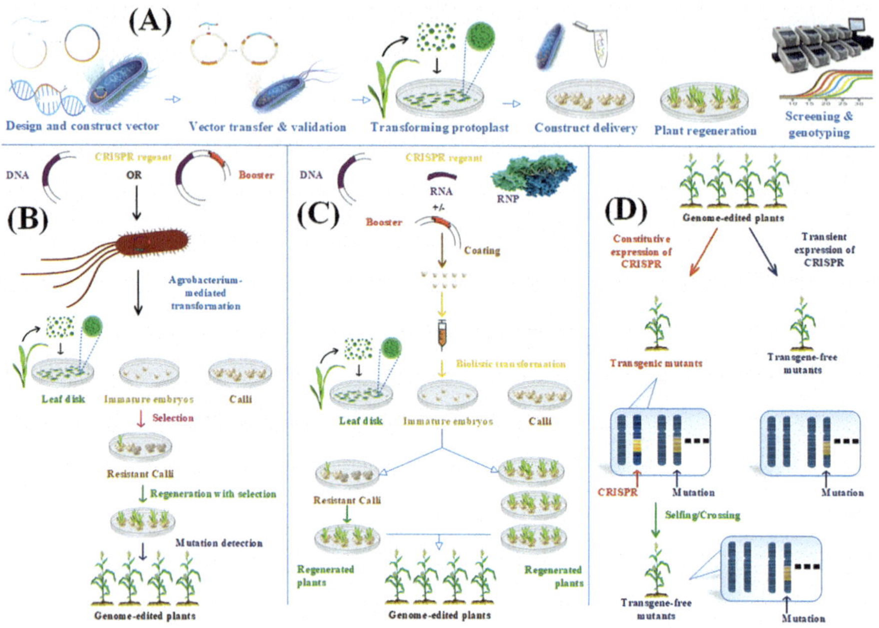

Fig. 8.2 Overall method of genome editing for plants. (Gao 2021)

genetically modified was significant. It involves transforming genetic traits required for carotenoid (precursors of vitamin A) production in the endosperm that lacks provitamin A due to a missing code gene (Qamar et al. 2020).

The U.S. Environmental Protection Agency and Food Inspection Agency have recognized "SmartStaxTM," a genetically engineered maize with eight stacked cry-genes: (Cry2Ab, Cry1A.105, Cry1F, Cry3Bb1, Cry34, Cry35Ab1, Cp4, and bar) (Wangila 2016). Moreover, transgenic Bt maize has the highest pest insects and herbicide resistance. African Agriculture Technology Foundation performed numerous initiatives with global organizations to generate water stress-resilient maize varieties in Africa through genetic engineering (Munawar et al. 2020). Utilizing the soil's carbon and boosting carbon dioxide absorption may increase cumulative carbon sequestration in agricultural soils (Nayak et al. 2019). Developing new types by using genetic engineering increased plant photosynthetic efficiency, deeper roots, lignin content quality, fertilizer use effectiveness, and water consumption effectiveness (Savy and Cozzolino 2022). Nitrous oxide generates 296 times more heat than CO_2 and may last hundred years. Optimizing nitrogen-enhanced crops may minimize fertilizer consumption and GHG emissions (Ussiri and Lal 2018). Because agriculture and fertilizer contribute 50% of U.S. GHG emissions. The latest advancement in transforming technologies limits soma clonal variability during tissue culturing. Using genetic modification, desirable alleles or genotypes may be transferred or programmed among plant and animal species (Sehgal and Khan 2020).

5.2 *Genome Editing (GE)*

The latest advancement in genetic engineering including genome-based understanding, efficient gene identification, and in vitro culturing methods enable for second-generation applications that rely on genetic modification and cis genesis (Ryngajłło et al. 2020). These technological innovations generate significance modern crop varieties to offset future agricultural sustainability issues (Suprasanna et al. 2021) and biotic genome editing (GE) technique presents a unique opportunity for disease prevention, insect control, resistance versus abiotic and biotic stressors, rapid crop breeding, and improved crop produce with lower production expense. Cutting-edge gene-editing approaches are categorized as "Oligonucleotide Directed Mutagenesis" (ODM) and "Site Directed Nucleases" (SDNs) (Sun et al. 2019; Verma et al. 2022). These enable genetic recombination, site modifications, and gene regulation management.

5.2.1 Oligonucleotide-Directed Mutagenesis

In "Oligonucleotide Directed Mutagenesis," a DNA fragment of 20–100 nucleotides is chemically produced and transferred to the plant genome at the target location via particle bombardment or polyethylene glycol (PEG)-assisted gene transformation (Munawar et al. 2020). Nevertheless, imposed mutation effectiveness is relatively low, around 0.05% at the target location (Cabrera-Ponce et al. 2019; Munawar et al. 2020).

5.2.2 Site-Directed Nucleases

Site-directed nucleases bond with particular DNA sequences (9–40 nucleotides). They split double strands to undertake deamination, demethylation, and acetylation processes (Ko et al. 2020). These biochemical events induce genetic suppression or genetic manipulation, altering bioactivity. Double-strand break (DSB) is the most significant biochemical process in biological systems. The DSB is restored by homologous recombination (HR) or non-homologous end joining (NHEJ) (Biehs et al. 2017). The HR is employed for massive DNA insertion and non-random modification. The NHEJ introduces randomized introduction at the target location to knockout genes or reduces gene expression (Amritha and Shah 2021).

Additional categories of site-directed nuclease are zinc finger nucleases (ZFNs), clustered regularly interspaced short palindromic repeats (CRISPR), and transcription activator-like effector nucleases are site-directed nucleases (TALENs).

5.2.2.1 Zinc Finger Nucleases (ZFNs)

Zinc finger nucleases are manufactured proteins used for gene targeting (Fig. 8.3; Eid and Mahfouz 2016). They have a DNA cleaving endonuclease constituent linked to zinc finger domains (Zarei et al. 2019). These are typically used to produce deletions or insertions in cleaved genomes. Zinc finger nucleases were the first target-specific protein tools in genetic modification (Gupta and Shukla 2017). Such synthetic restriction enzymes link zinc finger DNA binding region to a DNA cleavage region to identify unique DNA regions in complicated genomes (Fig. 8.2). Using the endogenous DNA restoration process, such elements might straight transform complicated species genome sequences. Zinc finger nucleases attach to DNA by recognizing 3 base-pairs (Vats et al. 2019; Chen et al. 2020). A specific zinc finger might require precision to connect a targeted genome, but unique constructions comprise of 3–4 fingers modifying an 18–24 kb ZFN target region (9–12 bp for each half) (Paschon et al. 2019). This length is enough to target particular regions in large genome sequences. The targeted site length is regulated by the context-specific synthesis of ZFN components (Wolter et al. 2019). Linking 2 DNA binding ZFNs to FokI (Restriction endonuclease) monomers creates a 5–7 bp spacer that helps FokI dimerize and generate DSBs. The approach induces focused DSBs using ZFNs that promote DNA repair through error-sensitive NHEJ and HDR (homology-directed repair) in cells. ZFNs have been applied to improve Arabidopsis (Saifaldeen

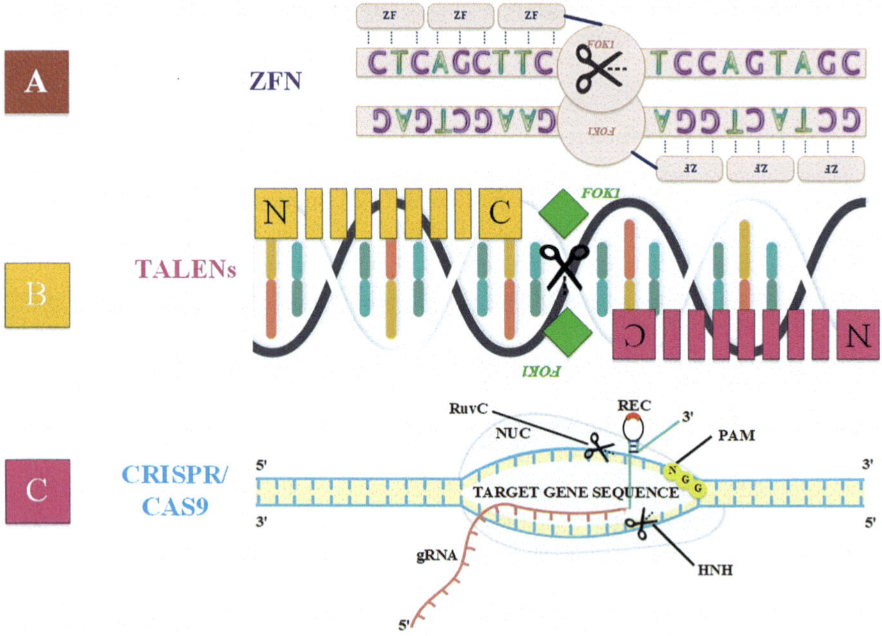

Fig. 8.3 Genome editing method's diagrammatic illustration: ZFN, TALENs, and CRISPR/Cas9. (Eid and Mahfouz 2016)

et al. 2020), tobacco (Oh et al. 2021), maize (Matres et al. 2021), oil crops such as rapeseed and soybean, flowering plants such as petunia, rice, and fruits varieties like figs and apples (Zhang et al. 2018). The ZFN-mediated transgenic integration was used for trait layering in maize, which allows for greater crop improvement (Chen et al. 2019). The ZFNs have little impact on climatic change, tolerance by modifying genes of the host plant, which are difficult to create and duplicate (Kausch et al. 2021).

5.2.2.2 CRISPR/Cas9 System

CRISPR/Cas9 is the third-generation genome editing technique (Fig. 8.3). The CRISPR/Cas technology is convenient, accurate, powerful, and inexpensive relative to TALENs and ZFNs (Wada et al. 2020). Developing a DNA binding motif by a target-sequence requires biochemistry and protein-engineering expertise, which may be reduced by using short RNA-sequences instead. CRISPR protects photosynthetic bacteria by targeting and destroying foreign DNA, such as viruses or plasmids (Glick and Patten 2022). CRISPR/Cas9 was first used to modify plant genomes in 2013 and is now a common GE technology. Effective CRISPR-Cas positioning requires matching-repetitions, crRNA processes, an operon of Cas genes that translate Cas protein sub-units, and invader DNA targeting spacers (Razzaq et al. 2019; Zess and Begemann 2021).

CRISPR/Cas-9 is another technique for modifying both RNA and DNA. CRISPR/Cas-9, a bacterial defensive mechanism, may be applied to crops, microbes, animals, and people (Zhang et al. 2021). Its uses to abiotic stressors are currently confined to *Arabidopsis*, although new successes in soybean and cocoa against drought and salt resistance (Drb2a/Drb2b genes) have been revealed (Tastan et al. 2020).

In subsequent years, several research shows changed genotypes in maize, wheat, soybeans, barley, potato, and tomato. This approach is also used in maize to knock off Zmzb7 genes to generate albino plants (Ansari et al. 2020). For overexpression, maize U3-promoter was used. Subsequently, agro-bacterium-mediated conversion of maize embryo yielded 31% mutant efficiency in T0 strains. CRISPR/Cas9 was used to knock down the maize waxy genes Wx1, which encodes for a granule linked to the amylose-producing enzyme GBSS (egranule-bound starch synthase). Wild-type (WT) seeds-starch has 75% amylopectin and 25% amylose, whereas Wx1 seed starch is 100% amylopectin (Leong et al. 2018; Li et al. 2019). Thus, the absence of amylose-generated rich amylopectin-maize or waxy-maize has improved digestibility and application in bioindustries generating food stabilizers, thickeners, papers, and adhesives (Lu et al. 2019).

5.2.2.3 Transcription Activator-Like Effector Nucleases (TALENs)

TALENs were discovered in a pathogen of the plant, Xanthomonas. Combinations of TALE repetitions and the FokI restriction enzyme address a single nucleotide instead of three at the target site, enabling TALENs accurate and flexible in target design while increasing the number of potential target sites (Fig. 8.3; Lino et al. 2018; Rothschild 2020). TAL (Transcription activator-like) effectors penetrate the nuclei and stimulate transcription by attaching to host genes promoter regions. 34-Residue tandem AARs (amino acid repeats) help TAL-effectors identify DNA (Cox et al. 2017).

First used to change OsSWEET14 promotor region in rice plants gene that affects vulnerability to bacterial blight. Because of an alteration in the promoter regions, "effector binding element" does not interact with the promotor, causing crop bacterial blight resistance (Oliva et al. 2019). TALEN was also employed to modify three homo alleles of the vulnerable MLO (Mildew Locus O) genes to induce powdery mildew resilience in wheat varieties. GL2 (GLABRA2) gene deletion in maize produced variants including shiny phenotype and reduced leaf waxes (Tyagi et al. 2021).

5.3 High-Density Genotyping and Resequencing

Maize, rice, durum wheat, and barley were improved by SNP (Single nucleotide polymorphisms) chips. In maize, chip-based genotyping developed by Cornell-CIMMYT (The International Maize and Wheat Improvement Center with Cornell University NY) led to three Illumina 1536-SNP chips that were shortly superseded by the Illumina Maize SNP50 Bead chip having 56,110 SNPs, 1 SNP/40 kb, and 19,540 genes with 2 SNPs/gene (Edriss et al. 2017). At an alternative level, genotyping may also be done using GBS. Illumina's next-generation sequencing technology has a straightforward, multiplexed technique for creating reduced representation libraries (Gurgul et al. 2019). GBS libraries were created by lowering genomic complexities using restriction enzymes that might reach sequence capture-inaccessible areas. In barley (Oregon Wolfe Barley) and maize (IBM) recombinant inbred line (RIL) populations, 25,000 and 200,000 sequence tags were identified, correspondingly (Aggarwal et al. 2021; Favre et al. 2021). Organisms without a full genome sequencing may construct a reference map around restriction sites during sample genotyping. By using GBS method, the CIMMYT Global Maize Program (CIMMYT-GMP) is using high-density genotyping to enhance distinctive features (Singh 2017). Numerous billion data points already have been created on significant germplasms. System optimization is improving SNP calls and reducing lost data points (Romay 2018). Also, in coming years, chip or GBS-based genotyping will rely on their cost for genotyping and associated data storage, analytics, and distribution systems. The genetic sequencing of B73 and Palomero, a Mexican popcorn landrace, is crucial milestones in maize genetic studies, having considerable significance for comprehending maize genome structure and evolvement and formulating ways to use genetic data in maize development (Roorkiwal et al. 2020). Palomero's genomes are 22% (140 Mb) smaller than B73's and contain many unknown sequence data, indicating a massive number of undiscovered genes. As sequencing grows more cost-effective and faster, resequencing is replacing genotyping. For evolutionary, biodiversity, and genomic investigations, maize and rice have undergone large-scale resequencing (Huanca-Mamani et al. 2018; Singh et al. 2022).

Maize haplotype map (HapMap) is the first of its kind. The genome of 27 maize inbred lines was characterized by substantially different haplotypes (Alekya et al. 2022). The specific marker-based mapping may be replaced with haplotype-based mapping to boost mapping power and find particular genotypes in genes or allele combinations at various loci that correlate to the identical targeted trait (Gupta et al. 2019). Additionally, to next-generation genotyping and sequencing, maize has various mapping populations as global maize genetic resources. The maize nested association mapping (NAM) population, with 5000 RILs, is a recent genetic resource (Nepolean et al. 2018; Tran et al. 2020). The NAM population integrates QTL (quantitative trait loci) mapping with association mapping to identify genes underlying complex characteristics. The NAM RILs (nested association mapping, recombinant inbred lines) represent global maize genetic variability, allowing researchers to explore genes related to numerous phenotypes, notably biotic, and abiotic stress resiliency (Scott et al. 2020; Wani et al. 2022).

5.4 *High-Throughput and Precision Phenotyping*

Developments in phenotyping help increase grain breeding efficiency by using high-throughput genotyping. Phenotyping has developed from relying primarily on direct assessment of target characteristics (e.g., grains production) during stresses in analytical breeding for abiotic resilient crops to remote phenotyping (Bhat et al. 2020). Remote sensing may imply whole-plant growth, water availability, or crop production. HTPP analyzes plant attributes to determine phenotypic traits. Such systems can forecast genotypic success in diverse climatic situations (particularly when "buried factors" are included in monitored investigatory environments (Ross-Tremblay 2020). Plant temperature, spectral reflectance, and canopy structure may be measured all across the life cycle of the crop. The CIMMYT International Maize Program is working to define field variance at significant phenotyping locations globally and enhance field-based maize germplasm phenotyping (Prasanna et al. 2021). This comprises non-destructive biomass calculation with NDVI (normalized difference vegetation index), soil moisture monitoring with neutron probes/TDR (time domain reflectometry), chlorophyll content with a SPAD (soil plant analysis development) meter, canopy behavior with infrared thermography, etc. (Fukuda et al. 2018; Brewer et al. 2022).

5.5 *Doubled Haploid Technology*

Doubled haploid (DH) technique helps to introduce unique germplasm into elite-breeding programs. DH plants are generated whenever haploid cells are subjected to chromosome duplication, whether naturally or through chemical modification, enabling a single round of recombination to produce a homozygous line (Blary and Jenczewski 2019; Kariyawasam 2021). DH promotes "forward breeding" and allows for an initial look at the possibilities of emerging lines, offering better information about their environmental adaptation prior to them being completely assessed and utilized for hybrids creation and commercial farming (Jiang et al. 2020; Maqbool et al. 2020). The maternal haploidy-based DH technique in maize may boost line development efficiency by lowering the duration of homozygosity from 7 to 2 seasons. In maize, spontaneous chromosomal duplication proved ineffective for breeders. In the past 10–15 years, commercialized maize breeding programs in North America, Europe, and most lately China have used DH technique (Andorf et al. 2019; Uliana Trentin et al. 2020). Eighty percent of commercialized maize breeding operations use DH technique. Since 2007, the CIMMYT International Maize Program has worked with the University of Hohenheim, Germany, to optimize DH technology for tropical/subtropical maize growing conditions (Chaikam and Prasanna 2020). CIMMYT Maize Program is phenotyping this DH population at different places globally for water stress, temperature, and poor nitrogen resistance (Trachsel et al. 2019). CIMMYT and IITA (The International Institute of Tropical Agriculture) built abiotic stress phenotyping and breeding networks for maize in eastern and southern Africa, including high-density genotyping and bioinformatics services. CIMMYT creates similar partnerships throughout Asia and Latin America, five genetic methods, and approaches for developing climate cereals, including inducement costs and rates of inducer line upkeep and seeds output (Gedil and Menkir 2019; Prasanna et al. 2021).

6 Future Maize Germplasm Adaptations

Within the past two decades, global firms have issued a variety of breeding products using molecular breeding, along with transgenic and marker-assisted selection (MAS) techniques (Prasanna et al. 2020). However, evidence concerning the size, nature, and scope of these initiatives is scant. Maize, barley, wheat, rice, and pearl millet have MAS-derived cultivars and advanced lines with better seed quality (Prasanna et al. 2021). MAS seems to have been effective in boosting seed quality and resistance to disease. Governmental maize organizations in underdeveloped nations have issued few MAS-bred cultivars. Five excellent maize lines in India are resistant to turcicum leaf blight and polysora rust (Ranganatha et al. 2021). MAS has shown its promise as a technique to help traditional genetic modification of crops, despite few examples of its usage in plant genetics. Still many effective implementations are likely to present but are exclusive to commercial-breeding global businesses (Boudry et al. 2021).

Biotechnology has improved grain productivity. CIMMYT began a maize breeding initiative in Southern Africa in 1997 to boost productivity in minimal resources, and water stress zones (Grote et al. 2021). Maize cultivars were designed for optimum, poor N, and drought-managed environments. All CIMMYT hybrids outperformed commercial checks. CIMMYT hybrids demonstrated a 40% production benefit over commercialized cultivars during severe water-deficit conditions (Worku et al. 2020). Water stress-resilient Maize for Africa has enhanced phenotyping capacity in ESA (Eastern and Southern Africa) from 6 to 35 ha since 2006 (funded by the Bill and Melinda Gates Foundation). It has contributed to raising maize's water stress resistance (Yahaya and Shimelis 2022). New drought-tolerant hybrids exhibited a 35% yield advantage over farmers' cultivars in ESA under low (3 t ha^{-1}) yield circumstances. The top variant (CZH0616) generated 26% more than the leading commercial cultivars. These findings show how maize modification may boost productivity under harsh conditions (Bekele 2020).

A 2 °C rise in temperature would reduce maize production beyond a 20% fall in moisture, according to The European Space Agency modeling (Rigden et al. 2020). Since CIMMYT started a maize drought-breeding program in Mexico in the 70s, maize germplasm was evaluated under a variety of conditions, such as a warmer temperature (Sheikh et al. 2017). This program's drought-tolerant maize exhibited a better heat tolerance due to indirect selection for heat exposure. In ESA, drought-tolerant maize improvement did not account for warm temperatures till the present (Nelimor et al. 2020). Analyzing maize strains from CIMMYT and IITA breeding projects in conditions with high temperatures found ESA maize germplasm was vulnerable to drought and heat stress at extremely high temperatures (Prasanna et al. 2021). These findings show that ESA's maize breeding system should include high-temperature conditions. Heat exposure resistance breeding involves the understanding impact of temperature on maize development and growth and genetic diversity for desired characteristics (Nelimor et al. 2020). Extreme heat reduces maize development period, absorptivity of light, and reproduction rate. Delaying sowing could be employed to evaluate for heat stress tolerance by exposing plants to extremely high temperatures between important development phases (Xia et al. 2021). By utilizing this strategy, researchers found considerable genetic diversity for heat tolerance in subtropical and tropical maize, showing conventional breeding might improve heat tolerance (Prasanna et al. 2021).

Maize production declines increased in Southern Africa over 30 °C during water stress. By the turn of the era, carbon dioxide might double (Muluneh 2020). Rising carbon dioxide level will accelerate photosynthetic activity in C3 plants, while in C4 plants, including maize, photosynthesis is peaked, and higher Carbon dioxide levels will make no influence (Éva et al. 2019; Tan et al. 2021). By partially closing stomata, increased Carbon dioxide lessens transpiration and enhances efficiency of water use. Partial stomatal closure may boost soil water availability and postpone water stress, but it might also raise canopy temperatures (Luan and Vico 2021). Higher temperature and water stress are predicted to rise in northern Namibia, northern Botswana, Southern Zimbabwe, and SSA (sub-Saharan Africa) and maize fields are likely to suffer from extreme heat amid water shortages. Another latest evidence on maize demonstrated that resistance to simultaneous water stress and extreme heat is genetically unique than resilience toward each condition separately (Nelimor et al. 2020; Malenica et al. 2021).

Twenty percent of manmade GHGs seems to be from agribusiness. Excessive fuel usage, farming equipment, fertilizer, and irrigation increase emissions of carbon dioxide (Panchasara et al. 2021). Conservation agriculture (CA) may decrease GHG emissions and enhance carbon sequestration, mitigating global warming (Balafoutis et al. 2017). SSA's (sub-Saharan Africa) maize cultivation techniques are rainfed, depend on the animal drive or human soil preparations, and consume relatively little fertilizers (Bationo et al. 2018). Therefore, in the territory, adaptability might be more crucial than remediation. Improved maize cultivars with enhanced resilience to high temperature and water stress will help SSA respond to climate change, but maize production remains among the smallest on the globe (Abegunde et al. 2019). Maize modification alone cannot boost present yields and mitigate future production losses from changing climate and increasing climatic fluctuation (Zougmoré et al. 2018). Growing period heat may not surpass the criteria for tropical and subtropical maize until 2050, but rising temperatures might increase plant evapotranspiration, rising plant's watering need, and limiting soil moisture availability to plants (Malhi et al. 2021). Enhanced soil moisture condition may shield crops from moisture stress during irregular rains and lessen drought intensity. CA reduces soil loss via tillage, conserving leftovers, and rotating crops. CA improves groundwater recharge, reduces evaporation, and reduces water overflow (Jovanovic et al. 2020; Bogati and Walczak 2022). CA increases soil water, according to much research. CA increased maize grain production by 1.8–2.7 times during moderate water stress. CA's greater level of soil water might shield maize productivity versus prolonged water stress situations (Hafez et al. 2021).

7 Farmers' Approaches for Lowering Climate Impacts

Climatic change and rising weather events have a smaller influence on producers' lifestyles since they are dynamic "agents" whose approach, and experiences determine advancement inside the limitations of their resources and knowledge (Rolnick et al. 2022). Farmers have devised adaptive mechanisms to deal with yearly fluctuations of rain. Changing climate may affect farmers' decisions and cultivation strategies (Abid et al. 2019; Rolnick et al. 2022). Detailed the kind of measures farmers use to move up in the world. By using this framework, the following are living development solutions for poor farmers facing environmental problems (Alreshidi 2019):

- Family enhances economic or physiological output of present growth dynamics, e.g., growers boost productivity by using external inputs such as irrigation and better breeds of crops they traditionally cultivate and by adopting water-conserving land management methods like the CA (Mustafa et al. 2019).
- The strategy allowed producers to increase revenue or avoid financial unpredictability by entering fresh or existing marketplaces. This includes innovative product farming and on-farm processing to increase value. In poor nations like SA,

this could involve shifting from maize to millet and sorghum. Considering smaller growth cycles and lower water needs, growers preferred maize (Reardon et al. 2019).

- Expanding their acreage or livestock increases farmers' earnings or assets. The extension could occur by restructuring, amassing of discarded farmland, or clearance of underutilized areas (Marewo 2020).
- Seasonal or long-term off-farm work helps farmers boost agricultural profitability. The earnings might be applied to agricultural or family expenses (Gautam 2017).
- Farmers depart agribusiness when they switch to some other agro-ecosystem or perhaps a nonfarming career. Climate change causes movement of people. Concerning farmers' adaptability approaches, plant genetics will serve a significant contribution (Altieri et al. 2015).

8 Challenges for Farmers' Access to Enhanced Germplasm

In certain circumstances, impoverished farmers have barriers to adopting modernized breeds, yet they normally cannot adopt better germplasm. However, poor farmers' limited property sizes hinder genuine poverty alleviation (Rodríguez et al. 2016). Adaptation is dispersed across a very limited area to affect earnings, although if better cultivars offer higher production. It is unclear whether impoverished farmers did not accept modern verities (Hayton 2020). Several poverty-related indicators correlate strongly with adaptation. Having access to credit and information is important in several research. Many researchers have emphasized land-owning sizes, although not everything, but mostly failed to correlate effects size with averaged or distribution of low holdings. Therefore, the middle classes land-owning impact is seldom analyzed (Combs et al. 2021; Foster and Rosenzweig 2022).

Adaptation is often linked to better results. Poor farmers do not have enough property to generate yield growth into high meaningful profit gains. This verifies the conclusions of those who ask what more property is needed to eliminate poverty given yield/income impacts (Gosnell et al. 2019). Even with small direct financial improvements, alternative paths are crucial. Small-holding poor workers may benefit from technology-driven labor supply. Limited areas of high-nutrient variety may offer essential nutrients, particularly for malnutrition populations (Alwang et al. 2019).

Studies demonstrate a considerable decrease in household poverty linked with adaptation, although when adaptation is prohibited. Considering the very minor direct impacts, such magnitudes are unlikely to be practical and indicate challenges with clear descriptions (Masten 2018). Farm productivity improvements are virtually exclusively responsible for short-term poverty reduction benefits. Property building or other variables may enable adopting growers to have greater wealth increases than the production benefit predicts, but the investigator must uncover the link (Fuglie et al. 2019; Jayne et al. 2019). Modest productivity increases in narrow sectors slow property buildup. The rate of poverty is a societal construction; usually, policymakers care about population-wide consequences (Turner et al. 2018). Direct revenue increases reduce poverty among farmers (producing a particular crop) and technology-adopting growers, according to surveys. Even huge declines in sub-population poverty are generally correlated with decreased countryside (national) poverty (Carter et al. 2019; Martey et al. 2020).

This chapter concluded: The amount of domestic economic consequences is substantially facilitated by the geographical range of innovation adaptation in the fields.

9 How Many Farmers Will Use Climate-Resistant Innovations in 2030?

Changing climate may intensify, exacerbate, or diversify various potential consequences. High temperatures, water shortages, high temperatures, floods, insects, herbs, and infection are important hazards to crop yields (Altieri and Nicholls 2017). A farmer's climate resilience is characterized by his/her ability to predict, withstand, and retrieve against environment disturbances, as well as adjust and become less susceptible and more robust (Appiah and Guodaar 2022). To ensure worldwide food security by 2030, it is important to determine what percentage of farms are climate-resilient. Small-scale farmers are especially susceptible to a changing environment because they have fewer assets and fewer opportunities for education, technology, wealth management, and support systems (Nyasimi et al. 2017). Some with less economic choices and those who are excessively skilled are less climate-resilient. Climatic adaptability depends on local climate hazards, agricultural techniques, and agricultural inputs (Table 8.2). Numerous contributors, including the UK and USA, are measuring resilience as a novel guidance efficiency criterion (Srivastav et al. 2021; Osumba et al. 2021).

Human nutrition will likely be generated by 750 million farms in 2030, an increase of 200 million small-holder farmlands (Singh 2019). A proportion must embrace climate-smart farming to secure food security worldwide despite increasing climatic stressors resulting from climate change. Adoption rates for many inventions have been approximately 1% every year during the preceding 15 years (Burchfield et al. 2020). Trying to recreate and adopt proven climate-smart techniques and technology can help small-holder farmers become more resilient. This entails establishing support from the public for inclusive public-private partnerships and effective extension programs to reach rural and deprived regions (Wally 2021).

10 Conclusions

Maize is among the three most essential cereals for food supply. Maize is a key element in animal feed in several industrialized and growing nations in Latin America and Asia. Rising requirements for animal supplies and altered diets in emerging

Table 8.2 Biotic and abiotic stress-resilient populations in Asia

Populations	Trait	Method/Strategy	Region	Reference
WE 2101, WE 2103, WE 2104, WE 2106, WE 2114, WE 2115, PAN 67, UH5051, UH5052, UH5053, MM3,	Early maturity, yield	Modern conventional methods	East Africa, Asia	Simtowe et al. (2019)
AMDROUT1-AMDROUT4, AMDROUT5 × 6, MARS7-MARS12	Drought, Genetic gains per year	Genomic selection	Asia	Vivek et al. (2017)
Hybrid TA5084	Yield	Gene introgression	Southeast Asia	Zoetbrood (2022)
MPS6, MPS5, MPS4, MPS3, MPS2, MPS1, and HSBC	Heat	RCGS (rapid-cycle genomic selection)	South Asia	Das et al. (2021)
GDRM-187	Extra-early maturity	Participatory plant breeding	Asia	Sahoo and Sanadya (2020)
BRS4154, MPS-A, MPS-B WLS, and WLCY	Waterlogging	Modified stratification method of phenotypic recurrent selection	Asia	Prasanna et al. (2021)

nations may quadruple maize consumption by 2050. Climate change complicates efforts to close the yield gap and increase output by breeding new high-yielding cultivars in many emerging economies, whose harvests stay low. Sustainable maize production improvement will need comprehensive techniques that help preserve existing production and boost crop productivity during unpredictable and dynamic environment. Asia's Green Revolution reveals that technology itself will not resolve the issue. Significant private- and public-sector investments and continuous political engagement are required to promote crop varieties and advances in seeds and feedstock supply chain networks. All of this will boost farmers' access to better seeds and equipment, along with crop management approaches that preserve soil quality and deal with extreme climate. Modern extensions and consulting mechanisms are required to help farmers understand, not just provide supplies. Small-holder maize producers in Asia, Latin America, and Africa will find extreme weather problematic.

The significant issues for the international scientific community are (a) to generate high-quality phenotypic data in breeding programs and integrating it with advanced methods and technologies, such as high-density genotyping data, doubled haploidy, and decision support tools, for the rapid evolution of sustainable and resilient germplasm; (b) to better understand the effects of climate change on cropping systems in various territories; and (c) to closely monitor climate crisis. Genetic engineering has lowered greenhouse gas emissions, sequestered carbon, generated cleaner fuels, developed stronger crop types, and lowered synthetic fertilizers usage. These methods have increased agricultural output and protected biodiversity and ecosystems. More research is needed to establish the vulnerability reduction advantages of climate-tolerant maize on critically poor farmers, although evidence is emerging out that it increases productivity, reduces uncertainties, and increases food and nutrition security. Developing seeds firms in LatAm, Asia, and SSA must be developed to be even more market-oriented and competitive to offer small farmers inexpensive climate-tolerant enhanced seeds.

References

Abdelmajid S, Mukhtar A, Baig MB, Reed MR (2021) Climate change, agricultural policy and food security in Morocco. In: Emerging challenges to food production and security in Asia, Middle East, and Africa. Springer, Cham, pp 171–196

Abdelrahman M, Burritt DJ, Tran L-SP (2018) The use of metabolomic quantitative trait locus mapping and osmotic adjustment traits for the improvement of crop yields under environmental stresses. Seminars in Cell and Dev Biol 83:86–94

Abegunde VO, Sibanda M, Obi A (2019) The dynamics of climate change adaptation in Sub-Saharan Africa: a review of climate-smart agriculture among small-scale farmers. Climate 7(11):132

Abid M, Scheffran J, Schneider UA, Elahi E (2019) Farmer perceptions of climate change, observed trends and adaptation of agriculture in Pakistan. Environ Manag 63(1):110–123

Afzal N, Afionis S, Stringer LC, Favretto N, Sakai M, Sakai P (2021) Benefits and trade-offs of smallholder sweet potato cultivation as a pathway toward achieving the sustainable development goals. Sustainability 13(2):552

Aggarwal S, Tolani P, Gupta S, Yadav AK (2021) Posttranslational modifications in systems biology. Adv Protein Chem Struct Biol 127:93–126

Alekya K, Krishna VA, Jagadeesh K, Pavani P, Mounika M, Srujana V (2022) Chapter-7 whole genome strategies for marker assisted selection in plant breeding. J Agric Biotech 75:95

Allen-Sader C, Thurston W, Meyer M, Nure E, Bacha N, Alemayehu Y, Stutt RO, Safka D, Craig AP, Derso E (2019) An early warning system to predict and mitigate wheat rust diseases in Ethiopia. Environ Res Lett 14(11):115004

Alreshidi E (2019) Smart sustainable agriculture (SSA) solution underpinned by internet of things (IoT) and artificial intelligence (AI). arXiv preprint arXiv:190603106

Altieri MA, Nicholls CI (2017) The adaptation and mitigation potential of traditional agriculture in a changing climate. Clim Chang 140(1):33–45

Altieri MA, Nicholls CI, Henao A, Lana MA (2015) Agroecology and the design of climate change-resilient farming systems. Agron Sustain Dev 35(3):869–890

Alwang J, Gotor E, Thiele G, Hareau G, Jaleta M, Chamberlin J (2019) Pathways from research on improved staple crop germplasm to poverty reduction for smallholder farmers. Agric Syst 172:16–27

Amritha P, Shah JM (2021) Can genetic engineering-based methods for gene function identification be eclipsed by genome editing in plants? A comparison of methodologies. Mol Genet Genomics 296(3):485–500

Andorf C, Beavis WD, Hufford M, Smith S, Suza WP, Wang K, Woodhouse M, Yu J, Lübberstedt T (2019) Technological advances in maize breeding: past, present and future. Theor Appl Genet 132(3):817–849

Ansari WA, Chandanshive SU, Bhatt V, Nadaf AB, Vats S, Katara JL, Sonah H, Deshmukh R (2020) Genome editing in cereals: approaches, applications and challenges. Int J Mol Sci 21(11):4040

Appiah DO, Guodaar L (2022) Smallholder farmers' perceptions and knowledge on climate variability and perceived effects in vulnerable rural communities in the Offinso Municipality, Ghana. Environ Dev 42:100691

Arunrat N, Sereenonchai S, Chaowiwat W, Wang C (2022) Climate change impact on major crop yield and water footprint under CMIP6 climate projections in repeated drought and flood areas in Thailand. Sci Total Environ 807:150741
Aryal JP, Sapkota TB, Khurana R, Khatri-Chhetri A, Rahut DB, Jat ML (2020) Climate change and agriculture in South Asia: adaptation options in smallholder production systems. Environ Dev Sustain 22(6):5045–5075
Ashraf A, Abd Rahman F, Abdullah N (2018) Poultry feed in Malaysia: an insight into the Halalan Toyyiban issues. In: Proceedings of the 3rd International Halal Conference (INHAC 2016). Springer, pp 511–531
Bailey R, Wellesley L (2017) Chokepoints and vulnerabilities in global food trade. Chatham House, London
Balafoutis A, Beck B, Fountas S, Vangeyte J, Van der Wal T, Soto I, Gómez-Barbero M, Barnes A, Eory V (2017) Precision agriculture technologies positively contributing to GHG emissions mitigation, farm productivity and economics. Sustainability 9(8):1339
Bationo A, Fening JO, Kwaw A (2018) Assessment of soil fertility status and integrated soil fertility management in Ghana. In: Improving the profitability, sustainability and efficiency of nutrients through site specific fertilizer recommendations in West Africa agro-ecosystems. Springer, Cham, pp 93–138
Behera BK, Rout PK, Behera S (2019) Move towards zero hunger. In: Move towards zero hunger. Springer, Singapore, pp 1–35
Bekele Y (2020) Genetic variation and association of yield and yield related traits in bread wheat (*Triticum aestivum* L.) genotypes under irrigation condition in south omo, southeren Ethiopia. Haramaya University, Harar
Ben Hassen T, El Bilali H (2022) Impacts of the Russia-Ukraine War on global food security: towards more sustainable and resilient food systems? Foods 11(15):2301
Benbi D (2018) Carbon footprint and agricultural sustainability nexus in an intensively cultivated region of Indo-Gangetic Plains. Sci Total Environ 644:611–623
Bhat JA, Deshmukh R, Zhao T, Patil G, Deokar A, Shinde S, Chaudhary J (2020) Harnessing high-throughput phenotyping and genotyping for enhanced drought tolerance in crop plants. J Biotechnol 324:248–260
Biehs R, Steinlage M, Barton O, Juhász S, Künzel J, Spies J, Shibata A, Jeggo PA, Löbrich M (2017) DNA double-strand break resection occurs during non-homologous end joining in G1 but is distinct from resection during homologous recombination. Mol Cell 65(4):671–684. e675
Blary A, Jenczewski E (2019) Manipulation of crossover frequency and distribution for plant breeding. Theor Appl Genet 132(3):575–592
Bloem S, de Pee S (2017) Developing approaches to achieve adequate nutrition among urban populations requires an understanding of urban development. Glob Food Sec 12:80–88
Bogati K, Walczak M (2022) The impact of drought stress on soil microbial community, enzyme activities and plants. Agronomy 12(1):189
Boudry P, Allal F, Aslam ML, Bargelloni L, Bean TP, Brard-Fudulea S, Brieuc MS, Calboli FC, Gilbey J, Haffray P (2021) Current status and potential of genomic selection to improve selective breeding in the main aquaculture species of International Council for the Exploration of the Sea (ICES) member countries. Aquac Res 20:100700
Brewer K, Clulow A, Sibanda M, Gokool S, Naiken V, Mabhaudhi T (2022) Predicting the chlorophyll content of maize over phenotyping as a proxy for crop health in smallholder farming systems. Remote Sens 14(3):518
Burchfield E, Matthews-Pennanen N, Schoof J, Lant C (2020) Changing yields in the Central United States under climate and technological change. Clim Chang 159(3):329–346
Cabrera-Ponce JL, Valencia-Lozano E, Trejo-Saavedra DL (2019) Genetic modifications of corn. In: Corn. Elsevier, Amsterdam, The Netherlands, pp 43–85
Campbell BM, Hansen J, Rioux J, Stirling CM, Twomlow S (2018) Urgent action to combat climate change and its impacts (SDG 13): transforming agriculture and food systems. Curr Opin Environ Sustain 34:13–20

Carter MR, Tjernström E, Toledo P (2019) Heterogeneous impact dynamics of a rural business development program in Nicaragua. J Dev Econ 138:77–98
Chaikam V, Prasanna B (2020) Doubled haploid technology for rapid and efficient maize breeding. In: Accelerated plant breeding, volume 1. Springer, pp 257–292
Chapagain D, Baarsch F, Schaeffer M, D'haen S (2020) Climate change adaptation costs in developing countries: insights from existing estimates. Clim Dev 12(10):934–942
Chen K, Wang Y, Zhang R, Zhang H, Gao C (2019) CRISPR/Cas genome editing and precision plant breeding in agriculture. Annu Rev Plant Biol 70(1):667–697
Chen S, Yao Y, Zhang Y, Fan G (2020) CRISPR system: discovery, development and off-target detection. Cell Signal 70:109577
Cirino G (2021) The impact of climate change events on asset classes: an empirical analysis to extract a climate change risk premium
Combs JG, Jaskiewicz P, Rau SB, Agrawal R (2021) Inheriting the legacy but not the business: when and where do family nonsuccessors become entrepreneurial? J Small Bus Manag 61(4):1–30
Cortés AJ, López-Hernández F (2021) Harnessing crop wild diversity for climate change adaptation. Genes 12(5):783
Cox KL, Meng F, Wilkins KE, Li F, Wang P, Booher NJ, Carpenter SC, Chen L-Q, Zheng H, Gao X (2017) TAL effector driven induction of a SWEET gene confers susceptibility to bacterial blight of cotton. Nat Commun 8(1):1–14
Das RR, Vinayan MT, Seetharam K, Patel M, Phagna RK, Singh SB, Shahi JP, Sarma A, Barua NS, Babu R (2021) Genetic gains with genomic versus phenotypic selection for drought and waterlogging tolerance in tropical maize (*Zea mays* L.). Crop J 9(6):1438–1448
Day L, Cakebread JA, Loveday SM (2021) Food proteins from animals and plants: differences in the nutritional and functional properties. Trends Food Sci Technol 119:428–442
Dempewolf H, Baute G, Anderson J, Kilian B, Smith C, Guarino L (2017) Past and future use of wild relatives in crop breeding. Crop Sci 57(3):1070–1082
Dilawari R, Kaur N, Priyadarshi N, Kumar B, Abdelmotelb K, Lal SK, Singh B, Tripathi A, Aggarwal S, Jat BS (2021) Genome editing: a tool from the vault of science for engineering climate-resilient cereals. In: Harsh environment and plant resilience. Springer, Cham, pp 45–72
Donovan J, Rutsaert P, Domínguez C, Peña M (2022) Capacities of local maize seed enterprises in Mexico: implications for seed systems development. Food Secur 14(2):509–529
Dowswell CR, Paliwal RL, Cantrell RP (2019) Maize in the third world. CRC press, Boca Raton
Edriss V, Gao Y, Zhang X, Jumbo MB, Makumbi D, Olsen MS, Crossa J, Packard KC, Jannink JL (2017) Genomic prediction in a large African maize population. Crop Sci 57(5):2361–2371
Eid A, Mahfouz MM (2016) Genome editing: the road of CRISPR/Cas9 from bench to clinic. Exp Mol Med 48(10):e265–e265. https://doi.org/10.1038/emm.2016.111
Elisante F (2020) Assessing the potentials of agricultural ecosystem pollination services to improve bean yield in smallholder farming systems
Éva C, Oszvald M, Tamás L (2019) Current and possible approaches for improving photosynthetic efficiency. Plant Sci 280:433–440
Favre F, Jourda C, Besse P, Charron C (2021) Genotyping-by-sequencing technology in plant taxonomy and phylogeny. In: Molecular plant taxonomy. Springer, New York, pp 167–178
Fischer R, Connor D (2018) Issues for cropping and agricultural science in the next 20 years. Field Crops Res 222:121–142
Fitting E (2021) 6. Importing corn, exporting labor: the neoliberal corn regime, GMOs, and the erosion of Mexican biodiversity. In: Food for the few. University of Texas Press, Austin, pp 135–158
Foster AD, Rosenzweig MR (2022) Are there too many farms in the world? labor market transaction costs, machine capacities, and optimal farm size. J Polit Econ 130(3):636–680
Fuglie K, Gautam M, Goyal A, Maloney WF (2019) Harvesting prosperity: technology and productivity growth in agriculture. World Bank Publications, Washington, DC

Fukuda A, Kondo K, Ikka T, Takai T, Tanabata T, Yamamoto T (2018) A novel QTL associated with rice canopy temperature difference affects stomatal conductance and leaf photosynthesis. Breed Sci 68:305–315. https://doi.org/10.1270/jsbbs.17129

Gao C (2021) Genome engineering for crop improvement and future agriculture. Cell 184(6):1621–1635

Gautam Y (2017) Seasonal migration and livelihood resilience in the face of climate change in Nepal. Mt Res Dev 37(4):436–445

Gedil M, Menkir A (2019) An integrated molecular and conventional breeding scheme for enhancing genetic gain in maize in Africa. Front Plant Sci 10:1430

Giller KE, Andersson J, Delaune T, Silva JV, Descheemaeker K, van de Ven G, Schut AG, van Wijk M, Hammond J, Hochman Z (2022) IFAD Research Series 83: the future of farming: who will produce our food?

Girma G, Nida H, Tirfessa A, Lule D, Bejiga T, Seyoum A, Mekonen M, Nega A, Dessalegn K, Birhanu C (2020) A comprehensive phenotypic and genomic characterization of Ethiopian sorghum germplasm defines core collection and reveals rich genetic potential in adaptive traits. Plant Genome 13(3):e20055

Glick BR, Patten CL (2022) Molecular biotechnology: principles and applications of recombinant DNA. John Wily & Sons, NW, Washington, USA

Gosnell H, Gill N, Voyer M (2019) Transformational adaptation on the farm: processes of change and persistence in transitions to 'climate-smart' regenerative agriculture. Glob Environ Change 59:101965

Grote U, Fasse A, Nguyen TT, Erenstein O (2021) Food security and the dynamics of wheat and maize value chains in Africa and Asia. Front Sustain Food Syst 4:617009

Gupta SK, Shukla P (2017) Gene editing for cell engineering: trends and applications. Crit Rev Biotechnol 37(5):672–684

Gupta PK, Kulwal PL, Jaiswal V (2019) Association mapping in plants in the post-GWAS genomics era. Adv Genet 104:75–154

Gurgul A, Miksza-Cybulska A, Szmatoła T, Jasielczuk I, Piestrzyńska-Kajtoch A, Fornal A, Semik-Gurgul E, Bugno-Poniewierska M (2019) Genotyping-by-sequencing performance in selected livestock species. Genomics 111(2):186–195

Hafez EM, Osman HS, Gowayed SM, Okasha SA, Omara AE-D, Sami R, Abd El-Monem AM, Abd El-Razek UA (2021) Minimizing the adversely impacts of water deficit and soil salinity on maize growth and productivity in response to the application of plant growth-promoting rhizobacteria and silica nanoparticles. Agronomy 11(4):676

Harms NE, Knight IA, Pratt PD, Reddy AM, Mukherjee A, Gong P, Coetzee J, Raghu S, Diaz R (2021) Climate mismatch between introduced biological control agents and their invasive host plants: improving biological control of tropical weeds in temperate regions. Insects 12(6):549

Hasegawa T, Fujimori S, Havlík P, Valin H, Bodirsky BL, Doelman JC, Fellmann T, Kyle P, Koopman JF, Lotze-Campen H (2018) Risk of increased food insecurity under stringent global climate change mitigation policy. Nat Clim Chang 8(8):699–703

Hayton B (2020) Vietnam: rising dragon. Yale University Press, London

Huanca-Mamani W, Arias-Carrasco R, Cárdenas-Ninasivincha S, Rojas-Herrera M, Sepúlveda-Hermosilla G, Caris-Maldonado JC, Bastías E, Maracaja-Coutinho V (2018) Long non-coding RNAs responsive to salt and boron stress in the hyper-arid Lluteno maize from Atacama Desert. Genes 9(3):170

Islam SMF, Karim Z (2019) World's demand for food and water: the consequences of climate change. In: Desalination-challenges and opportunities, London, UK. IntechOpen, pp 57–84

Jada KS, van den Berg M (2022) Identifying effective message-framing techniques in behavior change communication for healthy diets: an experimental study of promoting biofortified maize adoption in Ethiopia. Appetite 178:106263

Jayne T, Haggblade S, Minot N, Rashid S (2019) Agricultural commercialization, rural transformation and poverty reduction: what have we learned about how to achieve this? Gates Open Res 3(678):678

Jiang S, Cheng Q, Yan J, Fu R, Wang X (2020) Genome optimization for improvement of maize breeding. Theor Appl Genet 133(5):1491–1502

Jiri O, Mafongoya PL, Chivenge P (2017) Building climate change resilience through adaptation in smallholder farming systems in semi-arid Zimbabwe. Int J Clim Change Strateg Manag 9:151–165

Jovanovic N, Pereira L, Paredes P, Pôças I, Cantore V, Todorovic M (2020) A review of strategies, methods and technologies to reduce non-beneficial consumptive water use on farms considering the FAO56 methods. Agric Water Manag 239:106267

Kariyawasam T (2021) Genetics and cell biology of Chlamydomonas zygote development. University of British Columbia, Vancouver

Katsenios N, Sparangis P, Chanioti S, Giannoglou M, Leonidakis D, Christopoulos MV, Katsaros G, Efthimiadou A (2021) Genotype× Environment interaction of yield and grain quality traits of maize hybrids in Greece. Agronomy 11(2):357

Kausch AP, Wang K, Kaeppler HF, Gordon-Kamm W (2021) Maize transformation: history, progress, and perspectives. Mol Breed 41(6):1–36

Khaki S, Khalilzadeh Z, Wang L (2019) Classification of crop tolerance to heat and drought—a deep convolutional neural networks approach. Agronomy 9(12):833

Khan MA, Akhtar MS (2015) Agricultural adaptation and climate change policy for crop production in Africa. In: Hakeem KR (ed) Crop production and global environmental issues. Springer International Publishing, Cham, pp 437–541

Khan MMH, Rafii MY, Ramlee SI, Jusoh M, Al-Mamun M (2021) Bambara groundnut (*Vigna subterranea* L. Verdc): a crop for the new millennium, its genetic diversity, and improvements to mitigate future food and nutritional challenges. Sustainability 13(10):5530

Khatun S (2019) Australianama: The South Asian Odyssey in Australia. Oxford University Press

Kinfe H, Tesfaye A (2018) Yield performance and adoption of released sorghum varieties in Ethiopia. Edelweiss Appl Sci Technol 2(1):46–55

Ko Y-S, Kim JW, Lee JA, Han T, Kim GB, Park JE, Lee SY (2020) Tools and strategies of systems metabolic engineering for the development of microbial cell factories for chemical production. Chem Soc Rev 49(14):4615–4636

Koopmans ME, Rogge E, Mettepenningen E, Knickel K, Šūmane S (2018) The role of multi-actor governance in aligning farm modernization and sustainable rural development. J Rural Stud 59:252–262

Krell N, Morgan B, Gower D, Caylor K (2021) Consequences of dryland maize planting decisions under increased seasonal rainfall variability. Water Resour Res 57(9):e2020WR029362

Kumari S, Saini S, Dhania G (2022) Biotechnological approaches for mitigation and adaptation of climate change. In: Plant stress mitigators. Springer, Singapore, pp 369–385

Leitner S, Pelster DE, Werner C, Merbold L, Baggs EM, Mapanda F, Butterbach-Bahl K (2020) Closing maize yield gaps in sub-Saharan Africa will boost soil N2O emissions. Curr Opin Environ Sustain 47:95–105

Leong K-YB, Chan Y-H, Abdullah WMANW, Lim S-HE, Lai K-S (2018) The CRISPR/Cas9 system for crop improvement: progress and prospects. Next Gener Plant Breed 129:129–145

Li J, Li Y, Ma L (2019) CRISPR/Cas9-based genome editing and its applications for functional genomic analyses in plants. Small Methods 3(3):1800473

Lino CA, Harper JC, Carney JP, Timlin JA (2018) Delivering CRISPR: a review of the challenges and approaches. Drug Deliv 25(1):1234–1257

Lu X, Chen J, Zheng M, Guo J, Qi J, Chen Y, Miao S, Zheng B (2019) Effect of high-intensity ultrasound irradiation on the stability and structural features of coconut-grain milk composite systems utilizing maize kernels and starch with different amylose contents. Ultrason Sonochem 55:135–148

Luan X, Vico G (2021) Canopy temperature and heat stress are increased by compound high air temperature and water stress and reduced by irrigation – a modeling analysis. Hydrol Earth Syst Sci 25(3):1411–1423

Makkar H (2018) Feed demand landscape and implications of food-not feed strategy for food security and climate change. Animal 12(8):1744–1754

Malenica N, Dunić JA, Vukadinović L, Cesar V, Šimić D (2021) Genetic approaches to enhance multiple stress tolerance in maize. Genes 12(11):1760

Malhi GS, Kaur M, Kaushik P (2021) Impact of climate change on agriculture and its mitigation strategies: a review. Sustainability 13(3):1318

Maqbool MA, Beshir A, Khokhar ES (2020) Doubled haploids in maize: development, deployment, and challenges. Crop Sci 60(6):2815–2840

Maqbool MA, Beshir Issa A, Khokhar ES (2021) Quality protein maize (QPM): importance, genetics, timeline of different events, breeding strategies and varietal adoption. Plant Breed 140(3):375–399

Marewo MK (2020) Fast track land reform and belonging: examining linkages between resettlement areas and communal areas in Zvimba District, Zimbabwe

Martey E, Etwire PM, Kuwornu JK (2020) Economic impacts of smallholder farmers' adoption of drought-tolerant maize varieties. Land Use Policy 94:104524

Masten AS (2018) Resilience theory and research on children and families: past, present, and promise. J Fam Theory Rev 10(1):12–31

Matres JM, Hilscher J, Datta A, Armario-Nájera V, Baysal C, He W, Huang X, Zhu C, Valizadeh-Kamran R, Trijatmiko KR, Capell T, Christou P, Stoger E, Slamet-Loedin IH (2021) Genome editing in cereal crops: an overview. Transgenic Res 30(4):461–498. https://doi.org/10.1007/s11248-021-00259-6

Mbanyele V, Mtambanengwe F, Nezomba H, Groot JC, Mapfumo P (2021) Combinations of in-field moisture conservation and soil fertility management reduce effect of intra-seasonal dry spells on maize under semi-arid conditions. Field Crops Res 270:108218

Mercure J-F, Paim M-A, Bocquillon P, Lindner S, Salas P, Martinelli P, Berchin II, de Andrade GJ, Derani C, de Albuquerque Junior C (2019) System complexity and policy integration challenges: the Brazilian energy-water-food nexus. Renew Sust Energ Rev 105:230–243

Mojid M (2020) Climate change-induced challenges to sustainable development in Bangladesh. In: IOP conference series: earth and environmental science, vol 1. IOP Publishing, p 012001

Mottaleb KA, Kruseman G, Erenstein O (2018) Evolving food consumption patterns of rural and urban households in developing countries: a Bangladesh case. Br Food J 120(2):392–408

Muluneh A (2020) Impact of climate change on soil water balance, maize production, and potential adaptation measures in the Rift Valley drylands of Ethiopia. J Arid Environ 179:104195

Mulungu K, Ng'ombe JN (2019) Climate change impacts on sustainable maize production in Sub-Saharan Africa: a review. In: Maize Prod Use. IntechOpen

Munawar S, Mustafa G, Khan MS, Joyia FA (2020) Role of biotechnology in climate resilient agriculture. In: Environment, climate, plant and vegetation growth. Springer, Cham, pp 339–365

Munaweera T, Jayawardana N, Rajaratnam R, Dissanayake N (2022) Modern plant biotechnology as a strategy in addressing climate change and attaining food security. Agric Food Secur 11(1):1–28

Mustafa M, Mayes S, Massawe F (2019) Crop diversification through a wider use of underutilised crops: a strategy to ensure food and nutrition security in the face of climate change. In: Sustainable solutions for food security. Springer, pp 125–149

Nayak AK, Rahman MM, Naidu R, Dhal B, Swain CK, Nayak AD, Tripathi R, Shahid M, Islam MR, Pathak H (2019) Current and emerging methodologies for estimating carbon sequestration in agricultural soils: a review. Sci Total Environ 665:890–912

Nazir H, Louis C, Jose S, Prakash J, Muthuswamy N, Buan ME, Flox C, Chavan S, Shi X, Kauranen P (2020) Is the H2 economy realizable in the foreseeable future? Part I: H2 production methods. Int J Hydrog Energy 45(27):13777–13788

Nelimor C, Badu-Apraku B, Tetteh AY, Garcia-Oliveira AL, N'guetta AS-P (2020) Assessing the potential of extra-early maturing landraces for improving tolerance to drought, heat, and both combined stresses in maize. Agronomy 10(3):318

Nepolean T, Kaul J, Mukri G, Mittal S (2018) Genomics-enabled next-generation breeding approaches for developing system-specific drought tolerant hybrids in maize. Front Plant Sci 9:361

Ngurumwa JP (2016) Contribution of smallholder maize production towards household food security in Babati District. Mzumbe University

Niazian M, Niedbała G (2020) Machine learning for plant breeding and biotechnology. Agriculture 10(10):436

Nyasimi M, Kimeli P, Sayula G, Radeny M, Kinyangi J, Mungai C (2017) Adoption and dissemination pathways for climate-smart agriculture technologies and practices for climate-resilient livelihoods in Lushoto, Northeast Tanzania. Climate 5(3):63

Ochami FA (2021) Effect of spacing and number of plants per hill on growth and yield of Sc duma 43 in the coastal lowlands. KeMU

Oh Y, Kim H, Kim S-G (2021) Virus-induced plant genome editing. Curr Opin Plant Biol 60:101992

Oladosu Y, Rafii MY, Abdullah N, Magaji U, Miah G, Hussin G, Ramli A (2017) Genotype× Environment interaction and stability analyses of yield and yield components of established and mutant rice genotypes tested in multiple locations in Malaysia. Acta Agric Scand B Soil Plant Sci 67(7):590–606

Oliva R, Ji C, Atienza-Grande G, Huguet-Tapia JC, Perez-Quintero A, Li T, Eom J-S, Li C, Nguyen H, Liu B (2019) Broad-spectrum resistance to bacterial blight in rice using genome editing. Nat Biotechnol 37(11):1344–1350

Oliveira GLT, McKay B, Plank C (2017) How biofuel policies backfire: misguided goals, inefficient mechanisms, and political-ecological blind spots. Energy Policy 108:765–775

Osumba JJ, Recha JW, Oroma GW (2021) Transforming agricultural extension service delivery through innovative bottom–up climate-resilient agribusiness farmer field schools. Sustainability 13(7):3938

Panchasara H, Samrat NH, Islam N (2021) Greenhouse gas emissions trends and mitigation measures in australian agriculture sector—A review. Agriculture 11(2):85

Paschon DE, Lussier S, Wangzor T, Xia DF, Li PW, Hinkley SJ, Scarlott NA, Lam SC, Waite AJ, Truong LN (2019) Diversifying the structure of zinc finger nucleases for high-precision genome editing. Nat Commun 10(1):1–12

Patel SK, Sharma A, Singh GS (2020) Traditional agricultural practices in India: an approach for environmental sustainability and food security. Energy Ecol Environ 5(4):253–271

Pichler M, Bhan M, Gingrich S (2021) The social and ecological costs of reforestation. Territorialization and industrialization of land use accompany forest transitions in Southeast Asia. Land Use Policy 101:105180

Prasanna B, Nair SK, Babu R, Gowda M, Zhang X, Xu Y, Olsen M, Chaikam V, Cairns JE, Zaman-Allah M (2020) Increasing genetic gains in maize in stress-prone environments of the tropics. In: Genomic designing of climate-smart cereal crops. Springer, Amsterdam, pp 97–132

Prasanna BM, Cairns JE, Zaidi P, Beyene Y, Makumbi D, Gowda M, Magorokosho C, Zaman-Allah M, Olsen M, Das A (2021) Beat the stress: breeding for climate resilience in maize for the tropical rainfed environments. Theor Appl Genet 134(6):1729–1752

Purushothaman S, Patil S (2019) Agrarian change and urbanization in Southern India. India studies in business and economics. Springer

Qamar S, Tantray AY, Bashir SS, Zaid A, Wani SH (2020) Golden rice: genetic engineering, promises, present status and future prospects. In: Rice research for quality improvement: genomics and genetic engineering. Springer, Singapore, pp 581–604

Raffini F, Bertorelle G, Biello R, D'Urso G, Russo D, Bosso L (2020) From nucleotides to satellite imagery: approaches to identify and manage the invasive pathogen Xylella fastidiosa and its insect vectors in Europe. Sustainability 12(11):4508

Ranganatha HM, Lohithaswa HC, Pandravada A (2021) Mapping and validation of major quantitative trait loci for resistance to Northern corn leaf blight along with the determination of the relationship between resistances to multiple foliar pathogens of maize (*Zea mays* L.). Front Genet 11:548407

Razzaq A, Saleem F, Kanwal M, Mustafa G, Yousaf S, Imran Arshad HM, Hameed MK, Khan MS, Joyia FA (2019) Modern trends in plant genome editing: an inclusive review of the CRISPR/Cas9 toolbox. Int J Mol Sci 20(16):4045
Reardon T, Echeverria R, Berdegué J, Minten B, Liverpool-Tasie S, Tschirley D, Zilberman D (2019) Rapid transformation of food systems in developing regions: highlighting the role of agricultural research and innovations. Agric Syst 172:47–59
Rigden A, Mueller N, Holbrook N, Pillai N, Huybers P (2020) Combined influence of soil moisture and atmospheric evaporative demand is important for accurately predicting US maize yields. Nat Food 1(2):127–133
Rodríguez JA, Giménez Thomsen C, Arenas D, Pagell M (2016) NGOs' initiatives to enhance social sustainability in the supply chain: poverty alleviation through supplier development programs. J Supply Chain Manag 52(3):83–108
Rolla AL, Nuñez MN, Guevara ER, Meira SG, Rodriguez GR, de Zárate MIO (2018) Climate impacts on crop yields in Central Argentina. Adaptation strategies. Agric Syst 160:44–59
Rolnick D, Donti PL, Kaack LH, Kochanski K, Lacoste A, Sankaran K, Ross AS, Milojevic-Dupont N, Jaques N, Waldman-Brown A (2022) Tackling climate change with machine learning. ACM Comput Surv 55(2):1–96
Romay MC (2018) Rapid, affordable, and scalable genotyping for germplasm exploration in maize. In: The maize genome. Springer Nature, NY, United States, pp 31–46
Roorkiwal M, Bharadwaj C, Barmukh R, Dixit GP, Thudi M, Gaur PM, Chaturvedi SK, Fikre A, Hamwieh A, Kumar S (2020) Integrating genomics for chickpea improvement: achievements and opportunities. Theor Appl Genet 133(5):1703–1720
Ross-Tremblay P (2020) Thou Shalt Forget: indigenous sovereignty, resistance and the production of cultural oblivion in Canada. University of London Press, London
Rothschild J (2020) Ethical considerations of gene editing and genetic selection. Gen Fam Med 21(3):37–47
Ryngajłło M, Jędrzejczak-Krzepkowska M, Kubiak K, Ludwicka K, Bielecki S (2020) Towards control of cellulose biosynthesis by Komagataeibacter using systems-level and strain engineering strategies: current progress and perspectives. Appl Microbiol Biotechnol 104(15):6565–6585
Sahoo S, Sanadya SK (2020) Climate smart plant breeding. Agriculture & Food: e-Newsletter 2:192–194
Saifaldeen M, Al-Ansari DE, Ramotar D, Aouida M (2020) CRISPR FokI dead Cas9 system: principles and applications in genome engineering. Cell 9(11):2518
Sanchez PA (2019) Properties and management of soils in the tropics. Cambridge University Press, Cambridge
Sanjana Reddy P (2017) Sorghum, *sorghum bicolor* (L.) Moench. Millets and sorghum: biology and genetic improvement. Wiley, Hoboken, pp 1–48
Savy D, Cozzolino V (2022) Novel fertilising products from lignin and its derivatives to enhance plant development and increase the sustainability of crop production. J Clean Prod 366:132832
Scott MF, Ladejobi O, Amer S, Bentley AR, Biernaskie J, Boden SA, Clark M, Dell'Acqua M, Dixon LE, Filippi CV (2020) Multi-parent populations in crops: a toolbox integrating genomics and genetic mapping with breeding. Heredity 125(6):396–416
Searchinger T, Waite R, Hanson C, Ranganathan J, Dumas P, Matthews E, Klirs C (2019) Creating a sustainable food future: a menu of solutions to feed nearly 10 billion people by 2050. Final report. WRI
Sehgal D, Khan T (2020) Plant tissue culture: beyond being a tool for genetic engineering. In: Environmental microbiology and biotechnology. Springer Nature, NY, United States, pp 175–200
Shahzad A, Ullah S, Dar AA, Sardar MF, Mehmood T, Tufail MA, Shakoor A, Haris M (2021) Nexus on climate change: agriculture and possible solution to cope future climate change stresses. Environ Sci Pollut Res 28(12):14211–14232
Shawa HC (2019) Influence of environmental conditions on nutritional value of quality protein maize. University of the Free State

Sheikh FA, Dar ZA, Sofi P, Lone AA (2017) Recent advances in breeding for abiotic stress (drought) tolerance in maize. Int J Curr Microbiol App Sci 6(4):2226–2243
Simtowe F, Marenya P, Amondo E, Worku M, Rahut DB, Erenstein O (2019) Heterogeneous seed access and information exposure: implications for the adoption of drought-tolerant maize varieties in Uganda. Agric Food Econ 7(1):1–23
Singh AK (2017) Discovery and role of molecular markers involved in gene mapping, molecular breeding, and genetic diversity. In: Plant bioinformatics, Springer Nature, NY, United States. Springer, pp 303–328
Singh P (2019) Feeding 1.7 billion. In: Foundation day and 26th general body meeting presidential address, National Academy of Agricultural Sciences NASC, DPS Marg, New Delhi, India, 5th June, 2019.
Singh B (2020) Cowpea: the food legume of the 21st century, vol 164. Wiley
Singh D, Chaudhary P, Taunk J, Singh CK, Singh D, Tomar RSS, Aski M, Konjengbam NS, Raje RS, Singh S (2021) Fab advances in Fabaceae for abiotic stress resilience: from 'omics' to artificial intelligence. Int J Mol Sci 22(19):10535
Singh A, Ramakrishna G, Kaila T, Saxena S, Sharma S, Gaikwad AB, Abdin M, Gaikwad K (2022) Next-generation sequencing technologies: approaches and applications for crop improvement. In: Genomics of cereal crops. Springer, New York, pp 31–94
Srivastav AL, Dhyani R, Ranjan M, Madhav S, Sillanpää M (2021) Climate-resilient strategies for sustainable management of water resources and agriculture. Environ Sci Pollut Res 28(31):41576–41595
Sun L, Ke F, Nie Z, Wang P, Xu J (2019) Citrus genetic engineering for disease resistance: past, present and future. Int J Mol Sci 20(21):5256
Suprasanna P, Saddhe A, Ghuge SA, Ingle KP (2021) New and novel genetic tools for improving crops. CABI Rev 16:1–15
Tan TL, Zulkifli NA, Zaman ASK, binti Jusoh M, Yaapar MN, Rashid SA (2021) Impact of photoluminescent carbon quantum dots on photosynthesis efficiency of rice and corn crops. Plant Physiol Biochem 162:737–751
Tanumihardjo SA, McCulley L, Roh R, Lopez-Ridaura S, Palacios-Rojas N, Gunaratna NS (2020) Maize agro-food systems to ensure food and nutrition security in reference to the sustainable development goals. Glob Food Sec 25:100327
Tastan C, Yaşar S, Tanyolaç M, Turgut K, Şireli U, Atak C, Haliloğlu K, Benlioğlu K, Taşkın K, Barlas N (2020) CRISPR-of-things: applications and challenges of the most popular gene editing tool in the fields of health, agriculture and environment. Int J Innov Approach Sci Res 4(4)
Trachsel S, Dhliwayo T, Gonzalez Perez L, Mendoza Lugo JA, Trachsel M (2019) Estimation of physiological genomic estimated breeding values (PGEBV) combining full hyperspectral and marker data across environments for grain yield under combined heat and drought stress in tropical maize (*Zea mays* L.). PLoS One 14(3):e0212200
Tran QH, Bui NH, Kappel C, Dau NTN, Nguyen LT, Tran TT, Khanh TD, Trung KH, Lenhard M, Vi SL (2020) Mapping-by-sequencing via MutMap identifies a mutation in ZmCLE7 underlying fasciation in a newly developed EMS mutant population in an elite tropical maize inbred. Genes 11(3):281
Tripathi AD, Mishra R, Maurya KK, Singh RB, Wilson DW (2019a) Estimates for world population and global food availability for global health. In: The role of functional food security in global health, Academic Press, Massachusetts, USA. Elsevier, pp 3–24
Tripathi L, Ntui VO, Tripathi JN (2019b) Application of genetic modification and genome editing for developing climate-smart banana. Food Energy Secur 8(4):e00168
Turner GM, Larsen KA, Candy S, Ogilvy S, Ananthapavan J, Moodie M, James SW, Friel S, Ryan CJ, Lawrence MA (2018) Squandering Australia's food security—The environmental and economic costs of our unhealthy diet and the policy Path We're On. J Clean Prod 195:1581–1599
Tyagi S, Kumar R, Kumar V, Won SY, Shukla P (2021) Engineering disease resistant plants through CRISPR-Cas9 technology. GM Crops Food 12(1):125–144

Uffelmann E, Huang QQ, Munung NS, de Vries J, Okada Y, Martin AR, Martin HC, Lappalainen T, Posthuma D (2021) Genome-wide association studies. Nat Rev Dis Primers 1(1):59. https://doi.org/10.1038/s43586-021-00056-9

Uliana Trentin H, Frei UK, Lübberstedt T (2020) Breeding maize maternal haploid inducers. Plants 9(5):614

Ussiri D, Lal R (2018) Nitrogen cycling and dynamics in terrestrial ecosystems. Soil and climate. CRC Press, Boca Raton, pp 183–218

Vats S, Kumawat S, Kumar V, Patil GB, Joshi T, Sonah H, Sharma TR, Deshmukh R (2019) Genome editing in plants: exploration of technological advancements and challenges. Cell 8(11):1386

Verma KK, Song X-P, Budeguer F, Nikpay A, Enrique R, Singh M, Zhang B-Q, Wu J-M, Li Y-R (2022) Genetic engineering: an efficient approach to mitigating biotic and abiotic stresses in sugarcane cultivation. Plant Signal Behav 17(1):2108253

Vivek B, Krishna GK, Vengadessan V, Babu R, Zaidi P, Kha LQ, Mandal S, Grudloyma P, Takalkar S, Krothapalli K (2017) Use of genomic estimated breeding values results in rapid genetic gains for drought tolerance in maize. Plant Genome 10(1):plantgenome2016.2007.0070

Volk GM, Byrne PF, Coyne CJ, Flint-Garcia S, Reeves PA, Richards C (2021) Integrating genomic and phenomic approaches to support plant genetic resources conservation and use. Plants 10(11):2260

Wada N, Ueta R, Osakabe Y, Osakabe K (2020) Precision genome editing in plants: state-of-the-art in CRISPR/Cas9-based genome engineering. BMC Plant Biol 20(1):1–12

Wade CM, Baker JS, Van Houtven G, Cai Y, Lord B, Castellanos E, Leiva B, Fuentes G, Alfaro G, Kondash A (2022) Opportunities and spatial hotspots for irrigation expansion in Guatemala to support development goals in the food-energy-water nexus. Agric Water Manag 267:107608

Wally D (2021) Exploring the application of ICTs and big data analytics on climate data in climate-smart agriculture to increase productivity for small-scale farmers: the case of Ghana. PhD dissertation, Aalborg University

Wangila DS (2016) Resistance management of western corn rootworm, Diabrotica virgifera virgifera LeConte, to corn traits in Nebraska. The University of Nebraska-Lincoln, Lincoln

Wani SH, Samantara K, Razzaq A, Kakani G, Kumar P (2022) Back to the wild: mining maize (Zea mays L.) disease resistance using advanced breeding tools. Mol Biol Rep 49:1–17

Wilson EO, Lovejoy TE, Hannah LJ (2019) Biodiversity and climate change. Yale University Press, London

Withers PJ, Rodrigues M, Soltangheisi A, De Carvalho TS, Guilherme LR, Benites VM, Gatiboni LC, De Sousa DM, Nunes RS, Rosolem CA (2018) Transitions to sustainable management of phosphorus in Brazilian agriculture. Sci Rep 8(1):1–13

Wolter F, Schindele P, Puchta H (2019) Plant breeding at the speed of light: the power of CRISPR/Cas to generate directed genetic diversity at multiple sites. BMC Plant Biol 19(1):1–8

Worku M, De Groote H, Munyua B, Makumbi D, Owino F, Crossa J, Beyene Y, Mugo S, Jumbo M, Asea G (2020) On-farm performance and farmers' participatory assessment of new stress-tolerant maize hybrids in Eastern Africa. Field Crops Res 246:107693

Wu W, Ma BL, Whalen JK (2018) Enhancing rapeseed tolerance to heat and drought stresses in a changing climate: perspectives for stress adaptation from root system architecture. Adv Agron 151:87–157

Xia Z, Zhang S, Wang Q, Zhang G, Fu Y, Lu H (2021) Effects of root zone warming on maize seedling growth and photosynthetic characteristics under different phosphorus levels. Front Plant Sci 12:746152

Xu Y, Li P, Zou C, Lu Y, Xie C, Zhang X, Prasanna BM, Olsen MS (2017) Enhancing genetic gain in the era of molecular breeding. J Exp Bot 68(11):2641–2666

Yahaya MA, Shimelis H (2022) Drought stress in sorghum: mitigation strategies, breeding methods and technologies—a review. J Agron Crop Sci 208(2):127–142

Yang C, Fraga H, van Ieperen W, Trindade H, Santos JA (2019) Effects of climate change and adaptation options on winter wheat yield under rainfed Mediterranean conditions in southern Portugal. Clim Chang 154(1):159–178

Yerlikaya BA, Ömezli S, Aydoğan N (2020) Climate change forecasting and modeling for the year of 2050. In: Environment, climate, plant and vegetation growth. Springer Nature, NY, United States, pp 109–122

Zarei A, Razban V, Hosseini SE, Tabei SMB (2019) Creating cell and animal models of human disease by genome editing using CRISPR/Cas9. J Gene Med 21(4):e3082

Zess E, Begemann M (2021) CRISPR-Cas9 and beyond: what's next in plant genome engineering. In Vitro Cell Dev Biol 57(4):584–594

Zhang Y, Massel K, Godwin ID, Gao C (2018) Applications and potential of genome editing in crop improvement. Genome Biol 19(1):1–11

Zhang D, Zhang Z, Unver T, Zhang B (2021) CRISPR/Cas: a powerful tool for gene function study and crop improvement. J Adv Res 29:207–221

Zoetbrood A (2022) Private sector involvement in climate adaptation: studying the role and level of engagement of the private sector in public-private partnerships for food security in the global south and its justice implications

Zougmoré RB, Partey ST, Ouédraogo M, Torquebiau E, Campbell BM (2018) Facing climate variability in sub-Saharan Africa: analysis of climate-smart agriculture opportunities to manage climate-related risks. Cah Agric 27(3):1

Chapter 9
Climate Resilience Technologies for Wheat Production

Zahoor Ahmad, Ejaz Ahmad Waraich, Celaleddin Barutçular, Aiman Hina, Asim Abbasi, Muhammad Mohiuddin, Inzamam Ul Haq, Muhammad Ashar Ayub, and Sundas Sana

Abstract Climate change has greatly influenced overall agricultural production. Pakistan is an agricultural economy, and agriculture is the source of income for the majority of the population. Unfortunately, agriculture is most sensitive to climatic irregularities because drought, salinity, and heat stress are major causes of yield decline in both major and minor crops. This chapter provides useful insights regarding strategically utilization of the most recent research findings to speed up the development of wheat genotypes having resistance against climatic irregularities. The recommended strategies discussed in this chapter develop a link between translational research and breeding strategies. This chapter also addresses research gaps which together are anticipated to enhance wheat yield particularly under heat and

Z. Ahmad (✉)
Department of Botany, Constituent Punjab College, University of Central Punjab, Bahawalpur, Pakistan

E. A. Waraich
Department of Agronomy, University of Agriculture Faisalabad, Faisalabad, Pakistan

C. Barutçular
Department of Field Crops, Faculty of Agriculture, Cukurova University, Adana, Turkey

A. Hina
Ministry of Agriculture (MOA), National Centre for Soybean Improvement, State Key Laboratory for Crop Genetics and Germplasm Enhancement, Nanjing Agricultural University, Nanjing, China

A. Abbasi · M. Mohiuddin
Department of Environmental Sciences, Kohsar University Murree, Murree, Pakistan

I. U. Haq
College of Plant Protection, Gansu Agricultural University, Lanzhou, China

M. A. Ayub
Institute of Agro-Industry and Environment, The Islamia University of Bahawalpur, Bahawalpur, Punjab, Pakistan

S. Sana
Department of Botany, Sub Campus Rahim Yar Khan, The Islamia University of Bahawalpur, Bahawalpur, Pakistan

M. Hasanuzzaman (ed.), *Climate-Resilient Agriculture, Vol 2*,
https://doi.org/10.1007/978-3-031-37428-9_9

drought stress. In addition to meet the demand of local growers, breeders must ensure the release of improved varieties having wider adaptation to support the efficacy of selection program. The genotype × environment interactions (GEI) are directly or indirectly linked to the existing biotic and abiotic stresses, soil nourishment status, agronomic practices as well as genetic composition of any crop. In order to enhance the efficiency of different breeding strategies, understanding the basis of GEI is prerequisite. Furthermore, modeling techniques that take into account environmental and genomic data could be used to locate genetic locus underlying stability across multiple locations and abiotic stress response. In addition, the technique of crop simulations will help to understand the fundamentals of GEIs and could be useful in predicting morphological characterization and growth trend in wheat genotypes when projected to various stressed of varying intensity. It is also possible to transfer genes between artificially hybridized and bread wheat by the backcrossing breeding method. By backcrossing with high-yielding wheat cultivars, the breeder could obtain an upraise of 10–40% in yield of synthetic wheat even in drought and stress conditions. By using recombinant inbred lines (RILS) and near-isogenic lines (NILS), 1RS translocation lines of rye have demonstrated high levels of tolerance and greater biomass production under heat and stress conditions. Similar to gene editing, gene cloning is also not very common in developing improved genotypes with better adaptation options against stress. Additionally, the use and regulation of genome edit technologies such as CRISPR/Cas9 system are still in ambiguity which has made it difficult to integrate such technologies into breeding program with an international focus.

Keywords Climate resilience · Wheat · Genomic technologies · Drought · Heat

1 Introduction

Climate change poses a significant threat to wheat production, as it leads to unpredictable weather patterns and increased incidence of extreme weather events such as droughts and floods. Different climate variables have varying impacts on different crops and regions. Recent years have seen an expansion of grain-sown area and increase in grain production due to increased heat caused by climate change (Pickson et al. 2020; Yang et al. 2015). However, high temperatures can negate the positive effects of increased rainfall and CO_2 concentrations on crop production in certain areas (Malhi et al. 2021; Dai et al. 2018). Climate change also negatively impacts grain production by increasing the occurrence of pests and diseases, shortening crop growth cycles, and increasing the frequency of extreme weather events (Wang et al. 2021; Chen et al. 2018).

To mitigate these impacts and ensure the continued productivity of wheat crops, farmers are turning to a range of climate resilience technologies for ensuring food security and sustainable agricultural practices in the face of increasing climate variability and extreme weather events. These technologies include a range of genetic, agronomic, and management practices that can help farmers adapt to changing conditions and protect their crops from the effects of drought, heat, and extreme weather (Bei et al. 2022; Belton et al. 2021; Jiang et al. 2020).

One technology that can help to increase the resilience of wheat crops to drought is the use of drought-tolerant wheat varieties. These varieties have been bred to have deeper roots, better water-use efficiency, and improved tolerance to high temperatures (Bapela et al. 2022). Crop rotation is another farming practice that can help to increase the resilience of wheat crops to the impacts of climate change. By rotating crops, farmers can improve soil health, reduce pest and disease pressures, and increase overall crop yields (Jalli et al. 2021). Climate-smart farming practices such as conservation agriculture, agroforestry, and the use of weather and climate information in crop management can also help to increase the resilience of wheat production to climate change. By adopting these practices, farmers can ensure that their crops are better able to withstand the impacts of extreme weather events and continue to produce high-quality wheat. Technologies such as precision irrigation and precision fertilization can help farmers use water and fertilizer more efficiently, which can be especially important in regions that are becoming drier or more water-stressed due to climate change. For example, a study found that precision irrigation led to higher wheat yields and reduced water use in a region that was experiencing drought (Djanaguiraman et al. 2020).

Climate-resilient wheat breeding using precision breeding, genome editing, and gene editing can also help to breed more resilient wheat varieties that can adapt to the changing climate. This chapter highlights these technologies and discusses their role to ensure that wheat crops are better able to withstand the impacts of extreme weather events and continue to produce high-quality wheat (Gul et al. 2022; Abbas 2022; Zhang et al. 2022). These practices are essential for ensuring food security and sustainable agricultural practices in the face of increasing climate variability and extreme weather events.

2 Climate Change Trends and Their Impact on Agricultural Productivity

Climate consists of complex structures with spatiotemporal dynamics. Naturally, the earth's sphere constantly remains in a variable and unstable state, experiencing various configurational dimensions which cause irregularities in climatic conditions. It is stated that the existing climate is different compared to the Mesozoic era and is continuously undergoing changes (Mac et al. 1998). Intergovernmental panel on climate change (IPCC) endorsed the side effects of climate change, namely, temperature fluctuations, increased water flow from ice caps, and glaciers causing them to melt faster and unexpected rise in sea levels around the globe (IPCC 2007). This whole scenario has become a burning issue of every debate; therefore, several think tanks and people from different schools of thought are sitting together to find applicable and universal preventive measures which can be adopted globally (IPCC 2007).

Studies showed that climatic changes may be caused due to natural or man-made factors, e.g., severe rainfall, deforestation, and extinction of natural flora and fauna due to urbanization. Sadly, all these factors are slowly damaging the natural ecosystem, destroying the natural habitat of animals, adding poisonous gases to human breath, and leading to an unhygienic place for living (Singh 2007; IPCC 2007). According to the facts stated by several studies, excessive emission of toxic gases from fuel consumption has damaged and reduced the greenhouse effect which is not only risking human lives but also impacting the national economy and food security.

Pakistan is an agricultural economy, and agriculture is the source of income for the majority of the population. Unfortunately, agriculture is most sensitive to climatic irregularities because drought, salinity, and heat stress are major causes of yield decline in both major and minor crops (Deschenes and Greenstone 2007; Timmins 2006; Schlenker and Roberts 2006; Ashenfelter and Storchmann 2006). Therefore, underdeveloped nations are more vulnerable to the losses and damages caused by climatic aberrations, 8–11% more compared to developed nations. Furthermore, this devastating situation has also increased the cases of depression, theft, and anxiety among such nations (Alteiri and Nicholls 2017; Lesk et al. 2016).

Climate change brings lots of inevitable disasters, among them food insecurity is top-ranked. Among the 17 Sustainable Development Goals (SDG), the first and second goals focus on no poverty and zero hunger, respectively. Considering these two goals, it is predicted to achieve zero hunger by the end of 2030 which seems impossible to attain under the current circumstances of climate change as to statistics around 815 million population are suffering from malnutrition (Richardson et al. 2018).

The agricultural sector always endures heavy production losses, with each degree rise in temperate. Data have shown an instant decline in the yield of both commercial and staple crops after a strong heat wave, flooding, or prolonged drought (Ito et al. 2018). The overall crop production is estimated to decline at a much faster rate than before because the temperature jumps from 2.6 °C to reached 4 °C (Rogelj et al. 2016). Moreover, the existing cropping pattern followed by the farming community is not enough strong to cope with the seventies of climate (Reckling et al. 2018).

Currently, the changes in rainfall patterns, temperature extremes, and increased insect pest infestation are paving the way for maximum crop production loss which will result in scarcity of basic food items (Dhanker and Foyer 2018; Campbell et al. 2016; Kang et al. 2009). These fluctuations in temperature and expected decline in crop yield are leading toward the onset of World War IV which will be on the issue of food security. Therefore, improving and securing food grain should be the fundamental objective of future studies. This chapter discusses and highlights the damages caused by climate change and tried to suggest mitigation strategies to cope with the calamities caused by climate irregularities.

2.1 Temperature

Temperature elevations are one of the best methods to explain climate change because temperature fluctuations or extremes affect almost aspect of human lives as well as the national economy of every country (Rasul et al. 2011). For example, during the plant growth cycle, temperature balance is very critical as it influences the majority of the important growing and reproductive metabolisms. In the case of wheat, rice, sorghum, barley, and oats crops, the flowering stage is very critical. Under stress conditions, slight temperature changes can cause delayed flowerings or permanent sterility (Barnabás et al. 2008; Winkel et al. 1997; Saini and Aspenal 1982). Similar studies reported about a 35–75% decrease in grain setting because of water scarcity (Sheoran and Saini 1996; Saini and Aspenal 1981). If the temperature rises to more than 35 °C, it causes hindrance in the process of photosynthesis (Griffin et al. 2004). In maize, heat stress is reported to impact antioxidant activity (Gong et al. 1997). According to a report, in Mexico, where *Zea mays* is the leading crop, it is predicted that in near future, overall crop production will decline by up to 25.7%, among them maize would face more yield losses (Hellin et al. 2014). Moreover, when both drought and heat stress are combined in cereal crops, it causes more yield loss and destruction (Wang and Huang 2004).

2.2 Drought

The quality of the grain may alter significantly as a result of the water availability; the changes vary depending on whether there is too much or not enough water. Drought-affected plants have a lower photosynthetic activity which has an impact on grain production (Jiang et al. 2009; Ahmad et al. 2022; Bukhari et al. 2022). Proteins must build up in the grain to produce high-quality grain; however, the availability of water, a suitable plant-growing environment, and nutrient conditions have a significant impact on this process (Rodrigues and Teixeira 2010; Ahmad et al. 2021). According to Kobayashi et al. (2018), over a 115-day cycle, wheat used an average of 347.2 mm of water or around 3.02 mm daily. The consumption was 0.70 mm day^{-1} during the establishment phase, 0.93 mm per day during the tillering phase, 2.21 mm day^{-1} during the booting phase, 3.74 mm per day during the blooming phase, and 2.12 mm day^{-1} during the grain maturation phase. Moreover, the critical times for water stress are the end of the tillering phase, the commencement of stem elongation, throughout head development, and the commencement of the blooming period (Sharma and Singh 2022; Yimere and Assefa 2022). Moreover, indeed, some tillers will not produce spikes if they experience water stress during the stem elongation period. Water scarcity during the milking stage of kernel growth also affects the yield of the wheat plant. Conversely, the vegetative development, which is often most impacted by the interruption of irrigation in the early heading stage, productivity was more vulnerable to stress during the emergence of the head

(Moreira and Cardoso 2009). In their study of water stress in four stages of development (tillering, booting, grain filling, and physiological maturity), Zulkiffal et al. (2021) found that water stress decreased grain yield by 22.7% during the tillering stage, 41.6% during booting stage, and 9.1% during grain filling stage. As a result, the impacts of low water availability vary depending on the plant's phenological stage, the length of the water stress, and its severity.

2.3 Rainfall

Current climatic changes such as droughts, floods, cold and heat waves, and abrupt rainfall patterns significantly affect the output of field crops particularly those grown in areas (Olayide et al. 2016; Rasul et al. 2002). Among these climatic variations, rainfall is vital ecological factors that significantly affect crop yield, growth, and normal development. Mostly, smallholder farmers are the ones which are greatly affected by the uncertain rainfall patterns (Mar et al. 2018; Ndamani and Watanabe 2015). Rainfall causes soil erosion which depletes soil nutrients and ultimately hinders crop growth and yield (Zike 2019). Preceding crop patterns and future rainfall sequences significantly diminish wheat leaf area index, plant height, crop growth, and root colonizing arbuscular mycorrhizal fungi (AMF) (Tataw et al. 2016). The decline in seasonal rainfall is usually associated with decreased soil humidity which exerts pressure on soil moisture resulting in decreased plant growth and yield. Moreover, population and damage potential of certain crop pathogens are also affected with soil and air moisture contents. An upsurge in soil and air moisture is usually correlated with increased rainfalls which provide a conducive environment for growth of certain disease-causing pathogens (Coakley et al. 1999).

3 Climate-Resilient Technologies for Enhancing Wheat Growth under Stresses

The recommended strategies discussed in this chapter develop a link between translational research and breeding strategies. This chapter will also address the existing research gaps which together are anticipated to enhance wheat yield particularly under different stresses, e.g., drought and heat (Reynolds et al. 2021).

3.1 Crop Design Targets by Using De Novo Genome Assembly

Breeders must ensure that their varieties offer enough wider adaptation to support the investment and scope of their selection program while also ensuring that they fulfill the local needs of growers. Hence, plant breeders must find ways to maintain a balance between crop yield and its adaption stability for a particular set of given ecological conditions, while ensuring that new crop cultivars should have the ability to withstand against multi environments. Hence, the process of crop improvement is very delicate and requires a lot of expertise, along with few difficulties. On top, selecting a suitable parent is a very crucial step in all breeding programs especially under multiple environments where environment interactions (GEIs) are an important factor to consider as well. The genotype × environment interactions (GEI) are directly or indirectly linked to the existing biotic and abiotic stresses, soil nourishment status, agronomic practices as well as genetic composition of any crop. In order to enhance the efficiency of different breeding strategies, understanding the basis of GEI is prerequisite (Reynolds et al. 2021).

3.2 Screening of Stress-Resistant Germplasm to Refine Breeding Targets

Although major portion of wheat cultivated area is under severe threat of drought and heat, various wheat breeding programs still lack efficient screening protocols of wheat cultivars and usually utilize quite general drought and heat stress criteria (Braun et al. 2010). Moreover, various breeding programs are vulnerable to global climatic changing patterns. During breeding schemes, some important factors cannot be ignored such as soil nutritional shortages or toxic effects, and biotic stressors, because if not correctly detected, it may compromise the genetic gain and understanding of molecular and biochemical mechanism of stress adaption (Mathews et al. 2011; Bagci et al. 2007).

As a result, elucidating the various stress profiles to which wheat must respond such as acquiring morphological and adaptive features and improved yield of a specific area. Screening of significant genotypes depicting resistance against drought and heat under severely stress affected areas along with diverse range of ecological conditions (Opare et al. 2018; Ramírez-Villegas et al. 2011). A few characteristics with maximum response toward GEIs could be used to draw information about the combined effect of phenological and physiological dataset as well as weather forecasts, soil type, and required agronomic practices could also be predicted (Reynolds et al. 2004). This method allows the identification of molecular markers that are related to tolerance to certain stress profiles, as well as the primary consequences of high temperatures and/or drought (Messmer et al. 2009). Selective trialing may go further in terms of exact phenotypic expression, critical physiological features, and accurate environmental characterization; however, it is only possible to do so at a limited range of locations based on the resources available. However, the results of this research may be utilized to improve the methods for data acquisition, which will allow the fundamental issues of GEI to be solved on a much broader scale.

3.3 Prediction of Phenological Wheat Growth under Drought and Heat Conditions

Genotype × environment interactions have strong impact on plant development and number of grains produced (Reynolds et al. 2020). The technique of crop simulations will help to understand the fundamentals of GEIs and could be useful in predicting morphological characterization and growth trend in wheat genotypes when projected to various stressed of varying intensity (Wallach et al. 2021). In crops, extreme temperature always influences the heading stage, grain filling duration whereas unavailability of water causes decline in nutrient intake of crop and eventually affects crop growth which is detrimental to crop production. Therefore, use of global girded crop models (GGCM) combining the crop and climate simulation models is suggested to lessen the crop production losses. The consortium was developed between three members, namely, University of Florida (UF), International Food Policy Research Institute (IFPRI), and CIMMYT. This consortium established the three global gridded crop models (GGCM) in wheat (NWheat, CROPSIMCERES, and CROPSIM) with the idea to increase the spatial modeling capability of wheat to assess any climatic irregularities (Pequeno et al. 2021; Hernandez-Ochoa et al. 2019; Gbegbelegbe et al. 2017). Similarly, another crop simulation technique, namely, the mink system is also reported to use in agriculture (Robertson 2017). Recently, CIMMYT's high-performance computer clusters have also been using this technology to execute analyses of time period between 1980 and 2010 to calculate the net worth of these 30 years and at the same time to predict future climatic conditions during 2040–2070 (Pequeno et al. 2021; Lopes et al. 2018; Asseng et al. 2002). It is hoped that the use of these simulation models would help in better defining of future breeding targets by considering the environmental factors, crop development, and growth directions and important introgression of important traits that contribute adaptation and stability.

3.4 Introgression of Climate-Resilient Genetic Materials from Landraces

The hexaploid wheat developed by hybridizing *Triticum durum* (AABB) × *Aegilops tauschii* (DD) has demonstrated the ability to survive under the extreme environmental condition which was contributed from the wild diploid ancestor *A. tauschii* (Zhang et al. 2018; Elbashir et al. 2017; Sohail et al. 2011; Trethowan and Mujeeb-Kazi 2008; Chevre et al. 1989; Kihara and Lilienfeld 1949). Therefore, without interfering with naturally occurring meiotic division, it is also possible to transfer genes between artificially hybridized and bread wheat by the backcrossing breeding method. By backcrossing with high-yielding wheat cultivars, the breeder can obtain a raise of 10–40% in yield of synthetic wheat even in drought and stress conditions (Cossani and Reynolds 2015; Lopes and Reynolds 2011; Narasimhamoorthy et al.

2006; del Blanco et al. 2001). To date, almost 85 synthetically developed wheat have been approved for general cultivation and are growing on more than 6% of field area in India (Aberkane et al. 2020). Wild relatives of bread wheat are blessed with the indispensable rich source of elite genetic materials conferring resistance against biotic and abiotic stresses and for improved yield. For instance, the transfer of the *Thinopyrum 7E* gene increases the 13% wheat yield (Reynolds et al. 2001).

Similarly, the introgression of *A. ventricosa* and 1RS translocation Rye is the most successful examples of working collections in CIMMYT during the early 1980s (Juliana et al. 2019; Sharma et al. 2009; Braun et al. 1998; Singh et al. 1998). By using recombinant inbred lines (RILS) and near-isogenic lines (NILS), 1RS translocation lines demonstrated high levels of tolerance and greater biomass production under heat and stress conditions (Sharma et al. 2018; Pinto and Reynolds 2015; Hoffmann 2008; Zarco-Hernandez et al. 2005; Ehdaie et al. 2003; Villareal et al. 1995; Schlegel and Meinel 1994; Ludlow and Muchow 1990). Conversely, 1RS was reported to decrease grain production under water deficit conditions which show the influence of genotype × environment interactions (Tahmasebi et al. 2015; Peake et al. 2011). Several translocation lines have been created overall, but because of their history and agronomically undesirable traits, they have not been used in wheat breeding (Hao et al. 2020; Kishii 2019; Friebe et al. 1996).

Almost 10% of the wild relatives that have been captured are likely to have been employed in interspecific crossing also, few of them are examined to assess genetic variation linked to a trait that may increase yield or climate resistance (Hao et al. 2020; Kishii 2019; Friebe et al. 1996). Any *Aegilops* species can be used to create new amphidiploids, which are diploid hybrids to discover novel sources of resistance in addition to creating synthetic hexaploid kinds of wheat. Additionally, cytogenetics that is currently being expedited by the marker is a very efficient and uncontroversial method for transferring genes between related species to create superior lines technology (King et al. 2017).

3.5 Application of Phenomics for Selection of Elite Parents

The foundation of plant breeding is phenotyping, and the effectiveness of using genomic technologies is based on how well and how pertinently phenotyping is done. Rigid phenotyping must support genetic and physiological understanding to speed genetic gain, particularly using new genetic material into new crossing schemes (Reynolds et al. 2020; Molero et al. 2019; Rebetzke et al. 2018). Current breeding stock is the most readily available and rich source of genetically diverse germplasm for multiple stress-tolerant/resistant/adaptive characteristics. It is interesting to note that choosing parents among advanced breeding lines does not yet typically involve comprehensive physiological or genetic dissection (Rai et al. 2018; Varshney et al. 2018; Crain et al. 2018). Hence, advances in field phenotyping have made it possible to choose adaptable features at a breeding level and have eliminated inherent biases associated with dependence on spatial evaluations.

Although these qualities can contribute to breeding for certain conditions, they are also prone to genotype × environment interactions (GEI), namely, crop yield (Reynolds and Langridge 2016; Richards 2006). It is obvious from the significant efforts of corporate and public sector breeders of adapting high-throughput phenotyping (HTP), for effective and accurate selection (Roitsch et al. 2019; Gaffney et al. 2015).

Use of distant remote sensing technologies may help to morphologically characterize crops at reproductive stages in an eco-friendly manner which could help to increase the effectiveness of HTP (Araus and Kefauver 2018; Araus and Cairns 2014). Moreover, high throughput is also considering for precise and accurate phenotyping of few traits which are difficult to estimate via HTP (Reynolds et al. 2020; Molero et al. 2019; Reynolds and Langridge 2016; Richards et al. 2010). Field phenotyping technologies have proliferated recently and yet only a few numbers have been demonstrated to be effective before and during breeding procedures (Araus and Kefauver 2018; Araus and Cairns 2014). Using various spectral reflectance indices, remote sensing may be used to quantify the expression of attributes conferring stress adaptability, namely, earliness, vigor, biomass production, pigmentation, and water intake capacity of plants.

In addition to performing geographically and temporal growth analyses, few new methods, namely, low altitude RGB (red, green, and blue) images have been employed to calculate vegetation cover and phenological assessment. Hence, more precise phenotyping protocols must be established to discriminate between pertinent traits during different phases of breeding schemes and translational research. These protocols must take into account considering other factors as well such as day time, crop development stage weather, and environmental conditions while collecting data along with mode of cultivation (Reynolds et al. 2020). Root vigor and depth play a pivotal role in dealing with heat and drought, making them essential traits. Substantial progress has been made in high-throughput phenotyping technologies that rely on imaging principles. However, it's worth noting that these technologies often demonstrate their effectiveness primarily in controlled environments. Excavating roots for visual inspection or measuring plant DNA extracted from soil samples remain the only effective screening method under field conditions, although they have a low processing capacity (Pinto and Reynolds 2015).

3.6 Prediction of Root Features by Developing Selection Index

Currently, in breeding populations, root capacity cannot be taken into serious consideration due to the lack of compatible screening technique with field. However, there is reason to believe that root capacity could be predicted by remote sensing methods thanks to a few examples (Pinto and Reynolds 2015; Lopes and Reynolds 2010). A strong association has been observed between canopy temperature (CT) with root mass under drought and heat stress conditions (Pinto and Reynolds 2015) which suggests the possibility to establish a screening protocol of root on the basis of remotely sensed characters, which could generate root index. This advancement could enhance the selection procedure for root characters and could be used in traditional wheat breeding schemes. In addition to CT, other features such as water index (WI), carbon isotope discrimination (CID), and oxygen isotope enrichment could be useful for extracting further necessary information (Gutierrez et al. 2010; Cabrera-Bosquet et al. 2009). Furthermore, shoot biomass and WI represent strong correlation which could facilitate effective assessment of root capacity (Babar et al. 2006).

Under different stress conditions, root capacity can also be measured by calculating WI, CT water scarcity conditions, and then simulating all these factors together with oxygen isotope enrichment and CID. This data modeling will give total root biomass, estimates of root index, and root capacity; these findings could have significant advantages for root research in general as well as other crops.

3.7 Association of Rhizosphere with Genotype to Determine Abiotic Stress

The diversification of soil microbiome is strongly influenced by the plant genotype as well as the environment (Latz et al. 2021). The genetic composition and population size of the rhizospheric microbiota are also directly influenced by genetic characteristics, e.g., root structure and composition, which vary greatly from one to another species (Sasse et al. 2018; Schweitzer et al. 2008). There is a lot of interest in determining the effect of rhizosphere microbiota in enhancing overall crop development and growth such as nutrient transport in wheat (Azarbad et al. 2018; Ahkami et al. 2017; Donn et al. 2015; Yang et al. 2009). For instance, new research indicates that plants under drought stress alter the genetics constitution of root exudates for enhanced activity of microbes. As a reward, this can produce favorable post-drought circumstances brought on by plants, such as increased nutrient availability, for plant regeneration. Therefore, it may be assumed that specific genotypes with modified exudation patterns following drought events produce a microbiota to develop better resistance to drought conditions (de Vries et al. 2019). It raises few more questions such as when used in a breeding scheme, whether rhizospheric microbiota could be influenced by the association with specific genotype under stress conditions. One study in wheat answered to such question by showing significant association between host plant and phyllosphere (Bradáová et al. 2019; Hafez et al. 2019; Sapkota et al. 2015).

3.8 Understanding of Genetic Mechanisms of Climate Resilience in Crops

For genetics studies, morphological characters such as plant height in wheat and soybean could be an under drought and heat stress conditions and provide useful dataset. These genetic mapping studies include nested association mapping (NAM) and genome-wide association studies (GWAS). Quantitative trait loci (QTL) mapping can be used to describe and compile information on both well-known and unknown genetic areas associated to heat and drought adaptation. To achieve this, genomic areas from various mapping studies are matched to pan-genomes or reference genomes of wheat that are readily available, enabling comparisons between research at the level of physical position. By reviewing these comparisons, one can then detect hotspot segments or clusters that contain markers associated with stress, indicating that these markers have a crucial and consistent role in a range of genetic backgrounds and contexts. The results of meta-analysis studies could be used to compute global P-value using summary statistics across several GWAS data (Soriano and Alvaro 2019; Acuna-Galindo et al. 2015; Griffiths et al. 2009). In single meta-analysis studies, the estimates of single nucleotide polymorphism (SNP) depicted false-positive effects by using larger sample size (Joukhadar et al. 2021; Montenegro et al. 2017; Evangelou and Ioannidis 2013). CIMMYT has already released high-quality 08 wheat lines, highly adaptable to stress, in addition to that, next-generation sequencing profiles of hundreds of important breeding lines developed by CIMMYT are now accessible for further exploration. The aforementioned cross-cutting genetic and agronomic resources will be essential ancillaries for manipulating genetic materials, enabling a novel method of identification of alleles and the search for highly valuable useful variations. The promise of epigenetics is not explored in this chapter because it will require more knowledge than is now accessible to translate (Varotto et al. 2020). Similar to gene editing, gene cloning is also not very common in developing improved genotypes with better adaptation options against stress. Additionally, the use and regulation of genome edit technologies such as CRISPR/Cas9 system are still in ambiguity which has made it difficult to integrate such technologies into breeding program with an international focus.

4 Conclusion

Climate change is bringing inevitable calamities to the earth's sphere. These changes are increasing threats to food security and national economy. In Pakistan, wheat is consumed as staple crop; therefore, it is very important not only to improve yield but also to develop climate-resilient wheat varieties. Although conventional breeding approaches have played significant role in the development of drought- and heat-resistant cultivars, still there is a dire need to take advantage of new technologies such as introgression of genes from wild species, identification of genomic spots harboring stable and elite QTLs, identification of candidate genes underlying these QTLs, and genetic manipulation of germplasm by using genome edit technologies to develop climate-smart wheat.

References

Abbas S (2022) Climate change and major crop production: evidence from Pakistan. Environ Sci Pollut Res 29:5406–5414

Aberkane H, Payne T, Kishii M, Smale M, Amri A, Jamora N (2020) Transferring diversity of goat grass to farmers' fields through the development of synthetic hexaploid wheat. Food Secur 12:1017–1033

Acuna-Galindo MA, Mason RE, Subramanian NK, Hays DB (2015) Meta-analysis of wheat QTL regions associated with adaptation to drought and heat stress. Crop Sci 55:477–492

Ahkami AH, Allen White R, Handakumbura PP, Jansson C (2017) Rhizosphere engineering: enhancing sustainable plant ecosystem productivity. Rhizosphere 3:233–243

Ahmad Z, Anjum S, Skalicky M, Waraich EA, Muhammad Sabir Tariq R, Ayub MA, Hossain A, Hassan MM, Brestic M, Sohidul Islam M, Habib-Ur-Rahman M (2021) Selenium alleviates the adverse effect of drought in oilseed crops camelina (*Camelina sativa* L.) and canola (*Brassica napus* L.). Molecules 26(6):1699

Ahmad Z, Barutçular C, Zia Ur Rehman M, Sabir Tariq RM, Afzal M, Waraich EA, Ahmad A, Iqbal MA, Bukhari MA, Ahmad K, El Sabagh A (2022) Pod shattering in canola reduced by mitigating drought stress through silicon application and molecular approaches-a review. J Plant Nutr 46(1):101–128

Altieri MA, Nicholls CI (2017) The adaptation and mitigation potential of traditional agriculture in a changing climate. Clim Change 140:33–45

Araus JL, Cairns JE (2014) Field high-throughput phenotyping: the new crop breeding frontier. Trends Plant Sci 19:52–61

Araus JL, Kefauver SC (2018) Breeding to adapt agriculture to climate change: affordable phenotyping solutions. Curr Opin Plant Biol 45:237–247

Ashenfelter O, Storchmann K (2006) Using a hedonic model of solar radiation to assess the economic effect of climate change: the case of Mosel Valley vineyards, NBER Working Paper, p 12380

Asseng S, Turner NC, Ray JD, Keating BA (2002) A simulation analysis that predicts the influence of physiological traits on the potential yield of wheat. Eur J Agron 17:123–141

Azarbad H, Constant P, Giard-Laliberté C, Bainard LD, Yergeau E (2018) Water stress history and wheat genotype modulate rhizosphere microbial response to drought. Soil Biol Biochem 126:228–236

Babar MA, Reynolds MP, van Ginkel M, Klatt AR, Raun WR, Stone ML (2006) Spectral reflectance to estimate genetic variation for in-season biomass, leaf chlorophyll, and canopy temperature in wheat. Crop Sci 46:1046–1057

Bagci SA, Ekiz H, Yilmaz A, Cakmak I (2007) Effects of zinc deficiency and drought on grain yield of field-grown wheat cultivars in Central Anatolia. J Agron Crop Sci 193:198–206

Bapela T, Shimelis H, Tsilo TJ, Mathew I (2022) Genetic improvement of wheat for drought tolerance: progress, challenges and opportunities. Plants 11:1331

Barnabás B, Jäger K, Fehér A (2008) The effect of drought and heat stress on reproductive processes in cereals. Plant Cell Environ 31:11–38

Bei G, Zhang S, Guo Y, Yanli L, Hu N, Liu J (2022) Study on meteorological disaster monitoring of field fruit industry by remote sensing data. Adv Meteorol 2022:1659053

Belton B, Win MT, Zhang X, Filipski M (2021) The rapid rise of agricultural mechanization in Myanmar. Food Policy 101:102095

Bradáčová K, Sittinger M, Tietz K, Neuhäuser B, Kandeler E, Berger N, Ludewig U, Neumann G (2019) Maize inoculation with microbial consortia: contrasting effects on rhizosphere activities, nutrient acquisition and early growth in different soils. Microorganisms 7:1–16

Braun HJ, Payne TS, Morgounov A, van Ginkel M, Rajaram S (1998) The challenge: one billion tons of wheat by 2020. In: Slinkard AE (ed) 9th international wheat genetics symposium. University Extension Press, University of Saskatchewan, Saskatoon, pp 33–40

Braun HJ, Atlin G, Payne T (2010) Multi-location testing as a tool to identify plant response to global climate change. In: Reynolds MP (ed) Climate change and crop production. CABI, Wallingford, pp 115–138
Bukhari MA, Yousaf M, Ahmad Z, Rafay M, Shah AN, Abbas A, Shah AA, Javed T, Afzal M, Ali S, Abdullah MI (2022) Enhancing drought stress tolerance in Camelina (*Camelina sativa* L.) through exogenous application of potassium. Physiol Plant 174(5):e13779
Cabrera-Bosquet L, Molero G, Nogués S, Araus JL (2009) Water and nitrogen conditions affect the relationships of Δ13C and Δ18O to gas exchange and growth in durum wheat. J Exp Bot 60:1633–1644
Campbell BM, Vermeulen SJ, Aggarwal PK, Corner-Dolloff C, Girvetz E, Loboguerrero AM, Ramirez-Villegas J, Rosenstock T, Sebastian L, Thornton PK (2016) Reducing risks to food security from climate change. Glob Food Sec 11:34–43
Chen Y, Zhang Z, Tao F (2018) Impacts of climate change and climate extremes on major crops productivity in China at a global warming of 1.5 and 2.0 C. Earth Syst Dynam 9:543–562
Chevre AM, Jahier J, Trottet M (1989) Expression of disease resistance genes in amphiploids wheats—*Triticum tauschii* (Coss.) Schmal. Cereal Res Commun 17:23–29
Coakley SM, Scherm H, Chakraborty S (1999) Climate change and plant disease. Annu Rev Phytopathol 37:399–426
Cossani CM, Reynolds MP (2015) Heat stress adaptation in elite lines derived from synthetic hexaploid wheat. Crop Sci 55:2719–2735
Crain J, Mondal S, Rutkoski J, Singh RP, Poland J (2018) Combining high-throughput phenotyping and genomic information to increase prediction and selection accuracy in wheat breeding. Plant Genome 11:170043
Dai A, Zhao T, Chen J (2018) Climate change and drought: a precipitation and evaporation perspective. Curr Clim Change Rep 4:301–312
de Vries FT, Williams A, Stringer F, Willcocks R, McEwing R, Langridge H, Straathof AL (2019) Changes in root-exudate-induced respiration reveal a novel mechanism through which drought affects ecosystem carbon cycling. New Phytol 224:132–145
del Blanco IA, Rajaram S, Kronstad WE (2001) Agronomic potential of synthetic hexaploid wheat-derived populations. Crop Sci 41:670–676
Deschenes O, Greenstone M (2007) The economic impacts of climate change: evidence from agricultural output and random fluctuations in weather. Am Econ Rev 97:354–385
Dhankher OP, Foyer CH (2018) Climate resilient crops for improving global food security and safety. Plant Cell Environ 41:877–884
Djanaguiraman M, Raine SR, Kravchenko AN (2020) Precision irrigation and crop diversification as strategies to mitigate water use and enhance grain production in a dryland cropping system. Agric Water Manag 237:104869
Donn S, Kirkegaard JA, Perera G, Richardson AE, Watt M (2015) Evolution of bacterial communities in the wheat crop rhizosphere. Environ Microbiol 17:610–621
Ehdaie B, Whitkus RW, Waines JG (2003) Root biomass, water-use efficiency, and performance of wheat–rye translocations of chromosomes 1 and 2 in spring bread wheat 'Pavon'. Crop Sci 43:710–717
Elbashir AAE, Gorafi YSA, Tahir ISA, Kim JS, Tsujimoto H (2017) Wheat multiple synthetic derivatives: a new source for heat stress tolerance adaptive traits. Breed Sci 67:248–256
Evangelou E, Ioannidis JP (2013) Meta-analysis methods for genome-wide association studies and beyond. Nature Rev Genet 14:379–389
Friebe B, Jiang J, Raupp WJ, McIntosh RA, Gill BS (1996) Characterization of wheat–alien translocations conferring resistance to diseases and pests: current status. Euphytica 91:59–87
Gaffney J, Schussler J, Löffler C, Cai W, Paszkiewicz S, Messina C, Groeteke J, Keaschall J, Cooper M (2015) Industry-scale evaluation of maize hybrids selected for increased yield in drought-stress conditions of the US corn belt. Crop Sci 55:1608–1618
Gbegbelegbe S, Cammarano D, Asseng S, Robertson R, Chung U, Adam M, Abdallah O, Payne T, Reynolds M, Sonder K, Shiferaw B, Nelson G (2017) Baseline simulation for global wheat production with CIMMYT mega-environment specific cultivars. Field Crops Res 202:122–135

Gong M, Chen SN, Song YQ, Li ZG (1997) Effect of calcium and calmodulin on intrinsic heat tolerance in relation to antioxidant systems in maize seedlings. Funct Plant Biol 24:371–379

Griffin JJ, Ranney TG, Pharr DM (2004) Heat and drought influence photosynthesis, water relations, and soluble carbohydrates of two ecotypes of redbud (*Cercis canadensis*). J Am Soc Hortic Sci 129:497–502

Griffiths S, Simmonds J, Leverington M, Wang Y, Fish L, Sayers L, Albert L, Orford S, Wingen L, Herry L, Faure S, Laurie D, Bilham L, Snape J (2009) Meta-QTL analysis of the genetic control of ear emergence in elite European winter wheat germplasm. Theor Appl Genet 119:383–395

Gul A, Xiumin W, Chandio AA, Rehman A, Siyal SA, Asare I (2022) Tracking the effect of climatic and non-climatic elements on rice production in Pakistan using the ARDL approach. Environ Sci Pollut Res 29:31886–31900

Gutierrez M, Reynolds MP, Klatt AR (2010) Association of water spectral indices with plant and soil water relations in contrasting wheat genotypes. J Exp Bot 61:3291–3303

Hafez E, Omara AED, Ahmed A (2019) The coupling effects of plant growth promoting rhizobacteria and salicylic acid on physiological modifications, yield traits, and productivity of wheat under water deficient conditions. Agronomy 9:1–16

Hao M, Zhang L, Ning S, Huang L, Yuan Z, Wu B, Yan Z, Dai S, Jiang B, Zheng Y, Liu D (2020) The resurgence of introgression breeding, as exemplified in wheat improvement. Front Plant Sci 11:252

Hellin J, Bellon MR, Hearne SJ (2014) Maize landraces and adaptation to climate change in Mexico. J Crop Improv 28:484–501

Hernandez-Ochoa IM, Pequeno DNL, Reynolds M, Babar MA, Sonder K, Milan AM, Hoogenboom G, Robertson R, Gerber S, Rowland DL, Fraisse CW, Asseng S (2019) Adapting irrigated and rainfed wheat to climate change in semi-arid environments: management, breeding options and land use change. Eur J Agron 109:125915

Hoffmann B (2008) Alteration of drought tolerance of winter wheat caused by translocation of rye chromosome segment 1RS. Cereal Res Commun 36:269–278

IPCC (2007) Climate change 2007: synthesis report, summary for policymakers: an assessment of the Intergovernmental Panel on Climate Change. (Draft prepared by Bernstein L, Bosch P, Canziani O, Chen Z, Christ R, and Davidson O). Intergovernmental Panel on Climate Change, p 22

Ito R, Vasconcelos HL, Feeley KJ (2018) Global climate change increases risk of crop yield losses and food insecurity in the tropical Andes. Glob Change Biol 24:592–602

Jalli M, Huusela E, Jalli H, Kauppi K, Niemi M, Himanen S, Jauhiainen L (2021) Effects of crop rotation on spring wheat yield and pest occurrence in different tillage systems: a multi-year experiment in Finnish growing conditions. Front Sustain Food Syst 5:647335

Jiang D, Yue H, Wollenweber B, Tan W, Mu H, Bo Y, Dai T, Jing Q, Cao W (2009) Effects of post-anthesis drought glutenin subunits and glutenin macropolymers content in wheat grain. J Agron Crop Sci 195:89–97

Jiang M, Hu X, Chunga J, Lin Z, Fei R (2020) Does the popularization of agricultural mechanization improve energy-environment performance in China's agricultural sector? J Clean Prod 276:124210

Joukhadar R, Thistlethwaite R, Trethowan R, Keeble-Gagnère G, Hayden MJ, Ullah S, Daetwyler HD (2021) Meta-analysis of genome-wide association studies reveal common loci controlling agronomic and quality traits in a wide range of normal and heat stressed environments. Theor Appl Genet. https://doi.org/10.1007/s00122-021-03809-y

Juliana P, Poland J, Huerta-Espino J, Shrestha S, Crossa J, Crespo-Herrera L, Toledo FH, Govindan V, Mondal S, Kumar U, Bhavani S, Singh PK, Randhawa MS, He X, Guzman C, Dreisigacker S, Rouse MN, Jin Y, Pérez-Rodríguez P, Montesinos-López OA, Singh D, Rahman MM, Marza FF, Singh RV (2019) Improving grain yield, stress resilience and quality of bread wheat using large-scale genomics. Nat Genet 51:1530–1539

Kang Y, Khan S, Ma X (2009) Climate change impacts on crop yield, crop water productivity and food security—a review. Prog Nat Sci 19:1665–1674

Kihara H, Lilienfeld F (1949) A new synthesized 6×–wheat. Hereditas 35:307–319
King J, Grewal S, Yang CY, Hubbart S, Scholefield D, Ashling S, Edwards KJ, Allen AM, Burridge A, Bloor C, Davassi A, da Silva GJ, Chalmers K, King IP (2017) A step change in the transfer of interspecific variation into wheat from *Amblyopyrum muticum*. Plant Biotechnol J 15:217–226
Kishii M (2019) An update of recent use of Aegilops species in wheat breeding. Front Plant Sci 10:585
Kobayashi BF, Valeriano TTB, Oliveira AF, de Santana MJ, de Morais Borges R (2018) Estimation of evapotranspiration of wheat crop grown in Uberaba, MG. Braz J Irrigated Agric 12:2923–2931
Latz MAC, Kerrn MH, Sørensen H, Collinge DB, Jensen B, Brown JKM, Madsen AM, Jørgensen HJL (2021) Succession of the fungal endophytic microbiome in wheat is dependent on tissue-specific interactions of host genotype and environment. Sci Total Environ 759:143804
Lesk C, Rowhani P, Ramankutty N (2016) Influence of extreme weather disasters on global crop production. Nature 529:84
Lopes MS, Reynolds MP (2010) Partitioning of assimilates to deeper roots is associated with cooler canopies and increased yield under drought in wheat. Funct Plant Biol 37:147–156
Lopes MS, Reynolds MP (2011) Drought adaptive traits and wide adaptation in elite lines derived from resynthesized hexaploid wheat. Crop Sci 51:1617–1626
Lopes MS, Royo C, Alvaro F, Sanchez-Garcia M, Ozer E, Ozdemir F, Karaman M, Roustaii M, Jalal-Kamali MR, Pequeno D (2018) Optimizing winter wheat resilience to climate change in rain fed crop systems of Turkey and Iran. Front Plant Sci 9:563
Ludlow MM, Muchow RC (1990) A critical evaluation of traits for improving crop yields in water-limited environments. Adv Agron 43:107–153
Mac MJ, Opler PA, Puckett Haecker CE, Doran PD (1998) Status and trends of the Nation's biological resources. U.S. Department of the Interior, U.S. Geological Survey, pp 89–93
Malhi GS, Kaur M, Kaushik P (2021) Impact of climate change on agriculture and its mitigation strategies: a review. Sustainability 13:1318
Mar S, Nomura H, Takahashi Y, Ogata K, Yabe M (2018) Impact of erratic rainfall from climate change on pulse production efficiency in Lower Myanmar. Sustainability 10:1–16
Mathews KL, Trethowan R, Milgate AW, Payne T, van Ginkel M, Crossa J, Delacy I, Cooper M, Chapman SC (2011) Indirect selection using reference and probe genotype performance in multi-environment trials. Crop Pasture Sci 62:313–327
Messmer R, Fracheboud Y, Bänziger M, Vargas M, Stamp P, Ribaut JM (2009) Drought stress and tropical maize: QTL-by-environment interactions and stability of QTLs across environments for yield components and secondary traits. Theor Appl Genet 119:913–930
Molero G, Joynson R, Pinera-Chavez FJ, Gardiner LJ, Rivera-Amado C, Hall A, Reynolds MP (2019) Elucidating the genetic basis of biomass accumulation and radiation use efficiency in spring wheat and its role in yield potential. Plant Biotechnol J 17:1276–1288
Montenegro JD, Golicz AA, Bayer PE, Hurgobin B, Lee H, Chan CKK, Visendi P, Lai K, Doležel J, Batley J, Edwards D (2017) The pangenome of hexaploid bread wheat. Plant J 90:1007–1013
Moreira DR, Cardoso VJM (2009) Water stress and productivity in *Triticum aestivum*. IAC 1003. Naturalia 32:13–19
Narasimhamoorthy B, Gill BS, Fritz AK, Nelson JC, Brown-Guedira GL (2006) Advanced backcross QTL analysis of a hard winter wheat × synthetic wheat population. Theor Appl Genet 112:787–796
Ndamani F, Watanabe T (2015) Influences of rainfall on crop production and suggestions for adaptation. Int J Agric Sci 5:2167–2447
Olayide OE, Tetteh IK, Popoola L (2016) Differential impacts of rainfall and irrigation on agricultural production in Nigeria: any lessons for climate-smart agriculture? Agric Water Manag 178:30–36
Opare P, Akintonde JO, Obeng-Ofori D, Nelson V (2018) Using climate analogue tools to explore and build smallholder farmer capacity for climate smart agriculture. AAS Open Res 1:24
Peake AS, Gilmour A, Cooper M (2011) The 1BL/1RS translocation decreases grain yield of spring wheat germplasm in low yield environments of north-eastern Australia. Crop Pasture Sci 62:276–288

Pequeno DNL, Hernandez-Ochoa IM, Reynolds MP, Sonder K, Molero-Milan A, Robertson R, Lopes MSS, Xiong W, Kropff M, Asseng S (2021) Climate impact and adaptation to heat and drought stress of regional and global wheat production. Environ Res Lett 16:54070

Pickson RB, He G, Ntiamoah EB, Li C (2020) Cereal production in the presence of climate change in China. Environ Sci Pollut Res 27:45802–45813

Pinto RS, Reynolds MP (2015) Common genetic basis for canopy temperature depression under heat and drought stress associated with optimized root distribution in bread wheat. Theor Appl Genet 128:575–585

Rai N, Bellundagi A, Kumar PKC, Thimmappa RK, Rani S, Sinha N, Krishna H, Jain N, Singh GP, Singh PK, Chand S, Prabhu KV (2018) Marker-assisted back-cross breeding for improvement of drought tolerance in bread wheat (*Triticum aestivum* L.). Plant Breed 137:514–526

Ramírez-Villegas J, Lau C, Köhler AK, Jarvis A, Arnell NP, Osborne TM, Hooker J (2011) Climate analogues: finding tomorrow's agriculture today. Working Paper 12. Copenhagen: CCAFS

Rasul G, Chaudhry QZ, Mahmood A, Hyder KW (2002) Effect of temperature rise on crop growth & productivity. Pak J Meteorol 8:53–62

Rasul G, Chaudhry QZ, Mahmood A, Hyder W (2011) Effect of temperature rise on crop growth and productivity. Pak J Meteorol 8:53–62

Rebetzke GJ, Jimenez-Berni J, Fischer RA, Deery DM, Smith DJ (2018) Review: high-throughput phenotyping to enhance the use of crop genetic resources. Plant Sci 282:40–48

Reckling M, Döring TF, Bergkvist G, Chmielewski F, Stoddard F, Watson C, Seddig S, Bachinger J (2018) Grain legume yield instability has increased over 60 years in long-term field experiments as measured by a scale-adjusted coefficient of variation. Asp Appl Biol 138:15–20

Reynolds M, Langridge P (2016) Physiological breeding. Curr Opin Plant Biol 31:162–171

Reynolds MP, Calderini DF, Condon AG, Rajaram S (2001) Physiological basis of yield gains in wheat associated with the LR19 translocation from *Agropyron elongatum*. In: Bedö Z, Láng L (eds) Wheat in a global environment. Springer, Dordrecht, pp 345–351

Reynolds MP, Trethowan R, Crosse J, Vargas M, Sayre KD (2004) Erratum to Physiological factors associated with genotype by environment interaction in wheat. Field Crops Res 85:253–274

Reynolds M, Chapman S, Crespo-Herrera L, Thimmappa RK, Rani S, Sinha N, Krishna H, Jain N, Singh GP, Singh PK, Chand S, Prabhu KV (2020) Breeder friendly phenotyping. Plant Sci 295:110396

Reynolds MP, Janet Lewis JM, Ammar K, Basnet BR, Crespo-Herrera L, Crosse J, Dhugga KS, Dreisigacker S, Juliana P, Karwat H, Kishii M, Krause MR, Langridge P, Lashari A, Mondal S, Payne T, Pequeno D, Pinto F, Sansaloni C, Schulthess U, Singh RP, Sonder K, Sukumaran S, Xiong W, Braun HJ (2021) Harnessing translational research in wheat for climate resilience. J Exp Bot 72:5134–5157

Richards RA (2006) Physiological traits used in the breeding of new cultivars for water-scarce environments. Agric Water Manag 80:197–211

Richards RA, Rebetzke GJ, Watt M, Condon AG, Spielmeyer W, Dolferus R (2010) Breeding for improved water productivity in temperate cereals: phenotyping, quantitative trait loci, markers and the selection environment. Funct Plant Biol 37:85–97

Richardson KJ, Lewis KH, Krishnamurthy PK, Kent C, Wiltshire AJ, Hanlon HM (2018) Food security outcomes under a changing climate: impacts of mitigation and adaptation on vulnerability to food insecurity. Clim Change 147:327–341

Robertson R (2017) MINK, details of a global gridded crop modeling system. Washington, DC: Int Food Res J 11:24–31

Rodrigues OR, Teixeira MCC (2010) Ecophysiological bases for maintaining wheat quality. Bottom Step: Embrapa Wheat

Rogelj J, Den Elzen M, Höhne N, Fransen T, Fekete H, Winkler H, Schaeffer R, Sha F, Riahi K, Meinshausen M (2016) Paris Agreement climate proposals need a boost to keep warming well below 2 °C. Nature 534:631

Roitsch T, Cabrera-Bosquet L, Fournier A, Ghamkhar K, Jiménez-Berni J, Pinto F, Ober ES (2019) Review: new sensors and data-driven approaches—a path to next generation phenomics. Plant Sci 282:2–10

Saini H, Aspenall D (1981) Effect of water deficit on sporogenesis in wheat (*Triticum aestivum* L.). Ann Bot 48:623–633
Saini H, Aspenall D (1982) Abnormal sporogenesis in wheat (*Triticum aestivum* L.) induced by short periods of high temperature. Ann Bot 49:835–846
Sapkota R, Knorr K, Jørgensen LN, O'Hanlon KA, Nicolaisen M (2015) Host genotype is an important determinant of the cereal phyllosphere mycobiome. New Phytol 207:1134–1144
Sasse J, Martinoia E, Northen T (2018) Feed your friends: do plant exudates shape the root microbiome? Trends Plant Sci 23:25–41
Schlegel R, Meinel A (1994) A quantitative trait locus (QTL) on chromosome arm 1RS of rye and its effect on yield performance of hexaploid wheat. Cereal Res Commun 22:7–13
Schlenker W, Roberts M (2006) Non-linear effects of weather on corn yields. Rev Agric Econ 28:391–398
Schweitzer JA, Bailey JK, Fischer DG, LeRoy CJ, Lonsdorf EV, Whitham TG, Hart SC (2008) Plant–soil microorganism interactions: heritable relationship between plant genotype and associated soil microorganisms. Ecology 89:773–781
Sharma V, Singh PK (2022) Feasibility of aqua crop model for predicting yield and biomass of wheat crop. Indian J Ecol 49:831–836
Sharma S, Bhat PR, Ehdaie B, Close TJ, Lukaszewski AJ, Waines JG (2009) Integrated genetic map and genetic analysis of a region associated with root traits on the short arm of rye chromosome 1 in bread wheat. Theor Appl Genet 119:783–793
Sharma A, Sheikh I, Kumar R, Kumar K, Vyas P, Dhaliwal HS (2018) Evaluation of end use quality and root traits in wheat cultivars associated with 1RS.1BL translocation. Euphytica 214:62
Sheoran IS, Saini HS (1996) Drought-induced male sterility in rice: changes in carbohydrate levels and enzyme activities associated with the inhibition of starch accumulation in pollen. Sex Plant Reprod 9:161–169
Singh S (2007) Climatology. Prayag Pustak Bhawan, Allahabad, pp 339–394
Singh RP, Huerta-Espino J, Rajaram S, Crossa J (1998) Agronomic effects from chromosome translocations 7DL.7AG and 1BL.1RS in spring wheat. Crop Sci 38:27–33
Sohail Q, Inoue T, Tanaka H, Eltayeb AE, Matsuoka Y, Tsujimoto H (2011) Applicability of Aegilops tauschii drought tolerance traits to breeding of hexaploid wheat. Breed Sci 61:347–357
Soriano JM, Alvaro F (2019) Discovering consensus genomic regions in wheat for root-related traits by QTL meta-analysis. Sci Rep 9:10537
Tahmasebi S, Heidari B, Pakniyat H, Dadkhodaie A (2015) Consequences of 1BL/1RS translocation on agronomic and physiological traits in wheat. Cereal Res Commun 43:554–566
Tataw JT, Baier F, Krottenthaler F, Pachler B, Schwaiger E, Wyhlidal S, Formayer H, Hosch J, Baumgarten A, Zaller JG (2016) Climate change induced rainfall patterns affect wheat productivity and agroecosystem functioning dependent on soil types. Ecol Res 31:203–212
Timmins C (2006) Endogenous land use and the Ricardian valuation of climate change. Environ Res Econ 33:119–142
Trethowan RM, Mujeeb-Kazi A (2008) Novel germplasm resources for improving environmental stress tolerance of hexaploid wheat. Crop Sci 48:1255–1265
Varotto S, Tani E, Abraham E, Krugman T, Kapazoglou A, Melzer R, Radanović A, Miladinović D (2020) Epigenetics: possible applications in climate-smart crop breeding. J Exp Bot 71:5223–5236
Varshney RK, Singh VK, Kumar A, Powell W, Sorrells ME (2018) Can genomics deliver climate-change ready crops? Curr Opin Plant Biol 45:205–211
Villareal RL, del Toro E, Mujeeb-Kazi A, Rajaram S (1995) The 1BL/1RS chromosome translocation effect on yield characteristic in a *Triticum aestivum* L. cross. Plant Breed 114:497–500
Wallach D, Palosuo T, Thorburn P, Hochman Z, Andrianasolo F, Asseng S, Basso B, Buis S, Crout N, Dumont B, Ferris R, Gaiser T, Gayler S, Hiremath S, Hoek S, Horan H, Hoogenboom G, Huang M, Jabloun M, Jansson PE, Jing Q, Justes E, Kersebaum KC, Launay M, Lewan E, Luo Q, Maestrini B, Moriondo M, Olesen JE, Padovan G, Poyda A, Priesack E, Pullens JWM, Qian B, Schütze N, Shelia V, Souissi A, Specka X, Shrivastava AK, Stella T, Streck T, Trombi G, Wallor E, Wang J, Weber TKD, Weihermüller L, de Wit A, Wöhling T, Xiao L, Zhao C, Zhu Y, Seidel SJ (2021) Multi-model evaluation of phenology prediction for wheat in Australia. Agric For Meteorol 298–299:108289

Wang Z, Huang B (2004) Physiological recovery of Kentucky bluegrass from simultaneous drought and heat stress. Crop Sci 44:1729–1736

Wang CJ, Wang R, Yu CM, Dang XP, Sun WG, Li QF, Wang XT, Wan JZ (2021) Risk assessment of insect pest expansion in alpine ecosystems under climate change. Pest Manag Sci 77:3165–3178

Winkel T, Renno JF, Payne W (1997) Effect of the timing of water deficit on growth, phenology and yield of pearl millet (*Pennisetum glaucum* (L.) R. Br.) grown in Sahelian conditions. J Exp Bot 48:1001–1009

Yang J, Kloepper JW, Ryu CM (2009) Rhizosphere bacteria help plants tolerate abiotic stress. Trends Plant Sci 14:1–4

Yang X, Chen F, Lin X, Liu Z, Zhang H, Zhao J, Li K, Ye Q, Li Y, Lv S (2015) Potential benefits of climate change for crop productivity in China. Agric For Meteorol 208:76–84

Yimere A, Assefa E (2022) Current and future irrigation water requirement and potential in the Abbay River Basin. Ethiopia Air Soil Water Res 15:11786221221097929

Zarco-Hernandez JA, Santiveri F, Michelena A, Javier Peña R (2005) Durum wheat (*Triticum turgidum*, L.) carrying the 1BL/1RS chromosomal translocation: agronomic performance and quality characteristics under Mediterranean conditions. Eur J Agron 22:33–43

Zhang D, Zhou Y, Zhao X, Lv L, Zhang C, Li J, Sun G, Li S, Song C (2018) Development and utilization of introgression lines using synthetic octaploid wheat (*Aegilops tauschii* × hexaploid wheat) as donor. Front Plant Sci 9:1113

Zhang H, Chandio AA, Yang F, Tang Y, Ankrah Twumasi M, Sargani GR (2022) Modeling the impact of climatological factors and technological revolution on soybean yield: evidence from 13-major provinces of China. Int J Environ Res Public Health 19:5708

Zike T (2019) Review on the effect of moisture or rain fall on crop production. Civil Environ Res 11:17–25

Zulkiffal M, Ahsan A, Ahmed J, Musa M, Kanwal A, Saleem M, Javaid MM (2021) Heat and drought stresses in wheat (*Triticum aestivum* L.): substantial yield losses, practical achievements, improvement approaches, and adaptive. J Plant Stress Physiol 3:12–19

Chapter 10
Improving Plant Nutrient Use Efficiency for Climate-Resilient Agriculture

Paramita Deb, Anwesha Mandal, Harendra, Suborna Roy Choudhury, Anupam Das, Swati Hazra, Arkabanee Mukherjee, Dibakar Ghosh, Shuvasish Choudhury, Subhas Chandra Santra, and Debojyoti Moulick

Abstract As per the UN prediction, the world population is to surpass 9 billion by 2050, so is the food demand by at least 70%. To meet this huge demand, the crop production must go hand in hand. It is anticipated that to meet the raised demand, the current production practices would invite anthropogenic stresses on our soil and biosphere. The leftover chemicals seep deep into the ecosystem risking human

---

Anupam Das and Debojyoti Moulick contributed equally with all other contributors.

P. Deb · A. Mandal · Harendra
GD Goenka University, Gurgaon, Haryana, India

S. R. Choudhury
Department of Agronomy, Bihar Agricultural University, Bhagalpur, Bihar, India

A. Das
Department of Soil Science and Agricultural Chemistry, Bihar Agricultural University, Bhagalpur, Bihar, India

S. Hazra
School of Agricultural Sciences, Sharda University, Greater Noida, Uttar Pradesh, India

A. Mukherjee
Indian Institute of Tropical Meteorology, Pune, Maharashtra, India

D. Ghosh
ICAR–Indian Institute of Water Management, Bhubaneswar, Odisha, India

S. Choudhury
Plant Stress Biology and Metabolomics Laboratory, Department of Life Science and Bioinformatics, Assam University Silchar, Silchar, Assam, India

S. C. Santra
Department of Environmental Science, University of Kalyani, Nadia, West Bengal, India

D. Moulick (✉)
Department of Environmental Science, University of Kalyani, Nadia, West Bengal, India

Plant Stress Biology and Metabolomics Laboratory, Department of Life Science and Bioinformatics, Assam University Silchar, Silchar, Assam, India

M. Hasanuzzaman (ed.), *Climate-Resilient Agriculture, Vol 2*,
https://doi.org/10.1007/978-3-031-37428-9_10

health and triggering environmental degradation. This is even worsened with the extreme weather events in recent years, affecting farm production and the availability of basic food grains at both the national and international levels. Apart from salinity, unpredicted severe droughts and floods hit the same region causing major problems for farmers, agricultural scientists, and extension workers. This leads to shortage of food grains and increase in price of food grains and inflation which in turn affects the farmers and poor communities the most. There is little chance to things getting better in the near future. Therefore, developing a climate-resilient agriculture (CRA) system is the need of the hour. CRA refers to combined use of natural resources in crop and livestock to attain sustainable productivity and income under varied climatic conditions. Most agricultural soils are deficient of more than one essential nutrients. Fertilisers and supplements are crucial in providing nutrients and high yields. It is estimated that performance of these fertilisers has been 50% or lower for N, less than 10% for P and 40% for K. Nutrient efficiency decreases due to leaching, run off, gas emission, and nutrient fixation in soil. To increase the nutrient efficiency, the best nutrient management strategies should be adopted, depending on the source of nutrients, crop requirements, nutrient level, placement of nutrients associated with the crop, as also climate, soil, and several other factors.

Keywords Nutrient use efficiency · Climate-resilient agriculture (CRA) · Climate change · Yield

1 Introduction

Global population has quintupled through the last century (1.6 billion in 1900, 8.0 billion in 2022) (UN 2017). There has been tremendous stress (biotic and abiotic or in combinations) on the agricultural sector to meet the sharp increase in food consumption. Recent studies estimated an increase of 110% in global food production by next three decades to meet the increased demand (Stewart et al., 2005; Tilman et al. 2011). Scientific estimates proposed an increase of 1.4% in annual agricultural production (OECD 2018), while others estimated a need of 60% more cereal production (FAO 2009) and doubling food demand (Glenn et al. 2008; Baulcombe et al. 2009; Tilman et al. 2011) by 2050 to maintain the required food demand with the proportional increase in population (Choudhury and Moulick 2022). To optimise the performance of our resource-limited planet, a significant improvement in nutrient use efficiency (NUE) coupled with better water use efficiency (WUE) is required to meet the elevated food requirement. Researches confirmed, to maintain the present level production with diverse conventional fertiliser application at a high rate for a longer period, resulted in low use efficiency, increased cost, and serious environmental issues (Wilson et al. 2008; Liu and Lal 2015).

There is evidence of 200–300% expansion in synthetic fertiliser consumption between 1970 and 2010 (FAOSTAT 2013; Smith et al. 2014). Researches confirmed almost half the world's population depends on nitrogenous fertilisers to sustain their production (Erisman et al. 2008; Smil 2002). China's increased nitrogen fertiliser production by a factor of 39 between 1963 and 2015 (Luo et al. 2018) and 46% additional urea production between 2003 and 2013 (Heffer and Prud'homme 2016) also supports the claim. An increased prediction in annual fertiliser demand (1.5% of N, 2.2% of P, and 2.4% of K) was reported by FAO between 2016 and 2020 (FAOSTAT 2017). A significant increase in fertiliser doses is required to achieve increase in global food production due to scarce water resources and limited arable lands. Heavy doses of fertiliser were applied for maintaining the current levels of grain production, resulting environmental pollution (Liu and Lal 2015). In the case of chemical fertilisers, low use efficiency is the prominent one, which not only increases the cost of production but also causes several environmental pollutions (Wilson et al. 2008). Overdose of nitrogen and phosphorus fertilisers deteriorates the groundwater quality and leads to eutrophication in aquatic ecosystems. Further, due to the low of fertiliser use efficiency of conventional fertilisers (30–35%, 18–20%, and 35–40% for N, P, and K, respectively; Subramanian et al. 2015), food production needs to be much more efficient than ever before (Chinnamuthu and Boopathi 2009).

The major threats put pressure on agriculture are climate change, environmental degradations, loss of biodiversity, urbanisation, pesticide residue build-up, and imbalanced fertiliser use. Most of the applied nutrients may not be bioavailable to plants due to various reasons, namely, leaching, photolysis, hydrolysis, fixation, and decomposition. But the maximum utilisation of applied fertilisers is crucial in the critical growth period of the crop of plants. Therefore, it is essential to reduce nutrient losses at some point of fertilisation and to boom the crop yield through exploring the latest technology. With the rising global food demand, depleting natural resources, coupled with regressing environmental conditions, attaining improved Nutrient use efficiency (NUE) and better crop yield have become difficult. Improved Nutrient use efficiency (NUE) helps plants to grow under reduced fertiliser input cost and reduced nutrient losses and increased crop yields. In many parts of the world's agricultural regions, N and P are limiting nutrients for crop production; hence, it is crucial that these nutrients are used well for cropping systems to remain economically viable. In the modern era because of high-yielding crop varieties, there is a much more demand of nutrients for plant which causes the multi nutrient deficiency in plants and as well as soil. Nutrient use efficiency (NUE) can be better explained as the quantity of nutrient the crop has uptake in a certain period of time as compared to the nutrients availability in soil or externally applied in same duration.

Finck (1992) gave propaganda about the nutrient uptake from mineral fertilisers:

(i) About 50–60% of the nitrogen (N) in mineral fertilisers is used up in the first year.

Table 10.1 Typical nutrient use efficiency levels for cereals

Measure	N	P_2O_5	K_2O	Interpretation
Partial factor productivity (kg grain/kg nutrient)	40–90	45–110	60–165	High values indicate restricted productivity due to nutrient supply; lower values indicate less responsive soils or excess fertiliser application.
Agronomic efficiency (kg grain/kg nutrient)	15–30	7–15	7–15	Less value indicates improved crop response with altered practices at less input costs.
Recovery efficiency (%)	40–65	15–25	30–50	Smaller values signify improved management can enhance efficiency or nutrient pilling in the soil.
Partial nutrient balance (kg nutrient/kg nutrient)	0.7–0.9	0.7–0.9	0.7–0.9	Lower values reflect that improved management practices might increase efficiency or soil fertility may be improving. A decline in soil fertility may be indicated by higher levels.

Source: Fixen et al. (2015)

(ii) In the first year, mineral fertiliser use P at a rate of 10–25% (on average 15%).
(iii) Over the ensuing decades, an additional 1–2% every year will be added.

About 50–60% of the K in mineral fertilisers is used up in the first year. A significant amount of the applied fertiliser is lost within a year of application. To maximise NUE, crop management and fertiliser application must be optimised. Nutrient use efficiency levels of N, P, and K for cereals crop is mentioned in Table 10.1. However, it will necessitate regulatory changes that encourage a rise in NUE at the grassroots level with emphasis on technology that can achieve higher congruence among crop consumption and N availability from all sources such as fertilisers, organic inputs, and native soil N (Cassman et al. 2002; Dobermann 2005).

Farm operations, fertiliser application in precise, should maximise nutrient use efficiency (NUE) to overcome the soil-plant competition. NUE can be explained as plant's ability in efficient absorption of nutrients from soil, their transportation, storage, and remobilisation. In recent times, new technologies and strategies have been developed to serve the age-old concern of improving NUE in crops (Dobermann 2005). These are mentioned below. The actual challenge in front of the scientific community is not only to produce enough quantity of food but also imparts desired quality traits while maintaining sustainability in near future (Moulick et al. 2016a, 2018d, 2020, 2021, 2022). In this chapter, we are going to assess the scenario of NUE in agricultural sector under the severe climatic fluctuation, operating globally.

2 Optimisation of Agronomic Practice

Soil physico-chemical properties and biological characteristics can be altered by agronomic practices that could help to improve nutrient availability under varying climatic conditions (Simpson et al. 2011). The rhizospheric site (root–soil interface) serves best for plant–soil–microorganism interactions and also acts as a control hub for nutrient transformation and plant uptake (Zhang et al. 2010). This plays critical role in increasing nutrient availability and minimising losses. Some useful management strategies to improve NUE with improved agronomic interventions or rhizospheric modification mentioned below (Wang and Shen 2019).

2.1 The 4R Nutrient Stewardship

This framework confirms nutrient application with right product, at right rate, in right time, and in right place. This signifies required nutrient application promoting growth and development on experimental basis under different conditions (right rate), ensuring nutrient availability in critical growth stages (right time); site-specific nutrient placement related to volume of roots (right placement) and using genuine fertiliser and organic resources (right source).

2.1.1 Soil Testing

Although soil testing is still one of the most effective methods for assessing the soil's ability to supply nutrients, precise calibration data are also required to be relevant for generating fertiliser recommendations.

2.2 Cereal–Legume Intercropping

This is a systemised cropping pattern to improve productivity and sustainability under different conditions with better NUE and crop productivity (Zhang et al. 2010; Ghosh et al. 2022a, b). In addition, biological processes encourage the use of symbionts, namely, rhizobacteria and mycorrhizae; on contrary, chemical processes include rhizosphere acidification and organic anion secretion (Dong et al. 2018). Furthermore, the intercropping increases available P in soil and phosphatase enzyme activity by rhizosphere acidification (Rojas dowing et al. 2017).

2.3 Using Effective Microbial Consortia

Arbuscular mycorrhizal fungi (AMF), a universally important symbiont, colonises the root of all important crops (approx. 80–90% of all known species) and behaves as "biofertilizers and bioprotectors" in regenerative agriculture (Bücking et al. 2012; Qiu and Wang 2006). AMF supports uptake and transport of vital nutrients, namely, N, P, K, S, Ca, Cu, and Zn from source (soil) to sink (host plants) using extraradical mycelium emerging from colonised roots into the soil (Soka and Ritchie

2014). AMF increases the P uptake upto 77% under low P supply as compared to 49% in high condition (Thingstrup et al. 2000). *Pseudomonas* (*Pseudomonas stutzeri*), *Bacillus* sp. (*B. flexus, B. globisporus, B. mucilaginosus, B. megaterium*), and endosymbiotic rhizobia (*Rhizobium leguminosarum, R. etli, R. yantingense, R. tropici*, etc.) are regarded as effective phosphate solubilisers (Igual et al. 2001). Phosphorus-solubilising bacteria (PSB) can also effectively increase the P fertiliser use efficiency, both organic and inorganic, using different mechanisms (Khan et al. 2013). Thus, we can conclude combining P and AMF solubilisers (Alam et al. 2002) and N fixers serve the best inoculants for enhancing nutrient use efficiency.

2.4 Regulating Soil pH

Soil pH, the determinant for nutrient solubility and availability to plants, also got consequences on plants (Saha et al. 2019; Choudhury et al. 2021a, 2022a). Most of the plant essential nutrients are available between specific pH range of 6.2 and 7.0. Any disturbance in this pH range leads to disrupted nutrient availability in soil. Application of pH regulatory agents (lime, gypsum, or both) effectively improves soil pH and Ca content. Liming, gypsum application, or mixing of both application help reducing the toxic effects of hydrogen, aluminium, and manganese, additionally improving soil biological activities, cation exchange capacity (CEC), nutrient (P, Ca, and Mg), availability, and soil structure. This also helps promoting N_2 fixation, stimulating nitrification, and decreases availability of certain nutrients (e.g. K, Mn, Zn, Fe, B, and Cu) (Fageria et al. 2008). The increased soil pH due to liming enhances microbial activity and communities resulting in rapid decomposition of soil organic matter and subsequent release of Fe and Al oxides (Takahashi and Dahlgren 2016).

2.5 Nitrification Inhibitor

Nitrogen stabilisers such as nitropyrene, dicyandiamide (DCD), n-butyl-thiophosphoric triamide (NBPT) decelerate nitrification or urease activity to reduce the fertiliser loss. The most popular choices for extensive agriculture usage are polymer-coated products that distribute nutrients under controlled environment.

2.5.1 Enhancing Fertiliser Use Efficiency Through Chromatin Modulation

NGR-5 (nitrogen-mediated tiller growth response 5 gene), which is based on PRC-2, represses the genes in rice that cause tillering and, as a result, increases the number of tillers. Nitrogen fertilisation stimulates the genome re-framing of H3K27me3 methylation. Proteasomal destruction influenced by NGR5 is a target of gibberellin-GID1. The economic yield of rice species is increased, and nitrogen fertilisation is decreased by altering the competitive linkage between NGR5, DELLA proteins, and GID1. By that way, sustainability is maintained by a rise in output and a decrease in input. P consumption at the farm level will be improved by using new upgraded technologies including precision fertilisation, polymer-coated fertilisers and nutrient recycling from domestic, agricultural, and industrial wastes (Cordell and White 2013).

2.6 Use Integrated Soil Nutrient Management Practice

Diverse management strategies have been implemented, taking into account the soil types, cropping patterns, and farmers' resources, in order to lessen the severity of soil fertility degradation. Researches revealed combine application of both organic and inorganic phosphorus betters crop yields in less fertile soils (Opala et al. 2010; Otinga et al. 2013, Noushahi et al. 2019; Ghosh et al. 2020a, b, 2022b). Best management practices (BMPs), such as application of fertiliser with amendment (lime), recommended crop rotations, accelerate organic matter content and checks erosion, pest, disease, and weeds infestation and contribute to improvement in yield and optimisation of nutrient use efficiency significantly (Fageria et al. 2008).

2.6.1 Foliar Spray

Over-fertilisation is a risk possessed by soil application of the nutrients inviting economic losses associated with degraded soil fertility. Report suggests foliar application under rainfed conditions improved P and N feeding in wheat crop maximising yield and growth, better soil fertility, efficient resource use with improved profits (Abrol et al. 2020).

2.6.2 Seed Priming

Seed priming hastens and promotes germination coupled with better plant growth, stress resistance, NUE, and WUE (Moulick et al. 2023a, b; Choudhury et al. 2022b; Moulick et al. 2019a, b, 2018c, 2017, 2016b; Sahoo et al. 2019). The experiment was designed to investigate the absorption and use of nitrogen by rice seed treatment with halo- and hydro-priming using nitrate salts in two consecutive years. It was noticed that seed priming with $Mg(NO_3)_2$, KNO_3, and distilled water consistently beat non-primed control sets in promoting seed germination and nitrate reductase enzyme activity in leaves, which promotes nitrogen accumulation in various plant parts. With the help of this seed priming methodology raise the amount of both soluble and insoluble nitrogen content in seed as well as the ability of seeds and straws to absorb nitrogen (kg ha^{-1}). Seed priming was also found beneficial in encouraging grain yield after applying kg^{-1} nitrogen in the soil. Apart from application of nitrate, priming seeds with beneficial elements/agents have shown positive effect in minimising salinity, drought, and toxic elements and in inducing phytotoxicity and accumulation in the edible part also (Saha et al. 2019; Sahoo et al. 2019; Moulick et al. 2018a, b, 2021).

2.6.3 Identifying the Superior Genotypes

To select and breed nutrient-efficient species or genotypes within a species gives the benefits of lowered fertiliser cost and less risk of contaminated soil and water. Plants have developed different physiological, morphological, biochemical, and molecular adaptive systems such as altered root structure and increased carboxylate exudation to overcome the nutrient deficiency stress (Aziz et al. 2011). The modified structures and strategies help to release or solubilise available nutrients from insoluble organic and other pools. Farmers currently require nutrient-efficient genotypes which can efficiently absorb and use nutrients. Therefore, cultivar selection is an effective strategy to deal with nutrient-deficient soil.

2.7 Modification of Root Morphology and Physiology

Plants' ability to acquire and absorb nutrients is strongly influenced by morphological characteristics such as root length, thickness, surface area, and volume (Bengough et al. 2011). These factors affect the capacity of the roots to penetrate deep into soil layers and tolerate extremes in temperature, moisture, toxicity, and element deficiency. They can also alter the rates of nitrogen uptake and the pH of the rhizosphere. The root architecture in terms of transporters, exudates, and frequently the presence of symbiotic connections such as mycorrhiza will be the primary determinant of efficient acquisition. Therefore, promoting early root establishment, high-affinity transporter systems, microbe attachment (mycorrhiza), root proliferation, and improved mechanisms for increasing nutrient bioavailability will improve NUE.

Incorporating genetic variation in root hair, root hair length in particular, probably can improve crop productivity and nutrient uptake efficiency. A deeper root with more aerenchyma tissues in the cortex can significantly contribute to efficient N uptake with lower carbon input in root growth (Postma and Lynch 2011). Additionally, this modified structure might improve water uptake efficiency and thus help to overcome drought stress and also observed in case of salinity (Hund et al. 2009; Moulick et al. 2020; Choudhury et al. 2022c). However, identifying key regulators, both sufficiently upstream and robust, for developing species with altered root growth to address nutrient deficiency is a complex and challenging process (Wissuwa et al. 2017).

2.7.1 Improving Translocation (Partitioning/Remobilisation)

Total dry matter accumulation influenced by rate of fertiliser application influences nutrient demand (uptake/utilisation) (Dobermann 2007). Improved and efficient nutrient utilisation from PGPR and (or) AMF restricts water resource contamination and also cuts off increasing fertiliser costs (Adesemoye et al. 2008). Effective nutrient remobilisation from senescent organs to fresh, emerging organs, such as immature leaves and developing seeds, is also another interesting area for crop NUE development (Stigter and Plaxton 2015). As the seed matures, the nutrients must be remobilised and translocated to the developing seed, to address the nutrient use efficiency problem (Gregersen 2011).

2.7.2 Improving Internal Utilisation

Different plant physiological systems and their reaction to elemental deficiency, tolerance, and toxicity as well as environmental factors control the nutrient uptake and use efficiency. High photosynthetic activity per unit of nutrient and more effective nutrient remobilisation from older to young leaves are typically attributed to efficient internal nutrient utilisation (Stigter and Plaxton 2015).

3 Application of Advanced Techniques

Advanced and improved techniques such as site-specific/real-time nitrogen management, slow-release/controlled-release fertiliser (SR/CRF), site-specific precision nutrient management and urease/nitrification inhibitor play significant role in reducing fertiliser loss and improving NUE (Xiang et al. 2008).

3.1 Remote Sensing

In recent times, remote sensing is gaining importance to obtain important data on precision farming. GIS (geographic information system) helps establishing the field management information system by processing, analysing, and trimming the data of soil and crops (Xiang et al. 2008).

3.2 Nanotechnology

As compared to conventional fertilisers, the presence of an essential plant nutrient element in the substrate or soil in the form of nanofertiliser allows for greater dissolving, quicker absorption, and faster assimilation by the plants. This has been shown for nutrients including N, P, K, Ca, Mg, Fe, Mn, Zn, Cu, and Mo (Ditta and Arshad 2016).

3.3 Nanofertilisers

A collection of low-cost technical solutions for improved plant development and preservation is known as nanoscience or nanotechnology. The distribution of insecticides and fertilisers in a regulated way with high site specificity and little collateral harm is made possible by the use of nanoparticles and nanocapsules (Nair et al. 2010). Unwanted leftovers are released into the environment as a result of the use of pesticides and fertilisers to speed up food production (Sekhon 2014). Due to their superior effectiveness, agronomic efficiency, and ability to remediate contaminated soil and groundwater, nanoparticles are considered to be a more desirable option to conventional sources of fertilisers. Application of numerous nanofertilisers has a stronger influence on improving crop output, reducing environmental contamination risk, and lowering fertilisation costs for crop production. NUE can therefore be improved by using nanofertilisers in crop fields. When used in the right amounts and concentrations, nanofertilisers improve crop growth and yield. However, going above a specific point can have an inhibitory effect on agricultural plants, which could result in slower crop growth and output. The science of nanotechnology has made it possible to use nanoscale or nanostructured materials as fertiliser carriers or controlled-release carriers in order to lower the expense of environmental protection (Chinnamuthu and Boopathi 2009). Nanofertilisers can be explained as the nanoparticles or nanomaterials that provide nutrition to plants or increase the efficiency of the activities of conventional fertilisers. Replacing conventional fertilisers with nanofertilisers are advantageous as they accommodate steady and controlled release of nutrients into soils, which put a stop to water pollution (Naderi and Danesh-Shahraki 2013; Singh et al. 2015). Being at a nascent stage, there are limited researches or systemic studies on the effects and benefits of micronutrient-containing nanofertiliser application under field situations (Liu and Lal 2015). Additionally, nanofertilisers can be paired with nanodevices to synchronise the release of fertiliser N and P with crop absorption, reducing undesired nutrient losses to the environment and eliminating unfavourable nutrient interactions with soil, water, and air (De Rosa et al. 2010). The increased availability of nutrients due to nanofertilisers increases the synthesis of chlorophyll, the rate of photosynthesis, the generation of dry matter, and ultimately the development of the entire plant. According to reports, the seeds treated with nano-TiO_2 resulted in plants that displayed more chlorophyll-a, more dry weight, and a higher photosynthetic rate compared to the plants made from untreated (Zheng et al. 2005). The presence of an essential plant nutrient element, namely, N, P, K, Ca, Mg, Fe, Mn, Zn, Cu, and Mo. in the substrate or soil in the form of nanofertiliser permits greater dissolving, faster absorption, and assimilation by the plants compared to standard fertilisers (Ditta and Arshad 2016). These nanofertilisers could be used in agriculture both to increase crop yield and minimising environmental pollution (Liu and Lal 2015). The nanofertilisers due to their nanosize can be used for nanocoatings, for example, sulphur nanocoating (≤ 100 nm layer), ensuring their controlled release, surface protection, and overall increasing its use efficiency (Brady and Weil 1996; Ditta et al. 2015). Nevertheless, nanofertilisers have been proved more efficient compared to the traditional fertilisers as they reduce nitrogen emissions, leaching loss, and long-term incorporation by soil microorganisms (Liu et al. 2006). The slow release helps to reduce the toxicity of over application and improves soil health (Suman et al. 2010). The biodegradable nanoencapsulated slow-release fertilisers which include urea, calcium phosphate, and potassium chloride encourage regulated release of NPK fertiliser sources (Corradini et al. 2010). Kaolin and polymeric biocompatible nanoparticles are also used for similar purposes (Wilson et al. 2008).

Nanofertilisers and nanocomposites can be utilised to boost the effectiveness of nutrient utilisation by limiting the nutrient ions from the fertiliser granules which may become fixed or lost in the environment from the fertiliser granules (Subramanian et al. 2008). Extensive studies had been undertaken to characterise nitrogen nanofertilisers (Subramanian and Rahale 2013; Manikandan and Subramanian 2014), phosphate nanofertilisers (Bansiwal et al. 2006; Adhikari 2011; Behnassi et al. 2011), potassium nanofertilisers (Subramanian and Sharmila Rahale 2012), sulphur nanofertilisers (Subramanian and Rahale 2012; Selva Preetha et al. 2014; Thirunavukkarasu 2014). Nanofertilisers could be categorised into four categories including macronutrients nanofertilisers, micronutrient nanofertilisers, nutrient-loaded nanofertilisers, and plant-growth-enhancing nanomaterials.

3.4 Macronutrient Nanofertilisers

The chemical components of macronutrient nanofertilisers, which include N, P, K, Mg, and Ca, can provide plants with these vital macronutrients. In general, these macronutrient fertilisers (primarily N and P fertilisers) are used in large quantities to increase food, fibre, and other essential commodity production. (Liu and Lal 2015). Researchers claim that micronutrient nanofertilisers might improve the bioavailability of precursory nutrients to plants under extreme conditions (Liu and Lal 2015).

3.5 Nanomaterial-Enhanced Fertilisers and Nanoparticulate Fertiliser Carriers

These nanomaterial-enhanced fertilisers, when combined with plant nutrients, can boost the plants' ability to absorb those nutrients and/or lessen the aftermath of conventional fertiliser. The most important example of this type is nutrient-augmented zeolites (Liu and Lal 2015). Extensive studies revealed that the use of different forms of nanoparticles can somewhat (and at low concentrations) improve plant development without having any essential plant nutrients (Liu and Lal 2015). Typical examples of this type include nanoparticles of SiO_2 (Roldugin et al. 2015; Asadzade et al. 2015), selenium (Zhang et al. 2015; Chen et al. 2015), copper (Jain et al. 2016a, b), iron (Delfani et al. 2014; Bakhtiari et al. 2015; Soliman et al. 2015), zinc (El-Kereti et al. 2013; Tarafdar et al. 2014; Asadzade et al. 2015; Soliman et al. 2015), phosphorus (Liu and Lal 2014; Sharonova et al. 2015), and carbon nanotubes (Yatim et al. 2015).

Nanofertilisers (particularly biological ones) enjoy the advantages of high nutrient use efficiency, reduced soil toxicity, and reduced environmental impact in addition to boosting bioavailability, decreasing soil absorption, also fixation, and increasing the of insoluble minerals in soil. Nanofertilisers may improve fertiliser effectiveness and soil nutrient absorption ratios during crop production while also conserving fertiliser resources. An excessive fertiliser release in soil could poison the ecosystem and upset its natural balance. When applied, fertilisers' nutrient supply to the soil used by plants can be extended by nanofertilisers for a longer run. Nanofertilisers can extend the time of nutrient supply of fertilisers into soil that the plants are using at the moment of delivery. And remaining nutrients are then turned into salts that are insoluble in the soil. Leaching and/or leaking of fertiliser nutrients into soil can be reduced with the use of nanostructured formulations (Cui et al. 2010; Solanki et al. 2015). Moreover, some emerging techniques such as application of computational tools, metabolomics, next-generation sequencing, potentials of plant's wild relatives have also been explored for their possible role in better understanding of plant's responses and enhancing greater NUE and imparting resilience against climate fluctuation-induced stress in crops (Choudhury et al. 2021a, b; Mazumder et al. 2022, 2021; Hossain et al. 2021a, b, 2022).

4 Impact of Different Strategies to Reduce the Greenhouse Gases Due to Agricultural Activities

Three crucial trace gases that have an impact on the climate are carbon dioxide (CO_2), methane (CH_4), and nitrous oxide (N_2O) (Fig. 10.1).

In soil microbial activity, root respiration, chemical decay processes, heterotrophic respiration of soil fauna, and fungus all contribute to the synthesis of GHGs (Chapuis-Lardy et al. 2007). According to Ludwig et al. (2001), associated emission flux rates are significantly influenced by soil water content (humidity), soil temperature, the availability of nutrients, pH value, and land cover-related factors. For microbial and plant respiratory processes, nutrient availability is a critical factor for plant respiratory process and microbial activity in soil. Soil's natural N and C content, atmospheric deposition, manure treatments, and fertiliser use all play a significant influence. Their organically rich soils are regarded as one of the major contributors to CO_2 and N_2O emissions, especially when used for agriculture

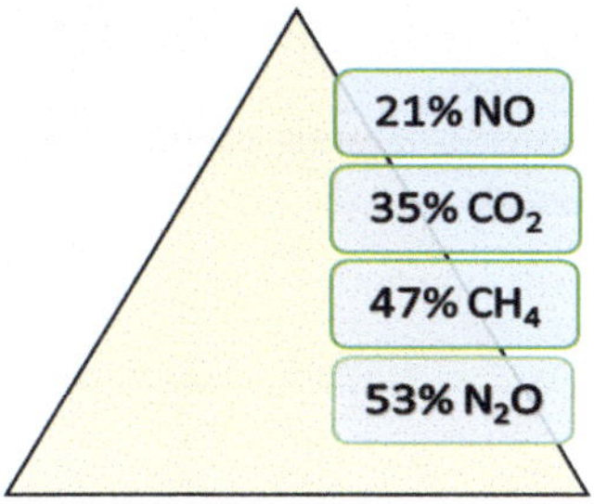

Fig. 10.1 Average greenhouse gases (GHGs) emissions from global soils (IPCC 2007)

purposes followed by drainage. High emission rates persist even after 20–30 years after agricultural activity has stopped, high emission rates persist. Also, emission rate is various fertiliser (nitrogenous fertiliser, phosphorous and potassic fertiliser) are different which is compiled in Table 10.2. According to Galloway et al. (2008), significant nitrification results in the loss of over 70% of nitrogen fertiliser, with an annual direct economic loss of about 81 million US dollars (urea costs between 0.54 and 0.80 US dollars per kilogramme) (Subbarao et al. 2012). According to estimates by 2050, the amount of nitrogen fertiliser used globally would double to 3.0108 t annually (Subbarao et al. 2012), and the amount of nitrogen lost from agricultural systems annually will increase to 6.15 t.

This will exacerbate issues with environmental pollution and raise the deficit of nitrogen fertiliser in agricultural systems. The biggest terrestrial carbon and nitrogen pools are found in the top metre of the earth's soil layer, where around 1500 Pg of total carbon and 136 (92–140) Pg of total nitrogen are stored despite their highly unequal distribution (of 3–4 orders of magnitude for carbon). This will make agricultural systems more susceptible to nitrogen fertiliser loss and make concerning environmental issues worse (Batjes 1996; Kutsch et al. 2009; Nieder and Benbi 2008; Schaufler et al. 2010; Schlesinger and Andrews 2000). The fluxes (flow rate per unit area) of CO_2, CH_4, and N_2O between 2000 and 2005 were nearly neutral. According to an analysis of the European terrestrial GHG balance (Schulze et al. 2009), the CO_2 sink of grasslands and forests entirely offsets agricultural CH_4 and N_2O emissions. The capacity of a soil to absorb atmospheric CH_4 is influenced by its qualities. Higher organic matter concentration in soils results in higher CH_4 uptake (1.49 mol CH_4 m^{-2} h^{-1}) than lower organic matter content soils (1.0 mol CH_4 m^{-2} h^{-1}). Additionally, the least amount of CH_4 is absorbed by soils with no vegetation: 0.52 mol CH_4 m^{-2} h^{-1} (Maljanen et al. 2004). Because of fertilising and tillage, using dried-up peat soils for agriculture increases CO_2 and N_2O emissions (Maljanen et al. 2007). Since soil degassing accounts for 35% of CO_2, 47% of CH_4, 53% of N_2O, and 21% of total annual emissions, GHG emissions from soils need to be adequately quantified for global budgets (IPCC 2007) even though different method of application of synthetic fertiliser and manure leads to N_2O emission from soil Table 10.3. According to Pilegaard et al. (2006), the C/N ratio is inversely associated with N_2O emissions, being greatest at a C/N-value of 11 (optimised decaying and humus development) and lowest at a C/N ratio of 30 (limited organic substances decaying) (Gundersen et al. 2012a, b). When paired with dryness and low pH

Table 10.2 Emission factors of different fertilisers from various countries

Fertiliser	Emission factor (kg CO_2-eq. kg of fertiliser^{-1})	Country	References
Emission factors for N fertiliser production			
Urea	1.3	Western Europe	Kongshaug (1998)
	1.6	Europe	Skowronska and Filipek (2014)
	1.9/2.7/5.5	Europe/Russia, USA/China	Brentrup et al. (2016)
	3.1	South-eastern United States	Albaugh et al. (2012)
	3.5	United Kingdom	Williams et al. (2010)
	4	Sweden and Western Europe	Davis and Haglund (1999)
Ammonium nitrate	6.2	Europe	Skowronska and Filipek (2014)
	6.5	United Kingdom	Elsayed et al. (2003)
	6.8	Western Europe	Kongshaug (1998)
	7	Sweden and western Europe	Davis and Haglund (1999)
	7.1	Netherlands	Kramer et al. (1999)
	7.2	United Kingdom	Williams et al. (2010)
	3.5/8/10.3	Europe/Russia/USA/China	Brentrup et al. (2016)
Emission factors for P fertiliser production			
Ammonium phosphate	1.3–1.8	Sweden and Western Europe	Davis and Haglund (1999)
	1.4/1.7/2.89	Europe/Russia, USA/China	Brentrup et al. (2016)
	6.4	South-eastern United States	Albaugh et al. (2012)
	7.8–8.9	China	Zhang et al. (2017)
Single super phosphate	0.6	United Kingdom	Williams et al. (2010)
	1	Sweden and Western Europe	Davis and Haglund (1999)
Triple superphosphate	0.4–0.54	Europe, Russia, USA, China	Brentrup et al. (2016)
	1	Sweden and western Europe	Davis and Haglund (1999)
	1	Brazil	da Silva and Kulay (2005)
	1.2	United Kingdom	Davis and Haglund (1999)
	1.6	Netherlands	Williams et al. (2010)
	1.6	Europe/Russia/USA/China	Skowronska and Filipek (2014)
Emission factors for K fertiliser production			
Potassium Chloride	0.14–0.25	China	Chen et al. (2018)

Table 10.3 N_2O emission occurs from manures and fertiliser under different methods of applications

Application method	Synthetic N	Manure/ slurry	Compost	Digestate	References
	Emission factor (kg CO_2/kg N)				
Broadcasting and incorporation	1.85	0.60–3.70	0.60	–	Akiyama et al. (2004)
Broadcasting and incorporation	5.36	1.37–3.78	1.50	–	Lopez-Fernandez et al. (2007)
Incorporation	9.66	6.88	0.33	–	Alluvione et al. (2010)
Subsurface	2.44	1.07	1.82	0.72	Meijide et al. (2009)
Broadcasting and incorporation	4.44	3.87	1.97–5.60	2.68	Meijide et al. (2009)
Broadcast	6.17	2.98	4.62	1.70	Vallejo et al. (2006)
Incorporation	0.09–0.36	0.27–0.33	–	0.15–0.30	Collins et al. (2011)
Subsurface	0.51	0.48–1.07	–	0.30	Baral et al. (2017)
Broadcast	0.51/2.06	0.89/3.67	–	0.42/1.19	Chantigny et al. (2007)
Broadcast & Subsurface	1.8	1.19–1.79	–	7.2–8.9	Saunders et al. (2012)
Subsurface	2.1	11.0–13.4	–	3.3–6.0	Lemke et al. (2012)
Broadcasting and incorporation	2.7	7.7–14.3	–	5.4	Bertora et al. (2008)
Side-dressed	4.62/26.61	10.51/16.27	–	8.3/17.6	Chantigny et al. (2010)
Incorporation	0.18–0.57	–	0.06–0.75	1.9–15.2	Verdi et al. (2018)

values, N_2O emission can be significantly reduced at C/N ratios of 20 (Gundersen et al. 2012a, b). The C:N ratio and CO_2 and CH_4 emissions, all three are positively correlated (Shi et al. 2014; Weslien et al. 2009). However, the availability of different electron donors such as $Fe^{3+,}$ Mn^{4+}, SO_4^{2-}, and NO_3^- may affect how much the CH_4 will be produced in soils (Achtnich et al. 1995; Dalal and Allen 2008; Fumoto et al. 2008; Kögel-Knabner et al. 2010).

Wetland ecosystems include peatland as a subtype. The organically rich soils are regarded as one of the major contributors to CO_2 and N_2O emissions, especially when used for agriculture purposes followed by drainage. High emission rates persist even after 20–30 years after cessation of agricultural activity. Compared to other types of land cover, these peatlands may operate as a sink for CH_4. These peatlands, as opposed to other types of land cover, might act as a sink for CH_4. Short-term, negligible CH_4 emission rates from drained peatland sites have been observed. They served as an annual sink for atmospheric CH_4. Application of different manures such as poultry manure and pig sully may increase the methane emission (Table 10.4); not only application of fertiliser, different types of crops also response

Table 10.4 Manure use in paddy field: CH_4 emission factors

Manure type	Emission factor (kg CO_2/kg of N)	Reference
Fowl manure	6.5	Yuan et al. (2017)
Poultry manure	23.5	Zhao et al. (2015)
Pig slurry/dairy slurry	0.9–46/0.35–60	Chadwick et al. (2000), Chen and Corson (2014)

Table 10.5 N fertiliser application: CH_4 emission factors

Crop type	Emission factor (kg CO_2/kg of N)	Reference
Rice	2.5	Yuan et al. (2017)
Rice	7.5	Das and Adhya (2014)
Rice, wheat	9.5	Yang et al. (2015), Zhao et al. (2015)

differently and increase methane emission with nitrogenous fertiliser (Table 10.5). The capacity of a soil to absorb atmospheric CH_4 is influenced by its qualities. Higher organic matter concentration in soils results in higher CH_4 uptake (1.49 mol CH_4 m^{-2} h^{-1}) than lower organic matter content soils (1.0 mol CH_4 m^{-2} h^{-1}). Additionally, the least amount of CH_4 is absorbed by soils with no vegetation: 0.52 mol CH_4 m^{-2} h^{-1} (Maljanen et al. 2004). Because of fertilising and tillage, using dried-up peat soils for agriculture increases CO_2 and N_2O emissions (Maljanen et al. 2007). In Finland, soil with barley crops on it released 7.52 mol CO_2 m^{-2} s^{-1}, compared to 3.66 mol CO_2 m^{-2} s^{-1} from surrounding bare soil.

By minimising unintentional losses, reducing greenhouse gas emissions, particularly N_2O, from agricultural land not only lessens their impact on the environment but also can enable farmers to use less fertiliser while maintaining output. The primary causes of the low nitrogen fertiliser utilisation rate are nitrate loss and nitrous oxide (N_2O) emissions brought on by nitrification and denitrification has significantly raised the earth's temperature. The studies assembled in Table 10.6 were conducted for a variety of crops and soil types, and no noticeable trend in terms of emissions appeared, regardless of crop rotation, soil type, temperature, country, application method, or climate.

4.1 *Strategies to Manage Greenhouse Gas Emissions*

The primary causes of the low nitrogen fertiliser utilisation rate are nitrate loss and nitrous oxide (N_2O) emissions brought on by nitrification and denitrification (Fig. 10.2). A greenhouse gas called N_2O has significantly raised the earth's temperature. With the addition of N fertilisers, soil respiration becomes less sensitive to soil temperature and more sensitive to soil moisture (Peng et al. 2011). Soil respiration falls during long-term N addition research (Bowden et al. 2004). In aerobic soil

Table 10.6 Nitrous oxide emission factors for N fertilisers under different crops and soil type

Country	Crop type	Emission factor (% N_2O/kg N)	Emission factor (kg CO_2 eq/kg N)	Soil type	References
Northeast China	Maize	0.03	0.09	Clay loam	Li et al. (2013)
North-western China	Silage corn	0.06	0.11–0.18	Sand	Kuang et al. (2018)
Pacific Northwest	Silage corn	0.03–0.12	0.09–0.36	Silt loam	Collins et al. (2011)
South China	Rice	0.14	0.45	Clay loam	Chen et al. (2013)
Denmark	Spring barley	0.17	0.51	Sandy loam	Baral et al. (2017)
Italy	None	0.06–0.19	0.11–0.57	Silt loam	Verdi et al. (2018)
South China	Wheat and maize	0.15–0.21	0.45–0.63	Sandy loam	Meng et al. (2005)
California	Almond	0.15–0.31	0.45–0.92	Sandy loam	Wolff et al. (2017)
United Kingdom	Ryegrass	0.1–0.4	0.30–1.19	Sandy clay loam/clay loam	Jones et al. (2005)
Japan	Rice	0.24	0.72	Sandy loam	Singla and Inubushi (2014)
Japan	Potato, sweet corn, winter wheat, sugar beet	0.26	0.77	Andisol	Koga (2013)
Japan, South China	Rice, wheat, and Pak Choi	0.3	0.89	Andisol, unspecified	Akiyama and Tsuruta (2003), Yuan et al. (2017)
France	Winter wheat	0.3–0.5	0.89–1.49	Silt	Gu et al. (2013)
North China Plain	Summer maize, Welsh onion, and winter wheat	0.48	1.43	Cambisol, sandy loam	Yao et al. (2017), Ding et al. (2013)
Mediterranean, India	Various, rice	0.5	1.49	Sandy clay loam	Das and Adhya (2014), Cayuela et al. (2017)
–	None	0.6	1.78	Silty clay loam	Saunders et al. (2012)
–	None	0.62	1.85	Sandy loam	Akiyama et al. (2004)
Northwest India	Oilseed crops, cereals, pulses, and millets	0.40–0.67	1.19–2.0	Sandy loam and Sandy clay loam	Jain et al. (2016a, b)
Eastern Canada	Timothy grass	0.17–0.69	0.51–2.06	Loam and sandy loam	Chantigny et al. (2007)
Northern Germany	None	0.68	2.03	Sandy loam	Senbayram et al. (2009)

(continued)

Table 10.6 (continued)

Country	Crop type	Emission factor (% N_2O/kg N)	Emission factor (kg CO_2 eq/kg N)	Soil type	References
China	Various vegetables	0.69	2.06	Various soil	Wang et al. (2018a, b)
Western Canada	Barley	0.7	2.09	Unspecified	Lemke et al. (2012)
Central Spain	Barley	0.82	2.44	Calcaric cambisol	Meijide et al. (2009)
–	None	0.9	2.68	Loam	Bertora et al. (2008)
China	Various	0.92	2.74	Various	Lu et al. (2006)
Germany, Global, China	Various	1	2.98	Various	Kaiser and Ruser (2000), Bouwman et al. (2002), Yang et al. (2015)
Xinjiang	Cotton	1.46	4.35	Calcaric fluvisol	Tao et al. (2018)
Central Spain	Maize	1.49	4.44	Sandy loam	Meijide et al. (2007)
Pennsylvania	Maize, alfalfa	1.6	4.77	Silt loam	Adviento-Borbe et al. (2010)
Central Spain	Maize, none	1.80	5.36	Sandy loam	Lopez-Fernandez et al. (2007)
Eastern Canada	Spring barley	1.1–2.1	3.28–6.26	Sandy loam, loam, silt loam	Zebarth et al. (2008)
Eastern Canada	Various	1.7	5.07	Various	Gregorich et al. (2005)
South-eastern China	Various vegetables	1.9	6.66	Silty clay loam	Zhang et al. (2016)
Central Spain	Potato	2.07	6.17	Clay loam	Vallejo et al. (2006)
North-western Italy	Corn	3.24	9.66	Silt loam	Alluvione et al. (2010)
Japan	Komatsuna	2.96–3.64	8.8–10.8	Sand	Singla et al. (2013)
Netherlands	None	4	6.3–11.9	Sand	Velthof et al. (2003)
United Kingdom	Various	6.5	1.2–19.4	Various	Dobbie and Smith (2003)
Northern United States	Maize	2–7	6.0–20.9	Fine and coarse loamy	McSwiney and Robertson (2005)
Eastern Canada	Corn	1.55–8.93	4.6–26.6	Clay, loam	Chantigny et al. (2010)
Global	Various	0.03–10	0.09–29.8	Various	Kim et al. (2013)
Japan	Corn	3.5–12.9	10.4–38.4	Andosol	Mukumbuta et al. (2017)

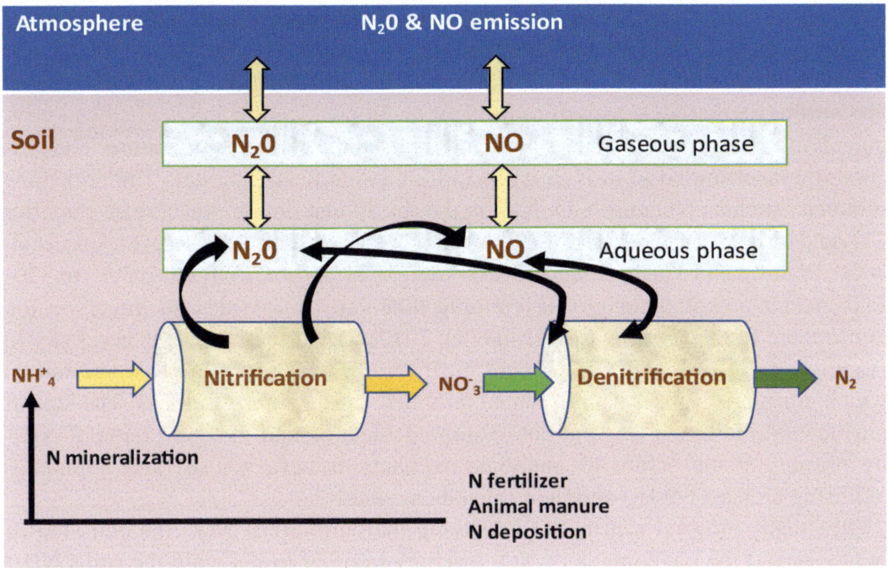

Fig. 10.2 Nitrous oxide emission from soil

conditions, the application of liquid manure (urea) increased N_2O emissions, whereas saturated soil conditions increased N_2O emissions from the application of NH_4^+ fertilisers (Tenuta and Beauchamp 2003). Since not all forms of nitrogen can be absorbed by plants, fertiliser application rates need to be adjusted to plant requirements in order to reduce N_2O emissions from agricultural lands. N levels that are unavailable to plants result in higher N_2O emissions (McSwiney and Robertson 2005). Increased N_2O emissions can be avoided by using controlled-release fertilisers or denitrification inhibitors (Shoji et al. 2001). To prevent increasing N_2O emissions, choosing the right type of fertiliser depends on the soil's water content (Sanz-Cobena et al. 2014). The tillage system also affects fertiliser applications. N_2O emissions under no-till and conservation tillage were higher when using urea; however, no differences could be seen when utilising urea-ammonium nitrate fertilisers (Venterea et al. 2005). However, there are conflicting results regarding how the tillage system affects N_2O emissions.

While some researchers found that no-till practise reduced N_2O emissions and that this was explained by lower soil temperatures (Six et al. 2002), others found that no-till increased N_2O emissions and explained this by higher microbial activity (Baggs et al. 2003). In no-till farming, increasing soil moisture causes higher N_2O emissions. Under no-till, increased C sequestration and CH_4 production cannot make up for increased N_2O emissions. Higher organic matter concentration in soils results in higher CH_4 uptake (1.49 mol CH_4 m^{-2} h^{-1}) than lower organic matter content soils (1.0 mol CH_4 m^{-2} h^{-1}). Additionally, the least amount of CH_4 is absorbed by soils with no vegetation: 0.52 mol CH_4 m^{-2} h^{-1} (Maljanen et al. 2004). Because of fertilising and tillage, using dried-up peat soils for agriculture increases CO_2 and N_2O emissions (Maljanen et al. 2007).

The primary causes of the low nitrogen fertiliser utilisation rate are nitrate loss and nitrous oxide (N_2O) emissions brought on by nitrification and denitrification. A greenhouse gas called N_2O has significantly raised the earth's temperature. The process of converting NH_4^+ to NO_3 will result in a significant amount of intermediate products, such as NO and N_2O. N_2O is the third most potent greenhouse gas after CO_2 and CH_4 (Gary and Lincoln 2000; Daniel et al. 2002; Ito et al. 2018; Anas et al. 2020; Walling and Vaneeckhaute, 2020; Wang et al. 2021a, b). It has a 298- to 310-fold greater potential for global warming than carbon dioxide and lingers in the atmosphere for a very long time (Liu et al. 2019). Because of this, it is necessary to employ certain techniques to prevent nitrification, which will increase the rate at which nitrogen fertiliser is utilised and minimise nitrogen fertiliser loss. The idea of nitrification inhibition was put out to stop nitrification and the conversion of NH_4^+ by preventing the action of ammonia-oxidising bacteria (AOB) and ammonia-oxidising archaea (AOA) and their related enzymes.

It actually slows down to NO_3, delaying the build-up of NO_3, the emission of N_2O gas, and the phenomena of NO_3 leaching in order to maintain the soil's NH_4^+ concentration as high as possible. Biological and industrial nitrification inhibitors (NIs) are the two primary categories of nitrification inhibitors currently available (BNIs). Common industrial nitrification inhibitors include dicyanamide (DCD), DMPP (3,4-dimethylpyrazole phosphate), and 2-chloro-6-(trichloromethyl) pyridine although their costs are higher and their duration of action in the soil is short.

4.2 *Biological Nitrification Inhibition (BNI)*

The term "biological nitrification inhibition" is used to describe the secretion of organic substances from plant roots that suppress nitrification and the natural molecules generated by plants are called biological nitrification inhibitors. Biological nitrification inhibitors play a significant role in a key element in enhancing nitrogen use and raising fertiliser quality. The biochemical nitrogen cycle of the globe is greatly impacted by BNI, which can also effectively raise the rate of nitrogen utilisation in soil and minimise nitrogen leaching and N_2O gas emissions (Ito et al. 2018; Liu et al. 2019) and increase crop yield. Zhang et al. (2011) observed that the total vegetable production treated with BNI was 163.2 t, which was 10.3 t greater than that of the sole application of urea. This finding demonstrated that the application of BNI greatly boosted agricultural productivity (Datta and Adhya 2014). According to the studies reported by Sun et al. (2016) and Cui et al. (2021), the use of BNI "*Nimin*" and *Xanthoderma lucidum* can greatly boost up rice yield in addition to reduced CH_4 and N_2O emissions. Inhibiting biological nitrification is thought to be a part of the adaptation strategy that enables plants to store and utilise nitrogen efficiently; nitrogen scarcity may have acted as a catalyst for the evolution of such inhibitors. Plant root discharge inhibitors: it is site-specific phenomenon that only

Table 10.7 Different plant species' roots release biological nitrogen inhibitors (BNI) which prevent nitrification

Sl. No.	Plant species	Total BNI released sum of four plants (ATU Day)	Specific BNI (ATU Day g^{-1} root dry wt.)
	Pasture grasses		
1	*Brachiaria humidicola (Rendle) Schweick*	51.5	13.4
2	*B. decumbens Stapf*	37.3	18.3
3	*Melinis minutiflora Beauv.*	21.4	3.8
4	*Panicum maximum Jacq.*	12.5	3.3
5	*Lolium perenne Lssp. Multiflorum (Lam.) Husnot*	13.5	2.6
6	*Andropogon gayanus Kunth*	11.7	7.7
7	*B. brizantha (A. Rich) Stapf*	6.8	2.0
	Cereal crops		
8	*Sorghum biocolour (L)Moench cv. Hybrid Sorgo*	26.1	5.2
9	*Penninsetum glaucum (L.) R. Br.cv.CIVT*	7.0	1.8
	Legume crops		
10	*Arachis hypogaea L. (var TMV2)*	9.4	2.5

restricted the area of the root that is exposed to NH_4^+. The nitrifying bacteria's activity may be increased by the soil NH_4^+ mineralisation of soil organic nitrogen and nitrogen fertilisers (Robinson 1963; Trenkel 2007). There are various plant species which is capable to release biological nitrification inhibitors from their root (Table 10.7). Additionally, one of the key directions for study is to generate more biological nitrification inhibitors with higher nitrification inhibitory potentiality using cutting-edge genetic and molecular methods by planting common crops in pastures, which will increase the usage of nitrogen (Oita et al. 2016).

5 Research Gap

Agricultural production is the result of numerous variables interacting each other. Through investigation of crop management, climate, and socioeconomic factors, and estimating the contribution of yield of yield-limiting factors in the yield gap. It is challenging to forecast how climate change would affect crop nutrient needs, nutrient availability, and soil nutrient cycling. Future climatic changes will not alter the fundamental concepts that have guided effective nutrient management in the past. Nutrient management in a changing climate is likely to follow the same guidelines as current practises. Plant scientists are on the verge of making significant strides in their understanding of nutrient use efficiency (NUE) and the creation of more nutrient-efficient plant types. However, a crucial component of the use and application of this strategy is that the methods must be evaluated in actual field condition. The traditional domains of plant breeding, crop physiology, and agronomy can increase their opportunities to research about genetic variations in NUE by using the new techniques in NUE and for establishing the link between genotype and phenotype. Nutrient use efficiency can be accelerated by two consecutive phenomenon, that is, efficiency of nutrient acquisition and nutrient utilisation. Modulations in these two perspectives could enhance nutrient use efficiency. The following scientific endeavours are required for such purposes:

(i) Development of resource-efficient crop cultivars specially modified root architecture
(ii) Development of control release fertiliser for preventing nutrient losses
(iii) Genetic manipulation of cultivars for higher nutrient acquisition

6 Conclusion

Achieving high NUE, higher crop productivity and environmental protection simultaneously have become a major challenge in global intensive agriculture. Various approaches comprising management practices, process-based rhizosphere management, and molecular and genetic modulation of plants for enhancing NUE are the keys for achieving higher nutrient use efficiency. Apart from the conventional fertilisation techniques, other methods such as control release fertilisation and use of nanoformulations uphold NUE and environmental safety. However, situation-specific modulations are indispensable for sustainable agriculture.

References

Abrol V, Singh AP, Kumar A, Chary R, Srinivasarao CH, Sharma P, Singh B, Salgotra S, Kapoor J, Dadhich H (2020) Effect of foliar application of nutrients on wheat (*Triticum aestivum*) crop performance, economics, resource use efficiency and soil properties under rainfed conditions. Indian J Agric Sci 90(1):138–141

Achtnich C, Bak F, Conrad R (1995) Competition for electron donors among nitrate reducers, ferric iron reducers, sulfate reducers, and methanogens in anoxic paddy soil. Biol Fertil Soil 19:65–72

Adesemoye AO, Torbert HA, Kloepper JW (2008) Enhanced plant nutrient use efficiency with PGPR and AMF in an integrated nutrient management system. Can J Microbiol 886:876–886. https://doi.org/10.1139/W08-081

Adhikari T (2011) Nano-particle research in soil science: micronutrients. In: Proceedings of the national symposium on 'applications of clay science: agriculture environment and industry', NBSS & LUP, Nagpur, 18–19 February 2011, pp 74–75

Adviento-Borbe M, Kaye J, Bruns M, McDaniel M, McCoy M, Harkcom S (2010) Soil greenhouse gas and ammonia emissions in long-term maize-based cropping systems. Soil Sci Soc Am J 74:1623–1634. https://doi.org/10.2136/sssaj2009.0446

Akiyama H, Tsuruta H (2003) Nitrous oxide, nitric oxide, and nitrogen dioxide fluxes from soils after manure and urea application. J Environ Qual 32:423–431. https://doi.org/10.2134/jeq2003.4230

Akiyama H, McTaggart IP, Ball BC, Scott A (2004) N_2O, NO, and NH_3 emissions from soil after the application of organic fertilizers, urea and water. Water Air Soil Pollut 156:113–129. https://doi.org/10.1023/b:Wate.0000036800.20599.46

Alam S, Khalil S, Ayub N, Rashid M (2002) In vitro solubilization of inorganic phosphate by phosphate solubilizing microorganisms from maize rhizosphere. Int J Agric Biol 4:1–6

Albaugh TJ, Vance ED, Gaudreault C, Fox TR, Allen HL, Stape JL, Rubilar RA (2012) Carbon emissions and sequestration from fertilization of pine in the Southeastern United States. For Sci 58:419–429. https://doi.org/10.5849/forsci.11-050

Alluvione F, Bertora C, Zavattaro L, Grignani C (2010) Nitrous oxide and carbon dioxide emissions following green manure and compost fertilization in corn. Soil Sci Soc Am J 74:384–395. https://doi.org/10.2136/sssaj2009.0092

Anas M, Liao F, Verma KK, Sarwar MA, Mahmood A, Chen ZL, Li Q, Zeng XP, Liu Y, Li YR (2020) Fate of nitrogen in agriculture and environment: agronomic, eco-physiological and molecular approaches to improve nitrogen use efficiency. Biol Res 53:47. https://doi.org/10.1186/s40659-020-00312-4

Asadzade N, Moosavi SG, Seghatoleslam MJ (2015) Effect of low irrigation and Zn and SiO_2 nano-fertilizers and conventional fertilizers on morphophysiological traits and seed yield of sunflower. Biol Forum – Int J 7(1):357–364. ISSN: 0975-1130

Aziz T, Ahmed I, Farooq M, Aamer MM, Sabir M (2011) Variation in phosphorus efficiency among *Brassica* cultivars I: internal utilization and phosphorus remobilization. J Plant Nutr 34:2006–2017. https://doi.org/10.1080/01904167.2011.610487

Baggs EM, Stevenson M, Pihlatie M, Regar A, Cook H, Cadisch G (2003) Nitrous oxide emissions following application of residues and fertiliser under zero and conventional tillage. Plant Soil 254:361–370

Bakhtiari M, Moaveni P, Sani B (2015) The effect of iron nanoparticles spraying time and concentration on wheat. Biol Forum – Int J 7(1):679–683. ISSN: 0975-1130

Bansiwal AK, Rayalu SS, Labhasetwar NK, Juwarkar AA, Devotta S (2006) Surfactant-modified zeolite as a slow-release fertilizer for phosphorus. J Agric Food Chem 54:4773–4779

Batjes NH (1996) Total carbon and nitrogen in the soils of the world. Eur J Soil Sci 47:151–163.

Baral KR, Labouriau R, Olesen JE, Petersen SO (2017) Nitrous oxide emissions and nitrogen use efficiency of manure and digestates applied to spring barley. Agric Ecosyst Environ 239:188–198. https://doi.org/10.1016/j.agee.2017.01.012

Baulcombe D, Crute I, Davies B, Dunwell J, Gale M, Jones J, Pretty J, Sutherland W, Toulmin C (2009) Reaping the benefits: science and the sustainable intensification of global agriculture. The Royal Society, London

Behnassi M, Shahid AS, D'Silva J (2011) Sustainable agricultural development. Springer Science Business Media, London, pp 171–184

Bengough AG, Mckenzie BM, Hallett PD, Valentine TA (2011) Root elongation, water stress, and mechanical impedance: a review of limiting stresses and beneficial root tip traits. J Exp Bot 62(1):58–68. https://doi.org/10.1093/jxb/erq350

Bertora C, Alluvione F, Zavattaro L, van Groenigen JW, Velthof G, Grignani C (2008) Pig slurry treatment modifies slurry composition, N_2O, and CO_2 emissions after soil incorporation. Soil Biol Biochem 40:1999–2006. https://doi.org/10.1016/j.soilbio.2008.03.021

Bouwman A, Boumans L, Batjes N (2002) Emissions of N_2O and NO from fertilized fields: summary of available measurement data. Global Biogeochem Cycles 16:6–13. https://doi.org/10.1029/2001GB001811

Bowden RD, Davidson E, Savage K, Arabia C, Steudler P (2004) Chronic nitrogen additions reduce total soil respiration and microbial respiration in temperate forest soils at the Harvard Forest. For Ecol Manag 196(1):43–56

Brady NC, Weil RR (1996) The nature and properties of soils. Prentice-Hall Inc., Upper Saddle River

Brentrup F, Hoxha A, Christensen B (2016) Carbon footprint analysis of mineral fertilizer production in Europe and other world regions. The 10th International Conference on Life Cycle Assessment of Food (LCA Food 2016)

Bücking H, Liepold E, Ambilwade P (2012) The role of the mycorrhizal symbiosis in nutrient uptake of plants and the regulatory mechanisms underlying these transport processes. In: Dhal NK, Sahu SC (eds) Plant science. Intechopen, London, pp 107–138. ISBN: 978-953-51-0905-1

Cassman KG, Dobermann A, Walters DT (2002) Agroecosystems, nitrogen-use efficiency, and nitrogen management. Ambio 31(2):132–140. https://doi.org/10.1579/0044-7447-31.2.132

Cayuela ML, Aguilera E, Sanz-Cobena A, Adams DC, Abalos D, Barton L, Ryals R, Silver WL, Alfaro MA, Pappa VA (2017) Direct nitrous oxide emissions in Mediterranean climate cropping systems: emission factors based on a meta-analysis of available measurement data. Agric Ecosyst Environ 238:25–35. https://doi.org/10.1016/j.agee.2016.10.006

Chadwick D, Pain B, Brookman S (2000) Nitrous oxide and methane emissions following application of animal manures to grassland. J Environ Qual 29:277–287. https://doi.org/10.2134/jeq2000.00472425002900010035x

Chantigny MH, Angers DA, Rochette P, B'elanger G, Mass'e D, Cote D (2007) Gaseous nitrogen emissions and forage nitrogen uptake on soils fertilized with raw and treated swine manure. J Environ Qual 36:1864–1872. https://doi.org/10.2134/jeq2007.0083

Chantigny MH, Rochette P, Angers DA, Bittman S, Buckley K, Masse D, Belanger G, Eriksen-Hamel N, Gasser MO (2010) Soil nitrous oxide emissions following band-incorporation of fertilizer nitrogen and swine manure. J Environ Qual 39:1545–1553. https://doi.org/10.2134/jeq2009.0482

Chapuis-Lardy L, Wrage N, Metay A, Chotte JL, Bernoux M (2007) Soils, a sink for N_2O. A review. Glob Change Biol 13:1–17

Chen X, Corson MS (2014) Influence of emission-factor uncertainty and farm characteristic variability in LCA estimates of environmental impacts of French dairy farms. J Clean Prod 81:150–157. https://doi.org/10.1016/j.jclepro.2014.06.046

Chen DJ, Jiang LN, Huang H, Toyota K, Dahlgren RA, Lu J (2013) Nitrogen dynamics of anaerobically digested slurry used to fertilize paddy fields. Biol Fertil Soils 49:647–659. https://doi.org/10.1007/s00374-012-0752-8

Chen J, Liu X, Wang C, Yin SS, Li XL, Hu WJ, Simon M, Shen ZJ, Xia Q, Chu CC, Peng XX (2015) Nitric oxide ameliorates zinc oxide nanoparticles-induced phytotoxicity in rice seedlings. J Hazard Mater 297:173–182. https://doi.org/10.1016/j.jhazmat.2015.04.077

Chen W, Geng Y, Hong J, Yang D, Ma X (2018) Life cycle assessment of potash fertilizer production in China. Resour Conserv Recycl 138:238–245. https://doi.org/10.1016/j.resconrec.2018.07.028

Chinnamuthu CR, Boopathi PM (2009) Nanotechnology and agroecosystem. Madras Agric J 96:17–31

Choudhury S, Moulick D (2022) Response of field crops to abiotic stress: current status and future prospects, 1st edn. CRC Press. https://doi.org/10.1201/9781003258063

Choudhury S, Moulick D, Mazumder MK (2021a) Secondary metabolites protect against metal and metalloid stress in rice: an *in silico* investigation using dehydroascorbate reductase. Acta Physiol Planta 43(1):1–10. https://doi.org/10.1007/s1173-020-03173-2

Choudhury S, Sharma P, Moulick D, Mazumder MK (2021b) Unrevealing metabolomics for abiotic stress adaptation and tolerance in plants. J Crop Sci Bioethanol 24(5):479–493. https://doi.org/10.1007/s12892-021-00102-8

Choudhary A, Senthil-Kumar M (2022) Drought attenuates plant defence against bacterial pathogens by suppressing the expression of CBP60g/SARD1 during combined stress. Plant Cell Environ 45:1127–1145. https://doi.org/10.1111/pce.14275

Choudhury S, Mazumder MK, Moulick D, Sharma P, Tata SK, Ghosh D, Ali MH, Siddiqui HM, Brestic M, Skalický M, Hossain A (2022a) A computational study of the role of secondary metabolites for mitigation of acid soil stress in cereals using dehydroascorbate and mono-dehydroascorbate reductases. Antioxidant 11(3):458. https://doi.org/10.3390/antiox11030458

Choudhury S, Moulick D, Mazumder MK, Pattnaik BK, Ghosh D, Vemireddy LR, Aldhahrani A, Soliman MM, Gaber A, Hossain A (2022b) An *in vitro* and *in silico* perspective study of seed priming with zinc on the phytotoxicity and accumulation pattern of arsenic in rice seedlings. Antioxidants 11(8):1500. https://doi.org/10.3390/antiox11081500

Choudhury S, Moulick D, Ghosh D; Soliman M, Alkhedaide A, Gaber A, Hossain, A (2022c) Drought-induced oxidative stress in pearl millet (Cenchrus americanus L.) at seedling stage: Survival mechanisms through alteration of morphophysiological and antioxidants activity. Life 12:1171. https://doi.org/10.3390/life12081171

Collins HP, Alva AK, Streubel JD, Fransen SF, Frear C, Chen S, Kruger C, Granatstein D (2011) Greenhouse gas emissions from an irrigated silt loam soil amended with anaerobically digested dairy manure. Soil Sci Soc Am J 75:2206–2216. https://doi.org/10.2136/sssaj2010.0360

Cordell D, White S (2013) For achieving phosphorus security. Agronomy 3:86–116

Corradini E, deMoura M, Mattoso LHC (2010) A preliminary study of the incorporation of NPK fertilizer into chitosan nanoparticles. Express Polym Lett 4(8):509–515. https://doi.org/10.3144/expresspolymlett.2010.64

Cui HX, Sun CJ, Liu Q, Jiang J, Gu W (2010) Applications of nanotechnology in agrochemical formulation, perspectives, challenges and strategies. In: International conference on Nanoagri, Sao Pedro, pp 28–33

Cui L, Li D, Wu Z, Xue Y, Xiao F, Zhang L, Song Y, Li Y, Zheng Y, Zhang J, Cui Y (2021) Effects of nitrification inhibitors on soil nitrification and ammonia volatilization in three soils with different pH. Agronomy 11(8):1674. https://doi.org/10.3390/agronomy11081674

Dalal RC, Allen DE (2008) Greenhouse gas fluxes from natural ecosystems. Turner review no. 18. Aust J Bot 56:369–407

Daniel JA, Luis ASS, Norman GH (2002) Molecular biology and biochemistry of ammonia oxidation by Nitrosomonas europaea. Arch Microbiol 178:250–255. https://doi.org/10.1007/s00203-002-0452-0

Das S, Adhya TK (2014) Effect of combine application of organic manure and inorganic fertilizer on methane and nitrous oxide emissions from a tropical flooded soil planted to rice. Geoderma 214:185–192. https://doi.org/10.1016/j.geoderma.2013.08.011

daSilva GA, Kulay LA (2005) Environmental performance comparison of wet and thermal routes for phosphate fertilizer production using LCA–A Brazilian experience. J Clean Prod 13:1321–1325. https://doi.org/10.1016/j.jclepro.2005.05.004

Datta A, Adhya TK (2014) Effects of organic nitrification inhibitors on methane and nitrous oxide emission from tropical rice paddy. Atmos Environ 92:533–545. https://doi.org/10.1016/j.atmosenv.2014.04.009

Davis J, Haglund C (1999) Life cycle inventory (LCI) of fertiliser production: fertiliser products used in Sweden and Western Europe. SIK Institutet for livsmedel och bioteknik, Goteborg

De Rosa MC, Monreal C, Schnitzer M, Walsh R, Sultan Y (2010) Nanotechnology in fertilizers. Nat Nanotechnol 5:91

Delfani M, Firouzabadi MB, Farrokhi N, Makarian H (2014) Some physiological responses of black-eyed pea to iron and magnesium nanofertilizers. Commun Soil Sci Plant Anal 45:530–540

Ding WX, Luo JF, Li J, Yu HY, Fan JL, Liu DY (2013) Effect of long-term compost and inorganic fertilizer application on background N_2O and fertilizer induced N_2O emissions from an intensively cultivated soil. Sci Total Environ 465:115–124. https://doi.org/10.1016/j.scitotenv.2012.11.020

Ditta A, Arshad M (2016) Applications and perspectives of using nanomaterials for sustainable plant nutrition. Nano technol Rev 5:209

Ditta A, Arshad M, Ibrahim M (2015) Nanoparticles in sustainable agricultural crop production: applications and perspectives. Nanotechnol Plant Sci:55–75. https://doi.org/10.1007/978-3-319-14502-0_4

Dobbie KE, Smith KA (2003) Nitrous oxide emission factors for agricultural soils in Great Britain: the impact of soil water-filled pore space and other controlling variables. Glob Change Biol 9:204–218. https://doi.org/10.1046/j.1365-2486.2003.00563.x
Dobermann AR (2005) Nitrogen use efficiency - state of the art. Agronomy & Horticulture Faculty Publications, p 316. http://digitalcommons.unl.edu/agronomyfacpub/316. Accessed 3 July 2018
Dobermann A (2007) Nutrient use efficiency – measurement and management. Agron Hortic 1442:1–28
Dong J, Hunt J, Delhaize E, Caixian T (2018) The impact of elevated CO2 on acid-soil tolerance of hexaploid wheat (Triticum aestivum L.) genotypes varying in organic anion efflux. Plant Soil 428:401–413. https://doi.org/10.1007/s11104-018-3687-5
El-Kereti MA, El-feky SA, Khater MS, Osman YA, El-sherbini EA (2013) ZnO nanofertilizer and He Ne laser irradiation for promoting growth and yield of sweet basil plant. Recent Pat Food Nutr Agric 5(3):169–181. https://doi.org/10.2174/2212798405666131112142517
Elsayed M, Matthews R, Mortimer N (2003) Carbon and energy balances for a range of biofuels options. Sheffield Hallam University, Resources research unit, United Kingdom
Erisman JW, Sutton MA, Galloway J, Klimont Z, Winiwarter W (2008) How a century of ammonia synthesis changed the world. Nat Geosci 1:636–639. https://doi.org/10.1038/ngeo325
Fageria NK, Baligar VC, Li YC (2008) The role of nutrient efficient plants in improving crop yields in the twenty first century. J Plant Nutri 31:1121–1157. https://doi.org/10.1080/01904160802116068
FAO (2009) FAOSTAT. FAO Statistics Division. Online at http://faostat.fao.org/
FAO (2013) FAOSTAT Database. Food and Agriculture Organization of the United Nations, Rome
FAO (2017) World fertilizer trends and outlook to 2020. Food and Agriculture Organization of United Nations, Rome
Finck A (1992) Fertilizers and their efficient use. In: Halliday DJ, Trenkel ME, Wichmann W (eds) IFA: world fertilizer use manual. International Fertilizer Industry Association, Paris
Fixen P, Brentrup F, Bruulsema T, Garcia F, Norton R, Zingore S (2015) Nutrient/fertilizer use efficiency: measurement, current situation and trends. In: Drechsel P, Heffer P, Magen H, Mikkelsen R, Wichelns D (eds) Managing water and fertilizer for sustainable agricultural intensification. International Fertilizer Industry Association (IFA), International Water Management Institute (IWMI), International Plant Nutrition Institute (IPNI), and International Potash Institute (IPI), pp 8–38
Fumoto T, Kobayashi K, Li C, Yagi K, Hasegawa T (2008) Revising a process-based biogeochemistry model (DNDC) to simulate methane emission from rice paddy fields under various residue management and fertilizer regimes. Glob Change Biol 14:382–402
Galloway JN, Townsend AR, Erisman JW, Bekunda M, Cai Z, Freney JR, Martinelli LA, Seitzinger SP, Sutton MA (2008) Transformation of the nitrogen cycle: recent trends, questions, and potential solutions. Science 320:889–892. https://doi.org/10.1126/science.1136674
Gary DB, Lincoln SD (2000) Inhibition of soil nitrifying bacteria communities and their activities by glucosinolate hydrolysis products. Soil Biol Biochem 32:1261–1269. https://doi.org/10.1016/S0038-0717(00)00043-2
Ghosh D, Brahmachari K, Brestic M, Ondrisik P, Hossain A, Skalicky M, Sarkar S, Moulick D, Dinda NK, Das A, Pramanick B (2020a) Integrated weed and nutrient management improve yield, nutrient uptake and economics of maize in the rice-maize cropping system of Eastern India. Agronomy 10(12):1906. https://doi.org/10.3390/agronomy10121906
Ghosh D, Brahmachari K, Skalicky M, Hossain A, Sarkar S, Dinda NK, Das A, Pramanick B, Moulick D, Brestic M, Raza MA (2020b) Nutrients supplementation through organic manures influence the growth of weeds and maize productivity. Molecules 25(21):4924. https://doi.org/10.3390/molecules25214924
Ghosh D, Brahmachari K, Sarkar S, Dinda NK, Das A, Moulick D (2022a) Impact of nutrient management in rice-maize-greengram cropping system and integrated weed management

treatments on summer greengram productivity. Indian J Weed Sci 54(1):25–30. https://doi.org/10.5958/0974-8164.2022.00004.1

Ghosh D, Brahmachari K, Skalický M, Roy D, Das A, Sarkar S, Moulick D, Brestič M, Hejnak V, Vachova P, Hassan MM, Hossain A (2022b) The combination of organic and inorganic fertilizers influence the weed growth, productivity and soil fertility of monsoon rice. PLoS One 17(1):e0262586. https://doi.org/10.1371/journal.pone.0262586

Glenn JC, Gordon TJ, Florescu E (2008) The millenium project: state of the future. World Federation of UN Associations, Washington, DC

Gregersen PL (2011) Senescence and nutrient remobilization in crop plants. In: First PB, Hawkesford MJ (eds) The molecular and physiological basis of nutrient use efficiency in crops. Wiley, Hoboken, p 83

Gregorich E, Rochette P, VandenBygaart A, Angers D (2005) Greenhouse gas contributions of agricultural soils and potential mitigation practices in Eastern Canada. Soil Tillage Res 83:53–72. https://doi.org/10.1016/j.still.2005.02.009

Gu J, Nicoullaud B, Rochette P, Grossel A, H'enault C, Cellier P, Richard G (2013) A regional experiment suggests that soil texture is a major control of N_2O emissions from tile-drained winter wheat fields during the fertilization period. Soil Biol Biochem 60:134–141. https://doi.org/10.1016/j.soilbio.2013.01.029

Gundersen P, Christiansen JR, Alberti G, Brüggemann N, Castaldi S, Gasche R, Kitzler B, Klemedtsson L, Lobo-do-Vale R, Moldan F, Rütting T, Schleppi P, Weslien P, Zechmeister-Boltenstern S (2012a) The response of methane and nitrous oxide fluxes to forest change in Europe. Biogeosciences 9:3999–4012

Gundersen P, Christiansen JR, Frederiksen P, Vesterdal L (2012b) Influence of hydromorphic soil conditions on greenhouse gas emissions and soil carbon stocks in a Danish temperate forest. For Ecol Manag 284:185–195

Heffer P, Prud'homme M (2016) Global nitrogen fertilizer demand and supply: trend, current level and outlook. International Nitrogen Initiative Conference, Melbourne, Australia

Hossain A, Ahmed S, Ahmad Z, Alam MJ, Moulick D, Saha B, Samanta S, Maitra S, Bhadra P, Bukhari MA, Aftab T (2021a) Next-generation genetic engineering tools for abiotic stress tolerance in plants. In: Frontiers in plant-soil interaction. Academic Press, pp 153–197. https://doi.org/10.1016/B978-0-323-90943-3.00008-0

Hossain A, Ahmad Z, Moulik D, Maitra S, Bhadra P, Ahmad A, Garai S, Mondal M, Roy A, Sabagh AE, Aftab T (2021b) Jasmonates and salicylates: mechanisms, transport and signalling during abiotic stress in plants. In: Jasmonates and salicylates signaling in plants. Springer, Cham, pp 1–29. https://doi.org/10.1007/978-3-030-75805-9_1

Hossain A, Maitra S, Pramanick B, Bhutia K, Ahmad Z, Moulik D, Syed MA, Shankar T, Adeel M, Hassan MM, Aftab T (2022) Wild relatives of plants as sources for the development of abiotic stress tolerance in plants. In: Plant perspectives to global climate changes. Academic Press, pp 471–518. https://doi.org/10.1016/B978-0-323-85665-2.00011-X

Hund A, Ruta N, Liedgens M (2009) Rooting depth and water use efficiency of tropical maize inbred lines, differing in drought tolerance. Plant Soil 318:311–325. https://doi.org/10.1007/s11104-08-9843-6

Igual JM, Valverde A, Cervantes E, Velázquez E (2001) Review article phosphate-solubilizing bacteria as inoculants for agriculture: use of updated molecular techniques in their study. Agronomie 21:561–568

IPCC (2007) In: Solomon S, Qin D, Manning M, Chen Z, Marquis M, Averyt KB, Tignor M, Miller HL (eds) Climate change: the physical science basis. Contribution of working group I to the fourth assessment report of the Intergovernmental panel on climate change. Cambridge University Press, Cambridge/New York

Ito A, Nishina K, Ishijima K, Hashimoto S, Inatomi M (2018) Emissions of nitrous oxide (N2O) from soil surfaces and their historical changes in East Asia: a model-based assessment. Prog Earth Planet Sci 5:55. https://doi.org/10.1186/s40645-018-0215-4

Jain R, Dominic D, Jordan N, Rene ER, Weiss S, van Hullebusch ED, Hübner R, Lens PNL (2016a) Preferential adsorption of Cu in a multi-metal mixture onto biogenic elemental selenium nanoparticles. Chem Eng J 284:917–992

Jain N, Arora P, Tomer R, Mishra SV, Bhatia A, Pathak H, Chakraborty D, Kumar V, Dubey D, Harit R (2016b) Greenhouse gases emission from soils under major crops in Northwest India. Sci Total Environ 542:551–561. https://doi.org/10.1016/j.scitotenv.2015.10.073

Jones S, Rees R, Skiba U, Ball B (2005) Greenhouse gas emissions from a managed grassland. Glob Planet Change 47:201–211. https://doi.org/10.1016/j.gloplacha.2004.10.011

Kaiser EA, Ruser R (2000) Nitrous oxide emissions from arable soils in Germany—an evaluation of six long-term field experiments. J Plant Nutr Soil Sci 163:249–259. https://doi.org/10.1002/1522-2624(200006)163:33.0.CO;2-Z

Khan MS, Ahmad E, Zaidi A, Oves M (2013) Functional aspect of phosphate-solubilizing bacteria: importance in crop production. Springer, Berlin/Heidelberg, pp 237–263. https://doi.org/10.1007/978-3-642-37241-4

Kim DG, Hernandez-Ramirez G, Giltrap D (2013) Linear and nonlinear dependency of direct nitrous oxide emissions on fertilizer nitrogen input: a meta-analysis. Agric Ecosyst Environ 168:53–65. https://doi.org/10.1016/j.agee.2012.02.021

Koga N (2013) Nitrous oxide emissions under a four-year crop rotation system in northern Japan: impacts of reduced tillage, composted cattle manure application and increased plant residue input. Soil Sci Plant Nutr 59:56–68. https://doi.org/10.1080/00380768.2012.733870

Kögel-Knabner I, Amelung W, Cao Z, Fiedler S, Frenzel P, Jahn R, Kalbitz K, Kölbl A, Schloter M (2010) Biogeochemistry of paddy soils. Geoderma 157:1–14

Kongshaug G (1998) Energy consumption and greenhouse gas emissions in fertilizer production. IFA Technical Conference, Marrakech, Morocco, 28 September–1 October, 1998, 18 pp.

Kramer KJ, Moll HC, Nonhebel S (1999) Total greenhouse gas emissions related to the Dutch crop production system. Agric Ecosyst Environ 72:9–16. https://doi.org/10.1016/S0167-8809(98)00158-3

Kuang W, Gao X, Gui D, Tenuta M, Flaten DN, Yin M, Zeng F (2018) Effects of fertilizer and irrigation management on nitrous oxide emission from cotton fields in an extremely arid region of Northwestern China. Field Crop Res 229:17–26. https://doi.org/10.1016/j.fcr.2018.09.010

Kutsch W, Bahn M, Heinemeyer A (2009) Soil carbon dynamics: an integrated methodology, xi. Cambridge University Press, Cambridge/New York, p 286

Lemke RL, Malhi SS, Selles F, Stumborg M (2012) Relative effects of anaerobically digested and conventional liquid swine manure, and N fertilizer on crop yield and greenhouse gas emissions. Agric Sci 3:799–805. https://doi.org/10.4236/as.2012.36097

Li LJ, Han XZ, You MY, Horwath WR (2013) Nitrous oxide emissions from Mollisols as affected by long-term applications of organic amendments and chemical fertilizers. Sci Total Environ 452:302–308. https://doi.org/10.1016/j.scitotenv.2013.03.002

Liu R, Lal R (2014) Synthetic apatite nanoparticles as a phosphorus fertilizer for soybean (*Glycine max*). Sci Rep 4:5686–5691

Liu R, Lal R (2015) Potentials of engineered nanoparticles as fertilizers for increasing agronomic productions. Sci Total Environ 514:131–139. https://doi.org/10.1016/j.scitotenv.2015.01.104

Liu F, Wen LX, Li ZZ, Yu W, Sun HY, Chen JF (2006) Porous hollow silica nanoparticles as controlled delivery system for water soluble pesticide. Mater Res Bull 41:2268–2275

Liu Y, Tang H, Muhammad A, Huang G (2019) Emission mechanism and reduction countermeasures of agricultural greenhouse gases – a review. Greenh Gases Sci Technol 9:160–174. https://doi.org/10.1002/ghg.1848

Lopez-Fernandez S, Diez J, Hernaiz P, Arce A, García-Torres L, Vallejo A (2007) Effects of fertiliser type and the presence or absence of plants on nitrous oxide emissions from irrigated soils. Nutr Cycling Agroecosyst 78:279–289. https://doi.org/10.1007/s10705-007-9091-9

Lu Y, Huang Y, Zou J, Zheng X (2006) An inventory of N_2O emissions from agriculture in China using precipitation-rectified emission factor and background emission. Chemosphere 65:1915–1924. https://doi.org/10.1016/j.chemosphere.2006.07.035

Ludwig J, Meixner FX, Vogel B, Förstner J (2001) Soil-air exchange of nitric oxide: an overview of processes, environmental factors, and modeling studies. Biogeochemistry 52:225–257

Luo Z, Hu S, Chen D (2018) A historical view of nitrogen metabolism and its driving forces in China's chemical industry: implications for cleaner production policy and practice. J Clean Prod 187:308–317. https://doi.org/10.1016/j.jclepro.2018.03.240

Maljanen M, Komulainen VM, Hytönen J, Martikainen PJ, Laine J (2004) Carbon dioxide, nitrous oxide and methane dynamics in boreal organic agricultural soils with different soil characteristics. Soil Biol Biochem 36:1801–1808

Maljanen M, Hytönen J, Mäkiranta P, Alm J, Minkkinen K, Laine J, Martikainen PJ (2007) Greenhouse gas emissions from cultivated and abandoned organic croplands in Finland. Boreal Environ Res 12:133–140

Manikandan A, Subramanian KS (2014) Fabrication and characterisation of nanoporous zeolite based N fertilizer. Afr J Agric Res 9:276–284

Mazumder MK, Sharma P, Moulick D, Tata S, Choudhury S (2021) Salicylic acid ameliorates zinc and chromium induced stress responses in wheat seedlings: a biochemical and computational analysis. Cereal Res Commun:1–12. https://doi.org/10.1007/s42976-021-00201-w

Mazumder MK, Moulick D, Choudhury S (2022) Iron (Fe^{3+})-mediated redox responses and amelioration of oxidative stress in cadmium (Cd^{2+}) stressed mung bean seedlings: a biochemical and computational analysis. J Plant Biochem Biotechnol 31(1):49–60. https://doi.org/10.1007/s13562-021-00654-4

McSwiney CP, Robertson GP (2005) Nonlinear response of N_2O flux to incremental fertilizer addition in a continuous maize (*Zea mays* L.) cropping system. Glob Change Biol 11:1712–1719. https://doi.org/10.1111/j.1365-2486.2005.01040.x

Meijide A, Díez JA, Sanchez-Martín L, Lopez-Fernandez S, Vallejo A (2007) Nitrogen oxide emissions from an irrigated maize crop amended with treated pig slurries and composts in a Mediterranean climate. Agric Ecosyst Environ 121:383–394. https://doi.org/10.1016/j.agee.2006.11.020

Meijide A, Garcia-Torres L, Arce A, Vallejo A (2009) Nitrogen oxide emissions affected by organic fertilization in a non-irrigated Mediterranean barley field. Agric Ecosyst Environ 132:106–115. https://doi.org/10.1016/j.agee.2009.03.005

Meng L, Ding WX, Cai ZC (2005) Long-term application of organic manure and nitrogen fertilizer on N_2O emissions, soil quality and crop production in a sandy loam soil. Soil Biol Biochem 37:2037–2045. https://doi.org/10.1016/j.soilbio.2005.03.007

Moulick D, Ghosh D, Santra SC (2016a) An assessment of some physicochemical properties and cooking characteristics of milled rice and associated health risk in two rice varieties of arsenic contaminated areas of West Bengal, India. Int J Agric Food Sci 6(2):44–55

Moulick D, Ghosh D, Santra SC (2016b) Evaluation of effectiveness of seed priming with selenium in rice during germination under arsenic stress. Plant Physiol Biochem 109:571–578. https://doi.org/10.1016/j.plaphy.2016.11.004

Moulick D, Santra SC, Ghosh D (2017) Seed priming with Se alleviate As induced phytotoxicity during germination and seedling growth by restricting As translocation in rice (*Oryza sativa* L cv IET-4094). Ecotoxicol Environ Safe 145:449–456. https://doi.org/10.1016/j.ecoenv.2017.07.060

Moulick D, Santra SC, Ghosh D (2018a) Effect of selenium induced seed priming on arsenic accumulation in rice plant and subsequent transmission in human food chain. Ecotoxicol Environ Saf 152:67–77. https://doi.org/10.1016/j.ecoenv.2018.01.037

Moulick D, Santra SC, Ghosh D (2018b) Rice seed priming with se: a novel approach to mitigate as induced adverse consequences on growth, yield and as load in brown rice. J Hazard Mater 355:187–196. https://doi.org/10.1016/j.jhazmat.2018.05.017

Moulick D, Santra SC, Ghosh D (2018c) Seed priming with Se mitigates As-induced phytotoxicity in rice seedlings by enhancing essential micronutrient uptake and translocation and reducing As translocation. Environ Sci Pollut Res 25(27):26978–26991. https://doi.org/10.1007/s11356-018-2711-x

Moulick D, Santra SC, Ghosh D (2018d) Consequences of paddy cultivation in arsenic-contaminated paddy fields of lower Indo-Gangetic plain on arsenic accumulation pattern and selected grain quality traits: a preliminary assessment. In: Mechanisms of arsenic toxicity and tolerance in plants. Springer, Singapore, pp 49–78. https://doi.org/10.1007/978-981-13-1292-2_3

Moulick D, Chowardhara B, Panda SK (2019a) Agroecotoxicological aspect of arsenic (As) and cadmium (Cd) on field crops and its mitigation: current status and future prospect. In: Plant-metal interactions. Springer, Cham, pp 217–246. https://doi.org/10.1007/978-3-030-20732-8_11

Moulick D, Santra SC, Ghosh D, Panda SK (2019b) An assessment of efficiency of zinc priming in rice (cv. MTU-7029) during germination and early seedling growth. In: Priming and pretreatment of seeds and seedlings. Springer, Singapore, pp 495–507. https://doi.org/10.1007/978-981-13-8625-1_24

Moulick D, Sarkar S, Awasthi JP, Ghosh D, Choudhury S, Tata SK, Bramhachari K, Santra SC (2020) Rice grain quality traits: neglected or less addressed? In: Rice research for quality improvement: genomics and genetic engineering. Springer, Singapore, pp 729–745. https://doi.org/10.1007/978-981-15-4120-9_29

Moulick D, Samanta S, Sarkar S, Mukherjee A, Pattnaik BK, Saha S, Awashy JP, Bhowmick S, Ghosh D, Samal AC, Mahanta S, Mazumder MK, Choudhury S, Bramhachari K, Biswas JK, Santra SC (2021) Arsenic contamination, impact and mitigation strategies in rice agro-environment: an inclusive insight. Sci Total Environ 800:149477. https://doi.org/10.1016/j.scitotenv.2021.149477

Moulick D, Ghosh D, Skalicky M, Gharde Y, Mazumder MK, Choudhury S, Biswas JK, Santra SC, Brestic M, Vachova P, Hossain A (2022) Interrelationship among rice grain arsenic, micronutrients content and grain quality attributes: an investigation from genotype× environment perspective. Front Environ Sci 180. https://doi.org/10.3389/fenvs.2022.857629

Moulick D, Ghosh D, Mandal J, Bhowmick S, Mondal D, Choudhury S, Santra SC, Vithanage M, Biswas JK (2023a) A cumulative assessment of plant growth stages and selenium supplementation on arsenic and micronutrients accumulation in rice grains. J Clean Prod 135764. https://doi.org/10.1016/j.jclepro.2022.135764

Moulick D, Bhutia KL, Sarkar S, Roy A, Mishra UN, Pramanick B, Maitra S, Shankar T, Hazra S, Skalicky M, Brestic M (2023b) The intertwining of Zn-finger motifs and abiotic stress tolerance in plants: current status and future prospects. Front Plant Sci 13:1083960. https://doi.org/10.3389/fpls.2022.1083960

Mukumbuta I, Shimizu M, Hatano R (2017) Mitigating global warming potential and greenhouse gas intensities by applying composted manure in cornfield: a 3-year field study in an andosol soil. Agriculture 7:13. https://doi.org/10.3390/agriculture7020013

Naderi MR, Danesh-Shahraki A (2013) Nanofertilizers and their roles in sustainable agriculture. Int J Agric Crop Sci 5(19):2229–2232

Nair R, Varghese SH, Nair BG, Maekawa T, Yoshida Y, Kumar DS (2010) Nanoparticulate material delivery to plants. Plant Sci 179:154–163. https://doi.org/10.1016/j.plantsci.2010.04.012

Nieder R, Benbi DK (2008) Carbon and nitrogen in the terrestrial environment. Springer Science, Dordrecht, p 430

Noushahi HA, Hussian M, Bilal M, Salim MA, Idrees F, Jawad M, Rizwan M, Atta B, Tanveer K (2019) Improving phosphorus use efficiency by agronomical and genetic means. World J Agric Sci 15(2):47–53. https://doi.org/10.5829/idosi.wjas.2019.48.54

Oita A, Malik A, Kanemoto K, Geschke A, Nishijima S, Lenzen M (2016) Substantial nitrogen pollution embedded in international trade. Nat Geosci 9(2):111–115. https://doi.org/10.1038/ngeo2635

Opala P, Othieno CO, Peter K (2010) Effects of combining organic materials with inorganic phosphorus sources on maize yield and financial benefits in western Kenya. Exp Agric 46(1):23–34. https://doi.org/10.1017/S0014479709990457

Otinga AN, Vanlauwe B, Okalebo JR, Six L, Emong'ole M, Pypers P, Merckx R, Njoroge R (2013) Partial substitution of phosphorus fertiliser by farmyard manure and its localised application increases agronomic efficiency and profitability of maize production. Field Crops Res 140:32–43. https://doi.org/10.1016/j.fcr.2012.10.003

Peng Q, Dong Y, Qi Y, Xiao S, He Y, Ma T (2011) Effects of nitrogen fertilization on soil respiration in temperate grassland in Inner Mongolia, China. Environ Earth Sci 62:1163–1171

Pilegaard K, Skiba U, Ambus P, Beier C, Brüggemann N, Butterbach-Bahl K, Dick J, Dorsey J, Duyzer J, Gallagher M, Gasche R, Horvath L, Kitzler B, Leip A, Pihlatie MK, Rosenkranz P, Seufert G, Vesala T, Westrate H, Zechmeister-Boltenstern S (2006) Factors controlling regional differences in forest soil emission of nitrogen oxides (NO and N_2O). Biogeosciences 3:651–661

Postma JA, Lynch JP (2011) Theoretical evidence for the functional benefit of root cortical aerenchyma in soils with low phosphorus availability. Ann Bot 107:829–841. https://doi.org/10.1093/aob/mcq199

Qiu B, Wang Y (2006) Phylogenetic distribution and evolution of mycorrhizas in land plants. Mycorrhiza 16:299–363. https://doi.org/10.1007/s00572-005-0033-6

Robinson JB (1963) Nitrification in a New Zealand grassland soil. Plant Soil 19:173–183. https://doi.org/10.1007/BF01350243

Rojas-downing MM, Nejadhashemi AP, Harrigan T, Woznicki SA (2017) Climate risk management climate change and livestock: impacts, adaptation, and mitigation. Clim Risk Manag 16:1–19. https://doi.org/10.1016/j.crm.2017.02.001

Roldugin VI, Fedotov MA, Folmanis GE, Kovalenko LV, Tananaevc IG (2015) Formation of aqueous colloidal solutions of selenium and silicon by laser ablation. Dokl Phys Chem 463(Part 1):161–164. https://doi.org/10.1134/S0012501615070064

Saud S, Wang D, Fahad S (2022) Improved nitrogen use efficiency greenhouse gas emissions in agricultural soils as producers of biological nitrification inhibitors. Front Plant Sci 13:854195. https://doi.org/10.3389/fpls.2022.854195

Saha B, Chowardhara B, Kar S, Devi SS, Awasthi JP, Moulick D, Tanti B, Panda SK (2019) Advances in heavy metal-induced stress alleviation with respect to exogenous amendments in crop plants. In: Priming and pretreatment of seeds and seedlings. Springer, Singapore, pp 313–332. https://doi.org/10.1007/978-981-13-8625-1_15

Sahoo S, Borgohain P, Saha B, Moulick D, Tanti B, Panda SK (2019) Seed priming and seedling pre-treatment induced tolerance to drought and salt stress: recent advances. In: Priming and pretreatment of seeds and seedlings, pp 253–263. https://doi.org/10.1007/978-981-13-8625-1_12

Sanz-Cobena A, Abalos D, Meijide A, Sanchez-Martin L, Vallejo A (2014) Soil moisture determines the effectiveness of two urease inhibitors to decrease N_2O emission. Mitig Adapt Strateg Glob Change. https://doi.org/10.1007/s11027-014-9548-5

Saunders OE, Fortuna AM, Harrison JH, Cogger CG, Whitefield E, Green T (2012) Gaseous nitrogen and bacterial responses to raw and digested dairy manure applications in incubated soil. Environ Sci Technol 46:11684–11692. https://doi.org/10.1021/es301754s

Schaufler G, Kitzler B, Schindlbacher A, Skiba U, Sutton MA, Zechmeister-Boltenstern S (2010) Greenhouse gas emissions from European soils under different land use: effects of soil moisture and temperature. Eur J Soil Sci 61:683–696. https://doi.org/10.1111/j.1365-2389.2010.01277.x

Schlesinger WH, Andrews JA (2000) Soil respiration and the global carbon cycle. Biogeochemistry 48:7–20

Schulze ED, Luyssaert S, Ciais P, Freibauer A, Janssens IA, Soussana JF, Smith P, Grace J, Levin I, Thiruchittampalam B, Heimann M, Dolman AJ, Valentini R, Bousquet P, Peylin P, Peters W, Rodenbeck C, Etiope G, Vuichard N, Wattenbach M, Nabuurs GJ, Poussi Z, Nieschulze J, Gash JHC (2009) Importance of methane and nitrous oxide for Europe's terrestrial greenhouse-gas balance. Nat Geosci 2:842–850. https://doi.org/10.1038/ngeo686

Sekhon BS (2014) Nanotechnology in agri-food production: an overview. Nanotechnol Sci Appl 7:31–53. https://doi.org/10.2147/NSA.S39406

Selva Preetha P, Subramanian KS, Rahale CS (2014) Sorption characteristics of nanozeolite based slow release sulphur fertilizer. Int J Dev Res 4:225–228

Senbayram M, Chen R, Mühling KH, Dittert K (2009) Contribution of nitrification and denitrification to nitrous oxide emissions from soils after application of biogas waste and other fertilizers. Rapid Commun Mass Spectrom 23:2489–2498. https://doi.org/10.1002/rcm.4067

Sharonova NL, Yapparov AK, Khisamutdinov NS, Ezhkova AM, Yapparov IA, Ezhkov VO, Degtyareva IA, Babynin EV (2015) Nanostructured water-phosphorite suspension is a new promising fertilizer. Nanotechnol Russ 10(7–8):651–661. https://doi.org/10.1134/S1995078015040187

Shi WY, Yan MJ, Zhang JG, Guan JH, Du S (2014) Soil CO2 emissions from five different types of land use on the semiarid Loess Plateau of China, with emphasis on the contribution of winter soil respiration. Atmos Environ 88:74–82

Shoji S, Delgado J, Mosier A, Miura Y (2001) Use of controlled release fertilizers and nitrification inhibitors to increase nitrogen use efficiency and to conserve air and water quality. Commun Soil Sci Plant Anal 32:1051–1070

Simpson RJ, Oberson A, Culvenor RA, Ryan MH, Veneklaas EJ, Lambers H, Lynch JP, Ryan PR, Delhaize E, Smith FA, Smith SE, Harvey PR, Richardson AE (2011) Strategies and agronomic interventions to improve the phosphorus-use efficiency of farming systems. Plant Soil 349(1):89–120. https://doi.org/10.1007/s11104-011-0880-1

Singh R, Shedbalkar UU, Wadhwani SA, Chopade BA (2015) Bacteriagenic silver nanoparticles: synthesis, mechanism, and applications. Appl Microbiol Biotechnol 99(11). https://doi.org/10.1007/s00253-015-6622-1

Singla A, Inubushi K (2014) Effect of biogas digested liquid on CH_4 and N_2O flux in paddy ecosystem. J Integr Agric 13:635–640. https://doi.org/10.1016/s2095-3119(13)60721-2

Singla A, Dubey SK, Iwasa H, Inubushi K (2013) Nitrous oxide flux from komatsuna (*Brassica rapa*) vegetated soil: a comparison between biogas digested liquid and chemical fertilizer. Biol Fertil Soils 49:971–976. https://doi.org/10.1007/s00374-013-0796-4

Six J, Feller C, Denef K, Ogle SM, Moraes Sa JC, de Albrecht A (2002) Soil organic matter, biota and aggregation in temperate and tropical soils—effects of no-tillage. Agronomy 22:755–775

Skowronska M, Filipek T (2014) Life cycle assessment of fertilizers: a review. Int Agrophys 28:101–110

Smil V (2002) Nitrogen and food production: proteins for human diets. Ambio 31:126–131. https://doi.org/10.1579/0044-7447-31.2.126

Smith P, Bustamante M, Ahammad H, Clark H, Dong H, Elsiddig EA, Haberl H, Harper R, House J, Jafari M (2014) Agriculture, forestry and other land use (AFOLU), climate change 2014: mitigation of climate change. In: Contribution of working group III to the fifth assessment report of the Intergovernmental panel on climate change. University Press, Cambridge

Soka G, Ritchie M (2014) Arbuscular mycorrhizal symbiosis and ecosystem processes: prospects for future research in tropical soils. Open J Ecol 4(1):11–22. https://doi.org/10.4236/oje.2014.41002

Solanki P, Bhargava A, Chhipa H, Jain N, Panwar J (2015) Nano-fertilizers and their smart delivery system. In: Rai M et al (eds) Nanotechnologies in food and agriculture. Springer, Cham, pp 81–101. https://doi.org/10.1007/978-3-319-14024-7_3

Soliman AS, El-feky SA, Darwish E (2015) Alleviation of salt stress on Moringa peregrina using foliar application of nanofertilizers. J Hortic For 2:36–47. https://doi.org/10.5897/JHF2014.0379

Stewart W, Dibb D, Johnston A, Smyth T (2005) The contribution of commercial fertilizer nutrients to food production. Agron J 97:1–6. https://doi.org/10.2134/agronj2005.0001

Stigter KA, Plaxton WC (2015) Molecular mechanisms of phosphorus metabolism and transport during leaf senescence. Plan Theory 4:773–798. https://doi.org/10.3390/plants4040773

Subbarao GV, Sahrawat KL, Nakahara K, Ishikawa T, Kishii M, Rao IM, Hash CT, George TS, Srinivasa Rao P, Nardi P, Bonnett D, Berry W, Suenaga K, Lata JC (2012) Biological nitrification inhibition-a novel strategy to regulate nitrification in agricultural systems. Adv Agron 114:249–302. https://doi.org/10.1016/B978-0-12-394275-3.00001-8

Subramanian KS, Rahale CS (2012) Ball milled nanosized zeolite loaded with zinc sulfate: a putative slow release Zn fertilizer. Int J Innov Hortic 1:33–40

Subramanian KS, Sharmila Rahale C (2013) Nano-fertilizers – synthesis, characterization and application. In: Adhikari T, Subba R (eds) Nanotechnology in soil science and plant nutrition. New India Publishing Agency, New Delhi

Subramanian KS, Paulraj C, Natarajan S (2008) Nanotechnological approaches in nutrient management. In: Chinnamuthu CR, Chandrasekaran B, Ramasamy C (eds) Nanotechnology applications in agriculture. TNAU technical bulletin. TNAU, Coimbatore, pp 37–42

Subramanian KS, Manikandan A, Thirunavukkarasu M, Rahale CS (2015) Nano-fertilizers for balanced crop nutrition. In: Ra M et al (eds) Nanotechnologies in food and agriculture. Springer International Publishing, Cham, pp 69–80. https://doi.org/10.1007/978-3-319-14024-7_3

Suman PR, Jain VK, Varma A (2010) Role of nanomaterials in symbiotic fungus growth enhancement. Curr Sci 99(9):1189–1191

Sun L, Lu Y, Yu F, Kronzucker HJ, Shi W (2016) Biological nitrification inhibition by rice root exudates and its relationship with nitrogen-use efficiency. New Phytol 212:646–656. https://doi.org/10.1111/nph.14057

Takahashi T, Dahlgren RA (2016) Nature, properties and function of aluminum-humus complexes in volcanic soils. Geoderma 263:110–121. https://doi.org/10.1016/j.geoderma.2015.08.032

Tao R, Wakelin SA, Liang Y, Hu B, Chu G (2018) Nitrous oxide emission and denitrifier communities in drip-irrigated calcareous soil as affected by chemical and organic fertilizers. Sci Total Environ 612:739–749. https://doi.org/10.1016/j.scitotenv.2017.08.258

Tarafdar JC, Raliya R, Mahawar H, Rathore I (2014) Development of zinc nanofertilizer to enhance crop production in pearl millet (*Pennisetum americanum*). Agric Res 3:257–262. https://doi.org/10.1007/s40003-014-0113-y

Tenuta M, Beauchamp EG (2003) Nitrous oxide production from granular nitrogen fertilizers applied to a silt loam soil. Can J Soil Sci 83:521–532. https://doi.org/10.4141/S02-062

Thingstrup I, Kahiluoto H, Jakobsen I (2000) Arbuscular mycorrhizal fungi at two levels of P fertilization phosphate transport by hyphae of field communities of arbuscular mycorrhizal fungi at two levels of P fertilization. Plant Soil 221:181–187. https://doi.org/10.1023/A:1004721626216

Thirunavukkarasu M (2014) Synthesis and evaluation of sulphur nano-fertilizers for groundnut. Ph.D. Thesis, Tamilnadu Agricultural University, Coimbatore, India

Tilman D, Balzer C, Hill J, Befort BL (2011) Global food demand and the sustainable intensification of agriculture. Biol Sci 108(50):20260–20264. https://doi.org/10.1073/pnas.1116437108

Trenkel MD (2007) Ullmann's agrochemicals, in Fertilizers, 3 (Chap. 4.4 and 4.5). Wiley-VCH Verlag GmbH & Co, KGaA, Weinheim

United Nations (2017) World population prospects: the 2017 revision, key findings and advance tables. Department of Economic and Social Affairs, Population Division. Working Paper No. ESA/P/WP/248

Vallejo A, Skiba UM, García-Torres L, Arce A, Lopez-Fernandez S, SanchezMartín L (2006) Nitrogen oxides emission from soils bearing a potato crop as influenced by fertilization with treated pig slurries and composts. Soil Biol Biochem 38:2782–2793. https://doi.org/10.1016/j.soilbio.2006.04.040

Velthof GL, Kuikman PJ, Oenema O (2003) Nitrous oxide emission from animal manures applied to soil under controlled conditions. Biol Fertil Soils 37:221–230. https://doi.org/10.1007/s00374-003-0589-2

Venterea RT, Burger M, Spokas KA (2005) Nitrogen oxide and methane emissions under varying tillage and fertilizer management. J Environ Qual 34:1467–1477

Verdi L, Mancini M, Ljubojevic M, Orlandini S, Dalla Marta A (2018) Greenhouse gas and ammonia emissions from soil: the effect of organic matter and fertilisation method. Ital J Agron 13:260–266. https://doi.org/10.4081/ija.2018.1124

Walling E, Vaneeckhaute C (2020) Greenhouse gas emissions from inorganic and organic fertilizer production and use: a review of emission factors and their variability. J Environ Manag 276:111211. https://doi.org/10.1016/j.jenvman.2020.111211

Wang L, Shen J (2019) Root/rhizosphere management for improving phosphorus use efficiency and crop productivity. Better Crops 13(1):36–39. https://doi.org/10.24047/BC103136

Wang G, Luo Z, Wang E, Zhang W (2018a) Reducing greenhouse gas emissions while maintaining yield in the croplands of Huang-Huai-Hai Plain, China. Agric For Meteorol 260:80–94. https://doi.org/10.1016/j.agrformet.2018.06.003
Wang X, Zou C, Gao X, Guan X, Zhang W, Zhang Y, Shi X, Chen X (2018b) Nitrous oxide emissions in Chinese vegetable systems: a meta-analysis. Environ Pollut 239:375–383. https://doi.org/10.1016/j.envpol.2018.03.090
Wang C, Amon B, Schulz K, Mehdi B (2021a) Factors that influence nitrous oxide emissions from agricultural soils as well as their representation in simulation models: a review. Agronomy 11:770. https://doi.org/10.3390/agronomy11040770
Wang X, Bai J, Xie T, Wang W, Zhang G, Yin S, Wang D (2021b) Effects of biological nitrification inhibitors on nitrogen use efficiency and greenhouse gas emissions in agricultural soils: a review. Ecotoxicol Environ Saf 220:112338. https://doi.org/10.1016/j.ecoenv.2021.112338
Weslien P, Kasimir Klemedtsson Å, Börjesson G, Klemedtsson L (2009) Strong pH influence on N_2O and CH_4 fluxes from forested organic soils. Eur J Soil Sci 60:311–320
Williams AG, Audsley E, Sandars DL (2010) Environmental burdens of producing bread wheat, oilseed rape and potatoes in England and Wales using simulation and system modelling. Int J Life Cycle Assess 15:855–868. https://doi.org/10.1007/s11367-010-0212-3
Wilson MA, Tran NH, Milev AS, Kannangara GSK, Volk H, Lu GQM (2008) Nanomaterials in soils. Geoderma 146:291–302. https://doi.org/10.1016/j.geoderma.2008.06.004
Wissuwa M, Heuer S, Gaxiola R, Schilling R, Herrera-estrella L, Damar L (2017) Improving phosphorus use efficiency: a complex trait with emerging opportunities. Plant J 90(5):868–885. https://doi.org/10.1111/tpj.13423
Wolff MW, Hopmans JW, Stockert CM, Burger M, Sanden BL, Smart DR (2017) Effects of drip fertigation frequency and N-source on soil N_2O production in almonds. Agric Ecosyst Environ 238:67–77. https://doi.org/10.1016/j.agee.2016.08.001
Xiang YAN, Ji-yun JIN, Ping HE, Ming-zao L (2008) Recent advances on the technologies to increase fertilizer use efficiency. Agric Sci China 7(4):469–479. https://doi.org/10.1016/S1671-2927(08)60091-7
Yang B, Xiong Z, Wang J, Xu X, Huang Q, Shen Q (2015) Mitigating net global warming potential and greenhouse gas intensities by substituting chemical nitrogen fertilizers with organic fertilization strategies in rice–wheat annual rotation systems in China: a 3-year field experiment. Ecol Eng 81:289–297. https://doi.org/10.1016/j.ecoleng.2015.04.071
Yao Z, Yan G, Zheng X, Wang R, Liu C, Butterbach-Bahl K (2017) Reducing N_2O and NO emissions while sustaining crop productivity in a Chinese vegetable-cereal double cropping system. Environ Pollut 231:929–941. https://doi.org/10.1016/j.envpol.2017.08.108
Yatim NM, Shaaban A, Dimin MF, Yusof F (2015) Statistical evaluation of the production of urea fertilizer-multiwalled carbon nanotubes using plackett burman experimental design. Procedia Soc Behav Sci 195:315–323
Yuan J, Sha ZM, Hassani D, Zhao Z, Cao LK (2017) Assessing environmental impacts of organic and inorganic fertilizer on daily and seasonal Greenhouse Gases effluxes in rice field. Atmos Environ 155:119–128. https://doi.org/10.1016/j.atmosenv.2017.02.007
Zebarth B, Rochette P, Burton D (2008) N_2O emissions from spring barley production as influenced by fertilizer nitrogen rate. Can J Soil Sci 88:197–205. https://doi.org/10.4141/CJSS06006
Zhang F, Shen J, Zhang J, Zuo Y, Li L, Chen X (2010) Rhizosphere processes and management for improving nutrient use efficiency and crop productivity: implications for China. Adv Agron 107:1–32. https://doi.org/10.1016/S0065-2113(10)07001-X
Zhang YY, Liu JF, Mu YJ, Pei SW, Lun XX, Chai FH (2011) Emissions of nitrous oxide, nitrogen oxides and ammonia from a maize field in the North China plain. Atmos Environ 45:2956–2961. https://doi.org/10.1016/j.atmosenv.2010.10.052
Zhang P, Ma Y, Zhang Z (2015) Interactions between engineered nanomaterials and plants: phytotoxicity, uptake, translocation, and biotransformation. Nanotech Plant Sci:77–99. https://doi.org/10.1007/978-3-319-14502-0_5

Zhang M, Li B, Xiong Z (2016) Effects of organic fertilizer on net global warming potential under an intensively managed vegetable field in southeastern China: a three-year field study. Atmos Environ 145:92–103. https://doi.org/10.1016/j.atmosenv.2016.09.024
Zhang F, Wang Q, Hong J, Chen W, Qi C, Ye L (2017) Life cycle assessment of diammonium- and monoammonium-phosphate fertilizer production in China. J Clean Prod 141:1087–1094. https://doi.org/10.1016/j.jclepro.2016.09.107
Zhao Z, Yue Y, Sha Z, Li C, Deng J, Zhang H, Gao M, Cao L (2015) Assessing impacts of alternative fertilizer management practices on both nitrogen loading and greenhouse gas emissions in rice cultivation. Atmos Environ 119:393–401. https://doi.org/10.1016/j.atmosenv.2015.08.060
Zheng L, Hong F, Lu S, Liu C (2005) Effect of nano-TiO_2 on strength of naturally aged seeds and growth of spinach. Biol Trace Elem Res 104:83–91. https://doi.org/10.1385/BTER:104:1:08

Chapter 11
Biochar for Plant Stress Tolerance for Climate-Resilient Agriculture

Kazim Ali, Hafiz Muhammad Mubasher, Umair Rasheed, Amjid Hussain, Ahmad Sher, and Abdul Qayyum

Abstract Biochar is also known as charcoal that is produced by the process of thermal degradation of biomass in anaerobic conditions such as pyrolysis. The liquid and gaseous biofuels are produced by pyrolysis. The pyrolysis platform is getting more and more attention all over the world due to the following four reasons: (i) renewable biofuels can be produced through pyrolysis; (ii) pyrolysis can be used to treat a lot of waste biomass and turn it into a fuel source; (iii) the addition of biochar to the soil results in the long-term sequestration of CO_2 that produced in the atmosphere, and (iv) the addition of biochar into the soil enhances the soil fertility and productivity of crops. Biochar is not widely used in agriculture at the moment because the agronomic importance of biochar in crop production, soil health, and mechanisms involved in soil fertility had not been yet widely determined and understood. On small farming, the biochar has direct effect on nutrient supply and it has also many indirect effects such as it improves the water and nutrient holding capacity, pH, conversion of P and S, cation exchange ability of soil, physical properties of soil, and population of soil microbes (mycorrhizal fungi). It also neutralizes the phytotoxic substances in the soil. The application of biochar in the soil increases the population of microbes in the rhizosphere through unidentified mechanisms, and it also enhances the populations of useful microorganisms that encourage growth of plant and resistance toward the biotic stresses. According to a few pieces of evidence, it has been seen that biochar also plays a significant role in the protection of plants against various foliar pathogens and soil-borne disease. There are many signs that show biochar enhances the ability of canopy to control the broad spectrum because it triggers responses in both the induced systemic resistance (ISR) and

K. Ali · H. M. Mubasher · U. Rasheed · A. Hussain
College of Agriculture, University of Layyah, Layyah, Pakistan

A. Sher (✉)
College of Agriculture, University of Layyah, Layyah, Pakistan

Department of Agronomy, Bahauddin Zakariya University, Multan, Pakistan
e-mail: ahmad.sher@bzu.edu.pk

A. Qayyum
Department of Agronomy, The University of Haripur, Haripur, Pakistan

M. Hasanuzzaman (ed.), *Climate-Resilient Agriculture, Vol 2*,
https://doi.org/10.1007/978-3-031-37428-9_11

systemic acquired resistance (SAR) pathways. This chapter shows that application of biochar in soil improves the relations between soil, plant, and microbes that may play a role in health of plant. One of the major advantages of biochar application to the soil potentially enhances the plant responses toward the disease.

Keywords Biochar · Plant stress tolerance · Biostimulate enzymes · Soil microbes · Phytotoxic substances

1 Introduction

Biochar can be described as a solid, organic waste that is derived by pyrolysis of biomass. Biochar can be used as soil amendments and has a profound impact on fertility of soils by altering the chemical physical and biological characteristics in the soil (Awad et al. 2018). The impact of biochar as a soil amendment improves the soil's quality and encourages the growth of plants, resulting in higher yields of crops. Biochar resource manufacturing process, soil quality, and condition and the kind of crop being planted could affect its efficiency. Most biochar is made from feedstocks like animal manures, agricultural wastes as well as paper materials. The significance of these wastes in the creation of biochar is a proven process to convert waste into beneficial and valuable substance (Brewer et al. 2014). Biochar production can be achieved on the small scale with an oven to a large scale with pyrolysis. Pyrolysis is a process that converts biomass into bio-oil, biochar, and syngas between 350 and 700 °C, without air (Varma et al. 2018). The chemical processes that make biochar that is solid include gasification and pyrolysis (Lehmann et al. 2011). Pyrolysis can be classified into two kinds that are slow pyrolysis as well as rapid pyrolysis. They depend on the time of residence as well as the rate at which they heat. Slow pyrolysis is more efficient at producing syngas, while faster pyrolysis creates more liquids and oils. Additionally, slow pyrolysis is able to facilitate more biochar production (36%) more than rapid pyrolysis (about 17%) as well as gasification (12%) (Uchimiya et al. 2011). Slow pyrolysis is also referred to as traditional carbonization. It creates biochar by heating the biomass. This is low for a lengthy period of time and this could require days (Cao et al. 2009). For fast pyrolysis, however, biochar is made at higher temperatures with a small residence time (11 s). The major difference between the slow and fast pyrolysis methods is the yields. The latter gives the highest yield of bio-oil, while the former yields better yields for biochar. When making biochar, the process starts with the drying of biomass in which the particles are further heated to allow volatile substances to release in the solid (Rhodes et al. 2008). The volatile substances that are produced can be carbon dioxide methane, carbon monoxide, as well as hydrogen or condensable organic substances like acetic acids and methanol. The polymerization and cracking processes during the phase of gas alter the spectrum of the product (Cetin et al. 2005).

In the past few years, there has been an increase in variety of risks to sustainable agriculture as a result of climate change, desertification and fertile soils, and insufficient, particularly stress from abiotic sources like the drought and heavy metals polycyclic aromatic hydrocarbons (PAH), persistent organic, and inorganic chemicals as well as. They have reduced soil fertility and affected development of plants, leading to an incredibly low agricultural production (Rajkumar et al. 2017). Plants have biochemical and physiological responses to deal with abiotic stressors and environmental stressors; the stressors whether as an individual or a combination (a heavy metal in drought, salinity and drought, and drought) have a significant effect on crop growth and the productivity of agriculture. For instance, the loss of crop due to drought India alone is believed to be more than 21% decrease in cereals. There is also a 5% decrease in the consumption of pulses and 33% decrease in sugarcane production, reduction of 29% in the production of citrus fruits, and 18% decrease in total production of food grains in 2012–2013 (Udmale et al. 2014). In the same way, heavy metals and salinity pose an extremely serious risk to crop production altering the plant's metabolic and physiological processes. Beyond the direct effects on the plants, climate change can also impact the growth of plants indirectly by altering the nature of soil and allowing for create new pathogens and pests.

2 Nutritional Composition of Biochar

Numerous research works since 1850–2011 have demonstrated the varying impacts of biochar on yields of plants. The majority of studies said positive effects, 20% claimed none while 30% were negative (Maddox 2013). According to earlier research, both types of biochar PLB1 and PLB2 that were used in our study contained the higher concentrations of the major micronutrients such as Zn, Mn, and Cu (Macdonald et al. 2014). The small ratio of carbon and nitrogen in the bio-solids biochar demonstrates that it could be a suitable resource to use as a fertilizer for N (Hossain et al. 2011) and the bio-solids were the most abundant in terms of overall Kjeldahl nitrogen (TKN) quantity among the investigated materials. However, there is large variability in bio-solids biochar's properties using different methods of processing. So, it is impossible to assume that all types of bio-solids biochar have similar effects on land application. Freitas et al. (2017) reported and tested in research that a variety of bio-solids and biochar samples have TKN concentrations ranging from 13,000 to 50,000.

Different methods can be used to make bio-solids. The various stabilization processes have significant effects on the labile P percentage of bio-solids (Brandt et al. 2004). In research, the conversion of bio-solids into biochar was analyzed; Freitas et al. (2017) found that bio-solids and biochar differ in their mineral and elemental content. The bio-solids biochar examined in this research had a TP level of 67,300 mg kg^{-1}, and the XRD method did not reveal any crystalline mineral phosphate. Bio-solids vary in characterization depending on the treatment process they use (Freitas et al. 2017; Silveira et al. 2019). The bio-solids source may alter the interactions of P.

3 Manufacturing of Biochar

Biochar is made by pyrolyzing biomass under anoxic or anaerobic conditions between 300 and 800 °C. Before pyrolysis, it was subjected to a variety of pretreatment procedures, such as crushing, sieving, and air-drying. The final product is made by processing the pyrolysis products. The process itself can be categorized as "slow," "medium," or "fast" pyrolysis based on the time and temperature of pyrolysis. Biochar is the primary product of slower pyrolysis (residence time greater than 1 h), as well as bio-oil and tiny amounts of syngas (including CO, H_2, $CH_{4,}$ and CO) are also produced. The yields of syngas, biochar, and bio-oil range from 13%, 12%, to 75%, respectively, during the fast pyrolysis (residence time less than 2 s) (Ahmad et al. 2014; Jin et al. 2016). Syngas can be reused as a clean fuel. Bio-oil, also known as biodiesel, is regarded as an eco-friendly alternative to diesel fuel due to its similar molecular weight and other properties to diesel fuel (El-Dalatony et al. 2017). So, when it is heated to high temperatures, all three products produced by biomass pyrolysis are highly recoverable. The physicochemical properties of biochar, which vary significantly depending on the feedstock used and production conditions, determine its adsorption and activation performance. Consequently, it is essential to comprehend the biochar production processes and control parameters that effects on the biochar's physicochemical characteristics.

4 Biomass Feedstock

The yield, elemental content, and microstructure of the biochar are all determined by the composition of the biomass. Higher the amount of lignin in the raw materials derived from plants more yield of biochar. Therefore the biochar that is made up of solid wastes including manure from animals and sludge is higher than those from crop residues as well as wood biomass, due to the nature of the material, which includes high amount of inorganic compounds. Enders et al. (2012) and Cantrell et al. (2012) found that metal content in feces of animals slows down the process of removing volatile substances to some extent. In addition, releases of volatile compounds aid in the formation of a fully ordered mesoporous or microstructure in biochar, thereby increasing the surface area as well as the number of locations that are active (Bruun et al. 2012). However, the surface structure and reaction properties of the biochar may be significantly impacted if the feedstock contains an excessive amount of mineral ash, even though the yield may rise. Additionally, the surface's functional groups may be reflected in the mineral content of the biochar. The properties of adsorption and activation in the biochar are closely linked to double-bond oxygen-, nitrogen-, and double-bond groups (Rajapaksha et al. 2016). The study found that high-temperature burning alters the elemental content of raw materials. The change is most obvious in the increase of C content and decreases in H and O content, both of which improve the aromatic properties and decrease the polarity of biochar. Biochar made from livestock manure and sewage sludge has a high S and N content, whereas biochar made from plant matter typically has a lower C and O content than biochar made from solid waste.

5 Role of Biochar in Plant Stress Tolerance

Biochar not only helps in plant growth and performance, but it also helps to protect them from the impacts of abiotic stresses like drought or salinity (Rizwan et al. 2016; Shaaban et al. 2018). According to recent research, the use of biochar is quickly reducing the negative effects of trace metals (Yuan et al. 2019). This approach is less expensive and has less impact on the environment (Younis et al. 2016). According to the most recent research, applying biochar to contaminated soil significantly enhances the antioxidant activity and boosts up plant growth. By absorbing soil contaminants within the plant, such as trace metals, biochar aids in the reduction of soil pollution (El-Naggar et al. 2020).

Although a lot of research has been done on the properties of biochar, the role that biochar plays in the removal of soil pollution and environmental sustainability has not been thoroughly explored (Yuan et al. 2019; Haider et al. 2022). The research studies concentrate on a specific kind of pollutant or the procedure for remediating it. Therefore, it is essential to highlight the pollutant removal induced by biochar in support of the actual efforts of soil remediation. This chapter examines recent developments regarding the application of biochar to the remediation of trace metal-contaminated soils. The environmental and agricultural effects of biochar's in the removal of trace metals from soil are discussed in greater detail in this study. Due to the toxicity of trace metals in plants, the potential effects of biochar on the production of reactive oxygen species (ROS) are also discussed.

6 Biochar Provides Abiotic Stress Tolerance

The biochar's reaction to various abiotic stresses (the heavy metal (HM) stress, drought stress, temperature stress, lodging stress) is described below in sub-headings.

6.1 Heavy Metal Stress

Biochar is a material that is negatively charged that has high cation exchange capacity (CEC) mineral elements and differs within functional groups. Biochar has ability to reduce the uptake of heavy metals in soil through the electrostatic effect, which is accompanied by adsorption precipitation and complexation. In particular, the electronegativity of biochar improves the attraction of negatively charged elements (Cu, Zn, Cr, Zn, etc.) and decreases the bioavailability of soil. In the same way, the large CEC of biochar could absorb heavy metals that are on their surface via the ion exchange mechanism (He et al. 2019). A number of researchers (Gonzaga et al. 2018) have reported previously that biochar is made from waste materials which dramatically reduce the bioavailability of heavy metals by the electrostatic attraction and adsorption process which in turn increases crop's growth and grain yield. In addition, the use of mineral-104-loaded biochar can precipitate metals, transforming to an insoluble form. In addition, the functional group found in biochar offers an excellent site of active binding of heavy metals that reduces the availability of metals and creates healthier conditions for plant growth. Biochar's effects on plant growth and resistance to stress in metal-contaminated soil.

6.2 Drought Stress

Drought is regarded as to be one of the most significant factors that negatively impact the growth of plants and their productivity. Biochar is being viewed as a tool for mitigation to alleviate the stress of drought and to encourage the growth of plants. When there is condition of drought, the plant suffers from severe water deficit because of the loss of soil's water retention capacity that significantly impacts the growth of plants and their metabolism. However, the use of biochar in the form of soil amendment can increase the ability to hold water and the amount of moisture in the soil by altering the soil's pore size distribution as well as the stability of the soil's aggregate (Tanure et al. 2019). The porous nature of biochar enhances the retention of water in soils through expanding the micropores distribution within the soil. Additionally, the application of biochar can increase the active substances in the osmotic system that reduce the loss of water from leaves as in addition, the cationic character of the biochar improves plant's ability to absorb water. Additionally, biochar decreases the density bulk of soil, which affects water infiltration process and helps to facilitate soil aeration as well as the process of microbial respiration, which in turn boosts the growth of plants in conditions of drought (Tanure et al. 2019). Recently, it was reported that the use of biochar can reduce some of the reactive components of thiobarbituric acid and electrolyte leakage from leaves. Furthermore, it boosts the antioxidant capacity of wheat that is grown in drought conditions which result in increased yield and stress tolerance of the plant (Abbas et al. 2018; El-Mageed et al. 2020). In a separate investigation, Hashem et al. (2019) found that biochar's amendment has increased the amount of osmotic-active chemicals like K^+ in the tissues of plants, resulting in an increase in the amount of water absorbed by plants which in turn increases the water content in the leaf under the stress of water condition and also protects the plant from losing water. In this instance, the increase in the quantity of osmotic-active substances may be related to the high cation content in biochar. In addition, it is the fact that the modification of biochar enhances the polyphenolic and proline-containing compounds, which is an antioxidant and cleanses drought-related ROS, and consequently enhances the plant's survival and yield (Afshar et al. 2014). The porous nature of biochar enhances the retention of water in soils by increasing the distribution of micropores in the soil. In addition, the use of biochar enhances the active substances in the osmotic system that reduces the loss of water from the leaves, as the cationic nature of the biochar improves plant's ability to absorb water. Additionally, biochar decreases the density bulk of soil, which alters the water absorption and aids in the aeration of soil and the process of microbial respiration, which in turn boosts the growth of plants in a condition of drought (Tanure et al. 2019). Recently, it was reported that the use biochar can reduce those reactive compounds of thiobarbituric acid and electrolyte leakage from leaves. Additionally, it enhances the antioxidant activities of wheat that is grown in drought conditions that resulted in improved resistance to stress and increased growth (Abbas et al. 2018).

6.3 Temperature Stress

Stress on temperature (heat or cold) impacts plant growth through reducing effectiveness of water of photosynthetic rate, photosynthetic efficiency, and conductance of the stomata. Furthermore, the increased temperatures trigger the production of ROS and lower antioxidant activity that triggers cell death, which in turn hinders the growth of plants (Bruno et al. 2020; Rajkumar et al. 2017). Biochar's use aids in protecting the plant by limiting the negative impacts caused by temperature stress in 119 conditions. Biochar will increase the osmotic adjustment and reduce loss of water. In addition, biochar enhances the antioxidant capacity of the plant, which assists the plant in reducing ROS accumulation caused by stress. It also improves the plant's growth (Abbas et al. 2018; El-Mageed et al. 2020). Yet, only a few studies have addressed the effect of biochar on temperature stress responses. For instance, Fahad et al. (2015) examined the impact of biochar application on rice under conditions of heat stress and found that the application of biochar reduces the negative impact of heat stress on plants by the increased photosynthesis and water efficacy of the plants. Similarly, Yuan et al. (2017) examined the effects of biochar application on the cold tolerance of rice seedlings and found that biochar application enhances the interaction between organic molecules with stress-related proteins, which improves the sensitivity of the rice seedling.

The stress of temperature (heat as well as cold) impacts on plant growth by reducing water efficiency, photosynthetic rates, and the conductance of stomata. Additionally, the alteration in temperatures triggers the production of ROS and reduces antioxidant activity, which causes cell death, which in turn hinders growth of the plant (Bruno et al. 2020; Rajkumar et al. 2017). Biochar's use is aid in protecting the plant by limiting the negative impacts that are induced by stress conditions at 119 °C. Biochar could increase osmotic adjustments and stop loss of water. Additionally, biochar increases the antioxidant capacity of the plant. This helps the plant reduce the ROS accumulation caused by stress and boosts the growth of the plant (Abbas et al. 2018; El-Mageed et al. 2020). But, only a handful of studies have explored the effect of biochar on temperature stress responses until now. For instance, Fahad et al. (2015) examined the impact of biochar application on rice under conditions of heat stress and found that the application of biochar reduces the negative impact of heat stress on plants by enhanced photosynthesis and increased water usage efficacy of the plants. Similarly, Yuan et al. (2017) examined the effects of biochar on the cold tolerance of rice seedlings. They discovered that the application of biochar improves the interaction between organic molecules with stress-related proteins, which enhances the tolerance of rice seedling.

6.4 Stress Caused by Plant Diseases

A number of studies have looked at the possibility of biochar soil amendments to affect the resistance of plants to pathogens causing disease. In relation to soil pathogens (Matsubara et al. 2002), although primarily concerned with the effects in the effects of AM fungal inoculations in the development of asparagus resistance to fusarium root rot, they also found the charcoal amendments could have an inhibitory effect on pathogen that is borne in soil *Fusarium* sp. They discovered that charcoal made from coconut fibers slowed the growth of fusarium root and crown rot and also the increase in AM the growth of asparagus seeds.

This confirms earlier findings that root lesions caused by *Fusarium* sp. asparagi and+ *F. proliferatum* the oxysporum were reduced when biochar made from ground hardwood was added to the asparagus field soil as compared to a control with no amendments (Elmer and Pignatello 2011). Additionally, biochar application enhanced colonization of AM in the asparagus root which in turn helped suppress the disease, even when they were supplemented with allelopathic substances which are known to inhibit AM colonization within aspen (Elmer and Pignatello 2011). These results support the notion that biochar could help in awarding against allelopathic effects through the adsorption and detoxification process of allelopathic agents, a process that was previously discovered by Wardle et al. (1998). Apart from detoxification of chemical agents, the reduction of soil pathogens may be the result of a number of mechanisms that include (i) supplying nutrients and enhancing their uptake and solubility nutrients. This helps improve the plant's growth and resistance and enhancement of soil microorganisms that cause pathogenic stress; (ii) stimulation of microbes, which offer an immediate defense against the soil's pathogenic organisms by antibiosis and parasitism, or competition; (iii) biochar-associated organic compounds can suppress the soil's sensitive microbiota, resulting in growth resistance of microbial populations; (iv) biochar could trigger the defense system of plants and elicitors could be either biochar-borne chemicals, or biochar-induced microorganisms.

Biochar lacks an indigenous population of microorganisms that could aid in disease prevention. It is initially sterile. However, biochar can impact the microbes and communities, as mentioned in the previous paragraphs, and these changes could be beneficial. Graber et al. (2010) identified a variety of biochar-derived compounds that are believed to negatively impact microbial growth and viability. They include ethylene glycol, propylene glycol, hydroxypropionic and butyric acids benzoic acid, o-cresol and benzoic acid as well as the quinones (resorcinol as well as hydroquinone) and 2-phenoxyethanol. The presence of low levels of these harmful substances could disrupt sensitive soil microbiota and lead to an increase in resistant microbe communities (Graber et al. 2010).

Biochar's potential to trigger resistance in plants against pathogens that cause disease is being investigated in different systems that deal with pathogens that cause foliar damage. The severity of disease caused by the necrotrophic (*Botrytis cinerea*) and biotrophic (*Oidiopsis sicula* (originally called by its teleomorphic term: *Leveillula taurica*)) plant pathogens that cause foliar damage in tomatoes and pepper (Elad et al. 2010) was significantly decreased in treatments with biochar amendments. The reduction in damage caused by broad mite (*Polyphagotarsonemus latus*) in plants of pepper treated with biochar was also noted (Elad et al. 2010).

6.5 *Stress Caused by Insecticides in the Plant*

In this chapter devoted to the phytopathological side of the addition of biochar to soil, it is remiss not to highlight the fact that biochar can adversely affect the effectiveness of pesticides applied to soil such as insecticides, fungicides, and herbicides, because of the large adsorption affinity and the capacity that biochar shows toward a variety of organic compounds. Additionally, the organic component that is soluble of biochar could create complexes with herbicides applied to soil and increase their downward transportation from the zone of soil (Cabrera et al. 2011). For the fungicide Pyrimethanil, it has been demonstrated that an increase in biochar content causes a gradual increase in adsorption (Yu et al. 2010). Biochar may inactivate pesticides due to the strong pesticide adsorption (Graber et al. 2011; Nag et al. 2011). Consequently, high amount of pesticides could be required to achieve the same degree of protection against pests such as in an investigation of the effectiveness of fumigants against the many nematodes (Graber et al. 2011).

Biochar with large surfaces area (specific surface area, also known as SSA) can be especially problematic for controlling pests because their adsorption power for various chemicals is typically much higher than that of lower SSA biochar (Wang et al. 2010; Yang et al. 2010). The SSA content of biochar typically rises with the increase in temperature of the pyrolysis. When soils were altered by 22% (52 T per ha) biochar with a high SSA biochar (SSA of 242 m^2/g), the control of the plant green the foxtail (*Setaria viridis*) was greatly reduced even when the maximum herbicide (S-metolachlor and sulfentrazone) labels were applied (Graber et al. 2011).

6.6 Stress Caused by Plant Microorganisms

According to a recent review by Lehmann et al. (2011), the evidence of soil biochar was increased which have significant effects on soil microorganisms. In most of the studies that were looked at in that review, the biochar-amended soils had more microbial biomass as compared to soil without biochar. Biochar additions can cause significant alterations to the composition of microbial community as well as enzymatic activity in many soils as well as the rhizosphere. In particular, the biochar amendment was typically characterized by increased number of members belonging to the Actinobacteria and Bacteroidetes phyla (Jesus et al. 2009; Kolton et al. 2011). Although there is not much information about how biochar influences microbial abundance as well as population structure widely acknowledged that soil microorganisms can be extremely influential on the productivity of plants, changes induced by biochar to soil microorganisms can contribute to "The Biochar Effect" (Graber et al. 2010).

Biochar addition to soils often leads to significant increases in mycorrhizal fungi and plant interaction (Warnock et al. 2007). For example, mycorrhizal colonization of wheat roots and grain yield significantly increased when biochar was applied in conjunction in conjunction with mineral fertilizer. The application of biochar as well as fertilizer increased mycorrhizal colonization within clover bioassay plant species, and it was found that biochar provides adequate conditions for mycorrhizal mycoflora to infiltrate the roots of plants. Solaiman et al. (2010) and Warnock et al. (2007) described four ways through which biochar can influence mycorrhizal growth and/or function: (i) alteration of soil's physicochemical characteristics; (ii) indirect effects on mycorrhizae due to impacts on other soil microbes; (iii) the interference of plant fungi in signaling and the detoxification of toxic chemicals by biochar; and (iv) providing refuge in the face of fungal invaders (Warnock et al. 2007). These mechanisms can also impact other soil-dwelling bacteria, including plant pathogens. Beyond the well-known roles in arbuscular mycorrhizal (AM) organisms in stimulating growth in plants, it is recognized that microorganisms of the rhizosphere in general and certain strains from groups like *Pseudomonas*, *Bacillus* as well as *Trichoderma* in particular may improve the growth of plants in a variety of cultivation systems. For instance, the growth increase was caused by the different species belonging to *Trichoderma* that are found in tomatoes tobacco, radish, and tomato (Windham et al. 1986) as well as by different species belonging to *Bacillus* (Kloepper et al. 2004) and in various crop species. There are not many studies that have looked at the growth of plant growth-promoting rhizobacteria/fungi (PGPR/F) in soils that have been amended with biochar. One of the most notable work cultures was bulk soil and rhizospheres taken from peppers that were mature that had their growth improved by biochar inclusions (Graber et al. 2010). Of the 20 distinct strains collected, phylogenetic characterization using the partial analysis of 16S rRNA genes found that eight of the strains shared high similarity in sequence and *Pseudomonas*, *Mesorhizobium*, *Brevibacillus,* and *Bacillus* strains that have been identified for their capability to act as growth-promoting agents (Graber et al. 2010).

However, this study, which is being the only one in the world is not definitively proving the role of biochar-stimulated bacteria in promoting plant growth, could suggest an important research direction that could aid in understanding this "Biochar effect." A variety of PGPR/F organisms have been found to boost plant health as well as plant growth, either directly tackling plant pathogens or by stimulating the plant's systemic resistance to pests and diseases (Kaewchai et al. 2009).

7 Conclusion and Future Perspectives

Soils contaminated with trace metals interrupt numerous physiological, biochemical and morphological processes that impose severe concerns after being absorbed and transported to various plants' parts. Most noticeably, too much production of ROS causes membrane degradation through lipid peroxidation and has negative effects on the cellular level. These processes are controlled by plants through the production of antioxidants, both enzymatic and non-enzymatic, which help the plants defend themselves against soil contamination stress.

On the other hand, metals like Cd, Al, Ni, Pb, Cr, and Hg cause oxidative stress, which causes the organelles of plants to completely break down. We attempted to explain the mechanisms that involve in the reclamation process and increased productivity crop in this chapter by addressing several studies that support the significant function of biochar amended in the rhizosphere in response to plant stress caused by trace metals. In addition, under stress, the use of biochar to plants can strengthen the immune system if done at the right level. The use of biochar in contaminated soils may be a promising way to increase agricultural productivity and protect plant biodiversity. However, future strategies must also conduct a thorough analysis of the best ways to promote, produce, and improve suitability of biochar. Its adsorption potential and sustainability act as the best tool for removing trace metals from contaminated agricultural soils.

References

Abbas T, Rizwan M, Ali S, Adrees M, Mahmood A, Zia-Urrehman M, Qayyum MF (2018) Biochar application increased the growth and yield and reduced cadmium in drought stressed wheat grown in an aged contaminated soil. Ecotoxicol Environ Saf 148:825–833

Afshar RK, Chaichi MR, Asareh MH, Hashemi M, Liaghat A (2014) Interactive effect of deficit irrigation and soil organic amendments on seed yield and flavonolignan production of milk thistle (*Silybum Marianum* L). Gaertn. Ind Crop Prod 58:166–172

Ahmad M, Rajapaksha AU, Lim JE, Zhang M, Bolan N, Mohan D, Ok YS (2014) Biochar as a sorbent for contaminant management in soil and water: a review. Chemosphere 99:19–33. https://doi.org/10.1016/j.chemosphere.2013.10.071

Awad YM, Lee SS, Kim KH, Ok YS, Kuzyakov Y (2018) Carbon and nitrogen mineralization and enzyme activities in soil aggregate-size classes: Effects of biochar, oyster shells, and polymers. Chemosphere 198: 40–48

Brandt RC, Elliott HA, Connor OGA (2004) Water-extractable phosphorus in biosolids: implications for land-based recycling. Water Environ Res 76(2):121–129. https://doi.org/10.2175/106143004x141645

Brewer CE, Chuang VJ, Masiello CA, Gonnermann H, Gao X, Dugan B, Davies CA (2014) New approaches to measuring biochar density and porosity. Biomass Bioenergy 66:176–185. https://doi.org/10.1016/j.biombioe.2014.03.059

Bruno LB, Karthik C, Ma Y, Kadirvelu K, Freitas H, Rajkumar M (2020) Amelioration of chromium and heat stresses in Sorghum bicolor by Cr6+ reducing-thermo tolerant plant growth promoting bacteria. Chemosphere 244(125521):125521. https://doi.org/10.1016/j.chemosphere.2019.125521

Bruun EW, Ambus P, Egsgaard H, Hauggaard-Nielsen H (2012) Effects of slow and fast pyrolysis biochar on soil C and N turnover dynamics. Soil Biol Biochem 46:73–79. https://doi.org/10.1016/j.soilbio.2011.11.019

Cabrera A, Cox L, Spokas KA, Celis R, Hermosín MC, Cornejo J, Koskinen WC (2011) Comparative sorption and leaching study of the herbicides fluometuron and 4-chloro-2-methylphenoxyacetic acid (MCPA) in a soil amended with biochars and other sorbents. J Agric Food Chem 59(23):12550–12560. https://doi.org/10.1021/jf202713q

Cantrell KB, Hunt PG, Uchimiya M, Novak JM, Ro KS (2012) Impact of pyrolysis temperature and manure source on physicochemical characteristics of biochar. Bioresour Technol 107:419–428

Cao X, Ma L, Gao B, Harris W (2009) Dairy-manure derived biochar effectively sorbs lead and atrazine. Environ Sci Tech 43(9): 3285–3291

Cetin E, Gupta R, Moghtaderi B (2005) Effect of pyrolysis pressure and heating rate on radiata pine char structure and apparent gasification reactivity. Fuel 84(10): 1328–1334

Elad Y, David D, Harel Y, Borenshtein M, Ben Kalifa H, Silber A, Graber ER (2010) Induction of systemic resistance in plants by biochar, a soil applied carbon sequestering agent. Phytopathology 100:913–921

El-Dalatony M, Salama ES, Kurade M, Hassan S, Oh S-E, Kim S, Jeon BH (2017) Utilization of microalgal biofractions for bioethanol, higher alcohols, and biodiesel production: a review. Energies 10(12):2110. https://doi.org/10.3390/en10122110

El-Mageed TA, Rady MM, Taha RS, Abd-El Azeam S, Simpson CR, Semida WM (2020) Effects of integrated use of residual sulfur-enhanced biochar with effective microorganisms on soil properties, plant growth and short-term productivity of (*Capsicum annuum)* under salt stress. Sci Horti 261:108930

Elmer WH, Pignatello JJ (2011) Effect of biochar amendments on mycorrhizal associations and Fusarium crown and root rot of asparagus in replant soils. Plant Dis 95(8):960–966. https://doi.org/10.1094/PDIS-10-10-0741

El-Naggar A, M-Lee H, Hur J, Lee YH, Igalavithana AD, Shaheen SM (2020) Biochar-induced metal immobilization and soil biogeochemical process: an integrated mechanistic approach. Sci Total Environ 698:134112. https://doi.org/10.1016/j.scitotenv.2019.134112

Enders A, Hanley K, Whitman T, Joseph S, Lehmann J (2012) Characterization of biochars to evaluate recalcitrance and agronomic performance. Bio-resour Technol 114:644–653

Fahad S, Hussain S, Saud S, Tanveer M, Bajwa AA, Hassan S, Huang J (2015) A biochar application protects rice pollen from high-temperature stress. Plant Physiol Biochem 96:281–287. https://doi.org/10.1016/j.plaphy.2015.08.009

Freitas AM, Nair VD, Harris WG, Mosquera-Losada MR, Nair PKR (2017) Phosphorus release behavior of biosolids and corresponding biochars in abstract retrieved from ASA, CSSA and SSSA international annual meetings

Gonzaga MIS, Mackowiak C, Quintão de Almeida A, Wisniewski A Jr, Figueiredo de Souza D, da Silva LI, Nascimento de Jesus A (2018) Assessing biochar applications and repeated (*Brassica juncea* L). production cycles to remediate Cu contaminated soil. Chemosphere 201:278–285. https://doi.org/10.1016/j.chemosphere.2018.03.038

Graber ER, Meller Harel Y, Kolton M, Cytryn E, Silber A, Rav David D, Tsechansky L, Borenshtein M, Elad Y (2010) Biochar impact on development and productivity of pepper and tomato grown in fertigated soilless media. Plant and soil 337: 481–496

Graber ER, Tsechansky L, Khanukov J, Oka Y (2011) Sorption, volatilization, and efficacy of the fumigant 1,3-dichloropropene in a biochar-amended soil. Soil Sci Soc Am J 75(4):1365–1373. https://doi.org/10.2136/sssaj2010.0435

Haider FU, Wang X, Farooq M, Hussain S, Cheema SA, Ain NU, Liqun C (2022) Biochar application for the remediation of trace metals in contaminated soils: implications for stress tolerance and crop production. Ecotoxicol Environ Saf 230(113165):113165. https://doi.org/10.1016/j.ecoenv.2022.113165

Hashem A, Kumar A, Al-Dbass AM, Alqarawi AA, Al-Arjani A-BF, Singh G, Abd Allah EF (2019) Arbuscular mycorrhizal fungi and biochar improves drought tolerance in chickpea. Saudi J Biol Sci 26(3):614–624. https://doi.org/10.1016/j.sjbs.2018.11.005

He L, Zhong H, Liu G, Dai Z, Brookes PC, Xu J (2019) Remediation of heavy metal contaminated soils by biochar: 133 mechanisms, potential risks and applications in China. Environ Pollut 252:846–855

Hossain MK, Strezov V, Chan KY, Ziolkowski A, Nelson PF (2011) Influence of pyrolysis temperature on production and nutrient properties of wastewater sludge biochar. J Environ Manag 92(1):223–228. https://doi.org/10.1016/j.jenvman.2010.09.008

Jesus da CE, Marsh TL, Tiedje JM, de S Moreira FM (2009) Changes in land use alter the structure of bacterial communities in Western Amazon soils. ISME J 3(9):1004–1011. https://doi.org/10.1038/ismej.2009.47

Jin Y, Liang XQ, He MM, Liu Y, Tian GM, Shi JY (2016) Manure biochar influence upon soil properties, phosphorus distribution and phosphatase activities: a microcosm incubation study. Chemosphere 142:128–135

Kaewchai S, Soytong K, Hyde KD (2009) Mycofungicides and fungal biofertilizers. Fungal Divers 38:25–50

Kloepper JW, Ryu CM, Zhang S (2004) Induced systemic resistance and promotion of plant growth by *Bacillus* spp. Phytopathology 94(11):1259–1266. https://doi.org/10.1094/PHYTO.2004.94.11.1259

Kolton M, MellerHarel Y, Pasternak Z, Graber ER, Elad Y, Cytryn E (2011) Impact of biochar application to soil on the root-associated bacterial community structure of fully developed greenhouse pepper plants. Appl Environ Microbiol 77(14):4924–4930. https://doi.org/10.1128/AEM.00148-11

Lehmann Johannes Rillig MC, Thies J, Masiello CA, Hockaday WC, Crowley D (2011) Biochar effects on soil biota – a review. Soil Biol Biochemist 43(9):1812–1836. https://doi.org/10.1016/j.soilbio.2011.04.022

Macdonald LM, Farrell M, Van Zwieten L, Krull ES (2014) Plant growth responses to biochar addition: an Australian soils perspective. Biol Fertil Soils 50(7):1035–1045. https://doi.org/10.1007/s00374-014-0921-z

Maddox N (2013) The promise [and uncertainties] of Biochar. CSA News 58(9):4–9. https://doi.org/10.2134/csa2013-58-9-1

Matsubara Y, Hasegawa N, Fukui H (2002) Incidence of Fusarium root rot in asparagus seedlings infected with arbuscular mycorrhizal fungus as affected by several soil amendments. Engei Gakkai Zasshi Engei Gakkai (Japan) 71(3):370–374. https://doi.org/10.2503/jjshs.71.370

Nag SK, Kookana R, Smith L, Krull E, Macdonald LM, Gill G (2011) Poor efficacy of herbicides in biochar-amended soils as affected by their chemistry and mode of action. Chemosphere 84(11):1572–1577. https://doi.org/10.1016/j.chemosphere.2011.05.052

Rajkumar M, Bruno LB, Banu JR (2017) Alleviation of environmental stress in plants: the role of beneficial Pseudomonas spp. Crit Rev Environ Sci Technol 47(6):372–407. https://doi.org/10.1080/10643389.2017.1318619

Rajapaksha AU, Chen SS, Tsang DC, Zhang M, Vithanage M, Mandal S, Gao B, Bolan NS, Ok YS (2016) Engineered/designer biochar for contaminant removal/immobilization from soil and water: potential and implication of biochar modification. Chemosphere 148: 276–291

Rizwan M, Ali S, Abbas T, Zia-Ur-Rehman M, Hannan F, Keller C, Al-Wabel MI, Ok YS (2016) Cadmium minimization in wheat: a critical review. Ecotoxicol Environ Saf 130:43–53

Rhodes AH, Carlin A, Semple KT (2008) Impact of black carbon in the extraction and mineralization of phenanthrene in soil. Environ Sci Tech 42(3): 740–745

Shaaban M, Van-Zwieten L, Bashir S, Younas A, Núñez-Delgado A, Chhajro MA (2018) Concise review of biochar application to agricultural soils to improve soil conditions and fight pollution. J Environ Manag 228:429–440. https://doi.org/10.1016/j.jenvman.2018.09.006

Silveira ML, O'Connor GA, Lu Y, Erickson JE, Brandani C, Kohmann MM (2019) Runoff and leachate phosphorus and nitrogen losses from grass-vegetated soil boxes amended with biosolids and fertilizer. J Environ Qual 48(5):1498–1506. https://doi.org/10.2134/jeq2019.03.0106

Solaiman ZM, Blackwell P, Abbott LK, Storer P (2010) Direct and residual effect of biochar application on mycorrhizal root colonisation, growth and nutrition of wheat. Soil Res 48(7):546. https://doi.org/10.1071/sr10002

Tanure MMC, da Costa LM, Huiz HA, Fernandes RBA, Cecon PR, Pereira Junior JD, da Luz JMR (2019) Soil water retention, physiological characteristics, and growth of maize plants in response to biochar application to soil. Soil Tillage Res 192:164–173. https://doi.org/10.1016/j.still.2019.05.007

Uchimiya M, Wartelle LH, Klasson KT, Fortier CA, Lima M (2011) Influence of pyrolysis temperature on biochar property and function as a heavy metal sorbent in soil. J Agric Food Chem 59:2501–2510

Udmale PD, Ichikawa Y, Kiem AS, Sudhindra N (2014) Drought impacts and adaptation strategies for 138 agriculture and rural livelihood in the Maharashtra State of India. Open Agric J 8:41–47

Varma AK, Shankar R, Mondal P (2018) A review on pyrolysis of biomass and the impacts of operating conditions on product yield, quality, and upgradation. In: Recent advancements in biofuels and bioenergy utilization, pp 227–259. https://doi.org/10.1007/978-981-13-1307-3_10

Wang H, Lin K, Hou Z, Richardson B, Gan J (2010) Sorption of the herbicide terbuthylazine in two New Zealand Forest soils amended with biosolids and biochars. J Soils Sediments 10(2):283–289. https://doi.org/10.1007/s11368-009-0111-z

Wardle DA, Zackrisson O, Nilsson M-C (1998) The charcoal effect in Boreal forests: mechanisms and ecological consequences. Oecologia 115(3):419–426. https://doi.org/10.1007/s004420050536

Warnock DD, Lehmann J, Kuyper TW, Rillig MC (2007) Mycorrhizal responses to biochar in soil – concepts and mechanisms. Plant Soil 300(1–2):9–20. https://doi.org/10.1007/s11104-007-9391-5

Windham MT, Elad Y, Baker T (1986) A mechanism for increased plant growth induced by *Trichoderma* spp. Phytopathology 76:518–521

Yang X-B, Ying G-G, Peng P-A, Wang L, Zhao J-L, Zhang L-J, He H-P (2010) Influence of biochars on plant uptake and dissipation of two pesticides in an agricultural soil. J Agric Food Chem 58(13):7915–7921. https://doi.org/10.1021/jf1011352

Younis U, Malik SA, Rizwan M, Qayyum MF, Ok YS, Shah MHR, Rehman RA (2016) Ahmad Biochar enhances the cadmium tolerance in spinach (*Spinacia oleracea*) through modification of Cd uptake and physiological and biochemical attributes. Environ Sci Pollut Res 23:21385–21394

Yu X, Pan L, Ying G, Kookana RS (2010) Enhanced and irreversible sorption of pesticide pyrimethanil by soil amended with biochars. J Environ Sci (China) 22(4):615–620. https://doi.org/10.1016/s1001-0742(09)60153-4

Yuan J, Meng J, Liang X, Yang E, Yang X, Chen W (2017) Organic molecules from biochar leacheates have a positive effect on rice seedling cold tolerance. Front Plant Sci 8:1624

Yuan P, Wang J, Pan Y, Shen B, Wu C (2019) Review of biochar for the management of contaminated soil preparation, application and prospect. Sci Total Environ 659:473–490

Chapter 12
Chitosan for Plant Growth and Stress Tolerance

Muhammad Saad Ullah, Athar Mahmood, Muhammad Mansoor Javaid, Maria Naqve, Safura Bibi, Zain Ul Abidin, Ikram ul Haq, and Shahid Raza Khan

Abstract The human population is increasing at an exponential rate and plants make the base of the food chain. All inhabitants including human beings are primarily dependent on plants, and the production of these plants or crops primarily depends on various environmental conditions prevailing around the atmosphere. Climatic change imposes many stresses on plants including biotic and abiotic stress with a drastic impact on the productivity of plants. Various synthetic chemicals cause irreparable environmental problems and may lead to the metabolic transmutation of microorganisms. Increase in plant productivity with minimal use of chemical fertilizers is direly needed. In this context, chitosan-based biopolymers are considered best in their action, which is commonly isolated from shells of shrimp, shellfish, and crabs, as well as cell wall of fungi. Chitosan has the ability to increase the growth of plants by stimulating their many growth parameters such as the uptake of nutrients, division, and elongation of cells. In short, chitosan inclusively increases plants' yielding capacity by activating its enzymes involved in various biochemical courses such as enzymes involved in protein production. Chitosan responds like a plant growth promotor, modifying various growth features, and has the ability to stop the growth of various kinds of plant pathogens. Chitosan is capable of alleviat-

M. Saad Ullah · M. Naqve · S. Bibi · Z. U. Abidin
Department of Botany, Faculty of Sciences, University of Agriculture, Faisalabad, Pakistan

A. Mahmood (✉)
Department of Agronomy, Faculty of Agriculture, University of Agriculture, Faisalabad, Pakistan
e-mail: athar.mahmood@uaf.edu.pk

M. M. Javaid (✉) · S. R. Khan
Department of Agronomy, College of Agriculture, University of Sargodha, Sargodha, Pakistan
e-mail: mansoor.javaid@uos.edu.pk

I. u. Haq
Department of Plant Breeding and Genetics, College of Agriculture, University of Sargodha, Sargodha, Pakistan

M. Hasanuzzaman (ed.), *Climate-Resilient Agriculture, Vol 2*,
https://doi.org/10.1007/978-3-031-37428-9_12

ing the negative impacts of various environmental stresses by activating the stress transduction mechanism with the help of primary and secondary messengers involved.

Keywords Chitosan · Climatic changes · Synthetic chemicals · Biopolymers · Abiotic stress

1 Introduction

Plants provide the base of the food chain and have the capacity to make food materials for consumption. Environmental factors are the main determinants that have a decisive role in the somatic and propagative growth of plants. Prevailing unsuitable environmental factors can cause various deficiencies and irregularities in plant growth, poor crop production, long-lasting damage, and even in severe conditions death of plants (Duque et al. 2013; Mahmood et al. 2022). Environmental stresses, both biotic and abiotic have remarkable effects on plant production. These stresses include heat, salinity, drought, and rising temperature while some biotic stresses induced by fungi, bacteria, viruses, and nematodes, may affect plant growth by disrupting the uptake of nutrients. The human population has increased vigorously and of course, there is a terrific rise in it day by day while resources are limited and under constant exploitation pressure. Due to a confluence of factors, comprising the burgeoning global hunger for sustenance, the swiftly shifting climatic conditions, the disquieting degeneration of fertile land, and the perpetually surging consumer craving for high-caliber, ecologically sound food products, there has been an upsurge in emphasis on the creation and promotion of environmentally friendly food items (Carvalho 2006; Wasaya et al. 2021). Today's major concern is to explore alternative biological methods that can fulfill the needs of the emerging world's population. But it clearly demands the vast use of chemical fertilizers, pesticides, and large-scale land consumption. However, these chemicals are problematic as they have a leaching or volatilization impact of considerable gathering inside the soil particles, organic molecules, or in the atmosphere, also cause metabolic transformation of microorganisms (Ali et al. 2020a; Padhan et al. 2021). Consequently, there exists a pressing necessity to discover alternative methodologies that facilitate the secure and efficient utilization of these chemicals (Shahzad et al. 2018). In recent times, the utilization of nanocarrier and nanosensor advancements has enraptured the scrutiny of erudite individuals hailing from diverse fields encompassed by the purview of plant sciences (Tanveer et al. 2012; Wani 2019). Biopolymer-based compounds exhibit significant activity against pathogens with minimal human toxicity and negligible environmental implications. The utilization of these materials can provide safeguarding against an extensive array of plant diseases and augment levels of agricultural productivity (Ali et al. 2020b; Hartmann and Six 2022).

Additionally, these substances showcase the prospect of augmenting the yield of a plethora of cultivated crops, via ameliorating and preempting the onerous reliance on chemical fertilizers and precarious agricultural techniques (Balal et al. 2016). Among the experienced biomaterials, the best outcomes were obtained using chitosan biopolymers, because of its unlimited biological features such as antibacterial and growth-regulating action, biodegradability, biocompatibility, and human nontoxicity. It has garnered exceptional eminence as one of the most extensively scrutinized substances in the domain of nanotechnology (de Sousa Victor et al. 2020).

Chitosan (CHT) is a good choice because of rich sources, environmental concerns, and the subsequent simple manufacturing process (Mujtaba et al. 2019). Many studies have exposed of chitosan's role in prompting plant resistance responses, comprising defense genes and enzymes, antioxidant enzymes, and total phenolics (Shahid et al. 2014; Chandra et al. 2015). It is also described that chitosan-created nanomaterials have prospective to excite plant growth and yield by promoting the growth of seedlings, photosynthesis, and also have the potential to increase nutrient uptake.

2 Chitosan

CHT is a chitin derivative, having a molecular weight of about 300 and 1000 kilo Daltons. The molecular mass varies, with the source from which it is derived (Balal et al. 2017). The chemical structure of Chitin comprises *1–4* linked *2-Acetamido-2-deoxy-β-D-Glucopyranose*. Its unique characteristics as biocompatibility, biodegradability, and mammalian nontoxic behavior are because of its straight structure (Teng et al. 2001). Chitosan is a copolymer of N-Acetyl-D-glucosamine and D-glucosamine. Chitin deacetylation is done through chemical hydrolysis or sometimes through enzymatic hydrolysis in the existence of certain enzymes including chitin deacetylase in the presence of strong alkaline conditions (Venkatesan and Kim 2010). Chitin is widely distributed and is supposed to be the world's second most abundant biopolymer after cellulose. It could be found in some invertebrates, crustaceans' shells, the cuticle of insects, sheaths of molds, and also traced in cell walls of green algae and yeast. Commercial or industrial-scale production of chitosan has two main sources: (1) from crustaceans and (2) from fungal mycelium (Kannan et al. 2010). The shells of a crustacean as crab and shrimp comprise 30–40% proteins, 30–50% $CaCO_3$, and 20–30% chitin (Ibrahim and El-Zairy 2015). The major sources of chitin and chitosan are insects (cuticle) crustaceans, squid (*Loligo abdominal wall*), and centric diatoms and mushrooms (*Mucor rouxi* and *Aspergillus nidulans*) (Kaczmarek et al. 2019). Chitosan is among the chief chitin derivatives produced in the prevalence of specific conditions (alkaline in presence of NaOH) by enzymatic hydrolysis (Fig. 12.1; Escorcia et al. 2009).

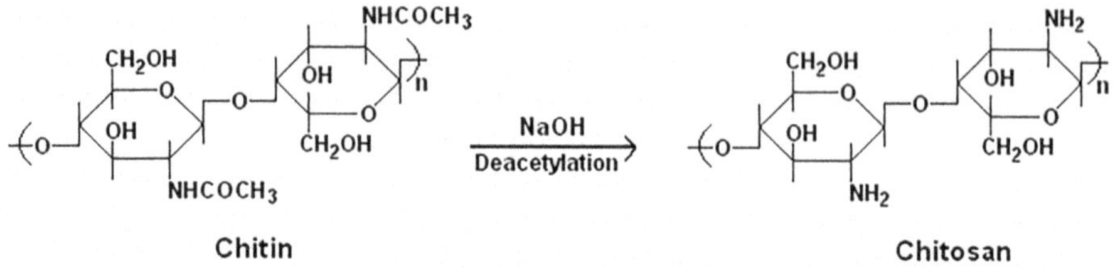

Fig. 12.1 Chitosan deacetylation from chitin (Escorcia et al. 2009)

2.1 Methods of Preparation of Chitosan

Chitosan is prepared by two following methods:

1. Chemical treatment
2. Enzymatic treatment

Chitosan is extracted from crustaceans by the chemical deacetylation procedure (Cheung et al. 2015). Demineralization, deproteinization, decolorization, and deacetylation are all phases of this process. In this procedure, mineral and protein residues are removed from crustacean shells by demineralizing them with HCl and deproteinizing with NaOH, respectively. Then, treatments with potassium manganese oxide eliminate molecules that are colored, leaving behind transparent chitin. The chitin is treated with alkaline to create chitosan. In the second technique, chitin deacetylase is used to generate chitosan enzymatically (Tsigos et al. 2000). We may get chitin deacetylase from different bacterial, fungal, and insect species (Zhao et al. 2010). The enzyme dissolves the N-acetamide bond found in the chitin molecule. Chitin is physically and chemically altered before being hydrolyzed because raw chitin is a poor substrate for enzymes. Enzymatic methods pose significant limitations in comparison to chemical methods, with their foremost drawback being the exorbitant cost of operation, particularly worrisome in the context of large-scale industrial production. This is primarily because the enzymes used for deproteinization and deacetylation are considerably more costly than their chemical counterparts. Furthermore, enzymatic methods exhibit lower efficiency than chemical methods due to their incapacity to achieve substantial reduction of the final 10% of residual proteins present in shells during the deproteinization process (Younes and Rinaudo 2015). Owing to the rapid growth and constant secretion of enzymes by microorganisms in reactors under the most favorable reaction conditions, the technique of fermentation has been devised as a remedy for the exorbitant cost of enzymes. The crux of the matter is that by harnessing the potential of fermentation processes, it is feasible to manufacture enzymes with greater efficiency and at a reduced cost in comparison to conventional methods. Furthermore, this approach facilitates the development of a more sustainable and ecologically sound mode of enzyme production (Younes and Rinaudo 2015). The diagrammatic demonstrations of chitosan formation by an enzymatic and chemical technique were shown in

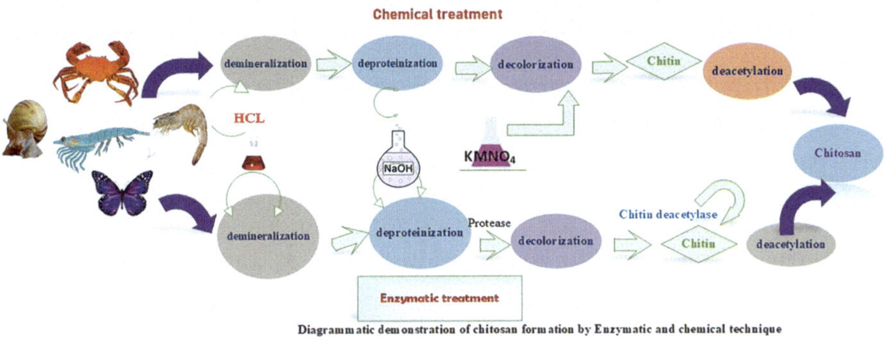

Fig. 12.2 Diagrammatic demonstration of chitosan formation by enzymatic and chemical technique (Madni et al. 2021)

Fig. 12.2 (Madni et al. 2021). The employment of the chemical method is prevalent due to its cost-efficiency and vast magnitude of yield in extracting chitosan from crustaceans. This methodology involves the use of chemicals to bearkdown the chitin present in crustacean shells, which is then deacetylated to produce chitosan. The process is economical and feasible for large-scale production, making it a popular choice in industrial applications (Cheung et al. 2015).

2.1.1 Chitosan Important Derivatives

Chitosan has many reactive amino side groups; these side groups increase the utility of chitosan and provide an opportunity to develop many chitosan by-products. Oligochitosan is an important water-soluble derivative of chitosan that can be established through many methods such as acid hydrolysis, oxidative degradation, and enzymatic hydrolysis (Kasaai et al. 2008). When CHT reacts with epoxide, hydroxyalkyl CHTs are formed. Ammonium tri methyl chitosan is another cationic derivative of water-soluble chitosan that is shaped due to the quaternization of the chitosan by reaction with CH_3I (methyl iodide) and NaOH (sodium hydroxide). N-carboxymethyl chitosan (CHT) is a water-soluble derivative of chitosan that finds wide application in diverse fields like food, medical, and gene therapy. Due to its high solubility and biocompatibility, it is extensively used in medical applications. Additionally, it is used as a food additive for improving texture, viscosity, and shelf life. N-carboxymethyl CHT shows promise for the development of new gene delivery systems owing to its excellent gene transfection properties, thus making it a popular choice in various industries (Khanjari et al. 2013). Another derivative of chitosan is N-methylene phosphonic chitosan, which demonstrates amphoteric properties, meaning that it can act as both an acid and a base. This unique feature gives it a wide range of potential applications, including use in water treatment and as a biomedical material.

3 Chitosan Role in Plant Growth

3.1 Chitosan Biopolymer Versus Plant Growth

CHT promotes the growth of various crops such as beans, potatoes, soybeans, radish cabbage, and other crops. When plant growth is promoted and allowed to grow luxuriously, yields also increase. Chitosan's major contribution is promotion of the growth rate of numerous types of plants (Ma et al. 2014). Described role of oligochitosan is that it can effectively stimulate the growth of wheat, germination capability, root length, seedling height, and strength that also enhances the activity of roots. It was also observed that early priming with chitosan causes good results as noted in case of pear millet, where it enhanced germination rate and seedling vigor. A similar enhancing effect was found in rapeseed where many growth attributes were seen positively enhanced when early seed soaking in chitosan was done. It improved sprouting rate, enhanced length, mass of hypocotyl and radicle were also increased (Sui et al. 2002). CHT molecules can lower water content in plants and boost many biological activities so overall play significant role in lowering the damage caused by stressful situations. Experiments on orchids were performed to conclude the effects of CHT on organogenesis chitosan-positive results were observed in this experiment even using very low concentrations (Nahar et al. 2012). Daikon radish, cabbage, sweet basil, soybean sprouts, and ornamental crops like gerbera and dendrobium orchids have all shown significant growth improvements in numerous studies using a variety of treatment methods, including in vitro, in vivo, soil application, pot application, and biofertilization (Choudhary et al. 2017). Chitosan and rhizobacteria together boost maize yields because they can spur plant development and function as organic fertilizer (Agbodjato et al. 2021). It is being used in freesia pot culture as a biostimulant (Salachna and Zawadzińska 2014). Chakraborty et al. (2020) show the modes of applications of chitosan and its promoting results on various crops (Fig. 12.3; Table 12.1). Even many crops, including cabbage (*Brassica oleracea*), bean sprouts, potatoes, strawberries, maize, and rice benefit from the growth-promoting effects of CHT. A CHT derivative called Chitogel has been discovered to promote vines' vegetative growth (Sani 2021). Tomato plants fed varying amounts of chitosan that showed greater shoot biomass, more flowers, and lower mycorrhization percentages (El Amerany et al. 2020). The spraying of CHT to foliage is known to boost the process of photosynthesis, which leads to improve overall plant growth and development (Rendina et al. 2019). CHT boosts agronomic properties of vegetables and improves mango tree fruit yield and growth. CHT is popular in agriculture for its ability to enhance plant growth, defend against diseases, and boost nutrient uptake, leading to sustainable and productive farming (Sajid et al. 2020). CHT application to leaves at the vegetative stage stimulates plant growth and development, boosting maize seed production (Mondal et al. 2013). CHT treatment to strawberry plants improves plant growth and productivity (Abdel-Mawgoud et al. 2010).

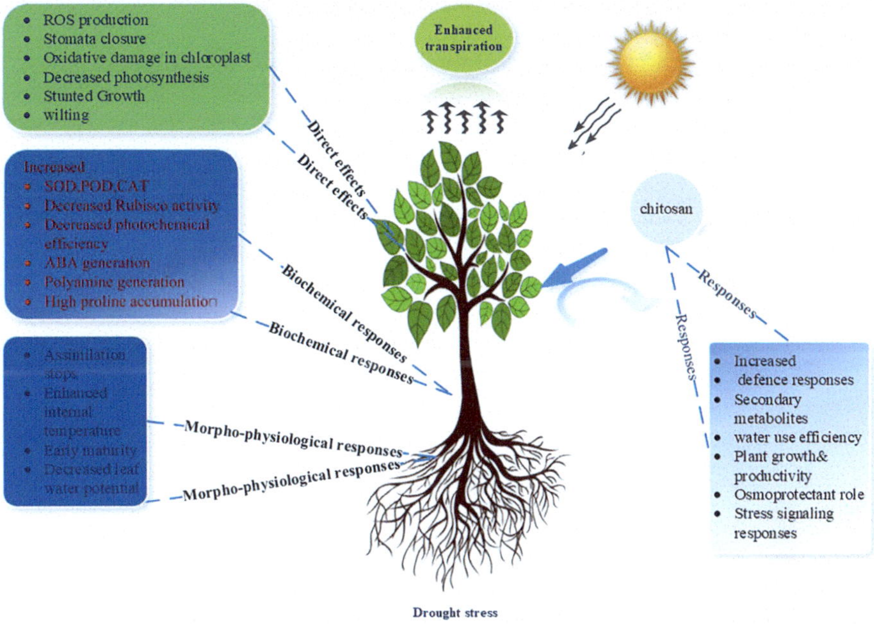

Fig. 12.3 Summary of chitosan-mediated disease alleviation in plants (Chakraborty et al. 2020)

3.2 *Chitosan Role in Plant Disease Suppression*

Plant development and productivity are negatively impacted by infections, diseases, and pests. To treat plant diseases or stop the spread of plant pathogens, a variety of synthetic chemicals, biocontrol agents, natural products (microorganisms, higher plants, and animals) are also utilized. These include viruses, nematodes, fungi, bacteria, and protozoa (Yoon et al. 2013). CHT is preferred and has certain benefits over other biocontrol agents since it has the ability to both control diseases and boost the capacity of the target plants to become resistant to them. Moreover, it also boosts the biodiversity in the rhizosphere. CHT has operative antimicrobial activity against an extensive variety of microorganisms, which include bacteria, fungi, viruses, and nematodes. Antimicrobial components have the capability to eliminate or restrain the growth of microorganisms. As the world battles the spread of infections and pathogens, these substances have gained significant importance. They function by targeting the essential components of microorganisms, leading to their demise or arrest. Antimicrobial agents can be obtained from natural or synthetic sources and are often used in producing antibiotics, disinfectants, and other medical interventions. Antimicrobial agents play a vital role in upholding public health and preventing drug-resistant strains from emerging (Reygaert 2018).

Table 12.1 Chitosan effects on plant growth and development

Plants species	Chitosan effects	Application method	Reference
Rice	Higher photosynthesis rate and increased plant growth	In vivo	Phothi and Theerakarunwong (2017)
Soybean	Increased plant growth	Soil application	Chibu et al. (2002)
Zea mays (corn)	Increased plant growth also grain weight	Biofertilization	Choudhary et al. (2017)
	Improved germination of seed	In vivo	Shao et al. (2005)
	Improved seed sprouting and staying power	In vivo	
Potato	Enlarged tuber size	In vivo	Falcón-Rodríguez et al. (2017)
	Amended growth and yield	In vitro and in vivo	Kowalski et al. (2006)
Solanum lycopersicum (Tomato)	Improved fruit productivity along its quality	In vivo	Sathiyabama and Charles (2015)
	Increased seed germination and vigor index	In vivo	Saharan et al. (2015)
Phaseolus vulgaris (Bean)	Increased leaf area, chlorophylls level, and carotenoids	In vitro	Santo Pereira et al. (2017)

3.2.1 Chitosan Antifungal Activity

In 1979, Allan and Hadwiger originally characterized CHT as a biofungicide. Subsequently, this substance has garnered significant recognition and has been thoroughly investigated in the field of plant protection research. The fungicidal utility of CHT has been seen in many molds and oomycetes (Pandey et al. 2018). The efficacy of chitosan has been observed to be highest in vitro for hindering the proliferation of numerous pathogenic fungi, including *Penicillium digitatum*, *Alternaria alternata*, *Botrytis cinerea*, and *Rhizopus stolonifera*. CHT performs through suppressing the growth of numerous pathogens in their different developmental stages, such as the hyphal growth stage, spore formation phase, spore viability germination, and at the end when the virulence factor is produced (Lemke et al. 2022). Chitosan is capable of quashing the growth of *Pythium aphanidermatum*. Against *Phytophthora capsici,* it causes disturbance in its endomembrane system, especially affecting its vacuoles' integrity (Chakraborty et al. 2020). The damping-off disease was successfully controlled by chi and its products (El-Mohamedya et al. 2019). Moreover, chitosan is being used as an antifungal mediator in the form of nanoparticles (Chouhan and Mandal 2021). Reports showed that root rot in wheat and kernel rot in maize are effectively controlled by chitosan (El-Gamal et al. 2021).

3.2.2 Chitosan Role as Antibacterial Agent

CHT derivatives are very effective against bacteria. Chitosan's antibacterial properties are often observed in human bacterial infections, which are frequently brought on by *Staphylococcus aureus*, some species of *Escherichia coli*, and occasionally other species of *Bacillus* (Shi et al. 2016). CHT is effective against various types of infections caused by bacteria and enhances plant strength against these infections. Chitosan efficiently controlled various plant pathogenic bacteria, like *Xanthomonas* spp. and *Pseudomonas* spp. (Toan et al. 2013). Chitosan inhibitory action and efficacy vary depending on the amount of chitosan material used, its molecular weight, the kind of bacteria utilized, the configuration of the bacterium's surface and cell walls, and the type of solvent used (Rabea and Steurbaut 2010). Chitosan has the best antibacterial activity against both gram-positive and gram-negative microorganisms. The presence of bi-charged cations on gram-positive bacteria's outer membrane helps to balance the central negative charges on LPS (lipopolysaccharide) molecules. Chitosan is thought to shift these divalent cations from the binding site, causing membrane disruption and cell death. Because chitosan binds to gram-positive bacteria more easily, the inhibitory result is more effective against these gram-positive bacteria as compared to gram-negative bacteria (Matica et al. 2019).

3.2.3 Antiviral Activity of CHT

Chitosan does not make up the virus's structural elements. Typically, chitin and similar polysaccharides are absent in viruses. Chitosan could only cure a small number of plant virus diseases, making it ineffective for use on plants (Iriti and Varoni 2015). It is yet unsure if CHT directly incapacitates viruses. Numerous investigations have demonstrated that CHT enhances the host's oversensitive reactions to contamination and stops viroids and viruses from completely multiplying in the plant (Badawy and Rabea 2011). Chitosan as a suppression response identified in *Tobacco mosaic* and necrosis viruses *X*, and peanut stunt viruses. According to Firmansyah (2017), CHT is also the best defense against the squash mosaic virus (SMV), which protects many plant species from both general and local infection. The impact of chitosan on several viruses is listed in Table 12.2 (Chakraborty et al. 2020).Nematocidal Activity of CHT

Numerous nematodes that function as pathogens in plants and cause a variety of illnesses that impact growth and development can be controlled by CHT (Kalaiarasan et al. 2006). The application of CHT to soils encourages the growth of chitinolytic bacteria that break down the chitin found in plant parasite tissues. Chitosan frequently decreases the ability of *Meloidogyne javanica* and *Heterodera schachtii* to lay eggs as well as the survivability of their larvae and adults. Chitosan has a significant quantity of nitrogen, so when it is applied, it produces large amounts of ammonia gas, which acts as a poisonous agent for nematodes and hinders their ability to survive. As a result, ammonia emission is lethal to nematodes (Maluin and Hussein 2020). By promoting full and selective confrontation of tomato plants with

Table 12.2 Chitosan effect on different viral infection (Chakraborty et al. 2020)

Plant species	Viruses	Effect	Inhibition rate (%)
Phaseolus vulgaris	TNV	Reduction in the number of local lesions	75–100
Phaseolus vulgaris	TNV	-----------------	75–100
Chenopodium quinoa	CNV	------------------	20–50
Chenopodium quinoa	ALMV-S	------------------	50–75
Phaseolus vulgaris	PSV	Reduction in number of systemically infected	75–100
Phaseolus vulgaris	ALMV	Plants	75–100
Pisum sativum (pea)	PSV	--------------	50–75
Pisum sativum		---------------	50–75

root-knot nematodes, CHT exemplifies elicitor activity. One of the stem nematodes, the pinewood nematodes (*Bursaphelen chusxylophilus*), was successfully controlled by chitosan-founded nanoparticles of avermectin (Liang et al. 2018).

4 Chitosan Application on Plant Responses with Distinct Position to Environmental Stresses

Plant stress is a state that exists around us and is unsuitable for plants to survive. Plants typically encounter and develop under less than ideal growing circumstances (Chadha et al. 2019). As they cause numerous abnormalities and shortages in growth patterns, crop yields, eternal damage, or death if they persist for a long time, beyond the plant tolerance boundaries, stress impacts are fatal for plants (Mehmood et al. 2018). Plant stresses are broadly classified into "abiotic" stresses that are triggered by nonliving causes (including drought, variations in temperature and salinity), and "biotic" stresses that are initiated by living organisms (such as microbes, insects, and plants).

4.1 Chitosan Effects on Abiotic Stresses

Drought, salt, and severe temperatures are the three main abiotic stressors that contribute to agricultural yield losses because they negatively affect plant growth, survival, and productivity (Javaid et al. 2018). Chitosan oligomers are good at helping plants to develop tolerance to a wide range of abiotic challenges, including water scarcity, heat stress, the predominance of salty environments, and the toxicity of heavy metals (Malerba et al. 2012). Although, how chitosan combats all of these factors is still not fully understood, it is evident from several papers that CHT triggered numerous defense-related functions in plants (Iriti and Faoro 2008). Application of chitin-built substances activates the defense system in plants because the cell membrane of plants has unique chitin receptors on its surface that perceive and create defense-involving responses. Chitin Elicitor-Binding Proteins (CEBiP), which have been identified from a variety of crops, can directly influence the course of gene expression to elicit an immune response. Similarly, chitosan-binding glycoprotein that belongs to the lectins family is removed from mustard leave (*Brassica campestris* L.) (Chen and Xu 2005). Vesicles were isolated from *Mimosa pudica* L. and *Cassia fasciculate* when analyzed by experts showed rapid activation of H-ATPase present on the plasma membrane, conforming to the importance of CHT receptors (Amborabé et al. 2008).

4.1.1 Effect on Drought Stress

Drought conditions are one of the obstacles to inclusive agricultural production (Wasaya et al. 2021). The quantity of CO_2 in the atmosphere has increased gradually since industrialization and has already surpassed 400 mol^{-1} (Xu et al. 2018; Javaid et al. 2022). This is a significant contributor to global warming, which alters the subsurface water table that plants can access and hence has an impact on unpredictable rainfall patterns. Drought has reduced the yield of many commercial crops globally as maize 75% (Kamara et al. 2003), wheat 22%, barley 50%, canola 30% (Masoud 2007). Reactive oxygen species (ROS) produce when drought shock boosts their production, and these ROS cause substantial damage to plants under stress by directly interacting with several macromolecules and peroxiding membrane lipids. Plant development is negatively impacted by drought stress, and its revenue is also affected (Yang et al. 2009). CHT treatment during drought conditions has the function of reducing the detrimental effects of stress by boosting the synthesis of antioxidant enzymes. By promoting root development, antioxidant enzyme synthesis boosts roots' capacity to absorb water. CHT application boosts plant nutrient absorption efficiency and soil fertility (El Amerany et al. 2020). Under controlled conditions, it also increases crop yield and other growth factors for a variety of crops, such as cowpea, potato, common beans, and wheat (Hidangmayum et al. 2019). CHT treatment in common beans impacts mineral accretion in the plant's body as well as the morphology of the roots and shoots. Drought stress is significant abiotic stress that restricts the functioning of many plants, damages their physiology, and alters their molecular and biochemical makeup (Salehi-Lisar and Bakhshayeshan-Agdam 2016).

In the instance of apples, chitosan was shown to be effective and produces adequate outcomes. Spraying chitosan on apple seedlings increased antioxidant activity, stopped electrolyte leakage, and preserved moisture levels for 35 days under persistent dry conditions. Increased antioxidant activity was observed in several other plants, resulting in changes such as higher H_2O_2 levels, improved root system growth, and a variety of other activities in potato, rice, grapes, and white clover (Chouhan and Mandal 2021). CHT controls stomata opening and the rate of

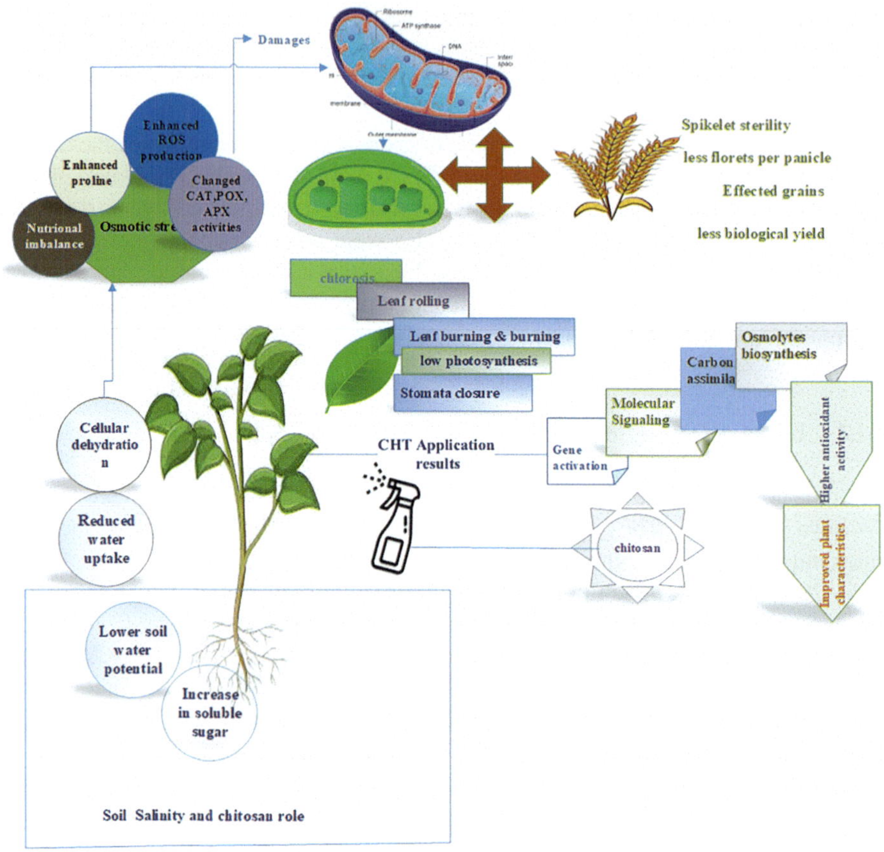

Fig. 12.4 The impacts of drought stress on plants and their corresponding adjustments (Seleiman et al. 2021)

transpiration in the stress phase by inducing ABA (abscisic acid) activities (Hidangmayum et al. 2019). CHT may therefore contribute as an antitransparent substance in several horticultural crops to help them to resist in stressful situation. The impacts of drought stress on plants and their corresponding adjustments are shown in Fig. 12.4 (Seleiman et al. 2021).

4.1.2 Influence on Biochemical Activities

Proline, an active osmoprotectant, controls osmotic pressure, quenches ROS, and maintains redox equilibrium under abiotic stresses. Leaf metabolic-free proline contents notably increase and leaf water potential decreases under extreme dry stress, reducing water loss (Ejaz et al. 2020). It raises the turgor of the leaves and assists in the delivery of water to them. MDA levels rise when there is a water shortage, which might lead to membrane leakage from the accumulation of free radicals. CHT functions in these situations as a constructive controller for osmotic regulation and eliminates the negative effects of drought stress conditions (Hidangmayum et al. 2019). Sara et al. (2012) reported that priming with chitosan decreased lipid peroxidation, removed ROS, and also helped to enhance membrane integrity in seedlings of beans, potatoes, thyme, and apples. In chitosan, an abundance of hydroxyl and amino groups are present that react with ROS and form firm, nontoxic macromolecular radicals. CHT can scavenge OH and O_2 radicals and possesses DNA-protective properties (Prashanth and Tharanathan 2007). When SOD level is enhanced, it promotes the manufacturing of malondialdehyde and diminishes lipid peroxidation in apple plantlets facing drought stress. Chitosan can also scavenge SOD and superoxide anion (Mudassir et al. 2018). When plants are treated with chitosan, various forms of carbohydrates including glucose, fructose, mannose, sorbitol, and myoinositol increase, even some extra perceived sugars also regulated (Li et al. 2017). These may improve drought resistance through an increase in osmotic modification and conservation of carbon stability in reaction to desiccation stress. CHT has been found to alleviate the impaired photosynthetic activity under drought stress in cowpea, chlorophyll, and carbohydrate contents that were increased when drenched by chitosan using 250 mg L^{-1}. Farouk and Amany (2012) reported that application of chitosan (at 250 mg L^{-1}) caused an increase in the thinness of the leaf blade through the increasing thickness of the mesophyll tissues, and water storage tissues. As a response of chitosan application, procambial activity of xylem tissues is stimulated during the differentiation phase of division, so it results as increased in xylem area for conductance.

4.2 Salinity Stress

One of the main biological issues impacting the quality and quantity of agricultural products is salt stress. Salinity stress is worsening day to day, affecting around 20% of the area that is farmed and more or less 30% of the irrigated area (Shrivastava and Kumar 2015). According to creation, there are typically two types of salinization: natural (caused by salty lakes and salts discharged from rocks and minerals), and manmade (by chemical contamination, unplanned crop rotation, inappropriate irrigation practices, and floods). It alone resulted in a 50–70% reduction in rice yield (Hussain et al. 2019). Based on the biology of the plant, different species have distinct salt sensitivity levels. According to previous research, halophytes are less vulnerable than glycophytes (Gupta et al. 2021). Yet many physiological processes in plants require sufficient salt, but excessive salt content has a negative impact on physiological development (Rasheed et al. 2022). Salinity retards the uptake of a number of major nutrients from soil because it alters the concentration around roots and interferes with the proper mode of diffusion and mass flow mechanisms (Shrivastava and Kumar 2015). According to Aravinthasamy et al. (2021), salty soils are defined as soils having an electrical conductivity of more than 4 dS m^{-1} at

25 °C. When there is a considerable amount of salt present, the external osmotic potential around the roots decreases, which might change the gradient and produce ex osmosis, and causes the plant to wilt. High salt and chloride ion concentrations cause toxicity near roots that affects the conductivity. Salt stress alters metabolic functions and encourages ROS, which interferes with cellular activities and results in oxidative stress (Ilangumaran and Smith 2017). Ion toxicity is the term used to describe the excessive buildup of MDA (malondialdehyde) in plants in salt conditions due to the peroxidation of membrane lipids. The application of chitosan might reduce these negative alterations. However, to mitigate these alterations, a low concentration is needed. Safflower and sunflower seeds treated with a small amount of CHT help to reduce the oxidative damage brought on by salt conditions (Hidangmayum et al. 2019). Both of these crops' enzymatic activity was altered by this treatment (Jabeen and Ahmad 2013). The harmful effects of salt stress on rice, maize, and mung bean were also observed to be mitigated by chitosan pretreatment in salinity, which resulted in increased antioxidant enzyme activities and reduced levels of malondialdehyde content (Ray et al. 2016). A hydroponic experiment carried out on wheat with 0.0625% oligochitosan gives progressive effects by a meaningfully cumulative antioxidant enzyme (POD, CAT, and SOD) during salt-prompted stress and may be capable to improve oxidative stress (Ma et al. 2012).

4.3 Effect on Biochemical Activities

Stress has an impact on a plant's metabolic processes, for example, by accumulating chlorophyllase and unstable protein complexes, which decrease the amount of chlorophyll in the leaf and cause the stomata to close (Pirasteh-Anosheh et al. 2016; Liaqat et al. 2019). This is further connected to low levels of carbon dioxide in leaves, which slow down photosynthesis and, as a result, reduced output and yield. When seeds are primed with oligochitosan (0.0625%), as was the case with wheat, the rates of photosynthesis and stomatal conductance both increase. This is due to stomatal conductance, which, along with a variety of other factors, is necessary for photosynthesis and depends on a protein involved in chlorophyll metabolism (Mahmood et al. 2022). Contrarily, several data indicate that even the administration of chitosan has little impact on the amount of internal CO_2, even if stomatal conductance is seen to be decreased. The reason for this is that the effectiveness of chitosan depends on several additional factors, including its molecular weight, the mode of application, the degree of deacetylation, and most importantly the responses of various plant species have to CHT (Joseph et al. 2020). Chitosan spraying sometime leads to decrease photosynthesis and constricted stomatal conductance without affecting internal CO_2 concentration. These contradictory findings could be attributed to the fact that the function of chitosan is dependent on a variety of factors such as application method, degree of deacetylation, molecular weight, and the perceptive abilities of different crops (Bhattacharya 2022). CHT role against alleviating drought and salinity stress has shown in Fig. 12.5 (Abdellatef et al. 2022).

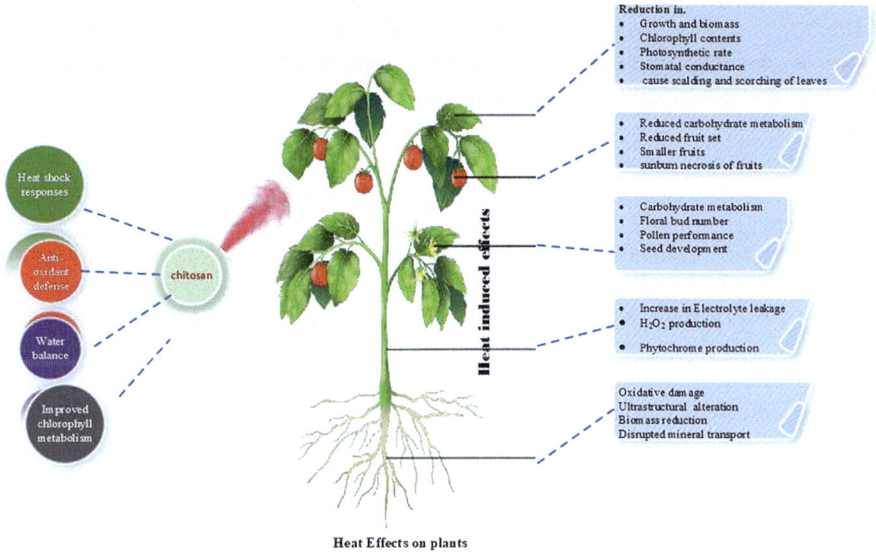

Fig. 12.5 Chitosan alleviating role against salinity and drought stress (Abdellatef et al. 2022)

4.4 Heavy Metal Stress

Anthropogenic activities that facilitate heavy metals' easy accessibility into plant bodies are a dominant contributor to their accumulation in soil. Typically, heavy metals are needed by plants to execute a variety of functions rather than posing serious risks by contaminating the whole food chain (Hassan et al. 2022). Plants utilize Cu, Fe, Mn, Ni, and Zn, but only in very small quantities; when these levels are exceeded, they become very hazardous (Chaffai and Koyam 2011). Chitosan's structure makes it suitable for carrying out a variety of crucial tasks. Chitosan couples amino and (OH) functional groups. These groups support CHT's ability to interact with many ions, nonnutrient elements, and heavy metals to create complexes. The symmetry of the metal adsorption rates on chitosan is Hg>Cu>Zn>As. ROS increases and alters the characteristics of cell membranes. In one instance, when *Brassica rapa* was grown hydroponically, metal stress was eliminated by using chitosan foliar applications of varying molecular weights to alleviate Cd harmful effects (Zhang et al. 2002). Similarly, a defensive result of cadmium (Cd) toxicity was shown in another experiment (Kamari et al. 2012). Silver (Ag), zinc (Zn), Cd, and lead (Pb) may all bind with chitosan, which provides evidence of metal accumulation in *Lolium multiflorum* L. and *Brassica napus* seed upon chitosan treatment. These findings open up a new path for mitigating phytotoxicity in the presence of heavy metals, leading one to the hypothesis that CHT can absorb toxicants in larger concentrations. It is also possible to increase the uptake and absorption of crucial minerals by plants. The running metabolism in the roots and shoots of plants alters when they are exposed to hazardous conditions, especially metal poisoning. Additionally, it disrupts intracellular processes, resulting in a variety of anomalies, as was seen in instances of Cd poisoning where it decreased stomata's conductivity. CHT was applied topically to the plant to reduce this toxicity, which eventually enhanced gas exchange and photosynthesis (Xu et al. 2017).

4.5 Heat Stress

Plants require an optimal temperature to flourish and yield well, but anthropogenic activities have caused disturbances in weather patterns. Climate change has become intensive for plant survival and a major threat to crop yield globally (Wang et al. 2018). Heat stress is essentially a situation that occurs when a temperature increase lasts long enough to cause enough permanent alterations in plants. Although some of them are well suited for their survival adaptations, it always has an impact on their growth and development traits (Wahid et al. 2007). Plants produce at their peak levels only under ideal environmental conditions. Denaturation of proteins interrupted enzyme activity, and fluidity of membrane lipids are all effects of temperatures that are excessively high. Additionally, the chloroplast contents in leaves are also impacted, and mitochondria may also be unable to adequately perform their functions. Consequently, it may be said that any change caused by stress, specifically heat stress, affects the order in which biochemical processes occur in plants. Long-term exposure to high temperatures damages cells at the cellular level, many of which are lethal to cells, causing them to die. High temperatures are hazardous as they harm lipids, proteins, and the fluidity of membranes. Continual exposure to high temperatures and extreme heat stress may cause cellular damage and cell death (Verma et al. 2020). Heat stresses effecting different plant aspects is shown in Fig. 12.5 (Ali et al. 2020b).

Heat stress is cited as a serious concern that frequently coexists with drought conditions. Under these situations, it is difficult to control and keep a record of because they change drastically with fluctuating conditions (McKersie and Lesheim 2013). Late-season sowing with chitosan, zinc, and humic acid mixture improves the crops' capacity to deal with heat stress; typically, foliar applications are recommended for this purpose (Ronga et al. 2021). However, it is confirmed that ABA has the capacity to activate genes associated with heat shock, such as ABF3 upregulation, which is responsible for boosting tolerance to heat stress. Therefore, it can be concluded that chitosan application lowers temperature stress by enhancing ABA action, which is responsible for closing the stomata in the best interest of avoiding plant (Bittelli et al. 2001).

5 Future Challenges and Perspective Outlooks

CHT is a significant, safe substance with a variety of mechanisms of action that benefits plant health. Its usage in agriculture can, at least in part, reduce the widespread consumption of chemical pesticides. This is enough evidence to prove that applying chitosan to plants have the expected positive effects concern with the concept of sustainable agriculture. It also results in resistance to numerous illnesses brought on by the presence of pathogens in them. Another plus point is that CHT is environment friendly, as it does not show any harmful impact. There has undoubtedly been a lot of work done, and many helpful aspects have been investigated, but more research is still needed because the mechanism underlying the restriction of pathogen development is not entirely known. Similarly, there is also a need to find out how plants acquire immunity against various stress conditions. Numerous derivatives used in agriculture had significant effects against worms, bacteria, and fungal assault, spurring the development of other derivatives. Overall, it can be said that using contemporary chitosan-based technology in agriculture has a bright future and will benefit crops in ways that are sustainable for the environment and persistent. Although there is yet no information on how chitosan is used to repel insects in important gardening crops, maybe it will be investigated shortly. Chitosan's usefulness against rising temperatures also needs to be investigated. Chitosan has a dual action: in addition to preventing pathogen development, it also changes the host plant's defensive mechanisms. The ideal operational concentrations and application techniques for CHT products are still unknown and not widely used.

References

Abdellatef MA, Elagamey E, Kamel SM (2022) Chitosan is the ideal resource for plant disease management under sustainable agriculture. https://doi.org/10.5772/intechopen.107958

Abdel-Mawgoud A, Tantawy A, El-Nemr M, Sassine Y (2010) Growth and yield responses of strawberry plants to chitosan application. Eur J Sci Res 39(1):170–177

Agbodjato NA, Mikpon T, Babalola OO, Dah-Nouvlessounon D, Amogou O, Lehmane H, Adoko MY, Adjanohoun A, Baba-Moussa L (2021) Use of plant growth promoting rhizobacteria in combination with chitosan on maize crop: promising prospects for sustainable, environmentally friendly agriculture and against abiotic stress. Agronomy 11(11):2205

Ali K, Tanveer A, Farooq N, Abbas T, Sarwar G, Nadeem MA, Hassan I, Javaid MM, Khan BA, Raza A (2020a) Seed polymorphism of *Rhyncosia capitata* (roth) dc. Enhance its tolerance to varying types and intensity of salt stresses. Pak J Weed Sci Res 26(4):393

Ali S, Rizwan M, Arif M, Ahmad R, Hasanuzzaman M, Ali DB, Hussain A (2020b) Approaches in enhancing thermotolerance in plants: an updated review. J Plant Growth Regul 39. https://doi.org/10.1007/s00344-019-09994-x

Amborabé B-E, Bonmort J, Fleurat-Lessard P, Roblin G (2008) Early events induced by chitosan on plant cells. J Exp Bot 59(9):2317–2324

Aravinthasamy P, Karunanidhi D, Subramani T, Roy PD (2021) Demarcation of groundwater quality domains using GIS for best agricultural practices in the drought-prone Shanmuganadhi River basin of South India. Environ Sci Pollut Res 28(15):18423–18435. https://doi.org/10.1007/s11356-020-08518-5

Badawy ME, Rabea EI (2011) A biopolymer chitosan and its derivatives as promising antimicrobial agents against plant pathogens and their applications in crop protection. Int J Carbohydr Chem 2011:1–29

Balal RM, Shahid MA, Javaid MM, Iqbal Z, Anjum MA, Garcia-Sanchez F, Mattson NS (2016) The role of selenium in amelioration of heat-induced oxidative damage in cucumber under high temperature stress. Acta Physiol Plant 38(6):1–14

Balal RM, Shahid MA, Javaid MM, Iqbal Z, Liu GD, Zotarelli L, Khan N (2017) Chitosan alleviates phytotoxicity caused by boron through augmented polyamine metabolism and antioxidant activities and reduced boron concentration in *Cucumis sativus* L. Acta Physiol Plant 39(1):1–15

Bhattacharya A (2022) Effect of low-temperature stress on germination, growth, and phenology of plants: a review. In: Physiological processes in plants under low temperature stress. Springer, Singapore, pp 1–106

Bittelli M, Flury M, Campbell GS, Nichols EJ (2001) Reduction of transpiration through foliar application of chitosan. Agric For Meteorol 107(3):167–175

Carvalho FP (2006) Agriculture, pesticides, food security and food safety. Environ Sci Pol 9(7–8):685–692

Chadha A, Florentine S, Chauhan BS, Long B, Jayasundera M, Javaid MM, Turville C (2019) Environmental factors affecting the germination and seedling emergence of two populations of an emerging agricultural weed: wild lettuce (*Lactuca serriola*). Crop Pasture Sci 70(8):709–717

Chaffai R, Koyama H (2011) Chapter 1 - Heavy Metal Tolerance in Arabidopsis thaliana. In: Kader J-C, Delseny M (eds) Advances in Botanical Research, vol 60. Academic Press, pp 1–49. https://doi.org/10.1016/B978-0-12-385851-1.00001-9

Chakraborty M, Hasanuzzaman M, Rahman M, Khan MAR, Bhowmik P, Mahmud NU, Tanveer M, Islam T (2020) Mechanism of plant growth promotion and disease suppression by chitosan biopolymer. Agriculture 10(12):624

Chandra S, Chakraborty N, Dasgupta A, Sarkar J, Panda K, Acharya K (2015) Chitosan nanoparticles: a positive modulator of innate immune responses in plants. Sci Rep 5:15195

Chen HP, Xu LL (2005) Isolation and characterization of a novel chitosan-binding protein from non-heading Chinese cabbage leaves. J Integr Plant Biol 47(4):452–456

Cheung RCF, Ng TB, Wong JH, Chan WY (2015) Chitosan: an update on potential biomedical and pharmaceutical applications. Mar Drugs 13(8):5156–5186

Chibu H, Shibayama H, Arima S (2002) Effects of chitosan application on the shoot growth of rice and soybean Japan J. Crop Sci 71(2):206–211

Choudhary RC, Kumaraswamy R, Kumari S, Sharma S, Pal A, Raliya R, Biswas P, Saharan V (2017) Cu-chitosan nanoparticle boost defense responses and plant growth in maize (*Zea mays* L.). Sci Rep 7(1):1–11

Chouhan D, Mandal P (2021) Applications of chitosan and chitosan-based metallic nanoparticles in agrosciences-a review. Int J Biol Macromol 166:1554–1569

de Sousa Victor R, da Marcelo Cunha Santos A, de Viana Sousa B, de Araújo Neves G, de Navarro Lima Santana L, Rodrigues Menezes R (2020) A review on chitosan's uses as biomaterial: tissue engineering, drug delivery systems and cancer treatment. Materials (Basel) 13(21). https://doi.org/10.3390/ma13214995

Duque G, Boersma D, Loza-Diaz G, Hassan S, Suarez H, Geisinger D, Suriyaarachchi P, Sharma A, Demontiero O (2013) Effects of balance training using a virtual-reality system in older fallers. Clin Interv Aging 8:257

Ejaz S, Fahad S, Anjum MA, Nawaz A, Naz S, Hussain S, Ahmad S (2020) Role of osmolytes in the mechanisms of antioxidant defense of plants. In: Sustainable agriculture reviews, vol 39. Springer, pp 95–117

El Amerany F, Rhazi M, Wahbi S, Taourirte M, Meddich A (2020) The effect of chitosan, arbuscular mycorrhizal fungi, and compost applied individually or in combination on growth, nutrient uptake, and stem anatomy of tomato. Sci Hortic 261:109015

El-Gamal NG, El-Mougy NS, Abdel-Kader MM, Ali Khalil MS (2021) Influence of inorganic salts and chitosan as foliar spray on wheat septoria leaf blotch disease severity under field conditions. Arch Phytopathol Plant Prot 54(13–14):836–849

El-Mohamedya R, Abd El-Aziz M, Kamel S (2019) Antifungal activity of chitosan nanoparticles against some plant pathogenic fungi in vitro. Agric Eng Int CIGR J 21:201–209

Escorcia D, Hernández D, Sánchez M, Benevente M (2009) Diseño y montaje de una planta piloto para la extracción de quitina y proteínas. Nexo Revista Científica 22(2):45–55

Falcón-Rodríguez AB, Costales D, Gónzalez-Peña D, Morales D, Mederos Y, Jerez E, Cabrera JC (2017) Chitosans of different molecular weight enhance potato (*Solanum* tuberosum L.) yield in a field trial. Span. J Agric Res 15(1):e0902–e0902

Farouk S, Amany AR (2012) Improving growth and yield of cowpea by foliar application of chitosan under water stress. Egypt J Biol 14:14–16
Firmansyah D (2017) Use of chitosan and plant growth promoting rhizobacteria to control squash mosaic virus on cucumber plants. IPB University Scientific Repository. http://repository.ipb.ac.id/handle/123456789/85776
Gupta S, Smith PMC, Boughton BA, Rupasinghe TWT, Natera SHA, Roessner U (2021) Inoculation of barley with Trichoderma harzianum T-22 modifies lipids and metabolites to improve salt tolerance. J Exp Bot 72(20):7229–7246. https://doi.org/10.1093/jxb/erab335
Hartmann M, Six J (2022) Soil structure and microbiome functions in agroecosystems. Nat Rev Earth Environ 4:1–15
Hassan MU, Nawaz M, Mahmood A, Shah AA, Shah AN, Muhammad F, Batool M, Rasheed A, Jaremko M, Abdelsalam NR (2022) The role of zinc to mitigate heavy metals toxicity in crops. Front Environ Sci 10:990223
Hidangmayum A, Dwivedi P, Katiyar D, Hemantaranjan A (2019) Application of chitosan on plant responses with special reference to abiotic stress. Physiol Mol Biol Plants 25(2):313–326
Hussain S, Shaukat M, Ashraf M, Zhu C, Jin Q, Zhang J (2019) Salinity stress in arid and semi-arid climates: effects and management in field crops. Clim Change Agric 13:201–226
Ibrahim H, El-Zairy E (2015) Chitosan as a biomaterial—structure, properties, and electrospun nanofibers. In: Concepts, compounds and the alternatives of anti bacterials, vol 1(1). IntechOpen, pp 81–101
Ilangumaran G, Smith DL (2017) Plant growth promoting rhizobacteria in amelioration of salinity stress: a systems biology perspective. Front Plant Sci 8:1768
Iriti M, Faoro F (2008) Abscisic acid is involved in chitosan-induced resistance to tobacco necrosis virus (TNV). Plant Physiol Biochem 46(12):1106–1111
Iriti M, Varoni EM (2015) Chitosan-induced antiviral activity and innate immunity in plants. Environ Sci Pollut Res Int 22(4):2935–2944
Jabeen N, Ahmad R (2013) The activity of antioxidant enzymes in response to salt stress in safflower (Carthamus tinctorius L.) and sunflower (Helianthus annuus L.) seedlings raised from seed treated with chitosan. J. Sci. Food Agric. 93 (7):1699-1705. J Sci Food Agric 93(7):1699–1705. https://doi.org/10.1002/jsfa.5953
Javaid MM, Florentine S, Ali HH, Weller S (2018) Effect of environmental factors on the germination and emergence of *Salvia verbenaca* L. cultivars (verbenaca and vernalis): an invasive species in semi-arid and arid rangeland regions. PLoS One 13(3):e0194319
Javaid MM, Wang X, Florentine SK, Ashraf M, Mahmood A, Li F-M, Fiaz S (2022) Effects on photosynthetic response and biomass productivity of *Acacia longifolia* ssp. longifolia under elevated CO_2 and water-limited regimes. Front Plant Sci 13:817730
Joseph B, Mavelil Sam R, Balakrishnan P, Maria JH, Gopi S, Volova T, Fernandes CMS, Thomas S (2020) Extraction of nano chitin from marine resources and fabrication of polymer nanocomposites: recent advances. Polymers 12(8):1664
Kaczmarek MB, Struszczyk-Swita K, Li X, Szczęsna-Antczak M, Daroch M (2019) Enzymatic modifications of chitin, chitosan, and chito oligosaccharides. Front Bioeng Biotechnol 7:24
Kalaiarasan P, Lakshmanan P, Rajendran G, Samiyappan R (2006) Chitin and chitinolytic biocontrol agents for the management of root knot nematode, Meloidogyne arenaria in groundnut (Arachis hypogaea L.) cv. Co3. Indian Indian J Nematol 36(2):181–186
Kamara A, Menkir A, Badu-Apraku B, Ibikunle O (2003) The influence of drought stress on growth, yield and yield components of selected maize genotypes. J Agric Sci 141(1):43–50
Kamari A, Ngah WSW, Ibrahim R, Hashim N, Isa IM, Mohamed A (2012) Sorption of Cd (II) and Pb (II) ions onto Shorea acuminata sawdust: Isotherm, kinetics and sorption mechanism studies. J Sci Math Lett 4(1):32–42
Kannan M, Nesakumari M, Rajarathinam K, Singh A (2010) Production and characterization of mushroom chitosan under solid-state fermentation conditions. Adv Biol Res 4(1):10–13
Kasaai MR, Arul J, Charlet G (2008) Fragmentation of chitosan by ultrasonic irradiation. Ultrason Sonochem 15(6):1001–1008

Khanjari A, Karabagias I, Kontominas M (2013) Combined effect of N, O-carboxymethyl chitosan and oregano essential oil to extend shelf life and control Listeria monocytogenes in raw chicken meat fillets. LWT-Food Sci Technol 53(1):94–99

Kowalski B, Terry FJ, Herrera L, Peñalver DA (2006) Application of Soluble Chitosan in vitro and in the Greenhouse to Increase Yield and Seed Quality of Potato Minitubers. Potato Res 49(3):167–176. https://doi.org/10.1007/s11540-006-9015-0

Lemke P, Jünemann L, Moerschbacher BM (2022) Synergistic antimicrobial activities of chitosan mixtures and chitosan–copper combinations. Int J Mol Sci 23(6):3345

Li Y, Wang N, Liu J, Hou X (2017) Demystifying neural style transfer. arXiv preprint arXiv:170101036

Liang W, Yu A, Wang G, Zheng F, Jia J, Xu H (2018) Chitosan-based nanoparticles of avermectin to control pine wood nematodes. Int J Biol Macromol 112:258–263

Liaqat A, Ihsan M, Rizwan M, Mehmood A, Ijaz M, Alam M, Abdullah M, Wajid M, Hussain R, Naeem M (2019) Inducing effect of chitosan on the physiological and biochemical indices of eggplant (*Solanum melongena* L.) genotypes under heat and high irradiance. Appl Ecol Environ Res 17(5):11273–11287

Ma CS, Deenick EK, Batten M, Tangye SG (2012) The origins, function, and regulation of T follicular helper cells. J Exp Med 209(7):1241–1253

Ma TY, Dai S, Jaroniec M, Qiao SZ (2014) Metal–organic framework derived hybrid Co3O4-carbon porous nanowire arrays as reversible oxygen evolution electrodes. J Am Chem Soc 136(39):13925–13931

Madni A, Kousar R, Naeem N, Wahid F (2021) Recent advancements in applications of chitosan-based biomaterials for skin tissue engineering. J Bioresour Bioprod 6(1):11–25

Mahmood A, Bibi S, Naqve M, Javaid MM, Zia MA, Jabbar A, Ud-Din W, Attia KA, Khan N, Al-Doss AA (2022) Physiological, biochemical, and yield responses of linseed (*Linum usitatissimum* L.) in α-Tocopherol-mediated alleviation of salinity stress. Front Plant Sci 13:867172

Malerba M, Crosti P, Cerana R (2012) Defense/stress responses activated by chitosan in sycamore cultured cells. Protoplasma 249(1):89–98

Maluin FN, Hussein MZ (2020) Chitosan-based agronano chemicals as a sustainable alternative in crop protection. Molecules 25(7):1611

Masoud SM (2007) The effects of water deficit during growth stages of canola (*Brassica napus* L.). Am Eurasian J Agric Environ Sci 2:417–422

Matica MA, Aachmann FL, Tøndervik A, Sletta H, Ostafe V (2019) Chitosan as a wound dressing starting material: antimicrobial properties and mode of action. Int J Mol Sci 20(23):5889

McKersie BD, Lesheim Y (2013) Stress and stress coping in cultivated plants. Springer Science and Business Media, Dordrecht

Mehmood A, Naeem M, Khalid F, Saeed Y, Abbas T, Jabran K, Sarwar MA, Tanveer A, Javaid MM (2018) Identification of phytotoxins in different plant parts of *Brassica napus* and their influence on mung bean. Environ Sci Pollut Res 25(18):18071–18080

Mondal M, Puteh A, Dafader N, Rafii M, Malek M (2013) Foliar application of chitosan improves growth and yield in maize. J Food Agric Environ 11(2):520–523

Mudassir A, Muhammad A, Javaid MM (2018) Enhancement in cotton growth and yield using novel growth promoting substances under water limited conditions. Pak J Bot 50(5):1691–1701

Mujtaba M, Morsi RE, Kerch G, Elsabee MZ, Kaya M, Labidi J, Khawar KM (2019) Current advancements in chitosan-based film production for food technology; A review. Int J Biol Macromol 121:889–904

Nahar SJ, Kazuhiko S, Haque SM (2012) Effect of polysaccharides including elicitors on organogenesis in protocorm-like body (PLB) of Cymbidium insigne in vitro. J Agric Sci Technol B 2(9B):1029

Padhan D, Rout PP, Kundu R, Adhikary S, Padhi PP (2021) Bioremediation of heavy metals and other toxic substances by microorganisms. In: Soil bioremediation: an approach towards sustainable technology. Wiley, pp 285–329

Pandey P, Verma MK, De N (2018) Chitosan in agricultural context-a review. Bull Environ Pharmacol Life Sci 7:87–96

Phothi R, Theerakarunwong CD (2017) Effect of chitosan on physiology, photosynthesis, and biomass of rice (*Oryza sativa* L.) under elevated ozone. Aust J Crop Sci 11(5):624–630

Pirasteh-Anosheh H, Ranjbar G, Pakniyat H, Emam Y (2016) Physiological mechanisms of salt stress tolerance in plants: an overview. In: Plant-environment interaction: responses and approaches to mitigate stress. Wiley, pp 141–160

Prashanth KH, Tharanathan R (2007) Chitin/chitosan: modifications and their unlimited application potential—an overview. Trends Food Sci Technol 18(3):117–131

RabEa EI, StEuRbaut W (2010) Chemically modified chitosans as antimicrobial agents against some plant pathogenic bacteria and fungi. Plant Prot Sci 46(4):149–158

Rasheed A, Raza A, Jie H, Mahmood A, Ma Y, Zhao L, Xing H, Li L, Hassan MU, Qari SH (2022) Molecular tools and their applications in developing salt-tolerant soybean (*Glycine max* L.) cultivars. Bioengineering 9(10):495

Ray SR, Javidul M, Bhuiyan H, Mahmud S, Tahjib-Ul-Arif M (2016) Chitosan Suppresses Antioxidant Enzyme Activities for Mitigating Salt Stress in Mungbean Varieties. IOSR J Agric Vet Sci 09:36-41

Rendina N, Nuzzaci M, Scopa A, Cuypers A, Sofo A (2019) Chitosan-elicited defense responses in *Cucumber mosaic* virus (CMV)-infected tomato plants. J Plant Physiol 234:9–17

Reygaert WC (2018) An overview of the antimicrobial resistance mechanisms of bacteria. AIMS Microbiol 4(3):482

Ronga D, Vitti A, Zaccardelli M, Pane C, Caradonia F, Cardarelli M, Colla G, Rouphael Y (2021) Root zone management for improving seedling quality of organically produced horticultural crops. Agronomy 11(4):630

Saharan V, Sharma G, Yadav M, Choudhary MK, Sharma S, Pal A, Raliya R, Biswas P (2015) Synthesis and in vitro antifungal efficacy of Cu–chitosan nanoparticles against pathogenic fungi of tomato Int. J Biol Macromol 75:346–353

Sajid M, Basit A, Ullah Z, Shah ST, Ullah I, Mohamed HI, Ullah I (2020) Chitosan-based foliar application modulated the yield and biochemical attributes of peach (*Prunus persica* L.) cv. Early Grand. Bull Natl Res Cent 44(1):1–11

Salachna P, Zawadzińska A (2014) Effect of chitosan on plant growth, flowering and corms yield of potted freesia. J Ecol Eng 15(3)

Salehi-Lisar SY, Bakhshayeshan-Agdam H (2016) Drought stress in plants: causes, consequences, and tolerance. In: Hossain MA, Wani SH, Bhattacharjee S, Burritt DJ, Tran L-SP (eds) Drought Stress Tolerance in Plants, Vol 1: Physiology and Biochemistry. Springer International Publishing, Cham, pp 1–16. https://doi.org/10.1007/978-3-319-28899-4_1

Sani MF (2021) Yield maximization of tomato (*Solanum lycopersicum* L.) by improving seedling characters using chiotsan raw material powder under silty clay loam soil. Department of Soil Science, Sher-e-bangla Agricultural University, Sher-e-Bangla Nagar, Dhaka

Santo Pereira AE, Silva PM, Oliveira JL, Oliveira HC, Fraceto LF (2017) Chitosan nanoparticles as carrier systems for the plant growth hormone gibberellic acid. Colloids Surf B: Biointerfaces 150:141–152

Sathiyabama M, Charles RE (2015) Fungal cell wall polymer-based nanoparticles in protection of tomato plants from wilt disease caused by *Fusarium oxysporum* f sp lycopersici. Carbohydr Polym 133:400–407

Sara K, Hossein A, Masoud SJ, Hassan M (2012) Effects of water deficit and chitosan spraying on osmotic adjustment and soluble protein of cultivars castor bean (Ricinus communis L.). J Stress Physiol Biochem 8(3):160–169

Seleiman MF, Al-Suhaibani N, Ali N, Akmal M, Alotaibi M, Refay Y, Dindaroglu T, Abdul-Wajid HH, Battaglia ML (2021) Drought stress impacts on plants and different approaches to alleviate its adverse effects. Plan Theory 10(2):259

Shahid MA, Balal RM, Javaid M, Aquel MA, Abbas T, Mattson N, Islam S (2014) Mitigation of stress induced by salinity and nickel in pea (*Pisum sativum* L.) by exogenous application of synthetic proline and proline-enriched *Lolium perenne* (L.) leaf extract. In: 2014 ASHS Annual Conference

Shahzad T, Ashraf M, Javaid MM, Waheed H, Abbas T, Sattar F (2018) Influence of field soil drought stress on some key physiological, yield and quality traits of selected newly-developed hexaploid bread wheat (*Triticum aestivum* L.) cultivars. Sains Malays 47(11):2625–2635

Shao C, Hu J, Song W, Hu W (2005) Effects of seed priming with chitosan solutions of different acidity on seed germination and physiological characteristics of maize seedling. J Zhejiang Univ (Agric Life Sci) 31(6):705–708

Shi S-F, Jia J-F, Guo X-K, Zhao Y-P, Chen D-S, Guo Y-Y, Zhang X-L (2016) Reduced Staphylococcus aureus biofilm formation in the presence of chitosan-coated iron oxide nanoparticles. Int J Nanomedicine 11:6499

Shrivastava P, Kumar R (2015) Soil salinity: a serious environmental issue and plant growth promoting bacteria as one of the tools for its alleviation. Saudi J Biol Sci 22(2):123–131

Sui G, Soohoo C, Affar EB, Gay F, Shi Y, Forrester WC, Shi Y (2002) A DNA vector-based RNAi technology to suppress gene expression in mammalian cells. Proc Natl Acad Sci 99(8):5515–5520

Tanveer A, Arshad MS, Ayub M, Javaid MM, Yaseen M (2012) Effect of temperature, light, salinity, drought stress and seeding depth on germination of *Cucumis melo var. agrestis*. Pak J Weed Sci 72(4):495–501

Teng WL, Khor E, Tan TK, Lim LY, Tan SC (2001) Concurrent production of chitin from shrimp shells and fungi. Carbohydr Res 332(3):305–316

Toan NV, Hanh TT, Thien PVM (2013) Antibacterial activity of chitosan on some common food contaminating microbes. Open Biomater J 4(1):1–5

Tsigos I, Martinou A, Kafetzopoulos D, Bouriotis V (2000) Chitin deacetylases: new, versatile tools in biotechnology. Trends Biotechnol 18(7):305–312

Venkatesan J, Kim S-K (2010) Chitosan composites for bone tissue engineering—an overview. Mar Drugs 8(8):2252–2266

Verma S, Kumar N, Verma A, Singh H, Siddique KHM, Singh NP (2020) Novel approaches to mitigate heat stress impacts on crop growth and development. Plant Physiol Reports 25(4):627–644. https://doi.org/10.1007/s40502-020-00550-4

Wahid A, Gelani S, Ashraf M, Foolad MR (2007) Heat tolerance in plants: an overview. Environ Exp Bot 61(3):199–223

Wang G, Wei Y, Qiao S, Lin P, Chen Y (2018) Generalized inverses: theory and computations, vol 53. Springer, Singapore

Wani RT (2019) Socioeconomic status scales modified Kuppuswamy and Udai Pareekh's scale updated for 2019. J Fam Med Prim Care 8(6):1846

Wasaya A, Abbas T, Yasir TA, Sarwar N, Aziz A, Javaid MM, Akram S (2021) Mitigating drought stress in sunflower (*Helianthus annuus* L.) through exogenous application of β-aminobutyric acid. J Soil Sci Plant Nutr 21(2):936–948

Xu P, Zheng G-W, Zong M-H, Li N, Lou W-Y (2017) Recent progress on deep eutectic solvents in biocatalysis. Bioresour Bioprocess 4(1):1–18

Xu Q, Wang X, Tang C (2018) The effects of elevated CO_2 and nitrogen availability on rhizosphere priming of soil organic matter under wheat and white lupin. Plant Soil 425(1):375–387. https://doi.org/10.1007/s11104-018-3601-1

Yoon M-Y, Cha B, Kim J-C (2013) Recent trends in studies on botanical fungicides in agriculture. Plant Pathol J 29(1):1

Younes I, Rinaudo M (2015) Chitin and chitosan preparation from marine sources. Structure, properties, and applications. Mar Drugs 13(3):1133–1174

Zhang M, Li XH, Gong YD, Zhao NM, Zhang XF (2002) Properties and biocompatibility of chitosan films modified by blending with PEG. Biomaterials 23(13):2641–2648. https://doi.org/10.1016/S0142-9612(01)00403-3

Zhao H, Zhang Y, Bradford PD, Zhou Q, Jia Q, Yuan F-G, Zhu Y (2010) Carbon nanotube yarn strain sensors. Nanotechnology 21(30):305502. https://doi.org/10.1088/0957-4484/21/30/305502

Chapter 13
Exogenous Application of Biostimulants and Commercial Utilization

Bushra Sarwar, Ahmad Sher, Muhammad Ijaz, Muhammad Irfan, and Sami Ul-Allah

Abstract Plant biostimulants are specialized goods that are used to boost crop productivity and swiftly spreading throughout the agricultural chemical and seed industries. Biostimulants are distinct from conventional crop inputs like fertilizers or pesticides can influence the plant growth and development through many channels from a single substance for influencing crop growth and development depending on both the timing and the location of application. However, there is wide variation in the effectiveness of biostimulants and little knowledge of the mechanisms underlying situations where variations are seen in field-tested experiments. These unidentified pathways might coincide with established indices of soil health, opening doors to untapped biostimulant potential regarding growth and development of the crop. Therefore, it is most frequently employed to provide the nutrients required to achieve the ideal yield at desired level, regardless of the fertilizer source and mode of administration used. The initial distinction between biostimulants and other agricultural inputs is in the adaptability of particular products in terms of the wanted response. The primary categories of crop biostimulants, known mechanisms of action, instances of their current field efficacy, and a future perspective are all addressed in this chapter.

Keywords Biostimulants · Application method · Yield level · Field efficacy · Soil health

---

B. Sarwar · M. Irfan
College of Agriculture, University of Layyah, Layyah, Pakistan

A. Sher (✉) · M. Ijaz
College of Agriculture, University of Layyah, Layyah, Pakistan

Department of Agronomy, Bahauddin Zakariya University, Multan, Pakistan
e-mail: ahmad.sher@bzu.edu.pk

S. Ul-Allah
College of Agriculture, University of Layyah, Layyah, Pakistan

Department of Plant Breeding and Genetics, Bahauddin Zakariya University, Multan, Pakistan

M. Hasanuzzaman (ed.), *Climate-Resilient Agriculture, Vol 2*,
https://doi.org/10.1007/978-3-031-37428-9_13

1 Introduction

Other categories are named for these products as plant biostimulators, probiotics, metabolic enhancers, and biofertilizers (Swift et al. 2018; Garca-Fraile et al. 2017). Plant biostimulants (PBs) are the most popular term used in beneficial manner for crop production. Although biostimulants have been used in commercial agriculture for many years, growers now have access to a much wider range of these compounds. Organic acids, seaweed extracts helpful microorganisms (fungus and bacteria), chitosan, and amino acid or protein hydrolysates are examples of goods that are frequently mentioned as biostimulants (Kauffman et al. 2007; du Jardin 2015; Halpern et al. 2015) and less popular but expanding categories include of concentrated enzymes, charcoal, and microbial extracts. The composition of individual products for the remaining biostimulants, which, more often than not, is not fully known, varies substantially with the exception of concentrated enzymes.

This information gap results from the biological origin of these goods, which contain several constituents. It is anticipated that the product's beneficial activity is caused by synergy among the elements rather than by the individual constituents acting alone (Yakhin et al. 2017; Bulgari et al. 2015). According to Yakhin et al. (2017), synergy among product ingredients makes it challenging to pinpoint the precise processes that trigger a crop response. As a result, a better method for defining biostimulant action should be based on their application and efficacy. Consequently, the remainder of this analysis will concentrate on applied biostimulants' agronomic effects on the production of row crops and their putative connections to soil health and is acknowledged.

Plant biostimulants (PBs) are not considered fertilizers or plant protection products because their main purposes are not to give nutrients or shield plants from pests and pathogens (Rouphael and Colla 2018). These items include elements and/or microbes that improve the availability of nutrients to plant roots and, subsequently, their uptake, promote the plant's ability to absorb nutrients, and, in some cases (Calvo et al. 2014), help the plant adapt to abiotic challenges. Less use of fertilizers is allowed by PBs that are frequently used in agriculture due to their effectiveness in enhancing plant nutrient uptake. Consequently, by potentially reducing the vast amounts of synthetic substances consumed by this activity, this could also help to increase the environmental sustainability of agriculture (Puglia et al. 2021). Starting with basic materials with incredibly various compositions and sources, it is possible to create compounds that have stimulatory effects on plants (Rouphael and Colla 2018). This is why various families of biostimulants have been established, including inorganic salts, complex organic materials, humic and fulvic substances, plant extracts and seaweed, chitosan, chitin derivatives, protein hydrolysates, amino acids and organic acids, animal/vegetable protein, and helpful microorganisms (yeast, filamentous fungi, and microalga bacteria like *Bacillus* and *Azotobacter* spp.) (Du Jardin 2015; La Torre et al. 2016). The beneficial impacts of biostimulants' various components may work in concert to enhance plant growth, productivity, yield, and quality. Because of this, it is still unclear how biostimulants work in general (Koleka et al. 2017). This is why a biostimulant is only considered to be such if it has been shown to boost plant nutrient uptake, production, and resilience to unfavorable environmental circumstances (Yakhin et al. 2017).

2 Role of PBs Coping the Toxic Effect of Compounds

Biostimulants can increase a tolerance of a crop to challenging environmental factors such as drought, UV radiation, high heat, and salinity by applying them sparingly to rhizosphere, plants, and seeds (Du Jardin 2015; Yakhin et al. 2017; Del Buono 2021). The production of chlorophyll and pigments, relative water content, leaf gas exchange or the activity of antioxidant enzymes, which control lipid membrane oxidation and water loss are a few essential biochemical physiological processes which are impacted by salt and drought stress in plants. According to recent studies, PBs may be able to mitigate these damages (Del Buono et al. 2020; Goñi et al. 2018). Furthermore, as biostimulated crops are more effective at getting and utilizing nutrients, PBs may enable a reduction in the usage of artificial fertilizers (Del Buono 2021).

3 Categories of Biostimulant

3.1 Seaweed Extracts

3.1.1 Proposed Mechanisms and Composition

The group of biostimulants known as seaweed extracts is made by processing several types of algae, most frequently macroalgae (seaweeds). Macroalgae are a renewable resource, and the species that are utilized to make the biostimulants (Ugarte et al. 2010) are meticulously watched to enable for continuous harvesting to keep the supply steady. The components of commercial products vary widely, depending on the species used, harvesting stage, and the specific extraction method used by each company (Goñi et al. 2016). Alkaline hydrolysis is the most widely used extraction technique while others include super-critical fluid, water-based, microwave, pressurized liquid extractions' ultrasound, acid hydrolysis, and enzyme (Shukla et al. 2019).

Seaweed extracts include betaines brassinosteroids, polyamines, and plant hormones in addition to the carbohydrates (Stirk et al. 2020). These chemicals work in concert to provide favorable effects in plants, including increased plant growth, resistance to biotic and abiotic stressors, and higher crop quality due to increased nutrient uptake. While it is easy to assess an interest crop response (nutrient absorption, plant growth, grain yield, etc.), it is exceedingly challenging to pinpoint the precise metabolite and mechanism at play in field experiments because of how the environment and agronomic techniques interact. Therefore, crop growth and yield response provide the best indicators of their treatment efficacy.

3.1.2 Efficacy and Field Application

The Roman Columella used seaweed extracts as organic manure amendments to their crops and as mulch throughout the first century, according to historical records (Newton 1951). The alleviation of alterations in environmental conditions regard as abiotic stress which mostly comprises of tolerance of water stress, is the targeted mode of action for seaweed extracts' foliar application of row crops. Additionally, there is a transformed emphasis on soil treatments to improve root development and root zone microbial activity. Phytohormones included in the product, stress reduction, and/or stimulation of plant metabolism have all been linked to increased nutrient uptake, grain yield, and plant growth when seaweed extracts are used (Calvo et al. 2014; González et al. 2013; Craigie 2011; Khan et al. 2009).

However, the absence of issued studies showing decreased yields and growth pattern do not imply that products enhance performance of crops if treated with seaweed extracts. Earlier studies concentrated on the differences between untreated and treated crop plants, and there is little information on how seaweed extract interacts with field conditions and other agronomic strategies.

3.2 Humic and Fulvic Acids

3.2.1 Proposed Mechanisms and Composition

The complex process of microbial breakdown of organic matter results in a wide range of different by-products along the course of degradation (Nardi et al. 2007). The final outcome of this route is traditionally thought to be soil organic matter (SOM), which is made up of refractory components that are not degraded due to resistance and is thought to contain stable chemical compounds collectively known as humus. These molecules are frequently categorized as humic acids (HA), fulvic acids (FA), and humin (acid insoluble and alkali), and the organic matter made in the soils is up to 60%. Humic acids are acid insoluble and alkali soluble (Lamar 2020). Traditional theories hold that these substances have degradation resistance, but a more recent theory holds that dynamically decomposition of organic matter is present and that substances that were once believed to be gone under stable reversible reactions. As a result, these substances have the potential to influence the soil microbiome (Lehmann and Kleber 2015).

3.2.2 Efficacy and Field Application

For many years, HA and FA have been employed as inputs in agricultural production, and their effects on communities of microbes, availability of nutrients, and growth of plants have been thoroughly researched (Celik et al. 2010; Jindo et al. 2016). Depending on the intended purpose, there are many different ways to use HA and FA. The two main uses in row crops are to improve nutrient uptake or amend the soil. It is frequently remarked that the performance of HA and FA treatments for enhancing grain production is not constant even applied in a comparison with common commercial fertilizers.

In addition, there are several accounts of applications of FA or HA at the field level having no advantages or even negative effects (Hartz and Bottoms 2010; de Santiago et al. 2010). Though, FA and HA can improve soil structure, preserve soil ammonium, and have an impact on soil biochemistry that is connected to nitrogen (N) and phosphorus (P) cycling. Alternative viewpoints for the HA and FA market might therefore be best centered on nitrogen management and the potential to enhance soil health.

3.3 Nitrogen-Fixing Bacteria

3.3.1 Known Mechanisms and Common Species

All living microbes require nitrogen, which is crucial for the creation of important substances including proteins and nucleic acids. The biggest source of readily available nitrogen is dinitrogen gas (N_2) in the atmosphere, but only a small number of bacteria (diazotrophs) can use nitrogenase to change N_2 into a bioactive form (NH_3). The metal cofactors of the three different nitrogenase enzyme complexes are vanadium-iron (V-Fe), iron-iron (Fe-Fe), and molybdenum-iron (Mo-Fe) (Zehr et al. 2003). Because not all microorganisms use all three nitrogenases, even if the Mo-Fe cofactor is the most prevalent, N fixation may be hampered by the availability of cofactor minerals (Vitousek and Howarth 1991).

Because oxygen irreversibly inhibits nitrogenase function, bacteria must find ways to safeguard the enzyme from oxygen while they are in aerobic settings. The creation of a heterocyst, which is common to cyanobacteria in aquatic settings, or a nodule are the two most prevalent ways for bacteria to isolate themselves from oxygen (Rhizobia–legume symbiotic association). In order to successfully integrate the diazotroph of interest into an agronomic system and maintain optimal product efficacy and biological nitrogen fixation (BNF), it is essential to understand how the diazotroph defends itself against high-oxygen concentrations.

3.3.2 Efficacy and Field Application

The possible replenishment of N to the growing crop, which reduces the requirement for N to be supplied as fertilizer, is the clear agronomic benefit for the usage of N-fixing microorganisms. Placement of these bacteria close to the growing crop's roots via in-furrow treatments or seed treatment is essential for maximizing their efficacy. In order to ensure a favorable crop response, the use of these bacteria will necessitate recommendations tailored to each particular farm to identify the suitable microorganism with the right administration method and with the necessary agronomic management.

3.4 Phosphorus-Solubilizing Bacteria (PSB)

3.4.1 Known Mechanisms and Common Species

Although P makes up only 0.1% of the water-soluble portion of soil content (w/w), this low availability makes fertilizer P necessary to meet plant need of nutrients for a particular crop yield (Sharma et al. 2013). Phosphorus that is applied to soils might become fixed there, making it unavailable for plant absorption while still contributing to the soil's P reservoir is only up to 90%. Microorganisms primarily achieve the solubilization of inorganic phosphates by the production of organic acids (Kalayu 2019) which can improve P availability in two different ways: (1) by preventing cations like Ca^{2+} (Calcium) and $Fe^{2+}/^{3+}$ (Ferrous/Ferric Iron) from fixing accessible P; and (2) releases mineral P-complexes by lowering the soil pH, particularly Ca (Walpola and Yoon 2012). Through the development of extracellular enzymes, organic phosphates can be hydrolyzed to increase the amount of soil-available P (Tarafdar et al. 2002). Although the mechanisms of P solubilization are well established, there is far less information available regarding the effectiveness of boosting those systems through inoculation or management to improve crop production.

3.4.2 Efficacy and Field Application

Numerous studies have been undertaken on phosphorus-solubilizing bacteria in both natural ecosystems and lab settings over a long period of time (Alori et al. 2017; Saeid et al. 2018). The phosphorus solubilizing microorganisms (PSM) have only lately been introduced as an agronomic input, hence their full commercialization potential has not yet been attained (Kalayu 2019). Because different soil types and agronomic activities (rotation, tillage, and fertilization) affect the amount and the source of soil P differently, it is essential to utilize the right microorganism to maximize P solubilization in the specific system. When three PSM strains were applied, wheat yield increased by 19–24% when compared to an uninoculated control, but by 33% when the three strains were co-applied (Turan et al. 2012). When used in conjunction with other agronomic methods, PSM in agronomic systems clearly has the ability to boost crop P uptake, crop grain yield, and soil-available P, the difficulty is to comprehend the species by environment interactions to maximize their usage. In addition, using PSM to provide P for crop growth rather than fertilizing with external P reduces P contamination of streams and can promote the establishment of more soil microorganisms.

3.5 *Arbuscular Mycorrhizal Fungi (AMF)*

3.5.1 Known Mechanisms and General Morphology

The symbiotic relationships between soil fungus and crop plants are widely recognized, and studies of these relationships have been conducted for a variety of crops, including rice, wheat, maize, and soybean (Mbodj et al. 2018; Sugiyama 2019). Due to the physical characteristics of vesicles and arbuscules that are created by these creatures, endophytic mycorrhizal fungi (AMF) are the most prevalent fungal/plant interaction. According to theory, these fungi coevolved with plant roots to enable adaptability to grow on dry land (Willis et al. 2013).

While genetic analysis enables the isolation and species-level differentiation of bacteria, AMF taxonomy frequently relies on the physical traits of the asexual resting spores. The synergistic link between soil bacteria and AMF and plants is another something that has been better understood as a result of advances in microbiological research, leading to the idea of potential co-inoculation (Miransari 2011). Future crop production will benefit from the understanding that AMF can interact with soil bacteria and impact how biostimulants, which support plant development and soil health, are produced or function.

3.5.2 Efficacy and Field Application

AMF's potential as applied biostimulants has been assessed by a number of research and reviews, with the main functions being to reduce the stress of saline conditions, defend against plant diseases, and boost nutrient availability and absorption, particularly connected to P (Plenchette et al. 2005). Increased uptake of Mg, P, N, Ca, and Kin maize in saline circumstances when AMF were present reduced salt stress (Lee et al. 2015). Additionally, stronger photosynthetic upregulation and a decrease in the generation of and in reaction to ROS were two additional ways that local AMF present alleviated salt stress more than inoculation with foreign species (Estrada et al. 2013).

4 Emerging Biostimulant Categories

4.1 *Enzymes*

A new class of biostimulants, which are pure enzymes, has been launched with the commercial usage of phosphatases in crop fields. Extracellular enzymes produced by organisms are found in soils; this is particularly evident in plants and microorganisms (Spiers and McGill 1979). These enzymes function as biological catalysts to speed up biochemical reactions, which are sometimes reliant on organic N or P chemicals in soils. Enzymes that can be applied to soil in cropping systems have recently been produced and purified by industrial manufacture of enzymes through microbial fermentation techniques (Nielsen et al. 2007).

Enzymes associated with the carbon (C) cycle are also of interest because, like phosphatases, they can catalyze the breakdown of residues and offer a potential tool for better management in high-organic matter systems like no-till or cover cropping. Hemicellulose and cellulose are two of the enzymes that are used to break down the polymers in plant tissues. These larger polymers become more hydrolyzable by microbial communities when they are broken down into smaller polymers or monomers. This decomposition may set off a series of events that speed up the mineralization of extra nutrients for upcoming crop absorption. The development of a perfect mixture with numerous enzymes targeting a certain organic component's disintegration and the release of a specific nutrient is theoretically possible.

4.2 *Biochar*

Pyrolysis, which is the thermochemical degradation of a fuel source without the addition of oxygen, produces biochar through high-heat processes (Weber and Quicker 2018). The final product is a highly carbonaceous substance with different properties depending on the source, the processing temperature range, and the processing duration (Leng and Huang 2018). Charcoal, which comes from woody biomass, is one of the most popular types of biochar. Hemicellulose, cellulose, and lignin are the main components of biomass, and when the structures degrade at different temperatures, the biochar's stability and activity change (Yang et al. 2007).

As a fuel source, construction material, filtering method, and most recently as an agricultural soil supplement, biochar is utilized in a wide range of industries. When applied to an agricultural land, biochar is extremely resistant to deterioration and functions as a steady carbon source. It can chelate with soil ions because it is porous and has a wide surface area. Higher plant productivity, improved soil treatment's ability to retain nutrients, and increased water holding capacity are all considered agricultural advantages of biochar (Biederman and Harpole 2013).

However, the potential of biochar as a long-term solution for increased crop production, improved soil health, and increased soil productivity is largely unexplored. Ongoing research into its use in conjunction with good strategies may lead to most direct biostimulant applications that are focused in growing season of plant by boosting the yield and plant growth.

5 The Biostimulant and Soil Health Potential

Numerous factors contribute to soil health, many of which are biologically mediated and can therefore be altered by the use of biostimulants. A few examples of the consistent procedures being developed by the USDA NRCS in collaboration with academic researchers across the United States and soil enzyme activity, total soil organic carbon, and soil respiration rates are considered as soil health evaluation and testing of indicators (USDA-NRCS 2021). These characteristics have the ability to simultaneously affect crop development and soil health since, as was previously said, they can be indications of biostimulant action in row crops. Many farmers use biostimulants in search of a yield response during the growing season, paying less attention to the possibility of long-term effects on their soils and repetitive applications over time.

A biostimulant may not produce a quick effect, but it has the ability to improve the health of the soil over time, increasing harvests in succeeding years. Long-term analyses of the effects of biostimulants on soils are, however, scarce. The possibility for greater carbon sequestration is in addition to a direct effect on soil biological activity for improved soil health.

Long-term studies of bioimpact stimulants on soil carbon soil health and nutrient cycling are therefore necessary. However, the long-term addition of C can modify the C:N ratios of soils, which may trap more N and lower crop performance. Although it might seem like the ideal solution, it will take time and a variety of techniques to fully realize and comprehend the value of using biostimulants to improve agronomic management for long-term improvements in soil health and crop yields.

6 Application Methods and Common Uses

Commercial biostimulants are often first applied to specialty crops since these crops frequently have better potential for profit per acre than row crops (Neill and Morgan 2021). Because specialty crops are often more vulnerable to environmental stressors (Kistner et al. 2018). Since there is no additional expense for application, including the biostimulant application with a current standard management practice gives the product a so-called "free ride." For instance, planting provides the chance to provide biostimulants via in-furrow or seed treatment to all planted acres.

Product compatibility with other agronomic inputs like pesticides and fertilizers is one of the biggest obstacles to the integration of biostimulants into a farmer system. There is a lack of understanding and a need for study defining the potential interactions that may happen during field application because there are so many distinctive chemistries and products on the market. Additionally, the reaction to biostimulants may vary depending on the stage of crop development at the time of application or the interactions with the climate, where extremes in precipitation and temperature may affect crop response.

7 Future Perspectives and Conclusion

The market for biostimulants faces major obstacles because of the countless potential uses for these substances. The intended aim of the application is frequently straightforward, even though all agronomic inputs (such as rotation, fertility, soil amendments, seed genetics, tillage, and pesticides) contain a variety of alternatives for product or method selection. For instance, the four distinct pesticide inputs herbicides, fungicides, insecticides, and nematodes each have a single-intended use that is the eradication of the consistent pest outbreak in form of nematodes, insects, and weeds of other microbial diseases. To stimulate signal pathways for mitigating abiotic stress, foliar spray at vegetative stages is considered to be best, however applying the same biostimulant, such seaweed extract, at planting may alter the microbial communities in the application zone. Growers have a wide range of options on the fertilizer market (Table 13.1).

The initial distinction between biostimulants and other agricultural inputs is in the adaptability of particular products in terms of the desired response. Boosting attention is being paid to increasing grain yield, which is frequently the outcome of more effective nutrient usage, and the primary research method for biostimulant use in row crop systems is now focusing on fertilizer recovery potential. While many biostimulants are intended to be applied to row crops in order to boost production, many products really produce these effects through having an impact on the biology of the root zone and the soil. A more thorough analysis of the impact of biostimulants on biological indicators and soil quality may uncover previously unrecognized advantages of their use. The use of biostimulants as a remedy for more sustainable practices and improved soil quality quantifiable yield increases as a result agronomic strategies are improved for public and governments sectors' awareness and the effects through which water quality is maintained and nutrient management is done.

Table 13.1 Role of plant biostimulants in mitigating heavy metal toxicity

Plant species	PB	Heavy metals	Recommended dose of PB	Results	Reference
Zea maize	Humic substances	Cr	4 mM C HA L^{-1}	Higher biomass production, higher stress signaling, and gene response in transcription, CAT, and proline increases	Canellas et al. (2020)
Zea maize	Silymarin-based biostimulant	Cd	0.24 g L^{-1}	Improves photosynthetic activity with enhanced antioxidant mechanism and expression of gene, restored hormonal homeostasis	Alharby et al. (2021)
Zea maize	Megafol	Metolachlor	2.5 L ha^{-1}	Enzymes (CAT, APX, GPX), lower levels of lipid membrane peroxidation, production, improved antioxidant activities increased germination, biomass	Panfili et al. (2019)
Helianthus annuus	Protein hydrolysates	Imazamox	3 L ha^{-1}	Improved plant growth, enhanced photosynthetic activity, increased chlorophyll contents, and improved stomatal conductance	Balabanova et al. (2016)
Glycine max	Fertiacyl Pòs	Glyphosate	0.4 L ha^{-1}	Slightly low yield losses and little symptoms of chlorosis and necrosis	Constantin et al. (2016)

References

Alharby HF, Al-Zahrani HS, Hakeem KR, Alsamadany H, Desoky ESM, Rady MM (2021) Silymarin-enriched bio-stimulant foliar application minimizes the toxicity of cadmium in maize by suppressing oxidative stress and elevating antioxidant gene expression. Biomol Ther 11:465

Alori ET, Glick BR, Babalola OO (2017) Microbial phosphorus solubilization and its potential for use in sustainable agriculture. Front Microbiol 8:971

Balabanova DA, Paunov M, Goltsev V, Cuypers A, Vangronsveld J, Vassilev A (2016) Photosynthetic performance of the imidazolinone resistant sunflower exposed to single and combined treatment by the herbicide imazamox and an amino acid extract. Front Plant Sci 7:1559

Biederman LA, Harpole WS (2013) Biochar and its effects on plant productivity and nutrient cycling: a meta-analysis. GCB Bioenergy 5:202–214

Bulgari R, Cocetta G, Trivellini A, Vernieri P, Ferrante A (2015) Bio-stimulants and crop responses: a review. Biol Agric Hortic 31:1–17

Calvo P, Nelson L, Kloepper JW (2014) Agricultural uses of plant bio-stimulants. Plant Soil 383:3–41

Canellas LP, Canellas NO, Irineu LESDS, Olivares FL, Piccolo A (2020) Plant chemical priming by humic acids. Chem Bio Technol Agric 7:12

Celik H, Katkat AV, Aşık BB, Turan MA (2010) Effect of foliar-applied humic acid to dry weight and mineral nutrient uptake of maize under calcareous soil conditions. Commun Soil Sci Plant Anal 42:29–38

Constantin J, de Oliveira RSE, Jr Gheno EA, Biffe DF, Braz GBP, Weber F, Takano HK (2016) Prevention of yield losses caused by glyphosate in soybeans with biostimulant. Afr J Agric Res 11:1601–1607

Craigi JS (2011) Seaweed extract stimuli in plant science and agriculture. J Appl Phycol 23:371–393

de Santiago A, Exposito A, Quintero JM, Carmona E, Delgado A (2010) Adverse effects of humic substances from different origin on lupin as related to iron sources. J Plant Nutr 33:143–156

Del Buono D (2021) Can bio-stimulants be used to mitigate the effect of anthropogenic climate change on agriculture? It is time to respond. Sci Total Environ 751:141763

Del Buono D, Regni L, Del Pino AM, Bartucca ML, Palmerini CA, Proietti P (2020) Effects of megafol on the olive cultivar 'Arbequina' grown under severe saline stress in terms of physiological traits, oxidative stress, antioxidant defences, and cytosolic Ca^{2+}. Front Plant Sci 11:603576

Du Jardin P (2015) Plant bio-stimulants: definition, concept, main categories and regulation. Sci Hortic 196:3–14

Estrada B, Aroca R, Barea JM, Ruiz-Lozano JM (2013) Native arbuscular mycorrhizal fungi isolated from a saline habitat improved maize antioxidant systems and plant tolerance to salinity. Plant Sci 201:42–51

García-Fraile P, Menéndez E, Celador-Lera L, Díez-Méndez A, Jiménez-Gómez A, Marcos-García M, Cruz-González XA, Martínez-Hidalgo P, Mateos PF, Rivas R (2017) Bacterial probiotics: a truly green revolution. In: Probiotics and plant health. Springer, Berlin/Heidelberg, pp 131–162

Goñi O, Fort A, Quille P, McKeown PC, Spillane C, O'Connell S (2016) Comparative transcriptome analysis of two Ascophyllum nodosum extract bio-stimulants: same seaweed but different. J Agric Food Chem 64:2980–2989

Goñi O, Quille P, O'Connell S (2018) Ascophyllum nodosum extract bio-stimulants and their role in enhancing tolerance to drought stress in tomato plants. Plant Physiol Biochem 126:63–73

González A, Castro J, Vera J, Moenne A (2013) Seaweed oligosaccharides stimulate plant growth by enhancing carbon and nitrogen assimilation, basal metabolism, and cell division. J Plant Growth Regul 32:443–448

Halpern M, Bar-Tal A, Ofek M, Minz D, Muller T, Yermiyahu U (2015) The use of bio-stimulants for enhancing nutrient uptake. In: Advances in agronomy, vol 130. Elsevier, Amsterdam, pp 141–174; ISBN 0065-2113

Hartz TK, Bottoms TG (2010) Humic substances generally ineffective in improving vegetable crop nutrient uptake or productivity. HortScience 45:906–910

Jindo K, Soares TS, Peres LEP, Azevedo IG, Aguiar NO, Mazzei P, Spaccini R, Piccolo A, Olivares FL, Canellas LP (2016) Phosphorus speciation and high-affinity transporters are influenced by humic substances. J Plant Nutr Soil Sci 179:206–214

Kalayu G (2019) Phosphate solubilizing microorganisms: promising approach as biofertilizers. Int J Agron:1–7. https://doi.org/10.1155/2019/4917256

Kauffman GL, Kneivel DP, Watschke TL (2007) Effects of a biostimulant on the heat tolerance associated with photosynthetic capacity, membrane thermostability, and polyphenol production of perennial ryegrass. Crop Sci 47:261–267

Khan W, Rayirath UP, Subramanian S, Jithesh MN, Rayorath P, Hodges DM, Critchley AT, Craigie JS, Norrie J, Prithiviraj B (2009) Seaweed extracts as bio-stimulants of plant growth and development. J Plant Growth Regul 28:386–399

Kistner E, Kellner O, Andresen J, Todey D, Morton LW (2018) Vulnerability of specialty crops to short-term climatic variability and adaptation strategies in the Midwestern USA. Clim Chang 146:145–158

Koleška I, Hasanagi'c D, Todorovi'c V, Murti'c S, Kloki'c I, Paradikovi'c N, Kukavica B (2017) Bio-stimulant prevents yield loss and reduces oxidative damage in tomato plants grown on reduced NPK nutrition. J Plant Interact 12:209–218

La Torre A, Battaglia V, Caradonia F (2016) An overview of the current plant bio-stimulant legislations in different European Member States. J Sci Food Agric 96:727–734

Lamar RT (2020) Possible role for electron shuttling capacity in elicitation of PB activity of humic substances on plant growth enhancement. In: The chemical biology of plant bio-stimulant, pp 97–121. John Wiley & Sons Ltd. N.J, USA

Lee Y, Krishnamoorthy R, Selvakumar G, Kim K, Sa T (2015) Alleviation of salt stress in maize plant by co-inoculation of arbuscular mycorrhizal fungi and *Methylobacterium coryza* CBMB20. J Korean Soc Appl Biol Chem 58:533–540

Lehmann J, Kleber M (2015) The contentious nature of soil organic matter. Nature 528:60–68

Leng L, Huang H (2018) An overview of the effect of pyrolysis process parameters on biochar stability. Bioresour Technol 270:627–642

Mbodj D, Effa-Effa B, Kane A, Manneh B, Gantet P, Laplaze L, Diedhiou AG, Grondin A (2018) Arbuscular mycorrhizal symbiosis in rice: establishment, environmental control and impact on plant growth and resistance to abiotic stresses. Rhizosphere 8:12–26

Miransari M (2011) Interactions between arbuscular mycorrhizal fungi and soil bacteria. Appl Microbiol Biotechnol 89:917–930

Nardi S, Muscolo A, Vaccaro S, Baiano S, Spaccini R, Piccolo A (2007) Relationship between molecular characteristics of soil humic fractions and glycolytic pathway and krebs cycle in maize seedlings. Soil Biol Biochem 39:3138–3146

Neill CL, Morgan KL (2021) Beyond scale and scope: exploring economic drivers of U.S. specialty crop production with an application to edamame. Front Sustain Food Syst 4:1–10

Newton L (1951) Seaweed utilization. Nature 167:1004

Nielsen PH, Oxenbøll KM, Wenzel H (2007) Cradle-to-gate environmental assessment of enzyme products produced industrially in Denmark by Novozymes A/S. Int J Life Cycle Assess 12:432

Panfili I, Bartucca ML, Marrollo G, Povero G, Del Buono D (2019) Application of a plant biostimulant to improve maize (*Zea mays*) tolerance to metolachlor. J Agric Food Chem 67:12164–12171

Plenchette C, Clermont-Dauphin C, Meynard JM, Fortin JA (2005) Managing arbuscular mycorrhizal fungi in cropping systems. Can J Plant Sci 85:31–40

Puglia D, Pezzolla D, Gigliotti G, Torre L, Bartucca ML, Del Buono D (2021) The opportunity of valorizing agricultural waste, through its conversion into bio-stimulants, biofertilizers, and biopolymers. Sustainability 13:2710

Rouphael Y, Colla G (2018) Synergistic biostimulatory action. Designing the next generation of plant bio-stimulants for sustainable agriculture. Front Plant Sci 9:1655

Saeid A, Prochownik E, Dobrowolska-Iwanek J (2018) Phosphorus solubilization by Bacillus species. Molecules 23:2897

Sharma SB, Sayyed RZ, Trivedi MH, Gobi TA (2013) Phosphate solubilizing microbes: sustainable approach for managing phosphorus deficiency in agricultural soils. Springer Plus 2:587

Shukla PS, Mantin EG, Adil M, Bajpai S, Critchley AT, Prithiviraj B (2019) Ascophyllum nodosum-based bio-stimulants: sustainable applications in agriculture for the stimulation of plant growth, stress tolerance, and disease management. Front Plant Sci 10:655

Spiers GA, McGill WB (1979) Effects of phosphorus addition and energy supply on acid phosphatase production and activity in soils. Soil Biol Biochem 11:3–8

Stirk WA, Rengasamy KRR, Kulkarni MG, van Staden J (2020) Plant bio-stimulants from seaweed: an overview. Chem Biol Plant Bio-stimulants 2:31–55

Sugiyama A (2019) The soybean rhizosphere: metabolites, microbes, and beyond—a review. J Adv Res 19:67–73

Swift R, Denton MD, Melino VJ (2018) Plant probiotics for nutrient acquisition by agriculturally important grasses: a comprehensive review of the science and the application. Annu Plant Rev Online: 2:1–47

Tarafdar JC, Yadav RS, Niwas R (2002) Relative efficiency of fungal intra-and extracellular phosphatases and phytase. J Plant Nutr Soil Sci 165:17–19

Turan M, Gulluce M, von Wirén N, Sahin F (2012) Yield promotion and phosphorus solubilization by plant growth–promoting rhizobacteria in extensive wheat production in Turkey. J Plant Nutr Soil Sci 175:818–826

Ugarte RA, Craigie JS, Critchley AT (2010) Fucoid flora of the rocky intertidal of the Canadian Maritimes: implications for the future with rapid climate change. In: Seaweeds and their role in globally changing environments. Springer, Berlin/Heidelberg, pp 69–90

USDA-NRCS Soil Health (2021) Available online: https://www.nrcs.usda.gov/wps/portal/nrcs/main/soils/health

Vitousek PM, Howarth RW (1991) Nitrogen limitation on land and in the sea: how can it occur? Biogeochemistry 13:87–115

Walpola BC, Yoon MH (2012) Prospectus of phosphate solubilizing microorganisms and phosphorus availability in agricultural soils: a review. African J Microbiol Res 6:6600–6605

Weber K, Quicker P (2018) Properties of biochar. Fuel 217:240–261

Willis A, Rodrigues BF, Harris PJC (2013) The ecology of arbuscular mycorrhizal fungi. Crit Rev Plant Sci 32:1–20

Yakhin OI, Lubyanov AA, Yakhin IA, Brown PH (2017) Bio-stimulants in plant science: a global perspective. Front Plant Sci 7:2049

Yang H, Yan R, Chen H, Lee DH, Zheng C (2007) Characteristics of hemicellulose, cellulose and lignin pyrolysis. Fuel 86:1781–1788

Zehr JP, Jenkins BD, Short SM, Steward GF (2003) Nitrogenase gene diversity and microbial community structure: a cross-system comparison. Environ Microbiol 5:539–554

Chapter 14
Cross Talk of Biostimulants with Other Signaling Molecules Under Abiotic Stress

Shruti Rohatgi, Riya Jain, Shivangi Mathur, Deeksha Singh, and Rajiv Ranjan

Abstract Abiotic stresses adversely affect plant growth, reproduction, deplete soil quality, and threaten food security. Plants mount extensive stress-specific reactions in response to abiotic stresses that involve signal transduction cascades, transcription of the appropriate responsive genes, accumulation of numerous transcripts, and metabolites specific to the stress, as well as coordinated biochemical and physiological adjustments tailored specifically to the stress. However, a plant's natural defenses are not always enough to assure plant survival in the presence of abiotic stress. Protein kinases that are similar to the *Saccharomyces* SNF1 and mammalian AMPK are involved in the main stress-signaling pathways, which suggest that stress signaling in plants is developed from energy sensing. To maintain cellular stability as well as ion-water equilibrium under stressful circumstances, stress signaling controls the expression of genes and the metabolism of proteins that are essential for ions and water transport. Biostimulants are nonnutrient chemicals or microorganisms that can enhance plant health and growth. They serve multiple purposes, including protection against the negative consequences of abiotic stress. In this chapter, we discuss the potential of various biostimulants on the activities of the plant defense system under stressful conditions and also their additional functions in controlling abiotic stresses. Biostimulants can resolve abiotic stresses, through the regulation of signaling molecules. These can provide long-term, cost-effective strategies that could raise agriculture productivity in the context of abiotic stress.

Keywords Abiotic stress · Signaling molecules · Biostimulants · Reactive oxygen species · Plant defense mechanism · Transcription: melatonin

Shruti Rohatgi and Riya Jain contributed equally with all other contributors.

S. Rohatgi · R. Jain · S. Mathur · D. Singh · R. Ranjan (✉)
Plant Biotechnology Lab, Department of Botany, Faculty of Science, Dayalbagh Educational Institute (Deemed to be University), Dayalbagh, Agra, India

M. Hasanuzzaman (ed.), *Climate-Resilient Agriculture, Vol 2*,
https://doi.org/10.1007/978-3-031-37428-9_14

1 Introduction

Plants are trying to survive in the environments where environmental conditions are continuously evolving negatively and frequently unfavorable for their growth and developments. These unfavorable environmental factors include biotic stress, like pathogen attack and herbivore assault, and abiotic stress, like heat, drought, frost, nutrient deficiencies, excess salt, or heavy metals like aluminum, cadmium, and arsenate in the soil. These significant environmental stresses, drought, salinity, and temperature affect the geographic diversity of plants in nature, restrict agriculture productivity, and adversely affect global food security. Climate change, which is expected to lead to an increase in the frequency of extreme weather, enhances the negative consequences of these abiotic stresses (Fedoroff et al. 2010). Practically everywhere in the world, drought, salt, unfavorable temperatures, and poor soil fertility are the most significant abiotic factors that restrict agricultural productivity. The majority of these issues, including drought and nutrient deficits, affect developing nations where agriculture is the primary source of income for rural residents (Verma and Deepti 2016). Actually, according to the FAO's "The State of Food and Agriculture 2007," only 3.5% of the world's land area is unaffected by environmental restrictions. Boyer predicted that yield losses brought on by unfavorable surroundings could reach 70% in 1982 (Boyer 1982; La Pena and Hughes 2007). Unfavorable stimuli can generate stress, which causes plants to respond by utilizing their energy to resist the stress rather than focusing on yield (Drobek et al. 2019).

To increase agricultural productivity and environmental sustainability, plant stress resistance must be improved. Crops with low-stress tolerance need excessive amounts of water and fertilizer, placing a heavy strain on the environment (Zhu 2016). In response to abiotic stresses, plants initiate a series of biochemical activities at the molecular level. Specific gene families are transcriptionally regulated to confer stress tolerance (Joshi et al. 2018). Based on how they function, these gene families are divided into three categories. The first category includes osmoprotective genes including heat shock proteins, late embryogenesis abundant (LEA) proteins, antioxidant enzymes, and osmoprotectants. The second category consists of the gene for ion transporters such as aquaporins and Na1/H1 channels. The third category consists of genes that control transcriptional regulation and signal perception, such as mitogen-activated protein kinases, salt-overly sensitive kinases (Ji et al. 2013), heat shock transcriptional factors, ethylene-responsive element-binding proteins, CBF/DREB, NAC, MYC/MYB, WRKY protein domains, and Cys2/His2 zinc-finger motifs (Umezawa et al. 2006).

Biostimulants are produced from several organic or inorganic materials and microbes that can increase the productivity and growth of plants and mitigate the impacts of abiotic stressors (Du Jardin 2015; Rouphael and Colla 2018). As of 2019, the EU regulation's definition of a biostimulants is "a substance that improves plant nutrition processes, regardless of the substance's nutrient content, with the primary objective of enhancing one or more of the following characteristics of the plant or the plant biota: (a) resistance to abiotic stress; (b) nutrients use efficiencies; (c) quality characteristics; or (d) presence of constrained soil nutrients or rhizosphere (García et al. 2020)." Biostimulants are natural or synthetic substances that can be applied to seeds, plants, and soil. To affect the growth of plants via improved resistance to abiotic stresses and enhance seed or grain quality and yield, these chemicals alter fundamental and structural processes. Biostimulants also decrease the requirement for fertilizers (Du Jardin 2015). Zhang and Schmidt (1997) introduced the term "biostimulant" to describe "materials that, in minute concentrations, enhance plant growth." It has been 25 years since researchers and commercial entities began to pay significant attention to biostimulants derived from natural sources (Herve 1994; Maini 2006; Zhang and Schmidt 1999; Crouch and Van Staden 1993; Sharma et al. 2014; Apone et al. 2010; Yakhin et al. 2016).

Biostimulants can be made up of both organic and inorganic substances and most of the constituents are still unknown. The omics technique, which includes determining the transcriptome, proteome, and metabolomic changes in treated plants, can be used to characterize the molecular mechanisms of action of biostimulants (Franzoni et al. 2022). Biostimulants are advanced agricultural resources which, based on the doses used, are placed between plant growth regulators and fertilizers. Exogenous applications of biostimulants are analogous to those of recognized plant hormones, the main ones of which are auxins, cytokinins, and gibberellins (Rouphael and Colla 2020; Yaronskaya et al. 2006). To alter the physiological processes of plants and increase productivity, biostimulants are progressively being used in production systems (Yakhin et al. 2017). Biostimulants can be produced chemically or organically composed of phytohormones and hormones precursor by plants. It can potentially promote growth, development, and adaptability to osmotic stress, salinity, and toxic chemicals like toxic aluminum when applied properly because it directly affects physiological processes in plants (Du Jardin 2012; Couto et al. 2012).

Biostimulants are categorized into two categories based on their formation: nonmicrobial plant biostimulants and microbial plant biostimulants. Nonmicrobial biostimulants are made from various organic matrices that have undergone various extraction methods, allowing the concentration of phytochemical compounds that promote crop development or increase a crop's resistance to harsh environmental circumstances (Mrid et al. 2021; Zulfiqar et al. 2020). It is extremely challenging to determine which substances are the most effective and essential for metabolic activities due to the complexity of the basic organic substances and, consequently, the nature of the final product. The substances that have a biostimulant effect can include chitin, chitosan, amino acids, vitamins, poly and oligosaccharides, nonnutritional minerals (like silicon and selenium), and traces of natural plant hormones (Mrid et al. 2021; Du Jardin 2015; Battacharyya et al. 2015; Xu and Geelen 2018).

Food waste and agricultural industrial by-products can also be used to make biostimulants. Utilizing by-products as a source of raw materials for the creation of biostimulants is one of the strategies used by the circular economy to make agriculture more sustainable (Xu and Geelen 2018; Franzoni et al. 2022). Microbial biostimulants, like plant growth-promoting rhizobium and arbuscular mycorrhizal fungi, are widely recognized as efficient and sustainable ways to ensure yield stability under low-input conditions, especially nitrogen and phosphorus deficit, as well as cutting-edge technology to increase crop tolerance to abiotic stresses, including extreme temperatures, salinity, and drought (Rouphael and Colla 2020).

Biostimulants when exogenously applied to these plants show physiobiochemical effects on plants. They regulate several cellular processes of plants like signal transduction, the activity of antioxidant enzymes, transcriptional regulation, and activation of signaling molecules which increases tolerance to abiotic stresses. The aim of this chapter is to increase plant growth and productivity under abiotic stress by the application of biostimulants and the activation of signaling molecules.

2 Types of Biostimulants

Biostimulants are compounds that are derived from various natural substances, such as bioactive compounds of plants, amino acids, and microorganisms (BorhannuddinBhuyan et al. 2020). Various authors have postulated several categories of biostimulants throughout the years depending on their source, mechanism of action, or primary component (Du Jardin 2015; Bulgari et al. 2015; Yakhin et al. 2017). The existing classification system is based on the origin of the raw materials, despite the fact that this decision does not always give accurate information about the product's biological activity. In accordance with Du Jardin, there are several different categories of biostimulants (Fig. 14.1), including fulvic and humic acids, seaweed and botanical extracts, N-containing compounds and protein hydrolysates, inorganic compounds, chitosan and other biopolymers, and beneficial microorganisms such as bacteria and fungi (Du Jardin 2012).

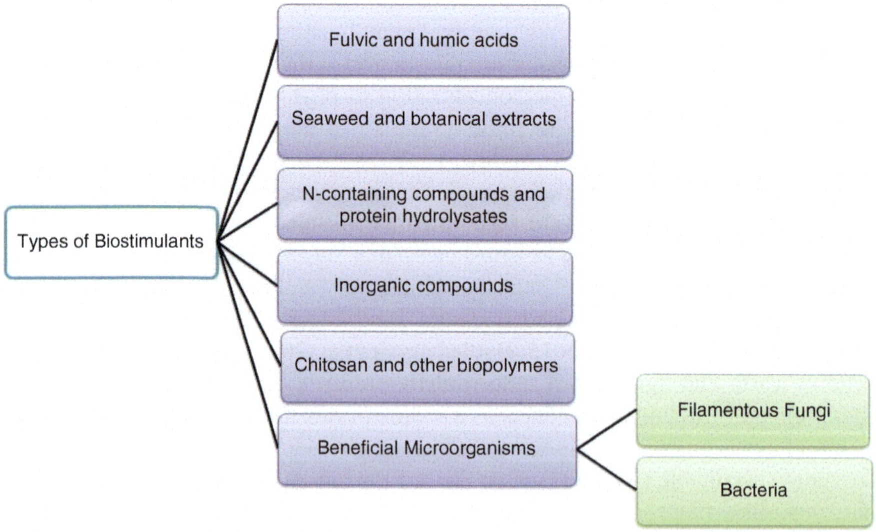

Fig. 14.1 Different types of biostimulants

2.1 Fulvic and Humic Acids

Humic substances (HS) are naturally occurring organic compounds that are mainly found in soil, and they are synthesized by plant and animal decomposition, microbial wastes, and the activity of microbes in the soil (Ukalska-Jaruga et al. 2021). HS is the group of heterogeneous compounds that were first divided into fulvic acids, humic acids, and humin according to their solubility as well as molecular weight. Consequently, interaction of td4r, md4r, and the plants' roots with microbes and soil organic matter leads to the production of humic substances and their complexes in the soil (Gautam et al. 2021). There are several natural sources of humic acids, including volcanic soils, peats, vermicompost, and mineral deposits (such as leonardite, an oxidized form of lignite) (Du Jardin 2012). Because of the long-standing capacity to affect the physical, chemical, biological, and physicochemical properties of soil, humic compounds play an important role in increasing soil fertility.

The effects of a variety of humic substance-based biostimulants are directed toward enhancing root nutrition in various ways. There are several reasons for this, such as an increase in macronutrient and micronutrient uptake, caused by the soil that contains poly-anionic humic substances, having a higher cation exchange capacity, and increased phosphorus availability caused by the interfering of HS with the precipitation of calcium phosphate. In addition, HS stimulates the H^+-ATPase of the plasma membrane to produce an electrochemical potential in the membrane that can be used to import nitrates and some other nutrients. Along with nutrition intake, ATPase's proton pumping through the plasma membrane also aids in cell wall thinning, cell expansion, and organ development (Jindo et al. 2012). The activity proposal of the biostimulation of HS also mentions stress management. The metabolism of phenylpropanoid is essential for biosynthetic pathways and several stress responses, as well as for the synthesis of phenolic chemicals. Humic substances of higher molecular mass have been shown to boost the activity of the key enzymes in this pathway in maize seedlings grown in a hydroponics system, showing that HS may affect the response to stress (Eyheraguibel et al. 2008; Olivares et al. 2017).

2.2 Seaweed and Botanical Extracts

Fresh seaweeds have long been used in agriculture as the main source of organic fertilizers and organic matter, but it has only recently been discovered that they also contain biostimulant properties. This leads to the commercial usage of seaweed extracts and refined compounds, such as alginates, carrageenans, and laminarin, as well as the by-products of their breakdown. Micronutrients, macronutrients, sterols, and N-containing substances like hormones and betaines are additional components that support plant development (Craigie 2011; Khan et al. 2009). Numerous of these chemicals are exclusive to their algal source, which explains why the research community and businesses are becoming more and more interested in these taxonomic families. *Ascophyllum, Fucus,* and *Laminaria* are the three primary genera that make up the phylum of brown algae, which contains the majority of the algal species. However, carrageenans are descended from red seaweeds, which represent a separate evolutionary branch. Khan et al. (2009) have identified the product names of around 20 seaweed products that have been used as biostimulants in recent times for better plant growth.

Seaweeds have an impact on plants and soil. They can be used as foliar treatments, hydroponic solutions, or on soils. Their polysaccharides aid in the aeration, water retention, and gel formation of soils. The fixation of heavy metals and soil remediation both benefit from the polyanionic compound's role in the exchange and fixation of cations. In suppressive soils, pathogen antagonists and bacteria that promote plant development are encouraged, which have positive impacts on the soil microflora. In addition to their various functions, macronutrients and micronutrients have nutritional effects on plants, which show that they operate as fertilizers. Hormonal impacts, which are considered to be key contributors to biostimulation activity on agricultural plants, harm the plant establishment, seed germination, and subsequent development and growth of the plant. Although bioassays and immunological methods have discovered auxins, cytokinins, gibberellins, abscisic acid, and other classes of hormone-like chemicals in seaweed extracts (Ghaderiardakani et al. 2019), there is scientific proof of the hormonal effects of the brown seaweed (*Ascophyllum nodosum*) extract, that is mostly explained by the up- and downregulation of the genes involved in hormone biosynthesis (Wally et al. 2013).

2.3 N-Containing Compounds and Protein Hydrolysates

Proteins from the by-products of agroindustry, such as animal wastes (such as epithelial tissues and collagen) and plant sources (crop leftovers), are hydrolyzed chemically and enzymatically to produce amino acids and peptide combinations (Du Jardin 2012; Calvo et al. 2014; Halpern et al. 2015). Single or combined molecules can also be made through chemical synthesis. Polyamines, betaines, and nonprotein amino acids are other nitrogenous compounds that are diverse in higher plants but little understood in terms of their ecological and physiological functions (Vranova et al. 2011). A unique instance of the derivatives of amino acids along with the well-known antistress benefits is *Glycine betaine* (Chen and Murata 2011). Case by case, it has been demonstrated that these substances serve a variety of functions as plant growth biostimulants (Du Jardin 2012; Calvo et al. 2014; Halpern et al. 2015). It modulates nitrogen absorption and assimilation and can have direct impacts on plants by controlling the structural genes and the enzymes that are involved in nitrogen assimilation as well as by affecting the signaling route for nitrogen acquisition in the roots. They also participate in the cross talk amongst carbon and nitrogen metabolisms by controlling the enzymes of the tri-carboxylic acid (TCA) cycle. Tissue hydrolysates and complex proteins have been found to have hormonal actions as well (Colla et al. 2014).

Some amino acids, including proline, have chelating properties that could protect plants from heavy metals while simultaneously assisting in the mobility and uptake of micronutrients. Scavenging of the free radicals by various nitrogenous chemicals, such as proline, betaine, and glycine, confers antioxidant action and helps to reduce environmental stress (Du Jardin 2015). There are several commercial goods available that are made from hydrolysates of protein from both animal and plant origin. In horticultural and agricultural crops, there have been varying reports of considerable gains in production and quality features (Calvo et al. 2014). In a recent assessment of the hydrolyzed proteins safety of animal origin, no ecotoxicity, phytotoxicity, or genotoxicity was noted based on the bioassays using the plants and yeasts traits as test organisms (Chaudhari 2017).

2.4 Inorganic Compounds

Beneficial elements are the inorganic substances that encourage plant development and may be necessary for some species but are not necessary for all plants (Brown et al. 2022). Cobalt, aluminum, silicon, sodium, and selenium are the five primary helpful elements. They are found in plants and soils as inorganic mineral salts and also as insoluble states as amorphous silica which is present in the gramineous species. These advantageous properties can manifest under specific environmental conditions, such as osmotic stress for sodium, and pathogen assault for selenium, or they might be fundamental, like the hardening of cell walls by the deposition of silica. Fungicides have been made from inorganic salts of helpful and necessary elements, such as chlorides, silicates, phosphates, and carbonates. These inorganic substances have an impact on osmotic pressure, pH, hormone signaling, redox homeostasis, and stress responses via enzymes such as peroxidases, while the exact mechanisms of action are still not entirely understood. More consideration should be given to their role as biostimulants of plant development, working on nutritional productivity and stress tolerance, which is separate from their fungicidal activity and their role as fertilizers and suppliers of nutrients (Deliopoulos et al. 2010; Du Jardin 2015).

2.5 Chitosan and Other Biopolymers

Deacetylated chitin is a biopolymer that is generated both naturally and artificially in the form of chitosan. Food, esthetics, pharmacy, and agriculture all employ polymers and oligomers of varying, regulated sizes. The physiological responses of oligomers of chitosan in plants are derived from their ability to bind a variety of cellular constituents, such as the components of the cell wall, plasma membrane, and DNA, as well as to bind particular receptors that are involved in the expression of defense genes, similar to the mode of action of elicitors in plant defense (El

Hadrami et al. 2010; Hadwiger 2013; Katiyar et al. 2015). The signaling mechanisms and receptors used by chitosan and chitin are different. Accumulation of H_2O_2 and the leakage of calcium ions into the cells have both been shown to be biological effects of chitosan binding to more or less particular receptors of the cell. These effects are anticipated to result in significant physiological changes since they are key participants in stress response signaling and the regulation of cell development. This hypothesis is supported by an examination of the proteomics or transcriptomics of plant cells that are treated with chitosan (Ferri et al. 2014; Povero et al. 2011). As a result, chitosan has been used in agriculture for many years, with a focus on protecting plants from fungal pathogens. However, applications of broader agriculture also concerned about the tolerance to abiotic stresses (such as salinity, drought, and cold stress), as well as quality characteristics referring to both primary and secondary metabolic activity. Chitosan induces closure of the stomata through an ABA-dependent mechanism, which contributes to the defense from environmental stress that is provided through this biostimulant (Iriti et al. 2009; Kannan 2019).

2.6 *Beneficial Microorganisms*

Microorganisms are obtained from plants, soil, and some other organic substances. They can increase crop yields either directly or indirectly depending on whether administered to the seeds or the soil (Castiglione et al. 2021). Microorganisms can influence agricultural crops directly by creating a symbiotic relationship (such as Mycorrhiza) or inadvertently by making nutrients more available to plants (Colla et al. 2015). This category mostly consists of – filamentous fungi, yeasts, and bacteria (Kour et al. 2019).

2.6.1 Filamentous Fungi

There are many ways that fungi can associate with a plant's roots, including parasitism or mutualism or in other words when both organisms thrive in close proximity to one another and form mutually beneficial relationships (Zamana et al. 2022). Since the beginning of land plants, fungi and plants have co-evolved, and the idea of a continuity of parasitism or mutualism is useful for describing the wide variety of connections that have emerged during evolutionary ages (Johnson and Graham 2013). Over 90% of all species of plants have symbiotic relationships with mycorrhizal fungi, which constitute a varied collection of taxa. Arbuscular-forming mycorrhiza (AMF) is a common endo-mycorrhiza that is associated with horticultural plants and crops, where *Glomeromycota* species develop arbuscules, branching structures built when fungi hyphae invade the cortex of plant roots (Bonfante and Genre 2010).

2.6.2 Bacteria

Bacteria participate in a variety of interactions with plants. (i) Similar to fungi, bacteria have continuity between parasitism and mutualism; (ii) bacterial habitats range from the cells interior to the soil; (iii) relationships can be temporary or long-lasting; and certain bacteria can even be transferred vertically through seeds; (iv) functions affecting plant life include participating in nutrient delivery, biogeochemical cycles, improved utilization of nutrients, induction of disease resistance, increased abiotic stress tolerance, and control of morphogenesis by the growth regulators of plants (Durán et al. 2021). Regarding the usage of biostimulants in agriculture, two primary categories should be considered inside this functional, taxonomic, and ecological diversity: (i) rhizobium-related mutualistic endosymbionts and (ii) rhizospheric plant growth-promoting rhizobacterium (PGPRs). The *Rhizobiu*m genus and its related taxa are sold as microbial inoculants or "bio-fertilizers" that assist plants in absorbing nutrients (Berg et al. 2014). Many functions of PGPRs affect various facets of plant life, including development and morphogenesis, growth and nutrition, reaction to abiotic and biotic stress, and collaboration with some other species in an agroecosystem (Philippot et al. 2013; Vacheron et al. 2013).

3 Role of Signaling Molecules Under Plant Defense Mechanism

The rapid changes in the global climate have led to an increase in the frequency and severity of abiotic stress on plants (Surabhi 2018; Ali et al. 2019). Plants frequently encounter various abiotic stresses that interfere with developmental processes and cellular membranes during their life processes (Jeandroz and Lamotte 2017; Aamer et al. 2018; Cheng et al. 2018; Ghaffari et al. 2019; Hasanuzzaman et al. 2019; Kerchev et al. 2020; Mohammadi et al. 2020). Unfavorable conditions such as salinity, cold, drought, pathogen attacks, wounding, heavy metals, excessive exposure to light, and nutritional restriction provide challenges (Fig. 14.2) to these plants (Okazaki and Saito 2014). To survive under these conditions, plants must recognize the environmental stresses and quickly exert several defense reactions, such as the peroxidation of lipids, overproduction of reactive oxygen species (ROS), accretion of osmolytes, and activation of antioxidant systems (Kazemi-Shahandashti and Maali-Amiri 2018; Yang and Guo 2018; Khan et al. 2019; Sharma et al. 2019; Ghosh et al. 2021).

An essential amino acid called proline is crucial for maintaining plant development and metabolism in the presence of abiotic stress. Numerous findings suggest a positive correlation between the accumulation of proline and plant tolerance to different abiotic stresses. It works like a molecular chaperone, an antioxidative defense molecule that acts as an antioxidant reactive oxygen species (ROS), and also has the signaling ability to activate particular gene activities that are necessary for the

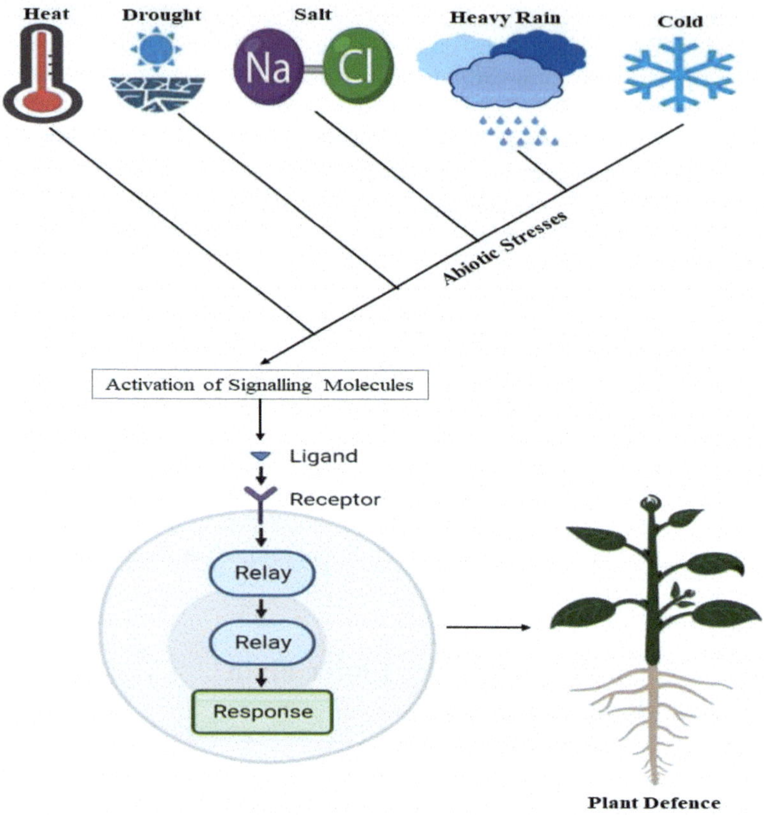

Fig. 14.2 Role of signaling molecules under abiotic stresses

recovery of plants from stresses due to its metal chelator capabilities. Furthermore, it serves as an osmoprotectant, a major carbon and nitrogen source, and is vital for growth and blooming of plants. Plant cells produce a large amount of proline to maintain water absorption, cellular homeostasis, osmoregulation, and redox equilibrium to repair cell structures and reduce oxidative damage. Numerous studies show that transgenic plants, especially those which overexpress genes that are designed for the accumulation of proline, are more tolerant to abiotic stress factors (Ghosh et al. 2022). Cells may react quickly to a variety of stimuli due to the reactive oxygen species (ROS), which are important signaling molecules. In plants, ROS plays an important role in the incorporation of various environmental signals, the stimulation of stress-response systems, and the detection of biotic and abiotic stresses. These processes support the development of plant resilience and defense mechanisms. Recent developments in the research of ROS signaling in plants also included the discovery of essential regulatory centers and ROS receptors that link this signaling to other critical stress-response processes for signal transduction and hormones, along with new functions for ROS in cell-to-cell signaling and organelle-to-organelle signaling (Mittler et al. 2022).

The primary components of biomembranes that can detect extracellular circumstances are lipids. Responses to numerous environmental stresses, including the change in temperature, drought, salinity, and pathogen attack, are involved in the signaling of lipids. Many lipids have been suggested to serve as signaling lipids, including fatty acid, lysophospholipid, diacylglycerol, phosphatidic acid, oxylipins, sphingolipid, N-acylethanolamine, and inositol phosphate. Research on these stress-induced lipid species has shown that the class of each lipid has unique biological significance, biosynthetic pathways, and signaling cascades that activate defense responses at the level of transcription. Lipids serve as mitigators of stress to lessen the severity of stresses in addition to their involvement in signaling. Under stressful conditions, it is common to witness the accelerated synthesis of particular lipids which accumulated in trace amounts under normal conditions of growth to counteract specific stresses. Recent findings have revealed that lipid remodeling leads to glucuronosyldiacylglycerol and oligogalactolipid accumulation. Interestingly, this process reduces freezing susceptibility and mitigates the detrimental effects of nutrition-depleting stress. Additionally, components generated from lipids such as cutin, suberin, and wax do not make up the lipid bilayer but help to prevent tissue damage and drought stress. These characteristics suggest that lipid-mediated protection from environmental stress aids in the survival of plants (Eyheraguibel et al. 2008; Okazaki and Saito 2014).

The plant's defense response to a pathogen attack is activated by salicylic acid (SA). According to Klessig et al. (2000), several potential SA signaling pathway components have been identified, including (i) the H2O2-scavenging ascorbate peroxidase and enzymes catalase, (ii) SA-induced protein kinases (SIPKs), (iii) a higher-affinity SA-binding proteins (SABP2), (iv) OBF/TGA family members of bZIP transcription factors, and (v) NPR1, an ankyrin repeat-containing protein required for SA signaling. Several defense genes, including the pathogenesis-related 1 gene (PR-1), have SA-responsive elements in their promoters which are physically bound by these bZIP factors (Klessig et al. 2000). Nitric oxide (NO) is another signaling molecule that activates defensive mechanisms in response to pathogen infection in recent research. Animal innate immunological and inflammatory responses have been demonstrated to be significantly influenced by NO (Schmidt and Walter 1994; Stamler 1994). The significance of NO in plants' defense against pathogens has just lately been studied, in contrast to the significant research on animal defense that has been done over the previous decade or more (Durner and Klessig 1999). At least three separate plant-pathogen systems have so far shown evidence of NO's involvement (Delledonne et al. 1998; Durner et al. 1998). After the infection of the tobacco mosaic virus, resistance but not susceptibility to the virus saw increases in the NO synthase (NOS)-like activities. The effects of NO on plant metabolism include modulating aconitase, guanylate cyclase, and mitogen-activated protein kinases (e.g., SIPK) (Klessig et al. 2000).

Due to its newly discovered functions in the control of plant development and growth as well as reactions to abiotic stresses, hydrogen sulfide (H_2S), a signaling molecule of gaseous nature in plants, has attracted a lot of interest in recent years. Recent studies have proposed that H_2S may contribute to signaling of plant defense by influencing the metabolic activity of glutathione, triggering the expression of defense- and pathogenesis-related (PR) genes, modifying the activity of enzymes by post-translational alteration, and interacting with the phytohormones like auxin, ethylene, and jasmonate (Choudhary et al. 2022).

Melatonin is a crucial phytohormone in the control of several plant functions, such as growth and stress response. Infections with pathogens seriously harm plants and lower agricultural output. According to recent studies, melatonin is crucial in preventing viral, bacterial, and fungal infections in plants and their postharvest fruits. In the interplay between plants and pathogens, reactive oxygen species, reactive nitrogen species, and melatonin create a complicated cycle that controls plant disease resistance. Additionally, melatonin interacts with additional phytohormones including auxin, salicylic acid, jasmonate, and abscisic acid to further activate the genes of plant defense. Melatonin is not only beneficial to plant immunity but also reduces pathogenicity. It is recognized as being vital in interactions between plants and pathogens due to the variety of methods through which it affects plant pathogenicity and immunity, emphasizing phytomelatonin as a crucial chemical in the plant immune function (Zeng et al. 2022).

4 Combined Action of Biostimulant and Signaling Molecules Under Abiotic Stress

Exogenous use of biostimulants enhances resistance to abiotic stress (Table 14.1) through controlling signaling molecules (e.g., nitric oxide, ROS, antioxidant enzymes, hydrogen peroxide) as shown in Fig. 14.3.

4.1 Relationship Between Melatonin and Nitric Oxide Under Abiotic Stress

Under biotic and abiotic stress conditions, melatonin, its derivatives, and precursors serve as powerful biostimulants, growth regulators, and antioxidants by scavenging ROS directly, delaying leaf senescence, undoing photosynthesis inhibition, and maintaining stable RNS and redox conditions (Debnath et al. 2019). Extracellular melatonin increases plant resistance to drought, high salt, heavy metals, harsh temperatures, acid rain, and infections while promoting growth of plants, antioxidant activity, and photosynthesis. Physiological functions are improved by melatonin in several ways. Melatonin delayed the usual production time in *Malus prunifolia* (apple) and shielded the new tissue from damage and pressures from the environment. Extracellular melatonin treatment under stressful circumstances increases indigenous melatonin and NO synthesis (Yin et al. 2013; Zhu et al. 2019).

Table 14.1 Comprehensive overview of biostimulants under abiotic stress

Sl. no.	Biostimulants	Plants	Abiotic stress	Physiological changes	References
1.	Brassinosteroids	*Phaseolus vulgaris* (French bean)	Heat stress	Improved NADPH oxidase activity, prevent photosynthetic machinery from oxidative stress	El-Bassiony et al. (2012)
2.	*Azospirillum brasilense*	*Triticum aestivum* (bread wheat)	Drought stress	Increase auxin content, induce root length, increased hydraulic conductivity.	Pereyra et al. (2012)
3.	Titanium dioxide and gibberellic acid	*Ocimum basilicum* (basil)	Drought stress	Increase CAT gene activity, enhance leaf relative water content, decrease peroxidation of lipids.	Kiapour et al. (2015)
4.	5-aminolevulinic acid	*Capsicum annuum* (chili)	Cold stress	Increase chlorophyll content, enhance glutamine synthetase and superoxide dismutase activity.	Korkmaz et al. (2010)
5.	Melatonin	*Solanum lycopersicum* (tomato)	Salinity and heat stress	Increase chlorophyll a and carotenoids, enhance glutamine synthetase enzyme activity and electron transport rate.	Martinez et al. (2018)
6.	Protein hydrolysates	*Lactuca sativa* (Lettuce)	Salinity stress	Increase glucosinolate, osmolytes, nitrogen metabolism, and reduce oxidative damage in plants.	Lucini et al. (2015)
7.	Protein hydrolysates	*Zea mays* (Maize)	Salinity stress	Increase antioxidant enzymes activity, enhance CAT gene, ascorbate peroxidase, and glutathione peroxidase enzyme activity.	Ertani et al. (2013)
8.	Nanoanatase	*Spinacia oleracea* (Spinach)	Ultra-violet radiation stress	Increase the activity of CAT gene, superoxide dismutase, and ascorbate peroxidase enzymes.	Lei et al. (2008)
9.	*Achromobacter piechaudii*	*Solanum lycopersicum* (Tomato)	Drought stress	Increase plant growth and yield and decrease ethylene content.	Mayak et al. (2004)
10.	Pepton 85/16	*Lactuca sativa* (Lettuce)	Cold stress	Increases relative growth rate and specific leaf area of plants.	Polo et al. (2006)

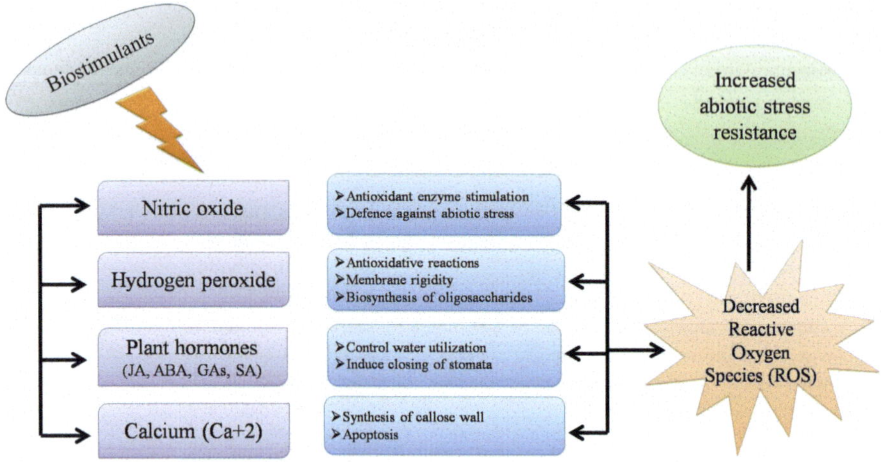

Fig. 14.3 Activation of signaling molecules by biostimulants under abiotic stresses

Melatonin works as a powerful antioxidant which upregulates the production of various antioxidative enzymes in the response to the abiotic stresses and raises the levels of Superoxide, peroxidase, glutathione s-transferase, glutathione reductase (GR), and sucrose (Foyer et al. 2001). Stress resistance is improved by antioxidant enzymes like glucose-6-phosphate dehydrogenase (G6PD) (Zhang et al. 2014). Additionally, melatonin can improve tolerance by changing the ratio of Na^+/K^+ (Kaya et al. 2020). In plant tissue, melatonin and NO have a similar method of action. The responses of abiotic stress involve NO as a crucial component. It interacts with the ROS to increase antioxidant capacity, induce the oxidative stress tolerance, and improve redox homeostasis (Correa-Aragunde et al. 2015). NO has a wide range of potential targets with which it may interact either indirectly or directly, changing the expression of genes and modulating protein activity (Zhu et al. 2019).

4.2 *Enhancement of Phytohormones and Bioactive Compounds Through Microbial Biostimulants to Combat Abiotic Stress*

Microbial biostimulants (MBs) enhance the development and growth of plants either indirectly or directly. These activities may be active concurrently or sequentially depending on the stage of plant growth and the surrounding environment. Gibberellic acids (GAs), abscisic acid (ABA), indole acetic acid (IAA), and cytokinins are a few well-known examples of phytohormones that are produced directly by plants. Other direct methods include biological nitrogen fixation and enhanced solubilization of mineral nutrients. One of the indirect processes is the creation of 1-amino-cyclopropane-1-carboxylic acids (ACC) deaminase. Other indirect mechanisms are the generation of siderophores, antifungal and antibacterial chemicals, antioxidant enzymes, exopolysaccharides (EPS), and biofilm development. A significant biotechnological strategy for reducing the negative impacts of abiotic stress on plants in various environmental contexts may be the development and implementation of MBs (Ali et al. 2022).

MBs mainly regulate the status of plant hormones by producing extracellular hormones and bioactive compounds which significantly lessen abiotic stress. The main plant hormone generated by MBs is auxin, which production takes place in several bacteria via different mechanisms. IAA, a signaling chemical generated by MBs and implicated in plant and microbe interactions, is well studied in both bacteria and fungi. Wheat roots grow more effectively when IAA-producing microbial biostimulant *Azospirillum brasilense* is present (Dobbelaere et al. 1999; Spaepen et al. 2008). Likewise, GA is a member of a significant family of phytohormones that is crucial for controlling the activity of leaves and root meristem, as well as cell proliferation and elongation. The development and resistance of plants to abiotic stimuli including heat, salt, heavy metals, and drought are improved by several GA-producing MBs (Choi et al. 2005; Kang et al. 2015; Backer et al. 2018).

4.3 Chitosan Activates Signaling Molecules Under Abiotic Stress

Chitosan is a natural biomolecule adapted from chitins that respond like an elicitor and a potential biostimulant in agriculture. Chitosan is a natural biomolecule adapted from chitins that respond like a biostimulant and elicitor in farming. Since it is nontoxic, biodegradable, and biocompatible, a wide range of applications may be possible. The use of secondary messengers in the stress transduction pathway improves physiological reactions to abiotic stresses and reduces their negative effects. Chitosan application increases antioxidant enzymes through hydrogen peroxide and nitric oxide signaling pathways and causes the formation of organic acids, polysaccharides, amino acids, and other metabolites that are necessary for osmotic adjustment, stress signaling, and mitochondrial biogenesis under stress (Hidangmayum et al. 2019).

This is employed as an antitranspirant substance via foliar spray in several crops, thus minimizing water usage ensures prevention from other adverse effects. Chitosan application reduces the negative effects brought on by water stress in cases of drought by enhancing the synthesis of antioxidant enzymes (Hidangmayum et al. 2019; Honglian et al. 2003; Yin et al. 2008). When chitosan is applied, defense responses are enhanced, activating oxidative burst, MAP-kinase, NO in the chloroplast, hydrogen peroxide through the octadecanoid pathway, and hypersensitive response (Rakwal et al. 2002; Lin et al. 2005; Hidangmayum et al. 2019; Pichyangkura and Chadchawan 2015).

4.4 Regulation of Reactive Oxygen Species by Biostimulants Under Abiotic Stress

Plant development and metabolism are directly impacted by humic acid (HA), a by-product of the plant biodegradation and microorganisms. Numerous crops have reported its growth-promoting benefits, which cannot be entirely attributable to hormone-like action (Muscolo et al. 2013). It has been discovered that humic acid reduces oxidative stress within plants. By depriving millet seedlings of watering during the three to five leaf stages, it was possible to boost seedling development and antioxidant qualities. In the study, millet leaves had higher levels of H_2O_2 and O_2 due to water stress; however, HA treatment lowered the pace at which oxygen radicals were produced and decreasing POD and SOD activity. Activities of photosystem I, photosystem II, and stomatal conductance were all enhanced by HA, which also assisted in reducing ROS formation and maintaining membrane integrity. These effects were shown even in the absence of antioxidant enzymes regulation (Shen et al. 2020).

Different biostimulants can contribute to the reduced ROS production in salt-damaged plants by improving their physiology. Again, the reduction of oxidative stress and ROS scavenging in plants under salinity is directly attributed to the increased antioxidant defense system contributed by biostimulant supplementation (Hasanuzzaman et al. 2021). According to reports, *P. putida* strain AKMP7, a thermotolerant bacterial biostimulant, improved root length and biomass under high temperatures. Additionally, according to experimental results, this bacterial strain reduces the production of ROS by enhancing the activity of the antioxidant enzymes SOD, CAT, and APX in wheat plants (Ali et al. 2011).

4.5 Impact of Biostimulants on Drought Stress by the Activation of Signaling Pathways

Biostimulants encourage the growth of root biomass, especially in soils with minimal levels of fertility and little availability of water. During a drought condition, at the beginning of the developmental stage, direct treatment of the seed promotes the recovery and growth of seedlings (Oliveira et al. 2015). When calcium and kinetin were used as biostimulants, they greatly increased the relative water content (RWC) and decreased the leakage of cellular electrolytes in soybean plants (Parihar et al. 2015). One of the efficient management techniques that can be accomplished through the use of biostimulants is the development of drought resistance in *Zea mays*, a plant that is sensitive to drought. RWC in leaves is improved, and there were fewer temperature variations between the surrounding atmosphere as well as the internal leaf environment due to the foliar spray of the Carbonsolo biostimulant, which contains 2% water-soluble nitrogen, 20% amino acids, 25% fulvic acids, and 50% humic acids (De Lacerda et al. 2010). By promoting suitable solutes (glycine, betaine, and proline) to reestablish the desired water potential gradients as well as its absorption in soil with decreased water availability, biostimulants increase plant resistance by encouraging root development despite shoot growth, enabling the plants to explore the deeper soil layers during the dry season (Aktar et al. 2015).

4.6 *Effect of Biostimulants on Antioxidant Enzymes*

The antioxidant defense system is directly impacted by biostimulants, particularly lycopene, phenolic compounds, and ascorbic acids. Antioxidant compounds, such as phenolic compounds and ascorbic acid, as well as antioxidant enzymes, such as superoxide dismutase (SOD), peroxidase (POX), and catalase (CAT), scavenge ROS, such as H_2O_2, OH, and $O_2^{\bullet-}$ (Jindo et al. 2012). Despite their effect on lycopene and ascorbate, protein hydrolysates are used as tomato biostimulants that have little effect on polyphenolic components (Ertani et al. 2013). The amount of antioxidant activity was increased in plants by the application of biostimulants. After the usage of the biostimulants, the fruit's antioxidant potential was stronger during the first season as compared to the second season. The changes in antioxidant capacities between the two seasons were found to be justified by climate change (Chen and Murata 2011). Antioxidative and enzymatic function in plants is positively impacted by chitosan-based nanoparticle biostimulants (CS-NP) and salicylic acid (SA). CAT, POX, and SOD activity in maize was enhanced by two levels after the 2 days of biostimulant administration, making them equivalent to plants treated with SA after various exposure durations (Ertani et al. 2013).

5 Conclusion and Future Prospective

Abiotic stress triggers a wide range of problems at the molecular and physiological levels in plants, which significantly lowers productivity globally. The generation of ROS is a natural response to all stresses, and understanding their signaling and destructive effect will aid in the development of crop varieties that are tolerant to stress. Agricultural researchers are constantly searching for effective strategies to reduce negative impacts on crop yields and regulate or preserve those yields through economic measures. This achievement was made possible by the identification of a variety of biostimulants that upregulate the activity of signaling molecules to maintain and enhance plant productivity under abiotic stress. Modern agriculture is switching to improved crop quality and yield by utilizing the natural stimulatory concept in plants. The use of biostimulants to increase abiotic stress tolerance is thus increasingly being recognized by the science community and the agricultural sector as a useful approach. The use of biostimulants may be a viable alternative to chemical fertilizers in promoting plant growth and increasing stress resistance.

Acknowledgment We are grateful to Director, Dayalbagh Educational Institute, Dayalbagh, Agrafor encouragement and kind support.

References

Aamer M, Muhammad UH, Li Z, Abid A, Su Q, Liu Y, Huang G (2018) Foliar application of glycinebetaine (GB) alleviates the cadmium (Cd) toxicity in spinach through reducing Cd uptake and improving the activity of anti-oxidant system. Appl Ecol Environ Res 16(6):7575–7583

Aktar M, Ali AZ, Rahman MK (2015) Application of nitrogen and boron on growth and nutrient contents of okra (*Abelmoschus esculentus* L.) grown on soil. J Asiatic Soc Bangladesh Sci 41(2):173–181

Ali SZ, Sandhya V, Grover M, Linga VR, Bandi V (2011) Effect of inoculation with a thermotolerant plant growth promoting *Pseudomonas putida* strain AKMP7 on growth of wheat (*Triticum* spp.) under heat stress. J Plant Interact 6(4):239–246

Ali S, Eum HI, Cho J, Dan L, Khan F, Dairaku K, Fahad S (2019) Assessment of climate extremes in future projections downscaled by multiple statistical downscaling methods over Pakistan. Atmos Res 222:114–133

Ali S, Moon YS, Hamayun M, Khan MA, Bibi K, Lee IJ (2022) Pragmatic role of microbial plant biostimulants in abiotic stress relief in crop plants. J Plant Interact 17(1):705–718

Apone F, Tito A, Carola A, Arciello S, Tortora A, Filippini L, Colucci G (2010) A mixture of peptides and sugars derived from plant cell walls increases plant defense responses to stress and attenuates ageing-associated molecular changes in cultured skin cells. J Biotechnol 145(4):367–376

Backer R, Rokem JS, Ilangumaran G, Lamont J, Praslickova D, Ricci E, Smith DL (2018) Plant growth-promoting rhizobacteria: context, mechanisms of action, and roadmap to commercialization of biostimulants for sustainable agriculture. Front Plant Sci 9:1473

Battacharyya D, Babgohari MZ, Rathor P, Prithiviraj B (2015) Seaweed extracts as biostimulants in horticulture. Sci Hortic 196:39–48

Berg G, Grube M, Schloter M, Smalla K (2014) Unraveling the plant microbiome: looking back and future perspectives. Front Microbiol 5:148

Bonfante P, Genre A (2010) Mechanisms underlying beneficial plant–fungus interactions in mycorrhizal symbiosis. Nat Commun 1(1):1–11

BorhannuddinBhuyan MHM, Mohsin SM, Mahmud JA, Hasanuzzaman M (2020) Use of biostimulants for improving abiotic stress tolerance in Brassicaceae plants. In: The plant family Brassicaceae. Springer, Singapore, pp 497–531

Boyer JS (1982) Plant productivity and environment. Science 218(4571):443–448

Brown PH, Zhao FJ, Dobermann A (2022) What is a plant nutrient? Changing definitions to advance science and innovation in plant nutrition. Plant Soil 476(1):11–23

Bulgari R, Cocetta G, Trivellini A, Vernieri P, Ferrante A (2015) Biostimulants and crop responses: a review. Biol Agric Hortic 31(1):1–17

Calvo P, Nelson L, Kloepper JW (2014) Agricultural uses of plant biostimulants. Plant Soil 383(1):3–41

Castiglione AM, Mannino G, Contartese V, Bertea CM, Ertani A (2021) Microbial biostimulants as response to modern agriculture needs: composition, role and application of these innovative products. Plan Theory 10(8):1533

Chaudhari PN (2017) Role of microbial bio-products in sustainable agriculture. Int J Innov Hortic 6(1):25–47

Chen TH, Murata N (2011) Glycine betaine protects plants against abiotic stress: mechanisms and biotechnological applications. Plant Cell Environ 34(1):1–20

Cheng C, Pei LM, Yin TT, Zhang KW (2018) Seed treatment with glycine betaine enhances tolerance of cotton to chilling stress. J Agric Sci 156(3):323–332
Choi WY, Rim SO, Lee JH, Lee JM, Lee IJ, Cho KJ, Kim JG (2005) Isolation of gibberellins-producing fungi from the root of several *Sesamum indicum* plants. J Microbiol Biotechnol 15(1):22–28
Choudhary AK, Singh S, Khatri N, Gupta R (2022) Hydrogen sulphide: an emerging regulator of plant defence signalling. Plant Biol 24(4):532–539
Colla G, Rouphael Y, Canaguier R, Svecova E, Cardarelli M (2014) Biostimulant action of a plant-derived protein hydrolysate produced through enzymatic hydrolysis. Front Plant Sci 5:448
Colla G, Rouphael Y, Di Mattia E, El-Nakhel C, Cardarelli M (2015) Co-inoculation of *Glomus intraradices* and *Trichoderma atroviride* acts as a biostimulant to promote growth, yield and nutrient uptake of vegetable crops. J Sci Food Agric 95(8):1706–1715
Correa-Aragunde N, Foresi N, Lamattina L (2015) Nitric oxide is a ubiquitous signal for maintaining redox balance in plant cells: regulation of ascorbate peroxidase as a case study. J Exp Bot 66(10):2913–2921
Couto CA, Peixoto CP, Vieira EL, Carvalho EV, Peixoto VAB (2012) Ação da cinetina, ácido indolbutírico e ácido giberélico na emergência do girassol sob estresse por alumínio. Comunicata Scientiae 3(3):206–209
Craigie JS (2011) Seaweed extract stimuli in plant science and agriculture. J Appl Phycol 23(3):371–393
Crouch IJ, Van Staden J (1993) Commercial seaweed products as biostimulants in horticulture. J Home Consum Hortic 1(1):19–76
De Lacerda CF, Costa RNT, Bezerra MA, Gheyi HR (2010) Estrategias de Manejo Para Uso de Água Salina na Agricultura; Embrapa Agroindústria Tropical—Capítulo em livro científico (ALICE): Fortaleza, Brazil
Debnath B, Islam W, Li M, Sun Y, Lu X, Mitra S, Qiu D (2019) Melatonin mediates enhancement of stress tolerance in plants. Int J Mol Sci 20(5):1040
Deliopoulos T, Kettlewell PS, Hare MC (2010) Fungal disease suppression by inorganic salts: a review. Crop Prot 29(10):1059–1075
Delledonne M, Xia Y, Dixon RA, Lamb C (1998) Nitric oxide functions as a signal in plant disease resistance. Nature 394(6693):585–588
Dobbelaere S, Croonenborghs A, Thys A, Vande Broek A, Vanderleyden J (1999) Phytostimulatory effect of *Azospirillum brasilense* wild type and mutant strains altered in IAA production on wheat. Plant Soil 212(2):153–162
Drobek M, Frąc M, Cybulska J (2019) Plant biostimulants: importance of the quality and yield of horticultural crops and the improvement of plant tolerance to abiotic stress—A review. Agronomy 9(6):335
Du Jardin P (2012) The science of plant biostimulants–a bibliographic analysis. Ad hoc study report
Du Jardin P (2015) Plant biostimulants: definition, concept, main categories and regulation. Sci Hortic 196:3–14
Durán D, Bernal P, Vazquez-Arias D, Blanco-Romero E, Garrido-Sanz D, Redondo-Nieto M, Martín M (2021) *Pseudomonas fluorescens* F113 type VI secretion systems mediate bacterial killing and adaption to the rhizosphere microbiome. Sci Rep 11(1):1–13
Durner J, Klessig DF (1999) Nitric oxide as a signal in plants. Curr Opin Plant Biol 2(5):369–374
Durner J, Wendehenne D, Klessig DF (1998) Defense gene induction in tobacco by nitric oxide, cyclic GMP, and cyclic ADP-ribose. Proc Natl Acad Sci 95(17):10328–10333
El Hadrami A, Adam LR, El Hadrami I, Daayf F (2010) Chitosan in plant protection. Mar Drugs 8(4):968–987
El-Bassiony AM, Ghoname AA, El-Awadi ME, Fawzy ZF, Gruda N (2012) Ameliorative effects of brassinosteroids on growth and productivity of snap beans grown under high temperature. Gesunde Pflanzen 64(4):175–182
Ertani A, Schiavon M, Muscolo A, Nardi S (2013) Alfalfa plant-derived biostimulant stimulate short-term growth of salt stressed *Zea mays* L. plants. Plant Soil 364(1):145–158

Eyheraguibel B, Silvestre J, Morard P (2008) Effects of humic substances derived from organic waste enhancement on the growth and mineral nutrition of maize. Bioresour Technol 99(10):4206–4212

Fedoroff NV, Battisti DS, Beachy RN, Cooper PJ, Fischhoff DA, Hodges CN, Zhu JK (2010) Radically rethinking agriculture for the 21st century. Science 327(5967):833–834

Ferri M, Franceschetti M, Naldrett MJ, Saalbach G, Tassoni A (2014) Effects of chitosan on the protein profile of grape cell culture subcellular fractions. Electrophoresis 35(11):1685–1692

Foyer CH, Theodoulou FL, Delrot S (2001) The functions of inter-and intracellular glutathione transport systems in plants. Trends Plant Sci 6(10):486–492

Franzoni G, Cocetta G, Prinsi B, Ferrante A, Espen L (2022) Biostimulants on crops: their impact under abiotic stress conditions. Horticulturae 8(3):189

García AL, García-Machado FJ, Borges AA, Morales-Sierra S, Boto A, Jiménez-Arias D (2020) Pure García organic active compounds against abiotic stress: a biostimulant overview. Front Plant Sci 11:575829

Gautam RK, Navaratna D, Muthukumaran S, Singh A, More N (2021) Humic substances: its toxicology, chemistry and biology associated with soil, plants and environment. In: Humic substances. IntechOpen, London, p 97

Ghaderiardakani F, Collas E, Damiano DK, Tagg K, Graham NS, Coates JC (2019) Effects of green seaweed extract on *Arabidopsis* early development suggest roles for hormone signalling in plant responses to algal fertilisers. Sci Rep 9(1):1–13

Ghaffari H, Tadayon MR, Nadeem M, Cheema M, Razmjoo J (2019) Proline-mediated changes in antioxidant enzymatic activities and the physiology of sugar beet under drought stress. Acta Physiol Plant 41(2):1–13

Ghosh UK, Islam MN, Siddiqui MN, Khan MAR (2021) Understanding the roles of osmolytes for acclimatizing plants to changing environment: a review of potential mechanism. Plant Signal Behav 16(8):1913306

Ghosh UK, Islam MN, Siddiqui MN, Cao X, Khan MAR (2022) Proline, a multifaceted signalling molecule in plant responses to abiotic stress: understanding the physiological mechanisms. Plant Biol 24(2):227–239

Hadwiger LA (2013) Multiple effects of chitosan on plant systems: solid science or hype. Plant Sci 208:42–49

Halpern M, Bar-Tal A, Ofek M, Minz D, Muller T, Yermiyahu U (2015) The use of biostimulants for enhancing nutrient uptake. Adv Agron 130:141–174

Hasanuzzaman M, Fujita M, Oku H, Islam MT (eds) (2019) Plant tolerance to environmental stress: role of phytoprotectants. CRC Press, Boca Raton

Hasanuzzaman M, Parvin K, Bardhan K, Nahar K, Anee TI, Masud AAC, Fotopoulos V (2021) Biostimulants for the regulation of reactive oxygen species metabolism in plants under abiotic stress. Cell 10(10):2537

Herve JJ (1994) Biostimulant, a new concept for the future and prospects offered by chemical synthesis and biotechnologies. Comptes Rendus de l'Academie d'Agriculture de France (France)

Hidangmayum A, Dwivedi P, Katiyar D, Hemantaranjan A (2019) Application of chitosan on plant responses with special reference to abiotic stress. Physiol Mol Biol Plants 25(2):313–326

Honglian G, Yuguang D, Xuefang B, Xiaoming Z (2003) Effects of active oxygen on suspended cotton cell culture by oligochitosan Zhongguo hai Yang yao wu. Chinese J Marine Drugs 22(1): 11–12

Iriti M, Picchi V, Rossoni M, Gomarasca S, Ludwig N, Gargano M, Faoro F (2009) Chitosan antitranspirant activity is due to abscisic acid-dependent stomatal closure. Environ Exp Bot 66(3):493–500

Jeandroz S, Lamotte O (2017) Plant responses to biotic and abiotic stresses: lessons from cell signaling. Front Plant Sci 8:1772

Ji H, Pardo JM, Batelli G, Van Oosten MJ, Bressan RA, Li X (2013) The salt overly sensitive (SOS) pathway: established and emerging roles. Mol Plant 6(2):275–286

Jindo K, Martim SA, Navarro EC, Pérez-Alfocea F, Hernandez T, Garcia C, Canellas LP (2012) Root growth promotion by humic acids from composted and non-composted urban organic wastes. Plant Soil 353(1):209–220

Johnson NC, Graham JH (2013) The continuum concept remains a useful framework for studying mycorrhizal functioning. Plant Soil 363(1):411–419

Joshi R, Singla-Pareek SL, Pareek A (2018) Engineering abiotic stress response in plants for biomass production. J Biol Chem 293(14):5035–5043

Kang SM, Khan AL, Waqas M, You YH, Hamayun M, Joo GJ, Lee IJ (2015) Gibberellin-producing *Serratia nematodiphila* PEJ1011 ameliorates low temperature stress in *Capsicum annuum* L. Eur J Soil Biol 68:85–93

Kannan R (2019) Biostimulant and its response in crop improvement: a review. Adv Agri Sci 16:151–173

Katiyar D, Hemantaranjan A, Singh B (2015) Chitosan as a promising natural compound to enhance potential physiological responses in plant: a review. Indian J Plant Physiol 20(1):1–9

Kaya C, Higgs D, Ashraf M, Alyemeni MN, Ahmad P (2020) Integrative roles of nitric oxide and hydrogen sulfide in melatonin-induced tolerance of pepper (*Capsicum annuum* L.) plants to iron deficiency and salt stress alone or in combination. Physiol Plant 168(2):256–277

Kazemi-Shahandashti SS, Maali-Amiri R (2018) Global insights of protein responses to cold stress in plants: signaling, defence, and degradation. J Plant Physiol 226:123–135

Kerchev P, van der Meer T, Sujeeth N, Verlee A, Stevens CV, Van Breusegem F, Gechev T (2020) Molecular priming as an approach to induce tolerance against abiotic and oxidative stresses in crop plants. Biotechnol Adv 40:107503

Khan W, Rayirath UP, Subramanian S, Jithesh MN, Rayorath P, Hodges DM, Prithiviraj B (2009) Seaweed extracts as biostimulants of plant growth and development. J Plant Growth Regul 28(4):386–399

Khan MI, Reddy PS, Ferrante A, Khan NA (2019) Plant signaling molecules: role and regulation under stressful environments. Woodhead Publishing, Cambridge

Kiapour H, Moaveni P, Habibi D (2015) Evaluation of the application of gibberellic acid and titanium dioxide nanoparticles under drought stress on some traits of basil (*Ocimum basilicum* L.). Int J Agron Agri Res 6(4):138–150

Klessig DF, Durner J, Noad R, Navarre DA, Wendehenne D, Kumar D, Silva H (2000) Nitric oxide and salicylic acid signaling in plant defense. Proc Natl Acad Sci 97(16):8849–8855

Korkmaz A, Korkmaz Y, Demirkıran AR (2010) Enhancing chilling stress tolerance of pepper seedlings by exogenous application of 5-aminolevulinic acid. Environ Exp Bot 67(3):495–501

Kour D, Rana KL, Yadav N, Yadav AN, Singh J, Rastegari AA, Saxena AK (2019) Agriculturally and industrially important fungi: current developments and potential biotechnological applications. In: Recent advancement in white biotechnology through fungi. Springer, Cham, pp 1–64

La Pena RD, Hughes J (2007) Improving Vegetable Productivity in a Variable and Changing Climate. J SAT Agric Res 4:666–675

Lei Z, Mingyu S, Xiao W, Chao L, Chunxiang Q, Liang C, Fashui H (2008) Antioxidant stress is promoted by nano-anatase in spinach chloroplasts under UV-B radiation. Biol Trace Elem Res 121(1):69–79

Lin W, Hu X, Zhang W, Rogers WJ, Cai W (2005) Hydrogen peroxide mediates defence responses induced by chitosans of different molecular weights in rice. J Plant Physiol 162(8):937–944

Lucini L, Rouphael Y, Cardarelli M, Canaguier R, Kumar P, Colla G (2015) The effect of a plant-derived biostimulant on metabolic profiling and crop performance of lettuce grown under saline conditions. Sci Hortic 182:124–133

Maini P (2006) The experience of the first biostimulant, based on amino acids and peptides: a short retrospective review on the laboratory researches and the practical results. Fertilitas Agrorum 1(1):29–43

Martinez V, Nieves-Cordones M, Lopez-Delacalle M, Rodenas R, Mestre TC, Garcia-Sanchez F, Rivero RM (2018) Tolerance to stress combination in tomato plants: new insights in the protective role of melatonin. Molecules 23(3):535

Mayak S, Tirosh T, Glick BR (2004) Plant growth-promoting bacteria that confer resistance to water stress in tomatoes and peppers. Plant Sci 166(2):525–530

Mittler R, Zandalinas SI, Fichman Y, Van Breusegem F (2022) Reactive oxygen species signalling in plant stress responses. Nat Rev Mol Cell Biol 23(10):663–679

Mohammadi H, Hazrati S, Ghorbanpour M (2020) Tolerance mechanisms of medicinal plants to abiotic stresses. In: Plant life under changing environment. Academic, Cambridge, pp 663–679

Mrid RB, Benmrid B, Hafsa J, Boukcim H, Sobeh M, Yasri A (2021) Secondary metabolites as biostimulant and bioprotectant agents: a review. Sci Total Environ 777:146204

Muscolo A, Sidari M, Nardi S (2013) Humic substance: relationship between structure and activity. Deeper information suggests univocal findings. J Geochem Explor 129:57–63

Okazaki Y, Saito K (2014) Roles of lipids as signaling molecules and mitigators during stress response in plants. Plant J 79(4):584–596

Olivares FL, Busato JG, de Paula AM, da Silva Lima L, Aguiar NO, Canellas LP (2017) Plant growth promoting bacteria and humic substances: crop promotion and mechanisms of action. Chem Biol Technol Agric 4(1):1–13

Oliveira FDAD, Medeiros JFD, Alves RDC, Lima LA, Santos STD, Régis LRDL (2015) Produção de feijão caupi em função da salinidade e regulador de crescimento. Revista Brasileira de Engenharia Agrícola e Ambiental 19:1049–1056

Parihar P, Singh S, Singh R, Singh VP, Prasad SM (2015) Effect of salinity stress on plants and its tolerance strategies: a review. Environ Sci Pollut Res 22(6):4056–4075

Pereyra MA, Garcia P, Colabelli MN, Barassi CA, Creus CM (2012) A better water status in wheat seedlings induced by *Azospirillum* under osmotic stress is related to morphological changes in xylem vessels of the coleoptile. Appl Soil Ecol 53:94–97

Philippot L, Raaijmakers JM, Lemanceau P, Van Der Putten WH (2013) Going back to the roots: the microbial ecology of the rhizosphere. Nat Rev Microbiol 11(11):789–799

Pichyangkura R, Chadchawan S (2015) Biostimulant activity of chitosan in horticulture. Sci Hortic 196:49–65

Polo J, Barroso R, Ródenas J, Azcón-Bieto J, Cáceres R, Marfà O (2006) Porcine hemoglobin hydrolysate as a biostimulant for lettuce plants subjected to conditions of thermal stress. HortTechnology 16(3):483–487

Povero G, Loreti E, Pucciariello C, Santaniello A, Di Tommaso D, Di Tommaso G, Perata P (2011) Transcript profiling of chitosan-treated *Arabidopsis* seedlings. J Plant Res 124(5):619–629

Rakwal R, Tamogami S, Agrawal GK, Iwahashi H (2002) Octadecanoid signaling component "burst" in rice (*Oryza sativa* L.) seedling leaves upon wounding by cut and treatment with fungal elicitor chitosan. Biochem Biophys Res Commun 295(5):1041–1045

Rouphael Y, Colla G (2018) Synergistic biostimulatory action: designing the next generation of plant biostimulants for sustainable agriculture. Front Plant Sci 9:1655

Rouphael Y, Colla G (2020) Biostimulants in agriculture. Front Plant Sci 11:40

Schmidt HH, Walter U (1994) NO at work. Cell 78(6):919–925

Sharma HS, Fleming C, Selby C, Rao JR, Martin T (2014) Plant biostimulants: a review on the processing of macroalgae and use of extracts for crop management to reduce abiotic and biotic stresses. J Appl Phycol 26(1):465–490

Sharma P, Sharma P, Arora P, Verma V, Khanna K, Saini P, Bhardwaj R (2019) Role and regulation of ROS and antioxidants as signaling molecules in response to abiotic stresses. In: Plant signaling molecules. Woodhead Publishing, Sawston, United Kingdom, pp 141–156

Shen J, Guo MJ, Wang YG, Yuan XY, Wen YY, Song XE, Guo PY (2020) Humic acid improves the physiological and photosynthetic characteristics of millet seedlings under drought stress. Plant Signal Behav 15(8):1774212

Spaepen S, Dobbelaere S, Croonenborghs A, Vanderleyden J (2008) Effects of *Azospirillum brasilense* indole-3-acetic acid production on inoculated wheat plants. Plant Soil 312(1):15–23

Stamler JS (1994) Redox signaling: nitrosylation and related target interactions of nitric oxide. Cell 78(6):931–936

Surabhi GK (2018) Update in root proteomics with special reference to abiotic stresses: achievements and challenges. J Protein Proteomics 9(1):31–35

Ukalska-Jaruga A, Bejger R, Debaene G, Smreczak B (2021) Characterization of soil organic matter individual fractions (fulvic acids, humic acids, and humins) by spectroscopic and electrochemical techniques in agricultural soils. Agronomy 11(6):1067

Umezawa T, Fujita M, Fujita Y, Yamaguchi-Shinozaki K, Shinozaki K (2006) Engineering drought tolerance in plants: discovering and tailoring genes to unlock the future. Curr Opin Biotechnol 17(2):113–122

Vacheron J, Desbrosses G, Bouffaud ML, Touraine B, Moënne-Loccoz Y, Muller D, Prigent-Combaret C (2013) Plant growth-promoting rhizobacteria and root system functioning. Front Plant Sci 4:356

Verma AK, Deepti S (2016) Abiotic stress and crop improvement: current scenario. Adv Plants Agric Res 4(4):345–346

Vranova V, Rejsek K, Skene KR, Formanek P (2011) Non-protein amino acids: plant, soil and ecosystem interactions. Plant Soil 342:31–48

Wally OS, Critchley AT, Hiltz D, Craigie JS, Han X, Zaharia LI, Prithiviraj B (2013) Regulation of phytohormone biosynthesis and accumulation in *Arabidopsis* following treatment with commercial extract from the marine macroalga *Ascophyllum nodosum*. J Plant Growth Regul 32(2):324–339

Xu L, Geelen D (2018) Developing biostimulants from agro-food and industrial by-products. Front Plant Sci 9:1567

Yakhin OI, Lubyanov AA, Yakhin IA (2016) Biostimulants in agrotechnologies: problems, solutions, outlook. Agrochem Herald 1:15–21

Yakhin OI, Lubyanov AA, Yakhin IA, Brown PH (2017) Biostimulants in plant science: a global perspective. Front Plant Sci 7:2049

Yang Y, Guo Y (2018) Elucidating the molecular mechanisms mediating plant salt-stress responses. New Phytol 217(2):523–539

Yaronskaya E, Vershilovskaya I, Poers Y, Alawady AE, Averina N, Grimm B (2006) Cytokinin effects on tetrapyrrole biosynthesis and photosynthetic activity in barley seedlings. Planta 224(3):700–709

Yin H, Bai X, Du Y (2008) The primary study of oligochitosan inducing resistance to *Sclerotinia sclerotiorum* on *Brassica napus*. J Biotechnol 136:S600–S601

Yin L, Wang P, Li M, Ke X, Li C, Liang D, Wu S, Ma X, Li C, Zou Y (2013) Exogenous melatonin improves Malus resistance to Marssonina apple blotch. J Pineal Res 54:426–434

Zamana SP, Kondratieva TD, Savich VI, Fedorovsky TG, Sokolov SA (2022) The influence of biostimulants on the qualitative composition of carrots. In: IOP conference series: earth and environmental science, vol 1045(1). IOP Publishing, p 012087

Zeng H, Bai Y, Wei Y, Reiter RJ, Shi H (2022) Phytomelatonin as a central molecule in plant disease resistance. J Exp Bot 73(17):5874–5885

Zhang X, Schmidt RE (1997) The impact of growth regulators on alpha-tocopherol status of water-stressed *Poa pratensis* L. Int Turfgrass Soc Res J 8:1364–2137

Zhang XH, Schmidt R (1999) Biostimulating turfgrasses. Grounds Maintenance 34:14–32

Zhang HJ, Zhang NA, Yang RC, Wang L, Sun QQ, Li DB, Guo YD (2014) Melatonin promotes seed germination under high salinity by regulating antioxidant systems, ABA and GA 4 interaction in cucumber (*Cucumis sativus* L.). J Pineal Res 57(3):269–279

Zhu JK (2016) Abiotic stress signaling and responses in plants. Cell 167(2):313–324

Zhu Y, Gao H, Lu M, Hao C, Pu Z, Guo M, Huang X (2019) Melatonin-nitric oxide crosstalk and their roles in the redox network in plants. Int J Mol Sci 20(24):6200

Zulfiqar F, Casadesús A, Brockman H, Munné-Bosch S (2020) An overview of plant-based natural biostimulants for sustainable horticulture with a particular focus on moringa leaf extracts. Plant Sci 295:110194

Chapter 15
Abiotic Stress Sensitivity and Adaptation in Field Crops

Sukamal Sarkar, Saikat Saha, Saswati Ghosh, Sujay Kumar Paul, Saikat Dey, Debojyoti Moulick, Subhas Chandra Santra, and Koushik Brahmachari

Abstract Agriculture production must be substantially increased in order to support the urgent requirement for food for the rapidly growing population. The depletion of natural resources combined with a hiking tendency in climatic variability has already harmed the worldwide agricultural production system. Numerous factors have a major role in determining how well crops are produced (climatic, agronomic, and soil nutrient status), and stress is defined as any unfavourable climatic situation that impairs plant development. Under today's changed scenarios of climate, abiotic stresses, such as drought, salinity, rising temperatures, submergence, and nutritional deficits, are becoming more common for crops. These stresses impede crop output by wreaking havoc on plant morphological processes, biochemical pathways, and physiological characteristics that are all intimately related to diverse plant development and yield phenomena. Abiotic stress is viewed as a multi-factorial event involving various processes in field crops. Numerous field crops, including cereals,

S. Sarkar · S. Dey
School of Agriculture and Rural Development, Ramakrishna Mission Vivekananda Educational and Research Institute, Narendrapur, West Bengal, India

S. Saha
Hooghly Krishi Vigyan Kendra, Bidhan Chandra Krishi Viswavidyalaya, Indian Council of Agricultural Research, Chinsurah, West Bengal, India

S. Ghosh
Department of Environmental Science, Kachrapara Collage, University of Kalyani, Kachrapara, West Bengal, India

S. K. Paul
Department of Agriculture, MITS Institute of Professional Studies, Bhujabala, Rayagada, Odisha, India

D. Moulick (✉) · S. C. Santra
Department of Environmental Science, University of Kalyani, Nadia, West Bengal, India

K. Brahmachari
Department of Agronomy, Faculty of Agriculture, Bidhan Chandra Krishi Viswa Vidyalaya, Mohanpur, Nadia, West Bengal, India

M. Hasanuzzaman (ed.), *Climate-Resilient Agriculture, Vol 2*,
https://doi.org/10.1007/978-3-031-37428-9_15

legumes, oilseeds, tuber crops, and other important cash crops, have seen drastic yield reduction due to abiotic stresses. Several abiotic stresses that plants experience, including temperature extremes, drought, salt, and heavy metals, will be discussed in this chapter. All of these are significant constraints on agricultural yield and sustainability. To address the challenges associated with elucidating the stress tolerance mechanism, this chapter also discusses several mitigating strategies involving improved breeding techniques, modern technology for field crop production, and the use of various nanotechnologies in conjunction with ICT interventions.

Keywords Abiotic stress · Stress types · Tolerance mechanism · Mitigating strategies

1 Introduction

Severe occurrences of abiotic stress (i.e. crop loss, temperature increase, soil salinity) significantly diminish agricultural production. The functioning of agroecosystems is significantly impacted by the rise in air temperature brought on by human-caused climate change. According to the Intergovernmental Panel on Climate Change (2007), if human-caused global warming continues at its present rate, there is probability of the earth's temperature to increase by at least 6.40 °C (Solomon et al. 2007). By the end of the twenty-first century, ocean position will have risen up to 59.0 cm as a result of glacier melting. In addition to altering the seasonal cycle, climate change raises the likelihood of new natural disasters such drought, flood, storms, and cyclones. Variations in temperature, moisture, and rainfall must negatively impact crop growth productivity because farming is primarily climate-dependent and sensitive to agroclimatic conditions (Yohannes 2016; Aryal et al. 2020; Choudhury and Moulick 2022).

An external event that negatively affects a plant's growth, development, or output is referred to as stress (Kitano 2002). A wide range of responses are triggered by stress, including altered gene expression, altered cellular metabolism, altered growth rates, and reduced agricultural production. Generally, manufacturing stress is a result of unanticipated changes in environmental conditions. Still, in stress-tolerant factory species, exposure to a given stress results in a time-dependent acclimatisation to that particular stress (Feist and Palon 2008). There are two types of plant stress: abiotic stress and biotic stress. Abiotic stress is quantifiable by physical or chemical processes, while biotic stress is quantifiable through natural units such as circumstances and insects. Certain pressures on crops harmed them in the same way that crops exhibit a variety of metabolic dysfunctions (Bemis and Torii 2007). Plants may recuperate from damage if the stress is modest or transitory; however, extreme stressors cause crop mortality by impeding blooming, seed conformation, and inducing anility. Stress-susceptible plants of a similar kind will be regarded similar (Feist and Palsson 2008).

Many different types of living things, most notably bacteria, fungi, nematodes, insects, arachnids, and weeds, can cause biotic stress in agricultural plants. The agents that cause biotic stress are directly reliant on their host for nourishment, which might result in the death of businesses. Significant biotic stress may occur as a result of pre- and post-harvest losses (Shavrukov et al. 2017). Despite the absence of an adaptive susceptible system, plants may compensate for biotic stressors via the evolution of some complex techniques (Fghire et al. 2015). The defense mechanism of plant that are activated in response to these shocks are genetically regulated by the restored inheritable law contained inside them. Hundreds of resistance genes against various biotic stressors are encoded from the industrial genome. Abiotic stress, which affects agricultural plants negatively due to non-living factors such as salinity, sunshine, warmth, cold, cataracts, and failure, is different from biotic stress (Shanker and Shanker 2016). The environment in which the crop thrives dictates the sort of biotic stress that can be evaluated on crop stores, as well as the capacity of the crop species to resist that form of stress. Numerous biotic stressors have an effect on photosynthesis, as biting insects diminish splint area and infectious agents lower photosynthesis rate per splint area.

Drought (water stress), flood (water logging), high heat, salinity in the soil atmosphere, and mineral toxicity can negatively impact the development, growth, production, and seed quality of a wide range of crops and other plant species (Browse and Xin 2001; Majumdar et al. 2023). Therefore, it is imperative to create crop types that can adapt to abiotic difficulties in order to ensure future food security and safety (Pearce 2001). A plant's primary defence against abiotic stress is root growth. If the soil carrying the nutrients is healthy and biologically distinct, there is a greater chance that it will survive harsh conditions (Volkov 2015). The shift of the Na/K ratio in the cytoplasm of the factory cell is one of the primary responses to abiotic stress (Flowers 2004).

2 Global Scenario of Yield Loss Due to Abiotic Stresses

Several academic articles predict that over the next 50–100 years, the earth's temperature will increase by 3–5 °C (Hall 2002). Temperature spikes and uneven rainfall, in combination with variations in rainfall and crop failure, are constantly taken into account. Anthropogenic conditioning, such as epidemics, unequal irrigation, and indiscriminate exploitation of natural resources, may contribute significantly to global abiotic stress (Schützendübel and Polle 2002). Under these circumstances, plants are probably exposed to abiotic stresses more frequently. Factory breeders have a responsibility to produce stress-tolerant cultivars in order to maintain food security and farmer safety. Molecular research will be conducted at the inheritable position in order to build mechanisms in shops that will assist them in dealing with various sorts of stress scenarios (Wuana and Okieimen 2011). Unless and until adaptive mechanisms against biotic and abiotic pressures are found, the shops will continue to be subjected to similar challenges, ultimately posing a serious threat to global agriculture.

According to many studies published, approximately 38% of the world's population, or up to 45% of the world's total arable land, experiences insufficient rainfall, and 3106 km^2 of land is afflicted by salt deposition (Flora et al. 2008). Additionally, 19.5% of irrigated agricultural land is classed as salty, and about 1% of global cultivable land is damaged by salinity (two million ha) and performs poorly (Takahashi et al. 2013). Throughout their growth and development, agricultural plants experience significant abiotic stress due to water shortages brought on by insufficient rainfall, cold temperatures, and salt build-up in the soil profile. Abiotic stress on agricultural plants is therefore a critical subject that will encompass world food security. Abiotic stresses are interconnected and have an effect on the connections between factory water and cellular and also whole crop genetic regulations, influencing both specific and ambiguous reactions that lead to a number of morphological, physiological, biochemical, and molecular changes that have a negative impact on the growth and development of crops.

3 Types of Abiotic Stresses in Plants

3.1 Heat Stress

Climate change has a greater impact on the agroecosystem and global crop production. During the past few decades, global warming has been creating a universal issue and leading to an enormous change in crop ecology and consequent yield loss (Mittler et al. 2012). Therefore, high temperatures are thought to be the main factor restricting crop development in general (Teskey et al. 2015; Ishimaru et al. 2016). Abiotic stress, particularly high-heat stress, has been identified when the extreme temperature imposed at their above threshold level over a certain period of time, imparts a detrimental harm to plant growth (Wahid et al. 2007). All phases of a plant's development could be affected by high-temperature stress, although some are more vulnerable than others, including the early seedling, flower blooming, gamete development, and grain filling stages (Wahid et al. 2007; Jagadish 2020). Plants exhibit a significant change in cell morphology and biochemical functions even when their ideal temperature for growth is exceeded by 5 °C. High temperatures at night increase the respiration capacity, which directly effects the energy consumption of the plant. It reduces photosynthetic functions, increases biomass production, and ultimately reduces the crop yield.

3.2 Cold Stress

Abiotic stress, whether high-temperature stress or chilling injury, both have a negative impact throughout the entire crop's growth and productivity. Extreme low temperature leads to significant changes in different physiological functions and reduces the rate of various plant metabolic activities (Ruelland et al. 2009). Kang and Saltveit (2002) showed that various field crops show susceptibility to cold injury, mostly at the root establishing and juvenile stages of the plant. Under cold stress, the percentage of germination is drastically reduced, resulting in yield loss (Kang and Saltveit 2002). Plant roots are very sensitive to cold injury, and it hinders root growth by altering the root morphology and root uptake pattern (Richner et al. 1996). Freezing injury generally occurs in the temperate regions. It reduces the water absorption capacity and deteriorates several enzymatic functions and cellular activities of the plant.

3.3 Salinity Stress

Soil salinity stress poses severe environmental threat to global agriculture and crop production. According to reports, saline soil covers 20% of the world's agricultural fields, diminishing crop productivity constantly (Van Schilfgaarde 1994; Munns 2002; Flowers 2004) and that after a few decades, soil salinity stress will deteriorate nearly half of the arable lands (Hasanuzzaman et al. 2014; Hossain 2019). Salinity stress in the plants occurs when the NaCl, Na_2SO_4, and other soluble salts are absorbed with water in excess through the rhizosphere and give rise to severe toxic effects throughout the plant body (Munns and Tester 2008). By modifying cell integrity, lowering osmotic potential, causing ion toxicity, and jeopardising the pattern of water intake, soil salinity stress has a detrimental effect on plants directly and ultimately results in plant cell death (Botella et al. 2007; Moulick et al. 2020a).

3.4 Stress Due to Acidity and Alkalinity

The severity of soil acidity is induced when the soil pH decreases to an extreme level (below 5.0 pH) and the accumulation of Al^{3+} and Mn^{2+} ions in higher concentrations in the soil. It has been reported that 40–50% of all cultivated land around the world is acidic (Von Uexkull and Mutert 1995). Most of the available plant nutrients are found in neutral soil, but in extreme acidic conditions, essential plant nutrients get fixed and become unavailable to the plant, which leads to multi-nutrient deficiencies and inhibits the growth of the plant (Hossain et al. 2023). Excess accumulation of Al^{3+} ions prevent root growth and affect the absorption capacity of nutrients and water of the plant (Hiscox and Isrealstam 1979). Alkaline toxicity to the plant occurs when an excess of $NaHCO_3$ and Na_2CO_3 are present in the soil solution (Fang et al. 2021). In addition to altering the integrity of the cell membrane and the osmotic adjustment, increased soil pH with an excess of carbonate and bicarbonate ions also impacts plant water absorption and photosynthetic capacity (Zhang et al. 2017; Kaiwen et al. 2020).

3.5 Stress Due to Contamination of Heavy Metal, Metalloids, and Non-Metal

Several geologic and anthropogenic activities, such as extensive agricultural practices, mining, and industrialisation, have caused environmental harm (Tiwari and Lata 2018). One of the biggest contributors to soil contamination is heavy metals, which are long-term deposits from mining, chemical manufacturing, municipal garbage, and sewage sludge. These activities lead to the deposition of heavy metals on the soil surface or leaching down into groundwater (Hakeem et al. 2015; Ozturk et al. 2015a, b; Basheer 2018). A reasonably dense metal, metalloid, or non-metal is considered to be hazardous if it has the ability to harm both the environment and the crop ecology (Asati et al. 2016).

Heavy metals, metalloids (substances in-between metals and non-metals), and non-metals are innate components of the environment, which arise in severe concentration due to various human-based activities. Some micro-elements such as boron, sodium, and chlorine are the vital micronutrients that are required in minute amounts to maintain various physiological and metabolic functions in plants, while some other elements including cadmium, arsenic, aluminium, and fluorine have no beneficial effects on plants and turn harmful when present in concentrations over their optimum levels. Generally, heavy metals cannot be decomposed and removed from the environment and some of them are immobile in nature, and they remain fixed where they are accumulated; however, movable metals are absorbed by the plant rhizosphere through diffusion and endocytosis (Ali and Jain 2004; Ashraf et al. 2010; Ali et al. 2015; Dehghani et al. 2016; Alharbi et al. 2018; Burakova et al. 2018). The most significant aspect that determines whether heavy metals actually restrict or stimulate plant growth is the concentration of those metals in the soil (Abolghassem et al. 2018). Heavy metal stresses impair the substitutional processes of metal ions in biomolecules and deactivate a number of crucial enzymes and other proteins. Thus, membrane integrity is compromised, which negatively affects fundamental plant metabolic processes as photosynthesis, respiration, and homeostasis (Hossain et al. 2012).

3.5.1 Arsenic

Having no positive effects on plants, arsenic (As) is a poisonous metalloid (Zhao et al. 2009). It is detrimental to the plants as well as consuming arsenic-affected crops leads to hyper/hypopigmentation and abnormal tissue growth on human skin (Panda et al. 2010). The inorganic arsenic in the soil is primarily absorbed by plants in the form of arsenite (As^{III}) and arsenate (As^{V}), which are readily available to the plant (Meharg and Hartley-Whitaker 2002). Inorganic sources of arsenic show comparatively high toxic effect than organic, while As^{III} is extremely hazardous than As^{V} along with notable amount of As accumulation can also be seen (Moulick et al. 2016a, 2018d, 2020b, 2021, 2022, 2023a). For agricultural soil, the allowable limit of As is 20.0 mg kg^{-1} of soil, and it is found to be toxic to sensitive crops at 5.0 mg L^{-1}in the soil (Garg and Singla 2011). It alters the plants' anatomical functions (Smith et al. 2010), decreases the biochemical reactions of plants, limits their metabolic processes, and eventually stunts their growth and development (Miteva 2002; Stoeva et al. 2003).

3.5.2 Boron

Boron is a helpful mineral for normal plant growth, but its toxicities and deficiencies are more common in most crops and environmental situations than any other micronutrient, which negatively impacts plant growth and development (Archana and Verma 2017). According to reports, boron shortage alters a number of physiological processes and biochemical processes, which in turn impacts how plants grow and develop normally (Dell and Huang 1997; Camacho-Cristobal et al. 2008). Yellowing of the leaves, a loss of strength, a delay in maturation, inappropriate fruit form and size, and a decreased yield are some of the most noticeable indicators of boron toxicity (Muntean 2009).

3.5.3 Sodium

A challenge to sustainable agriculture is posed by salt stress, also known as salinity stress, which is an excess of salt deposition in the soil and is regarded as one of the main environmental evils (Sahoo et al. 2019; Iqbal et al. 2020; Moulick et al. 2020a). Sodium salts such as $NaHCO_3$ and Na_2CO_3 are causing alkaline toxicity to the plant (Fang et al. 2021). Therefore, the stresses affect the osmotic adjustment, reduce photosynthetic capacity, and alter the integrity of cell membrane and water uptake (Zhang et al. 2017; Kaiwen et al. 2020). Globally, the stresses of soil salinity and alkalinity both seriously impair plant growth and crop yield (Wang et al. 2018).

3.5.4 Chlorine

Chloride (Cl^-) is the monovalent anion of chlorine that is beneficial nutrient for some plants. Although, when it reaches the required amount in the plant, it becomes toxic. Excessive chloride stress affects the plant physiological functions, disrupts their biosynthetic processes, inhibits the photosynthetic capacity, therefore suppresses the root and shoot growth, necrotic lesions appear on surface and margin of leaves, and induces opposed effect between anions (Ahmed et al. 2020).

3.5.5 Aluminium

One of the main issues with field crops is aluminium (Al) stress, which is toxic and detrimental under acidic conditions (Saha et al. 2019; Fan et al. 2022). It is estimated that 40–50% of the world's cultivated lands are occupied by acidic soil (Von Uexkull and Mutert 1995). Roots of the field crops have been found to be highly susceptible to Al toxicity (Ryan et al. 1993). Thus, root growth is extremely restricted by Al toxicity and needs to be identified immediately after contamination (Barcelo and Poschenrieder 2002). The Al^{3+} toxicity will reduce the grain yield and can even cause the death of the plant.

3.5.6 Cadmium

Cadmium (Cd) is an unessential heavy metal and one of the utmost pollutants which harms the environments (Zhao et al. 2010; Chowardhara et al. 2019a, b). Common anthropogenic activities that release Cd include significant industrialisation, municipal trash, excessive use of phosphatic fertilisers, and chemicals (Gupta and Gupta 1998; Xue et al. 2009) and deposited to the soil, leading to a terrible issue for agricultural land. It has been reported that 100 mg of Cd kg^{-1} of soil is acceptable for agricultural land (Salt et al. 1995). When the percentage of Cd present above their optimum limits, they become toxic and cause negative effect to plant such as leaf discoloration, discourage the plant growth, root rot and eventually destroy the plant (Mohanpuria et al. 2007; Guo et al. 2008). Excessive cadmium accumulation induces the iron deficiency, which affects the chlorophyll formation (Ghori et al. 2019).

3.5.7 Fluoride

Fluoride (F^-) is the most phytoavailable toxic non-metal that is known for its negative potentiality, even at very small amount to the plant (Choudhary et al. 2019). In general, F^- is released into the environment from parental rock, volcanic eruption, and bricks producing industries (Symonds et al. 1988). Fluoride is one of the most dangerous phytotoxicants found in common air pollutants (Haidouti et al. 1993). It was observed that higher amount of fluorine in acidic soil, enhancing the aluminium and reducing the phosphate uptake, decrease the crop productivity (Elrashidi et al. 1998). Inside the cell membrane, aluminium and fluoride are separated from each other to exert their toxic effects (Kinraide 1997). Fluoride is entered into the plant through vascular tissue and accumulates at the leaf edges (Threshow 1970). Amongst the all-toxic pollutants, gaseous fluorides are the major toxic and hazardous pollutants to the farmland (McCune and Weinstein 1971). Generally, F^- toxicity is observed in the plant during germination to early vegetative stages and interferes with the plant's morphology, physiology and metabolic activities (Panda 2015). The scorching and burning of foliage are the most prevalent apparent signs of F^- toxicity to plants. When a high concentration of F^- is absorbed and translocated to the shoots, it often leads to cell death (Miller 1993).

4 Field Crops Sensitivity of Abiotic Stress

Increased human population combined with diminishing land resources has resulted in increased pressure on limited natural resources, which has been exacerbated by climate change, which affects plant growth and development by destroying biomass (Moulick et al. 2023c; Wallace et al. 2003). It is a significant impediment to agricultural productivity. Abiotic stress is caused by the interaction of organisms with their physical environment (Vijayalakshmi 2015). Abiotic factors such as water shortage or excess, high or low temperatures, submergence of low-lying regions, salt, nutrient deficiency, pollution, and wind commonly influence crops (Pereira 2016). The technological development, policy ramifications, social behaviour, patterns of land and water use, level of economic development, and cultural diversity are depending on the population's structure (Dar et al. 2021). Numerous models based on multiple regression analysis are employed to determine the influence of varying agro-environmental situations on yields (Spitters et al. 1989). However, based on statistical models, it is difficult to discriminate between crop production differences caused directly or indirectly by abiotic stress (Cho and Oki 2012). This chapter will go through how different abiotic stresses affect field crops.

4.1 *Cereal Crops*

Abiotic stress has significantly affected the production of rice, wheat, and maize, three of the most important cereal crops in terms of daily calcium and protein consumption (Dar et al. 2021). Wheat and maize yields have decreased by 5.5 and 3.8%, respectively, since 1980, as a result of higher temperature and rainfall (Borlaug Institute for South Asia 2015). The yield loss in rice, wheat, and maize induced by drought and heat stress is shown in Fig. 15.1. About half of the world's rice-growing regions are affected by droughts of varied severity (Pandey and

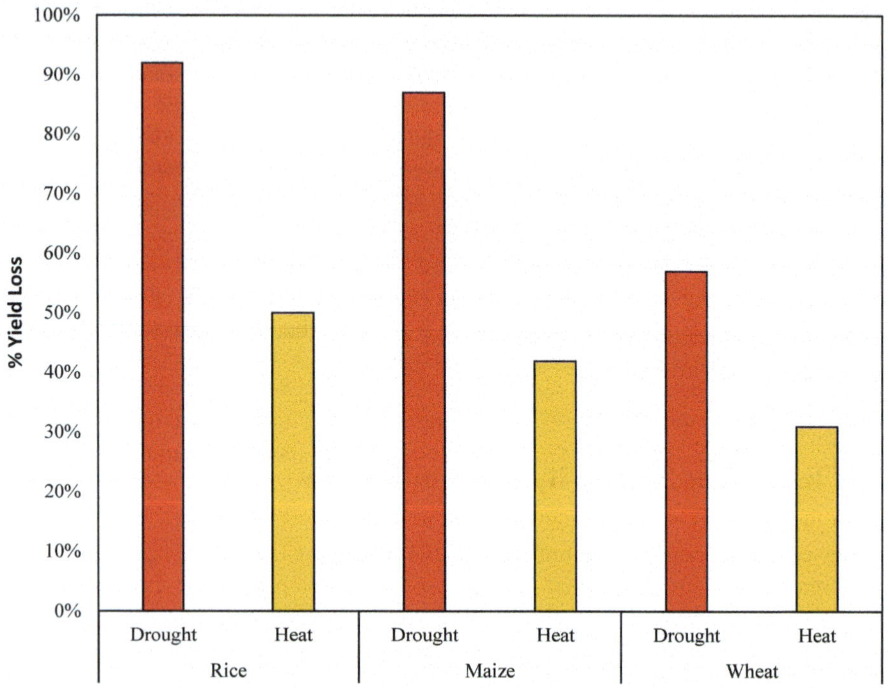

Fig. 15.1 Yield loss in major cereals due to moisture and temperature stress. (Source: Modified after Pandey and Bhandari 2008)

Bhandari 2008), and these conditions are projected to deteriorate further in the near future. The shift from vegetative to reproductive phases is mostly influenced by environmental factors. However, the unpredictability and varied timing of abiotic stress throughout critical developmental phases severely reduce grain production potential.

Abiotic stress during the early phases of pollen generation in rice and wheat has the biggest impact on flower and grain growth (Dolferus et al. 2011). Rice's vegetative growth and yield are adversely affected by drought (Ahadiyat et al. 2014). Low-moisture conditions have a detrimental effect on rice by limiting floret emergence, decreasing grain test weight, and ultimately reducing production. In contrast to wheat and barley, rice panicles have an open structure that is readily impacted by abiotic stress (Bennett et al. 1973). Figure 15.2 illustrates the plant's reaction to a water deficiency scenario schematically. Additionally, Chakrabarti et al. (2013) revealed that rising temperature had a significant effect on wheat crop productivity (Fig. 15.3).

Climate change, irregular rainfall, and sea level rise all contribute to an increase in flooding intensity. Rice can endure floods to a certain amount because to the existence of aerenchymatic tissue, but not for more than 7 days (Dar et al. 2021). Submergence-related mortality and degradation of rice plants are dependent on the flood's water quality, depth, and length (Mishra et al. 2010).

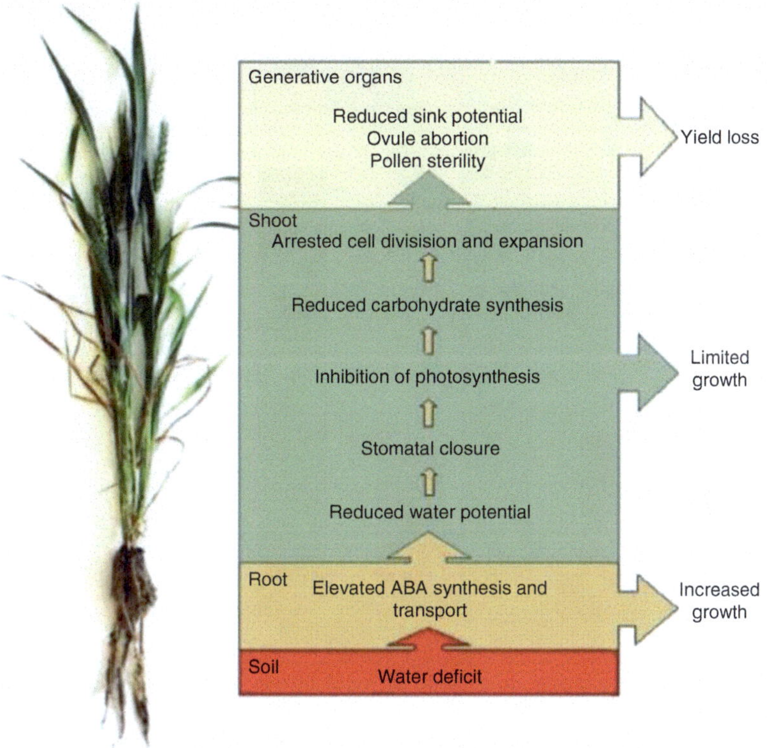

Fig. 15.2 Response of plant to water-deficient conditions. (Modified after Barnabás et al. 2008)

Rice is affected by low temperatures at all developmental stages. It decreases germination percentage, vegetative development, and ultimately yield by causing mayhem on rice throughout the blooming stage (Dar et al. 2021). In rice, prolonged exposure to cold stress results in leaf necrosis and chlorosis as a result of reduced photosynthetic activity and oxidative damage (Xiao et al. 2011). Low temperatures and drought stress during the initial microspore stage of pollen generation make self-pollinating rice and wheat produce sterile pollen (Ji et al. 2010; Oliver et al. 2005). However, since maize is a monoecious plant, drought stress is most severe during the reproductive stage (Dolferus et al. 2011). In maize, the incidence of kernel abortion rises during the pollination phases owing to dryness (Hidangmayum et al. 2018). Increased atmospheric CO_2 levels have been explored in laboratory and field environments in the absence of other limiting variables. In response to elevated CO_2 levels, photosynthetic activity increased, stomatal aperture decreased, and total biomass increased (Cho and Oki 2012).

Barley is reasonably resistant to abiotic stress of varying degrees (Munns and Tester 2008). Abiotic stress has a substantial impact on plant growth and development, reproductive time, and ultimately barley yield (Kazan and Lyons 2016).

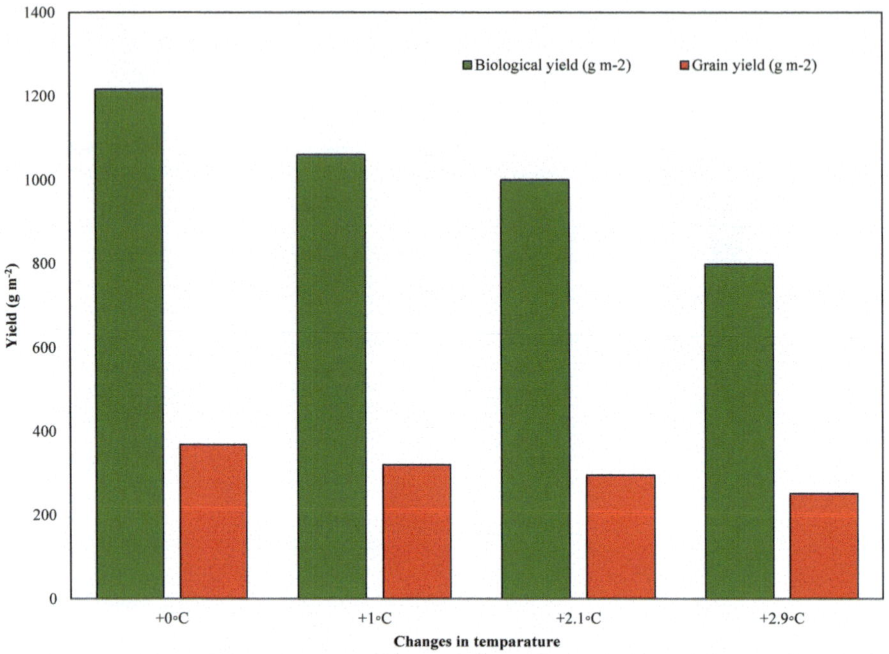

Fig. 15.3 Effect of increasing temperature on yield of wheat crop. (Source: Modified after Chakrabarti et al. 2013)

Drought circumstances prior to tillering and blooming considerably affect barley output by reducing the number of tillers, the number of spikes and grains per plant, and the weight of individual grains (Samarah 2005).

Grains exhibit a variety of phenotypic features when subjected to abiotic stress (Fig. 15.2). Rolling of leaves is seen in drought-prone plants (Plaza-Wüthrich et al. 2016). By contrast, stress has no effect on the length of cells in the internode. In finger millet, a reduction in plant height and lodging of the main stem were reported (Plaza-Wüthrich et al. 2016). Teff (*Eragrostis teff*) plants, which are produced in Ethiopia as a staple crop, also shown a detrimental impact of water scarcity (Fig. 15.4).

4.2 Legumes

Following the Poaceae family of crops, Leguminosae are an important element of the human diet. While legumes can adapt to a broad variety of agroclimatic conditions, the incidence of abiotic stress may be damaging to the crop's growth and development (Kang et al. 2017). The majority of environmental disturbances result in a rise in reactive oxygen species (ROS).

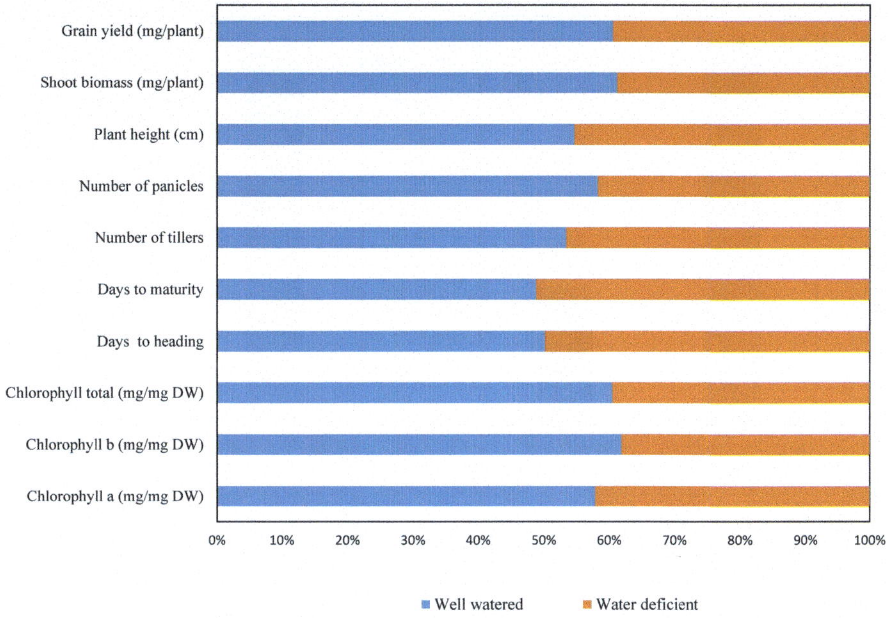

Fig. 15.4 Different traits of Tef plants (Poaceae family) grown under normal and stress situation in Africa. (Source: Modified after Plaza-Wüthrich et al. 2016)

Water stress damages legumes because it prevents photosynthesis, cell growth, and carbon metabolism (Araujo et al. 2015). In soybean, grown under controlled conditions, it has been demonstrated that drought has an impact on flowering, pod sterility, pod growth, yield, and individual seed weight (Desclaux et al. 2000). However, moderate dryness has been reported to aid faba bean in restricting vegetative growth and increasing yield and harvest index (Grashoff 1990). Stomata are affected by water stress, which decreases their number and increases their size (Araujo et al. 2015). Water stress causes root thinning in field peas, which scouts deeper for water supplies, but the overall root surface area is decreased (Benjamin and Nielsen 2006). Legumes, including cowpea and groundnut, can alter their root architecture in reaction to variations in moisture levels, allowing them to adapt to stressful situations (Pandey et al. 1984). In drought-tolerant pea cultivars, the accumulation of soluble sugars and proline caused by drought helps to maintain turgor pressure (Araujo et al. 2015). Pulses, on the other hand, such as pigeon pea and cowpea, can endure severe drought owing to their versatility. When faced with dryness, pigeon pea develops deep penetrating roots and osmoregulates in the leaves, while cowpea sheds its reproductive systems (Odeny 2007). Khan et al. (2019) have observed that drought stress dramatically decreased chlorophyll fluorescence and relative water content (percentage) in chickpea cv. Punjab Noor 2009 (Fig. 15.5).

Legume development and growth are significantly hampered by salt stress. It was estimated that near about 80 million acres of fertile land are directly or indirectly affected by soil salinity (Munns and Tester 2008). Pulses, on the other hand,

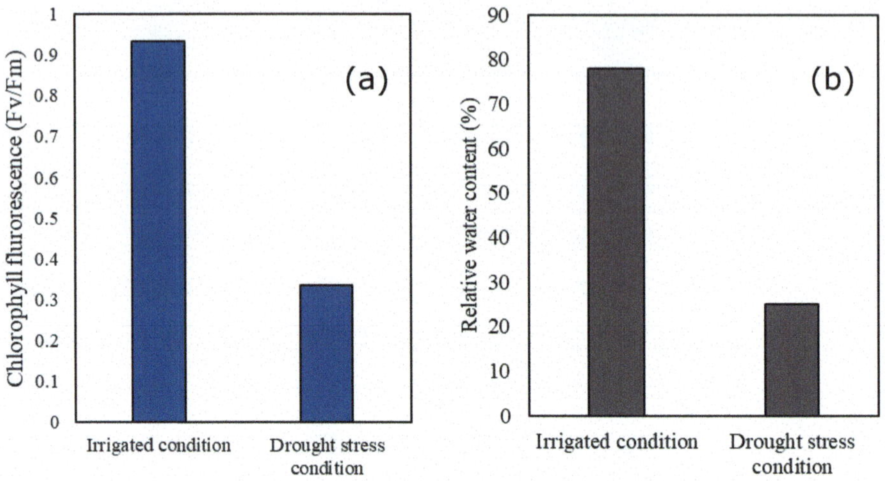

Fig. 15.5 Effect of irrigated and drought conditions on chlorophyll fluorescence (Fv/Fm) and relative water content (%) in chickpea variety Punjab Noor 2009. (Source: Modified after Khan et al. 2019)

are more susceptible to salt than grains and oilseeds (Hidangmayum et al. 2018). Salt stress impacts plant roots immediately, but moisture stress brought on by osmotic stress affects plant development over time (Munns 2005). Different degrees of salt tolerance and adaptation are noticed in legume crops such as soybean (Araujo et al. 2015). Adaptable genotypes exhibit increased root development in saline soil, but some exhibit sodium and chlorine ion compartmentalisation (Luo et al. 2005). Sodium ions are maintained primarily in roots, whereas chlorine ions reach shoots, impairing growth and, at deadly levels, resulting in plant death (Sanchez et al. 2011). Pulses build up too much salt, which accelerates the rate at which the leaves and stems turn anthocyanin-coloured, reducing seed germination and seedling establishment (Kumar et al. 2016). Due to salinity-induced moisture stress, pulse growth and development, particularly photosynthesis, are harmed. It occurs in conjunction with an increase in photorespiration (Araujo et al. 2015). Due to salt stress, the primary root of peas is reduced and the number of lateral roots increases. In peas exposed to drought or salt stress, structural changes such as an increase in epicuticular waxes to reduce transpiration have been observed (Sanchez et al. 2011). In mung bean, yellow mosaic disease, stem- and pod-borers, and salt stress all contributed to yield losses of 80–100%, especially during the rainy season (Sehrawat et al. 2015). Naher and Alam (2010) reported similar findings about the influence of salt on mung bean (Fig. 15.6).

Temperature is critical in the symbiotic association between legumes and rhizobium in soil (Araujo et al. 2015). Changes in temperature have an effect on crop growth and production potential as well. Temperatures over 25 °C have a significant effect on cool-season legumes such as chickpea, field pea, lentil, and faba bean, whilst temperatures above 35–40 °C have an effect on rainy season crops (Kumar

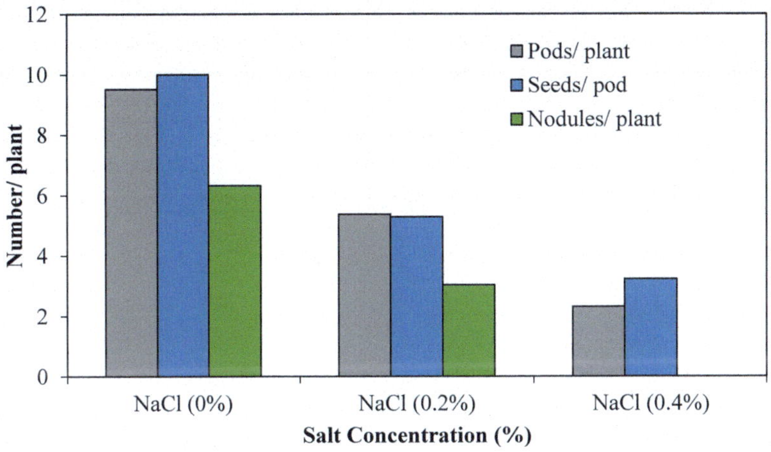

Fig. 15.6 Effect of different salinity concentrations in growth of mung bean cv. Bari Mung 5. (Source: Modified after Naher and Alam 2010)

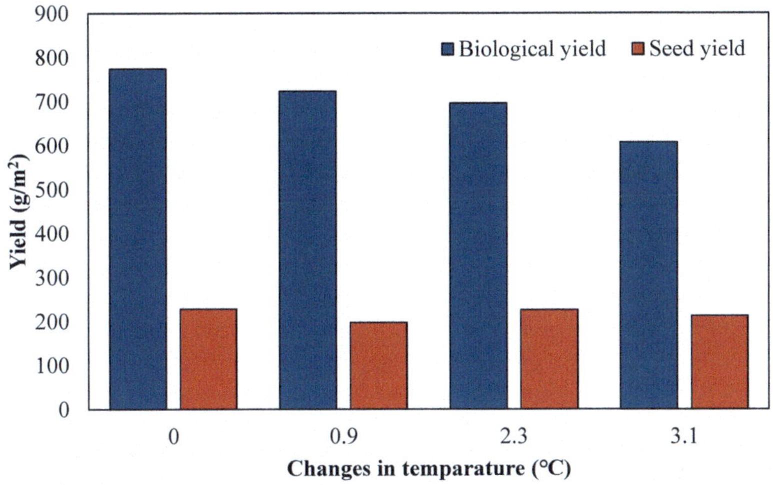

Fig. 15.7 Effect of increasing temperature on yield of chickpea. (Source: Modified after Chakrabarti et al. 2013)

et al. 2016). Similarly, terminal drought stress affects seed yields by 26–61% in chickpea (Kumar et al. 2016), and by 40–55% in pigeon pea (Kumar et al. 2016; Nam et al. 2001). According to Rainey and Griffiths (2005), heat stress significantly reduces yields of common bean (*Phaseolus vulgaris* L.) and peanut (*Arachis hypogea* L.). Flowering, seed setting, and pod development are all critical phases that are significantly influenced by changes in moisture and temperature (Araujo et al. 2015). Chakrabarti et al. (2013) also investigated the detrimental effect of temperature increases on chickpea yields (Fig. 15.7).

Pulses respond differently when exposed to suboptimal temperatures depending on their source. When temperatures drop to 0–15 °C, tropical and subtropical pulses are unable to function, although cold pulses can (Araujo et al. 2015). According to Antolin et al. (2005), growing pulses at temperatures as low as 20 °C lead to a two-fold increase in stem and root dry weight without affecting leaf dry weight. Pollen sterility occurs at colder temperatures over 20 °C, which inhibits the germination and emergence of cowpeas (Araujo et al. 2015).

4.3 Oilseeds

Oilseeds play a critical role in Indian agriculture and cuisine. Even with dramatic increases in output, there has always been a gap between demand for oilseeds and vegetable oils and supply (Chauhan et al. 2021). In India, rainy season groundnut, soybean, sunflower, niger, castor, and sesame are farmed. Oilseeds including linseed, safflower, and rapeseed-mustard are widely available throughout the Rabi season. In the pre-Kharif or summer season, oil seeds such as sesame, castor, sunflower, and peanut are also cultivated. Abiotic stress is often caused by cultivating oilseeds primarily under rainfed conditions, climate change, and growing area of salty soils.

Depending on the length and stage of the crop, drought or water scarcity conditions can affect oilseed crops to varied degrees. Drought initially affects the germination rate of seeds (Alqudah et al. 2011). In soybean, late-stage water deficits have a limiting influence on photosynthetic activity (El Sabagh et al. 2019). Under stress, a decrease in nitrogen transfer from roots to shoots was found, owing mostly to inadequate absorption (Hussain et al. 2008). Drought lowers stem elongation, leaf area, blooming, and achene filling in sunflower (Baldini et al. 2000). Soybeans grown during a drought have different fatty acid profiles and include more soluble carbohydrates and proline (Alqudah et al. 2011). While the majority of oilseeds, such as sunflower, sesame, and safflower, can tolerate mild drought, water stress, and high temperatures from blooming to grain/seed filling have a detrimental effect on the crops, namely, decreasing yield and oil content (Hussain et al. 2008). Oilseeds such as soybean, canola, and flax exhibit a drop in protein and oil content when the temperature rises (Carrera et al. 2009; Mirshekari et al. 2012; El Sabagh et al. 2019). However, drought has resulted in a decrease in oil content and an increase in protein content in safflower (Amini et al. 2014). Depending on the severity and stage of the crop, drought may have an impact on the number of seeds per siliqua, the number of siliquae per plant, the weight of a thousand seeds, the yield of seeds, the oil content of the seeds, and the output of oil from oilseeds (Nasri et al. 2008). Under conditions of drought, the sunflower's oleic, linoleic, and palmitic acid contents rise (El Sabagh et al. 2019). However, abiotic stress during flowering and seed development decreased both saturated (palmitic and stearic acid) and unsaturated fatty acids (oleic, linoleic, and linolenic acid) in safflower (Fernández-Cuesta et al. 2014). Drought-affected plants develop early, and their low linoleic acid content is a result of the shorter time required for oleic acid to be converted to linoleic acid. Dryness in the mid-season has little influence on groundnut, whereas drought in the late season increases stearic and oleic acid levels and decreases total oil and linoleic acid contents of groundnut seeds (Dwivedi et al. 1996).

Salinity stress has the potential to have a deleterious effect on oilseeds throughout their life cycle, from germination through reproductive maturity (Nayidu et al. 2013). Salinity stress is detrimental to crops because it causes physiological drought, ion excess, oxidative damage, and nutritional imbalance. Salinity hinders seedling growth and height development and lowers germination rates in the early stages of canola development (Sinaki et al. 2007). *Brassica* species vary in their tolerance for salinity stress, with *B. napus* being the most tolerant, followed by *B. carinata* and *B. juncea*, which can also withstand stress (Kumar et al. 2016). *B. nigra* and *B. rapa* are very salt tolerant (Kumar et al. 2016). The quality of seed oil extracted from oilseeds is directly proportional to the concentration and percentage of palmitic, oleic, stearic, and linoleic acids (Nayidu et al. 2013). Bergman et al. (2006) discovered that salt stress alters the proportions of palmitic, oleic, stearic, and linoleic acids. Di Caterina et al. (2007) observed that irrigation with salty water increases oleic acid content while decreasing linoleic acid levels in sunflower. Simultaneously, oilseed crop quality and productivity are negatively impacted by salt stress (Nayidu et al. 2013). Mensah et al. (2006) also observed that increased salinity has a detrimental effect on the pod and seed production per plant of rainy season groundnut.

Climate change causes crops to develop quicker, hence shortening their length, yet heat during the early and blooming phases may be very detrimental to the crop (Lobell and Gourdji 2012). Heat stress impairs the growth and development of oilseeds. In *Brassica* species, excessive heat stress results in poor seedling germination and development (Deol et al. 2003). CO_2 levels are gradually growing, and it has been shown that elevated CO_2 speeds up photosynthesis and also promotes rapid and healthy development (Lobell and Gourdji 2012). A temperature range of 25–35 °C is ideal for the growth of oilseeds (Kaya et al. 2006). Numerous research on soybeans have shown that increased CO_2 concentration has a fertilising impact and increases crop water use efficiency; however, the beneficial effect of raised CO_2 diminishes at very high temperatures (Lobell and Gourdji 2012). Proline content is increased by heat stress. Temperatures over 27 °C during blooming may result in floral sterility, reducing the quantity of flowers, seed size, and ultimate economic output (Morrison and Stewart 2002). Environmental extremes may have a significant effect on the growth and development of a crop, depending on the length of their exposure, the area of the crop, and its stage of development (Shah et al. 2016). Spring and winter temperatures are required for several reproductive phases such as gametogenesis, pollination, fertilisation, and embryogenesis (Angadi et al. 2000). Low-temperature stress damages leaf tissues and causes bleaching, withering, and, in severe situations, plant death. Thus, drought, salt stress, temperature extremes, and excessive CO_2 concentration all have a significant impact on the morphophysiological processes of oilseeds at various phases of crop development.

4.4 Tuber Crops

Tuber crops such as potatoes, sweet potatoes, cassava, and yam are critical components of the human diet and contribute significantly to global food security. Abiotic stressors such as severe heat, salt, and drought are unavoidable. Current circumstances necessitate producing crops that can withstand the challenges associated with changing environmental conditions.

After rice and wheat, the potato (*Solanum tuberosum* L.) is the world's third most significant food crop. Potato, which originates in South America's Andes, is subject to heat and drought stress (Singh et al. 2019). Tuber development requires an optimal night temperature of 17 °C; however, if the temperature surpasses this level, vegetative growth accelerates significantly, lowering tuber output (Setayesh et al. 2017). Sweet potato plants have decreased growth, stomatal conductance, nitrate reductase activity, and stomatal closure when water is scarce (Ravi and Indira 1999). Temperature increases have a physiological effect on tuber crops, slowing sprouting, increasing the frequency of tiny tubers, and resulting in uneven, bitter, poisonous, and fractured tubers (Demirel et al. 2017). Increased soil temperature results in dark discoloration of tubers, tuber rot, and sprout damage (Singh et al. 2019). Between 14 °C and 22 °C is the optimal temperature range for tuberisation and reduced haulm growth (Demirel et al. 2017). Above that, potato tuber growth and net assimilation both contribute to tuber yield reduction (Singh et al. 2019). Water deficit circumstances have a profound effect on sweet potato during the tuber initiation stage, depending on the volume of available soil moisture or the severity of the stress (Ravi and Indira 1999). Drought during the tuber development stage results in tuber lignification. Under water-scarce circumstances, the chlorophyll concentration of sweet potato leaves drops, that is, total chlorophyll, chlorophyll a, and chlorophyll b (Ravi and Indira 1999).

Around 20% of the world's cultivable land is significantly impacted by salinity (Sairam and Tyagi 2004). Tuber crops such as potato, which are moderately salt-sensitive, may flourish at soil salinity levels as high as 1.7 dS m^{-1} and irrigation water salinity levels as low as 1.1 dS m^{-1} (Chourasia et al. 2021). When exposed to salt, leaf and shoot tissues collect sodium and chloride ions, impairing metabolic function by lowering photosynthetic activity, protein synthesis, and inactivating enzymes (Tavakkoli et al. 2010). Exposure of plants to saltwater environments during their first phases of sprouting, emergence, and development results in yield losses of between 21% and 59% (Levy 1992). Salinity extremes impair sugar translocation, reducing microtuberisation, rooting capacity, root number, root diameter, and root length (Faried et al. 2016). Reduced leaf number, leaf area, and leaf development are found in tuber crops exposed to salt, resulting in leaf abscission and senescence (Chourasia et al. 2021). An increased sodium ion concentration in saline conditions hinders the absorption of other crucial elements such as potassium and magnesium (Ishikawa and Shabala 2019). Salt stress in tuber crops results in osmotic stress, which accumulates abscisic acid in the leaves, slowing photosynthesis, shutting stomata, and lowering photoassimilates translocation (Chourasia et al.

2021). However, low-salt concentrations in cassava have facilitated development and resulted in an increase in total starch and total protein in shoots rather than roots (Cheng et al. 2018).

Flooding tuber crops reduces the oxygen concentration in the root zone, impairing tuber development (Ravi and Indira 1999). Flooding increases haulm growth and decreases tuber number, size, and diameter throughout the formative phases. However, if excessive moisture stress happens during the crop's first stages, the crop rebounds from the initial shock and resumes normal development (Ravi and Indira 1999). During floods, tuber crops produce antioxidants in their leaves, which aid them in coping with the adverse conditions.

4.5 Sugar Crops

Abiotic stresses such as drought, salt, excessive heat or cold, and a lack of nitrogen do not affect sugar crops such as sugarcane and sugar beet very much. They possess adaptive morphological, physiological, and biochemical mechanisms that enable them to withstand mild stress situations (Narwade et al. 2016). At the moment, climate conditions are undergoing a dramatic shift, increasing the probability of extreme weather events. Any kind of stress during crucial development phases has a detrimental effect on crops. Unlike cereals, sugar crops, on the other hand, have a lower risk of complete crop failure owing to stress at key periods (Ober and Rajabi 2010).

Temperatures between 17 °C and 25 °C are ideal for sugar beet (Ober and Rajabi 2010), while temperatures between 18 °C and 40 °C are ideal for sugarcane (Narwade et al. 2016). Sugar beet is a reasonably resistant plant to drought and heat (Ober and Rajabi 2010). Moisture is vital in sugarcane throughout the key phases of tillering, internode elongation, and grand development, exerting a significant influence on crop growth. Sugarcane sett germination is influenced by soil moisture content, with both excess and deficiency reducing seminal root growth (Narwade et al. 2016). The stomata's ability to open and close and the activity of various enzymes are all influenced by moisture stress (Naik et al. 1993). Water availability in the soil has an effect on cell division, cell enlargement, and leaf growth in plants (Cosgrove 2000). Water stress has a detrimental influence on tiller development, dry matter accumulation, net photosynthesis, cane height, cane weight, internodal length, nutrient availability, root activity and production, and juice quality in sugarcane. Tillers in their infancy are particularly vulnerable to drought conditions (Naik et al. 1993). In drought-stricken sugarcane, a longer rope-like root system is visible, but in irrigated sugarcane, the root mass is less than 50 cm deep (Venkataramana and Naidu 1982). When sugar beet is subjected to both heat and water stress, it exhibits a decreased transpiration rate (Schubert et al. 2004). In tropical and subtropical climates, water stress and drought may result in yield losses of 50–70% owing to poor growth and tiller death (Narwade et al. 2016). Withholding irrigation during maturity, on the other hand, aids in limiting growth and enhancing photoassimilation (up to an 18% rise in sucrose) in sugarcane (Robertson and Donaldson 1998). High-moisture content in the soil and light soil with reduced nitrogen availability result in early blooming, thereby decreasing the sugar concentration (Narwade et al. 2016). Robertson et al. (1999) observed that withholding irrigation for 5 months before to maturity diminishes leaf and stalk extension, dramatically decreases the leaf area index (from 1.8 to 0.9), and induces leaf senescence in sugarcane. However, it has no effect on sucrose production.

Temperatures in the atmosphere and on the ground are increasing as a result of climate change. Due to increased mineralisation in the soil, this temperature rise boosts sugar plant nutrition uptake (Narwade et al. 2016). However, as the temperature rises beyond 25 °C, sugar beet leaf development, photosynthesis, and sugar accumulation are reduced mostly owing to higher respiration rate losses (Dambrosio et al. 2006). Critical factors in determining the degree of cell damage include the duration and intensity of heat stress (Narwade et al. 2016). Despite ambient water in the soil, a rapid reduction in the water content of the leaf tissues of sugarcane during heat stress was observed (Wahid and Close 2007). Internode growth, net absorption rate, and total biomass were all dramatically reduced by high-temperature stress, although tillering was improved (Wahid 2007). Heat stress results in protein denaturation, desiccation, dehydration, and membrane instability (Chalker-Scott 2002).

The sugar beet plant has a good tolerance for the stress caused by soil salt (Ober and Rajabi 2010). Sugarcane is fairly susceptible to salt, with a 1.4 dS m^{-1} threshold value (Tiwari et al. 2015). Sugarcane's seed germination stage is the most tolerant; however, salt stress during the shoot development stage is most detrimental (Plaut et al. 2000). However, compared to soils with no sodium, sugar beet grows successfully at low-salt concentrations (Khavari-Nejad et al. 2008). However, sugar beet's germination and early shoot development phases are salinity-sensitive (Jamil et al. 2006). In sugarcane subjected to lengthy periods of salinity stress, yellowing of leaves and salt damage to leaves were found (Gomathi and Thandapani 2005). Sugar beet cultivation on salty soils resulted in a decrease in water absorption and cell growth (Ober and Rajabi 2010). Sugarcane shoots undergo a decrease in tillering, internode narrowing, shoot height reduction, and photosynthesis reduction in salty and sodic soils (Tiwari et al. 2015). Critical nutrient partitioning in sugarcane plants exposed to extended salinity revealed an increase in the concentration of hazardous elements (Na, Cl, B, and Mo) inside plant tissue (Gomathi and Thandapani 2005). Crops cultivated on salty soils did not always result in a decrease in juice quality. The juice quality was mostly determined by the genotypes used, while tolerant genotypes (Co 94012 and Co 99004) exhibited minor quality degradation (Tiwari et al. 2015). When soil extracts had an electrical conductivity of 13.7 dSm^{-1}, sugar beet yields were reduced by up to 50% (Morillo-Velarde and Ober 2006). Sugarcane showed an increase in non-sugar solids and salts in salty soils and a decrease in sucrose %, brix, and purity (Tiwari et al. 2015). While a low-salt content in the soil has little or no impact on sugar beet production, salt build-up causes stress, severely reducing output (Ober and Rajabi 2010).

Sugarcane sprouted stubble buds poorly at low temperatures, had a low amount of reducing sugars, decreased photosynthesis, and decreased sucrose-phosphate synthase activity (Tiwari et al. 2015). Frost damage to young sugar beet plants is also a problem (Cary 1975). Under freezing circumstances, the majority of above-ground immature parts die, and regeneration occurs through below-ground secondary shoots. The degradation of sugarcane juice quality is mostly determined by the length of cold stress, the genotypes planted, and the degree of freezing (Tiwari et al. 2015).

Waterlogging is a serious stressor that affects enormous swaths of cultivable land in India and around the globe. Sugarcane is prone to waterlogging during the first 3–4 months of growth, but it is fairly resistant afterwards (Carter and Floyd 1975). Sugarcane roots get darkened when they are flooded, nutrition absorption and root respiration are impeded, and the whole subterranean root system becomes clogged (Gomathi and Thandapani 2005). However, adventitious roots that grow above the water layer have bigger intercellular gaps than original roots and provide the plant with the essential oxygen (Van Der Heyden et al. 1998). Flooding during the tillering stage results in tiller mortality, decreased tillering, elongation, and decreased leaf area (Gomathi and Thandapani 2005). Waterlogging reduces transpiration, photosynthetic rate, effective leaf area, leaf number, total chlorophyll content, and growth rate, while increasing respiration rate (Gomathi and Chandran 2009). Sugarcane afflicted by waterlogging might have yield reductions of up to 15–45% (Gomathi and Thandapani 2005).

4.6 *Other Important Field Crops*

Cotton is salt tolerant to a degree and may also adjust to moisture stress via stomatal control. Drought is deleterious throughout the early phases of development, and it also impairs the ultimate output during blooming. Low light stress is another issue that impedes cotton growth and development during the monsoon (Sabesh 2022). Cotton needs a cool night temperature, yet high temperatures cause blooming to be delayed and enzyme activity and protein synthesis to be decreased (Tavakkoli et al. 2010). Cotton's early stages are sensitive to salt and waterlogging, which impairs leaf development, root growth, and ultimate production (Sabesh 2022).

Jute cultivated in hot and humid climates exhibits a range of morphophysiological and biochemical changes in response to varied stress conditions (Rahman et al. 2021). In *Corchorus olitorius*, salinity decreased root length, fresh weight, fresh weight per plant, plant height, and number of leaves per plant (Tareq et al. 2018). When *C. olitorius* was exposed to a salt level of 175 mM NaCl, we found a drop in net photosynthesis, stomatal conductance, transpiration rate, pods/plant, and seeds/pods, as well as a decrease in pods plant^{-1} and seeds pods^{-1} (Yakoub et al. 2019). Rahman et al. (2021) observed that when *C. olitorius* and *C. capsularis* were exposed to acute moisture stress, they decreased plant height (35–50%), leaf number/plant (30%), leaf area (25–67%), stem diameter (16–42%), and yield (50–80%). While flooding with up to 10 cm of standing water resulted in a rise in adventitious root development, an increase in aerenchyma tissue in the adventitious root, a drop in fibre length (11–43%), a reduction in fibre strength (12–55%), and a 20–75% yield loss in jute (Rahman et al. 2021). Field circumstances for crop development are very complicated, necessitating the presence of abiotic stressors, and several stress situations may also arise, wreaking havoc on crops. Tobacco is a Solanaceae family plant. Salt stress in tobacco results in ion build-up at hazardous concentrations, impairing mineral nutrient uptake, and oxidative damage (Sun et al. 2020). The quality, dry weight, chlorophyll content, and yield of tobacco leaves are all negatively impacted by drought throughout the crop's growing stage, whereas the nitrogen and nicotine contents are positively impacted (Su et al. 2017). Whereas tobacco demonstrated decreased relative water content, transpiration rate, chlorophyll rate, and photosynthesis rate under flooding circumstances, as well as improved stomatal resistance and peroxidase activity in leaves.

5 Role of Ascorbate Glutathione Pathway in Maintaining ROS Homeostasis in Plants Under Abiotic Stress Condition

5.1 Generation of ROS

Aerobic activities will inevitably produce reactive oxygen species (Hasanuzzaman et al. 2019). The metabolism and ion balance of plants are significantly impacted by abiotic stresses. Plants accumulate ROS as a result of ion imbalance brought on by abiotic stress (Ahmad et al. 2009; Choudhury et al. 2021a, 2022a; b; c). Halliwell and Gutteridge (2015) discovered that in response to abiotic stress, plant cells create an excessive amount of ROS, including superoxide, hydrogen peroxide, hydroxyl radicals, and singlet oxygen (Halliwell and Gutteridge 2015). In plant cells, reactive oxygen species are mostly formed in the peroxisomes and chloroplasts (Ahmad et al. 2009). In the presence of light, photorespiration generates reactive oxygen species, while in the absence of light, mitochondria do so (Moller 2001). In plant cell, non-cyclic electron transport chains at complexes I and III of mitochondria, as well as an excess of membrane bound NADPH oxidases, produce superoxide radicals (Banerjee and Roychoudhury 2017). Due to the short life span of the superoxide radical, it does not cause much damage to plant cells. The cell membrane breaks down and lipid peroxidation occurs, when it becomes a hydroxyl radical (Halliwell 2006; Moulick et al. 2016b, 2017, 2018c). When there is too much photosynthetically active light or photooxidative stress, singlet oxygen species are generated in the chloroplast. Singlet oxygen causes oxidation of proteins, nucleic acids, and lipids in plant cells, leading to growth inhibition of plants (Krieger-Liszkay et al. 2008). Either univalent reduction or over-energising the electron transport chain produces hydrogen peroxide (Banerjee and Roychoudhury 2017). When ROS such as H_2O_2 is present in a higher concentration, it damages cellular macromolecules (Moller 2001). One of the most toxic ROS, hydroxyl radical, is produced when Fe^{2+} and Fe^{3+} act as catalysis in the Fenton reaction at neutral pH and when there is an excess of photosynthetically active light (Banerjee and Roychoudhury 2017). Excess accumulation of these ions is toxic for cells and causes plant cell death.

5.2 *ROS Scavenging Mechanisms (Enzymatic and Non-Enzymatic Means)*

Plants have special mechanisms for bearing abiotic stress and they comprise both enzymatic and non-enzymatic processes (Hasanuzzaman et al. 2019). The concentration and balance of ROS and ROS scavenging systems determine the degree of damage to the plant system. The plant experiences abiotic stress if the ROS concentration is higher than the level of scavenging ions (Hasanuzzaman et al. 2019). Non-enzymatic antioxidants generally comprise ascorbate, glutathione, flavonoids, tocopherols, carotenoids, etc. (Gill and Tuteja 2010). Ascorbate peroxidase, monodehydroascorbate reductase, glutathione peroxidase, dehydroascorbate reductase, glutathione reductase, superoxide dismutase, peroxiredoxin, catalase, and glutathione S-transferase are among the enzymes that act as antioxidants (Hasanuzzaman et al. 2019). All organisms either produce ascorbate in their bodies or get it as food from outside (Hasanuzzaman et al. 2019). Ascorbate biosynthesis is one of the earliest known life processes and helps in cell wall formation in plants. Glutathione is another important antioxidant, which comprises glutamine, cysteine, and glycine.

5.3 *The Role of the Ascorbate Glutathione Pathway in Abiotic Stress-Induced ROS Homeostasis in Plants*

Ascorbate and glutathione are an intrinsic part of the plant antioxidative system that catalases ROS in plants (Foyer and Noctor 2011). Plants accumulate antioxidants such as ascorbate and glutathione under abiotic stress conditions for maintaining ROS homeostasis in plants. The presence of ascorbate and glutathione has a detoxifying effect on ROS produced during abiotic stress in plants (Lopez-Delacalle et al. 2021). Higher ascorbate concentrations have frequently demonstrated that plant cells can withstand oxidative stress. Ascorbate peroxidase, dehydroascorbate reductase, monodehydroascorbate reductase, and glutathione reductase are the four enzymes that make up the ascorbate glutathione cycle, also known as the *Asada-Halliwell* pathway (Costa et al. 2021). They help in ROS scavenging and also to save the plant from possible damage from abiotic stress conditions (Hasanuzzaman et al. 2019; Mazumder et al. 2022a, b). Ascorbate acts as an electron donor in the ascorbate glutathione cycle and performs hydrogen peroxide detoxification by converting hydrogen peroxide into water, oxygen, and monodehydroascorbate (Foyer and Noctor 2011). This monodehydroascorbate is spontaneously converted to dehydroascorbate or to ascorbate by the action of monodehydroascorbate reductase

(Costa et al. 2021). Ascorbate and glutathione enhance osmotic regulation, water balance in plants, photosynthesis and other growth and yield attributes of plants (Lopez-Delacalle et al. 2021). The ascorbate glutathione cycle is an important ROS scavenging system that is mainly located in the cytoplasm and other cell organelles. Abiotic stress conditions result in a rise in hydrogen peroxide concentration and a fall in ascorbate and glutathione concentration.

6 Mitigation Strategies

6.1 Genetical and Breeding Improvement

To expedite breeding for desired genotypes, high-throughput phenotyping techniques, marker-assisted selection, next-generation sequencing, metabolomic profiling and genetic editing are applied (Choudhury et al. 2021b; Hossain et al. 2021e, f). The micronutrients, oils, phenolics, and other chemicals which modern cereal crop cultivars are enriched with are reviewed (Rasmussen 2020; Hossain et al. 2021e, g, h; Hossain et al. 2022). In recent years, the CRISPR/Cas9 system has become the most popular and commonly utilised method for altering the genome because of its ease of use and high efficacy. With the CRISPR/CAS9 system and its derivatives, new cultivars of field crops with increased yield potential are being developed (Araus et al. 2008).

For centuries, stress resistance cultivars of field crops development have relied on the technique of cultivating the plants that are capable of coping with a wide range of environmental challenges. Traditional breeding has resulted in several cultivars that are resistant to a variety of abiotic stresses (Hossain et al. 2021e). Conventional breeding, on the other hand, requires a lot of time and effort. An improved outcome may be achieved by selecting the best genotypes under stress conditions. Various abiotic stressors may affect the performance of different genotypes. However, it is possible that just one genotype will provide a high yield in a very demanding growing environment. More than one stress may sometimes improve the production potential of a plant. To cope with combined stress, these genotypes are critical. A genotype may sacrifice its ultimate output in order to survive in a stressful setting. Selecting abiotic stress-tolerant/resistant cultivars should thus prioritise yield potential.

Plants' resilience to abiotic stress may be improved by selecting a suitable habitat. The testing setting should resemble the natural habitat in which the plant is being grown. Indirect selection of genotypes may be advantageous when selection and target environment are comparable (Anbessa et al. 2010; Cormier et al. 2013; Hossain et al. 2021f). The degree of stress should be exerted proportionately. Better resistant cultivars may be developed if testing is subjected to greater strain stress. Some abiotic stresses, such as heat and drought, have a synergistic impact on each other. The best breeding method for selecting germplasm that promotes stress resistance in plants may involve researching the relationships between genotype and environment. Crop's ability to withstand a variety of environmental stresses is a particularly nuanced characteristic. Using Genome-Wide Association Studies (GWAS), researchers across the globe are attempting to elucidate the biological mechanisms that underlie these complicated features (GWAS) (Araus et al. 2002). In most cases, it requires a large collection of diverse germplasm to perform high-resolution gene mapping.

One of the most often used methods to identify genomic areas linked with target characteristics is QTL mapping. QTL mapping uses both the segregating and fixed populations from two parents to find genomic areas of interest. About 500 QTLs were reported in over 20 investigations looking for the QTL for salt tolerance (Azam et al. 2015; Hassan et al. 2018; Rasmussen 2020). Plant tissue culture-based approaches have been used in recent decades to address plant stress tolerance. The use of genetic engineering in crop development, particularly in response to varied stressors, is very effective. When compared to conventional breeding methods, it is a much faster solution (Hossain et al. 2021f). Particle bombardment and *Agrobacterium*-mediated gene delivery are two common strategies for introducing desired genes into plant cells in order to enhance certain features (Tal 1994). Breeders and biotechnologists used one or combination of those technologies to develop suitable stress resistance cultivars. Modern plant breeding tools have significantly curtailed down the time required for development and release a variety of a particular crop for location(s).

6.2 Agronomic Interventions

Mitigation of various biotic stresses in field crops by adoption of best or suitable management practices is quite a complex process. Stress mitigation through agronomic intervention requires holistic and whole-farm approaches (Hossain et al. 2021f). Abiotic stresses on field crops have greatly grown as a result of climate change and global warming (Sapkota et al. 2015; Hossain et al. 2021f). Abiotic stress in field crops is often classified as heat stress (high or low temperature); moisture stress (excess or less); and stress due to contamination of metals, metalloids, and non-metals (Meena et al. 2017; Moulick et al. 2016a, 2017; 2018a, b, c, 2019, 2023a, b, c; Hossain et al. 2021f; Moulick et al. 2022). In addition to reducing biotic pressures, agronomic management techniques including early planting, effective feeding and irrigation management, and the use of climate-smart technology can help increase wheat yield and production efficiency.

One of the most crucial elements that impacts the crop yield is the ideal sowing timing. Delays in sowing of the field crops beyond the optimal sowing window can negatively impact crop yield (Balwinder-Singh et al. 2016; Pal et al. 2017). Each site has an ideal planting date, and sowing before or after this time causes the yield to decrease (Sarkar et al. 2020a). Agrometeorological factors such as air temperature, solar radiation, and seasonal distribution of precipitation also play an important role in determining the yield of the field crops (Chen et al. 2013). Selecting the optimal crop sowing window is one of the most important agronomic measures to ensure safe seed germination, stable seedling formation, and yield maximisation (Shah et al. 2020). Sowing wheat beyond the optimal sowing window can result in suboptimal production, even in favourable weather conditions. Winter cereals should be sown as soon as possible to avoid poor seed germination, vegetative development, and reduced tillering capacity caused by harsh growing circumstances and altered crop microclimates (Dreccer et al. 2013; Shah et al. 2020). Shah et al. (2020) reported that the grain yield of winter wheat in China decreased by 1% because the optimum sowing date was shifted every day, which could be due to plant growth inhibition and yield-attributable traits. These gains might result from healthy photosynthesis that builds up in leaves and then spreads to economic parts (cereals) (Balwinder-Singh et al. 2015; Gou et al. 2017). In reality, it is quite difficult to suggest a standard sowing window for crops grown in the various wheat-growing regions of the world using the current cropping systems. Selecting the optimal sowing window for a site depends largely on agroclimatic conditions, existing cropping systems, and ultimately the ability of farmers to adapt to changing agronomic practices.

The source, timing, methods, and amount of phytonutrient application are the key factors behind the efficacy of phytonutrients in reducing abiotic stress in field crops. The wise utilisation of organic and inorganic plant nutrients is one of the most crucial elements of increasing abiotic stress resistance. Organic manures are more profitable than inorganic manures when exposed to high temperatures. Nitrogenous fertilisers have a lower volatilisation loss when administered in the organic form under high-temperature stress. Additionally, organic fertilisers boosted crop yields under conditions of temperature stress. A key agronomic input is straw mulching because it helps wheat plants that have been exposed to high temperatures by reducing moisture stress (Prasad 2005; Samui et al. 2020). When irrigation water is in short supply (Araya et al. 2019; Mondal et al. 2020a), organic plant nutrients can be used to increase wheat crop and water productivity. Additionally, it has been demonstrated in numerous papers that straw mulching reduces soil warmth, increases seed germination, and improves plant growth in field crops (Qin et al. 2020). Furthermore, it has been demonstrated that the majority of bulky organic manures contribute to moisture conservation in wheat under moisture stress. Organic fertilisers have been shown to boost the soil's capacity to retain water under water-stress conditions. Additionally, numerous studies have demonstrated that the use of organic manures in water-stressed conditions not only increases plant water use efficiency but also increases N availability in the root zone of wheat and potatoes (Wang et al. 2017; Brahmachari et al. 2019; Goswami et al. 2020). Nitrogenous fertilisers are critical for the development and productivity of all field crops. Effective nitrogen management has been shown to possibly decrease drought stress in wheat by preserving several metabolic processes (Agami et al. 2018; Hossain et al. 2021a). Additionally, an adequate supply of nitrogenous fertilisers might stimulate plant development, increase water use efficiency (WUE), and therefore mitigate the harmful effects of drought stress (Hossain et al. 2021a, b, d). According to reports, proper use of inorganic nitrogen fertilisers and irrigation water significantly boosts wheat output and WUE while also reducing the effects of drought stress (Zhao et al. 2020; Hossain et al. 2021c).

In the crop production system, managing irrigation is essential for reducing osmotic stress brought on by heat stress (Rashid et al. 2009; Rana et al. 2017; Bell et al. 2018). Irrigation water application is critical not only for increasing output but also for reducing numerous abiotic challenges. Field crop reproductive and grain production stages are typically when moisture stress is most likely to occur. Plant root systems are critical for controlling moisture and nutrient absorption patterns; partitioning of photosynthates is highly dependent on the interaction between the soil moisture and the root system. Modified irrigation water delivery techniques, such as ridge-furrow planting in conjunction with plastic film mulching, greatly enhance moisture and nutrient absorption, resulting in less drought stress (Goswami et al. 2020; Aryal et al. 2020). Efficient nutrient management, especially external nitrogenous fertiliser supply, is inextricably linked to irrigation management in order to control crop stress in relation to wheat output. Thus, optimising irrigation and nitrogen management may considerably increase photosynthetic capacity and production potential in wheat while also assisting in the reduction of various abiotic stressors (Wang et al. 2017; Hossain et al. 2021d). Additionally, by implementing a drip irrigation system, moisture stress was alleviated and WUE was improved in wheat crops grown under rainfed conditions (Samui et al. 2020; Mondal et al. 2020b). By regulating photosynthetic efficacy and antioxidant capacity, exogenous applications of osmotic balancing solutions such as glycine betaine and growth regulators such as methyl jasmonate, abscisic acid, and salicylic acid can successfully reduce WUE and drought stress in winter wheat under restricted irrigation conditions (Agarwal et al. 2005; Javadipour et al. 2019).

Seed priming and dipping of seedlings in various osmoregulatory and bioactive substitutes are one of the important agronomic interventions to mitigate the stress due to contamination of metals, metalloids, and non-metals substances (Wei et al. 2017; Saha et al. 2019; Sau et al. 2019; Yadav et al. 2020). It was reported that efficient seed priming techniques can successfully mitigate the negative impact of the metals, metalloids, and non-metals' contamination as well as play a crucial role to improve the growth and yield of the field crops (Moulick et al. 2016b, 2018a, b, c, 2021, 2023b, c Saha et al. 2019; Sahoo et al. 2019).

Seeds immersed in selenium (Se) solutions were shown to boost a variety of metabolic activities. When it came to rapid radical emergence in many crops, Se-soaked seeds outperformed unprimed seeds. Until now, seed priming technique has been effective in conferring resistance to a number of environmental challenges on a broad variety of crops (Moulick et al. 2021). Moulick et al. (2016a, b) published the first evidence of successfully using seed priming method to minimise As-induced phytotoxicity in rice. Their findings showed that priming rice seeds with Se not only significantly reduces As content when compared to unprimed seeds grown under similar As-stress conditions, but also mitigates As-induced phytotoxicity during germination and early seedling development.

6.3 Application of Nanotechnologies

Nanotechnology is seen as a developing area for overcoming abiotic stressors in various field crops, ensuring production, and addressing contemporary food security and food system concern (Hossain et al. 2021g; Moulick et al. 2023c; Tyagi et al. 2023). Nanotechnology has the ability to revolutionise agriculture by detecting and measuring soil fertility, soil moisture content, nutritional levels in plants, as well as temperature and disease pests in field crops (Hossain et al. 2021g). Additionally, these nanotools aid in monitoring crop development in real time and provide critical data for precision farming. Nanosensors and other sensing technologies provide vital information on the best times for planting and harvesting wheat as well as the best times to apply various agrochemicals (Panda et al. 2003; Hossain et al. 2021e). Numerous studies have demonstrated the crucial role that nanotechnology plays in reducing the negative impacts of abiotic stress on field crops (Liu et al. 2017; Hossain et al. 2021e, g). Additionally, in cereals, the multi-walled carbon nanotube (MWCNT) significantly influences seed germination and plant growth regulation (Lipşa et al. 2020). Globally, crop output is significantly impacted by drought, which also significantly reduces grain yield. It has been demonstrated that adding analcite nanoparticles to soil in drought-prone areas boosted wheat germination and growth development characteristics (Khan et al. 2017; Gutiérrez et al. 2017; Mahanta et al. 2023).

Silicon (Si) is regarded as a helpful element in plants that are subjected to abiotic stress. Si is not regarded as an essential nutrient in plants, but its high rate of accumulation in monocots through the root system encourages the ability to withstand water stress by raising root hydraulic conductivity (Hossain et al. 2021e, f). Salinity is a well-known environmental danger to wheat production. Salinity has been shown to have a variety of effects on seedlings of field crops, ranging from decreased germination percentages to uneven absorption of various nutrients owing to root development inhibition (Sarangi et al. 2015; Mondal et al. 2021a, b). It is found that soil application of green copper nanoparticles NPs (25 and 50 mg kg^{-1}) substantially reduces the oxidative stress Cr translocation while considerably increasing winter wheat plant development and productivity (Broadley et al. 2010; Hossain et al. 2021g). However, in wheat cultivated on salinity-stressed soil, the use of NPs improved both plant growth and germination performance (Panda et al. 2003; Hossain et al. 2021f). By modifying antioxidant enzymatic activity, seed priming with Ag NPs has been found to help wheat that is suffering from salt stress. Additionally, it is noted that varying dosages of Ag NPs may alleviate a variety of symptoms associated with salt stress in wheat (Hossain et al. 2021f).

6.4 Use of IoTs and Modern Agro-Tools

Hybrid intelligence strategies based on the Internet of Things (IoT) assist in identifying, detecting, and forecasting plant stress in agricultural production (Das et al. 2019). It conducts a comparative analysis of several machine learning (ML) approaches for observing and analysing various plant stress data (Galieni et al. 2020). Qualitative approaches, such as fluorescence, thermography, and VIS/NIR reflectance, allow a non-invasive assessment of the stressors' influence on plants, even over wide areas in a short period of time. Modern remote sensing methods such as satellite imaging and microwave remote sensors are increasingly being utilised to measure the drought stress of field crops (Olivera-Guerra et al. 2020; Ghosh et al. 2021). Abdulridha et al. (2019) effectively used hyperspectral imaging and machine learning to identify citrus canker disease using a UAV-based remote sensing approach. According to Puengsungwan and Jiraserccamomkul (2018), the chlorophyll fluorescence (ChF) approach may be an effective instrument for determining the leaf stress characteristics of lettuce.

Modern crop simulation models, such as APSIM and DSSAT, may also be useful decision support tools for water scheduling in rainfed and irrigated wheat farming systems (Gaydon et al. 2017; Sarkar et al. 2020b, 2022). The most effective cropping window for maximising yield while minimising abiotic stress in cereals may be predicted using crop simulation models that assess the complex connections between the plant-soil-atmosphere continuum and farmers' management practices (Zhao et al. 2014; Sarkar et al. 2020a). The diverging rice-wheat cropping system in the Indo-Gangetic Plains may benefit from the use of APSIM, a dynamic cropping system simulation model (Balwinder-Singh et al. 2015; Sarkar et al. 2022). Additionally, utilising the APSIM Oryza and wheat models (Bai and Tao 2017), was able to simulate a rice-wheat rotation system successfully in order to maximise crop cultivar and sowing time, water and nitrogen usage efficiency, and environmental effect.

7 Conclusion and Prospects

Over the past few decades, research trends regarding abiotic stress have been oriented around the identification, documentation (biomonitoring), and, to some extent, modeling the adverse impact of abiotic stress, often focusing on a single stressor. On the other hand, in applied research domain, the major focus was on developing suitable mitigation strategies, considering a single stressor. In reality, field condition crops usually encounter multiple stress (biotic and/or abiotic) in different combinations. Obviously, there exists a significant knowledge gap in terms of crop's response to multiple stress doses. In the era of drastic climatic fluctuations operating globally, consequences of climate change along with emerging stressors should be addressed with priority. Policymaking bodies should formulate long-term mitigation options and encourage strategic research initiatives to minimise the gap between the basic research and the applied research.

References

Abdulridha J, Batuman O, Ampatzidis Y (2019) UAV-based remote sensing technique to detect citrus canker disease utilizing hyperspectral imaging and machine learning. Remote Sens 11(11):1373

Abolghassem E, Ding Y, Farzad M, Xie Y (2018) Antioxidant response of bamboo (*Indocalamus latifolius*) as affected by heavy metal stress. J Elem 23:341–352

Agami RA, Alamri SAM, Abd El-mageed TA, Abousekken MSM, Hashem M (2018) Role of exogenous nitrogen supply in alleviating the deficit irrigation stress in wheat plants. Agric Water Manag 210:261–270

Agarwal S, Sairam RK, Srivastava GC, Meena RC (2005) Changes in antioxidant enzymes activity and oxidative stress by abscisic acid and salicylic acid in wheat genotypes. Biologia Plant 49(4):541–550

Ahadiyat YR, Hidayat P, Susanto U (2014) Drought tolerance, phosphorus efficiency and yield characters of upland rice lines. Emirates J Food Agric 26(1):25–34

Ahmad P, Jaleel CA, Azooz MM, Nabi G (2009) Generation of ROS and non-enzymatic antioxidants during abiotic stress in plants. Bot Res Int 2(1):11–20

Ahmed M, Hasanuzzaman M, Raza MA, Malik A, Ahmad S (2020) Plant nutrients for crop growth, development and stress tolerance. In: Roychowdhury R, Choudhury S, Hasanuzzaman M, Srivastava S (eds) Sustainable agriculture in the era of climate change. Springer, Cham, pp 43–92. https://doi.org/10.1007/978-3-030-45669-6_3

Alharbi OML, Basheer AA, Khattab RA, Ali I (2018) Health and environmental effects of persistent organic pollutants. J Mol Liq 263:442–453

Ali I, Jain CK (2004) Advances in arsenic speciation techniques. Int J Environ Anal Chem 84(12):947–964

Ali I, Al-Othman ZA, Sanagi MM (2015) Green synthesis of iron Nano-impregnated adsorbent for fast removal of fluoride from water. J Mol Liq 211:457–465

Alqudah AM, Samarah NH, Mullen RE (2011) Drought stress effect on crop pollination, seed set, yield and quality. In: Lichtfouse E (ed) Alternative farming systems, biotechnology, drought stress and ecological fertilization. Springer, Cham, pp 193–213

Amini H, Arzani A, Karami M (2014) Effect of water deficiency on seed quality and physiological traits of different safflower genotypes. Turk J Biol 38(2):271–282

Anbessa Y, Juskiw P, Good A, Nyachiro J, Helm J (2010) Selection efficiency across environments in improvement of barley yield for moderately low nitrogen environments. Crop Sci 50(2):451–457

Angadi SV, Cutforth HW, Miller PR, McConkey BG, Entz MH, Brandt SA, Volkmar KM (2000) Response of three Brassica species to high temperature stress during reproductive growth. Can J Plant Sci 80(4):693–701

Antolín MC, Hekneby M, Sánchez-Díaz M (2005) Contrasting responses of photosynthesis at low temperatures in different annual legume species. Photosynthetica 43:65–74

Araújo SS, Beebe S, Crespi M, Delbreil B, González EM, Gruber V, Lejeune-Henaut I, Link W, Monteros MJ, Prats E & Rao I (2015) Abiotic stress responses in legumes: strategies used to cope with environmental challenges. Crit Rev Plant Sci 34(1–3):237–280

Araus JL, Slafer GA, Reynolds MP, Royo C (2002) Plant breeding and drought in C3 cereals: what should we breed for? Ann Bot 89(7):925–940

Araus JL, Slafer GA, Royo C, Serret MD (2008) Breeding for yield potential and stress adaptation in cereals. Crit Rev Plant Sci 27(6):377–412

Araya A, Prasad PVV, Gowda PH, Afewerk A, Abadi B, Foster AJ (2019) Modeling irrigation and nitrogen management of wheat in northern Ethiopia. Agric Water Manag 216:264–272
Archana NP, Verma P (2017) Boron deficiency and toxicity and their tolerance in plants: a review. J Glob Biosci 6:4958–4965
Aryal JP, Rahut DB, Sapkota TB, Khurana R, Khatri-Chhetri A (2020) Climate change mitigation options among farmers in South Asia. Environ Dev Sustain 22(4):3267–3289
Asati A, Pichhode M, Nikhil K (2016) Effect of heavy metals on plants: an overview. Int J App Inno Eng Manag 5:56–66
Ashraf M, Ozturk M, Ahmad MSA (2010) Plant adaptation and phytoremediation. Springer, Cham, p 481
Azam F, Chang X, Jing R (2015) Mapping QTL for chlorophyll fluorescence kinetics parameters at seedling stage as indicators of heat tolerance in wheat. Euphytica 202(2):245–258
Bai H, Tao F (2017) Sustainable intensification options to improve yield potential and eco-efficiency for rice-wheat rotation system in China. Field Crops Res 211:89–105
Baldini M, Giovanardi R, Vannozzi G (2000) Effect of different water availability on fatty acid composition of the oil in standard and high oleic sunflower hybrids. In: Proceedings of the 15th international Sunflower conference, A
Balwinder-Singh HE, Sudhir-Yadav GDSS, Gaydon DS (2015) Options for increasing the productivity of the rice–wheat system of north-West India while reducing groundwater depletion. Part 1. Rice variety duration, sowing date and inclusion of mungbean. Field Crops Res 173:68–80
Balwinder-Singh HE, Gaydon DS, Eberbach PL (2016) Evaluation of the effects of mulch on optimum sowing date and irrigation management of zero till wheat in Central Punjab, India using APSIM. Field Crops Res 197:83–96
Banerjee A, Roychoudhury A (2017) Abiotic stress, generation of reactive oxygen species, and their consequences. In: Singh VP, Singh S, Tripathi DK, Prasad SM, Chauhan DK (eds) Reactive oxygen species in plants. https://doi.org/10.1002/9781119324928.ch2
Barceló J, Poschenrieder C (2002) Fast root growth responses, root exudates, and internal detoxification as clues to the mechanisms of aluminium toxicity and resistance: a review. Environ Exp Bot 48(1):75–92
Barnabás B, Jäger K, Fehér A (2008) The effect of drought and heat stress on reproductive processes in cereals. Plant Cell Environ 31(1):11–38
Basheer AA (2018) Chemical chiral pollution: impact on the society and science and need of the regulations in the 21st century. Chirality 30:402–406. https://doi.org/10.1002/chir.22808
Bell R, Haque M, Jahiruddin M, Rahman M, Begum M, Miah M, Islam M, Hossen M, Salahin N et al (2018) Conservation agriculture for rice-based intensive cropping by smallholders in the eastern gangetic plain. Agriculture 9(1):5
Bemis SM, Torii KU (2007) Autonomy of cell proliferation and developmental programs during Arabidopsis aboveground organ morphogenesis. Dev Biol 304(1):367–381
Benjamin JG, Nielsen DC (2006) Water deficit effects on root distribution of soybean, field pea and chickpea. Field Crops Res 97(2–3):248–253
Bennett MD, Finch RA, Smith JB, Rao MK (1973) The time and duration of female meiosis in wheat, rye and barley. Proc R Soc Lond B 183(1072):301–319
Bergman JW, Riveland NR, Flynn CR, Carlson GR, Wichman DM, Kephart KD (2006) Registration of "Montola 2004" safflower. Crop Sci 46(4):1818–1819. https://doi.org/10.2135/cropsci2005.12-0494
Borlaug Institute for South Asia (2015) Major accomplishments 2012–2014, BISA report series 1. Borlaug Institute for South Asia, New Delhi, p 38
Botella MA, Rosado A, Bressan R, Hasegawa PM (2007) Plant adaptive responses to salinity stress. In: Jenks MA, Hasegawa PM (eds) Plant abiotic stress. Blackwell Publishing, pp 37–70
Brahmachari K, Sarkar S, Sahoo D, Samui I, Nanda MK, Ghosh A, Mainuddin M (2019) Foliar nutrient management on potato grown under zero tillage and mulching in coastal saline soil of West Bengal, India. Proceedings 36(1):8

Broadley MR, Alcock J, Alford J et al (2010) Selenium biofortification of high-yielding winter wheat (*Triticum aestivum* L.) by liquid or granular Se fertilisation. Plant Soil 332:5–18. https://doi.org/10.1007/s11104-009-0234-4

Browse J, Xin Z (2001) Temperature sensing and cold acclimation. Curr Opin Plant Biol 4(3):241–246

Burakova EA, Dyachkova TP, Rukhov AV, Tugolukov EN, Galunin EV, Tkachev AG, Basheer AA, Ali I (2018) Novel and economic method of carbon nanotubes synthesis on a nickel magnesium oxide catalyst using microwave radiation. J Mol Liq 253:340–346

Camacho-Cristóbal JJ, Rexach J, González-Fontes A (2008) Boron in plants: deficiency and toxicity. J Integr Plant Biol 50(10):1247–1255

Carrera C, Martínez MJ, Dardanelli J, Balzarini M (2009) Water deficit effect on the relationship between temperature during the seed fill period and soybean seed oil and protein concentrations. Crop Sci 49(3):990–998

Carter CE, Floyd JM (1975) Inhibition of sugarcane yields by high water table during dormant season. In: Proceedings of the-American society of sugar cane technologists United States, vol 4. Proceedings – American Society of Sugar Cane Technologists (USA), pp 14–18

Cary JW (1975) Factors affecting cold injury of sugarbeet seedlings 1 (*Beta vulgaris* L.). Agron J 67(2):258–262

Chakrabarti B, Singh SD, Kumar V. et al. (2013) Growth and yield response of wheat and chickpea crops under high temperature. Ind J Plant Physiol 18:7–14. https://doi.org/10.1007/s40502-013-0002-6

Chalker-Scott L (2002) Do anthocyanins function as osmoregulators in leaf tissues? Adv Bot Res 37:103–127

Chauhan J, Choudhury P, Singh K (2021) Production, varietal improvement programme and seed availability of annual oilseeds in India: current scenario and future prospects. Indian Society of Oilseeds Research, pp 1–18

Chen C, Huang J, Zhu L, Shah F, Nie L, Cui K, Peng S (2013) Varietal difference in the response of rice chalkiness to temperature during ripening phase across different sowing dates. Field Crops Res 151:85–91

Cheng YE, Dong MY, Fan XW, Nong LL, Li YZ (2018) A study on cassava tolerance to and growth responses under salt stress. Environ Exp Bot 155:429–440

Cho J, Oki T (2012) Application of temperature, water stress, CO_2 in rice growth models. Rice (N Y) 5(1):10

Choudhary S, Rani M, Devika OS, Patra A, Singh RK, Prasad SK (2019) Impact of fluoride on agriculture: a review on it's sources, toxicity in plants and mitigation strategies. Int J Chem Stud 7:1675–1680

Choudhury S, Moulick D (2022) Response of field crops to abiotic stress: current status and future prospects, 1st edn. CRC Press. https://doi.org/10.1201/9781003258063

Choudhury S, Moulick D, Mazumder MK (2021a) Secondary metabolites protect against metal and metalloid stress in rice: an in silico investigation using dehydroascorbate reductase. Acta Physiol Planta 43(1):1–10. https://doi.org/10.1007/s1173-020-03173-2

Choudhury S, Sharma P, Moulick D, Mazumder MK (2021b) Unrevealing metabolomics for abiotic stress adaptation and tolerance in plants. J Crop Sci Biotechnol 24(5):479–493. https://doi.org/10.1007/s12892-021-00102-8

Choudhury S, Mazumder MK, Moulick D, Sharma P, Tata SK, Ghosh D, Ali HM, Siddiqui MH, Brestic M et al (2022a) A computational study of the role of secondary metabolites for mitigation of acid soil stress in cereals using dehydroascorbate and mono-Dehydroascorbate reductases. Antioxidants (Basel) 11(3):458. https://doi.org/10.3390/antiox11030458

Choudhury S, Moulick D, Mazumder MK, Pattnaik BK, Ghosh D, Vemireddy LR, Aldhahrani A, Soliman MM, Gaber A, Hossain A (2022b) An in vitro and in silico perspective study of seed priming with zinc on the phytotoxicity and accumulation pattern of arsenic in rice seedlings. Antioxidants (Basel) 11(8):1500. https://doi.org/10.3390/antiox11081500

Choudhury S, Moulick D, Ghosh D, Soliman M, Alkhedaide A, Gaber A, Hossain A (2022c) Drought-induced oxidative stress in pearl millet (*Cenchrus americanus* L.) at seedling stage: survival mechanisms through alteration of morphophysiological and antioxidants activity. Life 12(8):1171. https://doi.org/10.3390/life12081171

Chourasia KN, Lal MK, Tiwari RK, Dev D, Kardile HB, Patil VU, Kumar A, Vanishree G, Kumar D et al (2021) Salinity stress in potato: understanding physiological, biochemical and molecular responses. Life 11(6):545

Chowardhara B, Borgohain P, Saha B, Awasthi JP, Moulick D, Panda SK (2019a) Assessment of phytotoxicity of zinc on Indian mustard (*Brassica juncea*) varieties during germination and early seedling growth. Ann Plant Soil Res 21(3):239–244

Chowardhara B, Borgohain P, Saha B, Awasthi JP, Moulick D, Panda SK (2019b) Phytotoxicity of Cd and Zn on three popular Indian mustard varieties during germination and early seedling growth. Biocatal Agric Biotechnol 21:101349. https://doi.org/10.1016/j.bcab.2019.101349

Cormier F, Faure S, Dubreuil P, Heumez E, Beauchêne K, Lafarge S, Praud S, Le Gouis J (2013) A multi-environmental study of recent breeding progress on nitrogen use efficiency in wheat (*Triticum aestivum* L.). Theor Appl Genet 126(12):3035–3048

Cosgrove DJ (2000) Expansive growth of plant cell walls. Plant Physiol Biochem 38(1–2):109–124

Costa JH, Roque ALM, Aziz S, Dos Santos CP, Germano TA, Batista MC, Thiers KLL, da Cruz Saraiva KD, Arnholdt-Schmitt B (2021) Genome-wide identification of ascorbate-glutathione cycle gene families in soybean (*Glycine max*) reveals gene duplication events and specificity of gene members linked to development and stress conditions. Int J Biol Macromol 187:528–543. https://doi.org/10.1016/j.ijbiomac.2021.07.103

Dambrosio N, Arena C, Desanto AV (2006) Temperature response of photosynthesis, excitation energy dissipation and alternative electron sinks to carbon assimilation in *Beta vulgaris* L. Environ Exp Bot 55(3):248–257

Dar MH, Bano DA, Waza SA, Zaidi NW, Majid A, Shikari AB, Ahangar MA, Hossain M, Kumar A, Singh US (2021) Abiotic stress tolerance-progress and pathways of sustainable rice production. Sustainability 13(4):2078

Das R, Banerjee M, De S (2019) Emerging trends in disruptive technology management for sustainable development. Chapman & Hall, p 138

Dehghani MH, Sanaei D, Ali I, Bhatnagar A (2016) Removal of chromium(VI) from aqueous solution using treated waste newspaper as a low-cost adsorbent: Kinetic modeling and isotherm studies. J Mol Liq 215:671–679

Dell B, Huang L (1997) Physiological response of plants to low boron. Plant Soil 193(2):103–120. https://doi.org/10.1023/A:1004264009230

Demİrel U, Çalişkan S, Yavuz C, Tindaş İ, Polgar Z, Vaszily Z, Cernák I, Çalişkan ME (2017) Assessment of morphophysiological traits for selection of heat-tolerant potato genotypes. Turk J Agric For 41:218–232

Deol JS, Shivanna KR, Prakash S, Banga SS, Robbelen G (2003) Enarthrocarpus lyratus-based cytoplasmic male sterility and fertility restorer system in *Brassica rapa*. Plant Breed 122(5):438–440

Desclaux D, Huynh TT, Roumet P (2000) Identification of soybean plant characteristics that indicate the timing of drought stress. Crop Sci 40(3):716–722

Di Caterina R, Giuliani MM, Rotunno T, De Caro A, Flagella Z (2007) Influence of salt stress on seed yield and oil quality of two sunflower hybrids. Ann Appl Biol 151(2):145–154

Dolferus R, Ji X, Richards RA (2011) Abiotic stress and control of grain number in cereals. Plant Sci 181(4):331–341

Dreccer MF, Chapman SC, Rattey AR, Neal J, Song Y, Christopher JJ, Reynolds M (2013) Developmental and growth controls of tillering and water-soluble carbohydrate accumulation in contrasting wheat (*Triticum aestivum* L.) genotypes: can we dissect them? J Exp Bot 64(1):143–160

Dwivedi SL, Nigam SN, Rao RCN, Singh U, Rao KVS (1996) Effect of drought on oil, fatty acids and protein contents of groundnut (*Arachis hypogaea* L.) seeds. Field Crops Res 48(2–3):125–133

El Sabagh A, Hossain A, Barutçular C, Gormus O, Ahmad Z, Hussain S, Islam MS, Alharby H, Bamagoos A et al (2019) Effects of drought stress on the quality of major oilseed crops: implications and possible mitigation strategies—a review. Appl Ecol Environ Res 17(2):4019–4043

Elrashidi MA, Persaud N, Baligar VC (1998) Effect of fluoride and phosphate on yield and mineral composition of barley grown on three soils. Commun Soil Sci Plant Anal 29(3–4):269–283

Fan J, Chen K, Xu J, Abm K, Chen Y, Chen L, Yan X (2022) Physiological effects induced by aluminium and fluoride stress in tall fescue (*Festuca arundinacea* Schreb). Ecotoxicol Environ Saf 231:113192

Fang S, Hou X, Liang X (2021) Response mechanisms of plants under saline-alkali stress. Front Plant Sci 12:667458

Faried HN, Ayyub CM, Amjad M, Ahmed R (2016) Salinity impairs ionic, physiological and biochemical attributes in potato. PAKJAS 53(1):17–25

Feist AM, Palsson BØ (2008) The growing scope of applications of genome-scale metabolic reconstructions using *Escherichia coli*. Nat Biotechnol 26(6):659–667

Fernández-Cuesta Á, Velasco L, Ruiz-Méndez MV (2014) Novel safflower oil with high γ-tocopherol content has a high oxidative stability. Eur J Lipid Sci Technol 116(7):832–836

Fghire R, Anaya F, Ali OI, Benlhabib O, Ragab R, Wahbi S (2015) Physiological and photosynthetic response of quinoa to drought stress. Chilean J Agric Res 75(2):174–183

Flora SJS, Mittal M, Mehta A (2008) Heavy metal-induced oxidative stress and its possible reversal by chelation therapy. Indian J Med Res 128(4):501–523

Flowers TJ (2004) Improving crop salt tolerance. J Exp Bot 55(396):307–319

Foyer CH, Noctor G (2011) Ascorbate and glutathione: the heart of the redox hub. Plant Physiol 155(1):2–18. https://doi.org/10.1104/pp.110.167569

Galieni A, D'Ascenzo N, Stagnari F, Pagnani G, Xie Q, Pisante M (2020) Past and future of plant stress detection: an overview from remote sensing to positron emission tomography. Front Plant Sci 11:609155

Garg N, Singla P (2011) Arsenic toxicity in crop plants: physiological effects and tolerance mechanisms. Environ Chem Lett 9(3):303–321

Gaydon DS, Balwinder-Singh WE, Wang E, Poulton PL, Ahmad B, Ahmed F, Akhter S, Ali I, Amarasingha R et al (2017) Evaluation of the APSIM model in cropping systems of Asia. Field Crops Res 204:52–75

Ghori N-H, Ghori T, Hayat MQ, Imadi SR, Gul A, Altay V, Ozturk M (2019) Heavy metal stress and responses in plants. Int J Environ Sci Technol 16(3):1807–1828

Ghosh A, Nanda MK, Sarkar D, Sarkar S, Brahmachari K, Ray K (2021) Assessing the agroclimatic potentiality in Indian Sundarbans for crop planning by analyzing rainfall time series data. J Agrometeorol 23(1):113–121

Gill SS, Tuteja N (2010) Reactive oxygen species and antioxidant machinery in abiotic stress tolerance in crop plants. Plant Physiol Biochem 48(12):909–930. https://doi.org/10.1016/j.plaphy.2010.08.016

Gomathi R, Chandran K (2009) Effect of water logging on growth and yield of sugarcane clones. Quarterly news letter, 294: ICAR-Sugarcane Breeding Institute, Laknow, India, pp 1–2

Gomathi R, Thandapani TV (2005) Salt stress in relation to nutrient accumulation and quality of sugarcane genotypes. Sugar Tech 7(1):39–47

Goswami S, Mondal R, Puste AM (2020) Influence of irrigation and tillage management on growth, yield and water-use efficiency of wheat (*Triticum aestivum*) in Gangetic plains in West Bengal. Indian J Agron 65:47–52

Gou F, van Ittersum MK, van der Werf W (2017) Simulating potential growth in a relay-strip intercropping system: model description, calibration and testing. Field Crops Res 200:122–142

Grashoff C (1990) Effect of pattern of water supply on *Vicia faba* L. 1. Dry matter partitioning and yield variability. Neth J Agric Sci 38(1):21–44

Guo J, Dai X, Xu W, Ma M (2008) Overexpressing *GSH1* and *AsPCS1* simultaneously increases the tolerance and accumulation of cadmium and arsenic in *Arabidopsis thaliana*. Chemosphere 72(7):1020–1026

Gupta UC, Gupta SC (1998) Trace element toxicity relationships to crop production and livestock and human health: implications for management. Commun Soil Sci Plant Anal 29(11–14):1491–1522

Gutiérrez TJ, Ponce AG, Alvarez VA (2017) Nano-clays from natural and modified montmorillonite with and without added blueberry extract for active and intelligent food nanopackaging materials. Mater Chem Phys 194:283–292

Haidouti C, Chronopoulou A, Chronopoulos J (1993) Effects of fluoride emissions from industry on the fluoride concentration of soils and vegetation. Biochem Syst Ecol 21(2):195–208

Hakeem KR, Sabir M, Ozturk M, Mermut A (2015) Soil remediation and plants: prospects and challenges. Elsevier, p 724

Hall JL (2002) Cellular mechanisms for heavy metal detoxification and tolerance. J Exp Bot 53(366):1–11

Halliwell B (2006) Oxidative stress and neurodegeneration: where are we now? J Neurochem 97(6):1634–1658. https://doi.org/10.1111/j.1471-4159.2006.03907.x

Halliwell B, Gutteridge JM (2015) Free radicals in biology and medicine. Oxford University Press

Hasanuzzaman M, Nahar K, Alam MM, Bhowmik PC, Hossain MA, Rahman MM, Prasad MN, Ozturk M, Fujita M (2014) Potential use of halophytes to remediate saline soils. Biomed Res Int 2014:589341

Hasanuzzaman M, Bhuyan MHMB, Anee TI, Parvin K, Nahar K, Mahmud JA, Fujita M (2019) Regulation of ascorbate-glutathione pathway in mitigating oxidative damage in plants under abiotic stress. Antioxidants 8(9):384. https://doi.org/10.3390/antiox8090384

Hassan FSC, Solouki M, Fakheri BA, Nezhad NM, Masoudi B (2018) Mapping QTLs for physiological and biochemical traits related to grain yield under control and terminal heat stress conditions in bread wheat (*Triticum aestivum* L.). Physiol Mol Biol Plants 24(6):1231–1243

Hidangmayum A, Singh A, Kumar V, Dwivedi P (2018) Abiotic stress responses in cereals and pulses crop and their agronomic practices to enhance tolerance. Eurasian J Biosci 12(2):487–493

Hiscox JD, Israelstam GF (1979) A method for the extraction of chlorophyll from leaf tissue without maceration. Can J Bot 57(12):1332–1334

Hossain S (2019) Present scenario of global salt affected soils, its management and importance of salinity research. Int Res J Biol Sci 1(1):1–3

Hossain MA, Piyatida P, da Silva JAT, Fujita M (2012) Molecular mechanism of heavy metal toxicity and tolerance in plants: central role of glutathione in detoxification of reactive oxygen species and methylglyoxal and in heavy metal chelation. J Bot 2012:1–37

Hossain A, Bhatt R, Sarkar S (2021a) Cost-effective and eco-friendly agricultural technologies in rice-wheat cropping systems for food and environmental security. In: Jhariya MK, Banerjee A, Meena RS, Kumar S, Raj A (eds) Sustainable intensification for agroecosystem services and management. Springer, Cham, pp 69–96

Hossain A, Mottaleb KA, Maitra S et al (2021b) Conservation agriculture improves soil health: major research findings from Bangladesh. In: Jayaraman S, Dalal RC, Patra AK, Chaudhari SK (eds) Conservation agriculture: a sustainable approach for soil health and food security. Springer, Singapore. https://doi.org/10.1007/978-981-16-0827-8_26

Hossain A, Sarkar S, Rahman MA (2021c) Ecological intensification for sustainable agriculture in South Asia. In: Jhariya MK, Meena RS, Banerjee A (eds) Ecological intensification of natural resources for sustainable agriculture. Springer, Cham, pp 171–213

Hossain A, Skalicky M, Brestic M, Maitra S, Ashraful Alam M, Syed MA, Hossain J, Sarkar S, Saha S et al (2021d) Consequences and mitigation strategies of abiotic stresses in wheat (*Triticum aestivum* L.) under the changing climate. Agronomy 11(2):241

Hossain A, Ahmed S, Ahmad Z, Alam MJ, Moulick D, Saha B, Samanta S, Maitra S, Bhadra P, Bukhari MA, Aftab T (2021e) Next-generation genetic engineering tools for abiotic stress tolerance in plants. In: Aftab T, Hakeem KR (eds) Frontiers in plant-soil interaction. Academic, pp 153–197. https://doi.org/10.1016/B978-0-323-90943-3.00008-0

Hossain A, Ahmad Z, Moulik D, Maitra S, Bhadra P, Ahmad A, Garai S, Mondal M, Roy A, Sabagh AE, Aftab T (2021f) Jasmonates and salicylates: mechanisms, transport and signalling during abiotic stress in plants. Jasmonates and salicylates signaling in plants. Springer, Cham, pp 1–29. https://doi.org/10.1007/978-3-030-75805-9_1

Hossain A, Skalicky M, Brestic M, Mahari S, Kerry RG, Maitra S, Sarkar S, Saha S, Bhadra P et al (2021g) Application of nanomaterials to ensure quality and nutritional safety of food. J Nanomater 2021:1–19

Hossain A, Pramanick B, Bhutia KL, Ahmad Z, Moulick D, Maitra S, Ahmad A, Aftab T (2021h) Emerging roles of osmoprotectant glycine betaine against salt-induced oxidative stress in plants: a major outlook of maize (*Zea mays* L.). In: Aftab T, Hakeem KR (eds) Frontiers in plant-soil interaction. Academic, pp 567–587. https://doi.org/10.1016/B978-0-323-90943-3.00015-8

Hossain A, Maitra S, Pramanick B, Bhutia KL, Ahmad Z, Moulik D, Syed MA, Shankar T, Adeel M, Hassan MM, Aftab T (2022) Wild relatives of plants as sources for the development of abiotic stress tolerance in plants. In: Aftab T, Hakeem KR (eds) Plant perspectives to global climate changes. Academic, pp 471–518. https://doi.org/10.1016/B978-0-323-85665-2.00011-X

Hossain A, Maitra S, Sarker S, Mahmud AA, Ahmad Z, Emon RM, Vemuri H, Malek MA, Alam MA, Rahman MA, Alam MJ, Jahan N, Bhadra P, Moulick D, Saha S, Skalicky M, Brestic M (2023) Aluminium stress in plants: consequences and mitigation mechanisms. In: Tripathi DK, Pandey S, Singh VP, Chauhan DK, Sharma S (eds) Beneficial chemical elements of plants: recent developments and future prospects. Wiley, pp 1–46

Hussain M, Malik MA, Farooq M, Ashraf MY, Cheema MA (2008) Improving drought tolerance by exogenous application of glycinebetaine and salicylic acid in sunflower. J Agron Crop Sci 194(3):193–199

IPCC (2007) Summary for policymakers. In: Solomon S (ed) The physical science basis, contribution of working group I to the fourth assessment report of the intergovernmental panel on climate change. Cambridge University Press

Iqbal S, Waheed Z, Naseem A (2020) Nanotechnology and abiotic stresses. In: Javad S (ed) Nanoagronomy. Springer, Berlin, pp 37–52

Ishikawa T, Shabala S (2019) Control of xylem Na^+ loading and transport to the shoot in rice and barley as a determinant of differential salinity stress tolerance. Physiol Plant 165(3):619–631

Ishimaru T, Xaiyalath S, Nallathambi J, Sathishraj R, Yoshimoto M, Phoudalay L, Samson B, Hasegawa T, Hayashi K et al (2016) Quantifying rice spikelet sterility in potential heat-vulnerable regions: field surveys in Laos and southern India. Field Crops Res 190:3–9. https://doi.org/10.1016/j.fcr.2015.08.006

Jagadish SVK (2020) Heat stress during flowering in cereals – effects and adaptation strategies. New Phytol 226(6):1567–1572

Jamil M, Lee DB, Jung KY, Ashraf M, Lee SC, Rha ES (2006) Effects of salt (NaCl) stress on germination and early seedling growth of four vegetables species. J Cent Eur Agric 7(2):273–282

Javadipour Z, Balouchi H, Movahhedi Dehnavi M, Yadavi A (2019) Roles of methyl jasmonate in improving growth and yield of two varieties of bread wheat (*Triticum aestivum*) under different irrigation regimes. Agric Water Manag 222:336–345

Ji X, Shiran B, Wan J, Lewis DC, Jenkins CL, Condon AG, Richards RA, Dolferus R (2010) Importance of pre-anthesis anther sink strength for maintenance of grain number during reproductive stage water stress in wheat. Plant Cell Environ 33(6):926–942

Kaiwen G, Zisong X, Yuze H, Qi S, Yue W, Yanhui C, Jiechen W, Wei L, Huihui Z (2020) Effects of salt concentration, pH, and their interaction on plant growth, nutrient uptake, and photochemistry of alfalfa (*Medicago sativa*) leaves. Plant Signal Behav 15(12):1832373

Kang HM, Saltveit ME (2002) Chilling tolerance of maize, cucumber and rice seedling leaves and roots are differentially affected by salicylic acid. Physiol Plant 115(4):571–576

Kang JS, Singh H, Singh G, Kang H, Kalra VP, Kaur J (2017) Abiotic stress and its amelioration in cereals and pulses: a review. Int J Curr Microbiol Appl Sci 6(3):1019–1045

Kaya MD, Okçu G, Atak M, Cıkılı Y, Kolsarıcı Ö (2006) Seed treatments to overcome salt and drought stress during germination in sunflower (*Helianthus annuus* L.). Eur J Agron 24(4):291–295

Kazan K, Lyons R (2016) The link between flowering time and stress tolerance. J Exp Bot 67(1):47–60

Khan MN, Mobin M, Abbas ZK, AlMutairi KA, Siddiqui ZH (2017) Role of nanomaterials in plants under challenging environments. Plant Physiol Biochem 110:194–209

Khan N, Bano A, Rahman MA, Guo J, Kang Z, Babar MA (2019) Comparative physiological and metabolic analysis reveals a complex mechanism involved in drought tolerance in chickpea (*Cicer arietinum* L.) induced by PGPR and PGRs. Sci Rep 9(1):2097

Khavari-Nejad RA, Najafi F, Khavari-Nejad S (2008) Growth and some physiological parameters of four sugar beet (*Beta vulgaris* L.) cultivars as affected by salinity. Pak J Biol Sci 11(10):1390–1393

Kinraide TB (1997) Reconsidering the rhizotoxicity of hydroxyl, sulphate, and fluoride complexes of aluminium. J Exp Bot 48(5):1115–1124

Kitano H (2002) Systems biology: a brief overview. Science 295(5560):1662–1664

Krieger-Liszkay A, Fufezan C, Trebst A (2008) Singlet oxygen production in photosystem II and related protection mechanism. Photosynth Res 98(1–3):551–564. https://doi.org/10.1007/s11120-008-9349-3

Kumar K, Solanki S, Singh SN, Khan MA (2016) Abiotic constraints of pulse production in India. In: Biswas SK, Kumar S, Chand G (eds) Disease of pulse crops and their sustainable management. Biotech Books, New Delhi, pp 23–39

Levy D (1992) The response of potatoes (*Solunum tuberosum* L.). Ann Appl Biol 120(3):547–555

Lipşa F-D, Ursu E-L, Ursu C, Ulea E, Cazacu A (2020) Evaluation of the antifungal activity of gold–chitosan and carbon nanoparticles on *Fusarium oxysporum*. Agronomy 10(8):1143

Liu S, Lu Y, Yang C, Liu C, Ma L, Dang Z (2017) Effects of modified biochar on rhizosphere microecology of rice (*Oryza sativa* L.) grown in As-contaminated soil. Environ Sci Pollut Res Int 24(30):23815–23824

Lobell DB, Gourdji SM (2012) The influence of climate change on global crop productivity. Plant Physiol 160(4):1686–1697

Lopez-Delacalle M, Silva CJ, Mestre TC, Martinez V, Blanco-Ulate B, Rivero RM (2021) Synchronization of proline, ascorbate and oxidative stress pathways under the combination of salinity and heat in tomato plants. Environ Exp Bot 183:104351. https://doi.org/10.1016/j.envexpbot.2020.104351

Luo Q, Yu B, Liu Y (2005) Differential sensitivity to chloride and sodium ions in seedlings of *Glycine max* and *G. soja* under NaCl stress. J Plant Physiol 162(9):1003–1012

Majumdar A, Dubey PK, Giri B, Moulick D, Srivastava AK, Roychowdhury T, Bose S, Jaiswal, MK (2023) Combined effects of dry-wet irrigation, redox changes and microbial diversity on soil nutrient bioavailability in the rice field. Soil Tillage Res 232, p.105752. https://doi.org/10.1016/j.still.2023.105752

Mahanta S, Shree J, Santra SC, Moulick D, Hossain A (2023) Deciphering of mycogenic nanoparticles by spectroscopic methods. In Myconanotechnology and Application of Nanoparticles in Biology (pp. 93–117). Academic Press. https://doi.org/10.1016/B978-0-443-15262-7.00003-6

Mazumder MK, Moulick D, Choudhury S (2022a) Iron (Fe^{3+})-mediated redox responses and amelioration of oxidative stress in cadmium (Cd^{2+}) stressed mung bean seedlings: a biochemical and computational analysis. J Plant Biochem Biotechnol 31(1):49–60. https://doi.org/10.1007/s13562-021-00654-4

Mazumder MK, Sharma P, Moulick D, Tata SK, Choudhury S (2022b) Salicylic acid ameliorates zinc and chromium-induced stress responses in wheat seedlings: a biochemical and computational analysis. Cereal Res Commun 50(3):407–418. https://doi.org/10.1007/s42976-021-00201-w

McCune DC, Weinstein LH (1971) Metabolic effects of atmospheric fluorides on plants. Environ Pollut 1(3):169–174

Meena KK, Sorty AM, Bitla UM, Choudhary K, Gupta P, Pareek A, Singh DP, Prabha R, Sahu PK et al (2017) Abiotic stress responses and microbe-mediated mitigation in plants: the omics strategies. Front Plant Sci 8:172

Meharg AA, Hartley-Whitaker J (2002) Arsenic uptake and metabolism in arsenic resistant and nonresistant plant species. New Phytol 154(1):29–43

Mensah JK, Akomeah PA, Ikhajiagbe B, Ekpekurede EO (2006) Effects of salinity on germination, growth and yield of five groundnut genotypes. Afr J Biotechnol 5(20):1973–1979

Miller GW (1993) The effect of fluoride on higher plants with special emphasis on early physiological and biochemical disorders. Fluoride 26:3–22

Mirshekari M, Amiri R, Nezhad HI, Noori SS, Zandvakili OR (2012) Effects of planting date and water deficit on quantitative and qualitative traits of flax seed. Am Eurasian J Agric Environ Sci 12(7):901–913

Mishra AK, Erickson KW, Harris JM, Hallahan CB, Uematsu H (2010) Determinants of farm household income diversification in the United States: Evidence from farm-level data (No. 320-2016-9993)

Miteva E (2002) Accumulation and effect of arsenic on tomatoes. Commun Soil Sci Plant Anal 33(11–12):1917–1926

Mittler R, Finka A, Goloubinoff P (2012) How do plants feel the heat? Trends Biochem Sci 37(3):118–125. https://doi.org/10.1016/j.tibs.2011.11.007

Mohanpuria P, Rana NK, Yadav SK (2007) Cadmium-induced oxidative stress influence on glutathione metabolic genes of *Camellia sinensis* (L.). O. Kuntze. Environ Toxicol 22(4):368–374

Moller IM (2001) Plant mitochondria and oxidative stress: electron transport, NADPH turnover, and metabolism of reactive oxygen species. Annu Rev Plant Physiol Plant Mol Biol 52(1):561–591. https://doi.org/10.1146/annurev.arplant.52.1.561

Mondal M, Skalicky M, Garai S, Hossain A, Sarkar S, Banerjee H, Kundu R, Brestic M, Barutcular C et al (2020a) Supplementing nitrogen in combination with rhizobium inoculation and soil mulch in peanut (*Arachis hypogaea* L.) production system: part II. Effect on phenology, growth, yield attributes, pod quality, profitability and nitrogen use efficiency. Agronomy 10(10):1513

Mondal M, Skalicky M, Garai S, Hossain A, Sarkar S, Banerjee H, Kundu R, Brestic M, Barutcular C et al (2020b) Supplementing nitrogen in combination with rhizobium inoculation and soil mulch in peanut (*Arachis hypogaea* L.) production system: part I. effects on productivity, soil moisture, and nutrient dynamics. Agronomy 10(10):1582

Mondal M, Biswas B, Garai S, Adhikary S, Bandyopadhyay PK, Sarkar S, Banerjee H, Brahmachari K, Maitra S et al (2021a) Raising climate-resilient embolden rice (*Oryza sativa* L.) seedlings during the cool season through various types of nursery bed management. Sustainability 13(22):12910

Mondal M, Garai S, Banerjee H, Sarkar S, Kundu R (2021b) Mulching and nitrogen management in peanut cultivation: an evaluation of productivity, energy trade-off, carbon footprint and profitability. Energy Ecol Environ 6(2):133–147

Morillo-Velarde R, Ober ES (2006) Water use and irrigation. In: Draycott P (ed) Sugar beet. Blackwell Publishing, pp 50–255

Morrison MJ, Stewart DW (2002) Heat stress during flowering in summer *Brassica*. Crop Sci 42(3):797–803

Moulick D, Ghosh D, Santra SC (2016a) An assessment of some physicochemical properties and cooking characteristics of milled rice and associated health risk in two rice varieties of arsenic contaminated areas of West Bengal, India. Int J Agric Food Sci 6(2):44–55

Moulick D, Ghosh D, Chandra Santra S (2016b) Evaluation of effectiveness of seed priming with selenium in rice during germination under arsenic stress. Plant Physiol Biochem 109:571–578. https://doi.org/10.1016/j.plaphy.2016.11.004

Moulick D, Santra SC, Ghosh D (2017) Seed priming with Se alleviate as induced phytotoxicity during germination and seedling growth by restricting as translocation in rice (*Oryza sativa* L. c.v. IET-4094). Ecotoxicol Environ Saf 145:449–456. https://doi.org/10.1016/j.ecoenv.2017.07.060

Moulick D, Santra SC, Ghosh D (2018a) Effect of selenium induced seed priming on arsenic accumulation in rice plant and subsequent transmission in human food chain. Ecotoxicol Environ Saf 152:67–77. https://doi.org/10.1016/j.ecoenv.2018.01.037

Moulick D, Santra SC, Ghosh D (2018b) Rice seed priming with Se: a novel approach to mitigate as induced adverse consequences on growth, yield and as load in brown rice. J Hazard Mater 355:187–196. https://doi.org/10.1016/j.jhazmat.2018.05.017

Moulick D, Santra SC, Ghosh D (2018c) Seed priming with Se mitigates As-induced phytotoxicity in rice seedlings by enhancing essential micronutrient uptake and translocation and reducing as translocation. Environ Sci Pollut Res Int 25(27):26978–26991. https://doi.org/10.1007/s11356-018-2711-x

Moulick D, Santra SC, Ghosh D (2018d) Consequences of paddy cultivation in arsenic-contaminated paddy fields of lower Indo-Gangetic plain on arsenic accumulation pattern and selected grain quality traits: a preliminary assessment. In: Sharma P, Jha AB, Dubey RS (eds) Mechanisms of arsenic toxicity and tolerance in plants. Springer, Singapore, pp 49–78. https://doi.org/10.1007/978-981-13-1292-2_3

Moulick D, Chowardhara B, Panda SK (2019) Agroecotoxicological aspect of arsenic (As) and cadmium (Cd) on field crops and its mitigation: current status and future Prospect. In: Srivastava S, Srivastava A, Suprasanna P (eds) Plant-metal interactions. Springer, Cham, pp 217–246. https://doi.org/10.1007/978-3-030-20732-8_11

Moulick D, Samanta S, Saha B, Mazumder MK, Dogra S, Panigrahi K, Banerjee S, Ghosh D, Santra SC (2020a) Salinity stress responses in three popular field crops belonging to Fabaceae Family: current status and future prospect. In: Hasanuzzaman M, Araújo S, Gill SS (eds) The plant family Fabaceae. Springer, Singapore, pp 519–541

Moulick D, Sarkar S, Awasthi JP (2020b) Rice grain quality traits: neglected or less addressed? In: Rice research for quality improvement: genomics and genetic engineering. Springer Nature Singapore Pvt Ltd., pp 729–745

Moulick D, Samanta S, Sarkar S, Mukherjee A, Pattnaik BK, Saha S, Awasthi JP, Bhowmick S, Ghosh D et al (2021) Arsenic contamination, impact and mitigation strategies in rice agro-environment: an inclusive insight. Sci Total Environ 800:149477

Moulick D, Ghosh D, Skalicky M, Gharde Y, Mazumder MK, Choudhury S, Biswas JK, Santra SC, Brestic M et al (2022) Interrelationship among rice grain arsenic, micronutrients content and grain quality attributes: an investigation from genotype × environment perspective. Front Environ Sci 10. https://doi.org/10.3389/fenvs.2022.857629

Moulick D, Bhutia KL, Sarkar S, Roy A, Mishra UN, Pramanick B, Maitra S, Shankar T, Hazra S, Skalicky M, Brestic M (2023a) The intertwining of Zn-finger motifs and abiotic stress tolerance in plants: current status and future prospects. Front Plant Sci 13:1083960. https://doi.org/10.3389/fpls.2022.1083960

Moulick D, Ghosh D, Mandal J, Bhowmick S, Mondal D, Choudhury S, Santra SC, Vithanage M, Biswas JK (2023b) A cumulative assessment of plant growth stages and selenium supplementation on arsenic and micronutrients accumulation in rice grains. J Cleaner Product:135764. https://doi.org/10.1016/j.jclepro.2022.135764

Moulick D, Hazra S, Mukherjee A, Sinha S, Mahanta S, Das A, Saha B, Niazi NK, Biswas JK (2023c) Application of nanotechnology in mitigating arsenic stress and accumulation in crops: where we are and where we are moving towards. In: Niazi NK, Bibi I, Aftab T (eds) Global arsenic Hazard: ecotoxicology and remediation. Springer, Cham, pp 247–270. https://doi.org/10.1007/978-3-031-16360-9_12

Munns R (2002) Comparative physiology of salt and water stress. Plant Cell Environ 25(2):239–250

Munns R (2005) Response of crops to salinity, 25–27 April. International salinity forum, Riverside 339 p

Munns R, Tester M (2008) Mechanisms of salinity tolerance. Annu Rev Plant Biol 59:651–681

Muntean DW (2009) Boron, the overlooked essential element. Soil and Plant Laboratory, Inc., 1648 p

Naher N, Alam AK (2010) Germination, growth and nodulation of mungbean (*Vigna radiata* L.) as affected by sodium chloride. Int J Sustain Crop Prod 5(2):8–11

Naik GR, Somashekhar R, Hiremeth SM (1993) Effect of water stress on growth and stomatal characteristics in sugarcane cultivars. Indian Sugar 43(8):645–649

Nam NH, Chauhan YS, Johansen C (2001) Effect of timing of drought stress on growth and grain yield of extra-short-duration pigeonpea lines. J Agric Sci 136(2):179–189. https://doi.org/10.1017/S0021859601008607

Narwade, Ajay V., Kiran Bhagat, D. Patil, Yogeshwar Singh, A. Kumari, Y. Ban, H. Thakare, and C. Singh. "Abiotic stress responses in sugarcane." Challenges and Prospective of Plant Abiotic Stress; Pasala, R., Ed (2016): 419–446.

Nasri M, Khalatbari M, Zahedi H, Paknejad F, Moghadam HRT (2008) Evaluation of micro and macro elements in drought stress condition in cultivars of rapeseed (*Brassica napus* L.). Am J Agric Biol Sci 3(3):579–583

Nayidu N, Bollina V, Kagale S (2013) Oilseed crop productivity under salt stress. In: Ahmad P, Azooz MM, Prasad MNV (eds) Ecophysiology and responses of plants under salt stress. Springer, Singapore, pp 249–265

Ober ES, Rajabi A (2010) Abiotic stress in sugar beet. Sugar Tech 12(3–4):294–298

Odeny DA (2007) The potential of pigeonpea (*Cajanus cajan* (L.) Millsp.) in Africa. Nat Resour Forum 31(4):297–305

Oliver SN, Van Dongen JT, Alfred SC, Mamun EA, Zhao X, Saini HS, Fernandes SF, Blanchard CL, Sutton BG et al (2005) Cold-induced repression of the rice anther-specific cell wall invertase gene *OSINV4* is correlated with sucrose accumulation and pollen sterility. Plant Cell Environ 28(12):1534–1551

Olivera-Guerra L, Merlin O, Er-Raki S (2020) Irrigation retrieval from Landsat optical/thermal data integrated into a crop water balance model: a case study over winter wheat fields in a semi-arid region. Remote Sens Environ 239:111627

Ozturk M, Ashraf M, Aksoy A, Ahmad MSA (2015a) Phytoremediation for green energy, 1st edn. Springer, Cham

Ozturk M, Ashraf M, Aksoy A, Ahmad MSA (2015b) Plants, pollutants and remediation, 1st edn. Springer, Cham

Pal R, Mahajan G, Sardana V, Chauhan BS (2017) Impact of sowing date on yield, dry matter and nitrogen accumulation, and nitrogen translocation in dry-seeded rice in North-West India. Field Crops Res 206:138–148

Panda D (2015) Fluoride toxicity stress: physiological and biochemical consequences on plants. Int J Biores Environ Agric Sci 1:70–84

Panda RK, Behera SK, Kashyap PS (2003) Effective management of irrigation water for wheat under stressed conditions. Agric Water Manag 63(1):37–56

Panda SK, Upadhyay RK, Nath S (2010) Arsenic stress in plants. J Agron Crop Sci 196(3):161–174

Pandey S, Bhandari H (2008) Drought: economic costs and research implications. In: Serraj R, Bennett J, Hardy B (eds) Drought frontiers in Rice: crop improvement for increased rainfed production. World Scientific Publishing, pp 3–17

Pandey RK, Herrera WAT, Pendleton JW (1984) Drought response of grain legumes under irrigation gradient: I. yield and yield components 1. Agron J 76(4):549–553. https://doi.org/10.2134/agronj1984.00021962007600040009x

Pearce R (2001) Plant freezing and damage. Ann Bot 87(4):417–424

Pereira A (2016) Plant abiotic stress challenges from the changing environment. Front Plant Sci 7:1123

Plaut Z, Meinzer FC, Federman E (2000) Leaf development, transpiration and ion uptake and distribution in sugarcane cultivars grown under salinity. Plant Soil 218(1):59–69

Plaza-Wüthrich S, Blösch R, Rindisbacher A, Cannarozzi G, Tadele Z (2016) Gibberellin deficiency confers both lodging and drought tolerance in small cereals. Front Plant Sci 7:643

Prasad R (2005) Rice-wheat cropping systems. Adv Agron 86:255–339

Puengsungwan S, Jiraserccamomkul K (2018) IoT-based stress detection for organic lettuce farms using chlorophyll fluorescence (ChF). In: 6th Global Wireless Summit, GWS 2018, pp 354–357
Qin C, Tang Z, Chen J, Chen X (2020) The impact of soil and water resource conservation on agricultural production- an analysis of the agricultural production performance in Zhejiang, China. Agric Water Manag 240:106268
Rahman K, Ahmed N, Raihan MRH, Nowroz F, Jannat F, Rahman M, Hasanuzzaman M (2021) Jute responses and tolerance to abiotic stress: mechanisms and approaches. Plan Theory 10(8):1
Rainey KM, Griffiths PD (2005) Differential response of common bean genotypes to high temperature. J Am Soc Hort Sci 130(1):18–23
Rana L, Banerjee H, Ray K, Sarkar S (2017) System of Wheat Intensification (SWI) – a new approach for increasing wheat yield in small holder farming system. J Appl Nat Sci 9(3):1453–1464
Rashid MH, Alam MM, Khan MAH, Ladha JK (2009) Productivity and resource use of direct-(drum)-seeded and transplanted rice in puddled soils in rice–rice and rice–wheat ecosystems. Field Crop Res 113(3):274–281
Rasmussen SK (2020) Molecular genetics, genomics, and biotechnology in crop plant breeding. Agronomy 10(3):439
Ravi V, Indira P (1999) Crop physiology of sweet potato. Hortic Res 23:277–338
Richner W, Soldati A, Stamp P (1996) Shoot-to-root relations in field-grown maize seedlings. Agron J 88(1):56–61
Robertson MJ, Donaldson RA (1998) Changes in the components of cane and sucrose yield in response to drying-off of sugarcane before harvest. Field Crops Res 55(3):201–208
Robertson MJ, Inman-Bamber NG, Muchow RC, Wood AW (1999) Physiology and productivity of sugarcane with early and mid-season water deficit. Field Crops Res 64(3):211–227
Ruelland E, Vaultier MN, Zachowski A, Hurry V (2009) Chapter 2: Cold signalling and cold acclimation in plants. In: Advances in botanical research, vol 49, pp 35–150. https://doi.org/10.1016/S0065-2296(08)00602-2
Ryan PR, Ditomaso JM, Kochian LV (1993) Aluminium toxicity in roots: an investigation of spatial sensitivity and the role of the root cap. J Exp Bot 44(2):437–446
Sabesh M (2022) Abiotic stresses in cotton: a physiological approach. CICR. Available from https://www.cicr.org.in/Downloads.html
Saha B, Chowardhara B, Kar S, Devi SS, Awasthi JP, Moulick D, Tanti B, Panda SK (2019) Advances in heavy metal-induced stress alleviation with respect to exogenous amendments in crop plants. In: Hasanuzzaman M, Fotopoulos V (eds) Priming and pretreatment of seeds and seedlings. Springer, Singapore, pp 313–332. https://doi.org/10.1007/978-981-13-8625-1_15
Sahoo S, Borgohain P, Saha B, Moulick D, Tanti B, Panda SK (2019) Seed priming and seedling pretreatment induced tolerance to drought and salt stress: recent advances. In: Hasanuzzaman M, Fotopoulos V (eds) Priming and pretreatment of seeds and seedlings. Springer, Singapore, pp 253–263. https://doi.org/10.1007/978-981-13-8625-1_12
Sairam RK, Tyagi A (2004) Physiology and molecular biology of salinity stress tolerance in plants. Curr Sci 86:407–421
Salt DE, Blaylock M, Kumar NPBA, Dushenkov V, Ensley BD, Chet I, Raskin I (1995) Phytoremediation: a novel strategy for the removal of toxic metals from the environment using plants. Biotechnology 13(5):468–474
Solomon, S. (2007, December). IPCC (2007): Climate change the physical science basis. In Agu fall meeting abstracts (Vol. 2007, pp. U43D-01)
Samarah NH (2005) Effects of drought stress on growth and yield of barley. Agron Sustain Dev 25(1):145–149
Samui I, Skalicky M, Sarkar S, Brahmachari K, Sau S, Ray K, Hossain A, Ghosh A, Nanda MK et al (2020) Yield response, nutritional quality and water productivity of tomato (*Solanum lycopersicum* L.) are influenced by drip irrigation and straw mulch in the coastal saline ecosystem of Ganges Delta, India. Sustainability 12(17):6779

Sanchez DH, Pieckenstain FL, Szymanski J, Erban A, Bromke M, Hannah MA, Kraemer U, Kopka J, Udvardi MK (2011) Comparative functional genomics of salt stress in related model and cultivated plants identifies and overcomes limitations to translational genomics. PLoS One 6(2):e17094

Sapkota TB, Jat ML, Aryal JP, Jat RK, Khatri-Chhetri A (2015) Climate change adaptation, greenhouse gas mitigation and economic profitability of conservation agriculture: some examples from cereal systems of indo-Gangetic Plains. J Integr Agric 14(8):1524–1533

Sarangi SK, Maji B, Singh S, Burman D, Mandal S, Sharma DK, Singh US, Ismail AM, Haefele SM (2015) Improved nursery management further enhances the productivity of stress-tolerant rice varieties in coastal rainfed lowlands. Field Crops Res 174:61–70

Sarkar S, Gaydon DS, Brahmachari K, Nanda MK, Ghosh A, Mainuddin M (2020a) Modelling yield and seasonal soil salinity dynamics in rice-grasspea cropping system for the coastal saline zone of West Bengal, India. PRO 36:146

Sarkar S, Ghosh A, Brahmachari K (2020b) Application of APSIM model for assessing the complexities of rice-based cropping systems of South-Asia. In: Maitra S, Pramanick B (eds) Advanced agriculture. New Delhi Publishers, pp 212–233

Sarkar S, Gaydon DS, Brahmachari K, Poulton PL, Chaki AK, Ray K, Ghosh A, Nanda MK, Mainuddin M (2022) Testing APSIM in a complex saline coastal cropping environment. Environ Model Softw 147:105239

Sau S, Pal B, Sarkar S, Sarkar T (2019) Influence of seed priming on germination and seedling vigour of wood apple (*Feronia limonia* Swingle). Int J Bio-Resour Stress Manag 10(2):128–136

Schubert S, Abd-El-Motagally F, Steffens D (2004) Heat stress can be compensated by irrigation in sugar beet. In: Jacobsen E-E, Jensen CR, Porter JR (eds) Proceedings of the 8th European society of agronomy congress

Schützendübel A, Polle A (2002) Plant responses to abiotic stresses: heavy metal-induced oxidative stress and protection by mycorrhization. J Exp Bot 53(372):1351–1365

Sehrawat N, Yadav M, Bhat KV, Sairam RK, Jaiwal PK (2015) Effect of salinity stress on mungbean [*Vigna radiata* (L.) Wilczek] during consecutive summer and spring seasons. J Agric Sci BGD 60(1):23–32

Setayesh R, Kafi M, Mehrjerdi MZ (2017) Low sensitivity to photoperiod may increase potato yield in short day through the maintenance of sink and source balance. Pak J Bot 49(3):929–933

Shah SH, Ali S, Hussain Z, Jan SA, Din J, Ali GM (2016) Genetic improvement of tomato (*Solanum lycopersicum*) with *AtDREB1A* gene for cold stress tolerance using optimized *Agrobacterium*-mediated transformation system. Int J Agric Biol 18(3):471–482

Shah F, Coulter JA, Ye C, Wu W (2020) Yield penalty due to delayed sowing of winter wheat and the mitigatory role of increased seeding rate. Eur J Agron 119:126120

Shanker AK, Shanker C (2016) Abiotic and biotic stress in plants—recent advances and future perspectives. IntechOpen. Available from http://www.intechopen.com/books/abiotic-and-biotic-stress-in-plants-recent-advances-and-future-perspectives

Shavrukov Y, Kurishbayev A, Jatayev S, Shvidchenko V, Zotova L, Koekemoer F, de Groot S, Soole K, Langridge P (2017) Early flowering as a drought escape mechanism in plants: how can it aid wheat production? Front Plant Sci 8:1950

Sinaki JM, Majidi E, Shirani-Rad AL, Moormohammadi G, Zarei G (2007) The effect of water deficit during growth stage of canola (Brassica napus L). Am Eurasian J Agric Environ Sci 2:417–422

Singh B, Kukreja S, Goutam U (2019) Impact of heat stress on potato (*Solanum tuberosum* L.): present scenario and future opportunities. J Hortic Sci Biotechnol 95(4):407–424

Smith SE, Christophersen HM, Pope S, Smith FA (2010) Arsenic uptake and toxicity in plants: integrating mycorrhizal influences. Plant Soil 327(1–2):1–21

Spitters CJT, van Keulen H, van Kraalingen DWG et al (1989) A simple and universal crop growth simulator SUCROS8 7. In: Rabbinge R et al (eds) Simulation and systems management in crop protection. Pudoc, Wageningen, pp 147–181

Stoeva N, Berova M, Zlatev Z (2003) Physiological response of maize to arsenic contamination. Biol Plant 46(3):449–452
Su X, Wei F, Huo Y, Xia Z (2017) Comparative physiological and molecular analyses of two contrasting flue-cured tobacco genotypes under progressive drought stress. Front Plant Sci 8:827
Sun H, Sun X, Wang H, Ma X (2020) Advances in salt tolerance molecular mechanism in tobacco plants. Hereditas 157(1):5
Symonds RB, Rose WI, Reed MH (1988) Contribution of Cl^- and F^- bearing gases to the atmosphere by volcanoes. Nature 334(6181):415–418
Takahashi D, Li B, Nakayama T, Kawamura Y, Uemura M (2013) Plant plasma membrane proteomics for improving cold tolerance. Front Plant Sci 4:90
Tal M (1994) In vitro selection for salt tolerance in crop plants: theoretical and practical considerations. Vitro Cell Dev Biol Plant 30(4):175–180
Tareq MZ, Hossen B, Biswas SK, Hoque ABMZ, Hasan MM (2018) Yield and quality of jute (*Corchorus capsularis* L.) vegetable seed as affected by salinity environment. J Agroecol Nat Resour Manag 5(30–33):595
Teskey, R., Wertin, T., Bauweraerts, I., Ameye, M., McGuire, M. A., & Steppe, K. (2015). Responses of tree species to heat waves and extreme heat events. Plant, cell & environment, 38(9):1699–1712
Tavakkoli E, Rengasamy P, McDonald GK (2010) High concentrations of Na^+ and cl^- ions in soil solution have simultaneous detrimental effects on growth of faba bean under salinity stress. J Exp Bot 61(15):4449–4459
Threshow M (1970) Environment and plant response. Department of Biology, University of Utah
Tiwari S, Lata C (2018) Heavy metal stress, signaling, and tolerance due to plant-associated microbes: an overview. Front Plant Sci 9:452
Tiwari AK, Lal M, Singh AK (2015) Current status of sugarcane research in India. Nova Publishers
Tyagi S, Yadav P, Chakraborty A, Majumdar A, Moulick D, Santra SC, Upadhyay MK, Sahoo U, Maitra S, Hossain A (2023) Myconanotechnologies: an approach towards sustainable agriculture. In Myconanotechnology and Application of Nanoparticles in Biology (pp. 139–157). Academic Press. https://doi.org/10.1016/B978-0-443-15262-7.00005-X
Van Der Heyden C, Ray JD, Nable R (1998) Effects of waterlogging on young sugarcane plants. Aust Sugarcane 2:28–30
van Schilfgaarde J (1994) Irrigation – a blessing or a curse. Agric Water Manag 25(3):203–219
Venkataramana S, Naidu KM (1982) Root growth during the formative phase in irrigated and water stressed sugarcane and its relationship with shoot development and yield. Indian J Plant Physiol 32:43–50
Vijayalakshmi D (2015) Abiotic stresses and its management in agriculture. In: Handbook on e-Velanmai for extension professionals. Springer, pp 361–388
Volkov V (2015) Salinity tolerance in plants. Quantitative approach to ion transport starting from halophytes and stepping to genetic and protein engineering for manipulating ion fluxes. Front Plant Sci 6:873
von Uexküll HR, Mutert E (1995) Global extent, development and economic impact of acid soils. Plant–soil interactions at low pH: principles and management Grundon, N.J, Raymet GE, Probert ME, eds.: Kluwer Academic Publishers 171:1–15
Wahid A (2007) Physiological implications of metabolite biosynthesis for net assimilation and heat-stress tolerance of sugarcane (*Saccharum officinarum*) sprouts. J Plant Res 120(2):219–228
Wahid A, Close TJ (2007) Expression of dehydrins under heat stress and their relationship with water relations of sugarcane leaves. Biol Plant 51(1):104–109
Wahid A, Gelani S, Ashraf M, Foolad MR (2007) Heat tolerance in plants: an overview. Environ Exp Bot 61(3):199–223
Wallace JS, Acreman MC, Sullivan CA (2003) The sharing of water between society and ecosystems: from conflict to catchment–based co–management. Philos Trans R Soc Lond Ser B Biol Sci 358(1440):2011–2026

Wang Y, Janz B, Engedal T, de Neergaard AD (2017) Effect of irrigation regimes and nitrogen rates on water use efficiency and nitrogen uptake in maize. Agric Water Manag 179:271–276

Wang H, Takano T, Liu S (2018) Screening and evaluation of saline– alkaline tolerant germplasm of rice (*Oryza sativa* L.) in soda saline–alkali soil. Agronomy 8(10):205

Wei L-X, Lv B-S, Li X-W, Wang M, Ma H, Yang H, Yang R, Piao Z, Wang Z et al (2017) Priming of rice (*Oryza sativa* L.) seedlings with abscisic acid enhances seedling survival, plant growth, and grain yield in saline-alkaline paddy fields. Field Crops Res 203:86–93

Wuana RA, Okieimen FE (2011) Heavy metals in contaminated soils: a review of sources, chemistry, risks and best available strategies for remediation. ISRN Ecol 2011:1–20

Xiao Y, Pan Y, Luo L, Zhang G, Deng H, Dai L, Liu X, Tang W, Chen L, Wang G (2011) Quantitative trait loci associated with seed set under high temperature stress at the flowering stage in rice (*Oryza sativa* L.). Euphytica 178(3):331–338

Xue D, Chen M, Zhang G (2009) Mapping of QTLs associated with cadmium tolerance and accumulation during seedling stage in rice (*Oryza sativa* L.). Euphytica 165(3):587–596

Yadav T, Kumar A, Yadav RK, Yadav G, Kumar R, Kushwaha M (2020) Salicylic acid and thiourea mitigate the salinity and drought stress on physiological traits governing yield in pearl millet-wheat. Saudi J Biol Sci 27(8):2010–2017

Yakoub ARB, Tlahig S, Ferchichi A (2019) Germination, growth, photosynthesis, and osmotic adjustment of Tossa jute (*Corchorus olitorius* L.) seeds under saline irrigation. Pol J Environ Stud 28(2):935–942

Yohannes HA (2016) Review on relationship between climate change and agriculture. J Earth Sci Clim Change 7:335

Zhang H, Liu XL, Zhang RX, Yuan HY, Wang MM, Yang HY, Ma HY, Liu D, Jiang CJ, Liang ZW (2017) Root damage under alkaline stress is associated with reactive oxygen species accumulation in rice (*Oryza sativa* L.). Front Plant Sci 8:1580

Zhao FJ, Ma JF, Meharg AA, McGrath SP (2009) Arsenic uptake and metabolism in plants. New Phytol 181(4):777–794

Zhao K, Liu X, Xu J, Selim HM (2010) Heavy metal contaminations in a soil-rice system: identification of spatial dependence in relation to soil properties of paddy fields. J Hazard Mater 181(1–3):778–787

Zhao G, Bryan BA, Song X (2014) Sensitivity and uncertainty analysis of the APSIM-wheat model: interactions between cultivar, environmental, and management parameters. Ecol Model 279:1–11

Zhao H, Zhang P, Wang Y, Ning T, Xu C, Wang P (2020) Canopy morphological changes and water use efficiency in winter wheat under different irrigation treatments. J Integr Agric 19(4):1105–1116

Chapter 16
Biostimulants for Plant Abiotic Stress Resistance and Climate-Resilient Agriculture

Anwesha Chatterjee and Harshata Pal

Abstract Agriculture is the science of growing crops and rearing animals for sustainable food production and enhanced livelihood. Crops growing under favourable conditions often go through different environmental stresses which can be biotic or abiotic. Plants synthesise protective compounds as a response to the abiotic stresses such as salt, temperature, drought and mineral regulation. These compounds are usually produced within the plant metabolic pathways. In modern agriculture, agronomists are focusing on the development of biological functional compounds or biostimulants which can contribute to the enhancement of plant abiotic stress tolerance and productivity in adverse environmental conditions. Biostimulant also regulates the nutritional quality of the crops making them fit to consume. Biostimulants ensure little or no use of chemical fertilisers. Although some of the biostimulant applications are widely practised by the farmers, some modern technologies, such as the use of microbial biostimulant, are still unknown to many. Therefore, in this chapter, we aim to give an overview of recent advances in the research related to the theory and practice of biostimulants which will impart knowledge among the young researchers, agricultural industries, farmers and business collaborators for better understanding of the molecular processes and implementation of biostimulants in growing different crops.

Keywords Abiotic stress tolerance · Biostimulant · Nutrient use efficiency (NUE) · Plant growth-promoting microorganism (PGPM) · Sustainable agriculture

1 Introduction

Agricultural sectors are broadening themselves by practising bio-based technologies integrated with circular economy (Colla and Rouphael 2015; Rouphael and Colla 2020a). Using Earth's limited resources in a sustainable manner without

---

A. Chatterjee · H. Pal (✉)
Amity Institute of Biotechnology, Amity University Kolkata, Kolkata, West Bengal, India
e-mail: hpal@kol.amity.edu

M. Hasanuzzaman (ed.), *Climate-Resilient Agriculture, Vol 2*,
https://doi.org/10.1007/978-3-031-37428-9_16

harming the environment and providing market to the farmers is the ultimate requirement for food security. Farmers should use improved novel technologies to uplift the crop yield and ensure resource use efficiency for food security without compromising with the soil pollution. Biostimulants are bio-based products which prove to be an essential alternative to synthetic fertilisers and establishes improved production of crops making them adaptable to the surrounding environmental conditions (Jägermeyr 2020). Plant biostimulants can be natural substances or chemical derivatives of substances such as humic and fulvic acids, protein hydrolysates, seaweed extracts and also diverse classes of fungi and bacteria. Some inorganic compounds also serve as plant biostimulants (Rouphael and Colla 2020b). Plant biostimulants enhanced plant resilience by overcoming nutrient deficiency and increasing uptake and assimilation of the nutrients thereby getting adopted in both organic and conventional cropping systems (De Pascale et al. 2017). Fruits, vegetables and cereals represent an important constituent of human diet due to their nutritional value. Scientists have documented that various agricultural practices including plant-biostimulant application can greatly influence the quality of vegetables (Rouphael et al. 2018). While it is also documented that biostimulant application in fruit crops is limited as compared to vegetables and cereals which might be due to the perennial nature of the species. Despite having the limitations, selected biostimulants have been found to be effective with respect to increased fruit yield and quality (Basile et al. 2020). Apart from the earlier mentioned types, biostimulants can also be developed from food or agro-industrial waste products. This paves the way for recycling the waste products as the by-products are used as raw materials for biostimulant production thereby utilising the circular bioeconomy strategy in achieving sustainable agriculture. The by-products suitable for biostimulant production should be free of pesticide residues, low cost for collection and storage of the products and finally adequate supply and efficient synergistic properties (Xu and Geelen 2018). Various high-throughput technologies are used to monitor the mode of action of biostimulants. It is documented that biostimulants directly take part on the plant metabolism by modifying the biochemical pathways and thereby improve the water and nutrient use efficiency (NUE) of the plants. They also stimulate plant morphology and restrain abiotic stresses by ameliorating plant primary and secondary metabolism (Bulgari et al. 2019a). According to researchers, the effectiveness of a biostimulant is not due to a single component but due to the synergistic effect of a combination of bioactive compounds. Scientists are aiming to develop second-generation plant biostimulants by studying the chemistry behind the synergistic properties in order to deliver novel biostimulant consortia (Rouphael and Colla 2018).

Intensive research and effective results of biostimulant application have led to the continuous expansion of biostimulant market. European counties such as France, Italy and Spain are leading in producing biostimulants. Research showed that biostimulant market may reach approximately USD 6.7 billion by the year 2030 (Traon et al. 2014; Biostimulants Market Size, Share & Trends Analysis Report 2022). The fundamental plant-biostimulant interaction has been depicted in Fig. 16.1.

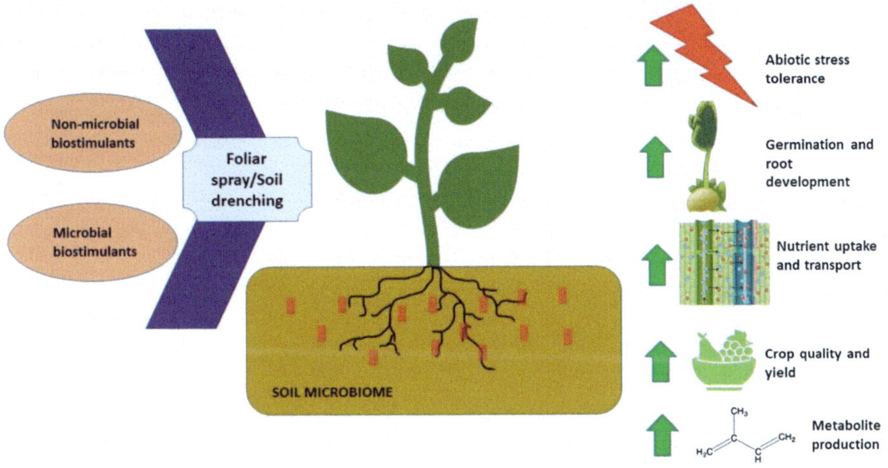

Fig. 16.1 Correlation between plant and biostimulants

2 Classification of Biostimulants

Scientists have classified biostimulants on the basis of their composition and mode of action. Some of the commercially available plant biostimulant and their contribution in agricultural science are listed in Table 16.1. In this chapter, we have classified biostimulant based on the origin of raw material (du Jardin 2015; Franzoni et al. 2022).

2.1 Humic Substances

Scientists have characterised and proved that humic substances have unique structures from which it is possible to understand how humic substances enhance root growth by interacting with the cell and regulate plant metabolism forming an ideal bioactive compound or biostimulant (García et al. 2019). The molecular size of the humic substances has shown variable intensities of growth in *Arabidopsis* and maize seedlings. Humic substances have been proved to be effective in lateral root emergence, root elongation and hairy root formation by interacting with the hormone signalling pathways. Humic substances also promote plant growth by their nutrient uptake and use efficiency (Nardi et al. 2021).

Table 16.1 List of some commercially available biostimulants that have been widely used in research over the years

Commercial name of the biostimulant	Components	Model plant	Effects	References
Powhumus® Huminbio Microsense Seed® Huminbio Microsense Bio® Fulvagra®	Humic substances	Cherry tomato	Positive effects on plant growth, fruit quality and yield, mineral content and antioxidant enzyme activity	Turan (2021)
Actiwave®	*Ascophillum nodosum*	*Camellia japonica* L.	Stimulated rooting when applied with gibberellic acid	Ferrante et al. (2013)
Trainer®	Plant-derived protein hydrolysates	Tomato	Positive effects on plant growth under water stress	Paul et al. (2019)
CycoFlow®	Sugar cane molasses and yeast extract	Tomato	Improved plant activities at high temperature	Francesca et al. (2020)
Megafol® and Viva®	Humic acids and amino acids/peptides	Tomato	Nutritional addition in fruits	Klokić et al. (2020)
Biozyme®	Sea weed extract	Soyabean	Enhanced plant growth and yield	Tandon and Dubey (2015)
Stimplex®	Sea weed extracts	Tomato	Higher lycopene and β-carotene in fruits	Sidhu et al. (2017)
Biological Fertilizer DC; Proradix®; RhizoVital® and the microbial consortia product (MCP) comprising of 12 different beneficial bacterial and fungal strains such as *Azotobacter vinlandii, Clostridium* sp., *Lactobacillus* sp., *Bacillus velezensis, B. subtilis* (SILo Sil® BS)*, B. thuringiensis, Pseudomonas fluorescens, Acetobacter, Enterococcus, Rhizobium japonicum, Nitrosomonas* and *Nitrobacter, Saccharomyces, Penicillium roqueforti, Monascus, Aspergillus oryzae, Trichoderma harzianum* (TRICHOSIL®) and algae extracts from *Arthrospira platensis* (Spirulina) and *Ascophyllum nodosum*	Microbial and algal consortia	Tomato	Increased the efficiency and reproducibility of the plant	Bradácová et al. (2019)

Retrosal®	Mixture of carboxylic acids, containing calcium oxide (CaO) and complexed by ammonium ligninsulphonate, Zinc (Zn) chelated by EDTA	Romaine lettuce	Stimulated growth of plants under salinity stress	Bulgari et al. (2019b)
Super Fifty® and Ecolicitor®	Sea weed extracts	Lettuce	Stimulated growth of plants under salinity stress	Guinan et al. (2013)
Asahi SL/Goemar Goteo	Combination of three phenolics and sea weed extract of *Ascophyllum nodosum*	Coriander	Enhanced plant performance under chilling stress	Pokluda et al. (2016)
ComCat®	*Lychnis viscaria*	Wheat and maize	Enhanced seedling growth	Van Der Watt and Pretorius (2013)
Kelpak®	*Ecklonia maxima*	Mung bean and Swiss chard	Promoted rooting in mung bean and enhanced overall growth in Swiss chard	Arthur et al. (2012)
Vivema Twin	Mixture of hydrolysable and condensed tannins obtained from waste materials	Tomato	Enhanced plant growth under salt stress	Campobenedetto et al. (2021b)
Expando®	Mineral elements, amino acids, vitamins and phytohormone	Tomato	Plant growth promotion and increase in total yield	Contartese et al. (2016)

(continued)

Table 16.1 (continued)

Commercial name of the biostimulant	Components	Model plant	Effects	References
TNC Bactorr^{S13} and Flortis Micorizze	Microbial consortia	Tomato	Plant growth promotion under salinity stress	Miceli et al. (2021)
AlgaeGreen®	Sea weed extract	Broccoli	Improved nutritional quality of the plant	Lola-luz et al. (2014)
Goëmar BM 86®	Sea weed extract	Apple	Improved fruit quality	Basak (2012)
Maxicrop®	Sea weed extract	Lettuce, cauliflower, barley, strawberry	Improved yield, frost resistance, insect resistance	Khan et al. (2009)
Soil-Life™ and Nutri-Life Platform®	Microbial consortia	Sugar cane	Enhanced plant growth	Berg et al. (2019)
ERANTHIS®	Sea weed and yeast extract	Tomato	Mitigated water stress effect on the plant	Campobenedetto et al. (2021a)
VIVA®	Mixture of polysaccharides, polypeptides, amino acids, selected humic acids and vitamins	Tomato	Increased plant biomass under drought stress	Petrozza et al. (2013)
BACSTIM®	A microbial consortium of five Bacilli strains	Maize	Growth promotion under drought stress	Nephali et al. (2021)
KIEM®	Lignin derivatives along with amino acids obtained from plants and molybdenum	Cucumber	Increased germination rate under heat stress	Campobenedetto et al. (2020)
Terra-Sorb®	Amino acid	Ryegrass plants	Recovery of plants from temperature stress	Botta (2013)

2.2 Protein Hydrolysates

Thermal and enzymatic hydrolysis of proteins from animal and plant sources gives rise to protein hydrolysates. Protein hydrolysates obtained from plant sources showed positive results on plant growth and development by interplaying a role in plant defence mechanisms (Colla et al. 2017). Foliar spraying of protein hydrolysates derived from plant sources on lettuce plant increased plant performance under salinity stress (Lucini et al. 2015). Protein hydrolysates when combined with microbial biostimulants showed enhanced growth and plant tolerance to alkalinity and salinity stress (Rouphael et al. 2017). Protein hydrolysates showed growth promotion in maize seedlings under nutrient and oxygen deficiency as well as high salinity (Trevisan et al. 2019).

2.3 Seaweed Extract and Phytochemicals

Brown, red and green multicellular marine algae belong to the class of seaweed extract and form an important source of plant fertiliser. Seaweed extracts have been used widely by the agronomists to stimulate plant growth. Scientists applied *Ascophyllum nodosum* extracts on seedless grapes which showed better quality fruits and higher yields (Norrie and Keathley 2006). Application of seaweed extract on broccoli plant showed a distinct increase in the yield and average weight of the curds. It also increased the macro- and micronutrient uptake and use in broccoli (Gajc-Wolska et al. 2012). The active mineral compounds such as laminarin, fucoidan, alginates and plant hormones are present in marine seaweed which contributes to the growth and development of plants (Battacharyya et al. 2015). Microalgae or seaweed extract production require less amount of renewable resources and are suitable for metabolic engineering (Chiaiese et al. 2018).

Bioactive compounds present in the plants taking part in various physiological activities also sometime serve as source of biostimulants (Bulgari et al. 2017). Scientists suspects that withanolides present in *Withania somnifera* might have biostimulant effect (Jonathan et al. 2015).

2.4 Chitosan

In agriculture, chitosan is considered to be an effective biostimulant for its plant protective and storage stability capacities. Chitosan interacts with plant defence mechanism and thereby regulates different biochemical pathways. The effectiveness of plant-chitosan interaction depends on its origin, degree of deacetylation and acetylation patterns (Stasinska-Jakubas and Hawrylak-Nowak 2022). Due to the film forming properties of chitosan, it is widely used in encapsulation. Chitosan with other agents forms binding substances and stimulates plant growth and gives protection against pathogens (Hadwiger 2013). Encapsulation with chitosan ensures proper water and gas permeability thereby stimulating germination (Korbecka-glinka et al. 2021). A study revealed that application of chitosan increased fruit yield and contents of carotenoids, anthocyanins, flavonoids and phenolics in strawberry plant (Rahman et al. 2018).

2.5 Inorganic Compounds

Inorganic compounds such as selenium, silicon, cobalt, aluminium and sodium are some of the beneficial elements that are responsible for carrying out growth in plants. These inorganic compounds are present in the soil as well as inside the plant in the form of salts and often cause cell wall strengthening (Kalam et al. 2020; Savvas and Ntatsi 2015).

2.6 Microorganisms

Microorganisms such as bacteria, fungi, yeast and microalgae are considered as biostimulants also known as plant growth-promoting microorganisms. They enhance nutrient and water uptake through soil and induce plant hormone biosynthesis. They are also known to enhance abiotic stress tolerance and production of volatile organic compounds (VOCs) that are responsible for carrying out several biological activities (Ruzzi and Aroca 2015). These organisms either establish a symbiotic relationship with the crop or increase the bioavailability of the nutrients to the plants for their proper growth and development (Colla et al. 2014). A study showed that combined application of bacteria such as *Azotobacter*, *Streptomyces* and fungal strains of *Trichoderma* increased nutrient uptake ability in tomato plants (Allaga et al. 2020). It has also been documented that combinational treatment of different fungal (Arbuscular mycorrhizal fungi) and two bacterial strains has enhanced water stress tolerance in tomato plants (Mannino et al. 2020). Inoculation of plant growth-promoting rhizobacteria (PGPR) strain *Bacillus subtilis* CBR05 in tomato plant improved the fruit quality with high contents of lycopene and carotenoids in the fruits (Chandrasekaran et al. 2019). In a study, it is shown that arbuscular mycorrhizal fungi *Rhizophagus intraradices* and also combination of *Rhizophagus intraradices* and *Funneliformis mosseae* increased synthesis of anthocyanins, vitamin c and polyphenols in saffron (Caser et al. 2018).

2.7 *Biostimulants from Waste*

Extracts from food waste, industrial waste, sewage treatments, aquaculture, manures and composts are a different class of biostimulant. Biostimulants extracted from waste products are known to improve plant production by interacting with several physiological processes giving rise to secondary metabolites (Xu and Geelen 2018). It has been reported that waste-based biostimulants enhance the activity of the enzyme phenylalanine ammonia lyase (PAL) (Ertani et al. 2011).

2.8 *Nanoparticles*

Nanoparticles are the newest class of plant biostimulants and have been reported to improve the quality of the plant products and make the plants tolerant to abiotic stress (Juarez-Maldonado et al. 2019). A study revealed that nano-titanium dioxide promoted photosynthesis in tomato leaves (Qi et al. 2013). Titanium dioxide nano-anatase application has shown decrease in the accumulation of superoxide radicals, hydrogen peroxide and malonyldialdehyde thereby increasing the superoxide dismutase, catalase, ascorbate peroxidase and guaiacol peroxidase in spinach under UV-B radiation (Lei et al. 2008). Plant-nanomaterial interaction enhances transport and nutrient uptake ability. Scientists showed that zinc oxide nanoparticles increased overall growth, chlorophyll and soluble protein in tomato (Raliya et al. 2015).

3 Production and Preparation of Biostimulants

Different raw materials and their processing and purification lead to the formulation of diverse classes of biostimulants. The technology for producing different types of biostimulant differs from each other. The methods frequently used for biostimulant production include cultivation, extraction, fermentation, hydrolysis cell rupture treatment and purification (Yakhin et al. 2017). Most of the research articles based on biostimulants did not clearly describe the composition of a biostimulant since the compounds are not yet characterised and mere presence of a compound do not confirm that to be a growth stimulator. Also, the substances in biostimulant raw materials differ with season, species and growth condition of the source organism. The high-throughput analytical techniques used in the process of identification of the biostimulant and its bioactive components include fingerprint techniques, ELISA, GC-MS, HPLC, Immunoblot, NMR, spectroscopy, molecular taxonomical characterisation, plate count methods, thermochemolysis and chromatography. Due to lack of knowledge about a bioactive compound in a biostimulant, the biostimulant composition changes over time and their functionality cannot be interpreted. Scientists perform high-throughput techniques to ensure the quality and stability of the product consistently (Shekhar et al. 2012). Yakhin et al. (2017) reviewed several studies and concluded that microbial biostimulants consist of auxin, cytokinins, betaines, gibberellins, amino acids, oligopeptides, peptidoglycans, lipopolysaccharides, melatonin, vitamins, siderophores, etc. Bacterial biostimulants are processed by means of cultivation, acid hydrolysis, alkali hydrolysis, enzymatic hydrolysis and fermentation while fungal biostimulants are cultivated by means of fermentation and lyophilisation. Seaweed extracts are obtained by acid and alkaline processing and extraction; alkaline hydrolysis; aqueous extraction; cell burst; cell rupture; freezing; alkaline and water extractions; cryoprocessing; enzyme-assisted extraction (EAE); fermentation; heated alkaline hydrolysis; microwave-assisted extraction (MAE); neutral extraction; pressurised liquid extraction (PLE); supercritical fluid extraction (SFE); ultrasound-assisted extraction (UAE). Humic substances and waste materials are extracted and undergo thermochemolysis for the formulation of biostimulants.

4 Mode of Action of Biostimulants

Every agricultural fertiliser or growth enhancer must demonstrate a specific mode of action for effective marketing and also for legal acceptance before selling the product. Biostimulants, due to their diverse nature do not rely on a specific biochemical target and their mode of action remains unidentified in most cases (Khan et al. 2009; Rathore et al. 2009; Paradiković et al. 2011). Ongoing researches suggest that the basic mechanism of action of biostimulant would be stimulating photosynthesis, enhancing nutrient uptake and regulating stress signalling pathways (Khan et al. 2009; Sharma et al. 2012). Protein-based biostimulants are widely studied as the researchers are able to identify the metabolic pathways targeted and eventually identify the mechanism of action (Nardi et al. 2016). The overall action of biostimulants after their application in plants include penetration, translocation and transformation; signal regulation and gene expression which finally positively effect the plant. Scientists reported two commonly investigated methods: radiolabelling amino acids and mathematical modelling in order to investigate the penetration and absorption of amino acid or peptide-based biostimulants into plant tissue. Mathematical modelling also revealed the transport of the bioactive compounds present in the biostimulants to distant tissues. Protein hydrolates enter plant tissue through diffusion of proteins. These biostimulants are soluble in water and other solvents which make them easier to penetrate into the interior of the plants. Sometimes additives such as surfactants are added to improve solubility of the biostimulants (Kolomaznı et al. 2012, Pecha et al. 2011).

After penetration and translocation, the biostimulants activate several signalling pathways which result in effecting productivity. When the biostimulant binds to its receptor, activates the secondary messengers such as lipids, sugars, and ions, which activate several cellular responses (Wang and Irving 2011). Biostimulant uptake occurs usually by membrane-mediated actions while enzyme-based biostimulants interact via 'Lock and key' mechanism. Molecular methods such as microarray, metabolomic, proteomic and transcriptomic analysis are done to identify the changes in gene expression (Jannin et al. 2012; Santaniello et al. 2013).

Protein-based biostimulants are known to enhance plant growth and development and defence mechanisms. Proteins contain a peptide site called 'cryptides' or 'cryptein' which has been isolated from marine organisms by protein hydrolysis by the researchers and they confirmed that it triggers plant defence mechanisms (Yakhin et al. 2017).

Hormones regulate metabolic processes during plant growth and development. It has been reported that biostimulants directly effect the plant hormonal status (Kurepin et al. 2014). Hormones could be present as an active compound in a biostimulant or application of biostimulant may activate the hormone synthesis pathways. Sometimes bioactive components present in a biostimulant may act as precursor of hormone synthesis (Paradiković et al. 2011). Therefore, hormones play an important role while understanding the mechanism of action of biostimulants.

5 Role of Biostimulants in Agriculture

According to scientists (Yakhin et al. 2017), biostimulants can be defined as 'a formulated product of biological origin that improves plant productivity as a consequence of the novel or emergent properties of the complex of constituents and not as a sole consequence of the presence of known essential plant nutrients, plant growth regulators, or plant protective compounds' in order to differentiate biostimulants from other growth enhancers. It has been reported that various biostimulants have increased plant growth and the possible mode of actions could be enhanced metabolism which directly stimulates growth and germination, increased nutrient absorption and biostimulants can also lower the effects of abiotic stress by interacting with the signalling cascades associated with the controlling of various abiotic stress such as salinity, heat, drought and chilling. The possible functions that a biostimulant targets are depicted in Fig. 16.2.

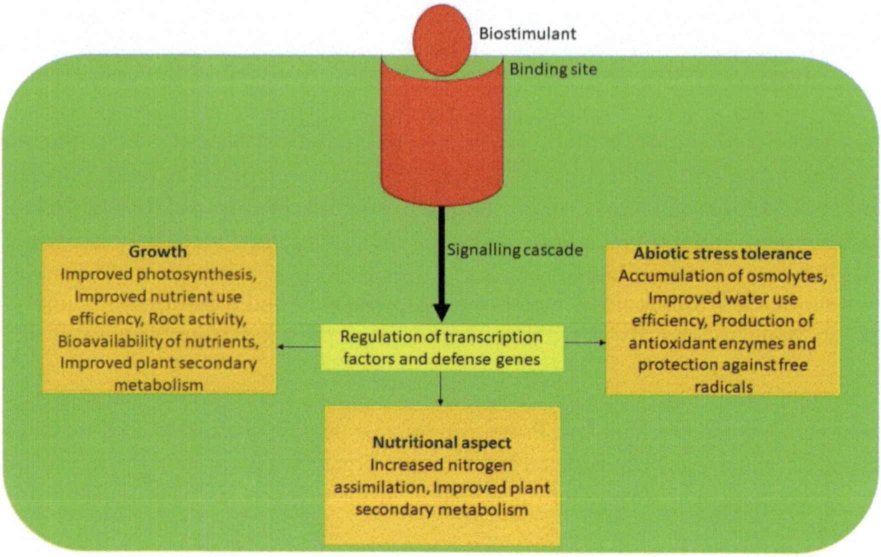

Fig. 16.2 Possible mechanism of actions of biostimulants

5.1 Crop Quality, Yield, Nutrient Uptake and Overall Physiological Performance

Agriculture is the science of production and maintaining of crops and livestock. Clinical researches have reported that fruit and vegetable consumption is the key to healthy and long human lives (Kyriacou and Rouphael 2017). Application of biostimulants is being currently suggested by various agronomists to enhance quality and yield of plants also making them tolerance to harsh environmental conditions. Scientists reported that seaweed extracts can be used as biostimulants as they optimised the chlorophyll biosynthesis pathway as a result producing intense coloured leaves (Khan et al. 2009) which is considered to be an essential parameter in crops for appealing the buyers. Studies have shown increased concentration of photosynthetic pigments in leafy vegetables when treated with biostimulants (Bulgari et al. 2014). Scientists reported the effects of using vegetal-based and protein hydrolysates that increased the antioxidant activities, metabolite contents and nutritional uptake in leafy vegetables, fruit and vegetable crops. Combined application of a number of commercial biostimulants improved fruit quality and metabolite accumulation (Soppelsa et al. 2019; Caruso et al. 2019; Gugała et al. 2019; Cozzolino et al. 2020). The synergistic effects of a number of commercial biostimulants whose primary ingredients are amino acids, organic acids and carbohydrates increased yield and improved fruit quality in pepper and tomato under stressful condition (Paradiković et al. 2011; Petrozza et al. 2013). Microbial biostimulants improve crop quality and yield by interacting with the metabolic pathways responsible for producing metabolites and antioxidant enzymes. Studies confirmed that plant growth-promoting strains of *Bacillus, Brevibacillus, Rhizobium, Pseudomonas, Azotobacter, Staphylococcus, Azospirillum and Kocuria* have significantly altered the quality of the products in terms of physical growth and abiotic stress tolerance (Yildirim et al. 2011; Fasciglione et al. 2012; Karlidag et al. 2013; Babu et al. 2015; Chandrasekaran et al. 2019). Brown seaweed extracts also stimulated plant growth and other metabolic activities in cabbage (Rengasamy et al. 2016). Apart from bacteria, arbuscular mycorrhizal fungi also function as biostimulant and have been reported to promote synthesis of metabolites and pigments and antioxidant enzymes (Caser et al. 2018; Caser et al. 2019). Several reports concluded that bioactive molecules present in the biostimulants based on vegetal extracts, seaweed extracts and smoke influenced the growth and photosynthetic pigment accumulation (Luziatelli et al. 2019; Kulkarni et al. 2019). Commercial biostimulants Goemar BM86 and Sesol have been reported to enhance macro- and micronutrients in broccoli along with increased leaf area, trunk diameter and total biomass (Gajc-Wolska et al. 2012). Moringa leaf extracts is being currently under examination. It has been reported that moringa leaf extracts when applied to the plants, it increases the vegetative growth and metabolite accumulation. It also regulates the stress responsive pathways and ensures improved growth under water stress (El Mageed et al. 2017; Abdel-rahman and Abdel-kader 2020).

Plant biostimulants increase the nutrient use efficiency which involves solubilisation of nitrogen, phosphorus, potassium and other essential macro- and micronutrients (De Pascale et al. 2017). Scientists showed that application of commercial biostimulants maintained plant growth under reduced nutrient supply (Papenfus et al. 2013; Koleška et al. 2017). The mechanism for nutrient uptake efficiency on biostimulant application is based on maintaining cell homeostasis and preventing oxidative stress. Studies showed that protein hydrolysates improved crop yield by enhancing nutrient uptake by the plants such as lettuce and baby spinach. It is also reported that nutrient uptake was associated with greater photosynthetic activity and secondary metabolite accumulation (Mola et al. 2019a, b, 2020). Application of a biostimulant containing both microbial and non-microbial products showed synergistic effects in growing lettuce. The yield increased under reduced nutrient supply and activation of various signalling pathways consequently improved the overall physical growth (Rouphael et al. 2020). The polysaccharide content in seaweed extracts increased iron uptake by plants thereby developing better root systems, increasing fruit production and leaf area (Vernieri et al. 2006; Spinelli et al. 2010). Improper nutrition often causes various diseases in plants. Biostimulant application may reduce the occurrence of the disease by enhancing nutrient uptake by plants. Scientists applied different biostimulant products that ensured increased yield and reduced the occurrence of Blossom-end rot in pepper caused due to calcium deficiency by efficient nutrient use (Parađiković et al. 2013).

5.2 *Production of Secondary Metabolites and Abiotic Stress Resistance*

The mechanism of stress tolerance by biostimulant is still under research. It is considered that the physiological effects caused in plants by biostimulants are due to the interaction of the bioactive molecules present in the biostimulant and the plant metabolic pathway. During stress, the plant metabolism gets activated which in turn helps the plant to adapt or recover themselves from the stress (Van Oosten et al. 2017). Biostimulants increase metabolites in plants which have antioxidant properties and play a key role in reduction of free radical accumulation during stressful situations (Staykov et al. 2021; Alharby et al. 2021). A number of recent studies based on the relationship between biostimulant and abiotic stress tolerance are listed in Table 16.2. Plant secondary metabolites are also directly related to the response to abiotic stress in plants. Biostimulants control the production and accumulation of the metabolites (Ashraf et al. 2018). The mode of action of biostimulants for the type of secondary metabolite it will produce depends on the type and severity of the stress. For example, a study showed the production of different classes of secondary metabolites under different intensities of salinity stress (Saia et al. 2021). Mode of action of biostimulants also depends on the type of the biostimulant. Scientists reported that microbial biostimulants accumulate secondary metabolites, namely, by overproducing reactive oxygen species (ROS) thereby activating the de novo pathway for antioxidant compound production (Ganugi et al. 2021). Humic substances interact with the plants differently depending upon their origin, doses, molecular size and structure (Nardi et al. 2021). In a study, protein hydrolysate increased carotene and lycopene production under normal water condition whereas, their level decreased under limited water availability (Francesca et al. 2021). Phenylpropanoid pathway is the major biochemical pathway that involves production of the secondary metabolites such as phenolic compounds. These phenolic compounds are required by the plants in order to adapt themselves to stressful conditions by activating the antioxidant potential. Studies show that biostimulants stimulate the production of phenolic compounds (Bulgari et al. 2017; Mola et al. 2019b). A study showed that amino acid and algal extract-based biostimulants increased the total phenolic acid content in Broccoli (Kałużewicz et al. 2017). Biostimulant application also promoted the production and function of anthocyanin and phenolic compounds in grapes (Kok 2021). Carotenoids are essential plant pigments taking part in photosynthesis. They also serve as nutraceuticals or antioxidant compounds. Scientists reported that carotenoid production in crops depends on the abiotic stress induction (Norshazila et al. 2017). A study revealed that biostimulant application on pepper increased the accumulation of carotenoids by activating the enzymes responsible for carrying out carotene biosynthesis (Barrajón-catalán et al. 2019). Application of biostimulants on lettuce also increased carotenoid accumulation on leaves (Mola et al. 2019a). Glucosinolates are an important class of secondary metabolites that play a major role in plant defence mechanism against stresses and also increase the nutritional quality of the plant (Hanschen and Rohn 2021).

Table 16.2 Some recent biostimulant-based researches revealing increased abiotic stress tolerance in agricultural products

Type of abiotic stress	Biostimulant	Model plant	Possible mechanism of action	References
Salinity stress	Diluted honey bee extract	Onion	Enhanced activity of plant antioxidative defence systems	Semida et al. (2019)
Salinity stress	*Azospirillum*	Lettuce	Enhanced expression of genes involved in antioxidant activity	Fasciglione et al. (2015)
Salinity and heavy metal stress	*Moringa oleifera* leaf extract	Beans	Increased activity of antioxidant enzymes	Howladar (2014)
Salinity stress	Liquorice root extract	Beans	Improvements in the activity of antioxidant defence systems	Rady et al. (2019)
Salinity stress	*Dunaliella salina* exopolysaccharides	Tomato	Activation and/or inhibition of various metabolic pathways including jasmonic acid-dependent pathways	El Arroussi et al. (2018)
Salinity stress	*Sargassum muticum* and *Jania rubens* (sea weeds)	Chickpea	Regulation of the four key amino acid (serine, threonine, proline and aspartic acid) metabolism	Latef et al. (2017)
Salinity stress	*Azospirillum brasilense* and *Pantoea dispersa*	Sweet pepper	The lower Cl^- accumulation in the mesophyll tissue may have improved reduction in the stomatal conductance	Del Amor and Cuadra-crespo (2012)
Drought stress	*Ascophyllum nodosum* (seaweed extract)	Spinach	Improved leaf water relations and maintained cell turgor pressure and stomatal conductance	Xu and Leskovar (2015)
Drought stress	*Azospirillum brasilense*	Tomato	Improved plant-water relationship through various mechanisms resulting in larger xylem vessel area of the stems, lower specific leaf area and a larger root system allowed a higher stem-specific conductivity	Romero et al. (2014)
Drought stress	Moringa leaf extract	Squash	Maintained higher relative water content, water use efficiency, osmoprotectants and lower electrolyte leakage	El Mageed et al. (2017)
Drought stress	*Ascophyllum nodosum* (sea weed extract)	Tomato	Maintenance of osmolytes and expression of tas14 dehydrin gene	Goñi et al. (2018)

(continued)

Table 16.2 (continued)

Type of abiotic stress	Biostimulant	Model plant	Possible mechanism of action	References
Drought stress	Fulvic acid	Tea	Regulation of starch, sucrose and phenylpropane activities, triterpenoid synthesis and heat shock proteins	Qiu et al. (2021)
Drought stress	Protein hydrolysate	Tomato	Increased accumulation of antioxidant molecules, carotenoids and lycopene	Francesca et al. (2021)
Heat stress	*Ascophyllum nodosum*	Tomato	Accumulation of soluble sugars and gene expression of protective heat shock proteins	Carmody et al. (2020)
Heat stress	*Ascophyllum nodosum* extracts and animal-based L-α amino acids	*Arabidopsis thaliana*	Activation of specific heat shock proteins, antioxidant systems and ROS scavengers	Cocetta et al. (2022)
Heat stress	Glutathione	Mung bean	Controlling antioxidant defence and methylglyoxal detoxification system	Nahar et al. (2015)
Heat stress	Ascorbic acid	Mung bean	Reduction of oxidative stress	Kumar et al. (2011)
Heat stress	Abscisic acid	Chickpea	Controlling the accumulation of osmolytes	Kumar et al. (2012)
Heat stress	Proline	Chickpea	Controlling the function of carbon enzymes and antioxidative pathways	Kaushal et al. (2011)
Chilling stress	Strains of *Arthrobacter, Flavimonas, Flavobacterium, Massilia, Pedobacter* and *Pseudomonas*	Tomato	Decreased membrane damage and activation of antioxidant enzymes and controlled proline synthesis in the leaves	Subramanian et al. (2016)
Chilling stress	*Pseudomonas* strains	Tomato	Enhanced antioxidant activity and improved expression of cold acclimation genes: LeCBF1andLeCBF3	Subramanian et al. (2015)
Chilling stress	5-aminolevulinic acid	Pepper	Enhanced relative water content, stomatal conductance and superoxide dismutase (SOD) enzyme activity and reduced membrane permeability	Korkmaz et al. (2010)
Chilling stress	Asahi SL/Goemar Goteo	Coriander	Increased L-ascorbic acid, phenolic concentration and antioxidant activity	Pokluda et al. (2016)

Biostimulant application manipulates the production and accumulation of glucosinolates in plants. Seaweed extracts elicited the crop quality by increasing the accumulation of glucosinolates in Broccoli (Hellín et al. 2018). A similar study showed that a fungal-based biostimulant promoted glucosinolate level in various leafy vegetables (Velasco et al. 2021). Apart from edible crops, biostimulants were also able to promote the secondary metabolite production in medicinal plants (Bernstein et al. 2019; Abeed et al. 2021). A study revealed that sago bagasse hydrolysate allowed to grow tomato plants with increased rate of germination and metabolite accumulation by controlling the expression in genes responsible for nitrogen and carbon metabolism (Kumar et al. 2019).

6 Global Market of Biostimulants

Biostimulants are newly discovered products and due to their effectivity and benefits, the industry for biostimulant is growing rapidly. According to research, the market value of biostimulants in Europe grew from €200 to €500 million in 2011–2013 with an annual growth potential of 10% or more. In Europe, France, Italy and Spain are leaders in biostimulant production. Biostimulant market in North America and USA valued $0.27 billion and $313.0 million in 2013 and 2014 respectively and growing each year. The biostimulants' market in the Asia-Pacific was valued at $0.25 billion in 2013 and India and China are considered to be contributing the poorest soil quality, increased soil degradation and use of chemicals in agricultural fields are persuading the need of healthy and sustainable alternatives to boost plant growth in agricultural sector. With increasing population and soil pollution, the demand of organic quality products is increasing thereby positively impacting the growth of global biostimulant market. In the year 2021, the global market size of biostimulants valued $2.79 billion and is expected that it will increase at the rate of 10.4% (compound annual growth rate) in the next 8 years. Europe is dominating the global biostimulant market from the year 2021 and has the share of over 37.5% of the total revenue. The second largest regional market in the year 2021 was achieved by Asia-Pacific with India and China having various production units and companies leading to rapid agricultural growth.

Although Central and South America is presently growing in a stagnant rate, agriculture-based country such as Brazil is having the scope of biostimulant market to grow exponentially in the coming years for a sustainable and eco-friendly agricultural practice. The biostimulant market growth and share depends on the crop type, active ingredients in the biostimulant product, the number of manufacturers and the location. Several companies are associated with research and development programmes and other marketing strategies such as high investments and forming joint ventures to increase the market. Some of the major companies playing an important role in the global biostimulant market are as follows: Isagro Group, BASF

SE, Biolchim SpA, Sapec Agro S.A., Platform Specialty Products Corp., Novozymes A/S, Valagro SpA, Italpollina SAP, Koppert B.V., Biostadt India Ltd. Apart from the geographical location, biostimulant market size can also be valued on the basis of crop type, biostimulant type and method of application. In 2021, row crops and cereals dominated the market with a revenue share of more than 59.5%. While turf and ornamental crops were in the second position. Based on the type of biostimulant, acid-based biostimulants played a significant role in the global biostimulant market. Seaweed extracts and microbial biostimulants are also gaining importance with time. In 2021, according to the method of application, foliar treatment influenced the market with a revenue share of more than 79.5%. This is because foliar treatment ensured faster nutrient use. Seed treatment is also gaining insight for its efficiency and cost-effectiveness due to limited labour requirement (Biostimulants Market Size, Share & Growth Report, 2030).

7 Limitations of Biostimulant

Due to unclear and unidentified mode of action of biostimulants, biostimulant market is often considered to be not based on science. Some studies reported the negative effects of biostimulants. For example, an amino acid-based biostimulant obtained from animal when applied on plant, it showed depression in growth whereas amino acid derived from plant showed positive effects (Cerdán et al. 2013). The complex nature of biological systems and biostimulants leads to ineffective responses. However, huge literature supports the fact that biostimulants are beneficial and are complementing the market growth thereby moving towards increased global food production. Development of biostimulants from broad range of raw materials, rigorous research and conferences and development of legal attributes is improving the market values. With growing market, the significant problem of biostimulant industry including highly complex and unidentified products is making it difficult to understand the mode of action and legal support is gained only on the basis of composition and not on the mode of action.

8 Conclusion

Despite having several beneficial applications, formulation of biostimulants is a complex procedure due the variable and heterogeneous nature of the starting materials and the final products. Therefore, it is difficult to assign a particular mode of action to a certain plant biostimulant. The efficiency of a biostimulant may vary when applied to different plant species with genetic variability. Also, the severity of environmental conditions or abiotic stress may compel the product to act differently. Moreover, it is recommended to consider that a biostimulant activity is the result of synergistic or antagonistic effect of all the components concerned with its production and application. Plant biostimulants are making notable offerings to sustainable crop management technology for resilient agriculture by replacing synthetic fertilisers. Although plant biostimulants are gathering importance and rapidly spreading its market, it is necessary to elucidate the molecular and physiological mechanisms of the biostimulants for a clear comprehensive knowledge among the agriculture practitioners. All the scientific outcomes that have been discussed in this chapter conclude that biostimulants are efficient growth-promoting compounds but cannot replace fertilisers completely on the present date. But the quantity of using chemical fertilisers can be reduced to some extent, thereby it can be considered as an initial step to reduce soil pollution besides sustainable agriculture. Scientists proposed that plant phenotypic, genomic and transcriptomic tools are opening new avenues for biostimulant formulation in order to amplify resource use efficiency and sustainable plant production (Briglia et al. 2019; Giovannini et al. 2020). The results obtained after biostimulant application both under greenhouse/laboratory or open field conditions are required to be validated depending on the guidelines proposed by the members of European Biostimulant Industry Council (EBIC) while justifying the plant-biostimulant potentiality (Ricci et al. 2019). Appropriate trials and data analyses depict plant performances and simultaneously acknowledge the quality of a plant biostimulant being suitable for commercial use.

References

Abdel-Rahman SSA, Abdel-Kader AAS (2020) Response of Fennel (*Foeniculum vulgare*, Mill) plants to foliar application of moringa leaf extract and benzyladenine (BA). S Afr J Bot 129:113–122

Abeed AHA, Ali M, Ali EF, Majrashi A, Eissa MA (2021) Induction of *Catharanthus roseus* secondary metabolites when *Calotropis procera* was used as bio-stimulant. Plan Theory 10:1623. https://doi.org/10.3390/plants10081623

Alharby HF, Al-zahrani HS, Hakeem KR, Alsamadany H, Desoky EM, Rady MM (2021) Silymarin-enriched biostimulant foliar application minimizes the toxicity of cadmium in maize by suppressing oxidative stress and elevating antioxidant gene expression. Biomol Ther 11:465. https://doi.org/10.3390/biom11030465

Allaga H, Bóka B, Poór P, Nagy VD, Szucs A, Stankovics I, Takó M, Manczinger L, Vágvölgyi C, Kredics L, Körmöczi P (2020) A composite bioinoculant based on the combined application of beneficial bacteria and fungi. Agronomy 10:220. https://doi.org/10.3390/agronomy10020220

Arthur G, Aremu AO, Moyo M, Stirk WA, van Staden J (2012) Growth-promoting effects of a seaweed concentrate at various pH and water hardness conditions. S Afr J Sci 109:1–6

Ashraf MA, Iqbal M, Rasheed R, Hussain I, Riaz M, Arif MS (2018) Environmental stress and secondary metabolites in plants: an overview. In: Plant metabolites and regulation under environmental stress, vol 8. Academic, pp 153–167

Babu AN, Jogaiah S, Ito S, Kestur A, Tran LP (2015) Improvement of growth, fruit weight and early blight disease protection of tomato plants by rhizosphere bacteria is correlated with their beneficial traits and induced biosynthesis of antioxidant peroxidase and polyphenol oxidase. Plant Sci 231:62–73

Barrajón-catalán E, Álvarez-martínez FJ, Borrás F, Pérez D, Herrero N, Ruiz JJ, Micol V (2019) Metabolomic analysis of the effects of a commercial complex biostimulant on pepper crops. Food Chem 310:125818. https://doi.org/10.1016/j.foodchem.2019.125818

Basak A (2012) The effect of seaweed product goemar bm 86 and pollinus, on the performance of chemical thinning using NAA and BA. Acta Hortic 932:109–115

Basile B, Rouphael Y, Colla G, Soppelsa S, Andreotti C (2020) Appraisal of emerging crop management opportunities in fruit trees, grapevines and berry crops facilitated by the application of biostimulants. Sci Hortic 267:109330. https://doi.org/10.1016/j.scienta.2020.109330

Battacharyya D, Babgohari MZ, Rathor P, Prithiviraj B (2015) Seaweed extracts as biostimulants in horticulture. Sci Hortic 196:39–48

Berg S, Dennis PG, Paungfoo-lonhienne C, Anderson J, Robinson N, Brackin R, Royle A, Dibella L, Schmidt S (2019) Effects of commercial microbial biostimulants on soil and root microbial communities and sugarcane yield. Biol Fertil Soils 56:565–580

Bernstein N, Gorelick J, Zerahia R, Koch S (2019) Impact of N, P, K, and humic acid supplementation on the chemical profile of medical cannabis (*Cannabis sativa* L). Front Plant Sci 10:736. https://doi.org/10.3389/fpls.2019.00736

Botta A (2013) Enhancing plant tolerance to temperature stress with amino acids: an approach to their mode of action. Acta Hortic 1009:29–36

Bradácová K, Florea AS, Bar-Tal A, Minz D, Yermiyahu U, Shawahna R, Kraut-cohen J, Zolti A, Erel R, Dietel K, Weinmann M, Zimmermann B, Berger N, Ludewig U, Neumann G, Posta G (2019) Microbial consortia versus single-strain inoculants: an advantage in PGPM-assisted tomato production. Agronomy 9:105. https://doi.org/10.3390/agronomy9020105

Briglia N, Petrozza A, Hoeberichts FA, Verhoef N, Povero G (2019) Investigating the impact of biostimulants on the row crops corn and soybean using high-efficiency phenotyping and next generation sequencing. Agronomy 9:761. https://doi.org/10.3390/agronomy9110761

Bulgari R, Podetta N, Cocetta G, Piaggesi A, Ferrante A (2014) The effect of a complete fertilizer for leafy vegetables production in family and urban gardens. Bulg J Agric Sci 20:1361–1367

Bulgari R, Morgutti S, Cocetta G, Negrini N, Farris S, Calcante A, Spinardi A, Ferrari E, Mignani I, Oberti R, Ferrante A (2017) Evaluation of borage extracts as potential biostimulant using a phenomic, agronomic, physiological, and biochemical approach. Front Plant Sci 8:935. https://doi.org/10.3389/fpls.2017.00935

Bulgari R, Franzoni G, Ferrante A (2019a) Biostimulants application in horticultural crops under abiotic stress conditions. Agronomy 9:306. https://doi.org/10.3390/agronomy9060306

Bulgari R, Trivellini A, Ferrante A (2019b) Effects of two doses of organic extract-based biostimulant on greenhouse lettuce grown under increasing NaCl concentrations. Front Plant Sci 9:1870. https://doi.org/10.3389/fpls.2018.01870

Campobenedetto C, Grange E, Mannino G, Arkel J, Van Beekwilder J, Karlova R, Garabello C, Contartese V, Bertea CM (2020) A biostimulant seed treatment improved heat stress tolerance during cucumber seed germination by acting on the antioxidant system and glyoxylate cycle. Front Plant Sci 11:836. https://doi.org/10.3389/fpls.2020.00836

Campobenedetto C, Agliassa C, Mannino G, Vigliante I, Contartese V, Secchi F, Bertea CM (2021a) A biostimulant based on seaweed (*Ascophyllum nodosum* and *Laminaria digitata*) and yeast extracts mitigates water stress effects on tomato (*Solanum lycopersicum* L.). Agriculture 11:557. https://doi.org/10.3390/agriculture11060557

Campobenedetto C, Mannino G, Beekwilder J, Contartese V, Karlova R, Bertea CM (2021b) The application of a biostimulant based on tannins affects root architecture and improves tolerance to salinity in tomato plants. Sci Rep 11:354. https://doi.org/10.1038/s41598-020-79770-5

Carmody N, Goñi O, Łangowski Ł, O'Connell S (2020) *Ascophyllum nodosum* extract biostimulant processing and its impact on enhancing heat stress tolerance during tomato fruit set. Front Plant Sci 11:807. https://doi.org/10.3389/fpls.2020.00807

Caruso G, Pascale S, De Cozzolino E, Cuciniello A, Cenvinzo V, Bonini P, Colla G, Rouphael Y (2019) Yield and nutritional quality of vesuvian piennolo tomato PDO as affected by farming system and biostimulant application. Agronomy 9:505. https://doi.org/10.3390/agronomy9090505

Caser M, Victorino ÍMM, Demasi S, Berruti A, Donno D, Lumini E, Bianciotto V, Scariot V (2018) Saffron cultivation in marginal alpine environments: how amf inoculation modulates yield and bioactive compounds. Agronomy 9:12. https://doi.org/10.3390/agronomy9010012

Caser M, Demasi S, Victorino ÍMM, Donno D, Faccio A, Lumini E, Bianciotto V, Scariot V (2019) Arbuscular mycorrhizal fungi modulate the crop performance and metabolic profile of saffron in soilless cultivation. Agronomy 9:232. https://doi.org/10.3390/agronomy9050232

Cerdán M, Sánchez-sánchez A, Jordá JD, Juárez M, Sánchez-andreu J (2013) Effect of commercial amino acids on iron nutrition of tomato plants grown under lime-induced iron deficiency. J Plant Nutr Soil Sci 176:859–866

Chandrasekaran M, Chun SC, Oh JW, Paramasivan M, Saini RK, Sahayarayan JJ (2019) *Bacillus subtilis* CBR05 for Tomato (*Solanum lycopersicum*) fruits in south Korea as a Novel Plant Probiotic Bacterium (PPB): Implications from total phenolics, flavonoids, and carotenoids content for fruit quality. Agronomy 9:838. https://doi.org/10.3390/agronomy9120838

Chiaiese P, Corrado G, Colla G, Kyriacou MC, Rouphael Y (2018) Renewable sources of plant biostimulation: microalgae as a sustainable means to improve crop performance. Front Plant Sci 9:1782. https://doi.org/10.3389/fpls.2018.01782

Cocetta G, Landoni M, Pilu R, Repiso C, Nolasco J, Alajarin M, Ugena L, Levy CCB, Scatolino G, Villa D, Ferrante A (2022) Priming treatments with biostimulants to cope the short-term heat stress response: a transcriptomic profile evaluation. Plan Theory 11:1130. https://doi.org/10.3390/plants11091130

Colla G, Rouphael Y (2015) Biostimulants in horticulture. Sci Hortic 196:1–2

Colla G, Rouphael Y, Mattia D, El-nakhel C, Cardarelli M (2014) Co-inoculation of *Glomus intraradices* and *Trichoderma atroviride* acts as a biostimulant to promote growth, yield and nutrient uptake of vegetable crops. J Sci Food Agric 95:1706–1715

Colla G, Hoagland L, Ruzzi M, Cardarelli M, Bonini P, Canaguier R, Rouphael Y (2017) Biostimulant action of protein hydrolysates: unraveling their effects on plant physiology and microbiome. Front Plant Sci 8:2202. https://doi.org/10.3389/fpls.2017.02202

Contartese V, Garabello C, Occhipinti A, Barbero F, Bertea CM (2016) Effects of a new biostimulant on gene expression and metabolic responses of tomato plants. Acta Hortic 1148:35–42

Cozzolino E, Giordano M, Fiorentino N, El-nakhel C, Pannico A, Di Mola I, Mori M, Kyriacou MC, Colla G, Rouphael Y (2020) Appraisal of biodegradable mulching films and vegetal-derived biostimulant application as eco-sustainable practices for enhancing lettuce crop performance and nutritive value. Agronomy 10:427. https://doi.org/10.3390/agronomy10030427

De Pascale S, Rouphael Y, Colla G (2017) Plant biostimulants: innovative tool for enhancing plant nutrition in organic farming. Eur J Hortic Sci 82:277–285

Del Amor FM, Cuadra-crespo P (2012) Plant growth-promoting bacteria as a tool to improve salinity tolerance in sweet pepper. Funct Plant Biol 39:82–90

du Jardin P (2015) Plant biostimulants: definition, concept, main categories and regulation. Sci Hortic 196:3–14

El Arroussi H, Benhima R, Elbaouchi A, Sijilmassi B, El Mernissi N, Aafsar A, Meftah-Kadmiri I, Bendaou N, Smouni A (2018) *Dunaliella salina* exopolysaccharides: a promising biostimulant for salt stress tolerance in tomato (*Solanum lycopersicum*). J Appl Phycol 30:2929–2941

El Mageed TAA, Semida WM, Rady MM (2017) Moringa leaf extract as biostimulant improves water use efficiency, physio-biochemical attributes of squash plants under deficit irrigation. Agric Water Manag 193:46–54

Ertani A, Schiavon M, Altissimo A, Franceschi C, Nardi S (2011) Phenol-containing organic substances stimulate phenylpropanoid metabolism in *Zea mays*. J Plant Nutr Soil Sci 174:496–503

Fasciglione G, Casanovas EM, Yommi A, Sueldo RJ, Barassi CA (2012) *Azospirillum* improves lettuce growth and transplant under saline conditions. J Sci Food Agric 92:2518–2523

Fasciglione G, Casanovas EM, Quillehauquy V, Yommi AK, Goñi MG, Roura SI, Barassi CA (2015) *Azospirillum* inoculation effects on growth, product quality and storage life of lettuce plants grown under salt stress. Sci Hortic 195:154–162

Ferrante A, Trivellini A, Vernier P, Piaggesi A (2013) Application of Actiwave® for improving the rooting of camellia cuttings. Acta Hortic 1009:213–218

Francesca S, Arena C, Mele BH, Schettini C, Ambrosino P, Barone A, Rigano MM (2020) The use of a plant-based biostimulant improves plant performances and fruit quality in tomato plants grown at elevated temperatures. Agronomy 10:363. https://doi.org/10.3390/agronomy10030363

Francesca S, Cirillo V, Raimondi G, Maggio A, Barone A, Rigano MM (2021) A novel protein hydrolysate-based biostimulant improves tomato performances under drought stress. Plan Theory 10:783. https://doi.org/10.3390/plants10040783

Franzoni G, Cocetta G, Prinsi B, Ferrante A, Espen L (2022) Biostimulants on crops: their impact under abiotic stress conditions. Horticulturae 8:189. https://doi.org/10.3390/horticulturae8030189

Gajc-Wolska J, Spiewski T, Grabowska A (2012) The effect of seaweed extracts on the yield and quality parameters of broccoli (*Brassica oleracea* var. *cymosa* L.) in open field production. Acta Hortic 1009:83–90

Ganugi P, Martinelli E, Lucini L (2021) Microbial biostimulants as a sustainable approach to improve the functional quality in plant-based foods: a review. Curr Opin Food Sci 41:217–223

García AC, Augusto T, De Castro VT, Santos LA, Carlos O, Tavares H, Castro RN, Luiz R, Berbara L, García-mina JM (2019) Structure–property–function relationship of humic substances in modulating the root growth of plants. Sci Rep 6:20798. https://doi.org/10.1038/srep20798

Giovannini L, Palla M, Agnolucci M, Avio L, Sbrana C, Turrini A, Giovannetti M (2020) Arbuscular mycorrhizal fungi and associated microbiota as plant biostimulants: research strategies for the selection of the best performing Inocula. Agronomy 10:106. https://doi.org/10.3390/agronomy10010106

Goñi O, Quille P, Connell SO (2018) *Ascophyllum nodosum* extract biostimulants and their role in enhancing tolerance to drought stress in tomato plants. Plant Physiol Biochem 126:63–73

Gugała M, Sikorska A, Zarzecka K, Findura P, Malaga-toboła U (2019) Chemical composition of winter rape seeds depending on the biostimulators used. Agronomy 9:716. https://doi.org/10.3390/agronomy9110716

Guinan KJ, Sujeeth N, Copeland RB, Jones PW, Brien NMO, Sharma HSS, Prouteau PFJ, O'Sullivan JT (2013) Discrete roles for extracts of *Ascophyllum nodosum* in enhancing plant growth and tolerance to abiotic and biotic stresses. Acta Hortic 1009:127–135

Hadwiger LA (2013) Multiple effects of chitosan on plant systems: solid science or hype. Plant Sci 208:42–49

Hanschen FS, Rohn S (2021) Advanced research on glucosinolates in food products. Foods 10:3148. https://doi.org/10.3390/foods10123148

Hellín P, Novo-Uzal E, Pedreño MA, Cava J, Garrido I, Molina MV, Fenoll J, Flores P (2018) Impact of plant biostimulants on the production of glucosinolates in broccoli. Acta Hortic 1194:99–104

Howladar SM (2014) A novel *Moringa oleifera* leaf extract can mitigate the stress effects of salinity and cadmium in bean (*Phaseolus vulgaris* L.) plants. Ecotoxicol Environ Saf 100:69–75

Jägermeyr J (2020) Agriculture's historic twin-challenge toward sustainable water use and food supply for all. Front Sustain Food Syst 4:35. https://doi.org/10.3389/fsufs.2020.00035

Jannin L, Arkoun M, Ourry A, Laîné P, Goux D, Garnica M, Fuentes M, Francisco SS, Baigorri R, Cruz F, Houdusse F, Garcia-Mina JM, Yvin JC, Etienne P (2012) Microarray analysis of humic acid effects on *Brassica napus* growth: involvement of N, C and S metabolisms. Plant Soil 359:297–319

Jonathan G, Rivka R, Avinoam S, Lumír H, Nirit B (2015) Hypoglycemic activity of withanolides and elicitated *Withania somnifera*. Phytochemistry 116:283–289

Juárez-Maldonado A, Ortega-Ortíz H, Morales-Díaz AB, González-Morales S, Morelos-Moreno Á, Cabrera-De la Fuente M, Sandoval-Rangel A, Cadenas-Pliego G, Benavides-Mendoza A (2019) Nanoparticles and nanomaterials as plant biostimulants. Int J Mol Sci 20:162. https://doi.org/10.3390/ijms20010162

Kalam O, Beom A, Park S, Mateja A, Ivan G, Sun K, Woo H, Park CH (2020) Silicon biostimulant enhances the growth characteristics and fortifies the bioactive compounds in common and Tartary buckwheat plant. J Crop Sci Biotechnol 24:51–59

Kałużewicz A, Gąsecka M, Spiżewski T (2017) Influence of biostimulants on phenolic content in broccoli heads directly after harvest and after storage. Folia Hortic 29(2):221–230

Karlidag H, Yildirim E, Turan M, Pehluvan M, Donmez F (2013) Plant growth-promoting Rhizobacteria mitigate deleterious effects of salt stress on strawberry plants (*Fragaria 3 ananassa*). HortScience 48:563–567

Kaushal N, Gupta K, Bhandhari K, Kumar S, Thakur P, Nayyar H (2011) Proline induces heat tolerance in chickpea (*Cicer arietinum* L.) plants by protecting vital enzymes of carbon and antioxidative metabolism. Physiol Mol Biol Plants 17:203–213

Khan W, Rayirath UP, Jithesh SSMN, Rayorath P, Hodges DM, Critchley AT, Craigie JS, Norrie J, Prithiviraj B (2009) Seaweed extracts as biostimulants of plant growth and development. J Plant Growth Regul 28:386–399

Klokić I, Koleška I, Hasanagić D, Murtić S, Bosančić B, Todorović V (2020) Biostimulants' influence on tomato fruit characteristics at conventional and low-input NPK regime. Acta Agric Scand Sect B Soil Plant Sci 70:233–240

Kok D (2021) Manipulating anthocyanin accumulation and phenolic compounds profile of cv. Trakya Ilkeren table grape (*V. vinifera* L.) by using foliar biostimulant and abscisic acid applications. Erwerbs-obstbau 63:15–21

Koleška I, Hasanagić D, Todorović V, Murtić S, Klokić I, Paradiković N, Kukavica B (2017) Biostimulant prevents yield loss and reduces oxidative damage in tomato plants grown on reduced NPK nutrition. J Plant Interact 12:209–218

Kolomaznı K, Pecha J, Friebrova V, Janacova D, Vasek V (2012) Diffusion of biostimulators into plant tissues. Heat Mass Transf 48:1505–1512

Korbecka-glinka G, Wiśniewska-wrona M, Kopania E (2021) The use of natural polymers for treatments enhancing sowing material. Polimery 66:11–20

Korkmaz A, Korkmaz Y, Demirkıran AR (2010) Enhancing chilling stress tolerance of pepper seedlings by exogenous application of 5-aminolevulinic acid. Environ Exp Bot 67:495–501

Kulkarni MG, Rengasamy KRR, Pendota SC, Gruzb J, Plačkovából L, Novákb O, Doležalb K, Van Stadena J (2019) Bioactive molecules derived from smoke and seaweed *Ecklonia maxima* showing phytohormone-like activity in *Spinacia oleracea* L. New Biotechnol 48:83–89

Kumar S, Kaur R, Kaur N, Bhandhari K, Kaushal N, Gupta K, Bains TS, Nayyar H (2011) Heat-stress induced inhibition in growth and chlorosis in mungbean (*Phaseolus aureus* Roxb.) is partly mitigated by ascorbic acid application and is related to reduction in oxidative stress. Acta Physiol Plant 33:2091–2101

Kumar S, Kaushal N, Nayyar H, Gaur P (2012) Abscisic acid induces heat tolerance in chickpea (*Cicer arietinum* L.) seedlings by facilitated accumulation of osmoprotectants. Acta Physiol Plant 34:1651–1658

Kumar S, Chinnannan K, Thamilarasana SKMS, Shanmuganathand R, Padikasan IA (2019) Biocatalysis and agricultural biotechnology enzymatically hydrolysed sago bagasse improves physiological, biochemical and molecular attributes of *Solanum lycopersicum*. Biocatal Agric Biotechnol 17:499–506

Kurepin LV, Zaman M, Pharis RP (2014) Phytohormonal basis for the plant growth-promoting action of naturally occurring biostimulators. J Sci Food Agric 94:1715–1722

Kyriacou MC, Rouphael Y (2017) Towards a new definition of quality for fresh fruits and vegetables. Sci Hortic 234:463–469

Latef AAHA, Srivastava AK, Saber H, Alwaleed EA, Tran LSP (2017) *Sargassum muticum* and *Jania rubens* regulate amino acid metabolism to improve growth and alleviate salinity in chickpea. Sci Rep 7:10537. https://doi.org/10.1038/s41598-017-07692-w

Lei Z, Mingyu S, Xiao W, Chao L, Chunxiang Q, Liang C, Hao H, Xiaoqing L, Fashui H (2008) Antioxidant stress is promoted by nano-anatase in spinach chloroplasts under UV-B radiation. Biol Trace Elem Res 121:69–79

Lola-luz T, Hennequart F, Gaffney M (2014) Effect on yield, total phenolic, total flavonoid and total isothiocyanate content of two broccoli cultivars (*Brassica oleraceae* var. *italica*) following the application of a commercial brown seaweed extract (*Ascophyllum nodosum*). Agric Food Sci 23:28–37

Lucini L, Rouphael Y, Cardarelli M, Canaguier R, Kumar P, Colla G (2015) The effect of a plant-derived biostimulant on metabolic profiling and crop performance of lettuce grown under saline conditions. Sci Hortic 182:124–133

Luziatelli F, Ficca AG, Colla G, Švecová EB, Ruzzi M (2019) Foliar application of vegetal-derived bioactive compounds stimulates the growth of beneficial bacteria and enhances microbiome biodiversity in lettuce. Front Plant Sci 10:60. https://doi.org/10.3389/fpls.2019.00060

Mannino G, Nerva L, Gritli T, Novero M, Fiorilli V, Bacem M, Bertea CM, Lumini E, Chitarra W, Balestrini R (2020) Effects of different microbial inocula on tomato tolerance to water deficit. Agronomy 10:170. https://doi.org/10.3390/agronomy10020170

Miceli A, Moncada A, Vetrano F (2021) Use of microbial biostimulants to increase the salinity tolerance of vegetable transplants. Agronomy 11:1143. https://doi.org/10.3390/agronomy11061143

Mola ID, Cozzolino E, Ottaiano L, Giordano M, Rouphael Y, Colla G, Mori M (2019a) Effect of vegetal- and seaweed extract-based biostimulants on agronomical and leaf quality traits of plastic tunnel-grown baby lettuce under four regimes of nitrogen fertilization. Agronomy 9:571. https://doi.org/10.3390/agronomy9100571

Mola ID, Ottaiano L, Cozzolino E, Senatore M, Giordano M, El-nakhel C, Sacco A, Rouphael Y, Colla G (2019b) Physiological, and qualitative responses of baby rocket leaves under diverse nitrogen conditions. Plan Theory 8:522. https://doi.org/10.3390/plants8110522

Mola ID, Cozzolino E, Ottaiano L, Nocerino S, Rouphael Y, Colla G, El-nakhel C, Mori M (2020) Nitrogen use and uptake efficiency and crop performance of baby spinach (*Spinacia oleracea* L.) and lamb's lettuce (*Valerianella locusta* L.) grown under variable sub-optimal n regimes combined with plant-based biostimulant application. Agronomy 10:278. https://doi.org/10.3390/agronomy10020278

Nahar K, Hasanuzzaman M, Alam M, Fujita M (2015) Exogenous glutathione confers high temperature stress tolerance in mung bean (*Vigna radiata* L.) by modulating antioxidant defense and methylglyoxal detoxification system. Environ Exp Bot 112:44–54

Nardi S, Pizzeghello D, Schiavon M, Ertani A (2016) Plant biostimulants: physiological responses induced by protein hydrolyzed- based products and humic substances in plant metabolism. Sci Agric 73:18–23

Nardi S, Schiavon M, Francioso O (2021) Chemical structure and biological activity of humic substances define their role as plant growth promoters. Molecules 26:2256. https://doi.org/10.3390/molecules26082256

Nephali L, Moodley V, Piater L, Steenkamp P, Buthelezi N, Dubery I, Burgess K, Huyser J, Tugizimana F (2021) A metabolomic landscape of maize plants treated with a microbial biostimulant under well-watered and drought conditions. Front Plant Sci 12:676632. https://doi.org/10.3389/fpls.2021.676632

Norrie J, Keathley JP (2006) Benefits of *Ascophyllum nodosum* marine-plant extract applications to 'Thompson seedless' grape production. Acta Hortic 727:243–248

Norshazila S, Othman R, Jaswir I, Yumi Zuhanis H (2017) Effect of abiotic stress on carotenoids accumulation in pumpkin plants under light and dark conditions. Int Food Res J 24:387–394

Papenfus HB, Kulkarni MG, Stirk WA, Finnie JF, Van Staden J (2013) Effect of a commercial seaweed extract (Kelpak®) and polyamines on nutrient-deprived (N, P and K) okra seedlings. Sci Hortic 151:142–146

Paradiković N, Vinković T, Vinković Vrček I, Žuntar I, Bojić M, Medić-Šarić M (2011) Effect of natural biostimulants on yield and nutritional quality: an example of sweet yellow pepper (*Capsicum annuum* L.) plants. J Sci Food Agric 91:2146–2152

Građiković N, Vinković T, Vrček IV, Tkalec M (2013) Natural biostimulants reduce the incidence of BER in sweet yellow pepper plants (*Capsicum annuum* L.). Agric Food Sci 22:307–317

Paul K, Sorrentino M, Lucini L, Rouphael Y, Bonini P, Begona M, Canaguier R, Reynaud H, Panzarova K, Colla G (2019) A combined phenotypic and metabolomic approach for elucidating the biostimulant action of a plant-derived protein hydrolysate on tomato grown under limited water availability. Front Plant Sci 10:493. https://doi.org/10.3389/fpls.2019.00493

Pecha J, Furst T, Kolomaznik K, Friebrova V, Svoboda P (2011) Protein biostimulant foliar uptake modeling: the impact of climatic conditions. AIChE 58:2010–2019

Petrozza A, Summerer S, Di Tommaso G, Di Tommaso D, Piaggesi A (2013) An evaluation of tomato plant root development and morpho-physiological response treated with VIVA® by image analysis. Acta Hortic 1009:155–160

Pokluda R, Sękara A, Jezdinský A, Kalisz A, Neugebauerová J, Grabowska A (2016) The physiological status and stress biomarker concentration of *Coriandrum sativum* L. plants subjected to chilling are modified by biostimulant application. Biol Agric Hortic 32:258–268

Qi M, Liu Y, Li T (2013) Nano-TiO_2 improve the photosynthesis of tomato leaves under mild heat stress. Biol Trace Elem Res 156:323–328

Qiu C, Sun J, Shen J, Zhang S, Ding Y, Gai Z, Fan K, Song L, Chen B, Ding Z, Wang Y (2021) Fulvic acid enhances drought resistance in tea plants by regulating the starch and sucrose metabolism and certain secondary metabolism. J Proteome 247:10433. https://doi.org/10.1016/j.jprot.2021.104337

Rady MM, Desoky EM, Elrys AS, Boghdady MS (2019) Can licorice root extract be used as an effective natural biostimulant for salt-stressed common bean plants? S Afr J Bot 121:294–305

Rahman M, Mukta JA, Sabir AA, Gupta DR, Mohi-ud-din M, Hasanuzzaman M, Miah G, Rahman M, Islam T (2018) Chitosan biopolymer promotes yield and stimulates accumulation of antioxidants in strawberry fruit. PLoS One 13:e0203769. https://doi.org/10.1371/journal.pone.0203769

Raliya R, Nair R, Chavalmane S, Wang W, Biswas P (2015) Mechanistic evaluation of translocation and physiological impact of titanium dioxide and zinc oxide nanoparticles on the tomato (*Solanum lycopersicum* L.) plant. Metallomics 7:1584–1594

Rathore SS, Chaudhary DR, Boricha GN, Ghosh A, Bhatt BP, Zodape ST, Patolia JS (2009) Effect of seaweed extract on the growth, yield and nutrient uptake of soybean (*Glycine max*) under rainfed conditions. S Afr J Bot 75:351–355

Rengasamy KRR, Kulkarni MG, Pendota SC, Staden JV (2016) Enhancing growth, phytochemical constituents and aphid resistance capacity in cabbage with foliar application of eckol – a biologically active phenolic molecule from brown seaweed. New Biotechnol 33:273–279

Ricci M, Tilbury L, Daridon B, Sukalac K (2019) General principles to justify plant biostimulant claims. Front Plant Sci 10:494. https://doi.org/10.3389/fpls.2019.00494

Romero AM, Vega D, Correa OS (2014) *Azospirillum brasilense* mitigates water stress imposed by a vascular disease by increasing xylem vessel area and stem hydraulic conductivity in tomato. Appl Soil Ecol 82:38–43

Rouphael Y, Colla G (2018) Synergistic biostimulatory action: designing the next generation of plant biostimulants for sustainable agriculture. Front Plant Sci 9:1655. https://doi.org/10.3389/fpls.2018.01655

Rouphael Y, Colla G (2020a) Biostimulants in Agriculture. Front Plant Sci 11:40. https://doi.org/10.3389/fpls.2020.00040

Rouphael Y, Colla G (2020b) Toward a sustainable agriculture through plant biostimulants: from experimental data to practical applications. Agronomy 10:1461. https://doi.org/10.3390/agronomy10101461

Rouphael Y, Cardarelli M, Bonini P, Colla G (2017) Synergistic action of a microbial-based biostimulant and a plant derived-protein hydrolysate enhances lettuce tolerance to alkalinity and salinity. Front Plant Sci 8:131. https://doi.org/10.3389/fpls.2017.00131

Rouphael Y, Kyriacou MC, Petropoulos SA, De Pascale S, Colla G (2018) Improving vegetable quality in controlled environments. Sci Hortic 234:275–289

Rouphael Y, Carillo P, Colla G, Fiorentino N, Sabatino L, El-nakhel C, Giordano M, Pannico A, Cirillo V, Shabani E, Cozzolino E, Lombardi N, Napolitano M, Woo SL (2020) Appraisal of combined applications of *trichoderma virens* and a biopolymer-based biostimulant on lettuce agronomical, physiological, and qualitative properties under variable N regimes. Agronomy 10:96. https://doi.org/10.3390/agronomy10020196

Ruzzi M, Aroca R (2015) Plant growth-promoting rhizobacteria act as biostimulants in horticulture. Sci Hortic 196:124–134

Saia S, Corrado G, Vitaglione P, Colla G, Bonini P, Giordano M, Stasio E, Di Raimondi G, Sacchi R, Rouphael Y (2021) An Endophytic fungi-based biostimulant modulates volatile and non-volatile secondary metabolites and yield of greenhouse basil (*Ocimum basilicum* L.) through variable mechanisms dependent on salinity stress level. Pathogens 10:797. https://doi.org/10.3390/pathogens10070797

Santaniello A, Giorgi FM, Di Tommaso D, Di Tommaso G, Piaggesi A, Perata P (2013) Genomic approaches to unveil the physiological pathways activated in *Arabidopsis* treated with plant-derived raw extracts. Acta Hortic 1009:161–174

Savvas D, Ntatsi G (2015) Biostimulant activity of silicon in horticulture. Sci Hortic 196:66–81

Semida WM, El-mageed TAA, Hemida K, Rady MM (2019) Natural bee-honey based biostimulants confer salt tolerance in onion via modulation of the antioxidant defence system. J Hortic Sci Biotechnol 94:632–642

Sharma SHS, Lyons G, McRoberts C, McCall D, Carmichael E, Andrews F, Swan R, McCormack R, Mellon R (2012) Biostimulant activity of brown seaweed species from Strangford Lough: compositional analyses of polysaccharides and bioassay of extracts using mung bean (*Vigno mungo* L.) and pak choi (*Brassica rapa chinensis* L.). J Appl Phycol 24:1081–1091

Shekhar SHS, Lyons G, McRoberts C, McCall D, Carmichael E, Andrews F, McCormack R (2012) Brown seaweed species from Strangford Lough: compositional analyses of seaweed species and biostimulant formulations by rapid instrumental methods. J Appl Phycol 24:1141–1157

Sidhu V, Nandwani D, Wang L, Wu Y (2017) A study on organic tomatoes: effect of a biostimulator on phytochemical and antioxidant activities. J Food Qual 2017:5020742. https://doi.org/10.1155/2017/5020742

Soppelsa S, Kelderer M, Casera C, Bassi M, Robatscher P, Matteazzi A, Andreotti C (2019) Foliar applications of biostimulants promote growth, yield and fruit quality of strawberry plants grown under nutrient limitation. Agronomy 9:483. https://doi.org/10.3390/agronomy9090483

Spinelli F, Fiori G, Noferini M, Sprocatti M, Costa G (2010) A novel type of seaweed extract as a natural alternative to the use of iron chelates in strawberry production. Sci Hortic 125:263–269

Stasinska-Jakubas M, Hawrylak-Nowak B (2022) Protective, biostimulating, and eliciting effects of chitosan and its derivatives on crop plants. Molecules 27:2801. https://doi.org/10.3390/molecules27092801

Staykov NS, Angelov M, Petrov V, Minkov P, Kanojia A, Guinan KJ, Alseekh S, Fernie AR, Sujeeth N, Gechev TS (2021) An *Ascophyllum nodosum*-derived biostimulant protects model and crop plants from oxidative stress. Metabolites 11:24. https://doi.org/10.3390/metabo11010024

Subramanian P, Mageswari A, Kim K, Lee Y, Sa T (2015) Psychrotolerant Endophytic *Pseudomonas* sp. Strains OB155 and OS261 induced chilling resistance in tomato Plants (*Solanum lycopersicum* Mill.) by Activation of their antioxidant capacity. Int Soc Mol Plant Microbe Interact 28:1073–1081

Subramanian P, Kim K, Krishnamoorthy R, Mageswari A, Selvakumar G, Sa T (2016) Cold stress tolerance in psychrotolerant soil bacteria and their conferred chilling resistance in tomato (*Solanum lycopersicum* Mill.) under low temperatures. PLoS One 11:e0161592. https://doi.org/10.1371/journal.pone.0161592

Tandon S, Dubey A (2015) Effect of biozyme (*Ascophyllum Nodosum*) biostimulant on growth and development of soybean [*Glycine Max* (L.) Merill]. Commun Soil Sci Plant Anal 46:845–858

Traon D, Amat L, Zotz F, Jardin PD (2014) A legal framework for plant biostimulants and agronomic fertiliser additives in the EU. Arcadia International

Trevisan S, Manoli A, Quaggiotti S (2019) A novel biostimulant, belonging to protein hydrolysates, mitigates abiotic stress effects on maize seedlings grown in hydroponics. Agronomy 9:28. https://doi.org/10.3390/agronomy9010028

Turan M (2021) Effect of biostimulants on yield and quality of cherry tomatoes grown in fertile and stressed soils. Hortic Sci 56:414–423

Van Der Watt E, Pretorius JC (2013) A triglyceride from *Lupinus albus* L. seed with bio-stimulatory properties. Afr J Biotechnol 35:5431–5443

Van Oosten MJ, Pepe O, De Pascale S, Silletti S, Maggio A (2017) The role of biostimulants and bioeffectors as alleviators of abiotic stress in crop plants. Chem Biol Technol Agric 4:5. https://doi.org/10.1186/s40538-017-0089-5

Velasco P, Rodríguez VM, Soengas P, Poveda J (2021) *Trichoderma hamatum* increases productivity, glucosinolate content and antioxidant potential of different leafy *Brassica* vegetables. Plan Theory 10:2449. https://doi.org/10.3390/plants10112449

Vernieri P, Borghesi E, Tognoni F, Ferrante A, Serra G, Piaggesi A (2006) Use of biostimulants for reducing nutrient solution concentration in floating system. Acta Hortic 718:477–484

Wang YH, Irving HR (2011) Developing a model of plant hormone interactions. Plant Signal Behav 6:494–500

Xu L, Geelen D (2018) Developing biostimulants from agro-food and industrial by-products. Front Plant Sci 9:1567. https://doi.org/10.3389/fpls.2018.01567

Xu C, Leskovar DI (2015) Effects of *A. nodosum* seaweed extracts on spinach growth, physiology and nutrition value under drought stress. Sci Hortic 183:39–47

Yakhin OI, Lubyanov AA, Yakhin IA, Brown PH (2017) Biostimulants in plant science: a global perspective. Front Plant Sci 7:2049. https://doi.org/10.3389/fpls.2016.02049

Yildirim E, Karlidag H, Turan M, Dursun A, Goktepe F (2011) Growth, nutrient uptake, and yield promotion of broccoli by plant growth-promoting Rhizobacteria with manure. Hortic Sci 46:932–936

Chapter 17
Soil Salinity and Sustainable Agriculture

Muhammad Nauman, Safura Bibi, Athar Mahmood ,
Muhammad Mansoor Javaid, Muhammad Azeem,
and Muhammad Ather Nadeem

Abstract Climate change is the major cause of environmental stresses, which badly affect agricultural crops. Abiotic stresses such as heat, drought, and soil salinization, among this soil salinity, are the main problem that is very dangerous to global food security and environmental sustainability. Soil salinity causes osmotic stress, led to an imbalance of nutrient uptake, and disturbs plant growth. It is necessary to overcome this problem through different approaches like the use of compost, UV radiations, nanotechnology, integrated agro-farming systems, salty farming combined with subsurface drainage, tolerant bacteria combined with cultivars of tolerant plants, and other emerging reclamation strategies. This chapter focuses on the strategies in order to restore agricultural sustainability and ensure global food security in the face of climate change leading to an increase in soil salinity.

Keywords Soil salinity · Sustainable · Strategies · Compost · Priming · Hormones

M. Nauman · S. Bibi
Department of Botany, Faculty of Sciences, University of Agriculture, Faisalabad, Pakistan

A. Mahmood (✉)
Department of Agronomy, Faculty of Agriculture, University of Agriculture, Faisalabad, Pakistan
e-mail: athar.mahmood@uaf.edu.pk

M. M. Javaid (✉) · M. A. Nadeem
Department of Agronomy, College of Agriculture, University of Sargodha, Sargodha, Pakistan
e-mail: mansoor.javaid@uos.edu.pk

M. Azeem
Biosaline Research Laboratories, Department of Botany, University of Karachi, Karachi, Pakistan

M. Hasanuzzaman (ed.), *Climate-Resilient Agriculture, Vol 2*,
https://doi.org/10.1007/978-3-031-37428-9_17

1 Introduction

Abiotic stress is the term used to describe how environmental factors negatively affect plants in a particular environment. Drought, salt, very low or high temperatures, and others are some of the stresses. Abiotic stress causes changes in a plant's morphology, physiology, biochemistry, and molecular structure that are harmful to its growth and productivity (Balal et al. 2016). Numerous genes that are activated by abiotic stress increase the concentration of a number of proteins and metabolites, some of which may have a role in the body's ability to defend itself against different stresses (Wasaya et al. 2021). Plants are directly affected by many environmental conditions, which result in changes to their morphological and anatomical characteristics (Tanveer et al. 2012; Tripathi et al. 2017). It has been stated that abiotic stress, such as water stress, flood, and heavy metal, causes about 50% of crop loss. Any external biotic (herbivore) or abiotic factor that decreases photosynthesis and makes it difficult for plants to turn energy into biomass is stress (Mustafa et al. 2022). The production of food must increase by 70% to feed the world expanding population. To meet the growing need for food, reducing agricultural losses caused by numerous environmental stresses is a serious challenge (Crist et al. 2017) Salinization reduces the growth of the crop and yield. The stability, development of bioenergy, and yield of basic food crops are significantly impacted by the primary abiotic stresses like excessive salt, water stress, cold, and heat up to 70%. Salt stress produces toxic minerals on plants such as Na^+ and Cl^-. It has been established that soil salinity predated agriculture and humans, but the issue has only recently arisen as a result of agricultural methods like poor irrigation (Shahzad et al. 2018; Giordano et al. 2021). Low soil water is one of major problems in the world. Among various effects of drought on plants, the first is the prevention of cell growth and division. This has an impact on normal physiological and chemical activities, such as ion absorption, respiration, photosynthesis, translocation, and nutrition metabolism. This inhibition can impact seed germination and the reduction of cell elongation during the early phases of plant growth. Lack of water inhibits leaf growth, which lowers photosynthetic area and pigments (chlorophyll *a* and *b*). Additionally, reactive oxygen species (ROS) production sets up oxidative damage in thylakoids. Chlorophyll and carotenoids in plants are affected, and the core complexes of photosystems I (PSI) and II (PSII) degrade as a result and inhibit crucial photochemical activities (Shahid et al. 2014; Guidi et al. 2017). Heat stress occurs at multiple time scales and with varying levels of severity and duration (Ishimaru et al. 2016). Heat stress has significant deleterious effects, especially on crops throughout the thermally sensitive developmental phases of early establishment, flowering, and gametogenesis (Jagadish 2020; Mahmood et al. 2022).

2 Sustainable Agriculture

Sustainable agriculture is a crucial component of world development (Janker et al. 2018). Three distinct views, such as those of economic stability, social stability, and environmental balance, must be used to approach the idea of sustainable agriculture (Farooq et al. 2019). Achieving the Sustainable Development Goals (SDG) depends on preserving and enhancing the sustainability of deteriorating irrigated drylands (UNDP, 2017). It has been documented that agronomic advances are beneficial for raising specific parameters (Devkota et al. 2015). However, because salinization impacts soils, crops, and the climate, a thorough assessment of innovations and technology is required to ascertain whether they have the potential to boost the sustainability of agricultural production (Chattha et al. 2017; Hopmans et al. 2021). Multiple sustainability indicators should be systematically quantified and compared, as various cropping systems, crops, and agronomic management techniques. This will give recommendations for adaptation and future steps to enhance the sustainability of deteriorating irrigated drylands. Additionally, in many situations, combining simulation, experimental, and multi-criteria techniques is required to increase sustainability. CA-based techniques (no-tillage, crop rotation, and residue retention) could improve the sustainability of crop production in salinity-affected irrigated drylands in Central Asia. These techniques were paired with water-saving nitrogen fertilizer rates and alternative wet and dry (AWD) irrigation techniques (Devkota et al. 2022; Mukhtar et al. 2022).

3 Soil Salinity

In order to feed the growing world population, which is expected to reach over 9.8 billion people in 2050, the soil is a crucial resource (UN 2020). Too much salt in the soil has made it saline, which is dangerous for both the environmental health and agriculture outputs (Díaz et al. 2021; Wei et al. 2021). Loss of soil fertility, changes to the properties of the soil, and detrimental effects on the environmental functions of the soil are all caused by saline buildup in the lower area or soil surface (Gorji et al. 2020). Saline–alkaline soils have high salt concentrations, pH levels, and Na concentrations. Salinized soils have electrical conductivity values of more than 4 dS m^{-1} and ESP greater than 15. Acidic soils have similar values of EC less than 4 dS m^{-1}, ESP greater than 15, and pH higher than 8.5, whereas alkaline soils have the opposite values of pH (Qamar et al. 2013; Seifi et al. 2020).

The main cause of the alkaline soil high pH is a high carbonate concentration (Decock et al. 2015). As a result, increased salinization levels cause the loss of the soil resources that are now accessible, which has an impact on agricultural growth and ecological well-being (Kumar et al. 2020). There are two main factors that contribute to soil salinity, i.e., natural (primary salinization) and anthropogenic activities (secondary salinization). The main natural sources of soil salinization are

Table 17.1 Distribution of salt-affected regions around the globe (Million ha) (Shahid et al. 2018)

Area	Sodic soil	Saline soil	Total	Percent
Africa	26.9	53.5	80.4	8.60
America	69.3	77.6	146.9	15.8
Australasia	340.0	17.6	357.6	38.4
Europe	22.9	7.8	30.8	3.30
Asia	121.9	94.7	316.5	33.9

constituent elements, physical or chemical weathering of minerals, and marine water intrusion (Ramos et al. 2020). Poor drainage conditions exacerbate the conditions of manmade salinization (Zain et al. 2017; Manasa et al. 2020).

Saline soils are soils with salt concentrations that impede plant growth (ECe > 4 dS m^{-1}). Salt stress refers to the effects that these salt accumulations have on the development, production, and seed quality of plants. Problems including improper agricultural land use, lack of rain, excessive evaporation, and poor drainage all cause the salinization of the land (Okur and Örçen 2020). Salinity stress is a crucial factor that lowers plant production. It is affecting 20% of irrigated soil and limiting food production by 30%. In salt-affected plants, nutritional issues and oxidative stress lead to hazardous levels of sodium and chlorine in the cell organelles and cytosol, which also have an impact on water uptake (Table 17.1; Machado and Serralheiro 2017).

The agriculture industry is essential to the nation's economy (Khan et al. 2020). This specific sector is responsible for producing close to 20% of the country's income. Its contribution to the GDP is 21%, and its employment as a whole is 43.7% of all employment. The country non-urban population, which makes up to 66% of the total population, depends on agriculture for at least some of their income (Abdullah et al. 2015; Elahi et al. 2022). 831 M ha of the world soil is influenced by salinity, of which 397 M ha is saline soil and 434 M ha is sodic soil (Kulshershthsa et al. 2022). Different natural or human-induced processes led to 4.5 M ha salt-affected areas in Pakistan (Aslam 2016). Natural processes include weathering of the original material and deposition of sea salt carried by rainfall and wind. As an alternative, human-induced activities such as irrigation using saltwater, a rise in the water table due to excessive irrigation, and poor drainage can also be considered. Because there is very little salt draining in these locations, salt builds up on the soil's surface, which is a serious danger. After soluble salts have been moved into the subsoil through leaching, a secondary consequence of soil salinity occurs where sodium is bonded to the soil due to the negative charges on clay (Zara et al. 2017; Leogrande and Vitti 2019).

According to the analysis, approximately 66.4% of the samples exhibited top 0–20 cm soil electrical conductivity (EC) values, and 72.8% showed sodium content (ESP) levels. Similar to this, 60.8% of the EC readings and 72% of the ESP values at soil depths of 20–40 cm exceeded safe limits. 56.8% of EC values and 79.2% of ESP values for soil depths between 40 and 60 cm exceeded the permissible limits (Solangi et al. 2019). The increased soluble salt content of the soil inhibits the uptake and metabolism of vital mineral nutrients for plants. The intake of vital nutrients such as P, K^+, N, and Ca^{2+} decreased due to particular ion toxicities brought on by increased salt uptake, or by excessive Na^+, Cl, or sulfate (SO_4^{-2}) ions (Yarami and Sepaskhah 2016).

3.1 *Effects of Salinity on Plant*

Salinity inhibits plant growth due to low osmotic potential of the soil solution and an inadequate intake of salt (Safdar et al. 2019). Plant growth is affected by salt in the soil and water for two reasons. Firstly, the presence of salt in the soil solution slows the plant's ability to absorb water, which in turn slows its rate of growth. Second, if the plant absorbs too much salt through the transpiration stream, the cells in the transpiring leaves are damaged, which could result in further growth reductions. This is called the salt-specific or ion-excess effect of salinity (Parihar et al. 2015). Plant growth may be adversely influenced by more acidity and toxicity of Na_2CO_3, HCO_3, and other anions in sodic soils. In the end, this harms plant nutrition and metabolism. High salinity and high EC of soil have a number of detrimental effects that result in plant cell dehydration, reduced growth, and even fatality in more sensitive plants (Syed et al. 2021). The reduced osmotic potential of the soil solution in saline soils has detrimental effects that can result in nutritional imbalances, ion toxicity, physiological drought, or a combination of all these issues. By building up salts in the shoots, it hinders the growth of plants and agricultural crops. Some distinct symptoms of salt-affected soils include restricted root growth, restricted flowering, marginal or leaf tip browning/burning, diminished vigor, and low crop yield (Sonon et al. 2015).

Osmotic stress, which is brought on by increased soil salinity, reduces the amount of water that plants need, resulting in physiological dryness. After these circumstances, the plant experiences ionic stress and its ionic balance degrades when an increase in Na^+ and Cl^- ions results from ionic stress in the medium, and they compete with K^+, Ca^{2+}, and Mg^{2+} leading to nutritional deficiency in plants. Osmotic and ionic stresses are the primary effects of salinity, whereas structural changes and the production of hazardous substances are the indirect effects (Fig. 17.1; Shahid et al. 2020).

Plants are impacted by salt stress due to toxicity brought on by osmotic pressures and ions. These impacts cause some unfavorable alterations in plants (Hussain et al. 2019). At the germination and seedling growth periods, plants are particularly sensitive. Many processes that are anticipated to take place naturally as the plant develops slow down or stop during these stages. Physiological dryness, sterility, stunted development, decreased leaf area, slow or absent blooming, disrupted membrane, production of the ROS, and decreased photosynthetic activity are a few more constraints that can occur. High salinity inhibits plant growth by reducing the fresh and dry weights of leaves and the shoot and root development (Hussain et al. 2019).

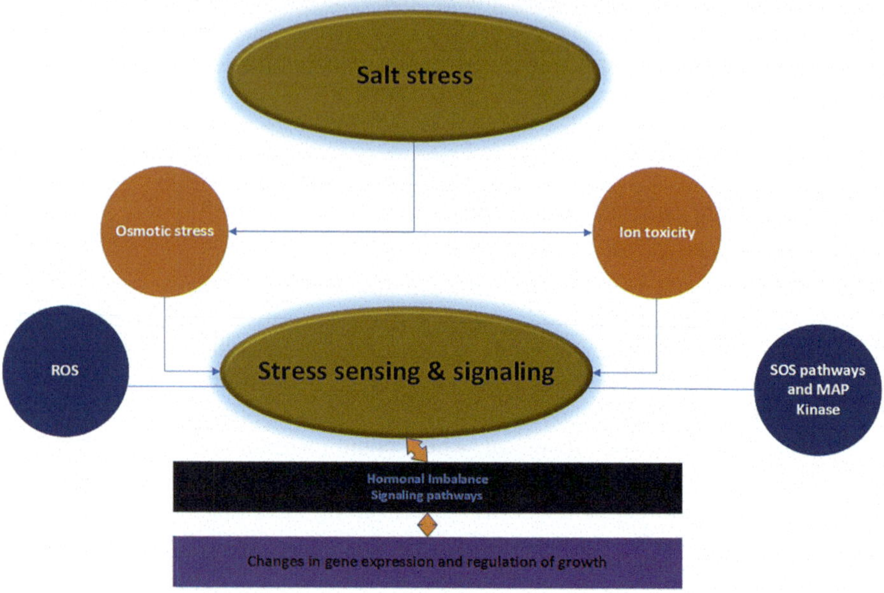

Fig. 17.1 Flow sheet diagram showing the effect of salt stress on plants (Hussain et al. 2019)

Although soluble salts are present in all soils, salty and sodic conditions cause an overabundance of salts in the root zone, which deteriorates the physical, chemical, and biological qualities of the soil. Salinity has a negative impact on plant growth, which is controlled by environmental factors such as the growing season, temperature, humidity, light, air pollution, and plant characteristics like growth stage, species, and variety. Soil factors include salinity level, moisture content, heating rate, and levels of heavy elements (Zamin et al. 2019).

4 Approaches to Mitigate Soil Salinity

4.1 Use of Compost

Municipal solid waste compost and green manure compost increase soil salinity but subsequently noticeably decrease it over time (El Azzouzi et al. 2019). Composting techniques for recovering salty soils have included a mix of sugarcane compost and green waste compost. Since there were salts that are dissolved in the compost and there was not enough flushing initially, this enhanced the soil EC over the first 100 days. The replacement of the Na^+ with Ca^{2+}, in soil exchange places and leaching of the solute, however, caused the EC to drop to 2.8 from 16.65 dS m^{-1} after 120 days, increasing the soil organic carbon content by 34.6% (Singh et al. 2019).

4.2 Phytoremediation

Phytoremediation is a widely used process in which we use plant species to reduce the salinity of the soil like *Lycium chinense*. Agroforestry and halophytic plants (*Sauda vera*) are available to overcome the salinity (Mukhopadhyay et al. 2021).

4.3 PGPB

Plant growth-promoting bacteria (PGPB) are presented in the soil rhizosphere, the area of soil around a plant root. Due to the presence of exudates of root, which serve as an attractant for a variety of organisms in soil plant, rhizosphere is rich in PGPB (Singh and Strong 2016). The improved plant growth induced by PGPB under stressful conditions is accompanied by a decrease in stress induced ethylene levels, which is mediated by antioxidative enzyme activity (Naing et al. 2021). Creating siderophores, phosphate solubilization, and nitrogen fixation to improve plant nutrition (Etesami and Beattie 2017). Exopolysaccharides are produced, which scavenge excess sodium ions and prevent their migration to plant leaves, thereby reducing sodium ion accumulation in plant roots (Qin et al. 2016; Etesami and Beattie 2017). Catalase (CAT) reduced glutathione (GR), superoxide dismutase (SOD), and ascorbate peroxidase (APX) are increased in plants to combat oxidative stress (Islam et al. 2016; Qin et al. 2016). The expression of ion transporters regulated a high sodium and potassium ratio and reduced ion toxicity (Fig. 17.2; Islam et al. 2016; Etesami et al. 2022; Mudassir et al. 2018).

4.4 Mycorrhiza

Mycorrhiza is recognized for assisting in the mobilization and solubilization of nutrients even in saline regions. There are reports on the positive effects of mycorrhiza, such as increased nutrient mobility and availability in soils (Chang et al. 2018). Furthermore, the vesicular–arbuscular mycorrhiza fungus (VAM) maintenance of a high K^+/Na^+ ratio in salty soils revealed its salt-tolerance mechanisms (Zahir et al. 2019). There are conflicting results that claim that rise in the concentration of salt in the medium reduced VAM colonization in wheat (Zafar et al. 2015; Zhu et al. 2016).

4.5 Priming

Physical techniques to boost plant growth have the potential to be used on a large scale and have environmental benefits. These techniques are an alternative to the present chemical or biological-based ones. Physical invigoration techniques are

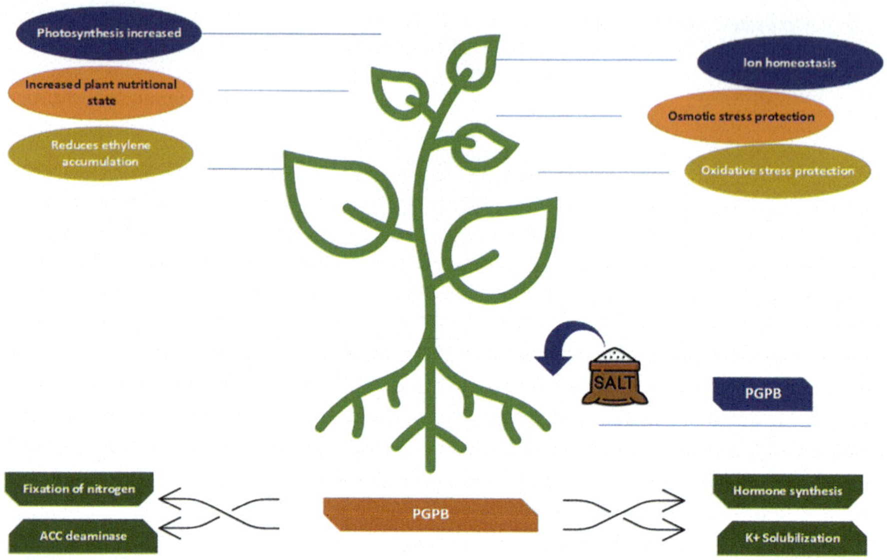

Fig. 17.2 Plant growth-promoting traits of PGPB for salinity stress tolerance in plants (Bhise and Dandge 2019)

also known as "physical priming." Physical invigoration techniques have been proven to be a successful strategy for creating new biotechnology-based solutions for the expanding seed industry (Araujo et al. 2016). According to several reports, effective pre-sowing seed treatments include the use of elements including temperature, ultraviolet (UV), and ionizing radiation (Araujo et al. 2016). In general, pre-sowing exposure to non-lethal concentrations of these physical agents has a good impact on plant health, which improves growth and yield in addition to encouraging germination. Breaking seed dormancy with pre-sowing temperature treatments is a common strategy that improves the general seed quality in batches (Liu and El-Kassaby 2015).

4.6 Use of Sulfur and Salicylic Acid

The formation of glycine betaine, photosynthetic, nitrogen fixation, proline metabolic activity, the control of the antioxidant defense system, and interactions among plants and water are only a few of the crucial physiological activities that salicylic acid controls in plants. As a result, it has been widely reported that salicylic acid protects plants from abiotic stresses (Khan et al. 2015; Wani et al. 2016). Numerous studies have demonstrated the role of salicylic acid in plant resistance to a range of abiotic stimuli, including drought, metals and metalloids, ozone, UV-B radiation, and temperature stress (Zhang et al. 2015) (Siboza et al. 2017). On the other hand, keeping the levels of plant mineral nutrients can aid plants in their ability to withstand stress (Jahan et al. 2020).

Soils contain sulfur in both organic and inorganic forms. Elementary sulfur or any of its form (sulfide, sulfate, thiosulfate) serves as a representation for the inorganically bound, which make up only 10–15% of the total sulfur. About 75–90% of all sulfur is organically bonded, which is found in organic substances such as amino acids, proteins, polypeptides, and others. Sulfur concentrations in soils are typically between 0.0% and 2.0% in humid regions, 1.0% in moorland soils, and up to 3.5% in marshland. The quantity of sulfur added by fertilizer and the content of organic matter are all important for the significant fluctuations in the total sulfur in soils (Morrison and Mojzsis 2021). The importance of the primary constituents and by-products of sulfur absorption has been well documented in plants' ability to withstand abiotic stressors, such as metals (Hussain et al. 2021), and chilling. Notably, in plants under stress, there is a strong relationship between salicylic acid and sulfur in terms of their physiological roles. For instance, the interaction outcomes of salicylic acid and sulfur significantly influenced plant growth, metabolism, and stress tolerance; exogenous SA was associated with increased GR activity and S/Cys-GSH concentration (Pal et al. 2014). Due to enhanced ATP-S and serine acetyltransferase (SAT) activity mediated by salicylic acid, cysteine and GSH levels were elevated (Nazar et al. 2015; Waheed et al. 2020).

4.7 Use of Nanofertilizer

The requirement to use excessive amounts of fertilizers has significantly increased in the current scenario of changing climate and expanding human population, but using chemical fertilizers excessively has led to the discharge of potentially dangerous compounds into the environment. In this regard, nanofertilizers have been suggested as a reliable solution to the issue (Yusefi-Tanha et al. 2020). According to reports, the use of nanofertilizers increases resistance to adverse environmental conditions, such as salinity (Adibah et al. 2020). *Zea mays* and *Arabidopsis thaliana* are subjected to saline regimes in order to investigate the impact of a betaine-rich nanofertilizer (50 mM). Under salt stress, enhanced growth characteristics and development were observed in both plant species. Another study found that tomato plants cultivated in salt benefited from the foliar application of commercial nanofertilizer based on the Lithovit VR standard (LITHO) to increase flower counts, leaf Ca^{2+}, and Mg^{2+} and decrease cellular leakage (Sassine et al. 2022). Zn-based nanofertilizer boosted growth and mitigated salt detrimental effects on plant and biomass in cotton. Si-based nanofertilizer improved the growth and output of cucumber plants, a potential worldwide vegetable crop (Yassen et al. 2017; Hussein and Abou-Baker 2018).

4.8 *Ultraviolet Radiation*

About 8–9% of all solar radiations correspond to the UV light component of the electromagnetic spectrum, which falls within the non-ionizing region (Coohill and Sutherland 1989). The amount of UV rising as a result of the stratospheric ozone layer thinning and research into the processes by which plants might defend themselves from this danger are growing as well. The molecular profile of these defensive responses, which are triggered by UV radiation in plants, could be utilized to develop new therapies to enhance plant sensitivity to abiotic restrictions like salinity. There is very little information on the subject of UV radiation's potential role in salinity relief (Ouhibi et al. 2014). During the investigation, two UV-C radiation doses were applied: 0.85 kJ/m² and 3.42 kJ/m². Fresh weights of roots and leaves decreased, K^+ ion absorption was hindered, and Na^+ ion concentrations increased in unprimed seeds under salt stress. UV-C therapy showed a dose-dependent impact, with its salt-mitigating effect being stronger at 0.85 kJ m^2 than at 3.42 kJ m^2. These impacts were decreased when plants were grown from UV-C primed seeds. The authors proposed UV-C priming as a quick and low-cost method for reducing stress brought on by NaCl in lettuce (Ouhibi et al. 2014). The production and concentration of phytochemicals are beneficial to health, the extending of the life span of fresh plant products, or the stimulation of plant defense against biotic stress Urban et al. 2016).

5 Conclusion

Different approaches like biological (PGPRs, fungi), chemical (NPs, sulfur), and phytohormone (salicylic acid) applications to the plants proved to be the best growth stimulator under abiotic stresses like salinity, which induced oxidative damages in plants. The medium applications of PGPRs to plants have been known to directly better plant growth and development. To address the needs of a world with increasing soil salinity problems, innovative ways for reducing saline in plants and promoting the production of crop under salinity stress are being developed. Different chemical, biochemical, and phytohormone applications are used in stress tolerance. The nanoparticles, sulfur, PGPRs, and salicylic acid enhanced plant growth under salt stress conditions. Different mechanisms are applied to tolerate the salt stress. Application in roots, priming, or foliar applications are highly effective in mitigating salt stress.

References

Abdullah DZ, Khan SA, Jebran K, Ali ABS (2015) Agricultural credit in Pakistan: past trends and future prospects. J Appl Environ Biol Sci 5(12):178–188

Adibah FSF, Jahan MS, Fatihah HNN (2020) Betaine-rich Nano fertilizer improves growth parameters of *Zea mays* var. saccharata and *Arabidopsis thaliana* under salt stress. Bulg J Agric Sci 26(1):177–185

Araujo SS, Paparella S, Dondi D, Bentivoglio A, Carbonera D, Balestrazzi A (2016) Physical methods for seed invigoration: advantages and challenges in seed technology. Front Plant Sci 7:646

Aslam M (2016) Agricultural productivity current scenario, constraints and future prospects in Pakistan. Sarhad J Agric 32(4):289–303

Balal RM, Shahid MA, Javaid MM, Iqbal Z, Anjum MA, Garcia-Sanchez F, Mattson NS (2016) The role of selenium in amelioration of heat-induced oxidative damage in cucumber under high temperature stress. Acta Physiol Plant 38:1–14

Bhise KK, Dandge PB (2019) Mitigation of salinity stress in plants using plant growth promoting bacteria. Symbiosis 79(3):191–204. https://doi.org/10.1007/s13199-019-00638-y

Chang W, Sui X, Fan X-X, Jia T-T, Song F-Q (2018) Arbuscular mycorrhizal symbiosis modulates antioxidant response and ion distribution in salt-stressed *Elaeagnus angustifolia* seedlings. Front Microbiol 9:652

Chattha M, Hassan M, Khan I, Chattha M, Munir H, Nawaz M, Mahmood A, Usman M, Kharal M (2017) Alternate skip irrigation strategy ensure sustainable sugarcane yield. J Anim Plant Sci 27:1604–1610

Coohill TP, Sutherland JC (1989) Free-electron lasers in ultraviolet photobiology. J Opt Soc Am B 6(5):1079–1082. https://doi.org/10.1364/JOSAB.6.001079

Costa SF, Martins D, Agacka-Mołdoch M, Czubacka A, Sousa Araújo SD (2018) Strategies to alleviate salinity stress in plants. In: Salinity responses and tolerance in plants, vol 1. Springer, pp 307–337

Crist E, Mora C, Engelman R (2017) The interaction of human population, food production, and biodiversity protection. Science 356(6335):260–264

Decock C, Lee J, Necpalova M, Pereira EP, Tendall DM, Six J (2015) Mitigating N 2 O emissions from soil: from patching leaks to transformative action. Soil 1(2):687–694

Devkota M, Martius C, Gupta R, Devkota K, McDonald A, Lamers J (2015) Managing soil salinity with permanent bed planting in irrigated production systems in Central Asia. Agric Ecosyst Environ 202:90–97

Devkota KP, Devkota M, Rezaei M, Oosterbaan R (2022) Managing salinity for sustainable agricultural production in salt-affected soils of irrigated drylands. Agric Syst 198:103390

Díaz FJ, Sanchez-Hernandez JC, Notario JS (2021) Effects of irrigation management on arid soils enzyme activities. J Arid Environ 185:104330

Elahi E, Khalid Z, Zhang Z (2022) Understanding farmers' intention and willingness to install renewable energy technology: a solution to reduce the environmental emissions of agriculture. Appl Energy 309:118459

El Azzouzi M, El Idrissi Y, Baghdad B, El Hasini S, Saufi H (2019) Assessment of trace metal contamination in periurban soils in the region of Kenitra-Morocco. Mediterr. J Chem 8(2):108–114

Etesami H, Beattie GA (2017) Plant-microbe interactions in adaptation of agricultural crops to abiotic stress conditions. In: Kumar V, Kumar M, Sharma S, Prasad R (eds) Probiotics and Plant Health. Springer Singapore, Singapore, pp 163–200. https://doi.org/10.1007/978-981-10-3473-2_7

Etesami H, Li Z, Maathuis FJM, Cooke J (2022) The combined use of silicon and arbuscular mycorrhizas to mitigate salinity and drought stress in rice. Environ Exp Bot 201:104955. https://doi.org/10.1016/j.envexpbot.2022.104955

Farooq M, Rehman A, Pisante M (2019) Sustainable agriculture and food security. In: Innovations in sustainable agriculture. Springer, pp 3–24

Giordano M, Petropoulos SA, Rouphael Y (2021) Response and defence mechanisms of vegetable crops against drought, heat and salinity stress. Agriculture 11(5):463

Gorji T, Yildirim A, Hamzehpour N, Tanik A, Sertel E (2020) Soil salinity analysis of Urmia Lake Basin using Landsat-8 OLI and Sentinel-2A based spectral indices and electrical conductivity measurements. Ecol Indic 112:106173

Guidi L, Tattini M, Landi M (2017) How does chloroplast protect chlorophyll against excessive light. Chlorophyll 21:21–36
Hopmans JW, Qureshi A, Kisekka I, Munns R, Grattan S, Rengasamy P, Ben-Gal A, Assouline S, Javaux M, Minhas P (2021) Critical knowledge gaps and research priorities in global soil salinity. Adv Agron 169:1–191
Hussain S, Bai Z, Huang J, Cao X, Zhu L, Zhu C, Khaskheli MA, Zhong C, Jin Q, Zhang J (2019) 1-Methylcyclopropene modulates physiological, biochemical, and antioxidant responses of rice to different salt stress levels. Front Plant Sci 10:124
Hussain SJ, Khan NA, Anjum NA, Masood A, Khan MIR (2021) Mechanistic elucidation of salicylic acid and sulphur-induced defence systems, nitrogen metabolism, photosynthetic, and growth potential of mungbean (*Vigna radiata*) under salt stress. J Plant Growth Regul 40(3):1000–1016
Hussein M, Abou-Baker N (2018) The contribution of nano-zinc to alleviate salinity stress on cotton plants. R Soc Open Sci 5(8):171809
Ishimaru T, Xaiyalath S, Nallathambi J, Sathishraj R, Yoshimoto M, Phoudalay L, Samson B, Hasegawa T, Hayashi K, Arumugam G (2016) Quantifying rice spikelet sterility in potential heat-vulnerable regions: field surveys in Laos and southern India. Field Crop Res 190:3–9
Islam F, Yasmeen T, Arif MS, Ali S, Ali B, Hameed S, Zhou W (2016) Plant growth promoting bacteria confer salt tolerance in *Vigna radiata* by up-regulating antioxidant defense and biological soil fertility. Plant Growth Regul 80(1):23–36
Jagadish SK (2020) Heat stress during flowering in cereals–effects and adaptation strategies. New Phytol 226(6):1567–1572
Jahan B, AlAjmi MF, Rehman MT, Khan NA (2020) Treatment of nitric oxide supplemented with nitrogen and sulfur regulates photosynthetic performance and stomatal behavior in mustard under salt stress. Physiol Plant 168(2):490–510
Janker J, Mann S, Rist S (2018) What is sustainable agriculture? Critical analysis of the international political discourse. Sustainability 10(12):4707
Khan MIR, Fatma M, Per TS, Anjum NA, Khan NA (2015) Salicylic acid-induced abiotic stress tolerance and underlying mechanisms in plants. Front Plant Sci 6:462
Khan I, Raza MA, Awan SA, Shah GA, Rizwan M, Ali B, Tariq R, Hassan MJ, Alyemeni MN, Brestic M (2020) Amelioration of salt induced toxicity in pearl millet by seed priming with silver nanoparticles (AgNPs): the oxidative damage, antioxidant enzymes and ions uptake are major determinants of salt tolerant capacity. Plant Physiol Biochem 156:221–232
Kumar P, Avtar R, Dasgupta R, Johnson BA, Mukherjee A, Ahsan MN, Nguyen DCH, Nguyen HQ, Shaw R, Mishra BK (2020) Socio-hydrology: a key approach for adaptation to water scarcity and achieving human well-being in large riverine islands. Prog Disaster Sci 8:100134
Leogrande R, Vitti C (2019) Use of organic amendments to reclaim saline and sodic soils: a review. Arid Land Res Manag 33(1):1–21
Liu Y, El-Kassaby YA (2015) Timing of seed germination correlated with temperature-based environmental conditions during seed development in conifers. Seed Sci Res 25(1):29–45
Machado RMA, Serralheiro RP (2017) Soil salinity: effect on vegetable crop growth. Management practices to prevent and mitigate soil salinization. Horticulturae 3(2):30
Mahmood A, Bibi S, Naqve M, Javaid MM, Zia MA, Jabbar A, Ud-Din W, Attia KA, Khan N, Al-Doss AA (2022) Physiological, biochemical, and yield responses of linseed (*Linum usitatissimum* L.) in α-tocopherol-mediated alleviation of salinity stress. Front Plant Sci 13:867172
Manasa M, Katukuri NR, Nair SSD, Haojie Y, Yang Z, Guo RB (2020) Role of biochar and organic substrates in enhancing the functional characteristics and microbial community in a saline soil. J Environ Manag 269:110737
Morrison PR, Mojzsis SJ (2021) Tracing the early emergence of microbial sulfur metabolisms. Geomicrobiol J 38(1):66-86. https://doi.org/10.1080/01490451.2020.1812773
Mudassir A, Muhammad A, Javaid MM (2018) Enhancement in cotton growth and yield using novel growth promoting substances under water limited conditions. Pak J Bot 50:1691–1701

Mukhopadhyay R, Sarkar B, Jat HS, Sharma PC, Bolan NS (2021) Soil salinity under climate change: challenges for sustainable agriculture and food security. J Environ Manag 280:111736

Mukhtar A, Awan MI, Sadaf S, Mahmood A, Javed T, Shah AN, Shabbir R, Alotaibi SS, Shah AA, Adamski R (2022) Sulfur enhancement for the improvement of castor bean growth and yield, and sustainable biodiesel production. Front Plant Sci 13:905738

Mustafa M, Szalai Z, Divéky-Ertsey A, Gál I, Csambalik L (2022) Conceptualizing multiple stressors and their consequences in agroforestry systems. Stresses 2:242–257

Naing AH, Maung TT, Kim CK (2021) The ACC deaminase-producing plant growth-promoting bacteria: influences of bacterial strains and ACC deaminase activities in plant tolerance to abiotic stress. Physiol Plant 173(4):1992–2012

Nazar R, Umar S, Khan N, Sareer O (2015) Salicylic acid supplementation improves photosynthesis and growth in mustard through changes in proline accumulation and ethylene formation under drought stress. S Afr J Bot 98:84–94

Okur B, Örçen N (2020) Soil salinization and climate change. In: Climate change and soil interactions. Elsevier, pp 331–350

Ouhibi C, Attia H, Rebah F, Msilini N, Chebbi M, Aarrouf J, Urban L, Lachaal M (2014) Salt stress mitigation by seed priming with UV-C in lettuce plants: growth, antioxidant activity and phenolic compounds. Plant Physiol Biochem 83:126–133

Pal DK, Wani SP, Sahrawat KL, Srivastava P (2014) Red ferruginous soils of tropical Indian environments: A review of the pedogenic processes and its implications for edaphology. CATENA 121:260–278. https://doi.org/10.1016/j.catena.2014.05.023

Palacios OA, Bashan Y, de Bashan LE (2014) Proven and potential involvement of vitamins in interactions of plants with plant growth-promoting bacteria—an overview. Biol Fertil Soils 50(3):415–432

Parihar P, Singh S, Singh R, Singh VP, Prasad SMJEs, research P (2015) Effect of salinity stress on plants and its tolerance strategies: a review. Environ Sci Pollut Res Int 22(3):4056–4075

Qamar R, Rehman A, Ali A, Ghaffar A, Mahmood A, Javeed HMR, Aziz M (2013) Growth and economic assessment of wheat under tillage and nitrogen levels in rice-wheat system. Am J Plant Sci 4:2083

Qin Y, Druzhinina IS, Pan X, Yuan Z (2016) Microbially mediated plant salt tolerance and microbiome-based solutions for saline agriculture. Biotechnol Adv 34(7):1245–1259

Ramos TB, Castanheira N, Oliveira AR, Paz AM, Darouich H, Simionesei L, Farzamian M, Gonçalves MC (2020) Soil salinity assessment using vegetation indices derived from Sentinel-2 multispectral data. application to Lezíria Grande, Portugal. Agric Water Manag 241:106387

Safdar H, Amin A, Shafiq Y, Ali A, Yasin R, Shoukat A, Hussan MU, Sarwar MI (2019) A review: impact of salinity on plant growth. Nat Sci 17(1):34–40

Sassine Y, Sajyan T, El Zarzour A, Abdelmawgoud A, Germanos M, Alturki S (2022) Integrative effects of biostimulants and salinity on vegetables: Contribution of bioumik and Lithovit®-urea50 to improve salt-tolerance of tomato. The Estonian University of Life Sciences, Tartu, Estonia, EMU DSpace

Seifi M, Ahmadi A, Neyshabouri M-R, Taghizadeh-Mehrjardi R, Bahrami H-A (2020) Remote and Vis-NIR spectra sensing potential for soil salinization estimation in the eastern coast of Urmia hyper saline lake, Iran. Remote Sens Appl Soc Environ 20:100398

Shahid MA, Balal RM, Javaid M, Aquel MA, Abbas T, Mattson N, Islam S (2014) Mitigation of stress induced by salinity and nickel in pea (*Pisum sativum* L.) by exogenous application of synthetic proline and proline-enriched *Lolium perenne* (L.) leaf extract. In: 2014 ASHS annual conference

Shahid SA, Zaman M, Heng L (2018) Soil salinity: historical perspectives and a world overview of the problem. In: Guideline for salinity assessment, mitigation and adaptation using nuclear and related techniques. Springer, pp 43–53

Shahid MA, Sarkhosh A, Khan N, Balal RM, Ali S, Rossi L, Gómez C, Mattson N, Nasim W, Garcia-Sanchez F (2020) Insights into the physiological and biochemical impacts of salt stress on plant growth and development. Agronomy 10(7):938

Shahzad T, Ashraf M, Javaid MM, Waheed H, Abbas T, Sattar F (2018) influence of field soil drought stress on some key physiological, yield and quality traits of selected newly-developed hexaploid bread wheat (*Triticum aestivum* L.) cultivars. Sains Malaysiana 47:2625–2635

Siboza XI, Bertling I, Odindo AO (2017) Enzymatic antioxidants in response to methyl jasmonate and salicylic acid and their effect on chilling tolerance in lemon fruit [Citrus limon (L.) Burm. F.]. Sci Hortic 225:659–667

Singh JS, Strong P (2016) Biologically derived fertilizer: a multifaceted bio-tool in methane mitigation. Ecotoxicol Environ Saf 124:267–276

Singh YP, Arora S, Mishra VK, Dixit H, Gupta RK (2019) Plant and soil responses to the combined application of organic amendments and inorganic fertilizers in degraded sodic soils of indo-gangetic plains. Commun Soil Sci Plant Anal 50(20):2640–2654

Solangi GS, Siyal A, Babar M, Siyal P (2019) Spatial analysis of soil salinity in the Indus River Delta, Pakistan. Eng Technol Appl Sci Res 9:4271–4275

Sonon LS, Saha U, Kissel DE (2015) Soil salinity testing, data interpretation and recommendations. University of Georgia Extension 1019

Syed A, Sarwar G, Shah SH, Muhammad S (2021) Soil salinity research in 21st century in Pakistan: its impact on availability of plant nutrients, growth and yield of crops. Commun Soil Sci Plant Anal 52(3):183–200

Tanveer A, Arshad MS, Ayub M, Javaid MM, Yaseen M (2012) Effect of temperature, light, salinity, drought stress and seeding depth on germination of Cucumis melo var. *agrestis*. Pak J Weed Sci Res 18:445–459

Tripathi DK, Singh S, Yadav V, Arif N, Singh S, Dubey NK, Chauhan DK (2017) Silicon: a potential element to combat adverse impact of UV-B in plants. In: UV-B radiation: from environmental stressor to regulator of plant growth. Wiley, pp 175–195

UN (2020) The sustainable development goals report 2020. United Nations. https://unstats.un.org/sdgs/report/2020/#sdg-goals

Urban L, Charles F, de Miranda MRA, Aarrouf J (2016) Understanding the physiological effects of UV-C light and exploiting its agronomic potential before and after harvest. Plant Physiol Biochem 105:1–11

Waheed H, Khan MA, Khan HMT, Javaid MM, Rahman FU, Aslam MM (2020) Enhancement of some key physiological, morphological and biochemical traits of watermelon induced by Trichoderma harzianum fungi. Semina Ciências Agrárias 41:2047–2060

Wani SH, Kumar V, Shriram V, Sah SK (2016) Phytohormones and their metabolic engineering for abiotic stress tolerance in crop plants. Crop J 4(3):162–176

Wasaya A, Abbas T, Yasir TA, Sarwar N, Aziz A, Javaid MM, Akram S (2021) Mitigating drought stress in sunflower (*Helianthus annuus* L.) through exogenous application of β-aminobutyric acid. Soil Sci Plant Nutr 21:936–948

Wei Y, Ding J, Yang S, Wang F, Wang C (2021) Soil salinity prediction based on scale-dependent relationships with environmental variables by discrete wavelet transform in the Tarim Basin. Catena 196:104939

Yarami N, Sepaskhah A (2016) Effect of irrigation water salinity, manure application and planting method on qualitative compounds of saffron (*Crocus sativus* L.). Int J Plant Prod 10(2):197–219

Yassen A, Abdallah E, Gaballah M, Zaghloul S (2017) Role of silicon dioxide nano fertilizer in mitigating salt stress on growth, yield and chemical composition of cucumber (*Cucumis sativus* L.). Int J Agric Res 22:130–135

Yusefi-Tanha E, Fallah S, Rostamnejadi A, Pokhrel LR (2020) Zinc oxide nanoparticles (ZnONPs) as a novel nanofertilizer: influence on seed yield and antioxidant defense system in soil grown soybean (*Glycine max* cv. Kowsar). Sci Total Environ 738:140240

Zafar M, Tanveer A, Majeed MA, Safdar ME, Ali HH, Javed MM (2015) Reducing herbicide dose in sugarcane by application of plant water extract. Herbologia 15. https://doi.org/10.5644/Herb.15.1.07

Zahir ZA, Nadeem SM, Khan MY, Binyamin R, Waqas MR (2019) Role of halotolerant microbes in plant growth promotion under salt stress conditions. In: Saline soil-based agriculture by halotolerant microorganisms. Springer, In, pp 209–253

Zain M, Khan I, Chattha M, Qadri R, Anjum S, Hassan M, Mahmood A, Ilyas M (2017) Foliar applied thiourea at different growth stages modulated late sown wheat. Pak J Sci 69:39

Zamin M, Khattak AM, Salim AM, Marcum KB, Shakur M, Shah S, Jan I, Fahad S (2019) Performance of *Aeluropus lagopoides* (mangrove grass) ecotypes, a potential turfgrass, under high saline conditions. Environ Sci Pollut Res 26(13):13410–13421

Zara Z, Iqbal J, Ayub K, Irfan M, Mahmood A, Khera RA, Eliasson B (2017) A comparative study of DFT calculated and experimental UV/Visible spectra for thirty carboline and carbazole based compounds. J Mol Struct 1149:282–298

Zhang Y, Xu S, Yang S, Chen Y (2015) Salicylic acid alleviates cadmium-induced inhibition of growth and photosynthesis through upregulating antioxidant defense system in two melon cultivars (*Cucumis melo* L.). Protoplasma 252(3):911–924

Zhu X, Song F, Liu S, Liu F (2016) Role of arbuscular mycorrhiza in alleviating salinity stress in wheat (*Triticum aestivum* L.) grown under ambient and elevated CO2. J Agron Crop Sci 202(6):486–496

Chapter 18
Mechanism and Approaches to Enhance Salt Stress Tolerance in Crop Plants

Saima Asghar, Safura Bibi, Asif Mukhtiar, Muhammad Anjum Zia, Maria Naqve, Athar Mahmood, Muhammad Mansoor Javaid, and Muhammad Azeem

Abstract Plants face different stresses in the environment, and among these environmental stresses, salinity is more devastating stress due to its negative impacts on crop plants. Salinity is stress that affects growth, physiology, and nutrient uptake in plants, which ultimately leads to food scarcity. Soluble salts decrease the water potential, resulting in an aqueous medium unavailable for the plant retarding plant development. Increased imbibition of generated seeds changes due to the decreased solute potential of the growing media, which causes ion toxicity. Increased concentration of reactive oxygen species (ROS) damages the lipid, protein, and nucleic acid; ultimately disrupts the cellular metabolism; and alters the enzymatic activities. The adverse effects due to the increased salinity levels can be mitigated via genetic diversity, osmoprotectant/osmolyte accumulation, hormonal regulation, and antioxidant mechanisms. This chapter elaborates on the impacts of salinity stress on morpho-physiological attributes of the plant and also describes different mechanisms and perspectives that can mitigate salt stress.

S. Asghar · S. Bibi · A. Mukhtiar · M. Naqve
Department of Botany, Faculty of Sciences, University of Agriculture, Faisalabad, Pakistan

M. A. Zia
Department of Biochemistry, Faculty of Sciences, University of Agriculture, Faisalabad, Pakistan

A. Mahmood (✉)
Department of Agronomy, Faculty of Agriculture, University of Agriculture, Faisalabad, Pakistan
e-mail: athar.mahmood@uaf.edu.pk

M. M. Javaid
Department of Agronomy, Faculty of Agriculture, University of Agriculture, Faisalabad, Pakistan

M. Azeem
Biosaline Research Laboratories, Department of Botany, University of Karachi, Karachi, Pakistan

M. Hasanuzzaman (ed.), *Climate-Resilient Agriculture, Vol 2*,
https://doi.org/10.1007/978-3-031-37428-9_18

Keywords Antioxidants · Reactive oxygen species · Salt stress · Ion homeostasis · Osmo-protectant

1 Introduction

The increased global population causes a rapid utilization of renewable and sustainable resources. The rapid increase in utilization of renewable resources due to increase in global population has been producing many problems for mankind. These problems include flooding, high temperature, salt stress, and water scarcity, which affect crop yield. Recent studies have shown that 20% of yield loss occurs because of salinity stress and low temperature, 40% due to elevated temperature, 17% due to drought stress, and 8% due to other factors. Salinity is classified as primary and secondary salinity. Various salts that are released by the weathering of rocks are the main causes of primary salinity. The anthropogenic activities, i.e., deforestation, overgrazing, intensive cropping, and irrigation, are considered the main cause of secondary salinity. The nature of Pakistan's soil is calcareous and alkaline with a low amount of organic matter (Mihoub et al. 2019). The soluble salts present in excess amounts in rhizospheric soil disturb plant growth, biochemistry, and physiology. The electrical conductivity of soil or irrigation water ranging from 4 dS m^{-1} or higher that is equal to 40 mM NaCl is referred to as saline condition (Ferreira et al. 2018). Globally, salinity has affected approximately 33% of total arable land (Mustafa et al. 2019), and about 0.2–0.4% of the cultivated area has been changed into barren land due to salinity (Zemni et al. 2022). The majority of crops that are utilized by humans as a food source are glycophytes sensitive to high concentrations of salt in the soil as compared to halophytes (salt-tolerant plants) in nature (Lombardi et al. 2022).

In the twenty-first century, the most emerging environmental issue is soil salinity. Increase in population demands more high food, so it is need of time to mitigate soil salinity to fulfill the need. The researchers are trying to discover different perspectives for the tolerance mechanisms against salinity stress in tolerant plants along with the interest to apply these mechanisms to plants that are sensitive to salinity (Shah et al. 2021). Recent studies have decoded major elements of the entire defensive mechanisms. This chapter is mainly based to discuss different mechanisms of salt tolerance.

About 30% of global agricultural productivity comes from 17% of irrigated land (Ortiz-Bobea et al. 2021). Approximately 6% land area constitutes 800 million hectares of dryland area that is unsuitable for cultivation due to salinity stress (Nanduri et al. 2019). Globally, 45 million hectares of irrigated area are also affected due to salinity stress (Carillo et al. 2011). According to another estimate, 831 × 106 hectares has been damaged by salts (Morris 2007). Furthermore, about one-half of the irrigated area (approximately 2.5 × 108 hectares) (Junpen et al. 2018) is affected by salinity. In 2050, approximately 50% of arable land area is expected to be vulnerable to salinity (Schneider and Asch 2020). Kaya-Okur et al. (2019) stated that annually 10% of land area is being destroyed around the world due to salt stress (Arora et al. 2018).

The geographical area of Pakistan is about 79.61 M ha (Syed et al. 2021), where 80% of which is irrigated through canal water. The sustainable agricultural practices of Pakistan are under continual threat due to many factors like change in water scarcity climate, drought, and soil salinity. Major factors among these abiotic stresses are the salinity and sodicity that cause soil degradation (Hailu and Mehari 2021; Ondrasek et al. 2011). In Pakistan, the potential of soil for crop production has declined due to salt stress. Soil erosion is mainly caused by the mismanagement of agricultural practices and improper cultivation practices. In Pakistan, almost 1.4 M ha of agricultural area has escaped the cultivation practices, which is equal to 25% irrigated area (Bhatt et al. 2019). In Pakistan, the growth and crop yield have declined notably due the salinity (Abideen et al. 2022).

2 Plant Responses to Salinity Stress

The crop plants under salinity stress express some symptoms like slow growth and low germination rate (Gamalero and Glick 2022). Moreover, salinity disturbs the concentration of gibberellic acid and abscisic acid, which lead to retard seed germination (Bomle et al. 2021; Feng et al. 2020). With the emergence of radicle and root hairs, the osmotic stress affects the water absorption by roots and its uptake toward the aerial plant parts: firstly, the osmotic stress in which plant growth is inhibited by less water absorption. Then, the inoculation of salt ions in the plant transpiration stream results in ion toxicity, which inhibits photosynthesis, disturbs homeostasis, and increases lipid peroxidation (Gupta et al. 2022; Abbaspour et al. 2021). Oxidative stress, a secondary effect, is caused by salinity, due to accumulation of the ROS in the cells. It affects the normal functioning of plants by peroxidation of lipids, proteins, and nucleic acids (Mushtaq et al. 2020). When the concentration of salts becomes high, it disturbs the mineral nutrition that causes the disruption of ions in the cell. The increased level of sodium results in the denaturation of proteins, modification in membrane solubility, and metabolism of ROS (Kaur et al. 2022).

2.1 *Osmotic Effect*

Increased salt concentrations impose stress on the plants either by enhancing toxic ions concentration or by high osmotic potential (Dourado et al. 2022; Joshi et al. 2022). The soluble salts retard water potential by making aqueous medium unavailable to the plants for absorption, and it is the main cause of the retardation for plant growth under salinity stress (Bukhari et al. 2019). A moderate level of osmotic stress does not affect root growth but has an adverse effect on aerial parts of crop plants (Yadav et al. 2020; Hosseinifard et al. 2022). The osmotic effect causes damage to cell, which is mainly dependent on various aspects like stress time duration, types of plant species, methods of the stress application, and type of cell and tissue (O'Flanagan et al. 2019; Zhang et al. 2019).

2.2 Ionic Effect

The toxicity of specific ion mainly occurs due to the accumulation of excess ions like sulfate, sodium, and chloride. It affects the crop from germination to maturity. The effect of specific ion in the lateral growth stages results in the failure of crops (Bhattacharya 2022). With reference to tolerance against salinity various, crops show different degrees of responses. Higher plants that are mostly agricultural crop plants are highly diplomatic to salinity stress. In the salinity stress environment, high levels of chloride and sodium ion combined with low potassium ion level were reported. The senescence is due to the aggregation of salt ions in the older leaves of the plant (Ayuso-Calles et al. 2021). The root fresh weight (RFW) and the shoot fresh weight (SFW) were reduced by up to 50% due to increased levels of Na^+ and Cl^- ion accumulation in the interior leaf sap (Shahzad et al. 2022).

2.3 Germination Effect

Salinity disturbs the germination process in many ways. The imbibition process of seeds alters with the decrease in the solute potential of growing media (Baha 2022; Uçarlı 2020) causing toxicity in the ionic concentrations, which disturbs the enzymatic mechanisms of metabolic processes of nucleic acid (Nowicka 2022) and changes the metabolism of proteins (Johnson et al. 2020), variation in hormonal balance (Giersch et al. 2020), and depletion of seed reserves (Carter 2019). Among cultivars and species, the germination rate and the germinated seed ratio vary due to salinity stress (Goro and Sinha 2020). Salinity stress deteriorates both halophytes and glycophytes mainly at the germination stage. Seed germination is the main stage required for the successful growth and development of plant seedlings; these seedlings are much more sensitive to salinity stress in comparison with the vegetative growth stages. Accumulation of toxic ions due to salinity causes a mineral imbalance in plants. Due to the reduction in the uptake of essential ions, the plant physiological activities get disturbed. A low level of salt stress promotes seed dormancy, while higher levels of salinity stress retard the germination of seed (Kataria et al. 2022).

To overcome these limitations of nutritional values, seeds developed mechanisms to attain their decreased water potential or various tolerance mechanisms to cope with the disturbance produced by the salt stress. Plant germination is disturbed in many ways due to salinity. Salnity leads to a decrease in soil osmotic potential, which declines the seed water imbibition (Mahpara et al. 2022). Due to the production of ionic toxicity that changes the enzymatic actions involved in the metabolic process of nucleic acid. Salt stress also affects the germination of seed including modifications in protein metabolism (Johnson and Puthur 2021). The seeds are more sensitive to salt stress because of the direct contact with topsoil (Chen et al. 2020). The germination rate of the seeds decreases due to the ionic stress acquired due to the accumulation of sodium chloride in the soil (Toderich et al. 2020). Due to low water potential, the water absorption capacity of seed is decreased; thus, salt stress produces deleterious effects on the development of embryo, resulting in late germination processes (El-Hendawy et al. 2019; Song et al. 2023). The time required for germination of the seed depends on the genotypes and strength of salinity stress. Salinity stress and seed germination are inverse to each other (Konuşkan et al. 2017; Bakhshandeh et al. 2020). Deleterious effects are pronounced on the seed size and germination index of the chickpea (Johal et al. 2022). The size and germination of seed show inverse relation under salinity stress (Farid et al. 2021).

2.4 Growth and Physiological Effect

The decreased growth rate is one of the primary effects of salinity stress in plants. Accumulation of salts in soil retards plant growth in two ways. First, it diminishes the water uptake that results in stunted growth. Second, it enters the transpirational stream, leads to damage to the leaves, and further reduces plant growth. The salinity stress reduces many physiological processes of the plant such as photosynthesis, photomorphogenesis, respiration, photoperiodism, seed germination, plant nutrition, circadian rhythms, and stomatal functioning nastic movements (Minorsky 2019). Under salinity, stress plants show different deviations from the normal function, which results in a reduction in crop yield. Plant–water relation is disturbed when high amounts of salts accumulate in the root of plants. Salinity stress retards the turgor pressure and osmotic potential of plants when the level of water taken up by the plants is low. A decrease in the uptake of water retards the mechanism of stomatal opening and cell division, which ultimately lowers photosynthesis and affects plant tissues (Nawaz et al. 2010; El Rasafi et al. 2022). Lowering of turgor pressure results in closing of stomata, which retards the gaseous exchange via the transpiration process (Liu et al. 2019). Salinity affects physiological activities like modifications occurring in the membrane permeability, leads to denaturing of membrane proteins (Rawat et al. 2021), and inhibits the photosynthetic phenomenon (Sforza et al. 2020). A lower rate of photosynthesis occurs because of a lowering in pigments and denaturation of enzymes involved in photosynthesis (Farghaly et al. 2015; Kreslavski et al. 2023).

2.5 *Oxidative Effect*

The greater production of ROS, for example, OH^- and O^{2-}, is the major outcome of salinity (Khan et al. 2022). ROS oxidatively damage the nucleic acids, proteins, and lipids, which ultimately disturb cellular metabolism. Production of ROS takes place due to decreased oxygen level, which disrupts the metabolic processes of the plants (Thorat et al. 2018; Sharma et al. 2020). Plants have distinctive ways to retrieve the ROS production as enzyme stimulation of the anti-oxidative pathway (Streyczek et al. 2022). During cell growth and normal body functioning, ROS are produced at a low level (Shin et al. 2020), but its production is increased during stress conditions (Bulgari et al. 2019). The osmotic effect regulates superoxide accumulation in the chloroplast, inhibits the required supply of CO_2 for photosynthesis, and disturbs the opening of stomata. This superoxide accumulation stimulates photooxidation and photoinhibition in plant cells (Abdelaal et al. 2022).

2.6 *Nutritional Imbalance*

The excessive production of sodium and chloride ions and low consumption of different minerals like Ca^{2+} and K^+ are the major causes of ion discrepancy. If K^+ and Na^+ ion ratio exceeds the normal range, it disturbs the proper functioning and inactivation of enzymes in the plants. The plant–water relation is disturbed when the deposition of salts is high; it results in the low consumption of major nutrients and limited uptake. Therefore, cellular metabolic activities and normal enzyme functioning are affected (Lennicke and Cochemé 2021). Excess sodium ion uptake results in a reduction in potassium ion uptake. Inside the plant, maintenance of Ca^{2+} is necessary and is a vital parameter under salinity stress. The K is known as a major important element for protein formation, osmoregulation, photosynthesis, and regulating turgor pressure of the plant cell. Potassium ions coupled with Ca^{2+} are essential for the proper functioning of the cell membrane and regulating the integrity of plant cell (Kohli et al. 2022).

3 Approaches to Mitigate Salinity Stress

Salinity affects the plant physiology and enzymatic reactions in many ways ultimately reducing the yield of crop plants. To minimize these losses, there are some approaches that mitigate the negative impacts of salinity. Some of these approaches are depicted here (Fig. 18.1).

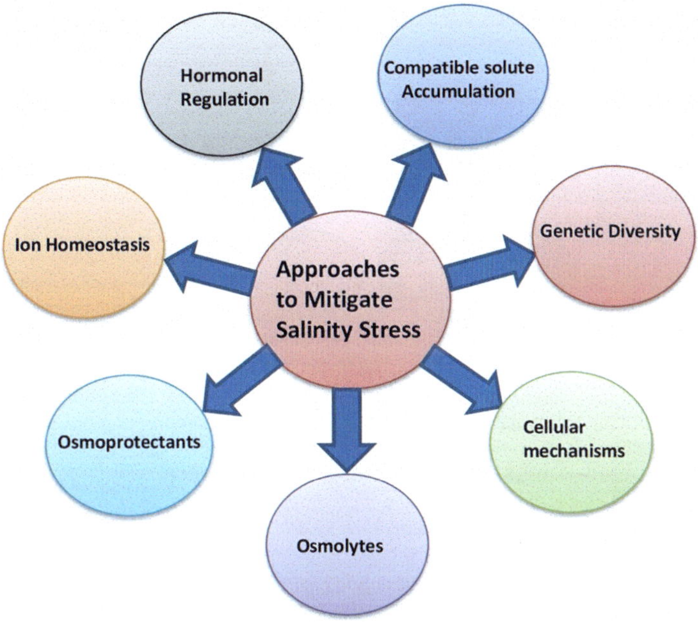

Fig. 18.1 Various approaches to mitigate salinity stress in plants. (Dhiman et al. 2021)

3.1 Salt Tolerance Through Genetic Diversity in Plants

Many plant genera show the phenomenon of salt tolerance against salt stress through genetic diversity (Sarabi et al. 2017). Many crop plants are found salt-sensitive, and some are discovered as hypersensitive in comparison with that of halophytes, which are native to saline environment. Specific morphological and anatomical adaptations or some avoidance mechanisms are present in various halophytes that have the potential to cope with extreme saline conditions (Aslam et al. 2011). However, these characteristics are suitable for crops in which gene introgression is difficult or impossible. Recent research has elaborated that most glycophytes and halophytes tolerate salinity (Lokhande and Suprasanna 2012). Cytotoxic ions of the saline environment, mainly sodium and chloride ions, result in the compartmentalization of the ions occurring in the vacuole that is used as the osmotic potential (Djanaguiraman and Prasad 2013). A similarity is present between many molecular entities mediating the ionic homeostasis process and signaling of salinity stress among the plants (Ji et al. 2013; Zhu et al. 2021).

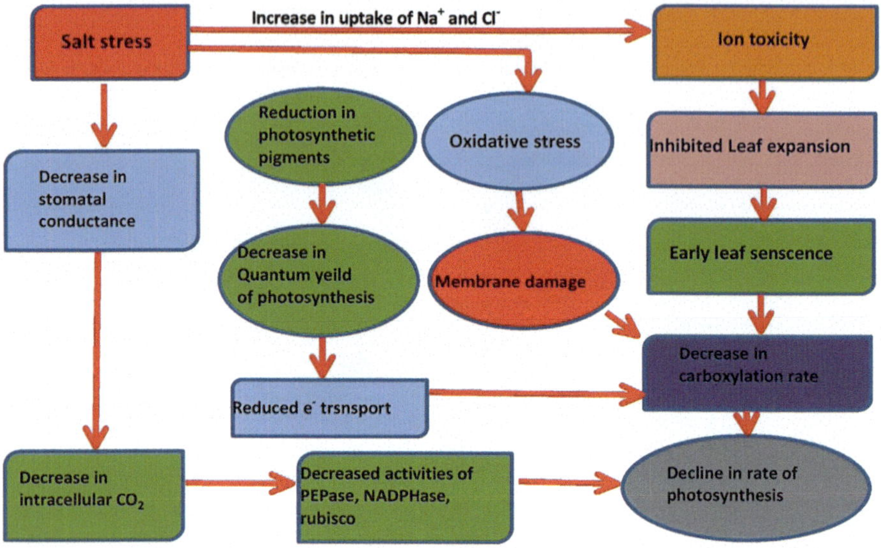

Fig. 18.2 Depicting the negative impacts of salinity on physiological and enzymatic processes in plants. (Khan et al. 2020)

3.2 *Cellular Mechanisms of Salt Stress Survival, Recovery, and Growth*

When the levels of salts are high, it can cause ion imbalance and hyperosmotic stress that cause cellular damage and secondary osmotic impacts (Singh et al. 2014). Consequently, plants survive by both tolerating and avoiding salt stress, i.e., few plants remain dormant during saline conditions, and some develop special mechanisms to cope with such saline environment (Acosta-Motos et al. 2017). The growth and developmental processes of the plants ceased when the turgor of plant cell wall falls below the threshold yield (Kutschera and Niklas 2013) (Fig. 18.2).

3.3 *Osmolytes and Osmoprotectant*

Osmotic adjustment is the cellular reaction to lowering the turgor of cell (Blum 2017). As discussed earlier, salt tolerance needs suitable accumulation of solutes in the organelles and cytosol where they perform different functions like osmotic protection and osmotic adjustment (Farouk 2011). Some suitable osmolytes are beneficial ions like K^+, but a greater number of them are organic solutes. In response to increased osmotic stress, the accumulation of suitable solutes is a universal process in all organisms as diverse as bacteria to animals and plants. Therefore, the accumulations of solutes are dependent on the type of organism and even the type of species. A broad range of organic compatible solutes include sugars (mainly sucrose and fructose), sugar alcohols (methylated inositol and glycerol), and complicated sugars (fructose, trehalose, and raffinose) (Conde et al. 2011; Yang et al. 2022). In plants, during stress conditions the abovementioned sugars act as metabolic signals by playing a crucial role in the detoxification of reactive oxygen species (ROS) and regulation of photosynthetic proficiency and cellular organization. Moreover, these sugars protect the plants by enhancing physiological responses, for example, by harmonizing antioxidant activity, regulating membrane integrity, and maintaining water efficiency under various abiotic stresses such as pesticide exposure (Koza et al. 2022).

Others include the derivatives of quaternary amino acids (glycine betaine, proline, betaine, and β-alanine betaine), sulfonium compounds, and tertiary amines (Hill et al. 2010; Jouyban 2012). Several osmolytes are assumed as osmoprotectants, as their range of concentration is not sufficient for the solute alignment. An amino acid, glycine betaine, involves in the protection of thylakoid and cell membrane processes when exposed to extreme saline conditions and even high temperature or freezing (Bhandari and Nayyar 2014; Shah et al. 2017; Fell et al. 2020). However, several osmoprotectants serve the function of tolerance mechanisms under saline environment in plants when they are expressed as transgene products (Omari Alzahrani 2021; Hossain and Islam 2022). An adaptive microbial activity of osmoprotectants is the hunt of ROS which are the results of ionic stress, hyperosmotic conditions, causing dysfunction of membrane and necrosis (Onaga and Wydra 2016; Guerrero-Rubio et al. 2020). A simple characteristic of compatible solutes is that they have the ability to deposit high salt concentration without disturbing the internal chemistry (Malda et al. 2013). The significant solute concentration has the ability to prevent the enzymatic functions that perform in saline environment. There is minimal effect on pH, luminal division of organelles, and net balance checking of cytosol due to these compounds. These solutes can be accumulated in the cell without affecting normal functioning (Yancey 2005).

3.4 Ion Homeostasis

The latest research has reported that molecular entities mediate the homeostasis of K^+ and Na^+ ions and also give an insight into the Ca^{2+} functioning for the adjustment of transport systems in plants. Recently, SOS, the stress signaling mechanism, was identified as a pivotal regulator for the salt tolerance mechanism and ionic homeostasis of the plants (Basu et al. 2021).

3.5 Hormonal Regulation

Phytohormones play a vital role in growth, development, and sustainability of the plant. To mitigate salt stress at different developmental stages, phytohormones like gibberellic acid, indoleacetic acid (IAA), jasmonic acid (JA), abscisic acid (ABA), ethylene, brassinosteroids, and salicylic acid are induced (Chitnis et al. 2014). ABA plays a vital role under salt stress conditions because of greater production in response to stress environment. Abscisic acid (ABA) helps the plant to cope with the deleterious impact of salinity stress on translocation of assimilates, growth, and development of photosynthesis and promotes the plant–water status through the guard cells (Toh et al. 2021). During salinity stress, increased abscisic acid production in the root and shoot takes place that suppresses the toxic effects of salinity stress on transportation of assimilates, photosynthesis, and growth (Ji et al. 2018). It is reported that ABA plays a crucial role during the cellular signaling process and regulates the expression of responsive genes for water and salt stress (Osakabe et al. 2014). Production of ABA has been associated with improving the contents of Ca^{2+}, K^{+}, sugars, and proline in root vacuole to cope with salt stress (Shabala and Pottosin 2014). Accumulation of ABA is suppressed under salinity stress, which means leaf and root tissues might have distinctive inducing mechanisms for ABA under water and salt stress (Ruiz-Sola et al. 2014; Redondo-Gómez et al. 2021). Ethylene is a gaseous hormone that promotes the development and growth of plant under the stress environment and interactions of ethylene signaling pathways with that of salt stress (Riyazuddin et al. 2020). Remarkable biosynthesis of ethylene was noticed in salt-tolerant genotypes of soybean in comparison with that of salt-sensitive genotypes under salt stress (Freitas et al. 2018).

JA has been observed as a regulator for the production of tolerance mechanism against saline condition. In *Arabidopsis*, the over-expression of TaAOC1 genes of wheat encoding the enzyme AOC enhances the levels of JAs that upregulate the mechanism of tolerance under salinity stress (Xiong et al. 2014; Pavlović et al. 2019). In tomato, a phytohormone, systemin regulates the production of JA and promotes the salt tolerance mechanism among dependent strategies of abscisic acid. Transgenic tomatoes express ProSystem, which shows enhanced biomass and stomatal conductance under saline conditions (Orsini et al. 2010; Bagues et al. 2021). Over-expression of numerous ethylene and JA-inductive genes was described for ordination of the salt tolerance mechanisms in several species without deleterious impact on plant growth reported (Kazan 2015) that can be considered initiating point for the genetic enhancement of tolerance under salt stress in the plants.

3.6 Compatible Solute Accumulation

Under stress, many plants enhance the cellular activity of osmotically active chemicals. Hydrophilic properties are present in compatible solutes, meaning they restore water on the protein complex surface, membrane, or protein while preventing to disturb the normal metabolic processes. In this way, they reduce the inhibition of enzyme activity caused by high ion concentration. Compatible solutes include amino acid and certain derivatives of amino acid, cyclic polyols, sugars, fructans, ectoine, quaternary amino acid, and sulfonium complexes (Singh et al. 2015). There is an increase in compatible solutes under osmotic stress, but trehalose does not show any response under osmotic stress while it is preventive even at minimum concentration and acts as osmoprotectants (Velázquez-Márquez et al. 2015; Slama et al. 2015). Low quantities of mannitol in chloroplast aid to decrease damage by creating hydroxyl radicals and high concentration of other suitable solutes to minimize the inhibitory responses of ions on enzyme function. The major purpose of suitable solutes is to stabilize proteins, membranes, and protein complexes and scavenge oxygen radicals. A frequent adaptive response to abiotic stressors is the accumulation of proline (Ghosh et al. 2022). The vacuoles, chloroplast, cytosol, and cytoplasm all contain high level of proline. It performs crucial functions such as detoxification of ROS, osmotic adjustment, antioxidant activity, and protein and protein complex stabilization (Lokhande et al. 2011; Bose et al. 2014). Likewise, Muchate et al. (2016) reported that the plants that are sensitive to salinity stress (EC of 16–20 ds m^{-1}) had significantly high levels of glycine betaine and proline than plants not exposed to salt stress. In salt-sensitive barley cultivars, proline and glycine butane were found to contribute significantly to cellular osmosis potential, maintaining the lower cytosolic potassium ions level (Zhonghua et al. 2011; Hao et al. 2021). The genes P5CS and proline-5 carboxylate reductase are joined to proline aggregation as a result of salinity stress (Sarkar et al. 2021). Excessive expression of gene P5CS resulted in twofold rises in proline concentration (Nounjan et al. 2012). Proline is a signaling or regulatory molecule that governs gene expression and maintains metabolite pools and redox equilibrium, in addition to its role in salt tolerance (Huang et al. 2022).

Glycine butane is located in chloroplast and performs a vital role in the prevention of protection of chloroplast, thylakoid membrane, osmotic adjustment, and photosynthetic system under salt stress (Shu et al. 2015). Excessive expression of gene BADH in rice resulted in significant glycine betaine buildup under salt stress (Hussain Wani et al. 2013). The total rise of polar sugars is also observed in response to salt stress. Sugars contribute around half of the overall osmotic potential in glycophytes exposed to salt environment (Rasool et al. 2013; Misbah et al. 2022). Trehalose, a deoxidizing sugar, is important for metabolic balance. Excessive expression of gene OsTPS1, which encodes this sugar, increased the capacity of rice cultivars in salt tolerance (Liew et al. 2011; Khan et al. 2015).

3.7 Antioxidant Defense

Crop plants possess the ability to produce strong ROS scavenging metabolism, involving proteolytic or nonproteolytic mitigating pathways to show resistance against the harmful impacts of ROS reproduction. Enzymes that are antioxidant including ascorbate peroxidase, glutathione peroxidase, glutathione-S-transferase, glutathione reductase, superoxide dismutase, catalase, ascorbate peroxidase, monodehydroascorbate reductase, guaiacol peroxidase, and nonproteolytic antioxidants like phenolic compounds, ascorbic acid, glutathione, non-enzymatic amino acids, carotenoids, and alpha-tocopherols also protect free radicals (Ravisankar et al. 2015). Superoxide dismutase is an important antioxidant enzyme that dismutase superoxide radicals to produce H_2O_2 as part of defense strategy against free radicals' toxicity. Superoxide dismutase (SOD) is divided on the basis of metal cofactor: Copper and zinc are the cofactors of Cu/Zn SOD, manganese is the cofactor of Mn-SOD, and iron is the cofactor of Fe-SOD, which are present in the mitochondria, chloroplast, peroxisomes, and cytosol. Catalases scavenge hydrogen peroxide (H_2O_2) during the fatty acids oxidation and play a role in the photorespiration processes under abiotic and salt stresses. Catalase enzyme is essential for the detoxification of ROS because it possesses the ability to directly convert hydrogen peroxide into water and carbon dioxide. Catalase has the highest turnover rate of any enzymes, converting six million H_2O_2 molecules into H_2O and carbon dioxide every minute (Gill and Tuteja 2010; Ighodaro and Akinloye 2018). Catalase activity decreases to varying degrees under salt stress in CAT-1, CAT-2, CAT-3, and CAT-4. The CAT-2 activity dropped by 45 percent, while CAT-3 activity dropped by 29 percent (Yamashita et al. 2007).

Ascorbate peroxidase (APX) serves the identical role of catalase, but instead of employing ascorbate as a reductant and catalase as the elimination of H_2O_2. APX can be found in the stroma and thylakoid, the microbody that includes peroxisomes, glyoxysomes, the mitochondria, and cytoplasm. Stromal APX, cytosolic APX, and mitochondrial APX are the different types. After salt stress, several studies found an excess in ascorbate peroxidase activity in plants like cotton, pea, and rice. Therefore, ascorbate peroxidase activity is reduced in halophytes and mangroves (Shackira and Puthur 2017). It has also been reported that greater MDAR activity in transgenic tobacco contributes to greater tolerance to salt and osmotic stresses. Ascorbic acid, a polar antioxidant, is found in plant tissues with meristem and photosynthetic cells (Gallie 2013). The NADPH-oxidized reaction of oxidized glutathione is catalyzed by the glutathione reductase enzyme. When comparing salt-stressed plants to control plants, glutathione reductase activity was identified to be higher in the aerial parts of plants that are under stress conditions (Pessarakli et al. 2015).

4 Conclusion and Future Perspectives

About 20% of crop yield is lost due to abiotic stress. Salinity stress in soil affects plants through oxidative, osmotic, and ionic stress. Increased concentration of salt in the soil has several negative results on the morpho-physiological parameters of the crop plants. Salinity stress hampers plant growth and germination at every developmental stage because salt stress affects protein synthesis, photosynthesis, and respiration. Salt tolerance mechanisms are via vacuole, sodium and potassium discrimination, osmotic adjustment, and salt inclusion and exclusion. The application of different mechanisms like salt tolerance through genetic diversity, cellular mechanisms, use of osmoprotectants and different osmolytes, ionic homeostasis, hormonal regulation, and antioxidant defensive mechanism is also discussed. It is concluded at the end of this chapter that despite various indices for the tolerance mechanisms against salinity stress, none of these can be a specific index in determining the level of tolerance against salt stress. In fact, all indices might be mandatory but are not abundant; therefore, further research is required for determining the efficient criteria for every plant species.

References

Abbaspour H, Pour FS, Abdel-Wahhab MA (2021) Arbuscular mycorrhizal symbiosis regulates the physiological responses, ion distribution and relevant gene expression to trigger salt stress tolerance in pistachio. Physiol Mol Biol Plants 27(8):1765–1778

Abdelaal K, Alsubeie MS, Hafez Y, Emeran A, Moghanm F, Okasha S, Omara R, Basahi MA, Darwish DBE, Ibrahim M (2022) Physiological and biochemical changes in vegetable and field crops under drought, salinity and weeds stresses: control strategies and management. Agriculture 12(12):2084

Abideen Z, Koyro HW, Hussain T, Rasheed A, Alwahibi MS, Elshikh MS, Hussain MI, Zulfiqar F, Mansoor S, Abbas Z (2022) Biomass production and predicted ethanol yield are linked with optimum photosynthesis in phragmites karka under salinity and drought conditions. Plan Theory 11(13):1657

Acosta-Motos JR, Ortuño MF, Bernal-Vicente A, Diaz-Vivancos P, Sanchez-Blanco MJ, Hernandez JA (2017) Plant responses to salt stress: adaptive mechanisms. Agronomy 7(1):18

Arora NK, Fatima T, Mishra I, Verma M, Mishra J, Mishra V (2018) Environmental sustainability: challenges and viable solutions. Environ Sustain 1:309–340

Aslam R, Bostan N, Nabgha-e-Amen MM, Safdar W (2011) A critical review on halophytes: salt tolerant plants. J Med Plant Res 5(33):7108–7118

Ayuso-Calles M, Flores-Félix JD, Rivas R (2021) Overview of the role of rhizobacteria in plant salt stress tolerance. Agronomy 11(9):1759

Bagues M, Zaghdoud C, Hafsi C, Boussora F, Triki T, Nagaz K (2021) Combined effect of deficit irrigation with saline water affects gas exchange, phytochemical profiles, antioxidant activities and grain yield of barley landraces "Ardhaoui" at heading stage. Plant Biosyst 155(3):436–446

Baha N (2022) Comparative effects of osmotic and salt stresses on germination and seedling growth of alfalfa: physiological responses involved. Agric Conspec Sci 87(4):311–319

Bakhshandeh E, Bradford KJ, Pirdashti H, Vahabinia F, Abdellaoui R (2020) A new halothermal time model describes seed germination responses to salinity across both sub-and supra-optimal temperatures. Acta Physiol Plant 42:1–15

Basu S, Kumar A, Benazir I, Kumar G (2021) Reassessing the role of ion homeostasis for improving salinity tolerance in crop plants. Physiol Plant 171(4):502–519

Bhandari K, Nayyar H (2014) Low temperature stress in plants: an overview of roles of cryoprotectants in defense. In: Physiological mechanisms and adaptation strategies in plants under changing environment, vol 1. Springer, New York, pp 193–265

Bhatt R, Kaur R, Ghosh A (2019) Strategies to practice climate-smart agriculture to improve the livelihoods under the rice-wheat cropping system in South Asia. In: Sustainable management of soil and environment. Springer, Singapore, pp 29–71

Bhattacharya A (2022) Effect of low-temperature stress on germination, growth, and phenology of plants: a review. In: Physiological processes in plants under low temperature stress. Springer, Singapore, pp 1–106

Blum A (2017) Osmotic adjustment is a prime drought stress adaptive engine in support of plant production. Plant Cell Environ 40(1):4–10

Bomle DV, Kiran A, Kumar JK, Nagaraj LS, Pradeep CK, Ansari MA, Alghamdi S, Kabrah A, Assaggaf H, Dablool AS (2021) Plants saline environment in perception with rhizosphere bacteria containing 1-aminocyclopropane-1-carboxylate deaminase. Int J Mol Sci 22(21):11461

Bose J, Rodrigo-Moreno A, Shabala S (2014) ROS homeostasis in halophytes in the context of salinity stress tolerance. J Exp Bot 65(5):1241–1257

Bukhari SAH, Peerzada AM, Javed MH, Dawood M, Hussain N, Ahmad S (2019) Growth and development dynamics in agronomic crops under environmental stress. In: Agronomic crops: volume 1: production technologies. Springer, Singapore, pp 83–114

Bulgari R, Franzoni G, Ferrante A (2019) Biostimulants application in horticultural crops under abiotic stress conditions. Agronomy 9(6):306

Carillo P, Annunziata MG, Pontecorvo G, Fuggi A, Woodrow P (2011) Salinity stress and salt tolerance. Abio Stress Plant Mech Adap 1:21–38

Carter E (2019) Seed and seedling dynamics of annual medic pastures in South Australia. In: Proceedings of the XIV international grassland congress. CRC Press, pp 447–450

Chen L, Liu L, Lu B, Ma T, Jiang D, Li J, Zhang K, Sun H, Zhang Y, Bai Z (2020) Exogenous melatonin promotes seed germination and osmotic regulation under salt stress in cotton (*Gossypium hirsutum* L.). PLoS One 15(1):e0228241

Chitnis VR, Gao F, Yao Z, Jordan MC, Park S, Ayele BT (2014) After-ripening induced transcriptional changes of hormonal genes in wheat seeds: the cases of brassinosteroids, ethylene, cytokinin and salicylic acid. PLoS One 9(1):e87543

Conde A, Chaves MM, Gerós H (2011) Membrane transport, sensing and signaling in plant adaptation to environmental stress. Plant Cell Physiol 52(9):1583–1602

Dhiman P, Rajora N, Bhardwaj S, Sudhakaran SS, Kumar A, Raturi G, Chakraborty K, Gupta OP, Devanna B, Tripathi DK (2021) Fascinating role of silicon to combat salinity stress in plants: an updated overview. Plant Physiol Biochem 162:110–123

Djanaguiraman M, Prasad PV (2013) Effects of salinity on ion transport, water relations and oxidative damage. In: Ecophysiology and responses of plants under salt stress. Springer, New York, pp 89–114

Dourado PRM, de Souza ER, Santos MA, Lins CMT, Monteiro DR, Paulino MKSS, Schaffer B (2022) Stomatal regulation and osmotic adjustment in sorghum in response to salinity. Agriculture 12(5):658

El Rasafi T, Oukarroum A, Haddioui A, Song H, Kwon EE, Bolan N, Tack FM, Sebastian A, Prasad M, Rinklebe J (2022) Cadmium stress in plants: a critical review of the effects, mechanisms, and tolerance strategies. Crit Rev Environ Sci Technol 52(5):675–726

El-Hendawy S, Elshafei A, Al-Suhaibani N, Alotabi M, Hassan W, Dewir YH, Abdella K (2019) Assessment of the salt tolerance of wheat genotypes during the germination stage based on germination ability parameters and associated SSR markers. J Plant Interact 14(1):151–163

Farghaly FA, Radi AA, Abdel-Wahab DA, Hamada AM (2015) Comparative study of alkaline and saline stresses on two oil-producing plants. Egypt J Exp Biol (Botany) 10(1):13–13

Farid M, Anshori MF, Musa Y, Iswoyo H, Sakinah AI (2021) Interaction of rice salinity screening in germination and seedling phase through selection index based on principal components. Chil J Agric Res 8(3):368–377

Farouk S (2011) Osmotic adjustment in wheat flag leaf in relation to flag leaf area and grain yield per plant. J Stress Physiol Biochem 7(2):117–138

Fell JB, Fischer JP, Baer BR, Blake JF, Bouhana K, Briere DM, Brown KD, Burgess LE, Burns AC, Burkard MR (2020) Identification of the clinical development candidate MRTX849, a covalent KRASG12C inhibitor for the treatment of cancer. J Med Chem 63(13):6679–6693

Feng Y-N, Cui J-Q, Zhou T, Liu Y, Yue C-P, Huang J-Y, Hua Y-P (2020) Comprehensive dissection into morpho-physiologic responses, ionomic homeostasis, and transcriptomic profiling reveals the systematic resistance of allotetraploid rapeseed to salinity. BMC Plant Biol 20(1):1–22

Ferreira JF, Sandhu D, Liu X, Halvorson JJ (2018) Spinach (*Spinacea oleracea* L.) response to salinity: nutritional value, physiological parameters, antioxidant capacity, and gene expression. Agriculture 8(10):163

Freitas VS, de Souza MR, Costa JH, de Oliveira DF, de Oliveira PS, de Castro ME, Freire RS, Prisco JT, Gomes-Filho E (2018) Ethylene triggers salt tolerance in maize genotypes by modulating polyamine catabolism enzymes associated with H_2O_2 production. Environ Exp Bot 145:75–86

Gallie DR (2013) L-ascorbic acid: a multifunctional molecule supporting plant growth and development. Scientifica 2013:795964

Gamalero E, Glick BR (2022) Recent advances in bacterial amelioration of plant drought and salt stress. Biology 11(3):437

Ghosh U, Islam M, Siddiqui M, Cao X, Khan M (2022) Proline, a multifaceted signalling molecule in plant responses to abiotic stress: understanding the physiological mechanisms. Plant Biol 24(2):227–239

Giersch GE, Charkoudian N, Stearns RL, Casa DJ (2020) Fluid balance and hydration considerations for women: review and future directions. Sport Med 50:253–261

Gill SS, Tuteja N (2010) Reactive oxygen species and antioxidant machinery in abiotic stress tolerance in crop plants. Plant Physiol Biochem 48(12):909–930

Goro MG, Sinha VB (2020) Seed germination responses for varying KNO_3 and $NaNO_3$ stress in Trifolium alexandrinum. L cultivars. Biocatal Agric Biotechnol 25:101618

Guerrero-Rubio MA, Escribano J, García-Carmona F, Gandía-Herrero F (2020) Light emission in betalains: from fluorescent flowers to biotechnological applications. Trends Plant Sci 25(2):159–175

Gupta A, Mishra R, Rai S, Bano A, Pathak N, Fujita M, Kumar M, Hasanuzzaman M (2022) Mechanistic insights of plant growth promoting bacteria mediated drought and salt stress tolerance in plants for sustainable agriculture. Int J Mol Sci 23(7):3741

Hailu B, Mehari H (2021) Impacts of soil salinity/sodicity on soil-water relations and plant growth in dry land areas: a review. J Nat Sci Res 12(3):1–10

Hao S, Wang Y, Yan Y, Liu Y, Wang J, Chen S (2021) A review on plant responses to salt stress and their mechanisms of salt resistance. Horticulturae 7(6):132

Hill R, Li C, Jones A, Gunn J, Frade P (2010) Abundant betaines in reef-building corals and ecological indicators of a photoprotective role. Coral Reefs 29:869–880

Hossain MT, Islam T (2022) Amelioration of salinity stress by bacillus species as promoters of plant growth in saline soil. In: Bacilli in agrobiotechnology. Springer, pp 199–208

Hosseinifard M, Stefaniak S, Ghorbani Javid M, Soltani E, Wojtyla Ł, Garnczarska M (2022) Contribution of exogenous proline to abiotic stresses tolerance in plants: a review. Int J Mol Sci 23(9):5186

Huang Q, Farooq MA, Hannan F, Chen W, Ayyaz A, Zhang K, Zhou W, Islam F (2022) Endogenous nitric oxide contributes to chloride and sulphate salinity tolerance by modulation of ion transporter expression and reestablishment of redox balance in *Brassica napus* cultivars. Environ Exp Bot 194:104734

Hussain Wani S, Brajendra Singh N, Haribhushan A, Iqbal Mir J (2013) Compatible solute engineering in plants for abiotic stress tolerance-role of glycine betaine. Curr Genomics 14(3):157–165

Ighodaro O, Akinloye O (2018) First line defence antioxidants-superoxide dismutase (SOD), catalase (CAT) and glutathione peroxidase (GPX): their fundamental role in the entire antioxidant defence grid. Alexandria J Med 54(4):287–293

Ji H, Pardo JM, Batelli G, Van Oosten MJ, Bressan RA, Li X (2013) The salt overly sensitive (SOS) pathway: established and emerging roles. Mol Plant 6(2):275–286

Ji X, Cheng J, Gong D, Zhao X, Qi Y, Su Y, Ma W (2018) The effect of NaCl stress on photosynthetic efficiency and lipid production in freshwater microalga—Scenedesmus obliquus XJ002. Sci Total Environ 633:593–599

Johal N, Batish D, Pal A, Chandel S, Pal M (2022) Investigating the effects of 2850 MHz electromagnetic field radiations on the growth, germination and Antioxidative defense system of chickpea (*Cicer arietinum* L.) seedlings. Russ J Plant Physiol 69(6):1–8

Johnson R, Puthur JT (2021) Seed priming as a cost effective technique for developing plants with cross tolerance to salinity stress. Plant Physiol Biochem 162:247–257

Johnson EC, Dammer EB, Duong DM, Ping L, Zhou M, Yin L, Higginbotham LA, Guajardo A, White B, Troncoso JC (2020) Large-scale proteomic analysis of Alzheimer's disease brain and cerebrospinal fluid reveals early changes in energy metabolism associated with microglia and astrocyte activation. Nat Med 26(5):769–780

Joshi S, Nath J, Singh AK, Pareek A, Joshi R (2022) Ion transporters and their regulatory signal transduction mechanisms for salinity tolerance in plants. Physiol Plant 174(3):e13702

Jouyban Z (2012) The effects of salt stress on plant growth. Tech J Eng Appl Sci 2(1):7–10

Junpen A, Pansuk J, Kamnoet O, Cheewaphongphan P, Garivait S (2018) Emission of air pollutants from rice residue open burning in Thailand, 2018. Atmosphere 9(11):449

Kataria S, Anand A, Raipuria RK, Kumar S, Jain M, Watts A, Brestic M (2022) Magnetopriming actuates nitric oxide synthesis to regulate phytohormones for improving germination of soybean seeds under salt stress. Cells 11(14):2174

Kaur H, Hussain SJ, Kaur G, Poor P, Alamri S, Siddiqui MH, Khan MIR (2022) Salicylic acid improves nitrogen fixation, growth, yield and antioxidant defence mechanisms in chickpea genotypes under salt stress. J Plant Growth Regul 41(5):2034–2047

Kaya-Okur HS, Wu SJ, Codomo CA, Pledger ES, Bryson TD, Henikoff JG, Ahmad K, Henikoff S (2019) CUT and tag for efficient epigenomic profiling of small samples and single cells. Nat Commun 10(1):1930

Kazan K (2015) Diverse roles of jasmonates and ethylene in abiotic stress tolerance. Trend Plant Sci 20(4):219–229

Khan MS, Ahmad D, Khan MA (2015) Utilization of genes encoding osmoprotectants in transgenic plants for enhanced abiotic stress tolerance. Electron J Biotechnol 18(4):257–266

Khan I, Raza MA, Awan SA, Shah GA, Rizwan M, Ali B, Tariq R, Hassan MJ, Alyemeni MN, Brestic M (2020) Amelioration of salt induced toxicity in pearl millet by seed priming with silver nanoparticles (AgNPs): the oxidative damage, antioxidant enzymes and ions uptake are major determinants of salt tolerant capacity. Plant Physiol Biochem 156:221–232

Khan MN, Li Y, Fu C, Hu J, Chen L, Yan J, Khan Z, Wu H, Li Z (2022) CeO_2 nanoparticles seed priming increases salicylic acid level and ROS scavenging ability to improve rapeseed salt tolerance. Global Chall 6(7):2200025

Kohli SK, Kaur H, Khanna K, Handa N, Bhardwaj R, Rinklebe J, Ahmad P (2022) Boron in plants: uptake, deficiency and biological potential. Plant Growth Regul:1–16. https://doi.org/10.1007/s10725-022-00844-7

Konuşkan Ö, Gözübenli H, Atiş İ, Atak M (2017) Effects of salinity stress on emergence and seedling growth parameters of some maize genotypes (*Zea mays* L.). Turk J Agric Food Sci Technol 5(12):1668–1672

Koza NA, Adedayo AA, Babalola OO, Kappo AP (2022) Microorganisms in plant growth and development: roles in abiotic stress tolerance and secondary metabolites secretion. Microorganisms 10(8):1528

Kreslavski VD, Khudyakova AY, Kosobryukhov AA, Balakhnina TI, Shirshikova GN, Alharby HF, Allakhverdiev SI (2023) The effect of short-term heating on photosynthetic activity, pigment content, and pro−/antioxidant balance of a. thaliana Phytochrome mutants. Plan Theory 12(4):867

Kutschera U, Niklas KJ (2013) Cell division and turgor-driven stem elongation in juvenile plants: a synthesis. Plant Sci 207:45–56

Lennicke C, Cochemé HM (2021) Redox metabolism: ROS as specific molecular regulators of cell signaling and function. Mol Cell 81(18):3691–3707

Liew LN, Shi J, Li Y (2011) Enhancing the solid-state anaerobic digestion of fallen leaves through simultaneous alkaline treatment. Bioresour Technol 102(19):8828–8834

Liu J, Hu T, Fang L, Peng X, Liu F (2019) CO_2 elevation modulates the response of leaf gas exchange to progressive soil drying in tomato plants. Agric For Meteorol 268:181–188

Lokhande VH, Suprasanna P (2012) Prospects of halophytes in understanding and managing abiotic stress tolerance. In: Environmental adaptations and stress tolerance of plants in the era of climate change, pp 29–56. https://doi.org/10.1007/978-1-4614-0815-4_2

Lokhande VH, Nikam TD, Patade VY, Ahire ML, Suprasanna P (2011) Effects of optimal and supra-optimal salinity stress on antioxidative defence, osmolytes and in vitro growth responses in *Sesuvium portulacastrum* L. Plant Cell Tissue Organ Cult 104(1):41–49

Lombardi T, Bertacchi A, Pistelli L, Pardossi A, Pecchia S, Toffanin A, Sanmartin C (2022) Biological and agronomic traits of the main halophytes widespread in the Mediterranean region as potential new vegetable crops. Horticulturae 8(3):195

Mahpara S, Zainab A, Ullah R, Kausar S, Bilal M, Latif MI, Arif M, Akhtar I, Al-Hashimi A, Elshikh MS (2022) The impact of PEG-induced drought stress on seed germination and seedling growth of different bread wheat (*Triticum aestivum* L.) genotypes. PLoS One 17(2):e0262937

Malda J, Visser J, Melchels FP, Jüngst T, Hennink WE, Dhert WJ, Groll J, Hutmacher DW (2013) 25th anniversary article: engineering hydrogels for biofabrication. Adv Mater 25(36):5011–5028

Mihoub A, Amin AE-EAZ, Naeem A, Bouhoun MD (2019) Improvement in phosphorus nutrition of wheat plants grown in a calcareous sandy soil by incorporating chemical phosphorus fertilizer with some selected organic substances. Acta Agric Slov 113(2):263–272

Minorsky PV (2019) The functions of foliar nyctinasty: a review and hypothesis. Biol Rev 94(1):216–229

Misbah N, Akram I, Saleem MA (2022) Influences induced by salinity stress on germination, growth and proline contents of maize (*Zea mays* L.). J Agric Food Environ Anim Sci 3(1):15–26

Morris JT (2007) Estimating net primary production of salt marsh macrophytes. In: Principles and standards for measuring primary production, pp 106–119. https://doi.org/10.1093/acprof:oso/9780195168662.003.0007

Muchate NS, Nikalje GC, Rajurkar NS, Suprasanna P, Nikam TD (2016) Plant salt stress: adaptive responses, tolerance mechanism and bioengineering for salt tolerance. Bot Rev 82:371–406

Mushtaq Z, Faizan S, Gulzar B (2020) Salt stress, its impacts on plants and the strategies plants are employing against it: a review. J Appl Biol Biotechnol 8(3):81–91

Mustafa G, Akhtar MS, Abdullah R (2019) Global concern for salinity on various agro-ecosystems. In: Salt stress, microbes, and plant interactions: causes and solution, vol 1, pp 1–19. https://doi.org/10.1007/978-981-13-8801-9_1

Nanduri KR, Hirich A, Salehi M, Saadat S, Jacobsen SE (2019) Quinoa: a new crop for harsh environments. In: Sabkha Ecosystems: volume VI: Asia/Pacific. Springer, Cham, pp 301–333

Nawaz K, Hussain K, Majeed A, Khan F, Afghan S, Ali K (2010) Fatality of salt stress to plants: morphological, physiological and biochemical aspects. Afr J Biotechnol 9(34):5475–5480

Nounjan N, Nghia PT, Theerakulpisut P (2012) Exogenous proline and trehalose promote recovery of rice seedlings from salt-stress and differentially modulate antioxidant enzymes and expression of related genes. J Plant Physiol 169(6):596–604

Nowicka B (2022) Heavy metal–induced stress in eukaryotic algae—mechanisms of heavy metal toxicity and tolerance with particular emphasis on oxidative stress in exposed cells and the role of antioxidant response. Environ Sci Pollut Res 29(12):16860–16911

O'Flanagan CH, Campbell KR, Zhang AW, Kabeer F, Lim JL, Biele J, Eirew P, Lai D, McPherson A, Kong E (2019) Dissociation of solid tumor tissues with cold active protease for single-cell RNA-seq minimizes conserved collagenase-associated stress responses. Genome Biol 20(1):1–13

Omari Alzahrani F (2021) Metabolic engineering of osmoprotectants to elucidate the mechanism (s) of salt stress tolerance in crop plants. Planta 253:1–17

Onaga G, Wydra K (2016) Advances in plant tolerance to abiotic stresses. Plant Genome 10(9):229–272

Ondrasek G, Rengel Z, Veres S (2011) Soil salinisation and salt stress in crop production. In: Abiotic stress in plants—mechanisms and adaptations. IntechOpen, London, pp 171–190
Orsini F, D'Urzo MP, Inan G, Serra S, Oh D-H, Mickelbart MV, Consiglio F, Li X, Jeong JC, Yun D-J (2010) A comparative study of salt tolerance parameters in 11 wild relatives of Arabidopsis thaliana. J Exp Bot 61(13):3787–3798
Ortiz-Bobea A, Ault TR, Carrillo CM, Chambers RG, Lobell DB (2021) Anthropogenic climate change has slowed global agricultural productivity growth. Nat Clim Chang 11(4):306–312
Osakabe Y, Yamaguchi-Shinozaki K, Shinozaki K, Tran LSP (2014) ABA control of plant macroelement membrane transport systems in response to water deficit and high salinity. New Phytol 202(1):35–49
Pavlović I, Mlinarić S, Tarkowská D, Oklestkova J, Novák O, Lepeduš H, Bok VV, Brkanac SR, Strnad M, Salopek-Sondi B (2019) Early *Brassica* crops responses to salinity stress: a comparative analysis between Chinese cabbage, white cabbage, and kale. Front Plant Sci 10:450
Pessarakli M, Haghighi M, Sheibanirad A (2015) Plant responses under environmental stress conditions. Adv Plants Agric Res 2(6):73
Rasool S, Hameed A, Azooz M, Siddiqi T, Ahmad P (2013) Salt stress: causes, types and responses of plants. In: Ecophysiology and responses of plants under salt stress. Springer, New York, pp 1–24
Ravisankar P, Reddy AA, Nagalakshmi B, Koushik OS, Kumar BV, Anvith PS (2015) The comprehensive review on fat soluble vitamins. IOSR J Pharm 5(11):12–28
Rawat N, Singla-Pareek SL, Pareek A (2021) Membrane dynamics during individual and combined abiotic stresses in plants and tools to study the same. Physiol Plant 171(4):653–676
Redondo-Gómez S, Mesa-Marín J, Pérez-Romero JA, López-Jurado J, García-López JV, Mariscal V, Molina-Heredia FP, Pajuelo E, Rodríguez-Llorente ID, Flowers TJ (2021) Consortia of Plant-Growth-Promoting Rhizobacteria isolated from halophytes improve response of eight crops to soil salinization and climate change conditions. Agronomy 11(8):1609
Riyazuddin R, Verma R, Singh K, Nisha N, Keisham M, Bhati KK, Kim ST, Gupta R (2020) Ethylene: a master regulator of salinity stress tolerance in plants. Biomol Ther 10(6):959
Ruiz-Sola MÁ, Arbona V, Gomez-Cadenas A, Rodriguez-Concepcion M, Rodriguez-Villalon A (2014) A root specific induction of carotenoid biosynthesis contributes to ABA production upon salt stress in Arabidopsis. PLoS One 9(3):e90765
Sarabi B, Bolandnazar S, Ghaderi N, Ghashghaie J (2017) Genotypic differences in physiological and biochemical responses to salinity stress in melon (Cucumis melo L.) plants: prospects for selection of salt tolerant landraces. Plant Physiol Biochem 119:294–311
Sarkar J, Chakraborty U, Chakraborty B (2021) High-temperature resilience in Bacillus safensis primed wheat plants: a study of dynamic response associated with modulation of antioxidant machinery, differential expression of HSPs and osmolyte biosynthesis. Environ Exp Bot 182:104315
Schneider P, Asch F (2020) Rice production and food security in Asian Mega deltas—a review on characteristics, vulnerabilities and agricultural adaptation options to cope with climate change. J Agron Crop Sci 206(4):491–503
Sforza E, Pastore M, Franke SM, Barbera E (2020) Modeling the oxygen inhibition in microalgae: an experimental approach based on photorespirometry. New Biotechnol 59:26–32
Shabala S, Pottosin I (2014) Regulation of potassium transport in plants under hostile conditions: implications for abiotic and biotic stress tolerance. Physiol Plant 151(3):257–279
Shackira A, Puthur JT (2017) Enhanced phytostabilization of cadmium by a halophyte—Acanthus ilicifolius L. Int J Phytoremediation 19(4):319–326
Shah LR, Husain M, Raja A (2017) Abiotic stress mechanism in herbaceous crops: an overview. J Pharm Innov 6(11): 339–346
Shah AN, Tanveer M, Abbas A, Fahad S, Baloch MS, Ahmad MI, Saud S, Song Y (2021) Targeting salt stress coping mechanisms for stress tolerance in *Brassica*: a research perspective. Plant Physiol Biochem 158:53–64

Shahzad B, Yun P, Shabala L, Zhou M, Sellamuthu G, Venkataraman G, Chen Z-H, Shabala S, Wu H (2022) Unravelling the physiological basis of salinity stress tolerance in cultivated and wild rice species. Funct Plant Biol 49(4):351–364

Sharma A, Kumar V, Shahzad B, Ramakrishnan M, Singh Sidhu GP, Bali AS, Handa N, Kapoor D, Yadav P, Khanna K (2020) Photosynthetic response of plants under different abiotic stresses: a review. J Plant Growth Regul 39:509–531

Shin J, Song M-H, Oh J-W, Keum Y-S, Saini RK (2020) Pro-oxidant actions of carotenoids in triggering apoptosis of cancer cells: a review of emerging evidence. Antioxidants 9(6):532

Shu S, Yuan Y, Chen J, Sun J, Zhang W, Tang Y, Zhong M, Guo S (2015) The role of putrescine in the regulation of proteins and fatty acids of thylakoid membranes under salt stress. Sci Rep 5(1):1–16

Singh M, Kumar J, Singh V, Prasad S (2014) Proline and salinity tolerance in plants. Biochem Pharmacol 3(6):e170

Singh M, Kumar J, Singh S, Singh VP, Prasad SM (2015) Roles of osmoprotectants in improving salinity and drought tolerance in plants: a review. Rev Environ Sci Biotechnol 14:407–426

Slama I, Abdelly C, Bouchereau A, Flowers T, Savouré A (2015) Diversity, distribution and roles of osmoprotective compounds accumulated in halophytes under abiotic stress. Ann Bot 115(3):433–447

Song KE, Hwang HR, Konvalina P, Jun WJ, Jung JW, Shim S (2023) Hydrogen peroxide ameliorates the adversities of drought stress during germination and seedling growth in sorghum (*Sorghum bicolor* L.). Agronomy 13(2):330

Streyczek J, Apweiler M, Sun L, Fiebich BL (2022) Turmeric extract (*Curcuma longa*) mediates anti-oxidative effects by reduction of nitric oxide, iNOS protein-, and mRNA-synthesis in BV2 microglial cells. Molecules 27(3):784

Syed A, Sarwar G, Shah SH, Muhammad S (2021) Soil salinity research in 21st century in Pakistan: its impact on availability of plant nutrients, growth and yield of crops. Commun Soil Sci Plant Anal 52(3):183–200

Thorat B, Bagkar T, Raut S (2018) Responses of rice under salinity stress: a review. IJCS 6(4):1441–1447

Toderich KN, Mamadrahimov AA, Khaitov BB, Karimov AA, Soliev AA, Nanduri KR, Shuyskaya EV (2020) Differential impact of salinity stress on seeds minerals, storage proteins, fatty acids, and squalene composition of new quinoa genotype, grown in hyper-arid desert environments. Front Plant Sci 11:607102

Toh S, Takata N, Ando E, Toda Y, Wang Y, Hayashi Y, Mitsuda N, Nagano S, Taniguchi T, Kinoshita T (2021) Overexpression of plasma membrane H+-ATPase in guard cells enhances light-induced stomatal opening, photosynthesis, and plant growth in hybrid aspen. Front Plant Sci 12:766037

Uçarlı C (2020) Effects of salinity on seed germination and early seedling stage. In: Abiotic Stress in Plants, vol 211. IntechOpen, pp 211–231. https://doi.org/10.5772/intechopen.91549

Velázquez-Márquez S, Conde-Martínez V, Trejo C, Delgado-Alvarado A, Carballo A, Suárez R, Mascorro J, Trujillo A (2015) Effects of water deficit on radicle apex elongation and solute accumulation in *Zea mays* L. Plant Physiol Biochem 96:29–37

Xiong H, Li J, Liu P, Duan J, Zhao Y, Guo X, Li Y, Zhang H, Ali J, Li Z (2014) Overexpression of OsMYB48-1, a novel MYB-related transcription factor, enhances drought and salinity tolerance in rice. PLoS One 9(3):e92913

Yadav S, Modi P, Dave A, Vijapura A, Patel D, Patel M (2020) Effect of abiotic stress on crops. Sustain Crop Prod 3. https://doi.org/10.5772/intechopen.88434

Yamashita K, Shiozawa A, Banno S, Fukumori F, Ichiishi A, Kimura M, Fujimura M (2007) Involvement of OS-2 MAP kinase in regulation of the large-subunit catalases CAT-1 and CAT-3 in *Neurospora crassa*. Genes Genet 82(4):301–310

Yancey PH (2005) Organic osmolytes as compatible, metabolic and counteracting cytoprotectants in high osmolarity and other stresses. J Exp Biol 208(15):2819–2830

Yang M, Fu Y, Zhao R, Koorem K, Li B, Siemann E, Yang Q (2022) The effects of light availability on plant-soil interactions and salinity tolerance of invasive tree species, *Triadica sebifera*. For Ecol Manag 506:119964

Zemni N, Slama F, Bouksila F, Bouhlila R (2022) Simulating and monitoring water flow, salinity distribution and yield production under buried diffuser irrigation for date palm tree in Saharan Jemna oasis (North Africa). Agric Ecosyst Environ 325:107772

Zhang S, Vijayavenkataraman S, Lu WF, Fuh JY (2019) A review on the use of computational methods to characterize, design, and optimize tissue engineering scaffolds, with a potential in 3D printing fabrication. J Biomed Mater Res B Appl Biomater 107(5):1329–1351

Zhonghua T, Yanju L, Xiaorui G, Yuangang Z (2011) The combined effects of salinity and nitrogen forms on *Catharanthus roseus*: the role of internal ammonium and free amino acids during salt stress. J Plant Nutr Soil Sci 174(1):135–144

Zhu X, Xie S, Tang K, Kalia RK, Liu N, Ma J, Bressan RA, Zhu J-K (2021) Non-CG DNA methylation-deficiency mutations enhance mutagenesis rates during salt adaptation in cultured *Arabidopsis* cells. Stress Biol 1(1):1–12

Chapter 19
Mechanisms and Approaches of Enhancing Drought Stress Tolerance in Crop Plants

Nono Carine Temegne, Esaïe Tsoata, Godswill Ntsomboh-Ntsefong, Atabong Paul Agendia, Francis Emmanuel Ngome, and Emmanuel Youmbi

Abstract Human accomplishments have been implicated in activating global warming, producing drought, which affects crops. Crops are more vulnerable to drought, and as a result, production is diminished. Producing varieties adapted to sustainable agriculture, in particular those more resilient to drought, is a challenge for geneticists. Based on a characterization of the climatic risk, a multidisciplinary approach combining ecophysiology and genetics can help define varietal ideotypes combining adaptive characters, which can be utilized in breeding. Selection aimed at better adaptation to water deficit relies both on the existence of natural diversity or diversity generated by hybridization and on the implementation of screening tests for adaptive traits. To reduce the impacts of water starvation on crops, it is also imperative to find a response machinery of crop drought tolerance. This mechanism comprises morpho-physiological, biochemical, cellular, and molecular processes. Enhancement of the architecture of roots, arrangement of leaves, osmotic equilibrium, relative H_2O content, and adjustment of stomata are reflected as the most important traits of water deficit tolerance in crops. Signaling and reactive oxygen species are vital machineries to cope with water starvation through Ca^{2+} and plant hormones. Microorganisms (fungi, bacteria) have a key function in improving drought tolerance. The transgenic techniques and exogenous substances are also vital for increasing drought tolerance.

Keywords Drought stress · Gene expression · Tolerant crop

N. C. Temegne (✉) · E. Tsoata · G. Ntsomboh-Ntsefong · A. P. Agendia · E. Youmbi
Department of Plant Biology, University of Yaoundé I, Yaoundé, Cameroon
e-mail: carine.temegne@facsciences-uy1.cm

F. E. Ngome
Institute of Agricultural Research for Development (IRAD), Yaoundé, Cameroon

M. Hasanuzzaman (ed.), *Climate-Resilient Agriculture, Vol 2*,
https://doi.org/10.1007/978-3-031-37428-9_19

1 Introduction

Climatic modification is a significant defy, which menaces the durable production of grain cereals. It generates several abiotic constraints as drought. Drought affects about 1.43 billion people in the world (The World Bank 2021). It is the main limitation of agricultural production. The overall reduction in the volume of rainfall can schematically result either from a diminution in the duration of the raining period, or from the more frequent occurrence of episodes when the rain is lacking throughout the cycle (Lacape et al. 2015). It is commonly accepted that climate modification is escorted by augmented menaces in rapports to inter-annual changeability and the occurrence of risky climatic happenings. The effects of epochs of drought on harvests vary according to the plants, particularly in their mode of development and reproduction, for example, critical stages of sensitivity differentiated between plants with continuous growth on one hand and cereals on the other hand (Lacape et al. 2015; Sandhu and Kumar 2017). Drought, whether episodic, sporadic, or permanent, is strongly linked to the loss of crop yield. Studies, carried out at the level of the plant, make it possible to understand the different strategies implemented during water deficit.

The mechanisms used by plants to adapt to drought can change with environmental conditions. Similarly, genotypes deemed to be tolerant may present different mechanisms of adaptation to water deficit. Water deficit tolerance mechanisms comprise the physiology of physical stature and biochemical, cellular, and molecular processes. Improved knowledge of crop water deficit mechanisms is provided by numerous research studies on water deficit carried out crosswise organizations on plant roles, plant breeding, molecular genetics, biotechnologies, and cell and molecular biology. They have aided researchers to plan better approaches to diminish crop return losses during water deficit period. These comprise several essential agronomic characters linked to water deficit resistance, the quantitative trait loci (QTL) identification for return in water deficit conditions, marker-assisted pyramiding of genetic areas, which raise harvest in water deficit period, improvement of proficient practices for genetic modification, entire sequencing, gloss of genomes, and syntonic researches (Sandhu and Kumar 2017).

2 Selection of Crops Tolerant to Water Deficit and Genetic Breeding

2.1 *Plant Adaptation Mechanisms*

Screening of cultivars and diffusion to producers of enhanced cultivars that are more suitable appear to be an appropriate solution simple to implement and diminish the influence of water starvation. This response by the "seed" must most often be reasoned in combination with an action on the technical itineraries and the conduct of the culture. In cultivated plants, the adaptive response to drought is classically described as involving three key approaches (Blum 1988):

- **Escape** by adjusting the growth cycle. The water deficit escape tactic assembles mechanisms permitting crops to achieve their cycles before land and crop drought occurs. Mechanisms supporting developing plasticity, fast morphological growth, and redeployment of preanthesis osmolytes to generative plant's structures permit crops to escape water deficit.
- **Avoidance** matches the crop's capability to evade dehydrating, classically by monitoring H_2O loss or preserving its absorption. Drought avoidance stratagem assembles mechanisms permitting crops to preserve moderately great tissue H_2O potential notwithstanding H_2O deprivation. Machineries for enhancing H_2O uptake, H_2O storing in cell, and reducing H_2O damage favor water deficit evading. This comprises, amid further machineries, augmented asset in roots, restricted leaf surfaces via coming loose of elder foliage, rolling of leaves, fruit production, and cuticles of leaves and stomatal closing.
- **Tolerance** or ability of the plant to overcome a deterioration in its hydric state (avoidance of dehydration), for example, by osmotic adjustment. The water deficit tolerance stratagem assembles mechanisms permitting crops to fight drought through mediocre cell H_2O potential. The crops' reply to cell drought governs their level of water deficit resistance. Growth of the effective antioxidant complex and preservation of turgor via osmotic regulation are amid the machineries permitting water deficit resistance.

2.2 *Suitable Parent Selection*

The pilot and essential phase of breeding programs implicates the selection of appropriate parents (donors). This selection from a great germplasm collection is a fundamental phase. The usage of a particular parent with exceptional features for a precise milieu might be conducted for the achievement of cultivars and feature improvement programs (Sandhu and Kumar 2009). The maximum of the old-fashioned parent have numerous unwanted characteristics and consequently are not appropriate for immediate utilization in breeding programs. Rice, for example, has unwanted traits like high crop height, weak harvest potential, and poor grain quality. However, they have a required water deficit resistance feature. Contrarily, new rice cultivars have appropriate characteristics like great harvest, medium height, and good grain type, but they are water deficit susceptible (Sandhu and Kumar 2009). Breeding for a chosen feature to acquire novel gene combinations needs the utilization of genetic variation, which is present in old-fashioned cultivars carrying wanted features and newly enhanced cultivars with great harvest potential (Ashraf and Akram 2009).

2.3 Toward Adaptation Trait Identification

Yield, an integrative feature at the crop and plot scales and over time, generally reveals feeble heritability and significant genotype×environment interaction components. It is therefore indispensable for breeding to decompose yield into individual components that are simpler, more heritable, and can be utilized as selection tests that means more facile to evaluate than field yield, by correlation with it, and can be implemented on huge numbers. In function of varieties, the machineries implicated, or the scales of observation, different physiological traits are suggested for application in screening characterizations comprising general indicators (stress indices, rolling leaves, etc.), traits related to photosynthesis and gas exchange, water intake and efficiency of utilization, H_2O status parameters, and osmotic balance (Lacape et al. 2015; Tsoata et al. 2016). Iqbal et al. (2021) stated that improved root architecture, structure of leaves, osmotic adjustment, H_2O relative content, and regulation of stomata are looked at as the most important characteristics of drought tolerance crops. Additionally, signaling and ROS are vital machineries for surviving dryness through Ca^{2+} and plant's hormones like abscisic acid (ABA) and brassinosteroid. Belowground organisms, like fungi and bacteria, have a crucial function in enhancing drought tolerance. The amount of distinguishing loci, transgenic approaches, and the usage of osmoprotectants are too essential for increasing drought resistance.

The creation of varieties that perform better in drought conditions requires the implementation of multidisciplinary methods. It is found on the evaluation of genetic changeability for harvest in drought circumstances or for the physiological characteristics implicated, within a support plant material. In marker-assisted selection (MAS) methods, plant material (segregating populations or diverse genetic resources) is subject to simultaneous characterization of genotype using DNA and phenotype markers (in stressed versus unstressed conditions). The link between the two characterizations makes it possible to identify the regions of the genome involved, or QTL, and to select the progenies indirectly. Current advances in molecular biology and genomics allow a significant modification of scale in terms of genetic dissection of characters, making it possible to analyze the variation in the expression of thousands of genes in large numbers of individuals' underneath stress (Blum 2011). When candidate genes are recognized, genetic transformation represents a possible way to verify the effect of genes, or even to mature merchantable transgenic cultivars (Varshney et al. 2011). Although such approaches, MAS or genetic transformation, applied to the improvement of plant tolerance to biotic and abiotic constraints are now widely developed by private sector seed companies, the known examples of applications are still few in number (Lacape et al. 2015).

3 Breeding for Improved Tolerance to Water Deficit

Several research structures have dedicated 10–30 years and more to mature drought-resistant varieties via traditional and modern techniques linked to secondary traits (Table 19.1), and examples include International Maize and Wheat Improvement Center (CIMMYT), International Crops Research Institute for the Semi-Arid Tropics (ICRISAT), and International Rice Research Institute (IRRI) (Mondal et al. 2020; Srivastava et al. 2022). Water deficit replies in crops are multifaceted physiological and biochemical processes and imply modifications in anatomical structures (Tsoata et al. 2016, 2017). These features are controlled by groups of genes in the genome. It is indispensable to find drought-related genes or proteins. An assortment of drought response-related proteins (OsMYBR57 in rice), phytohormonal regulators (salicylic, jasmonic, and abscisic acids, ethylene, auxin, gibberellin, brassinosteroids, peptide, etc.), signaling genes, and protective enzymes participate to drought resistance in crops (Yang et al. 2022a, b).

Breeders should emphasize these features when selecting drought-tolerant plant cultivars. Wild species of selected plants like tomato, chili, pumpkin, and onion can be used to create drought-tolerant cultivars. Further tools that are able to be used for drought breeding comprise concealing, marker-assisted selection (simple sequence repeat, cleaved amplified polymorphic sequences, single nucleotide polymorphisms, etc.), plant modification, phenotyping, and conventional breeding methods (Kumar et al. 2022). Advanced methods like RNA interfering, grouped constantly interspaced small palindromic replications, or transgenic gene silencing offer an enormous possibility for misuse (Rajput et al. 2021). So, climate-resilient breeding is the time required and should be made possible by establishing appropriate methods rendering to the requisite of drought conditions.

Identification of genomic regions participating in drought tolerance aids in developing cultivars adapted to rain-fed areas via marker-assisted breeding. Selamat and Nadarajah (2021) found it in rice. Marker-assisted backcross breeding (MABB) has allowed novel chances for the introgression of regions governing drought resistance via cautious QTL identification and adequate mapping works. Selection for a well-built root system could increase water deficit resistance in rice, and these characteristics could be fruitfully transferred to popular cultivars by MABB. Therefore, MABB was carried out to introgress root feature QTLs from a deep-rooted, drought-tolerant japonica cultivar, CT9993, into common cultivars. These QTLs for basal root thickness, grain returns, and root tensile strength on chromosome 4 and penetrated root thickness on chromosome 9 of CT9993 were introgressed into the common cultivars by crossing and the offspring were carefully chosen for forefront analysis by the QTL flanking markers.

Table 19.1 Some conventional breeding developed at the research institute for drought tolerance and secondary traits associated

Crops	Commercial/ register name	Methods	Secondary traits	Research structure	Area	References
Oryza sativa L.	MR219	MAB	High yield	Universiti Kebangsaan Malaysia/IRRI	Asia	Shamsudin et al. (2016)
O. sativa L.	IR64, Vandana	Genetic variations, conventional and molecular approaches (MABB, QTL introgression)	High yield	IRRI	South and Southeast Asia, Africa	Kumar et al. (2014)
Hordeum vulgare L.	Lenetah	Multiple crosses $F_{5\text{-}6}$ line	High yield	Agricultural Research Service Aberdeen/Idaho Agric. Exp. Station	America	Obert et al. (2008)
Phaseolus vulgaris L.	CO46348	Single cross	High yield, best seed quality, resistance to rust (*Uromyces appendiculatus*)	Colorado Agricultural Experiment Station (CAES)/Univ. of Idaho/ USDA_ARS	America	Brick et al. (2008)
Triticum aestivum L.	NEO1643	Multiple crosses $F_{3\text{-}4}$ line	High grain yield (non-irrigated production)	Nebraska Agric. Exp. Station/ USDA_ARS	America	Baenziger et al. (2008)
T. aestivum L.	Ripper	Multiple crosses $F_{3\text{-}4}$ line	High yield, superior milling and bread-baking quality	CAES/Colorado State Univ./USDA_ARS	America	Haley et al. (2007)
T. aestivum L.	Jinmai 50	Multiple crosses $F_{2\text{-}5}$ line	Early maturity, vigor, cold tolerance, high yield	WBIG of Cotton Res. Inst. of Shanxi Agric. Sci. Acad. China	Asia	Xinglai et al. (2006)
H. vulgare L.	Giza 126	Multiple crosses	High yield	Field Crops Research Institute, Giza, Egypt	Africa	Noaman et al. (1995)
Cicer arietinum L.	FLIP 87-59C	Cross	High yield	ICARDA/ICRISAT	Asia	Singh et al. (1996)
C. arietinum L.	ICC 4958	Collection	High yield, short duration, terminal branching growth habit	Legumes Program — ICRISAT/ICARDA (India and Syria)	Asia	Saxena et al. (1993)

4 Morphological Mechanisms to Water Deficit Adaptation

4.1 *Root Morphology and Plasticity in Relation to Drought Resistance*

The root system fixes the crops in the land and gets H_2O and minerals from the rhizosphere and regulates lateral root emergence. Roots adjust their configuration in reply to water deficit to upsurge infiltration, sharing, and interaction with the land for enhanced H_2O and nutrient uptake (Rongsawat et al. 2021). These adaptations guarantee required nutrition and H_2O uptake, upholding crop physiological functions, and production throughout water deficit. Numerous works were carried out on relationships amid root characters and crop outputs under water deficit. Pinpointing these characteristics in diverse crops in reply to water deficit will help in the selection of water starvation-resistant varieties.

Fonta et al. (2022) stated that the malleability of root phenotypes in Azucena cultivar (vs IR64 cultivar) participates to its water deficit resistance by decreasing the metabolic fee of land exploration and enhancing the effectiveness of H_2O transportation.

The basic function of abscisic acid is to change the root system, its development shape, and restraint its development. So, ABA-lacking mutants (aba2-1 and aba3-1) show an improved level of lateral root emergence (Deak and Malamy 2005). Fascinatingly, ABA insensitive (abi4) demonstrates an improved no. of lateral roots (Quiroz-Figueroa et al. 2010). Consequently, as a substitute for abscisic acid rates, abscisic acid gesturing is too implicated in lateral root formation. ABA is implicated in preserving root meristem and root growth. Drought controls root cell integrity and elongation by checking respiratory burst oxidase homolog gene expression through ROS oozes (Zhang et al. 2014).

4.2 *Microorganisms and Tolerance to Drought*

Rhizobacteria liberate substances like plants hormone (ACC deaminase, auxins, cytokinin, ethylene), volatile compounds (acetic acid), siderophores, osmolytes (sugars, proline, amino acids), and exopolysaccharides, which enhance crops drought tolerance by maintaining higher land dampness content (Shoaib et al. 2022). Exopolysaccharides secreted by *Azotobacter* sp., *Bacillus* sp., *Klebsiella* sp., *Paenibacillus* sp., *Pseudomonas* sp., and *Sphingomonas* sp. facilitate land water conservation and porousness, rise land aggregation, create a protective capsule around roots, and enhance water and nutrient uptake during the water deficit period (Costa et al. 2018). Butanediol produced by *P. chlororaphis* mediated stomatal closing in *Arabidopsis* (Cho et al. 2008; Shoaib et al. 2022). Cytokinin exsuded by *P. fluorescens* improved drought response on *Solanum lycopersicum* L. (Mekureyaw et al. 2022).

Arbuscular mycorrhizal fungi (AMF) raise H_2O deficit tolerance and decrease its vulnerability through enhancing H_2O, P-uptake, soil-C, and photosynthetic rate; extending the land rhizosphere via extra radical hyphae; improving soil stability via glomalin secretion, which favors soil aggregation; and augmenting apoplastic H_2O transportation and root hydraulic conductivity (Smith and Read 2010; Tsoata et al. 2015; Temegne 2018; Verbruggen et al. 2021). Furthermore, they sporadically control lateral root development like in *Citrus limon* L. (Liu et al. 2022). AMF impact root hydraulic characteristics and improve crop water deficit tolerance by monitoring the gene, which controls plasma membrane proteins (PIP). So, in common bean, AMF were intensely linked to the PIP2 protein control and diminished hydraulic conductivity underneath drought (Aroca et al. 2007). AMF surge antioxidant enzyme actions, decreasing oxidative stress under water deficit, like bringing more H_2O_2 effluxes of the tap and lateral roots of Bitter orange (Huang et al. 2017).

5 Physiological Mechanisms to Water Deficit Adaptation

5.1 Stomatal Closure

Stomatal closure occurs a few minutes subsequently crop sensitivity of a drought. Stomata are tiny apertures in the epidermis of crop shoot, constituted of guard cells and the turgor state that regulates the pore size (Xu et al. 2022). They permit crops to enhance gas exchanges (carbon dioxide, oxygen, and H_2O vapor) with the air reliant on the environmental surroundings. In drought, the decrease in H_2O loss by transpiration linked to stomatal closing allows crops to maintain their H_2O equilibrium. Nonetheless, this helpful influence derives at the expense of a decrease in the entry of carbon dioxide in the leaves and an upsurge of temperature of leaves, since the extra sun energy eliminated through open stomata below ideal irrigating milieus is no longer dissipated. In low carbon dioxide/oxygen ratio and leaf warming, photorespiration is favored at the outlay of photosynthesis leading to a reduction in carbohydrate production (Zhou et al. 2007).

Physiological procedures that control stomata role have been determined. They include complex relations with external and internal elements. Abscisic acid is closely involved in monitoring the aperture and closure of stomata in reply to modifications in cell water status, while it is still not clear by what means cell H_2O shortage provokes ABA synthesis. The pointer might be a weakened pressure of cells, membrane changes, solute contents, or cellular barrier pressure (Muhammad et al. 2022).

In guard cells of stomata, abscisic acid is accountable for the upsurge in the calcium concentration that stimulates membrane anion canals and vacuolar potassium canals. The anion canals permit the discharge of anions out of the cells (Cl^-, NO_3^-), while the vacuolar potassium canals discharge the vacuolar potassium in the cytoplasm. This leads to a plasma membrane depolarization that neutralizes inner-rectifying potassium (potassium in) canals, stimulates outward rectifying potassium (potassium out) canals, and conducts potassium efflux out of the guard cells. The anions efflux and potassium from guard cells participate to damage guard cell turgor that conducts to the closing of stomata. Abscisic acid deters the H^+/ATPases, the potassium in canals, and the importation of chloride, malate, and nitrate in the cytosol that are the machineries implicated in the stomatal aperture (Kim et al. 2010).

Apart from ABA, other plant hormones like auxin, brassinosteroids, ethylene, gibberellin, jasmonic acid, salicylic acid, or cytokinins are implicated in this machinery over a multifaceted interaction with ABA gesturing path. Jasmonate and brassinosteroids were revealed to bring stomatal closing and deter stomatal aperture underneath drought (Wang et al. 2021a, b; Iqbal et al. 2021; Xu et al. 2022).

Non-stomata restrictions might be as well implicated in the reduced photosynthesis level perceived underneath drastic drought. They principally relate to the reduced synthesis of the enzymes implicated in C incorporation and photophosphorylation in low humidity environments (Bota et al. 2004) and to lesser circulation of carbon dioxide through the mesophyll of leaves. A polemic exists on whether drought reduces photosynthesis principally via stomatal closing or via metabolic weakening (Bota et al. 2004). The efforts to reply to this query first of all displayed that the weakening of the Calvin cycle does not restrain photosynthesis until that water deficit is extremely drastic. Furthermore, the relative cell H_2O content at which a photosynthetic metabolism is reduced will be intensely specie reliant on (Bota et al. 2004). Crassulacean acid metabolism and C4 crops specifically have settled machineries to enhance their photosynthetic machines underneath short H_2O availability in comparison with the C3 crops (Hamim 2005).

5.2 *Defense Against Oxidative Damages*

In drought, crop submission to sunlight intensity surpasses their ability of carbon dioxide absorption below low stomatal conductance that conducts to an extreme ROS secretion. The PS (photosystem) II action is decreased to match an existing carbon substrate when the carboxylation is reduced. The surplus of electrons from the photophosphorylation electron sequence is relocated to carbon dioxide at PSI in the Mehler reaction, conducting the liberation of ROS like O_2^- and H_2O_2 inside the chloroplasts. Moreover, an improved oxygenase activity of the RuBisCO enzyme underneath drought leads to ROS liberation inside the peroxisomes (Apel and Hirt 2004). ROS is able to engender main oxidative troubles in crops, and their harmful impacts comprise DNA deterioration, amino acid, protein oxidation, and lipid peroxidation. They disrupt cellular membrane and enzymes, disturbing the roles of cells. Therefore, crop endurance underneath drought is determined by an increase in a proficient antioxidant system, with enzymatic (superoxide dismutase, catalase, ascorbate peroxidase, etc.) and non-enzymatic (amino acid, pigments, and polyphenols) constituents (Muhammad et al. 2022; Apel and Hirt 2004).

5.3 *Osmotic Adjustment*

To reduce water loss, crops can dynamically gather organic composites (osmoprotectants); it is the osmotic adjustment. Osmoprotectants are very soluble with weak molecular weight and non-noxious in great cytosolic contents. They comprise a large assortment of compounds that hinge on the type of drought, the crops, and the cultivars. Proteins and amino acids (aspartic acid, proline, etc.), quaternary amines (glycine betaine, alanine betaine, etc.), polyols and sugars (D-pinitol, mannitol, etc.), organic acids (malic acid, citric acid, etc.), hydrophilic proteins (LEA, heat shock proteins, etc.), and ions (Ca^{2+}, K^+, Cl^-, etc.) are among the crop osmoprotectants (Wang et al. 2004; Tunnacliffe and Wise 2007).

Osmoprotectant permits cell to lessen their osmotic potential, so preserving H_2O consummation and cellular turgor underneath drought. These molecules safeguard the proteins, the cellular membranes, and the metabolic mechanism to fight dryness, by cooperating with H_2O and avoiding harmful interactions of molecules (Sanders and Arndt 2012).

6 Molecular Mechanisms to Water Deficit Adaptation

Three main groups of molecular mechanisms of drought adaptation are well-known: mechanisms implicated in signal detecting and transduction, those implicated in transcriptional regulation, and those implicated in protection machineries, comprising osmotic adjustment, drought-responsive gene expression, protein and membrane defenses, and H_2O and ion uptake (Fig. 19.1).

6.1 *Signal Detecting*

To have suitable physiological and molecular replies, crops need to perceive the drought via particular receptors like AHK1/ATHK1 osmosensor (Tran et al. 2007) and ABA receptors (Table 19.2).

6.2 *Signal Transduction*

After discernment of drought signal by membrane receptors, secondary messengers (Ca signals, ROS, proteins kinases, protein phosphatases) are implicated in the gesture transduction to the cytosol and nucleus.

Ca is one of the most essential secondary messengers in reply to extracellular incitements in crops. It has a vital function in drought signaling. The Ca^{2+} content in

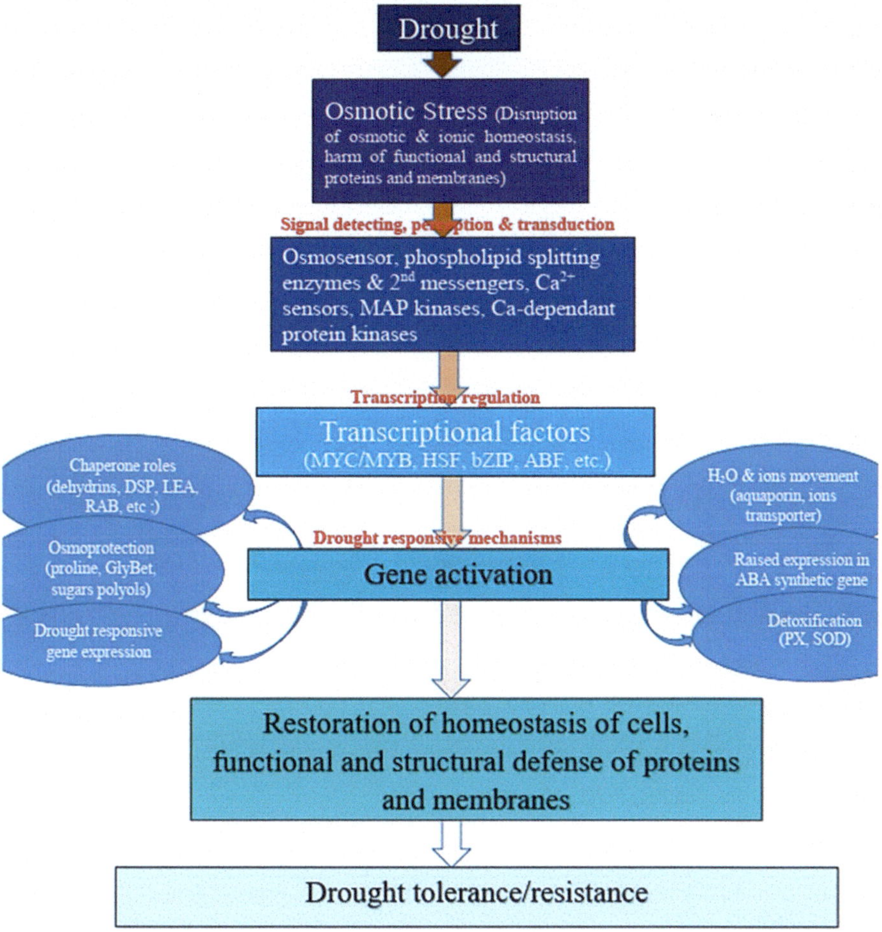

Fig. 19.1 Crop molecular reactions to drought. *ABA:* Abscisic acid, *bZIP:* basic leucine zipper domain, *DSP:* desiccation stress protein, *GlyBet:* glycine-betaine, *HSF:* heat shock factors, *LEA:* late embryogenesis abundant, *MAP:* mitogen-activated protein, *MYC/MYB:* murine c-myc/murine c-Myb proteins, *RAB:* Ras-associated binding proteins, *SOD:* superoxide dismutase. Adapted or modified from Wang et al. (2003), Albert (2017), and Muhammad et al. (2022)

the cell is well-adjusted by the existence of Ca^{2+} similar to vacuoles, endoplasmic reticulum, mitochondria, and cell walls. In reply to drought, the Ca^{2+} canals sited in the vacuole and endoplasmic reticulum are stimulated by abscisic acid via the intermediary of the secondary messenger inositol-1,4,5-triphosphate (IP3), conducting to the Ca^{2+} discharge in the cytoplasm. The Ca^{2+} is able to be returned back to the vacuoles and ER via the stimulation of the Ca^{2+}ATPase and H^+/Ca^{2+}antiporters when the cell's environment becomes ideal again.

The disparities in the Ca^{2+} cytoplasmic content are well-known by Ca^{2+}-sensors or Ca^{2+}-binding proteins that are principally *calmodulins*, *calmodulin-like proteins*,

Table 19.2 Selected genes/transporters for enhanced drought resistance via controlling ABA signaling way in crops

Receptors/genes / transporters	Roles	Plants	References
VaPYL4	Improved survival rate, fresh weight of seedlings and siliques, POD	*Arabidopsis*	Ren et al. (2022)
MdPYL9	Increased photosynthetic rate, RWC	*Malus domestica*	Yang et al. (2022a, b)
TaPYL4	Enhanced growth features, regulated (+) stomata movement, osmoprotectant biosynthesis, and root architecture	*T. aestivum* L.	Zhang et al. (2022)
ZmbZIP33	Augmented chlorophyll concentration and root length	*Arabidopsis*; *Zea* mays L.	Cao et al. (2021)
AtSAUR32	Involved in the abscisic acid mediation gesture transduction	*Arabidopsis*	He et al. (2021)
AtBBD1	Augmented proline concentration	*Arabidopsis*	Huque et al. (2021)
OsMADS23	Promoted ABA and proline biosynthesis	*O. sativa* L.	Li et al. (2021)
SlGRAS4	Modulated ROS homeostasis	*S. lycopersicum*	Liu et al. (2021)
OsSWEET13	Raised the sucrose concentration in leaf and root tissues and in phloem sap	*O. sativa* L.	Mathan et al. (2021)
DcABF3	Enhanced stomatal density, root length, and germination percentage	*Arabidopsis*	Wang et al. (2021a, b)
GmCIPK2	Improved drought resistance and hairy roots	*Arabidopsis*, *Glycine max*	Xu et al. (2021)

RWC relative water content, *POD* peroxidase

calcineurin B-like proteins, and Ca^{2+}-dependent protein kinases (CDPK). These Ca sensors can bring the manifestation of drought response genes (Tuteja and Mahajan 2007).

The ROS are unmistakably implicated like 2nd messengers amid drought receptors and the downstream gesturing cascade. While Ca^{2+} gesturing is mainly regulated in crops by packing and discharge, ROS gesturing is regulated by secretion and scavenging. Two diverse loops are stated to be implicated in the ROS gesture transduction path. A common protection reply is able to be stimulated to delete ROS; however, a step-up reply can be stimulated to improve ROS gestures through the NADPH oxidases action, which can dynamically produce H_2O_2 (Kwak et al. 2003; Muhammad et al. 2022). H_2O_2 can be like ROS implicated in the drought-reply gesture transduction ways, due to its relatively steady configuration and its readiness to diffuse from one cell part to another. The identified downstream actions controlled by H_2O_2 are Ca^{2+} deployment via stimulation of Ca^{2+} porous canals in the plasma membrane, protein phosphorylation via MAPK/CDPK cascades, and gene control (Chen et al. 2021; Mansoor et al. 2022).

Protein kinase is implicated in catalyzing the phosphorylation of other proteins corresponding to the addition of a PO_4^{3-} ion. On the contrary, protein phosphatase catalyzes the dephosphorylation of protein that signifies the subtraction of a PO_4^{3-} cluster through hydrolysis. Phosphorylation/dephosphorylation machineries are implicated in the stimulation and deactivation of enzyme action and in controlling protein–protein exchanges inside gesturing nets. Among kinase proteins, CDPK-stimulated MAPK and sucrose non-fermentation 1-related kinase (SnrK) (Chen et al. 2021; Mansoor et al. 2022) were stated to be implicated in the transduction of drought gestures from the plasma membrane to the nucleus. The MAPK protein is implicated in a controlling cascade constituted of principally three mechanisms, a MAPK kinase kinase, a MAPK kinase, and a MAPK, linked to everyone further via events of phosphorylation (Chen et al. 2021; Mansoor et al. 2022). SnrK proteins of clusters 2 and 3 have an important function in the abscisic acid signaling path via their contribution to the inhibition of the H^+ATPase and in the stimulation of anion and K^+ in canals implicated in stomatal closing machinery (Chen et al. 2021; Mansoor et al. 2022). Among phosphatase proteins, two clusters are generally characterized in collected works and termed conferring to their capacity to bind particular proteins rests: serine/threonine protein phosphatases and tyrosine-specific protein phosphatases. Many phosphatases, among the tyrosine-specific protein phosphatase, are implicated in the abscisic acid signalization path (Schweighofer et al. 2004).

6.3 *Transcriptional Regulator*

Crop genomes hold numerous transcription factors, and the majority have their place in limited vast multi-gene groups. Single affiliates of an identical group frequently reply in a different way to numerous abiotic stress stimuli (drought); nonetheless, many stresses reactive genes might distribute the identical transcription factors since significant overlays amid the drought signaling pathways. Naturally, the transcription factors implicated in reply to drought are well-known conferring to their sensitivity to abscisic acid, describing two kinds of transcriptional paths, the abscisic acid facilitated paths and the non-abscisic acid facilitated paths; however, there are proofs for multifaceted cross-talk amid sorts of paths (Golldack et al. 2014).

6.4 *Genes Induced in Reaction to Drought*

Important analogous studies of gene expressions were piloted in crops by the improvement of the cDNA microarrays and the RNA sequencing expertise, permitting the selection of genes diversely overexpressed in reply to drought in wheat (Chu et al. 2021). Certain of these genes were analyzed in reverse genetic research studies to clarify their role; however, a huge number of genes still have unidentified roles. Three main gene families defined in collected works for their participation in reply to drought are heat shock and molecular chaperon proteins, LEA and dehydrins, and aquaporins. Nonetheless, as drought disturbs cell growth and prime and secondary metabolites, several other genes have their manifestation adjusted underneath drought conditions.

The inactivity of proteins and enzymes is typically produced in crops exposed to drought. Upholding proteins in their serviceable conformations, averting the accumulation of denatured proteins, and rejecting worthless peptides are vital. It is the function of the molecular chaperones, specifically the heat shock proteins (HSPs) that are up-controlled underneath drought. Among the five main families of HSPs, the small HSPs are the most predominant in crops. Combined with their participation in membrane and protein defense, HSPs cooperate with numerous signaling compounds and with further drought response mechanisms like osmotic regulation and depollution machineries. Cross-talk machineries with HSPs, chaperon proteins, and other drought-reactive machineries will be discovered to make available a more comprehensive understanding of crop reply to drought (Wang et al. 2004).

LEA and dehydrins are proteins significantly produced in reply to drought for the duration of the last phase of embryogenesis in dehydrated kernels. Their roles are still generally unidentified. It was proposed that they might be implicated in binding H_2O, ion requisitioning, macromolecule, and membrane steadiness (Tunnacliffe and Wise 2007).

Aquaporins are proteins permitting H_2O and lesser neutral solutes to travel transversely through cellular membranes. Conferring to amino acid sequence resemblance, they are grouped into four subgroups (Chaumont et al. 2005). Aquaporins have been presented to regulate the widespread H_2O transportations from the roots to the leaves for the duration of transpiration flux, the transportation of solutes in the phloem, the closing or opening of the leaf stomata, the leaf movement, and the cytosolic homeostasis. They are stimulated via phosphorylation/dephosphorylation machineries brought by CDPK. Acid pH and free Ca^{2+} decrease the H_2O porousness of the aquaporins (Albert 2017).

6.5 Histone Changes and DNA Methylation

Following transcriptional controls via precise transcription factors, epigenetic procedures intensely impact the effectiveness of gene manifestation in reply to water deficit (Sun et al. 2022). Epigenetic procedures comprise histone changes, DNA demethylation, and non-encrypting DNA-linked gene and transferrable component silencing machineries. Crucial enzymes implicated in demethylation machineries were described (Sahu et al. 2013). Small RNAs were as well stated to have a significant function in epigenetic control in reply to drought through transcriptional gene silencing via RNA-assisted DNA methylation (Khraiwesh et al. 2010). Moreover, among other plant hormones, abscisic acid might be implicated in controlling epigenetic changes (Chinnusamy and Zhu 2009). Nevertheless, exhaustive information on the particular machineries, which triggered epigenetic control below disclosure in the environment, is only gradually developing. Little is recognized about the mitotic and meiotic heritability of histone change and DNA methylation (Albert 2017).

7 Biotechnologies for Enhancing Water Deficit Tolerance

Climate modification exacerbates a lesser favorable growth environment. It leads to an important rise in constraints generated by abiotic stresses like drought, limiting crop returns. Enhancement of varieties should be a path of solution. Breeding programs are presently set up to encounter novel challenges. In this situation, crop breeding in association with molecular biology and biotechnology presents prodigious prospects for enhancement of worldwide farming. The progresses in plant molecular biology and biotechnology comprising tissue culture have been bearing up in the past.

Disentanglement of the genomes of numerous plants, the capacity to decipher many genetic networks below diverse conditions, and the arrival of new-generation sequencing tools have unlocked completely novel ways for improving the genetic potential of crop cultivars. Molecular marker helped breeding leads to enhanced cultivars in many cultivated plants via marker-assisted selection. Current developments in genomics, transcriptomics, proteomics, and bioinformatics promise concrete keys to improved drought tolerance. Bright usage of these tools is projected to help in the identification of genes and genomic areas accountable for controlling numerous essential characters that can be exploited for enhanced plant output and defense.

7.1 Plant Cell Tissue and Organ Culture

The usage of tools of biotechnologies founded on in vitro crop tissue culture and genetic engineering has permitted us to address the serious difficulties of cultivated plant enhancement for durable farming. In vitro regeneration is a main tool in transgenic creation; meanwhile, progresses in molecular genetics, e.g., gene Nover expression, gene suppression, promoter analysis, and TDNA tagging, necessitate proficient modification schemes. Effective tissue culture is consequently an important stage, necessary for the authentication and utilization of information produced by these great tools. Application of procedures for regeneration is consequently an essential prerequisite for genetic modification and other tissue culture derivative methods to produce diversity like somaclonal disparity, in vitro mutagenesis, doubled haploid culture, and extensive hybridization (Singh et al. 2022).

7.2 Somaclonal Variation and In Vitro Mutagenesis

Tissue culture produces a varied sort of genetic variant in crops that may be integrated into crop breeding programs. Somaclonal variance is well-definite as the genetic and phenotypic difference amid clonally propagated crops of a sole contributor clone. Genetic differences happen in undistinguishable cells, lonely protoplasts, callus, tissues, and morphological characters of redeveloped crops. The reason for disparity is habitually credited to modifications in the chromosome number and structure. Usually, the word somaclonal variance is utilized for genetic instability existing amid all types of cells/crops gained from cells cultured in vitro. It is well-defined as genetic modifications, which ascend in vitro amid clonal regenerants and their matching donor crops (Leva and Rinaldi 2017). Crops redeveloped from tissue and cell cultures express genetic dissimilarity for qualitative and quantitative characters. Somaclonal disparity produced by the procedure of tissue culture is too named tissue culture-induced disparity to precisely delimit the activating milieu. The happening of unregulated and unstructured dissimilarity for the period of the culture procedure is an unforeseen and regularly unwanted happening when crops are micro-propagated at the profitable scales. Nevertheless, apart from these harmful impacts, its utility in plant breeding through the formation of new variants has been widely stated (Bhatia 2015).

The induced somaclonal disparity may be utilized for the genetic operation of plants with polygenic characters. The novel cultivars resulting from in vitro tissue culture might reveal drought tolerance and enhancement in quality and better return. Somaclonal variants are able to be detected utilizing numerous methods that are largely registered as morpho-physiological, biochemical, and molecular discovery methods. There are two principal methods for the loneliness of somaclonal variants: concealing and cell choice. Concealing implicates the scrutiny of a huge amount of cells or redeveloped crops for the discovery of variant entities. Modification for many characters may be far-off more effortlessly lonely from cell cultures than from the entire crop population. This is because a great amount of cells might be effortlessly and successfully concealed for mutant characters. Concealing several crops would be more problematic, normally tedious. Mutants might be efficiently chosen to enhance the adaptation to drought. Several works have stated that the in vitro culture or associated with mutagenesis, and physicochemical or biological mediators, might be used to upsurge genetic variability and mutants, as a possible basis of novel commercial varieties. In vitro culture milieus may be mutagenic and crops redeveloped from organ cultures, callus, protoplasts, and through somatic embryogenesis occasionally reveal phenotypic and genotypic disparities. It is vital to underline that tissue culture upsurges the effectiveness of mutagenic ministrations and permits the management of huge populations and fast replication of chosen variants. The impacts brought by the drought in the crop cultivated in vitro and in vivo environments are similars. So, the in vitro method may be utilized as a substitute for farm assessments for learning the common influence of drought on crop growth and development. The greatest broadly utilized technique for the screening of genotypes resistant to drought is the choice constraint method. This is founded on the in vitro culture of crop cells, tissues, or organ medium complemented by choosy agents, permitting choosing, and redeveloping crops using desired features (Bhatia 2015).

In vitro choice permits to gain the time necessary for regenerating drought-tolerant lines. Mutants got from somaclonal variants may be successfully chosen for adapting drought resistance in several plants. Selection has been gainfully and broadly used for the seclusion of cell clones, which generate greater amounts of certain mutants. In the cell screening method, appropriate pressure is used to allow privileged survival or development of variant cells. Screening approaches have been effectively established for the salvage of genotypes tolerant against drought. When the screening pressure permits only the mutant cells to subsist or split, it is named positive screening. In negative selection, the wild-type cells split usually and then are destroyed by a counter-choice agent. The mutant cells are incapable to split, and as an outcome of that they are avoided as counter-choice agent. These cells are afterward saved by the exclusion of the counter-choice agent. This method has been carried out utilizing a number of crop material parts (callus, suspension cultures, somatic embryos, leaf and stem cultures, etc.) that have been selected for disparity in their aptitude to resist fairly great concentrations of salt in the medium of culture (Bhatia 2015).

7.3 Transgenic Approaches

Transgenic methods are among the existing instruments for crop enhancement programs found on methods of biotechnologies. Currently, several machineries and gene groups that lead to enhanced output and adaptation to drought are documented. These gene groups may be used in new mixtures, expressed ectopically, or conveyed to plants in which they do not logically happen. Hence, the opportunity to change the plant accessions with genes from any biological source is a very great instrument for molecular crop breeding. Achievements in genetic enhancement of drought have implicated the use of lone or limited genes implicated in signaling or controlling paths or which encrypt enzymes implicated in these paths like osmolytes or well-matched solutes, antioxidants, molecular chaperones or osmoprotectants, and H_2O and ion transporters. The weakness of this method is the presence of several interconnecting genes implicated, and endeavors to enhance plant drought resistance over the use of a single or a little of them are frequently linked with further frequently unwanted, pleiotropic, and phenotypic modifications. More proficient and reproducible regeneration and transformation methods can be established in main cultivated plants that are able to ease the transference of suitable genes. For the progress of transgenic tolerance against drought, isolation, and description of new genes from bacteria (*Bacillus marisflavi* CRDT-EB-1) and higher plants, host facilitated signal transduction in reply to stress, works on physiological, genetic, and molecular mechanisms underlying drought resistance, and identification, adequate mapping and cloning of genes accountable for imparting drought resistance can be prolonged in future (Gowtham et al. 2021). The advancement of transgenic crops utilizing instruments of biotechnologies has come to be essential in drought stress biology. Workings on genetics and molecular methods have revealed that most of the drought-resistant characters are multigenic. Hence, to enhance drought resistance many drought-related genes require to be transferred. Lately, the use of lone transcription elements has delivered a similar influence as the manipulation of manifold genes. This has come to be a hopeful method to acquire drought to develop resistant plants.

ABA controls the adaptive reply of crops to drought through varied physiological and developmental procedures. The abscisic acid bio-synthetic path has been intensely investigated, and several of the vital enzymes implicated in abscisic acid synthesis have been utilized in transgenic crops for enhancing drought resistance (Muhammad et al. 2022). Transgenic crops expressing the genes implicated in ABA production illustrated augmented resistance to drought (Ren et al. 2022). Likewise, several additional mechanisms implicated in crop defense against drought implicates the up control of well-matched osmoprotectants. Their principal role is to preserve cellular turgor, nonetheless they are as well implicated in escaping oxidative destruction and overseeing over and done with direct steadying of membranes and proteins. Several genes implicated in the production of these solutes have been identified for their prospect in engineering crop drought resistance. The usage of genes associated with solute production has been fruitfully utilized in creating drought-resistant cultivated plants, and the relocation of glycine-betaine intermediates has enhanced the water starvation-resistance of transgenic crops in several examples. Proline is an amino acid that happens extensively in higher plants and is usually amassed in huge quantities in reply to drought (Tsoata et al. 2017). The osmoprotectant function of proline has been proved in many plants by over-expressing genes implicated in proline synthesis (Wang et al. 2022). Further methods efficaciously established in an assortment of cultivated plants to get drought-tolerant crops by transgenesis have been the use of transcription factors (TFs), late embryogenesis abundant (LEA) proteins, and antioxidant proteins (Tunnacliffe and Wise 2007; Hu et al. 2022).

7.4 *Nanotechnologies for Enhancing Drought Tolerance in Crops*

Nanomaterials are used in global positioning and geological information technology and remote detection systems using satellite television aimed at PC imaging fields. They aid in assessing much localized environmental states, defining whether plants are developing at optimum effectiveness or pinpointing the precise type and position of dilemmas of crops and their environment. When a water deficit is sensed, an adjustment of the watering rates can be switched on immediately. Nanoparticles like ZnO have been established to be very effective in mitigating the impact of water starvation in crops (Dimkpa et al. 2020). Carbon nanotubes have stimulated water transportation in dry and arid climates, improving plant biomass harvests and grain crop yields, comprising corn (Tiwari et al. 2014). In addition, Ag nanoparticles improved the development of soybean seedlings below inundation by modulating the monitoring of protein attribute for misfolded proteins in the endoplasmic reticulum (Hashimoto et al. 2020). Further studies are needed to investigate the function of nanomaterials in mitigating the harmful impacts of drought and the underlying mechanisms that allow crops to deal with the harsh impacts of suboptimal situations.

8 Conclusion and Future Perspectives

Traditional breeding techniques alongside the suitable implementation of molecular markers for marker-assisted breeding, gene pyramiding, marker-assisted foreground and background selections, and latest methods of genotyping by sequencing may speed up the procedure of reaching improved crop cultivars in forthcoming years. So, information, precocity of cautionary schemes and farming of elevated yielding, great attribute, and drought-resistant cultivars might deliver a key to the problem of water deficit. Selection and introduction of appropriate characters which taper the break amid projected and real harvest; understanding accurate morphophysio-molecular mechanisms of water deficit resistance; and scheming a standard screening process for a huge population might help for the growth of drought-resistant crops cultivars. Implementing suitable approaches might help breeders' effort to create better water starvation-resistant cultivars. Conventional and marker-assisted breeding stratagems built on the usage of water deficit-resistant parents, and the advancement of appropriate plotting populations to recognize QTLs genes touching harvest might lead to return enhancement and steadiness during water deficit. Replies to drought implicate numerous genes and gene functions regulating water deficit resistance. The expression of water deficit resistance-associated characters and the network of water starvation-associated genes have been studied. The use of phytohormones and osmoprotectants for crop development, engineering for water deficit tolerance, and high-throughput new tools might be beneficial machineries in recognizing genes to enhance return during water deficit.

References

Albert E (2017) Genetic and genomic determinants of response to water deficit in tomato (*Solanum lycopersicum*) and impact on fruit quality. PhD thesis, University of Avignon, 174 p

Apel K, Hirt H (2004) Reactive oxygen species: metabolism, oxidative stress, and signal transduction. Annu Rev Plant Biol 55:373–399. https://doi.org/10.1146/annurev.arplant.55.031903.141701

Aroca R, Porcel R, Ruiz-Lozano JM (2007) How does arbuscular mycorrhizal symbiosis regulate root hydraulic properties and plasma membrane aquaporins in *Phaseolus vul-*

garis under drought, cold or salinity stresses? New Phytol 173(4):808–816. https://doi.org/10.1111/J.1469-8137.2006.01961

Ashraf M, Akram NA (2009) Improving salinity tolerance of plants through conventional breeding and genetic engineering: an analytical comparison. Biotechnol Adv 27:744–752. https://doi.org/10.1016/j.biotechadv.2009.05.026

Baenziger PS, Beecher B, Graybosch RA, Ibrahim AMH, Baltensperger DD, Nelson LA (2008) Registration of 'NEO1643' wheat. J Plant Registr 2:36–42. https://doi.org/10.3198/jpr2007.06.0327crc

Bhatia S (2015) Application of plant biotechnology. In: Bhatia S, Sharma K, Dahiya R, Bera T (eds) Modern applications of plant biotechnology in pharmaceutical sciences. Academic, pp 157–207. https://doi.org/10.1016/B978-0-12-802221-4.00005-4

Blum A (1988) Plant breeding for stress environments. CRC Press, Boca Raton, 231 p. https://doi.org/10.1201/9781351075718

Blum A (2011) Drought resistance–is it really a complex trait? Funct Plant Biol 38(10):753–757. https://doi.org/10.1071/FP11101

Bota J, Medrano H, Flexas J (2004) Is photosynthesis limited by decreased Rubisco activity and RuBP content under progressive water stress? New Phytol 162(3):671–681. https://doi.org/10.1111/j.1469-8137.2004.01056.x

Brick MA, Ogg JB, Singh SP, Schwartz HF, Johnson JJ, Pastor-Corrales MA (2008) Registration of drought-tolerant, rust-resistant, high-yielding pinto bean germplasm line CO46348. J Plant Registr 2:120–134. https://doi.org/10.3198/jpr2007.06.0359crg

Cao L, Lu X, Wang G, Zhang Q, Zhang X, Fan Z, Cao Y, Wei L, Wang T, Wang Z (2021) Maize ZmbZIP33 is involved in drought resistance and recovery ability through an abscisic acid-dependent signaling pathway. Front Plant Sci 12:442. https://doi.org/10.3389/fpls.2021.629903

Chaumont F, Moshelion M, Daniels MJ (2005) Regulation of plant aquaporin activity. Biol Cell 97:749–764. https://doi.org/10.1042/bc20040133

Chen X, Ding Y, Yang Y, Song C, Wang B, Yang S, Guo Y, Gong Z (2021) Protein kinases in plant responses to drought, salt, and cold stress. J Integr Plant Biol 63:53–78. https://doi.org/10.1111/jcpb.13061

Chinnusamy V, Zhu J (2009) Epigenetic regulation of stress responses in plants. Curr Opin Plant Biol 12(2):133–139. https://doi.org/10.1016/j.pbi.2008.12.006

Cho SM, Kang BR, Han SH, Anderson AJ, Park J-Y, Lee Y-H, Cho BH, Yang K-Y, Ryu C-M, Kim YC (2008) 2R, 3R-butanediol, a bacterial volatile produced by *Pseudomonas chlororaphis* O6, is involved in induction of systemic tolerance to drought in *Arabidopsis thaliana*. Mol Plant-Microbe Interact 21:1067–1075. https://doi.org/10.1094/MPMI-21-8-1067

Chu C, Wang S, Paetzold L, Wang Z, Hui K, Rudd JC, Xue Q, Ibrahim AMH, Metz R, Johnson CD, Rush CM, Liu S (2021) RNA-seq analysis reveals different drought tolerance mechanisms in two broadly adapted wheat cultivars 'TAM 111' and 'TAM112'. Sci Rep 11:4301. https://doi.org/10.1038/s41598-021-83372-0

Costa OYA, Raaijmakers JM, Kuramae EE (2018) Microbial extracellular polymeric substances: ecological function and impact on soil aggregation. Front Microbiol 9:1636. https://doi.org/10.3389/fmicb.2018.01636

Deak KI, Malamy J (2005) Osmotic regulation of root system architecture. Plant J 43:17–28. https://doi.org/10.1111/J.1365-313x.2005.02425.x

Dimkpa CO, Andrews J, Sanabria J, Bindraban PS, Singh U, Elmer WH, Gardea-Torresdey JL, White JC (2020) Fertiliser, and zinc oxide nanoscale and bulk particles on wheat performance and grain nutrient accumulation. Sci Total Environ 722:137808. https://doi.org/10.1016/j.scitotenv.2020.137808

Fonta JE, Giri J, Vejchasarn P, Lynch JP, Brown KM (2022) Spatiotemporal responses of rice root architecture and anatomy to drought. Plant Soil 479:443–464. https://doi.org/10.1007/s11104-022-05527-w

Golldack D, Li C, Mohan H, Probst N (2014) Tolerance to drought and salt stress in plants: unraveling the signaling networks. Front Plant Sci 5. https://doi.org/10.3389/fpls.2014.00151

Gowtham HG, Duraivadivel P, Ayusman S, Sayani D, Gholap SL, Niranjana SR, Hariprasad P (2021) ABA analogue produced by *Bacillus marisflavi* modulates the physiological response of *Brassica juncea* L. under drought stress. Appl Soil Ecol 159:103845. https://doi.org/10.1016/j.apsoil.2020.103845

Haley SD, Johnson JJ, Peairs FB, Quick JS, Stromberger JA, Clayshulte SR (2007) Registration of 'Ripper' wheat. J Plant Registr 1:1–6. https://doi.org/10.3198/jpr2006.10.0689crc

Hamim (2005) Photosynthesis of C3 and C4 species in response to increased CO_2 concentration and drought stress. Hayati J Biosci 12(4):131–138. https://doi.org/10.1016/S1978-3019(16)30340-0

Hashimoto T, Mustafa G, Nishiuchi T, Komatsu S (2020) Comparative analysis of the effect of inorganic and organic chemicals with silver nanoparticles on soybean under flooding stress. Int J Mol Sci 21(4):1300. https://doi.org/10.3390/ijms21041300

He Y, Liu Y, Li M, Lamin-Samu AT, Yang D, Yu X, Izhar M, Jan I, Ali M, Lu G (2021) The Arabidopsis SMALL AUXIN UP RNA32 protein regulates ABA-mediated responses to drought stress. Front. Plant Sci 12:259. https://doi.org/10.83389/fpls.2021.625493

Hu Y, Chen X, Chen X (2022) Regulatory network established by transcription factors transmits drought stress signals in plants. Stress Biol 2:26. https://doi.org/10.1007/S44154-022-00048-z

Huang Y-M, Zou Y-N, Wu Q-S (2017) Alleviation of drought stress by mycorrhizas is related to increased root H_2O_2 efflux in trifoliate orange. Sci Rep 7:42335. https://doi.org/10.1038/srep42335

Huque AKM, So W, Noh M, You MK, Shin JS (2021) Overexpression of AtBBD1, *Arabidopsis* bifunctional nuclease, confers drought tolerance by enhancing the expression of regulatory genes in ABA-mediated drought stress signaling. Int J Mol Sci 22:2936. https://doi.org/10.3390/ijms22062936

Iqbal S, Wang X, Mubeen I, Kamran M, Kanwal I, Diaz GA, Abbas A, Parveen A, Atiq MN, Alshaya H, El-Abedin TKZ, Fahad S (2021) Phytohormones trigger drought tolerance in crop plants: outlook and future perspectives. Front Plant Sci 12:799318. https://doi.org/10.3389/fpls.2021.799318

Khraiwesh B, Arif MA, Seumel GI, Ossowski S, Weigel D, Reski R, Frank W (2010) Transcriptional control of gene expression by microRNAs. Cell 140:111–122. https://doi.org/10.1016/j.cell.2009.12.023

Kim T-H, Bohmer M, Hu H, Nishimura N, Schroeder JI (2010) Guard cells signal transduction network: advances in understanding abscisic acid CO_2, and Ca^{2+} signalling. Ann Rev Plant Biol 61:561–591. https://doi.org/10.1146/annurev-arplant-042809-112226

Kumar A, Dixit S, Ram T, Yadaw RB, Mishra KK, Mandal NP (2014) Breeding high-yielding drought-tolerant rice: genetic variations and conventional and molecular approaches. J Exp Bot 65:6265–6278. https://doi.org/10.1093/jxb/eru363

Kumar S, Kamboj D, Mishra CN, Sing GP (2022) Prospects of molecular markers for wheat improvement in postgenomic era. In: Sharma P, Yadav D, Gaur RK (eds) Bioinformatics in agriculture. Academic, pp 323–340. https://doi.org/10.1016/B978-0-323-89778-5.00014-3

Kwak JM, Mori IC, Pei ZM, Leonhardt N, Torres MA, Dangl JL, Bloom RE, Bodde S, Jones JD, Schroeder JI (2003) NADPH oxidase *AtrbohD* and *AtrbohF* genes function in ROS-dependent ABA signaling in Arabidopsis. EMBO J 22:2623–2633. https://doi.org/10.1093/emboj/cdg277

Lacape J-M, Loison R, Foncéka D (2015) Amélioration de l'adaptation des cultures à la sécheresse en zone de savane africaine. In: Torquebiau E (ed) Changement climatique et agricultures du monde. Quae Edition, Versailles, pp 63–74

Leva A, Rinaldi LMR (2017) Breeding genetics and biotechnology. In: Thomas B, Murray BG, Murphy DJ (eds) Encyclopedia of applied plant sciences, vol 2, 2nd edn. Elsevier, pp 468–473. https://doi.org/10.1016/B9478-0-12-394807-6.00150-7

Li X, Yu B, Wu Q, Min Q, Zeng R, Xie Z, Huang J (2021) OsMADS23 phosphorylated by SAPK9 confers drought and salt tolerance by regulating ABA biosynthesis in rice. PLoS Genet 17:e1009699. https://doi.org/10.1371/journal.pgen.1009699

Liu Y, Wen L, Shi Y, Su D, Lu W, Cheng Y, Li Z (2021) Stress-responsive tomato gene SlGRAS4 function in drought stress and abscisic acid signaling. Plant Sci 304:110804. https://doi.org/10.1016/j.plantsci.2020.110804

Liu C, Guo X, Wu X, Dai F, Wu Q (2022) The comprehensive effects of *Rhizophagus intraradices* and P on root system architecture and p transportation in *Citrus limon* L. Agriculture 12:317. https://doi.org/10.3390/agriculture12030317

Mansoor S, Ali Wani O, Lone JK, Manhas S, Kour N, Alam P, Ahmad A, Ahmad P (2022) Reactive oxygen species in plants: from source to sink. Antioxidants 11:225. https://doi.org/10.3390/antiox11020225

Mathan J, Singh A, Ranjan A (2021) Sucrose transport in response to drought and salt stress involves ABA-mediated induction of OsSWEET13 and OsSWEET15 in rice. Physiol Plant 171:620–637. https://doi.org/10.1111/ppl.13210

Mekureyaw MF, Pandey C, Hennessy RC, Nicolaisen MH, Liu F, Nybroe O, Roitsch T (2022) The cytokinin-producing plant beneficial bacterium *Pseudomonas fluorescens* G20-18 primes tomato (*Solanum lycopersicum*) for enhanced drought stress responses. J Plant Physiol 270:153629. https://doi.org/10.1016/J.JPLPH.2022.153629

Mondal S, Dutta S, Crespo-Herrera L, Huerta-Espino J, Braun HJ, Singh RP (2020) Fifty years of semi-dwarf spring wheat breeding at CIMMYT: grain yield progress in optimum, drought and heat stress environments. Field Crops Res 250:107757. https://doi.org/10.1016/j.fcr.2020.107757

Muhammad AM, Waseem M, Jakada BH, Okal EJ, Lei Z, Saqib HSA, Yuan W, Xu W, Zhang Q (2022) Mechanisms of abscisic acid-mediated drought stress responses in plants. Int J Mol Sci 23(3):1084. https://doi.org/10.3390/ijms23031084

Noaman MM, El Sayad AA, Asaad FA, El Sherbini AM, El Bawab AO, El Moselhi MA (1995) Registration of 'Giza 126' barley. Crop Sci 35:710. https://doi.org/10.2135/cropsci1995.0011183x003500060037x

Obert DE, Evans CP, Wesenberg DM, Windes JM, Erickson CA, Jackson EW (2008) Registration of 'Lenetah' spring barley. J Plant Registr 2:85–97. https://doi.org/10.3198/jpr2008.01.0021crc

Quiroz-Figueroa F, Rodríguez-Acosta A, Salazar-Blas A, Hernández-Domínguez E, Campos ME, Kitahata N, Asami T, Galaz-Avalos RM, Cassab GI (2010) Accumulation of high levels of ABA regulates the pleiotropic response of the nhr1 Arabidopsis mutant. J Plant Biol 53:32–44. https://doi.org/10.1007/s12374-009-9083-1

Rajput M, Choudhary K, Kumar M, Vivekanand V, Chawade A, Ortiz R, Pareek N (2021) RNA interference and CRISPR/Cas gene editing for crop improvement: paradigm shift towards sustainable agriculture. Plan Theory 10(9):1914. https://doi.org/10.3390/plants10091914

Ren C, Kuang Y, Lin Y, Guo Y, Li H, Fan P, Li S, Liang Z (2022) Overexpression of grape ABA receptor gene VaPYL4 enhances tolerance to multiple abiotic stresses in *Arabidopsis*. BMC Plant Biol 22:271. https://doi.org/10.1186/s12870-022-03663-0

Sahu PP, Pandey G, Sharma N, Puranik S, Muthamilarasan M, Prasad M (2013) Epigenetic mechanisms of plant stress responses and adaptation. Plant Cell Rep 32:1151–1159. https://doi.org/10.1007/s00299-013-1462-x

Sanders GJ, Arndt SK (2012) Osmotic adjustment under drought conditions. In: Aroca R (ed) Plant responses to drought stress. Springer, Berlin/Heidelberg, pp 199–229. https://doi.org/10.1007/978-3-642-32653-0_8

Sandhu N, Kumar A (2017) Bridging the rice yield gaps under drought: QTLs, genes, and their use in breeding programs. Agronomy 7(2):27. https://doi.org/10.3390/agronomy7020027

Saxena NP, Krishnamurthy L, Johansen C (1993) Registration of a drought-resistant chickpea germplasm. Crop Sci 33(6):1424. https://doi.org/10.2135/cropsci1993.0011183x003300060088x

Schweighofer A, Hirt H, Meskiene I (2004) Plant PP2C phosphatases: emerging functions in stress signaling. Trends Plant Sci 9:236–243. https://doi.org/10.1016/j.tplants.2004.03.007

Selamat N, Nadarajah KK (2021) Meta-analysis of quantitative traits loci (QTL) identified in drought response in rice (*Oryza sativa* L.). Plan Theory 10(4):716. https://doi.org/10.3390/plants10040716

Shamsudin NAA, Swamy BPM, Ratnam W, Cruz MTS, Raman A, Kumar A (2016) Marker assisted pyramiding of drought yield QTLs into a popular Malaysian rice cultivar, MR219. BMC Genet 17:30. https://doi.org/10.1186/s12863-016-0334-0

Shoaib M, Banerjee BP, Hayden M, Kant S (2022) Roots' drought adaptive traits in crop improvement. Plan Theory 11:2256. https://doi.org/10.3390/plants11172256

Singh KB, Omar M, Saxena MC, Johansen C (1996) Registration of FLIP 87-59C, a drought tolerant chickpea germplasm line. Crop Sci 36:1–2. https://doi.org/10.2135/cropsci1996.0011183x003600020054x

Singh KP, Indoliya Y, Agrawal L, Awasthi S, Deeba F, Dwivedi S, Chakrabarty D, Shirke PA, Pandey V, Singh N, Dhankher OP, Barik SK, Tripathi RD (2022) Genomic and proteomic responses to drought stress and biotechnological interventions for enhanced drought tolerance in plants. Curr Plant Biol 29:100239. https://doi.org/10.1016/J.CPB.2022.100239

Smith SE, Read DJ (2010) Mycorrhizal symbiosis. Academic, Cambridge, p 800

Srivastava RK, Yadav OP, Kaliamoorthy S, Gupta SK, Serba DD, Choudhary S, Govindaraj M, Kholová J, Murugesan T, Satyavathi CT, Gumma MK, Singh RB, Bolam S, Gupta R, Varshney RK (2022) Breeding drought-tolerant pearl millet using conventional and genomic approaches: achievements and prospects. Front Plant Sci 13:781524. https://doi.org/10.3389/fpls.2022.781524

Sun M, Yang Z, Liu L, Duan L (2022) DNA methylation in plant responses and adaptation abiotic stresses. Int J Mol Sci 23:6910. https://doi.org/10.3390/ijms23136910

Temegne NC (2018) Improvement in the performances of Voandzou (*Vigna subterranea* (L.) Verdc.) in response to phosphate deficiency through chemical and biological fertilization. PhD thesis, University of Yaounde I, Faculty of Science (Cameroon), 151 p

The World Bank (2021) An EPIC response: innovative governance for flood and drought risk management. https://www.worldbank.org/en/topic/water/publication/an-epic-response-innovative-governance-for-flood-and-drought-risk-management

Tiwari D, Dasgupta-Schubert N, Cendejas LV, Villegas J, Montoya LC, Garcia SB (2014) Interfacing carbon nanotubes (CNT) with plants: enhancement of growth, water and ionic nutrient uptake in maize (*Zea mays*) and implications for nanoagriculture. Appl Nanosci 4(5):577–591. https://doi.org/10.1007/S13204-013-0236-7

Tran L-SP, Urao T, Qin F, Maruyama K, Kakimoto T, Shinozaki K, Yamaguchi-Shinozaki K (2007) Functional analysis of AHK1/ATHK1 and cytokinin receptor histidine kinases in response to abscisic acid, drought, and salt stress in Arabidopsis. Proc Natl Acad Sci U S A 104:20623–20628

Tsoata E, Njock SR, Youmbi E, Nwaga D (2015) Early effects of water stress on some biochemical and mineral parameters of mycorhizal *Vigna subterranea* (L.) Verdc. (Fabaceae) cultivated in Cameroon. IJAAR 7(2):21–35

Tsoata E, Temegne NC, Youmbi E (2016) Analysis of early physiological criteria to screen four Fabaceae plants for their tolerance to water stress. Int J Recent Sci Res 7(11):14334–14338

Tsoata E, Temegne CN, Youmbi E (2017) Analysis of early biochemical criterion to screen four Fabaceae plants for their tolerance to drought stress. Int J Curr Res 9(1):44568–44575

Tunnacliffe A, Wise MJ (2007) The continuing conundrum of the LEA proteins. Naturwissenschaften 94:791–812. https://doi.org/10.1007/s00114-007-0254-y

Tuteja N, Mahajan S (2007) Calcium signaling network in plants: an overview. Plant Signal Behav 2:9–85. https://doi.org/10.4161/PSB.2.2.4176

Varshney RK, Bansal KC, Aggarwal PK, Datta SK, Craufurd PQ (2011) Agricultural biotechnology for crop improvement in a variable climate: hope or hype? Trends Plant Sci 16(7):363–371. https://doi.org/10.1016/j.tplants.2011.03.004

Verbruggen E, Struyf E, Vicca S (2021) Can arbuscular mycorrhizal fungi speed up carbon sequestration by enhanced weathering? Plants People Planet 3:445–453. https://doi.org/10.1002/PPP3.10179

Wang W, Vinocur B, Altman A (2003) Plant responses to drought, salinity and extreme temperatures: towards genetic engineering for stress tolerance. Planta 218:1–14

Wang W, Vinocur B, Shoseyov O, Altman A (2004) Role of plant heat-shock proteins and molecular chaperones in the abiotic stress response. Trends Plant Sci 9:244–252. https://doi.org/10.1016/j.tplants.2004.03.006

Wang X, Li Q, Xie J, Huang M, Cai J, Zhou Q, Dai T, Jiang D (2021a) Abscisic acid and jasmonic acid are involved in drought priming-induced tolerance to drought in wheat. Crop J 9(1):120–132. https://doi.org/10.1016/j.cj.2020.06.002

Wang Y-H, Que F, Li T, Zhang R-R, Khadr A, Xu Z-S, Tian Y-S, Xiong A-S (2021b) DcABF3, an ABF transcription factor from carrot, alters stomatal density and reduces ABA sensitivity in transgenic *Arabidopsis*. Plant Sci 302:110699. https://doi.org/10.1016/j.plantsci.2020.110699

Wang Z, Yang Y, Yadav V, Zhao W, He Y, Zhang X, Wei C (2022) Drought-induced proline is mainly synthesized in leaves and transported to roots in watermelon under water deficit. Hortic Plant J 8(5):615–626. https://doi.org/10.1016/j.hpj.2022.06.009

Xinglai P, Sangang X, Qiannying P, Yinhong S (2006) Registration of 'Jinmai 50' wheat. Crop Sci 46:983–995. https://doi.org/10.2135/cropsci2005.0217

Xu M, Li H, Liu Z-N, Wang X-H, Xu P, Dai S-J, Cao X, Cui X-Y (2021) The soybean CBL-interacting protein kinase, GmCIPK2, positively regulates drought tolerance and ABA signaling. Plant Physiol Biochem 167:980–989. https://doi.org/10.1016/j.plaphy.2021.09.026

Xu B-Q, Wang J-J, Peng Y, Huang H, Sun L-L, Yang R, Suo L-N, Wang S-H, Zhao W-C (2022) SIMYC2 mediates stomatal movement in response to drought stress by repressing SICHS1 expression. Plant Sci 13:952758. https://doi.org/10.3389/fpls.2022.952758

Yang J, Wang M, Zhou S, Xu B, Chen P, Ma F, Mao K (2022a) The ABA receptor gene MdPYL9 confers tolerance to drought stress in transgenic apple (*Malus domestica*). Environ Exp Bot 194:104695. https://doi.org/10.1016/j.envexpbot.2021.104695

Yang L, Chen Y, Xu L, Wang J, Qi H, Guo J, Zhang L, Shen J, Wang H, Zhang F, Xie L, Zhu W, Lü P, Qian Q, Yu H, Song S (2022b) The OsFTIP6-OsHB22-OsMYBR57 module regulates drought response in rice. Mol Plant 15(7):1227–1242. https://doi.org/10.1016/j.molp.2022.06.003

Zhang C, Bousquet A, Harris JM (2014) Abscisic acid and lateral root organ defective/modulate root elongation via reactive oxygen species in *Medicago truncatula*. Plant Physiol 166:644–658. https://doi.org/10.1104/pp.114.248542

Zhang Y, Zhao Y, Li T, Ni C, Han L, Du P (2022) TaPYL4, an ABA receptor gene of wheat, positively regulates plant drought adaptation through modulating the osmotic stress-associated processes. BMC Plant Biol 22:423. https://doi.org/10.1186/s12870-022-03799-z

Zhou Y, Lam HM, Zhang J (2007) Inhibition of photosynthesis and energy dissipation induced by water and high light stresses in rice. J Exp Bot 58(5):1207–1217. https://doi.org/10.1093/jxb/erl291

Chapter 20
Conferring Plant Tolerance to Drought and Salinity by the Application of Biochar

Ayesha Batool, Maria Naqve, Muhammad Anjum Zia, Athar Mahmood, Sahar Mumtaz, and Muhammad Mansoor Javaid

Abstract A variety of organic waste feedstock, such as municipal sewage sludge and agricultural waste, can be used to create the carbon-rich material known as biochar. Biochar has gained popularity owing to its unique properties, which include stable structure, cation exchange capacity, a sizable specific surface area, and high carbon content. With different types of feedstocks, biochar possesses different physiochemical characteristics. Pyrolysis, hydrothermal carbonization, and gasification are the basic methods used frequently to produce biochar. Biochar can be remolded by gas purging, acid, steam, alkali, carbonaceous material, oxidizing agent, and metal ions. The environmental application fields determine the modification techniques to use. It is crucial to investigate the biochar's impact on soil microbes in order to identify the main reason for increased soil fertility under different soil and feedstock conditions. The emission of polycyclic aromatic hydrocarbons (PAHs) and heavy metals into the environment needs to be considered when using biochar to remediate the environment. Due to its numerous distinctive qualities, biochar is an effective, economical, and sustainable material to combat stress-induced toxicity in plants.

A. Batool · M. Naqve (✉)
Department of Botany, Faculty of Sciences, University of Agriculture, Faisalabad, Pakistan
e-mail: maria.naqve@uaf.edu.pk

M. A. Zia
Department of Biochemistry, Faculty of Sciences, University of Agriculture, Faisalabad, Pakistan

A. Mahmood
Department of Agronomy, Faculty of Agriculture, University of Agriculture, Faisalabad, Pakistan

S. Mumtaz
Department of Botany, Division of Science and Technology, University of Education, Lahore, Pakistan

M. M. Javaid
Department of Agronomy, College of Agriculture, University of Sargodha, Sargodha, Pakistan

M. Hasanuzzaman (ed.), *Climate-Resilient Agriculture, Vol 2*,
https://doi.org/10.1007/978-3-031-37428-9_20

Keywords Biochar · Plant stress · Pyrolysis · Feedstock · Hydrothermal carbonization

1 Introduction

The carbon-rich product, also known as biochar, is created from organic feedstock using specific thermal combustion techniques and little oxygen (Joseph et al. 2010; Mahmood et al. 2022). Two most commonly used organic wastes that can be utilized as feedstock are agricultural waste and municipal solid waste (Wang and Wang 2019). Adsorption is the main technique used by biochar to eliminate organic and heavy metal pollutants. Biochar's adsorption capacity is directly correlated to the physiochemical properties, which include cation exchange capacity, surface area, functional groups, and pore size distribution. However, physiochemical properties of biochar are dependent on the conditions of preparation (Lonappan et al. 2016). In general, biochar manufactured at high temperature has more carbon content because volume of the micropores is increased by the elimination of volatile organic compounds at high temperatures (Zhang et al. 2017). However, as the temperature rises, the yield of biochar declined (Xue et al. 2019).

The solid by-product of biomass pyrolysis known as biochar, most commonly used as charcoal, has been manufactured and used for many thousand years. Biochar has a wide range of uses, including the generation of heat and power, the purification of flue gases, metallurgical uses, use in agricultural and animal husbandry, building materials, and even medical applications. It has grown in favor in recent years as a replacement for fossil carbon carriers in a number of these applications in an effort to reduce greenhouse gas emissions (Weber and Quicker 2018).

While some of the biomass is decomposed during carbonization, a significant amount of its carbon content is kept. The product's characteristics change; it becomes more carbonaceous and, as a result, easier to substitute in technological procedures (Weber and Quicker 2018). Depending on the required char characteristics, feedstock and carbonization conditions are selected. Torrefaction, or pyrolysis at temperatures up to 300 °C, greatly enhances a number of raw biomass problematic qualities. These include mechanical characteristics such as grindability (Deng et al. 2009). They are frequently a limiting factor in co-firing and co-gasification applications, which involve processing biomass using machinery intended for coal.

As a soil additive, biochar has been used to increase soil fertility and sequester carbon. Fast pyrolysis and gasification both produce solids with higher carbon content as a by-product, but the quality is frequently insufficient for many uses. For example, the solid product of gasification may contain hazardous chemicals such as polycyclic aromatic hydrocarbons, while chars from flash pyrolysis may frequently have comparably low carbon concentrations (Weber and Quicker 2018).

1.1 Production of Biochar

The solid by-product of biomass pyrolysis is biochar. Since several thousand years ago, charcoal has been manufactured and used. It can be manufactured by using woody feedstocks (Quicker and Weber 2016). When a fuel is thermochemically broken down at high temperatures without the aid of outside oxygen, this method is called pyrolysis. The first step in the procedure is to dry the biomass. Further heating of the particle causes the solid to release volatile compounds. These volatile compounds (acetic acid and methanol) can result in condensable organic compounds or long-lasting gases (e.g., CO_2, CO, CH_4, and H_2).

Cracking and polymerization are examples of subsequent reactions that can change the entire product spectrum in the gas phase. The three distinct products are tar, one or more liquid phases, solid residue, and permanent gases. Due to the competition between these reactions, it leads toward dissimilar products. Depending on the required product of the procedure, few basic rules can be adapted in order to maximize the outcome of required by-product. The objective of fast pyrolysis is to produce liquid oil. In order to prevent the condensable volatiles from polymerizing into char or cracking into light gases, it is crucial to cool them as soon as they are released from the solid feedstock. The biomass is quickly warmed up to the reaction temperature within seconds. Gases are extinguished to finish subsequent reactions. Seventy-five percent of the dry matter in the feedstock is the liquid yield.

1.2 Biochar Preparation

The pyrolysis states, such as reactor type, residence time, temperature, and heating rate, as well as the many feedstock kinds, have a notable effect on the chemical and physical characteristics of biochar. Slow pyrolysis, microwave-assisted pyrolysis, traditional carbonization, flash carbonization, and fast pyrolysis are the basic thermochemical practices used to produce biochar (Xiu et al. 2017). It has been observed that biochar manufactured at excessive temperatures between 600 and 700 °C has few H and O functional groups, but shows an extreme aromatic nature with well-orderly C layers and has little ion exchange properties because of the oxygenation and dehydration of biomass (Lago et al. 2021). A higher assemblage of C=C and C–H functional groups can be found in the biochar manufactured at lower temperatures ranging from 300 to 400 °C, which is also having a variety of organic properties, comprising of cellulose-type and aliphatic structures (Ko et al. 2015). Because of its complicated and diverse chemical and physical makeup, biochar has significant characteristics for pollutant withdrawal through surface assimilation (Rajapaksha et al. 2016).

1.2.1 Feedstocks

Biomass is the raw material for microwave-assisted pyrolysis and conventional pyrolysis. The amount of lignin, cellulose, and hemicellulose tells the amount of bio-oil, gas, and biochar in pyrolysis by-products. It has been observed that raw material having more lignin manufactures the large amount of biochar when pyrolyzed at average temperature of 500 °C. Consequently, the selection of pyrolysis raw material might be affected by the required balance between the different pyrolysis outcomes, that is, bio-oil, gas, and biochar. Biomass is a biological material that is alive or has recently been alive and has the potential to produce energy. Biomass is expected to get more popularity among primary energy sources in the twenty-first century. Due to its carbon neutrality, renewable nature, accessibility, and low sulfur content, biomass has the capability to act as an excellent fossil fuel replacement. Changing from remnant fuels to biomass as an origin of energy will help to cope with the current energy crisis as well as reduce pollution and global warming. Municipal solid waste, residues, and virgin resources are the three main types of biomass resources. Among the virgin resources, seeds from cereal and oil crops as well as forests are included. Residential or nonresidential wastes are included in municipal solid waste (MSW). These types of biomasses have the potential to be promising future energy sources with the right conversion method. Agricultural residues, municipal solid wastes, and forest residues are being used as a successful substitute for microwave pyrolysis processes.

1.2.2 Conventional Pyrolysis

In this process of heating system, heat is transferred to biomass by an external source by the process of convention, radiation, and conduction. The raw materials' temperature is highest at the start of this process and gradually drops as it moves toward the center. The moderate heating rate and the prolonged fumes residence hour are the two key variables in the conventional pyrolysis process (Fisher et al. 2012). This process has been used to manufacture charcoal for a long time. The yield and effectiveness of this pyrolysis are influenced by a few well-known variables. These parameters are related to the pressure and moisture content of the feedstock (Wang et al. 2013). The highest temperature experienced during a typical pyrolysis process is referred to as the "peak temperature." According to reports, the fixed-content characteristics of the manufactured biochar increase as the peak temperature rises. A study of the pyrolysis of various biomass raw materials, that is, cane bagasse, eucalyptus wood, and radiata pine (Godwin et al. 2019), revealed that rise in operating pressure results in the reduction of whole surface size of the char. A slow pyrolysis of Miscanthus produced similar results, which was reported by Wafiq et al. (2016). They perceived a remarkable drop in the char's BET surface size from 161.7 to 0.137 $m^2\ g^{-1}$. Researchers claim that these results are a result of the tar deposits obstructing the char pores under intense pressure. Recent research showed that high moisture content increases charcoal yields at high pressures.

Because of this, it is preferable to use the majority of biomass with excessive vapor contents for producing biochar through conventional pyrolysis (Ghodake et al. 2021). This pyrolysis method depends on the system's convection current and biomass thermal conductivity in addition to being relatively slow and inefficient (Foong et al. 2020).

1.2.3 Microwave-Assisted Pyrolysis

Instead of just heating, microwave pyrolysis of biomass involves energy conversion (Fig. 20.1; Kalt et al. 2020). In this method, dielectric heating transforms electromagnetic energy into thermal energy (Capodaglio and Callegari 2018). Additionally, rather than coming from an outside source, heat is produced throughout the biomass in bulk. In contrast to conventional pyrolysis, feedstock's interior temperature is higher than its exterior and surface temperatures. The most successful technique for enhancing and accelerating chemical reactions is microwave-assisted pyrolysis. Compared to other thermochemical processes, the chemical reactions are completed more quickly and efficiently due to heat transfer profile (Patel et al. 2021). This technique has the advantage over other pyrolysis methods because it saves energy by accelerating chemical reactions and cutting down residence time.

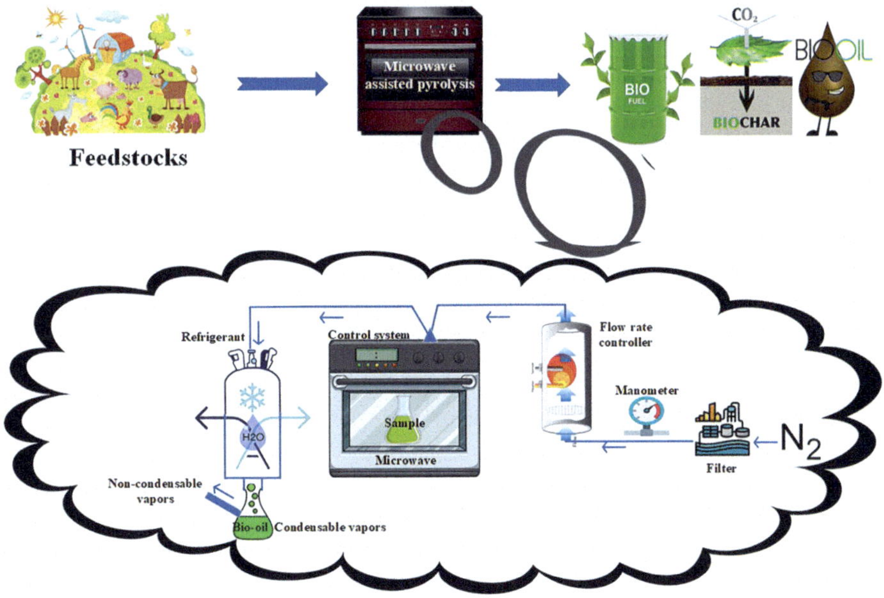

Fig. 20.1 Schematic diagram for microwave-assisted pyrolysis from feedstock to products. (Kalt et al. 2020)

2 Plant Stress and Its Consequence

Abiotic stress is the umbrella term for any environmental factor that can restrict plant growth and development. Plant scientists and agronomists have been focusing on plant abiotic stress from last 50 years because it has threatened global food security due to the exponential growth in human populations. Scientists started estimating crop losses caused by environmental constraints and finding out the causes of these losses. Numerous reports mentioning crop loss as a result of various abiotic stresses have been published (Van Ittersum et al. 2013; Zandalinas et al. 2018; Javaid et al. 2022). Some of these stresses have an impact on cultivated lands for a long time, which eventually affects crop yield and variety. Despite the fact, it is hard to determine the impact of abiotic stresses on agriculture lands. Mohanta et al. (2017) based their findings on FAO reports, concluding that approximately 96.5% of agricultural land is affected by abiotic stresses globally. The estimated percentages of stress contributing toward crop loss were later broken down by him (Imran et al. 2021). A variety of unfavorable factors such as salinity, cold, drought, and heat can impact both farmland plants and crop, making it a continuously changing habitat (Raza et al. 2019).

Numerous pests, parasites, and pathogens can infect plants and cause biotic stress. Necrotrophic (kill host cells by secreting toxins) and biotrophic are the two types of fungal parasites that rely upon feeding on living host cells. Biotic stress causes different symptoms on plants, that is, plant cankers, spots on leaves, and vascular wilting (Markell et al. 2015). The main causes of soil-borne illnesses that lead to nutrient deficiencies, stunted growth, and wilting are nematodes, which feed on plant tissue (Khan and Sharma 2020). Similar to bacteria, viruses can harm an area or an entire population resulting in stunted growth and chlorosis (Garcia-Ruiz 2019). On the other hand, insects and mites also cause harm to plants by feed or laying eggs there. Insects also behave as transporters of bacteria and viruses (Tanveer et al. 2012; Berg et al. 2017). Plants contain a progressive immune system to overcome these stresses (Iqbal et al. 2021). Plants contain defense system in the form of physical barriers such as cuticles, wax, and trichomes to fight against insects and pathogens. Moreover, plants produce chemicals to protect themselves from pathogens (Ali et al. 2013; Zaynab et al. 2019).

3 Biochar and Drought Stress

It has been widely reported that using biochar is advantageous when there is little water available (Ali et al. 2017). After applying biochar, drought-stressed plants grew and produced more biomass. In the case of maize (Tayyab et al. 2018) and okra, plant height and leaf area of plant increased by biochar under drought stress (Zheng et al. 2021). Similar to this, adding biochar to field-grown wheat increased its biomass under semi-arid Mediterranean regions (Ali et al. 2017). Fruit yield and quality of drought-stressed tomato plants can be improved by the application of biochar (Gavili et al. 2018). Adding biochar to sandy soil can improve growth of tomato seedlings and help to develop resistance to wilting (Mulcahy et al. 2013). Application of biochar to silty soil resulted in clear increase in tomato growth in comparison to control tomatoes (Semida et al. 2019). Tayyab et al. (2018) investigated that adding biochar to sandy loam soil can enhance tomato leaf quality. After receiving a 1% biochar treatment from straw gasification, barley grown in coarse sand exhibited increased growth of shoots and roots. However, when there was a water shortage in sandy loam soil, the gasification of biochar had no impact on the growth of barley (Hansen et al. 2016). Studies show that treating plants with biochar under drought condition improved their capacity to produce oxygen (Gogoi et al. 2019; Hassan et al. 2021).

4 Biochar and Salt Stress

The increased stomatal conductance brought on by biochar is frequently linked to the enhanced plant growth facing salt stress. Recent research has shown that adding biochar to saline soils increased stomatal conductance and density in plants such as herbaceous plants, wheat, and tomatoes (Dahlawi et al. 2018). Roots in soil treated with biochar become slightly resistance to osmotic condition because of improved soil properties, Na^+ binding, and increased soil moisture (Patel et al. 2017). Then roots cause reduction in the amount of ABA produced, which resulted in an increase in leaf growth and stomatal conductance. Biochar remarkably decreases the concentration of Na^+ in potato xylem sap in comparison to control potatoes, while significantly increases the concentrations of K^+ and ratio of Na^+ to K^+ (Akhtar et al. 2015; Hassan et al. 2019). Similarly, biochar resulted in decrease of Na^+ uptake in lettuce facing salt stress (Mehdizadeh et al. 2020). In another study, the uptake of Na^+ and K^+ under salt condition is also impacted by the application of biochar in maize (Akram et al. 2016; Gogi et al. 2018).

5 Addition of Biochar to Saline Soils

Addition of biochar to saline soils reportedly improves photosynthesis, biomass, and growth of plants. For instance, biochar that is made from hardwood and softwood resulted in increased tuber yield, shoot biomass, and root length in potatoes facing salt stress (Ali et al. 2017). Similarly, maize grow in soils with high level of exchangeable sodium and soluble salts produced more biomass when rice hull biochar was added to the soil (Dahlawi et al. 2018).

The biochar (*Fagus grandifolia* pyrolyzed at 378 °C at 50 t ha^{-1}) used in a study allowed herbaceous species, *Prunella vulgaris* and *Abutilon theophrastis*, to grow and produce more biomass when subjected to salt stress (Honermeier et al. 2013;

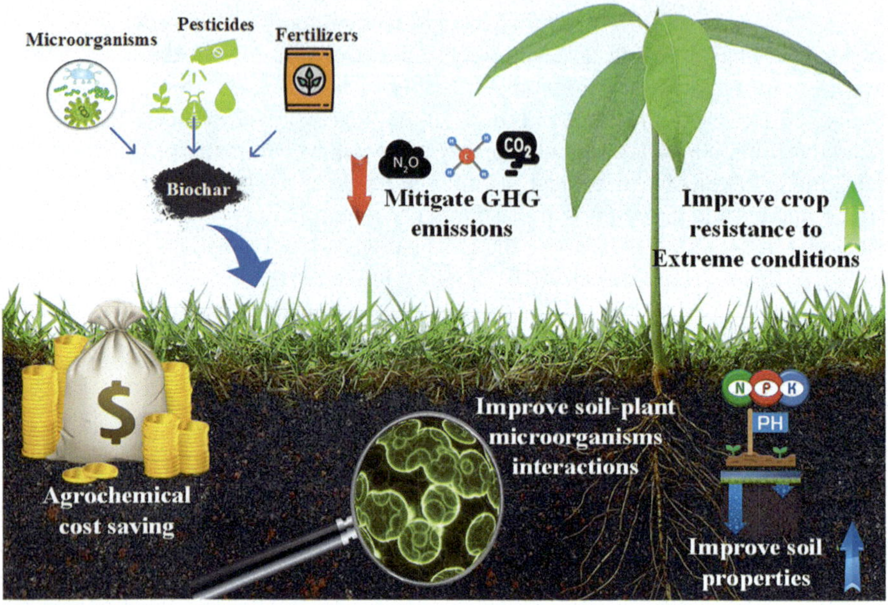

Fig. 20.2 Key role of biochar in improving soil-plant-microorganisms interactions and mitigating greenhouse gases. (Tayyab et al. 2018)

Thomas et al. 2013). Biochar had no apparent effect on photosynthetic carbon gain or chlorophyll fluorescence and WUE in both species facing salt stress. Response of biochar to salt condition varied depending on the type of plant. Furthermore, using composted biochar could increase biomass production in saline environment. In an experiment of field study for 2 years, PS and BPC treatments in saline soil led to improvements in yield of maize grains, plant height, photosynthetic pigments, plant density, leaf area index, and root length (Singh et al. 2020). Similar to this, when exposed to composted biochar, two halophyte species produced more biomass (You et al. 2021). The addition of biochar and soil microbes improves plant biomass and growth in comparison to only biochar treatments under salinity stress (Fig. 20.2; Tayyab et al. 2018).

6 Conclusion

Biochar application is thought to be a method of carbon sequestration to refine the characteristics of soil. Manufacturing of biochar takes place by heating the organic waste in incomplete or full absence of oxygen and resulted into a solid waste called biochar. The beneficial impacts of biochar on soil's biological, chemical, and physical properties have already been studied. Nutrient retention, improvement in soil cation exchange capacity, water holding capacity, as well as increase in enzyme activities of soil and diversity of microbial communities are directly linked with biochar application. Moreover, beneficial effects caused by biochar on soil modifications are much entangled. By modifying the biochemical characteristics, physiological processes, and efficiency of photosynthesis, as well as enhancing the activities of antioxidant enzymes in plants, biochar supplementation can enhance a plant's ability to withstand abiotic stress conditions.

References

Akhtar SS, Andersen MN, Naveed M, Zahir ZA, Liu F (2015) Interactive effect of biochar and plant growth-promoting bacterial endophytes on ameliorating salinity stress in maize. Funct Plant Biol 42:770–781

Akram MS, Shahid M, Tariq M, Azeem M, Javed MT, Saleem S, Riaz S (2016) Deciphering *Staphylococcus sciuri* SAT-17 mediated anti-oxidative defense mechanisms and growth modulations in salt stressed maize (*Zea mays* L.). Front Microbiol 7:867

Ali H, Tanveer A, Nadeem M, Asghar H, Javaid M (2013) Germination ecology of *Rhynchosia capitata*: an emerging summer weed in Asia. Planta Daninha 31:249–257

Ali S, Rizwan M, Qayyum MF, Ok YS, Ibrahim M, Riaz M, Arif MS, Hafeez F, Al-Wabel MI, Shahzad AN (2017) Biochar soil amendment on alleviation of drought and salt stress in plants: a critical review. Environ Sci Pollut Res 24:12700–12712

Berg G, Koberl M, Rybakova D, Muller H, Grosch R, Smalla K (2017) Plant microbial diversity is suggested as the key to future biocontrol and health trends. FEMS Microbiol Ecol 93:fix050

Capodaglio AG, Callegari A (2018) Feedstock and process influence on biodiesel produced from waste sewage sludge. J Environ Manag 216:176–182

Dahlawi S, Naeem A, Rengel Z, Naidu R (2018) Biochar application for the remediation of salt-affected soils: challenges and opportunities. Sci Total Environ 625:320–335

Deng J, Wang G, Kuang J, Zhang Y, Luo Y (2009) Pretreatment of agricultural residues for co-gasification via torrefaction. J Anal Appl Pyrolysis 86:331–337

Fisher E, Dupont C, Darvell L, Commandre JM, Saddawi A, Jones J, Grateau M, Nocquet T, Salvador S (2012) Combustion and gasification characteristics of chars from raw and torrefied biomass. Bioresour Technol 119:157–165

Foong SY, Liew RK, Yang Y, Cheng YW, Yek PNY, Mahari WAW, Lee XY, Han CS, Vo DVN, Van-Le Q (2020) Valorization of biomass waste to engineered activated biochar by microwave pyrolysis: progress, challenges, and future directions. Chem Eng J 389:124–401

Garcia-Ruiz H (2019) Host factors against plant viruses. Mol Plant Pathol 20:1588–1601

Gavili EA, Moosavi A, Moradi-Choghamarani F (2018) Cattle manure biochar potential for ameliorating soil physical characteristics and spinach response under drought. Arch Agron Soil Sci 64:1714–1727

Ghodake GS, Shinde SK, Kadam AA, Saratale RG, Saratale GD, Kumar M, Palem RR, AL-Shwaiman HA, Elgorban AM, Syed A (2021) Review on biomass feedstocks, pyrolysis mechanism and physicochemical properties of biochar: state-of-the-art framework to speed up vision of circular bioeconomy. J Clean Prod 297:126–645

Godwin PM, Pan Y, Xiao H, Afzal MT (2019) Progress in preparation and application of modified biochar for improving heavy metal ion removal from wastewater. J Bioresour Bioprod 4:31–42

Gogi MD, Arif MJ, Mahmood A, Chatha U, Niaz S, Khan I, Hassan MU, Chatha MB, Usman M (2018) Effect of abiotic factors and leaf-quality of sugarcane on the seasonal occurrence of *Pyrilla perpusilla* walker (homoptera: lophopidae) and *Epipyrops melanoleuca* fletcher (lepidoptera: epipyropidae). Int J Sugar 120:230–235

Gogoi N, Sarma B, Mondal SC, Kataki R, Garg A (2019) Use of biochar in sustainable agriculture. In: Innovations in sustainable agriculture. Springer, pp 501–528

Hansen V, Hauggaard-Nielsen H, Petersen CT, Mikkelsen TN, Muller-Stover D (2016) Effects of gasification biochar on plant-available water capacity and plant growth in two contrasting soil types. Soil Tillage Res 161:1–9

Hassan M, Chattha M, Chattha M, Mahmood A, Sahi S (2019) Chemical composition and methane yield of sorghum as influenced by planting methods and cultivars. JAPS 29:251–259

Hassan MU, Chattha MU, Khan I, Chattha MB, Barbanti L, Aamer M, Iqbal MM, Nawaz M, Mahmood A, Ali A (2021) Heat stress in cultivated plants: nature, impact, mechanisms, and mitigation strategies – a review. Int J Plant Biol 155:211–234

Honermeier B, Ali S, Leschhorn B, Mahmood A, Ijaz M, Russo M, Shafiee-Hajiabad M, Ullah H, Zeller S (2013) Cultivation of medicinal and spice plants in Germany. A review. Int J Agric Biol 15:1379–1388

Imran QM, Falak N, Hussain A, Mun BG, Yun BW (2021) Abotic stress in plants; stress perception to molecular response and role of biotechnological tools in stress resistance. Agronomy 11:1579

Iqbal Z, Iqbal MS, Hashem A, Abd-Allah EF, Ansari MI (2021) Plant defense responses to biotic stress and its interplay with fluctuating dark/light conditions. Front Plant Sci 12:631–810

Javaid MM, Florentine SK, Ashraf M, Mahmood A, Sattar A, Wasaya A, Li FM (2022) Photosynthetic activity and water use efficiency of *Salvia verbenaca* L. under elevated CO_2 and water-deficit conditions. Int J Plant Biol 208:536–551

Joseph SD, Camps-Arbestain M, Lin Y, Munroe P, Chia CH, Hook J, van Zwieten L, Kimber S, Cowie A, Singh BP, Lehmann J, Foidl N, Smernik RJ, Amonette JE (2010) An investigation into the reactions of biochar in soil. Soil Res 48(7):501–515. https://doi.org/10.1071/SR10009

Kalt G, Lauk C, Mayer A, Theurl MC, Kaltenegger K, Winiwarter W, Erb KH, Matej S, Haberl H (2020) Greenhouse gas implications of mobilizing agricultural biomass for energy: a reassessment of global potentials in 2050 under different food-system pathways. Environ Res Lett 15:34–66

Khan MR, Sharma RK (2020) Fusarium-nematode wilt disease complexes, etiology and mechanism of development. Indian Phytopathol 73:615–628

Ko JK, Kim Y, Ximenes E, Ladisch MR (2015) Effect of liquid hot water pretreatment severity on properties of hardwood lignin and enzymatic hydrolysis of cellulose. Biotechnol Bioeng 112:252–262

Lago BC, Silva CA, Melo LCA, de Morais EG (2021) Predicting biochar cation exchange capacity using Fourier transform infrared spectroscopy combined with partial least square regression. Sci Total Environ 794:148762

Lonappan L, Rouissi T, Das RK, Brar SK, Ramirez AA, Verma M, Surampalli RZ, Valero RJ (2016) Adsorption of methylene blue on biochar microparticles derived from different waste materials. Waste Manag 49:537–544

Mahmood S, Ahmed W, Alatalo JM, Mahmood M, Imtiaz M, Ditta A, Ali EF, Abdelrahman H, Slaný M, Antoniadis V, Rinklebe J, Shaheen SM, Li W (2022) Herbal plants- and rice straw-derived biochars reduced metal mobilization in fishpond sediments and improved their potential as fertilizers. Sci Total Environ 826:154043. https://doi.org/10.1016/j.scitotenv.2022.154043

Markell SG, Harveson RM, Block CC, Gulya TJ (2015) Sunflower diseases. In: Sunflower. Elsevier, pp 93–128

Mehdizadeh L, Moghaddam M, Lakzian A (2020) Amelioration of soil properties, growth and leaf mineral elements of summer savory under salt stress and biochar application in alkaline soil. Sci Hortic 267:109–319

Mohanta TK, Bashir T, Hashem A, Abd-Allah EF (2017) Systems biology approach in plant abiotic stresses. Plant Physiol Biochem 121:58–73

Mulcahy DN, Mulcahy DL, Dietz D (2013) Biochar soil amendment increases tomato seedling resistance to drought in sandy soils. J Arid Environ 88:222–225

Patel A, Khae P, Patra D (2017) Biochar mitigates salinity stress in plants. In: Plant adaptation strategies in changing environment. Springer, pp 153–182

Patel HK, Joshi MP, Kalaria RK (2021) Biochar: a futuristic tool to remove heavy metals from contaminated soils. In: Biochar and its application to bioremediation. Springer, pp 231–258

Quicker P, Weber K (2016) Biokohle. Herstellung, Eigenschaften und Verwendung von Biomassekarbonisaten, 1 edn. Springer Vieweg Wiesbaden, Germany. https://doi.org/10.1007/978-3-658-03689-8

Rajapaksha AU, Chen SS, Tsang DC, Zhang M, Vithanage M, Mandal S, Gao B, Bolan NS, Ok YS (2016) Engineered/designer biochar for contaminant removal/immobilization from soil and water: potential and implication of biochar modification. Chemosphere 148:276–291

Raza A, Razzaq A, Mehmood SS, Zou X, Zhang X, Lv Y, Xu J (2019) Impact of climate change on crops adaptation and strategies to tackle its outcome: a review. Plan Theory 8:34

Semida WM, Beheiry HR, Setamou M, Simpson CR, Abd-El-Mageed TA, Rady MM, Nelson SD (2019) Biochar implications for sustainable agriculture and environment: a review. S Afr J Bot 127:333–347

Singh B, Guldhe A, Shahi D, Bauddh K (2020) Biochar amendment in agricultural soil for mitigation of abiotic stress. In: Ecological and practical applications for sustainable agriculture. Springer, p 305

Tanveer A, Arshad MS, Ayub M, Javaid MM, Yaseen M (2012) Effect of temperature, light, salinity, drought stress and seeding depth on germination of *Cucumis melo* var. agrestis. J Agron Crop Sci 18(6):445–459

Tayyab M, Islam W, Khalil F, Ziqin P, Caifang Z, Arafat Y, Hui L, Rizwan M, Ahmad K, Waheed S (2018) Biochar: an efficient way to manage low water availability in plants. Appl Ecol Environ Res 16:2565–2583

Thomas SC, Frye S, Gale N, Garmon M, Launchbury R, Machado N, Melamed S, Murray J, Petroff A, Winsborough C (2013) Biochar mitigates negative effects of salt additions on two herbaceous plant species. J Environ Manag 129:62–68

Van-Ittersum MK, Cassman KG, Grassin P, Wolf J, Tittonell P, Hochma Z (2013) Yield gap analysis with local to global relevance – a review. Field Crop Res 143:4–17

Wafiq A, Reichel D, Hanafy M (2016) Pressure influence on pyrolysis product properties of raw and torrefied Miscanthus: role of particle structure. Fuel 179:156–167

Wang J, Wang S (2019) Preparation, modification and environmental application of biochar: a review. J Clean Prod 227:1002–1022

Wang L, Skreiberg O, Gronli M, Specht GP, Antal MJ (2013) Is elevated pressure required to achieve a high fixed-carbon yield of charcoal from biomass. The importance of particle size. Energy Fuel 27:2146–2156

Weber K, Quicker P (2018) Properties of biochar. Fuel 217:240–261

Xiu S, Shahbazi A, Li R (2017) Characterization, modification and application of biochar for energy storage and catalysis: a review. Trends Renew Energy 3:86–101

Xue Y, Wang C, Hu Z, Zhou Y, Xiao Y, Wang T (2019) Pyrolysis of sewage sludge by electromagnetic induction: biochar properties and application in adsorption removal of Pb (II), Cd (II) from aqueous solution. Waste Manag 89:48–56

You X, Yin S, Suo F, Xu Z, Chu D, Kong Q, Zhang C, Li Y, Liu L (2021) Biochar and fertilizer improved the growth and quality of the ice plant (*Mesembryanthemum crystallinum* L.) shoots in a coastal soil of Yellow River Delta. China Sci Total Environ 775:144–893

Zandalinas SI, Mittler R, Balfagon D, Arbona V, Gomez-Cadenas A (2018) Plant adaptations to the combination of drought and high temperatures. Physiol Plant 162:2–12

Zaynab M, Fatima M, Sharif Y, Zafar MH, Ali H, Khan KA (2019) Role of primary metabolites in plant defense against pathogens. Microb Pathog 137:103–728

Zhang X, Gao B, Zheng Y, Hu X, Creamer AE, Annable MD, Li Y (2017) Biochar for volatile organic compound (VOC) removal: sorption performance and governing mechanisms. Bioresour Technol 245:606–614

Zheng J, Wang S, Wang R, Chen Y, Siddique KH, Xia G, Chi D (2021) Ameliorative roles of biochar-based fertilizer on morpho-physiological traits, nutrient uptake and yield in peanut (*Arachis hypogaea* L.) under water stress. Agric Water Manag 257:107–112

Chapter 21
Accumulation and Toxicity of Arsenic in Rice and Its Practical Mitigation

Asanka Tennakoon, G. A. H. Galahitigama, P. W. M. Tharindi, T. Suthajini, N. P. M. Abeysinghe, Salinda Sandamal, Karthika Rajendran, Adel A. Hadifa, Ayman El Sabagh, and Disna Ratnasekera

Abstract Arsenic (As) poisoning in agroecosystem is a major concern globally because rice is a main food source for a vital community in As-polluted areas. This effort is notable for its thorough examination of an extensive series of topics, including the health concerns allied with As disclosure; As sources in soil-rice systems; As toxicity symptoms at various stages of rice growth; As uptake, metabolism, and detoxification; and strategies to reduce As bioaccumulation. Moreover, significant efforts are being made to reduce As accumulation. The effectiveness of mitigation strategies has been thoroughly investigated, with a focus on soil amendments, irrigation management, electrokinetic remediation, exogenous application of chemicals and hormones, phytoremediation, bioremediation, transgenic variety development, and other agronomic techniques. Furthermore, among the various methods currently in use, biotechnology may be a good strategy for reducing As accretion in rice grains. Molecular engineering could be a feasible method for identifying the genes involved in the As metabolic route in plants. Despite this, the majority of these novel approaches are still being researched. As a result, this study explores into As accessibility in paddy soil, the mechanism of plant As uptake, the effects on humans and plants, and remediation approaches for mitigating As accretion.

Keywords Heavy metal · Phytotoxicity · Reactive oxygen species · Remediation technique · Oxidative stress

A. Tennakoon · T. Suthajini
Department of Biosystems Technology, Faculty of Technology, Eastern University, Chenkalady, Sri Lanka

G. A. H. Galahitigama · P. W. M. Tharindi · N. P. M. Abeysinghe
Department of Export Agriculture, Faculty of Agricultural Sciences, Sabaragamuwa University of Sri Lanka, Balangoda, Sri Lanka

M. Hasanuzzaman (ed.), *Climate-Resilient Agriculture, Vol 2*,
https://doi.org/10.1007/978-3-031-37428-9_21

1 Introduction

Arsenic is a noxious metalloid, and highly cancer-causing substance that accumulates in food chains over time and can be found in almost every region of the world (Wang et al. 2019; Yang et al. 2022). As accumulation in terrestrial soils is caused by both human and natural processes. The main human actions that lead to As accumulation in soil include metal excavating and smelting, application of As-containing agrochemicals, food additives, and wood stabilizers (Yu et al. 2017). Exposure to humans causes a wide range of other acute and long-lasting health issues; namely, skin lesions, diabetes, cancer, and liver issues (Shrivastava et al. 2020). The primary source of As exposure in humans is contaminated groundwater used for drinking (Majumdar and Bose 2018). Furthermore, consumption of packaged foods has recently been shown to exacerbate this problem (Fahad et al. 2016).

Plant phytotoxicity, on the other hand, which results in abridged root expansion, chlorosis in leaves, shrinkage, and necrosis in aerial plant parts, etc., also poses concerns to human health (Carbonell-Barrachina et al. 1998). As buildup inhibits plant metabolic activities, develops leaves scorching, prevents root-to-shoot transfer, decreases biomass accumulation, and lowers fertility and total yield (Garg and Singla 2011). Additionally, paddy grown in soils with higher than 60 mg kg^{-1} of As concentration show phytotoxic symptoms as poor growth, brown acnes, and burning foliage (Bakhat et al. 2017). According to reports, rice production in traditional paddy fields in Bangladesh decreased significantly from 7.5 to 2.5 t ha^{-1} as a result of higher As in the soil (Stoeva and Bineva 2003). When As enters the cells, it causes significant damage to the cells because of the production of reactive oxygen species (Gupta et al. 2013). For nutrition, most people depend on rice, which is

S. Sandamal
State Key Laboratory of Systematic and Evolutionary Botany, Institute of Botany, Chinese Academy of Sciences, Beijing, China

University of Chinese Academy of Sciences, Beijing, China

K. Rajendran
VIT School of Agricultural Innovations and Advanced Learning (VAIAL), Vellore Institute of Technology (VIT), Vellore, Tamil Nadu, India

A. A. Hadifa
Department of Rice Research, Field Crops Research Institute, Agricultural Research Center (ARC), Giza, Egypt

A. El Sabagh
Department of Field Crops, Faculty of Agriculture, Siirt University, Siirt, Turkey

Department of Agronomy, Faculty of Agriculture, Kafrelsheikh University, Kafr Elsheikh, Egypt

D. Ratnasekera (✉)
Department of Agricultural Biology, Faculty of Agriculture, University of Ruhuna, Matara, Sri Lanka
e-mail: disnar@agbio.ruh.ac.lk

grown on As-polluted paddy soils. Additionally, it has been revealed that because rice is grown in flooded environments, the plants can accumulate as much as many folds more As than crops such as maize and wheat (Shrivastava et al. 2020). As a result, Asia, where rice consumption is frequently higher, would have a larger risk of human As poisoning (Dave et al. 2013).

Several types of As are present in the environment. Paddy fields contain more As (III) than As (V) under flooded paddy fields conditions (Meharg and Jardine 2003). Organic As is generally less toxic to plants than inorganic As (Abedi and Mojiri 2020). As availability in soil impacted by various factors, comprising but not limited to pH, redox potential, dissolved organic carbon, organic matter, and biotic factors). However, As (III) and As (V) penetrate the rice plant through distinct pathways. Aquaporin channels (OsPIP2;7, OsPIP2;6), may allow As (III) entry, but activation of phosphate transporter proteins (PHTs) may be necessary for As (V) to enter the roots of rice plants. Lsi1 proteins have a role in the silicic acid transport system, which rice plants use to absorb organic As compounds such as dimethylarsinic acid (DMA) and monomethylarsonic acid (MMA) (Suriyagoda et al. 2018). Some genes including OsPT2, OsPIP1;2, and OsPHT1;1 are also involved in this procedure of As translocation from the root to the shoot. Genes such as OsPTR7 and OsNPF8;1 are crucial for DMA accumulation and translocation in rice grains (Abedi and Mojiri 2020).

Because As bioaccumulation poses negative health hazards to humans and other organisms, many approaches have been developed to minimize As accretion in the ecosystem and living cells. Among them, adapting agronomic approaches can be applied effectively (Mitra et al. 2017). Changing irrigation schedules, nutrient management, and amendment application are all highly successful on a trial-and-error basis (Mitra et al. 2017). Intermittent irrigation, sprinkler irrigation, and alternate wetting-drying (AWD) are three water management methods widely regarded as crucial for reducing As bioaccumulation (Upadhyay et al. 2020). The absorption and movement of As in paddy plants was also decreased by inorganic fertilizers such as S, Fe, P, and Si (Mitra et al. 2017). Soil amendments, including organic matter (Suriyagoda et al. 2018), biochar (Kamali et al. 2022), and modified biochar (Lin et al. 2017; Qiao et al. 2018; Irshad et al. 2022a, b), nano-based amendments (Li et al. 2019; Wu et al. 2020) have shown positive outcomes in reducing As buildup in plants. Many hormones applied to plant leaves, or foliar application, reduced As buildup (Ramesh et al. 2015). Microbial (Hare et al. 2017) and plant-based (Mitra et al. 2017) bioremediation strategies are both viable options for reducing bioaccumulation in rice. In addition to these methods, many cutting-edge technologies are now being used. Biotechnological methods, electrokinetic remediation, and batch and column operations are some examples of technologies (Chen et al. 2017; Ali 2018; Zhang et al. 2022). Their ability to reduce plant As buildup has been studied. The majority of these cutting-edge methods, however, are still in the research stage. As a result, this study delves into As' accessibility in paddy soil, the mechanism of plants' As absorbtion, the effects on humans and plants, and remediation options for reducing As accumulation.

2 Health Risk of As Exposure

Given the trend of population expansion and the increasing demand for rice, farmers' increased use of pesticides to safeguard their crops may have an impact on the level of trace (metalloids) in soils (Okamura et al. 2002). Trace (metalloids) contamination on a paddy field has grown significantly in importance as a major environmental problem in recent years due to its bioaccumulation in the ecosystem and nonbiodegradability (Hojsak et al. 2015). Trace (metalloids), one of the most dangerous substances, has repeatedly been considered a potential rice contaminant due to the tremendous potential harm for both the environment and human health (Hojsak et al. 2015). In nature, organic and inorganic trace elements (metalloids) are two distinct categories (Hite 2013). Inorganic As poses a greater threat than organic As owing to the presence of pentavalent inorganic As compound that dissolves in water to generate a weak acid and produces arsenate (Grund et al. 2008), which then contaminates groundwater and causes harm to humans (Chung et al. 2014). Environmental heavy metals in their inorganic forms may undergo gradual decomposition into less hazardous substances via chemical and biological processes (Ayangbenro and Babalola 2017).

Millions of people have suffered from chronic diseases as a result of drinking polluted water over the last several decades in the globe (Bundschuh et al. 2012; McClintock et al. 2012; Chakraborti et al. 2018). One of the most important sources of inorganic As exposure is rice consumption in humans (Chatterjee et al. 2010; Kumarathilaka et al. 2019), as soil and rice pollution is well-documented worldwide (Jahan et al. 2021). Rice provides for over 50% of the average person's daily As intake, making it a significant source of As exposure (Kumarathilaka et al. 2019; Rahman et al. 2021). According to the recent risk assessments, it has been found that consuming As-contaminated rice and foods made with rice increases the threat to human health (Signes-Pastor et al. 2016, 2017). Inorganic As is toxic to humans when it is present in water and food due to interactions with protein sulfhydryl groups and phosphate substitution in ATP (Garbinski et al. 2019). Persistent low-level As exposure has been linked to changes in the cardiovascular system, the digestive tract (Rahman et al. 2009), the nervous system (Mochizuki et al. 2019; Naujokas et al. 2013), the immune system, the developmental processes, metabolic issues, diabetes (Nizam et al. 2013), and liver fibrosis (Das et al. 2012). Chronic As exposure (>10 g/L) has been associated with skin (Leonardi et al. 2012), lung (Smith et al. 2022), bladder (Meliker et al. 2010), and prostate (Bulka et al. 2016) cancers in humans because of As-induced oxidative stress, suppression of DNA repair, signal transduction pathway disruption, and chromosomal abnormalities (WHO 2017; Engwa et al. 2019). As is extremely toxic and can result in coma and death (Hu et al. 2021).

Dietary arsenic (As) is a serious public health issue worldwide. In humans, organ damage is a common complication of As poisoning. As may enter the body in a variety of routes, including ingesting (by food) it, breathing it in (inhalation), absorbing it through the skin (skin penetration), or/and contacting it with the mucous membranes. Its toxicokinetics are determined by the type and duration of contact, the physicochemical properties (pKa and solubility) of the substance, the route obtained, and the living beings affected. Between 70% and 90% of ingested inorganic As (iAs (III) and iAs (V)) is absorbed by the body through the digestive tract. The blood then carries it to the liver, kidneys, bladder, lungs, and other organs, muscles, and nerves. Most absorbed and distributed inorganic As is stored in the liver. Urine is the primary route for its excretion (Watanabe and Hirano 2013).

The metabolism of intestinal microbiota increases As bioavailability, systemic distribution, and toxicity. Similarly, where inorganic As and its metabolites are concentrated in the body is determined by the protein composition of different tissues and the extent to which they were exposed. As a consequence of their binding to the sulfhydryl groups of keratin proteins, trivalent arsenicals accumulate in hair, skin, and nails (Watanabe and Hirano 2013). First, it binds to keratin and buildups in hair and nails, which are early indicators of skin irregularities. Keratosis, a skin condition, is frequently one of the first signs of As poisoning. Recent reports from Asian countries including India, Nepal, and Bangladesh have described cases of melanosis, squamous cell carcinoma (SQCC), and keratosis associated with Bowen's syndrome (Chakraborti et al. 2016; Hu et al. 2021). The amount of monomethylarsonate (MMA) and the amount of dimethylarsonate (DMA) in an As species directly correlate with the risk of bladder, lung, and skin cancer (Chervona et al. 2012; Eckstein et al. 2017).

Chronic As exposure weakens cellular and humoral immunity, which is harmful to the immune system and may promote immunosuppression (Biswas et al. 2008; Dangleben et al. 2013). Similar study indicates that ongoing exposure raises serum levels of immunoglobulin IgA, IgG, and IgE, which may hasten the development of respiratory conditions such as pneumonia, allergic bronchitis, and chronic obstructive pulmonary disease (Islam et al. 2007). Chronic As exposure may also cause glomerulonephritis, acute tubular necrosis, hypercalciuria, albuminuria, nephrocalcinosis, and renal papillae necrosis (Roggenbeck et al. 2016). Both Chronic Kidney Disease (CKD) and the onset of incipient nephropathy have been linked to As-induced nephron damage (Robles-Osorio et al. 2015).

Quansah et al. (2015) reported that the people who consume water contaminated with As have an increased incidence of spontaneous abortion, stillbirth, and neonatal and infant death. Placental insufficiency problems caused by chronic inorganic As exposure include preterm birth and intrauterine growth retardation. Inorganic As has also been discovered in meals made of rice that are taken by infants and young children, who are particularly susceptible to the negative consequences of As consumption (Signes-Pastor et al. 2016; Islam et al. 2017).

Chronic As toxicity has been linked to many health problems, particularly cancer (kidney, lung, liver, bladder, prostate), cardiovascular diseases, skin rashes and cancer (hyperkeratosis), catarrhal changes, nutritional and gastrointestinal disturbance, diabetes mellitus, abnormal pregnancy outcomes, nervous system disorders (neuropathies, psychosis, impaired intellectual function), and musculoskeletal disorders. Additionally, some of the most common acute As toxicities include hypotension, acute tubular necrosis, organ failure (heart, liver and renal failure), nausea, convulsions, pulmonary edema, diarrhea, vomiting, delirium, dehydration, lethargy, abdominal pain, and a burning sensation in the stomach and esophagus (Rahaman et al. 2021). Risks associated with chronic As exposure in rice may be reduced through a combination of agronomic and biotechnological methods, careful dietary management, and increased public awareness of the issue.

3 As Sources in Soil-Rice Systems

As may contaminate rice-soil systems in a variety of ways, including both natural and anthropogenic sources (Afzal et al. 2018). Geothermal and volcanic activity and weathering of soils and rocks are among natural processes that exacerbate the As buildup. Industrial waste, medications, feed additives, metal mining, pesticides, fertilizers, and irrigation with As-contaminated groundwater are only a few examples of the anthropogenic factors that cause an alarming rate of As accumulation in the environment (Suriyagoda et al. 2018). Natural As pollution is caused by geochemical pathways including alkali desorption, sulphide oxidation, reductive dissolution, and geothermal activity (Afzal et al. 2018). As accumulates in paddy soils as a result of irrigation with As-contaminated groundwater. Previous studies have shown a favorable association between the amount of As in irrigation water and As buildup in irrigated soil (Afzal et al. 2018). South and southeast Asia has been severely affected with the natural As contamination in the ground water.

The parent material's composition as well as the current geological circumstances can both have an impact on As levels in soil. As is frequently associated with sulfur in nature (Awasthi et al. 2017). As exists in three different forms—inorganic, organic, and gaseous—each with its own set of characteristics. Trimethylarsine oxide (TMAO), monomethylarsonic acid, dimethylarsenic acid, and other organic forms of As, as well as inorganic forms such as arsenate As (V) and arsenite As (III), are produced by biomethylation in paddy soils (Zheng et al. 2013). Additionally, paddy soils may contain organic forms of As, such as, arsenolipids, dimethylarsinylriboside, arsenobetaine (AB), arsenosugars, and arsenocholine (AC), even though in much lesser concentrations (Suriyagoda et al. 2018). Arsine gas is the gaseous state of As. As is apparently more hazardous in its inorganic forms than its organic forms. The maximum permissible concentration of As in soil in the United States has been set at 24 mg/kg by the EPA. As (III) accounts for 87–94% of total As in anaerobic conditions in aquatic and soil habitats (Khan et al. 2022). It is common for As (III) to exist as H_3AsO_3 and As (V) to exist as H_2AsO_3 or $HAsO_4{}^{2-}$ at neutral pH (Suriyagoda et al. 2018). It has a strong affinity for hydroxides, oxyhydroxides, and iron and manganese oxides present in soil due to its negative charge (Saifullah et al. 2018). Despite organic trimethylarsine (TMA) being recognized as nontoxic, arsenite (III) is thought to be more mobile and hazardous than arsenate (V) (Zheng et al. 2013). In comparison to other As forms, As (III) and DMA are thought to be more soluble and bioavailable. As forms are transformed by oxidation, reduction, methylation, and demethylation. The physical, chemical, and biological features of the soil, in contrast, have a significant impact on determining the As content (Suriyagoda et al. 2018).

4 Toxicity Symptoms of As at Different Growth Stages of Rice

The assimilation of As has a significant negative impact on rice plant growth, and As stimulates the normal metabolic process. It frequently manifests as a wide range of symptoms at various phases of development, resulting in stunted growth and decreased productivity. Because plant hormone controls and regulates the growth, development of plant, and abiotic stress responses, when stress is applied to rice plants, the level and distribution of endogenous phytohormone are altered. Indeed, Indole-3-acetic acid (IAA) has been related to the responses of the rice plants to environmental stress (Choudhury et al. 2011).

4.1 *Seedling Stage*

Germination is one of the most susceptible phases and is easily disrupted by abiotic problems. In general, a rice plant that has accumulated As exhibits the following symptoms during the germination phase: slower elongation, low dry matter, shorter roots and shoots, fewer leaves, narrower leaf blades, and lower levels of chlorophyll (Shaibur et al. 2009). The statement recently expressed is supported by numerous studies. Shaibur et al. (2006) revealed a 43% and 33% decline in shoot height and root height, respectively. Additionally, leaf blade size and the number of leaves also decline. Furthermore, when As intake increases, leaf number and width of the leaf blade decrease, which affects the dry matter yield (Shaibur et al. 2006). Begum and Mondal (2019) investigated the negative effects of As at 5 ppm or greater in nutrient solutions on seed germination. The length of the root and shoot is also reduced by 40% in seeds treated with As (Bastías and Beldarrain 2016). According to a prior study, treatments with 50 μM and 100 μM As caused a 70% decrease in root length because roots are more sensitive to As than shoots at the seedling stage of rice plant (Mlangeni et al. 2020).

4.2 *Vegetative Stage*

Arsenic accumulation generally limits plant height, root length, and leaf number, similar to the seedling stage. Furthermore, it has a direct impact on the biomass and chlorophyll synthesized during the vegetative development stage (Shaibur et al. 2009). Using higher As concentrations (2–8 mg/L) alters the rice plant height from 91.1–84.1 cm to 79.2–63.8 cm (Mecwan Neha et al. 2018). According to Begum and Mondal (2019), the symptoms of chlorosis are caused by an increase in As (V) and As (III) concentrations as well as a consistent decrease in chlorophyll content. When the Fe plaque forms on the surface of the rice root, the chlorotic symptoms become more pronounced (Mitra et al. 2017). In addition to chlorophyll, other pigments such as xanthophyll and carotene are also drastically diminished in the rice plant (Choudhury et al. 2011). Root symptoms such as yellowing of the roots and a decline in root vigor were occasionally noticed when rice plants were subjected to 60 mg/Kg (Bhattacharya et al. 2021).

4.3 Reproductive Stage

Shaibur et al. (2009) reported the following symptoms for the As accumulation during reproductive growth: Reduced tiller number, poor grain filling, less productive tillers, and yield loss due to sterility. Das et al. (2013) recorded productivity declines of 55.5% and 54.8% during the tillering and harvesting periods, respectively. Above 60 mg kg^{-1} As concentration, a significant production decline of approximately 80% was observed (Bhattacharya et al. 2021). Abedin et al. (2002), on the other hand, demonstrated a low grain yield and found a significant correlation between grain yield and arsenate treatment. Nevertheless, the application of arsenate did not significantly decline the straw yield.

5 Factors Determining As Mobility and Bioavailability

The physiochemical parameters and biological properties of soil, as well as the affinity of metal soil particles, are the primary determinants of metal interaction in soil (Liu et al. 2018; Srivastava et al. 2021). The presence of metal ions in the soil also alters the microclimate of the root rhizosphere. Environmental and soil factors have been linked to bioavailability and bioaccumulation of As in rice (Mesa et al. 2017). Consequently, it is critical to understand the factors that influence As mobilization.

5.1 As Speciation

Arsenic discussed previously like arsenate and arsenite are the commonly available forms of As, whereas monomethyl and dimethyl arsenic are organic As species found in soil and groundwater (Mitra et al. 2017). However, the selective assimilation of specific As forms by paddy roots affects the state of the soil in paddy fields (Han et al. 2017), with inorganic species being absorbed much faster than organic species (Liu et al. 2018). Arsenate (V) predominates in submerged environments, whereas As (III) predominates in soil under aerobic conditions. According to Han et al. (2017), arsenate, monomethyl arsonic acid, arsenite, and dimethyl arsenic acid are the compounds that are more bioavailable to As in rice plants.

5.2 Soil pH

The pH of the soil has a potential influence on plant mobility, bioavailability, and accumulation (Roy et al. 2015). Because soil pH has a direct impact on As leaching and speciation, it also has a direct impact on bioavailability. As absorption is facilitated by low pH because it makes the molecule that binds As, Fe-oxyhydroxide, more soluble (Liu et al. 2018). In addition, a high pH raises the negative surface charges and encourages As desorption from Fe oxide and enhances As mobilization and buildup (Liu et al. 2018). A negative correlation between the As concentration in rice grains and soil pH was reported by Yao et al. (2021), in contrast to the positive correlation observed for wheat and pepper.

5.3 Organic Matter

Soil organic matter and As buildup are highly correlated (Huang et al. 2021). It has been demonstrated that rice plants contained high levels of As in soils that were rich in organic matter. It guarantees that organic matter in the soil regulates the movement of As (Liu et al. 2018). Organic matter, according to Syu et al. (2019), increased the release of As into the soil, which caused rice plants to absorb a large amount of As. The degree of increase is determined by the decomposition capacity and pace of organic matter application.

5.4 Soil Texture

The increased surface area and decreased particle size of the soil both influence the reduced toxicity of As. Because silt and clay soil have finer textures than sandy soil, clay soil has five times less phototoxicity than sandy soil (Liu et al. 2018). But, under acidic pH, Yao et al. (2021) found substantial As accumulation in rice crops grown in soils which have higher clay content. The small particle size of clay provides an increased specific surface area to bind a notable quantity of As, and an increase in pH makes it easier for As that has been absorbed on clay particles to leak out, hence increasing the amount of As that has been absorbed and accumulated on clay particles (Roy et al. 2015).

5.5 Genotype Variation

Plant genotype is another factor which influences the As accumulation in rice fields (Fu et al. 2017). When compared to TD71 and Yinjingruanzhan, IAPAR9 and Nanyangzhan acquire more iAs and ToAs (Bastías and Beldarrain 2016). According to study conducted by Bastías and Beldarrain (2016), the Nanyangzhan genotype includes more DMA than any other As species. Tolerance in shoot systems and grain As accumulation were studied by Murugaiyan et al. (2021), who found that the genotypes WTR1-BRRI dhan 69, NPT-IR68552-55-3-2, OM997, and GSR IR1-5-Y4-S1-Y1 were the most As-tolerant species (Cantamessa et al. 2020). Results from the research conducted by Fernández-Baca et al. (2021) discussed that the amount of inorganic As in rice grains varied, suggesting that genotype variation influences iAs accumulation (Ma et al. 2018).

5.6 *Iron Plaque and Radial Oxygen Loss*

Radial oxygen loss (ROL) refers to the transport of oxygen from the aerenchyma of rice roots into the rhizosphere. It is determined by the soil oxygen content and the depth of saturated soil (Karimi et al. 2009). As concentration is negatively related to ROL rates (Fu et al. 2017). This results in the precipitation of iron oxides or hydroxides on the root surface, as ferric iron Fe (III) is produced from ferrous iron Fe (II). Iron plaque is a term used to describe this phenomenon. The As content in rice tissue is affected by iron plaque, which is a prime sink for As (III) oxidizing bacteria (Karimi et al. 2009).

5.7 *Microbial Contribution for the As Availability in Paddy Soils*

The toxicity and availability of As in rice soil are significantly impacted by the biological, chemical, and physical characteristics of the soil. In As reduction, oxidation, methylation, and volatilization, microbes are therefore crucial (Chen et al. 2019). Although As is regarded as a harmful substance, the majority of living things have evolved a resistance to it or a mechanism to use it in their physiological functions. Microbes have the capacity to oxidize As (III) to As (V) and reduce its toxicity. Furthermore, As (III) contributes electrons to bacterial metabolism as an electron donor (Zheng et al. 2013). According to recent research, the As (III) concentration is greater than 80% in paddy soils due to this microbial activity. According to reports, bacteria, anaerobic archaea, halophiles, and many cyanobacteria and algae influence As biomethylation in paddy soils. The methylcobalamin-dependent non-enzymatic methylation of As has been attributed to anaerobic prokaryotes and cyanobacteria. This As methylation is crucial to the As cycling process (Huang et al. 2016). The methanogenic archaea and sulfate-reducing bacteria (SRB) were recently discovered to be actively involved in the methylation and demethylation of As (Chen et al. 2019).

The incorporation of organic matter potentially alters the diversity of microorganisms in the soil. Dissimilatory iron-reducing bacteria (DIRB) are abundant in paddy soils, according to Zeng et al. (2013). If the activity of these bacteria is reduced, ferric oxide may become available in the soil and As may become immobilized (Yi et al. 2019). The activity of ferrous iron-oxidizing microorganisms may reduce As bioavailability in rice plants by increasing the adsorption of more As ions to ferric minerals under flooding conditions by oxidizing ferrous ions.

When symbiotically associated with rice plant roots, arbuscular mycorrhizal fungus (AMF) can enhance the plant's tolerance to As. Rice roots may be prevented from absorbing As (V), As (III), and MMA by mycorrhiza (Hussain et al. 2021). According to some studies, mycorrhizal roots release signaling molecules that downregulate the Lsi1 gene (Si transporter), which is responsible for As uptake and transportation (Abedi and Mojiri 2020). Furthermore, bacteria play a unique role in the release of As sequestrated by an iron plaque in the rice rhizosphere during the decomposition process shortly after harvesting. The soil solution and groundwater table may become contaminated as a result of this process, which may quickly raise the As percentage in paddy soils (He et al. 2020).

5.8 Effect of Redox Potential

As mobility is highly dependent on soil redox conditions. The oxygen consumption by rice roots and microorganisms in flooded paddy fields may cause a quick depletion of the soil's available oxygen, leading to a reduced condition (Bakhat et al. 2017). These biological transformations may result in a decrease in soil redox potential (Yamaguchi et al. 2011). In oxidized state, As (V) is dominant and has a higher affinity for Fe-oxy-hydroxide, reducing mobility and limiting bioavailability of rice roots. In contrast, under anaerobic/reducing conditions, As (III) exhibits higher mobility and bioavailability (Afzal et al. 2018).

Through their participation in the reactions, microorganisms can significantly contribute to redox processes. As (III) and As (V) both adsorb into Fe, Al, or Mn oxides and clay minerals during the adsorption and desorption processes that control the mobility of As in the soil. Among these, As (V) and As (III) have a high affinity for iron (hydr)oxides (FeOOH) (Afzal et al. 2018). Because the redox condition may affect the behavior of the FeOOH, the redox condition of ferrous is crucial in determining the As behavior in the soil (Bakhat et al. 2017).

6 As Uptake, Metabolism, and Detoxification

Since As is a class I carcinogen and a factor in a variety of human diseases, its accumulation in soils and subsequent entry into human diets pose a significant threat to human health. Rice is the staple food of South Asians. It has a higher As uptake tendency than other cereals because it is grown in submerged soil, which causes reductive dissolution of iron oxyhydroxide minerals and a subsequent decrease in As adsorbed on soil minerals, thereby increasing dissolved concentrations and plant availability (Biswas et al. 2018; Muehe et al. 2019). As is absorbed by rice plants from the soil and water systems; it ultimately culminates in the cooked grain (Fig. 21.1). As accumulation in rice roots is dependent on soil-water concentration levels and plant absorption caused by ROL and the formation of Fe plaques in the rhizosphere and on root surfaces.

Arsenate and inorganic phosphate share the same phosphate transport pathway in higher plants and have chemical structures that are very similar, according to physiological and electrophysiological research. As (V) absorption in *Arabidopsis thaliana* is mediated by phosphate transporters (PHTs), such as AtPHT1;1, AtPHT1;4, AtPHT1;5, AtPHT1;8, and AtPHT1;9 (Cao et al. 2017; Fontenot et al. 2015; Remy et al. 2012; Shin et al. 2004). DiTusa et al. (2015) discovered PvPht, which belongs to the phosphate transporter 1 (Pht1) family, in *Pteris vittata*. This gene boosted As (V) absorption and accumulation due to its strong combined ability for As (V). There are 13 OsPT (OsPT1 to OsPT13) genes that consist of rice phosphate transporter belonging gene family (Paszkowski et al. 2002). Among these genes, the roles of OsPT1, OsPT4, and OsPT8 were discovered (Cao et al. 2017; Kamiya et al. 2013; Wang et al. 2016; Ye et al. 2017). When certain phosphate genes are overexpressed, As (V) is taken up more readily, but when these genes are knocked down, As (V) is taken up less readily and accumulates less. When the OsPT4 gene was overexpressed, the amount of As in the roots and shoots went up by 40–66%. It also caused to raise the As levels in the straw and grain by 22–30%. As (V) sensitivity increased at high As (V) levels, knocking down the gene with CRISPR/Cas-9-reduced As accumulation in roots by 17–30%. As (V) either undergoes reducing to As (III) by arsenate reductases such as OsHAC1;1, OsHAC1;2, and OsHAC4 whenever it enters the cell (Shi et al. 2016; Xu et al. 2017) or is fed into the xylem by phosphate transporters (Mendoza-Cózatl et al. 2011; Wu et al. 2011). Furthermore, high-affinity phosphate transporter proteins involve As (V) absorption and transport in Oryza species. There are a few different types of aquaglyceroporins and a Nodulin 26-like Intrinsic Protein (NIP) of aquaporin channels that are responsible for arsenite (III) transport in plants. OsNIP2;1 (OsLsi1), a silicon transporter in rice roots, is responsible for the uptake of arsenite (III), while OsNIP2;1 (OsLsi2) is responsible for its efflux to the xylem (Li et al. 2009a, b; Ma et al. 2008; Mosa et al. 2012).

Both exodermal and endodermal cells have the Lsi1 and Lsi2 proteins in their plasma membranes (Ma et al. 2007; Ma and Yamaji 2006). Because of the bidirectional action of the Lsi1 protein channel, a portion of the As (III) taken up by root cells is released into the rhizosphere (Zhao et al. 2010). In contrast to the OsNIP2;1 and OsNIP2;2 transporters, several other NIPs, such as OsNIP1;1, OsNIP3;1, and OsNIP3;2, have been linked to As (III) efflux (Bienert et al. 2008; Ma et al. 2008; Sun et al. 2018). Aquaporin subfamilies such as plasma membrane intrinsic proteins (OsPIP2;4, OsPIP2;6, and OsPIP2;7), natural resistance-associated macrophage protein (OsNRAMP1), Small Basic Intrinsic Proteins (SIPs), Tonoplast Intrinsic Proteins (TIPs), and uncategorized intrinsic proteins (XIPs) are associated with the transport of As (III) (Maurel et al. 2015; Mosa et al. 2012; Tiwari et al. 2014).

The metabolism of As begins with the reduction of arsenate As (V) to As (III). Both enzymatic and nonenzymatic processes may have contributed to the decline. As (V) is transformed to As (III) nonenzymatically by oxidizing two reduced glutathione (GSH) molecules into oxidized GSH, which may immediately be recycled into two GSH molecules via GSH reductase. Furthermore, this process is too slow to effectively diminish As (V). During enzymatic activities, arsenate reductase transforms As (V) directly to As (III) (Foyer et al. 2011; Mukhopadhyay et al. 2000; Mukhopadhyay and Rosen 2002). Chao et al. (2014) found a High Arsenic Content 1 (HAC1) gene employing genome-wide association (GWAS) mapping which may transform As (V) to As (III) and enable the efflux of As (III) into the soil, limiting its accumulation in roots and transport to the shoot in Arabidopsis. Numerous arsenate reductases in Oryza, such as OsHAC1;1, OsHAC1;2, and OsHAC4, have been identified and functionally characterized (Shi et al. 2016; Xu et al. 2017; Fig. 21.1). The role of these proteins in arsenite efflux from the root to the soil is well recognized. When these genes overexpressed, enhanced arsenite efflux into the soil, and decreased As accumulation in paddy, their knockout reduced As (III) efflux and enhanced As accumulation in paddy shoots and grains.

As (III) levels rise; the plant metabolizes the residual As (III) by building complexes with sulfhydryl-rich molecules. GSH concentrations in plant cytosol often rise with exposure. It substantially influences As tolerance since it is a precursor of phytochelatin (PC) biosynthesis, in addition to being a limiting factor for high phytochelatin synthases (PCS) synthesis (Guo et al. 2008). Phytochelatin is produced from glutathione by phytochelatin synthases, and ABCC1/ABCC2 transporters subsequently transport it into vacuoles when it forms complexes with As (Kumar and Trivedi 2018). Two PCS genes in Oryza species, OsPCS1 and OsPCS2, contribute to catalyzing PC biosynthesis (Yamazaki et al. 2018; Fig. 21.1). Additionally, As (III) and GSH are conjugated to form the As (III)-GSH complex, which is catalyzed by glutathione S-transferases (GSTs) enzymes, with GSH acting as a substrate (Kumar and Trivedi 2018). Once As (III)-GSH and As (III)-PC complexes have been formed in the cytosol, vacuolar transporters bring the complexes into the vacuoles, where they are sequestered and detoxified (Verbruggen et al. 2009). Vacuolar transport of these nonprotein thiol-independent As (III) complexes is mediated by C-type ATP-binding cassette (ABC) transporters. OsABCC1, an ABCC1 orthologue from *A. thaliana*, is involved in As (III)-PC sequestration within rice vacuoles. The transporter gene (OsABCC1) was knocked out, resulting in significantly higher accumulation of As in paddy grains. The phloem is polluted with As due to a failure of nodal vasculature to properly compartmentalize the substance (Song et al. 2014). The regulation of GSH homeostasis is also a crucial part of the detoxification process for As (III). Glutathione is produced in the plastids and transported into the cytoplasm by chloroquine resistance transporters (CRT). OsCLT1, a gene encoding a CRT such as transporter 1 in rice, was mutated, lowering GSH levels in the cytoplasm, and thus PCs. The Osclt1 mutant accumulates less As in its roots and more or equally in its shoots when exposed to As because its PC2 levels are much lower than those of the wild species (Yang et al. 2016).

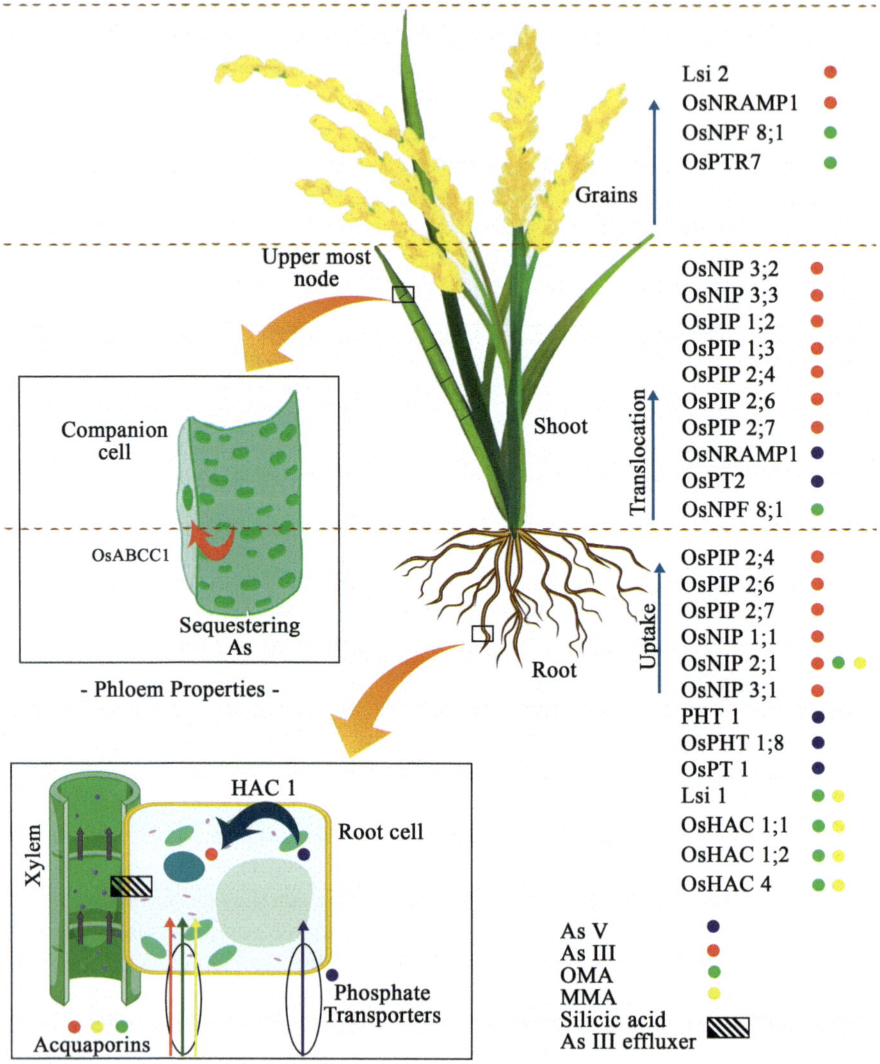

Fig. 21.1 Genes involved in As uptake and translocation in rice

7 Strategies to Mitigate As Bioaccumulation in Rice

Various methods have been investigated to mitigate As uptake by rice plants. However, Awasthi et al. (2017) identified three key approaches for discouraging As absorption by plants: (a) sustainable agronomic practices; (b) molecular changes to the As transporters involved in uptake, and (c) lowering As mobility by using various soil amendments. However, each method has its own restrictions, rewards, and drawbacks (Srivastava et al. 2018; Upadhyay et al. 2020). Thus, an integrated approach involving multiple techniques would yield effective results in preventing As accumulation in grains. The following discussion is primarily focused on various techniques that can be utilized to limit As bioaccumulation in grains in relation to the three primary strategies stated above.

7.1 Agronomic Practices to Mitigate As Accumulation

The primary objective of agronomic strategies is to reduce As absorption via application of various techniques (Abedi and Mojiri 2020). Water management and irrigation methods, nutrient management, application of soil amendments, electrokinetic remediation, exogenous application of hormones and chemicals, phytoremediation, and bioremediation are major agronomic practices that could be utilized (Srivastava et al. 2018, 2019; Huhmann et al. 2019; Seyfferth et al. 2019; Upadhyay et al. 2020).

7.1.1 Water Management and Irrigation Methods

Rice fields are frequently flooded during their vegetative phase in the traditional rice cultivation method (Srivastava et al. 2019) because rice is a water-dependent, semi-aquatic crop. However, under flooding circumstances, anaerobic conditions are being produced; As mobility is enhanced as a result of the reductive termination of Fe-(oxyhydro) oxides (Majumder and Bose 2018), which changes the redox states of the soil. This situation eventually promotes As (V) conversion into As (III), whereas As (III) concentration in the soil increases rather than As (V) encourages plant availability of As (III) which is more noxious compared to As (V) (Majumdar et al. 2019). As a result, water management is critical in rice cultivation. Furthermore, by managing water supplementation in rice cultivation, higher yields with minimal As accumulation in grains are guaranteed (Upadhyay et al. 2020).

To mitigate As bioaccumulation, numerous water management strategies have been investigated, including intermittent irrigation, sprinkler irrigation, and alternate wetting-drying (AWD). Periodical irrigation allows for minor As translocation from environment to plants (Mlangeni et al. 2020). Furthermore, during the dry period, sporadic irrigation ensures fewer lowering circumstances, which reduces As accumulation in plants (Majumdar et al. 2020). However, a prolonged dry phase reduces As accumulation in grains while decreasing final yield and productivity (Upadhyay et al. 2020). Avoiding ground water irrigation and allowing rain-fed cultivation is another effective way to reduce As bioaccumulation (Upadhyay et al. 2020). When compared to other irrigation techniques, this is a simple and effective process. Lower As was absorbed into the rice plant due to rain water irrigation (Sharma et al. 2014). Moreover, Majumdar et al. (2020) recommended that "rizi-pisciculture" (A way of cultivating rice coculture with fishes) modifies the As bioavailability in rice fields. Thus, As could be absorbed by fish and converted to less hazardous compounds such as "arsenobetaine" (Majumdar et al. 2020).

7.1.2 Nutrient Management

Fertilizer application is critical for maximum yield in rice cultivation. Farmers use a variety of fertilizers and mineral nutrients to increase yield. Several minerals, including S, Fe, P, and Si, have been shown to reduce As absorption and movement in rice plants (Mitra et al. 2017).

Iron (Fe) is a significant nutrient compound that reduces As absorption in rice (Nath et al. 2014). Thus, iron application encourages the development of Fe plaque around the roots, which lessens As absorption and precipitates both Fe and As (Bakhat et al. 2017). Fe plaque formation in the root zone of rice plant reduces As absorption and encourages As sequence in the root zone, reducing As uptake and translocation (Syu et al. 2014). Fe-oxide, which acts as a sink for As, has been used to grow rice, and the amount of As in the seeds has been observed to be abridged by 47% (Matsumoto et al. 2015; Mitra et al. 2017). Simultaneously, Fe application would reduce As-arbitrated oxidative stress in paddy plants, which is considered an added benefit (Nath et al. 2014).

Phosphorous (P), in addition to Fe, is essential for the uptake of As by plants as well as the solubility of As in soil. Evidence from earlier studies on the incorporation of P into soil shows that decreasing the amount of As in Fe plaque causes the concentration of As in the soil to upsurge, which improves the bioavailability of As (Smith et al. 2002; Bogdan and Schenk 2009). However, phosphorous competes with the same As transporter and reduces As uptake once a P concentration reaches a threshold level (Pigna et al. 2010; Rahaman et al. 2011). Thus, the amount of As accumulated in paddy plants relies on the As:P ratio in the soil.

Silicon (Si) is also regarded as a micronutrient for rice plants (Tavakkoli et al. 2011). Si is available to rice plants as mono silicic acid, and soil physicochemical properties influence the Si bioavailability. The transporter gene "NIPs," which is thought to be the same transporter for As, is involved in Si absorption and transportation; therefore, silicon application would reduce As absorption by paddy plants (Bogdan and Schenk 2008). According to early studies, applying Si to shoots reduced As uptake and accumulation while improving antioxidant defense responses (Li et al. 2009a, b; Fleck et al. 2013; Tripathi et al. 2013).

A soil application of sulphur (S) also reduced the absorption and translocation of As in rice plants (Zhang et al. 2011; Dixit et al. 2015). By promoting the development of Fe plaques, increasing the desorption of As (V) from Fe plaques, and preventing As movement into cells, sulphur application would limit As absorption and translocation (Mitra et al. 2017). Beside these pathways, sulphur metabolism is used in As detoxification to safeguard plants from As in the soil (Finnegan and Chen 2012). The detoxification process begins at the cellular level with the production of low molecular mass sulfur-rich compounds such as glutathione and phytochelatin, which subsequently interact with As (III) and are deposited in vacuoles (Song et al. 2010). However, it has been noted that applying a higher concentration of sulphur hindered the movement of As from roots to shoot (Dixit et al. 2016). Additionally, the use of sulphate fertilizers when growing paddy increases the chemistry of As in reducing environments, promoting the accumulation of insoluble As-sulphide (Signes-Pastor et al. 2007). Moreover, it was noted that the development of Fe and

Mn plaques in the soil by the application of elemental sulfur and gypsum highly abridged the rice seeds As level by further 39.1% (Zhang et al. 2021a, b). Furthermore, the As in rice roots decreased when nitrogen fertilizer was applied (Srivastava et al. 2019).

7.1.3 Reduce the Mobility of As Through Adding Amendments

7.1.3.1 Application of Biochar

Several eco-friendly immobilizing agents for reducing As mobility in rice plants have recently been investigated (Chen et al. 2016). Biochar application has produced the best results, particularly in terms of reducing As mobility in soil. Biochar is made through a gradual pyrolysis process that uses organic materials such as wood chips, straw, paddy husk, and so on. Previous research findings showed that applying biochar drastically altered the physicochemical properties of soil, while immobilizing heavy metals such as As (Kamali et al. 2022). Interestingly, the material utilized to prepare the biochar indicated that it had an impact on the movement of As in soil. Since biochar has a larger capacity for adsorption, it retains As on its surfaces (Khan et al. 2013). However, the majority of research have demonstrated that the final outcome of applying biochar to mitigate As depends on the amount applied, the temperature at which biochar is prepared by pyrolysis, and the surrounding environment (Amen et al. 2020). For an instance, applying more biochar at a higher dosage increased the As mobility in potted paddy soil (Beesly et al. 2013). Additionally, sorption on biochar is limited because of its limited surface area and high negative charge (Amen et al. 2020).

Zerovalent iron (ZVI), which adsorbs relatively high As ions in the soil, significantly decreased the As absorption by rice plants when soil amendments containing Fe were applied (Qiao et al. 2018; Mlangeni et al. 2020). In general, Fe is effective at removing As; consequently, Fe-modified biochar will assimilate more As than biochar alone (Yang et al. 2022). The application of iron-modified biochar in As-contaminated paddy soils lessened the As concentration in rice seeds by altering the soil As fraction (Wu et al. 2022). Increased surface area and increased Fe functional groups on biochar surfaces as a result of surface alterations would effectively remove As from polluted soil (Premarathna et al. 2019). Moreover, use of hematite, manganese salts, and KOH-mediated modified biochar created positive results (Cope et al. 2014; Hu et al. 2015). Concurrently, it was discovered that iron-impregnated biochar reduced the bioaccumulation of As in rice (Yin et al. 2017). Furthermore, soil treatment with ferromanganese oxide-modified biochar (FMBC) reduced As accumulation in paddy seeds (Lin et al. 2017). Similarly, bismuth-coated biochar inhibited As accumulation in paddy soils under anaerobic environments (Zhu et al. 2019). Recent research on As remediation in paddy soil identified the use of goethite-modified biochar (GMBC) to reduce the movement and bioavailability of As in paddy soils (Zhu et al. 2020; Irshad et al. 2022a, b; Table 21.1). Herath et al. (2020) elaborated that rice husk biochar modified with Si fertilizer (Si-RHBC) decreased the quantity of As (III) in rhizosphere.

7.1.3.2 Nano-Based Approaches

With the recent advances in science, nanotechnological methods for lowering the buildup of trace metals in rice plant have become widespread worldwide. In particular, metal oxides were used initially, but owing to poor adsorption capacities, nano-metal oxides have been used instead (Mawia et al. 2021). Iron (hydro) oxides, alumina, titanium dioxide, copper oxide, and zinc oxide nanoparticles evaluated to eliminate trace metal(loid)s from polluted areas (Lata and Samadder 2016). For instance, Wang et al. (2018b) and Ma et al. (2020) reported that ZnO-nanoparticle-treated hydroponically grown rice plants reduced As (III) and As (V) in the roots (Table 21.1). Furthermore, nanostructured α-MnO_2 (Li et al. 2019), SiO_2 (Liu et al. 2014), and TiO_2 (Wu et al. 2020) reduced the availability of As in rice plants

Table 21.1 Different soil amendments used to mitigate As bioaccumulation in paddy

Amendment type	Decreasing level of As uptake	Remarks	References
Goethite-modified biochar (GMBC)	Roots sequestered the As by 174%	Increased the functions of soil peroxidase (POD) and catalase (CAT)	Irshad et al. (2022a, b)
Ferromanganese oxide-modified biochar (FMBC)	68.9–78.3% in rice grains	Enhanced the Fe and Mn plaque content	Lin et al. (2017)
Rice husk biochar modified with Si (Si-RHBC)	76% reduced in rhizosphere and 16% reduction in paddy straw	Oxidation of As (III) adsorbed on Si-based ferrihydrite to As (V) by aioA gene activation	Herath et al. (2020)
Rice husk biochar modified with nano-montmorillonite clay (NMRHBC)	Reduced As (III) by 73% in root zone of the rice	As concentration of grains not changed significantly	Herath et al. (2020)
ZnO nanoparticles (ZnO-NPs)	Total As in rice roots reduced significantly from the rhizosphere	ZnO-NPs applied with 100 mg/liter concentration	Wang et al. (2018b)
	Decreased total As accumulation in rice roots by 39.5%		Ma et al. (2020)
CuO nanoparticles (CuO-NPs)	Total As concentration reduced in the rice plants	CuO-NPs applied with 100 mg/liter concentration	Wang et al. (2019)
α-MnO_2 nanoparticles (α-MnO_2-NPs)	As accumulation in total plant including grains has been reduced	Application rate was 0.2–2.0 w/w	Li et al. (2019)
SiO_2 nanoparticles (SiO_2-NPs)	Declined the total As in the grains by 28% and 40% for pot	The oxidative destruction was reduced in roots and simultaneously antioxidant concentration augmented	Liu et al. (2014)
TiO_2 nanoparticles (TiO_2-NPs)	Reduced As bioaccumulation in rice seedlings by 40–90%	As adsorption to TiO_2 particles lower the As concentration in the media	Wu et al. (2020)

(Table 21.1). By modifying the microbial activities in soil, the addition of nano zerovalent iron-modified biochar (nZVI-BC) effectively reduced As pollution in mine soil (Fan et al. 2020).

7.1.3.3 Application of Organic Matter

Incorporation of soil organic matter increases microbial development and promotes the closure of Fe-oxyhydroxides, which enhance the As mobility (Ma et al. 2018). Similarly, it has been suggested that the chemical conformation of soil organic matter may affect the mobility of As in soil (Suriyagoda et al. 2018). A number of studies have demonstrated and identified negative relationship between organic matter and paddy seed As accumulation (Rahaman et al. 2011; Kar et al. 2013). Due to the formation of an organo-As complexes, organic matter has strong sorption affinity for As (Paikaray et al. 2005). However, dissolved organic carbon produced by decomposing organic matter would reduce As bioavailability in root zone due to As leaching deeper into the soil (Farooq et al. 2010).

Due to their potential for reuse, numerous organic and inorganic waste constituents have been identified as an excellent soil adjustments (Arco-Lázaro et al. 2016). As remediation techniques, organic amendments are considered vital and environmentally friendly (Beesley et al. 2014). However, contradictory impacts of organic compounds on As dynamics have been described, and this probably depends on the chemical composition of the soil (Niazi et al. 2011; Arco-Lázaro et al. 2016). Rendering to Das et al. (2008), the application of farmyard manure in paddy cultivation significantly decreased the As content in soil. Contradictorily, Jia et al. (2013) found that incorporation of biogas slurry into soil significantly improved the As bioaccumulation in paddy plants.

7.1.3.4 Electrokinetic Remediation

Electrokinetic remediation (EKR) is regarded as the novel technology for the removal of As from fine-grained soils due to its fundamental functioning principles, electromigration, and electroosmotic flow (Zhang et al. 2022). EKR is primarily dependent on soil particle size, redox potential, and As chemical speciation (Xu et al. 2021). By applying inorganic acids or reductive chemicals, As dissolution and release into the soil solution would be accelerated (Lee et al. 2016). Dithionite and thiosulfate are effective reductive agents, but dithionite is unstable in acidic environments (Kim and Baek 2015). Different concentrations of biogas slurry and sucrose have been used effectively to enhance As migration ability, resulting in a significant level of As removal from paddy soil (Zhang et al. 2022). However, this technology is still in the research phase, and field trials must be carried out before it can be used in open field cultivation.

7.1.3.5 Exogenous Application of Chemicals and Hormones

In addition, numerous soil amendments, foliar application of inorganic substances, and hormones may reduce As bioaccumulation in plants by regulating the physiological processes of plants (Srivastava et al. 2020; Mawia et al. 2021). Co-usage of auxin and selenium to As-exposed rice seedlings, for example, increased chlorophyll production and lipid synthesis (Pandey and Gupta 2015). This might be related to Se and auxin's synergistic actions on the γ-Aminobutyric acid (GABA) bypass path in plants (Ramesh et al. 2015). Exogenous NO application has been shown to protect plants from As by decreasing As movement from root to shoot and modulating As-involved regulatory networks and detoxification (Singh et al. 2016a; Singh et al. 2017). Si foliar spray increases plants' growth and protects against trace metal stresses such as As (Chen et al. 2018). Foliar application of Si, for example, abridged total As amount in As (V) and As (III)-exposed rice plants up to 67% and 78% respectively (Dwivedi et al. 2020). Simultaneously, Zhang et al. (2020) discovered that foliar Si application throughout the tillering phase considerably condensed As content of rice husk and grain. Furthermore, Syu et al. (2020) found that supplementing As-contaminated soils with CaO_2, increased grain yield, while decreasing As accretion in brown rice by 25–45% in flooded paddy soil. Thus, As bioaccumulation can be reduced with chemicals and hormones.

7.1.3.6 Phytoremediation

Phytoremediation is commonly used to accumulate harmful compounds from contaminated soil and aqueous media. It is a promising method for removing As from polluted paddy soil without negatively impacting soil quality (Mecwan Neha et al. 2018). Plants use a variety of mechanisms in phytoremediation, including phytostabilization, phytoextraction, phytoporation, and phytodegradation (Bhattacharya et al. 2021). Phytoextraction, in general, allows As to be absorbed and stored in plant shoots and leaves, whereas phytostabilization causes As to be immobilized in soil by plant roots (Liu et al. 2018). Because of its low cost, ability to be used in situ, lack of secondary pollution, and preservation of soil quality, it is the most efficient and commonly used remediation method for recovering As. However, the extended time required is unavoidable (Ma et al. 2018). Due to the significance of rice consumption in the human diet, phytoremediation of As in paddy fields will favor human health.

The use of hyperaccumulators in contaminated fields may gradually decline the amount of As in rice plant tissues. Hyperaccumulators have a number of desirable characteristics, including a rapid growth rate, a high biomass content, and a good potential for As accumulation. Furthermore, bioaccumulation and translocation factors should be greater than one for good hyperaccumulators (Srivastava et al. 2021). Because it has been difficult to find plants with all of the desired characteristics, a microbial association has been used to enhance biomass and plant growth, resulting in effective As accumulation (Mesa et al. 2017). Hyperaccumulators also have effective vacuole sequestration, quick translocation to shoots, and rapid root uptake. *Pteris vittate* has been used for As phytoremediation for the past 20 years because it concurs with the aforementioned statement. It has also been considered as a model plant to study how As is removed through photosynthesis (Han et al. 2017).

This study examines some previous studies on phytoremediation, which is used to lessen As accumulation in rice plants. Chinese brake fern (*Pteris vittala* L.), *Pityrogramma calomelanos*, and *Mimosa pudica* are a few of the recognized hyper-accumulators (Mecwan Neha et al. 2018). According to Mandal et al. (2012) study, Chinese brake fern has the capability for the As phytoextraction, which results in a 52% reduction in the amount of As in rice grains. *Pteris vittata* is the most researched and favored As-hyperaccumulating plant because of its unique characteristics, including its vigorous root system, increased biomass content, and fast growth rate. A study by Praveen et al. (2017) found that when rice was planted beside three accumulators including *Phragmites australis*, *Vetiveria zizanioides*, and *Pteris vitatta*, the total As concentration in the rice shoot, root, husk, and grain decreased. The As content in grains was found to be within the WHO-recommended range (1 mg kg^{-1}).

Additional studies have suggested remediation by intercropping with accumulators. In the intercropping of rice plant with aquatic vegetables, water spinach (*Ipomoea aquatica* Forsk) outperformed water celery (*Oenanthe javanica*) and Guangdong white arrowhead (*Sagittaria sagittifolia* L.) (Huang et al. 2021). Furthermore, Indian mustard and sunflower have the potential to be hyperaccumulators due to their large biomass concentration (Mecwan Neha et al. 2018). Phosphorous is the crucial element that encourages the buildup of As in contaminated soil. The phosphorus application is reported by Fu et al. (2017), and it may cause the plant *Pteris* to remove As.

Interactions between plants and microbes may hasten phytoremediation. The use of As-tolerant bacterial strains such as *Staphylococcus arlettae* (NBRIEAG-6), *Staphylococcus* sp. (NBRIEAG-8), and *Brevibacillus* sp. (NBRIEAG-9) in paddy fields on rice plant growth promotion and As uptake was studied by Singh et al. (2016). They found that there was a 30–40% reduction in As uptake compared to the non-inoculated plant.

7.1.3.7 Bioremediation

It is crucial to take the essential actions to get rid of or limit the As accumulation in rice in order to assure food safety. Reduced As uptake, translocation, or sequestration by biological agents throughout the plant is made possible via bioremediation (Panthri and Guptha 2019).

As was previously mentioned, soil microorganisms are crucial for As transport through paddy soil and rice plant absorption (Table 21.2). They assist the processes of sorption, precipitation, methylation, mineralization, immobilization, redox, and volatilization (Panthri and Guptha 2019). Several soil microorganisms including *Bacillus*, *Rhodococcus*, and *Halobacterium* species have been shown to sorb As via

Table 21.2 Different microorganisms tested in As bioremediation in rice plants

Name of the Microorganisms	Type	Effects on As accumulation in rice plants	References
Piriformospora indica	Agaricomycetes rhizospheric fungus	Increased As immobilization in roots, translocation of iron to shoots, and increase the As stress tolerance	Ghorbani et al. (2021)
D. hansenii	Yeast strain	Under As stress modulates APX and SOD activity to scavenge reactive oxygen species and lower the grain As content by 40% Enhances the development and nutrient condition	Kaur et al. (2020)
Kocuria flava AB402 *Bacillus vietnamensis* AB403	Rhizospheric microorganism	Reduce As absorption, and accumulation, in rice grains and increase seedling growth under As stress	Mallick et al. (2018)
Iron-reducing bacteria (FeRB), sulphate reducing bacteria (SRB)	Rhizospheric microorganism	Control the amount of As in the rhizosphere	Dai et al. (2020)
Arbuscular mycorrhizal fungi (AMF)	Soil fungi	Inhibits the uptake and accumulation of AS (III) transporter by reducing its expression	Chen et al. (2013)
Chlorella vulgaris and *Nannochloropsis* sp.	Microalgae	Enhance As accumulation in root and shoots	Upadhyay et al. (2016)
Anabaena azotica	Cyanobacteria	The movement and accumulation of several As species reduce by 43–70% in rice	Wang et al. (2018a, b)
Leonardite + *Bacillus pumilus*, *Pseudomonas* sp., *Bacillus thuringiensis*	Endophytic bacteria	Reduce As content and accumulation of grain As	Dolphen and Thiravetyan (2019)

their cell surface, which is a temperature and pH-dependent process (Mitra et al. 2017). *Bacillus subtilis* contributes to As detoxification by forming Fe hydroxides on its cell surface via the inner-sphere complex formation. Recent research has discovered new rhizospheric bacterial strains that act as plant growth inducers and As remediators. The bacterial strains EA 106, BBAU/MMM1 TWD-2 and D54 prevent As from building up in roots and lower its concentration and also transport and accumulation in aerial plant components (Hare et al. 2017).

Through bio-oxidation processes, *Leptospirillum ferrooxidans*, *Thiobacillus caldus*, *A. ferrooxidans*, and *Thiobacillus ferrooxidans* oxidize Cu_3AsS_4, FeAsS, and As_4S_4 (Bakhat et al. 2017). It is more effective than abiotic oxidation and is known as bioleaching. Microbes convert ferrous to ferric ions, and the following chemical oxidation of sulfides by Fe results in As-containing sulfide minerals. Arsenic solubilization in soil solution does, however, increase as a result of reductive dissolution (Bakhat et al. 2017).

Along with the aforementioned bacteria, the AMF, which live in symbiosis with rice plant roots, also perform a protective role by sending out a signal that inhibits the expression of the *OsLsi1* and *OsLsi2* genes, which are thought to be the mediating genes for As (III) transport. The AMF assists the rice plant in maintaining its natural biomass and yield even when exposed to As without accumulating As in the grains (Mitra et al. 2017).

Owing to their ability, introducing and encouraging the growth of these microorganisms into As-contaminated soil could help to reduce As stress on plants via the As methylation and volatilization process. It has been demonstrated that *Saccharomyces cerevisiae* that has been genetically altered with the soil fungus (*Westerdykella aurantiaca*) gene WaarsM exhibits significant As volatilization and methylation (Khan et al. 2022). One of the most innovative agricultural methods is thought to be the use of rhizospheric bacteria to reduce As stress in rice.

Planting rice with As accumulating plants such as *Pteris Vittata* has also been shown to be a successful method. However, there is a possibility that the accumulated As will enter the food chain because the plant does not use it for metabolism (Khan et al. 2022).

7.2 Transgenic Rice Production for the As Tolerance

Transgenic techniques have increased As tolerance in rice plants. The upregulation of specific genes, such as the phytochelatin synthase gene (PCS), which produces chelators to detoxify As accumulated in plants. The PCS gene (CdPCS1) isolated from *Ceratophyllum demersum* was introduced into rice plants to reduce As accumulation in shoots and roots (Shri et al. 2015). The arsenite-thiol complex has been shown to scavenge stabilized As forms in plant parts other than grain. The effective efflux of As (III) in rice grains was investigated in the ACR3 gene isolated from yeast cells (Roychowdhury et al. 2018). The *ArsM* gene isolated from *Westerdykella aurantiaca* soil fungus is used to create transgenic rice capable of producing monomethylarsonic acid MMA (V) and DMA (V) that can bind inorganic As via As (III)-S-adenosyl methionine methyl transferase (Panthri and Guptha 2019). Due to glutaredoxin's ability to decrease As (V) and control As (III) levels in cells, two rice glutaredoxins (OsGrx C7 and OsGrxC2.1) were crucial in As-stress response, lowering As (III) accumulation. To exploit the genes that produce low As rice, novel gene editing tools such as CRISPR/Cas9 can target the genes that are responsible for As uptake and transport within the plant tissues (Chen et al. 2017).

7.3 Breeding Rice for As Tolerance

Rice cultivars differ in their potential for accumulating As in grains and other plant parts. Several studies have shown that rice cultivars perform differently in As accumulation when grown in similar soil conditions due to genetic variability (Khanam and Nayak 2021). More oxygen-releasing rice types may be able to keep the redox potential high in the root rhizosphere and increase the amount of Fe complexes on root surface while decreasing As accumulation. It has been discovered that rice genotypes with red bran in the seed accumulate more than genotypes with brown bran in the seed (Shakoor et al. 2019). The investigations show that japonica rice varieties with less As in the grains have a strong potential to contribute to the breeding program for As-tolerant rice (Norton et al. 2008). Additionally, the World Rice Core Collection reported the existence of a rice cultivar with increased grain DMAV content (Panthri and Guptha 2019). A cross between low As rice accessions exhibits low As accumulation among 1/4th of the second-generation plants which has demonstrated great potential for As tolerance rice breeding (Farooq et al. 2016). To increase As tolerance in rice by lowering As accumulation in rice grains, it is crucial to adopt low As variants in the current rice germplasm.

8 Conclusion and Future Works

As biomagnification has caused severe health issues, as well as prominent negative effect on plant development. Since rice has been identified as a primary dietary food source of As, scientists have investigated a variety of mechanisms to mitigate As absorption in paddy plants. Thus, to diminish As accumulation in rice, modern agronomic, bioremediation, biotechnological, and nanotechnological approaches are used globally. However, molecular and physiological mechanisms relating to As absorption, translocation, and accumulation in *Oryza sativa* must still be elucidated in order to understand the whole picture. As a result, it is critical to use molecular, physiological, multi-omics, and bioinformatics approaches to understand As absorption, transport, and resistance in rice. Moreover, most of the studies were conducted in controlled laboratory conditions, which may not be representative of field conditions. Therefore, more extensive field-level experiments are needed to obtain reliable results. Additionally, investigations regarding As behavior under combined stress situations are required for investigation. Arsenic accumulation in seeds will ultimately diminish grain quality due to nutritional imbalance, and research into this issue should be expanded.

References

Abedi T, Mojiri A (2020) Arsenic uptake and accumulation mechanisms in rice species. Plan Theory 9(2):129

Abedin MDJ, Cresser MS, Meharg AA, Feldmann J, Cotter-Howells J (2002) Arsenic accumulation and metabolism in rice (*Oryza sativa* L.). Environ Sci Technol 36(5):962–968

Afzal B, Hussain I, Farooqi A (2018) Arsenic in paddy soils and potential health risk. In: Hashmi MZ, Varma A (eds) Environmental pollution of paddy soils. Springer, Cham, pp 151–163

Ali I (2018) Microwave assisted economic synthesis of multi walled carbon nanotubes for arsenic species removal in water: batch and column operations. J Mol Liq 271:677–685

Amen R, Bashir H, Bibi I, Shaheen SM, Niazi NK, Shahid M, Hussain MM, Antoniadis V, Shakoor MB, Al-Solaimani SG, Wang H, Bundschuh J, Rinklebe J (2020) A critical review on arsenic removal from water using biochar-based sorbents: the significance of modification and redox reactions. Chem Eng J 396:125195

Arco-Lázaro E, Agudo I, Clemente R, Bernal MP (2016) Arsenic(V) adsorption-desorption in agricultural and mine soils: effects of organic matter addition and phosphate competition. Environ Pollut 216:71–79

Awasthi S, Chauhan R, Srivastava S, Tripathi RD (2017) The journey of arsenic from soil to grain in rice. Front Plant Sci 8:1007

Ayangbenro A, Babalola O (2017) A new strategy for heavy metal polluted environments: a review of microbial biosorbents. Int J Environ Res Public Health 14(1):94

Azeh Engwa G, Udoka Ferdinand P, Nweke Nwalo FN, Unachukwu M (2019) Mechanism and health effects of heavy metal toxicity in humans. In: Karcioglu O, Arslan B (eds) Poisoning in the modern world – new tricks for an old dog? IntechOpen. https://doi.org/10.5772/intechopen.82511

Bakhat HF, Zia Z, Fahad S, Abbas S, Hammad HM, Shahzad AN, Abbas F, Alharby H, Shahid M (2017) Arsenic uptake, accumulation and toxicity in rice plants: possible remedies for its detoxification: a review. Environ Sci Pollut Res 24(10):9142–9158

Bastías JM, Beldarrain T (2016) Arsenic translocation in rice cultivation and its implication for human health. Chilean J Agric Res 76(1):114–122

Beesley L, Marmiroli M, Pagano L, Pigoni V, Fellet G, Fresno T, Vamerali T, Bandiera M, Marmiroli N (2013) Biochar addition to an arsenic contaminated soil increases arsenic concentrations in the pore water but reduces uptake to tomato plants (*Solanum lycopersicum* L.). Sci Total Environ 454–455:598–603

Beesley L, Inneh OS, Norton GJ, Moreno-Jimenez E, Pardo T, Clemente R, Dawson JJC (2014) Assessing the influence of compost and biochar amendments on the mobility and toxicity of metals and arsenic in a naturally contaminated mine soil. Environ Pollut 186:195–202

Begum M, Mondal S (2019) Relative toxicity of arsenite and arsenate on early seedling growth and photosynthetic pigments of rice. Curr J Appl Sci Technol:1–5. https://doi.org/10.9734/cjast/2019/v33i430087

Bhattacharya S, Sharma P, Mitra S, Mallick I, Ghosh A (2021) Arsenic uptake and bioaccumulation in plants: a review on remediation and socio-economic perspective in Southeast Asia. Environ Nanotechnol Monit Manag 15:100430

Bienert GP, Thorsen M, Schüssler MD, Nilsson HR, Wagner A, Tamás MJ, Jahn TP (2008) A subgroup of plant aquaporins facilitate the bi-directional diffusion of $As(OH)_3$ and $Sb(OH)_3$ across membranes. BMC Biol 6(1):26

Biswas R, Ghosh P, Banerjee N, Das J, Sau T, Banerjee A, Roy S, Ganguly S, Chatterjee M, Mukherjee A, Giri A (2008) Analysis of T-cell proliferation and cytokine secretion in the individuals exposed to arsenic. Hum Exp Toxicol 27(5):381–386

Biswas A, Biswas S, Das A, Roychowdhury T (2018) Spatial variability and competing dynamics of arsenic, selenium, iron and bioavailable phosphate from ground water and soil to paddy plant parts. Groundw Sustain Dev 7:328–335

Bogdan K, Schenk MK (2008) Arsenic in rice (*Oryza sativa* L.) related to dynamics of arsenic and silicic acid in paddy soils. Environ Sci Technol 42(21):7885–7890

Bogdan K, Schenk MK (2009) Evaluation of soil characteristics potentially affecting arsenic concentration in paddy rice (*Oryza sativa* L.). Environ Pollut 157(10):2617–2621

Bulka CM, Jones RM, Turyk ME, Stayner LT, Argos M (2016) Arsenic in drinking water and prostate cancer in Illinois counties: An ecologic study. Environ Res 148:450–456

Bundschuh J, Litter MI, Parvez F, Román-Ross G, Nicolli HB, Jean J-S, Liu C-W, López D, Armienta MA, Guilherme LRG, Cuevas AG, Cornejo L, Cumbal L, Toujaguez R (2012) One century of arsenic exposure in Latin America: a review of history and occurrence from 14 countries. Sci Total Environ 429:2–35

Cantamessa S, Massa N, Gamalero E, Berta G (2020) Phytoremediation of a highly arsenic polluted site, using *Pteris vittata* L. and *Arbuscular mycorrhizal* fungi. Plan Theory 9(9):1211

Cao Y, Sun D, Ai H, Mei H, Liu X, Sun S, Xu G, Liu Y, Chen Y, Ma LQ (2017) Knocking out *OsPT4* gene decreases arsenate uptake by rice plants and inorganic arsenic accumulation in rice grains. Environ Sci Technol 51(21):12131–12138

Carbonell AA, Aarabi MA, DeLaune RD, Gambrell RP, Patrick WH (1998) Bioavailability and uptake of arsenic by wetland vegetation: effects on plant growth and nutrition. J Environ Sci Health A 33(1):45–66

Chakraborti D, Rahman MM, Ahamed S, Dutta RN, Pati S, Mukherjee SC (2016) Arsenic groundwater contamination and its health effects in Patna district (capital of Bihar) in the middle Ganga plain, India. Chemosphere 152:520–529

Chakraborti D, Singh S, Rahman M, Dutta R, Mukherjee S, Pati S, Kar P (2018) Groundwater arsenic contamination in the Ganga river basin: a future health danger. Int J Environ Res Public Health 15(2):180

Chao DY, Chen Y, Chen J, Shi S, Chen Z, Wang C, Danku JM, Zhao F-J, Salt DE (2014) Genome-wide association mapping identifies a new arsenate reductase enzyme critical for limiting arsenic accumulation in plants. PLoS Biol 12(12):e1002009

Chatterjee D, Halder D, Majumder S, Biswas A, Nath B, Bhattacharya P, Bhowmick S, Mukherjee-Goswami A, Saha D, Hazra R, Maity PB, Chatterjee D, Mukherjee A, Bundschuh J (2010) Assessment of arsenic exposure from groundwater and rice in Bengal Delta Region, West Bengal, India. Water Res 44(19):5803–5812

Chen Y, Xu W, Shen H, Yan H, Xu W, He Z, Ma M (2013) Engineering arsenic tolerance and hyperaccumulation in plants for phytoremediation by a PvACR3 transgenic approach. Environ Sci Technol 47(16):9355–9362

Chen Z, Wang Y, Xia D, Jiang X, Fu D, Shen L, Wang H, Li QB (2016) Enhanced bioreduction of iron and arsenic in sediment by biochar amendment influencing microbial community composition and dissolved organic matter content and composition. J Hazard Mater 311:20–29

Chen Y, Han YH, Cao Y, Zhu YG, Rathinasabapathi B, Ma LQ (2017) Arsenic transport in rice and biological solutions to reduce arsenic risk from rice. Front Plant Sci 8:268

Chen R, Zhang C, Zhao Y, Huang Y, Liu Z (2018) Foliar application with nano-silicon reduced cadmium accumulation in grains by inhibiting cadmium translocation in rice plants. Environ Sci Pollut Res 25(3):2361–2368

Chen C, Li L, Huang K, Zhang J, Xie WY, Lu Y, Dong X, Zhao FJ (2019) Sulfate-reducing bacteria and methanogens are involved in arsenic methylation and demethylation in paddy soils. ISME J 13(10):2523–2535

Chervona Y, Arita A, Costa M (2012) Carcinogenic metals and the epigenome: understanding the effect of nickel, arsenic, and chromium. Metallomics 4(7):619

Choudhury B, Chowdhury S, Biswas AK (2011) Regulation of growth and metabolism in rice (*Oryza sativa* L.) by arsenic and its possible reversal by phosphate. J Plant Interact 6(1):15–24

Chung JY, Yu SD, Hong YS (2014) Environmental source of arsenic exposure. J Prev Med Public Health 47(5):253–257

Cope CO, Webster DS, Sabatini DA (2014) Arsenate adsorption onto iron oxide amended rice husk char. Sci Total Environ 488–489:554–561

Dai J, Tang Z, Jiang N, Kopittke PM, Zhao F-J, Wang P (2020) Increased arsenic mobilization in the rice rhizosphere is mediated by iron-reducing bacteria. Environ Pollut 263:114561
Dangleben NL, Skibola CF, Smith MT (2013) Arsenic immunotoxicity: a review. Environ Health 12(1):73
Das DK, Sur P, Das K (2008) Mobilisation of arsenic in soils and in rice (*Oryza sativa* L.) plants affected by organic matter and zinc application in irrigation water contaminated with arsenic. Plant Soil Environ 54(1):30–37
Das N, Paul S, Chatterjee D, Banerjee N, Majumder NS, Sarma N, Sau TJ, Basu S, Banerjee S, Majumder P, Bandyopadhyay AK, States JC, Giri AK (2012) Arsenic exposure through drinking water increases the risk of liver and cardiovascular diseases in the population of West Bengal, India. BMC Public Health 12(1):639
Das I, Ghosh K, Das DK, Sanyal SK (2013) Assessment of arsenic toxicity in rice plants in areas of West Bengal. Chem Speciat Bioavailab 25(3):201–208
Dave R, Singh PK, Tripathi P, Shri M, Dixit G, Dwivedi S, Chakrabarty D, Trivedi PK, Sharma YK, Dhankher OP, Corpas FJ, Barroso JB, Tripathi RD (2013) Arsenite tolerance is related to proportional thiolic metabolite synthesis in rice (*Oryza sativa* L.). Arch Environ Contam Toxicol 64(2):235–242
DiTusa SF, Fontenot EB, Wallace RW, Silvers MA, Steele TN, Elnagar AH, Dearman KM, Smith AP (2015) A member of the phosphate transporter 1 (Pht1) family from the arsenic-hyperaccumulating fern *Pteris vittata* is a high-affinity arsenate transporter. New Phytol 11. https://doi.org/10.1111/nph.13472
Dixit G, Singh AP, Kumar A, Singh PK, Kumar S, Dwivedi S, Trivedi PK, Pandey V, Norton GJ, Dhankher OP, Tripathi RD (2015) Sulfur mediated reduction of arsenic toxicity involves efficient thiol metabolism and the antioxidant defense system in rice. J Hazard Mater 298:241–251
Dixit G, Singh AP, Kumar A, Mishra S, Dwivedi S, Kumar S, Trivedi PK, Pandey V, Tripathi RD (2016) Reduced arsenic accumulation in rice (*Oryza sativa* L.) shoot involves sulfur mediated improved thiol metabolism, antioxidant system and altered arsenic transporters. Plant Physiol Biochem 99:86–96
Dolphen R, Thiravetyan P (2019) Reducing arsenic in rice grains by leonardite and arsenic–resistant endophytic bacteria. Chemosphere 223:448–454
Dwivedi S, Kumar A, Mishra S, Sharma P, Sinam G, Bahadur L, Goyal V, Jain N, Tripathi RD (2020) Orthosilicic acid (OSA) reduced grain arsenic accumulation and enhanced yield by modulating the level of trace element, antioxidants, and thiols in rice. Environ Sci Pollut Res 27(19):24025–24038
Eckstein M, Eleazer R, Rea M, Fondufe-Mittendorf Y (2017) Epigenomic reprogramming in inorganic arsenic-mediated gene expression patterns during carcinogenesis. Rev Environ Health 32(1–2):93–103
Fahad S, Hussain S, Saud S, Hassan S, Chauhan BS, Khan F, Ihsan MZ, Ullah A, Wu C, Bajwa AA, Alharby H, Amanullah, Nasim W, Shahzad B, Tanveer M, Huang J (2016) Responses of rapid viscoanalyzer profile and other rice grain qualities to exogenously applied plant growth regulators under high day and high night temperatures. PLoS One 11(7):e0159590
Fan J, Chen X, Xu Z, Xu X, Zhao L, Qiu H, Cao X (2020) One-pot synthesis of nZVI-embedded biochar for remediation of two mining arsenic-contaminated soils: arsenic immobilization associated with iron transformation. J Hazard Mater 398:122901
Farooq SH, Chandrasekharam D, Berner Z, Norra S, Stüben D (2010) Influence of traditional agricultural practices on mobilization of arsenic from sediments to groundwater in Bengal delta. Water Res 44(19):5575–5588
Farooq MA, Islam F, Ali B, Najeeb U, Mao B, Gill RA, Yan G, Siddique KHM, Zhou W (2016) Arsenic toxicity in plants: cellular and molecular mechanisms of its transport and metabolism. Environ Exp Bot 132:42–52
Fernández-Baca CP, McClung AM, Edwards JD, Codling EE, Reddy VR, Barnaby JY (2021) Grain inorganic arsenic content in rice managed through targeted introgressions and irrigation management. Front Plant Sci 11:612054
Finnegan PM, Chen W (2012) Arsenic toxicity: the effects on plant metabolism. Front Physiol 3:182

Fleck AT, Mattusch J, Schenk MK (2013) Silicon decreases the arsenic level in rice grain by limiting arsenite transport. J Plant Nutr Soil Sci 176(5):785–794
Fontenot EB, Ditusa SF, Kato N, Olivier DM, Dale R, Lin W-Y, Chiou TJ, Macnaughtan MA, Smith AP (2015) Increased phosphate transport of *Arabidopsis thaliana* Pht1;1 by site-directed mutagenesis of tyrosine 312 may be attributed to the disruption of homomeric interactions: Pht1;1 Y312 is key to oligomerization and function. Plant Cell Environ 38(10):2012–2022
Foyer CH, Noctor G, Hodges M (2011) Respiration and nitrogen assimilation: targeting mitochondria-associated metabolism as a means to enhance nitrogen use efficiency. J Exp Bot 62(4):1467–1482
Fu J-W, Liu X, Han Y-H, Mei H, Cao Y, de Oliveira LM, Liu Y, Rathinasabapathi B, Chen Y, Ma LQ (2017) Arsenic-hyperaccumulator *Pteris vittata* efficiently solubilized phosphate rock to sustain plant growth and As uptake. J Hazard Mater 330:68–75
Garbinski LD, Rosen BP, Chen J (2019) Pathways of arsenic uptake and efflux. Environ Int 126:585–597
Garg N, Singla P (2011) Arsenic toxicity in crop plants: physiological effects and tolerance mechanisms. Environ Chem Lett 9(3):303–321
Ghorbani A, Tafteh M, Roudbari N, Pishkar L, Zhang W, Wu C (2021) *Piriformospora indica* augments arsenic tolerance in rice (*Oryza sativa*) by immobilizing arsenic in roots and improving iron translocation to shoots. Ecotoxicol Environ Saf 209:111793
Grund SC, Hanusch K, Wolf HU (2008) Arsenic and arsenic compounds. In: Ullmann's encyclopedia of industrial chemistry. Wiley-VCH Verlag GmbH and Co. KGaA, Weinheim. https://doi.org/10.1002/14356007.a03_113.pub2
Guo J, Dai X, Xu W, Ma M (2008) Overexpressing GSH1 and AsPCS1 simultaneously increases the tolerance and accumulation of cadmium and arsenic in *Arabidopsis thaliana*. Chemosphere 72(7):1020–1026
Gupta DK, Inouhe M, Rodríguez-Serrano M, Romero-Puertas MC, Sandalio LM (2013) Oxidative stress and arsenic toxicity: role of NADPH oxidases. Chemosphere 90(6):1987–1996
Han Y-H, Liu X, Rathinasabapathi B, Li H-B, Chen Y, Ma LQ (2017) Mechanisms of efficient As solubilization in soils and As accumulation by As-hyperaccumulator *Pteris vittata*. Environ Pollut 227:569–577
Hare V, Chowdhary P, Baghel VS (2017) Influence of bacterial strains on *Oryza sativa* grown under arsenic tainted soil: accumulation and detoxification response. Plant Physiol Biochem 119:93–102
He S, Wang X, Wu X, Yin Y, Ma LQ (2020) Using rice as a remediating plant to deplete bioavailable arsenic from paddy soils. Environ Int 141:105799
Herath I, Zhao F-J, Bundschuh J, Wang P, Wang J, Ok YS, Palansooriya KN, Vithanage M (2020) Microbe mediated immobilization of arsenic in the rice rhizosphere after incorporation of silica impregnated biochar composites. J Hazard Mater 398:123096
Hite AH (2013) Arsenic and rice: a call for regulation. Nutrition 29(1):353–354
Hojsak I, Braegger C, Bronsky J, Campoy C, Colomb V, Decsi T, Domellöf M, Fewtrell M, Mis NF, Mihatsch W, Molgaard C, van Goudoever J, ESPGHAN Committee on Nutrition (2015) Arsenic in rice: a cause for concern. J Pediatr Gastroenterol Nutr 60(1):142–145
Hu X, Ding Z, Zimmerman AR, Wang S, Gao B (2015) Batch and column sorption of arsenic onto iron-impregnated biochar synthesized through hydrolysis. Water Res 68:206–216
Hu Y, Xiao T, Zhang A (2021) Associations between and risks of trace elements related to skin and liver damage induced by arsenic from coal burning. Ecotoxicol Environ Saf 208:111719
Huang K, Chen C, Zhang J, Tang Z, Shen Q, Rosen BP, Zhao F-J (2016) Efficient arsenic methylation and volatilization mediated by a novel bacterium from an arsenic-contaminated paddy soil. Environ Sci Technol 50(12):6389–6396
Huang SY, Zhuo C, Du XY, Li HS (2021) Remediation of arsenic-contaminated paddy soil by intercropping aquatic vegetables and rice. Int J Phytoremediation 23(10):1021–1029
Huhmann B, Harvey CF, Uddin A, Choudhury I, Ahmed KM, Duxbury JM, Ellis T, van Geen A (2019) Inversion of high-arsenic soil for improved rice yield in Bangladesh. Environ Sci Technol 53(7):3410–3418

Hussain MM, Bibi I, Niazi NK, Shahid M, Iqbal J, Shakoor MB, Ahmad A, Shah NS, Bhattacharya P, Mao K, Bundschuh J, Ok YS, Zhang H (2021) Arsenic biogeochemical cycling in paddy soil-rice system: interaction with various factors, amendments and mineral nutrients. Sci Total Environ 773:145040

Irshad MK, Ibrahim M, Noman A, Shang J, Mahmood A, Mubashir M, Khoo KS, Ng HS, Show PL (2022a) Elucidating the impact of goethite-modified biochar on arsenic mobility, bioaccumulation in paddy rice (*Oryza sativa* L.) along with soil enzyme activities. Process Saf Environ Prot 160:958–967

Irshad MK, Noman A, Wang Y, Yin Y, Chen C, Shang J (2022b) Goethite modified biochar simultaneously mitigates the arsenic and cadmium accumulation in paddy rice (*Oryza sativa* L.). Environ Res 206:112238

Islam LN, Nurun Nabi AHM, Rahman MM, Zahid MSH (2007) Association of respiratory complications and elevated serum immunoglobulins with drinking water arsenic toxicity in human. J Environ Sci Health A 42(12):1807–1814

Islam S, Rahman MM, Rahman MA, Naidu R (2017) Inorganic arsenic in rice and rice-based diets: health risk assessment. Food Control 82:196–202

Jahan I, Abedin MA, Islam MR, Hossain M, Hoque TS, Quadir QF, Hossain MI, Gaber A, Althobaiti YS, Rahman MM (2021) Translocation of soil arsenic towards accumulation in rice: magnitude of water management to minimize health risk. Water 13(20):2816

Jia Y, Sun G-X, Huang H, Zhu Y-G (2013) Biogas slurry application elevated arsenic accumulation in rice plant through increased arsenic release and methylation in paddy soil. Plant Soil 365(1/2):387–396

Kamali M, Sweygers N, Al-Salem S, Appels L, Aminabhavi TM, Dewil R (2022) Biochar for soil applications-sustainability aspects, challenges and future prospects. Chem Eng J 428:131189

Kamiya T, Islam R, Duan G, Uraguchi S, Fujiwara T (2013) Phosphate deficiency signaling pathway is a target of arsenate and phosphate transporter OsPT1 is involved in As accumulation in shoots of rice. Soil Sci Plant Nutr 59(4):580–590

Kar S, Das S, Jean J-S, Chakraborty S, Liu C-C (2013) Arsenic in the water–soil–plant system and the potential health risks in the coastal part of Chianan Plain, Southwestern Taiwan. J Asian Earth Sci 77:295–302

Karimi N, Ghaderian SM, Raab A, Feldmann J, Meharg AA (2009) An arsenic-accumulating, hypertolerant brassica, *Isatis capadocica*. New Phytol 184(1):41–47

Kaur J, Anand V, Srivastava S, Bist V, Tripathi P, Naseem M, Nand S, Anshu, Khare P, Srivastava PK, Bisht S, Srivastava S (2020) Yeast strain *Debaryomyces hansenii* for amelioration of arsenic stress in rice. Ecotoxicol Environ Saf 195:110480

Khan S, Chao C, Waqas M, Arp HPH, Zhu Y-G (2013) Sewage sludge biochar influence upon rice (*Oryza sativa* L) yield, metal bioaccumulation and greenhouse gas emissions from acidic paddy soil. Environ Sci Technol 47(15):8624–8632

Khan Z, Thounaojam TC, Upadhyaya H (2022) Arsenic stress in rice (*Oryza sativa*) and its amelioration approaches. Plant Stress 4:100076

Khanam R, Nayak AK (2021) Arsenic accumulation in rice: current status and future management options. Indian Farming 71(04):50–52

Kim EJ, Baek K (2015) Enhanced reductive extraction of arsenic from contaminated soils by a combination of dithionite and oxalate. J Hazard Mater 284:19–26

Kumar S, Trivedi PK (2018) Glutathione S-transferases: role in combating abiotic stresses including arsenic detoxification in plants. Front Plant Sci 9:751

Kumarathilaka P, Seneweera S, Ok YS, Meharg A, Bundschuh J (2019) Arsenic in cooked rice foods: assessing health risks and mitigation options. Environ Int 127:584–591

Lata S, Samadder SR (2016) Removal of arsenic from water using nano adsorbents and challenges: a review. J Environ Manag 166:387–406

Lee J-C, Kim EJ, Kim H-W, Baek K (2016) Oxalate-based remediation of arsenic bound to amorphous Fe and Al hydrous oxides in soil. Geoderma 270:76–82

Leonardi G, Vahter M, Clemens F, Goessler W, Gurzau E, Hemminki K, Hough R, Koppova K, Kumar R, Rudnai P, Surdu S, Fletcher T (2012) Inorganic arsenic and basal cell carcinoma in areas of Hungary, Romania, and Slovakia: a case–control study. Environ Health Perspect 120(5):721–726

Li R-Y, Ago Y, Liu W-J, Mitani N, Feldmann J, McGrath SP, Ma JF, Zhao F-J (2009a) The rice aquaporin Lsi1 mediates uptake of methylated arsenic species. Plant Physiol 150(4):2071–2080

Li RY, Stroud JL, Ma JF, McGrath SP, Zhao FJ (2009b) Mitigation of arsenic accumulation in rice with water management and silicon fertilization. Environ Sci Technol 43(10):3778–3783

Li B, Zhou S, Wei D, Long J, Peng L, Tie B, Williams PN, Lei M (2019) Mitigating arsenic accumulation in rice (*Oryza sativa* L.) from typical arsenic contaminated paddy soil of Southern China using nanostructured α-MnO_2: pot experiment and field application. Sci Total Environ 650:546–556

Lin L, Gao M, Qiu W, Wang D, Huang Q, Song Z (2017) Reduced arsenic accumulation in indica rice (*Oryza sativa* L.) cultivar with ferromanganese oxide impregnated biochar composites amendments. Environ Pollut 231:479–486

Liu C, Wei L, Zhang S, Xu X, Li F (2014) Effects of nanoscale silica sol foliar application on arsenic uptake, distribution and oxidative damage defense in rice (*Oryza sativa* L.) under arsenic stress. RSC Adv 4(100):57227–57234

Liu L, Li W, Song W, Guo M (2018) Remediation techniques for heavy metal-contaminated soils: principles and applicability. Sci Total Environ 633:206–219

Ma JF, Yamaji N (2006) Silicon uptake and accumulation in higher plants. Trends Plant Sci 11(8):392–397

Ma JF, Yamaji N, Mitani N, Tamai K, Konishi S, Fujiwara T, Katsuhara M, Yano M (2007) An efflux transporter of silicon in rice. Nature 448:5

Ma JF, Yamaji N, Mitani N, Xu X-Y, Su Y-H, McGrath SP, Zhao F-J (2008) Transporters of arsenite in rice and their role in arsenic accumulation in rice grain. Proc Natl Acad Sci U S A 105(29):9931–9935

Ma J, Lei E, Lei M, Liu Y, Chen T (2018) Remediation of arsenic contaminated soil using malposed intercropping of *Pteris vittata* L. and maize. Chemosphere 194:737–744

Ma X, Sharifan H, Dou F, Sun W (2020) Simultaneous reduction of arsenic (As) and cadmium (Cd) accumulation in rice by zinc oxide nanoparticles. Chem Eng J 384:123802

Majumdar A, Bose S (2018) A glimpse on uptake kinetics and molecular responses of arsenic tolerance in rice plants. In: Hasanuzzaman M, Nahar K, Fujita M (eds) Mechanisms of arsenic toxicity and tolerance in plants. Springer, Singapore, pp 299–315

Majumdar A, Upadhyay MK, Kumar JS, Sheena, Barla A, Srivastava S, Jaiswal MK, Bose S (2019) Ultra-structure alteration via enhanced silicon uptake in arsenic stressed rice cultivars under intermittent irrigation practices in Bengal delta basin. Ecotoxicol Environ Saf 180:770–779

Majumdar A, Kumar JS, Sheena, Bose S (2020) Agricultural water management practices and environmental influences on arsenic dynamics in rice field. In: Srivastava S (ed) Arsenic in drinking water and food. Springer, Singapore, pp 425–443

Mallick I, Bhattacharyya C, Mukherji S, Dey D, Sarkar SC, Mukhopadhyay UK, Ghosh A (2018) Effective rhizoinoculation and biofilm formation by arsenic immobilizing halophilic plant growth promoting bacteria (PGPB) isolated from mangrove rhizosphere: a step towards arsenic rhizoremediation. Sci Total Environ 610–611:1239–1250

Mandal A, Purakayastha TJ, Patra AK, Sanyal SK (2012) Phytoremediation of arsenic contaminated soil by *Pteris Vittata* L. II. Effect on arsenic uptake and rice yield. Int J Phytoremediation 14(6):621–628

Matsumoto S, Kasuga J, Taiki N, Makino T, Arao T (2015) Inhibition of arsenic accumulation in Japanese rice by the application of iron and silicate materials. Catena 135:328–335

Maurel C, Boursiac Y, Luu D-T, Santoni V, Shahzad Z, Verdoucq L (2015) Aquaporins in plants. Physiol Rev 95(4):1321–1358

Mawia AM, Hui S, Zhou L, Li H, Tabassum J, Lai C, Wang J, Shao G, Wei X, Tang S, Luo J, Hu S, Hu P (2021) Inorganic arsenic toxicity and alleviation strategies in rice. J Hazard Mater 408:124751

McClintock TR, Chen Y, Bundschuh J, Oliver JT, Navoni J, Olmos V, Lepori EV, Ahsan H, Parvez F (2012) Arsenic exposure in Latin America: biomarkers, risk assessments and related health effects. Sci Total Environ 429:76–91

Mecwan Neha V, Yagnik B, Solanki H (2018) A review on phytoremediation of arsenic-contaminated soil. Int Res J Environ Sci 7(4):27–36

Meharg AA, Jardine L (2003) Arsenite transport into paddy rice (*Oryza sativa*) roots. New Phytol 157(1):39–44

Meliker JR, Slotnick MJ, AvRuskin GA, Schottenfeld D, Jacquez GM, Wilson ML, Goovaerts P, Franzblau A, Nriagu JO (2010) Lifetime exposure to arsenic in drinking water and bladder cancer: a population-based case–control study in Michigan, USA. Cancer Causes Control 21(5):745–757

Mendoza-Cózatl DG, Jobe TO, Hauser F, Schroeder JI (2011) Long-distance transport, vacuolar sequestration, tolerance, and transcriptional responses induced by cadmium and arsenic. Curr Opin Plant Biol 14(5):554–562

Mesa V, Navazas A, González-Gil R, González A, Weyens N, Lauga B, Gallego JLR, Sánchez J, Peláez AI (2017) Use of endophytic and rhizosphere bacteria to improve phytoremediation of arsenic-contaminated industrial soils by autochthonous *Betula celtiberica*. Appl Environ Microbiol 83(8):e03411–e03416

Mitra A, Chatterjee S, Moogouei R, Gupta D (2017) Arsenic accumulation in rice and probable mitigation approaches: a review. Agronomy 7(4):67

Mlangeni AT, Perez M, Raab A, Krupp EM, Norton GJ, Feldmann J (2020) Simultaneous stimulation of arsenic methylation and inhibition of cadmium bioaccumulation in rice grain using zero valent iron and alternate wetting and drying water management. Sci Total Environ 711:134696

Mochizuki H, Phyu KP, Aung MN, Zin PW, Yano Y, Myint MZ, Thit WM, Yamamoto Y, Hishikawa Y, Thant KZ, Maruyama M, Kuroda Y (2019) Peripheral neuropathy induced by drinking water contaminated with low-dose arsenic in Myanmar. Environ Health Prev Med 24(1):23

Mosa KA, Kumar K, Chhikara S, Mcdermott J, Liu Z, Musante C, White JC, Dhankher OP (2012) Members of rice plasma membrane intrinsic proteins subfamily are involved in arsenite permeability and tolerance in plants. Transgenic Res 21(6):1265–1277

Muehe EM, Wang T, Kerl CF, Planer-Friedrich B, Fendorf S (2019) Rice production threatened by coupled stresses of climate and soil arsenic. Nat Commun 10(1):4985

Mukhopadhyay R, Rosen BP (2002) Arsenate reductases in prokaryotes and eukaryotes. Environ Health Perspect 110(suppl 5):745–748

Mukhopadhyay CK, Mazumder B, Fox PL (2000) Role of hypoxia-inducible factor-1 in transcriptional activation of ceruloplasmin by iron deficiency. J Biol Chem 275(28):21048–21054

Murugaiyan V, Ali J, Frei M, Zeibig F, Pandey A, Wairich A, Wu L-B, Murugaiyan J, Li Z (2021) Identification of promising genotypes through systematic evaluation for arsenic tolerance and exclusion in rice (*Oryza sativa* L.). Front Plant Sci 12:753063

Nath S, Panda P, Mishra S, Dey M, Choudhury S, Sahoo L, Panda SK (2014) Arsenic stress in rice: redox consequences and regulation by iron. Plant Physiol Biochem 80:203–210

Naujokas MF, Anderson B, Ahsan H, Aposhian HV, Graziano JH, Thompson C, Suk WA (2013) The broad scope of health effects from chronic arsenic exposure: update on a worldwide public health problem. Environ Health Perspect 121(3):295–302

Niazi NK, Singh B, Shah P (2011) Arsenic speciation and phytoavailability in contaminated soils using a sequential extraction procedure and xanes spectroscopy. Environ Sci Technol 45(17):7135–7142

Nizam S, Kato M, Yatsuya H, Khalequzzaman MD, Ohnuma S, Naito H, Nakajima T (2013) Differences in urinary arsenic metabolites between diabetic and non-diabetic subjects in Bangladesh. Int J Environ Res Public Health 10(3):1006–1019

Norton GJ, Lou-Hing DE, Meharg AA, Price AH (2008) Rice–arsenate interactions in hydroponics: whole genome transcriptional analysis. J Exp Bot 59(8):2267–2276

Okamura H, Piao M, Aoyama I, Sudo M, Okubo T, Nakamura M (2002) Algal growth inhibition by river water pollutants in the agricultural area around Lake Biwa, Japan. Environ Pollut 117(3):411–419

Paikaray S, Banerjee S, Mukherji S (2005) Sorption of arsenic onto Vindhyan shales: role of pyrite and organic carbon. Curr Sci 88(10):1580–1585

Pandey C, Gupta M (2015) Selenium and auxin mitigates arsenic stress in rice (*Oryza sativa* L.) by combining the role of stress indicators, modulators and genotoxicity assay. Hazard Mater 287:384–391

Panthri M, Gupta M (2019) Plausible strategies to reduce arsenic accumulation in rice. In: Advances in rice research for abiotic stress tolerance. Woodhead Publisher/Elsevier, pp 371–384

Paszkowski U, Kroken S, Roux C, Briggs SP (2002) Rice phosphate transporters include an evolutionarily divergent gene specifically activated in arbuscular mycorrhizal symbiosis. Proc Natl Acad Sci U S A 99(20):13324–13329

Pigna M, Cozzolino V, Giandonato Caporale A, Mora ML, Di Meo V, Jara AA, Violante A (2010) Effects of phosphorus fertilization on arsenic uptake by wheat grown in polluted soils. J Soil Sci Plant Nutr 10(4):428–442

Praveen A, Mehrotra S, Singh N (2017) Rice planted along with accumulators in arsenic amended plots reduced arsenic uptake in grains and shoots. Chemosphere 184:1327–1333

Premarathna KSD, Rajapaksha AU, Sarkar B, Kwon EE, Bhatnagar A, Ok YS, Vithanage M (2019) Biochar-based engineered composites for sorptive decontamination of water: a review. Chem Eng J 372:536–550

Qiao J, Liu T, Wang X, Li F, Lv Y, Cui J, Zeng X, Yuan Y, Liu C (2018) Simultaneous alleviation of cadmium and arsenic accumulation in rice by applying zero-valent iron and biochar to contaminated paddy soils. Chemosphere 195:260–271

Quansah R, Armah FA, Essumang DK, Luginaah I, Clarke E, Marfoh K, Cobbina SJ, Nketiah-Amponsah E, Namujju PB, Obiri S, Dzodzomenyo M (2015) Association of arsenic with adverse pregnancy outcomes/infant mortality: a systematic review and meta-analysis. Environ Health Perspect 123(5):412–421

Rahaman S, Sinha AC, Mukhopadhyay D (2011) Effect of water regimes and organic matters on transport of arsenic in summer rice (*Oryza sativa* L.). J Environ Sci 23(4):633–639

Rahaman MDS, Rahman MDM, Mise N, Sikder MDT, Ichihara G, Uddin MDK, Kurasaki M, Ichihara S (2021) Environmental arsenic exposure and its contribution to human diseases, toxicity mechanism and management. Environ Pollut 289:117940

Rahman MM, Ng JC, Naidu R (2009) Chronic exposure of arsenic via drinking water and its adverse health impacts on humans. Environ Geochem Health 31(S1):189–200

Rahman MM, Alauddin M, Alauddin ST, Siddique AB, Islam MDR, Agosta G, Mondal D, Naidu R (2021) Bioaccessibility and speciation of arsenic in children's diets and health risk assessment of an endemic area in Bangladesh. J Hazard Mater 403:124064

Ramesh SA, Tyerman SD, Xu B, Bose J, Kaur S, Conn V, Domingos P, Ullah S, Wege S, Shabala S, Feijó JA, Ryan PR, Gilliham M (2015) GABA signalling modulates plant growth by directly regulating the activity of plant-specific anion transporters. Nat Commun 6(1):7879

Remy E, Cabrito TR, Batista RA, Teixeira MC, Sá-Correia I, Duque P (2012) The Pht1;9 and Pht1;8 transporters mediate inorganic phosphate acquisition by the *Arabidopsis thaliana* root during phosphorus starvation. New Phytol 195(2):356–371

Robles-Osorio ML, Sabath-Silva E, Sabath E (2015) Arsenic-mediated nephrotoxicity. Ren Fail 37(4):542–547

Roggenbeck BA, Banerjee M, Leslie EM (2016) Cellular arsenic transport pathways in mammals. J Environ Sci 49:38–58

Roy M, Giri AK, Dutta S, Mukherjee P (2015) Integrated phytobial remediation for sustainable management of arsenic in soil and water. Environ Int 75:180–198

Roychowdhury R, Khan MDH, Choudhury S (2018) Arsenic in Rice: An overview on stress implications, tolerance and mitigation strategies. In: Hasanuzzaman M, Nahar K, Fujita M (eds) Plants under metal and metalloid stress. Springer, Singapore, pp 401–415

Saifullah, Dahlawi S, Naeem A, Iqbal M, Farooq MA, Bibi S, Rengel Z (2018) Opportunities and challenges in the use of mineral nutrition for minimizing arsenic toxicity and accumulation in rice: a critical review. Chemosphere 194:171–188
Seyfferth AL, Amaral D, Limmer MA, Guilherme LRG (2019) Combined impacts of Si-rich rice residues and flooding extent on grain As and Cd in rice. Environ Int 128:301–309
Shaibur MR, Kitajima N, Sugawara R, Kondo T, Huq SMI, Kawai S (2006) Physiological and mineralogical properties of arsenic-induced chlorosis in rice seedlings grown hydroponically. Soil Sci Plant Nutr 52(6):691–700
Shaibur MR, Kitajima N, Sugawara R, Kondo T, Huq SMI, Kawai S (2009) Effect of arsenic on phytosiderophores and mineral nutrition of barley seedlings grown in iron-depleted medium. Soil Sci Plant Nutr 55(2):283–293
Shakoor MB, Riaz M, Niazi NK, Ali S, Rizwan M, Arif MS, Arif M (2019) Recent advances in arsenic accumulation in rice. In: Advances in rice research for abiotic stress tolerance. Elsevier, pp 385–398
Sharma AK, Tjell JC, Sloth JJ, Holm PE (2014) Review of arsenic contamination, exposure through water and food and low-cost mitigation options for rural areas. Appl Geochem 41:11–33
Shi S, Wang T, Chen Z, Tang Z, Wu Z, Salt DE, Chao D-Y, Zhao F-J (2016) OsHAC1;1 and OsHAC1;2 function as arsenate reductases and regulate arsenic accumulation. Plant Physiol 172(3):1708–1719
Shin H, Shin H-S, Dewbre GR, Harrison MJ (2004) Phosphate transport in *Arabidopsis*: Pht1;1 and Pht1;4 play a major role in phosphate acquisition from both low- and high-phosphate environments. Plant J 39(4):629–642
Shri M, Dave R, Diwedi S, Shukla D, Kesari R, Tripathi RD, Trivedi PK, Chakrabarty D (2015) Heterologous expression of *Ceratophyllum demersum* phytochelatin synthase, CdPCS1, in rice leads to lower arsenic accumulation in grain. Sci Rep 4(1):5784
Shrivastava A, Barla A, Majumdar A, Singh S, Bose S (2020) Arsenic mitigation in rice grain loading via alternative irrigation by proposed water management practices. Chemosphere 238:124988
Signes-Pastor A, Burló F, Mitra K, Carbonell-Barrachina AA (2007) Arsenic biogeochemistry as affected by phosphorus fertilizer addition, redox potential and pH in a West Bengal (India) soil. Geoderma 137(3–4):504–510
Signes-Pastor AJ, Carey M, Meharg AA (2016) Inorganic arsenic in rice-based products for infants and young children. Food Chem 191:128–134
Signes-Pastor AJ, Vioque J, Navarrete-Muñoz EM, Carey M, García de la Hera M, Sunyer J, Casas M, Riaño-Galán I, Tardón A, Llop S, Amorós R, Amiano P, Bilbao JR, Karagas MR, Meharg AA (2017) Concentrations of urinary arsenic species in relation to rice and seafood consumption among children living in Spain. Environ Res 159:69–75
Singh N, Srivastava S, Rathaur S, Singh N (2016) Assessing the bioremediation potential of arsenic tolerant bacterial strains in rice rhizosphere interface. J Environ Sci 48:112–119
Singh AP, Dixit G, Kumar A, Mishra S, Singh PK, Dwivedi S, Trivedi PK, Chakrabarty D, Mallick S, Pandey V, Dhankher OP, Tripathi RD (2016a) Nitric oxide alleviated arsenic toxicity by modulation of antioxidants and thiol metabolism in rice (*Oryza sativa* L.). Front Plant Sci 6:1272
Singh AP, Dixit G, Kumar A, Mishra S, Kumar N, Dixit S, Singh PK, Dwivedi S, Trivedi PK, Pandey V, Dhankher OP, Norton GJ, Chakrabarty D, Tripathi RD (2017) A protective role for nitric oxide and salicylic acid for arsenite phytotoxicity in rice (*Oryza sativa* L.). Plant Physiol Biochem 115:163–173
Smith E, Naidu R, Alston AM (2002) Heavy metals in the environment. J Environ Qual 31:7
Smith NK, Keltie E, Sweeney E, Weerasinghe S, MacPherson K, Kim JS (2022) Toenail speciation biomarkers in arsenic-related disease: a feasibility study for investigating the association between arsenic exposure and chronic disease. Ecotoxicol Environ Saf 232:113269
Song W-Y, Park J, Mendoza-Cózatl DG, Suter-Grotemeyer M, Shim D, Hörtensteiner S, Geisler M, Weder B, Rea PA, Rentsch D, Schroeder JI, Lee Y, Martinoia E (2010) Arsenic tolerance in

Arabidopsis is mediated by two ABCC-type phytochelatin transporters. Proc Natl Acad Sci U S A 107(49):21187–21192

Song W-Y, Yamaki T, Yamaji N, Ko D, Jung K-H, Fujii-Kashino M, An G, Martinoia E, Lee Y, Ma JF (2014) A rice ABC transporter, OsABCC1, reduces arsenic accumulation in the grain. Proc Natl Acad Sci 111(44):15699–15704

Srivastava S, Srivastava S, Bist V, Awasthi S, Chauhan R, Chaudhry V, Singh PC, Dwivedi S, Niranjan A, Agrawal L, Chauhan PS, Tripathi RD, Nautiyal CS (2018) *Chlorella vulgaris* and *Pseudomonas putida* interaction modulates phosphate trafficking for reduced arsenic uptake in rice (*Oryza sativa* L.). J Hazard Mater 351:177–187

Srivastava S, Pathare VS, Sounderajan S, Suprasanna P (2019) Nitrogen supply influences arsenic accumulation and stress responses of rice (*Oryza sativa* L.) seedlings. J Hazard Mater 367:599–606

Srivastava S, Suprasanna P, Tripathi RD (2020) Safeguarding rice from arsenic contamination through the adoption of chemo-agronomic measures. In: Srivastava S (ed) Arsenic in drinking water and food. Springer, Singapore, pp 411–424

Srivastava S, Shukla A, Rajput VD, Kumar K, Minkina T, Mandzhieva S, Shmaraeva A, Suprasanna P (2021) Arsenic remediation through sustainable phytoremediation approaches. Fortschr Mineral 11(9):936

Stoeva N, Bineva T (2003) Oxidative changes and photosynthesis in oat plants grown in as-contaminated soil. Bulg J Plant Physiol 29(1–2):87–95

Sun S-K, Chen Y, Che J, Konishi N, Tang Z, Miller AJ, Ma JF, Zhao F-J (2018) Decreasing arsenic accumulation in rice by overexpressing OsNIP1;1 and OsNIP3;3 through disrupting arsenite radial transport in roots. New Phytol 219(2):641–653

Suriyagoda LDB, Dittert K, Lambers H (2018) Arsenic in rice soils and potential agronomic mitigation strategies to reduce arsenic bioavailability: a review. Pedosphere 28(3):363–382

Syu C-H, Lee C-H, Jiang P-Y, Chen M-K, Lee D-Y (2014) Comparison of As sequestration in iron plaque and uptake by different genotypes of rice plants grown in As-contaminated paddy soils. Plant Soil 374(1/2):411–422

Syu C-H, Wu P-R, Lee C-H, Juang K-W, Lee D-Y (2019) Arsenic phytotoxicity and accumulation in rice seedlings grown in arsenic-contaminated soils as influenced by the characteristics of organic matter amendments and soils. J Plant Nutr Soil Sci 182(1):60–71

Syu C-H, Yu C-H, Lee D-Y (2020) Effect of applying calcium peroxide on the accumulation of arsenic in rice plants grown in arsenic-elevated paddy soils. Environ Pollut 266:115140

Tavakkoli E, Lyons G, English P, Guppy CN (2011) Silicon nutrition of rice is affected by soil pH, weathering and silicon fertilisation. J Plant Nutr Soil Sci 174(3):437–446

Tiwari M, Sharma D, Dwivedi S, Singh M, Tripathi RD, Trivedi PK (2014) Expression in Arabidopsis and cellular localization reveal involvement of rice NRAMP, OsNRAMP1, in arsenic transport and tolerance: OsNRAMP1 in arsenic transport and tolerance. Plant Cell Environ 37(1):140–152

Tripathi P, Tripathi RD, Singh RP, Dwivedi S, Goutam D, Shri M, Trivedi PK, Chakrabarty D (2013) Silicon mediates arsenic tolerance in rice (*Oryza sativa* L.) through lowering of arsenic uptake and improved antioxidant defence system. Ecol Eng 52:96–103

Upadhyay AK, Singh NK, Singh R, Rai UN (2016) Amelioration of arsenic toxicity in rice: comparative effect of inoculation of *Chlorella vulgaris* and *Nannochloropsis* sp. on growth, biochemical changes and arsenic uptake. Ecotoxicol Environ Saf 124:68–73

Upadhyay MK, Majumdar A, Suresh Kumar J, Srivastava S (2020) Arsenic in rice agro-ecosystem: solutions for safe and sustainable rice production. Front Sustain Food Syst 4:53

Verbruggen N, Hermans C, Schat H (2009) Mechanisms to cope with arsenic or cadmium excess in plants. Curr Opin Plant Biol 12(3):364–372

Wang P, Zhang W, Mao C, Xu G, Zhao F-J (2016) The role of OsPT8 in arsenate uptake and varietal difference in arsenate tolerance in rice. J Exp Bot 67(21):6051–6059

Wang X, Sun W, Zhang S, Sharifan H, Ma X (2018a) Elucidating the effects of cerium oxide nanoparticles and zinc oxide nanoparticles on arsenic uptake and speciation in rice (*Oryza sativa*) in a hydroponic system. Environ Sci Technol 52(17):10040–10047

Wang Y, Li YQ, Lv K, Cheng JJ, Chen XL, Ge Y, Yu XY (2018b) Soil microalgae modulate grain arsenic accumulation by reducing dimethylarsinic acid and enhancing nutrient uptake in rice (*Oryza sativa* L.). Plant Soil 430(1–2):99–111

Wang X, Sun W, Ma X (2019) Differential impacts of copper oxide nanoparticles and Copper(II) ions on the uptake and accumulation of arsenic in rice (*Oryza sativa*). Environ Pollut 252:967–973

Watanabe T, Hirano S (2013) Metabolism of arsenic and its toxicological relevance. Arch Toxicol 87(6):969–979

World Health Organization (2017) WHO methods and data sources for global burden of disease estimates 2000–2015. Global Health Estimates Technical Paper. WHO/HIS/IER/GHE/2017.1. Accessed 11 July 2022

Wu C, Ye Z, Shu W, Zhu Y, Wong M (2011) Arsenic accumulation and speciation in rice are affected by root aeration and variation of genotypes. J Exp Bot 62(8):2889–2898

Wu F, Fang Q, Yan S, Pan L, Tang X, Ye W (2020) Effects of zinc oxide nanoparticles on arsenic stress in rice (*Oryza sativa* L.): germination, early growth, and arsenic uptake. Environ Sci Pollut Res 27(21):26974–26981

Wu K, Wu C, Jiang X, Xue R, Pan W, Li W-C, Luo X, Xue S (2022) Remediation of arsenic-contaminated paddy field by a new iron oxidizing strain (*Ochrobactrum* sp.) and iron-modified biochar. J Environ Sci 115:411–421

Xu J, Shi S, Wang L, Tang Z, Lv T, Zhu X, Ding X, Wang Y, Zhao F, Wu Z (2017) OsHAC4 is critical for arsenate tolerance and regulates arsenic accumulation in rice. New Phytol 215(3):1090–1101

Xu Y, Lu Q, Li J, Wan L, Chen S, Lu Y (2021) Effect of humus on the remediation of arsenic-contaminated soil by electrokinetic technology. Environ Technol Innov 21:101297

Yamaguchi N, Nakamura T, Dong D, Takahashi Y, Amachi S, Makino T (2011) Arsenic release from flooded paddy soils is influenced by speciation, Eh, pH, and iron dissolution. Chemosphere 83(7):925–932

Yamazaki S, Ueda Y, Mukai A, Ochiai K, Matoh T (2018) Rice phytochelatin synthases Os PCS 1 and Os PCS 2 make different contributions to cadmium and arsenic tolerance. Plant Direct 2(1):e00034

Yang J, Gao M, Hu H, Ding X, Lin H, Wang L, Xu J, Mao C, Zhao F, Wu Z (2016) OsCLT1, a CRT-like transporter 1, is required for glutathione homeostasis and arsenic tolerance in rice. New Phytol 211(2):658–670

Yang X, Shaheen SM, Wang J, Hou D, Ok YS, Wang SL, Wang H, Rinklebe J (2022) Elucidating the redox-driven dynamic interactions between arsenic and iron-impregnated biochar in a paddy soil using geochemical and spectroscopic techniques. J Hazard Mater 422:126808

Yao B-M, Chen P, Zhang H-M, Sun G-X (2021) A predictive model for arsenic accumulation in rice grains based on bioavailable arsenic and soil characteristics. J Hazard Mater 412:125131

Ye Y, Li P, Xu T, Zeng L, Cheng D, Yang M, Luo J, Lian X (2017) OsPT4 contributes to arsenate uptake and transport in rice. Front Plant Sci 8:2197

Yi XY, Yang YP, Yuan HY, Chen Z, Duan GL, Zhu YG (2019) Coupling metabolisms of arsenic and iron with humic substances through microorganisms in paddy soil. J Hazard Mater 373:591–599

Yin D, Wang X, Peng B, Tan C, Ma LQ (2017) Effect of biochar and Fe-biochar on Cd and As mobility and transfer in soil-rice system. Chemosphere 186:928–937

Yu HY, Wang X, Li F, Li B, Liu C, Wang Q, Lei J (2017) Arsenic mobility and bioavailability in paddy soil under iron compound amendments at different growth stages of rice. Environ Pollut 224:136–147

Zhang J, Zhao Q-Z, Duan G-L, Huang Y-C (2011) Influence of sulphur on arsenic accumulation and metabolism in rice seedlings. Environ Exp Bot 72(1):34–40

Zhang S, Geng L, Fan L, Zhang M, Zhao Q, Xue P, Liu W (2020) Spraying silicon to decrease inorganic arsenic accumulation in rice grain from arsenic-contaminated paddy soil. Sci Total Environ 704:135239

Zhang D, Du G, Zhang W, Gao Y, Jie H, Rao W, Jiang Y, Wang D (2021a) Remediation of arsenic-contaminated paddy soil: effects of elemental sulfur and gypsum fertilizer application. Ecotoxicol Environ Saf 223:112606

Zhang Q, Zou D, Zeng X, Li L, Wang A, Liu F, Wang H, Zeng Q, Xiao Z (2021b) Effect of the direct use of biomass in agricultural soil on heavy metals activation or immobilization? Environ Pollut 272:115989

Zhang C, Hao Y, Zhang K, Chen S, Yang Z (2022) Effect of biogas slurry and sucrose addition on electrokinetic removal of arsenic from paddy soil. Int J Environ Sci Technol. https://doi.org/10.1007/s13762-022-04006-0

Zhao F-J, Ago Y, Mitani N, Li R-Y, Su Y-H, Yamaji N, McGrath SP, Ma JF (2010) The role of the rice aquaporin Lsi1 in arsenite efflux from roots. New Phytol 186(2):392–399

Zheng R, Sun G, Zhu Y (2013) Effects of microbial processes on the fate of arsenic in paddy soil. Chin Sci Bull 58(2):186–193

Zhu N, Qiao J, Yan T (2019) Arsenic immobilization through regulated ferrolysis in paddy field amendment with bismuth impregnated biochar. Sci Total Environ 648:993–1001

Zhu S, Zhao J, Zhao N, Yang X, Chen C, Shang J (2020) Goethite modified biochar as a multifunctional amendment for cationic Cd(II), anionic As(III), roxarsone, and phosphorus in soil and water. J Clean Prod 247:119579

Chapter 22
Mechanism and Approaches to Enhancing Heat Stress Tolerance in Crop Plants

Maira Tanveer, Athar Mahmood , Bushra Sarfraz, Muhammad Anjum Zia, Muhammad Mansoor Javaid, Safura Bibi, Maria Naqve, Muhammad Ather Nadeem, Muhammad Azeem, and Abdul Jabbar

Abstract Rising temperature is a major agricultural issue throughout most of the world. Heat stress has harmful effects on plants that vary depending on the species and affects their growth and development. Major tolerance mechanism tools include proteins, ion transporters, osmoprotectants, proteins, and antioxidants. High temperatures may also have adverse effects, gas exchange parameters, photorespiration, evapotranspiration, and membrane stability, in later stages. Plants have many adaptations, avoidance, and tolerance methods to reduce the damaging effects of high temperatures. Plant species have developed a broad collection of adaptation methods for harsh environments, such as heat shock protein induction (HSPs). All of the main HSPs serve important roles in reducing protein aggregation and misfolding. Plants resist heat stress by generating antioxidants and stress proteins, and moreover sustaining membrane permeability. Recent advancements in omics techniques have increased our understanding of the molecular pathways producing heat stress. Choosing the right planting time, seed priming, nutrients, and water management are all important approaches for enhancing plant behavior under severe HS conditions due to climate change.

M. Tanveer · B. Sarfraz · S. Bibi · M. Naqve · A. Jabbar
Department of Botany, Facility of sciences, University of Agriculture, Faisalabad, Pakistan

A. Mahmood (✉)
Department of Agronomy, Faculty of Agriculture, University of Agriculture, Faisalabad, Pakistan
e-mail: athar.mahmood@uaf.edu.pk

M. A. Zia
Department of Biochemistry, University of Agriculture, Faisalabad, Pakistan

M. M. Javaid (✉) · M. A. Nadeem
Department of Agronomy, College of Agriculture, University of Sargodha, Sargodha, Pakistan
e-mail: mansoor.javaid@uos.edu.pk

M. Azeem
Biosaline Research Laboratories, Department of Botany, University of Karachi, Karachi, Pakistan

M. Hasanuzzaman (ed.), *Climate-Resilient Agriculture, Vol 2*,
https://doi.org/10.1007/978-3-031-37428-9_22

Keywords Heat stress · Climate change · Crop tolerance

1 Introduction

Heat stress is an environmental stress which plants experience, and it initiates the synthesis of various heat shock proteins (HSPs). Currently, an increase in global average temperature is brought by global warming, causing severe problems in agriculture across the world, with a predictable loss of agricultural output amounting to countless dollars annually. Heat shocks inactivate essential enzymes, disrupt the protein synthesis, and damage the membrane, ultimately harming the crop growth. High temperatures prevent plants from growing and cause oxidative stress, which lowers crop yields and nutrient quality. Heat stress has a significant influence on their activities because plants are sessile creatures that cannot move to more favorable conditions (Javaid et al. 2022a, b, c). Heat stress affects germination, growth, development, and reproduction, leading to reduced crop productivity. As a result, in order to survive under stressful situations, plants display a wide range of morphological, physiological, and biochemical reactions (Janská et al. 2010; Tanveer et al. 2020). To achieve good productivity under challenging circumstances, specific field management techniques can be used. Abiotic variables are the primary determinants of yield production in crop plants (Sihi et al. 2022). High temperatures, salt, drought, and heavy metal stress, among other conditions, all have an impact on agricultural plant development (Waqas et al. 2019). The aforementioned pressures affect 90% of agricultural production and can result in up to 70% productivity loss for major food crops (Onaga and Wydra 2021). Integration of crop yield and climate change models estimated that, important crops, including rice, wheat, and maize, would continue to lose productivity, which might have an adverse effect on food security (Mehmood et al. 2018b; Hussain et al. 2020). HSPs support the transmission of stress signals, repair damaged proteins and membranes, and protect photosynthesis. It is widely acknowledged that the heat tolerance adaptive approach promotes the expression of various HSPs. Heat shock responses are regulated by transcriptional and translational levels. Heat shock element (HSE) and a cis-acting DNA region are required for heat-induced transcription (Asthir et al. 2015). In the presence of heat stress, such advancements enable plant growth and development. However, resistance level for heat stress varied between and within the species. There is a great deal of variation both within and across species, offering chances to genetically enhance crop heat stress resistance. There have been some successful efforts to develop heat-tolerant genotypes using traditional plant breeding techniques (Wahid et al. 2007). Plants often produce heat shock proteins, osmoprotectants, and antioxidants in response to high temperatures (Mudassir et al. 2018; Khan and Shahwar 2020).

2 Loss of Productivity Due to Heat Stress

Drought conditions are linked to heat stress and affected the growth, development and ultimately reduce the farm income. At a 40% water reduction, maize yield was decreased by up to 40% and wheat by 21% (Daryanto et al. 2016). An important crop in Africa is cowpea where productivity reductions might vary from 34% to 68% depending on when drought stress develops (Farooq et al. 2017; Chadha et al. 2019).

Morphological, physiological, and biochemical changes occurred under stressful conditions that decreased crop development and growth (Wahid et al. 2007). High temperatures have a variety of negative effects on crops, such as poor germination, leaf senescence, drastically reduced pollen viability, and therefore, fewer grains with smaller grain sizes (Ugarte et al. 2007; Waheed et al. 2020). High-temperature stress damages the protein and cell membrane along with disruption of the protein synthesis and inactivates major enzymes (Smertenko et al. 1997). In general, the effects of heat stress are more pronounced throughout reproductive stage, which leads to reduction of the fertilization and causing flower abortion. The gas exchange parameters, water relation, and cell membrane stability are mainly affected by unfavorable temperatures. Many studies have indicated yield reductions in several crops due to heat stress (Table 22.1).

3 Plant Response to Heat Stress

In natural environment, both light and temperature have an impact on plants. However, when they are exposed to a temperature for a long time more than their optimum range, it disrupts the whole plant's life cycle. The main difficulties for plants are extreme environmental conditions; they all have an adverse effect on their development and growth. Examples of these situations include diverse biotic and abiotic stressors (Onyekachi et al. 2019; Waheed et al. 2019). Plants have developed an astounding variety of methods for reducing their adverse environmental impacts because they are sessile and so unable to move to more favorable environments

Table 22.1 Effect of drought and heat stress on yield reduction of different crops

Crop	Stress	Yield reduction	References
Maize	Drought, heat	62–86%	Gabaldón-Leal et al. (2016)
Rice	Drought, heat	51–89%	Sun et al. (2022)
Wheat	Heat	54%	Akter and Rafiqul Islam (2017)
Soybean	Drought	44–69%	Cohen et al. (2021)
Chickpea	Drought	42–66%	Gaur et al. (2013)
Sunflower	Drought	58%	Van der Merwe et al. (2015)
Canola	Heat	28%	Lohani et al. (2022)
Potato	Drought	11%	Tang et al. (2018)

(Thornton et al. 2014). Moreover, it has been shown that plant cells have sophisticated systems to react to several stressors, including HSR, which has a positive effect which might boost agricultural production under conditions of climate change. Extremely high temperatures can cause devastating damage to crop and reduce crop production (Javaid et al. 2020b; Elahi et al. 2022). The heat response to stress generates a significant amount of reactive oxygen species (ROS), which causes oxidative stress as a secondary stress (de Castro Cavallari et al. 2019).

3.1 Seed Germination and Plant Growth Under Heat Stress Conditions

The percentage of plant development is influenced by a major factor known as heat stress. Every species has a different temperature tolerance level (Khan and Shahwar 2020). Heat stress has been associated to a decrease in germination percentages, which has a negative influence on the emergence, growth, and development of a wide range of plant species (Kumar et al. 2011). Effects of high temperature on plant germination are elaborated in Fig. 22.1. Visual stress symptoms such as fruit and leaf discoloration and senescence can also be caused by high temperature (Vollenweider and Günthardt-Goerg 2005).

The seed germination inhibition takes place by HS by disrupting enzymes' activity involved in starch breakdown and promoting the abscisic acid (ABA) production (Essemine et al. 2010). Heat stress affects plant phenology and plant life cycle. For example, above the optimum level, slight temperature rise (1–2 °C) shortened the duration of grain filling in cereals, resulting in lower productivity (Hassan et al. 2021). Protein denaturation at high heat stress caused cell death in particular regions

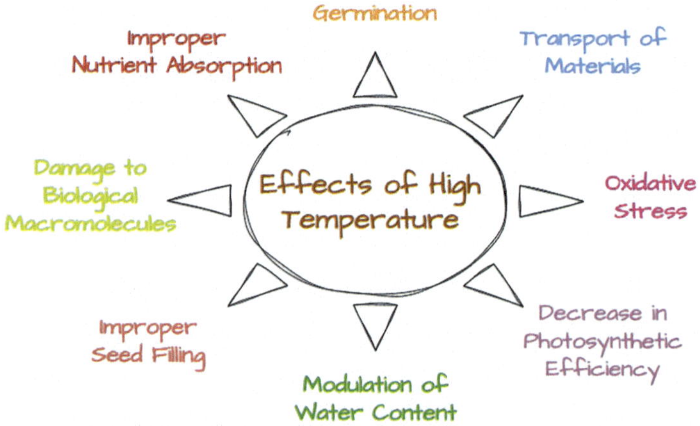

Fig. 22.1 Various effects of heat stress on growth and development of plants. (Kumari et al. 2020)

and tissues. As a result, both severe and mild heat stress cause plant death, fruit and flower abortion, and leaf falling.

A rise in temperature from the thresholds stimulates many mechanisms that affected the rate of development and shortening growing seasons, potentially reducing yield (Stöckle et al. 2018). In the embryo, temperatures beyond 37 °C prevented the production of proteins, which inhibited germination. Also, coleoptile development stopped entirely at 45 °C (Khan and Shahwar 2020).

3.2 *Nutrients Absorption Influenced by Heat Stress*

High temperatures inhibit plant development by reducing nutrient and water absorption (Huang et al. 2012; Shahid et al. 2015). It also influences the source-sink connections between plants, production of enzymes, and nutrients imbalance in plants (Rennenberg et al. 2006). Heat stress reduces protein levels involved in assimilation and absorption of nutrients as well as the total protein content of roots. That is why, because of global warming, increased heat stress may lower crop production and nutritional quality. In future, plants are supposed to experience, more often, a hotter and longer season of abrupt heat stress (such as heat waves) because temperature increases the chances of global warming, which will have an unfavorable effect on plant function (Fatima et al. 2020).

Heat stress has a significant impact just because it typically affects the concentration of important nutrients in plants and, in certain cases, the overall nutritional contents (Soares et al. 2019). Heat stress affected key enzymes involved in food metabolisms, involved in nitrate and ammonium absorption as well as nutrient uptake (Lopez-Delacalle et al. 2020). In addition, the nutrient uptake ratio is lowered down by heat stress, because of a decrease in root hair surface and nutrient absorption (Calleja-Cabrera et al. 2020). The few prior studies have demonstrated that heat stress can be damaging, although it is unknown if heat stress has immediate or long-term effect on root systems (Lipiec et al. 2013).

3.3 *Yield Response to Heat Stress*

Heat and drought have a number of damaging effects on agriculture production, including reduced leaf photosynthesis and increased leaf senescence (Wang et al. 2013). The causes of the decreased yield in high temperatures are reduced reproduction components like pollen embryo and impaired meiosis (Pandey and Rastogi 2019). Abiotic and biotic stresses cause more than 50% of crop losses. Crop production is significantly decreased by HS during the grain filling process. The lower rate of seed filling ultimately reduced the potential for yield (Sehgal et al. 2018).

Heat stress also harms nutritional value of cereals and oilseeds because it significantly drops the levels of carbohydrates, protein, and oil (Wahid et al. 2007).

Sorghum studies demonstrate that during the flowering period, long exposure to high temperatures (above 36–38 °C) for 10–15 days inhibited pollen germination, prevented fertilization, and caused flower abortion.

3.4 *Impact of Heat Stress on Partitioning Assimilation*

Plant development and final production, because of heat stress, reduced source-sink activities (Hassan et al. 2021). However, significant differences in assimilate partitioning have been mentioned under heat stress (Yang et al. 2002). Heat stress influences plant source-sink interactions by decreasing carbon absorption and partitioning in plants (Hassan et al. 2021). Previous investigations showed that heat exhaustion resulted in callus development (Scharf et al. 2012). However, in the current investigation, it seems that heat stress impacted the "sink strength," or the capacity of early flower buds to absorb water to move around while absorbing under HS; the movement of assimilates from leaf area to uptake is also disrupted (Kaur et al. 2021). Wheat photosynthetic rates were highest between 20 and 30 °C. Temperatures above 30 °C inhibited photosynthesis and caused assimilates to be transferred from the leaf area to the reproductive organs (Abrol and Ingram 1996; Mehmood et al. 2018a). Heat stress reduced water-soluble carbohydrate movement from stem to grain and caused a meaningful decrease in overall production (Hassan et al. 2021).

3.5 *Biochemical Responses Under Heat Stress Conditions*

Among major environmental stress, heat stress has negative impact on crop growth and production (Beltagi et al. 2016). Temperature-sensitive biochemical processes are abundant in plant growth and development. Plant responses to high temperatures vary according to plant type, temperature intensity, and temperature duration (Fahad et al. 2015) (Fig. 22.2).

3.6 *Impact of Heat Stress on Plant Physiology*

Temperature rises initially promote plant growth, photosynthesis, respiration, and enzyme activity; but, at a certain point, these factors tend to decline as a result of modifications to the plant's cellular state, membrane fluidity, and organelle properties (Khan and Shahwar 2020). In a variety of developmental phases, plants may be exposed to various types of stress, and in different tissues, their stress response mechanisms may vary as elaborated in Fig. 22.3 (Queitsch et al. 2000). It also reduces the crop's relative water contents and leads to damage of cell structures and associated proteins.

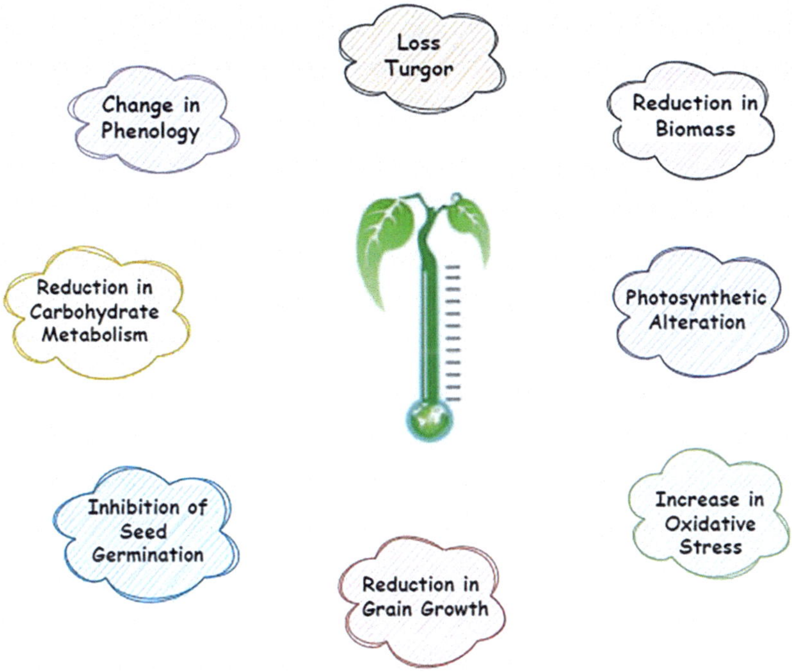

Fig. 22.2 Various physiological responses to heat stress. (dos Santos et al. 2022)

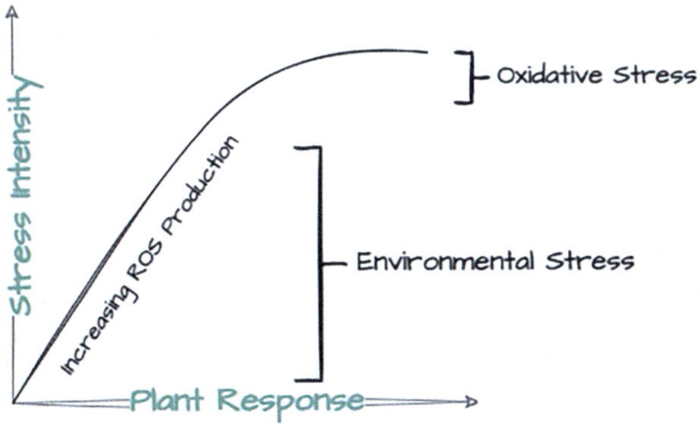

Fig. 22.3 Response of a plant system to increased stress in the context of ROS generation and oxidative stress. (Mullineaux and Baker 2010)

3.6.1 Photosynthesis

High temperatures reduced photosynthesis, which influences the production and development of plants (Javaid et al. 2020a). Photosynthesis rate is heat-dependent physiological process. Photosynthetic pigments, glycolic metabolism, electron transport system, CO_2 fixation pathways, and photosystems all played critical roles in photosynthesis efficiency. Photosynthetic capacity may be reduced when any of these components get damaged (Ashraf and Harris 2013). The rate of photosynthetic activity was reduced by heat stress (5 days, 33 °C) in rice plants. The 15–30% closure of stomata was the reason for the decrease in photosynthesis (Pandey and Rastogi 2019). Photosynthesis includes a particularly complex physiological mechanism in plants, especially gas exchange parameters (Ashraf and Harris 2013; Javaid et al. 2022a). High temperatures change intercellular CO_2 concentration and leaf stomatal conductance. Stomata closure may also contribute to reduced photosynthesis by influencing intercellular CO_2 concentration (Greer and Weedon 2012; Javaid et al. 2022c).

High temperatures also affect plant water relations, decreasing water potential and leaf area, which reduced photosynthetic activity (Netondo et al. 2004). Long-term heat stress also caused a deficiency of carbohydrate reservoirs that disrupts the photosynthesis activity (Abbas et al. 2015; Hassan et al. 2021). Heat stress altered the activity of several enzymes and distribution of energy due to disrupting inactivating oxygen-evolving enzymes and changing regeneration rate of RuBP (Sharma et al. 2020).

3.6.2 Respiration

The rate of respiration increases dramatically as the temperature rises, reaching around 40–50 °C. Nevertheless, beyond 50 °C, respiration decreased significantly because of respiratory process damage (Hassan et al. 2021). Moreover, heat stress causes an increase in ROS production while decreasing ATP production (Zaboli et al. 2019). During heat stress, wheat leaf area respiration increased significantly, resulting in a considerable yield reduction (Almeselmani et al. 2012). Heat stress degrades kinetics of CO_2 and O_2 in Rubisco (Cossani and Reynolds 2012). Heat stress increased O_2 and CO_2 solubility in the leaves of crops, lowering the effectiveness of CO_2 concentration mechanisms. Rubisco is an important enzyme that is well known to catalyze photosynthesis and photorespiration (Keys 1986). The mesophyll CO_2 concentration limits the carboxylase activity under HT conditions. Heat stress reduced photosynthetic efficiency due to CO_2 loss from photorespiration (Hassan et al. 2021).

3.6.3 Damage Due to Oxidative Stress

High temperatures are well recognized to cause oxidative stress in plants, which is due to an imbalance between photosynthesis and respiration brought by heat (Khan and Shahwar 2020). Proteins, membranes, lipids, and DNA are all damaged because of oxidative stress. The plasma membrane is disrupted at different levels by high temperature stress, which also generates activated oxygen species (Das and Roychoudhury 2014; Hassan et al. 2021). Membrane lipids and pigments peroxide cause cellular damage by destroying membrane permeability and function. ROS damaged the photosynthetic machinery and other cellular components, which caused hindrance in metabolic activity and reduced metabolic flow activities, which affect plant growth and production and can cause cell death (Aamer et al. 2021; Ul Hassan et al. 2021).

Like other abiotic stimuli, accumulation of unwanted and damaging ROS species occurs due to heat stress (Javaid et al. 2020b; Mahmood et al. 2022). Different physiological problems in plants are brought on by ROS. For instance, DNA, proteins, lipids, photosynthetic pigments, and even all components of the cell are reacted by OH radicals (Anjum et al. 2015). Plant plasma membrane surfaces accumulate ROS because of prolonged high temperature. These circumstances lead to accumulating the ROS in cells which triggered the cell death (Petrov et al. 2015).

3.6.4 Water Loss

High temperatures are typically linked to less water availability for the plants, which inhibits photosynthesis and depletes the reserves because respiration still requires substrates (Scafaro et al. 2021). This results in a reduction in nutrients supplied to the reproductive organs and a reduction in cellular activity. When water availability is high, plants typically try to regulate water contents in tissues regardless of temperature. Nevertheless, a temperature rise can be harmful when water is scarce (Wahid et al. 2007; Rehman et al. 2022).

Although when water is abundant in the soil, fast decrease of water contents in the tissue of leaves of sugarcane occurred when exposed to high temperatures (Wahid 2007). Due to gas exchange processes like transpiration and photosynthesis, this loss of water is higher throughout the day than at night. Under heat stress conditions, more water is lost during the day due to increased transportation rate than at night, causing stress in snap bean plants (Hassan et al. 2021).

3.6.5 Reproductive Development

Although heat stress affects all plant tissues, reproductive organs are most susceptible. For example, a few degrees higher in temperature during the blooming stage of plants caused a significant loss of grain (Jagadish et al. 2021). Heat stress leads to abrupt floral buds and flower abortion, though sensitivity varies greatly within and among plant species and varieties (Arnao and Hernández-Ruiz 2020). This sterility under abiotic stress conditions inhibits the growth of endosperm and pollen tubes, lowers ovule viability, causes abnormalities in style and stigma positions, causes fewer pollen grains, and lowers the pollen fertility rate (Wahid et al. 2007; Mahmood et al. 2022). The aforementioned conditions cause a decrease in the quantity of pollen on the stigma during the blooming stage (Wu et al. 2020).

4 Adaptation of Plants to Heat Stress

4.1 Avoidance Mechanism and Tolerance Mechanism

The primary processes by which plants respond to heat stress are tolerance and avoidance mechanisms, which are employed in several ways. Plants adopt short-term avoidance or assimilation strategies to deal with high temperatures, such as shifting the orientation of their leaves, cooling through transpiration, or modifying their membrane lipid compositions. Frequent heat-induced plant aspects include stomata shutting, less water loss, and bigger xylem vessels (Fig. 22.4; Srivastava et al. 2012).

Early maturity in a variety of agricultural plants is often linked to decreased yield losses in high temperatures (Adams et al. 2001; Rodríguez et al. 2005). Many plants showed varying degrees of leaf rolling in response to the temperature. The physiological purpose of leaf rolling in maintaining adaptation potential was to boost the efficiency of water metabolism in wheat leaf tissues (Wahid et al. 2007; Naqve et al. 2021).

4.2 Antioxidant Defense

Antioxidant defense system is reliant on the effectiveness of the thermotolerance system (Maestri et al. 2002). When plants are subjected to high temperature stress, they increase the activity of antioxidant enzymes and also accumulate more nonenzymatic compounds. However, most of the time, even in vulnerable genotypes, these increased activities are insufficient to help plants tolerate stress (Almeselmani et al. 2009). As a

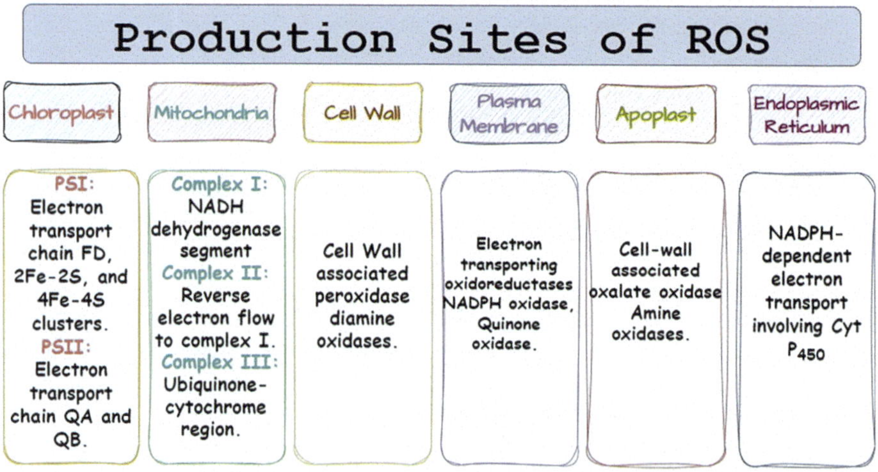

Fig. 22.4 Reactive oxygen species (ROS) production sites in plants. (Hasanuzzaman et al. 2013)

defense mechanism against the damaging effects of ROS, plants have a variety of enzymatic and nonenzymatic antioxidants (Wahid et al. 2007; Nawaz et al. 2022).

Some of the enzymatic antioxidants are as follows:

- Catalase enzyme (CAT)
- Dehydroascorbate reductase enzyme (DHAR)
- Monodehydroascorbate reductase enzyme (MDHAR)
- Superoxide dismutase enzyme (SOD)
- Glutathione peroxidase enzyme (GPX)
- Glutathione S-transferase (GST) and glutathione reductase (GR)

Some of the nonenzymatic antioxidants are as follows:

- Glutathione (GSH)
- Carotenoids
- Anthocyanins
- Ascorbate (AsA)

4.3 Reactive Oxygen Species in Heat Tolerance

Plant hormones can induce thermotolerance that is induced by plant hormones through ROS production, altering the redox signaling and production of NADPH oxidases (Mittler and Blumwald 2015; Mhamdi and Van Breusegem 2018). Hydroxyl radicals and superoxide anions are ionic forms while singlet oxygen and hydrogen peroxide are molecular states of ROS. Recent research has shown that ROS is a significant signal for regulating pathogen infection response and environmental stressors (Mittler et al. 2004).

4.4 Production Sites of ROS in Plants

During different processes like respiration and photosynthesis, ROS are produced in mitochondria, chloroplast, peroxisomes, and other cell sites. ROS are always produced because of electron leakage from chloroplasts, plasma membranes, and mitochondrial electron transport systems. ROS are a particular class of free radical and reactive molecule produced by oxygen (Asada 1987; Hassan et al. 2022).

4.5 Heat Shock Proteins

F. Ritossa, an Italian scientist, discovered heat shock proteins in Drosophila melanogaster in early 1960s while investigating high-temperature stress (Ritossa 1962). HSPs play a significant role in the life cycle of plants because their function goes well beyond protection against biotic and abiotic stressors. However, HSPs were initially identified based on their capacity to tolerate high temperatures (Vierling 1991). HSPs appear to be well-suited for their functions in protein folding, assembly, translocation, and degradation during normal cellular growth and development (Park and Seo 2015).

4.5.1 Classification of HSPs

Heat shock proteins are found in almost all species, including humans, microorganisms, plants, and mammals (Haslbeck and Vierling 2015). HSP families have molecular sizes ranging from 10 to 100 kDa and are present in a variety of cellular compartments (Jee 2016). HSPs are classified into the following subfamilies based on their location and function (Table 22.2).

- Small HSPs
- HSP100
- HSP90
- HSP70
- HSP60

5 Approaches to Enhance Heat Tolerance

Surviving to heat stress is a built-in mechanism in plants; however, heat tolerance capability differs by species and even within species (Poór et al. 2021). New generation molecular breeding, marker-assisted selection (MAS), genome editing approaches, and precision breeding are effective tools in the study of heat tolerance (Singh et al. 2019).

Thermo tolerance is a multigenic characteristic that is regulated by the interaction of genotype and environment affecting the heat tolerance in plants and is

Table 22.2 Location and function of heat shock proteins

Classification	Location	Function	References
Small HSPs	Cytosol, nucleus	Prevent wrong protein aggregation	Bakthisaran et al. (2015)
HSP100	Cytoplasm, nucleus, mitochondria	Thermo tolerance, facilitates reactivation of proteins	Tustumi et al. (2022)
HSP90	Cytoplasm	Prevents from aggregation, facilitates signal transduction	Priya et al. (2019)
HSP70	Cytosol, endoplasmic reticulum	Protection against stress, protein refolding	Lubkowska et al. (2021)
HSP60	Mitochondria, cytosol	Prevent aggregation of denatured proteins, help in protein folding	Deng et al. (2019)

regulated by multigenic characteristics interactions. Traditional breeding is time-consuming and a laborious procedure for developing heat-tolerant species (Rauf et al. 2016). Assistance of MAS significantly accelerates plant breeding programs in terms of heat tolerance development.

5.1 Biotechnological Approaches in Enhancing Heat Stress Tolerance

Biotechnology has recently made significant contributions to the discovery of genes that are involved in heat tolerance in maize and tomato crops (Hassan et al. 2021). Changes in transcriptional and translational activity can control HS tolerance in plants. During the stressful phase, transcription is necessary to facilitate the necessary translational processes (Hassan et al. 2021). Such plant activities change significantly during seed germination, as well as during heat and drought stress (Khaleghi et al. 2019).

The major biotechnology techniques used to increase heat stress resistance in crops are genetic transformation and MAS.

5.1.1 Genetic Transformation

The development and production of essential crops for agriculture are significantly hampered by heat stress, which occurs when high temperatures permanently harm plant development or function. To create "climate-smart" crops, it is essential to understand biochemical, physiological, and molecular processes of thermotolerance as the world's population expands and temperatures rise. Heat stress tolerance in plants offers comprehensive, multidisciplinary overview of the most recent findings in this significant area of research (Akter and Rafiqul Islam 2017). This work investigates heat stress, its effect on agricultural plants, and numerous methods to modify tolerance levels. It presents contributions from an international team of plant scientists and researchers. Due to the detrimental effects it has on crop growth and output, high temperatures have gained international attention. Stress-related proteins are abruptly expressed more when there is heat stress, and these proteins help plants defend themselves and offer tolerance. When plants experience heat stress, antioxidant enzymes and HSPs are crucial. Repression of regular cellular protein synthesis and stimulation of HSPs production are characteristics of the heat shock response (Boopathy et al. 2022). Under HT stress, numerous enzymatic and nonenzymatic antioxidants are upregulated, cell membrane stability is maintained, a variety of suitable solutes are produced, and hormonal changes take place. Plants are protected from damaging radicals that may otherwise harm lipophilic proteins by ROS, which involves numerous mechanisms such as glutathione peroxidase, Fenton reactions, and Haber Weiss (Asthir 2015). The quantitative trait loci and identified genes imparting heat tolerance will be helpful in a marker-assisted breeding program. These approaches will identify relevant genetic variables that may be exploited in plant engineering to increase heat tolerance (Nguyen et al. 2018). This review examines various protective mechanisms for HT tolerance. The impact of abiotic stress on plant productivity has been mitigated to a limited extent by conventional plant breeding. This might be because qualities controlled by numerous genes located at QTL are complex (Monostori et al. 2017). Moreover, traditional breeding for better heat resistance qualities has been successful. For instance, Becker et al. (2016) created the "Ripper" type of wheat, which is drought-resistant. Under Colorado, this cultivar did well with greater grain yields in nonirrigated circumstances. Additionally, the cultivar offers excellent milling and bread-baking qualities. The maize cultivars created by Badu-Apraku and Yallou (2009) produced more grain per acre than the control varieties when planted in drought-stricken areas. These maize varieties also did well when exposed to biotic stress.

5.1.2 Marker-Assisted Selection (MAS)

Because HS is complicated and phenotypic selection for HS tolerance is difficult, MAS is a crucial strategy for enhancing heat tolerance in crops (Hassan et al. 2021). Therefore, using MAS, it is important to identify the genes, either at the level of the entire plant or of each component, that are in control of heat tolerance. Recently, multiple genetic markers have been discovered that diverse abiotic stressors. Using this approach, a significant number of QTLs with significant effects on HS at various plant development stages have been identified and reported (Simko et al. 2021).

Many plants in large populations exhibit varying degrees of adaptability to a new environmental condition, and using this ability, plants change their phenotypic, morphological, physiological, and genetic orientation in response to heat stress (Gantait et al. 2019). It is necessary to identify, select, and employ plant populations in breeding programs that demonstrate sustainable growth under changing environmental conditions (Gilliham et al. 2017). By using only phenotypical observations in conventional breeding programs, genotypes are chosen without knowledge of the choice of linked genes. To speed up the selection of plant populations with traits that improve their resistance to abiotic stress, molecular breeding approaches, in particular marker-assisted selection, provide a significant potential (Hasan et al. 2015). Increased accuracy through the discovery and use of DNA markers (spread across the genome) is connected to features of interest. Furthermore, because environmental conditions cannot alter molecular markers, which are readily observable at any stage of plant development, it is possible to screen for stress-tolerant plant populations even at the seedling stage without wasting time, space, or precious resources. The quantitative fingerprinting, germplasm screening, and marker-assisted selection techniques have all been used (Gantait et al. 2019).

5.2 *Osmic Approaches to Enhance Heat Stress Tolerance*

5.2.1 Planting Duration

A management strategy to achieve maximum production under various heat stress conditions is planting duration (Hassan et al. 2021). Crop yield under HS is improved by choosing an appropriate planting or sowing period that prevents high temperature at anthesis and grain filling. Additionally, sowing timing influences crop growth and development, making choice of appropriate sowing timing a crucial strategy to achieve optimum yield under such challenging circumstances (Briak and Kebede 2021). Crops should be planted at an appropriate period to alleviate the damaging impacts of high temperatures to increase the yield potential and this approach increases yield potential under elevated temperature conditions (Nyakane 2019).

5.2.2 Nutrient and Water Management

Optimal nutrient availability protects plant integrity by assisting with a variety of physiological functions. Nitrogen and magnesium are necessary elements for photosynthesis. Phosphorus is a basic requirement in the production of energy, whereas potassium is needed for enzyme activation and gas exchange (Tränkner et al. 2018). Several authors demonstrate that silicon and potassium enhance the protective effects of several nutrients against abiotic stresses (Ahmad et al. 2019). ROS can be produced at both high and low temperatures, which is terrible for plants (Munjal 2019). Additionally, selenium (Se) protects plants from damaging effects of ROS by serving a structural role in the synthesis of glutathione peroxidase (Chauhan et al. 2019). An essential micronutrient that supports maintaining membrane health is zinc. Permeability and an ideal Zn supply mitigated the harmful effects of heat stress (Hassan et al. 2021). They are all essential for the body to function. This suggests that sufficient mineral feeding is required to keep plants performing well under stress (Teklić et al. 2021).

5.2.3 Seed Priming

Priming of the seed is an effective method for promoting crop stand establishment all over the world (Farooq et al. 2019). Seed priming greatly enhanced seed germination and emergence, which leads to improved growth and development. There is proof that seeds with normal or low vigor can be primed to increase the growth, size, and number of roots (Pawar and Laware 2018). Seed priming improved crop production against abiotic stressors, for example, heat stress, drought, and pests (Prodhan et al. 2022).

6 Conclusion

Heat stress has become a serious concern for agricultural production across the world because it has a considerable influence on plant growth, development, and output. Surprisingly, the frequency and duration of high temperatures determine the extent to which this occurs in different climatic zones. Heat stress, on the other hand, has the largest impact on reproductive development. As a result, the presence of HS throughout the reproductive stage is associated with the most severe decreases in final production. Moreover, excessive heat causes oxidative stress, damaged photosynthetic machinery, reduced photosynthesis, decreased ATP production, and increased membrane permeability. In high-temperature conditions, plants accumulate a variety of metabolites as well as metabolic pathways and activities such as antioxidants, osmoprotectants, and heat shock proteins (HSPs). These changes highlight the importance of physiological and molecular studies in understanding the mechanisms underlying stress responses. Under heat stress, plants may accumulate a variety of thermoprotectants, phytohormones, and signaling molecules that contribute to high temperatures.

References

Aamer M, Bilal Chattha M, Mahmood A, Naqve M, Hassan MU, Shaaban M, Rasul F, Batool M, Rasheed A, Tang H, Chuan Z, Shao J, Huang G (2021) Rice residue-based biochar mitigates N_2O emission from acid red soil. Agronomy 11(12):2462

Abbas T, Balal RM, Shahid MA, Pervez MA, Ayyub CM, Aqueel MA, Javaid MM (2015) Silicon-induced alleviation of NaCl toxicity in okra (*Abelmoschus esculentus*) is associated with enhanced photosynthesis, osmoprotectants and antioxidant metabolism. Acta Physiol Plant 37:1–15

Abrol YP, Ingram KT (1996) Effects of higher day and night temperatures on growth and yields of some crop plants. In: Global climate change and agricultural production direct and indirect effects of changing hydrological, pedological and plant physiological processes. Wiley, Chichester, pp 124–140

Adams S, Cockshull K, Cave C (2001) Effect of temperature on the growth and development of tomato fruits. Ann Bot 88(5):869–877

Ahmad P, Ahanger MA, Alam P, Alyemeni MN, Wijaya L, Ali S, Ashraf M (2019) Silicon (Si) supplementation alleviates NaCl toxicity in mung bean [*Vigna radiata* (L.) Wilczek] through the modifications of physio-biochemical attributes and key antioxidant enzymes. J Plant Growth Regul 38:70–82

Akter N, Rafiqul Islam M (2017) Heat stress effects and management in wheat. A review. Agron Sustain Dev 37:1–17

Almeselmani M, Deshmukh P, Sairam R (2009) High temperature stress tolerance in wheat genotypes: role of antioxidant defence enzymes. Acta Agron Hung 57(1):1–14

Almeselmani M, Deshmukh P, Chinnusamy V (2012) Effects of prolonged high temperature stress on respiration, photosynthesis and gene expression in wheat (*Triticum aestivum* L.) varieties differing in their thermotolerance. Plant Stress 6(1):25–32

Anjum NA, Sofo A, Scopa A, Roychoudhury A, Gill SS, Iqbal M, Lukatkin AS, Pereira E, Duarte AC, Ahmad I (2015) Lipids and proteins—major targets of oxidative modifications in abiotic stressed plants. Environ Sci Pollut Res 22:4099–4121

Arnao M, Hernández-Ruiz J (2020) Melatonin in flowering, fruit set and fruit ripening. Plant Reprod 33:77–87

Asada K (1987) Production and scavenging of active oxygen in chloroplasts. In: Kyle DJ, Osmond CB, Arntzen CJ (eds) Photoinhibition. Elsevier, Amsterdam, pp 227–287

Ashraf M, Harris PJ (2013) Photosynthesis under stressful environments: an overview. Photosynthetica 51:163–190

Asthir B (2015) Protective mechanisms of heat tolerance in crop plants. J Plant Interact 10:202–210

Asthir B, Thapar R, Bains NS, Farooq M (2015) Biochemical responses of thiourea in ameliorating high temperature stress by enhancing antioxidant defense system in wheat. Russ J Plant Physiol 62 (6):875–882. https://doi.org/10.1134/S1021443715050040

Badu-Apraku B, Yallou CG (2009) Registration of Striga-Resistant and Drought-Tolerant Tropical Early Maize Populations TZE-W Pop DT STR C4 and TZE-Y Pop DT STR C4. J Plant Regist 3(1):86–90. https://doi.org/10.3198/jpr2008.06.0356crg

Bakthisaran R, Tangirala R, Rao CM (2015) Small heat shock proteins: role in cellular functions and pathology. Biochim Biophys Acta, Proteins Proteomics 1854(4):291–319

Becker SR, Byrne PF, Reid SD, Bauerle WL, McKay JK, Haley SD (2016) Root traits contributing to drought tolerance of synthetic hexaploid wheat in a greenhouse study. Euphytica 207:213–224

Boopathy LRA, Jacob-Tomas S, Alecki C, Vera M (2022) Mechanisms tailoring the expression of heat shock proteins to proteostasis challenges. J Biol Chem 298:101796

Briak H, Kebede F (2021) Wheat (*Triticum aestivum*) adaptability evaluation in a semi-arid region of Central Morocco using APSIM model. Sci Rep 11(1):23173

Calleja-Cabrera J, Boter M, Oñate-Sánchez L, Pernas M (2020) Root growth adaptation to climate change in crops. Front Plant Sci 11:544

Chadha A, Florentine S, Chauhan BS, Long B, Jayasundera M, Javaid MM, Turville C (2019) Environmental factors affecting the germination and seedling emergence of two populations of an emerging agricultural weed: wild lettuce (*Lactuca serriola*). Crop Pasture Sci 70(8):709–717

Chauhan R, Awasthi S, Srivastava S, Dwivedi S, Pilon-Smits EA, Dhankher OP, Tripathi RD (2019) Understanding selenium metabolism in plants and its role as a beneficial element. Crit Rev Environ Sci Technol 49(21):1937–1958

Cohen I, Zandalinas SI, Fritschi FB, Sengupta S, Fichman Y, Azad RK, Mittler R (2021) The impact of water deficit and heat stress combination on the molecular response, physiology, and seed production of soybean. Physiol Plant 172(1):41–52

Cossani CM, Reynolds MP (2012) Physiological traits for improving heat tolerance in wheat. Plant Physiol 160(4):1710–1718

Daryanto S, Wang L, Jacinthe P-A (2016) Global synthesis of drought effects on maize and wheat production. PLoS One 11(5):e0156362

Das K, Roychoudhury A (2014) Reactive oxygen species (ROS) and response of antioxidants as ROS-scavengers during environmental stress in plants. Front Environ Sci 2:53

de Castro Cavallari F, Leal CLV, Zvi R, Hansen PJ (2019) Effects of melatonin on production of reactive oxygen species and developmental competence of bovine oocytes exposed to heat shock and oxidative stress during in vitro maturation. Zygote 27(3):180–186

Deng Y, Hu Z, Chai Z, Tang YZ (2019) Molecular cloning of heat shock protein 60 (Hsp60) and 10 (Hsp10) genes from the cosmopolitan and harmful dinoflagellate *Scrippsiella trochoidea* and their differential transcriptions responding to temperature stress and alteration of life cycle. Mar Biol 166:1–14

dos Santos TB, Ribas AF, de Souza SGH, Budzinski IGF, Domingues DS (2022) Physiological responses to drought, salinity, and heat stress in plants: a review. Stress 2(1):113–135

Elahi E, Khalid Z, Tauni MZ, Zhang H, Lirong X (2022) Extreme weather events risk to crop-production and the adaptation of innovative management strategies to mitigate the risk: a retrospective survey of rural Punjab, Pakistan. Technovation 117:102255

El-Beltagi HS, Ahmed OK, Hegazy AE (2016) Protective effect of nitric oxide on high temperature induced oxidative stress in wheat (*Triticum aestivum*) callus culture. Not Sci Biol 8(2):192–198

Essemine J, Ammar S, Bouzid S (2010) Impact of heat stress on germination and growth in higher plants: physiological, biochemical and molecular repercussions and mechanisms of defence. J Biol Sci 10(6):565–572

Fahad S, Hussain S, Matloob A, Khan FA, Khaliq A, Saud S, Hassan S, Shan D, Khan F, Ullah N (2015) Phytohormones and plant responses to salinity stress: a review. Plant Growth Regul 75:391–404

Farooq M, Gogoi N, Barthakur S, Baroowa B, Bharadwaj N, Alghamdi SS, Siddique KH (2017) Drought stress in grain legumes during reproduction and grain filling. J Agron Crop Sci 203(2):81–102

Farooq M, Usman M, Nadeem F, ur Rehman H, Wahid A, Basra SM, Siddique KH (2019) Seed priming in field crops: potential benefits, adoption and challenges. Crop Pasture Sci 70(9):731–771

Fatima Z, Ahmed M, Hussain M, Abbas G, Ul-Allah S, Ahmad S, Ahmed N, Ali MA, Sarwar G, Iqbal P (2020) The fingerprints of climate warming on cereal crops phenology and adaptation options. Sci Rep 10(1):1–21

Gabaldón-Leal C, Webber H, Otegui ME, Slafer GA, Ordóñez R, Gaiser T, Lorite I, Ruiz-Ramos M, Ewert F (2016) Modelling the impact of heat stress on maize yield formation. Field Crop Res 198:226–237

Gantait S, Sarkar S, Verma SK (2019) Marker-assisted selection for abiotic stress tolerance in crop plants. In: Molecular plant abiotic stress: biology and biotechnology. Wiley, pp 335–368

Gaur PM, Jukanti AK, Samineni S, Chaturvedi SK, Basu PS, Babbar A, Jayalakshmi V, Nayyar H, Devasirvatham V, Mallikarjuna N, Krishnamurthy L, Laxmipathi Gowda CL (2013) Climate change and heat stress tolerance in chickpea. In: Climate change and plant abiotic stress tolerance, pp 837–856. https://doi.org/10.1002/9783527675265.ch31

Gilliham M, Able JA, Roy SJ (2017) Translating knowledge about abiotic stress tolerance to breeding programmes. Plant J 90(5):898–917. https://doi.org/10.1111/tpj.13456

Greer DH, Weedon MM (2012) Modelling photosynthetic responses to temperature of grapevine (*Vitis vinifera* cv. Semillon) leaves on vines grown in a hot climate. Plant Cell Environ 35(6):1050–1064

Hasan MM, Rafii MY, Ismail MR, Mahmood M, Rahim HA, Alam MA, Ashkani S, Malek MA, Latif MA (2015) Marker-assisted backcrossing: a useful method for rice improvement. Biotechnol Biotechnol Equip 29(2):237–254

Hasanuzzaman M, Nahar K, Alam MM, Roychowdhury R, Fujita M (2013) Physiological, biochemical, and molecular mechanisms of heat stress tolerance in plants. Int J Mol Sci 14 (5):9643–9684

Haslbeck M, Vierling E (2015) A first line of stress defense: small heat shock proteins and their function in protein homeostasis. J Mol Biol 427(7):1537–1548

Hassan MU, Chattha MU, Khan I, Chattha MB, Barbanti L, Aamer M, Iqbal MM, Nawaz M, Mahmood A, Ali A (2021) Heat stress in cultivated plants: nature, impact, mechanisms, and mitigation strategies—a review. Plant Biosyst 155(2):211–234

Hassan MU, Ghareeb RY, Nawaz M, Mahmood A, Shah AN, Abdel-Megeed A, Abdelsalam NR, Hashem M, Alamri S, Thabit MA (2022) Melatonin: a vital pro-tectant for crops against heat stress: mechanisms and prospects. Agronomy 12(5):1116

Huang B, Rachmilevitch S, Xu J (2012) Root carbon and protein metabolism associated with heat tolerance. J Exp Bot 63(9):3455–3465

Hussain S, Huang J, Huang J, Ahmad S, Nanda S, Anwar S, Shakoor A, Zhu C, Zhu L, Cao X, Jin Q, Zhang J (2020) Rice production under climate change: adaptations and mitigating strategies. In: Fahad S, Hasanuzzaman M, Alam M et al (eds) Environment, climate, plant and vegetation growth. Springer, Cham, pp 659–686. https://doi.org/10.1007/978-3-030-49732-3_26

Jagadish SK, Way DA, Sharkey TD (2021) Plant heat stress: concepts directing future research. Plant Cell Environ 44(7):1992–2005

Janská A, Maršík P, Zelenková S, Ovesná J (2010) Cold stress and acclimation–what is important for metabolic adjustment? Plant Biol 12(3):395–405

Javaid MM, Waheed H, Nazami N, Ashraf M, Li F-M, Tanveer A (2020a) Response of chickpea to foliar supply of Hoagland's solution under rain-fed conditions. Semina Cienc Agrar 41:3053–3066

Javaid M, Zia A, Waheed H, Nargis J, Shahid A, Aziz A, Wasaya A (2020b) Effect of isoproturon with and without adjuvants on photosynthetic attributes of wheat and its associated weeds. Planta Daninha 38. https://doi.org/10.1590/S0100-83582020380100024

Javaid MM, Florentine SK, Ashraf M, Mahmood A, Sattar A, Wasaya A, Li FM (2022a) Photosynthetic activity and water use efficiency of *Salvia verbenaca* L. under elevated CO2 and water-deficit conditions. J Agron Crop Sci 208(4):536–551

Javaid MM, Mahmood A, Alshaya DS, AlKahtani MD, Waheed H, Wasaya A, Khan SA, Naqve M, Haider I, Shahid MA (2022b) Influence of environmental factors on seed germination and seedling characteristics of perennial ryegrass (*Lolium perenne* L.). Sci Rep 12(1):9522

Javaid MM, Wang X, Florentine SK, Ashraf M, Mahmood A, Li F-M, Fiaz S (2022c) Effects on photosynthetic response and biomass productivity of *Acacia longifolia* ssp. *longifolia* under elevated CO2 and water-limited regimes. Front Plant Sci 13:817730

Jee H (2016) Size dependent classification of heat shock proteins: a mini-review. J Exerc Rehabil 12(4):255

Kaur H, Ozga JA, Reinecke DM (2021) Balancing of hormonal biosynthesis and catabolism pathways, a strategy to ameliorate the negative effects of heat stress on reproductive growth. Plant Cell Environ 44(5):1486–1503

Keys A (1986) Rubisco: its role in photorespiration. Philos Trans R Soc 313(1162):325–336

Khaleghi A, Naderi R, Brunetti C, Maserti BE, Salami SA, Babalar M (2019) Morphological, physiochemical and antioxidant responses of *Maclura pomifera* to drought stress. Sci Rep 9(1):19250

Khan Z, Shahwar D (2020) Role of heat shock proteins (HSPs) and heat stress tolerance in crop plants. In: Roychowdhury R, Choudhury S, Hasanuzzaman M, Srivastava S (eds) Sustainable agriculture in the era of climate change. Springer, Cham, pp 211–234. https://doi.org/10.1007/978-3-030-45669-6_9

Kumar B, Verma SK, Singh H (2011) Effect of temperature on seed germination parameters in Kalmegh (*Andrographis paniculata* Wall. ex Nees.). Ind Crop Prod 34(1):1241–1244

Kumari P, Rastogi A, Yadav S (2020) Effects of heat stress and molecular mitigation approaches in orphan legume, chickpea. Mol Biol Rep 47:4659–4670

Lipiec J, Doussan C, Nosalewicz A, Kondracka K (2013) Effect of drought and heat stresses on plant growth and yield: a review. Int Agrophys 27(4):463

Lohani N, Singh MB, Bhalla PL (2022) Short-term heat stress during flowering results in a decline in canola seed productivity. J Agron Crop Sci 208(4):486–496

Lopez-Delacalle M, Camejo DM, García-Martí M, Nortes PA, Nieves-Cordones M, Martínez V, Rubio F, Mittler R, Rivero RM (2020) Using tomato recombinant lines to improve plant tolerance to stress combination through a more efficient nitrogen metabolism. Front Plant Sci 10:1702

Lubkowska A, Pluta W, Strońska A, Lalko A (2021) Role of heat shock proteins (HSP70 and HSP90) in viral infection. Int J Mol Sci 22(17):9366

Maestri E, Klueva N, Perrotta C, Gulli M, Nguyen HT, Marmiroli N (2002) Molecular genetics of heat tolerance and heat shock proteins in cereals. Plant Mol Biol 48:667–681

Mahmood A, Bibi S, Naqve M, Javaid MM, Zia MA, Jabbar A, Ud-Din W, Attia KA, Khan N, Al-Doss AA (2022) Physiological, biochemical, and yield responses of linseed (*Linum usitatissimum* L.) in α-tocopherol-mediated alleviation of salinity stress. Front Plant Sci 13:867172

Mehmood A, Tanveer A, Javed MM, Nadeem MA, Naeem M, Abbas T (2018a) Estimation of economic threshold level of alligator weed (*Alternanthera philoxeroides* (Mart.) Griseb.) to tackle grain quality and yield losses in rice. Arch Agron Soil Sci 64(2):208–218

Mehmood A, Naeem M, Khalid F, Saeed Y, Abbas T, Jabran K, Sarwar MA, Tanveer A, Javaid MM (2018b) Identification of phytotoxins in different plant parts of *Brassica napus* and their influence on mung bean. Environ Sci Pollut Res 25:18071–18080

Mhamdi A, Van Breusegem F (2018) Reactive oxygen species in plant development. Development 145:dev164376

Mittler R, Blumwald E (2015) The roles of ROS and ABA in systemic acquired acclimation. Plant Cell 27(1):64–70

Mittler R, Vanderauwera S, Gollery M, Van Breusegem F (2004) Reactive oxygen gene network of plants. Trends Plant Sci 9(10):490–498

Monostori I, Szira F, Tondelli A, Arendas T, Gierczik K, Cattivelli L, Galiba G, Vagujfalvi A (2017) Genome-wide association study and genetic diversity analysis on nitrogen use efficiency in a Central European winter wheat (*Triticum aestivum* L.) collection. PLoS One 12(12):e0189265

Mudassir MA, Hussain SZ, Khan M, Asma ST, Iqbal Z, Huma Z, Ullah N, Zhang H, Ansari TM, Hussain I (2018) Polyacrylamide exotemplate-assisted synthesis of hierarchically porous nanostructured TiO_2 macrobeads for efficient photodegradation of organic dyes and microbes. RSC Adv 8(52):29628–29636

Mullineaux PM, Baker NR (2010) Oxidative stress: antagonistic signaling for acclimation or cell death. Plant Physiol 154(2):521–525

Munjal R (2019) Oxidative stress and antioxidant defense in plants under high temperature. In: Reactive oxygen, nitrogen and sulfur species in plants: production, metabolism, signaling and defense mechanisms. Wiley, pp 337–352

Naqve M, Shahbaz M, Naseer M, Mahmood A (2021) Alpha tocopherol application as seed priming alters antioxidative defense system of okra against salinity stress. Pol J Environ Stud 30(5):4143–4152

Nawaz M, Hassan MU, Chattha MU, Mahmood A, Shah AN, Hashem M, Alamri S, Batool M, Rasheed A, Thabit MA (2022) Trehalose: a promising osmo-protectant against salinity stress—physiological and molecular mechanisms and future prospective. Mol Biol Rep 49:1–17

Netondo GW, Onyango JC, Beck E (2004) Sorghum and salinity: I. Response of growth, water relations, and ion accumulation to NaCl salinity. Crop Sci 44(3):797–805

Nguyen HC, Lin KH, Ho SL, Chiang CM, Yang CM (2018) Enhancing the abiotic stress tolerance of plants: from chemical treatment to biotechnological approaches. Physiol Plant 164(4):452–466

Nyakane N (2019) Use of magnetic Field nutrient solution amended with mycorrhizal and ratio of calcium to magnesium on yield and oil quality of rose geranium (*Pelargonium Graveolens* L.)

Onaga G, Wydra K (2021) Mechanisms of plant abiotic stress response and tolerance

Onyekachi OG, Boniface OO, Gemlack NF, Nicholas N (2019) The effect of climate change on abiotic plant stress: a review. In: Abiotic biotic stress plants. IntechOpen, p 17

Pandey AK, Rastogi A (2019) Mechanisms of heat stress tolerance in plants: a physiological approach. J Pharmacogn Phytochem 8(5):1732–1737

Park C-J, Seo Y-S (2015) Heat shock proteins: a review of the molecular chaperones for plant immunity. Plant Pathol J 31(4):323

Pawar V, Laware S (2018) Seed priming a critical review. Int J Sci Res Biol Sci 5:94–101

Petrov V, Hille J, Mueller-Roeber B, Gechev TS (2015) ROS-mediated abiotic stress-induced programmed cell death in plants. Front Plant Sci 6:69

Poór P, Nawaz K, Gupta R, Ashfaque F, Khan MIR (2021) Ethylene involvement in the regulation of heat stress tolerance in plants. Plant Cell Rep 2021:1–24

Priya M, Dhanker OP, Siddique KH, HanumanthaRao B, Nair RM, Pandey S, Singh S, Varshney RK, Prasad PV, Nayyar H (2019) Drought and heat stress-related proteins: an update about their functional relevance in imparting stress tolerance in agricultural crops. Theor Appl Genet 132:1607–1638

Prodhan MM, Sarker U, Hoque MA, Biswas MS, Ercisli S, Assouguem A, Ullah R, Almutairi MH, Mohamed HR, Najda A (2022) Foliar application of GA3 stimulates seed production in cauliflower. Agronomy 12(6):1394

Queitsch C, Hong S-W, Vierling E, Lindquist S (2000) Heat shock protein 101 plays a crucial role in thermotolerance in *Arabidopsis*. Plant Cell 12(4):479–492

Rauf S, Al-Khayri JM, Zaharieva M, Monneveux P, Khalil F (2016) Breeding strategies to enhance drought tolerance in crops. In: Advances in plant breeding strategies: agronomic, abiotic and biotic stress traits. Springer, pp 397–445

Rehman S, Chattha MU, Khan I, Mahmood A, Hassan MU, Al-Huqail AA, Salem MZ, Ali HM, Hano C, El-Esawi MA (2022) Exogenously applied trehalose augments cadmium stress tolerance and yield of mung bean (*Vigna radiata* L.) grown in soil and hydroponic systems through reducing cd uptake and enhancing photosynthetic efficiency and antioxidant defense systems. Plants 11(6):822

Rennenberg H, Loreto F, Polle A, Brilli F, Fares S, Beniwal R, Gessler A (2006) Physiological responses of forest trees to heat and drought. Plant Biol 8(05):556–571

Ritossa F (1962) A new puffing pattern induced by temperature shock and DNP in Drosophila. Experientia 18(12):571–573

Rodríguez M, Canales E, Borrás-Hidalgo O (2005) Molecular aspects of abiotic stress in plants. Biotecnol Apl 22(1):1–10

Scafaro AP, Fan Y, Posch BC, Garcia A, Coast O, Atkin OK (2021) Responses of leaf respiration to heatwaves. Plant Cell Environ 44(7):2090–2101

Scharf K-D, Berberich T, Ebersberger I, Nover L (2012) The plant heat stress transcription factor (Hsf) family: structure, function and evolution. Biochim Biophys Acta Gene Regul Mech BBA-Gene Regul Mech 1819(2):104–119

Sehgal A, Sita K, Siddique KH, Kumar R, Bhogireddy S, Varshney RK, Hanumantha Rao B, Nair RM, Prasad PV, Nayyar H (2018) Drought or/and heat-stress effects on seed filling in food crops: impacts on functional biochemistry, seed yields, and nutritional quality. Front Plant Sci 9:1705

Shahid MA, Balal RM, Pervez MA, Abbas T, Aqueel MA, Javaid M, Garcia-Sanchez F (2015) Foliar spray of phyto-extracts supplemented with silicon: an efficacious strategy to alleviate the salinity-induced deleterious effects in pea (*Pisum sativum* L.). Turk J Bot 39(3):408–419

Sharma A, Kumar V, Shahzad B, Ramakrishnan M, Singh Sidhu GP, Bali AS, Handa N, Kapoor D, Yadav P, Khanna K (2020) Photosynthetic response of plants under different abiotic stresses: a review. J Plant Growth Regul 39:509–531

Sihi S, Bakshi S, Maiti S, Nayak A, Sengupta DN (2022) Analysis of DNA polymerase λ activity and gene expression in response to salt and drought stress in *Oryza sativa* Indica rice cultivars. J Plant Growth Regul 41(4):1499–1515

Simko I, Jia M, Venkatesh J, Kang B-C, Weng Y, Barcaccia G, Lanteri S, Bhattarai G, Foolad MR (2021) Genomics and marker-assisted improvement of vegetable crops. Crit Rev Plant Sci 40(4):303–365

Singh B, Salaria N, Thakur K, Kukreja S, Gautam S, Goutam U (2019) Functional genomic approaches to improve crop plant heat stress tolerance. F1000 Res 8(1721):1721

Smertenko A, DRÁBER P, Viklický V, Opatrný Z (1997) Heat stress affects the organization of microtubules and cell division in *Nicotiana tabacum* cells. Plant Cell Environ 20(12):1534–1542

Soares JC, Santos CS, Carvalho SMP, Pintado MM, Vasconcelos MW (2019) Preserving the nutritional quality of crop plants under a changing climate: importance and strategies. Plant Soil 443(1):1–26. https://doi.org/10.1007/s11104-019-04229-0

Srivastava S, Pathak AD, Gupta PS, Shrivastava AK, Srivastava AK (2012) Hydrogen peroxide-scavenging enzymes impart tolerance to high temperature induced oxidative stress in sugarcane. J Environ Biol 33(3):657

Stöckle CO, Higgins S, Nelson R, Abatzoglou J, Huggins D, Pan W, Karimi T, Antle J, Eigenbrode SD, Brooks E (2018) Evaluating opportunities for an increased role of winter crops as adaptation to climate change in dryland cropping systems of the US Inland Pacific Northwest. Clim Chang 146:247–261

Sun Q, Zhao Y, Zhang Y, Chen S, Ying Q, Lv Z, Che X, Wang D (2022) Heat stress may cause a significant reduction of rice yield in China under future climate scenarios. Sci Total Environ 818:151746

Tang R, Niu S, Zhang G, Chen G, Haroon M, Yang Q, Rajora OP, Li X-Q (2018) Physiological and growth responses of potato cultivars to heat stress. Botany 96(12):897–912

Tanveer A, Khan MA, Ali HH, Javaid MM, Raza A, Chauhan BS (2020) Influence of different environmental factors on the germination and seedling emergence of *Ipomoea eriocarpa* R. Br Crop Protect 130:105070

Teklić T, Parađiković N, Špoljarević M, Zeljković S, Lončarić Z, Lisjak M (2021) Linking abiotic stress, plant metabolites, biostimulants and functional food. Ann Appl Biol 178(2):169–191

Thornton PK, Ericksen PJ, Herrero M, Challinor AJ (2014) Climate variability and vulnerability to climate change: a review. Glob Chang Biol 20(11):3313–3328

Tränkner M, Tavakol E, Jákli B (2018) Functioning of potassium and magnesium in photosynthesis, photosynthate translocation and photoprotection. Physiol Plant 163(3):414–431

Tustumi F, Agareno GA, Galletti RP, da Silva RBR, Quintas JG, Sesconetto LA, Szor DJ, Wolosker N (2022) The role of the heat-shock proteins in esophagogastric cancer. Cell 11(17):2664

Ugarte C, Calderini DF, Slafer GA (2007) Grain weight and grain number responsiveness to pre-anthesis temperature in wheat, barley and triticale. Field Crop Res 100(2–3):240–248

Ul Hassan M, Rasool T, Iqbal C, Arshad A, Abrar M, Abrar MM, Habib-ur-Rahman M, Noor MA, Sher A, Fahad S (2021) Linking plants functioning to adaptive responses under heat stress conditions: a mechanistic review. J Plant Growth Regul 2021:1–18

Van der Merwe R, Labuschagne MT, Herselman L, Hugo A (2015) Effect of heat stress on seed yield components and oil composition in high-and mid-oleic sunflower hybrids. S Afr J Plant Soil 32(3):121–128

Vierling E (1991) The roles of heat shock proteins in plants. Annu Rev Plant Biol 42(1):579–620

Vollenweider P, Günthardt-Goerg MS (2005) Diagnosis of abiotic and biotic stress factors using the visible symptoms in foliage. Environ Pollut 137(3):455–465

Waheed H, Javaid MM, Shahid A, Ali HH, Nargis J, Mehmood A (2019) Impact of foliar-applied Hoagland's nutrient solution on growth and yield of mash bean (*Vigna mungo* L.) under different growth stages. J Plant Nutr 42(10):1133–1141

Waheed H, Khan MA, Khan HMT, Javaid MM, Rahman FU, Aslam MM (2020) Enhancement of some key physiological, morphological and biochemical traits of watermelon induced by *Trichoderma harzianum* fungi. Semina Cienc Agrar 41(5 suppl 1):2047–2060

Wahid A (2007) Physiological implications of metabolite biosynthesis for net assimilation and heat-stress tolerance of sugarcane (*Saccharum officinarum*) sprouts. J Plant Res 120:219–228

Wahid A, Gelani S, Ashraf M, Foolad MR (2007) Heat tolerance in plants: an overview. Environ Exp Bot 61(3):199–223

Wang P, Sun X, Li C, Wei Z, Liang D, Ma F (2013) Long-term exogenous application of melatonin delays drought-induced leaf senescence in apple. J Pineal Res 54(3):292–302

Waqas MA, Kaya C, Riaz A, Farooq M, Nawaz I, Wilkes A, Li Y (2019) Potential mechanisms of abiotic stress tolerance in crop plants induced by thiourea. Front Plant Sci 10:1336

Wu C, Cui K, Tang S, Li G, Wang S, Fahad S, Nie L, Huang J, Peng S, Ding Y (2020) Intensified pollination and fertilization ameliorate heat injury in rice (*Oryza sativa* L.) during the flowering stage. Field Crop Res 252:107795

Yang J, Sears R, Gill B, Paulsen G (2002) Genotypic differences in utilization of assimilate sources during maturation of wheat under chronic heat and heat shock stresses. Euphytica 125:179–188

Zaboli G, Huang X, Feng X, Ahn DU (2019) How can heat stress affect chicken meat quality: a review. Poult Sci 98(3):1551–1556

Chapter 23
Mechanisms and Responses to Enhancing Pollutants Stress Tolerance in Crop Plants

Arun Kumar Maurya

Abstract Plants are very robust and tolerant group of organisms in the nature. They face all environmental changes in static conditions by evolving and developing morpho-anatomical structures and physio-biochemical mechanisms. Various man-made substances acting as pollutants have entered into the environment and causing pollution have posed danger and risk to the ecosystem health and badly affected all organisms including the crop plants. The nature of pollutants is solid (paper, tin, heavy metals, toxic metals, metalloids, plastics); liquid (oil spills, pesticides, detergents); gases (CO, CH_4, NOx, SO_2, SF_6, CFC and PFC); and radiations (ozone, X-rays, UV-radiation, radioactive radiation). These pollutants are present in various concentrations and amounts and have different impacts on crop plants. The effects are visible ultimately as decline in production and productivity of crops. Plants have evolved diverse mechanisms in responses to pollutants to counteract the pollution stress. It can be seen as bioaccumulation, biomass partitioning, detoxification, modification, phytochelation, biotransformation, exclusion, increased hormone, and metabolites synthesis. Some of these mechanisms are general in nature, while others are species-specific, depending upon the type and nature of the pollution stress that shows various responses in relation with nature and type of pollutants and interaction with plant. These responses are visible at morphological, anatomical, physiological, biochemical, and molecular levels. These responses are an outcome of tolerance capacity of the plant. The pollutant and plant interaction further give rise to clues to help plant to evolve against the pollutant. The diverse and ever-rising threat of pollutant has made human being conscious to regulate, control, and protect the ecosystem. It has pushed to invent and use novel tools and techniques that help to overcome or clean up the environment. These innovative techniques are nanotechnology, bioremediation, use of unmanned aerial vehicle, and modern biotechnological tools. They help to develop tolerant crops by either classical or modern biotechnology tools to meet the rising need of human population across globe.

A. K. Maurya (✉)
Department of Botany, Multanimal Modi College, Ghaziabad, Uttar Pradesh, India
e-mail: akmaurya@mmcmodinagar.ac.in

M. Hasanuzzaman (ed.), *Climate-Resilient Agriculture, Vol 2*,
https://doi.org/10.1007/978-3-031-37428-9_23

Keywords Antioxidants · Heavy metals · Oxidative stress · Ozone · Pollution

1 Introduction

Pollution is threatening every organism on earth in current scenario. Diverse nature and types of pollutants are released by man-made activities into the environment ranging from solid, liquid, gas, and radiation including noise. Pollutants are affecting air, water, soil, and biological domain in various permutation, combination, and intensity. Plants are an integral component of biological domain of ecosystem and therefore are not intact from effects of pollutants. These pollutants affect the ecosystem and develop stress on plants which have been classified as abiotic stress. It influences morphological, anatomical, physiological, biochemical, and molecular aspects of plants including crop plants. Current scenario of climate change, global warming, and environmental pollution are becoming lethal combination for inducing various abiotic and biotic stress leading to sharp decline in plant growth, development, and survival. Therefore, there is great need to act swiftly to mitigate such multifactorial stress and find ways to augment the tolerance of crops (Zandalinas et al. 2021).

Most of the pollutants causing environmental pollution and affecting crop plants are man-made and released by human being into ecosystem knowingly or unknowingly. The pollutants can be classified as per their nature affecting crop plants as follows:

1. Air pollutants (CO, CH_4, NOx, SO_2, SF_6, VOx, CFC, and PFC)
2. Water pollutants (oil spills, pesticides, detergents, sewage waste)
3. Soil pollutants (e-waste, pesticide, heavy metals, metalloids)
4. Radiations (ozone, X-rays, UV lights, radioactive pollutants)
5. Pollutants having potential to pollute air, water, and soil.

Due to rapidly rising urbanization and industrialization, release of diversified pollutants causing great threat and resulting into great impacts on crop plant physiology and consequent effects are visible as crop loss. Depending upon factors and their interactions such as geographic zone, climatic conditions, anthropogenic activities, and time develops complex and variable air pollution compositions (Mishra et al. 2021). Air pollution considers harmful, reactive, gaseous fraction, and particulate matter which are above certain limit. Gases include carbon-based pollutants, carbon monoxides and dioxide (CO, CO_2), methane (CH_4), and volatile organic compounds (VOCs), nitrogenous pollutants include ammonia (NH_3), ammonium (NH_4^+), dinitrogen tetroxide (N_2O_4), nitrogen mono- and dioxide (NO, NO_2); sulfur-based (sulphur dioxide (SO_2)); ozone (O_3), mercuric vapours (Hg), chlorine (Cl_2), and fluorides (HF, SiF_6, CF_4 and F_2). Their interactions between and among them and with other atmospheric gaseous molecules can form other, even more harmful secondary compounds. Key gaseous organic carbon forms causing atmospheric pollution belong to VOCs that include benzene, ethylbenzene, hexane (C_6H_6), xylenes (BTEX), and phenol vapours (Molina and Segura 2021).

Air pollutants are released from point as well as non-point pollution sources. They show great impact and spread rapidly in comparison with soil and water pollutants. Air pollutants have widespread and greater impacts on the morpho-physio and biochemical changes in crop's plants. It is important to note that all air pollutants are not harmful; some have beneficial roles in plant growth and yields, but most of them modify physiological and morphological processes. To minimize negative impacts of air pollutants on crop plants, there is greater need for developing better understanding about interaction and uptake of air pollutants vis-à-vis response shown by crop plants for making policy and programme to protect crop plants (Shrestha et al. 2022). Heavy metals (HMs) and polycyclic aromatic hydrocarbon (PAHs) are another group of toxic atmospheric pollutants whose effects are dependent upon different environmental conditions such as soil pH, temperature, and plants' physiological and genetic status (Shrestha et al. 2022).

Ozone is a gaseous, highly reactive, triatomic oxygen molecule and prominently found in stratosphere, but considered as phytotoxic air pollutant when present in troposphere, causing decline in several important agricultural crops yield threatening food security (Shang et al. 2022). The fall of solar UV-B radiation in Antarctica peninsula is found to be enhanced many days due to ozone depletion which affects plants' photosynthesis by impairment in the upper mesophyll cells as observed in *Colobanthus quitensis* (Kunth) Bartl. and *Deschampsia antarctica* Desv., which is associated with light-independent enzymatic reaction and consequently, biomass depletion is observed (Xiong and Day 2001). High concentration of ozone exposure to crops causes substantial reductions in yield as ozone enters into leaves through stomata and induces formation of ROS which develops oxidative stress, consequently reduction in photosynthesis and accelerated senescence are observed (Montes et al. 2022). An estimate of yield losses attributed to ozone between 2014 to 2017 for key crops across China showed production losses are 34–91 million metric tonnes (mt/yr) with highest losses in Henan provinces (Wang et al. 2022a). Looking into the gravity of matter, China has made commitment to cut down greenhouse gases (GHG) emission from different sectors by rigorous planning for mitigation including agricultural activities (open burning of crop residues, rice farming, change in cropland, and associated emissions, machinery uses, nitrogen fertilizer, and pesticide production) for ensuring carbon neutrality by 2060 (Liang et al. 2021).

Ozone pollution affects vital plant-pollinator system by disrupting floral visuals, olfactory perception of volatile communication signals, learning memories, and behaviour of pollinators. It has been experimentally validated by adding air with ozone that has caused increased reduction in floral volatile concentrations with distance from volatile sources, degradation of floral scents, their ratios showing negative impacts on pollination, and pollinator relationships (Farré-Armengol et al. 2016). In addition to that ozone-induced oxidative stress, disturbance in olfactory coding in the honeybee brain was observed that link with the pollinators' olfaction and foraging behaviours in addition to a possible long-term harmful effect on pollinator services (Démares et al. 2022).

Soil pollution is another category of pollution causing great economic loss and environmental degradation. Key sources of soil pollutants are fertilizers, pesticides, solid waste containing heavy metals, e-wastes, and plastics. The excessive application of nitrogen (N) fertilizers has given rise to numerous environmental problems such as excessive atmospheric N_2O, ammonia (NH_3) volatilization from crop lands, groundwater contaminations, and soil acidification which in turn give rise to economic losses and serious threat to human health (Gou et al. 2022) along with NO emission which has greater role in climate forcing and air pollution (Wang et al. 2022a, b, c). The magnitude and timing of ammonia emission are dependent upon agricultural practices adopted, climatic conditions, and edaphic factors. An estimate from China showed that total ammonia emission from fertigated cropland amounts to 4.3 Tg NH_3 Yr^{-1} in 2017 with an overall ammonia emission factor of 12% where vegetable, maize, and rice were found to be largest emitter. Apart from that, more emission hotspot is reported in South China and temporally multiple emission peaks reported in a year (Wang et al. 2021a).

Many HMs and metalloids are reported to be present in nature and classified in three categories as per toxicity, such as ultra-highly toxic (Cd, Hg, Pb, Cu, Zn); highly toxic (Mo, Mn, Fe); medium toxic (Ni, Co); and low level toxic (Sr, Zr). HMs such as Cu, Fe, Zn, Mn, Co, and Se are required by organisms in small quantities, essential for metabolic and key enzymatic functions playing important oxido-reductive roles. Rest of the HMs (Pb, Cd, Hg, and As) which do not play any role in life cycle of an organism induce negative impacts.

The pollutants released into atmosphere tend to fall on earth surface through rain or settlement over the period and cause soil and water pollution. Similarly, the soil pollutants have tendency to pollute surface water bodies, running, and ground water as they gradually enter into these sources by surface bodies by percolation or run off and leaching. Some soil pollutants have tendency to volatilize or become so small that in dried condition move to atmospheric domain as dust or particulate matter and cause air pollution. Thus, pollutants circulate in ecosystem among ecosystems and contaminate soil, water, and air. Some of these pollutants gain entry into biological systems from ecosystem through absorption by plants into their body which are eaten by herbivores. This way they become part of food chain and show biomagnification phenomenon.

2 Mechanisms Enhancing Pollutant Stress Tolerance

Pollution stress is a global problem whose effects are visible on crop plants too. To overcome this challenge, crop plants have devised several mechanisms and strategies to adopt the stress conditions (Fig. 23.1). These mechanisms are as follows:

1. Bioaccumulation
2. Biomass partitioning

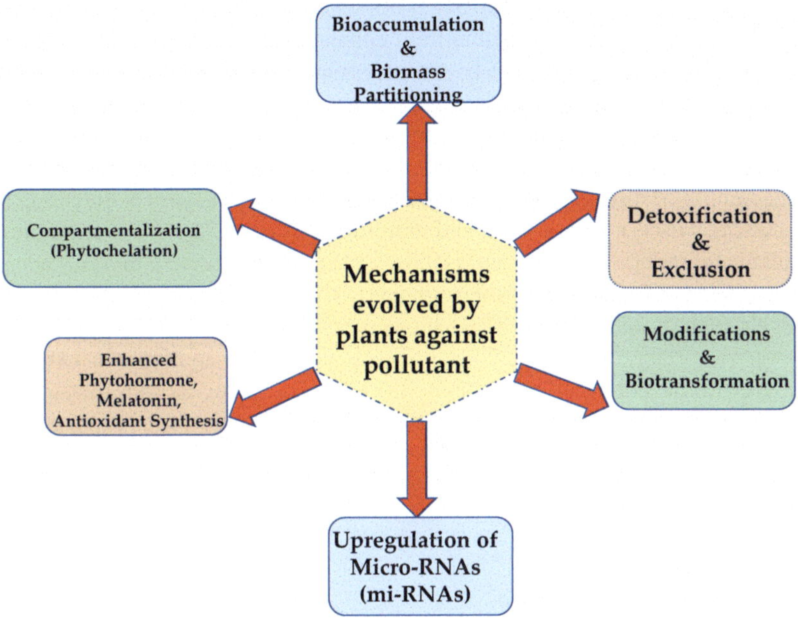

Fig. 23.1 Various mechanisms evolved by plants for pollutant stress tolerance

3. Exclusion
4. Detoxification
5. Modification
6. Phytochelation
7. Enhanced phytohormones
8. Biotransformation
9. Enhanced antioxidant synthesis
10. Melatonin synthesis
11. MicroRNAs (miRNAs)

2.1 *Bioaccumulation*

Plants have evolved mechanism of bioaccumulation for tolerance of several pollutants. It is an increased pollutant concentration in certain tissues, organelles of terrestrial, or aquatic organisms following uptake from the ambient environmental medium. Bioaccumulation processes in plant are governed by uptake rate, assimilation efficiency, and efflux of pollutants. Among plant species, *Sorghum bicolor* and *Sedum alfredii* are found to potential accumulator as hyperaccumulator of HMs having great value to be utilized for phytoremediation (Mishra et al. 2021; Ge et al. 2022). The plant, *Sedum alfredii*, a perennial herb, Cd/Zn hyperaccumulator, native to China, grows in abandoned Pb/Zn/Cd mines and highly tolerant to Cu, Mn, Ni, and Pb. There are two key issues that limit enhanced HMs tolerance are small plant biomass and slow growth rate of hyperaccumulator plants and consequently phytoremediation potential.

Phytoremediation technique involving plant growth-promoting bacteria (PGPB) is also used to clean up HMs contaminated soils such as *Solanum nigrum*, a Cd hyperaccumulator rhizospheric region led to isolation of *Bacillus* sp. PGP15 (Zhang et al. 2022a). HMs' accumulation and amplification take place by consumption of such plant and induce health hazard (Hlihor et al. 2022) as seen in high Cd accumulation in grains of wheat (*Triticum aestivum* L.). External application of protective chemicals such as spermidine (SPD) showed protective role in Cd + Pb combined stress on rice (*Oryza sativa* L.) seedlings grown under hydroponic experiment where spermidine enhanced tolerance to the HMs and many morpho-physiological traits (Gu et al. 2022). Mitigation strategies such as the omics, functional gene, and other strategies for uncovering Cd stress have been discussed in detail (Zhou and Li 2022).

2.2 *Biomass Partitioning*

Biomass partitioning, though obscure, is an in vivo strategy of plants to tackle pollution problems by petitioning their biomass between different organs. Three possible drivers identified for biomass partitioning are reduced plant size combined with allometric scaling between organs, early decrease in root surface affecting water uptake, and increased biomass allocation to root system compensating for lower soil resource acquisition (Delerue et al. 2022). Biomass allocation strategies in relation with Pb were observed in bamboo tissues where reduced biomass allocation was observed in new bamboo growth, whereas rhizome received increasing biomass allocation; consequently Pb accumulation trend was seen as rhizomes/old stems > new roots/old roots > new leaves > new stems among various tissue (Cai et al. 2021).

2.3 *Exclusion*

Plants eliminate toxic material (metal, metalloids, organic materials) by employing various strategies such as exclusion, detoxification, modifications, or by accumulating diverse metabolites. Metalloids such as arsenic (As) are removed by glutathione (GSH) and phytochelatins (PC) like metabolites having thiol-containing groups. The GSH and PC act as chelators conjugates with As in cytoplasm, transport, and sequester them in vacuole. Anthocyanin, flavonoid, and a secondary metabolite have also been found to provide As tolerance by following the same strategy adopted by using GSH & PC. Exogenous application or enhanced endogenous concentration of GSH by use of hormone conferred As stress tolerance in plants (Ahammed and Yang 2022). Therefore, this can be chosen as a strategy to minimize the impact of As tolerance. Current climate changed scenario, inducing the intense drying-rewetting cycle affecting soil microbial community composition and function as these factors influence the enzyme responses (Dehydrogenase (DHA) and alkaline phosphomonoesterase (ALP)) and perturbation caused due to HMs. It helps in making predictions related with microbial function in soils contaminated with HMs and climate change (Tan et al. 2023).

2.4 Detoxification

Crop plants employ the detoxification mechanism for getting rid of their ill effects of HMs or other pollutant molecules by employing immobilization in vacuole or cell walls. Contrary to that, the ex vivo mechanism is employed by alfalfa (*Medicago sativa* L.) and sorghum (*Sorghum bicolor*) where pollutants are detoxified externally by action of tyrosinase, laccase, or peroxidase like enzymes exudated through their root systems. They cause polymerization of pollutants in humic acid in soil, making it immobilized and inaccessible for enzymatic detoxification. Apart from polymerization, the enzymes also oxidize the compounds (phenols, PAH by using H_2O_2 as electron accepter) into easily degradable compounds making indirect detoxification in the external environment. If the pollutant gains entry into plant body, immobilization strategy is the main detoxification mechanism. The enzyme cytochrome P450 monooxygenase detoxifies organic compounds by using electron from NADPH that activates molecular oxygen (O_2) and inserts a single oxygen atom into the substrate making them hydrophilic and helping to detoxify easily by catalysing diverse reactions involving variety of natural compounds. The enzyme has ability to oxidize aromatic compounds such as when plant *A. thaliana* was exposed to PAH (phenanthrene) by incorporating two hydroxyl groups in the substrate.

Another ex vivo strategy is based on degradation of pollutant in soil by root-stimulated soil microbial enzymes and oxygen transport observed in alfalfa, fescue grass (*Festuca arundinacea*), maize (*Zea mays* L.), switch grass (*Panicum virgatum*), sorghum, and soyabean. Plants also get rid of HMs or pollutants either by excretion through use of diverse transporters such as aquaporins, pumps, or by chelation using low-molecular-weight molecules such as glutathione, metallothioneins, and phytochelatins which facilitate sequestration of the HMs or pollutants into vacuoles. The activation of cell signalling mechanism induces gene expression of genes involved in the HMs transportaion. Plant hormones also help overcoming HMs stress through their own signalling as well as activating other signalling pathways that induce enzymatic and non-enzymatic antioxidant defence responses helping to scavenge ROS/RNS and supporting cellular activities, improving quality and productivity crops (Noor et al. 2022).

2.5 Modification

Modification is another strategy or mechanism to get rid of ill effects of pollutants or HMs. It is an efficient strategy, where backbone of chemical is undisturbed, but proteins posttranslational modification (PTM) occurs that help to regulate protein function. In response to Cd stress in turnip seedlings grown under hydroponic condition, 547 succinylated sites on 256 protein were revealed which participated in various biological processes that can help developing Cd acclimation and tolerant crops (Li et al. 2022).

2.6 Phytochelation

Potentially toxic elements such as As, Cr, Cd, and Pb can be cleaned up by eco-friendly and cheaper technique called phytoremediation, used for ecological restoration by plant-soil feedback (Zhu et al. 2022), but not yet achieved its full potential due to poor understanding of molecular mechanisms at proteomics level (Alsafran et al. 2022). Detailed discussions are made about proteomics and metabolomics *vis-à-vis* HMs chelation, their toxicity, tolerance, and partitioning in food crops for better health (Anani et al. 2022).

Rising industrialization and urbanization leads to increase and accumulation of metals, metalloids, persistent compounds type pollutants, and plastics in environment and greatly affects highly productive ecosystem like coastal and wetlands. To improve such ecosystem, halophiles growing in coastal area and having high tolerance to meta(loid) toxicity stress can be used for phytoremediation in coastal area like *Kosteletzkya pentacarpos* (L.). Ledebour (Malvaceae) is found as a suitable species for controlling HMs contamination (Zhou et al. 2021).

As it is known that Cd is a nonessential HM, human uses are releasing Cd in environment causing high pollution leading to absorption into the body causing damage to kidneys. It occurs by damaging proximal tubule of nephrons that dysfunction electron transport chain in mitochondria, electron leakage, and consequent ROS generation. NADPH oxidase also got impaired and generated ROS developing oxidative stress which in turn damages DNA, proteins, and lipids, triggering epithelial cell death and ultimately making kidney nonfunctional (Yan and Allen 2021). Research progress of HMs effects especially about Cd in plants is greatly discussed in relation with its uptake, transport, and accumulation along with the strategies to get rid of HMs (Yang et al. 2021b). Cd-contaminated soils have been phytoremediated by using Cd hyperaccumulator accumulating in root and xylem cell wall of willow species (*Salix matsudana* var. *umbraculifera* Rehd). The upregulation of differently expressed genes and proteins along with phenylpropanoid synthesis helped in upregulation of lignin content in roots by enhanced synthesis of lignin synthesizing enzymes (Yu et al. 2022). Salinity stress and HM stress (Cu) tolerance by halophilic *Brachybacterium muris* were observed by inducing increased amount of unsaturated fatty acid and associated content improving membrane stability along with transporter and antioxidants content (Liu et al. 2022).

Chemical pesticides are one of the organic pollutants and environmental hazards to the biosphere, especially in the agricultural ecosystem. Phytoextraction is one of the eco-friendly, sustainable solutions of phytoremediation that can be used as a method for soil clean-up with the possibility of re-use of extracted metals through phyto-mining. Phyto-mining or phytoextraction technique helps in environmental clean-up by re-use of extracted metals using plant species having hyperaccumulating property. Usually, plants have limited ability to accumulate HMs or pollutants that limit phyto-mining. It can be improved by the process known as priming (Karalija et al. 2022). Silicon- (Si) based fertilizers are helpful improving soil quality and alleviated stress induced by abiotic (salt and drought) and/or biotic factors and remediated HMs such as Al, As, Cd, Cr, and Zn (Zhao et al. 2022a, b, c). Similarly, sulfur amendments made on HMs phytoextraction from contaminated soils have shown positive response (Zakari et al. 2021).

Phytoremediation involving phytoextraction involves uptake and transport of HMs into aerial part of plants making the processes environmental friendly method to reduce the HMs load in soil. It is done by efficient and wide variety of transporters and their regulators in the plant system for developing hyperaccumulator through transgenic or gene editing techniques (Yang et al. 2022a).

2.7 *Enhanced Phytohormones Synthesis*

Phytohormones or plant growth regulators (PGR) are signalling biomolecules, act in nanomolar concentration, diverse chemical structure, and physicochemical properties, and get highly modulated during pollution or HMs stress (Kosakivska et al. 2022). Widespread industrial uses of Cd have led to major environmental problem contaminating water and soil. Plants show Cd toxicity in seed germination, water and mineral uptake, and photosynthesis metabolic alteration because of oxidative stress. Cd stress modulates PGR such as abscisic acid (ABA), auxins, brassinosteroids, cytokinin's (CK), gibberellins, jasmonic acid, and NO. ABA is found to be involved in strengthening against HMs tolerance by improving osmolyte and antioxidant levels (Kumar et al. 2022). Modern techniques have helped to develop Cd-tolerant plants by limiting Cd bioavailability and toxicity for long-term management of Cd-polluted soils by identification of QTLs, CRISPAR/Cas9, and functional genomics (Zulfiqar et al. 2022).

2.8 *Biotransformation*

The phytoremediation technique under bioremediation is a low-cost, eco-friendly method, especially where pollutants have hazardous effect on the health of non-targeted organisms, and uses macro- or micro-algae for removal of pollution and waste water treatment (Hejna et al. 2022). Recently, emerging domains such as algomics, potential characterization of biodegrading enzymes, and genetic engineering can be combined to develop strategies for tapping the benefits of biotransformation (Manzi et al. 2022).

2.9 Enhanced Antioxidant Synthesis

Antioxidants are a group of low MW molecules such as AsA, GSH, α-tocopherol (vitamin E), phenolic compounds, flavonoids, alkaloids, and nonprotein amino acids. They help to regulate ROS generation and their scavenging. Enzymes such as superoxide dismutase (SOD), catalase (CAT), peroxidase (POX), polyphenol oxidase (PPO), ascorbate peroxidase (APX), monodehydroascorbate reductase (MDHAR), dehydroascorbate reductase (DHAR), glutathione reductase (GR), glutathione peroxidase (GPX), glutathione S-transferase (GST), thioredoxin reductase (TRX), and peroxiredoxin (PRX) are also involved in regulating ROS. CAT, APX, GPX, or AsA-GSH cycle helps in the conversion of harmful H_2O_2 into water. AsA-GSH cycle is a major antioxidant defence cycle and is also known as Asada-Halliwell Cycle (AHC). AsA and GSH are the two most abundant soluble antioxidants found in plants. (Hasanuzzaman et al. 2020). These molecules are generated, and enzymes get activated during stress induced by pollutants. Hg stress induced ion leakage, but overexpression of AsA-GSH led to Hg tolerance by improved photosynthesis, higher antioxidant level, and high osmolyte accumulation in tomato plant. Such strategies can lead to develop Hg-resistant tomato plants and can be applied to reclaim Hg-contaminated soils (Bashir et al. 2022).

2.10 Melatonin Synthesis

Melatonin (MT) (N-acetyl-5-methoxytryptamine), a pleiotropic, is now considered as plant growth regulator, signalling, and biostimulator molecule that improves plant growth by enhancing HMs tolerance. MT helps positively against HMs by improving redox and nutrient homeostasis, primary and secondary metabolism, and osmotic balance in plant cell (Hoque et al. 2021). Exogenous supplementation of MT relieved Cd-induced oxidative stress in *Brassica napus* growing under Cd stress (Menhas et al. 2022) and Cd and Pb tolerance when supplemented with NO by enhancing Ca^{++} and K^+ uptake and increased exudation of organic acid into the rhizosphere (Imran et al. 2022).

2.11 MicroRNAs (miRNAs)

MicroRNAs are small non-coding RNA and found to be involved in regulation of HMs stress along with other abiotic stress. Metal-regulated miRNAs and their target genes are part of a complex regulatory network and coordinate plant responses to HMs stress through antioxidant functions, hormone signals, metal transporters, root growth, and transcription factors (TF) (Yang et al. 2022a).

3 Pollutant Stress *Vis-à-Vis* Responses Shown by Crop Plants

3.1 Impact of Pollutants on the Morphological Characteristics

As the pollutants show diversity in nature, their responses on plant morphology are also variable. Consequently, plants have also evolved various strategies to overcome pollutants stress including HMs inducing cell death, chlorosis, physio-biochemical disturbances, epinasty, necrosis, stunted growth, and reduced biomass (Sandalio et al. 2001; Ortega-Villasante et al. 2005; Singh et al. 2020). Chromium (VI and III) concentrations causing Cr stress ranging from 0 to 100 ppm elevate putrescine level up to 10 time and their accumulation was quicker in Cr IV and reached 3000–5000 ppm in leaves (dry wt). Initially, it led to reduced root growth, chlorosis, and after that, reduced shoot growth, low water content, and induction of leaf chitinase activity occur (Hauschild 1993). Rice is susceptible to arsenic (As) contamination under paddy soil conditions and depends upon uptake that in turn is dependent upon the rice genotypes where fine genotype was greatly affected than coarse genotype (Niazi et al. 2022). Ozone as air pollutant causes enhanced leaf senescence, leading to a loss of green leaf area and consequently affecting grain productivity (Soja and Soja 1995). Microplastics (MPs) pollution effects were studied in Radish and Broccoli where it was found that MPs affect plant biomass, tissue composition, and root traits (López et al. 2022). Deciduous tree species *Callistemon citrinus* in petrochemical polluted area and high temperature condition in South of Iran showed decrease in characters such as leaf area, length, and breadth of blade (Seyyednejad et al. 2015). Industrial and agricultural activities along with rapid urbanization are seen as key contributor for HMs stress problem (Hou et al. 2020).

3.2 Effect of Pollutants on the Physio-biochemical and Molecular Changes

Pollutants enter into the plant body, start interacting with cellular components, and influence their physio-biochemistry. It is known that HMs inactivate biomolecules by blocking key functional groups or displace metal ions associated with them that leads to induction of ROS (Ortega-Villasante et al. 2005). Cu stress negatively affects seed germination, chlorophyll biosynthesis, antioxidant synthesis, and mineral nutrition that leads to plant growth and productivity (Mir et al. 2021).

Chromium (Cr) accumulation in Cr-sensitive variety of Chickpea crops led to reduction of yield which was associated with negatively regulated traits (chlorophyll content, enzyme activities, nutrient content) and upregulated traits such as proline and antioxidants levels along with reduced Cr uptake in tolerant cultivars (Singh et al. 2022). Glutamate, cysteine, and proline are the amino acids that confer protections against Cr(VI) which causes decrease in expression of glutamate receptors that hampers protection against HMs (López-Bucio et al. 2022). Rice is very susceptible for As contamination which is one of the causes to decline in production of this important crop; shows upregulation of antioxidant level, high pigment content, and H_2O_2 in tolerant varieties (Niazi et al. 2022).

Plants employ detoxification, chelation, compartmentalization, excretion, and export using transporters and elevated level of antioxidant system including NO that are the key mechanisms to overcome pollutants and HMs at the physio-biochemical and molecular level. Both enzymatic and non-enzymatic mechanisms are employed by plants for NO generation (Corpas and Barroso 2013) which is oscillatory in nature during HMs stress and helps overcoming by induction of antioxidant level that lowers or eliminates the effects of oxidative stress generated by ROS or reactive nitrogen species (RNS).

NO is a small, gaseous, bioactive signalling molecules, involved in various regulatory physiological responses in plants such as flowering (Guo et al. 2003), senescence (Crawford 2006), root development (Gupta and Kaiser 2010), pollen tube growth (Wang et al. 2012), posttranslational modification such as S-nitrosylation that regulates gene transcription (Mengel et al. 2013), gene expression, the mobilization of second messenger (Astier and Lindermayr, 2012), programmed cell death (PCD) (Wang et al. 2013, Lin et al. 2012), seed germination, stomatal movements (Fan and Liu. 2012), seed dormancy (Arc et al. 2013), and induced during HMs stress (Cu) in mung bean (Gaur et al. 2021). The plant hormone abscisic acid (ABA) also plays key roles in mitigating HMs stress (Bücker-Neto et al. 2017). There are ample reports that crosstalk between NO and ABA signalling molecules occurs in higher plants (Prakash et al. 2019). NO also modulates metal transporters (NIP, NRAMP, ABC and iron transporters, stress-related genes such as CytP450, GSTs, GRXs, TFs, amino acid, hormone(s), signalling and secondary metabolism genes involved in As detoxification (Singh et al. 2017), *Cicer* sp. (Yadav and Mani 2019) and Cr tolerance (Maurya and Sinha 2020), and in Arsenate (AsV)-induced stress in *Vicia faba* L. (Siddiqui et al. 2021).

Cadmium (Cd) is highly hazardous, toxic to living being even in low concentration, triggers oxidative stress *in Glycine max* (L.) roots (Pérez-Chaca et al. 2014), geno-cytotoxicity and lipid peroxidation in *Allium sativum* and *Vicia faba* (Ünyayar et al. 2006), and negatively affects growth and development of plants. Cd effects are being mitigated by use of the Fe-fortification that provides tolerance by upregulating antioxidant level in lentil (Bansal et al. 2021), by use of Si and iron nanoparticles (Fe-NPs) (Ahmad et al. 2021a). Cadmium stressed plants synthesised proline, expression of proline transporters and malondialdehyde in response to exogenous

ABA application in pea plants (Zdunek-Zastocka et al. 2021), epigenetic modification in root meristems cell nuclei of *Vicia faba* (Zabka et al. 2021), decreasing NO generation, improving iron homoeostasis and induction of HY1 in *Arabidopsis* (Han et al. 2014), change in growth and cell cycle gene expression in suspension-culture cells of soybean (Sobkowiak and Deckert 2003), and DNA methylation (Chakrabarti and Mukherjee 2021). Cd induced NO that led to PCD in roots of lupine plants (Arasimowicz-Jelonek et al. 2012), pea (*Pisum sativum* L.) roots, shown by imaging of ROS and NO accumulation in vivo (Rodriguez-Serrano et al. 2006). Stress induced by Cd increased total soluble sugars, reducing sugars but decreased non-reducing sugars, concomitant increase in sucrose degrading enzyme, acid invertase, sucrose synthase, and decrease in sucrose phosphate synthase in rice (Verma and Dubey 2001). Exogenous supplementation of NO helped mitigating Hg stress (Ahmad et al. 2021a, b), co-inoculation of *Rhizobia* and arbuscular mycorrhiza fungi (AMF), which helped in coping Cd stress (Wang et al. 2021a, b).

Morphological, biomass, and gene expression associated with antioxidant activities showed negative effects in response to Cd and Hg^{2+} in *Lolium perenne* (Cruz et al. 2021). Tolerant *Hydrangea strigosa* plant upregulated SOD, POD, CAT, and AsA content in root and released root exudates containing sucrose, glycolic acid, and nonanoic acid, which showed potential to remediate Pb from rhizospheric region making it fit for rhizoremediation (Jin et al. 2022).

Indium (I), an element having great application in electronic industry, is an emerging component of e-waste and great environmental threat. It gets accumulated in root, restricting growth and perturbation of phosphate, magnesium, and iron homeostasis (Cheah et al. 2022), and activates HMs-tolerant genes and phosphate starvation responses including phosphate-regulated transcription factor and transporters in roots. Exogenously applied phosphate alleviated I toxicity by reducing its uptake (Cheah et al. 2022). Deciduous tree species *Callistemon citrinus* growing in high temperature conditions in petrochemical area of South Iran, in comparison to unpolluted regions, where VOC released from plants led to rise in soluble carbohydrates, proline, chlorophyll a and b, and carotenoids (Seyyednejad et al. 2015). Microplastics released in both aquatic and land ecosystems showed downregulation of glucosinolate (GSL), anthocyanin content, but upregulation of MDA contents affecting growth parameters (López et al. 2022). Radiation pollutant ozone causes chlorophyll fluorescence and decline in plastoquinone's (PQ), affecting plant development in wheat (Soja and Soja 1995) (Fig. 23.2)

4 E-Waste, Indium (In) Pollution, and Responses of Crop Plants

E-waste is a new challenge of the present century posing great health and environmental hazards. An estimate reports that around 50 million tonnes of e-waste are generated annually across the globe. The effects of e-waste are seen as infertility, endocrine, hormonal, and neurological problems in both animals and human

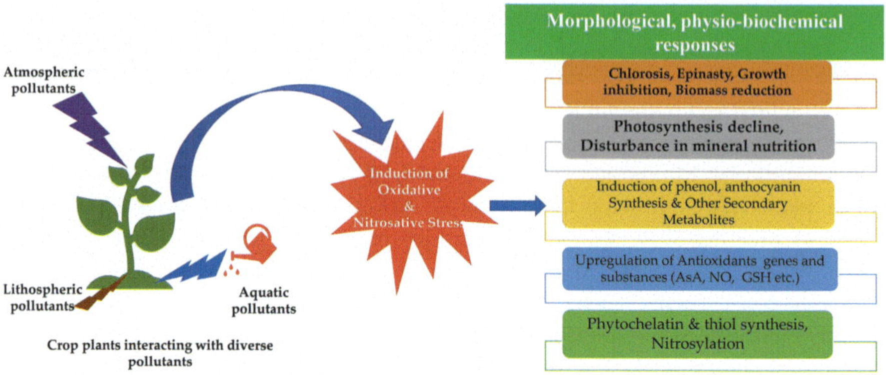

Fig. 23.2 Diverse physio-biochemical responses shown by plants against pollutant stress

(especially children's) beings. E-waste disposal and management greatly vary across the globe in developing and developed countries. Only a small fraction (15%) is recycled annually and the rest are discarded unplanned as landfilling, burned, illegally traded, or thrown in nearby areas. Such practice leads to pollution of all components of nature, namely soil, air, and water. Such a situation has kept India in the 3rd largest e-waste contributor position. This e-waste is discarded without following proper disposal mechanism and almost 80% of this e-waste is dumped or burned into environment making situation worse (Forti et al. 2020).

Indium is a post-transitional metal placed in group 13 of periodic table, used mainly in electronics and related industries for making devices such as flat-panel display screens, touch panels, semiconductors, and heat reflectors. Indium demand for making such products has been pushed from 662 mt to 760 mt (~15% increase) within a decade (2011–19) (USGS 2020). Parallelly, an increase of ~111% e-waste was reported within a decade due to huge consumption of electronic devices. Apart from electronic waste, Indium is also released into the environment by coal burning, zinc smelters, semiconductor fabrication, and e-waste recycling sites, leading to environmental pollution (White and Shine 2016). The free released Indium gets entry into plants consumed by human being through soil or water taken up by vegetables like cabbage, garlic, lettuce and water spinach, Vietnamese vegetables, and beets, posing health risk after their consumption. Key staple crops such as rice and wheat show high accumulation of Indium in the roots and some reports show translocation into grains (Cheah et al. 2022). The reports suggest excess Indium accumulation in plant tissues that restricts plant growth and delays development and changes in plant morphology. It is also seen causing breakdown of root outer layer and root elongation inhibition on Indium exposure in cow pea (Kopittke et al. 2009). Indium also inhibited root and shoot growth in *Arabidopsis*; the disrupted synthesis of transporters involved in phosphate uptake and translocation (Chang et al. 2020a). Similar effects were seen in rice (*Oryza sativa* L.) with additional symptoms like leaf necrosis and the acute perturbation of phosphorus, magnesium, and iron homeostasis (Cheah et al. 2022). Treatment of C_3 and C_4 plants with In_2O_3 nanoparticles

(NPs) significantly reduced growth and photosynthesis, induced oxidative damage (H_2O_2, lipid peroxidation), and impaired P and Fe homeostasis, particularly in the young leaves of C_4 plants (Shabbaj et al. 2022). Upon exposure to Indium, plant shows induction of oxidative-nitrosative stress, anthocyanization in the leaves and petioles, stimulates organic acid (citrate) that has overcoming effects (Chang et al. 2020a, b), generation of Nitric Oxide (NO), and upregulation of *S*-nitrosoglutathione, *S*-nitrosoglutathione reductase activity, and glutathione contents in wheat (Zhao et al. 2022a, b, c). It is also reported that the effects of Indium on the growth and uptake by wheat plants grown in soils contaminated by Indium depend upon Al and acidic pH (Syu et al. 2021). Indium is mostly restricted by plants in roots by inducing accumulation with small translocation into the grains that showed no adverse effects on human health, but further studies are required to ascertain the risk factor (Chang et al. 2020b). The above ground accumulation of Ga and In was found low in perennial ryegrass (Lolium perenne L.) and likely to be accumulated in soil from where it reaches to our food directly or indirectly (Jensen et al. 2018). Low doses exposure (25–2000 ppm) of indium had either no or minor effects on root elongation, total chlorophyll, and MDA development in *Arabidopsis* (Ma et al. 2013). Using bioremediation technology, several organisms have potential to develop biofilm attached to e-waste and promote substrate bioavailability, metabolite transfer, and cell viability that accelerates bioleaching and degradation (Bharathi et al. 2022).

5 Soil Microorganisms/PGPR and Pollution Stress

Soil microorganisms play great role in HMs tolerance and removal and are deeply discussed (Mao et al. 2022). Research constrains like limited screening of strains, focus on single strain, non-evaluation of pathogenicity, and low field experiments are needed to be overcome to get maximum output from this domain. It has been reported that cloned cytochrome P450 gene was upregulated and downregulated by exogenous brassinolide (BR) by Cd stress in hybrid poplar. It led to Cd and brassinolide accumulation and reduced oxidative stress (Tian et al. 2022). Priming is an effective, eco-friendly, cost-effective, and value-added technique in which plant or plant parts are traeted with novel compounds which helps in better performance that can be exploited (Wiszniewska 2021) such as seed priming resulted in improved seed germination, reduced seedling emergence time, shortened crop time, enhanced stress tolerance, and grain production. Priming uses extra-terrestrial physical agents, ionization radiation (X-rays, Gamma rays), nonionizing radiation (ultrasonic waves, magnetic fields, microwaves, infra-red radiation) and ensured enhanced production over conventional methods. (Bera et al. 2022).

Microbes also promote Se uptake and accumulation in crops and have been suggested as biofortification agent (Yang et al. 2021a, b). Metal-resistant plant growth-promoting bacteria (PGPB) promote biotic, abiotic, and alters metal bioavailability in soils reducing toxicity in plants and making them a better phytoremediation agent (Ma et al. 2022; Wang et al. 2022a, b, c).

6 Plastic Pollution, Mechanisms, and Responses

Microplastic pollution (MP) is an outcome of modern society and has become nuisance for ecosystem (Cholewińska et al. 2022) and induce cellular toxicity in mammals (Banerjee and Shelver, 2021). Micro (~5 mm) and nano plastic (~1000 nm) are the types of MP causing serious water and soil pollution. These varieties of plastics make soil more porous and retain more water, but no correlation with stability or increase in microbial diversity. These MPs have changed soil dynamics by increasing evapotranspiration, organic carbon, CO_2 flux, water saturation, nitrogen content, and microbial biomass, but decreased the N_2O, flux, water use efficiency, soil bulk density, and soil microbial diversity (Zhang et al. 2022a, b, c).

Plants are acting as sink for MP and NP as they are adsorbed on their surface and only NP sometime escapes to enter into plant body, inducing toxic effects, photosynthesis, and oxidative stress (Yin et al. 2021). It was reported that maize was more sensitive than rice and wheat in terms of phytotoxic effects within the Poaceae family, while wheat and lettuce were less sensitive to microplastics exposure (Zhang et al. 2022b). The hermetic effects of NP to *Lactuca sativa* showed that NP effects led to decline in water content, osmotic potential, the decreased stomatal limitation (Ls) and A/Ci, and increased mesophyll and low carboxylation efficiency leading to decline in photosynthesis as a non-stomatal limitation. NP also increased trapping fluxes; absorption with a decrease in electron fluxes that caused photosystem reaction centres' disruption (Yildiztugay et al. 2022). In addition to the fact of plastic accumulation in living organisms, there is a great and urgent need to minimize the influx of microplastics into the environment (Cholewińska et al. 2022).

7 Modern Techniques and Strategies for Pollution Mitigation and Crop Improvement

There are various technologies that are getting applied for clean-up of the environment. These can be discussed in the following.

7.1 Nanotechnology

Nanotechnology is an emerging high value technology having great potential in crop management and sustainable agriculture. Nanotechnology helps the element to bring it to a nano level and develops NPs that show better, efficient, and enhanced performance in comparison to their original atomic size of element. Application of nanotechnology in agriculture can be myriad, but currently seen as nano-fertilizers, nano-pesticides, and herbicides which are mainly responsible for agricultural pollution. With certain limitations, fungal organisms are found as potential sources for synthesis of biogenic NPs due to their enhanced bioavailability, bioactivity, and high tolerance capacity for heavy metals in crop fields (Sonawane et al. 2022). Diverse applications of nanoscale zero-valent iron particles (NZVI) are reported in a variety of industries, but negligently released into environment and taken up by living organisms (plants, soil animals, and microorganisms) which poses risk and induces toxic effects (Zhang et al. 2022a, b, c).

7.2 Bioremediation

Pollutants or HMs-tolerant plant species are chosen for clean-up polluted soils or xenobiotic management present in the environment by phyto- or bioremediation. Such technology is augmented by using plant bio-stimulants (PB) making agriculture sustainable by minimizing impacts of pollutants and enhancing phytoremediation efficiency of plant species. The PBs are derived from amino acid, humic substances inorganic salts, microbes, protein hydrolysate, plant extracts, and seaweeds (Bartucca et al. 2022). Widespread household application of formaldehyde (HCHO) as gaseous or liquid form as disinfectants, adhesives, and wood-based furniture is leading to undesirable emissions becoming health hazard and environmental problems. Formaldehyde can be easily removed by phytoremediation technique using microbes contributing in making clean and healthy environment to achieve the UN sustainable development goals (Peng et al. 2022).

7.3 Unmanned Aerial Vehicle (UAV) Technology

Unmanned aerial vehicle (UAV) technology is a very recent innovation which has gained diverse applications in short span of time. UAV is helpful for environmental management because of its cost effectiveness especially in large farmland with rugged terrain as it does't require specific take-off platforms. Pesticide application through UAV in large farmland areas with rugged terrain have contributed for better, efficient, and effective application with minimizing pesticide waste and environmental pollution (Hu et al. 2022).

7.4 cDNA Microarray

Plants are facing diverse environmental stress including pollutants. cDNA microarray analysis is one of the modern tools used to detect changes in gene expression induced by pollutants including ozone, heavy metals, salinity, and temperature. cDNA microarray technology is very sensitive, reliable, and specific and has been applied in ozone-induced stress gene expression analysis (Calatayud and Barreno

2001). Environmental stressors such as acid rain, drought, high and low temperature, salinity, and UV-B radiation show shifts in gene expression and are found to be distinct from those in response to ozone. Diagnosis of stress such as drought, salinity, UV-B, low temperature, high temperature, and acid rain and ozone have been made using cDNA microarray analaysis in Arabidopsis. (Tamaoki et al. 2004)

8 Conclusion and Future Perspective

Pollution stress is inducing diverse negative impacts on the growth and development of crop plants in the current scenario. To cope up with the impacts, diverse defence mechanisms have been evolved by plants such as bioaccumulation, resource partitioning, modification, detoxification, compartmentalization, which help plants to either neutralize the effects in vivo or excrete outside to get rid of toxic pollutants. Plants in response to these pollutants develop oxidative or nitrosative stress which further induces synthesis of antioxidant chemicals, secondary metabolites, and various signalling molecules such as NO. Nitric oxide acts as cytoprotective or cytotoxic molecule, depending upon the cellular redox condition. The plants have also evolved several morpho-anatomical modifications as well as physio-biochemical adaptations such as antioxidant synthesis, proline, ascorbic acid, and lignification. These modification or changes help in survival of plants. The pollution stress has become critical issue and major concern in current scenario. To overcome this problem, various strategies and techniques have been evolved to clean up pollutants or carry out the process that harms less to the environment. These techniques include bioremediation including phytoremediation, nanotechnology, augmentation by use of beneficial chemicals, use of unmanned aerial vehicles (UAV), and cDNA microarray. These techniques are hoped to clean up the polluted environment to bring it back to natural and sustainable level.

References

Ahammed GJ, Yang Y (2022) Anthocyanin-mediated arsenic tolerance in plants. Environ Pollut 292(Pt B):118475. https://doi.org/10.1016/j.envpol.2021.118475

Ahmad A, Yasin NA, Khan WU, Akram W, Wang R, Shah AA, Akbar M, Ali A, Wu T (2021a) Silicon assisted ameliorative effects of iron nanoparticles against cadmium stress: attaining new equilibrium among physiochemical parameters, antioxidative machinery, and osmoregulators of *Phaseolus lunatus*. Plant Physiol Biochem 166:874–886

Ahmad P, Alyemeni MN, Wijaya L, Ahanger MA, Ashraf M, Alam P, Paray BA, Rinklebe J (2021b) Nitric oxide donor, sodium nitroprusside, mitigates mercury toxicity in different cultivars of soybean. J Hazard Mater 408:124852. https://doi.org/10.1016/j.jhazmat.2020.124852

Alsafran M, Usman K, Ahmed B, Rizwan M, Saleem MH, Al Jabri H (2022) Understanding the phytoremediation mechanisms of potentially toxic elements: a proteomic overview of recent advances. Front Plant Sci 13:881242. https://doi.org/10.3389/fpls.2022.881242

Anani OA, Abel I, Olomukoro JO, Onyeachu IB (2022) Insights to proteomics and metabolomics metal chelation in food crops. J Proteins Proteom 2022:1–15

Arasimowicz-Jelonek M, Floryszak-Wieczorek J, Deckert J, Rucinska-Sobkowiak R, Gzyl J, Pawlak-Sprada S, Abramowski D, Jelonek T, Gwozdz EA (2012) Nitric oxide implication in cadmium-induced programmed cell death in roots and signaling response of yellow lupine plants. Plant Physiol Biochem 58:124–134

Arc E, Sechet J, Corbineau F, Rajjou L, Marion-Poll A (2013) ABA crosstalk with ethylene and nitric oxide in seed dormancy and germination. Front Plant Sci 4:63. https://doi.org/10.3389/fpls.2013.00063

Astier J, Lindermayr C (2012) Nitric oxide-dependent posttranslational modification in plants: an update. Int J Mol Sci 13(11):15193–15208

Banerjee A, & Shelver WL (2021) Micro- and nanoplastic induced cellular toxicity in mammals: A review. Sci Total Environ 755(Pt 2), 142518. https://doi.org/10.1016/j.scitotenv.2020.142518

Bansal R, Priya S, Dikshit HK, Jacob SR, Rao M, Bana RS, Kumari J, Tripathi K, Kumar A, Kumar S, H M Siddique K (2021) Growth and antioxidant responses in iron-biofortified lentil under cadmium stress. Toxics 9(8):182. https://doi.org/10.3390/toxics9080182

Bartucca ML, Cerri M, Del Buono D, Forni C (2022) Use of biostimulants as a new approach for the improvement of phytoremediation performance-a review. Plants (Basel) 11(15). https://doi.org/10.3390/plants11151946

Bashir S, Jan N, Wani UM, Raja V, John R (2022) Co-over expression of Ascorbate Glutathione pathway enzymes improve mercury tolerance in tomato. Plant Physiol Biochem 186:170–181

Bera K, Dutta P, Sadhukhan S (2022) Seed priming with non-ionizing physical agents: plant responses and underlying physiological mechanisms. Plant Cell Rep 41(1):53–73

Bharathi SD, Dilshani A, Rishivanthi S, Khaitan P, Vamsidhar A, Jacob S (2022) Resource recycling, recovery, and xenobiotic remediation from e-wastes through biofilm technology: a review. Appl Biochem Biotechnol. https://doi.org/10.1007/s12010-022-04055-8

Bücker-Neto L, Paiva A, Machado RD, Arenhart RA, Margis-Pinheiro M (2017) Interactions between plant hormones and heavy metals responses. Genet Mol Biol 40(1 suppl 1):373–386

Cai X, Jiang M, Liao J, Yang Y, Li N, Cheng Q, Li X, Song H, Luo Z, Liu S (2021) Biomass allocation strategies and Pb-enrichment characteristics of six dwarf bamboos under soil Pb stress. Ecotoxicol Environ Saf 207:111500. https://doi.org/10.1016/j.ecoenv.2020.111500

Calatayud A, Barreno E (2001) Chlorophyll a fluorescence, antioxidant enzymes and lipid peroxidation in tomato in response to ozone and benomyl. Environ Pollut 115(2):283–289

Chakrabarti M, Mukherjee A (2021) Investigating the underlying mechanism of cadmium-induced plant adaptive response to genotoxic stress. Ecotoxicol Environ Saf 209:111817. https://doi.org/10.1016/j.ecoenv.2020.111817

Chang HF, Wang SL, Lee DC, Hsiao SS, Hashimoto Y, Yeh KC (2020a) Assessment of indium toxicity to the model plant *Arabidopsis*. J Hazard Mater 387:121983. https://doi.org/10.1016/j.jhazmat.2019.121983

Chang HF, Yang PT, Lin HW, Yeh KC, Chen MN, Wang SL (2020b) Indium uptake and accumulation by rice and wheat and health risk associated with their consumption. Environ Sci Technol 54(23):14946–14954

Cheah BH, Liao PC, Lo JC, Wang YT, Tang IC, Yeh KC, Lee DY, Lin YF (2022) Insight into the mechanism of indium toxicity in rice. J Hazard Mater 429:128265. https://doi.org/10.1016/j.jhazmat.2022.128265

Cholewińska P, Moniuszko H, Wojnarowski K, Pokorny P, Szeligowska N, Dobicki W, Polechoński R, Górniak W (2022) The occurrence of microplastics and the formation of biofilms by pathogenic and opportunistic bacteria as threats in aquaculture. Int J Environ Res Public Health 19(13). https://doi.org/10.3390/ijerph19138137

Corpas FJ, Barroso JB (2013) Nitro-oxidative stress vs oxidative or nitrosative stress in higher plants. New Phytol 199(3):633–635

Crawford NM (2006) Mechanisms for nitric oxide synthesis in plants. J Exp Bot 57(3):471–478

Cruz Y, Villar S, Gutiérrez K, Montoya-Ruiz C, Gallego JL, Delgado MDP, Saldarriaga JF (2021) Gene expression and morphological responses of *Lolium perenne* L. exposed to cadmium ($Cd^{(2+)}$) and mercury ($Hg^{(2+)}$). Sci Rep 11(1):11257. https://doi.org/10.1038/s41598-021-90826-y
Delerue F, Scattolin M, Atteia O, Cohen GJV, Franceschi M, Mench M (2022) Biomass partitioning of plants under soil pollution stress. Commun Biol 5(1):365. https://doi.org/10.1038/s42003-022-03307-x
Démares, F., Gibert, L., Creusot, P., Lapeyre, B., & Proffit, M. (2022). Acute ozone exposure impairs detection of floral odor, learning, and memory of honey bees, through olfactory generalization. The Science of the total environment, 827, 154342. https://doi.org/10.1016/j.scitotenv.2022.154342
Farré-Armengol, G., Filella, I., Llusia, J., & Peñuelas, J. (2016). Bidirectional Interaction between Phyllospheric Microbiotas and Plant Volatile Emissions. Trends in plant science, 21(10), 854–860. https://doi.org/10.1016/j.tplants.2016.06.005
Fan QJ, Liu JH (2012) Nitric oxide is involved in dehydration/drought tolerance in P*oncirus trifoliata* seedlings through regulation of antioxidant systems and stomatal response. Plant Cell Rep 31(1):145–154
Forti V, Baldé CP, Kuehr R, Bel G. The Global E-waste Monitor 2020: Quantities, flows and the circular economy potential. United Nations University (UNU)/United Nations Institute for Training and Research (UNITAR)–co-hosted SCYCLE Programme, International Telecommunication Union (ITU) & International Solid Waste Association (ISWA), Bonn/Geneva/Rotterdam
Gaur S, Kumar J, Prasad SM, Sharma S, Bhat JA, Sahi S, Singh VP, Tripathi DK, Chauhan DK (2021) Silicon and nitric oxide interplay alleviates copper induced toxicity in mung bean seedlings. Plant Physiol Biochem 167:713–722
Ge J, Tao J, Zhao J, Wu Z, Zhang H, Gao Y, Tian S, Xie R, Xu S, Lu L (2022) Transcriptome analysis reveals candidate genes involved in multiple heavy metal tolerance in hyperaccumulator *Sedum alfredii*. Ecotoxicol Environ Saf 241:113795. https://doi.org/10.1016/j.ecoenv.2022.113795
Gou Z, Liu G, Wang Y, Li X, Wang H, Chen S, Su Y, Sun Y, Ma NL, Chen G (2022) Enhancing N uptake and reducing N pollution via green, sustainable N fixation-release model. Environ Res 214(Pt 4):113934. https://doi.org/10.1016/j.envres.2022.113934
Gu J, Hu C, Jia X, Ren Y, Su D, He J (2022) Physiological and biochemical bases of spermidine-induced alleviation of cadmium and lead combined stress in rice. Plant Physiol Biochem 189:104–114
Guo FQ, Okamoto M, Crawford NM (2003) Identification of a plant nitric oxide synthase gene involved in hormonal signaling. Science 302(5642):100–103
Gupta KJ, Kaiser WM (2010) Production and scavenging of nitric oxide by barley root mitochondria. Plant Cell Physiol 51(4):576–584
Han B, Yang Z, Xie Y, Nie L, Cui J, Shen W (2014) Arabidopsis HY1 confers cadmium tolerance by decreasing nitric oxide production and improving iron homeostasis. Mol Plant 7(2):388–403. https://doi.org/10.1093/mp/sst122
Hasanuzzaman M, Bhuyan MHMB, Zulfiqar F, Raza A, Mohsin SM, Mahmud JA, Fujita M, Fotopoulos V (2020) Reactive oxygen species and antioxidant defense in plants under abiotic stress: revisiting the crucial role of a universal defense regulator. Antioxidants 9(8):681. https://doi.org/10.3390/antiox9080681
Hauschild MZ (1993) Putrescine (1,4-diaminobutane) as an indicator of pollution-induced stress in higher plants: barley and rape stressed with Cr(III) or Cr(VI). Ecotoxicol Environ Saf 26(2):228–247
Hejna M, Kapuścińska D, Aksmann A (2022) Pharmaceuticals in the aquatic environment: a review on eco-toxicology and the remediation potential of algae. Int J Environ Res Public Health 19(13). https://doi.org/10.3390/ijerph19137717
Hlihor RM, Roşca M, Hagiu-Zaleschi L, Simion IM, Daraban GM, Stoleru V (2022) Medicinal plant growth in heavy metals contaminated soils: responses to metal stress and induced risks to human health. Toxics 10(9). https://doi.org/10.3390/toxics10090499

Hoque MN, Tahjib-Ul-Arif M, Hannan A, Sultana N, Akhter S, Hasanuzzaman M, Akter F, Hossain MS, Sayed MA, Hasan MT, Skalicky M, Li X, Brestič M (2021) Melatonin modulates plant tolerance to heavy metal stress: morphological responses to molecular mechanisms. Int J Mol Sci 22(21):11445. https://doi.org/10.3390/ijms222111445

Hou D, O'Connor D, Igalavithana AD, Alessi DS, Luo J, Tsang DCW, Sparks DL, Yamauchi Y, Rinklebe J, Ok YS (2020) Metal contamination and bioremediation of agricultural soils for food safety and sustainability. Nat Rev Earth Environ 1:366–381

Hu H, Kaizu Y, Huang J, Furuhashi K, Zhang H, Li M, Imou K (2022) Research on methods decreasing pesticide waste based on plant protection unmanned aerial vehicles: a review. Front Plant Sci 13:811256. https://doi.org/10.3389/fpls.2022.811256

Imran, M., Khan, A. L., Mun, B. G., Bilal, S., Shaffique, S., Kwon, E. H., Kang, S. M., Yun, B. W., & Lee, I. J. (2022). Melatonin and nitric oxide: Dual players inhibiting hazardous metal toxicity in soybean plants via molecular and antioxidant signaling cascades. Chemosphere, 308 (Pt 3), 136575. https://doi.org/10.1016/j.chemosphere.2022.136575

Jensen H, Gaw S, Lehto NJ, Hassall L, Robinson BH (2018) The mobility and plant uptake of gallium and indium, two emerging contaminants associated with electronic waste and other sources. Chemosphere 209:675–684

Jin J, Song Z, Zhao B, Zhang Y, Wang R (2022) Physiological and metabolomics responses of *Hydrangea macrophylla* (Thunb.) Ser. and *Hydrangea strigosa* Rehd. to lead exposure. Ecotoxicol Environ Saf 243:113960. https://doi.org/10.1016/j.ecoenv.2022.113960

Karalija E, Selović A, Bešta-Gajević R, Šamec D (2022) Thinking for the future: Phytoextraction of cadmium using primed plants for sustainable soil clean-up. Physiol Plant 174(4):e13739. https://doi.org/10.1111/ppl.13739

Kopittke PM, McKenna BA, Blamey FPC, Wehr JB, Menzies NW (2009) Metal-induced cell rupture in elongating roots is associated with metal ion binding strengths. Plant Soil 322:303–315

Kosakivska IV, Vedenicheva NP, Babenko LM, Voytenko LV, Romanenko KO, Vasyuk VA (2022) Exogenous phytohormones in the regulation of growth and development of cereals under abiotic stresses. Mol Biol Rep 49(1):617–628

Kumar, S., Shah, S. H., Vimala, Y., Jatav, H. S., Ahmad, P., Chen, Y., & Siddique, K. H. M. (2022). Abscisic acid: Metabolism, transport, crosstalk with other plant growth regulators, and its role in heavy metal stress mitigation. Frontiers in plant science, 13, 972856. https://doi.org/10.3389/fpls.2022.972856

Li X, Yang D, Yang Y, Jin G, Yin X, Zheng Y, Xu J, Yang Y (2022) Quantitative succinyl-proteome profiling of turnip (*Brassica rapa* var. rapa) in response to cadmium stress. Cell 11(12). https://doi.org/10.3390/cells11121947

Liang D, Lu X, Zhuang M, Shi G, Hu C, Wang S, Hao J (2021) China's greenhouse gas emissions for cropping systems from 1978-2016. Sci Data 8(1):171. https://doi.org/10.1038/s41597-021-00960-5

Lin A, Wang Y, Tang J, Xue P, Li C, Liu L, Hu B, Yang F, Loake GJ, Chu C (2012) Nitric oxide and protein S-nitrosylation are integral to hydrogen peroxide-induced leaf cell death in rice. Plant Physiol 158(1):451–464

Liu S, Liu X, Shi Y, Zhuang S, Chen Q (2022) The adaptive mechanism of halophilic *Brachybacterium muris* in response to salt stress and its mitigation of copper toxicity in hydroponic plants. Environ Pollut:120124. https://doi.org/10.1016/j.envpol.2022.120124

López MD, Toro MT, Riveros G, Illanes M, Noriega F, Schoebitz M, García-Viguera C, Moreno DA (2022) *Brassica* sprouts exposed to microplastics: effects on phytochemical constituents. Sci Total Environ 823:153796. https://doi.org/10.1016/j.scitotenv.2022.153796

López-Bucio JS, Ravelo-Ortega G, López-Bucio J (2022) Chromium in plant growth and development: toxicity, tolerance and hormesis. Environ Pollut 312:120084. https://doi.org/10.1016/j.envpol.2022.120084

Ma C, Chhikara S, Xing B, Musante C, Clarkson JC, Dhankher OP (2013) Physiological and molecular response of *Arabidopsis thaliana* (L.) to nanoparticle cerium and indium oxide exposure. ACS Sustain Chem Eng 1(7):768–778

Ma Y, Wang Y, Shi XJ, Chen XP, Li ZL (2022) Mechanism and application of plant growth-promoting bacteria in heavy metal bioremediation. Huan Jing Ke Xue 43(9):4911–4922

Manzi HP, Abou-Shanab RAI, Jeon BH, Wang J, Salama ES (2022) Algae: a frontline photosynthetic organism in the microplastic catastrophe. Trends Plant Sci 27(11):1159–1172. https://doi.org/10.1016/j.tplants.2022.06.005

Mao Y, Tan H, Wang M, Jiang T, Wei H, Xu W, Jiang Q, Bao H, Ding Y, Wang F, Zhu C (2022) Research progress of soil microorganisms in response to heavy metals in rice. J Agric Food Chem 70(28):8513–8522

Maurya AK, Sinha D (2020) Nitric oxide as a savior molecule against stress induced by chromium and cadmium. Int J Plant Environ Sci 6(04):253–263

Mengel A, Chaki M, Shekariesfahlan A, Lindermayr C (2013) Effect of nitric oxide on gene transcription–*S*-nitrosylation of nuclear proteins. Front Plant Sci 4:293. https://doi.org/10.3389/fpls.2013.00293

Menhas S, Yang X, Hayat K, Ali A, Ali EF, Shahid M, Shaheen SM, Rinklebe J, Hayat S, Zhou P (2022) Melatonin enhanced oilseed rape growth and mitigated Cd stress risk: a novel trial for reducing Cd accumulation by bioenergy crops. Environ Pollut 308:119642. https://doi.org/10.1016/j.envpol.2022.119642

Mir AR, Pichtel J, Hayat S (2021) Copper: uptake, toxicity and tolerance in plants and management of cu-contaminated soil. Biometals 34(4):737–759

Mishra D, Kumar S, Mishra BN (2021) An overview of morpho-physiological, biochemical, and molecular responses of sorghum towards heavy metal stress. Rev Environ Contam Toxicol 256:155–177

Molina L, Segura A (2021) Biochemical and metabolic plant responses toward polycyclic aromatic hydrocarbons and heavy metals present in atmospheric pollution. Plan Theory 10(11). https://doi.org/10.3390/plants10112305

Montes CM, Demler HJ, Li S, Martin DG, Ainsworth EA (2022) Approaches to investigate crop responses to ozone pollution: from $O_{(3)}$-FACE to satellite-enabled modeling. Plant J 109(2):432–446

Niazi NK, Hussain MM, Bibi I, Natasha, Shahid M, Ali F, Iqbal J, Shaheen SM, Abdelrahman H, Akhtar W, Wang H, Rinklebe J (2022) The significance of eighteen rice genotypes on arsenic accumulation, physiological response and potential health risk. Sci Total Environ 832:155004. https://doi.org/10.1016/j.scitotenv.2022.155004.

Noor I, Sohail H, Sun J, Nawaz MA, Li G, Hasanuzzaman M, Liu J (2022) Heavy metal and metalloid toxicity in horticultural plants: tolerance mechanism and remediation strategies. Chemosphere 303:135196. https://doi.org/10.1016/j.chemosphere.2022.135196

Ortega-Villasante C, Rellán-Álvarez R, Del Campo FF, Carpena-Ruiz RO, Hernández LE (2005) Cellular damage induced by cadmium and mercury in *Medicago sativa*. J Exp Bot 56:2239–2251

Peng WX, Yue X, Chen H, Ma NL, Quan Z, Yu Q, Wei Z, Guan R, Lam SS, Rinklebe J, Zhang D, Zhang B, Bolan N, Kirkham MB, Sonne C (2022) A review of plants formaldehyde metabolism: implications for hazardous emissions and phytoremediation. J Hazard Mater 436:129304. https://doi.org/10.1016/j.jhazmat.2022.129304

Pérez-Chaca MV, Rodriguez-Serrano M, Molina AS, Pedranzani HE, Zirulnik F, Sandalio LM, Romero-Puertas MC (2014) Cadmium induces two waves of reactive oxygen species in *Glycine max* (L.) roots. Plant Cell Environ 37:1672–1687

Prakash, V., Singh, V. P., Tripathi, D. K., Sharma, S., & Corpas, F. J. (2019). Crosstalk between nitric oxide (NO) and abscisic acid (ABA) signalling molecules in higher plants. Environmental and Experimental Botany, 161, 41–49. https://doi.org/10.1016/j.envexpbot.2018.10.033

Rodriguez-Serrano M, Romero-Puertas MC, Zabalza A, Corpas FJ, Gomez M, Del Rio LA, Sandalio LM (2006) Cadmium effect on oxidative metabolism of pea (*Pisum sativum* L.) roots. Imaging of reactive oxygen species and nitric oxide accumulation in vivo. Plant Cell Environ 29:1532–1544

Sandalio LM, Dalurzo HC, Gómez M, Romero-Puertas MC, Del Río LA (2001) Cadmium-induced changes in the growth and oxidative metabolism of pea plants. J Exp Bot 52:2115–2126

Seyyednejad SM, Niknejad M, Yusefi M (2015) The effect of air pollution on some morphological and biochemical factors of *Callistemon citrinus* in petrochemical zone in south of Iran. Asian J Plant Sci 8:562–565

Shabbaj II, AbdElgawad H, Balkhyour MA, Tammar A, Madany MMY (2022) Elevated CO_2 differentially mitigated oxidative stress induced by indium oxide nanoparticles in young and old leaves of C_3 and C_4 crops. Antioxidants 11(2):308. https://doi.org/10.3390/antiox11020308

Shang B, Fu R, Agathokleous E, Dai L, Zhang G, Wu R, Feng Z (2022) Ethylenediurea offers moderate protection against ozone-induced rice yield loss under high ozone pollution. Sci Total Environ 806:151341. https://doi.org/10.1016/j.scitotenv.2021.151341

Shrestha RK, Shi D, Obaid H, Elsayed NS, Xie D, Ni J, Ni C (2022) Crops' response to the emergent air pollutants. Planta 256(4):80. https://doi.org/10.1007/s00425-022-03993-1

Siddiqui MH, Alamri S, Mukherjee S, Al-Amri AA, Alsubaie QD, Al-Munqedhi B, Ali HM, Kalaji HM, Fahad S, Rajput VD, Narayan OP (2021) Molybdenum and hydrogen sulfide synergistically mitigate arsenic toxicity by modulating defense system, nitrogen and cysteine assimilation in faba bean (*Vicia faba* L.) seedlings. Environ Pollut 290:117953. https://doi.org/10.1016/j.envpol.2021.117953

Singh PK, Indoliya Y, Chauhan AS, Singh SP, Singh AP, Dwivedi S, Tripathi RD, Chakrabarty D (2017) Nitric oxide mediated transcriptional modulation enhances plant adaptive responses to arsenic stress. Sci Rep 7:3592. https://doi.org/10.1038/s41598-017-03923-2

Singh D, Sharma NL, Singh CK, Sarkar SK, Singh I, Dotaniya ML (2020) Effect of chromium (VI) toxicity on morpho-physiological characteristics, yield, and yield components of two chickpea (*Cicer arietinum* L.) varieties. PLoS One 15(12):e0243032. https://doi.org/10.1371/journal.pone.0243032

Singh D, Singh CK, Singh D, Sarkar SK, Prasad SK, Sharma NL, Singh I (2022) Glycine betaine modulates chromium (VI)-induced morpho-physiological and biochemical responses to mitigate chromium toxicity in chickpea (*Cicer arietinum* L.) cultivars. Sci Rep 12(1):8005. https://doi.org/10.1038/s41598-022-11869-3

Sobkowiak R, Deckert J (2003) Cadmium-induced changes in growth and cell cycle gene expression in suspension-culture cells of soybean. Plant Physiol Biochem 41:767–772

Soja G, Soja AM (1995) Ozone effects on dry matter partitioning and chlorophyll fluorescence during plant development of wheat. Water Air Soil Pollut 85:1461–1466. https://doi.org/10.1007/BF00477187

Sonawane H, Shelke D, Chambhare M, Dixit N, Math S, Sen S, Borah SN, Islam NF, Joshi SJ, Yousaf B, Rinklebe J, Sarma H (2022) Fungi-derived agriculturally important nanoparticles and their application in crop stress management - prospects and environmental risks. Environ Res 212:113543. https://doi.org/10.1016/j.envres.2022.113543

Syu CH, Chen LY, Lee DY (2021) The growth and uptake of gallium (Ga) and indium (in) of wheat seedlings in Ga- and In-contaminated soils. Sci Total Environ 759:143943. https://doi.org/10.1016/j.scitotenv.2020.143943

Tamaoki M, Matsuyama T, Nakajima N, Aono M, Kubo A, Saji H (2004) A method for diagnosis of plant environmental stresses by gene expression profiling using a cDNA microarray. Environ Pollut 131(1):137–145. https://doi.org/10.1016/j.envpol.2004.01.008

Tan X, He J, Nie Y, Ni X, Ye Q, Ma L, Megharaj M, He W, Shen W (2023) Climate and edaphic factors drive soil enzyme activity dynamics and tolerance to cd toxicity after rewetting of dry soil. Sci Total Environ 855:158926. https://doi.org/10.1016/j.scitotenv.2022.158926

Tian F, Han C, Chen X, Wu X, Mi J, Wan X, Liu Q, He F, Chen L, Yang H, Zhong Y, Qian Z, Zhang F (2022) PscCYP716A1-mediated brassinolide biosynthesis increases cadmium tolerance and enrichment in poplar. Front Plant Sci 13:919682. https://doi.org/10.3389/fpls.2022.919682

Ünyayar S, Çelik A, Çekiç FO, Gözel A (2006) Cadmium-induced genotoxicity, cytotoxicity and lipid peroxidation in *Allium sativum* and *Vicia faba*. Mutagenesis 21:77–81

USGS (2020) Mineral commodity summaries 2020 U.S. Government Publishing Office; Telecommunication Union (ITU) & International Solid Waste Association (ISWA),the circular

economy potential. United Nations University (UNU)/United Nations Institute for Training and Research (UNITAR) – co-hosted SCYCLE Programme, International UK
Verma S, Dubey R (2001) Effect of Cadmium on soluble sugars and enzymes of their metabolism in rice. Biol Plant 44:117–123. https://doi.org/10.1023/A:1017938809311
Wang YH, Li XC, Zhu-Ge Q, Jiang X, Wang WD, Fang WP, Chen X, Li XH (2012) Nitric oxide participates in cold-inhibited *Camellia sinensis* pollen germination and tube growth partly via cGMP in vitro. PLoS One 7(12):e52436. https://doi.org/10.1371/journal.pone.0052436
Wang Y, Loake GJ, Chu C (2013) Cross-talk of nitric oxide and reactive oxygen species in plant programmed cell death. Front Plant Sci 4:314. https://doi.org/10.3389/fpls.2013.00314
Wang C, Cheng K, Ren C, Liu H, Sun J, Reis S, Yin S, Xu J, Gu B (2021a) An empirical model to estimate ammonia emission from cropland fertilization in China. Environ Pollut 288:117982. https://doi.org/10.1016/j.envpol.2021.117982
Wang X, Fang L, Beiyuan J, Cui Y, Peng Q, Zhu S, Wang M, Zhang X (2021b) Improvement of alfalfa resistance against Cd stress through rhizobia and arbuscular mycorrhiza fungi co-inoculation in Cd-contaminated soil. Environ Pollut 277:116758. https://doi.org/10.1016/j.envpol.2021.116758
Wang Y, Narayanan M, Shi X, Chen X, Li Z, Natarajan D, Ma Y (2022a) Plant growth-promoting bacteria in metal-contaminated soil: current perspectives on remediation mechanisms. Front Microbiol 13:966226. https://doi.org/10.3389/fmicb.2022.966226
Wang Y, Wild O, Ashworth K, Chen X, Wu Q, Qi Y, Wang Z (2022b) Reductions in crop yields across China from elevated ozone. Environ Pollut 292:118218. https://doi.org/10.1016/j.envpol.2021.118218
Wang Y, Yao Z, Zheng X, Subramaniam L, Butterbach-Bahl K (2022c) A synthesis of nitric oxide emissions across global fertilized croplands from crop-specific emission factors. Glob Chang Biol 28(14):4395–4408
White SJO, Shine JP (2016) Exposure potential and health impacts of indium and gallium, metals critical to emerging electronics and energy technologies. Curr Environ Health Rep 3(4):459–467
Wiszniewska A (2021) Priming strategies for benefiting plant performance under toxic trace metal exposure. Plan Theory 10(4). https://doi.org/10.3389/fpls.2022.919682
Xiong FS, Day TA (2001) Effect of solar ultraviolet-B radiation during springtime ozone depletion on photosynthesis and biomass production of Antarctic vascular plants. Plant Physiol 125(2):738–751
Yadav BS, Mani A (2019) Analysis of bHLH coding genes of *Cicer arietinum* during heavy metal stress using biological network. Physiol Mol Biol Plants 25:113–121
Yan LJ, Allen DC (2021) Cadmium-induced kidney injury: oxidative damage as a unifying mechanism. Biomol Ther 11(11). https://doi.org/10.3390/biom11111575
Yang D, Hu C, Wang X, Shi G, Li Y, Fei Y, Song Y, Zhao X (2021a) Microbes: a potential tool for selenium biofortification. Metallomics 13(10). https://doi.org/10.1093/mtomcs/mfab054
Yang GL, Zheng MM, Tan AJ, Liu YT, Feng D, Lv SM (2021b) Research on the mechanisms of plant enrichment and detoxification of cadmium. Biology 10(6). https://doi.org/10.3390/biology10060544
Yang Y, Huang J, Sun Q, Wang J, Huang L, Fu S, Qin S, Xie X, Ge S, Li X, Cheng Z, Wang X, Chen H, Zheng B, He Y (2022a) microRNAs: key players in plant response to metal toxicity. Int J Mol Sci 23(15):8642. https://doi.org/10.3390/ijms23158642
Yang Z, Yang F, Liu JL, Wu HT, Yang H, Shi Y, Liu J, Zhang YF, Luo YR, Chen KM (2022b) Heavy metal transporters: functional mechanisms, regulation, and application in phytoremediation. Sci Total Environ 809:151099. https://doi.org/10.1016/j.scitotenv.2021.151099
Yildiztugay E, Ozfidan-Konakci C, Arikan B, Alp FN, Elbasan F, Zengin G, Cavusoglu H, Sakalak H (2022) The hormetic dose-risks of polymethyl methacrylate nanoplastics on chlorophyll a fluorescence transient, lipid composition and antioxidant system in *Lactuca sativa*. Environ Pollut 308:119651. https://doi.org/10.1016/j.envpol.2022.119651
Yin L, Wen X, Huang D, Du C, Deng R, Zhou Z, Tao J, Li R, Zhou W, Wang Z, Chen H (2021) Interactions between microplastics/nanoplastics and vascular plants. Environ Pollut 290:117999. https://doi.org/10.1016/j.envpol.2021.117999

Yu M, Zhuo R, Lu Z, Li S, Chen J, Wang Y, Li J, Han X (2022) Molecular insights into lignin biosynthesis on cadmium tolerance: morphology, transcriptome and proteome profiling in Salix matsudana. J Hazard Mater 441:129909. https://doi.org/10.1016/j.jhazmat.2022.129909

Żabka A, Winnicki K, Polit JT, Wróblewski M, Maszewski J (2021) Cadmium (ii)-induced oxidative stress results in replication stress and epigenetic modifications in root meristem cell nuclei of *Vicia faba*. Cell 10(3):640. https://doi.org/10.3390/cells10030640

Zakari S, Jiang X, Zhu X, Liu W, Allakonon MGB, Singh AK, Chen C, Zou X, Akponikpè PBI, Dossa GGO, Yang B (2021) Influence of sulfur amendments on heavy metals phytoextraction from agricultural contaminated soils: a meta-analysis. Environ Pollut 288:117820. https://doi.org/10.1016/j.envpol.2021.117820

Zandalinas SI, Fritschi FB, Mittler R (2021) Global warming, climate change, and environmental pollution: recipe for a multifactorial stress combination disaster. Trends Plant Sci 26(6):588–599

Zdunek-Zastocka E, Grabowska A, Michniewska B, Orzechowski S (2021) Proline concentration and its metabolism are regulated in a leaf age dependent manner but not by abscisic acid in pea plants exposed to cadmium stress. Cell 10(4):946. https://doi.org/10.3390/cells10040946

Zhang, Y., Cai, C., Gu, Y., Shi, Y., & Gao, X. (2022a). Microplastics in plant-soil ecosystems: A meta-analysis. Environmental pollution (Barking, Essex : 1987), 308, 119718. https://doi.org/10.1016/j.envpol.2022.119718

Zhang Y, Zhao S, Liu S, Peng J, Zhang H, Zhao Q, Zheng L, Chen Y, Shen Z, Xu X, Chen C (2022b) Enhancing the phytoremediation of heavy metals by combining hyperaccumulator and heavy metal-resistant plant growth-promoting bacteria. Front Plant Sci 13:912350. https://doi.org/10.3389/fpls.2022.912350

Zhang S, Yi K, Chen A, Shao J, Peng L, Luo S (2022c) Toxicity of zero-valent iron nanoparticles to soil organisms and the associated defense mechanisms: a review. Ecotoxicology 31(6):873–883

Zhao H, Qian R, Liang X, Ou Y, Sun C, Lin X (2022a) Indium induces nitro-oxidative stress in roots of wheat (*Triticum aestivum*). J Hazard Mater 428:128260. https://doi.org/10.1016/j.jhazmat.2022.128260

Zhao K, Yang Y, Zhang L, Zhang J, Zhou Y, Huang H, Luo S, Luo L (2022b) Silicon-based additive on heavy metal remediation in soils: toxicological effects, remediation techniques, and perspectives. Environ Res 205:112244. https://doi.org/10.1016/j.envres.2021.112244

Zhao W, Teng M, Zhang J, Wang K, Zhang J, Xu Y, Wang C (2022c) Insights into the mechanisms of organic pollutant toxicity to earthworms: advances and perspectives. Environ Pollut 303:119120. https://doi.org/10.1016/j.envpol.2022.119120

Zhou M, Li Z (2022) Recent advances in minimizing cadmium accumulation in wheat. Toxics 10(4):187. https://doi.org/10.3390/toxics10040187

Zhou M, Lutts S, Han R (2021) *Kosteletzkya pentacarpos*: a potential halophyte candidate for phytoremediation in the meta(loid)s polluted saline soils. Plan Theory 10(11). https://doi.org/10.3390/plants10112495

Zhu SC, Zheng HX, Liu WS, Liu C, Guo MN, Huot H, Morel JL, Qiu RL, Chao Y, Tang YT (2022) Plant-soil feedbacks for the restoration of degraded mine lands: a review. Front Microbiol 12:751794. https://doi.org/10.3389/fmicb.2021.751794

Zulfiqar U, Jiang W, Xiukang W, Hussain S, Ahmad M, Maqsood MF, Ali N, Ishfaq M, Kaleem M, Haider FU, Farooq N, Naveed M, Kucerik J, Brtnicky M, Mustafa A (2022) Cadmium phytotoxicity, tolerance, and advanced remediation approaches in agricultural soils; a comprehensive review. Front Plant Sci 13:773815. https://doi.org/10.3389/fpls.2022.773815

Chapter 24
Phytohormones as Stress Mitigator in Plants

Zain Ul Abidin, Athar Mahmood, Safura Bibi, Muhammad Mansoor Javaid, Muhammad Anjum Zia, Muhammad Saad Ullah, Muhammad Azeem, Muhammad Ather Nadeem, and Bilal Ahmad Khan

Abstract Climatic change causes many stresses to crops such as drought, salinity, heat, cold, and heavy metal. Abiotic factors in plants have gained global attention and many strategies have been carried out to alleviate these factors in plants. Several protective mechanisms are developed by plants to alleviate stress. These mechanisms include synthesis of phytohormones (organic substances) which are signaling molecules produced within plants at very low concentration. These include JA (jasmonic acid), melatonin, abscisic acid, polyamines, IAA (indole acetic acid), SA (salicylic acid), ethylene, PGPR-mediated phytohormones, brassinosteroids, and GA (gibberellic acid). Some new growth regulators such as strigolactones have been discovered in plants. Phytohormones are plant growth regulators at different stages of development. Extreme conditions, such as stress, initiate signaling pathways of phytohormones that orchestrate adaptive responses even at very low concentrations. This hormonal initiation regulates series of reactions which helps plants to grow in suboptimal growth phases. During abiotic stress conditions, phytohormones' application helps plants to produce more osmolytes such as polyamines, proline, and others which protect cellular machinery of plants. This chapter deals with understanding the role and mechanism of phytohormones in plants for mitigation of stress.

Keywords Phytohormones · Abiotic stresses · Strigolactones · Cytokinins · Ethylene

---

Z. Ul Abidin · S. Bibi · M. Saad Ullah
Department of Botany, Faculty of Sciences, University of Agriculture, Faisalabad, Pakistan

A. Mahmood (✉)
Department of Agronomy, Faculty of Agriculture, University of Agriculture, Faisalabad, Pakistan
e-mail: athar.mahmood@uaf.edu.pk

M. Mansoor Javaid (✉) · M. A. Nadeem
Department of Agronomy, College of Agriculture, University of Sargodha, Sargodha, Pakistan
e-mail: mansoor.javaid@uos.edu.pk

M. Hasanuzzaman (ed.), *Climate-Resilient Agriculture, Vol 2*,
https://doi.org/10.1007/978-3-031-37428-9_24

1 Introduction

World population level is rising up continuously, making it possible to increase by more than 2.4 billion by 2050 (DESA 2015). So, increase in production as well as quality of food is needed up to 70% by 2050 to fulfill the need of population. For this, there is need to expand the production of crops such as cereals, grains, and pulses. The production of crops mainly depends upon the environmental conditions (extracellular components and developmental signals), but with increase in the population the number of industries is increasing every day. This increase is destroying the nature and is a big cause of climate disturbance. Environmental stresses are the main factors limiting production of crops. Among various climatic changes, drought stress, logging of water, heat stress, low temperature, heavy metals stress, and salinity are harmful for plants growth (Abbas et al. 2019; Sabagh et al. 2022).

Environmental stress conditions, such as heat and cold stress, have extreme effect on plants production (Gull et al. 2019). Abiotic stress is induced in plants by chemical or physical environment, while biotic stress is induced by some disease-causing bacteria, fungi, viruses, insects, etc. These fungi, bacteria, nematodes, and viruses are the living organisms sometimes known as pathogens which cause biotic stress in plants. These pathogens cause disturbance in uptaking the nutrients which lead to the death of plants (de Oliveira and Bell 2022).

As plants cannot move, they have to face these drastic conditions. In these unfavorable conditions, different types of stresses such as salinity, drought, and high temperature can co-occur. Different defensive mechanisms are developed by plants under stress. Plants have five defensive metabolites which protect them from extreme environment. These include cuticles, unsaturated fatty acids, oxylipin precursor, scavengers of reactive species, and molecular chaperones (He et al. 2018; Aqueel et al. 2015). The metabolic acclimatization response, in our opinion, is best viewed as a "Facultative metabolic adaptation" (Table 24.1; Rhodes and Nadolska-Orczyk 2001; Asif et al. 2012).

Except for these defensive metabolites, some plant hormones are recognized that are helpful to plants in promoting growth under changing environment, called phytohormones. Phytohormones are signal molecules that function at very less

M. A. Zia
Department of Biochemistry, Faculty of Sciences, University of Agriculture, Faisalabad, Pakistan

M. Azeem
Biosaline Research Laboratories, Department of Botany, University of Karachi, Karachi, Pakistan

B. A. Khan
Department of Agronomy, College of Agriculture, University of Sargodha, Sargodha, Pakistan

Department of Soil and Crop Sciences, Texas A&M University, College Station, TX, USA

Table 24.1 Facultative metabolic adaptations of plants to different stresses (Rhodes and Nadolska-Orczyk 2001)

Stress	Facultative metabolic adaptation	Proposed function(s)
Heat	Heat shock proteins' induction	HSPs can interact with proteins that become unstable due to heat stress, may act as molecular chaperones, and confer thermotolerance
Heavy metals	Copper induces the production of metallothioneins while cadmium activates phytochelatin synthase	Cadmium can be chelated and detoxifies by phytochelatin. Similarly, copper can be chelated and detoxifies by metallothioneins
Salinity Water deficits	Osmotic adjustment (OA); active accumulation of solutes, including compatible organic solutes (sugars, polyols, amino acids, and onium compounds)	OA might help in limiting dehydration and maintaining turgor; organic solutes can act as osmo-protectants and cryoprotectants. Additionally, certain organic solutes can also act as antioxidants

concentration and act as chemical messengers having vital role in exposure of responses toward stress (Aziz et al. 2019; Zhao et al. 2021).

2 Abiotic Stresses

Plants face several extreme circumstances that have effect on crops production. Abiotic stresses are interconnected and can occur in form of osmotic stress as well as homeostasis (Khalid et al. 2019). Under drastic conditions, several genes present in plants change their expression patterns. This change in genes pattern has effect on the growth rate and crop production. That is the reason to spot the genes that cause abiotic stress. This will be helpful in clarifying mechanism of stresses by identification of genes. Abiotic stress affecting plants includes drought, salinity, heavy metals, heat, and low temperature stress described below.

2.1 Drought Stress

Production of crops is restricted by drought stress. It refers to lack of rainfall that causes a shortage of water in groundwater. Lack of groundwater lowers the water table, which can hinder plant growth and survival. Drought stress can cause deleterious effects on plants by interfering with some activities such as assimilation of carbon, gaseous exchange of leaf, turgor pressure, ionic balance, activity of enzymes, elongation of stem, size of leaf, and root proliferation (Babar et al. 2009; Ali et al. 2020).

Most of human nutrition is mainly provided through three different crops. Those three crops are grains, tubers, and legumes. Rice, maize, and wheat are main three grains. Some other grains which are consumed on daily basis are oats, great millet, and cereal rye which are accounting about 50% of proteins and 56% dietary energy consumed on land. Drought stress causes increase in the osmotic pressure of leaf, decrease in transpiration, turgor pressure in leaves as well as decreased hydraulic conductivity and rate of sap flow in roots. Generally, drought stress has impact on plant-water ratio, which makes plants unsteady and affects plants growth (Yadav et al. 2021). Physiologically, drought stress minimizes photosynthesis, closing of stomata, and chlorophyll content reduction (Javaid and Tanveer 2014; Wahab et al. 2022). By affecting enzyme activity and Calvin cycle, drought stress influences production of starch in plants (Uarrota et al. 2018).

2.2 *Salinity Stress*

Salinity has adverse effects on crops and limits agricultural productivity. Salinity stress is total salts concentration in soil mainly sodium or potassium ions (Saberi Riseh et al. 2021). According to Javaid et al. (2012) and Gupta and Huang (2014), salt stress has affected about 20% of agricultural land. Soil salinity usually retards the metabolism of plants by reducing plants' ability to absorb water (Akhtar 2019; Javaid et al. 2020). Salinity impairs plants' morphological, biochemical, and physiological processes, which involve the absorption of water, nutrients, and germination of seeds. Salinity stress can be ionic or osmotic stress. In both conditions, salinity has harmful impact on plants. The osmotic pressure of plant cell is increased with increase of salts present in soil solution because of salinity stress, which limits the nutrient and water absorption like Ca^{2+}, K^{+} by plants from soil (Chaudhry and Sidhu 2021; Javaid et al. 2022).

2.3 *Heat Stress*

Increase in temperature from optimum level which can harm plants is heat stress (Hu et al. 2020). Mostly, higher plants are ranked as mesophilic having optimum growth at 10 °C to 30 °C. 37 °C to 42 °C or above temperature is explained as extreme high temperature for model plant (*Arabidopsis thaliana)* (Javaid and Tanveer 2013; Desaint et al. 2021). ROS excessive accumulation leading to oxidative stress, denaturation of proteins leading to improper proteins folding as well as alterations in lipid membranes leading to impaired permeability are physiological and biochemical results of heat stress. Heat stress has bad impact on plant growth and there is big decrease in yield of crops due to heat stress as 42% yield loss is found in maize, 31% in wheat, and 50% in rice (Fahad et al. 2017).

An adaptive strategy has been developed by plants to acquire tolerance toward heat. Plants have developed an adaptive mechanism to have tolerance against heat stress. Upon sensing heat stress, metabolism is changed, and antioxidants are increased by plants. This results in homeostasis. The accretion and expressing of heat shock proteins (HSPs) is increased as a chaperone to save the protein from irreversible heat damage. Thus, in reaction to an increase in temperature, the cellular signaling pathways are triggered to coordinate physio-biochemical processes (Li et al. 2021b; Mahmood et al. 2022b).

2.4 Cold Stress

Along with drought stress, cold stress is most damaging environmental condition for plants. Cold stress is divided into chilling stress. Temperate plants have cold acclimation capacity as they can tolerate freezing temperature, while subtropical and tropical plants are very tactful to chilling stress, so, they do not have the ability of cold acclimation. Temperate climate plants can withstand cold stress occurring in early winter and spring. Moreover, they can also tolerate the environmental temperature changes. Rice, maize, soybeans, potatoes, and tomatoes as well as some other important crops are very delicate to cold stress and cannot adapt to it (Chinnusamy et al. 2007; Theocharis et al. 2012).

2.5 Heavy Metal Stress

Growth of plants and productivity is extremely affected by some heavy metals present in the atmosphere. These heavy metals cause stress in plants. Elements with densities above 5 g cm^{-3} are known as heavy metals and 53 elements are categorized as heavy metals from 93 elements occurring in nature. Among them, there are also some biological trace elements of heavy metals, needed for plants development. They are 17 in number. But high content of these micronutrients has very harmful impacts on plants. Some of them are mercury, silver, cadmium, lead, chromium, and aluminum (Tanveer et al. 2012; Mousavi et al. 2022).

These metals enter in the plants through roots and disturb plants' metabolism like photosynthesis, mineral nutrition, and absorption of water. To counteract the toxicity of heavy metals, plants use different processes. These processes include exclusion, segregation, formation of chelating metals, osmoprotectors biosynthesize and storage, and others (Wani et al. 2018).

3 Biotic Stress

Pathogens are the agents that cause diseases in plants through biotic stress other than abiotic stress. These pathogens include bacteria, viruses, nematodes, and fungi. Biotic stress is also induced in plants by pests. Climatic changes are main source of these pests and other disease-causing agents. It is recognized that increase in temperature is a big cause of spreading of pathogens. Fungi are biotic organisms that are main threat for plants than other biotic organisms. About 8100 fungi species are identified as pathogens. Likewise, viruses are also known as a big threat to plants other than fungi worldwide. Wilting of plants, leaf spot, root rot, and seed damage are the symptoms that are caused by different types of pathogens (Anderson et al. 2004). Pathogens infect plants parts such as flowers and leaves. By this way, they provide a source to other bacteria and viruses to harm other plants. Some other wild species of plants are present that grow on their own known as weeds. These weeds reduce the plants' growth directly or indirectly. Weeds are fast growing and high yielding plants; due to this reason, they dominate over other plants and environment (Tanveer et al. 2013; Saddique et al. 2018).

4 Phytohormones

Phytohormones are natural substances and are chemically synthesized as growth regulators. The term "phytohormones" was introduced by Went and Timan in 1937 and is used as the title of their book (Gill and Patranabis 2021). Phytohormones have role not only in the enhancement of growth of plants, but they also have role in producing tolerance of heat stress in plants and increasing yield (Torres et al. 2017). Plant life cycle passes through different circumstances like stress. Plant growth under salinity is reduced by ions which have effect on metabolism of plants or may be due to relationship of these ions with water. Under salinity stress, different techniques were used to increase the growth of plants. To date, nine different types of phytohormones were recognized; the first hormone discovered was auxin, likewise, salicylates, cytokinin, gibberellin, brassinosteroids, jasmonates, abscisic acid, and strigolactones were identified. Abscisic acid, salicylates, ethylene, and jasmonates have pivotal role in producing tolerance against stresses in plants. Indole acetic acid is believed to be useful as a potential prodrug for cancer therapy in combination with horseradish peroxidase (HRP) (Wan et al. 2022).

5 Categories of Phytohormones

Mainly, there are five naturally occurring phytohormones. These are auxin, cytokinins, ethylene, abscisic acid, and gibberellic acid (Biswas et al. 2018). All these phytohormones are categorized in two classes as growth promoter and growth inhibitors.

5.1 Growth Promoters

Auxin, gibberellins, and cytokinins have been given the status of growth promoters in plants.

5.1.1 Auxin

Auxins are endogenous regulators having vital role in the synthesis of roots and shoots as well as their relative growth. In plants, indole acetic acid is produced by tryptophan and its indole precursor dependently or independently. Auxins have role in plants' defensive mechanisms toward high salinity conditions. In thale cress, *Oryza sativa*, and soybean plant, it is investigated that auxin activates the process of transcription of different genes (Zhang et al. 2021). Reactive oxygen species, homeostasis, architecture of roots and ABA- (abscisic acid) sensitive genes are modulated by auxin which increases the drought-resistant phenotype (Jogawat et al. 2021).

As roots are the main organs in plants which sense the heavy metal stress, it is needed to determine the interconnection between the stimuli of heavy metal stress and auxin homeostasis. Heavy metals modulate auxin concentration and distribution pattern, resulting in disruption homeostasis of auxin in plants (Fattorini et al. 2017). When plants face metalloids, transcript level of auxin genes is changed, which maintains homeostasis through rearranging auxin active pool by its deactivation, mortification, and transport for better adaptation under stress. The graphical representation of auxin as heavy metal stress mitigator is shown in Fig. 24.1 (Wang et al. 2015; Singh et al. 2021).

5.1.2 Gibberellins

More than 100 years ago, gibberellins (also known as gibberellic acid) were found. Since then, about 100 types of gibberellins are identified. All about 136 gibberellins have been discovered based on their structure (Bon et al. 2018). They are responsible for cell elongation and inducing enzyme synthesis in barley. GAs (gibberellic acids) are diterpenes that are synthesized from acetic and mevalonic acid via the isoprenoid pathway. An alternative pathway for gibberellic acid biosynthesis has

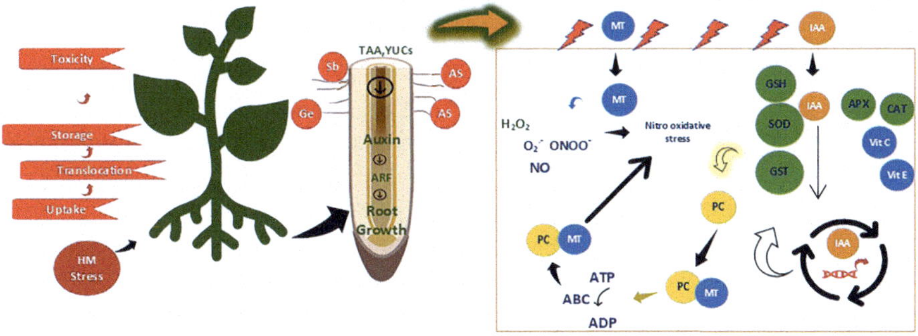

Fig. 24.1 Schematic diagram representing role of auxin as heavy metal stress mitigator in plants (Singh et al. 2021)

Table 24.2 Role and mechanism of gibberellins in different plants

Plant	Role and mechanism	References
Rice	Regulates proteins (salt regulated proteins)	Wen et al. (2010)
Linseed	Enhancement of growth, physio-biochemical attributes, amino acid content, and antioxidant enzyme activity	Khan et al. (2010)
Wheat	Induces inflection of ion and phytohormone homeostasis	Iqbal and Ashraf (2013)

been proposed, in which the precursor is pyruvate rather than mevalonic acid (Fàbregas and Fernie 2022).

Reduction in plant height was interconnected to short internodes. A change or lack of GA was critical in the evolution of semidwarf and varieties of *Triticum aestivum* and *Oryza sativa* with high yield during Green Revolution, which enhanced the yield of both grain (Plaza-Wüthrich et al. 2016). Gibberellins also have role in increasing plant resistance toward heavy metal stress by regulating various plant processes. In *Vicia faba* (L.), GA administration helps restore lead (Pb) and Cd-induced mitotic indices (Saini et al. 2021). In addition, by stimulating the mobilization of electrons in *H. vulgare*, the redox homeostasis is regulated by GA (Table 24.2). Gibberellic acid involvement in redox balance helps plants adapt to less than optimal growing conditions (Mark et al. 2016).

By controlling several salt-regulated proteins, GA reduces the NaCl-induced growth suppression in rice by showing a minor impact of gibberellins on salinity in *O. sativa*. Under salt stress, the enhancement in the production was due to ionic and hormonal homeostasis induced by GA priming. The treatment of GA and $CaCl_2$ combination have reduced the impact of salt stress by accelerating the growth rate, metabolic framework, and amino acid in flaxseed. Application of gibberellins has also minimized salt stress in other plants (Fahad et al. 2015). It is reported that gibberellins increase plant nitrogen use efficiency and chlorophyll content to alleviate salinity (Criado et al. 2017). Gibberellins enhance protein content, RNA content, and mitotic activity in plants which produce tolerance against cadmium stress (Niharika et al. 2021).

5.1.3 Cytokinins

Among the growth regulators, cytokinins have key function in promoting division of cells, chlorophyll biosynthesis, and modifying apical dominance in plants. CK (Cytokinin) members are considered important regulators that strongly influence plant growth (Hai et al. 2020). Cytokinins are present in benzylaminopurine form, but they are very expensive. Due to specific spatial and timing expression of cytokinin metabolism, application or manipulation of exogenous CK can have effect on stress tolerance of plants, making endogenous cytokinin application or manipulation more difficult to achieve with respect to overall stress (Yu et al. 2021).

Table 24.3 Cytokinins effect on plants pathogens

Pathogen type	Pathogens	Host plants	Cytokinins' sources	Effects of CK	References
Fungi	*Erysiphe graminis f.* sp. *Tritici*	*T. aestivum*	Exogenous trans-zeatin	Immunity induced by CK and cytokinin-induced susceptibility	Babosha (2009)
Insects, pests	Gypsy moth (*Lymantria dispar*)	*Populus*	Exogenous application of benzylaminopurine (BAP) 100 μM, (Totally 25 ml per plant)	Daunt insect feeding, delay larval development, or reduce weight gained by insect larvae, wound-inducible accumulation of JA (jasmonic acid) and LNA (linolenic acid)	Dervinis et al. (2010)
Bacteria	*P. syringae*	*A. thaliana*	*P. fluorescens*	*P. fluorescens* G20-18 produced by CK is used as biocontrol against infection caused by *P. syringae*	Großkinsky et al. (2016)

Cytokinins' application effect on plant stress resistance may be positive or negative. Cytokinins have important function in reduction of stress in plants caused due to bacteria, fungi, and other pests. Cytokinins' effect producing tolerance in plants against biotic stress-causing agents like pathogens is given in Table 24.3 (Akhtar et al. 2020). Cytokinins' application under abiotic stress conditions can directly slow down leaf aging by scavenging free radicals (Hönig et al. 2018). *Solanum melongena* have shown enhancement in tolerating the salt-induced stress on cytokinins application, which enhanced photosynthesis, roots/shoots biomass accumulation, width of stem, and malondialdehyde content (Wu et al. 2014; Weller et al. 2021).

5.2 *Growth Inhibitors*

5.2.1 Abscisic Acid

In plants, abscisic acid hormone is related with primary response to stress. For more than 40 years, abscisic acid biosynthesis' major components are identified by molecular genetics and biochemical applications. Abscisic acid is stored rapidly when plants go through stress such as salt stress. When the plants face adverse conditions, then ABA (Abscisic acid) level modulation in tissues and cells is critical for balancing the protection and other development processes. Different processes are involved in controlling abscisic acid level in plants. These include transport, synthesis as well as degradation, and metabolism (Chen et al. 2020; Iqbal et al. 2021). On the behalf of concentration, plant to plant as well as climatic conditions, abscisic acid has different effects on the root growth of plants. Abscisic acid's high

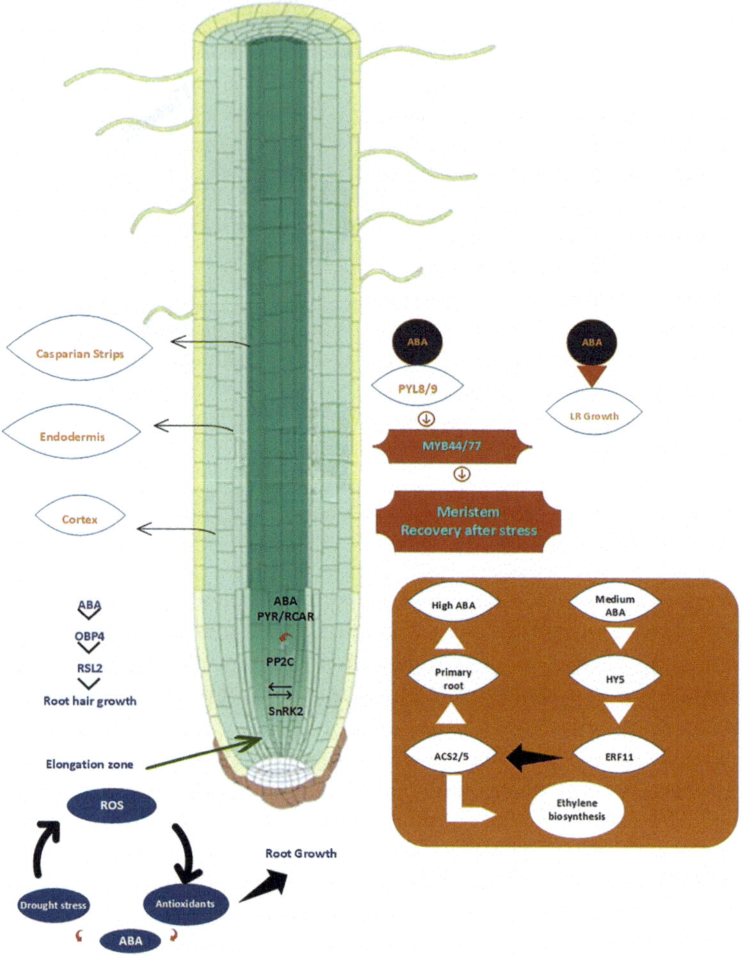

Fig. 24.2 Role of abscisic acid in plants during drought stress (Ali et al. 2020)

concentration causes inhibition of root formation; however, low content stimulates them (Rowe et al. 2016; Mahmood et al. 2022a).

When the abscisic acid is applied to plants exogenously, then it is confirmed that formation of root hairs is negatively regulated except ABA-insensitive mutant. It controls the root hairs' formation by DOF-type transcriptional regulator which is OBP4. It binds with RSL2 gene and causes inhibition of its expression (Fig. 24.2; Rymen et al. 2017). Exogenous ABA treatment minimizes the falling of leaves as well as enhances the salinity tolerance in citrus plant and common bean (Mahmood et al. 2012; Gupta et al. 2021).

5.2.2 Ethylene

Ethylene is present in gaseous form. Less amount of ethylene in plants facilitates the initiation of protective signals, while high concentration in cucumber and wheat appeared to be inhibitory at the time of postharvest storage (Riyazuddin et al. 2020). HSPs (heat shock proteins) have important functions in minimizing impacts of heat stress by maintaining other proteins and helping them to refold. This can be possible by the application of protective hormones such as ethylene (Poór et al. 2022). On the application of abscisic acid, seed germination process is inhibited, but by application of ethylene and gibberellins hormones, this abscisic acid is inhibited and germination process is started again (Ullah et al. 2015; Verma et al. 2016).

A substance is used to replace ethylene which has same properties and function as ethylene and is known as "Ethephon." For instance, under salt stress condition to examine the ethylene function, ethephon was applied to seeds of *Medicago sativa* (Zain et al. 2017; Wang et al. 2020). Ethylene is found to relieve water lodging stress through the development of cross-disordered aerenchyma cells that oscillate with abscisic acid, gibberellins, auxins, and kinetin (Shimamura et al. 2014). Ethylene is formed by two enzymes. These enzymes are 1-Aminocyclopropane-1-Carboxylic acid synthase (ACS) and 1-Aminocyclopropane-1-Carboxylic acid oxidase (ACO). Under normal conditions, ethylene is synthesized by Met. ACS convert *S*-adenosyl-methionine (SAM) into 1-aminocyclopropane-1-carboxylic acid (ACC). Then 1-aminocyclopropane-1-carboxylic acid is oxidized to synthesize ET through an enzyme ACO (Fig. 24.3; Thao et al. 2015).

5.3 Other Stress-Related Phytohormones

Salicylic acid, jasmonate, melatonin, strigolactones, brassinosteroids, polyamines, and PGPR-mediated phytohormones are some other stress-mitigating hormones.

5.3.1 Salicylic Acid

Salicylic acid has pivotal role in different processes occurring in plants and protecting plants from various stresses. In plants, salicylic acid is present mostly in powdered form. It is a vital phytohormone which has important role in enhancing photosynthetic activity by regulating carbohydrate metabolism in plants (Arif et al. 2020). Salicylic acids' protective functions include regulation of ROS and antioxidant enzymes (Sharma et al. 2020). Salicylic acid application is studied in different crops by using different methods such as foliar application on maize, soaking seeds of wheat, or by rooting medium of wheat (Ahmad et al. 2021). Under several abiotic stress conditions, salicylic acid showed increased growth of plants, photosynthetic

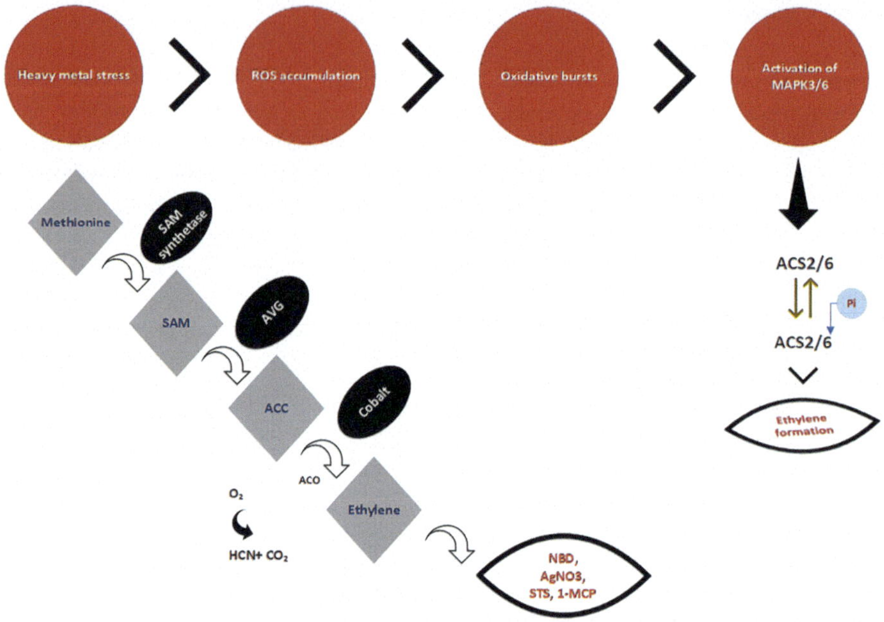

Fig. 24.3 Biosynthesis of ethylene under normal conditions and heavy metal stress (Thao et al. 2015)

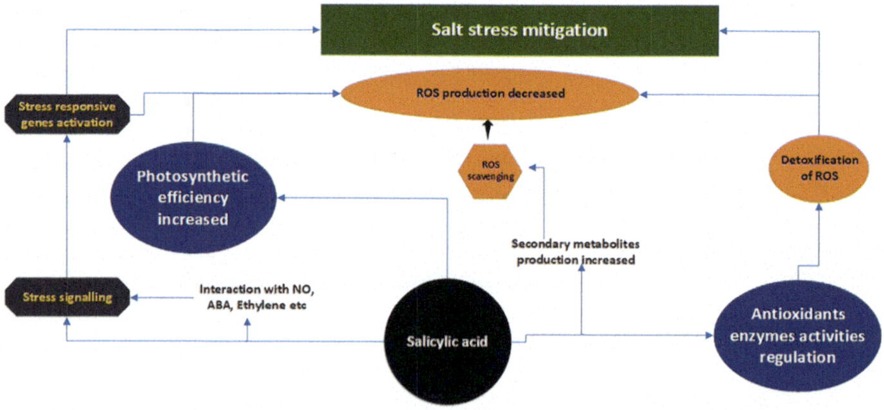

Fig. 24.4 Mechanism of action of salicylic acid in alleviation of abiotic stress (Hasanuzzaman et al. 2017)

activity, and minimized ROS production shown in two pea varieties (Fig. 24.4; Hasanuzzaman et al. 2017). It is investigated that application of salicylic acid with trehalose mitigates drought stress in sweet basil by enhancing the antioxidant enzymatic activity (Zulfiqar et al. 2021).

5.3.2 Jasmonates

Plants show different physiological and biochemical responses to environmental conditions. These responses are exhibited by the application of jasmonic acid (JA). Jasmonates are combination of jasmonic acid and its derivatives (Ruan et al. 2019). Many reports indicate that exogenous jasmonic acid can reduce salt toxicity by stimulating antioxidant activity. Foliar application of jasmonic acid on leaves of various crops minimizes negative impact of salinity and increases crop productivity and yields by inducing enzymes that scavenge ROS and absorb ions. It is identified that formation of terpenoid, alkaloid, and antioxidant (secondary metabolites) is regulated by jasmonates. Secondary metabolites' production is increased in tomato by activation of jasmonates. Activation of signaling pathway and mechanism of jasmonic acid under stress climate is shown in Table 24.4; Fig. 24.5 (Farhangi-Abriz and Ghassemi-Golezani 2018).

5.3.3 Melatonin

In phytohormones, melatonin is a remarkable and strong hormone which has severe effect on dehydration occurring in plants. Melatonin is a natural growth hormone. That is why melatonin is applied to plants such as medicinal plants, vegetable plants, beauty plants, and fruiting trees to mitigate the effects of drought (Moustafa-Farag et al. 2020). In many studies, melatonin is recognized as antiviral compound in animals. In plants with bacterial interactions, melatonin is used as a strong anti-bacterial compound against different pathogens. As an example, in rice plants BLS (bacterial leaf streak) caused by Xoo (*Xanthomonas oryzae* pv. *oryzae*) is reduced by 17% by foliar application of melatonin (Chen et al. 2019). It is studied that melatonin enhances the mechanism of tolerance in mustard greens under salinity (Park et al. 2021). Melatonin also has role in the production of other growth regulators in plants, which helps them to tolerate different stresses. It is tested at different concentrations in different literatures (Table 24.5; Shafi et al. 2021). Plant stress is regulated directly or indirectly through melatonin. Directly, melatonin inhibits accumulation of ROS and nitrogen species, and indirectly it affects stress responsive pathway.

Table 24.4 Jasmonates' exogenous and endogenous regulatory mechanism against environmental stresses

Types	Stresses	Species of plants	Jasmonates	Role of protection	References
Abiotic stress	Heavy metal Nickel	*Glycine max*	Exogenous (JA)	Expression of genes, antioxidant enzyme activation, osmolytes increased	Sirhindi et al. (2016)
Biotic stress	*Fusarium culmorum*	*Triticum aestivum*	Exogenous (MeJA)	Significantly minimized the concentration of H_2O_2 and peroxidation of lipids	Brenya et al. (2020)

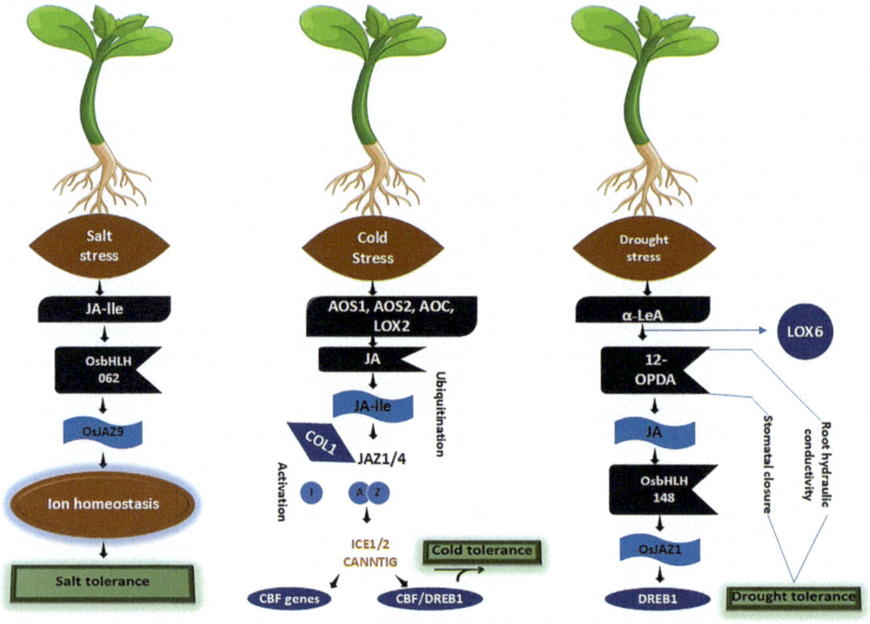

Fig. 24.5 Signaling pathway of JA in response to different abiotic stresses (Wang et al. 2021)

Table 24.5 Melatonin role in production of other growth regulators under stressed conditions

Stress types	Dose of melatonin	Species of plants	Plant growth regulators	Responses	References
Low temperature stress	150 µM	*Citrullus lunatus*	JA accumulation	MeJA and H_2O_2 level enhanced. Displaying role in grafting induced cold tolerance by melatonin	Li et al. (2021a)
Salinity stress	10 µM	*Arabidopsis thaliana*	ABA	Melatonin induces salt tolerance via upregulating of abscisic acid response gene, induces antioxidant defensive system, minimizes the high production of ROS induced by salt	Shukla et al. (2021)
Cd (Heavy metal)	100 µM	*Triticum aestivum*	NO scavenger (cPTIO)	Enhanced growth characteristics, reduced oxidative stress, decreased NO (nitric oxide) concentrations	Kaya et al. (2019)

5.3.4 Strigolactones

Recently, some new plant hormones have been discovered and studied. These are known as strigolactones. They are originally identified as host-derived stimulants for germinating the weeds of witch weeds and *Orobanche* genera (Saeed et al. 2017). Furthermore, strigolactones were recognized as signal molecules that have pivotal role in stimulating symbiosis between plant roots and AMF (Banerjee and Bhadra 2020). Higher secretion of strigolactones has role in stimulating germination of seeds of giant witch weeds more than 100 times (Crosino and Genre 2022).

In rice plants, strigolactones is studied to regulate the antioxidant defensive mechanism for the reduction of arsenic stress (Mostofa et al. 2021). Recently, strigolactones' function in plant microorganisms' interaction in rhizosphere and nodule formation has been studied. Strigolactones regulate root architecture. In rhizosphere, mycorrhizal association between rhizobia and AMF (arbuscular mycorrhizal fungi) is established by strigolactones which promote parasitic interaction (Faizan et al. 2020).

5.3.5 Brassinosteroids

Another class of phytohormones are brassinosteroids (BRs). They are member of polyhydroxysteroids class. Brassinosteroids can stimulate many processes such as germination of seeds, division in cells, theca growth, pollen tube formation and differentiation of tracheal elements, polarization of cell membrane, leaf senescence, and death. It has been identified that brassinosteroids are present in all parts of plants such as fruit, seed, root, shoot, flowers, and pollens (Manghwar et al. 2022). It has also been shown that after administration of 24-epibrasinolide, level of lead in *Beta vulgaris* is 50% less than plants with only metal treatment as this hormone reduces metal absorption (Rajewska et al. 2016).

Two types of gene regulation patterns are exhibited on application with exogenous brassinosteroids and exposing to heat: (I) Brassinosteroids rescue development protein expression, which is suppressed by heat stress. (II) Brassinosteroids induce high level of protecting protein than simply heat stress. BRs have also positive impacts on cereals crops such as wheat under stressed conditions shown in Table 24.6 (Kothari and Lachowiec 2021).

Table 24.6 Effect of brassinosteroids on cereal plants under normal and stress condition

Compounds	Treatment method	Quantity	Plant species	Type of stress	References
24-epibrassinolide	Foliar application	0.026 μM 0.05 μM 0.07 μM	*Triticum aestivum*	Salt	Shahbaz et al. (2008)
28-homobrassinolide	Foliar application	101 μM 202 μM 404 μM	*Triticum aestivum*	Salt	Eleiwa et al. (2011)

5.3.6 Polyamines

Polyamine is a ubiquitous aliphatic polycationic amine. Plants' growth and development is increased by application of polyamines. In addition, they regulate many different processes such as stress in plants. They can do this because of having character of free radical destruction, accommodation of osmotic level, and improvement of cell membrane integration (Spormann et al. 2021). It is recognized that polyamines have role in tolerating stresses such as salt and drought (Hassan et al. 2018).

Mostly used polyamines in plants are putrescines, spermine, and spermidines (Saha et al. 2015). Particularly, high concentration of polyamines under different stressed conditions exhibits two types of effects. At one side, application of polyamines was associated with tolerance of plants to different stresses because of role in inactivation of free radicals, while on the other side polyamines have been identified to reduce plant's ability to tolerate stressed condition due to high level of H_2O_2 because of polyamine catabolism. Polyamines also have been identified as protectors against different abiotic factors in plants (Alcázar et al. 2020). In tomatoes, application of putrescine results in better cold tolerance (Song et al. 2014).

5.3.7 GPR-Mediated Phytohormones

Sometimes, plants growth is stimulated by some bacteria known as PGPR (plant growth promoting rhizobacteria). They colonize with roots of plants. PGPR are potential candidates for plant protection through colonization in the rhizosphere and production of metabolites of antimicrobial (antagonists). Phytohormones which are generated by these types of bacterial communities provide plants' growth and immunization through regulating hormones (Maheshwari et al. 2015). Crops yield is increased by the treatment of PGPR under different types of stressed conditions through induction of different properties. They also decreased need of fertilizers as well as contribute to sustainability of agricultures. PGPR reduces both type of stress and stimulate plants' growth by various methods (Fig. 24.6; Niranjana and Hariprasad 2014).

PGPR have two types. These are extracellular PGPR (ePGPR) and intracellular PGPR (iPGPR). Extracellular PGPR exists in root cortex and intracellular PGPR usually exists in special structures of roots. Bacterial genera of the ePGPR class are *Agrobacteria, Arthrobacters, Azotobacter, Azospirillum, Bacillus, Burkholderia, Caulobacter, Chromobacterium, Erwinia, Flavobacterium, Micrococcus, Pseudomonas, and Serratia* (More et al. 2022). iPGPR belongs to the Rhizobiaceae family, which includes *Allorhizobium* and *Bradyrhizobium* (Subhash et al. 2021). PGPR alleviates stress through symbiosis of rhizobia and legumes. The ability to fixing nitrogen depends on types of bacteria (Etesami 2022).

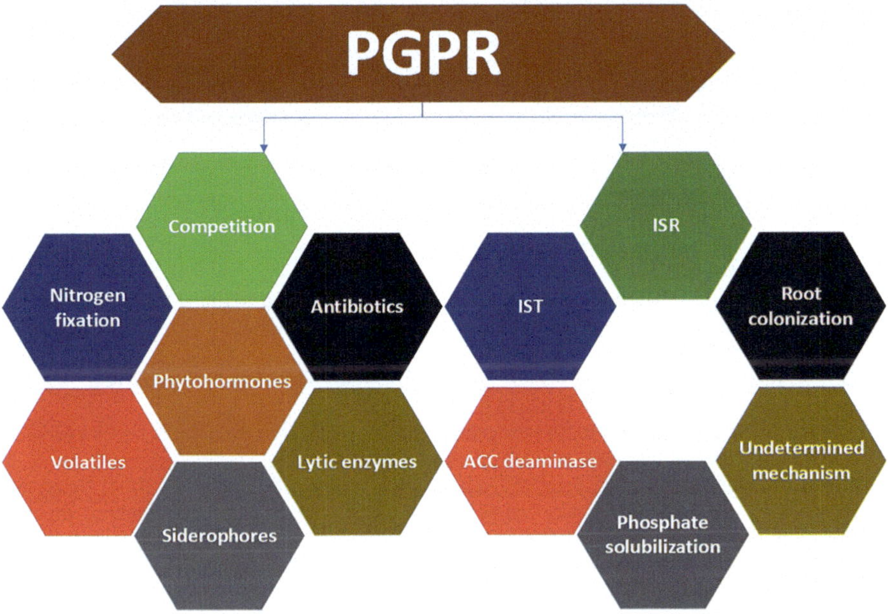

Fig. 24.6 Mechanism of mitigation of stresses and growth promotion by PGPR (Niranjana and Hariprasad 2014)

6 Conclusion and Future Perspective

Plants face many types of stress conditions during germination and growth processes. These unfavorable conditions have bad impacts on the plants which cause decrease in number of tillers, growth, and yield. To overcome these stresses, plants develop some growth regulators known as phytohormones (PHs). These hormones are involved in regulating integrity of the plants to endure with stress. Phytohormones stimulate genes expressions in plants which help them to mitigate stresses. Auxin, gibberellin, and cytokinin help plant in promoting growth. Under abiotic stress conditions, jasmonates, abscisic acid, ethylene, polyamines as well as salicylic acid impart tolerance to plants. Melatonin is observed in mitigation of plants' diseases caused by some bacteria. Strigolactones and melatonin promote arbuscular mycorrhizal association in between plants and fungi. Enhancement in plants' growth and crop production is investigated by application of these plant hormones. But more detailed investigation is needed about role of these phytohormones in tissue culture techniques, breaking dormancy, setting of fruits quality, and many other processes as they all are still unclear.

References

Abbas T, Nadeem MA, Tanveer A, Matloob A, Zohaib A, Safdar ME, Ali HH, Farooq N, Javaid MM, Tabassum T (2019) Herbicide mixtures and row spacing effects on Fenoxaprop resistant Phalaris minor in wheat. Int J Agric Biol 22(4):840

Ahmad A, Aslam Z, Naz M, Hussain S, Javed T, Aslam S, Raza A, Ali HM, Siddiqui MH, Salem MZ (2021) Exogenous salicylic acid-induced drought stress tolerance in wheat (*Triticum aestivum* L.) grown under hydroponic culture. PLoS One 16(12):e0260556

Akhtar MS (2019) Salt stress, microbes, and plant interactions: mechanisms and molecular approaches: Volume 2. Springer

Akhtar SS, Mekureyaw MF, Pandey C, Roitsch T (2020) Role of cytokinins for interactions of plants with microbial pathogens and pest insects. Front Plant Sci 10. https://doi.org/10.3389/fpls.2019.01777

Alcázar R, Bueno M, Tiburcio AF (2020) Polyamines: small amines with large effects on plant abiotic stress tolerance. Cell 9(11):2373

Ali S, Hayat K, Iqbal A, Xie L (2020) Implications of abscisic acid in the drought stress tolerance of plants. Agronomy 10(9):1323

Anderson PK, Cunningham AA, Patel NG, Morales FJ, Epstein PR, Daszak P (2004) Emerging infectious diseases of plants: pathogen pollution, climate change and agrotechnology drivers. Trends Ecol Evol 19(10):535–544

Aqueel MA, AbM R, Balal RM, Shahid MA, Mustafa I, Javaid MM, Leather SR (2015) Tritrophic interactions between parasitoids and cereal aphids are mediated by nitrogen fertilizer. Insect Sci 22(6):813–820

Arif Y, Sami F, Siddiqui H, Bajguz A, Hayat S (2020) Salicylic acid in relation to other phytohormones in plant: a study towards physiology and signal transduction under challenging environment. Environ Exp Bot 175:104040. https://doi.org/10.1016/j.envexpbot.2020.104040

Asif T, Javaid M, Abdul K, Abbas R, Ahsan A (2012) Allelopathic effect of Echinochloa crus-galli on field crops. Herbologia 13(2):9–17

Aziz A, Ashraf M, Asif M, Safdar M, Shahzad S, Javaid M, Akhtar N, Waheed H, Nadeem M, Ali10 S (2019) Impact of mulching materials on weeds dynamics, soil biological properties and lettuce (*Lactuca sativa* L.) productivity. Int J Bot Stud 4(4):128–134.

Babar BH, Tanveer A, Tahir M, Aziz A, Auh A, Nadeem MA, Javaid MM (2009) Allelopathic potential of wild onion (*Asphodelus tenuifolius*) on the germination and seedling growth of chickpea (*Cicer arietinum*). Weed Biol Manage 9(2):146–151

Babosha AV (2009) Regulation of resistance and susceptibility in wheat–powdery mildew pathosystem with exogenous cytokinins. J Plant Physiol 166(17):1892–1903

Banerjee P, Bhadra P (2020) Mini-review on Strigolactones: newly discovered plant hormones. Bioresour Biotech Res Comm 13(3)

Biswas S, Shivaprakash M, Maina C (2018) Biosynthesis of phytohormones by potassium solubilising bacteria isolated from banana rhizosphere. Int J Agric Environ Biotechnol 11(3):497–501

Bon DJYD, Mander LN, Lan P (2018) Syntheses of gibberellins A15 and A24, the key metabolites in gibberellin biosynthesis. J Org Chem 83(12):6566–6572. https://doi.org/10.1021/acs.joc.8b00876

Brenya E, Chen Z-H, Tissue D, Papanicolaou A, Cazzonelli CI (2020) Prior exposure of Arabidopsis seedlings to mechanical stress heightens jasmonic acid-mediated defense against necrotrophic pathogens. BMC Plant Biol 20(1):548. https://doi.org/10.1186/s12870-020-02759-9

Chaudhry S, Sidhu GPS (2021) Climate change regulated abiotic stress mechanisms in plants: a comprehensive review. Plant Cell Rep 1:1–31

Chen X, Sun C, Laborda P, He Y, Zhao Y, Li C, Liu F (2019) Melatonin treatments reduce the pathogenicity and inhibit the growth of *Xanthomonas oryzae* pv. oryzicola. Plant Pathol 68(2):288–296. https://doi.org/10.1111/ppa.12954

Chen K, Li G-J, Bressan RA, Song C-P, Zhu J-K, Zhao Y (2020) Abscisic acid dynamics, signaling, and functions in plants. J Integr Plant Biol 62(1):25–54. https://doi.org/10.1111/jipb.12899

Chinnusamy V, Zhu J, Zhu J-K (2007) Cold stress regulation of gene expression in plants. Trends Plant Sci 12(10):444–451

Criado MV, Veliz CG, Roberts IN, Caputo C (2017) Phloem transport of amino acids is differentially altered by phosphorus deficiency according to the nitrogen availability in young barley plants. Plant Growth Regul 82(1):151–160. https://doi.org/10.1007/s10725-017-0247-6

Crosino A, Genre A (2022) Peace talks: symbiotic signaling molecules in arbuscular mycorrhizas and their potential application. J Plant Interact 17(1):824–839

de Oliveira TS, Bell RW (2022) Introduction to subsoil constraints for crop production. In: Oliveira TS, Bell RW (eds) Subsoil constraints for crop production. Springer International Publishing, Cham, pp 1–10. https://doi.org/10.1007/978-3-031-00317-2

Dervinis C, Frost CJ, Lawrence SD, Novak NG, Davis JM (2010) Cytokinin primes plant responses to wounding and reduces insect performance. J Plant Growth Regul 29(3):289–296. https://doi.org/10.1007/s00344-009-9135-2

DESA U (2015) United Nations, Department of Economic and Social Affairs, Population Division (UN DESA), world population prospects: the 2015 revision, key findings and advance tables. Working Paper No. ESA/P/WP. 241, United Nations, Department of Economics.

Desaint H, Aoun N, Deslandes L, Vailleau F, Roux F, Berthomé R (2021) Fight hard or die trying: when plants face pathogens under heat stress. New Phytol 229(2):712–734

EL Sabagh A, Islam MS, Hossain A, Iqbal MA, Mubeen M, Waleed M, Reginato M, Battaglia M, Ahmed S, Rehman A, Arif M, Athar H-U-R, Ratnasekera D, Danish S, Raza MA, Rajendran K, Mushtaq M, Skalicky M, Brestic M, Soufan W, Fahad S, Pandey S, Kamran M, Datta R, Abdelhamid MT (2022) Phytohormones as growth regulators during abiotic stress tolerance in plants. Front Agron 4. https://doi.org/10.3389/fagro.2022.765068

Eleiwa ME, Bafeel SO, Ibrahim S (2011) Influence of brassinosteroids on wheat plant (*Triticum aestivum* L.) production under salinity stress conditions. I-Growth parameters and photosynthetic pigments. Aust J Basic Appl Sci 5(5):58–65

Etesami H (2022) Root nodules of legumes: a suitable ecological niche for isolating non-rhizobial bacteria with biotechnological potential in agriculture. Curr Res Biotechnol

Fàbregas N, Fernie AR (2022) The reliance of phytohormone biosynthesis on primary metabolite precursors. J Plant Physiol 268:153589

Fahad S, Nie L, Chen Y, Wu C, Xiong D, Saud S, Hongyan L, Cui K, Huang J (2015) Crop plant hormones and environmental stress. Sustain Agric Rev 1:371–400

Fahad S, Bajwa AA, Nazir U, Anjum SA, Farooq A, Zohaib A, Sadia S, Nasim W, Adkins S, Saud S, Ihsan MZ, Alharby H, Wu C, Wang D, Huang J (2017) Crop production under drought and heat stress: plant responses and management options. Front Plant Sci 8. https://doi.org/10.3389/fpls.2017.01147

Faizan M, Faraz A, Sami F, Siddiqui H, Yusuf M, Gruszka D, Hayat S (2020) Role of strigolactones: Signalling and crosstalk with other phytohormones. Open Life Sci 15(1):217–228

Farhangi-Abriz S, Ghassemi-Golezani K (2018) How can salicylic acid and jasmonic acid mitigate salt toxicity in soybean plants? Ecotoxicol Environ Saf 147:1010–1016

Fattorini L, Ronzan M, Piacentini D, Della Rovere F, De Virgilio C, Sofo A, Altamura M, Falasca G (2017) Cadmium and arsenic affect quiescent centre formation and maintenance in Arabidopsis thaliana post-embryonic roots disrupting auxin biosynthesis and transport. Environ Exp Bot 144:37–48

Gill A, Patranabis S (2021) Phytohormones as potential anticancer agents. Int J Res Appl Sci Biotechnol 8(3):37–43

Großkinsky DK, Tafner R, Moreno MV, Stenglein SA, García de Salamone IE, Nelson LM, Novák O, Strnad M, van der Graaff E, Roitsch T (2016) Cytokinin production by Pseudomonas fluorescens G20-18 determines biocontrol activity against Pseudomonas syringae in *Arabidopsis*. Sci Rep 6(1):23310. https://doi.org/10.1038/srep23310

Gull A, Lone AA, Wani NUI (2019) Biotic and abiotic stresses in plants. In: Abiotic biotic stress plants:1–19

Gupta B, Huang B (2014) Mechanism of salinity tolerance in plants: physiological, biochemical, and molecular characterization. Int J Genomics 2014:701596. https://doi.org/10.1155/2014/701596

Gupta D, Singh G, Tiwari S, Patel A, Fatima A, Dubey A, Naaz N, Pandey J, Prasad SM (2021) Salt stress toxicity amelioration by Phytohormones, synthetic product, and nutrient amendment practices. Physiology of salt stress in plants: perception, signalling, omics and tolerance mechanism, pp. 198–228

Hai NN, Chuong NN, Tu NHC, Kisiala A, Hoang XLT, Thao NP (2020) Role and regulation of cytokinins in plant response to drought stress. Plan Theory 9(4):422

Hasanuzzaman M, Nahar K, Bhuiyan TF, Anee TI, Inafuku M, Oku H, Fujita M (2017) Salicylic acid: an all-rounder in regulating abiotic stress responses in plants. Phytohormones-Signaling Mechanisms and crosstalk in plant development and stress responses 16:31–75.

Hassan FAS, Ali EF, Alamer KH (2018) Exogenous application of polyamines alleviates water stress-induced oxidative stress of Rosa damascena miller var. trigintipetala Dieck. S Afr J Bot 116:96–102. https://doi.org/10.1016/j.sajb.2018.02.399

He M, He C-Q, Ding N-Z (2018) Abiotic stresses: general defenses of land plants and chances for engineering multistress tolerance. Front Plant Sci 9:1771

Hönig M, Plíhalová L, Husičková A, Nisler J, Doležal K (2018) Role of cytokinins in senescence, antioxidant defence and photosynthesis. Int J Mol Sci 19(12):4045

Hu S, Ding Y, Zhu C (2020) Sensitivity and responses of chloroplasts to heat stress in plants. Front Plant Sci 11:375

Iqbal M, Ashraf M (2013) Gibberellic acid mediated induction of salt tolerance in wheat plants: growth, ionic partitioning, photosynthesis, yield and hormonal homeostasis. Environ Exp Bot 86:76–85. https://doi.org/10.1016/j.envexpbot.2010.06.002

Iqbal M, Iqbal MM, Saghir A, Mahmood A, Akram M, Husnain H, Shahid M, Ahmad S, Raza A, Hussain A (2021) Performance of early and late planting cotton genotypes under agro-ecological conditions of Multan, Punjab, Pakistan. Pak J Agric Res 34(3):569

Javaid MM, Tanveer A (2013) Optimization of application efficacy for POST herbicides with adjuvants on three-cornered jack (*Emex australis* Steinheil) in wheat. Weed Technol 27(3):437–444

Javaid M, Tanveer A (2014) Germination ecology of E mex spinosa and E mex australis, invasive weeds of winter crops. Weed Res 54(6):565–575

Javaid M, Tanveer A, Ahmad R, Yaseen M, Khaliq A (2012) Optimizing activity of herbicides at reduced rate on Emex spinosa campd. with adjuvants. Planta Daninha 30:425–435

Javaid M, Zia A, Waheed H, Nargis J, Shahid A, Aziz A, Wasaya A (2020) Effect of isoproturon with and without adjuvants on photosynthetic attributes of wheat and its associated weeds. Planta Daninha 38

Javaid MM, Florentine SK, Ashraf M, Mahmood A, Sattar A, Wasaya A, Li FM (2022) Photosynthetic activity and water use efficiency of *Salvia verbenaca* L. under elevated CO_2 and water-deficit conditions. J Agron Crop Sci 208(4):536–551

Jogawat A, Yadav B, Lakra N, Singh AK, Narayan OP (2021) Crosstalk between phytohormones and secondary metabolites in the drought stress tolerance of crop plants: a review. Physiol Plant 172(2):1106–1132

Kaya C, Okant M, Ugurlar F, Alyemeni MN, Ashraf M, Ahmad P (2019) Melatonin-mediated nitric oxide improves tolerance to cadmium toxicity by reducing oxidative stress in wheat plants. Chemosphere 225:627–638. https://doi.org/10.1016/j.chemosphere.2019.03.026

Khalid MF, Hussain S, Ahmad S, Ejaz S, Zakir I, Ali MA, Ahmed N, Anjum MA (2019) Impacts of abiotic stresses on growth and development of plants. In: Plant tolerance to environmental stress. CRC Press, pp 1–8

Khan MN, Siddiqui MH, Mohammad F, Naeem M, Khan MMA (2010) Calcium chloride and gibberellic acid protect linseed (*Linum usitatissimum* L.) from NaCl stress by inducing antioxidative defence system and osmoprotectant accumulation. Acta Physiol Plant 32(1):121–132

Kothari A, Lachowiec J (2021) Roles of Brassinosteroids in mitigating heat stress damage in cereal crops. Int J Mol Sci 22(5):2706

Li H, Guo Y, Lan Z, Xu K, Chang J, Ahammed GJ, Ma J, Wei C, Zhang X (2021a) Methyl jasmonate mediates melatonin-induced cold tolerance of grafted watermelon plants. Hortic Res 8. https://doi.org/10.1038/s41438-021-00496-0

Li N, Euring D, Cha JY, Lin Z, Lu M, Huang L-J, Kim WY (2021b) Plant hormone-mediated regulation of heat tolerance in response to global climate change. Front Plant Sci 11. https://doi.org/10.3389/fpls.2020.627969

Maheshwari DK, Dheeman S, Agarwal M (2015) Phytohormone-producing PGPR for sustainable agriculture. In: Maheshwari DK (ed) Bacterial metabolites in sustainable agroecosystem. Springer International Publishing, Cham, pp 159–182. https://doi.org/10.1007/978-3-319-24654-3_7

Mahmood A, Javaid MM, Aziz M, Shahzad AN, Rehman A, Honermeier B (2012) Yield and quality response of sorghum hybrids to different planting densities and sowing times. Int J Agric Appl Sci 4(1)

Mahmood A, Bibi S, Naqve M, Javaid MM, Zia MA, Jabbar A, Ud-Din W, Attia KA, Khan N, Al-Doss AA (2022a) Physiological, Biochemical, and yield responses of linseed (*Linum usitatissimum* L.) in α-Tocopherol-mediated alleviation of salinity stress. Front Plant Sci 13

Mahmood A, Wang X, Ali A, Awan MI, Ali L, Fiaz S, Naqve M, Hassan MU, Javaid MM, Wang H (2022b) Bio-methane production from maize with varying nitrogen levels and harvesting times under semi-arid conditions of Pakistan. Pol J Environ Stud 31(3)

Manghwar H, Hussain A, Ali Q, Liu F (2022) Brassinosteroids (BRs) role in plant development and coping with different stresses. Int J Mol Sci 23(3):1012

Mark C, Zór K, Heiskanen A, Dufva M, Emnéus J, Finnie C (2016) Monitoring intra- and extracellular redox capacity of intact barley aleurone layers responding to phytohormones. Anal Biochem 515:1–8. https://doi.org/10.1016/j.ab.2016.09.011

More N, Verma A, Bharagava RN, Kharat AS, Gautam R, Navaratna D (2022) Sustainable development in agriculture by revitalization of PGPR. In: Bioremediation. CRC Press, pp 127–142

Mostofa MG, Ha CV, Rahman MM, Nguyen KH, Keya SS, Watanabe Y, Itouga M, Hashem A, Abd-Allah EF, Fujita M, Trans L-SP (2021, 1815) Strigolactones modulate cellular antioxidant defense mechanisms to mitigate arsenate toxicity in rice shoots. Antioxidants 10(11)

Mousavi S, Brodie G, Payghamzadeh K, Raiesi T, Srivastava A (2022) Lead bioavailability in the environment, its exposure, and effects. J Adv Environ Health Res 10(1):1–14

Moustafa-Farag M, Mahmoud A, Arnao MB, Sheteiwy MS, Dafea M, Soltan M, Elkelish A, Hasanuzzaman M, Ai S (2020) Melatonin-induced water stress tolerance in plants: recent advances. Antioxidants 9(9):809

Niharika SNB, Singh A, Khare S, Yadav V, Bano C, Yadav RK (2021) Mitigating strategies of gibberellins in various environmental cues and their crosstalk with other hormonal pathways in plants: a review. Plant Mol Biol Rep 39(1):34–49. https://doi.org/10.1007/s11105-020-01231-0

Niranjana SR, Hariprasad P (2014) Understanding the mechanism involved in PGPR-mediated growth promotion and suppression of biotic and abiotic stress in plants. In: Future challenges in crop protection against fungal pathogens:59.

Park H-S, Kazerooni EA, Kang S-M, Al-Sadi AM, Lee I-J (2021) Melatonin enhances the tolerance and recovery mechanisms in *Brassica juncea* (L.) Czern. under saline conditions. Front Plant Sci 12. https://doi.org/10.3389/fpls.2021.593717

Plaza-Wüthrich S, Blösch R, Rindisbacher A, Cannarozzi G, Tadele Z (2016) Gibberellin deficiency confers both lodging and drought tolerance in small cereals. Front Plant Sci 7. https://doi.org/10.3389/fpls.2016.00643

Poór P, Nawaz K, Gupta R, Ashfaque F, Khan MIR (2022) Ethylene involvement in the regulation of heat stress tolerance in plants. Plant Cell Rep 41(3):675–698. https://doi.org/10.1007/s00299-021-02675-8

Rajewska I, Talarek M, Bajguz A (2016) Brassinosteroids and response of plants to heavy metals action. Front Plant Sci 7:629

Rhodes D, Nadolska-Orczyk A (2001) Plant stress physiology. e LS

Riyazuddin R, Verma R, Singh K, Nisha N, Keisham M, Bhati KK, Kim ST, Gupta R (2020) Ethylene: a master regulator of salinity stress tolerance in plants. Biomol Ther 10(6):959

Rowe JH, Topping JF, Liu J, Lindsey K (2016) Abscisic acid regulates root growth under osmotic stress conditions via an interacting hormonal network with cytokinin, ethylene and auxin. New Phytol 211 (1):225–239. doi:https://doi.org/https://doi.org/10.1111/nph.13882.

Ruan J, Zhou Y, Zhou M, Yan J, Khurshid M, Weng W, Cheng J, Zhang K (2019) Jasmonic acid signaling pathway in plants. Int J Mol Sci 20(10):2479

Rymen B, Kawamura A, Schäfer S, Breuer C, Iwase A, Shibata M, Ikeda M, Mitsuda N, Koncz C, Ohme-Takagi M, Matsui M, Sugimoto K (2017) ABA suppresses root hair growth via the OBP4 transcriptional regulator. Plant Physiol 173(3):1750–1762. https://doi.org/10.1104/pp.16.01945

Saberi Riseh R, Ebrahimi-Zarandi M, Tamanadar E, Moradi Pour M, Thakur VK (2021) Salinity stress: toward sustainable plant strategies and using plant growth-promoting rhizobacteria encapsulation for reducing it. Sustainability 13(22):12758

Saddique M, Kamran M, Shahbaz M (2018) Differential responses of plants to biotic stress and the role of metabolites. In: Plant metabolites and regulation under environmental stress. Elsevier, pp. 69–87

Saeed W, Naseem S, Ali Z (2017) Strigolactones biosynthesis and their role in abiotic stress resilience in plants: a critical review. Front Plant Sci 8. https://doi.org/10.3389/fpls.2017.01487

Saha J, Brauer EK, Sengupta A, Popescu SC, Gupta K, Gupta B (2015) Polyamines as redox homeostasis regulators during salt stress in plants. Front Environ Sci 3. https://doi.org/10.3389/fenvs.2015.00021

Saini S, Kaur N, Pati PK (2021) Phytohormones: key players in the modulation of heavy metal stress tolerance in plants. Ecotox Environ Safety 223:112578

Shafi A, Singh AK, Zahoor I (2021) Melatonin: role in abiotic stress resistance and tolerance. In: Plant growth regulators. Springer, pp 239–273

Shahbaz M, Ashraf M, Athar H-U-R (2008) Does exogenous application of 24-epibrassinolide ameliorate salt induced growth inhibition in wheat (*Triticum aestivum* L.)? Plant Growth Regul 55(1):51–64. https://doi.org/10.1007/s10725-008-9262-y

Sharma A, Sidhu GPS, Araniti F, Bali AS, Shahzad B, Tripathi DK, Brestic M, Skalicky M, Landi M (2020) The role of salicylic acid in plants exposed to heavy metals. Molecules 25(3):540

Shimamura S, Yoshioka T, Yamamoto R, Hiraga S, Nakamura T, Shimada S, Komatsu S (2014) Role of abscisic acid in flood-induced secondary aerenchyma formation in soybean (*Glycine max)* hypocotyls. Plant Prod Sci 17(2):131–137

Shukla MR, Bajwa VS, Freixas-Coutin JA, Saxena PK (2021) Salt stress in Arabidopsis thaliana seedlings: role of indoleamines in stress alleviation. Melatonin Res 4(1):70–83

Singh H, Bhat JA, Singh VP, Corpas FJ, Yadav SR (2021) Auxin metabolic network regulates the plant response to metalloids stress. J Hazard Mater 405:124250

Sirhindi G, Mir MA, Abd-Allah EF, Ahmad P, Gucel S (2016) Jasmonic acid modulates the physio-biochemical attributes, antioxidant enzyme activity, and gene expression in Glycine max under nickel toxicity. Front Plant Sci 7. https://doi.org/10.3389/fpls.2016.00591

Song Y, Diao Q, Qi H (2014) Putrescine enhances chilling tolerance of tomato (*Lycopersicon esculentum* Mill.) through modulating antioxidant systems. Acta Physiol Plant 36(11):3013–3027. https://doi.org/10.1007/s11738-014-1672-z

Spormann S, Soares C, Teixeira J, Fidalgo F (2021) Polyamines as key regulatory players in plants under metal stress—a way for an enhanced tolerance. Ann Appl Biol 178(2):209–226

Subhash AP, Nandana M, Sivapriya S, Anith K (2021) Microbial inoculants for mineral nutrient solubilization and mobilization. Biota Res Today 3(11):957–960

Tanveer A, Arshad MS, Ayub M, Javaid MM, Yaseen M (2012) Effect of temperature, light, salinity, drought stress and seeding depth on germination of Cucumis melo var. agrestis. Pak J Weed Sci Res 18(4)

Tanveer A, Mumtaz K, Javaid M, Chaudhry M, Balal R, Khaliq A (2013) Effect of ecological factors on germination of horse purslane (*Trianthema portulacastrum*). Planta Daninha 31:587–597

Thao NP, Khan MIR, Thu NBA, Hoang XLT, Asgher M, Khan NA, Tran L-SP (2015) Role of ethylene and its cross talk with other signaling molecules in plant responses to heavy metal stress. Plant Physiol 169(1):73–84. https://doi.org/10.1104/pp.15.00663

Theocharis A, Clément C, Barka EA (2012) Physiological and molecular changes in plants grown at low temperatures. Plant 235(6):1091–1105. https://doi.org/10.1007/s00425-012-1641-y

Torres CA, Sepúlveda G, Kahlaoui B (2017) Phytohormone interaction modulating fruit responses to photooxidative and heat stress on apple (*Malus domestica* Borkh.). Front Plant Sci 8:2129

Uarrota VG, Stefen DLV, Leolato LS, Gindri DM, Nerling D (2018) Revisiting carotenoids and their role in plant stress responses: from biosynthesis to plant signaling mechanisms during stress. In: Antioxidants and antioxidant enzymes in higher plants. Springer, pp 207–232

Ullah H, Mahmood A, Awan MI, Honermeier B (2015) Effect of row spacing and seed rate on fruit yield, essential oil and composition of anise (*Pimpinella anisum* L.). Pak J Agric Sci 52(2)

Verma V, Ravindran P, Kumar PP (2016) Plant hormone-mediated regulation of stress responses. BMC Plant Biol 16(1):86. https://doi.org/10.1186/s12870-016-0771-y

Wahab A, Abdi G, Saleem MH, Ali B, Ullah S, Shah W, Mumtaz S, Yasin G, Muresan CC, Marc RA (2022) Plants' physio-biochemical and Phyto-hormonal responses to alleviate the adverse effects of drought stress: a comprehensive review. Plan Theory 11(13):1620

Wan X, Yin J, Yan Q, Hu H, Zheng T, Chai Y, Pan W, Gao Y, Li N, Tang B (2022) Sustained-release nanocapsule based on a 3D COF for long-term enzyme prodrug therapy of cancer. Chem Comm 58(39):5877–5880. https://doi.org/10.1039/D2CC00373B

Wang R, Wang J, Zhao L, Yang S, Song Y (2015) Impact of heavy metal stresses on the growth and auxin homeostasis of Arabidopsis seedlings. Biometals 28(1):123–132. https://doi.org/10.1007/s10534-014-9808-6

Wang Y, Diao P, Kong L, Yu R, Zhang M, Zuo T, Fan Y, Niu Y, Yan F, Wuriyanghan H (2020) Ethylene enhances seed germination and seedling growth under salinity by reducing oxidative stress and promoting chlorophyll content via ETR2 pathway. Front Plant Sci 11. https://doi.org/10.3389/fpls.2020.01066

Wang Y, Mostafa S, Zeng W, Jin B (2021) Function and mechanism of Jasmonic acid in plant responses to abiotic and biotic stresses. Int J Mol Sci 22(16):8568

Wani W, Masoodi KZ, Zaid A, Wani SH, Shah F, Meena VS, Wani SA, Mosa KA (2018) Engineering plants for heavy metal stress tolerance. Rendi Lincei Sci Fis Nat 29(3):709–723

Weller S, Florentine S, Javaid MM, Welgama A, Chadha A, Chauhan BS, Turville C (2021) *Amaranthus retroflexus* L.(Redroot Pigweed): Effects of elevated CO_2 and soil moisture on growth and biomass and the effect of radiant heat on seed germination. Agronomy 11(4):728

Wen F-p, Zhang Z-h, Bai T, Xu Q, Pan Y-h (2010) Proteomics reveals the effects of gibberellic acid (GA3) on salt-stressed rice (*Oryza sativa* L.) shoots. Plant Sci 178(2):170–175. https://doi.org/10.1016/j.plantsci.2009.11.006

Wu X, He J, Chen J, Yang S, Zha D (2014) Alleviation of exogenous 6-benzyladenine on two genotypes of eggplant (*Solanum melongena* Mill.) growth under salt stress. Protoplasma 251(1):169–176. https://doi.org/10.1007/s00709-013-0535-6

Yadav B, Jogawat A, Gnanasekaran P, Kumari P, Lakra N, Lal SK, Pawar J, Narayan OP (2021) An overview of recent advancement in phytohormones-mediated stress management and drought tolerance in crop plants. Plant Gene 25:100264

Yu Y, Li Y, Yan Z, Duan X (2021) The role of Cytokinins in plant under salt stress. J Plant Growth Regul. https://doi.org/10.1007/s00344-021-10441-z

Zain M, Khan I, Chattha M, Qadri R, Anjum S, Hassan M, Mahmood A, Ilyas M (2017) Foliar applied thiourea at different growth stages modulated late sown wheat. Pak J Sci 69(1):39

Zhang H, Yu Z, Yao X, Chen J, Chen X, Zhou H, Lou Y, Ming F, Jin Y (2021) Genome-wide identification and characterization of small auxin-up RNA (SAUR) gene family in plants: evolution and expression profiles during normal growth and stress response. BMC Plant Biol 21(1):1–14

Zhao D, Wang H, Chen S, Yu D, Reiter RJ (2021) Phytomelatonin: an emerging regulator of plant biotic stress resistance. Trends Plant Sci 26(1):70–82

Zulfiqar F, Chen J, Finnegan PM, Younis A, Nafees M, Zorrig W, Hamed KB (2021) Application of Trehalose and salicylic acid mitigates drought stress in sweet basil and improves plant growth. Plan Theory 10(6):1078

Chapter 25
Role of Plant Extracts and Biostimulant in Mitigating Plant Drought and Salinity Stress

Shaimaa I. M. Elsayed, Ali S. Sabra, and Elsayed A. Omer

Abstract Phytoconstituents influence biochemical and enzymatic processes by serving as substrates, cofactors, and inhibitors. Numerous phytochemicals have been utilized to cure a variety of illnesses and stop the spread of disease. They offer protection to plants from both biotic and abiotic stressors. It is important to improve their goods to acquire the best concentrations and identify acceptable alternative sources in order to save costs because the presence of bioactive chemicals varies in quantity. In this regard, many classes of components in plant extracts have been identified, including alkaloids, polysaccharides, polyphenols, fatty alcohols, terpenoids, indole, pyridine, essential oils, and phytosterols. Plant extracts and biostimulants, which are bioactive substances produced from plants, aid in the pro-ecological production of strategically important crops. In response to environmental stress, a variety of nitrogenous substances present in protein hydrolysates, including betaines, amino acids, polyamines, nonprotein amino acids, hormones, and others, act as biostimulants of plant growth, metabolic production, and recovery. Utilizing natural nanoparticles to minimize the negative effects of environmental pressures on plants is a relatively new area. Agriculture may employ nanoparticles in a lucrative and environmentally responsible way to deal with scarce water resources and to strengthen crops' tolerance to drought stress. A seaweed (macroalgae) extract can boost plant growth and tolerance to harmful environmental factors like salt and drought by triggering biochemical and genetic pathways that give the plant resilience. The study of and application of phytochemicals as abiotic stress mitigators has significantly increased with the development of contemporary technologies. The utilization of plant extracts and biostimulants in their natural form or in nanometric form as environmentally benign external factors to aid plants in withstanding the detrimental effects of environmental conditions that occur in drought or salinity is discussed and described in this chapter.

S. I. M. Elsayed (✉)
Horticultural Crops Technology Department, National Research Centre, Dokki, Cairo, Egypt
e-mail: si.el-sayed@nrc.sci.eg

A. S. Sabra · E. A. Omer
Medicinal and Aromatic Plants Researches Department, National Research Centre, Dokki, Cairo, Egypt

M. Hasanuzzaman (ed.), *Climate-Resilient Agriculture, Vol 2*,
https://doi.org/10.1007/978-3-031-37428-9_25

Keywords Drought · Salinity · Plant extracts · Biostimulants · Nanoparticles

1 Introduction

Due to results of world climate change, important crops have produced less in many regions of the world due to variable rainfall, rising temperatures, droughts, flooding, and salinized soils (Ahmad et al. 2022). Abiotic stress is a condition triggered by elements including smog, dryness, salt, bright light, and high temperatures. Abiotic stress causes multiple physiological and molecular changes in cells, including a decrease in photosynthetic activity, an increase in the manufacture of reactive oxygen species (ROS), turgor loss, changes in membrane composition and fluidity, an increase in solute concentration and accumulation, and the accumulation of specific osmolytes (Bartwal et al. 2013).

One of the abiotic stresses is drought that affects agricultural systems most adversely. Water scarcity brought on by the drought reduces moisture availability, which in turn restricts crop development and yield. Plants eventually suffer oxidative damage from an overabundance of ROS because of decreased stomatal opening, rapid oxygen (O_2) photoreduction in the chloroplast, and enhanced photorespiration triggered by water shortage stress (Iqbal et al. 2020). In response to drought stress, plants produce the phytohormone abscisic acid (ABA), modify gene expression, shut their stomata, and preserve their osmotic equilibrium (Gong et al. 2020). Flooding or waterlogging, on the other hand, occurs when too much water is retained in the soil as a result of prolonged, heavy precipitation, insufficient drainage, or other factors.

Water logging frequently causes hypoxia, anoxia, reduced gaseous exchange, increased lipid peroxidation, and ROS buildup in plant roots. Furthermore, changes in soil pH cause considerable losses of important nutrients, salt formation, and an increase in heavy metal availability, all of which are adverse to plant growth (Fukao et al. 2019).

Furthermore, besides decreasing a plant's ability to absorb water, salt stress also damages biological membranes, as a result of ionic imbalance, oxidative damage, and nutritional imbalance; slows cell division and growth; decreases the rate of photosynthesis; alters a plant's lipid metabolism; and alters the yield characteristics. Agricultural production has fallen as a result of the salinity of the soil. When plants that are salt-stressed stop absorbing water, conditions that resemble drought start to appear. Salinity decreases stomatal conductance, damages photosynthetic enzymes, and disrupts the photosystem (PS) in plants, all of which result in the manufacture of ROS (Hasanuzzaman et al. 2018). Around one-fifth of irrigation regions are affected by salinity (Morton et al. 2019).

These elements together pose a grave danger to the ability of millions of people to produce enough food to consume, especially in light of the fight over water resources. Finding ways to improve plants' capacity to adapt to difficult circumstances like salt and drought is vital. Plant growth and productivity can be increased while reducing the negative effects of these stresses by using tolerant genotypes, best cultivation management practices, genetic engineering, adding microorganisms that promote plant growth, using biostimulants and plant extracts, and inoculating plants with these agents. This chapter focuses on expanding the use of biostimulants to minimize the damaging results from salt and drought on plants. This covers the use of synthetic compounds, bacterial and fungal stimulants, nanoparticle-coated plant extracts, and other extracts. A biostimulant is any chemical or microbe that is administered to plants in modest quantities to improve their uptake of nutrients, growth, productivity, and quality in addition to their tolerance to abiotic challenges. On this phrase, there is conflicting literature. Humic acids, seaweed, plant extracts, chitosan, biopolymers, as well as advantageous bacteria and fungus are a few of these biostimulants (Du-Jardin 2015). This definition will be used in this chapter.

2 Plant Extracts

The use of environmentally friendly inputs in crop production is a trend in environmentally friendly agriculture. Exogenous application of plant extracts as biostimulants is among them (Pourhadi et al. 2018). Different plant components' extracts are employed in a variety of ways to improve crop output. Depending on the technique of treatments, the concentration, the plant species, and the extract, different plant parts (roots, shoots, seeds) gave a varied reaction (Chojnacka et al. 2015). When exogenously applied, the chemical compositions of various plant components change how the plant grows (Rouphael et al. 2018).

Besides nutrients and antifungal and antibacterial compounds that shield the plant from biotic and abiotic stress, they may also include phytohormones. Different bioactive substances and secondary metabolites, such as nitriles, flavonoids, alkaloids, and antioxidants, are found in plants (Ahmed et al. 2016)

Extraction is a procedure that uses traditional methods and particular solvents to separate soluble bioactive components from undesirable insoluble elements or to remove biological active substances from plants, microbes, or marine life.

Solvent extraction, sometimes referred to as liquid-liquid extraction or solid-liquid extraction, can be used to carry out the extraction. Both extraction techniques are based on the separation of compounds according to their relative solubility in two distinct immiscible liquids or solid matter components.

When two immiscible liquids are combined for the purpose of extracting liquid from liquid, the solvent is used to separate one or more components. It entails removing a substance from one liquid phase and placing it in another.

In solid-liquid extraction, the active agent and inert matrix, which make up a solid substance's soluble components, are extracted using a solvent in accordance with the solid-liquid extraction principle. The extract from the extraction materials may be a liquid or a solid. Similar to how oil in oil seeds is integrated into cells, caffeine in coffee may also be a thin dispersion on a solid object.

The extraction processes are applied in the following fields:

- The chemical industry (inorganic, organic)
- Chemical metallurgy, and other fields (i.e., extraction of metallic salts from diluted brines)
- Nuclear engineering (i.e., reprocessing of nuclear fuels)
- Ecological technology (waste water treatment)
- Food science (edible oil refining)
- The pharmaceutical sector (extraction of agents)
- Using a plant stress-tolerance spray

2.1 Plant Sample Pre-extraction Preparation

The first step in researching medicinal plants is plant samples preparation in order to preserve the biomolecules in the plants before extraction. Plant samples such as leaves, barks, roots, fruits, and flowers can be obtained from either living or dried plant material. Grinding and drying of plant materials, for example, have an impact on the retention of phytochemicals in the final extracts.

2.2 The Extraction Solvent

Menstruum is the name given to the extraction agent utilized with plants or other sources. The kind of plant, the plant organ that has to be extracted, the make-up of the bioactive components, and the accessibility of the solvent are all factors that affect the solvent choice. While water, methanol, and ethanol are examples of polar solvents used in the extraction of polar compounds, hexane and dichloromethane are examples of nonpolar solvents (Pandey and Tripathi 2014; Sasidharan et al. 2011; Altemimi et al. 2017). To separate liquid from liquid, it is customary to combine two immiscible solvents, such as water and dichloromethane, water and ether, or water and hexane.

Due to its more polarity and immiscibility with these organic solvents, water is present in every combination. The item that will be extracted via liquid-liquid extraction should be soluble in an organic solvent, but not in water to help in separation (Majekodunmi 2015). There is also a classification of the polarity of the solvents used in extraction, with water being the most polar and n-hexane being the least polar (Sasidharan et al. 2011; Pandey and Tripathi 2014; Altemimi et al. 2017). The following 11 alternative extraction solvents are given in Table 25.1 in order of increasing polarity, according to Das et al. (2010) and Pandey and Tripathi (2014).

The selected solvent is added to the whole extract during fractionation in ascending polarity order, starting with water, which has the largest polarity, and moving from n-hexane, which has the least polarity (Das et al. 2010; Pandey and Tripathi 2014). When selecting five solvents to utilize during fractional process, the

Table 25.1 The main organic extraction solvents in ascending polarity order

Solvents	Formula	Polarity
n-Hexane	C_6H_{14}	0.009
Petroleum ether	C_6H_{14}	0.117
Diethyl ether	$C_2H_5O_2$	0.117
Ethyl acetate	$C_4H_8O_2$	0.228
Chloroform	$CHCl_3$	0.259
Dichloromethane	CH_2Cl_2	0.309
Acetone	C_3H_6O	0.355
n-Butanol	$C_4H_{10}O$	0.586
Ethanol	C_2H_5OH	0.654
Methanol	CH_3OH	0.762
Water	H_2O	1.000

conventional approach is to use two solvents with low polarity (n-hexane, chloroform), with two medium polarity (dichloromethane, n-butanol), and with one the largest polarity (water).

2.3 *Solvents Used for Preparing Extractions*

1. *Water* is employed as the most polar solvent to extract a range of polar compounds. It has a wide range of dissolving properties, is inexpensive, harmless, flammable, and has high polarity. It takes a lot of heat to concentrate the extract, and doing so might cause hydrolysis and promote the development of bacteria and mould (Das et al. 2010; Tiwari et al. 2011).
2. *Alcohol* may extract polar secondary metabolites and has a polar property. It is also miscible with water. It possesses self-preservative qualities at concentrations greater than 20%; however, low quantities are safe.
3. *Chloroform* works well as a nonpolar solvent for extracting substances including terpenoids, flavonoids, lipids, and oils. It has the advantages of being colorless, smelling pleasant, and being soluble in alcohols (Cowan 1999; Tiwari et al. 2011; Pandey and Tripathi 2014).
4. *Ether* is a nonpolar solvent that may be used to extract substances including fatty acids, terpenoids, coumarins, and alkaloids. It is characterized with low boiling point, tastelessness, and miscibility with water. Additionally, it is an actual stable substance that does not react with metals, acids, or bases. Negative aspects include its extreme volatility and flammability in nature (Cowan 1999; Tiwari et al. 2011; Pandey and Tripathi 2014)
5. *Ethyl acetate* has a medium polarity, low toxicity, and has the ability to extract many biological molecules, both polar and nonpolar, such as steroids, terpenes, waxes, and carotenoids.

6. *Ionized water:* a specific extraction solvent that is highly polar and remarkably heat stable is ionized water (green solvent). It can sustain a liquid state up to 3000 °C and is appropriate for high-temperature applications. With water and other solvents, it is extremely miscible, and it performs well for polar chemical extraction. Because it absorbs and transmits microwave radiation, it is appropriate for microwave-assisted extraction. Because it is extremely polar, nonflammable, and nontoxic, it is suitable for liquid-liquid extraction (Bhan 2017).

2.4 Commonly Used Methods in the Extraction

Procedures for extracting natural materials are frequently viewed as bottlenecks in analytical techniques and extract preparation. Several attempts to replace traditional extraction processes with new ones have been attempted in recent years. Each method seeks to increase the amount of extracted components while minimizing the amount of extraction solvent needed and the sample preparation time.

2.4.1 Maceration

It is an extractive method that involves many processes in order to extract essential oils and active chemicals from plant sources. In a closed jar, the powdered material is mixed with the solvent of choice (menstruum). The solvent is then strained out, followed by crushing the solid marc residue left over from the extraction process, to recover the largest amount of occluded solution. Chemical is often combined and shook if it is in a container to ensure proper extraction. Following extraction, filtration or decantation is used to separate the micelle and marc.

Then in a water bath or an oven, the menstruum and micelle are separated by evaporation (Doughari 2012; Ujang et al. 2013; Pandey and Tripathi 2014; Azwanida, 2015; Majekodunmi 2015; Ingle et al. 2017). This methodology is useful and more suitable for thermo labile plant material.

2.4.2 Ultrasonic-Assisted Extraction

Ultrasonic frequencies larger than 20 kHz, which have the ability to break cell walls and enhance solvent migration into cells, are utilized in this procedure to extract bioactive components from matrix. The plants are homogenized, coupled with a solvent or a mixture of solvents, and then incubated in a water bath in the sonication bath depending on the polarity of the analytes (Altemimi et al. 2017). The extraction of thermo labile chemicals at low temperatures is made possible by this simple approach.

2.4.3 Microwave-Assisted Extraction

Microwaves, which are nonionizing electromagnetic waves with frequencies (electromagnetic radiations) ranging from 0.3 to 300 GHz, are used in this technique to get bioactive compounds from plants. The magnetic and electrical fields contained in this electromagnetic radiation boost the efficiency of analytes extraction. Extracts produced by microwave-assisted extraction (MAE) showed improved antioxidant activity by applying a number of methodologies, including ferric reducing antioxidant influence (FRAP), oxygen radical absorbance capacity (ORAC), and total phenolics (Altemimi et al. 2017). This method has an advantage over conventional extraction methods since it consumes less solvent and less time.

2.4.4 Accelerated Solvent Extraction

A technique for extracting different components from a challenging solid or half solid sample matrix is called accelerated solvent extraction (ASE). The ASE sample preparation technique extracts a wide range of bioactive compounds using organic solvents at high pressures of up to 1500 psi. This makes it feasible to extract desired analytes from complicated matrices more successfully. The solvent is retained in a liquid form even after it has boiled due to the high pressure, which aids analytes in diffusing into the solvent and increases recovery. In an automated system, after combining plant powder with filler material like sand in the extraction cells, the solvent or mixture of solvents penetrates the plant material under high pressure in an accelerator solvent extractor. This utilizes less solvent and produces high-quality extract in comparably less time than conventional extraction techniques. The recovery of volatile and semi-volatile chemicals has excellent potential with accelerated solvent extraction.

2.5 *Use of Plant Extracts in Combating Abiotic Stresses*

Plant extracts (PEs), which come from plants and don't include any artificial chemicals, are a green and environmentally acceptable method for reducing harmful environmental stresses (Ahmad et al. 2022). Plant extracts are made from the full plants of several medicinal and fragrant plants, including their roots, leaves, stems, flowers, and fruits. They contain a variety of bioactive substances in addition to minerals, pigments, and proteins, including phenolic acids, flavonoids, anthocyanins, tannins, sterols, fatty acids, glucosinolates, terpenoids, polysaccharides, and phytohormones; however, the content will differ according to the plant part which utilized to prepare the extract (Ahmad et al. 2022). The advantageous effects that extract have on plants may result from the interaction of several distinct chemicals or chemical families.

2.6 Application Methods

Plants are given bioactive extracts through a number of methods, such as foliar spraying, soil-based additions, and seed priming. To enhance their absorption and penetration to the leaves through the stomata, they might be used with wetting agent and some substances (Yakhin et al. 2017; Ahmad et al. 2022). The efficiency of these extracts depends on the plants used type, the degree or duration of stress, and differences in the chemical makeup of the area where the plant was cultivated. As a result, it is challenging to predict a single method of use for plant extracts. Nevertheless, a number of metabolic changes that occur in plants after treatment might be viewed as pathways for providing resistance.

2.7 Ways of Action

Plant performance and development characteristics are improved by exogenous application of Plant extracts (PEs) via modifying physiological and biochemical processes, primary and secondary metabolic pathways, and signal transduction pathways (Ahmad et al. 2022). Applying PEs to stressed plants improves phenotypic parameters such as plant weight, root size, leaf area, leaves number, thickness of leaks, number of fruits, seeds, and yield (Rady and Mohamed 2015; Suryaman et al. 2021). Better photosynthetic activity, water relations, pigment accumulation, and nutrient absorption might all contribute to this improvement (Ahmad et al. 2022). Plants under salt stress experience decreased photosynthetic rate and transpiration rate. ROS begin to build up and produce oxidative stress as a result of an imbalance in the water balance inside cells and a disruption in the electron transport system. These ROS target macromolecules like proteins, lipids, and nucleic acids, which have an influence on the metabolism of the cells (Munns and Tester 2008). Plants respond by enzymatic and nonenzymatic antioxidants becoming active. Nonenzymatic antioxidants such as ascorbate, glutathione, phenolics, and flavonoids are crucial parts of plant extracts. PEs may be more successful in reducing stress-related conditions because they contain more of these chemicals; however, more study on a wide range of plants is necessary to ascertain if the reaction of plants is dose-dependent and what the crucial threshold level is.

Phytohormones including ascorbic acid (ASA), jasmonic acid (JA), salicylic acid (SA), and ethylene, which are known to activate different signal transduction pathways in plants in response to biotic and abiotic stress, are present in plant extracts in addition to bioactive chemicals (Ahmad et al. 2022). After being exposed to stress, these phytohormones are synthesized and activate signalling pathways that cause the activation of genes relevant to stress. This hypothesis was supported by evidence from the *Arabidopsis* plant, where the activation of ABA, SA, and auxin-induced signal pathways by Moringa leaves extract (MLE) increased plant tolerance to salt (Brazales-Cevallos et al. 2022; Ahmad et al. 2022). Another opinion is that the presence of PEs makes it easier for certain solutes to accumulate, such as proline (Pro), glycine betaine (GB), sugar alcohols, and polyamines, which are necessary for osmotic adjustment in stressful circumstances.

2.8 Application Under Drought Stress

Many plant extracts have been used exogenously as foliar sprays to increase plant growth. According to research by Rady and Mohamed (2015), the exogenous application of moringa seed extract as a foliar spray enhanced *Capsicum annuum* L.'s antioxidant capabilities and seed germination during heavy metal stress. Several studies have demonstrated how effectively PEs may improve plant performance and development in the presence of abiotic stresses. By altering biochemical traits and the antioxidant system, neem tree (*Azadirachta indica*) leaf extract, for example, lessened the deleterious effects of water restriction on quinoa (*Chenopodium quinoa*) (Naz et al. 2022). Neem extract used at 5% concentration increased the synthesis of appropriate solutes like Pro and GB as well as antioxidant enzyme activities and the prevention of oxidative damage in this regard. Spraying *Salix babylonica* hydrolyzed extract on *Houttuynia cordata* plants promoted growth, increased osmolyte solution accumulation, and elevated antioxidant enzyme activities, supporting the application of this extract as a secure protector and growth regulator during drought deficiency (Ri and Ri 2022). Foliar application of biostimulants and natural extracts, such as propolis and maize grain extract, improved growth characteristics, including photosynthetic rate, pigment content, gas exchange parameters (RWC), membrane stability, and accumulation of compatible solutes under stressful conditions. Furthermore, this therapy increased the activity of antioxidants that are not enzymes, such as Pro, glutathione, ascorbate, and tocopherol, as well as antioxidant enzymes, such as catalase, ascorbate peroxidase, and peroxidase (POD), and superoxide dismutase (SOD) (Desoky et al. 2021). In a field test, the quinoa plants were given exogenous growth stimulants like tryptophan and banana peel extract. Tryptophan was inferior to banana extract for increasing quinoa production and quality. Exogenous treatment resulted in increases in growth indices, indole acetic acid, pigments, phenolic compounds, antioxidant enzyme activities, and nutritional properties of seeds (Bakry et al. 2016). PEs can be used with other stimulants to create results that are more pronounced because of their synergistic effects. When rhizobacteria and MLE were combined, drought-stressed wheat plants produced more growth (Lalarukh et al. 2022a). This beneficial effect was demonstrated by increases in fresh weight, dry weight, shoot and root length, photosynthetic apparatus, total soluble sugars, proteins, and minerals (calcium, phosphorus, potassium, and nitrogen). Indicators of oxidative stress such as malondialdehyde (MDA) and hydrogen peroxide (H_2O_2) were also reported to be decreasing (Lalarukh et al. 2022b). The application of *Brassica* water extract (BWE) at 2% improved the drought resistance of two wheat cultivars by increasing morphological traits, such as fresh and dry weights, as well as antioxidant enzyme activities, such as SOD,

CAT, and POD, an increase in chlorophyll and carotenoid contents, and a reduction in oxidative damage by free radicals (Khaliq et al. 2022).

Elsayed et al. (2022) investigated the effects of *Ocimum basillicum* (Basil) leaves extract (BLE) on seedling development, chemical makeup, and various qualities for essential oil of *Eucalyptus citriodora*. The findings showed that whereas proline and MDA levels fell under the same treatment, vegetative growth metrics and photosynthetic pigments rose in plants that received six-day irrigation. All root features, total sugar content in all organs, and antioxidant enzyme activity (polyphenol oxidase (PPO), SOD, POD, and CAT) were increased by irrigation intervals every 8 days, whereas leaf area showed a large rise in plants that received irrigation every 4 days. Except for Pro and MDA, which exhibited the lowest value at the same rate, *E. citriodora* plants sprayed with BLE at a rate of 20% generated the greatest value for all vegetative development metrics and all chemical compositions. The optimum method for producing fresh leaves for essential oils was irrigation every six or eight days with a BLE spray of 20 to 40 percent. The primary component of *E. citriodora* fresh leaves essential oil is citronellal, which was increased to 71.54% by watering every 8 days with the application of 40% BLE. The oil's monoterpenoid content ranged from 92.59 to 97.93%. The main monoterpenes in oil showed moderate to strong antioxidant activity.

O. sanctum L. (*O. tenuiflorum*, Tulsi) has been utilized for thousands of years in Ayurveda because of its wide range of therapeutic characteristics. Tulsi leaves are substantially abundant in antioxidant components, according to Pandeya et al. (2016). Its leaf extract is therefore thought to effectively scavenge ROS when given to plants, avoiding oxidative damage under drought stress. In order to determine the effectiveness of an aqueous leaf extract of *O. sanctum* against drought stress in two rice genotypes grown in a glass house, the current study was carried out in the fall of 2013 and spring of 2014 seasons. Here, we demonstrate how the treatment of *Ocimum* affected a number of morpho-physiological (chlorophyll fluorescence, leaf rolling score, leaf tip burn, number of senesced leaves, and total dry matter) and biochemical (Pro, MDA, and SOD activity) parameters in both seasons. When applied to rice plants under drought stress, *Ocimum* extract boosted the expression of dehydrin genes while decreasing the expression of aquaporin genes. Application of *Ocimum* leaf extract under drought stress is thus recommended as a viable technique to reduce drought stress in an affordable, practical, and environmentally friendly way.

Hanafy (2017) looked at the effectiveness of foliar spraying of leaf extract of *M. oleifera* on *Glycine max* plants to reduce drought stress (cv. Giza 111). Drought stress caused significant drops in growth indices including shoot and root length, fresh and dried shoot and root weight, and photosynthetic pigments like chlorophyll a, chlorophyll b, carotenoids, and total pigments. Additionally, it significantly increased oxidative damage (lipid peroxidation), osmolyte compounds (Pro, total soluble sugars, and total phenols), non-enzymatic antioxidants (ascorbic acid, tocopherol, and reduced glutathione), enzymatic antioxidants (glutathione reductase (GR), SOD, and APX), and oxidative damage (lipid peroxidation). Furthermore, as compared to either the control plants or the drought-stressed plants, foliar spraying with MLE improved all the aforementioned metrics. It appeared that MLE might improve the tested plant's resistance to drought stress.

When applied topically to drought-stressed soybean plants, MLE considerably enhanced both physiochemical characteristics and plant growth indices. The improvement in growth could be attributed to its function as a growth enhancer because it is rich in protein, which is necessary for the formation of protoplasm as well as vitamin C, and important nutrients like potassium, calcium, and magnesium. In addition, MLE serves as good sources of natural antioxidant compounds like ascorbic acid, flavonoids, phenolics, and carotenoids, which make it an excellent plant growth enhancer.

It possesses growth-promoting chemicals like auxins and cytokinins (particularly zeatin) that can increase cell division, multiplication, and enlargement as well as trigger the synthesis of chlorophyll. To sustain plants' ability to photosynthesize when under water stress, chlorophyll concentration must be preserved. In the current study, drought stress significantly lowers total pigment, carotenoids, and chlorophyll (chlorophyll a, b) concentrations compared to control. This decline may be linked to photo-oxidation or chlorophyll breakdown via the synthesis of proteolytic enzymes like chlorophyllase, degeneration of the chloroplast, or stomatal closure. It may be thought of as a classic indication of oxidative stress. On the other hand, a potential drop in stomata conductance would result in less photosynthetic activity caused by the inhibitory impact of less water content in leaves. The chlorophyll content of *G. max* leaves was enhanced concurrently by foliar MLE treatment. By applying Moringa extract topically, plants may produce more phyto-regulator cytokinins or their levels may be connected with MLE's levels of zeatin, a cytokinin that delays the aging of leaves and the loss of chlorophyll. The nutritional potentialities of several macro elements, including Mg, a component of chlorophyll that would be responsible for causing an increase in the levels of chlorophyll a, b in *G. max* plants, are also highly concentrated in moringa leaves. On the other hand, spraying plants with MLE caused soybean plants to have significantly higher GSH, ascorbic acid, and tocopherol concentrations than the comparable controls. MLE is a good source of minerals, amino acids, phenolics, soluble sugars, and certain antioxidants such as free Pro and ascorbate. As a result, ascorbic acid (ASA), tocopherol, and reduced glutathione concentrations were produced as antioxidants, which allowed soybean plants to withstand drought stress. In contrast to drought-stressed plants and control plants, foliar spraying with MLE extract decreased the level of MDA in the leaves of soybean plants. A substantial amount of calcium is included in *M. oleifera* extract, which can protect membranes from harm and leakage, thus preserving membrane integrity under severe drought.

2.9 Application Under Salinity Stress

Due to the presence of polyphenol chemicals, application of *Rosmarinus officinalis* extract proved effective in reducing the negative effects of salt on apple seedlings (Mahmoudand and Dahab 2018). The roselle *(Hibiscus sabdariffa)* plants' growth and production were enhanced by the exogenous application of aloe extract and potassium silicate, whether under full irrigation or water deficiency situations

(Abdou et al. 2022). Under salty irrigation circumstances, seed priming with maize grain extract (MGE) and zeatin increased wheat development and yield by enhancing the hormonal contents and boosting antioxidant components, such as ascorbate, glutathione, and tocopherol, therefore minimizing oxidative damage (Alharby et al. 2020).

The potential protective effects of a watery extract of *O. basilicum* leaves (OLE) on bean plants watered with various salt levels were explored(0.0, 50, 100 or 150 mM NaCl). In order to evaluate the cytotoxic effects of salt on *Vicia faba* plants, oxidative stress markers such as lipid peroxidation was measured. On bean plants grown under salinity stress in the presence or absence of a watery extract of *O. basilicum* leaves, growth parameters, antioxidant enzyme activity (CAT, POD, and APX), organic solutes (soluble sugars, soluble proteins, and Pro), and ions content (Na, K^+, Ca^{2+}, and Mg^{2+}) were also measured (Abou-Alhamd and Loutfy 2020). The statistics revealed that, in comparison to control plants, salinity stress caused a considerable rise in the activity of antioxidant enzymes, organic solutes, lipid peroxidation, and ions content. *Ocimum* leaves extract (OLE) considerably reduced the negative effects of salt on *V. faba* plants, while also increasing the activity of antioxidant enzymes, osmolytes (soluble sugar and protein), and ions content (K^+, Ca^{2+}, and Mg^{2+}), which may have been an induced defense mechanism against salinity stress. NaCl stress caused changes in the growth of *V. faba* by affecting the physiological and biochemical behaviors. In addition to lowering Na^+ and MDA concentrations, treatment with OLE altered development by raising organic solutes (soluble sugars, soluble proteins, and Pro), critical minerals (K^+, Ca^{2+}, and Mg^{2+}), and the activity of antioxidant enzymes (CAT, POD, and APX). Therefore, it might be inferred that OLE plays a part in reducing the negative effects of NaCl stress. As a result, we suggest that spraying OLE onto the leaves of *V. faba* plants may be a good way to increase their resistance to NaCl stress. Additionally, this approach is simple to use, eco-friendly, and might be approved by farmers for usage in the field to boost plant growth and productivity.

Sweet basil (*O. basilicum* L. cv. cispum) plants planted with or without salt stress were irrigated using watery leaf extract in order to investigate the effectiveness of *M. oleifera* as biostimulants. A liter of distilled water was used to homogenize 200 g of fresh moringa leaves before being filtered and made into the following dilutions, 2.5, 5.0, 10, and 20%. Malondialdehyde and Pro levels in stressed basil had risen, but with the use of MLE, they considerably decreased. MLE extract increased the basil leaf area by 60% under salt stress when compared to control basil; growth metrics such as shoot length, shoot fresh weight, shoot dry weight, number of branches, and root length and dry weight all increased. Basil plants after *M. oleifera* treatment had significantly higher levels of anthocyanin, total carbohydrates, and SOD activity. The different chemical structure of the *Moringa* species highlighted the resistance mechanisms in basil. With the simultaneous watering of basil utilizing Moringa, our hypothesis was that the effectiveness would even grow significantly stronger (Hassanein et al. 2019).

Even though the damask rose (*Rosa damascena* Mill. var. trigintipetala Dieck) is a commercially significant fragrant plant, salt stress has severely stunted its growth and development. According to reports, MLE contributes to the ability to cope with salt stress (Hassan et al. 2021). However, nothing is known about the effects of MLE on fragrant plants, particularly damask rose under salinity. Therefore, the purpose of this study was to find out how MLE affects the reduction of salt stress in damask roses. Rose plants were given a treatment of 200 mmol L^{-1} NaCl, and at 1:30 MLE which was applied topically (v:v). The growth characteristics, chlorophyll content, RWC, Pro content, and membrane stability index under salinity were all improved by foliar spraying with MLE. When MLE were administered, less MDA and H_2O_2 accumulated as well. Additionally, under stress, MLE treatment significantly improved the antioxidant enzyme activity (CAT and SOD), total phenols, and radical scavenging activity. Together, MLE possessed an effective antioxidant defense mechanism that captured ROS and provided significant protection against oxidative damage brought on by salt. The fact that these treatments increased RWC and chlorophyll content can be used to explain how salt stress's negative effects on growth were lessened by MLE. The contribution of MLE to enhancing the salinity-induced tolerance has been demonstrated in diverse crops (Rehman et al. 2014; Rady and Mohamed 2015; Ali et al. 2018). MLE can be used as a biostimulant and plant growth enhancer since it includes a variety of phytohormones, flavonoids, amino acids, antioxidants, and vital nutrients (Taiz and Zeiger 2010; Hassan and Fetouh 2019).

3 Nanoparticles-Coated Extracts

The use of natural nanoparticles in reducing the negative effects of environmental stressors on plants is a new field. Utilizing nanoparticles in agriculture is a productive and environmentally beneficial way to deal with scarce water resources and to make crops more resilient to drought stress (Dowom et al. 2022).

3.1 Application Under Drought Stress

In a pot experiment, wheat plants were treated with iron oxide nanoparticles (Fe_3O_2-NPs) made from ginger and cumin seeds in order to increase their resistance to drought stress (Noor et al. 2022). The germination rate, biomass, pigment concentration, Pro content, soluble sugars, antioxidant enzyme activities, such as SOD, POX, and APX, as well as a decrease in lipid peroxidation and electrolyte leakage were all improved by cumin nanoparticles at 1.2 mM concentration more than ginger particles. Similarly, a recent study looked at how application of Chitosan nanoparticles (CNPs) affected the physiological, phytochemical, and biochemical responses of *Salvia abrotanoides* plants growing under drought stress (Dowom et al. 2022). By increasing RWC, photosynthetic pigments, phenolics, flavonoids, soluble sugars, proteins, suitable solutes, and antioxidant activities like SOD, PPO, and Guaiacol peroxidase (GPX), CNPs treatment reduced the deleterious impacts of drought. There were also observed anatomical alterations, including an increase in stomata density following treatment. In contrast, the elicitor reduced oxidative damage as seen by a decline in electrolyte leakage (Dowom et al. 2022). Another recent study used papaya fruit extract and Plant growth-promoting rhizobacteria (PGPR) to apply green produced zinc oxide nanoparticles (ZnO-NPs) to reduce the combined effects of heat and drought stress on wheat plants (Azmat et al. 2022). Through improvements in growth indices, biomass, photosynthetic pigments, nutrients, and accumulation of soluble sugars, protein, and indole acetic acid, treated plants subjected to single or combination shocks demonstrated greater tolerance. Proline content and antioxidant enzyme activities such as SOD, POX, CAT, APX, GR, and dehydroascorbate reductase (DHAR) rose when ZnO-NPs and rhizobacteria were applied together. Meanwhile, following application, a decrease in oxidative damage was seen as a decline in electrolyte leakage, MDA, and H_2O_2, and this finally led to an improvement in tolerance (Azmat et al. 2022). In addition to enhancing plant development properties, biostimulants also promote the accumulation of secondary metabolites (Giglou et al. 2022). In this study, when chitosan biostimulants (Kitoplus at 1% and chitosan-coated iron oxide nanoparticles at 10 μM) were applied, mint plants under drought stress displayed an increase in menthone content. The main component of the essential oil, menthol, rose in stressed plants following 0.5% Kitoplus treatment. This shows that using biostimulants and chitosan nanoparticles to increase the synthesis of essential oils under drought stress was successful (Giglou et al. 2022).

3.2 *Application Under Salinity Stress*

The detrimental effects of salt on plants have been lessened using nano molecules. In a recent research, wheat plants were subjected to salt stress and received foliar applications of fertilizer containing nano-silicon and nano-chitosan (Hajihashemi and Kazemi 2022). In this study, the use of these compounds raised antioxidant enzyme activity, which can snuff out ROS and lessen lipid peroxidation while also enhancing plant development. After application, osmotic adjustment-related molecules such as Pro, GB, carbohydrates, and free amino acids increased. In a research experiment on two wheat (*Triticum aestivum*) cultivars, foliar application of nano-Zn particles, in particular at 50 mg L^{-1}, promoted the buildup of osmolytes and enhanced nutrient absorption, hence inducing salt tolerance and enhancing physiological parameters (Lalarukh et al. 2022b). However, this treatment resulted in a decrease in plant pigments as well as enzymatic and nonenzymatic antioxidants, indicating that plants used alternate toleration mechanisms. In a recent study, polysaccharides crude nanoparticles extract was made from the waste of Taif rose petals (*R. damascene*), in order to lessen the negative effects of salt and stress on eggplants (Alabdallah and Alluqmani 2022). By raising the chlorophyll content, enhancing the antioxidant enzyme activities (SOD, CAT, POX, GR), boosting the membrane stability, and minimizing oxidative damage, the administration of this extract benefitted stressed plants (Alabdallah and Alluqmani 2022). In a field experiment, the effectiveness of applying Magnesium nanoparticles (MgO-NP) alone or in conjunction with effective microorganisms (Ems) was investigated on *Ipomoea batatas* under salt stress. As a result of greater concentrations of osmolytes such as Pro, free amino acids, soluble sugars, as well as accumulation of phenolic compounds and carotenoids, the combined application increased the ability to withstand osmotic stress. Both enzymatic and nonenzymatic antioxidants have higher activity (Abd El-Mageed et al. 2022). The improvement in ionic homeostasis brought on by the K^+ and Na^+ ion balance finally resulted in greater growth and yield. Under in vitro salt stress, green biosynthesized silver nanoparticles (AgNPs) with spherical shapes and sizes of 9 and 30 nm were employed to counteract the negative effects of salinity on *Maerua oblongifolia* shoots (Shaikhaldein et al. 2022). Application of AgNPs improved the pigments' content and the antioxidant enzyme activities as well as reduced the oxidative damage; therefore, AgNPs application was found to be an effective way for enhancing salinity tolerance. Zinc oxide nanoparticles (ZnO NPs) synthesized using plant extracts were found to improve the shoot growth and reduce salt toxicity of *S. officinalis* exposed to salinity stress (Alenezi et al. 2022). The effect was dose-dependent and most of the best effects were at 10 mg L^{-1} concentration. At this concentration, shoot number, fresh weight and shoot dry weight, and total chlorophyl were increased. Also, the contents of Pro, antioxidant enzymes, such as CAT, SOD, GR, were enhanced as well. Therefore, the green biosynthesized ZnO NPs can be considered an effective nano-fertilizer for plants under salinity stress.

All these studies suggest that application of nanoparticles can be an effective strategy in mitigating the adverse effects of drought and salinity and maintaining sustainable crop production.

4 Biostimulants

Biostimulants are now defined as substances that can help plants develop and improve nutrient absorption while they are under abiotic stress (Colla and Rouphael 2015; Du-Jardin 2015). In response to environmental stress, it has been shown that a variety of nitrogenous compounds found in protein hydrolysates, such as betaines, amino acids, polyamines, nonprotein amino acids, hormones, and others, serve as biostimulants of plant growth, metabolic production, and recovery (Vranova et al. 2011; Colla et al. 2015). It has been discovered that phytohormones like auxins, gibberellins, and cytokinins, as well as polyamines, sterols, and antioxidant enzymes that are naturally abundant in plant extracts, can effectively relieve the stress of plants. They controlled endogenous stress-responsive genes and activated hormone-producing genes (Craigie 2011; Bhattacharyya et al. 2015).

The use of various chemical compounds to chelate macro- or micronutrients, microbial plant growth regulators, and plant-based biostimulants has been demonstrated to be effective in promoting plant growth and production under stressful environments (Ali et al. 2020). Without being fertilizers, herbicides, or soil-improving agents, biostimulants are advantageous for plant development (Calvo et al. 2014; Du-Jardin 2015; Halpern et al. 2015). Biostimulants include helpful fungi and microorganisms like plant growth-promoting rhizobacteria (PGRs), which are both of synthetic or natural origin (Halpern et al. 2015). One method for enhancing plant growth and output under various environmental circumstances involves the exogenous application of biostimulants in a variety of ways. The term "biostimulants" was first used in the literature that is now available by Kauffman et al. (2007). "Materials other than fertilizers that stimulate plant development when is administered in modest quantities" is how they characterized biostimulants. Exogenous delivery of these biochemicals by various routes (seed soaking, foliar spraying, or soil treatment) has been shown to be beneficial in enhancing plant growth and output under stressful circumstances. These biostimulants include plant-based biostimulants that are useful in enhancing plant growth and output under stressful conditions, plant growth regulators of microbial origin, and the chelation of macro- or micronutrients with various organic compounds (Ali et al. 2020). Chitosan, SA, silicon, poly-γ-glutamic acid, −aminobutyric acid, melatonin, Pro, GB, brassinosteroids, etc. are some of these chemical stimulants.

4.1 Application Under Drought Stress

Chitin, a substance found in the cell walls of fungus and a key component of the exoskeletons of insects and crustaceans, is the source of chitosan. Chitosan is widely used in the agricultural and pharmaceutical industries (Ali et al. 2020). Deacetylated chitin can be generated naturally by plants or artificially in the form of chitosan (Du-Jardin 2015). This polycationic can attach to certain cell receptors that are engaged in defensive mechanisms against biotic invasion as well as to cellular elements including DNA, plasma membrane, and cell wall (Du-Jardin 2015). By boosting the buildup of H_2O_2 as a signalling molecule, exogenous application of chitosan enhances the plant antioxidant defense responses in several crops. A variety of secondary metabolites, including polyphenolics, lignin, flavonoids, and phytoalexins, are produced in various crops, including tomato (El-Tantawy 2009), wheat (Zeng and Luo 2012), and rice as a result of the elevated concentration of H_2O_2 (Pongprayoon et al. 2013). Utilizing chitosan can modify the accumulation of bioactive molecules and provide drought tolerance. In a recent study, it was discovered that seed priming and chitosan foliar spraying at a concentration of 0.25% increased the accumulation of hydroxycinnamic acid and hydroxybenzoic acid derivatives, including ferulic acid, p-coumaric acid, o-coumaric acid, gallic acid, and salicylic acid (SA), in kidney beans, *Phaseolus vulgaris*, that were experiencing drought (Manoj et al. 2022). This was largely due to improved antioxidant activity following chitosan treatment. SA, a signaling molecule that affects how plants react to environmental difficulties like salt, drought, metal toxicity, etc., and how their metabolism works, is a cheap and effective drug for minimizing the negative impacts of environmental stressors on plants (Damalas and Koutroubas 2022). Plantago ovata plants were less affected by mild drought stress thanks to SA, in particular at 0.8 mM, because it enhanced the cell membrane stability, water consumption efficiency, soluble sugar content, Pro content, total phenolics, and catalase enzyme activity. As a result, seed output was increased (Roumani et al. 2022). Another ecologically advantageous elicitor in agriculture, poly-glutamic acid (PGA), may lessen the negative effects of environmental stressors on plants by altering plant growth, development, and the microbial community in the rhizosphere (Ma et al. 2022). Exogenous PGA treatment improved the dry weight of maize, root development, concentrations of soluble sugar and Proline, as well as the photosynthetic rate in drought-stressed *Zea mays* plants. Additionally, application increased the activity of the soil enzyme urease and increased the number of beneficial bacteria, such as *Actinobacteria*, *Chloroflexi*, *Firmicutes*, *Alphaproteobacteria*, and *Deltaproteobacteria*, in the rhizosphere (Ma et al. 2022). Plants can become more drought-tolerant thanks to the discovery of melatonin and the signaling molecule Ca^{2+} (Cisse et al. 2022). By improving growth traits, photosynthetic rate, stomatal conductance, transpiration rate, water use efficiency, pigment content, and activating antioxidant enzyme activities, such as POD, SOD, and CAT, as well as nonenzymatic antioxidants, exogenous administration of Ca^{2+} combined with melatonin reduced the detrimental effects of drought stress on the woody tree *Dalbergia odorifera*. This may suggest that melatonin and calcium work in concert to help woody plants endure drought (Cisse et al. 2022). On turf grass, gamma-aminobutyric acid (GABA), either alone or in conjunction with Proline, was treated. Greater turf quality, a darker green color, higher leaf RWC, and longer stolons were all maintained in treated bentgrass, which improved turf grass performance under dry conditions (Chapman et al. 2022). Another investigation on licorice (*Glycyrrhiza glabra*) discovered that root mycorrhiza inoculation and silicon treatment at 300 ppm enhanced plant performance under drought stress by increasing Proline levels and altering the amino acid composition (Haghighi et al. 2022).

4.2 Application Under Salinity Stress

The treatment with SA even at low quantities (0.5 mM) increased growth parameters and reduced the detrimental effects of salinity on *R. damascena* in recent research (Omidi et al. 2022). It became clear from this study that SA treatment improved the growth traits, pigment content, antioxidant systems both enzymatic and nonenzymatic, and the activation of genes involved to the stress response, conferring salt tolerance in rose. When subjected to salt stress, SA was found to increase *Celosia argentea* growth (Gholamzadeh et al. 2022). The physiological and morphological properties of plants were enhanced by foliar application of SA at a concentration of 1 mM, including increased photosynthetic rate, pigment, and Pro levels. This suggests that SA may be utilized as a workable method to lessen salt stress. Chickpea (*Cicer arietinum*) seedlings were primed with salicylic acid at 0.5 mM to increase the growth rate, biomass, capacity to fix nitrogen, antioxidant activity, and reduce oxidative damage, which enhanced overall plant salt tolerance (Kaur et al. 2022). In a second experiment, priming seeds with salicylic acid significantly reduced the detrimental effects of salinity on pea (*Pisum sativum*) seedlings, even at higher salt concentrations up to 300 mM NaCl. By exogenously administering salicylic acid at 0.75 mM and $CaCl_2$ at 50 mM, red bean (*P. calcaratus*) performance and growth characteristics were enhanced through increased pigment content, decreased sugars, a significant decrease in oxidative damage, and an increase in the scavenging capacity of free radicals, with the exogenous supply of salicylic acid and Ca^{2+}; it is therefore feasible to improve salt tolerance by controlling the antioxidant machinery (Mahdavian 2022).

Seed priming with salicylic acid at 1 mM was a successful tactic for minimizing the detrimental effects of salinity on plants in maize plants under salt stress. The treated plants had increased root dry matter, cob yield, leaf RWC, and Pro content when compared to the untreated plants (Islam et al. 2022). A separate chemical, triacontanol, has demonstrated potential in lessening the damage that salt causes to plants. Triacontanol was applied topically to plants in a pot experiment at a concentration of 75 M, which significantly enhanced growth metrics and physio-biochemical traits including pigment content, gas exchange capacity, and antioxidant enzyme activities like SOD and CAT. On the other hand, after the application of triacontanol at 75 μM, oxidative damage parameters including MDA and H_2O_2 were reduced (Sarwar et al. 2022). In agricultural production, tetrapyrrol, a regulator of 5-aminolevulinic acid (ALA) is used to control plant development and improve their resistance to stress (Wu et al. 2022). Raising the amount of heme in the leaves by exogenous spraying of ALA on cucumber seedlings' leaves boosted growth and physiological parameters, proving that heme is essential for plants' responses to salt stress. Application of melatonin at a greater dosage (150 M) resulted in improved photosynthetic metrics, relative water content, and elemental contents, such as K^+, Ca^{2+}, and P, while diminishing Na^+ and Cl contents. The expression of genes related to defense systems enhanced, whereas MDA and H_2O_2, markers of oxidative stress, were decreased (Sheikhalipour et al. 2022). Brassinosteroids are a class of phytohormones that regulate plant growth and development. Brassinosteroids, such as brassinolide (BL), were found to improve grape vine seedlings' tolerance to salt in salinity conditions in a field trial (El-Banna et al. 2022). The capacity of BL to reduce the effects of salinity was attributed to an increase in photosynthetic pigments, preservation of ion homeostasis, improvement of the K^+/Na^+ ratio, improvement of water relations, and an increase in antioxidant enzyme activities (El-Banna et al. 2022). Nitric oxide and silicon can be used to lessen salt's detrimental effects on plants; when treated weekly or every 14 days, nitric oxide and silicon were found to promote the growth and productivity of pepper plants under salt stress (Badem and Soylemez 2022). Exogenous dose of sodium nitroprusside, a nitric oxide donor, reduced the damaging effects of salt on marigold plants (Barzin et al. 2022). Application increased plant growth and performance by raising photosynthetic activity and pigment content, Pro and glutathione accumulation, boosting K^+/Na^+ ratio, and activating antioxidant enzymes involved in defense mechanisms.

Examples of small-molecular-weight substances known as appropriate solutes that build up when plants get used to environmental problems like drought, salt, and extreme temperatures are pro and GB. These chemicals do not alter metabolism and are essential for osmotic correction (Sakamoto and Murata 2002). The capacity of cells to respond to stimuli is enhanced by the knowledge that Pro, GB, polyamines, and sugars (manitol, sorbitol, galactinol, and trehalose) regulate the osmotic pressure of cells and preserve their integrity. Numerous phytohormones, such as (ABA) cytokinin, ethylene, (JA), SA, and melatonin, contribute to the buildup of these osmoprotectants in response to environmental stresses (Hossain et al. 2022).

Numerous studies have shown how effective these appropriate solutes are in assisting stressed plants to restore their osmotic equilibrium. Exogenous applications of 400, 200, and 100 mg L^{-1} of GB, Pro, and SA respectively, enhanced the physiological traits and development of potato genotypes subjected to salt stress; consequently, they develop a greater tolerance for salinity (Zaki and Radwan, 2022). Similar to this, foliar application of pro increased the accumulation of carotenoids and phenolic compounds in acclimated *Aloe vera* plants, which preserved the efficiency of the photosystem II (PS II) and improved ionic homeostasis, water relations, Pro accumulation, and antioxidant system. The improvement, nevertheless, was only seen in plants that had acclimated to salt (Nakhaie et al. 2022). The physiological damage brought on by salt and dehydration was also reduced by foliar treatments of GB and ABA in two types of turf grass; however, the effects differed across the two species and were more prominent in Kentucky bluegrass (*Poa pratensis*). MDA and electrolyte leakages showed that application was beneficial in reducing oxidative damage and it also increased the activities of antioxidant enzymes such as APX, POD, and SOD, which boosted growth and performance (Yang et al. 2012).

5 Microbial Biostimulants

5.1 Bacterial Biostimulants

Microbial biostimulants are cutting-edge technologies that can guarantee excellent nutritional content in agricultural produce while overcoming the negative impacts resulting from environmental changes. A microbe from one of four distinct genera *Azotobacter* spp., Mycorrhizal fungi, *Rhizobium* spp., and *Azospirillum* spp. is what makes up a microbial plant biostimulant (Castiglione et al. 2021).

5.1.1 Application Under Drought Stress

Applying helpful microbes is thought to be a successful tactic for improving plant tolerance to environmental challenges (Osman et al. 2022). Due to the activation of genes related to defense mechanisms, including *Capsicum annuum* dehydrin, small heat shock protein, Pro synthesizing enzymes, and transcriptional factors, as well as the activation of POD and glutathione peroxidase enzymes, pepper plants treated with *Bacillus butanolivorans* KJ40 showed better tolerance to drought stress (Kim et al. 2022). Additionally, the phytochemical profile changed following the microbial treatment, with an increase in luteolin and catechin content and a decrease in capsaicin, dihydrocapsaicin, and naringenin concentration (Kim et al. 2022). Different *Bacilli s*trains used as microbial biostimulants changed the metabolism of amino acids, hormones, phenolics, steviol glycosides, and oxylipins, enhancing the growth and drought resilience of maize plants (Othibeng et al. 2022). Rhizobacteria and MLE were used to lessen the negative effects of water shortage on wheat plants (Lalarukh et al. 2022a). *Pseudomonas aeruginosa*, a microbe that fixes nitrogen, was injected into stressed plants, and MLE was sprayed on them. The combined approach greatly boosted photosynthetic activity, plant growth metrics, and the accumulation of critical ions including calcium, nitrogen, phosphorus, and potassium as well as soluble carbohydrates and proteins (Lalarukh et al. 2022a).

In a greenhouse trial, inoculating *Glycyrrhiza uralensis* plants with a bacterial solution containing *Bacillus amyloliquefaciens* FZB42 at a concentration of 108 CFU/ml boosted plant growth and survival (Yue et al. 2022). The Pro and sucrose content of the plants increased following the inoculation, and they also accumulated compatible solutes that improved osmotic adjustment and increased antioxidant enzymes (CAT). The accumulation of pharmacologically significant substances, such as flavonoids, liquiritin, and glycyrrhizic acid, was also accelerated by bacterial inoculation. After vaccination, JA levels increased as well, which is engaged in defense mechanisms against various stressors. The use of biofertilizers, such as Nitroxin, Super Nitro Plus, and Biophosphorus, reduced the negative effects of drought *on Dracocephalum kotschyi* plants and enhanced most of physiological parameters, such as RWC, Pro accumulation, anthocyanins, and flavonoids contents, and also activated antioxidant enzyme activities, such as SOD, GPX, CAT, and APX (Cham et al. 2022).

5.1.2 Application Under Salinity Stress

Utilizing endophytic bacteria is one of the strategies for reducing salinity's negative effects on plants (Omer et al. 2022). In this study, inoculating isolates from *A. brasilense* and *P. geniculate* into salt-stressed flax (*Linum usitatissimum*) either separately or together improved the growth characteristics and increased the content of chlorophylls and carotenoids, soluble sugars, protein, and total phenolics, ascorbic acid, and potassium. Additionally, immunization was linked to an increase in antioxidant enzymes including SOD, POX, and APX as well as a decrease in oxidative stress markers like sodium, (H_2O_2) and Malondialdehyde (MDA). This may suggest that endophytes' inoculation might be a novel strategy for reducing the stress caused by salt on plants (Omer et al. 2022). Due to enhanced accumulation of pigments, soluble sugars, protein, and enzymatic and nonenzymatic antioxidants, halotolerant *B. safensis* PM22 was reported to promote growth and increase salt tolerance in maize plants. Consequently, these bacteria may be used as a biofertilizer to help plants function better in salinity-prone environments (Azeem et al. 2022). To lessen the detrimental effects of salt on plants, rhizobia can be injected into the plants. Under salt stress, lima bean plants were injected with Plant growth-promoting bacteria (PGPB), which promoted growth and induced negative impacts on growth and productivity (Lopes et al. 2022).

In this investigation, plants infected with rhizobia strains and *A. baldaniorum* strain had higher growth, root length, biomass, sum of modules, and pigment values than plants that weren't inoculated (Lopes et al. 2022). By enhancing growth parameters like RWC and photosynthetic pigments, as well as by reducing oxidative damage markers like MDA and H_2O_2 and electrolyte leakage, priming wheat seeds with growth-promoting rhizobacteria strains like *Bacillus* and *Pseudomonas* reduced the detrimental effects of salinity stress on plants (Haroon et al. 2022). The inoculated plants displayed increased enzymatic activity and salt tolerance-granting (*SOS1*, *SOS4*) gene expression (Haroon et al. 2022). The underlying mechanisms involved changing the expression of genes involved in ion homeostasis, which resulted in an increase in K^+, Ca^{2+}, and P accumulation and a decrease in Na^+. Similar to this, there was an increase in the expression of antioxidant enzymes involved in phenylpropanoid production and glutathione metabolism. Surprisingly, inoculation caused the Pro concentration to decrease because the endophytic bacteria increased the Pro catabolism gene, which decreased the accumulation of Pro (Yue et al. 2022).

5.2 *Fungal Biostimulants*

Arbuscular mycorrhiza fungi (AMF) are fungi that live in suggestion with plant roots to help plants survive environmental conditions including salt and drought (Borde et al. 2017). By helping growth, improving food availability, photosynthetic rate, preserving osmotic balance, and improving water usage efficiency, AMF helps plants better withstand these challenges. Additionally, these fungi affect the antioxidant machinery, and the majority of research found that inoculated plants with the fungus had increased antioxidant enzyme activity (Borde et al. 2017).

5.2.1 Application Under Drought Stress

Arbuscular mycorrhizal fungus immunization of olive trees dramatically increased growth (shoot length, leaf area, and leaf number), RWC, stomatal conductance, and Pro and sugar levels, highlighting the crucial function of AMF in plant adaptation to water deficiency circumstances (Aganchich et al. 2022). In a separate study, AMF inoculation significantly enhanced growth in the medicinal woody plant *Cinnamomum migao* by raising levels of soluble carbs, protein, and Pro as well as the photosynthetic rate, antioxidant activities like SOD, CAT, and POX. The MDA concentration was reduced concurrently with the injection of arbuscular mycorrhizal fungus, which is a marker of the oxidative damage caused by dryness (Yan et al. 2022). By using biofertilizers (mycorrhiza, nitrozist, and phosphozist) as well as seaweed extract, two wheat cultivars cultivated in water-deficit situations had their physiological features, such as root volume, leaf RWC, membrane stability index, and photosynthetic pigment content, enhanced (Vafa et al. 2022). As a result, grain production eventually rose and the scarce water supply become more resilient.

5.2.2 Application Under Salinity Stress

The development of taro (*Colocasia esculenta*) plantlets following AMF inoculation was increased by higher levels of protein, Pro and GB, phenolics, and antioxidant capability. Due to this advancement, the plantlets were able to establish themselves more successfully during acclimatization (Baltazar-Bernal et al. 2022). By employing endophytes fungus as biofertilizer, plants' resilience to abiotic stresses can be improved. It was looked into if introducing the endophytic fungus (*Stremphylium lycopersici*) to maize plants may mitigate the negative effects of salt (Ali et al. 2022a). In addition to boosting the activity of antioxidant enzymes like CAT and APX, the endophytic contact increased the amount of Indole acetic acid (IAA), phenolics, and flavonoids. (Ali et al. 2022a). In addition to enhancing the activity of antioxidant enzymes like CAT and APX, the endophytic contact increased the amount of Indole acetic acid (IAA), phenolics, and flavonoids. The altered cationic contents had increased Ca^{2+}, K^{+}, and Mg^{2+} contents, while decreasing Na^{+} contents improved the K^{+}/Na^{+} ratio and increased plant tolerance.

6 Macroalgae, Microalgae, and Fern Extracts

Plant-derived biostimulants, such as seaweed (macroalgae) extract, have been found to increase plant productivity and resistance to adverse environmental conditions like salt and drought (Sujeeth et al. 2022). These biostimulants stimulate the genetic and molecular mechanisms that provide the plant resilience. There is a lot of interest in using these biostimulants as an effective and environmentally benign way to promote tolerance in sustainable agricultural production systems (Sudiro et al. 2022). Seaweed extract is a rich source of biostimulating ingredients that can help to lessen the harmful effects of drought and increase tolerance (Tinte et al. 2022). Microalgae, which are tiny, unicellular organisms that live in freshwater and the ocean, may also carry out the photosynthesis process (Kumar et al. 2022). These microalgae include biostimulant materials, particularly carotenoids, which encourage plant growth and production while also improving the resilience of the plants to harsh environmental conditions and the sustainability of agriculture. They may be used as a foliar treatment, a soil supplement, or a seed primer (Kumar et al. 2022). They do this by improving crop quality, stimulating root growth, and improving the efficiency with which water is utilized.

6.1 Application Under Drought Stress

Of a current study, seaweed extract treatment changed the primary and secondary metabolism in maize plants under drought stress, including the buildup of phenylpropanoid chemicals, fatty acid metabolism, and amino acid metabolism (Tinte et al. 2022). Such substances promote drought resistance by raising growth parameters, reducing oxidative damage, enhancing antioxidant activities, and improving nutrient intake (Tinte et al. 2022). Common bean (*P. vulgaris*) cultivars subjected to drought stress had their production increased by foliar application of seaweed extract at a rate of 4 L ha^{-1} (Ziaei and Pazoki 2022). The enhanced protein, TSS, RWC, chlorophyll content, CAT, and SOD activities were all credited with the growth enhancement. According to Ziaei and Pazoki (2022), the fatty acids profile differed at different water regimes, with drought deficit causing a decrease in saturated fatty acids (SFAs) and monounsaturated fatty acids (MUFAs) and an increase in polyunsaturated fatty acids (PUFAs), which may be more important for plants' ability to adapt to these difficult conditions. In this research, cowpea seeds were primed with seaweed extract, vermicompost leachate, and smoke water to reduce drought stress by modifying tolerance mechanisms, such as an increase in suitable solutes and photosynthetic pigments (Voko et al. 2022). Depending on the stimulant, the response changed; for instance, seaweed extract and leachable vermicompost promoted the production of shoots and nodules, whereas smoke-water enhanced shoot height, root length, and bloom number. However, the levels of phenolic and flavonoid compounds were decreased as a result of the usage of biostimulants in this study (Voko et al. 2022). In a recent study, the effects of rubbing Ascophyllum nodosum extract on okra (*Abelmoschus esculentus*) leaves under a water deficit were examined (Ali et al. 2022b). The concentration of Pro, anthocyanins, and chlorophyll was increased, as well as the activity of antioxidant enzymes, by spraying the extract at 0.3%. (APX, POX, CAT).

The foliar spray improved okra performance and drought resistance (Ali et al. 2022b). In a related study, the effects of *Kappaphycu salvarezii* seaweed extract on alleviating drought stress in maize plants were examined (Trivedi et al. 2022). Improvements in growth, yield, and antioxidant enzyme activity were noted after using seaweed extract. This led to an increase in water tolerance because of the extract's active ingredients, including GB, choline chloride, and zeatin, as well as the development of the antioxidant system (Trivedi et al. 2022). In a field experiment, the application of seaweed extract topically to sugarcane plants in Brazil promoted growth, improved stalk production, sugar accumulation, boosted antioxidant machinery, and decreased oxidative damage to counteract the negative effects of drought (MDA). To decrease the effects of drought, the scientists suggested using seaweed extract as a foliar biostimulant to sugarcane (Jacomassi et al. 2022).

Rice plants were exposed to brown algae (*Dictyota dichotoma*) extract under salt and drought stress (El-Katony et al. 2020). Both aqueous extract and powder exhibited positive benefits in reducing the impacts of salt and drought stress on rice plants; however, the growth improvement was more pronounced in response to algal extract than in reaction to powder and under drought stress than under salinity stress. In stressed plants, application increased phenolics, flavonoids, sterols, hormones, and vitamins (El-Katony et al. 2020).

6.2 Application Under Salinity Stress

Tomato plants were exposed to benefit importantly from the use of extracts of algal derivatives as a defense against the osmotic stress brought on by salinization (Di-Stasio et al. 2020). The extracts of these biostimulants promoted plant adaptability to these challenging environments by improving water relationships and water usage efficiency (Di-Stasio et al. 2020). *Azolla* extract application reduced the negative effects of salt on cotton plants (Ibrahim 2022). The extract enhanced cotton plant growth, yield, and chemical make-up. The greatest findings for physio-biochemical properties, such as pigments, total soluble sugars, total phenolics, carotenoids, protein, amino acids, and antioxidant activities (SOD, CAT, POX, GR), were obtained when seed soaking and foliar spray were combined (Ibrahim 2022).

7 Conclusions and Future Perspectives

It's crucial to remember that relying alone on one technique might not be enough to entirely counteract the effects of salt on plants or to preserve a sustainable yield. Under these unfavorable conditions, a variety of integrated treatments, including microbial inoculation, fertilizers, soil amendments, crop rotation, chemical stimulants, plant extracts, and farm management strategies, may be used to achieve the sustainability of production. Use of standardized extracts and botanicals is strongly encouraged if you wish to acquire consistent and conclusive results. This evaluation acknowledged the efficacy of plant extracts and other biostimulants in decreasing some of the damaging effects of salt and drought on plants. Numerous studies at the molecular, proteome, and metabolomics levels are needed to close the information gap regarding the mechanisms of tolerance and mode of action of these extracts in drought and salinity-stressed plants. It is crucial to consider how the extraction process affects the number of bioactive compounds in plant extract. The efficiency of extracting bioactive substances will depend on the extraction selectivity and limitations of each method. Therefore, the key to increasing the value of plant extracts and the added value of applying such extracts to the plant in the issue of abiotic stress mitigation will be the development of new technology, as well as improving and promoting the existing method. Plant-derived bioactive compounds known as plant extracts and biostimulants support the pro-ecological production of strategically essential crops. Although plant extracts have been shown to have positive benefits, they are rarely employed in conventional growing methods. Farmers anticipate higher cultivation expenses as well as a decline in plant quality and quantity, which would harm crop profitability, as a result of their ignorance about the uses and applications of biostimulants. In order to get the greatest and best quality yields, it is also necessary to choose the right biostimulant for a particular plant type from among the many preparations that are readily accessible. The market necessitates the creation of preparations that are user-friendly, versatile, and may be coupled with other agents. Utilizing plant extracts and biostimulants for commercial purposes would minimize the quantity of mineral fertilizers released into the environment, lowering soil, water, and air pollution. In light of global warming, this is very significant. Although the recently created bio preparation technologies may significantly aid in environmental preservation, their primary focus is on sustainable horticulture and agricultural production in order to provide low-cost, easily available, high-quality food. Plant kinds, application techniques, and environmental conditions, as well as the raw material and method used to generate them, all affect how effectively plant extracts and biostimulants work. It is important to emphasize the positive effects of mixing microorganisms and plant hydrolysates on agricultural plant development and yield. The key advantages include an improvement in crop quality and performance, no adverse effects on humans, animals, or the environment, a rise in the variety of advantageous microbes, and better soil qualities. The usage of biostimulants in horticulture and agriculture will require solutions that are suited for the location and time of year. It will be necessary to create stewardship strategies for maximizing their usage, as well as monitoring methods for the efficiency of biostimulants. When making decisions on the farm, the long-term consequences of ecological services and biogeochemical cycles should be assessed. Producing companies for biostimulants will be expected to contribute to complete strategies at the agro system, farm, and landscape levels, of which biostimulants are simply one element. Stakeholders, farmers, public research, and regulatory organizations will need to be involved if lucrative and sustainable plant productions are to benefit from the advantages that biostimulants may provide. More than ever, government action is required to harmonize policies and rules, provide a strong framework for risk assessment that complies with the proportionality principle, and avoid data duplication across rules.

References

Abd El-Mageed TA, Gyushi MA, Hemida KA, El-Saadony MT, Abd El-Mageed SA, Abdalla H, Abu Qamar SF, El-Tarabily KA, Abdelkhalik A (2022) Coapplication of effective microorganisms and nanomagnesium boosts the agronomic, physio-biochemical, osmolytes, and antioxidants defenses against salt stress in *Ipomoea batatas*. Front Plant Sci 13:883274. https://doi.org/10.3389/fpls.2022.883274

Abdou NM, El-Saadony FM, Roby MH, Mahdy HA, El-Shehawi AM, Elseehy MM, El-Tahan AM, Abdalla H, Saad AM, Abou Sreea AIB (2022) Foliar spray of potassium silicate, aloe extract composite and their effect on growth and yielding capacity of roselle (*Hibiscus sabdariffa* L.) under water deficit stress conditions. Saudi J Biol Sci. https://doi.org/10.1016/j.sjbs.2022.02.033

Abou-Alhamd FM, Loutfy N (2020) *Ocimum basilicum* leaf extract induces salinity stress tolerance in Faba bean plants. Egypt J Bot 60(3):681–690

Aganchich B, Wahbi S, Yaakoubi A, El-Aououad H, Bota J (2022) Effect of arbuscular mycorrhizal fungi inoculation on growth and physiology performance of olive tree under regulated deficit irrigation and partial rootzone drying. S Afr J Bot 148:1–10

Ahmad A, Blasco B, Martos V (2022) Combating salinity through natural plant extracts based biostimulants: a review. Front Plant Sci 13:1665. https://doi.org/10.3389/fpls.2022.862034

Ahmed MB, Zhou JL, Ngo HH, Guo W, Chen M (2016) Progress in the preparation and application of modified biochar for improved contaminant removal from water and waste water. Bioresour Technol 21(4):836–851

Alabdallah NM, Alluqmani SM (2022) The synthesis of polysaccharide crude nanoparticles extracts from Taif rose petals and its effect on eggplant seedlings under drought and salt stress. J King Saud Univ Sci 34(5):102055. https://doi.org/10.1016/j.jksus.2022.102055

Alenezi NA, Al-Qurainy F, Tarroum M, Nadeem M, Khan S, Salih AM, Shaikhaldein HO, Alfarraj NS, Gaafar ARZ, Al-Hashimi A, Alansi S (2022) Zinc oxide nanoparticles (ZnO NPs), biosynthesis, characterization and evaluation of their impact to improve shoot growth and to reduce salt toxicity on *Salvia officinalis* in vitro cultivated. PRO 10(7):1273. https://doi.org/10.3390/pr10071273

Alharby HF, Alzahrani YM, Rady MM (2020) Seeds pretreatment with zeatin or maize grain-derived organic biostimulant improved hormonal contents, polyamine gene expression, and salinity and drought tolerance of wheat. Int J Agric Biol 24(4):714–724. https://doi.org/10.17957/IJAB/15.1491

Ali EF, Hassan FAS, Elgimabi M (2018) Improving the growth, yield and volatile oil content of *Pelargonium graveolens* L. Herit by foliar application with moringa leaf extract through motivating physiological and biochemical parameters. S Afr J Bot 119:383–389. https://doi.org/10.1016/j.sajb.2018.10.003

Ali Q, Shehzad F, Waseem M, Shahid S, Hussain A, Haider ZM, Habib N, Hussain SM, Javed M, Perveen R (2020) Plant based biostimulants and plant stress responses. In: Plant ecophysiology and adaptation under climate change: mechanisms and perspectives I, Holder. Springer Nature, Singapore, pp 185–202. https://doi.org/10.1007/978-981-15-2156-0_22

Ali R, Gul H, Rauf M, Arif M, Hamayun M, Khilji SA, Ud-Din A, Sajid ZA, Lee IJ (2022a) Growth-promoting endophytic fungus (*Stemphylium lycopersici*) ameliorates salt stress tolerance in maize by balancing ionic and metabolic status. Front Plant Sci 13:565–890. https://doi.org/10.3389/fpls.2022.890565

Ali J, Jan I, Ullah H, Ahmed N, Alam M, Ullah R, El-Sharnouby M, Kesba H, Shukry M, Sayed S, Nawaz T (2022b) Influence of *Ascophyllum nodosum* extract foliar spray on the physiological and biochemical attributes of okra under drought stress. Plants 11(6):790. https://doi.org/10.3390/plants11060790

Altemimi A, Lakhssassi N, Baharlouei A, Watson DG, Lightfoot DA (2017) Phytochemicals: extraction, isolation, and identification of bioactive compounds from plant extracts. Plants 6(4):42. https://doi.org/10.3390/plants6040042

Azeem MA, Shah FH, Ullah A, Ali K, Jones DA, Khan MEH, Ashraf A (2022) Biochemical characterization of halotolerant *bacillus safensis* pm22 and its potential to enhance growth of maize under salinity stress. Plan Theory 11(13):1721

Azmat A, Tanveer Y, Yasmin H, Hassan MN, Shahzad A, Reddy M, Ahmad A (2022) Coactive role of zinc oxide nanoparticles and plant growth promoting rhizobacteria for mitigation of synchronized effects of heat and drought stress in wheat plants. Chemosphere 297:133982. https://doi.org/10.1016/j.chemosphere.2022.133982

Azwanida NN (2015) A review on the extraction methods use in medicinal plants, principle, strength and limitation. Med Aromat Plants 4(3):1000196

Badem A, Söylemez S (2022) Effects of nitric oxide and silicon application on growth and productivity of pepper under salinity stress. J King Saud Univ Sci 34(6):1–11. https://doi.org/10.1016/j.jksus.2022.102189

Bakry BA, Ibrahim FM, Abdallah MMS, El-Bassiouny HMS (2016) Effect of banana peel extract or tryptophan on growth, yield and some biochemical aspects of quinoa plants under water deficit. Int J Pharmtech Res 9(8):276–287

Baltazar BO, Spinoso JL, Mancilla-Álvarez E, Bello-Bello JJ (2022) Arbuscular mycorrhizal fungi induce tolerance to salinity stress in taro plantlets (*Colocasia esculenta* l. schott) during acclimatization. Plants 11(13):1780. https://doi.org/10.3390/plants11131780

Bartwal A, Mall R, Lohani P, Guru SK, Arora S (2013) Role of secondary metabolites and brassinosteroids in plant defense against environmental stresses. J Plant Growth Regul 32(1):216–232. https://doi.org/10.1007/s00344-012-9272-x

Barzin G, Kazemi MM, Entezari M (2022) The interaction effects of NaCl stress and sodium nitroprusside on growth, physiological and biochemical responses of *Calendula officinalis* L. Biologia 77(6):2081–2091. https://doi.org/10.1007/s11756-022-01068-w

Battacharyya D, Babgohari MZ, Rathor P, Prithiviraj B (2015) Seaweed extracts as biostimulants in horticulture. Sci Hortic 196:39–48

Bhan M (2017) Ionic liquids as green solvents in herbal extraction. Int J Adv Res Dev 2:10–12

Borde M, Dudhane M, Kulkarni M (2017) Role of arbuscular mycorrhizal fungi (AMF) in salinity tolerance and growth response in plants under salt stress conditions. In: Mycorrhiza eco-physiology, secondary metabolites, nanomaterials, pp 71–86. https://doi.org/10.1007/978-3-319-57849-1_5

Brazales DK, Romero JY, Vences MÁ, Torres M, Aviles NY, Sohlenkamp C, Serrano M (2022) Transcriptional characterization of the biostimulant effect of *Moringa oleifera* leaf extracts using *Arabidopsis thaliana* as a model. S Afr J Bot 144:250–256. https://doi.org/10.1016/j.sajb.2021.09.011

Calvo P, Nelson L, Kloepper JW (2014) Agricultural uses of plant biostimulants. Plant Soil 383:3–41. https://doi.org/10.1007/s11104-014-2131-8

Castiglione AM, Mannino G, Contartese V, Bertea CM, Ertani A (2021) Microbial biostimulants as response to modern agriculture needs: composition, role and application of these innovative products. Plants 10(8):1533

Cham R, Abtahi SA, Jafarinia M, Yasrebi J (2022) Physiological responses of *Dracocephalum kotschyi Boiss* to drought stress and Bio-fertilizers. S Afr J Bot 148:180–189. https://doi.org/10.1016/j.sajb.2022.04.008

Chapman C, Rossi S, Yuan B, Huang B (2022) Differential regulation of amino acids and nitrogen for drought tolerance and post stress recovery in creeping bent grass. J Am Soc Hortic Sci 147(4):208–215. https://doi.org/10.21273/JASHS05215-22

Chojnacka K, Michalak I, Dmytryk A, Wilk R, Górecki H (2015) Innovative natural plant growth biostimulants. In: Shishir S, Pant KK (eds) Advances in fertilizer technology II. Studium Press LLC, pp 452–489

Cisse EHM, Zhang LJ, Pu YJ, Miao LF, Li DD, Zhang J, Yang F (2022) Exogenous Ca^{2+} associated with melatonin alleviates drought-induced damage in the woody tree *Dalbergia odorifera*. J Plant Growth Regul 4(16):2359–2374. https://doi.org/10.1007/s00344-021-10449-5

Colla G, Rouphael Y (2015) Biostimulants in horticulture. Sci Hortic 196:1–2. https://doi.org/10.1016/j.scienta.2015.10.044

Colla G, Nardi S, Cardarelli M, Ertani A, Lucini L, Canaguier R, Rouphael Y (2015) Protein hydrolysates as biostimulants in horticulture. Sci Hortic 196:28–38

Cowan MM (1999) Plant products as antimicrobial agents. Clin Microbiol Rev 2(4):564–582. https://doi.org/10.1128/CMR.12.4.564

Craigie JS (2011) Seaweed extract stimuli in plant science and agriculture. J Appl Phycol 23:371–393. https://doi.org/10.1007/s10811-010-9560-4

Damalas CA, Koutroubas SD (2022) Exogenous application of salicylic acid for regulation of sunflower growth under abiotic stress: a systematic review. Biologia 77(7):1685–1697. https://doi.org/10.1007/s11756-022-01020-y

Das K, Tiwari RK, Shrivastava DK (2010) Techniques for evaluation of medicinal plant products as antimicrobial agents: current methods and future trends. J Med Plants Res 4(2):104–111

Desoky ESM, Elrys AS, Mansour E, Eid RS, Selem E, Rady MM, Ali EF, Mersal GA, Semida WM (2021) Application of biostimulants promotes growth and productivity by fortifying the antioxidant machinery and suppressing oxidative stress in faba bean under various abiotic stresses. Sci Hortic 288:110340

Di-Stasio E, Cirillo V, Raimondi G, Giordano M, Esposito M, Maggio A (2020) Osmo-priming with seaweed extracts enhances yield of salt-stressed tomato plants. Agronomy 10(10):1559. https://doi.org/10.3390/agronomy10101559

Doughari JH (2012) Phytochemicals: extraction methods, basic structures and mode of action as potential chemotherapeutic agents. In: Rao V (ed) Phytochemicals - a global perspective of their role in nutrition and health. ISBN: 978-953-51-0296-0.

Dowom SA, Karimian Z, Ei Dehnavi MM, Samiei L (2022) Chitosan nanoparticles improve physiological and biochemical responses of *Salvia abrotanoides* (Kar.) under drought stress. BMC Plant Biol 22(1):364. https://doi.org/10.1186/s12870-022-03689-4

Du-Jardin P (2015) Plant biostimulants: definition, concept, main categories and regulation. Sci Hortic 196:3–14. https://doi.org/10.1016/j.scienta.2015.09.021

El-Banna MF, AL-Huqail AA, Farouk S, Belal BE, El-Kenawy MA, Abd El-Khalek AF (2022) Morpho-physiological and anatomical alterations of salt-affected Thompson seedless grapevine (*Vitis vinifera* L.) to Brassinolide spraying. Horticulturae 8(7): 568. https://doi.org/10.3390/horticulturae8070568

El-Katony TM, Deyab MA, El-Adl MF, Ward FME (2020) The aqueous extract and powder of the brown alga *Dictyota dichotoma* (Hudson) differentially alleviate the impact of abiotic stress on rice (*Oryza sativa* L.). Physiol Mol Biol Plants 26(6):1155–1171. https://doi.org/10.1007/s12298-020-00805-2

Elsayed SIM, Mazhar AMA, El-Sayed SM, Said AM (2022) Improvement the drought tolerance of *Eucalyptus citriodora* seedling by spraying basil leaves extract and its influence on growth, volatile oil components and some enzymatic activity. Egypt J Chem 65(12):619–635. https://doi.org/10.21608/EJCHEM.2022.127566.5662

El-Tantawy EM (2009) Behavior of tomato plants as affected by spraying with chitosan and aminofort as natural stimulator substances under application of soil organic amendments. Pak J Biol Sci 12(17):1164–1173. https://doi.org/10.3923/pjbs.2009.1164.1173

Fukao T, Barrera-Figueroa BE, Juntawong P, Peña-Castro JM (2019) Submergence and waterlogging stress in plants: a review highlighting research opportunities and understudied aspects. Front Plant Sci 10:340. https://doi.org/10.3389/fpls.2019.00340

Gholamzadeh AA, Mousavi-Fard S, Nejad RA (2022) Morphological and physiological characteristics for evaluation of salicylic acid effects on *Celosia argentea* L. under salinity stress. Iran. J Plant Physiol 12(1):4027–4037. https://doi.org/10.30495/IJPP.2022.689078

Giglou MT, Giglou RH, Esmaeilpour B, Azarmi R, Padash A, Falakian M, Śliwka J, Gohari G, Lajayer HM (2022) A new method in mitigation of drought stress by chitosan-coated iron oxide nanoparticles and growth stimulant in peppermint. Ind Crop Prod 187:115286. https://doi.org/10.1016/j.indcrop.2022.115286

Gong Z, Xiong L, Shi H, Yang S, Herrera-Estrella LR, Xu G, Zhu JK (2020) Plant abiotic stress response and nutrient use efficiency. Sci China Life Sci 63:635–674. https://doi.org/10.1007/s11427-020-1683-x

Haghighi TM, Saharkhiz MJ, Kavoosi G, Jowkar A (2022) Monitoring amino acid profile and protein quality of Licorice (*Glycyrrhiza glabra* L.) under drought stress, silicon nutrition and mycorrhiza inoculation. Sci Hortic 295:110808

Hajihashemi S, Kazemi S (2022) The potential of foliar application of nano-chitosan-encapsulated nano-silicon donor in amelioration the adverse effect of salinity in the wheat plant. BMC Plant Biol 22:148. https://doi.org/10.1186/s12870-022-03531-x

Halpern M, Bar-Tal A, Ofek M, Minz D, Muller T, Yermiyahu U (2015) The use of biostimulants for enhancing nutrient uptake. In: Donald L (ed) Advanced in agronomy, vol 130. Academic Press, pp 141–174. https://doi.org/10.1016/bs.agron.2014.10.001

Hanafy SR (2017) Using *Moringa oleifera* leaf extract as a biofertilizer for drought stress mitigation of *Glycine max* L. Plants Egypt J Bot 57(2):281–292. https://doi.org/10.21608/ejbo.2017.596.1027

Haroon U, Khizar M, Liaquat F, Ali M, Akbar M, Tahir K, Batool SS, Kamal A, Chaudhary HJ, Munis MFH (2022) Halotolerant plant growth-promoting rhizobacteria induce salinity tolerance in wheat by enhancing the expression of SOS genes. J Plant Growth Regul 41(6):2435–2448. https://doi.org/10.1007/s00344-021-10457-5

Hasanuzzaman M, Oku H, Nahar K, Bhuyan MHM, Mahmud JA, Baluska F, Fujita M (2018) Nitric oxide-induced salt stress tolerance in plants: ROS metabolism, signaling, and molecular interactions. Plant Biotechnol Rep 12:77–92. https://doi.org/10.1007/s11816-018-0480-0

Hassan FAS, Fetouh MI (2019) Does moringa leaf extract have preservative effect improving the longevity and postharvest quality of gladiolus cut spikes? Sci Hortic 250:287–293

Hassan F, Al-Yasi H, Ali E, Alamer K, Hessini K, Attia H, El-Shazly S (2021) Mitigation of salt-stress effects by moringa leaf extract or salicylic acid through motivating antioxidant machinery in damask rose. Can J Plant Sci 101(2):157–165. https://doi.org/10.1139/CJPS-2020-0127

Hassanein AR, Abdelkader FA, Faramawy MH (2019) Moringa leaf extracts as biostimulants-inducing salinity tolerance in the sweet basil plant. Egypt J Bot 59(2):303–318

Hossain A, Pamanick B, Venugopalan VK, Ibrahimova U, Rahman MA, Siyal AL, Maitra S, Chatterjee S, Aftab T (2022) Emerging roles of plant growth regulators for plants adaptation to abiotic stress–induced oxidative stress. In: Emerging plant growth regulators in agriculture. Academic Press, pp 1–72. https://doi.org/10.1016/B978-0-323-91005-7.00010-2

Ibrahim AA (2022) Alleviation the adverse effect of the salinity on cotton plant by using azolla extract. Egypt J Chem 65(7):155–164

Ingle KP, Deshmukh AG, Padole DA, Dudhare MS, Moharil MP, Khelurkar VC (2017) Phytochemicals: extraction methods, identification, and detection of bioactive compounds from plant extracts. J Pharm Phytoch 6:32–36

Iqbal MS, Singh AK, Ansari MI (2020) Effect of drought stress on crop production. In: Rakshit A, Singh HB, Singh AK, Singh US, Fraceto L (eds) New frontiers in stress management for durable agriculture. Springer, Singapore, pp 35–47. https://doi.org/10.1007/978-981-15-1322-0_3

Islam AT, Ullah H, Himanshu SK, Tisarum R, Cha-um S, Datta A (2022) Effect of salicylic acid seed priming on morpho-physiological responses and yield of baby corn under salt stress. Sci Hortic 304:11304

Jacomassi LM, de Oliveira VJ, Oliveira MP, Momesso L, de Siqueira GF, Crusciol CAC (2022) A seaweed extract-based biostimulant mitigates drought stress in sugarcane. Front Plant Sci 13:865291. https://doi.org/10.3389/fpls.2022.865291

Kauffman GL, Kneivel DP, Watschke TL (2007) Effects of a biostimulant on the heat tolerance associated with photosynthetic capacity, membrane thermostability and polyphenol production of perennial ryegrass. Crop Sci 47(1):261–267

Kaur H, Hussain SJ, Kaur G, Poor P, Alamri S, Siddiqui MH, Khan MIR (2022) Salicylic acid improves nitrogen fixation, growth, yield and antioxidant defence mechanisms in chickpea genotypes under salt stress. J Plant Growth Regul 41(5):2034–2047

Khaliq A, Ibrahim MU, Hussain S, Ul Z, Haq M, Al-Huqail AA, Nawaz M, Ali B, Khan F, Ali HM, Siddiqui MH (2022) The hormetic effects of a brassica water extract triggered wheat growth and antioxidative defense under drought stress. Appl Sci 12(9):45–82

Kim ST, Yoo SJ, Weon HY, Song J, Sang MK (2022) *Bacillus butanolivorans* KJ40 contributes alleviation of drought stress in pepper plants by modulating antioxidant and polyphenolic compounds. Sci Hortic 301:1111

Kumar S, Korra T, Singh UB, Singh S, Bisen K (2022) Microalgal based biostimulants as alleviator of biotic and abiotic stresses in crop plants. In: Singh HB, Vaishnav A (eds) New and future developments in microbial biotechnology and bioengineering, pp 195–216. https://doi.org/10.1016/B978-0-323-85577-8.00013-5

Lalarukh I, Al-Dhumri SA, Al-Ani LKT, Hussain R, Al-Mutairi KA, Mansoora N, Amjad SF, Abbas MH, Abdelhafez AA, Poczai P, Meena KR (2022a) A combined use of rhizobacteria and moringa leaf extract mitigates the adverse effects of drought stress in wheat (*Triticum aestivum* L). Front Microbiol 13:813–415

Lalarukh I, Zahra N, Al Huqail AA, Amjad SF, Al-Dhumri SA, Ghoneim AM, Alshahri AH, Almutari MM, Alhusayni FS, Al-Shammari WB, Poczai P (2022b) Exogenously applied ZnO nanoparticles induced salt tolerance in potentially high yielding modern wheat (*Triticum aestivum* L.) cultivars. Environ Technol Innov 7:102–799

Lopes ÁL, Setubal IS, Neto VP, Zilli JE, Rodrigues AC, Bonifacio A (2022) Synergism of Bradyrhizobium and *Azospirillum baldaniorum* improves growth and symbiotic performance in lima bean under salinity by positive modulations in leaf nitrogen compounds. Appl Soil Ecol 180:104–603

Ma H, Li P, Liu X, Li C, Zhang S, Wang X, Tao X (2022) Poly-γ-glutamic acid enhanced the drought resistance of maize by improving photosynthesis and affecting the rhizosphere microbial community. BMC Plant Biol 22(1):11

Mahdavian K (2022) Effect of salicylic acid and calcium chloride on lipid peroxidation and scavenging capacity of radical of red bean (*Phaseolus calcaratus* L.) under salt stress. Int J Hortic Sci Technol 9(1):55–72

Mahmoudand RA, Dahab AA (2018) Response of apple seedlings grown under saline conditions to natural plant extracts. Biosci Res 15(2):589–601

Majekodunmi SO (2015) Review of extraction of medicinal plants for pharmaceutical research. MRJMMS 3:521–527

Manoj BS, Gupta M, Jeelani MI, Gupta S (2022) Chitosan augments bioactive properties and drought resilience in drought-induced red kidney beans. Food Res Int 159:111–597

Morton MJL, Awlia M, Al-Tamimi N, Saade S, Pailles Y, Negrão S, Tester M (2019) Salt stress under the scalpel—dissecting the genetics of salt tolerance. Plant J 97:148–163

Munns R, Tester M (2008) Mechanisms of salinity tolerance. Annu Rev Plant Biol 59:651–681. https://doi.org/10.1146/annurev.arplant.59.032607.092911

Nakhaie A, Habibi G, Vaziri A (2022) Exogenous proline enhances salt tolerance in acclimated Aloe vera by modulating photosystem II efficiency and antioxidant defense. S Afr J Bot 47:1171–1180

Naz H, Akram NA, Ashraf M, Hefft DI, Jan BL (2022) Leaf extract of neem (Azadirachta indica) alleviates adverse effects of drought in quinoa (*Chenopodium quinoa* Willd.) plants through alterations in biochemical attributes and antioxidants. Saudi. J Biol Sci 29(3):1367–1374

Noor R, Yasmin H, Ilyas N, Nosheen A, Hassan MN, Mumtaz S, Khan N, Ahmad A, Ahmad P (2022) Comparative analysis of iron oxide nanoparticles synthesized from ginger (*Zingiber officinale*) and cumin seeds (*Cuminum cyminum*) to induce resistance in wheat against drought stress. Chemosphere 292:133–201

Omer AM, Osman MS, Badawy AA (2022) Inoculation with Azospirillum brasilense and/or Pseudomonas geniculata reinforces flax (*Linum usitatissimum*) growth by improving physiological activities under saline soil conditions. Bot Stud 63(1):15. https://doi.org/10.1186/s40529-022-00345-w

Omidi M, Khandan-Mirkohi A, Kafi M, Zamani Z, Ajdanian L, Babaei M (2022) Biochemical and molecular responses of Rosa damascena mill. cv. Kashan to salicylic acid under salinity stress. BMC Plant Biol 22(1):373

Osman HS, Rady A, Awadalla A, Omara AED, Hafez EM (2022) Improving the antioxidants system, growth, and sugar beet quality subjected to long-term osmotic stress by phosphate solubilizing bacteria and compost tea. Int J Plant Prod 16(1):119–135

Othibeng K, Nephali L, Myoli A, Buthelezi N, Jonker W, Huyser J, Tugizimana F (2022) Metabolic circuits in sap extracts reflect the effects of a microbial biostimulant on maize metabolism under drought conditions. Plants 11(4):510. https://doi.org/10.3390/plants11040510

Pandey A, Tripathi S (2014) Concept of standardization, extraction, and pre-phytochemical screening strategies for herbal drug. J Pharmacogn Phytochem 2(5):15–119

Pandeya V, Ansarib MW, Tulac S, Sahooc RK, Bainsa GK, Tutejae N, Shuklaa A (2016) *Ocimum sanctum* leaf extract induces drought stress tolerance in rice. Plant Signal Behav 11(5):1–9

Pongprayoon W, Roytrakul S, Pichayangkura R, Chadchawan S (2013) The role of hydrogen perox-ide in chitosan-induced resistance to osmotic stress in rice (*Oryza sativa* L.). Effect of chitosan on plant growth, flowering and corms yield of potted freesia. Plant Growth Regul 70(2):159–173

Pourhadi M, Badi HN, Mehrafarin A, Omidi H, Hajiaghaee R (2018) Phytochemical and growth responses of *Mentha piperita* to foliar application of biostimulants under greenhouse and field conditions. Herba Polonica 64(2):1–12

Rady MM, Mohamed GF (2015) Modulation of salt stress effects on the growth, physio-chemical attributes and yields of *Phaseolus vulgaris* L. plants by the combined application of salicylic acid and *Moringa oleifera* leaf extract. Sci Hortic 193:105–113

Rehman H, Nawaz MQ, Basra SMA, Afzal I, Yasmeen A, Hassan FU (2014) Seed priming influence on early cropgrowth, phenological development and yield performance of linola (*Linumusita tissimum* L.). J Integr Agric 13:990–996

Ri K, Ri J (2022) Hydrolysis of willow (*Salix babylonica* L.) Extract alleviates drought effects on Houttuynia Thunb. Theor Exp Plant Physiol 34(1):71–81

Roumani A, Biabani A, Karizaki AR, Alamdari EG (2022) Foliar salicylic acid application to mitigate the effect of drought stress on isabgol (*Plantago ovata* forssk). Biochem Syst Ecol 104:104453. https://doi.org/10.1016/j.bse.2022.104453

Rouphael Y, Giordano M, Cardarelli M, Cozzolino E, Mori M, Kyriacou M, Colla G (2018) Plant- and seaweed-based extracts increase yield but differentially modulate nutritional quality of greenhouse spinach through biostimulant action. Agronomy 8(7):126. https://doi.org/10.3390/agronomy8070126

Sakamoto A, Murata N (2002) The role of glycine betaine in the protection of plants from stress: clues from transgenic plants. Plant Cell Environ 25(2):163–171

Sarwar M, Anjum S, Alam MW, Ali Q, Ayyub CM, Haider MS, Ashraf MI, Mahboob W (2022) Triacontanol regulates morphological traits and enzymatic activities of salinity affected hot pepper plants. Sci Rep 12(1):3736

Sasidharan S, Chen Y, Saravanan D, Sundram KM, Yoga Latha L (2011) Extraction, isolation and characterization of bioactive compounds from plants extracts. Afr J Tradit Complement Altern Med 8:1–10

Shaikhaldein HO, Al-Qurainy F, Nadeem M, Khan S, Tarroum M, Salih AM, Alansi S, Al-Hashimi A, Alfagham A, Alkahtani J (2022) Assessment of the impacts of green synthesized silver nanoparticles on *Maerua oblongifolia* shoots under *in vitro* salt stress. Materials 15(14):4784

Sheikhalipour M, Mohammadi SA, Esmaielpour B, Zareei E, Kulak M, Ali S, Nouraein M, Bahrami MK, Gohari G, Fotopoulos V (2022) Exogenous melatonin increases salt tolerance in bitter melon by regulating ionic balance, antioxidant system and secondary metabolism-related genes. BMC Plant Biol 22(1):380

Sudiro C, Guglielmi F, Hochart M, Senizza B, Zhang L, Lucini L, Altissimo A (2022) A Phenomics and metabolomics investigation on the modulation of drought stress by a biostimulant plant extract in tomato (*Solanum lycopersicum*). Agronomy 12(4):764

Sujeeth N, Petrov V, Guinan KJ, Rasul F, O'Sullivan JT, Gechev TS (2022) Current insights into the molecular mode of action of seaweed-based biostimulants and the sustainability of seaweeds as raw material resources. Int J Mol Sci 23:54–76

Suryaman M, Sunarya Y, Istarimila I, Fudholi A (2021) Effect of salinity stress on the growth and yield of mungbean (*Vigna radiata* (L.) R. Wilczek) treated with mangosteen pericarp extract. Biocatal Agric Biotechnol 36:102–132

Taiz L, Zeiger E (2010) Plant physiology, 5th edn. Sinauer Associates, Sunderland, p 782

Tinte MM, Masike K, Steenkamp PA, Huyser J, van der Hooft JJ, Tugizimana F (2022) Computational metabolomics tools reveal metabolic reconfigurations underlying the effects of biostimulant seaweed extracts on maize plants under drought stress conditions. Metabolites 12(6):487

Tiwari P, Kumar B, Kaur M, Kaur G, Kaur H (2011) Phytochemical screening and extraction: a review. Int Pharm Sci 1:98–106

Trivedi K, Anand KG, Kubavat D, Ghosh A (2022) Role of *Kappaphy cusalvarezii* seaweed extract and its active constituents, glycine betaine, choline chloride, and zeatin in the alleviation of drought stress at critical growth stages of maize crop. J Appl Phycol 34(3):1791–1804

Ujang ZB, Subramaniam T, Diah MM, Wahid HB, Abdullah BB, Rashid AA, Appleton D (2013) Bioguided fractionation and purification of natural bioactive obtained from *Alpiniaconchigera* water extract with melanin inhibition activity. J Biomater Nanobiotechnol 4:265–272

Vafa ZN, Sohrabi Y, Mirzaghaderi G, Heidari G (2022) Soil microorganisms and seaweed application with supplementary irrigation improved physiological traits and yield of two dryland wheat cultivars. Front Plant Sci 13. https://doi.org/10.3389/fpls.2022.855090

Voko MP, Kulkarni MG, Ngoroyemoto N, Gupta S, Finnie JF, Van-Staden J (2022) Vermicompost leachate, seaweed extract and smoke-water alleviate drought stress in cowpea by influencing phytochemicals, compatible solutes and photosynthetic pigments. Plant Growth Regul 97(2):327–342

Vranova V, Rejsek K, Skene KR, Formanek P (2011) Non-protein amino acids: plant, soil and ecosystem interactions. Plant Soil 342(1):31–48

Wu Y, Li J, Wang J, Dawuda MM, Liao W, Meng X, Yuan H, Xie J, Tang Z, Lyu J, Yu J (2022) Heme is involved in the exogenous ALA-promoted growth and antioxidant defense system of cucumber seedlings under salt stress. BMC Plant Biol 22(1):329

Yakhin OI, Lubyanov AA, Yakhin IA, Brown PH (2017) Biostimulants in plant science: a global perspective. Front Pant Sci 7:2049

Yan Q, Li X, Xiao X, Chen J, Liu J, Lin C, Guan R, Wang D (2022) Arbuscular mycorrhizal fungi improve the growth and drought tolerance of *Cinnamomum migao* by enhancing physio-biochemical responses. Ecol Evol 12(7):9091

Yang Z, Yu J, Merewitz E, Huang B (2012) Differential effects of abscisic acid and glycine betaine on physiological responses to drought and salinity stress for two perennial grass species. J Am Soc Hortic Sci 137(2):96–106

Yue Z, Chen Y, Wang Y, Zheng L, Zhang Q, Liu Y, Hu C, Chen C, Ma K, Sun Z (2022) Halotolerant *Bacillus altitudinis* WR10 improves salt tolerance in wheat via a multi-level mechanism. Front Plant Sci 13:941388

Zaki HE, Radwan KS (2022) The use of OSM regulators and antioxidants to mitigate the adverse impacts f salinity stress in diploid and tetraploid potato genotypes (*Solanum* spp.). Chem Biol Technol Agric 9(1):19

Zeng D, Luo X (2012) Physiological effects of chitosan coating on wheat growth and activities of protective enzyme with drought tolerance. Open J Soil Sci 2:282–288

Ziaei M, Pazoki A (2022) Foliar-applied seaweed extract improves yield of common bean (*phaseolus vulgaris* l.) cultivars through changes in biochemical and fatty acid profile under irrigation regimes. J Soil Sci Plant Nutr 22:2969–2979. https://doi.org/10.1007/s42729-022-00860-6

Chapter 26
Secondary Metabolism and Its Role in Enhancing Drought Stress Tolerance

Rasha Fouad, Hend Fouad, Ahmed H. El-Desoky, and Elsayed A. Omer

Abstract Plants produce vivid secondary metabolites that are crucial for environmental adaptation and management of abiotic stress such as salinity, drought, temperature, etc. In addition to their different roles in plant life, secondary metabolites are essential for protecting plants from biotic and abiotic stresses. There is a large difference in secondary metabolites levels and kinds depending on the plant species in response to stress. Secondary metabolites are diverse phytochemical classes found in the plant kingdom. These secondary metabolites are divided into three main groups: phenolic compounds (flavonoids and phenylpropanoids), nitrogen-containing compounds (glucosinolates and alkaloids), and terpenes. Secondary metabolites levels in plants may change in response to stress. To combat negative physiological changes caused by stress, plants produce secondary metabolites, which are by-products of primary metabolites. Secondary metabolites with great specificity in reducing the impacts of abiotic stressors include flavonoids, carotenoids, and isoprenoids. To counteract the negative impacts of abiotic stresses, plants undergo fast changes in their secondary metabolites biosynthesis and accumulation. Consequently, there is a significant relationship between secondary metabolites and plants' tolerance to withstand abiotic stress. Several chemical substances like abscisic acid, methyl jasmonate, polyamines, salicylic acid, brassinosteroids, and melatonin had been used as exogenous application to improve plant tolerance against abiotic stress. The aim of metabolic engineering is to increase the production of secondary metabolites and so increase the ability of plants to withstand diverse environmental stresses. In order to provide abiotic stress tolerance, metabolic engineering has been employed successfully in many plants to increase the levels of secondary metabolites. In this chapter, we discuss and narrate recent research and trends in the field secondary metabolites and drought stress issue in

R. Fouad (✉) · H. Fouad · E. A. Omer
Department of Medicinal and Aromatic Plants Research, National Research Centre, Giza, Egypt
e-mail: rf.ahmad@nrc.sci.eg

A. H. El-Desoky
Department of Pharmacognosy, National Research Centre, Giza, Egypt

M. Hasanuzzaman (ed.), *Climate-Resilient Agriculture, Vol 2*,
https://doi.org/10.1007/978-3-031-37428-9_26

addition to the metabolic engineering activities to increase the secondary metabolites synthesis as a stress-response mechanism in plants to make them more stress-tolerant.

Keywords Secondary metabolism · Abiotic stress · Drought · Metabolic engineering

1 Introduction

Drought is considered as the most dangerous abiotic stress, especially in areas that are dry and semidry (Bray 2001; Rahdari and Hoseini 2012; Rana et al. 2013; Zlatev and Lidon 2012). Drought may cause a severe food crisis due to the growing world population, which is anticipated to reach 10 billion by 2050. Climate change could worsen this scenario by making water shortages more frequent, more severe, and last longer. Therefore, it must be urgent to deepen our comprehension of the complex mechanisms underpinning drought resistance and create premier more drought-resistant crop varieties while maintaining the same level of agronomic and quality parameters (Hossain et al. 2016).

Water serves a crucial function in plant growth, evolution, and metabolism, as it makes up between 80% and 95% of most plants fresh biomass (Hirt and Shinozak 2004; Salehi-lisar et al. 2012). When there are insufficient water supplies for the plants, water stress takes place. A great loss of water and a reduction in the available water absorbed by the plants occur as a result of different weather conditions. There is a great loss of water because of the atmospheric conditions which also lead to a significant reduction in water absorbed by plants (Sharma and Kumar 2021). Additionally, there are numerous factors that contribute to water deficiency in plants, such as extremes in temperatures, low rainfall, high light density, salinity, and more. Further, in several situations, plants are unable to absorb water even though there is sufficient water in the soil. This kind of drought stress is physiological and known as pseudodrought (Arbona et al. 2013; Ashraf et al. 2009; Salehi-lisar et al. 2012). Many negative alterations in the physiological, morphological, ecological, biochemical, and molecular processes and features of plants are mostly caused by drought stress, as it is considered a multidimensional stress (Bhargava and Sawant 2013; Farooq et al. 2009; Kumar et al. 2021; Shao et al. 2008), which decrease photosynthesis (McKay et al. 2003), reduce cell turgor pressure (Taiz and Zeiger 2006), and decline cell elongation and division (Bal et al. 2010). In the root zone during drought stress, there is a great decrement in the growth of root tissues that inhibits the nutrient uptake and nitrogen transport and its metabolism as well as it causes a reduction in some compounds like ammonium transporter, glutamine synthetase, and nitrate reductase by the roots and their movement to the target sites from roots to shoots (Sharma and Kumar 2021). The main causes of drought and plant reactions are illustrated in Fig. 26.1. Drought stress also can negatively affect gene expression, quality, and quantity of plant growth and yields which may lead to plant death (Cattivelli et al. 2008; Prasad et al. 2008; Jaleel et al. 2009; Zlatev and

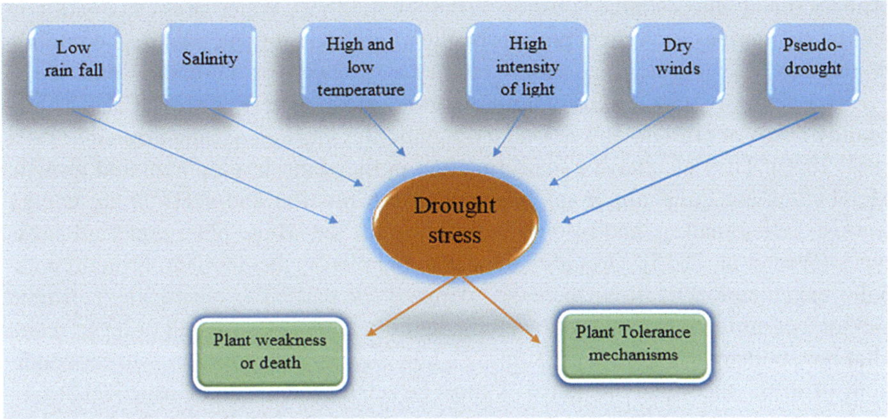

Fig. 26.1 Drought-causing factors and plant behavior

Lidon 2012; Nezhadahmadi et al. 2013; Chen et al. 2022). According to previous reports, drought stress dramatically reduces 70% of the essential crops production (Kaur et al. 2008; Mantri et al. 2012; Bromham et al. 2013).

A plant reacts to a water deficiency based on the plant species, its age, its stage of development, and the duration of the water shortage and its severity (Madhava Rao et al. 2006). All plants, at different levels, have various defense mechanisms to tolerate these stresses, but it varies from plant to plant (Bieniasz 2004; Cseke et al. 2016). There are numerous approaches in plants to tolerate drought stress, like morphological, physiological, biochemical, ecological, developmental, and molecular mechanisms (Madhava Rao et al. 2006; Ashraf et al. 2009; Salehi-lisar et al. 2012).

Two keys premeditated responses of plants to stress effects could be noticed; coping skills or creating stress tolerance by the help of their metabolites as well as avoiding stress effect by physiologically inactive phases (Aertsen and Michiels 2004). Plant metabolites are very important for stresses tolerance in plants. They have many tasks, as antioxidants or defense against pathogens, signals or regulators, and well-suited solutes. The metabolites piling up in plants frequently arise under stress from different signalling molecules or elicitors because it is established that growth factors such as temperature, lightning, and providing nutritions improved the buildup of diverse natural product (Ballhorn 2011; Rini Vijayan and Raghu 2020; Chen et al. 2022). Metabolites are a plant's arsenal of adaptable building blocks for consumption and structure. They enable plants to devise strategies to defeat adversaries, displace rivals, and withstand environmental stresses. Responses of plants to stress are powerful and contain multidimensional interaction between many different regulation levels, such as expression of genes and regulation of metabolism for morphological and physiological adaption (Chen et al. 2022). Plants produce two types of metabolites, primary and secondary metabolites.

Secondary metabolites are typically synthesized to satisfy specific requirements, whereas primary metabolites most probably serve the same biological function across all species. Secondary metabolites' productivity in plants is induced by use of altered synthesis pathways employing primary metabolites or by using the same primary metabolite's source substrates (Pott et al. 2019). Water stress adaptation in plants is a complex biological process that involves changing metabolites' composition and gene expression patterns (Urano et al. 2009). It is a common practice to assess a plant resistance to drought stress by evaluating at how well it can sustain healthy levels of primary and secondary metabolites and defense mechanisms (Patel et al. 2020). They also have a main function in the plant development and growth. Metabolites are quite important in membrane formation and scaffolding, energy storage, cell signaling, and resource allocation for the whole plant under all stress types (Wen et al. 2015). Drought stress disrupts plants' metabolism through metabolic enzymes inactivation, increased demand for particular chemicals, substrate lack or a combination of these preceding, and many more causes. In order to retain vital metabolism and adapt by assuming a new steady-state amid the current conditions of stress, the metabolic network must be rebuilt (Obata and Fernie 2012).

As a result of the abiotic and biotic stress, there is variability in the secondary metabolites accumulation. This variability leads to the secondary metabolites biosynthesis (Zhi-lin et al. 2007). Environmental stresses target the synthesis of the secondary metabolites and affect the functioning of plant species. In this way, abiotic and biotic stresses play a critical role in the secondary metabolites synthesis in the plant. Plants under stress exhibit an increment biosynthesis and secondary metabolites accumulation. Plants develop a defense mechanism against stress through these secondary metabolites synthesis (Sharma and Kumar 2021). Metabolites like phenolic compounds, antioxidants, phytohormones, amino acids, compatible solutes, polyamines, and pathogen-related proteins are remarkable for plant growth, defense, and stress tolerance (Chen et al. 2022). Secondary metabolites are considered as matchless sources for flavors, industrial biochemicals and pharmaceutical products (Ravishankar and Venkataraman 1990; Ramakrishna and Ravishankar 2011; Ravishankar and Rao 2000). Plant secondary metabolites also are responsible for the certain colors, odors, and tastes in plants (Bennett and Wallsgrove 1994). The physiological tolerance of plants for adaptation is another key role for secondary metabolites (Seigler 1998). Compounds that play no important part in the preservation of a plant's life cycle are frequently known in plants as secondary metabolites (Seigler 1998; Ramakrishna and Ravishankar 2011), but they play a major role in defense and adaptation to the environmental and in coping stress conditions (Ramakrishna and Ravishankar 2011).

Plants create several secondary metabolites that are very reliant on the growing conditions and have an impact on the metabolic pathways that lead to the accumulation of related natural products. Flavonoids, phenolics, polyamines, abscisic acid (ABA), salicylic acid (SA), nitric oxide, and jasmonates (JA) are taking part in plants' response to stress (Tuteja and Sopory 2008). The physiological and morphological stages of the plants have a major impact on the production of these components, which is frequently low (Rao and Ravishankar 2002). Many reactions occur when plants are exposed to drought stress which are common in plants. Cellular dehydration caused by this stress leads to osmotic stress, which pulls water out of the cytoplasm and into vacuoles (Ramakrishna and Ravishankar 2011). Recently, secondary metabolites have recently drawn more attention due to their significant applicability in a variety of disciplines. In particular, their roles and uses in cosmetics, nutrition, medicine, and plant tolerance to stress are more notable (He and Giusti 2010; Ngo et al. 2013).

2 Secondary Metabolites and Drought Stress Protection

In general, secondary metabolites are not necessary for plant cells to function normally, but they are crucial for defending plants against biotic and abiotic stress. In response to stresses, considerable variation in secondary metabolites levels and types are produced by different plant species (Akula and Ravishankar 2011; Kusano et al. 2015; Verma and Shukla 2015). According to the estimation, the plant kingdom contains more than 100,000 secondary metabolites. These secondary metabolites are primarily categorized into three categories: nitrogen-containing compounds (alkaloids and glucosinolates), phenolics (flavonoids and phenylpropanoids), and terpenes (Fang et al. 2011; Zagorchev et al. 2013; Wink 2015). Secondary metabolites levels in plants may change in response to drought stress, ion toxicity, and other stressors (Akula and Ravishankar 2011). To combat negative physiological changes caused by stress, plants produce secondary metabolites, which are derived from their primary metabolites (Kinghorn 1994; Ramakrishna and Ravishankar 2011). Under drought stress, the induction of secondary metabolites can control redox homoeostasis, ion transport, enzyme activity, and the stiffness and turgidity of tissues and cells (Seki et al. 2007; Rodziewicz et al. 2014). There are many secondary metabolites synthesized in plant tissues, which help in protecting the plant under water stress conditions and increase the capacity of drought stress tolerance (Verma and Shukla 2015). Extreme drought stress causes a significant decrease in the activity of photosynthesis and consistently a huge increment in the secondary metabolites concentration, which includes different metabolites with relation to the species (Zobayed et al. 2007). Drought stress leads to oxidative stress and offers a great increment in the flavonoid and phenolic levels (Larson 1988). In contrast, water stress conditions also decrease the production of carotenoids, saponins, and chlorophyll *a* and *b* (Anjum et al. 2003). The accumulation of anthocyanin occurs with drought stress at very low temperature (Soliz-Guerrero et al. 2002). The concentrations of different macronutrients, essential oils, proline, and carbohydrates are affected significantly by water scarcity conditions (Khalid 2006). Plant tissues become more resistant to drought when it contains high anthocyanins content (Chalker-Scott 1999). For instance, the purple chilli may withstand stress of drought better than the green chilli (Bahler et al. 1991). Plants can be protected by flavonoids while they are under drought stress (Sharma and Kumar 2021). Environmental and abiotic factors affected saponin content (Szakiel et al. 2010). Plant reproductive organs synthesize saponins, which enhance the chemical defense and response of the plant to the different environmental conditions (Lin et al. 2009).

3 Secondary Metabolites Under Drought Stress

The main method conserving cell osmotic potential under drought stress in plants is metabolic regulation. Secondary metabolites of the plant show very important role in the ability to withstand drought, as they are involved in many metabolic pathways (Kumar et al. 2021). Secondary metabolites exhibit an extensive range of protective mechanisms and signals as a result of biotic and abiotic stresses. The understanding of secondary metabolites in plant stress physiology may also be advanced through study on the biosynthetic processes of secondary metabolites in plants. Therefore, it is very important to promote effective and creative methods to comprehend the significance of secondary metabolites in enhancing plant resistance under drought (Yadav et al. 2021). Some examples explaining the role of secondary metabolites in the tolerance of plants to water stress are discussed below and are illustrated in Fig. 26.2.

3.1 Glucosinolates

Glucosinolates, cyanogenic glucosides, indole alkaloids, nonprotein amino acids, and phytoalexins are the main sulfur and nitrogen-containing secondary metabolites. It was discovered in many plant species that these secondary metabolites

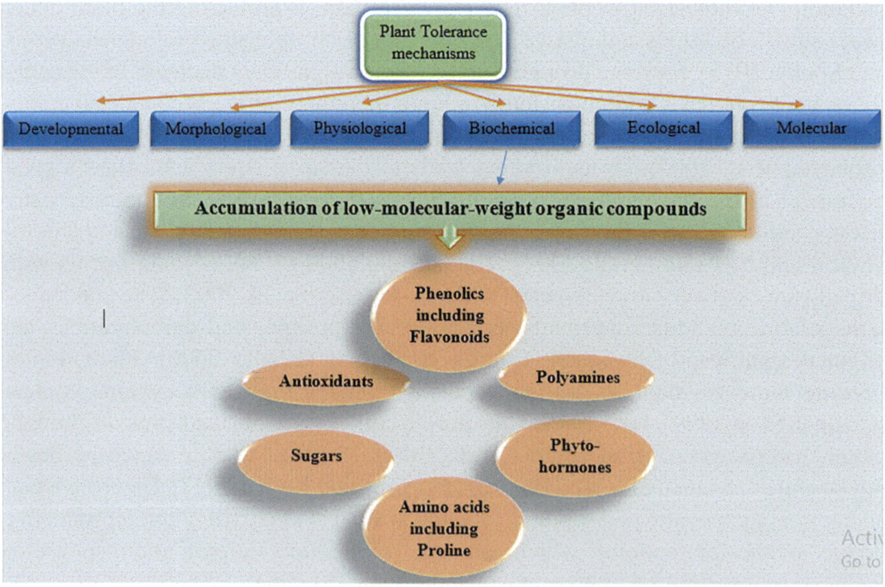

Fig. 26.2 Plant tolerance mechanisms and the main secondary metabolites' synthesis as a result of drought stress

modified as a response to drought (Kleinwachter and Selmar 2015). Glucosinolates are major secondary metabolites distributed in sixteen families within the order Brassicales. They are divided into three divisions based on the amino acid precursor: aromatic, indolic, and aliphatic glucosinolates (Sønderby et al. 2010; Abuelsoud et al. 2016). Glucosinolates are very important in the defense mechanism and in improving plants' immunity under drought conditions (Katz et al. 2015). The enzyme myrosinase breaks glucosinolates into thiohydroximate-O-sulfonates which can rearrange it and make isothiocyanates, nitriles, and related compounds (Bednarek et al. 2009). Sharma and Kumar et al. (2021) stated that the reduction in glucosinolates resulted in lower drought tolerance while the plants which have great concentration of glucosinolates generally tolerate drought stress.

Studies on the impact of drought on glucosinolate content have been conducted in a variety of plants, but with maladjusted results. According to several studies, drought stress increases the quantity of glucosinolates, notably aliphatic glucosinolates, while indolic glucosinolates were either unaffected or decreased (Radovich et al. 2005; Zhang et al. 2008; Schreiner et al. 2009; Mewis et al. 2012; Tong et al. 2014). In *Brasica juncea* leaves, reducing soil water content for 21 days from 18% to 6% increased aliphatic glucosinolates from 13.6 to 16.7 mg g^{-1}, whereas the indolic glucosinolates content was unaltered (Tong et al. 2014). Moreover, López-Berenguer et al. (2008) found that the glucosinolates in *B. oleracea* var. *italica* were increased by 1.49 μmol g^{-1} dry weight day^{-1} at low drought stress when the relative water content was less than 82%. On the other hand, Robbins et al. (2005) and Khan et al. (2010 and 2011) stated that glucosinolates in *B. oleracea* were not affected by drought stress as their content didn't change or even reduced. It was noticed in Chinese cabbage under drought stress that transcripts of glucosinolates production increased as a result of glucosinolates stomatal closure (Eom et al. 2018). The explanation given by Schreiner et al. (2009) and Khan et al. (2011) for the conflicting results was the various genotypical responses to water stress. However, Schreiner et al. (2009) have concluded that glucosinolates metabolism is related to the soil water content and have a role in the adjustment of osmotic, but several studies have stated that the physiological function played by glucosinolates response to drought remains unclear. Numerous investigations connected abscisic acid synthesis to the accumulation of glucosinolates caused by drought, showing a connected role in root-to-shoot transmission (Tong et al. 2014). Further results clarified a connection between glucosinolates and phytohormones network in response to drought stress (Malitsky et al. 2008; Morant et al. 2010; Chen et al. 2012). Generally, the role of glucosinolates and the role which they play on a molecular level in drought stress are still unclear. Hence, either glucosinolates are playing part in osmotic regulation, serving as antioxidants, or interconnection with hormone signaling; this still needs more evidence (Abuelsoud et al. 2016).

3.2 Phenolics

Phenolics are one of the main secondary metabolites in every part of plants (Lone et al. 2020). It is very crucial to accumulate phenolics to eliminate the negative effect caused in plants under water stress (Naikoo et al. 2019). Numerous metabolic investigations, particularly those done on the *Arabidopsis* plant, demonstrate that flavonoids accumulate under drought stress which provides protection against drought stress and enable plants to endure harsh conditions (Nakabayashi et al. 2014). Phenolics are aromatic compounds synthesized by phenylpropanoid pathway (Ghimire et al. 2017). Phenolics are aromatic substances which are formed by the mevalonic acid pathway or by shikimic acid pathway (in higher plants, fundamentally) (Krzyzanowska et al. 2010). Phenolics play a crucial function in plant-defenses mechanisms to pathogens and abiotic stress like salinity, drought, and UV (Lattanzio 2013; Sharma et al. 2019). An enzyme in the shikimic acid pathway called 5-enolpyruvylshikimate-3-phosphate synthase is involved in the aromatic amino acids synthesis like tyrosine and phenylalanine, the precursors in the biosynthesis of phenylpropanoid and so help in the defense mechanism (Ghimire et al. 2017). Under drought stressful conditions, phenolic compounds are highly accumulated in plants for survival by controlling the water ions flux (Mirás-Avalos and Intrigliolo 2017; González-Chavira et al. 2017). By causing the stomatal closure, these phenolic compounds prevent the plants from losing any more water (Yadav et al. 2021). Phenolics can regulate drought stress by acting as antioxidants (Nichols et al. 2015). The buildup of phenolics in the cytoplasm can detoxify the hurtful molecules of hydrogen peroxide generated in the plant by water stress (Hernandez et al. 2009; Sharma and Kumar 2021). Salicylic acid boosts plant antioxidant defenses and enhances cell metabolism (Hein et al. 2016). During drought stress, ferulic acid (a potent antioxidant) can enhance the integrity of cell membranes (Hura et al. 2012; Piasecka et al. 2017). Under abiotic stress, the flavonoids' quantity varies and could be related to the generation of reactive oxygen species (ROS) (Havaux and Kloppstech 2001). Drought increases phenolic acids due to cell walls lignification and amino acids production to maintain cell osmotic equilibrium (Ayaz et al. 2000).

3.3 Flavonoids

A major large group of naturally occurring phenolics are called flavonoids, such as flavones, flavanones, isoflavones, flavonols, chalcones, anthocyanidins, and proanthocyanidins (Mierziak et al. 2014; Panche et al. 2016). Flavonoids are paramount phytochemicals which play many novel roles in plants (Gould and Lister 2006). Under stressful conditions, plants acquire sophisticated enzymatic and nonenzymatic antioxidant defense mechanisms to prevent high ROS concentrations during their growth. Moreover, oxidative harm results from an imbalance between antioxidant defenses and ROS production (Doke 1997; Dat et al. 2000; Mahalingam and Fedoroff 2003). Increased syntheses of flavonoids have been noticed as a response to drought stress (Hernández et al. 2004). Plants that are suffering oxidative stress from drought produce ROS as a result. Commonly adapted natural substances like flavonoids and phenolics enable the plant to sweep ROS (Treml and Smejkal 2016). The position of hydroxyl groups, double carbon bonds, and modifications such as glycosylation, prenylation, and methylation influence the antioxidant properties of flavonoids (Rice-Evans et al. 1997). After the oxidation of flavonoids, there is a reconversion of flavonoids by ascorbic acid into primary metabolites, which increases the plant tolerance capacity against water stress conditions (Hernandez et al. 2009).

Flavonoids production is increased as a result of a variety of abiotic and biotic stresses, like defense mechanisms against cold, drought, and pathogens (Tattini et al. 2004; Olsen et al. 2009; Agati et al. 2011) and therefore this increment in flavonoids content may assist plants in preventing oxidative damage brought on by stress (Tattini et al. 2004; Olsen et al. 2009; Agati et al. 2011; Di Ferdinando et al. 2012). Due to the existence of molecules like the dihydroxy B-ring-substituted flavonoid glycosides, certain flavonoids can function as antioxidants and scavenge ROS when plants are exposed to oxidative stress (Di Ferdinando et al. 2012).

It has been observed that Artemisia and *Hypericum brasiliense* plants have high concentrations of artemisinin, quercetin, rutin, and betulinic acid (Verma and Shukla 2015). In addition, flavonoids were found with high content under drought in *Glechoma longituba* and high amounts of several macronutrients, proline, and essential oils were also produced significantly under water stress in some plants like *Ocimum americanum* and *O. basilicum* (Ashraf et al. 2018; Zhang et al. 2012). Caser et al. (2018) mentioned that phenolics and flavonoids increased in *Salvia sinaloensis* fern under drought stress; also Caser et al. (2019) found 139% total phenols and 101% favonoids in 50% water capacity as compared to 100% in *S. dolomitica*. It was stated that rice cultivar Q8 showed the greatest tolerance between twenty with 65.3 mg phenolics and 37.8 mg flavonoids comparing to Q2 (most susceptible to drought) which recorded 33.9 mg total phenols and 27.4 mg total flavonoids (Quan et al. 2016). Under water stress conditions, quercetin and oil decreased in *Sesamum indicum*. In contrast, total content of phenolics and favonoids as well as antioxidant activity increased in the same plant under water stress (Kermani et al. 2019). Drought-mediated synthesis of natural products such as isoprenoids, alkaloids, and total phenolics in medicinal and aromatic plants may be considered as an important tactic to increase the output of these worthy products (Kleinwachter and Selmar 2014, 2015). Jaafar et al. (2012) stated that in plants under water stress, flavonoids and total phenolics are produced more abundantly overall as well as in higher concentrations. Contrarily, total flavonoids content was almost the same whether the plants were raised under drought-stressed or well-watered circumstances.

However, drought stress conditions increased furoquinones concentrations, but total furoquinones content decreased in *S. miltiorrhiza* plants growing under water stress comparing to plants that receive adequate water (Liu et al. 2011).

3.4 Carotenoids

Nonvolatile terpenoids in plants include chlorophyll, abscisic acid, and carotenoids. During drought stress, carotenoids have the ability to act as antioxidants in plants. They serve as light collectors, scavengers of singlet oxygen species and triplate state chlorophylls, dissipators of surplus damaging energy during stressful conditions, and membrane stabilizers (Uarrota et al. 2018). Carotenoids are remarkable antioxidants; playing a direct role in reducing the harm effect of ROS resulted from drought stress (Niyogi et al. 1997; Treutter 2006). In many higher plants, carotenoids are significant indicators of drought tolerance (Farooq et al. 2009). Under stress conditions, carotenoids participate in regulating pigment system by maintaining function of chlorophyll and protecting photosynthesis against singlet oxygen (Huseynova et al. 2016). The main line of defense and protection against singlet oxygen is composed of β-carotene and α-tocopherol (Krieger-liszkay 2005). Under drought conditions, differential response of terpenoids including carotenoids was noticed in shoots and roots. Levels of terpenoids under water stress reduced in shoots while increased in roots (Kleine and Müller 2014). Sorghum root exudates were found to impede hydroxyphenyl pyruvate dioxygenase activity, which led to a deficit of plastoquinone and, thus, interfered with carotenoids biosynthesis (Meazza et al. 2002). Romagni et al. (2000) found that usnic acid inhibits the PDS enzyme's ability to convert phytoene into carotenoids. Reactive oxygen species such as superoxide radical, singlet oxygen ($^1O_2$), hydrogen peroxide, and hydroxyl radicals are produced in greater quantities during drought and waterlogging conditions (Ashraf 2012; Choudhury et al. 2013). Triplet chlorophylls, which are the sources of $^1O_2$ in leaves, can be directly quenched by the potent antioxidants known as carotenoids, as well as detoxify many types of ROS (Ramel et al. 2012). Through the direct $^1O_2$-oxidation of β-carotene and zeaxanthin and the dissipation of the extremely energetic molecular oxygen ($^1O_2$-$^3O_2$ transition), carotenoids also contribute to the quenching of $^1O_2$ (Brunetti et al. 2015). Under moderate drought, the content of carotenoids was found to decrease but slightly increase under drought stress (Sudrajat et al. 2015). Likewise, total carotenoids content in alpine plants and olive trees was increased (Doupis et al. 2013; Ramalho et al. 2014; Buchner et al. 2017). Xanthophyll cycle activation may be related to carotenoids' increment as a response to water stress (Ma et al. 2013; Mattos and Moretti 2016; Schweiggert et al. 2017). Increased content of zeaxanthin under water stress has been reported in vine berries (Savoi et al. 2016) and African eggplants (Mibei et al. 2016). The transformation of violaxanthin into antheraxanthin enables the light's surplus energy to disperse (Buchner et al. 2017). The enhancement of plant tolerance to various abiotic stresses could be also a result of the rise in zeaxanthin levels (Wu et al. 2015).

Carotenoids play a significant role in this regard for plant survival. To better understand the role of carotenoids in all types of plants stress, more direct research is required.

3.5 Terpenoids

The building blocks of terpenoids are terpenes. Monoterpenes, diterpenes, triterpenes, tetraterpenes, polyterpenes, and sterols are subgroups of terpenes, which are hydrocarbons. Isoprenoids or terpenoids are the names for compounds derived from terpenes. Terpenoids can be found in both volatile and nonvolatile forms. Additionally, nonvolatile terpenoids in plants include chlorophyll, carotenoids, and abscisic acid. These terpenes and their biosyntheses are produced during drought stress conditions (Turtola et al. 2003; Ormeno et al. 2007). Isopentenyl diphosphate and its allylic isomer (dimethylallyl diphosphate) are used to synthesize terpenoids (McGarvey and Croteau 1995). Plants synthesize these compounds by two distinct separate pathways, the plastidic methylerythritol phosphate and the cytosolic mevalonic acid pathways (Dudareva et al. 2013). Plants can store terpenoids in specific structures (Gershenzon and Croteau 1991). For instance, conifers have resinous ducts that collect terpenoids (Wu and Hu 1997) and terpenoids can account in conifer species for 1–2% of leaves' dry matter (Blanch et al. 2009).

Under low to moderate drought stress, the plants accumulate carbon that stimulates the synthesis of terpenoids and helps in the defense mechanism of plants against acute drought stress or water deficiency conditions (Llusià and Peñuelas 1998). Terpenoids' functions are decreased by increase in the time of drought stress; prolonged drought stress can decline the functioning of terpenoids whether seasonal mild to moderate water stress increases the stimulation and management of the terpenoids and helps in protecting plants under abiotic stress. Therefore, for terpenoids stimulation, severe drought stress conditions serve as a determining factor that can determine the relative reduction in the synthesis of terpenoids. There is no direct link between terpenoid synthesis and carbon assimilation shown in some plants. The carbon assimilation increases by increased drought stress and terpenoids' content depleted by prolonged drought stress and high temperature over a long time (Staudt et al. 2017; Sharma and Kumar 2021).

Several studies that showed the relationship between drought stress and terpenes concluded the following: (i) terpenoid synthesis can mitigate the impacts of drought stress by quenching reducing power, minimizing oxidative stress, and avoiding photoinhibition (Nowak et al. 2012; Selmar and Kleinwächter 2013), (ii) diterpenes promote ROS scavenging system that increases the resistance to drought (Munne-Bosch et al. 2001), (iii) drought stress was shown to activate several genes' participation in the synthesis of terpenoids (Ma et al. 2014), (iv) when there is a lack of water, phenyl-1,2-ethanediol is converted to cumin aldehyde, which may be debated in defense mechanisms (Rebey et al. 2012), (v) the level of terpenoids differs in shoot and root under drought stress, as it decreases in the shoot whereas increases in roots (Kleine and Müller 2014), (vi) the ability to reduce water stress was found to be associated with drought-induced production of carotenoids, glucosides, monoterpenes, and terpenoids (Liu et al. 2014; Savoi et al. 2016), (vii) several kinds of monoterpenes, camphor (a terpenoid), and (E, E)-α-farnesene (a sesquiterpene) are increased under water stress (Loreto et al. 2014), and (viii) otherwise, Delfne et al.

(2005) concluded that monoterpenes yields are negatively affected with water stress. Moreover, drought stress was found to affect the transcription of genes participating in the synthesis of flavonoids and terpenes (Payton et al. 2011; Zhang et al. 2014).

Terpenoids are thought to have a major impact on the ability to defend against abiotic and biotic stress (Loreto and Schnitzler 2010; Loreto et al. 2014). Terpenoids assist in conifer defense against diseases and herbivores (Martin et al. 2003, Gershenzon and Dudareva 2007; Loreto et al. 2014). Although terpenoids play a variety of well-known roles in stress reduction (Vickers et al. 2009), less investigations have been done on how different stress forms affect the quantity of terpenoids in plants. Several earlier studies investigated drought effects on the terpenoid content in conifers leaves and wood. The results were contradictory in *Pinus taeda*, *Picea abies*, and *Pinus sylvestris* nonetheless as drought led to increase in the content of monoterpenoids (Hodges and Lorio 1975; Turtola et al. 2003), while content of some monoterpenoids even in *P. taeda* (the same species) was decreased (Gilmore 1977). Similar to this, drought raised the monoterpenes' content in coniferous and deciduous plants needles such as *P. abies*, *P. halepensis*, *Quercus ilex*, *Q. coccifera*, *Rosmarinus officinalis*, and *S. officinalis* (Kainulainen et al. 1992; Llusià and Peñuelas 1998; Llusià et al. 2006; Nowak et al. 2012), but had no impact on monoterpenes' content of other species (*Cistus albidus*) (Llusià and Peñuelas 1998). Furthermore, it has been demonstrated that the terpenoid concentration of several provenances of the same conifer species can differ significantly (Von Rudloff 1972; Manninen et al. 2002) and it is unclear whether various provenances adapted to certain site conditions react to a given environmental stress in the same way. Although terpenoids in roots may be influenced equally by abiotic stress, the results of this study on the effects of drought stress on the terpenoid contents of shoots are ambiguous and this dearth of understanding on drought-induced changes in terpenoids of root is in stark contrast.

3.6 Phytohormones Under Drought Stress

Phytohormones and other secondary metabolites are the structures which act towards a main function in plants overcoming the stressed conditions and surviving in that environmental stress. Various phytohormones like jasmonic acid, abscisic acid, ethylene, cytokinins, SA, and brassinosteroids (BRs) induce the accumulation of osmolytes in plant tissues that help plants to survive under the conditions of water stress (Sharma and Kumar 2021). Phytohormones perform a major role in the physiological and biochemical mechanisms of the plants and provide a protective shield to tolerate and protect the plants from the stressful environment (Fang et al. 2011). Drought stress conditions cause changes in different plant growth hormones syntheses and alter the functioning and behavior of compounds like gibberellins, SA, abscisic acid, jasmonic acid, auxins, and cytokinin (Sharma and Kumar 2021). For example, with drought conditions, abscisic acid (the major growth hormone) is immediately produced and starts its synthesis against drought. Abscisic acid is first synthesized in roots parts and then moved to the leaves and branches for acting. ABA is important for stomatal aperture regulation, other growth activities, and expression of ABA-responsive genes (Sharma and Kumar 2021). It is generally known that under drought stress, abscisic acid concentration rises significantly. Prior researches studied the chemical reactions series involving carotenoids for abscisic acid biosynthesis and catabolism (Schwartz et al. 2003; Schroeder and Nambara 2006; Sridha and Wu 2006; Ikegami et al. 2009). Furthermore, salicylic acid content was increased significantly due to water deficiency (Munne-Bosch and Penuelas 2003).

To overcome different abiotic and biotic stress, phytohormones play the function of chemical messengers in response to stress. Plant growth hormones help in activating the different plant physiological, biochemical, and developmental processes when they get released from water stress. These plant processes contain aggregation of osmolytes, stomatal closure and enhancement growth of shoot and root to cope with drought stress (Sharma and Kumar 2021).

4 Enhancing the Natural Products Synthesis and Accumulation in Drought-Stressed Plants

Shikimic acid, mevalonic acid, phenylpropanoic acid, and methylerythritol phosphate are among the pathways for secondary metabolites (Banerjee and Sharkey 2014; Akhi et al. 2021). Plants under drought stress produce ROS, which leads to oxidative stress. Secondary metabolites mainly scavenge ROS to shield plant cells from lipid peroxidation, in addition to having a substantial impact on other defense-related activities (Treml and Smejkal 2016). Under stressful circumstances, important phenylpropanoid pathway enzymes such as phenylalanine ammonia-lyase (PAL), 4-coumaroyl CoA ligase (4CL), and coumarate-4- hydroxylase (C4H) are implicated (Bartwal et al. 2013). Secondary metabolites' accumulation enhances plants' tolerance to stress by adjusting the plant's biochemical and physiological parameters (Kabera et al. 2014). Plants generate more natural compounds and build up them faster when stressed by drought. The production and accumulation of secondary phytoproducts can be significantly influenced by nutrient supply, light, and temperature (Falk et al. 2007; Gershenzon 1984).

Drought conditions can lead to a variety of interactions between different factors; water shortages and below-freezing temperatures are accompanied together, lower water availability and high light intensity are associated mostly with increased temperatures. All these factors induce drought stress and may also entail in the soil's higher salt concentrations. Different studies demonstrated that exposing plants to drought stress can accumulate higher secondary metabolites content compared to well-watered plants. Several researches suggested that different kinds of natural products share the elevation in concentrations caused by drought stress. Corresponding increase was noticed for both simple and complicated phenolic as well as for the different terpene classes. Additionally, the concentration of compounds that contain nitrogen, like cyanogenic glucosides, alkaloids, and glucosinolates is positively influenced by water stress, too. It should be considered that plant growth decreases under drought stress; therefore, even without increasing the natural products' total amount, the reduction in biomass could have merely increased their concentration in fresh or dry weight (Kleinwächter and Selmar 2015). Many studies revealed that water stress conditions result in massive or slight increases in the secondary metabolites content such as essential oil and phenolics as well as benzylisoquinoline alkaloids (jatrorrhizine, berberine, and palmatine) as explained by Xia et al. (2007), Petropoulos et al. (2008), Nowak et al. (2010), and Manukyan (2011) and they attributed these results to the reduction in the growth of plants under drought, thus the final content of the secondary metabolites is extremely decreased compared to the well-watered plants. Kleinwächter and Selmar (2015) found that drought stress conditions considerably increase the content of plant secondary products in almost all of the examined plants except a small number of instances that described a similar rise in the natural products total content of per plant. This could be a result of the absence of information on these plants' biomass or the fact that the reduction in biomass caused by stress considerably compensates for the rise in concentrations of pertinent natural secondary metabolites.

5 Improvement of Drought Stress Tolerance by Exogenous Treatments with Metabolites

5.1 *Abscisic Acid*

The generation of abscisic acid (ABA) in roots and its transportation to the plant leaves are well known to have a main role in controlling plant water content under water stress (Zhang and Davies 1990). It was also observed that ABA played a command function in modulating the plant's adaptability to drought stress by enhancing oxygen scavenging effectiveness, organizing essential enzymes expression, and boosting sugar accumulation (Jiang and Zhang 2001; Liu et al. 2013), stomata closure controlling, and limiting leaf expansion, that way prohibiting the drying of leaf tissues and increasing the likelihood of survival under a protracted drought (Liu et al. 2005). Foliar spraying of abscisic acid improved the tolerance of plants to water stress. Shoot height as well as shoot and root dry weights increased under the conditions of drought stress by the addition of abscisic acid (Wei et al. 2015). Increment of ABA content in the root may preserve root development and enhance root hydraulic conductivity (Liu et al. 2005). Glutathione and ascorbate concentrations in leaves and roots increase significantly as a response to abscisic acid foliar application under drought stress (Li et al. 2014). Abscisic acid increased lipid peroxidation due to rising content of reducing sugars, anthocyanin, H_2O_2, polyamines, cysteine, soluble proteins, proline, photosynthetic pigments, malondialdehyde, ascorbic acid, and antioxidant enzyme activity (Hussain et al. 2010; Iqbal et al. 2010; Ruan et al. 2012). Administration of absicic acid (ABA) reduced seed weight, seed setting, and number of seeds in each spike marginally. The better performance of seedlings derived from plants' seeds that are treated with ABA under temperature stress may be attributed to high level of endogenous ABA and the improvement in antioxidant enzyme activities (Li et al. 2014), which can be anticipated to increase resistance to drought as well.

The transcript levels of the genes that code for the Ascorbate-Glutathione (AsA-GSH) pathway, also known as Asada–Halliwell pathway cycle enzymes (these genes showed various patterns of expression in the leaf, root, and stomatal conductance for reducing water loss), were also temporally modulated by abscisic acid in ABA-treated plants (Li et al. 2014).

5.2 *Methyl Jasmonate*

The methyl ester of jasmonic acid (Methyl jasmonate (MeJA)) has an advantageous effect on plants to withstand drought when applied exogenously. MeJA application reduces the detrimental effects of drought on growth by promoting photosynthesis rate, stomatal conductance, accumulation of osmolytes, and water use efficiency by enhancing antioxidant activities retarding leaves senescence (Munemasa et al. 2007; Ma et al. 2014; Sheteiwy et al. 2018; Sadeghipour 2018). Jasmonates have been linked with secondary metabolites' accumulation which comprise a part of the defensive reaction (van der Fits and Memelink 2000). Application of methyl jasmonate results in increased POD, CAT, SOD, GR, and APX activities that detoxify hydrogen peroxide collectively along with higher soluble sugar and proline content that improves drought tolerance (Wu et al. 2012). MeJA raised the content of shikonin and its derivatives, red naphthoquinonone, and boosted endogenous IAA biosynthesis in plants (Grsic et al. 1999).

5.3 *Salicylic acid*

Salicylic acid (SA) is an endogenous phenolic growth regulator that contributes in controlling physiological activities, including photosynthesis, growth, and other metabolic activities. Numerous studies back up the prominent role of SA in regulating how plants react to diverse abiotic and biotic stressors. Additionally, SA may lead to many different metabolic responses in plants as well as alter plant-water interactions (Mohamed et al. 2020). Salicylic acid application may enhance plant response to drought stress by: (i) boosting or maintaining plant growth, antioxidant activity, and Rubisco activity (Munne-Bosch and Penuelas 2003; Bechtold et al. 2010; Khokon et al. 2011; Ying et al. 2013), (ii) enhancing the plant relative water content, the rate of photosynthetic activity, the proline content, and greater activity of CAT and SOD comparing with control plants (Ying et al. 2013; Nazar et al. 2015), (iii) lowering various indicators of oxidative stress, including electrolyte leakage and lipid peroxidation, indicating that SA can help maintain the integrity of the membrane to some extent, (iv) acting like an ROS scavenger (Kang et al. 2013), (v) controlling stomatal closure that was inhibited by salicylhydroxamic acid's activity, a cell wall peroxidase inhibitor that suggested SA may have triggered stomatal closure by way of the ROS generated by cell wall peroxidases (Khokon et al. 2011; Miura et al. 2013), (vi) increasing nitrogen amounts and the activity of nitrate reductase (Singh and Usha 2003), and (vii) enhancing soluble sugars, protein content, and the stability index of the membrane coupled with grain production and yield (Khan et al. 2012).

5.4 Brassinosteroids

Brassinosteroids (a group of steroidal plant substances that are present in nature) have a vast range of biological activities (Clouse and Sasse 1998). Brassinosteroids are important for plant growth and development and they can also organize the survival of plants grown under stress conditions (Planas-Riverola et al. 2019). Brassinosteroids' shortage led to an increment in the plants drought sensitivity. It was noticed that endogenous BRs such as typhasterol and 28-norbrassinolide were greater in drought-tolerant plants compared to sensitive plants (Tůmová et al. 2018).

Foliar spraying of brassinosteroids reduced the negative effects of drought in a variety of plants by: (i) enhancing growth, photosynthetic activities, and ROS scavenging systems that improve yield, photosynthesis, pollen viability, and germination (Hu et al. 2013; Khamsuk et al. 2018; Hu et al. 2019; Kaya et al. 2019), (ii) increasing protein content, improving ROS scavenging system, and reducing lipid peroxidation (Anjum et al. 2011), (iii) prompting ABA level (Yuan et al. 2010; Jangid and Dwivedi 2017), (iv) modifying proline, soluble sugars, and antioxidant activity (Zhang et al. 2016), (v) boosting up terpenoids, abscisic acid, phenolics, and ethylene levels, and (vi) encouraging polyphenoloxidase enzymes and peroxidase activities, which are engaged in polyphenols metabolism (Fujioka and Yokota 2003).

5.5 Polyamines

Polyamines are small aliphatic organic molecules with two or more primary amino groups. Putrescine (diamine), spermine (tetramine), and spermidine (triamine) are the major polyamines found in plants (Wang et al. 2003; Alcázar et al. 2006; Yamaguchi et al. 2007; Kusano et al. 2007; Minguet et al. 2008; Alcázar et al. 2010). Metabolism of polyamines is linked with the production of ethylene, which could be important in responding to stress (Zapata et al. 2004). Polyamines are involved in stomatal closure, in addition to their osmoregulatory functions (Liu et al. 2000). Polyamines also function as free reactive oxygen radical's scavengers, as they are part of the antioxidant system (Das and Misra 2004; Kuznetsov et al. 2007). Polyamines are accumulated intensively in numerous plants when subjected to abiotic stress (Alcázar et al. 2006, 2010; Evans and Malmberg 1989; Hussain et al. 2011). Stress-tolerant plants have a great ability to promote polyamine biosynthesis under water stress (Gill and Tuteja 2010). While stress tolerance decreased due to the biosynthesis of polyamine inhibition, the simultaneous application of exogenous polyamines reverses this effect (Kumar et al. 2008).

Exogenous foliar application of polyamines is highly effective in regulating plant growth and increasing plant tolerance to different stresses. Also, plants treated with seed priming and polyamines foliar application under water stress showed more drought tolerance compared to not treated plants (Farooq et al. 2009).

5.6 Melatonin

Melatonin (N-acetyl-5-methoxytryptamine), a neurohormone generated by the pineal gland, has been found latterly in kingdom of plantae (Arnao 2006; Ramakrishna et al. 2009). It is an eco-friendly chemical molecule with strong antioxidant activity and scavenging reactive nitrogen species (RNS) and ROS (Manchester et al. 2015; Arnao and HernandezRuiz 2015; Zhang et al. 2015). Drought tolerance of both drought-sensitive and drought-tolerant plants is significantly increased by melatonin pretreatment. Melatonin application under drought stress conditions leads to better water conservation in leaves, stability in chlorophyll content, decreased electrolyte loss, and higher performance in photosynthesis (Li et al. 2015). The resulting oxidative stress from drought stress was reduced and leaf senescence was delayed by adding melatonin to the soil (Wang et al. 2013).

6 Secondary Metabolites' Engineering to Improve Plants Against Abiotic Stress

Plants produce bright secondary metabolites that are crucial for environmental adaptation and management of abiotic stresses like drought, temperature, and salinity. Secondary metabolites with high expertise in lessening abiotic stress effects include isoprenoids, carotenoids, and flavonoids. In order to combat the negative effect of abiotic stress, the biosynthesis and accumulation of these secondary metabolites change quickly under stress. As a result, secondary metabolites and plants' abilities to withstand drought stress are strongly correlated. The promise of metabolic engineering is to improve the secondary metabolites' production and thereby increase the capacity of plants to withstand various environmental stresses. Many plants have been successfully engineered to contain higher amounts of secondary metabolites, which confer the tolerance of drought stress. In this part of the chapter, a deep discussion will be presented about metabolic engineering to increase the secondary metabolites' synthesis and production that plants produce to protect them from stress and make them more resilient to it.

Abiotic stress including salinity, temperature, and drought is one of the most important difficulties for plant growth and production in arid environments. Abiotic stress, particularly drought resulting from climate changes and global warming, can result in continuous decrease in vegetative cover of Saharan and sub-Saharan areas which can negatively affect the production of economic crops (Verma and Deepti 2016) and threaten the stability of food supplies (WHO 2020). Several downstream pathways including those encoding secondary metabolites biosynthesis are stimulated by water stress. During drought, plants tend to prevent water loss that is first detected by roots as a result of dry soil. Then, cell to cell signaling will transfer this stimulus via roots to leaves through xylem, inducing secondary metabolites biosynthesis via controlling systemic phytohormone signaling (Tardieu et al. 2014; Khan et al. 2015; Yadav et al. 2021; Jogawat et al. 2021). The produced secondary metabolites can induce stomatal shutdown, alleviate oxidative stress by improving phenolic compounds' production (Anjum et al. 2017; Ibrahim et al. 2019), and produce bitter materials to protect against herbivores or guard against drought-induced microbial infections.

6.1 *Biosynthesis of Secondary Metabolites*

Primary plant metabolites are the main metabolites necessary for the growth and function of the plant likes carbohydrates, amino acids, organic acids, proteins, and lipids, whose metabolism is mainly connected by tricarboxylic acid cycle (TCA cycle) and glycolysis. However, secondary metabolites do not fundamentally participate in any metabolic process and are formed for several purposes in plants such as interindividual communication, attraction, defense, detoxification and storage of nitrogen or phosphorus, or inhibition of competitors. Moreover, stress conditions can switch the plant's metabolic nature toward the biosynthesis of secondary metabolites that could be needed to mitigate the hazards of stress and allow plant survival and improve its tolerance. According to origin and biosynthetic modules, secondary metabolites have three main scaffolds. The first one is produced by shikimic acid pathway and contains phenolics (such as phenolic acids), flavonoids, lignans, stilbenes, coumarins, and tannins; the second is originating from isoprene units via mevalonate pathway and contains mainly terpenes like mono-, di-, sester-, and triterpenes, carotenoids, majority of volatile oils components, and sterols; and the third is nitrogen-containing compounds like glucosinolates and alkaloids (Krzyzanowska et al. 2010; Nunnery et al. 2010; Jamwal et al. 2018; Isah 2019).

6.2 Secondary Metabolites Perform Several Important Functions for the Plant Under Drought Stress

Since water is the essential component in plant growth and metabolism and as it is the main vehicle for the transport of metabolites and nutrients to the plant's different parts and it is the main medium for all metabolic reactions, drought stress decreases both the turgor pressure and water potential. It subsequently modifies a number of different physiological processes, such as the biosynthesis of secondary metabolites (Ashraf et al. 2018), which can aid in the development of the plants' resistance to water stress (Verma and Shukla 2015). Plant oxidative stress rises in response to drought leading to ROS production. ROS can damage many cellular organelles by the oxidation of lipids and release of free radicles that can damage proteins and alter the functions of cell membrane affecting vital properties of the cell such as osmosis (Yang et al. 2018). Phenolic substances such as flavonoids, tannins, stilbenes, phytoalexins, and coumarins are well-known antioxidants with the ability to scavenge ROS that can alleviate oxidative stress which plants produce in excess in responding to increased oxidative stress (Treml and Smejkal 2016). Some other metabolites can promote the cellular ROS scavenging machinery to speed up the elimination of ROS such as *Salvia officinalis* diterpene (Munne-Bosch et al. 2001). Moreover, several secondary metabolites can increase energy production and adjust osmotic regulation in the plant by promoting glutamic acid-mediated proline biosynthesis (Qu et al. 2019).

In numerous instances, it was shown that water stress caused an increase in the secondary metabolites levels. Total phenolic content of *Trachyspermum ammi* was increased under drought conditions (Azhar et al. 2011). *Artemisia* and *Hypericum brasiliense* produced higher amounts of artemisinin, quercetin, rutin, and betulinic acid under drought (Verma and Shukla 2015). *Glechoma longituba* displayed increased level of flavonoids under drought stress, However, drought stress caused considerable variations in the essential oil content of *O. americanum and O. basilicum plants* (Zhang et al. 2012; Ashraf et al. 2018). *S. sinaloensis* subjected to 50% water deprivation displayed substantial increment in the flavonoids and polyphenols levels (Caser et al. 2018). Similar results were observed in *S. officinalis* and *S. dolomitica* (Caser et al. 2019). Artificially induced drought stress in seedlings of *Carum copticum* by mannitol altered the concentration of thymol from 18.7% to 20.8% and γ-terpinene from 20.0% to 24.6% in the volatile oil (Razavizadeh and Komatsu 2018). Two edible ferns showed increased resistance to drought at early stages in terms of relative leaf water decrement and a noticeable increment in phenolics, flavonoids, proanthocyanidins, enzyme activity (POD, SOD, CAT, and APX), and osmotic substances (Yanlin et al. 2019). Significant increase in total polyphenolics, flavonoids, and antioxidant activity was noticed in *Sesamum indicum* L. under water stress conditions, while a decrement in quercetin and the oil content of sesamin was observed in the same plant under the same conditions (Kermani et al. 2019). In addition, the medicinal plants' production of the important secondary metabolites under drought can be adopted as a novel approach (Kleinwachter and Selmar 2014; Kleinwachter and Selmar 2015).

6.3 Terpenoids and Drought Stress

Terpenes are mainly composed of the condensation of two or more five-carbons isoprene units via the mevalonate pathway. They can be categorized based on the building blocks number into monoterpenes, sesqui-, di-, sester-, tri-, tetra-, polyterpenes, and sterols. Lower molecular weight terpenes are the main components of plants' volatile oils. Whereas higher molecular weight terpenes such as the triterpene alcohols α-amyrin and β-amyrin are of the main components of plant waxes, tetraterpenes can be plant pigments such as β-carotene or precursors of vitamin A that serve a main function in the antioxidant system of the plant. Triterpenes and saponins can constitute the aglycone part of saponins or cardiac glycosides that have several biological activities.

Mediterranean plants growing in arid regions of North Africa and pinewood plants are producing a wide variety of volatile terpenes including monoterpene, diterpene, and sesquiterpene, whose production is stimulated under drought stress (Turtola et al. 2003; Ormeno et al. 2007). Sometimes terpene production can be increased by agricultural treatment such as spraying lemon grass by SA that enhanced the citral content, which is the main terpene responsible for the lemon odor and helped to withstand drought (Idrees et al. 2010). The yields of monoterpenes in spearmint and rosemary were decreased by drought (Delfine et al. 2005). Under drought, biosynthetic genes involved in terpene biosynthesis were observed to be overexpressed in *Cistus creticus* (Ma et al. 2014). Similarly, drought induced the genes transcription encoding the flavonoids and terpene biosynthesis in cotton and potato (Payton et al. 2011; Zhang et al. 2014). Cumin triterpenes were found to be structurally altered by drought. Furthermore, drought induces the conversion of phenyl-1,2-ethanediol to the antioxidant compound cumin aldehyde (Rebey et al. 2012). Interestingly, organ-specific response of terpene accumulation was observed in response to drought. While the level of terpenes was high in roots, it was low in shoots (Kleine and Müller 2014). Drought stress and herbivores' attack in rapeseed led to an increase in several monoterpeses like camphor and sesquiterpenes such as (*E*, *E*)-α-farnesene (Loreto et al. 2014).

6.4 Phenylpropanoids and Drought Stress

Phenylpropanoids are aromatic compounds derived from phenolic acids such as coumarins, flavonoids, lignins, tannins, etc. (Jamwal et al. 2018). Flavonoids and phenolic acid can make an important contribution to drought tolerance via ameliorating oxidative stress as previously discussed. Lignins are the main cell wall strengthening compounds with polymeric phenolic nature; they are the main structural support substances in plants and their biosynthetic genes were discovered to be overexpressed in various plants under drought. Meanwhile, several drought-resistant cultivars were engineered to produce higher amounts of lignin (Ma et al. 2014; Zhang et al. 2014; Quan et al. 2016; Savoi et al. 2016; Li et al. 2020). Accumulation of flavonoids in *Arabidopsis* was proven to be increased during drought and was found to be useful to fight drought hazards via metabolomic and transcriptomic studies (Nakabayashi et al. 2014).

Genetic modification in cells to increase the yield and production rate of a specific metabolite by remodeling or deregulating metabolic networks is known as metabolic engineering (Stephanopoulos 1999). Initially, metabolic engineering strategies relied on engineering a single enzyme regarded as rate limiting. Nowadays, the ability to transfer all or part of a biosynthetic pathway to particular cell systems is made possible through metabolic engineering (Naqvi et al. 2010). There have been numerous attempts over the last 30 years to manipulate carotenoids, monoterpene, sesquiterpene, diterpene, or triterpene syntheses (Pribat et al. 2013).

Scientific developments in the domains of molecular biology have facilitated and improved our understanding of the regulation and biosynthesis of secondary metabolites in plants. The secondary metabolites' production in the plant can be increased by metabolic engineering, which has been confirmed to be important and advantageous. In order to create plants that are resistant to environmental stresses, it is more effective to engineer numerous enzymatic processes to the same or various pathways of metabolites. Metabolic engineering is getting more significant as a means of giving plants special abilities to cope with challenging and shifting environmental conditions. Comprehensive analyses of genes and metabolic pathways are necessary to completely understand the precise functions that genes and pathways of metabolites play during abiotic stress and to enhance abiotic stress tolerance through metabolic engineering (Ganjewala et al. 2019). They come to the conclusion that additional such efforts are required to promote the plant tolerance in industrially significant crops exposed to various abiotic stresses, and they offer several extremely helpful examples for researchers to create stress-tolerant transgenic plants. Applications of genome/transcription sequencing implementation in crops that are sentient to stress could be led to the identification of novel genes for secondary metabolic pathways involved in the tolerance to abiotic stress. Additionally, the overexpression or repression of genes linked to abiotic stress may be facilitated by the cutting-edge CRISPER-Cas9 gene editing technology. This method is now used to create genetically altered sweet potatoes that can withstand abiotic stress. Although the most straightforward method for changing gene expressions is metabolic engineering, genetically modified plants have numerous complicated problems that primarily affect the general public. Numerous earlier researches have supplied insightful data on plants' abilities to manage abiotic stress in an efficient and timely manner (Lopez-Arredondo et al. 2016). The metabolic engineering of isoprenoid, flavonoids, and carotenoids biosynthesis will be covered in the section that follows.

6.5 Carotenoids

Carotenoids, like flavonoids, are experts at mediating abiotic stresses; as a result, metabolic engineering has revealed tremendous potential in plants' abiotic stress tolerance. In spite of that, only little efforts have been made to date to manipulate the biosynthesis of carotenoids in few plants, containing sweet potatoes, tobacco, and *Arabidopsis*. However, because sweet potatoes are a highly worthy industrial crop and an excellent antioxidant carotenoids source, the most extensive work on metabolic engineering of the biosynthesis of carotenoids has been done on this plant (Chen et al. 2011; Wang et al. 2016). Other metabolites derived from carotenoids, such as ABA, are well-known for controlling a number of procedures, including the response of plants to drought. Small soluble protein receptors known as PYR/PYL/RCAR (pyrabactin resistance1/PYR1-like/regulatory component of ABA receptor) are found in the cytoplasm and nucleus of plant cells and bind to ABA (Ganjewala et al. 2019). According to studies, constituent ABA receptor overexpression in plants may greatly increase their ability to withstand drought. Exact modulation of the activity of one or more ABA receptors is essential to increase drought tolerance without decreasing plant production. However, overexpressing ABA may have the potential to negatively impact yield in non-stressed plants. Using an engineered PYR1 ABA-receptor that can be activated by nonherbicidal agrochemicals and inducers of ABA responses, transgenic *Arabidopsis* with drought tolerance has been produced. (Lopez-Arredondo et al. 2016).

6.6 Isoprenoids

In all aspects of life, the biosynthesis of isoprenoid is a well-known biochemical pathway. It can generally be broken down into three stages. The universal C5 building blocks iso-pentenyl diphosphate (IPP) and dimethylallyl diphosphate (DMAPP) are synthesized in the early stage via the mevalonate (MVA) and methylerythritol phosphate (MEP) pathways. In order to create prenyl diphosphates of various lengths, the isoprene units linearly condense during the central stage. Prenyl diphosphates are distributed to various branches of the pathway during the late stage, where they are further altered to produce final products such as sterols, carotenoids, etc. (Vranová et al. 2013). All necessary components for isoprenoid biosynthesis are present in natural terpenoid producers, including the complex subcellular organization of this metabolic pathway and storage capacity at the subcellular or tissue level. Additionally, plants' virtually limitless carbon reserves, guaranteed by photosynthesis, make them a practical substitute for industrial incubators for the microorganisms' production of isoprenoids. The study of plant isoprenoid metabolism is a rapidly developing field of study (Pribat et al. 2013). Plants' physiological adaptation to high salinity and low temperature is aided by isoprenoids. Isoprenoids in alfalfa play significant functions during drought (Soto et al. 2011). Alfalfa is particularly sensitive to abiotic stress, so its yield is 50% lower when subjected to osmotic stress. The acetate-MVA pathway's thiolase-II enzyme, also known as MsAACT, has been cloned to control the biosynthesis of isoprenoid during salinity stress adaptation in alfalfa (Soto et al. 2011). In salt-stress conditions, transgenic alfalfa overexpressing MsAACT1 generated more squalene without affecting 3-hydroxy-3-methylglutaryl-CoA reductase (HMGR) activity. Mevastatin and vitamin C (a natural antioxidant) have been shown to inhibit the acetate-MVA pathway, proposing that this pathway is essential for the abiotic stress-induced antioxidant system in plants.

6.7 Flavonoids

A number of researchers have successfully tried to engineer the pathways that produce flavonoids (Butelli et al. 2008 and 2012; Nakabayashi et al. 2014). Numerous trials have shown a connection between plant tolerance to UV-B radiation and flavonoid accumulation (Stracke et al. 2010; Kusano et al. 2011). However, it is still uncertain how exactly flavonoids help the body deal with UV-B radiation. According to Butelli et al. (2008, 2012), it is possible to modify the genes expression that make flavonoids in *Arabidopsis* and other plants. Thirty-five genes have been found to date that encode enzymes or transcriptionally engaged in the anthocyanins and flavonoids production (Yonekura-Sakakibara et al. 2008). Metabolome and transcriptome profiles' analysis of wild-type during drought and oxidative stresses, single MYB12/PFG1 or MYB75/PAP1 overexpressors (which produce flavonol glycosides or anthocyanin pigment, respectively), double MYB12 and PAP1 overexpressors, and transparent testa4 as a flavonoid-deficient mutant and flavonoid-deficient MYB12 or PAP1 overexpressing lines (obtained by crossing tt4 and the specific MYB over expresser) have disclosed the flavonoids piling up. This overabundance was the main factor in the mutant's increased tolerance to these stresses. (Nakabayashi et al. 2014).

The anthocyanins cumulation in plants is also related to the ability of the plant to withstand salinity. Kim et al. (2012) displayed good anthocyanin gathering in *Brassica napus* by overexpressing *A. thaliana* dihydroflavonol-4-reductase (AtDFR) to significantly improve the tolerance of salinity. The development of *Brassica* species resistance to stress is highly eligible, given the continuously growing cultivation area worldwide. It may be highly beneficial in other commercially significant plants, such as *Brassica*, to modify and/or confer salinity and drought stress resistance through the overexpression of the *AtDFR* gene. It is not yet obvious how the cumulation of flavonoids increases plant stress tolerance because of the limited knowledge we presently have of the biochemical and molecular pathways underpinning their occurrence in stressed plants.

7 Conclusion and Future Prospects

The demand for food increases due to the steady increase in the population. On the contrary, arable lands and available water suffer from a decrease as resulted from global water stress factors and an increment in the causes of environmental pressures, which constitute an increasing challenge for food production. The primary global factor limiting crop yield is drought. Secondary metabolites serve many functions in plant life, including protecting plants from abiotic and biotic stress. There is significant variation in both secondary metabolites levels and types as a response to stress across plant species. Nearly 100,000 secondary metabolites are thought to exist in the kingdom of plantae. The three main divisions of secondary metabolites are phenolics (flavonoids and phenylpropanoids), terpenes, and nitrogen-containing compounds containing nitrogen (alkaloids and glucosinolates). Secondary metabolite levels in plants can change as a result of abiotic stress and other stressors. To resist the harmful physiological changes brought on by stress, plants develop secondary metabolites, which are derivatives of their primary metabolites. In order to thrive and tolerate stress, plants have developed a complex network of defense mechanisms. One well-known method of plant resistance to drought stress is the buildup of appropriate solutes to enable osmotic adaptation. Osmolytes can build up in high concentrations due to their low molecular weight without interfering with metabolism or cellular components. Solute accumulation activates soil water uptake and raises cellular osmotic pressure. During water stress, they might also take the place of water molecules surrounding membranes, proteins (including enzymes), and nucleic acids. By dislodging water molecules surrounding cellular components and preventing the instability of vital macromolecules, compatible solutes may inhibit cellular components from interacting with ions (at high concentrations). It is evident that foliar spraying of several chemicals such as ABA, ABA, MeJA, SA, melatonin, Pas, and BRs can increase the tolerance of plants experiencing water stress. As a result, secondary metabolites and plants' ability to withstand drought stress are strongly correlated. Drought resistance could be achieved by modifying the SMs' biosynthesis pathway genes. The promise of metabolic engineering is to raise the secondary metabolites' productivity and thereby increase the capacity of plants to withstand various environmental stresses. Many plants have been successfully engineered to contain higher secondary metabolites content, which confer drought stress tolerance. The quality and quantity of different crops are adversely impacted by intense drought conditions, leading to higher food prices. Therefore, new environmentally sustainable agricultural practices must be followed to overcome these issues.

References

Abuelsoud W, Hirschmann F, Papenbrock J (2016) Sulfur metabolism and drought stress tolerance in plants. In: Hossain MA, Bhattacharjee S, Tran L-SP, Wani SH, Burritt DJ (eds) drought stress tolerance in plants. Physiology and biochemistry. Springer Nature, pp 227–249

Aertsen A, Michiels CW (2004) Stress and how bacteria cope with death and survival. Crit Rev Microbiol 30:263–273

Agati G, Biricolti S, Guidi L, Ferrini F, Fini A, Tattini M (2011) The biosynthesis of flavonoids is enhanced similarly by UV radiation and root zone salinity in *L. vulgare* leaves. J Plant Physiol 168:204–212

Akhi MZ, Haque MM, Biswas MS (2021) Role of secondary metabolites to attenuate stress damages in plants. In: Waisundara V (ed) Antioxidants. IntechOpen, London

Akula R, Ravishankar GA (2011) Influence of abiotic stress signals on secondary metabolites in plants. Plant Signal Behav 6:1720–1731

Alcázar R, Marco F, Cuevas JC, Patrón M, Ferrando A, Carrasco P, Tiburcio AF, Altabella T (2006) Involvement of polyamines in plant response to abiotic stress. Biotechnol Lett 28:1867–1876

Alcázar R, Altabella T, Marco F, Bortolotti C, Reymond M, Koncz C, Carrasco P, Tiburcio AF (2010) Polyamines: Molecules with regulatory functions in plant abiotic stress tolerance. Planta 231:1237–1249

Anjum F, Yaseen M, Rasool E, Wahid A, Anjum S (2003) Water stress in barley (*Hordeum vulgare* L.) I. effect on morphological characters. Pak J Agri Sci 40(1-2):43–44

Anjum S, Wang L, Farooq M, Hussain M, Xue L, Zou C (2011) Brassinolide application improves the drought tolerance in maize through modulation of enzymatic antioxidants and leaf gas exchange. J Agron Crop Sci 197(3):177–185

Anjum SA, Ashraf U, Tanveer M, Khan I, Hussain S, Shahzad B, Zohaib A, Abbas F, Saleem MF, Ali I (2017) Drought induced changes in growth, osmolyte accumulation and antioxidant metabolism of three maize hybrids. Front Plant Sci 8:69. https://doi.org/10.3389/fpls.2017.00069

Arbona V, Manzi M, de Ollas C, Gómez-Cadenas A (2013) Metabolomics as a tool to investigate abiotic stress tolerance in plants. Int J Mol Sci 14:4885–4911

Arnao MB, Hernandez-Ruiz J (2006) The physiological function of melatonin in plants. Plant Signal Behav 3:89–95

Arnao MB, Hernandez-Ruiz J (2015) Functions of melatonin in plants: a review. J Pineal Res 59:133–150

Ashraf MA (2012) Waterlogging stress in plants: a review. Afr J Agri Res 7:1976–1981

Ashraf M, Ozturk M, Athar HR (2009) Salinity and water stress, improving crop efficiency. Springer, The Netherlands, p 243

Ashraf MA, Iqbal M, Rasheed R, Hussain I, Riaz M, Arif MS (2018) Environmental stress and secondary metabolites in plants: an overview. In: Ahmad P, Ahanger MA, Singh VP, Tripathi DK, Alam P, Alyemen MN (eds) . Elsevier, Plant metabolites and regulation under environmental stress, pp 153–167

Ayaz F, Kadioglu A, Turgut R (2000) Water stress effects on the content of low molecular weight carbohydrates and phenolic acids in *Ctenanthe setosa* (Rosc.) Eichler. Can J Plant Sci 80:373–378

Azhar N, Hussain B, Ashraf MY, Abbasi KY (2011) Water stress mediated changes in growth, physiology and secondary metabolites of desi ajwain (*Trachyspermum ammi* L.). Pak J Bot 43:15–19

Bahler BD, Steffen KL, Orzolek MD (1991) Morphological and biochemical comparison of a purple-leafed and a green-leafed pepper cultivar. HortScience 26:736

Bal W, Kozlowski H, Robbins R, Pettit LD (2010) Improving drought tolerance by exogenous application of glycine betaine and salicylic acid in sunflower. J Agron Crop Sci 194:193–199

Ballhorn DJ (2011) Constraints of simultaneous resistance to a fungal pathogen and an insect herbivore in lima bean (*Phaseolus lunatus* L.). J Chem Ecol 37:141–144

Banerjee A, Sharkey TD (2014) Methylerythritol 4-phosphate (MEP) pathway metabolic regulation. Nat Prod Rep 31:1043–1055
Bartwal A, Mall R, Lohani P, Guru SK, Arora S (2013) Role of secondary metabolites and brassinosteroids in plant defense against environmental stresses. J Plant Growth Regul 32:216–232
Bechtold U, Lawson T, Mejia-Carranza J, Meyer RC, Brown IR, Altmann T, Ton J, Mullineaux PM (2010) Constitutive salicylic acid defences do not compromise seed yield, drought tolerance and water productivity in the Arabidopsis accession C24. Plant Cell Environ 33:1959–1973
Bednarek P, Piślewska-Bednarek M, Svatoš A, Schneider B, Doubský J, Mansurova M, Humphry M, Consonni C, Panstruga R, Sanchez-Vallet A, Molina A, SchulzeLefert P (2009) A glucosinolate metabolism pathway in living plant cells mediates broad-spectrum antifungal defense. Science 323(5910):101–106
Bennett RN, Wallsgrove RM (1994) Secondary metabolites in plant defence mechanisms. New Phytol 127:617–633
Bhargava S, Sawant K (2013) Drought stress adaptation: metabolic adjustment and regulation of gene expression. Plant Breed 132:21–32
Bieniasz PD (2004) Intrinsic immunity: a front-line defense against viral attack. Nat Immunol 5:1109–1115
Blanch JS, Peñuelas J, Sardans J, Llusià J (2009) Drought, warming and soil fertilization effects on leaf volatile terpene concentrations in Pinus halepensis and Quercus ilex. Acta Physiol Plant 31:207–218
Bray EA (2001) Plant response to water-deficit stress. Encycl Life Sci. https://doi.org/10.1002/9780470015902.a0001298.pub2
Brunetti C, Guidi L, Sebastiani F, Tattini M (2015) Isoprenoids and phenylpropanoids are key components of the antioxidant defense system of plants facing severe excess light stress. Environ Exp Bot 119:54–62
Bromham L, Saslis-Lagoudakis CH, Bennett TH, Flowers TJ (2013) Soil alkalinity and salt tolerance: adapting to multiple stresses. Biol Lett 9(5):20130642
Buchner O, Roach T, Gertzen J, Schenk S, Karadar M, Stöggl W, Miller R, Bertel C, Neuner G, Kranner I (2017) Drought affects the heat-hardening capacity of alpine plants as indicated by changes in xanthophyll cycle pigments, singlet oxygen scavenging, α-tocopherol and plant hormones. Environ Exp Bot 133:159–175
Butelli E, Titta L, Giorgio M et al (2008) Enrichment of tomato fruit with health-promoting anthocyanins by expression of select transcription factors. Nat Biotechnol 26:1301–1308
Butelli E, Licciardello C, Zhang Y, Liu J, Mackay S, Bailey P, Reforgiato-Recupero G, Martin C (2012) Retrotransposons control fruit-specific, cold-dependent accumulation of anthocyanins in blood oranges. Plant Cell 24:1242–1255
Caser M, D'Angiolillo F, Chitarra W, Lovisolo C, Ruffoni B, Pistelli L, Pistelli L, Scariot V (2018) Ecophysiological and phytochemical responses of *Salvia sinaloensis* fern to drought stress. Plant Growth Regul 84:383–394
Caser M, Chitarra W, D'Angiolillo F, Perrone I, Demasi S, Lovisolo C, Pistelli L, Pistelli L, Scariot V (2019) Drought stress adaptation modulates plant secondary metabolite production in *Salvia dolomitica* Codd. Ind Crop Prod 129:85–96
Cattivelli L, Rizza F, Badeck FW, Mazzucotelli E, Mastrangelo AM, Francia E, Marè C, Tondelli A, Stanca AM (2008) Drought tolerance improvement in crop plants: an integrated view from breeding to genomics. Field Crop Res 105:1–14
Chalker-Scott L (1999) Environmental significance of anthocyanins in plant stress responses. Photochem Photobiol 70:1–9
Chen X, Han H, Jiang P, Nie L, Bao H, Fan P, Lv S, Feng J, Li Y (2011) Transformation of blycopene cyclase genes from *Salicornia europaea* and *Arabidopsis* conferred salt tolerance in Arabidopsis and Tobacco. Plant Cell Physiol 52(5):909–921
Chen Y, Pang Q, He Y, Zhu N, Branstrom I, Yan X, Chen S (2012) Proteomics and metabolomics of *Arabidopsis responses* to perturbation of glucosinolate biosynthesis. Mol Plant 5(5):1138–1150

Chen D, Mubeen B, Hasnain A, Rizwan M, Adrees M, Naqvi SAH, Iqbal S, Kamran M, El-Sabrout AM, Elansary HO, Mahmoud EA, Alaklabi A, Sathish M, Ud Din GM (2022) Role of promising secondary metabolites to confer resistance against environmental stresses in crop plants: current scenario and future perspectives. Front Plant Sci 13:Article 881032

Choudhury S, Panda P, Sahoo L, Panda SK (2013) Reactive oxygen species signaling in plants under abiotic stress. Plant Signal Behav 8:e23681

Clouse SD, Sasse JM (1998) Brassinosteroids: essential regulators of plant growth and development. Annu Rev Plant Physiol Plant Mol Biol 49:427–451

Cseke LJ, Kirakosyan A, Kaufman PB, Warber S, Duke JA, Brielmann HL (2016) Natural products from plants, 2nd edn. CRC Press, Boca Raton FL, p 632

Das KC, Misra HP (2004) Hydroxyl radical scavenging and singlet oxygen quenching properties of polyamines. Mol Cell Biochem 262:127–133

Dat J, Vandenabeele S, Vranova E, Van Montagu M, Inze D, Van Breusegem F (2000) Dual action of the active oxygen species during plant stress responses. Cell Mol Life Sci 57:779–795

Delfine S, Loreto F, Pinelli P, Tognetti R, Alvino A (2005) Isoprenoids content and photosynthetic limitations in rosemary and spearmint plants under water stress. Agric Ecosyst Environ 106:243–252

Di Ferdinando M, Brunetti C, Fini A, Tattini M (2012) Flavonoids as antioxidants in plants under abiotic stresses. In: Ahmad P, Prasad M (eds) Abiotic stress responses in plants. Springer, New York, pp 159–179

Doke N (1997) The oxidative burst: Role in signal transduction and plant stress. In: Scandalios SG (ed) Oxidative stress and the molecular biology of antioxidant defences. Cold Spring Harbor Laboratory Press, New York, pp 785–813

Doupis G, Bertaki M, Psarras G, Kasapakis I, Chartzoulakis K (2013) Water relations, physiological behavior and antioxidant defence mechanism of olive plants subjected to different irrigation regimes. Scient Hortic 153:150–156

Dudareva N, Klempien A, Muhlemann JK, Kaplan I (2013) Biosynthesis, function and metabolic engineering of plant volatile organic compounds. New Phytol 198(1):16–32

Eom SH, Baek S-A, Kim JK, Hyun TK (2018) Transcriptome analysis in Chinese cabbage (*Brassica rapa* ssp. pekinensis) provides the role of glucosinolate metabolism in response to drought stress. Molecules 23:1186

Evans PT, Malmberg RL (1989) Do polyamines have roles in plant development? Annu Rev Plant Physiol Plant Mol Biol 40:235–269

Falk KL, Tokuhisa, JG, Gershenzon J (2007) The effect of sulfur nutrition on plant glucosinolate content: physiology and molecular mechanisms. Plant Biol 9:573–581

Fang X, Yang C, Wei Y, Ma Q, Yang L, Chen X (2011) Genomics grand for diversified plant secondary metabolites. Plant Divers Resour 33:53–64

Farooq M, Wahid A, Kobayashi N, Fujita D, Basra SMA (2009) Plant drought stress: effects, mechanisms and management. Agron Sustain Dev 29:185–212

Fujioka S, Yokota T (2003) Biosynthesis and metabolism of brassinosteroids. Annu Rev Plant Biol 54:137–164

Ganjewala D, Gurminder K, Nidhi S (2019) Metabolic engineering of stress protectant secondary metabolites to confer abiotic stress tolerance in plants. In: Singh SP, Upadhyay SK, Pandey A, Kumar S (eds) Molecular approaches in plant biology and environmental challenges. Springer, pp 207–227

Gershenzon J (1984) Changes in the levels of plant secondary metabolites under water and nutrient stress. Recent Adv Phytochem 18:273–320

Gershenzon J, Croteau R (1991) Terpenoids. In: Rosenthal GA, Berenbaum MR (eds) Herbivores: Their Interactions with Secondary Plant Metabolites, 2nd Edition. Academic Press, Orlando, Florida, pp 165–219

Gershenzon J, Dudareva N (2007) The function of terpene natural products in the natural world. Nat Chem Biol 3:408–414

Ghimire BK, Seong ES, Yu CY, Kim SH, Chung IM (2017) Evaluation of phenolic compounds and antimicrobial activities in transgenic *Codonopsis lanceolata* plants via overexpression of the γ-tocopherol methyltransferase (γ-tmt) gene. S Afr J Bot 109:25–33
Gill SS, Tuteja N (2010) Polyamines and abiotic stress tolerance in plants. Plant Signal Behav 5:26–33
Gilmore AR (1977) Effects of soil moisture stress on monoterpenes in loblolly pine. J Chem Ecol 3:667–676
González-Chavira MM, HerreraHernández MG, Guzmán-Maldonado H, Pons-Hernández JL (2017) Controlled water deficit as abiotic stress factor for enhancing the phytochemical content and adding-value of crops. Sci Hortic 234:354–360
Gould KS, Lister C (2006) Flavonoid functions in plants. In: Andersen ØM, Markham KR (eds) Flavonoids: chemistry, biochemistry and applications. CRC Press, Boca Raton, pp 397–441
Grsic S, Kirchheim B, Piepe K, Fritsch M, Hilgenberg W, Ludwig-Mueller J (1999) Induction of auxin biosynthetic enzymes by jasmonic acid and in clubroot diseased Chinese cabbage plants. Physiol Plant 105:521–531
Havaux M, Kloppstech K (2001) The protective functions of carotenoid and flavonoid pigments against excess visible radiation at chilling temperature investigated in *Arabidopsis npq* and tt mutants. Planta 213:953–966
He J, Giusti MM (2010) Anthocyanins: natural colorants with health-promoting properties. Annu Rev Food Sci Technol 1:163–187
Hein JA, Sherrard ME, Manfredi KP, Abebe T (2016) The fifth leaf and spike organs of barley (*Hordeum vulgare* L.) display different physiological and metabolic responses to drought stress. BMC Plant Biol 16:248
Hernández I, Alegre L, Munné-bosch S (2004) Drought-induced changes in flavonoids and other low molecular weight antioxidants in *Cistus clusii* grown under Mediterranean field conditions. Tree Physiology 24:1303–1311
Hernández I, Alegre L, Van Breusegem F, Munné-Bosch S (2009) How relevant are flavonoids as antioxidants in plants? Trends Plant Sci 14(3):125–132
Hirt H, Shinozaki K (2004) Plant responses to abiotic stress. Springer, Germany, p 315
Hodges JD, Lorio PL (1975) Moisture stress and composition of xylem oleoresin in loblolly pine. Forest Sci 21:283–290
Hossain MA, Bhattacharjee S, Tran L-SP, Wani SH, Burritt DP (2016) Drought stress tolerance in plants, physiology and biochemistry. Springer Nature, p 538
Hu WH, Yan XH, Xiao YA, Zeng JJ, Qi HJ, Ogweno JO (2013) 24-Epibrassinosteroid alleviate drought-induced inhibition of photosynthesis in *Capsicum annuum*. Sci Hortic 150:232–237
Hu WH, Yan XH, He Y, Xi R (2019) 24-epibrassinolide alleviate drought-induced photoinhibition in *Capsicum annuum* via upregulation of AOX pathway. Sci Hortic 243:484–489
Hura T, Hura K, Dziurka K, Ostrowska A, Baczek-Kwinta R, Grzesiak M (2012) An increase in the content of cell wall-bound phenolics correlates with the productivity of triticale under soil drought. J Plant Physiol 169:1728–1736
Huseynova IM, Rustamova SM, Aliyeva DR, Babayev HG, Aliyev JA (2016) Photosynthesis, antioxidant protection and drought tolerance in plants. In: Hossain MA, Bhattacharjee S, Tran L-SP, Wani SH, Burritt DJ (eds) Drought stress tolerance in plants, physiology and biochemistry. Springer Nature, pp 349–378
Hussain S, Saleem M, Ashraf M, Cheema M, Haq M (2010) Abscisic acid, a stress hormone helps in improving water relations and yield of sunflower (*Helianthus annuus* L.) hybrids under drought. Pak J Bot 42(3):2177–2189
Hussain SS, Ali M, Ahmad M, Siddique KHM (2011) Polyamines: natural and engineered abiotic and biotic stress tolerance in plants. Biotechnol Adv 29:300–311
Ibrahim W, Zhu Y-M, Chen Y, Qiu C-W, Shuijin Z, Wu F (2019). Genotypic differences in leaf secondary metabolism, plant hormones and yield under alone and combined stress of drought and salinity in cotton genotypes. Physiol Plant 165(2):343–355
Idrees M, Khan MMA, Aftab T, Naeem M, Hashmi N (2010) Salicylic acid-induced physiological and biochemical changes in lemongrass varieties under water stress. J Plant Interact 5:293–303

Ikegami K, Okamoto M, Seo M, Koshiba T (2009) Activation of abscisic acid biosynthesis in the leaves of *Arabidopsis thaliana* in response to water deficit. J Plant Res 122(2):235–243
Iqbal S, Bano A, Ilyas N (2010) Drought and abscisic acid (ABA) induced changes in protein and pigment contents of four wheat (*Triticum aestivum* L.) accessions. J Agric Res 48(1):1–13. https://doi.org/10.21608/ZJAR.2021.165605
Isah T (2019) Stress and defense responses in plant secondary metabolites production. Biol Res 52:39
Jaafar HZ, Ibrahim MH, Fakri NFM (2012) Impact of soil field water capacity on secondary metabolites, phenylalanine ammonia lyase (PAL), maliondialdehyde (MDA) and photosynthetic responses of Malaysian kacip fatimah (*Labisia pumila* Benth). Molecules 17(6):7305–7322
Jaleel CA, Manivannan P, Wahid A, Farooq M, Somasundaram R, Panneerselvam R (2009) Drought stress in plants: a review on morphological characteristics and pigments composition. Int J Agric Biol 11:100–105
Jamwal K, Bhattacharya S, Puri S (2018) Plant growth regulator mediated consequences of secondary metabolites in medicinal plants. J Appl Res Med Aromat Plant 9:26–38
Jangid KK, Dwivedi P (2017) Physiological and biochemical changes by nitric oxide and brassinosteroid in tomato (*Lycopersicon esculentum* mill.) under drought stress. Acta Physiol Plant 39(3):73
Jiang M, Zhang J (2001) Effect of abscisic acid on active oxygen species, antioxidative defence system and oxidative damage in leaves of maize seedlings. Plant Cell Physiol 42(11):1265–1273
Jogawat A, Yadav B, Chhaya LN, Singh AK, Narayan OP (2021) Crosstalk between phytohormones and secondary metabolites in the drought stress tolerance of crop plants: a review. Physiol Plant 172:1106–1132
Kabera JN, Semana E, Mussa AR, He X (2014) Plant secondary metabolites: Biosynthesis, classification, function and pharmacological properties. J Pharm Pharmacol 2:377–392
Kainulainen P, Oksanen J, Palomäki V, Holopainen JK, Holopainen T (1992) Effect of drought and waterlogging stress on needle monoterpenes of *Picea abies*. Can J Bot 70:1613–1616
Kang GZ, Li GZ, Liu GQ, Xu W, Peng XQ, Wang CY, Zhu YJ, Guo TC (2013) Exogenous salicylic acid enhances wheat drought tolerance by influence on the expression of genes related to ascorbate-glutathione cycle. Biol Plant 57:718–724
Katz E, Nisani S, Yadav BS, Woldemariam MG, Shai B, Obolski U, Ehrlich M, Shani E, Jander G, Chamovitz DA (2015) The glucosinolate breakdown product indole-3-carbinol acts as an auxin antagonist in roots of *Arabidopsis thaliana*. Plant J 82(4):547–555
Kaur G, Kumar S, Nayyar H, Upadhyaya HD (2008) Cold stress injury during the pod-filling phase in chickpea (Cicer arietinum L.): effects on quantitative and qualitative components of seeds. J Agron Crop Sci 194(6):457–464
Kaya C, Ashraf M, Wijaya L, Ahmad P (2019) The putative role of endogenous nitric oxide in brassinosteroid-induced antioxidant defence system in pepper (*Capsicum annuum* L.) plants under water stress. Plant Physiol Biochem 143:119–128
Kermani SG, Saeidi G, Sabzalian MR, Gianinetti A (2019) Drought stress infuenced sesamin and sesamolin content and polyphenolic components in sesame (*Sesamum indicum* L.) populations with contrasting seed coat colors. Food Chem 289:360–368
Khalid KA (2006) Influence of water stress on growth, essential oil, and chemical composition of herbs [*Ocimum* sp.]. Int Agrophys 20(4):289–296
Khamsuk O, Sonjaroon W, Suwanwong S, Jutamanee K, Suksamrarn A (2018) Effects of 24-epibrassinolide and the synthetic brassinosteroid mimic on chili pepper under drought. Acta Physiol Plant 40(6):106
Khan MA, Ulrichs C, Mewis I (2010) Influence of water stress on the glucosinolate profile of *Brassica oleracea* var. *italica* and the performance of *Brevicoryne brassicae* and *Myzus persicae*. Entomol Exp Appl 137(3):229–236
Khan MAM, Ulrichs C, Mewis I (2011) Water stress alters aphid-induced glucosinolate response in *Brassica oleracea* var. *italica* differently. Chemoecology 21(4):235–242
Khan SU, Asghari B, Gurmani A (2012) Abscisic acid and salicylic acid seed treatment as potent inducer of drought tolerance in wheat (*Triticum aestivum* L.). Pak J Bot 44:43–49

Khan MIR, Asgher M, Fatma M, Per TS, Khan NA (2015) Drought stress vis a vis plant functions in the era of climate change. Clim Chan Environ Sustain 3:13–25

Khokon MDAR, Okuma E, Hossain MA, Munemasa S, Uraji M, Nakamura Y, Mori IC, Murata Y (2011) Involvement of extracellular oxidative burst in salicylic acid-induced stomatal closure in *Arabidopsis*. Plant Cell Environ 34:434–443

Kim SH, Ahn YO, Ahn MJ, Lee HS, Kwak SS (2012) Down-regulation of β-carotene hydroxylase increases β-carotene and total carotenoids enhancing salt stress tolerance in transgenic cultured cells of sweet potato. Phytochem 74:69–78

Kinghorn AD (1994) The discovery of drugs from higher plants. Biotechnol 26:81–108

Kleine S, Müller C (2014) Drought stress and leaf herbivory affect root terpenoid concentrations and growth of *Tanacetum vulgare*. J Chem Ecol 40:1115–1125

Kleinwachter M, Selmar D (2014) Influencing the product quality by applying drought stress during the cultivation of medicinal plants. In: Ahmad P, Wani MR (eds) . Springer, Physiological mechanisms and adaptation strategies in plants under changing environment, pp 57–73

Kleinwachter M, Selmar D (2015) New insights explain that drought stress enhances the quality of spice and medicinal plants: potential applications. Agron Sustain Dev 35:121–131

Krieger-Liszkay A (2005) Singlet oxygen production in photosynthesis. J Exp Bot 56(411):337–346

Krzyzanowska J, Czubacka A, Oleszek W (2010) Dietary phytochemicals and human health. In: Giardi MT, Rea G, Berra B (eds) Bio-farms for nutraceuticals. Springer, pp 74–98

Kumar V, Giridhar P, Chandrashekar A, Ravishankar GA (2008) Polyamines influence morphogenesis and caffeine biosynthesis in in vitro cultures of *Coffea canephora* P ex Fr. Acta Physiol Plant 30:217–223

Kumar M, Patel MK, Kumar N, Bajpai AB, Siddique KHM (2021) Review metabolomics and molecular approaches reveal drought stress tolerance in plants. Int J Mol Sci 22:9108

Kusano T, Yamaguchi K, Berberich T, Takahashi Y (2007) Advances in polyamine research in 2007. J Plant Res 120:345–350

Kusano M, Tohge T, Fukushima A, Kobayashi M, Hayashi N, Otsuki H, Kondou Y, Goto H, Kawashima M, Matsuda F, Niida R, Matsui M, Saito K, Fernie AR (2011) Metabolomics reveals comprehensive reprogramming involving two independent metabolic responses of *Arabidopsis* to UV-B light. Plant J 67:354–369

Kusano M, Yang Z, Okazaki Y, Nakabayashi R, Fukushima A, Saito K (2015) Using metabolomic approaches to explore chemical diversity in rice. Mol Plant 8:58–67

Kuznetsov V, Shorina M, Aronova E, Stetsenko L, Rakitin V, Shevyakova N (2007) NaCl and ethylenedependent cadaverine accumulation and its possible protective role in the adaptation of the common ice plant to salt stress. Plant Sci 172:363–370

Larson RA (1988) The antioxidants of higher plants. Phytochemistry 27(4):969–978

Lattanzio V (2013) Phenolic compounds: Introduction. In: Ramawat K, Mérillon JM (eds) Natural products. Springer, pp 1543–1580

Li X, Cai J, Liu F, Dai T, Cao W, Jiang D (2014) Exogenous abscisic acid application during grain filling in winter wheat improves cold tolerance of offspring's seedlings. J Agron Crop Sci 200:467–478

Li C, Tan DX, Liang D, Chang C, Jia D, Ma F (2015) Melatonin mediates the regulation of ABA metabolism, free-radical scavenging, and stomatal behaviour in two *Malus* species under drought stress. J Exp Bot 66:669–680

Li X-L, Xiang L, Wang X-H, Qin P, Zhang M-S, Ren M-J (2020) Biotic and abiotic stress-responsive genes are stimulated to resist drought stress in purple wheat. J Integr Agric 19:33–50

Lin JT, Chen SL, Liu SC, Yang DJ (2009) Effect of harvest time on saponins in Yam (*Dioscorea pseudojaponica* Yamamoto). J Food Drug Anal 17:116–122

Liu Q, Zhang Y, Shouyi C (2000) Plant protein kinase genes induced by drought, high salt and cold stresses. Chin Sci Bull 45:1153–1157

Liu F, Jensen CR, Andersen MN (2005) A review of drought adaptation in crop plants: changes in vegetative and reproductive physiology induced by ABA-based chemical signals. Aust J Agric Res 56:1245–1252

Liu H, Wang X, Wang D, Zou Z, Lianga Z (2011) Effect of drought stress on growth and accumulation of active constituents in *Salvia miltiorrhiza* Bunge. Ind Crop Prod 33:84–88

Liu Z, Bao H, Cai J, Han J, Zhou L (2013) A novel thylakoid ascorbate peroxidase from Jatrophacurcas enhances salt tolerance in transgenic tobacco. Int J Mol Sci 15(1):171–185

Liu F, Xu G, Wu X, Ding Q, Zheng J, Zhang R, Gao Y (2014) Effect of drought stress and re-watering on emissions of volatile organic compounds from *Rosmarinus offcinalis*. J Zhejiang A&F Univ 31:264–271

Llusià J, Peñuelas J (1998) Changes in terpene content and emission in potted Mediterranean woody plants under severe drought. Can J Bot 76:1366–1373

Llusià J, Peñuelas J, Alessio GA, Estiarte M (2006) Seasonal contrasting changes of foliar concentrations of terpenes and other volatile organic compound in four dominant species of a Mediterranean shrubland submitted to a field experimental drought and warming. Physiol Plant 127:632–649

Lone R, Shuab R, Kamil AN (2020) Plant phenolics in sustainable agriculture volume 1. Springer Nature, p 594

Lopez-Arredondo D, Gonzalez-Morales SI, Bello-Bello E, Alejo-Jacuinde G, Herrera L (2016) Engineering food crops to grow in harsh environments. F1000Res 4(F1000 Faculty Rev), p 651

López-Berenguer C, Martínez-Ballesta MC, García-Viguera C, Carvajal M (2008) Leaf water balance mediated by aquaporins under salt stress and associated glucosinolate synthesis in broccoli. Plant Sci 174(3):321–328

Loreto F, Schnitzler JP (2010) Abiotic stresses and induced BVOCs. Trends Plant Sci 15:154–166

Loreto F, Pollastri S, Fineschi S, Velikova V (2014) Volatile isoprenoids and their importance for protection against environmental constraints in the Mediterranean area. Environ Exp Bot 103:99–106

Ma G, Zhang L, Matsuta A, Matsutani K, Yamawaki K, Yahata M, Wahyudi A, Motohashi R, Kato M (2013) Enzymatic formation of citraurin from cryptoxanthin and zeaxanthin by carotenoid cleavage dioxygenase4 in the flavedo of citrus fruit. Plant Physiol 163:682–695

Ma D, Sun D, Wang C, Li Y, Guo T (2014) Expression of favonoid biosynthesis genes and accumulation of favonoid in wheat leaves in response to drought stress. Plant Physiol Biochem 80:60–66

Madhava Rao KV, Raghavendra AS, Janardhan Reddy KJ (2006) Physiology and molecular biology of stress tolerance in plants. Springer, p 345

Mahalingam R, Fedoroff N (2003) Stress response, cell death and signalling: the many faces of reactive oxygen species. Physiol Plant 119:56–68

Malitsky S, Blum E, Less H, Venger I, Elbaz M, Morin S, Eshed Y, Aharoni A (2008) The transcript and metabolite networks affected by the two clades of *Arabidopsis glucosinolate* biosynthesis regulators. Plant Physiol 148(4):2021–2049

Manchester LC, Coto-Montes A, Boga JA, Andersen LP, Zhou Z, Galano A, Vriend J, Tan DX, Reiter RJ (2015) Melatonin: an ancient molecule that makes oxygen metabolically tolerable. J Pineal Res 59(4):403–419

Manninen AM, Tarhanen S, Vuorinen M, Kainulainen P (2002) Comparing the variation of needle and wood terpenoids in scots pine provenances. J Chem Ecol 28:211–228

Manukyan A (2011) Effect of growing factors on productivity and quality of lemon catmint, lemon balm and sage under soilless greenhouse production: I. Drought stress. Med Aromat Plant Sci Biotechnol 5(2):119–125

Mantri N, Patade V, Penna S, Ford R, Pang E (2012) Abiotic stress responses in plants: present and future. In: Ahmad P, Prasad MNV (eds) Abiotic stress responses in plants: metabolism, productivity and sustainability. Springer, New York, pp 1–19

Martin DM, Gershenzon J, Bohlmann J (2003) Induction of volatile terpene biosynthesis and diurnal emission by methyl jasmonate in foliage of Norway spruce. Plant Physiol 132:1586–1599

Mattos LM, Moretti CL (2016) Oxidative stress in plants under drought conditions and the role of different enzymes. Enzym Eng 5:136

McGarvey DJ, Croteau R (1995) Terpenoid metabolism. Plant Cell 7:1015–1026

McKay JK, Richards JH, Mitchell-Olds T (2003) Genetics of drought adaptation in *Arabidopsis thaliana*: I. Pleiotropy contributes to genetic correlations among ecological traits. Mol Ecol 12:1137–1151

Meazza G, Scheffler BE, Tellez MR, Rimando AM, Nanayakkara NPD, Khan IA, Abourashed EA, Romagni JG, Duke SO, Dayan FE (2002) The inhibitory activity of natural products on plant *p-hydroxyphenylpyruvate dioxygenase*. Phytochemistry 59:281–288

Mewis I, Khan MA, Glawischnig E, Schreiner M, Ulrichs C (2012) Water stress and aphid feeding differentially influence metabolite composition in *Arabidopsis thaliana* (L.). PLoS One 7(11):e48661

Mibei E, Ambuko J, Giovannoni J, Onyango A, Owino W (2016) Carotenoid profiling of the leaves of selected African eggplant accessions subjected to drought stress. Food Sci Nutr 5:113–122

Mierziak J, Kostyn K, Kulma A (2014) Flavonoids as important molecules of plant interactions with the environment. Molecules 19:16240–16265

Minguet EG, Vera-Sirera F, Marina A, Carbonell J, Blazquez MA (2008) Evolutionary diversification in polyamine biosynthesis. Mol Biol Evol 25:2119–2128

Mirás-Avalos JM, Intrigliolo DS (2017) Grape composition under abiotic constrains: water stress and salinity. Front Plant Sci 8:851

Miura K, Okamoto H, Okuma E, Shiba H, Kamada H, Hasegawa PM, Murata Y (2013) SIZ1 deficiency causes reduced stomatal aperture and enhanced drought tolerance via controlling salicylic acid-induced accumulation of reactive oxygen species in Arabidopsis. Plant J 73(1):91–104

Mohamed HI, El-Shazly HH, Badr (2020) A role of salicylic acid in biotic and abiotic stress tolerance in plants. In: Lone R (ed) Plant phenolics in sustainable agriculture. Springer Nature, pp 533–554

Morant M, Ekstrøm C, Ulvskov P, Kristensen C, Rudemo M, Olsen CE, Hansen J, Jørgensen K, Jørgensen B, Møller BL, Bak S (2010) Metabolomic, transcriptional, hormonal, and signaling crosstalk in superroot2. Mol Plant 3(1):192–211

Munemasa S, Oda K, Watanabe-Sugimoto M, Nakamura Y, Shimoishi Y, Murata Y (2007) The coronatine-insensitive 1 mutation reveals the hormonal signaling interaction between abscisic acid and methyl jasmonate in *Arabidopsis* guard cells. Specific impairment of ion channel activation and second messenger production. Plant Physiol 143(3):1398–1407

Munne-Bosch S, Penuelas J (2003) Photo and antioxidative protection and a role for salicylic acid during drought and recovery in field-grown *Phillyrea angustifolia* plants. Planta 217(5):758–766

Munne-Bosch S, Mueller M, Schwarz K, Alegre L (2001) Diterpenes and antioxidative protection in drought-stressed *Salvia offcinalis* plants. J Plant Physiol 158:1431–1437

Naikoo MI, Dar MI, Raghib F, Jaleel H, Ahmad B, Raina A, Khan FA, Naushin F (2019) Role and regulation of plants phenolics in abiotic stress tolerance: an overview. Plant Signal Mol:157–168

Nakabayashi R, Yonekura-Sakakibara K, Urano K, Suzuki M, Yamada Y, Nishizawa T, Matsuda F, Kojima M, Sakakibara H, Shinozaki K, Michael AJ, Tohge T, Yamazaki M, Saito K (2014) Enhancement of oxidative and drought tolerance in *Arabidopsis* by overaccumulation of antioxidant flavonoids. Plant J 77:367–379

Naqvi S, Farré G, Sanahuja G, Capell T, Zhu C, Christou P (2010) When more is better: multigene engineering in plants. Trends Plant Sci 15:48–56

Nazar R, Umar S, Khan N, Sareer O (2015) Salicylic acid supplementation improves photosynthesis and growth in mustard through changes in proline accumulation and ethylene formation under drought stress. S Afr J of Bot 98:84–94

Nezhadahmadi A, Hossain Prodhan Z, Faruq G (2013) Drought tolerance in wheat. Sci World J 2013:1–12

Ngo LT, Okogun JI, Folk WR (2013) 21st century natural product research and drug development and traditional medicines. Nat Prod Rep 30:584–592

Nichols SN, Hofmann RW, Williams WM (2015) Physiological drought resistance and accumulation of leaf phenolics in white clover interspecific hybrids. Environ Exp Bot 119:40–47

Niyogi KK, Björkman O, Grossman AR (1997) The roles of specific xanthophylls in photoprotection. – Proc Nat Acad Sci 94:14162–14167

Nowak M, Manderscheid R, Weigel H-J, Kleinwächter M, Selmar D (2010) Drought stress increases the accumulation of monoterpenes in sage (*Salvia officinalis*), an effect that is compensated by elevated carbon dioxide concentration. J Appl Bot Food Qual 83:133–136

Nowak M, Kleinwächter M, Manderscheid R, Weigel HJ, Selmar D (2012) Drought stress increases the accumulation of monoterpenes in sage (*Salvia officinalis*), an effect that is compensated by elevated carbon dioxide concentration. J Appl Bot Food Qual 83:133–136

Nunnery JK, Mevers E, Gerwick WH (2010) Biologically active secondary metabolites from marine cyanobacteria. Curr Opin Biotechnol 21:787–793

Obata T, Fernie AR (2012) The use of metabolomics to dissect plant responses to abiotic stresses. Cell Mol Life Sci 69:3225–3243

Olsen KM, Slimestad R, Lea US, Brede C, Lovdal T, Ruoff P, Verheul M, Lillo C (2009) Temperature and nitrogen effects on regulators and products of the flavonoid pathway: experimental and kinetic model studies. Plant Cell Environ 32:286–299

Ormeno E, Mevy J, Vila B, Bousquet-Melou A, Greff S, Bonin G, Fernandez C (2007) Water deficit stress induces different monoterpene and sesquiterpene emission changes in Mediterranean species. Relationship between terpene emissions and plant water potential. Chemosphere 67:276–284

Panche AN, Diwan AD, Chandra SR (2016) Flavonoids: an overview. J Nutr Sci 5:e47. https://doi.org/10.1017/jns.2016.41

Patel MK, Kumar M, Li W, Luo Y, Burritt DJ, Alkan N, Tran L-SP (2020) Enhancing salt tolerance of plants: from metabolic reprogramming to exogenous chemical treatments and molecular approaches. Cells 9:2492

Payton P, Kottapalli KR, Kebede H, Mahan JR, Wright RJ, Allen RD (2011) Examining the drought stress transcriptome in cotton leaf and root tissue. Biotechnol Lett 33:821–828

Petropoulos SA, Daferera D, Polissiou MG, Passam HC (2008) The effect of water deficit stress on the growth, yield and composition of essential oils of parsley. Sci Hortic-Amsterdam 115:393–397

Piasecka A, Sawikowska A, Kuczynska A, Ogrodowicz P, Mikołajczak K, Krystkowiak K, Gudys K, Guzy-Wróbelska J, Krajewski P, Kachlicki P (2017) Drought-related secondary metabolites of barley (*Hordeum vulgare* L.) leaves and their metabolomic quantitative trait loci. Plant J 89:898–913

Planas-Riverola A, Gupta A, Betegón-Putze I, Bosch N, Ibañes M, Caño-Delgado AI (2019) Brassinosteroid signaling in plant development and adaptation to stress. Development 146(5):dev151894

Pott DM, Osorio S, Vallarino JG (2019) From central to specialized metabolism: An overview of some secondary compounds derived from the primary metabolism for their role in conferring nutritional and organoleptic characteristics to fruit. Front Plant Sci 10:835

Prasad PVV, Staggenborg S, Ristic Z (2008) Impacts of drought and/or heat stress on physiological, developmental, growth, and yield processes of crop plants. Adv Agric Syst Model Ser 1:301–355

Pribat A, Boureau L, Mortain-Bertrand A, Bert LS, Rolin D, Teyssier E, Gallusci P (2013) Metabolic engineering of isoprenoid biosynthesis. In: Ramawat K, Mérillon J-M (eds) Natural products. Springer, pp 2813–2851

Qu X, Wang H, Chen M, Liao J, Yuan J, Niu G (2019) Drought stress–induced physiological and metabolic changes in leaves of two oil tea cultivars. J Am Soc Hortic Sci 144:439–447

Quan NT, Anh LH, Khang DT, Tuyen PT, Toan NP, Minh TN, Bach DT, Ha PTT, Elzaawely AA, Khanh TD (2016) Involvement of secondary metabolites in response to drought stress of rice (*Oryza sativa* L.). Agriculture 6:23

Radovich TJ, Kleinhenz MD, Streeter JG (2005) Irrigation timing relative to head development influences yield components, sugar levels and glucosinolate concentrations in cabbage. J Am Soc Hortic Sci 130(6):943–949

Rahdari P, Hoseini SM (2012) Drought stress: a review. Intl J Agron Plant Prod 3:443–446
Ramakrishna A, Ravishankar GA (2011) Influences of abiotic stress signals on secondary metabolites in plants. Plant Signal Behav 6:1720–1731
Ramakrishna A, Giridhar P, Ravishankar GA (2009) Indoleamines and calcium channels influence morphogenesis in *in vitro* cultures of *Mimosa police* L. Plant Signal Behav 12:1136–1141
Ramalho JC, Zlatev ZS, Leitão AE, Pais IP, Fortunato AS, Lidon FC (2014) Moderate water stress causes different stomatal and non-stomatal changes in the photosynthetic functioning of Phaseolus vulgaris L. genotypes. Plant Biol 16:133–146
Ramel F, Birtic S, Cuine S, Triantaphylides C, Ravanat JL, Havaux M (2012) Chemical quenching of singlet oxygen by carotenoids in plants. Plant Physiol 158:1267–1278
Rana RM, Rehman SU, Ahmed J, Bilal M (2013) A comprehensive overview of recent advances in drought stress tolerance research in wheat (*Triticum aestivum* L.). Asian J Agric Biol 1:29–37
Rao SR, Ravishankar GA (2002) Plant cell cultures: chemical factories of secondary metabolites. Biotechnol Adv 20:101–153
Ravishankar GA, Rao SR (2000) Biotechnological production of phytopharmaceuticals. J Biochem Mol Biol Biophys 4:73–102
Ravishankar GA, Venkataraman LV (1990) Food applications of plant cell cultures. Curr Sci 57:381–383
Razavizadeh R, Komatsu S (2018) Changes in essential oil and physiological parameters of callus and seedlings of *Carum copticum* L. under *in vitro* drought stress. J Food Meas Charact 12:1581–1592
Rebey IB, Jabri-Karoui I, Hamrouni-Sellami I, Bourgou S, Limam F, Marzouk B (2012) Effect of drought on the biochemical composition and antioxidant activities of cumin (*Cuminum cyminum* L.) seeds. Ind Crop Prod 36:238–245
Rice-Evans CA, Miller N, Paganga G (1997) Antioxidant properties of phenolic compounds. Trends Plant Sci 2(4:152–159
Rini Vijayan KP, Raghu AV (2020) Methods for enhanced production of metabolites under in vitro conditions. In: Sukumaran ST, Sugathan S, Abdulhameed S (eds) Plant metabolites: methods, applications and prospects. Springer, pp 111–140
Robbins RJ, Keck A, Banuelos G, Finley JW (2005) Cultivation conditions and selenium fertilization alter the phenolic profile, glucosinolate and sulforaphane content of broccoli. J Med Food 8(2):204–214
Rodziewicz P, Swarcewicz B, Chmielewska K, Wojakowska A, Stobiechi M (2014) Influence of abiotic stresses on plant hormone and metabolome changes. Acta Physiol Plant 36:1–19
Romagni J, Duke S, Dayan F (2000) Inhibition of plant asparagine synthetase by monoterpene cineoles. Plant Physiol 123:725–732
Ruan Y, Dong S, Liu L, Sun C, Wang L, Guo Q, Gai Z (2012) Effect of exogenous abscisic acid (ABA) on the physiological characteristics during soybean flowering under drought stress. Soybean Sci 31(3):385–394
Sadeghipour O (2018) Drought tolerance of cowpea enhanced by exogenous application of methyl jasmonate. Int J Mod Agric 7(4):51–57
Salehi-lisar SY, Motafakkerazad R, Hossain MM, Rahman IMM (2012) Water stress in plants: causes, effects and responses. In: IMM R, Hasegawa H (eds) Water stress. InTechOpen, pp 1–14
Savoi S, Wong DC, Arapitsas P, Miculan M, Bucchetti B, Peterlunger E, Fait A, Mattivi F, Castellarin SD (2016) Transcriptome and metabolite profiling reveals that prolonged drought modulates the phenylpropanoid and terpenoid pathway in white grapes (*Vitis vinifera* L.). BMC Plant Biol 16:67
Schreiner M, Beyene B, Krumbein A, Stützel H (2009) Ontogenetic changes of 2-propenyl and 3- indolylmethyl glucosinolates in *Brassica carinata* leaves as affected by water supply. J Agric Food Chem 57(16):7259–7263
Schroeder JI, Nambara E (2006) A quick release mechanism for abscisic acid. Cell 126(6):1023–1025

Schwartz SH, Qin X, Zeevaart JA (2003) Elucidation of the indirect pathway of abscisic acid biosynthesis by mutants, genes, and enzymes. Plant Physiol 131(4):1591–1601

Schweiggert R, Ziegler J, Metwali E, Mohamed F, Almaghrabi O, Kadasa N, Carle R (2017) Carotenoids in mature green and ripe red fruits of tomato (*Solanum lycopersicum* L.) grown under different levels of irrigation. Arch Biol Sci 69:305–314

Seigler DS (1998) Plant Secondary Metabolism. Springer, p 759. https://doi.org/10.1007/978-1-4615-4913-0

Seki M, Umezawa T, Urano K, Shinozaki K (2007) Regulatory metabolic networks in drought stress responses. Curr Opin Plant Biol 10:296–302

Selmar D, Kleinwächter M (2013) Stress enhances the synthesis of secondary plant products: the impact of the stress-related over-reduction on the accumulation of natural products. Plant Cell Physiol 54:817–826

Shao HB, Chu LY, Jaleel CA, Zhao CX (2008) Water-deficit stress-induced anatomical changes in higher plants. C R Biol 331:215–225

Sharma M, Kumar P (2021) Role of secondary metabolites and phytohormones in drought stress tolerance. J Food Process Preserv 8(3):1917–1946

Sharma A, Shahzad B, Rehman A, Bhardwaj R, Landi M, Zheng B (2019) Response of phenylpropanoid pathway and the role of polyphenols in plants under abiotic stress. Molecules 24:2452

Sheteiwy MS, Gong D, Gao Y, Pan R, Hu J, Guan Y (2018) Priming with methyl jasmonate alleviates polyethylene glycol-induced osmotic stress in rice seeds by regulating the seed metabolic profile. Environ Exp Bot 153:236–248

Singh B, Usha K (2003) Salicylic acid induced physiological and biochemical changes in wheat seedlings under water stress. Plant Growth Regul 39:137–141

Solíz-Guerrero JB, De Rodriguez DJ, Rodríguez-García R, Angulo-Sánchez JL, Méndez-Padilla G (2002) Quinoa saponins: concentration and composition analysis. In: Janick J, Whipkey A (eds) Trends in new crops and new uses. Proceedings of the Fifth National Symposium, pp 110–114

Sønderby IE, Geu-Flores F, Halkier BA (2010) Biosynthesis of glucosinolates-gene discovery and beyond. Trends Plant Sci 15(5):283–290

Soto G, Stritzler M, Lisi C, Alleva K, Pagano ME, Ardila F, Mozzicafreddo M, Cuccioloni M, Angeletti M, Ayub ND (2011) Acetoacetyl-CoA thiolase regulates the mevalonate pathway during abiotic stress adaptation. J Exp Bot 62:5699–5711

Sridha S, Wu K (2006) Identification of AtHD2C as a novel regulator of abscisic acid responses in *Arabidopsis*. Plant J 46(1):124–133

Staudt M, Bourgeois I, Al Halabi R, Song W, Williams J (2017) New insights into the parametrization of temperature and light responses of mono-and sesquiterpene emissions from Aleppo pine and rosemary. Atmos Environ 152:212–221

Stephanopoulos G (1999) Metabolic fluxes and metabolic engineering. Metab Eng 1:1–11

Stracke R, Jahns O, Keck M, Tohge T, Niehaus K, Fernie AR, Weisshaar B (2010) Analysis of production of flavonol glycosides-dependent flavonol glycoside accumulation in *Arabidopsis thaliana* plants reveals MYB11-, MYB12- and MYB111-independent flavonol glycoside accumulation. New Phytol 188:985–1000

Sudrajat DJ, Siregar IZ, Khumaida N, Siregar UJ, Mansur I (2015) Adaptability of white jabon (*Anthocephalus cadamba* MIQ.) seedling from 12 populations to drought and waterlogging. Agrivita 37:130–143

Szakiel A, Paczkowski C, Henry M (2010) Influence of environmental abiotic factors on the content of saponins in plants. Phytochem Rev 2:25

Taiz L, Zeiger E (2006) Plant physiology, 4th edn. Sinauer Associates Inc, p 700

Tardieu F, Parent B, Caldeira CF, Welcker C (2014) Genetic and physiological controls of growth under water deficit. Plant Physiol 164:1628–1635

Tattini M, Galardi C, Pinelli P, Massai R, Remorini D, Agati G (2004) Differential accumulation of flavonoids and hydroxycinnamates in leaves of *Ligustrum vulgare* under excess light and drought stress. New Phytol 163:547–561

Tong Y, Gabriel-Neumann E, Ngwene B, Krumbein A, George E, Platz S, Rohn S, Schreiner M (2014) Topsoil drying combined with increased sulfur supply leads to enhanced aliphatic glucosinolates in *Brassica juncea* leaves and roots. Food Chem 152:190–196
Treml J, Smejkal K (2016) Flavonoids as potent scavengers of hydroxyl radicals. Compr Rev Food Sci Food Saf 15:720–738
Treutter D (2006) Significance of flavonoids in plant resistance: a review. Environ Chem Lett 4:147–157
Tůmová L, Tarkowská D, Řehoˇrová K, Markova H, Kocˇová M, Rothova O, Čečetka P, Holá D (2018) Drought-tolerant and drought-sensitive genotypes of maize (*Zea mays* L.) differ in contents of endogenous brassinosteroids and their drought-induced changes. PLoS One 13(5):e0197870
Turtola S, Manninen A-M, Rikala R, Kainulainen P (2003) Drought stress alters the concentration of wood terpenoids in scots pine and Norway spruce seedlings. J Chem Ecol 29:1981–1995
Tuteja N, Sopory SK (2008) Chemical signaling under abiotic stress environment in plants. Plant Signal Behav 3:525–536
Uarrota V, Stefen Deivid, Leolato L, Medeiros G, Nerling D (2018) Revisiting carotenoids and their role in plant stress responses: from biosynthesis to plant signaling mechanisms during stress. In: Gupta Dk, Palma JM, Corpas FJ (eds) Antioxidants and Antioxidant Enzymes in Higher Plants. Springer, pp 207–232
Urano K, Maruyama K, Ogata Y, Morishita Y, Takeda M, Sakurai N, Suzuki H, Saito K, Shibata D, Kobayashi M, Yamaguchi-Shinozaki K, Shinozaki K (2009) Characterization of the ABA-regulated global responses to dehydration in Arabidopsis by metabolomics. Plant J 57:1065–1078
van der Fits L, Memelink J (2000) ORCA3, a jasmonate responsive transcriptional regulator of plant primary and secondary metabolism. Science 289:295–297
Verma A, Deepti S (2016) Abiotic stress and crop improvement: current scenario. Adv Plants Agric Res 4:00149
Verma N, Shukla S (2015) Impact of various factors responsible for fluctuation in plant secondary metabolites. J Appl Res Med Aromat Plants 2:105–113
Vickers CE, Possell M, Cojocariu CI, Velikova VB, Laothawornkitkul J, Ryan A, Mullineaux PM, Hewitt CN (2009) Isoprene synthesis protects transgenic tobacco plants from oxidative stress. Plant Cell Environ 32:520–531
Von Rudloff E (1972) Chemosystematic studies in the genus Pseudotsuga. I. Leaf oil analysis of the coastal and rocky mountain varieties of the Douglas fir. Can J Bot 50:1025–1040
Vranová E, Coman D, Gruissem W (2013) Network analysis of the MVA and MEP pathways for isoprenoid synthesis. Annu Rev Plant Biol 64:665–700
Wang HY, Huang YC, Chen SF, Yeh KW (2003) Molecular cloning, characterization and gene expression of a water deficiency and chilling induced proteinase inhibitor I gene family from sweet potato (*Ipomoea batatas* Lam.) leaves. Plant Sci 165:191–203
Wang P, Sun X, Li C, Wei Z, Liang D, Ma F (2013) Long-term exogenous application of melatonin delays drought-induced leaf senescence in apple. J Pineal Res 54:292–302
Wang H, Wang H, Shao H, Tang X (2016) Recent advances in utilizing transcription factors to improve plant abiotic stress tolerance by transgenic technology. Front Plant Sci 7:67
Wei L, Wang L, Yang Y, Wang P, Guo T, Kang G (2015) Abscisic acid enhances tolerance of wheat seedlings to drought and regulates transcript levels of genes encoding ascorbate-glutathione biosynthesis. Front Plant Sci 6:458
Wen W, Li K, Alseekh S, Omranian N, Zhao L, Zhou Y, Xiao Y, Jin M, Yang N, Liu H, Florian A, Li W, Pan Q, Nikoloski Z, Yan J, Fernie AR (2015) Genetic determinants of the network of primary metabolism and their relationships to plant performance in a maize recombinant inbred line population. Plant Cell 27:1839–1856
WHO (2020) A report about Drought. Retrieved from: https://www.who.int/health-topics/drought#tab=tab_1

Wink M (2015) Modes of action of herbal medicines and plant secondary metabolites. Medicines 2:251–286

Wu H, Hu Z-H (1997) Comparative anatomy of resin ducts of the Pinaceae. Trees11:135–143

Wu H, Wu X, Li Z, Duan L, Zhang M (2012) Physiological evaluation of drought stress tolerance and recovery in caulifl ower (*Brassica oleracea* L.) seedlings treated with methyl jasmonate and coronatine. J Plant Growth Regul 31(1):113–123

Wu J, Ji J, Wang G, Wu G, Diao J, Li Z, Chen X, Chen Y, Luo L (2015) Ecotopic expression of the *Lyciumbarbarum* β-carotene hydroxylase gene (chyb) enhances drought and salt stress resistance by increasing xanthophyll cycle pool in tobacco. Plant Cell Tiss Organ Cult 121:559–569

Xia L, Yang W, Xiufeng Y (2007) Effects of water stress on berberine, jatrorrhizine and palmatine contents in amur corktree seedlings. Acta Ecol Sin 27(1):58–64

Yadav B, Jogawat A, Rahman MS, Narayan OP (2021) Secondary metabolites in the drought stress tolerance of crop plants: a review. Gene Rep 23:1–14

Yamaguchi K, Takahashi Y, Berberich T, Imai A, Takahashi T, Michael AJ, Kusano T (2007) A protective role for the polyamine spermine against drought stress in *Arabidopsis*. Biochem Biophys Res Commun 352:486–490

Yang L, Wen K-S, Ruan X, Zhao Y-X, Wei F, Wang Q (2018) Response of plant secondary metabolites to environmental factors. Molecules 23:762

Yanlin W, Gao S, He X, Li Y, Li P, Zhang Y, Chen W (2019) Growth, secondary metabolites and enzyme activity responses of two edible fern species to drought stress and rehydration in Northeast China. Agron 9(3):137. https://doi.org/10.3390/agronomy9030137

Ying Y, Yue Y, Huang X, Wang H, Mei L, Yu W, Zheng B, Wu J (2013) Salicylic acid induces physiological and biochemical changes in three Red bayberry (*Myric rubra*) genotypes under water stress. Plant Growth Regul 71:181–189

Yonekura-Sakakibara K, Tohge T, Matsuda F, Nakabayashi R, Takayama H, Niida R, WatanabeTakahashi A, Inoue E, Saito K (2008) Comprehensive flavonol profiling and transcriptome coexpression analysis leading to decoding gene-metabolite correlations in Arabidopsis. Plant Cell 20:2160–2176

Yuan GF, Jia CG, Li Z, Sun B, Zhang LP, Liu N, Wang Q-M (2010) Effect of brassinosteroids on drought resistance and abscisic acid concentration in tomato under water stress. Sci Hortic 126(2):103–108

Zagorchev L, Seal CE, Kranner I, Odjakova M (2013) A central role for thiols in plant tolerance to abiotic stress. Int J Mol Sci 14:7405–7432

Zapata PJ, Serrano M, Pretel MT, Amoros A, Botella MA (2004) Polyamines and ethylene changes during germination of different plant species under salinity. Plant Sci 167:781–788

Zhang J, Davies W (1990) Changes in the concentration of ABA in xylem sap as a function of changing soil water status can account for changes in leaf conductance and growth. Plant Cell Environ 13:277–285

Zhang H, Schonhof I, Krumbein A, Gutezeit B, Li L, Stützel H, Schreiner M (2008) Water supply and growing season influence glucosinolate concentration and composition in turnip root (*Brassica rapa* ssp. *rapifera* L.). J Plant Nutr Soil Sci 171(2):255–265

Zhang Y, Shen YY, Wu XM, Wang JB (2016) The basis of pod dehiscence: anatomical traits of the dehiscence zone and expression of eight pod shatter-related genes in four species of Brassicaceae. Biol Plant 60(2):343–354

Zhang L, Wang Q, Guo Q, Chang Q, Zhu Z, Liu L, Xu H (2012) Growth, physiological characteristics and total favonoid content of *Glechoma longituba* in response to water stress. J Med Plant Res 6:1015–1024

Zhang N, Liu B, Ma C, Zhang G, Chang J, Si H, Wang D (2014) Transcriptome characterization and sequencing-based identification of drought-responsive genes in potato. Mol Biol Rep 41:505–517

Zhang N, Sun Q, Zhang H, Cao Y, Weeda S, Ren S, Guo YD (2015) Roles of melatonin in abiotic stress resistance in plants. J Exp Bot 66:647–656

Zhi-lin Y, Chuan-chao D, Lian-qing C (2007) Regulation and accumulation of secondary metabolites in a plant-fungus symbiotic system. Afr J Biotechnol 6(11):1266–1271
Zlatev Z, Lidon FC (2012) An overview on drought induced changes in plant growth, water relations and photosynthesis. Emir J Food Agric 24:57–72
Zobayed SMA, Afreen F, Kozai T (2007) Phytochemical and physiological changes in the leaves of St. John's wort plants under a water stress condition. Environ Exp Bot 59(2):109–116

Chapter 27
Seed Priming for Abiotic Stress Tolerance

Kazim Ali, Hafiz Muhammad Mubasher, Ahmad Sher, Abdul Sattar, and Abdul Manaf

Abstract Seed priming is relatively a quick and easy way to boost germination, yield in many crops of field and vigor index, especially those cultivated in harsh situations of field environment. Priming of seed is best known to start the usual processes of metabolism in the early stages of growth, precisely germination, before the process of radicle protrusion. Plants experience a range of stresses as they grow and expand. Utilizing a number of methods, such as osmo-, magnetic, chemical, biological, hydro-, matrix-, and hormonal priming, a specific physiological growth stage in the seed is induced before the germination. Early vigor from seed priming causes crops to emerge more uniformly, rapidly, and with a better crop stand. Therefore, seed priming is a very smart, useful, and practical alternative that makes it simpler for plants to defend themselves from various stress situations while seedling establishment is taking place. This chapter discusses recent developments in our knowledge of how seed priming affects numerous molecular, biochemical, physiological processes during seed germination and seedling establishment as well as how it helps to control a plant's resistance to abiotic stimuli. Moreover, in the current state of study research, the types and functions of priming (in drought, heat, and metal stress) were explained.

Keywords Seed priming · Abiotic stress · Tolerance · Hydropriming · Biochemical mechanisms

---

K. Ali · H. M. Mubasher
College of Agriculture, University of Layyah, Layyah, Pakistan

A. Sher (✉) · A. Sattar
College of Agriculture, University of Layyah, Layyah, Pakistan

Department of Agronomy, Bahauddin Zakariya University, Multan, Pakistan
e-mail: ahmad.sher@bzu.edu.pk

A. Manaf
Department of Agronomy, PMAS–Arid Agriculture University, Rawalpindi, Pakistan

M. Hasanuzzaman (ed.), *Climate-Resilient Agriculture, Vol 2*,
https://doi.org/10.1007/978-3-031-37428-9_27

1 Introduction

The most significant ones of all the abiotic factors that have a negative impact on crop production are salinity, drought, and severe temperatures (Thakur et al. 2010; Jaleel et al. 2009; Jakab et al. 2005). These stresses of abiotic are to blame to limit or delay the germination of seeds and the early growth of seedlings (Almansouri et al. 2001). During the critical germination period, plants react swiftly to changes in their surroundings. It is controlled by biochemical, molecular, and physiological processes relevant to the basic development of embryo (Lutts et al. 2016; Bewley et al. 2013). Various biochemical, physiological, and molecular processes within basic unit of cells are altered by abiotic stresses (Xiong and Zhu 2002). This will upset the vital balance of the metabolism in plant cells, which can result in oxidative stress, membrane damage, decreased cellular respiration, and the production of oxygen species which are reactive (ROS) (Suzuki and Mittler 2006).

By increasing the rate of emergence and germination uniformity, seed priming is also a viable method for optimizing the effectiveness of seeds, resulting in faster and better seedling growth (Sohail et al. 2018; Castaares and Bouzo 2018; Basra et al. 2004; Cramer 2002). Many seed priming strategies have been lately developed to improve seedling growth and seed germination. Seed priming using a variety of chemicals is a promising method for increasing plant stress tolerance. It is possible to prime seeds using a variety of chemicals, organic, ions, hormones, compounds, and antioxidants (Masondo et al. 2018; Ibrahim 2016; Jisha et al. 2013; Eskandari 2013; Nawaz et al. 2013; Ghassemi-Golezai and Esmaeilpour 2008). Environmental conditions, plant type, seed lot and vigor, and priming technique all play a significant role in the success of seed priming (Maiti and Pramanik 2013; Farooq et al. 2013; Corbineau and Come 2006; Parera and Cantliffe 1994). By stimulating many of the biochemical, physiological, and molecular mechanisms involved in the early stages of germination, seed priming also boosts abiotic stress tolerance systems that enhance growth both during and after seedling establishment. Seed priming starts a number of biochemical, physiological, systemic, cellular, and molecular changes in plants that encourage plant growth when they are subjected to abiotic stress (Pal et al. 2017; Ibrahim 2016; Lutts et al. 2016; Chen and Arora 2013; Siri et al. 2013; Eisvand et al. 2010; Varier et al. 2010). Seed priming raises the uniformity and germination rate by osmotic adjustment, metabolic repair while ingestion, and activation of metabolic pathways of pre-germinative in the primed seeds. Consequences for early reserve mobilization include DNA replication processing and repair processes (Lutts et al. 2016). Energy metabolism is among the numerous germination-related processes that reserve mobilization. Seed priming also encourages the weakening, respiration, and expansion of embryonic cells in the endosperm (Chen and Arora 2011; Varier et al. 2010; Pandita et al. 2007).

Seed priming makes better the germination by regulating pathways of DNA repair, deteriorating catalase, enzymes, and other antioxidant-scavenging enzymes, as well as the production of phospholipids and sterols and the de novo synthesis of proteins and nucleic acids (Rajjou et al. 2012; Chen 2011; Afzal et al. 2002).

Additionally, improved elongation, cell division, and abiotic stress response proteins may all aid in the production of healthy seedlings from seeds that are being primed (Ibrahim 2016). Furthermore, seed priming increases specific abiotic stress reactions linked to defense mechanisms. In late embryogenesis, rich protective proteins like heat shock proteins (HSPs) and GPX, SOD, and CAT accumulate more, membrane pumps of efflux are activated, peroxiredoxin-encoding genes are increased, and antioxidant-scavenging enzymes like SOD, CAT, and GPX are enhanced (Chen 2011; Catusse et al. 2011; Varier et al. 2010; Li et al. 2005). Seed priming controls how effectively organisms withstand stress by increasing post-translational modification of the potential for protein synthesis and controlling the ideal translational turnover ratio (Kubala et al. 2015).

2 Role of Seed Priming in Abiotic Stress Tolerance in Plants

Plants used as crops commonly encounter these stresses during their lifetime in basic natural conditions, which have a negative impact on their germination and output (Hussain et al. 2018). Several stresses, including salt, drought, heat, and other climate-related issues, are affecting crop output. To combat the adversities of these stresses, plants alter a number of biochemical and physiological processes. It is very well known that most of these stressors reduce absorption during the first few days after germination. Abiotic stresses that reduce water availability also impact cell elongation, which hinders embryo growth and causes poor seedling emergence. Both in regular and stressful conditions, the appropriate stand has been developed successfully using the practical method of seed priming (Hussain et al. 2016a; Jafar et al. 2012). The benefits of the process of priming under various stresses in different crops have been highlighted in recent articles (Hussain et al. 2018; Hussain et al. 2016a, b; Jisha et al. 2013).

Due to enhanced energy metabolism, OA, quick cellular defenses, larger embryos, and increased enzyme activity, the primed seeds generated plants with a strong head start and greater stress tolerance (Jisha et al. 2013). Seeds primed with PEG, as compared to the ones that are non-primed seeds, showed better process of germination, vigor, and relative water content, according to Pant and Bose (2016). Salicylic acid (SA) + hydrogen peroxide (H_2O_2) dramatically increased seed priming during cold stress; it was closely related to metabolite buildup, hormone metabolism, ROS detoxification, and efficiency of energy delivery (Li et al. 2017). Seed priming in crop of tobacco increased the plant's capacity to endure cold conditions during seedling and germination growth by modulating the antioxidant defense in tissues of plant (Xu et al. 2011). According to Guan et al. (2009), seed priming enhanced the development and germination of crop of maize seedlings under the conditions of low temperature. Seed priming was suggested by Elkoca et al. (2007) for improved seedling germination and robust growth of crop of chickpea under low temperature stress (hydro- and osmo-priming). According to Srivastava et al. (2010), a variety of seed priming techniques can help Indian mustard grow under salinity and water stress. Kumar et al. (2016) claim that hydro- and halo-priming ($Mg(NO_3)_2$ and $Ca(NO_3)_2$) can improve wheat growth, length of seedling, levels of soluble sugar, and amylase activity to decrease the negative influence of heavy metal ($HgCl_2$) stress.

3 Types of Seed Priming

Seed priming has following types:

3.1 Hydropriming

Hydropriming involves process of soaking the seeds basically in water, then again drying them to their previous and original content of moisture, and then planting them. This approach is economical and kind to the environment. Prior to sowing, the practice of seed wetting and drying of surface on a farm is known as on-farm priming, also known as hydropriming (Harris et al. 2005). Farmers in marginal tropical environments with few resources might find use for this method. The main drawbacks of hydropriming, however, are uneven levels of seed hydration brought on by unchecked water uptake and asynchronous emergence caused by variations in the activation of pathways of metabolism within seeds. Adjusting the hydration parameters including treatment duration, water volume, and temperature, used in hydropriming, is essential for ensuring consistent seed scheduled emergence and hydration. Commercial priming known as drum priming, adaption of basically hydropriming, begins by rotating seeds in a drum (Rowse 1996). The seeds are then gradually watered by the addition of H_2O in the form of vapor. Drum priming is more effective than traditional hydropriming because it offers regulated seed imbibition. After hydropriming maize seeds for 12 h, two proteins, embryonic protein globulin-1 and DC-8, which may be important candidates for the markers of protein for seed vigor and priming effect, changed in amount in the profiles of protein in the embryos (Gong et al. 2013). The actin isoform and cytosolic glyceraldehyde 3-phosphate dehydrogenase, two proteins that are involved in seed dehydration and function in the germination and priming processes, have been linked in previous studies to the priming and germination processes (Gallardo et al. 2001).

3.2 Osmo- and Halopriming

Osmo-priming involves soaking seeds in liquids with water potential that is low, such as mannitol, glycerol, polyethylene glycol, sorbitol, or inorganic salts like, KH_2PO_4, $MgSO_4$, $CaCl_2$, KCl, KNO_3, NaCl, and K_3PO_4 (Yacoubi et al. 2013).

Halopriming is the term used to describe the use of inorganic salts as priming agents. Early period of germination is activated due to the progressive seed imbibition, which occurs when seeds are dipped in osmotic solutions with water potential that is low (Di Girolamo and Barbanti 2012). Additionally, by producing osmolytes and/or signaling molecules, halopriming promotes seed metabolism and aids in germination. Potassium nitrate (KNO_3) was used to prime tomato seeds, Lara et al. (2014) found that this hastened germination is because the soaked nitrate gets in the seed embryo's metabolism by the help of the pathway of nitrate reductase enzyme, producing the molecule that will signal the nitrite/nitric oxide. Sahin et al. (2011) found that sodium chloride priming of tomato seeds caused the accumulation of osmoregulating defense molecules (such as proline (Pro) and anthocyanin) and the activation of the antioxidative enzyme system.

3.3 Solid Matrix Priming

Because of the expensive nature of osmotic agents and the aeration process' technical issues, osmo-priming has some limitations (Paparella et al. 2015). Osmo-priming, which includes regulating seed water intake, is an alternate technique to solid matrix priming. As method includes mixing and incubating seeds with a moist solid water carrier for a set amount of time, solid matrix priming is economical. The seeds are then taken out of the matrix, cleaned, and dried once again. Seeds can slowly hydrate when placed in a solid media, initiating the natural process that happens when seeds are planted in earth (McDonald 2000).

3.4 Hormone Priming

This involves the application of growth regulators of plants that can directly affect seed germination and includes kinetin, ethylene, polyamines, salicylic acid abscisic acid, auxin, and gibberellin.

3.5 Biological Priming

Biological priming entails ingesting seeds along with microorganisms that encourage plant growth (Callan et al. 1990). Mentioned priming technique protects seedlings from soil- and seed-borne diseases and also speeds up and uniforms germination.

3.6 *Chemical Priming*

In order to boost abiotic stress tolerance and germination rate, above-mentioned method of priming involves treating seed by specific synthetic or natural compounds (melatonin, polyamines, hydrogen sulfide, hydrogen peroxide, nitric oxide).

3.7 *Ultraviolet Priming*

Low doses of UV radiation are applied to seedlings during UV priming to increase plant immune to abiotic and biotic stress, increase yield, and increase quality. According to Dhanya Thomas and Puthur (2017), seedlings of a variety of crops, including wheat, rice, maize, cucumber, and cowpea, grew more quickly and produced more biomass and fresh and dry weight when exposed to UV radiation (0.004–4 Wm^2). Additionally, the ROS-scavenging system's activities increased, which made plants more resistant to abiotic stress.

4 Seed priming Methods and Agents

A range of these priming techniques, which basically are much used and developed, can be used for the purpose of speeding up and synchronizing germination of seed (Lutts et al. 2016; Nawaz et al. 2013). Osmo-conditioning, osmo-priming, and halopriming are the top three methods of priming. To allow for seed imbibition while preventing radicle protrusion, primed seeds are dipped in low-water potential aerated solutions. In order to produce water potential low solutions, a variety of agents are used (Chen and Arora 2011). The secret to the efficiency of seed priming is allowing precisely the right amount of water to support metabolic functions while preventing the cell from further expanding and developing. Osmo-key priming's objective basically is to reduce the oxidative stress caused by the ROS while increasing the duration of the entry of water. Therefore, when developing a method, it is crucial to take the water potential of osmotic agent into account (Paparella et al. 2015). Similar to a prolonged early seed imbibition, the osmo-priming process starts the slow evolution required for pre-germinative metabolism activities. To better understand how seeds shift from a dry, physiologically dormant state to a hydrated, active state, it is especially helpful to employ osmo-priming as a model (Chen and Arora 2011).

Many organic or man-made materials have been investigated in various crops as seed primers. Seed priming agents frequently include polyethylene glycol, inorganic salts, nutrients, pure water, and bio-stimulants (Masondo et al. 2018; Hussain et al. 2015; Butler et al. 2009). It also has been told that seed priming with polyamines, particular plant growth regulators, and other organic sources increases the quality of seeds in a variety of crops. Agents used in priming come in varieties and levels of effectiveness. As a result, each plant species must be optimized for the solutions of priming (Horii et al. 2007).

4.1 Polyethylene Glycol

The most widely utilized chemical to limit water potential in primed seed is polyethylene glycol (PEG), which has a molecular size between 6000 and 8000 mw, is harmless, and prevents it from accessing the desired seed (Thomas et al. 2000). By osmo-priming with help of PEG, the antioxidant system is improved and the seed's germination potential is raised, both of which increase the seed's resistance to the abiotic stresses (Mouradi et al. 2016; Chen and Arora 2011). It has been demonstrated that priming sweet corn seeds at 0.5 MPa for two days, tomato seeds only at 1.25 MPa for two straight days, and onion seeds at maximum 1.5 MPa for seven straight days all increases germination (Dorna et al. 2013; Govinden-Soulange and Levantard 2008).

In Moroccan alfalfa seed that was undergoing germination under a water shortage, seed priming with the help of PEG at maximum 0.6 MPa for straight 24 hours at the temperature 25 °C increased the basic activities of CAT and POD and lowered the MDA level and leakage of electrolyte (Mouradi et al. 2016). Additionally, this priming enhances the proline Pro-concentration in the seedlings of sunflower under stress of salt (Moghanibashi et al. 2013). Seed priming using PEG enhances the quality of sweet peppers and repaired damaged membranes, which increased the synthesis of the antioxidant system that eliminates ROS and decreased lipid peroxidation, according to Siri et al. (2013). Additionally, PEG priming reduced the production of antioxidant enzymes in response to the stress caused by nano-ZnO (Salah et al. 2015).

4.2 Inorganic Salts

Inorganic salts like acetic acid, $CaCl_2$, KNO_3, $NaNO_3K_3PO_4$, $MgSO_4$, $MnSO_4$, $CuSO_4$, $ZnSO_4$, KH_2PO, and $MgCl_2$ as well as organic compounds like glycerol, sorbitol, Pro, mannitol, butenolide, putrescine, chitosan, choline, and paclobutrazol are also used to control water potential in seed (Paparella et al. 2015; Farooq et al. 2005).

Seed priming with inorganic salts affects the mobilization of organic compounds to various sections of the embryo and enhances the antioxidant enzyme activities involved in seed germination. When compared to hydro-primed plants, seeds primed with $ZnSO_4$ had stronger CAT, SOD, dismutase, and POD activity (Aboutalebian and Nazari 2017). By boosting the antioxidant levels in the *Medicago sativa* seedlings, this therapy also reduced the effects of drought stress (Fallah et al. 2018).

NaCl seed priming increased Pro and soluble carbohydrate content, antioxidant enzyme activity, and seed membrane damage protection (Farhoudi et al. 2011). Activating the gibberellin biosynthetic gene and genes that weaken the endosperm cap increased the quantity of gibberellin in the tomato seed, which boosted germination ability (Nakaune et al. 2012). Amaranth seeds primed with NaCl resulted in a decrease in MDA generation, an increase in cell membrane stability, and an improvement in salinity tolerance via encouraging K^+ and Ca^{2+} accumulation (Omami 2005). By seed priming with sodium nitroprusside, which also increased the activity of antioxidant enzymes and levels of ascorbic acid, Pro, and total phenolics, the adverse effects of salt stress were lessened (Ali et al. 2017). Calcium nitrate seed priming increased germination rates and seedling emergence when temperatures weren't perfect (Batista et al. 2016). After seed priming with KNO_3, acid phytase and phosphatase activity was elevated in the roots, cotyledons, and shoots of lettuce seeds, increasing the seeds' capacity to tolerate salt (Nasri et al. 2011). The positive effect of calcium chloride priming on improving salinity tolerance may be due to the antioxidant system being activated and enhanced proline Pro build up in cucumber seedlings (Joshi et al. 2013). Priming sunflower seeds with KNO_3 improved their tolerance to salt by enhancing the activity of the POX enzyme and decreasing the seedling MDA level (Farhoudi 2012). Seed priming with KNO3 solution decreased peroxidase activity in *Silybum marianum* (Zavariyan et al. 2015). Pigeon pea seeds treated with $CaC1_2$ or KNO_3 showed improvements in soluble sugars, free amino acids, and proteins during germination under salt stress. In pepper seedlings grown from primed seeds, protein synthesis, water content, soluble sugar, polyphenols, soluble proteins, carotenoid content, and bulk were all improved with KCl, NaCl, and $CaCl_2$ (Aloui et al. 2017).

4.3 Fertilizers

To increase the effectiveness of seed priming, various fertilizers are utilized. This is due to the fact that seeds of soaking okra in a one diammonium phosphate or superphosphate solution boost growth and accelerated growth of seedling (Shah et al. 2011). Instead of using clean water to prime seedlings, dissolved nutrients are employed in seed priming with micronutrients (Arif et al. 2005). The status of mineral-nutrient of the plants is crucial for increasing plant abiotic stress tolerance (Marschner 1995).

4.4 Plants Growth Regulators

Hormonal priming is the term for the process of prepping seeds with phytohormones or plant growth agents. Crop emergence and germination are improved when seeds are primed with plant growth regulators (Carvalho et al. 2011; Gao et al.

2002). Hormones of plants play a critical part in the growth and physiology of plants with the help of producing and conveying various forms of vital signals within and between the cells; under abiotic stresses, the levels of endogenous of phytohormones significantly change (Iqbal and Ashraf 2013). Hormones of plants are essential for controlling molecular and physiological reactions to stresses of abiotic, which is necessary for overcoming unfavorable conditions of environment (Pandey 2017; Ku et al. 2018). Phytohormone priming of seed can enhance germination of seed by activating enzymes like protease and amylase that hydrolyze parts of protein and starch into simple forms that are edible by embryos (Miransari and Smith 2014). Abiotic stress also lowers the basic levels of most of the growth upregulators; however, the exogenous hormone therapy also be able to substitute for their absence (Babu et al. 2012). In response to salinity stress, plant tissues produce more abscisic acid (ABA) and jasmonic acid (JA) and less auxin, salicylic acid cytokinin, and gibberellins. Exogenous process of plant growth regulators can be utilized to undo these changes, which is the first step towards slowing the loss in plant growth due to stress. Use of the seed priming with a range of plant growth regulators, like JA and salicylic acid, has increased abiotic stress tolerance, seed germination, and seedling establishment (Javid et al. 2011; Krantev et al. 2008). Additionally, above mentioned treatments maintain steady levels of the substance like cytokinin and IAA in tissues of plant, which stimulate cell growth (Sakhabutdinova et al. 2003). Hormone priming of seeds of maize improved resistance to oxidative stress, which in turn raised resistance to abiotic stress (Afzal et al. 2008; Farooq et al. 2008).

4.5 Polyamines

Plant development and growth are known to be positively impacted by polyamines (PAs) (Watson and Malmberg 1998). They consist of putrescine and spermidine. The control of several processes of metabolism, like synthesis of protein, transcription, RNA modification, and regulation of enzyme concentration, as well as a number of plant stress responses has been linked to polyamines, which are tiny, ubiquitous molecules (Takahashi and Kakehi 2010). By maintaining membranes beneath high stress conditions and protecting plants from diverse abiotic challenges, polyamines act as cellular defenders since they bind to cations and can be combined with anionic membrane components like phospholipids (Kusano et al. 2008). Exogenous polyamines' (PA) impact on germination of seed is influenced by the kind, concentration, and dormancy state of the embryo (Farooq et al. 2011). Different crops' germination and seedling vigor were improved by priming of seed with PAs (Khan et al. 2021; Afzal et al. 2009).

Hence, by reducing oxidative damage, adjusting antioxidant, osmolytes, and photosynthetic systems, ensuring the antioxidant enzymes are active, and lowering hydrogen peroxide and MDA levels, spermine-primed seeds increased rice seedlings' ability to withstand salt stress (Paul and Aryadeep 2016). By increasing the activity of enzymes, the starch metabolism and the ascorbate-glutathione cycle, as well as the removal of ROS, reduction of lipid peroxidation, and strengthening of cell membrane stability, priming of seed with spermidine (Spd) increased stress tolerance of water in white clover (Li et al. 2014). By upregulating the genes encoding the PA biosynthesis-related enzymes, increasing the quantity of glycine betaine, flavonoids, total phenolics, and also strengthening the antioxidant enzyme activities, seed priming with Spd improved rice's ability to withstand chilling (Sheteiwy et al. 2017).

4.6 Ascorbic Acid

By increasing catalase and peroxidase activity, seed priming of ascorbic acid lessened stress of salt in pumpkin seedlings and decreased the activity of the corresponding enzymes, which is presumably due to the neutralization of free radicals brought on by stress of salt (Fazlali et al. 2013). However, under conditions of water constraint, ascorbic acid priming of rapeseed seeds increased growth rates, root and length of shoot, seedling vigor, and POX and CAT activity (Razaji et al. 2014). Under drought conditions, ascorbic acid-primed wheat seed resulted in increased seedling germination, development, and moisture stature (Farooq et al. 2013).

4.7 Beta-aminobutyric Acid

It is well known that plants can also be effectively trained to tolerate the abiotic stress by beta-amino butyric acid (BABA). This is accomplished via interacting with a variety of many hormones like ethylene, salicylic acid, and abscisic acid, or by creating a cascade of signaling pathways controlled by H_2O_2 (Zhong et al. 2014). The lipid peroxidation of cell membrane is decreased, and activity of enzymes, basically antioxidant enzymes, is increased; priming rice grains with BABA reduced the amount of MDA and increased the resistance of seedlings of rice to salt stress (Jisha and Puthur 2016). Furthermore, priming of the seed by beta-aminobutyric acid led to a rise in total carbohydrate, proline, and proteins; a combination of accumulation and growth in activity of nitrate reductase. It also boosted activity of enzymes antioxidant ones and decreased the amount of MDA in green grams (Jisha and Puthur 2016).

4.8 Aminolaevulinic Acid

After enhancing the activity of antioxidant enzymes, priming of seed by using 5-aminolevulinic acid enhanced seedling growth and germination (Kanto et al. 2015). The seed treatment by using 5-aminolevulinic acid enhanced the chilling endurance in 2 rice cultivars by enhancing the use of the antioxidant enzyme activities (APX, SOD, and GPX), boosting the substance of the glycine betaine, flavonoids, total phenolics, and increasing expression of the genes encoding PA biosynthesis-related enzymes (Sheteiwy et al. 2017).

4.9 Glycine Betaine

By boosting SOD enzyme activity, raising Pro content, and decreasing MDA content, salt tolerance was improved by glycine betaine (GB) priming pepper seeds (Korkmaz and Sirikci 2011). The prevention of damage of cell membrane the positive effects of priming of glycine betaine on enhancing tolerance of salt in the safflower may be attributed to decreasing lipid oxidation of membrane and raising homeostasis of ion (Alasvandyari et al. 2017). By enhancing activities of enzyme antioxidant, lowering water level, and reducing damages of cell membrane, seed treatment with GB also controls the ability of cotton seedlings to withstand cold stress (Cheng et al. 2018). Increases in Pro, IAA, excitation of antioxidant enzymes, dissolved fructose intake, a decrease in MDA and water, in addition to the stimulation of antioxidant enzymes, can all be achieved by seed soaking to counteract the damaging effects caused by water stress applied on the canola plants (Dawood et al. 2014).

4.10 Melatonin

Using seed priming by enhancing the working of enzymes, basically antioxidant enzymes, raising total phenolic, relative water, and Pro content, and decreasing sodium material, membrane relative electrolyte leakage, and lipid peroxidation level, melatonin reduced the impacts of salt stress on maize (Jiang et al. 2016). By raising the concentration of potassium, CAT, IAA, total carbohydrate, pigments, and total phenolic, as well as K^+/Na^+ and Ca^{2+}/Na^+ ratios, seed priming with melatonin reduced salt stress in faba bean seedlings (Dawood and EL-Awadi 2015).

4.11 Chitosan

Priming known as solid matrix with chitosan under saline stress increases seedling growth and germination (Sen and Mandal 2016). Priming known as chitosan improved the germination of wheat seeds also under osmotic stress seedling vigor (Hameed et al. 2014). Priming of seed with chitosan increases germination of seed and the seedlings' vigor because of synthesis and the mobilization of food resources, activating of specific enzymes, and the synthesis of RNA and DNA, in addition to the triggering of genes' germination response (Sadeghi et al. 2011). Chitosan priming of seed reduced the average germination time under low temperature stress, enhanced seedling vigor and CAT and POX activity, and lowered MDA ingredients and relative plasma membrane permeability, and higher levels of soluble sugars and Pro (Guan et al. 2009).

5 Seed Priming-Induced Abiotic Stress Tolerance

Complex processes that include a variety of adjustments, including the activation of particular genes, increased antioxidative activities, brief rises in the plant levels of regulator, protective proteins, inhibition of consuming of energy pathways, and also accumulation of osmolytes, are all part of the process of adapting to abiotic stresses (Bartels and Sunkar 2005). Abiotic stress conditions can benefit many plants from the exogenous applications of certain compounds through priming of seed to enhance seed seedling establishment and germination. The practical physiological technique for helping different plant species endure harmful abiotic conditions including drought, salt, and cold is seed priming (Hussain et al. 2016b; Khan et al. 2012). Under abiotic stress circumstances, seed priming's positive effects may be more apparent (Masondo et al. 2018; Ibrahim 2016; Chen et al. 2013).

There are presumably two methods through which seed priming achieves abiotic stress resistance. In the first method, pre-germination metabolic mechanisms such as enhanced energy metabolism, early seed food mobilization, elongation of embryonic cells, and breakdown of the endosperm are activated by seed priming (Chen and Arora 2011; Sun et al. 2010). In the second technique, seed priming subjects seeds to biotic pressures, which decrease radicle protrusion but promote stress responses such as osmotic adjustment, enzyme activation, and cross-tolerance to abiotic stressors. In germinating primed seeds, these stress coping mechanisms create a "priming memory" that may be called upon after stress exposure to promote greater stress tolerance (Ibrahim 2016; Bruce et al. 2007). Pre-germination stress exposure known as "seed priming" enables seeds to better withstand subsequent environmental stressors (Tanou et al. 2012). Prior exposure to some elements during the seed priming process gives plants the ability to withstand future stress exposure. By increasing the capacity for enzymatic and nonenzymatic antioxidants, accumulating essential transcription factors and dormant signaling proteins, and triggering of the epigenetic alterations and the development of defense reactions to epigenetic alterations through jasmonate, it enables an accelerated defense mechanism against harmful environments (Kasote et al. 2019; Tanou et al. 2012).

Most crucial defense mechanism for sprouting primed seeds to minimize ROS formation from abiotic stimuli is the antioxidant system (Banerjee and Roychoudhury 2018; Ibrahim 2016). Priming of seed raises the levels of ascorbic acid and the activity of the following antioxidant enzymes including SOD, GR, POX, and CAT. Through decrease in hydrogen peroxide and superoxide production, these actions add to the scavenging of ROS and seed protection (Kasote et al. 2019; Pal et al. 2017; Paparella et al. 2015). Additionally, anthocyanin, a photoprotective pigment, is encouraged to accumulate in seeds through seed priming, which enhances the ROS elimination and promotes protection of plant (Banerjee and Roychoudhury 2016).

Numerous plants have demonstrated the bountiful effects of the process of seed priming under stress of salt, including tomato (Cuartero et al. 2006), watermelon (*Cucumis melo* L.), (Sivritepe et al. 2003), hot pepper (Khan et al. 2009a), lettuce (Nasri et al. 2011), and okra (Dkhil et al. 2014). Carrot (Pill and Finch-Savage 1988), watermelon (Demir and Oztokat 2003), the muskmelon (Nascimento 2003), and also asparagus all benefited from seed priming when exposed to high temperatures (Bittencourt et al. 2004). At 1.0 and 1.2 MPa, carrot seed priming in PEG 6000 seed improved the germination and emergence of seedling (Pereira et al. 2009). Advantage of process of seed priming on seed of onion's performance at low temperatures might be due to the boost in the activity of endo-mannanase in the seeds' anterior to lateral root extrusion, an enzyme involved in the breakdown of endosperm cell walls. By raising the maximal germination temperature, seed priming reduced thermos inhibition in lettuce (Schwember and Bradford 2010). Additionally, for maize hybrids considered to be sensitive to cold temperatures, seed priming may be favorable (Hacisalihoglu et al. 2018).

5.1 Drought Stress

Water stress has the ability to impact early seedling growth and seed germination (Ahmad et al. 2009). Seeds of famous Chinese cabbage, also known as (*Brassica rapa*), were pretreated by the use of two hundred mM of potassium nitrate, also two hundred mM of urea, according to Yan (2015) who found that this encouraged early seedling growth while the plants were under the stress of drought. By controlling the basic function of peroxidase, superoxide dismutase, and catalase as well as the amounts of soluble sugar and Pro, seed priming treatments improved drought tolerance. These results suggested that under moisture-stress conditions, seed priming can also be employed to increase Chinese cabbage implantation and early seedling growth. Early vigor in plants is caused by seed priming: another method of dealing with abiotic stress.

Using magnetically treated seeds to grow chickpea plants and subjecting them to a moisture deficit of about 0.2 MPa resulted in improved water consumption, efficiency of radiation use, and biomass (Mridha et al. 2016). Due to better soil-water connections, magneto priming in maize enabled the seedlings to tolerate moisture stress (Anand et al. 2012). Under drought stress, priming of seed with hormone of salicylic acid and methyl jasmonate or the chemical of (paclobutrazol) elicitors boosted the development of the rice drought-responsive (RD2 and RD1) genes (of the AP2/ERF family) and the total phenolic content (Samota et al. 2017). Under controlled and stress of drought conditions, plants grown from the seeds that were primed outperformed those plants grown by seeds that were not primed in terms of development and growth.

Sunflower (*Helianthus annuus* L.) seeds were pretreated with ascorbic acid by Fatemi (2014). who found that pretreated seeds exhibited improved germination, seed durability, and growth rates than untreated seeds under PEG 6000-induced minimal potentials for water (2 to 4 bar). Ascorbic acid administered exogenously proved successful in reducing the consequences of stress caused by drought. Also similar results were observed when vitamin C or ascorbic acid was used to prime safflower seeds, boosting the antioxidant system of both nonenzymatic and enzymatic substances to minimize impacts of drought caused stress. Additionally, in primed seeds, the amylase or amylase capabilities were higher, which caused a larger starch degradation and an increase in sugar levels. In comparison to unprimed seeds, this one was followed by an improvement in germination speed, respiration, seed viability, and establishment of seedling (Amooaghaie and Nikzad 2013). Li et al. (2014a, b) found out that during stress of water in the very early stages involved in the germination of seed, chemicals boost the expression of the -amylase gene such as the chemical of polyamines.

Soybean plants that had pretreatment with H_2O_2 showed increased levels of the galactinol and the myo-inositol and when subjected to drought stress. They are thought to act as osmoprotectants and ROS scavengers. D-myo-inositol 3-phosphate synthase 2 and Galactinol synthase that are the crucial enzymes needed to make oligosaccharides (GolS), exhibit increased gene expression that is related to the buildup of galactinol and myo-inositol, respectively. Ishibashi et al. (2011) and Sghaier-Hammami et al. (2010) suggested that DC-8 plays a part in the defense of structures of embryonic cellular against drought stress and tolerance to desiccation. PEG priming increased AP2/EREBP regulon, a part of hormone, sugar, and redox signaling that is used to communicate with abiotic stresses, in *Arabidopsis* and *Medicago* seeds during germination (Maia et al. 2011; Dietz et al. 2010).

5.2 Heat Stress

Crop plants' metabolic processes are adversely affected by extreme temperature stress. Researchers from several fields employ seed priming methods to protect crops from temperature stress. Wheat that was planted late has lower yields due to heat stress. Osmoprimed late-sown wheat increased tiller numbers, biological yield, and harvest index to mitigate the negative effects of high temperatures (Mustafa et al. 2017). While, on the other hand, many crops' development and yield are constrained by chilling stress. Low temperatures cause membrane damage, impaired cellular respiration, and higher ROS levels, which all contribute to poor and unpredictable germination (Xing and Rajashekar 2001). Key physiological functions are disrupted as a result of damage that is induced by low temperature to organelles of cell, restriction of cell growth and division, as well as plant tissues suffering from metabolic imbalance, which are its primary symptoms, ultimately appear as altered seed shape (Ruelland et al. 2009). However, Hussain et al. (2016a) investigated how different methods for priming seeds that include hydropriming, redox priming, osmo-priming, hormonal priming, and chemical priming affect rice plants' ability to tolerate cold. Cold temperatures influenced a wide range of physiological processes in unprimed rice, notably postponed sprouting and poor germination rate. The mobilization and respiration of the seed reserve's associated metabolic processes were slowed down. Rice seedlings under control had higher levels of lipid peroxidation and H_2O_2 buildup. By triggering the production of glutathione and free radicals, as well as the activity of enzyme of antioxidant systems such as peroxidase, oxidative dismutase, and catalase Pro, seed priming protects the seedlings from oxidative stress. Melatonin-primed corn seeds exhibit greater germination tolerance to stress caused by chilling (Kołodziejczyk et al. 2016).

5.3 Heavy Metal Stress

Heavy metals such as cadmium, lead, copper, mercury, and chromium are known to cause severe environmental hazards in our day and age. The buildup of heavy metals in cropland is a major cause for concern since it could negatively impact crop development, soil health, environmental safety, and the quality and viability of food markets (Nagajyoti et al. 2010). By the contamination of the water, air, and soil, the plants' metabolism has an impact on the heavy metals' redistribution in the crust of earth and also in living things. According to researcher (Barceló 1990; Nagajyoti et al. 2010) plants that are grown on soils and contaminated with heavy metals exhibit metabolic changes, reduced growth, altered water relations, and heavy metal accumulation in edible plant sections.

However, studies show that utilizing plant hormones to prime seeds can significantly reduce the uptake of heavy metals in plant tissues while improving crop production under heavy metal stress. As an illustration, under cadmium stress, Gibberellic acid was used to prepare pigeon pea seeds, ethylene, abscisic acid, and cytokinine (each with a capacity of 10 or 100 M). Priming of seed with either hormone of plant at both doses resulted in an increase in the germinating rate index, overall regeneration of final germination percentage, time, and duration to fifty percent germination (Sneideris et al. 2015). Investigation found that priming of seed with 0.1 mM gibberellic acid also improved photosynthesis during the light phase and capacity of antioxidant in *Trifolium repens* L. in the soil contaminated with heavy metals (Galhaut et al. 2014). Under calcium chloride stress, salicylic acid seed priming of safflower seeds increased germination rates, dried seedling weight, and particular leaves mass (Jam et al. 2012).

6 Physiological and Biochemical Responses of Seed Priming to Abiotic Stress Tolerance

Crop stand establishment depends significantly on the time between the planting of seeds and the emergence of seedlings (Hubbard et al. 2012). Priming of seed has been implemented to boost rates of germination and regularity (Jisha et al. 2013; Di Girolamo and Barbanti 2012). After the seed has been re-desiccated, various germination-promoting processes are basically activated by the process of seed priming (Asgedom and Becker 2001). Because the imbibition lag is less with primed seeds, germination is typically more rapid and uniform. Fundamental mechanisms of seed priming initially include the enzyme activation, improved metabolism of germination, enhanced restore mechanisms, and osmotic correction (Hussain et al. 2015; Farooq et al. 2006). Physiological, metabolic, molecular, and cellular changes caused by priming of seed improve performance of seed and promote earlier as well as more uniform germination (Manonmani et al. 2014; Siri et al. 2013).

The activation of the priming memory is a hallmark of phase I. In this stage, stress hormone genes are activated, and proteins are produced, DNA and mitochondria are repaired, and signaling pathways are controlled. The most important phase is phase II, during which the main cellular and metabolic processes rise and germination begins. During this stage, the embryo synthesizes a number of proteins, including those involved in ROS signaling and scavenging, stress response, and storage, and employing both current and DNA, newly generated messenger ribonucleic acids (mRNA) and mitochondria are also created and repaired. The embryonic axis elongates, considerable reserve material mobilization takes place, the absorption of water and oxygen increases, and the radicle eventually emerges from the seed coat to finish germination and start the post-germination stage (Rosental et al. 2014; Chen and Arora 2013).

The advantages of process of seed priming are primarily involved in modulation of germination metabolic processes within the seed required for activation of enzyme, activities of antioxidant, nucleic acid (RNA and DNA) and repair processes, protein synthesis, breaking seed dormancy, translation, transcription, and DNA replication, and metabolism of germination, repairing damaged parts of seed, and reducing leakage of metabolite. Increasing ATP and the number of mitochondria, priming seeds improves germination performance, vigorous growth, and abiotic stress tolerance by triggering a number of physiological, biochemical, and metabolic changes (Wojtyla et al. 2016; Paparella et al. 2015).

One benefit of priming of seed is activation of genes and proteins related to division of cell, cell membrane transcription, modification, capacity translation, mobilization, oxidant stress reaction, and the water ways, where membrane and DNA repair is superior to ordinary ingestion (Kubala et al. 2015; Bewley et al. 2013). The enhanced process of germination caused by seed priming is attributed to a greater protein abundant supply engaged in oxidative stress tolerance, focused protein denaturation, posttranslational processing capabilities, separation or devastation of dormancy blocks, cell wall repair, premature development of embryo, adjustment of cells surrounding embryo, and an overall enhancement of pre-germination metabolism (Giri and Schillinger 2003).

7 Conclusion and Future Prospective

The main issues in agriculture include abiotic pressures like salinity, drought, and severe temperatures. These pressures frequently trigger various cellular responses and signaling pathways, as well as various types of damage. The use of seed priming can help crops combat the negative impacts of abiotic stressors. Seed priming improves plant defense-related responses and cell signaling pathways. It has been discovered that seed priming causes certain physiologic, biochemical, and molecular alterations. The advantages of seed priming programe include the activation of specific genes involved in enzyme synthesis, the use of stored energy, the restoration of organ of mitochondria and cell membranes, the production as well as maintenance of nucleic acids, and mobilization of mitochondria and reserve food. DNA transcription, translation, and replication are all controlled by seed priming. Designing proper and efficient priming methods and substances for many different plants will be crucial for increasing abiotic plant tolerance for stress in a basic sustainable manner. Research in future should also focus on the physiological, biochemical, and molecular changes caused through seed treatment under different abiotic conditions.

References

Afzal I, Basra SMA, Ahmad N, Cheema MA, Warraich EA, Khaliq A (2002) Effect of priming and growth regulator treatment on emergence and seedling growth of hybrid maize (*Zea mays*). Int J Agric Biol 4:303–306

Afzal I, Basra SMA, Shahid M, Farooq M, Saleem M (2008) Priming enhances germination of spring maize (*Zea mays* L.) under cool conditions. Seed Sci Technol 36:497–503

Afzal I, Ashraf S, Qasim M, Basra SMA, Shahid M (2009) Does halopriming improve germination and seedling vigour in marigold. Seed Sci Technol 37:436–445

Aboutalebian MA, Nazari S (2017) Seedling emergence and activity of some antioxidant enzymes of canola (Brassica napus) can be increased by seed priming. J Agric Sci 155(10):1541–1552

Ahmad S, Ahmad R, Ashraf MY, Ashraf M, Waraich EA (2009) Sunflower (Helianthus annuus L.) response to drought stress at germination and seedling growth stages. Pak J Bot 41:647–654

Alasvandyari F, Mahdavi B, Hosseini SM (2017) Glycine betaine affects the antioxidant system and ion accumulation and reduces salinity-induced damage in safflower seedlings. Arch Biol Sci 69(1):139–147

Ali Q, Daud MK, Haider MZ, Ali S, Rizwan M, Aslam N, Noman A, Iqbal N, Shahzad F, Deeba F, Ali I, Zhu SJ (2017) Seed priming by sodium nitroprusside improves salt tolerance in wheat (*Triticum aestivum* L.) by enhancing physiological and biochemical parameters. Plant Physiol Biochem 119:50–58

Almansouri M, Kinet JM, Lutts S (2001) Effect of salt and osmotic stresses on germination in durum wheat (*Triticum durum*). Plant Soil 231:243–254

Aloui H, Aymen EM, Chérif H (2017) Seed priming to improve seedling growth of pepper cultivars exposed to salt concentrations. Int J Veg Sci 23(6):489–507

Amooaghaie R, Nikzad K (2013) The role of nitric oxide in priming-induced low-temperature tolerance in two genotypes of tomato. Seed Sci Res 23:123–131

Anand A, Nagarajan S, Verma APS, Joshi DK, Pathak PC, Bhardwaj J (2012) Pre-treatment of seeds with static magnetic field ameliorates soil water stress in seedlings of maize (*Zea mays* L.). Ind J Biochem Biophys 49:63–70

Arif M, Ali S, Shah A, Javed N, Rashid A (2005) Seed priming maize for improving emergence and seedling growth. Sarhad J Agri 21:17–20

Asgedom H, Becker M (2001) Effects of seed priming with nutrient solutions on germination, seedling growth and weed competitiveness of cereals in Eritrea. In Proc Deutscher Tropentag 2001, University of Bonn and ATSAF. Magrraf Publishers Press, Weickersheim, p 282

Babu MA, Singh D, Gothandam KM (2012) Effect of salinity on growth, hormones and mineral elements in leaf and fruit of tomato cultivar PKM. J Anim Plant Sci 22(1):159–164

Banerjee A, Roychoudhury A (2016) Group II late embryogenesis abundant (LEA) proteins structural and functional aspects in plant abiotic stress. Plant Growth Regul 79(1):1–17

Banerjee A, Roychoudhury A (2018) Seed priming Technology in the Amelioration of salinity stress in plants. In: Rakshit A, Singh H (eds) Advances in seed priming. Springer, Singapore

Barceló J, Poschenrieder C (1990) Plant water relations as affected by heavy metal stress. A review J. Plant Nutr 13:1–37

Bartels D, Sunkar R (2005) Drought and salt tolerance in plants. Crit Rev Plant Sci 24:23–58

Basra SMA, Farooq M, Hafeez K, Ahmad N (2004) Osmohardening a new technique for rice seed invigoration. Int Rice Res Notes 29:80–81

Batista TB, Cardoso ED, Binotti FFS, Costa E, Sa ME (2016) Priming and stress under high humidity and temperature on the physiological quality of *Brachiaria brizantha* cv. MG-5 seeds. Acta Sci 38(1):123–127

Bewley JD, Bradford KJ, Hilhorst HWM, Nonogaki H (2013) Seeds physiology of development, germination and dormancy, 3rd edn. Springer, New York

Bittencourt MLC, Dias DCFS, Dias LAS, Araujo EF (2004) Efeito do condicionamentoosmotico das sementesnagerminaçao e no crescimento das plantulas de aspargo. Revista Brasileira de Sementes 26:50–56

Bruce TJA, Matthes MC, Napier JA, Pickett JA (2007) Stressful memories of plants evidence and possible mechanisms. Plant Sci 173:603–608

Butler L, Hay F, Ellis R, Smith R, Murray T (2009) Priming and re-drying improve the survival of mature seeds of Digitalis purpurea during storage. Ann Bot 103:1261–1270

Callan NW, Mathre DE, Miller JB (1990) Bio-priming seed treatment by biological control of Pythium ultimum pre-emergence damping-off in sh2 sweet corn. Plant Disease 74:368

Carvalho RF, Piotto FA, Schmidt D, Peters LP, Monteiro CC, Azevedo RA (2011) Seed priming with hormones does not alleviate induced oxidative stress in maize seedlings subjected to salt stress. Sci Agric 68:598–602

Castañares JL, Bouzo CA (2018) Effect of different priming treatments and priming durations onmelon germination behavior under suboptimal conditions. Open Agric 3(1):386–392

Catusse J, Meinhard J, Job C, Strub JM, Fischer U, Pestova E, Westohoff P, Van Dorselaer A, Job D (2011) Proteomics reveals potential biomarkers of seed vigor in sugar beet. Proteomics 11:1569–1580

Chen K, Arora R (2011) Dynamics of the antioxidant system during seed osmopriming, post-priming germination, and seedling establishment in spinach (*Spinacia oleracea*). Plant Sci 180:212–220

Chen K, Arora R (2013) Priming memory invokes seed stress-tolerance. Environ Exp Bot 94:33–45

Chen K, Fessehaie A, Arora R (2013) Aquaporin expression during seed osmopriming and post-priming germination in spinach. Biol Plant 57:193–198

Cheng C, Pei LM, Yin TT, Zhang KW (2018) Seed treatment with glycine betaine enhances tolerance of cotton to chilling stress. J Agri Sci:1–10

Corbineau F, Come D (2006) Priming: a technique for improving seed quality. Seed Test Int 132:38–40

Cramer GR (2002) Sodium-calcium interactions under salinity stress in Läuchli a, Lüttge u, salinity Environment-plants-molecules. Kluwer Academic Publishers, Dordrecht, pp 205–227

Cuartero J, Bolarin MC, Asins MJ, Moreno V (2006) Increasing salt tolerance in tomato. J Exp Bot 57:1045–1058

Dawood MG, El-Awadi ME (2015) Alleviation of salinity stress on (*Viciafaba* L) plants via seed priming with melatonin. Acta Biol Colom 20:223–235

Dawood MG, Taie HAA, Nassar RMA, Abdelhamid MT, Schmidhalter U (2014) The changes induced in the physiological, biochemical and anatomical characteristics of Vicia faba by the exogenous application of proline under seawater stress. South Afr J Bot 93:54–63

Demir I, Oztokat C (2003) Effect of salt priming on germination and seedling growth at low temperatures in watermelon seeds during development. Seed Sci Technol 31:765–770

Dhanya Thomas TT, Puthur JT (2017) UV radiation priming: a means of amplifying the inherent potential for abiotic stress tolerance in crop plants. Environ Exp Bot 138:57–66

Di Girolamo G, Barbanti L (2012) Treatment conditions and biochemical processes influencing seed priming effectiveness. Ital J Agron 7:8–18

Dietz KJ, Vogel MO, Viehhauser A (2010) AP2/EREBP transcription factors are part of gene regulatory networks and integrate metabolic, hormonal and environmental signals in stress acclimation and retrograde signaling. Protoplasma 245:3–14

Dkhil T, Zaman N, Rima R, Nahar K (2014) Yield response of okra to Seed priming with different chemicals. J Hort Sci 1:10–14

Dorna H, Jarosz M, Szopinska D, Szulc I, Rosinska A (2013) Germination vigour and health of primed Allium cepa L. seeds after storage. Acta Sci Pol Hortorum Cultus 12:43–58

Eisvand HR, Tavakkol-Afshari R, Sharifzadeh F, Maddah-Aref H, Hesamzadeh-Hejaz SM (2010) Effects of hormonal priming and drought stress on activity and isozyme profiles of antioxidant enzymes in deteriorated seed of tall wheat grass. Seed Sci Technol 38:280–297

Elkoca E, Haliloglu K, Esitken A, Ercisli S (2007) Hydro-and osmo-priming improve chickpea germination. Acta Agric Sci Sect B Soil Plant Sci 57(3):193–200

Eskandari H (2013) Effects of priming technique on seed germination properties, emergence and-field performance of crops. A review. Int J Agron Plant Prod 4:454–458

Fallah S, Malekzadeh S, Pessarakli M (2018) Seed priming improves seedling emergence and reduces oxidative stress in Nigella sativa under soil moisture stress. J Plant Nutr 41(1):29–40

Farhoudi R, Saeedipour S, Mohammadreza D (2011) The effect of NaCl seed priming on salt tolerance, antioxidant enzyme activity, proline and carbohydrate accumulation of muskmelon (*Cucumis melo* L.) under saline condition. Afr J Agric Res 6(6):1363–1370

Farhoudi R (2012) Evaluation effect of KNO_3, seed priming on seedling growth and cell membrane damage of sunflower (*Helianthus annus*) under salt stress. Am Eurasian J Agric Environ Sci 12(3):384–388

Farooq M, Basra SMA, Saleem BA, Nafees M, Chishti SA (2005) Enhancement of tomato seed germination and seedling vigor by osmopriming. Pak J Agric 42(3–4):36–41

Farooq M, Basra SMA, Khalid A, Tabassum R, Mehmood T (2006) Nutrient homeostasis reserves metabolism and seedling vigor as affected by seed priming in coarse rice. Can J Bot 84:1196–1202

Farooq M, Basra SMA, Rehman H, Hussain M (2008) Seed priming with polyamines improves the germination and early seedling growth in fine rice. J New Seeds 9:145–155

Farooq M, Aziz T, Rehman H, Rehman A, Cheema SA, Aziz T (2011) Evaluating surface drying and re-drying for wheat seed priming with polyamines effects on emergence early seedling growth and starch metabolism. Acta Physiol Plant 33:1707–1713

Farooq M, Irfan M, Aziz T, Ahmad I, Cheema SA (2013) Seed priming with ascorbic acid improves drought resistance of wheat. J Agron Crop Sci 199:12–22

Fatemi SN (2014) Germination and seedling growth in primed seeds of sunflower under water stress. Annu Res Rev Biol 4(23):3459–3469

Fazlali R, Asli DE, Moradi P (2013) The effect of seed priming by ascorbic acid on bioactive compounds of naked seed pumpkin (*Cucurbita pepo* var. *styriaca*) under salinity stress. Int J Farm All Sci 2(17):587–590

Galhaut L, Lespinay A, Walker DJ, Bernal MP, Correal E, Lutts S (2014) Seed priming of (*Trifolium repens* L.) improved germination and early seedling growth on heavy metal contaminated soil. Water Air Soil Pollution 225:1–15

Gallardo K, Job C, Groot SP, Puype M, Demol H, Vandekerckhove J, Job D (2001) Proteomic analysis of Arabidopsis seed germination and priming. Plant Physiol 126:835–848

Gao YP, Bonham-Smith PC, Gusta LV (2002) The role of peroxiredoxin antioxidant and calmodulin in ABA-primed seeds of Brassica napus exposed to abiotic stresses during germination. J Plant Physiol 159:951–958

Ghassemi-Golezani K, Esmaeilpour B (2008) The effect of salt priming on the performance of differentially matured cucumber (*Cucumis sativus*) seeds. Notulae Bot Hortic Agro bot Cluj Napoca 36(2):67–70

Giri GS, Schillinger WF (2003) Seed priming winter wheat for germination, emergence and yield. Crop Sci 43:2135–2141

Gong F, Wu X, Wang W (2013) Comparative proteomic identification of embryo proteins associated with hydropriming induced rapid-germination of maize seeds. Plant Omics J 6(5):333–339

Govinden-Soulange J, Levantard M (2008) Comparative studies of seed priming and pelleting on percentage and meantime to germination of seeds of tomato (Lycopersicon esculentum Mill). Afr J Agric Res 3(10):725–731

Guan YJ, Hu J, Wang XJ, Shao CX (2009) Seed priming with chitosan improves maize germination and seedling growth in relation to physiological changes under low temperature stress. J Zhejiang Univ Sci B 10(6):427–433

Hacisalihoglu G, Kantanka S, Miller N, Gustin JL, Settles AM (2018) Modulation of early maize seedling performance via priming under sub optimal temperatures. PLoS One 13(11)

Hameed A, Sheikh MA, Hameed A, Farooq T, Basra SMA, Jamil A (2014) Chitosan seed priming improves seed germination and seedling growth in Wheat (*Triticum aestivum* L.) under osmotic stress induced by polyethylene glycol. Philip Agric Sci 97(3):294–299

Harris D, Breese WA, Kumar Rao JVDK (2005) The improvement of crop yield in marginal environments using on-farm seed priming nodulation, nitrogen fixation and disease resistance. Aust J Agric Res 56:1211–1218

Horii A, McCue P, Shetty K (2007) Seed vigour studies in corn, soybean and tomato in response to fish protein hydrolysates and consequences on phenolic-linked responses. Bioresour Technol 98:2170–2177

Hubbard M, Germida J, Vujanovic V (2012) Fungal endophytes improve wheat seed germination under heat and drought stress. Botany 90:137–149

Hussain S, Zheng M, Khan F, Khaliq A, Fahad S, Peng S, Huang J, Cui K, Nie L (2015) Benefits of rice seed priming are offset permanently by prolonged storage and the storage conditions. Sci Rep 5:1–12

Hussain S, Khan F, Cao W, Wu L, Geng M (2016a) Seed priming alters the production and detoxification of reactive oxygen intermediates in rice seedlings grown under sub-optimal temperature and nutrient supply. Front Plant Sci 7:439

Hussain S, Khan F, Hussain HA, Nie L (2016b) Physiological and biochemical mechanisms of seed priming-induced chilling tolerance in Rice cultivars. Front Plant Sci 7:116

Hussain HA, Hussain S, Khaliq A, Ashraf U, Anjum SA, Men S, Wang L (2018) Chilling and-drought stresses in crop plants implications cross talk and potential management opportunities. Front Plant Sci 9:393

Ibrahim EA (2016) Seed priming to alleviate salinity stress in germinating seeds. J Plant Physiol 192:38–46

Iqbal M, Ashraf M (2013) Gibberellic acid mediated induction of salt tolerance in wheat plants: Growth, ionic partitioning, photosynthesis, yield and hormonal homeostasis. Environ Exp Bot 86:76–85

Ishibashi Y, Yamaguchi H, Yuasa T, Inoue IM, Arima S, Zheng SH (2011) Hydrogen peroxide spraying alleviates drought stress in soybean plants. J Plant Physiol 168:1562–1567

Jafar MZ, Farooq M, Cheema MA, Afzal I, Basra SMA, Wahid MA, Aziz T, Shahid M (2012) Improving the performance of wheat by seed priming under saline conditions. J Agron Crop Sci 198(1):38–45

Jakab G, Ton J, Flors V, Zimmerli L, Metraux JP, Mauch-Mani B (2005) Enhancing Arabidopsis salt and drought stress tolerance by chemical priming for its abscisic acid responses. Plant Physiol 139(1):267–274

Jaleel CA, Riadh K, Gopi R, Manivannan P, Ines J, Al-Juburi HJ, Chang-Xing Z, Hong-Bo S, Panneerselvam R (2009) Antioxidant defense responses physiological plasticity in higher plants under abiotic constraints. Acta Physiol Plant 31(3):427–436

Jam BJ, Shekari FARID, Azimi MR, Zangani ESMAEIL (2012) Effect of priming by salicylic acid on germination and seedling growth of safflower seeds under $CaCl_2$ stress. Int J Agri Res 2:1097–1105

Javid MG, Sorooshzadeh A, Moradi F, Sanavy SAMM, Allahdadi I (2011) The role of phytohormones in alleviating salt stress in crop plants. Aust J Crop Sci 5(6):726–734

Jiang X, Li H, Song X (2016) Seed priming with melatonin effects on germination and seedling growth in maize under salinity stress. Pak J Bot 48(4):1345–1352

Jisha KC, Puthur JT (2016) Seed priming with beta-amino butyric acid improves abiotic stress tolerance in rice seedlings. Rice Sci 23(5):242–254

Jisha KC, Vijayakumari K, Puthur JT (2013) Seed priming for abiotic stress tolerance: an overview. Acta Physiol Plant 35(5):1381–1396

Joshi N, Jain A, Arya K (2013) Alleviation of Salt stress in Cucumis sativus L. through seed priming with calcium chloride. Indian J App Res 3:22–25

Kanto U, Jutamanee K, Osotsapar Y, Chai W, Jattupornpong S (2015) Promotive effect of priming with 5-aminolevulinic acid on seed germination capacity, seedling growth and antioxidant enzyme activity in rice subjected to accelerated ageing treatment. Plant Prod Sci 18(4):443–454

Kasote DM, Lee JHJ, Jayaprakasha GK, Patil BS (2019) Seed priming with iron oxide nanoparticles modulate antioxidant potential and defense-linked hormones in watermelon seedlings. ACS Sustain. Chem Eng 7:5142–5151

Khan HA, Ayub CM, Pervez MA, Bilal RM, Shahid MA, Ziaf K (2009a) Effect of seed priming with NaCl on salinity tolerance of hot pepper (*Capsicum annuum* L.) at seedling stage. Soil Environ 28:81–87

Khan HA, Pervez MA, Ayub CM, Ziaf K, Bilal RM, Shahid MA, Akhtar MA (2009b) Hormonal priming alleviates salt stress in hot pepper (Capsicum annuum L.). Soil Environ 28:130–135

Khan HA, Ziaf K, Amjad M, Iqbal Q (2012) Exogenous application of polyamines improves germination and early seedling growth of hot pepper. Chilean J Agric Res 72(3):429–433

Khan M, Muhammad MA, Nawab A, Saeed K, Faiq A, Zia Ud D (2021) Effect of seed Priming techniques and different durations on germination and seedling growth of Wheat (*Triticum aestivum* L.). Plant Sci J 10(1):17–27

Kołodziejczyk, I, Dzitko K., Szewczyk R, Posmyk, M M (2016) Exogenous melatonin expediently modifies proteome of maize (*Zea mays* L.) embryo during seed germination. Acta Physiologiae Plantarum, 38 pp 1–18.

Korkmaz A, Sirikci R (2011) Improving salinity tolerance of germinating seeds by exogenous application of glycinebetaine in pepper. Seed Sci Technol 39:377–388

Krantev A, Yordanova R, Janda T, Szalai G, Popova L (2008) Treatment with salicylic acid decreases the effect of cadmium on photosynthesis in maize plants. J Plant Physiol 165:920–931

Ku YS, Sintaha M, Cheung MY, Lam HM (2018) Plant hormone signaling crosstalks between biotic and abiotic stress responses. Int J Mol Sci 19(10):3206

Kubala S, Wojtyla Ł, Quinet M, Lechowska K, Lutts S, Garnczarska M (2015) Enhanced expression of the proline synthesis gene *P5CSA* in relation to seed osmopriming improvement of *Brassica napus* germination under salinity stress. J Plant Physiol 183:1–12

Kumar M, Pant B, Mondal S, Bose B (2016) Hydro and halo priming: influenced germination responses in wheat Var-HUW-468 under heavy metal stress. Acta Physiol Plant 38:217

Kusano T, Berberich T, Tateda C, Takahashi Y (2008) Polyamines essential factors for growth and survival. Planta 228:367–381

Lara TS, Lira JMS, Rodrigues AC, Rakocevic M, Alvarenga AA (2014) Potassium nitrate priming affects the activity of nitrate reductase and antioxidant enzymes in tomato germination. J Agric Sci 6(2):72–80
Li F, Wu X, Tsang E, Cutler AJ (2005) Transcriptional profling of imbibed Brassica napus seed. Genomics 86:718–730
Li X, Cai J, Liu F, Dai T, Cao W, Jiang D (2014) Physiological, proteomic and transcriptional responses of wheat to combination of drought or waterlogging with late spring low temperature. Funct Plant Biol 41:690–703. https://doi.org/10.1071/FP13306
Li Z, Peng Y, Zhang XQ, Pan MH, Ma X, Huang LK (2014) Exogenous spermidine improves water stress tolerance of white clover (Trifolium repens L.) involved in antioxidant defence gene expression and proline metabolism. Plant Omics 7:517–526
Li Z, Xu J, Gao Y, Wang C, Guo G, Luo Y, Huang Y, Hu W, Sheteiwy MS, Guan Y, Hu J (2017) The synergistic priming effect of exogenous salicylic acid and H_2O_2 on chilling tolerance enhancement during maize (*Zea mays* L.) seed germination. Front Plant Sci 8:1153
Lutts S, Benincasa P, Wojtyla L, Kubala S, Pace R, Lechowska K, Quinet M, Garnczarska M (2016) Seed priming new comprehensive approaches for an old empirical technique. In: Araujo S, Balestrazzi A (eds) New challenges in seed biology – basic and translational research driving seed technology. Intech Open Publisher, London
McDonald MB (2000) Seed priming. In Black M Bewley JD (eds) Seed technology and its biological basis. Sheffield Academic Press Sheffield pp 287–325
Maia J, Dekkers BJW, Provart NJ, Ligterink W, Hilhorst HWM (2011) The re-establishment of desiccation tolerance in germinated Arabidopsis thaliana seeds and its associated transcriptome. PLoS One 6:291
Maiti R, Pramanik K (2013) Vegetable seed priming a low cost simple and powerful techniques for farmers livelihood. Int J Bioresour Stress Manag 4(4):475–481
Manonmani V, Begum MAJ, Jayanthi M (2014) Halo priming of seeds. Res J Seed Sci 7:1–13
Marschner H (1995) Mineral nutrition of higher plants, 2nd edn. Academic Press, New York
Masondo NA, Kulkarni MG, Finnie JF, Van Staden J (2018) Influence of biostimulants-seed priming on Ceratotheca triloba germination and seedling growth under low temperatures, low osmotic potential and salinity stress. Ecotoxicol Environ Saf 147:43–48
Miransari M, Smith DL (2014) Plant hormones and seed germination. Envir Exper Bot 99:110121
Moghanibashi M, Karimmojeni H, Nikneshan P (2013) Seed treatment to overcome drought and salt stress during germination of sunflwer (*Helianthus annuus* L.). J Agrobiol 30:89–96
Mouradi M, Bouizgaren A, Farissi M, Makoudi B, Kabbadj A, Very AA, Sentenac H, Qaddoury A, Ghoulam C (2016) Osmo priming improves seeds germination growth antioxidant responses and membrane stability during early stage of Moroccan alfalfa populations under water deficit. Chilean J Agric Res 76(3):265–272
Mridha N, Chattaraj S, Chakraborty D, Anand A, Aggarwal P, Nagarajan S (2016) Pre-sowing static magnetic field treatment for improving water and radiation use efficiency in chickpea (*Cicer arietinum* L.) under soil moisture stress. Bioelectromagnetics 37:400–408
Mustafa H, Mahmood T, Ullah A, Sharif A, Nafees A, Nadeem M, Ali R (2017) Role of seed priming to enhance growth and development of crop plants against biotic and abiotic stresses. Bull Biol Alli Sci Res Sect Plant Sci 2:1–11
Nagajyoti PC, Lee KD, Sreekanth TVM (2010) Heavy metal, occurrence and toxicity for plants. A review. Environ Chem Lett 8:199–216
Nakaune M, Hanada A, Yin YG, Matsukura C, Yamaguchi S (2012) Molecular and physiological dissection of enhanced seed germination using short-term low-concentration salt seed priming in tomato. Plant Physiol Biotechnol 52:28–37
Nascimento WM (2003) Muskmelon seed germination and seedling development in response to seed priming. Sci Agricola 60(1):71–75
Nasri N, Kaddour R, Mahmoudi H, Baatour O, Bouraoui N, Lachaâl M (2011) The effect of osmopriming on germination, seedling growth and phosphatase activities of lettuce under saline condition. Afr J Biotechnol 10:14366–14372

Nawaz J, Hussain M, Jabbar A, Nadeem GA, Sajid M, Subtain MU, Shabbir I (2013) Seed priming a technique. Int J Agric Crop Sci 6(20):1373–1381
Omami EN (2005) Response of Amaranth to Salinity Stress. PhD Dissertation University of Pretoria Pretoria 235
Pal A, Ali MA, Pal AK (2017) Effect of seed priming on reserve mobilization, water uptake and anti-oxidative enzyme activities in germinating seeds of groundnut under salinity stress. Int J Agric Sci 9(36):4542–4545
Pandey GK (2017) Mechanism of hormone signaling under stress, vols 1 and 2. Wiley-Blackwell, Hoboken
Pandita VK, Anand A, Nagarajan S (2007) Enhancement of seed germination in hot pepper following presowing treatments. Seed Sci Technol 35:282–290
Pant B, Bose B (2016) Mitigation of the influence of PEG-6000 imposed water stress on germination of halo primed rice seeds. Int J Environ Agric Biotech 9:275–281. https://doi.org/10.5958/2230-732X.2016.00036.X
Paparella S, Araújo SS, Rossi G, Wijayasinghe M, Carbonera D, Balestrazzi A (2015) Seed priming state of the art and new perspectives. Plant Cell Rep 34(8):1281–1293
Parera CA, Cantliffe DJ (1994) Pre-sowing seed priming. Hortic Rev 16:109–141
Paul S, Aryadeep R (2016) Seed priming with spermine ameliorates salinity stress in the germinated seedlings of two rice cultivars differing in their level of salt tolerance. Trop Plant Res 3(3):616–633
Pereira MD, Dias DCFS, Dias LAS, Araujo EF (2009) Primed carrot seeds performance under water and temperature stress. Sci Agric 66:174–179
Pill WG, Finch-Savage WE (1988) Effects of combining priming and plant growth regulator treatments on the synchronization of carrot seed germination. Ann Appl Biol 113:383–389
Rajjou L, Duval M, Gallardo K, Catusse J, Bally J, Job C, Job D (2012) Seed germination and vigor. Annu Rev Plant Biol 63:507–533
Razaji A, Farzanian M, Sayfzadeh S (2014) The effects of seed priming by ascorbic acid on some morphological and biochemical aspects of rapeseed (Brassica napus L.) under drought stress condition. Int J Biosci 4:432–442
Rosental L, Nonogaki H, Fait A (2014) Activation and regulation of primary metabolism during seed germination. Seed Sci Res 24:1–15
Rowse HR (1996) Drum-priming—A non-osmotic method of priming seeds. Seed Sci Technol 24:281–294
Ruelland E, Vaultier MN, Zachowski A, Hurry V (2009) Cold signaling and cold acclimation in plants. Adv Bot Res 49:35–150
Sadeghi H, Khazaei F, Yari L, Sheidaei S (2011) Effect of seed osmopriming on seed germination behavior and vigor of soybean (*Glycine max* L.). J Agric Biol Sci 6(1):39–43
Sahin FI, Iseri OD, Haberal M (2011) NaCl priming improves salinity response of tomato (*Lycopersium esculentum* Mill.) at seedling stage. Curr Opin Biotechnol 122:46–47
Sakhabutdinova AR, Fatkhutdinova DR, Bezrukova MV, Shakirova FM (2003) Salicylic acid prevents the damaging action of stress factors on wheat plants. Bulg J Plant Physiol:314–319
Salah SM, Yajing G, Dongdong C, Jie L, Amir N, Qijuan H, Weimin H, Mingyu N, Jin H (2015) Seed priming with polyethylene glycol regulating the physiological and molecular mechanism in rice (Oryza sativa L.) under nano-ZnO stress. Sci Rep 5:14278
Samota MK, Sasi M, Awana M, Yadav OP, Mithra ASV, Tyagi A, Singh A (2017) Elicitor-induced biochemical and molecular manifestations to improve drought tolerance in rice (*Oryza sativa* L.) through seed-priming. Frontiers in Plant Science 8:934
Schwember AR, Bradford KJ (2010) A genetic locus and gene expression pattern associated with the priming effect on lettuce seed germination at elevated temperature. Plant Mol Biol 73:105–118
Sen SK, Mandal P (2016) Solid matrix priming with chitosan enhances seed germination and seedling invigoration in mung bean under salinity stress. J Centrl Europ Agric 17(3):749–762

Sghaier-Hammami B, Jorrin Novo J, Gargouri-Bouzid R, Drira N (2010) Abscisic acid and sucrose increase in the protein content in date palm somatic embryos, causing changes in 2-DE profile. Phytochemistry 71:1223–1236

Shah AR, Ara N, Shafi G (2011) Seed priming with phosphorus increased germination and yield of okra. Afr J Agric Res 6(16):3859–3876

Sheteiwy M, Shen H, Xu J, Guan Y, Song W, Hu J (2017) Seed polyamines metabolism induced by seed priming with spermidine and 5-aminolevulinic acid for chilling tolerance improvement in rice (*Oryza sativa* L.) seedlings. Environ Exp Bot 137:58–72

Siri B, Vichitphan K, Kaewnaree P, Vichitphan S, Klanrit P (2013) Improvement of quality, membrane integrity and antioxidant systems in sweet pepper (*Capsicum annuum* Linn) seeds affected by osmopriming. Aust J Crop Sci 7(13):2068–2073

Srivastava AK, Lokhande VH, Patade VY, Suprasanna P, Sjahril R, D'Souza SF (2010) Comparative evaluation of hydro-, chemo-, and hormonal-priming methods for imparting salt and PEG stress tolerance in Indian mustard (*Brassica juncea* L.) Acta Physiol Plant 32:1135–1144

Sivritepe N, Sivritepe HO, Eris A (2003) The effects of NaCl priming on salt tolerance in melon seedlings grown under saline conditions. Sci Hortic 97:229–237

Sneideris LC, Gavassi MA, Campos ML, Damiao DAV, Carvalho RF (2015) Effects of hormonal priming on seed germination of pigeon pea under cadmium stress. Anais da Academia Brasileira Ciencias 87(3):1847–1852

Sohail SA, Chaurasia AK, Bara BM (2018) Effect of different seed priming methods on germination and vigour of Kabuli Chickpea (*Cicer kabulium* L.) seeds. Int J Curr Microbiol Appl Sci 7(8):1396–1404

Sun YY, Sun YJ, Wang MT, Li XY, Guo X, Hu R Jun MA (2010) Effects of seed priming on germination and seedling growth under water stress in rice. Acta Agron Sinica 36(11):1931–1940

Suzuki N, Mittler R (2006) Reactive oxygen species and temperature stresses a delicate balance between signaling and destruction. Physiol Plant 126:45–51

Thakur P, Kumar S, Malik, JA, Berger, J D, Nayyar H (2010) Cold stress effects on reproductive development in grain crops: an overview. Environ Exp Bot 67(3):429–443

Takahashi T, Kakehi J (2010) Polyamines ubiquitous polycations with unique roles in growth and stress responses. Annals Bot 105:1–6

Tanou G, Filippou P, Belghazi M, Diamantidis G, Fotopoulos V, Molassiotis A (2012) Oxidative and nitrosative-based signaling and associated post-translational modifications orchestrate the acclimation of citrus plants to salinity stress. Plant J 72:1365–1313

Thomas UC, Varughese K, Thomas A, Sadanandan S (2000) Seed priming for increased vigour viability and productivity of upland rice. Leisa India 4:14

Varier A, Vari AK, Dadlani M (2010) The sub cellular basis of seed priming. Curr Sci 99:450–456

Watson MB, Malmberg RL (1998) Arginine decarboxylase (polyamine synthesis) mutants of Arabidopsis thaliana exhibit altered root growth. Plant J 13:231–239

Wojtyla Ł, Lechowska K, Kubala S, Garnczarska M (2016) Molecular processes induced in primed seeds increasing the potential to stabilize crop yields under drought conditions. J Plant Physiol 203:116–126

Xing W, Rajashekar CB (2001) Glycinebetaine involvement in freezing tolerance and water stress is Arabidopsis thaliana. Environ Exp Bot 46:21–28

Xiong L, Zhu JK (2002) Molecular and genetic aspects of plant responses to osmotic stress. Plant Cell Environ 25: 131–139

Xu S, Hu J, Li Y, Ma W, Zheng Y, Zhu S (2011) Chilling tolerance in *Nicotiana tabacum* induced by seed priming with putrescine. Plant Growth Regul 63:279–290

Yacoubi R, Job C, Belghazi M, Chaibi W, Job D (2013) Proteomic analysis of the enhancement of seed vigour in osmoprimed alfalfa seeds germinated under salinity stress. Seed Sci Res 23:99–110

Yan M (2015) Seed priming stimulate germination and early seedling growth of Chinese cabbage under drought stress. South Afr J Bot 99:88–92

Zavariyan A, Rad M, Asghari M (2015) Effect of seed priming by potassium nitrate on germination and biochemical indices in (*Silybum marianum* L.) under salinity stress. Int J Life Sci 9:23–29

Zhong YP, Wang B, Yan JH, Cheng LJ, Yao LM, Xiao L, Wu TL (2014) DL-β-amino butyric-acid induced resistance in soybean against *Aphis glycines* Matsumura (*Hemiptera: Aphididae*). PLoS One 9(1):1–11

Chapter 28
Advances in Biotechnological Tools and Their Impact on Global Climate Change and Food Security

Zafar Iqbal, Asad Azeem, Sami Ul-Allah, Ahmad Sher, Muhammad Qadir Ahmad, Bilal Haider, and Muhammad Asghar

Abstract Climate change exists in a variety of forms such as temperature fluctuation, changing rainfall pattern, flood, and disease which impose negative impact on crop production and are alarming for food security. Climatic variations chiefly affect the agroecology in a number of ways, but the effects vary along with latitudes and longitudes. As agricultural production is dependent on climate and environment, food security is threatened by climate change. The effects of climate change on agriculture and food security may be minimized by adopting advanced agricultural and biotechnological techniques. This chapter summarizes the advanced research work on the role of modern biotechnology to ensure the global food security.

Keywords Modern biotechnology · Food security · Climate change

The original version of the chapter has been revised. A correction to this chapter can be found at https://doi.org/10.1007/978-3-031-37428-9_44

Z. Iqbal · B. Haider · M. Asghar
College of Agriculture, University of Layyah, Layyah, Pakistan

A. Azeem
Department of Plant Breeding and Genetics, University of Agriculture, Faisalabad, Pakistan

S. Ul-Allah (✉)
College of Agriculture, University of Layyah, Layyah, Pakistan

Department of Plant Breeding and Genetics, Bahauddin Zakariya University, Multan, Pakistan
e-mail: samipbg@bzu.edu.pk

A. Sher
College of Agriculture, University of Layyah, Layyah, Pakistan

Department of Agronomy, Bahauddin Zakariya University, Multan, Pakistan

M. Q. Ahmad
Department of Plant Breeding and Genetics, Bahauddin Zakariya University, Multan, Pakistan

M. Hasanuzzaman (ed.), *Climate-Resilient Agriculture, Vol 2*,
https://doi.org/10.1007/978-3-031-37428-9_28

1 Introduction

Climate change refers to long-term shifts in temperature and weather patterns. Normally, these changes are natural, but since the 1800s, the activities of humans have been the major source of climate change and give rise to heat-trapping gases (Kolawole & Okonkwo 2022). The burning of fossil fuels produces greenhouse gas which acts like a blanket that is wrapped around the earth and traps the heat and increases the earth's temperature (Khan 2018). The major greenhouse gases that affect the climate are carbon dioxide and methane. These gases are produced by using gasoline and coal which can be used for heating and cutting trees, and the clearing of land indirectly generates carbon dioxide.

It is clear that climate change in the world is caused by human activities and this is a growing danger to society. The activities include production of greenhouse gases, deforestation, industrialization, and uncontrolled urbanization as designated in Fig. 28.1. Greenhouse gases (methane, water vapors, CO_2, chlorofluorocarbons, and nitrous oxides) have a key role in the global climate cycles. The earth traps heat with sun rays; some of the heat is accumulated while the remaining of that energy is transferred back. Greenhouse gases in the environment lead to increasing global temperature and cause global warming (Ali 2021).

Agriculture is linked with food security and also with the variations in climate. Agriculture is responsible for a larger portion of global greenhouse gas emissions. However, when the productivity of agriculture increases, less carbon dioxide is emitted which produces more food. Agriculture is also an important carbon sink,

Fig. 28.1 Major factors that contribute to the climate change

and it absorbs an equal amount of carbon dioxide as it emits; therefore, in most of the times it neutralizes its negative impacts.

Increase in urbanization due to ever-increasing population has resulted in deforestation and clearing of agricultural lands. Deforestation has a major effect on climate change as it increases global warming. According to the earth's day network, deforestation is the second major contributor to the greenhouse gases of the world (Malhi et al. 2021). The important participation in climate change is the diversification of human activities like fossil fuels burning for the production of electricity, heat, and transport purpose that emits more amount of carbon dioxide which increases global warming and plays key role in climate change. Livestock in the aspect of agriculture also contributes to the climate change. According to studies, livestock around the globe is responsible for 50% of annual global greenhouse gas emissions (Beer et al. 2000; Fróna et al. 2021). The greenhouse gases released from animal agriculture such as methane and nitrous oxide are the causes of temperature fluctuation and overall climate change.

Biotechnology is a relatively new field, and it has revolutionized many biology-related fields. Especially, agriculture biotechnology has opened new scenarios to produce food and fiber. Agriculture scientists are using biotechnological techniques and tools to cope with biotic and abiotic stresses developed due to the changes in weather and climate. Therefore, major objective of this chapter is to analyze the detrimental effects of climate change on food production and food security and potential role of biotechnological tools to cope with this impact.

2 Effect of Climate Change on Food Security

Agriculture is the base of food security. Global food security depends upon global agricultural production. Climate change is recognized as a universal truth all around the world which shows adverse effects on agriculture production, water resources, animal and human health, forest system, and social and economic sectors as designated in Fig. 28.2. The major economic effect of climate change is on agriculture that is because of the size and susceptibility of this sector. An increase in temperature and warming causes danger to agriculture in developing countries normally because many growing areas in the low latitudes already induce climates that are too warm. The increasing food demand due to the increase in the size of the population results in intensive agriculture practices that include unprecedented use of agrochemicals, generation of livestock, and expansion of resources of water. This has further disturbed the situation by producing greenhouse gas (GHG) and resulting in natural resources pollution. Forests decrease the amount of CO_2 present in atmosphere, but due to an increase in the rate of deforestation, this ability has been decreased (Kamilaris et al. 2018).

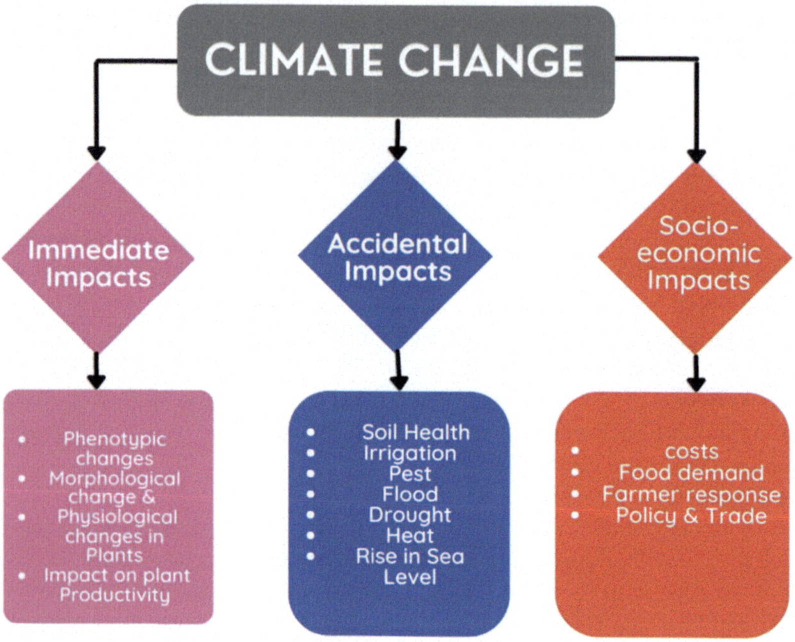

Fig. 28.2 Impacts of climate change on crop productivity and food demand

Food security and climate change are connected in different ways because the change in climate causes biotic and abiotic stresses that have a dangerous impact on the agricultural production and ultimately on food production. The earth and its agriculture are being impacted by climate change in many ways, like changes or shifts in annual rainfall, waves of heat, changes in sea level, change or shift in atmospheric CO_2, effects on ozone level and pests' modification, weeds, and microbes' modifications (Wolff et al. 2021). All of these changes have a negative impact on the food production of the world, so these shifts have driven the attention of all the scientists around the globe.

Biotechnology has played a very important role in managing the negative impacts of climate change. This includes but not limited to incorporation of resistance against biotic and abiotic stresses caused by climate change and increasing the production potential of food and fiber crops. Here in this chapter, role of modern techniques of biotechnology to mitigate the effects of climate change on food production and food security has been discussed.

In the situation of climate change adaptation and reduction, biotechnology has responded positively toward decreasing the sensitivity of the natural and human systems to address the effects of climate change. Here we can see how biotechnology plays a positive role in changing the weather conditions and food security, and how it decreases the negative aspects of climate change on the environment.

3 Agricultural Biotechnology and Climate-Resilient Food Production

Agricultural biotechnology can be defined as "The practical applications of living organisms and their compounds or products in an agricultural sector via utilizing the biotechnological techniques". The techniques used in agricultural biotechnology are tissue culture, marker-assisted breeding, and genetic engineering. Tissue culture is the growth of plant cells or tissues on the specific nutrient media. Under optimum conditions, the whole plant can be regenerated from a single tissue. In molecular-assisted breeding, DNA markers are utilized to locate genes and identify desirable traits by understanding the functions of the genes they carry. Molecular markers increase the efficiency of genes and understand the genetic diversity and relationships between the plant species.

Producing climate-resilient crops can improve food safety and security evenly. Changes in climate create huge problems for agricultural production due to newly emerged biotic and abiotic threats. The question raised in this sense is that what are climate-resilient crops? The answer to this question is that these are the crops that are tolerant to pests, biotic and abiotic effects of the environment, and they are able to enhance the yields of crops under stress conditions such as flooding, heat, freezing, drought, and salinity.

3.1 Decrease of Greenhouse Gases

Deforestation and the utilization of fertilizers increase emission of greenhouse gases; biotechnology provides a solution to decrease the greenhouse gases' emission and plays a positive role to reduce the detrimental role of change in climate by using environment-friendly fuels and decreasing the use of inorganic fuels. Biofilm production from traditional and genetically modified organisms helps to decrease the adverse effects of greenhouse gas emission (Shahzad et al. 2021).

Climate changes and population growth enhance the food demands for people, so the goal of attaining food security for generations is most important. Biotechnology can cause changes in crops, so these crops stand in different stress situations caused by the climate change which are difficult or impossible to attain by using conventional breeding techniques.

3.2 Functional Genomics

Genomics is the complete set of genetic instructions which is provided by DNA. The genomics is of two types, that is, structural genomics and functional genomics. The aim of functional genomics is to determine how the genetic components of living organism work together to produce a specific phenotype. Functional genomics focuses on expression of genes and their products in a particular context like, at developmental stage or during disease. Resistance in other crops can be developed by using the mixture of traditional breeding as well as transgenic and genome editing technologies (Tesfahun et al. 2018).

3.3 Selection of Molecular Markers Related to Stresses

Molecular marker techniques enable plant breeders to select plants based on their markers like their genotype rather than their phenotype (observable phenotype). This technique is marker-assisted selection (MAS). Molecular markers are gaining importance in plant breeding. MAS has many benefits which include enhancing the plant breeding process, enhancing efficiency, and reducing cost. In MAS, molecular markers (DNA fragments) are linked with crop production under various stress conditions; they either are linked with tolerance or with susceptibility.

In phenotypic selection, there may be various problems linked with environmental variations and human error which may influence the results. Moreover, for phenotypic selection, researchers have to wait till the maturity of the crop, but in MAS selection, crop genotypes can be selected even at early growth stages based on their molecular markers profile. There are many types of molecular markers, for example, RAPD, SNPs, SSR, ISSR, RFLP, and AFLP, some of which are PCR based and some are not PCR based. A list of molecular markers linked with different climate stress conditions in different crops is presented in Table 28.1.

Table 28.1 Markers associated with various crop stresses

Markers	Crop	Stress	Reference
RAPD	*Solanum lycopersicum*	High temperature	
SRAP	*Solanum lycopersicum*	High temperature	Comlekcioglu et al. (2010)
EST–SSR	*Manihot esculenta Cranz*	Drought	Wang et al. (2017)
SNP	*Zea mays*	Salt	Cui et al. (2015)
SNP	*Hordeum vulgare*	Salt	Sbei et al. (2014)
SSR and AFLP	*Triticum aestivum*	High temperature	Mohammadi et al. (2008)
SSR	*Triticum aestivum*	Drought	Dolferus et al. (2019)
SSR	*Triticum aestivum*	Drought	Ahmad et al. (2014)
SSR	*Triticum aestivum*	Cold stress	Wainaina et al. (2018)
SSR	*Oryza sativa*	Drought	Babu et al. (2003)
SFP and SSR	*Oryza sativa*	High temperature	Xiao et al. (2011)
SSR	*Gossypium hirsutum*	Salt stress	Saeed et al. (2014)
RFLP	*Sorghum bicolor*	Drought	Kebede et al. (2001)
QTL	*Citrus reshni* *Poncirus trifoliata*	Salt	Raga et al. (2016)

3.4 Impact of Biotechnology on Biofuel Consumption

Agriculture sector is one of the major consumers of fuel; hence, the price of fuel affects the prices of the food and it affects food security. Increasing energy prices and decreasing fossil fuel reserves have developed the interest in the conversion of biomass into the production of biofuel. Biofuels are developed from renewable resources of energy, and they are environment-friendly and also have the ability to reach more than half of the demand of the world for transportation fuels. Biofuels also decrease the dependence on imported petroleum, decrease the emission of greenhouse gas, enhance the economy of the region and also provide jobs, and increase the demand for bioproducts (Alshammari et al. 2015).

For biofuel production, the term synthetic biology is applied in which biofuel is produced by developing more efficient enzymes and their products which break down the solid mass; this process directly produces biofuel. Biofuel such as ethanol is produced from food crops and biomass through different processes such as biochemical and thermochemical process. First-generation biofuel is derived from crops such as oil seeds, sugar crops, and cereals. The techniques which are used to develop first-generation biofuels are well developed. Industrial biotechnology has many benefits which help to protect the environment and also meet global climate change (Praveena et al. 2018).

3.5 Impact of Biotechnology on Use of Fertilizers

Biofertilizers are derived from living resources which are applied to seed, root, and soil to increase the ability of nutrients and also enhance the soil health. Synthetic fertilizers have adverse effects on the crops and also on the environment, so biofertilizers are needed to decrease the impacts of synthetic fertilizers. This helps in safeguarding the health of soil and also enhances the quality of products of crops (Harfouche et al. 2019). The benefits of biofertilizers are as follows:

- Biofertilizers are cost effective.
- Enhance the growth of plants.
- Provide safety against drought.
- Improve the yield of crops.
- Biofertilizers are environment-friendly.
- Enhance the fertility of soil.

Biofertilizers are available in many forms, some of which are given below.

- For leguminous crops, *Rhizobium* fertilizer is used.
- For nonleguminous crops, *Azotobacter* is used.
- For sugarcane, *Acetobacter* is used.
- For low land and paddy rice, blue green algae are used.

Different methods are used to apply biofertilizers, some are listed below.

(a) Seedling root dip method: This method is used for rice crops.
(b) Treatment of seeds: The seeds are soaked in a nitrogen mixture and then sown.
(c) Treatment of soil: Biofertilizers and compost are mixed and then this mixture is spread on the soil where seeds are sown (Sheikh et al. 2019).

3.6 Role of Biotechnology in Biotic Stress

Biotic stress can be defined as the damage or harmful effect that is caused by a living organism on another living organism. The living organism that causes damage can be bacteria, virus, parasites, fungi, harmful insects, weeds, and cultured or local plants. Basically, the living organisms that cause biotic stress in plants directly decrease the nutrients in their host and as a result the plant dies (Arora et al. 2019).

The effort or competition for the survival of organisms in the environment is a natural process. The microbes compete with the plants for their survival, while on the other hand plants are combating with the microorganisms for their survival (Saavoss et al. 2019). Plants do not have an adaptive immune system, but they develop a different mechanism for their survival. Plants have genetic codes which are stored in them genetically; they use these codes as their defense mechanism for their survival. There are hundreds of resistant genes which are coded in plants' genomes against biotic stress. Different biotic stresses impact photosynthesis like the insects which chew the leaves decrease the leaf area and infection of viruses also decreases the leaf area for photosynthesis.

The organisms that cause stress in plants affect the plants in many ways, and as a result, decrease the productivity of plants. However, many mechanisms or techniques have been developed to decrease the effect of this stress through research. Biotic stress in plants can be diminished by studying the genetic mechanism of organisms that cause stress in plants. Genetically modified plants have been produced, and these plants have developed resistance against the stress-causing organisms (Hazra et al. 2020).

Insect pest tolerance decreases many crops of plants. The use of chemicals to control the pests causes harmful effects on the plants and chemicals also damage the environment. Bt (Cry) gene has been extracted from the bacteria "*Bacillus thuringiensis*." This gene is proved effective in controlling different lepidopteran insects. There are many *Bt* crops developed, most important which are *Bt* cotton and *Bt* corn; both of which contribute to food security.

The diseases in plants that are caused by viral infections are main source of yield loss and they are also dangerous to the crop production. Different genetically modified varieties of plants have been produced which show resistance against viral infections. For example: genetically modified papaya has been produced which shows resistance against ring spot virus. In fruit crops, protein that proves lethal against viruses has been introduced in plant tissues and this provides resistance against viral infections. The use of chemicals to control the viral infections in plants is costly to the farmers and also dangerous for plants' health and are not eco-friendly. Therefore, different strategies have been developed to overcome the viral infections through RNA silencing and resistance gene defense mechanism.

4 Conclusion

Climate change is a big threat to global food security. Although the climate change may be beneficial in some respects or for some areas, but overall, it negatively affects the agricultural food production due to rise in temperature, unusual rainfall and floods, and drought spells. Biotechnology is a relatively new field, especially in agriculture, and it can be helpful to mitigate the hazardous effects of climate change. With advances in biotechnology, climate-resilient crops and genotypes can be screened with environment-specific molecular markers and transgenic crops can be produced with traits adoptable to the changing climate. Although biotechnology has opened many ways to improve food production in the changing climate, interdisciplinary projects are required to ensure the food security in changing climate.

References

Ahmad MQ, Khan SH, Khan AS, Kazi AM, Basra SM (2014) Identification of QTLs for drought tolerance traits on wheat chromosome 2A using association mapping. Int J Agric Biol 16:862–870

Ali U, Jing W, Zhu J, Omarkhanova Z, Fahad S, Nurgazina Z, Khan ZA (2021) Climate change impacts on agriculture sector: a case study of Pakistan. Ciência Rural 51

Alshammari F (2015) Biochemical profile of bacillus thuringiensis strains obtained from middle Tennessee. Doctoral dissertation, Tennessee State University

Arora NK (2019) Impact of climate change on agriculture production and its sustainable solutions. Environ Sustain 2:95–96

Babu RC, Nguyen BD, Chamarerk V, Shanmugasundaram P, Chezhian P, Jeyaprakash P, Ganesh SK, Palchamy A, Sadasivam S, Sarkarung S, Wade LJ (2003) Genetic analysis of drought resistance in rice by molecular markers: association between secondary traits and field performance. Crop Sci 43:1457–1469

Beer J, Mende W, Stellmacher R (2000) The role of the sun in climate forcing. Quat Sci Rev 19:403–415

Comlekcioglu N, Şimşek ÖZHAN, Boncuk M, Aka-Kacar YILDIZ (2010) Genetic characterization of heat tolerant tomato (*Solanum lycopersicon*) genotypes by SRAP and RAPD markers. Genet Mol Res 9:2263–2274

Cui D, Wu D, Somarathna Y, Xu C, Li S, Li P, Zhang H, Chen H, Zhao L (2015) QTL mapping for salt tolerance based on snp markers at the seedling stage in maize (*Zea mays* L.). Euphytica 203:273–283

Dolferus R, Thavamanikumar S, Sangma H, Kleven S, Wallace X, Forrest K, Rebetzke G, Hayden M, Borg L, Smith A, Cullis B (2019) Determining the genetic architecture of reproductive stage drought tolerance in wheat using a correlated trait and correlated marker effect model. G3-Genes Genom Genet 9:473–489

Fróna D, Szenderák J, Harangi-Rákos M (2021) Economic effects of climate change on global agricultural production. Nat Conserv 44:117–139

Harfouche AL, Jacobson DA, Kainer D, Romero JC, Harfouche AH, Mugnozza GS, Moshelion M, Tuskan GA, Keurentjes JJ, Altman A (2019) Accelerating climate resilient plant breeding by applying next-generation artificial intelligence. Trends Biotechnol 37:1217–1235

Hazra A, Bej S, Mondal A, Murmu NC, Banerjee P (2020) Discerning detection of mutagenic biopollutant TNP from water and soil samples with transition metal-containing luminescence metal–organic frameworks. ACS Omega 5:15949–15961

Kamilaris A, Anton A, Blasi AB, Prenafeta-Boldú FX (2018) Assessing and mitigating the impact of livestock agriculture on the environment through geospatial and big data analysis. Int J Sustain Agric Manag Inform 4:98–122

Kebede H, Subudhi PK, Rosenow DT, Nguyen HT (2001) Quantitative trait loci influencing drought tolerance in grain sorghum (*Sorghum bicolor* L. Moench). Theor Appl Genet 103:266–276

Khan MA, Tahir A (2018) Economic effects of climate change on agriculture productivity by 2035: a case study of Pakistan (No. 2058-2018-5277).

Kolawole E, Okonkwo W (2022) Impacts of climate change on environment and the remedies. Int J Weather Clim Chang Conserv Res 8:1–9

Malhi GS, Kaur M, Kaushik P (2021) Impact of climate change on agriculture and its mitigation strategies: a review. Sustainability 13:1318

Mohammadi VA, Zali AA, Bihamta MR (2008) Mapping QTLs for heat tolerance in wheat. J Agric Sci Technol 10:261–267

Praveena SM, Shaifuddin SNM, Sukiman S, Nasir FAM, Hanafi Z, Kamarudin N, Ismail THT, Aris AZ (2018) Pharmaceuticals residues in selected tropical surface water bodies from Selangor (Malaysia): occurrence and potential risk assessments. Sci Total Environ 642:230–240

Raga V, Intrigliolo DS, Bernet GP, Carbonell EA, Asins MJ (2016) Genetic analysis of salt tolerance in a progeny derived from the citrus rootstocks Cleopatra mandarin and trifoliate orange. Tree Genet Genom 12:1–16

Saavoss M (2019) How might cellular agriculture impact the livestock, dairy, and poultry industries? Choices 34:1–6

Saeed M, Wangzhen G, Tianzhen Z (2014) Association mapping for salinity tolerance in cotton (*Gossypium hirsutum* L.) germplasm from US and diverse regions of China. Aust J Crop Sci 8:338–346

Sbei H, Sato K, Shehzad T, Harrabi M, Okuno K (2014) Detection of QTLs for salt tolerance in Asian barley (*Hordeum vulgare* L.) by association analysis with SNP markers. Breed Sci 64:378–388

Shahzad A, Ullah S, Dar AA, Sardar MF, Mehmood T, Tufail MA, Shakoor A, Haris M (2021) Nexus on climate change: agriculture and possible solution to cope future climate change stresses. Environ Sci Pollut Res 28:14211–14232

Sheikh Mohamed M, Sakthi Kumar D (2019) Application of nanotechnology in genetic improvement in crops. In Nanosci Sustain Agric:3–24

Tesfahun W (2018) Retracted article: climate change mitigation and adaptation through biotechnology approaches: a review. Cogent Food Agric 4:1512837

Wainaina CM, Makihara D, Nakamura M, Ikeda A, Suzuki T, Mizukami Y, Nonoyama T, Doi K, Kikuta M, Samejima H, Menge DM (2018) Identification and validation of QTLs for cold tolerance at the booting stage and other agronomic traits in a rice cross of a Japanese tolerant variety, Hananomai, and a NERICA parent, WAB56-104. Plant Prod Sci 21:132–143

Wang B, Guo X, Zhao P, Ruan M, Yu X, Zou L, Yang Y, Li X, Deng D, Xiao J, Xiao Y (2017) Molecular diversity analysis, drought related marker-traits association mapping and discovery of excellent alleles for 100-day old plants by EST-SSRs in cassava germplasms (*Manihot esculenta Cranz*). PLoS One 12:e0177456

Wolff NH, Zeppetello LRV, Parsons LA, Aggraeni I, Battisti DS, Ebi KL, Game ET, Kroeger T, Masuda YJ, Spector JT (2021) The effect of deforestation and climate change on all-cause mortality and unsafe work conditions due to heat exposure in Berau, Indonesia: a modelling study. Lancet Planet Health 5:e882–e892

Xiao Y, Pan Y, Luo L, Zhang G, Deng H, Dai L, Liu X, Tang W, Chen L, Wang GL (2011) Quantitative trait loci associated with seed set under high temperature stress at the flowering stage in rice (*Oryza sativa* L.). Euphytica 178:331–338

Chapter 29
Biotechnological Attributes of Bio-stimulants for Relieving Abiotic Stress

Proma Ghosh and Harshata Pal

Abstract Abiotic stress has become an integral part of agriculture worldwide due to climate change, natural calamity, and pollution. Several unusual environmental parameters such as heat, cold, freezing, draught, waterlogging, light intensity, UV radiation, metal toxicity, and nutrient deficiency are collectively called abiotic Stress. Less than 10% of world's arable land can be termed as "stress-free land." Draught, salinity, and heavy metal are the major abiotic stresses affecting the crop yield by more than 50%. Abiotic stresses can influence the alteration of morphological, biochemical, and metabolic pathways by generating reactive oxygen species (ROS), influencing ionic toxicity, and affecting the plant growth and productivity. The rapid change in environmental factors is overruling the adaptive potential of the plants. To cope with the loss of agricultural yield, researchers are getting the bio-stimulant as most promising way. Bio-stimulants can be active molecules (except pesticide and chemical fertilizer), or plant growth-promoting microbes/PGPM (bacteria or fungus) that can contribute to the plant metabolism in stress tolerance and enhance the production. Many researches have been done, but scientists are still in search for more specific and effective bio-stimulant for sustainable agriculture to feed the growing population. In this chapter, we discuss the different bio-stimulants and the pathways they act on to relieve the abiotic stress of the plant and increase the food production.

Keywords Abiotic stress · Bio-stimulant · PGPM · ROS · Sustainable agriculture

1 Introduction

To achieve sustainable development worldwide, United Nation announced several goals, one of them is to reach zero hunger by 2030. To achieve this, food production should be doubled as food demand is increasing heavily (Tilman et al. 2011).

P. Ghosh · H. Pal (✉)
Amity Institute of Biotechnology, Amity University Kolkata, Kolkata, West Bengal, India
e-mail: hpal@kol.amity.edu

M. Hasanuzzaman (ed.), *Climate-Resilient Agriculture, Vol 2*,
https://doi.org/10.1007/978-3-031-37428-9_29

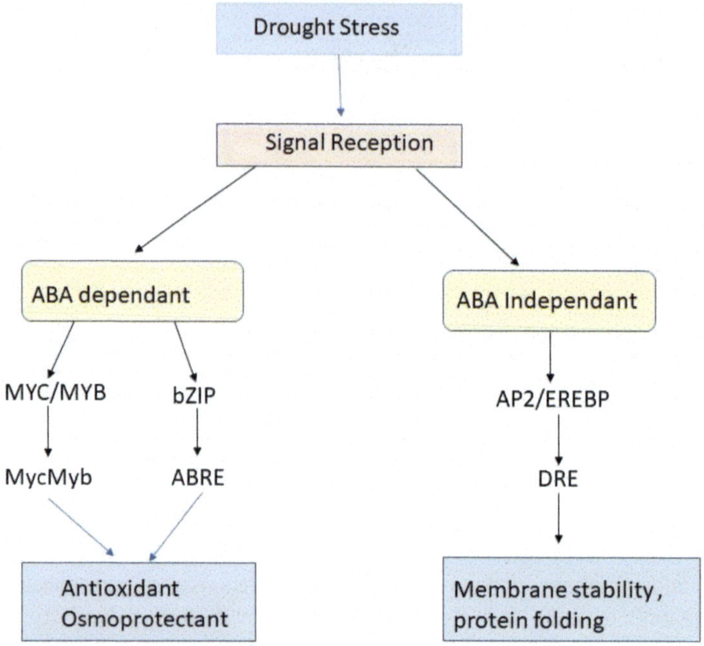

Fig. 29.1 Effect of draught stress in plant

Climate change, pollution, and excessive use of chemical fertilizer are affecting the crop production due to increase in abiotic stress such as heat, draught, salinity, and heavy metal. Among the abiotic stresses, draught stress affects the productivity of the land maximum. Second comes salinity which affects almost 10% of the land worldwide. Drought stress causes cell dehydration, inhibition of cell division, and enhancement reactive oxygen species (ROS) generation which in turn cause osmotic stress and oxidative stress in plants (Vurukonda et al. 2016). This is depicted in Fig. 29.1.

Salinity stress induces physiological drought condition as it changes the osmolarity of the soil and the plant cell. In turn, it also causes ionic toxicity inducing leaf senescence (Kaushal and Wani 2016). In Fig. 29.2, this has been shown by schematic.

By reducing water potential, plants accumulate proline and betaine derivatives to counteract the negative effects of salt stress. Additional water can be obtained from the environment due to the buildup of suitable osmolytes engaged in osmoregulation, hence reducing the immediate impact of a water scarcity. An imbalance of cellular ions caused by high salt concentration leads to ionic toxicity, osmotic stress, and the generation of active oxygen species (Rouhier et al. 2008).

As the heavy metals such as Pb, Cr, and As interact with the electron transport chain of chloroplast and mitochondrial membrane, heavy metal stress in plant causes high amount of ROS production as well as lipid peroxidation and biological macromolecule breakdown in plants (Anjum et al. 2017; Carrasco-Gil et al. 2012;

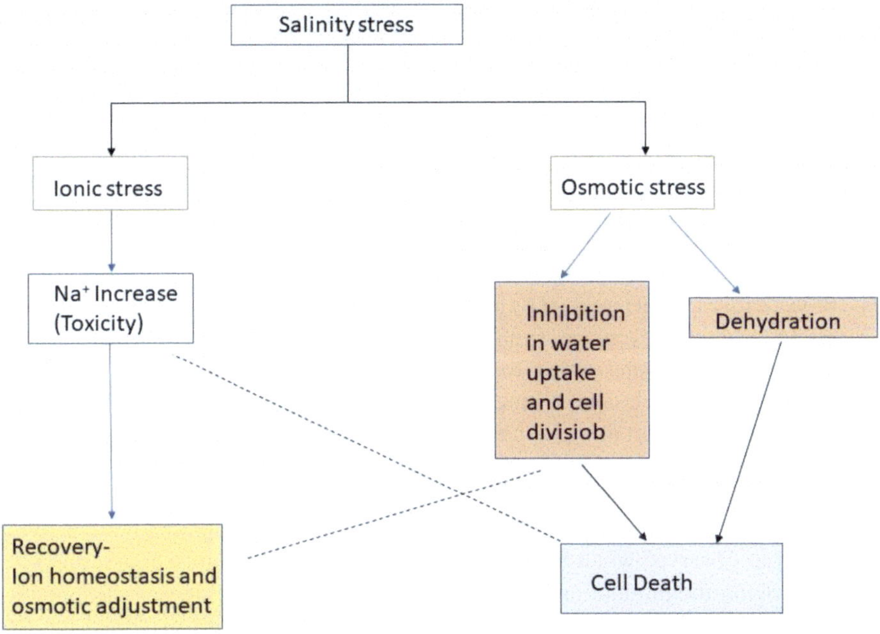

Fig. 29.2 Effect of salt stress on plant

Venkatachalam et al. 2017). It also acts as genotoxic material as it causes DNA damage and chromosomal abbreviation in plant cells (Aslam et al. 2017; Arya et al. 2013).

Plants can sense a very low level of stress signals in their environment. To mitigate negative effects of stressors, plant systems have undergone several biochemical evolutions (Singh et al. 2019). Over the past few decades, great progress has been made in understanding the mechanism underlying plant resistance/tolerance to individual biotic and/or abiotic stresses. However, plants in nature simultaneously handle multiple interacting loads that typically occur at the same time (Ma et al. 2020).

Bio-stimulants are the active compounds or microbes that can induce growth in plants, helping in decreasing dependence on chemical fertilizer for agriculture. The current definition of plant BSs by the EU regulation (2019) is: "A product that stimulates plant nutrition processes independently of the product's nutrient content, with the sole aim of improving one or more of the following characteristics of the plant or the plant rhizosphere: (a) nutrient use efficiency; (b) tolerance to abiotic stress; (c) quality traits; or (d) availability of confined nutrients in the soil or rhizosphere." Bio-stimulant covers a wide range of compounds from amino acids or amines to biopolymers (Yakhin et al. 2017). The current classification of bio-stimulants done by "Du Jardin" (2015) divides all bio-stimulants into different groups – humic and fulvic acids, seaweed and botanical extracts, protein hydrolysates and N-containing compounds, chitosan and other biopolymers, inorganic compounds, and beneficial fungi and bacteria (Bulgari et al. 2019). Here we have discussed about some biotechnological discoveries of bio-stimulants which enhance growth and production in plants during stressed condition.

2 Activity of Bio-stimulants in Stress Tolerance

2.1 Humic and Fulvic Acids

The bioactivity of humic acid (HA) helps reduce fertilization rates, improve nutrient utilization efficiency, replace synthetic plant regulators, improve fruit quality, increase, or reduce water stress tolerance and disease incidence improves early growth and flowering. Usage of humic acid as a bio-stimulant in horticultural crops is emerging as an important sustainable technology that can be integrated with other agricultural practices to increase the productivity of the cultivation system. It is more efficient, and has less negative impact on the environment (Canellas et al. 2015). Enhancement in nitrate reductase and phenylalanine ammonia lyase activity in leaves was observed when humates were applied in urban lettuce cultivation system, improving the production (Hernandez et al. 2016). Glucokinase, phosphoglucose isomerase, aldolase, and pyruvate kinase are enzymes involved in the metabolism of glucose. HS had a deleterious impact on their activity. The hydrolysis of sucrose into hexose, a substrate available to developing cells, was encouraged by increased invertase activity (Pizzeghello et al. 2001). Since enzymes involved in N assimilation were often boosted by HS, it is probable that these metabolites can be employed to prolong growth and enhance N metabolism when total carbohydrate content and sugar reduced after the application of humates. Therefore, in maize treated with HS, substantial net photosynthesis rates could be seen (Rouhier et al. 2008).

In summary, numerous HA-based formulations can be used as biologically active natural substances to advance sustainable agriculture.

2.2 Chitosan and Other Biopolymers

Chitosan and its oligomers (oligochitosans), the second most abundant polymer after cellulose, are linear polysaccharides formed from β-(1-4)-linked D-glucosamine and N-acetyl-D-glucosamine (Hemantaranjan 2015). Chitosan is taken out of the cell walls of fungus and the shells of crustaceans such as shrimp and crab. Chitin, a copolymer comprising N-acetyl-D-glucosamine and D-glucosamine, can be converted into chitosan by removing more than 80% of the acetyl groups (Mehrafarin 2017). Chitosan enhances plant potency, reduces the negative effects of adverse conditions, and stimulates plant growth. Chitosan influences various physiological responses such as plant immunity and defense mechanisms that involve various enzymes such as phenylalanine ammonium lyase, polyphenol oxidase and tyrosine ammonia lyase, and antioxidant enzymes such as superoxide dismutase, catalase, peroxide against adverse conditions (Hemantaranjan 2015). In addition to its environmental friendliness and effects on biotic and abiotic stress, chitosan has attracted a great deal of interest.

Regarding stress tolerance, chitosan has been reported to increase the stress tolerance of maize, improving antioxidant in draught condition (Rabêlo et al. 2019). Metabolomic study done on chitosan-treated sweet clover revealed that this biopolymer enhanced the production of different osmoprotectants that are involved in antioxidant defense and stress signaling (Li et al. 2017). Increased endogenous chitosans (CTS) content through exogenous application of CTS effectively alleviates dehydration-induced leaf senescence, growth inhibition, and cellular damages. The complete analysis of metabolomic and transcriptomics profile revealed that CTS enhanced aminoalkanoic and carbohydrate metabolism, energy production and conversion, the AsA-GSH and TCA cycles, and therefore the GABA shut pathway, as manifested by improved accumulation of abundant metabolites. Many of those metabolites are known to be involved in antioxidant defense, stress signaling, and energy production. The changes in metabolites and genes induced by CTS provide the evidence for the beneficial role of CTS in draught resistance in plant (Li et al. 2017).

Seaweed extract, a protective product based on algae extract, effectively improved sugarcane stalk yield and quality under drought stress in this study. SWE also enhanced the biometric parameters of sugarcane under drought stress, with taller and thicker stalks and, consequently, higher stalk and sugar yields and a metabolically stronger plant (Jacomassi et al. 2022).

Among the notable hormones, abscisic acid (ABA), jasmonate (JA), ethylene, and plant hormone (CK) are incontestable to mediate plant defense responses against abiotic stresses (Nakashima and Yamaguchi-Shinozaki 2013). One among the quickest responses to abiotic stress is the accumulation of ABA, which may be a key regulator within the activation of plant cellular adaptation to drought and salt stresses (Chen et al. 2013). The accumulated levels of ABA facilitate the binding to its receptor to initiate signal transduction, resulting in cellular responses to stresses (Ng et al. 2014; Sah et al. 2016). Additionally, several different salt-stress-related pathways, reminiscent of amino acid accumulation and Ca signaling, also are physiologically coupled to the ABA communication pathway (Yang et al. 2017).

Direct inoculation of lipochitooligosaccharides (LCOs) and/or thuricin-17 peptide, compounds defined as bacterial signals, can shield flowers in opposition to one of a kind abiotic stress (Nazari and Smith 2020). For instance, while soybean seed had been treated with each compound, they had more proof against excessive salt stress (Subramanian et al. 2016b).

2.3 Beneficial Fungi and Bacteria

Plant growth-promoting microorganisms, including plant growth-promoting bacteria (PGPB), rhizobia, and arbuscular mycorrhizal fungi (AMF), are found in free-living soil, rhizosphere/root surface (e.g., root bacteria and ectomycorrhizal fungi), or in plant beneficial tissues (e.g., endophytic bacteria, endomycorrhizal fungi, and AMF) (Ma et al. 2019). The role of PGPMs in plant growth, nutrient uptake, and biocontrol activity is well established. Despite the differences between these types of microorganisms, these PGPM strains are able to colonize the rhizosphere soil or the rhizosphere of plants and respond to abiotic stresses (drought, salinity, temperature extremes, etc.) and it can protect plants from biological stresses (such as B. plant pathogens) (Ma et al. 2019) as well as promote plant establishment and growth through the same plant growth-promoting mechanisms, including direct and indirect pathway. In general, direct mechanisms include enhanced nutrient uptake (nitrogen fixation, siderophore sequestration, potassium, phosphate solubilization, etc.), plant hormone synthesis (auxins, cytokinins, ABA, GA, etc.) (Ma et al. 2019), and induction of ACC deaminase with volatile or nonvolatile compounds.

Nowadays, use of microbes as bio-stimulants and bio control agent is increasing. Several researches are going on to get specific microbes for specific stress tolerance. Among all fungi, *Trichoderma* is widely used as biocontrol agent. Recent study also shows that this not only acts as a biocontrol agent but also shows plant resistance, plant growth, and development resulting in an increase in crop production (Ghazanfar et al. 2018). *Trichoderma*, an imperfect fungus, can be isolated from natural soil, decaying material, etc. *Trichoderma* was first introduced as biocontrol agent in early 1930s; later, scientists discovered that it could control fungal infection in different plant parts' life leaf and seed (Ghazanfar et al. 2018). *Trichoderma hamatum,* a fungal endophyte, isolate DIS 219b, was found to delay the onset of the drought response in the cocoa plant (*Theobroma cacao*); it may be due to changes in gene expression related to stomatal conductance, net photosynthesis, and green fluorescence emission (Ma et al. 2019).

Plant growth-promoting bacteria (PGPB) mainly reside in rhizospheric soil. Many of them also have endophytic characters. One of the most diverse groups of PGPB in rhizosphere is *Bacillus* (Ghosh et al. 2003). Recent works have shown that several species of Bacilli promote plant growth associated with their capacity to use ACC. Thus, a search for *Bacillus* was carried out as endophytes of seeds of different tomato varieties (*Lycopersicum esculentum Mill.*). *subtilis HYT-12-1* was isolated and found to both induce plant growth and possess ACC deaminase activity. The different isolates reported in that work also showed that other mechanisms of plant growth promotion were operative in those bacteria including the production of IAA, siderophores, and the solubilization of phosphates and nitrogen fixation.

Typically, plants exposed to high amount of stress increase the production of ethylene. PGPB associated with the plant cell help in degrading the precursor of ethylene, i.e., ACC (1-aminocyclopropane-1-carboxylate). This in turn lowers the

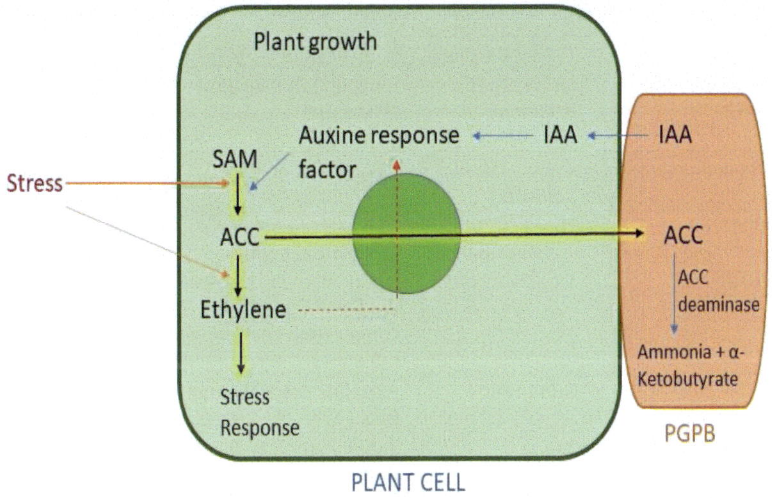

Fig. 29.3 PGPB inhibiting ethylene production by ACC deaminase alleviating stress in plant cell

production of ethylene, causing growth promotion and survival of the plant in stress condition. PGPB also produce IAA which in turn promote cell division and growth in plants. This has been depicted in Fig. 29.3.

By sequencing the 16S rDNA marker, representative isolates were determined to be closely related to five *Bacillus* species: *B. subtilis* (Xu et al. 2014). A much salt-tolerant Bacillus strain with ACC deaminase activity was isolated from Thai saline soil. The authors did not explain whether the strain was isolated from the tomato rhizosphere or from the bulk soil. However, it showed an interesting result that the strain demonstrates the ability to use ACC as the sole carbon source under saline conditions (up to 0.6 M NaCl). Tomato plant (Lycopersicon esculentum Mill. cv. Seeda) with several newly isolated bacterial strains (one after the other) increased the percentage of seeds that germinated, the root length of tomato plants, and the dry weight of the seedlings. Four of the five newly isolated strains were characterized as *Bacillus licheniformis* B2r, *B. vietnamensis* Apa, and *B. licheniformis* by homology to the 16S ribosomal RNA gene (Chookietwattana and Maneewan 2012). Salt tolerance mechanisms of endophytes are very much similar to plant growth-promoting rhizospheric bacteria (PGPR). Like PGPR, endophytes produce ROS scavengers such as glutathione, ascorbate, and tocopherol and the enzymes include superoxide dismutases (SOD), catalases (CAT), ascorbate- or thiol-dependent peroxidases (APX), glutathione reductases (GR), dehydroascorbate reductases (DHAR), and monodehydroascorbate reductases (MDHAR) (Rouhier et al. 2008). A wide range of PGPB is listed in the Table 29.1.

Table 29.1 List of PGPB used in mitigation of abiotic stress

No.	Bacteria	Benefitted plant	Growth-promoting activity	References
1	*Bacillus subtilis* BERA 7	Chickpea (*Cicer arietinum* cv. Giza 1)	Induction of osmoprotectant production	Abd-Allah et al. (2018)
2	*Pseudomonas spp*	Canola (*Brassica napus L.*)	IAA production	Akhgar et al. (2014)
3	*Bacillus amyloliquefaciens* NBRISN13	Rice (*Oryza sativa* L. indica var. Narayan)	IAA and siderophores production, highly competitive in rhizosphere, lipid, and fatty acid composition change under salt stress	Nautiyal et al. (2013)
4	*Rhizobium* spp.	Sunflower	Alleviate draught stress by EPS	Alami et al. (2000)
5	*Pseudomonas fluorescens*	Maize	Salinity stress alleviation. ACCd, increased NPK uptake maize	Nadeem et al. (2013)
6	*Achromobacter xylosoxidans*	*Phragmites australis, Ipomoea aquatica, and Vetiveria zizanioides*	Catechol and phenol degradation	Ho et al. (2012)
7	*Kocuria flava Bacillus vietnamensis*	*Oryza sativa*	The isolates showed significant reduction in As(III) uptake and increase in rice seedling growth in As-stressed area	Mallick et al. (2018)
8	*Methylobacterium oryzae*	*Acacia farnesiana*	Plants associated with *M. oryzae* showed growth in increased As-concentration without any reduction in biomass and chlorophyll	Alcántara-martínez et al. (2018)
9	*Rhizoglomus intraradices, Glomus etunicatum*	*Triticum aestivum*	AM colonization helped the host plant to overcome As-induced P deficiency. Increase of nutrients (N, P, S), lowered lipid peroxidation and H_2O_2	Sharma et al. (2017)
10	*Trichoderma* sp.	*Helianthus annuus*	Siderophore production, IAA, ACC deaminase activity, and phosphate solubilization. Inoculated plants in As and Pb-amended soil showed higher biomass production than uninoculated plant	Govarthananm et al. (2018)

(continued)

Table 29.1 (continued)

No.	Bacteria	Benefitted plant	Growth-promoting activity	References
11	*Pseudomonas putida*	*Triticum aestivum*	Phytochelatin synthase gene modified to overcome Cd, Hg, stress	Yong et al. (2014)
12	*Glomus mosseae*	*Solanum nigrum*	Cd stress tollerance	Luo et al. (2011)
13	*B. licheniformis B2r*	*Solanum lycopersicum* (tomato)	Alleviate some of the detrimental effects of salt stress through production of ACC deaminase and gibberellin and phosphate solubilization	Chookietwattana and Maneewan (2012)
14	*Flavobacterium glaciei*	*Solanum lycopersicum*	Alleviate chilling or cold stress by increasing super oxide dismutase, proline accumulation, decreasing lipid peroxidation	Subramanian et al. (2016a)
15	*Arthrobacter protophormiae*	*Pisum sativum*	Tolerance to salt and polyethylene glycol enhanced rhizobial nodulation and mycorrhizal colonization	Barnawal et al. (2014)
16	*Pseudomonas* sp. *A3R3*	*Alyssum serpyllifolium*	Alleviate Ni stress through production of IAA, siderophores, ACCD, and solubilization of P; excreted cellulase and pectinase. Increased the biomass of *B. juncea* and Ni content in *A. serpyllifolium*; showed high level of colonization in tissue interior of both plant species	Ma et al. (2011)
17.	Microbacterium sp. NCr-8, Arthrobacter sp. NCr-1,	*Noccaea caerulescens, Thlaspi perfoliatum*	Ni stress tolerance by production of IAA, siderophores, and ACCD. Ni translocation enhanced	Visioli et al. (2014)
18	*Bacillus thuringiensis GDB-1*	*Alnus firma*	Tolerance to As, Cu, Cd, Ni, Pb, and Zn through production of IAA, siderophores, ACCD, and phosphate solubilization	Babu et al. (2013)

3 Conclusion

Taking present scenario about the climate change into account, the hardest problem to attain zero hunger is the sustained crop loss due to abiotic stresses (mainly salinity, draught, and heavy metal stress) that is increasing day by day due to anthropogenic interventions such as excessive use of chemical fertilizer, pollution, and urbanization. To overcome this problem, bio-stimulants are being explored worldwide. Much progress has been done before and many strategies have been applied previously to successfully overcome or tolerate the abiotic stresses. But it is needed to find out more eco-friendly and economic approach to get control of this highly evoking problem. Bio-stimulants are the wide range of biological compounds and microorganisms that help plant to alleviate stress and increase growth and productivity. Still specific, eco-friendly and economical bio-stimulants need to be identified for specific stresses. Here we have briefly discussed the different classes of bio-stimulants that are discovered and their mechanism of action to alleviate stress is detected.

References

Abd-Allah EF, Alqarawi AA, Hashem A, Radhakrishnan R, Al-Huqail AA, Al-Otibi FON, Malik JA, Alharbi RI, Egamberdieva D (2018) Endophytic bacterium *Bacillus subtilis* (BERA 71) improves salt tolerance in chickpea plants by regulating the plant defense mechanisms. J Plant Interact 13:37–44

Akhgar AR, Arzanlou M, Bakker PAHM, Hamidpour M (2014) Characterization of 1-Aminocyclopropane-1-carboxylate (ACC) deaminase-containing pseudomonas spp. in the rhizosphere of salt-stressed canola. Pedosphere 24:461–468

Alami Y, Achouak W, Marol C, Heulin T (2000) Rhizosphere soil aggregation and plant growth promotion of sunflowers by an exopolysaccharide-producing *rhizobium* sp. strain isolated from sunflower roots. Appl Environ Microbiol 66:3393–3398

Alcántara-Martínez N, Figueroa-Martínez F, Rivera-Cabrera F, Gutiérrez-Sánchez G, Volke-Sepúlveda T (2018) Science of the Total environment an endophytic strain of *Methylobacterium* sp. increases arsenate tolerance in *Acacia farnesiana* (L.) Willd: a proteomic approach. Sci Total Environ 625:762–774

Anjum SA, Ashraf U, Khan I, Tanveer M, Shahid M, Shakoor A, Wang L (2017) Phyto-toxicity of chromium in maize: oxidative damage, Osmolyte accumulation, anti-oxidative defense and chromium uptake. Pedosphere 27:262–273

Arya SK, Basu A, Mukherjee A (2013) Lead induced genotoxicity and cytotoxicity in root cells of *Allium cepa* and *Vicia faba*. Nucleus 56:183–189

Aslam R, Bhat TM, Choudhary S, Ansari MYK (2017) An overview on genotoxicity of heavy metal in a spice crop (*Capsicum annuum* L.) in respect to cyto-morphological behaviour. Caryologia 70:42–47

Babu AG, Kim J, Oh B (2013) Enhancement of heavy metal phytoremediation by *Alnus firma* with endophytic *bacillus thuringiensis* GDB-1. J Hazard Mater 250–251:477–483

Barnawal D, Bharti N, Maji D, Chanotiya CS, Kalra A (2014) ACC deaminase containing *Arthrobacter protophormiae* induces NaCl stress tolerance through reduced ACC oxidase activity and ethylene production resulting in improved nodulation and mycorrhization in *Pisum sativum*. J Plant Physiol 171:884–894

Bulgari R, Franzoni G, Ferrante A (2019) Biostimulants application in horticultural crops under abiotic stress conditions. Agronomy 9:306

Canellas LP, Olivares FL, Aguiar NO, Jones DL, Nebbioso A, Mazzei P, Piccolo A (2015) Humic and fulvic acids as biostimulants in horticulture. Sci Hortic 196:15–27

Carrasco-Gil S, Estebaranz-Yubero M, Medel-Cuesta D, Millán R, Hernández LE (2012) Influence of nitrate fertilization on hg uptake and oxidative stress parameters in alfalfa plants cultivated in a hg-polluted soil. Environ Exp Bot 75:16–24

Chen X, Zhu W, Azam S, Li H, Zhu F, Li H, Hong Y, Liu H, Zhang E, Wu H, Yu S, Zhou G, Li S, Zhong N, Wen S, Li X, Knapp SJ, Ozias-Akins, Varshney PRK, Liang X (2013) Deep sequenc-

ing analysis of the transcriptomes of peanut aerial and subterranean young pods identifies candidate genes related to early embryo abortion. Plant Biotechnol J 11:115–127
Chookietwattana K, Maneewan K (2012) Selection of efficient salt-tolerant bacteria containing ACC deaminase for promotion of tomato growth under salinity stress. Soil Environ 31:30–36
Ghazanfar MU, Raza M, Raza W (2018) Trichoderma as potential biocontrol agent, its exploitation. In: Agriculture: a review. Plant Protection 112:53–59
Ghosh S, Penterman JN, Little RD, Chavez R, Glick BR (2003) Three newly isolated plant growth-promoting bacilli facilitate the seedling growth of canola, *Brassica campestris*. Plant Physiol Biochem 41:277–281
Govarthananm M, Mythili R, Selvankumar T, Kamala-Kannan S, Kim H (2018) Myco-phytoremediation of arsenic- and lead-contaminated soils by *Helianthus annuus* and wood rot fungi, *Trichoderma* sp. isolated from decayed wood. Ecotoxicol Environ Saf 151:279–284
Hemantaranjan DKA (2015) Chitosan as a promising natural compound to enhance potential physiological responses in plant: a review. Indian J Plant Physiol 20:1–9
Hernandez OL, Calderín A, Huelva R, Guridi F, Aguiar NO, Olivares FL, Hernandez OL, Calderín A, Huelva R, Martínez-balmori D, Guridi F, Hernandez OL, Calderín A, Huelva R (2016) Humic substances from vermicompost enhance urban lettuce production. Agron Sustain Dev 35:225–232. https://doi.org/10.1007/s13593-014-0221-x
Ho YN, Mathew DC, Hsiao SC, Shih CH, Chien MF, Chiang HM, Huang CC (2012) Selection and application of endophytic bacterium *Achromobacter xylosoxidans* strain F3B for improving of phenolic pollutants. J Hazard Mater 219–220:43–49
Jacomassi LM, Viveiros JO, Oliveira MP, de Siqueira GFML, Crusciol CAC (2022) A seaweed extract-based biostimulant mitigates drought stress in sugarcane. Front Plant Sci 13:865291
Kaushal M, Wani SP (2016) Rhizobacterial-plant interactions: strategies ensuring plant growth promotion under drought and salinity stress. Agric Ecosyst Environ 231:68–78
Li Z, Zhang Y, Zhang X, Merewitz E, Peng Y, Huang L, Yan Y (2017) Metabolic pathways regulated by chitosan contributing to drought resistance in white clover. J Proteome Res 16(8):3039–3052
Luo SL, Chen L, Chen JL, Xiao X, Xu TY, Wan Y, Rao C, Liu CB, Liu YT, Lai C, Zeng GM (2011) Analysis and characterization of cultivable heavy metal-resistant bacterial endophytes isolated from Cd-hyperaccumulator *Solanum nigrum* L. and their potential use for phytoremediation. Chemosphere 85(7):1130–1138
Ma Y, Rajkumar M, Luo Y, Freitas H (2011) Inoculation of endophytic bacteria on host and non-host plants — effects on plant growth and Ni uptake. J Hazard Mater 195:230–237
Ma Y, Vosátka M, Freitas H (2019) Editorial: Beneficial microbes alleviate climatic stresses in plants. Front Plant Sci 10:1–2
Ma Y, Dias MC, & Freitas H (2020) Drought and salinity stress responses and microbe-induced tolerance in plants. Front Plant Sci 11(November). https://doi.org/10.3389/fpls.2020.591911
Mallick I, Bhattacharyya C, Mukherji S, Dey D, Chandra S, Kumar U, Ghosh A (2018) Effective rhizoinoculation and biofilm formation by arsenic immobilizing halophilic plant growth promoting bacteria (PGPB) isolated from mangrove rhizosphere: a step towards arsenic rhizoremediation. Sci Total Environ 610–611:1239–1250
Mehrafarin A (2017) Review overview on chitosan as a valuable ingredient and biostimulant. Pharm Ind Agric Prod 15:83–91
Nadeem SM, Zahir ZA, Naveed M, Nawaz S (2013) Mitigation of salinity-induced negative impact on the growth and yield of wheat by plant growth-promoting rhizobacteria in naturally saline conditions. Ann Microbiol 63:225–232
Nakashima K, Yamaguchi-Shinozaki K (2013) ABA signaling in stress-response and seed development. Plant Cell Rep 32:959–970
Nautiyal CS, Srivastava S, Chauhan PS, Seem K, Mishra A, Sopory SK (2013) Plant growth-promoting bacteria *Bacillus amyloliquefaciens* NBRISN13 modulates gene expression profile of leaf and rhizosphere community in rice during salt stress. Plant Physiol Biochem 66:1–9
Nazari M, Smith DL (2020) A PGPR-produced bacteriocin for sustainable agriculture: a review of thuricin 17 characteristics and applications. Front Plant Sci 11:916

Ng LM, Melcher K, Teh BT, Xu HE (2014) Abscisic acid perception and signaling: structural mechanisms and applications. Acta Pharmacol Sin 35:567–584

Pizzeghello D, Nicolini G, Nardi S (2001) Hormone-like activity of humic substances in Fagus sylvaticae forests. New Phytol 151(3):647–657. https://doi.org/10.1046/j.0028-646x.2001.00223.x

Rabêlo VM, César P, Bressanin LA, Teixeira D, Oliveira C, Karam D, Doriguetto AC (2019) The foliar application of a mixture of semisynthetic chitosan derivatives induces tolerance to water deficit in maize, improving the antioxidant system and increasing photosynthesis and grain yield. Sci Rep 9(1):8164

Rouhier N, Koh CS, Gelhaye E, Corbier C, Favier F, Didierjean C, Jacquot JP (2008) Redox based anti-oxidant systems in plants: biochemical and structural analyses. Biochim Biophys Acta 1780:1249–1260

Sah SK, Reddy KR, Li J (2016) Abscisic acid and abiotic stress tolerance in crop plants. Front Plant Sci 7:1–26

Sharma S, Anand G, Singh N, Kapoor R (2017) Arbuscular mycorrhiza augments arsenic tolerance in wheat (*Triticum aestivum* L.) by strengthening antioxidant defense system and thiol metabolism. Front Plant Sci 8:906

Singh D, Singh VK, & Singh AK (2019) Endophytic microbes: Prospects and their application in abiotic stress management and phytoremediation. Seed Endophytes: Biology and Biotechnology, 313–333. https://doi.org/10.1007/978-3-030-10504-4_15

Subramanian P, Kim K, Krishnamoorthy R, Mageswari A, Selvakumar G, Sa T (2016a) Cold stress tolerance in psychrotolerant soil bacteria and their conferred chilling resistance in tomato (*solanum lycopersicum* Mill.) under low temperatures. PLoS One 11:1–17

Subramanian S, Ricci E, Souleimanov A, Smith DL (2016b) A proteomic approach to lipo-chitooligosaccharide and thuricin 17 effects on soybean germination unstressed and salt stress. PLoS One 11:1–19

Tilman D, Balzer C, Hill J, Befort BL (2011) Global food demand and the sustainable intensification of agriculture. Proc Natl Acad Sci U S A 108:20260–20264

Venkatachalam P, Jayalakshmi N, Geetha N, Sahi SV, Sharma NC, Rene ER, Sarkar SK, Favas PJC (2017) Accumulation efficiency, genotoxicity and antioxidant defense mechanisms in medicinal plant *Acalypha indica* L. under lead stress. Chemosphere 171:544–553

Visioli G, Egidio SD, Vamerali T, Mattarozzi M, Maria A (2014) Chemosphere culturable endophytic bacteria enhance Ni translocation in the hyperaccumulator *Noccaea caerulescens*. Chemosphere 117:538–544. https://doi.org/10.1016/j.chemosphere.2014.09.014

Vurukonda SSKP, Vardharajula S, Shrivastava M, SkZ A (2016) Enhancement of drought stress tolerance in crops by plant growth promoting rhizobacteria. Microbiol Res 184:13–24

Xu M, Sheng J, Chen L, Men Y, Gan L, Guo S, Shen L (2014) Bacterial community compositions of tomato (*Lycopersicum esculentum* Mill.) seeds and plant growth promoting activity of ACC phytoremediation deaminase producing *Bacillus subtilis* (HYT-12-1) on tomato seedlings. World J Microbiol Biotechnol 30:835–845

Yakhin OI, Lubyanov AA, Yakhin IA, Brown PH (2017) Biostimulants in plant science: a global perspective. Front Plant Sci 7:2049

Yang Z, Wang Y, Wei X, Zhao X, Wang B, Sui N (2017) Transcription profiles of genes related to hormonal regulations under salt stress in sweet sorghum. Plant Mol Biol Report 35:586–599

Yong X, Chen Y, Liu W, Xu L, Zhou J, Wang S, Chen P, Ouyang P, Zheng T (2014) Enhanced cadmium resistance and accumulation in *pseudomonas putida* KT2440 expressing the phytochelatin synthase gene of *Schizosaccharomyces pombe*. Lett Appl Microbiol 58:255–261

Chapter 30
Biotechnological Techniques for Sustainable Waste Management

Mujahid Farid, Sheharyaar Farid, Muhammad Arslan Aslam, Khadija Azhar, Khadija Tariq, Zainab Akbar Ali, Zunera Ijaz, Zoya Sarfraz, Muqaddas Khalid, and Mohid Ahmad

Abstract The environment is considered the most significant component for man and other biotic units for their survival. The level of environmental sustainability reflects the fundamental value of both biotic and abiotic components. To achieve the goal of sustainability, the best comprehensive technique is recycling the waste and other by-products to make the waste more beneficial. It will also help the relationship between biotic and abiotic components to preserve a visual and vigorous equilibrium that distinguishes a perfect environment. Biotechnology is considered an evolving science for the sustainability of the environment. The conventional treatment methods required a high amount of cost and energy in most cases to remove the pollutants from the contaminated area. However, most of these processes only change the state of pollutants rather than eliminate them. On the other hand, biological methods include the degradation of the contaminants. In biological treatment methods, pollutants can be fixed, purified, and detached by using mainly microorganisms. Due to the comparatively low cost and the dissimilarity of work development, the bioremediation procedure has been most extensively used all over the world. Environmental biotechnology has been in existence for a while, but with the emergence of new methods such as contemporary microbiology and molecular genetics, researchers can now address the more pressing environmental issues such as detoxifying hazardous waste over the use of microbes. Exponential population growth has resulted in a decrease in the natural ecosystem, and disruptions to the

---

M. Farid (✉) · K. Azhar · K. Tariq · Z. A. Ali · Z. Ijaz · Z. Sarfraz · M. Khalid · M. Ahmad
Department of Environmental Sciences, University of Gujrat, Hafiz Hayat Campus, Gujrat, Pakistan

S. Farid
School of Natural Sciences and Ryan Institute, University of Galway, Galway, Ireland

M. A. Aslam
Department of Energy, Environment and Climate Change, Asian Institute of Technology (ATI), Pathum Thani, Thailand

Faculty of Science and Technology, University of the Basque Country, Basque, Spain

M. Hasanuzzaman (ed.), *Climate-Resilient Agriculture, Vol 2*,
https://doi.org/10.1007/978-3-031-37428-9_30

stability of natural cycles have a detrimental effect on both humans and other living systems. Biotechnology seems to be the most sustainable method of nourishing the environment, despite the obstacles posed by the increase in population and the ensuing problems with pollution.

Keywords Environmental sustainability · Biotechnology · Waste treatment · Trends in biotechnology

1 Introduction

The environment is the basic element for the survival of humans and the biotic component. The purpose of environmental sustainability may be characterized as the capability to meet the fundamental necessities of society today without compromising the demands of future generations. When it comes to the management of waste, it may be linked to managing natural resources to ensure that their current and future useful uses are not affected (Englande and Jin 2006). A healthy environment offers the basic goods to all the living organisms present in the ecosystem; however, the natural resources are under stress due to the rapid increase in the population. The world's population is predicted to grow by around ten billion people by the time 2050, putting an enormous additional burden on the planet's resources (Maja and Ayano 2021). The world's policymakers now place a significant emphasis on sustainable development. The strategy to achieve the goal of sustainable development must be comprehensive, cost-effective, and determined to protect public health and the environment (Sachs et al. 2019).

Among the wide spectrum of technologies that could help us achieve sustainability, the best way for sustaining the environment is to yield back the entire waste constituent in a recyclable means so that waste becomes beneficial. However, biotechnological techniques are eco-friendly and particularly take an important place in food production, pollution prevention, and bioremediation (Kurade et al. 2021). Biotechnological technologies are used for adapting the development or modification of living organisms, chemical processes, and products that are already present in the environment. Additionally, as a complement to traditional chemical technologies, they are employed in the industrial sector to improve environmental performance (Singh et al. 2018).

Environmental biotechnology refers to the usage of microorganisms to improve the quality of the environment. Due to the rapid increase in the population, the natural ecosystem is under pressure and causes instability in the natural cycle. These tools can help to produce new and innovative methods for sustainable and more efficient ways to produce conventional products having no harmful impact on the natural environment through cell modification in living organisms (Chen et al. 2005; Ezeonu et al. 2012).

Biotechnological techniques seem to be the most effective technology in the industrial sector, especially in the process of production compared to the other traditional chemical processing methods as they are more environmentally friendly. The necessity for innovative, cost-effective, and sustainable biological techniques of pollution treatment has been driven by numerous issues with traditional methods of pollutant treatment such as incineration or landfills. Human activities in the environment include several chemicals mixture in the process of renovating the natural goods in his environment into other forms suitable for his feeding. In the manufacturing procedure, man generates problems either knowingly or unknowingly (Ezeonu et al. 2012; Ghahari et al. 2021).

Microorganisms are added to the substrates that produce the desired products for the industrial sector such as biofiltration, bio-detergent, bioleaching (biomining), bioremediation, biocatalyst, biomass fuel generation, and biomonitoring. These are the several biotechnological tools that help to sustain the environment by controlling pollution (Ghahari et al. 2021). This chapter will provide an overall comprehensive overview of the different biotechnological techniques for sustainable waste management along with the development and modification of these methods with time.

2 Tools for Environmental Sustainability

2.1 Biofuels

There is an obvious need for the world to shift its energy dependency from fossil fuels to renewable energy resources to combat the accelerating rate of climate change due to the emission of greenhouse gases (GHGs) into the atmosphere. The study by Liu et al. (2021) mentioned that according to the US environmental protection agency, the US alone was responsible to emit around 6.67 gigatons of GHGs into the environment in the year 2018. Due to the environmental regulations to minimize the emission of GHG, and the recent advancement in fermentation processes and crop engineering, the production of biodiesel and bioethanol has become affordable and environmentally friendly replacements for fuels based on petroleum (Osman et al. 2021). Biofuels are made from the cellulose in the wood, grasses, and non-edible parts of plants that can ominously diminish greenhouse gas emissions in contrast to fossil fuels. The types of the biofuel can be differentiated on the basis of different key factors including the variety of feedstock and the conversion process (Fig. 30.1) (Jeswani et al. 2020).

Biofuel technology has developed remarkable improvements through multiple generations. The main issue with the first generation of biofuels is that the feedstocks for it are food crops such as sugar cane, which compete with food production as they required soil, water, and fertilizer. The use of pesticides and genetically engineered crops is further constrained by strict rules and regulation regarding their

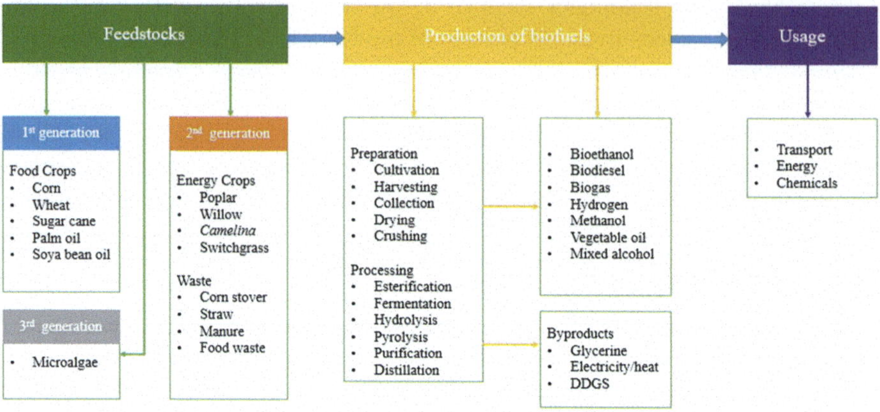

Fig. 30.1 An overview of the raw materials and manufacturing procedures for various biofuels. (Adapted from Jeswani et al. (2020))

usage (Liu et al. 2021). Moreover, to overcome these drawbacks, second-generation biofuels are made from the residues of the non-edible plant called lignocellulose, which is the most prevalent type of biomass on Earth (Isikgor and Becer 2015). Due to the economic benefits and the net carbon footprint, second-generation biofuels are very attractive. The production of biofuels through the transesterification of microalgae is generally known as the third generation of biofuels. However, the production of biofuels from second and third generations is referred to as advanced biofuels (Jeswani et al. 2020). As compared to the other generations of biofuels, the production of the fuel in microalgae-based biofuels occurs in one single organism which is the more efficient and direct process with zero energy demand (Wijffels and Barbosa 2010).

2.2 *Bioremediation*

The term "bioremediation" refers to the removal of pollutants from different environmental media through plants or microbes. The elimination of pollutants by using plants from water or soil occurs due to the absorption process by the roots and their accumulation in the leaves of plants. However, the microorganism can also eliminate or detoxify the pollutants that are present in the environment (Khalid et al. 2017). There are two major types of bioremediations which include phytoremediation and microorganism remediation.

The term "phytoremediation" is also divided into different types. The process by which the plant roots translocate and accumulate the contaminants from soil or water in their shoots which are then harvested is known as phytoextraction. Phytoextraction involves five major mechanisms, including the mobilization of contaminates in the rhizosphere and, the absorption of pollutants through plant roots

(Ali et al. 2013; Muthusaravanan et al. 2018). Moreover, rhizofiltration, also known as the name for phytofiltration, is the process of absorbing contaminants into the root zone or adsorbing them onto plant roots from a solution. The degree of success of phytofiltration depends upon the understating of pollutants heterogeneity and the correlation of all pollutants and nutrients (Dhanam 2017; Khan et al. 2019).

On the other hand, the term "phytostabilization" is the mechanism that involves the usage of plants having the ability to reduce the bioavailability of contaminants through the prevention of their seepage into the groundwater and their addition to the food chain through a variety of mechanisms such as adsorption by plant roots (Sarwar et al. 2017). However, phytostabilization does not offer a long-term remedy for contamination because it only works to minimize contamination of surrounding media/areas rather than reducing pollutant concentrations (Khalid et al. 2017).

Phytovolatilization is another type of phytoremediation technique, which uses plants to absorb pollutants, convert them into volatile substances, and release them into the atmosphere either in their original form or with slight modifications because of the plant's biochemical and evaporation forces (Khan et al. 2019). Moreover, phytodegradation, often referred to as phytotransformation, is the decomposition of pollutants absorbed by plants via internal biological pathways. Most plants can use their metabolic processes or enzymes to convert the ingested pollutants into less hazardous chemicals (da Conceição Gomes et al. 2016; Muthusaravanan et al. 2018).

The key process in microorganism remediation of a contaminated site is to inactivate and limit the availability of contaminants. Microorganisms are unable to degrade different inorganic pollutants such as heavy metals. However, they are capable to transform into another form because of the changes in their chemical and physical properties (Ashraf et al. 2019). Through their metabolic reactions, microorganisms contribute to the degradation of the targeted pollutant by functioning as biocatalysts and accelerating the process. Different factors can affect the optimization of the microbial remediation process including the presence of microbe's ability to degrade the pollutants and different environmental conditions such as temperature, soil type, pH, and redox potential (Mishra et al. 2021).

Though bioremediation is a potential approach for cleaning up the contaminated sites, there are some difficulties such as it requires more time as compared to the other treatment technologies. For most species, environmental factors might not always be appropriate. Moreover, using invasive plant species as hyperaccumulators may have an impact on the diversity of native flora. Similarly, the sustainability of biological treatment methods could be compromised by the exogenous application of microbes (Ashraf et al. 2019; Gerhardt et al. 2017).

2.3 *Biomining*

The general term "biomining" can be described as the methods for extracting and recovering metals from ores and waste products that rely on biological systems, primarily prokaryotic microorganisms. The presence of metals in the environment is a great concern as it can cause several harmful effects on the ecosystem and human health. (Johnson 2014). In the modern era, the concept of biomining refers to combined two-step biological mechanisms including bioleaching followed by biosorption to extract and recover the metals from different secondary sources such as e-waste, mining, and industrial waste (Kucuker and Kuchta 2018).

Several types of electronic wastes have a significant number of metals including rare earth elements having a high potential for economic development. The transformation to a greener, low-carbon economy relies significantly on crucial elements, such as elements of rare earth (REEs) and valuable metals, which are the basic components of all electronic waste materials; thus their extraction is significantly important (Tuncuk et al. 2012).

In previous studies, several traditional treatment techniques such as hydrometallurgical, mechanical, and biometallurgical methods have been suggested to recover the metals. However, these conventional processes have a significant environmental impact and required a larger scale of operation as compared to biomining which is considered the most sustainable and eco-friendly method to extract and recover the metals from the waste material (Kucuker and Kuchta 2018).

Biomining has been well recognized in the domain of recovering metals from refractory gold ores and low-grade and polymetallic base metal ores. In the present era, environmental and energy restrictions become more regulated and there will be a significant need to recycle old, abandoned wastes which frequently contain higher concentrations of metals, in addition, to recovering and reprocess metals from electronic as well as other metallic wastes (Johnson 2014).

2.4 *Biosensor*

Environmental pollution is currently the most serious problem we are facing, constantly growing and causing significant harm to the planet. The improvement in the standard of living causes different environmental problems such as rises in CO_2 concentration. Moreover, the soil ecosystem is also affected by the presence of different contaminants including pesticides and non-biodegradable substances. In addition to this man-made pollution, organic pollutants like bacteria and viruses are also polluting water sources (Wang et al. 2014).

The comprehensive analysis and continuous monitoring of the pollutants are important and ensure the safety of human health. Different types of biological agents are already in place for the detection and quantification of pollutants in all the environmental compartments. Moreover, classical analytical techniques like spectrophotometer and chromatography are costly, skill-intensive, and required more time. Thus, there is a need to develop more sustainable approaches with new advanced techniques. Therefore, biosensors appear to be a good option as a systematic tool for this (Hassani et al. 2017).

A biosensor is an integrated tool consisting of two parts: a bioreceptor (biological component) and a transducer. The first component, which is known as a bioreceptor, responds to the targeted compound by producing a physiochemical reaction, then the other component transducer transforms these physiochemical impulses into an electrical signal which provides an electronic display. These biosensors provide a real-time series signal with high sensitivity, and therefore used to quantify different biomaterial specimens and in other variety of the applications such as in the management of the fermentation process and environmental quality control (Negi and Choephel 2020).

There are different ways (biological component and transducer) by which a biosensor can be categorized into various types. Based on the biological component (bioreceptor), it could be an enzyme, antibody, or aptamer. However, transducers can also be used to classify a biosensor which could be electrochemical, optical, piezoelectric, electrical, or colorimetric. Despite the development of biosensors for the identification of environmental analytes, there are very few commercially available (Negi and Choephel 2020; Turner 2013).

3 Trends in Biotechnology

3.1 *Modern Biological Strategies*

- Deterioration typically means that a product is altered, but not continuously to the degree that it is broken down or transformed. Many decay demonstrations rely on proof that a single chemical is out of place, without determining whether or not additional products are being produced (Rochkind-Dubinsky et al. 1986). New harmful goods can be created through degradation. Right biodegradation is the metabolism of a substance to innocent products, to put it simply (Focht 1988). Determining the paths of degradation as well as the acceptable products and quantities is crucial. More inexperienced biodegradation approaches can be added by scientists and engineers with a better understanding of microbial ecology. An ideal all-natural strategy for research and development to break down a pollutant might consist of the following:
- Identifying the optimal organism with offshoot abilities.
- Describing the circumstances that permit the microbe to exist and its traits.
- Identifying the pollutant's metabolic route and any related or important cellular products.
- Identifying and describing the pathway's enzymes.
- Describing the setting for the therapy. If genetic engineering is used, the following significant steps are crucial:

Identifying the genes that control the pathways and the enzymes, and then modifying those genes to increase substrate diversity, stability, or degradation costs. In some cases, sequencing the genes of hobbyists may also provide some hints for how to control gene products to break down persistent substances (Kelly M, 1987,

Director, technology staff, Office of Solid Waste and Emergency Response, U.S. Environmental Protection Agency, personal communication).

3.2 Metabolic Pathway Design

To create effective metabolic pathways, three methods are being employed in the lab (the first two are more frequently used):

- Chemostats and other lab systems, in which organisms are cultivated under long-term selection circumstances to encourage the organisms to metabolize novel substrates.
- In vivo genetic transfers, whereby an herbal genetic technique is used to recruit the gene of a helpful enzyme from one organism into a pathway of another creature.
- Recombinant DNA era, in which genes are transferred using in vitro techniques to a new host to produce a whole new pathway (Pritchard PH, 1987, Acting Branch Chief, Microbial Ecology and Biotechnology. U.S. Environmental Protection Agency, Sabine Island, Gulf Breeze, personal communication).

The era of recombinant DNA allows for the most precise gene alteration, but it also necessitates extensive knowledge of one's ancestry and subsequent study and development. It is usually possible to complete selective pressure and in vivo transfer without extensive primary research. Positive examples are the many years of exposure to creosote and pentachlorophenol (PCP) waste, which has put selection pressure on organisms to develop the ability to metabolize novel substrates (Timmis et al. 1988).

The modified *Pseudomonas* can effectively break down 2,4,5-T, one of Agent Orange's active ingredients. A plasmid (an additional chromosomal DNA unit) in this situation has the genes necessary to produce one or more enzymes that break down the chemical. Environmental application must be the goal of modifying the plasmid so that it might be injected and maintained in a variety of host species that can occur on harmful internet domains (Ghosal et al. 1985).

3.3 Genetic Enhancement of Organisms

One strategy makes use of the era of recombinant DNA to logically construct pathways that can break down xenobiotic substances. These pathways can be constructed in two different ways: by redesigning existing pathways or by creating completely new pathways from enzymes or enzyme fragments (Ghosal et al. 1985). Patchwork assembly is the name for the later strategy. Many see the benefits of adopting native organisms' natural routes wherever feasible as well as the use of recombinant DNA creation to speed up processes for resistant substances. Work on molecular biological strategies to train several resistant command kilos is ongoing. For instance, it is believed that a mechanism exists that converts Dichlorobenzene (DCB), a suitable product, from Dichlorodiphenyltrichloroethane (DDT), one of the most persistent pesticides in the environment (Rochkind-Dubinsky et al. 1986). In many cases, it could be desirable to do without the oxygen need (Focht 1998). To create creatures that can paint without oxygen, basic recombinant technology research is required (Odelson et al. 1987; Swank 1998). Other research facilities are developing genetic strategies to speed up the removal of hazardous metals, which damage many soils and waste streams. This technique can recover priceless metals (Brierley 1982; Torma 1986; Wichlacz 1986).

3.4 Microbial Physiology and Ecology

In comparison to microbial biochemistry and genetics, microbial ecology and frame structure research and development are far less progressed. The use of organisms in the environment to reduce trash and pollution is severely impacted by these chemicals. Of all soil organisms, only 1–10% are known or have been grown (Check 1978). Even for bacteria that have been identified, very little is known about the whole range of reactions that may occur in any given organism, much less how those reactions are connected and managed. The relationship between a creature and its environment, as well as with other species, is thought about even much less. In nutrient enrichment and bioaugmentation, an understanding of frame form and ecology is crucial. Otherwise, the biosystem's performance and result cannot be taken into consideration (Raymond R, 1987, Sr, President, Biosystems, Chester Township, personal communication). In nutrient enrichment and bioaugmentation, an understanding of frame form and ecology is crucial. Otherwise, the biosystem's performance and result cannot be taken into consideration (Raymond R, 1987, Sr, President, Biosystems, Chester Township, personal communication).

3.5 Microbial Communities

Microbes exist in mixed microbial enterprises rather than being isolated within the environment. On occasion, microbial communities can destroy contaminants that a single cell could not. If the circumstances are right, an organism community may complete a series of responses. One reaction that changes benzoate into acetate, hydrogen, and carbon dioxide, for instance, and another that turns hydrogen and carbon dioxide into methane are just a few examples of how the dichlorination process previously mentioned is seen (Dolfing and Tiedje 1987). In another instance, enrichment using an analog chemical—a compound with a structure identical to the pollutant but without the chlorine attached—causes the exquisite bacteria to grow and promotes deterioration before being consumed by different, unreachable species. The analog chemical should be updated by developing new genotypes that would function well with the local organisms in real-world settings. At least one lab is investigating this method (Focht 1988). The cost of treatment will significantly increase if oxygen and vitamins are added, both in terms of the price of the raw material and the cost of adding and combining the additives. Significant financial savings should result from reducing the need for chemicals.

4 Bioengineering and Waste Treatment Processes for Energy Generation

4.1 Geothermal Energy

Energy production through the geothermal process is a sustainable and renewable resource and can minimize the risk of global warming (Soltani et al. 2021). At the end of the twentieth century, the focus in the field of renewable energy production has been placed on creating ecologically friendly and financially feasible solutions for disposing of geothermal waste (Premuzic et al. 1997).

Solid waste is trash composed of both inorganic and organic substances that are no longer valuable and should be managed to preserve the environment (Nanda and Berruti 2021). Geothermal processes can produce waste, particularly solid waste material. According to a study by Pasqualetti and Dellinger (1988) the major sources of solid waste in the US from the field of geothermal are drilling mud remnants, drill cuttings, and chemical waste from power plants and cooling towers.

This geothermal waste includes several types of contaminants including heavy metals. However, it is already acknowledged that many methods, such as oxidation, reduction, precipitation, and adsorption on the surface, can be used by microbes to engage with metallic materials. These processes provided the basis to create innovative methods that use biochemical processes to concentrate and eliminate hazardous metals from wastes and transform them into by-products that are appropriate for the environment and regulatory requirements (Premuzic et al. 1991).

4.2 Microbial Fuel Cells

Water and power conservation poses a worldwide problem due to several circumstances, including environmental changes, population growth, and rising commercial and home energy demand (Asim et al. 2021). Microbial fuel cells (MFCs) are a novel technology that has been evolving to tackle these issues. Microbial fuel cells may produce energy as well as they may also improve the removal efficacy of various types of contaminants from wastewater (Gude 2016a).

Microbial fuel cells utilize the electrochemical catalytic properties of microorganisms, to generate energy through the oxidation of both substrates organic and inorganic. Ideally, practically all substances which can be decomposed by microbes can be used as feedstock in MFCs, including simple chemical compounds, sewage from metropolitan areas, agricultural residues, and different industrial effluents (Gude 2016b).

All these MFCs are made up of two chambers including the anode and cathode. These chambers are kept separate through a proton exchange membrane. Organic feedstocks are oxidized through the biocatalyst at the anode, which ultimately releases electrons and protons. The proton exchange membrane (PEM) transports the protons to the cathode compartment, whereas the external circuit conveys the electrons. However, on the cathode side, both electrons and protons react while oxygen is simultaneously reduced to water (Rahimnejad et al. 2015).

The following categories list the primary MFC applications created in recent decades (Rahimnejad et al. 2015):

- Production of bioelectricity
- Biohydrogen generation
- Effluent treatment
- Biosensor

5 Waste Hierarchy

The expanding human population, economic expansion, fast urbanization, rising incomes, and increased demand for products and services all contribute to a rise in solid waste production (Minghua et al. 2009). From 1900 to 2010, the pace of global solid waste output increased from less than 0.3 Mt per day to more than 3.5 Mt per day and this rate is expected to be doubled by 2025 and tripled by 2100 (Hoornweg and Bhada-Tata 2012). The World Bank estimates that the yearly rate of municipal garbage production worldwide has reached around 2.01 billion tons and will increase to about 3.4 billion tons by 2050 (Kaza 2018). Garbage management challenges have gotten worse because of this sudden rise in waste generation (Iqbal et al. 2020). One of the biggest problems facing local governments in this century is managing solid waste.

Solid waste management accounts for a larger share of the recurring budgets (20–50%) of cities across the world (UN 2010). Municipalities have been managing solid waste since the mid-1900s, when uncollected trash and inadequate sanitation were first identified as the root causes of infectious illnesses (UN 2010). Humans and the ecology suffer when municipal solid waste is not managed properly.

Burning garbage and open dumping cause air pollution with hazardous fumes and soil contamination with leachates (Iqbal et al. 2020). The linear economic model, sometimes known as the "take, make, and discard" pattern, has achieved an unparalleled degree of growth as the globe transitions to an urban future, but it has also placed significant pressure on trash creation and resource supply concerns on the anthroposphere. Urban solid waste is widely believed to need to be handled to reduce any possible harm to the environment and public health. Another widely held belief is that contemporary civilizations should strive to be sustainable and pursue zero-waste society in accordance with circular economy concepts (Kijak and Moy 2004; Gamberini et al. 2013; VanEwijk and Stegemann 2014).

Policymakers and technical experts have adopted suitable methods for managing municipal solid waste and preserving the environment because of the ongoing promotion of environmental management and the accomplishment of sustainable development goals (Patil 2012). In this sense, the necessity of recycling and reusing material wastes has grown over time (Umar et al. 2016). The concept of waste hierarchy (WH) is well-known and has been utilized in various ways in waste management strategies. However, the first hierarchy for waste management was presented in the Dutch parliament in 1979 by politician Ad Lansink, whose name is derived from the Dutch phrase "Ladder van Lansink." The "Ladder of Lansink" is a framework that outlines the preferred order of waste management and resource conservation options, with "reduce" being the most favored and "landfill" the least. Over time, this approach has evolved into the widely recognized waste hierarchy, which is a crucial aspect of waste legislation both within the EU and globally. Unlike the 3Rs framework of the circular economy, the waste hierarchy specifically addresses the priority of waste processing through a five-stage pyramid, starting with "prevention" as the most desirable option and ending with "disposal" as the least preferable.

5.1 Sustainable Solid Waste Management Hierarchy

The ideas of "waste management hierarchy" serve as the foundation for sustainable solid waste management today (Yakubu and Zhou 2018). The hierarchy can be shown in a variety of ways, but the overall idea is to move waste management "up the hierarchy" such that reduction, reuse, and recycling (the "3Rs") are closer to the "top" and garbage is diverted from disposal, which is at the "bottom." The "3Rs"—reduce, reuse, and recycle—are referred to as the "waste hierarchy" and are used to categorize waste management techniques according to how desirable they are for minimizing trash. The waste hierarchy seeks to create the least amount of trash while maximizing the practical advantages from items.

Waste minimization is the priority in the hierarchy of waste management. Reusing old materials, traveling to supermarkets or local stores with shopping baskets, and other practices are all part of waste reduction. Reduced costs for disposal and treatment, preservation of natural resources, and the resolving of environmental issues are some advantages that might result from trash reduction (Sasikumar and Krishna 2009). In eight states in Malaysia, the government has mandated waste separation at source since 2015. This action was made to lessen the amount of garbage that was transported to disposal facilities since some of them have surpassed their carrying capacity and there isn't enough room in cities to build new ones (Sabariah et al. 2018).

The second strategy, waste reuse, is repurposing or reusing outdated products that still have value or function (Agamuthu 2004). Residents may help minimize the amount of trash produced by reusing old goods that still have value or giving them to others rather than tossing them away, according to Bees and William (2017). Many times, households have started the recycling process without recognizing it. For instance, discarded polythene bags are recycled, and persons in need are given old furniture or clothing.

The next step in the waste management hierarchy is recycling, which some homes do every day. There are several ways to recycle home solid waste for future use. Paper, aluminum, steel, glass, and plastic are all recyclable materials.

The least favored choice in the hierarchy of waste management is disposal. Open dumps or sanitary landfills can be used for disposal. The latter is superior and better for the environment. Since open garbage dumping pollutes the ecosystem, experts recommend using landfills as the destination for waste disposal.

6 Applications of Biotechnology in Waste Management Approaches

6.1 Toxicity Characterizations (Biomarkers)

Toxicity characterization techniques aim to define what is "toxic in toxic portions." Toxicity plays a sizable feature inside the strength of mind of water excellent-based necessities and effects on organic treatment plants. Polychlorinated dibenzo-p-dioxins and dibenzofurans are called dioxins and are a source of concern in the environment due to their biological build-up and harmful health effects. Incineration, namely waste material burning, is a major source of these persistent pollutants (Fiedler et al. 1990) and global interests have concentrated on methods for monitoring their emission from industrial operations.

The development of enzymatic biomarkers, the study of endocrine disruptors, and the screening of suspected carcinogens using gene assay are all current areas of attention that will continue in future.

The strain of a yeast with the dioxin receptor of human and a reporter gene has been created. In the presence of polycyclic aromatic hydrocarbons (PAHs), the reporter gene activates the receptor. The toxic reactions to dioxin, polychlorinated biphenyls (PCBs), benzopyrene, hexachlorobenzene, and certain PAHs may be mediated by this strain. Monitoring of PAHs and dioxin (or dioxin-like) chemicals in environmental samples is another use for this strain. It is now being used by researchers at the University of Tennessee to demonstrate bioremediation efforts (Miller III et al. 2004).

The development of a reporter gene assay technique using recombinant budding yeast that expresses human and rat AhR and Arnt proteins is another development (Kawanishi et al. 2013). To increase substance influx and prevent its efflux, this gene has been further modified by being removed from yeast assay strains that encode cell wall mannoproteins and ATP-binding cassette transporters which resulted in the development of an assay method for these yeast protoplasts that increased the detection limit 40-fold and decreased the assay's time by 40% (Kawanishi et al. 2003).

Over the past 10 years, specific disciplines in toxicology and environmental technical knowledge have arisen around the study of endocrine disruptive chemical compounds (EDCs) or xenobiotics with hormone interest. To evaluate the health of water bodies, monitoring environmental estrogens has therefore become more important (Kase et al. 2018). Semi-low-throughput in vitro tests have been developed because of research on the molecular mechanisms of EDC, and they may be used to quickly ascertain if environmental substances, compounds, or residues influence hormone levels (Bolger et al. 1998). In evaluating estrogenic activity, the use of in vitro reporter gene assays has been found to be useful because they provide a quantitative and integrated evaluation of ER-active contaminants. The assays measure the effect of the contaminants in terms of 17-estradiol equivalents (EEQs), taking into account the complexity of environmental mixtures which may contain both known and unknown compounds (Mehinto et al. 2018).

In this work, 33 surface (SW) and wastewater (WW) samples were collected from throughout Europe and tested using the newly developed EASZY assay, which employs transgenic *cyp19a1b-GFP* zebrafish (Danio rerio) embryos to measure in vivo estrogenic activity. A considerable and concentration-dependent up-regulation of the ER-regulated *cyp19a1b* gene expression in the developing brain was discovered in 18 of the 33 SW and WW samples, which caused estrogenic responses in the EASZY test (Brion et al. 2019).

The purpose of an aquatic pathobiology studies software is to discover using fish and one-of-a-kind non-mammals in fundamental danger assessment studies as environmental and biomedical fashions for carcinogenesis and reproductive/developmental consequences. The Japanese medaka (*Oryzias latipes*) is used in studies as a biological model for evaluating the carcinogenic and developmental effects of mixtures of hydrocarbons, thanks to support from the US Environmental Protection Agency. The US Department of Defense funds research on using the Japanese medaka as a biomedicine to examine the impact of chemical weapons on fetal and developmental outcomes. Additionally, the US Department of Energy encourages the study of wild fish as indicators of exposure to various chemical pollutants in marshes typical of the Mississippi River Basin (Hartley 2002).

6.2 Evaluations of Ecotoxicity

For sustainable development to become a reality, the device's eco-integrity must be upheld. Ecotoxicological evaluations are therefore more prevalent in industry. To maintain the status quo of water pleasant, based needs, evaluation of toxicity both acute and persistent, direct, and oblique, brief, and lengthy term is necessary. The fate and behavior of pollutants within the environment, risk assessment, and risk management are crucial aspects of the opinions. The future standard will include extensive ecological laboratory profiles on trash and goods to assist predict destiny and impact. To determine biodegradability and bacterial toxicity, BASF Ludwigshafen, Germany, uses more than 18 analytical tests. For sustainable development to be a reality, the eco-integrity of the device must be maintained. Therefore, ecotoxicological analyses are more common in business. Evaluation of acute and persistent, direct, and indirect, short- and long-term, and acute and persistent toxicity is required to preserve the status quo of water pleasant, based demands. The fate and behavior of pollutants within the environment, risk assessment, and risk management are crucial aspects of the opinions. The future standard will include extensive ecological laboratory profiles on trash and goods to assist predict destiny and impact. In this regard, the German company BASF Ludwigshafen uses more than 18 analytical tests to characterize features such as biodegradability, bacterial toxicity, and ecotoxicology (Strotmann and Weisbrodt 1994).

6.3 Treatment Trends

It is necessary to reduce the chronic organics and toxic compounds in wastewater to acceptable levels to limit the ecotoxicological impacts on receiving waterways. The following techniques can be used to improve organic treatment in the chemical and petrochemical industry: pre-treatment using chemical and/or advanced oxidation techniques to reduce toxicity and improve biodegradability; multi-degree treatment for moderate energy wastewaters, more effective precise contaminant removal, and reduced shock loadings; VOCs reduction via acclimation of biomass and implementation of biofilters for off-gasoline treatment; and selector designs (Eckenfelder and Englande 1998). It has been discovered that the immediate most SOUR (SOURim) is a greater degree of microbial reactivity to transient loadings and toxic/inhibitory inputs (Shamas and Englande 1992).

Further research is necessary to enhance the practical applications of this approach. The 16S ribosomal DNA polymerase chain reaction (PCR) is a technique that amplifies distinct DNA fragments to provide sufficient genetic material for identification using methods such as gel electrophoresis. Denaturing gradient gel electrophoresis (DGGE) is a method that separates and identifies DNA fragments by creating a denaturing gradient. This results in the separation of DNA into distinct bands, which can then be sequenced for identification. The PLFA and PCR-DGGE strategies have been used to determine the amount of potential biomass and the types of microorganisms present in various aquatic environments, deep subsurface environments, and marine environments. This has been demonstrated through studies of these environments (Hiorns et al. 1997; Methé et al. 1998; Gonzalez and Moran 1997; Zwart et al. 1998; Berner and Berner 1987; Ringelberg et al. 1997; Crump and Baross 1998).

Future installations of organic treatment methods will need to be designed in a variety of ways and reactor layouts to achieve the highest treatment effectiveness while requiring the least amount of space and power. A Bio Hoch reactor, which is defined under the terminology "progressive organic therapy," is an example of a format that achieves such goals (Trobisch 1992).

6.4 Residual Management

The management of residuals or bio-solids is becoming increasingly important in the context of conventional waste management and sustainable reuse. For sustainable processes to become viable options for waste control, they must be transformed into "value-added products," where the value of the final product exceeds the cost of processing. Some potential value-added products from properly treated bio-solids include engineered soil, ornamental horticultural fertilizer, turf grass, synthetic coal, and activated carbon. The US market for these products has been documented by Englande and Reimers (2001).

7 Ways of Biotechnology That Make the World Sustainable

7.1 Role of Biotechnology in Sustainability

Biotechnology refers to a group of legal technologies that are important to many different industrial industries (Zeng et al. 2018). For the commercial production of biotechnology goods and the transportation of its facilities, a fourth castigation—differently designated as biochemical, bioprocess, and biotechnology engineering—is required. None of the several techniques that biotechnology offers are used in all industrial divisions (Zenget al. 2018). Many nations are now expressing and implementing cogent strategies for harnessing biotechnology for industrial regeneration, job creation, and social improvement after realizing its tactical importance.

7.2 Bio Pulping

Managing wood chips with lignin-humiliating fungus before pulping is known as bio pulping. An experimental process called bio pulping has been thoroughly studied, mostly as a pre-treatment before mechanically pulping wood. By promoting the dispersion and effectiveness of chemicals during the heating of wood chips for breaking the cellulose fibers from the lignin before bio pulping substantially facilitates subsequent mechanical and chemical pulping. Therefore, bio pulping reduces the need for energy and chemicals, improves paper quality, and lessens the impact of pulp manufacturing on the environment (Patil 2012).

7.3 Bioenergy and Fuels

The production of fuels through biotechnology continues to intrigue many people. Examples of biofuels include bioethanol (Wyman 2018; Roehr 2001), biogas, biodiesel (Graboski and McCormick 1998), and biohydrogen (Nandi and Sengupta 1998). Except for biohydrogen, the other biofuels have a history of established or emerging commercial or pilot experimental application. Despite extensive research (Aden et al. 2002) on the bioconversion of lignocellulosic biomass to sugars for fermentation to ethanol, it is still rigid. The bioconversion of starch to sugars to produce bioethanol is more efficient and popular.

Similarly, gasoline ethanol produced from cane and beet sugar processing leftovers has been in use for a long time. Another widely utilized method is anaerobic digestion of organic waste to methane. The conventional manufacture of bioethanol has already been significantly sped up by modern biotechnology. For instance, the cost of the primary feedstock is reduced by the higher yielding naturally adapted corn; the starch found in genetically modified corn is more easily converted to sugars using enzymatic bioconversion than natural corn starch. Microbial enzymes have been engineered for enhanced stability and the ability to rapidly convert starch to fermentable sugars. Additionally, microorganisms have been designed to tolerate higher levels of toxic ethanol and undergo faster fermentation. These advancements, along with others, will make the production of bioethanol more economically feasible in the future.

7.4 Biofertilizers and Soil Inoculants

As a low-cost and secure alternative to chemical fertilizers that are used to deliver nitrogen, phosphorous, potassium, sulfur, and other inorganic nutrients necessary for crop development, biofertilizers and inoculants are gaining popularity (Ashraf et al. 2019). Rhizobia bacteria that fix nitrogen were the foundation of the first generation of biological fertilizers, which were undoubtedly present in the root nodules of legume plants. These bacteria fix nitrogen in the atmosphere so that the plant has access to attackable nitrogen. By enhancing plants' interest in conventional fertilizers, microbial inoculants can be employed as a companion to them. It is predicted that increased usage of biofertilizers will significantly contribute to reducing the pollution, energy use, and resource consumption associated with traditional fertilizer use.

7.5 *Biopesticides*

Crop safety, managing weeds, controlling insects, treating seeds, preventing algae growth in swimming pools, and protecting wood and textiles are all areas where pesticides are employed (Waxman 1998). Any tiny biological manager produced by microorganisms for use in controlling rodent pests, weeds, and insects is referred to as a "biopesticide." Biopesticides are packaged, handled, stored, and applied in the same ways as old-fashioned pesticides. Although biopesticides have achieved some spectacular things, their effectiveness has been questioned (Auld and Louise 1995). In 2000, sales of bio pesticides were around US$160 million. Of this, sales of Bt products accounted for more than 90% of the total (Vega et al. 1999).

Less than 2% of the world's pesticide market is now occupied by biopesticides, but it is expected that this share will rise significantly over time. Compared to traditional chemical pesticides, biopesticides often have less negative environmental effects overall, are more highly targeted, do not leave harmful residues, and lower the chance of resistance development in the target species (Hart and Pimentel 2002). Plant fungal infections have been treated using biofungicides in both the phylloclade and rhizosphere. Certain soil-borne plant diseases have been successfully managed by seed coatings containing *Bacillus* and *Pseudomonas* species (Weisz et al. 1984).

7.6 *Plastics and Other Polymers*

Since the 1920s, researchers have been aware of the presence of biodegradable polymers in bacteria, such as polyhydroxy alkanoic acids (PHA). For a very long time, bio plastics were ignored due to the costs associated with manufacturing them and the availability of flexible, affordable polymers made from petrochemicals. Renewing interest in biologically produced polymers is being driven by concerns over the permanence of petrochemical plastics in the environment (Simpson et al. 2019). A microbe was created by the Japan Institute of Physics and Chemical Research to manufacture up to 96% of its dry weight in biodegradable plastic (Lenz 1995). There are now many different plastics available, but no plastic biopolymers. Their manufacturing and usage are environmentally sustainable, even though they

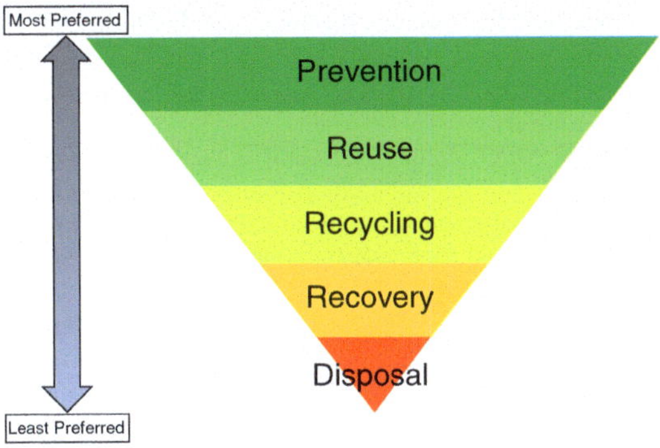

Fig. 30.2 The hierarchy of waste management

are still somewhat pricey. Improved biopolymers and microbial strains capable of generating them are being developed.

8 Conclusions and Recommendations

The production of waste is directly linked with the overall population and consumption globally. It is necessary to utilize the available resources optimally as the production of waste drastically increases due to the rapid increase in human population and industrial activities. Preventing the negative effects of waste through proper waste management is equally vital. To fulfil the requirements of the existing population and establish a planet that is resource-rich for the future, alternative resources must be used. Because of environmental and economic benefits, the applications of biotechnological tools in agriculture and industrial sectors are essential to treat different types of waste materials. These technologies can improve the quality of the processing method and the quality of the product in a cost-effective way. Due to various modifications in the microbiological processes, now it is possible to evaluate various wastes around us. These tools are not only used for the treatment of various pollutants, but, on the other hand, they can also provide different renewable energy resources. The modifications from conventional treatment technologies to cost-effective and sustainable methods are the biggest achievements of the twenty-first century (Fig. 30.2).

References

Aden A, Ruth M, Ibsen K, Jechura J, Neeves K, Sheehan J, Wallace B, Montague L, Slayton A, Lukas J (2002) Lignocellulosic biomass to ethanol process design and economics utilizing co-current dilute acid prehydrolysis and enzymatic hydrolysis for corn stover. United States. https://doi.org/10.2172/15001119

Agamuthu P, Fauziah HS, Simon C (2004) Municipal solid waste management in Malaysia - possibility of improvement? Malaysian J Sci 23:61–70

Ali H, Khan E, Sajad MA (2013) Phytoremediation of heavy metals—concepts and applications. Chemosphere 91(7):869–881

Ashraf S, Ali Q, Zahir ZA, Ashraf S, Asghar HN (2019) Phytoremediation: environmentally sustainable way for reclamation of heavy metal polluted soils. Ecotoxicol Environ Saf 174:714–727

Asim AY, Mohamad N, Khalid U, Tabassum P, Akil A, Lokhat D, Siti H (2021) A glimpse into the microbial fuel cells for wastewater treatment with energy generation. Desalin Water Treat 214:379–389

Auld BA, Louise M (1995) Constraints in the development of bioherbicides. Weed Technol 9:638–652

Bees AD, Williams ID (2017) Explaining the differences in household food waste collection and treatment provisions between local authorities in England and Wales. Waste Manage 70:222–235

Berner EK, Berner RA (1987) Marginal marine environments: estuaries. In: The global water cycle: geochemistry and environment. Prentice-Hall, Englewood Cliffs

Bolger R, Nestich S, Wiese T, Ervin K, Checovich W (1998) Rapid screening of environmental chemicals for estragon receptor binding capacity. Environ Health Perspect 106:551–557

Brierley CL (1982) Microbiological mining. Sci Am 247(2):44–53. https://doi.org/10.1038/scientificamerican0882-44

Brion F, De Gussem V, Buchinger S, Hollert H, Carere M, Porcher JM, Piccini B, Féray C, Dulio V, Könemann S, Simon E, Werner I, Kase R, Aït-Aïssa S (2019) Monitoring estrogenic activities of waste and surface waters using a novel in vivo zebrafish embryonic (EASZY) assay: comparison with in vitro cell-based assays and determination of effect-based trigger values. Environ Int 130:104896. https://doi.org/10.1016/j.envint.2019.06.006

Check W (1978) The reemergence of microbial ecology. Mosaic 18(4):24–35

Chen W, Mulchandani A, Deshusses MA (2005) Environmental biotechnology: challenges and opportunities for chemical engineers. AICHE J 51(3):690–695

Crump BC, Baross JA (1998) Particle-attached bacteria and heterotrophic plankton in the Columbia River estuary. Mar Ecol Prog Ser 138:265–273

da Conceição Gomes MA, Hauser-Davis RA, de Souza AN, Vitória AP (2016) Metal phytoremediation: general strategies, genetically modified plants and applications in metal nanoparticle contamination. Ecotoxicol Environ Saf 134:133–147

Dhanam S (2017) Strategies of bioremediation of heavy metal pollutants toward sustainable agriculture. In: Sustainable agriculture towards food security, pp 349–358

Dolfing J, Tiedje JM (1987) Growth yield increase linked to reductive dechlorination in a defined 3-chlorobenzoate degrading methanogenic coculture. Arch Microbiol 149:102–105. https://doi.org/10.1007/BF00425073

Eckenfelder WW, Englande AJ (1998) Innovative biological treatment for sustainable development in the chemical industries. In: Proceedings of IAWQ 19th Biennial International Conference, Vancouver, 16–20 June, pp 95–102

Englande AJ, Jin G (2006) Application of biotechnology in waste management for sustainable development: an overview. Manag Environ Qual 17(4):467–477. https://doi.org/10.1108/14777830610670526

Englande AJ, Reimers RS (2001) Bio solids management – sustainable development status and future direction. Water Sci Technol 44:41–46

Ezeonu CS, Tagbo R, Anike EN, Oje OA, Onwurah INE (2012) Biotechnological tools for environmental sustainability: prospects and challenges for environments in Nigeria—a standard review. Biotechnol Res Int 450802:1. https://doi.org/10.1155/2012/450802

Fiedler H, Hutzinger O, Timms CW (1990) Dioxins: sources of environmental load and human exposure. Toxicol Environ Chem 29:157–234

Focht DD (1988) Performance of biodegradative microorganisms in soil: xenobiotic chemicals as unexploited metabolic niches. Environmental biotechnology: reducing risks from environmental chemicals through biotechnology, 15–29
Focht DD (1998) Professor, Department of Soil and Environmental Sciences, University of California, Riverside, personal communication
Gamberini R, Del Buono D, Lolli F, Rimini B (2013) Municipal solid waste management: identification and analysis of engineering indexes representing demand and costs generated in virtuous Italian communities. Waste Manag 33(11):2532–2540. https://doi.org/10.1016/j.wasman.2013.06.003
Gerhardt KE, Gerwing PD, Greenberg BM (2017) Opinion: taking phytoremediation from proven technology to accepted practice. Plant Sci 256:170–185
Ghahari S, Ghahari S, Ghahari S, Nematzadeh GA, Sarma H (2021) Environmental biotechnology: toward a sustainable future. In: Biotechnology for sustainable environment, pp 1–31
Ghosal D, You I, Chatterjee DK, Chakrabarty AM (1985) Microbial degradation of halogenated compounds. Science 228(4696):135–142
Gonzalez JM, Moran MA (1997) Numerical dominance of a group of marine bacteria in the α-subclass of the class Proteobacteria in coastal seawater. Appl Environ Microbiol 63:4237–4242
Graboski MS, McCormick RL (1998) Combustion of fat and vegetable oil derived fuels in diesel engines. Prog Energy Combust Sci 24:125–164. https://doi.org/10.1016/S0360-1285(97)00034-8
Gude VG (2016a) Microbial fuel cells for wastewater treatment and energy generation. Micro Electro Fuel Cell:247–285. https://doi.org/10.1016/B978-1-78242-375-1.00008-3
Gude VG (2016b) Wastewater treatment in microbial fuel cells–an overview. J Clean Prod 122:287–307
Hart K, Pimentel D (2002) Public health and costs of pesticides. Encycl Pest Manag 1:677–679. https://doi.org/10.1201/noe0824706326.ch313
Hartley P (2002) Interpersonal communication, 2nd edn. Routledge. https://doi.org/10.4324/9780203019719
Hassani S, Momtaz S, Vakhshiteh F, Maghsoudi AS, Ganjali MR, Norouzi P, Abdollahi M (2017) Biosensors and their applications in detection of organophosphorus pesticides in the environment. Arch Toxicol 91(1):109–130
Hiorns WD, Methe BA, Nierzwicki-Bauer SA, Zehr JP (1997) Bacterial diversity in Adirondack Mountain lakes as revealed by 16S rRNA gene sequences. Appl Environ Microbiol 63:2957–2960
Hoornweg D, Bhada-Tata P (2012) What a Waste : A Global Review of Solid Waste Management. Urban development series; knowledge papers no. 15. World Bank, Washington, DC. http://hdl.handle.net/10986/17388
Iqbal A, Liu X, Chen G (2020) Municipal solid waste: review of best practices in the application of life cycle assessment and sustainable management techniques. Sci Total Environ 729:138622
Isikgor FH, Becer CR (2015) Lignocellulosic biomass: a sustainable platform for the production of bio-based chemicals and polymers. Polym Chem 6(25):4497–4559
Jeswani HK, Chilvers A, Azapagic A (2020) Environmental sustainability of biofuels: a review. Proc R Soc A: Math Phys Eng Sci 476(2243):20200351. https://doi.org/10.1098/rspa.2020.0351
Johnson DB (2014) Biomining—biotechnologies for extracting and recovering metals from ores and waste materials. Curr Opin Biotechnol 30:24–31. https://doi.org/10.1016/j.copbio.2014.04.008
Kase R, Javurkova B, Simon E, Swart K, Buchinger S, Könemann S, Escher BI, Carere M, Dulio V, Ait-Aissa S, Hollert H, Valsecchi S, Polesello S, Behnisch P, di Paolo C, Olbrich D, Sychrova E, Gundlach M, Schlichting R, Leborgne L (2018) Screening and risk management solutions for steroidal estrogens in surface and wastewater. TrAC Trend Analy Chem 102:343–358. https://doi.org/10.1016/j.trac.2018.02.013
Kawanishi M, Sakamoto M, Ito A, Kishi K, Yagi T (2003) Construction of reporter yeasts for mouse aryl hydrocarbon receptor ligand activity. Mutat Res 540:99–105

Kawanishi M, Ohnisi K, Takigami H, Yagi T (2013) Simple and rapid yeast reporter bioassay for dioxin screening: evaluation of the dioxin-like compounds in industrial and municipal waste incineration plants. Environ Sci Pollut Res 20:2993–3002

Kaza S (2018) What a waste 2.0 a global snapshot of solid waste management to 2050. The World Bank, Washington, D.C.

Kelly M (1987) Director, technology staff, Office of Solid Waste and Emergency Response, U.S. Environmental Protection Agency, personal communication

Khalid S, Shahid M, Niazi NK, Murtaza B, Bibi I, Dumat C (2017) A comparison of technologies for remediation of heavy metal contaminated soils. J Geochem Explor 182:247–268

Khan I, Iqbal M, Shafiq F (2019) Phytomanagement of lead-contaminated soils: critical review of new trends and future prospects. Int J Environ Sci Technol 16(10):6473–6488

Kijak R, Moy DA (2004) Decision support framework for sustainable waste management. J Ind Ecol 8(3):33–50. https://doi.org/10.1162/1088198042442 30

Kucuker MA, Kuchta K (2018) Biomining – biotechnological systems for the extraction and recovery of metals from secondary sources. Global NEST J 20(4):737–742. https://doi.org/10.30955/GNJ.002692

Kurade MB, Ha YH, Xiong JQ, Govindwar SP, Jang M, Jeon BH (2021) Phytoremediation as a green biotechnology tool for emerging environmental pollution: a step forward towards sustainable rehabilitation of the environment. Chem Eng J 415:129040. https://doi.org/10.1016/j.cej.2021.129040

Lenz RW (1995) Biodegradable polymers and plastics in Japan: Research, development, and applications. Japanese Technology Evaluation Center, JTEC/WTEC Program Loyola College in Maryland, Baltimore, Maryland.

Liu Y, Cruz-Morales P, Zargar A, Belcher MS, Pang B, Englund E, Dan Q, Yin K, Keasling JD (2021) Biofuels for a sustainable future. Cell 184(6):1636–1647. https://doi.org/10.1016/j.cell.2021.01.052

Maja MM, Ayano SF (2021) The impact of population growth on natural resources and farmers' capacity to adapt to climate change in low-income countries. Earth Syst Environ 5(2):271–283. https://doi.org/10.1007/s41748-021-00209-6

Mehinto AC, Kroll KJ, Jayasinghe BS, Lavelle CM, VanDervort D, Adeyemo OK, Bay SM, Maruya KA, Denslow ND (2018) Linking in vitro estrogenicity to adverse effects in the inland silverside (*Menidia beryllina*). Environ Toxicol Chem 37(3):884–892. https://doi.org/10.1002/etc.4024

Methé BA, Hiorns WD, Zehr JP (1998) Contrasts between marine and freshwater bacterial community composition: analyses of communities in Lake George and six other Adirondack lakes. Limno and Oceano 43(2):368–374. https://doi.org/10.4319/lo.1998.43.2.0368

Miller CA III, Martinat M, Hyman L (2004) Assessment of aryl hydrocarbon receptor complex interactions using pBEVY plasmids: expression vectors with bi-directional promoters for use in Saccharomyces cerevisiae. Nucleic Acids Res 26:3577–3583

Minghua Z, Xiumin F, Rovetta A, Qichang H, Vicentini F, Bingkai L, Giusti A, Yi L (2009) Municipal solid waste management in Pudong new area, China. Waste Manag 29:1227–1233

Mishra M, Singh SK, Kumar A (2021) Chapter 5 - Environmental factors affecting the bioremediation potential of microbes. In: Kumar A, Singh VK, Singh P, Mishra EC (eds) Food science, technology and nutrition. Woodhead Publishing, pp 47–58. https://doi.org/10.1016/B978-0-12-821199-1.00005-5

Muthusaravanan S, Sivarajasekar N, Vivek JS, Paramasivan T, Naushad M, Prakashmaran J, Gayathri V, Al-Duaij OK (2018) Phytoremediation of heavy metals: mechanisms, methods and enhancements. Environ Chem Lett 16(4):1339–1359. https://doi.org/10.1007/s10311-018-0762-3

Nanda S, Berruti F (2021) Municipal solid waste management and landfilling technologies: a review. Environ Chem Lett 19(2):1433–1456. https://doi.org/10.1007/s10311-020-01100-y

Nandi R, Sengupta S (1998) Microbial production of hydrogen: an overview. Crit Rev Microbiol 24(1):61–84. https://doi.org/10.1080/10408419891294181

Negi NP, Choephel T (2020) Biosensor: an approach towards a sustainable environment. In: Mohsin M, Naz R, Ahmad A (eds) Nanobiosensors for agricultural, medical and environmental applications. Springer Singapore, pp 43–62. https://doi.org/10.1007/978-981-15-8346-9_3
Odelson DA, Rasmussen JL, Smith CJ (1987) Extrachromosomal systems and gene transmission in anaerobic bacteria. Plasmid 17:87–109
Osman AI, Qasim U, Jamil F, Al-Muhtaseb AH, Jrai AA, Al-Riyami M, Al-Maawali S, Al-Haj L, Al-Hinai A, Al-Abri M, Inayat A, Waris A, Farrell C, Maksoud MIAA, Rooney DW (2021) Bioethanol and biodiesel: bibliometric mapping, policies and future needs. Renew Sust Energy Rev 152:111677. https://doi.org/10.1016/j.rser.2021.111677
Pasqualetti MJ, Dellinger M (1988) Hazardous waste from geothermal energy: a case study. J Energy Develop 13(2):275–295. http://www.jstor.org/stable/24807777
Patil YD (2012) Sustainable development & construction industry. SSRN Electro J:2004957. https://doi.org/10.2139/ssrn.2004957
Premuzic ET, Lin MS, Zhu H, Gremme AM (1991) Selectivity in metal uptake by stationary phase microbial populations. Arch Environ Contam Toxicol 20:234–240
Premuzic ET, Lin MS, Jin JZ, Hamilton K (1997) Geothermal waste treatment biotechnology. Energy Source 19(1):9–17. https://doi.org/10.1080/00908319708908828
Pritchard PH (1987) Acting Branch Chief, Microbial Ecology and Biotechnology. U.S. Environmental Protection Agency, Sabine Island, Gulf Breeze, FL, personal communication.
Rahimnejad M, Adhami A, Darvari S, Zirepour A, Oh SE (2015) Microbial fuel cell as new technology for bioelectricity generation: a review. Alex Eng J 54(3):745–756. https://doi.org/10.1016/j.aej.2015.03.031
Raymond R (1987) Sr, President, Biosystems, Chester Township, PA, personal communication
Ringelberg DB, Sutton S, White DC (1997) Biomass, bioactivity and biodiversity: microbial ecology of the deep subsurface: analysis of ester-linked phospholipid fatty acids. FEMS Microbiol Rev 20:371–377
Rochkind-Dubinsky ML, Sayler GS, Blackburn JW (1986) Microbiological decomposition of chlorinated aromatic compounds. [online] Available at: https://www.osti.gov/biblio/7162255. Accessed 22 Nov 2022.
Roehr M (2001) The biotechnology of ethanol: classical and future applications. Wiley-VCH, Chichester, p 232
Sabariah B, Norfadzrin FIMN, Mohamad NI, Aswanimiyuni A, Haryani H (2018) Effect of probiotics in fermented palm kernel meal (FPKM) and total mixed ration (TMR) to improve milk production in Mafriwal dairy cattle of Malaysia Veterinary Institute (IVM) farm. Malays J Veter Res 9(1): 45–51
Sachs JD, Schmidt-Traub G, Mazzucato M, Messner D, Nakicenovic N, Rockström J (2019) Six transformations to achieve the sustainable development goals. Nature Sustain 2(9):805–814. https://doi.org/10.1038/s41893-019-0352-9
Sarwar N, Imran M, Shaheen MR, Ishaque W, Kamran MA, Matloob A, Rehim A, Hussain S (2017) Phytoremediation strategies for soils contaminated with heavy metals: modifications and future perspectives. Chemosphere 171:710–721
Sasikumar K, Krishna SG (2009) Solid waste management. PHI Learning, New Delhi
Shamas JY, Englande AJ (1992) Use of the immediate maximum specific oxygen uptake rate as an activated sludge process control parameter. Water Sci Technol 25:123–132
Simpson JD, Samuel AS, Kristofer J, Thurecht GS (2019) Engineered polymeric materials for biological applications: overcoming challenges of the bio–nano interface. Polym 11(9):1441. https://doi.org/10.3390/polym11091441
Singh RP, Singh PK, Gupta R, Singh RL (2018) Biotechnological tools to enhance sustainable production. In: Singh RL, Mondal SA (eds) . Woodhead Publishing, pp 19–66. https://doi.org/10.1016/B978-0-12-812160-3.00002-7

Soltani M, Moradi Kashkooli F, Souri M, Rafiei B, Jabarifar M, Gharali K, Nathwani JS (2021) Environmental, economic, and social impacts of geothermal energy systems. Renew Sustain Energy Rev 140:110750. https://doi.org/10.1016/j.rser.2021.110750

Strotmann UJ, Weisbrodt W (1994) Wastewater treatment and integrated environmental protection at the BASF AG in Ludwigshafen, Germany. Water Sci Technol 29:185–192

Swank R (1998) Environmental research laboratory. U.S. Environmental Protection Agency, Athens, personal communication

Timmis KN, Rojo F, Ramos JL (1988) Prospects for laboratory engineering of bacteria to degrade pollutants. In: Omenn GS (ed) Environmental biotechnology. Basic Life Sci 45:61–79. Springer, Boston. https://doi.org/10.1007/978-1-4899-0824-7_4

Torma AE (1986) Biohydrometallurgy as an emerging technology. In: Proceedings of the biotechnology and bioengineering symposium 16, p 49

Trobisch KH (1992) Recent developments in the treatment of chemical wastewater in Europe. Water Sci Technol 26:319–332

Tuncuk A, Stazi V, Akcil A, Yazici EY, Deveci H (2012) Aqueous metal recovery techniques from e-scrap: hydrometallurgy in recycling. Miner Eng 25(1):28–37. https://doi.org/10.1016/j.mineng.2011.09.019

Turner APF (2013) Biosensors: sense and sensibility. Chem Soc Rev 42(8):3184–3196

Umar UA, Shafiq N, Malakahmad A, Nuruddin MF, Khamidi MF, Salihi IU (2016) Impact of construction waste minimization at construction site: case study. J Technol 78(5–3):33–40

UN (2010) Solid waste management in the world's cities: water and sanitation in the world's cities. Earthscan, Washington, D.C.

VanEwijk S, Stegemann JA (2014) Limitations of the WH for achieving absolute reductions in material throughput. J Clean Prod 132:122–128. https://doi.org/10.1016/j.jclepro.2014.11.05

Vega FE, Jackson MA, McGuire MR (1999) Germination of conidia and blastospores of the fungal entomopathogen Paecilomyces fumosoroseus on the cuticle of Bemisia argentifolii and on two types of agar. Mycopathol 147:33–35

Waxman MF (1998) Agrochemical and Pesciticide Safety Handbook. Lewis, Boca Raton

Wang X, Lu X, Chen J (2014) Development of biosensor technologies for analysis of environmental contaminants. Trend Environ Anal Chem 2:25–32

Weisz JR, Rothbaum FM, Blackburn TC (1984) Standing out and standing in: the psychology of control in America and Japan. Am Psychol 39(9):955–969. https://doi.org/10.1037/0003-066X.39.9.955

Wichlacz PL (1986) Practical aspects of genetic engineering for the mining and mineral industries. In: Biotechnology bioengineering symposium 16. Wiley, New York

Wijffels RH, Barbosa MJ (2010) An outlook on microalgal biofuels. Science 329(5993):796–799

Wyman C (2018) Handbook on bioethanol. Routledge 1:1–18

Yakubu Y, Zhou J (2018) Novel approach to quantify municipal solid waste management hierarchy based on analytical hierarchy. Int J Environ Sci Technol 16:1897–1908. https://doi.org/10.1007/s13762-018-1768-x

Zeng L, Li X, Fan S, Zhang M, Yin Z, Tadé M, Liu S (2018) Insight into MoS2 synthesis with biophotoelectrochemical engineering and applications in levofloxacin elimination. ACS Appl Energy Mater 1(8):3752–3762

Zwart G, Hiorns WD, Methe BA, Van Agterveld MP, Huismans R, Nold SC, Zehr JP, Laanbroek HJ (1998) Nearly identical 16S rRNA sequences recovered from lakes in North America and Europe indicate the existence of clades of globally distributed freshwater bacteria. Syst Appl Microbiol 21:546–556

Chapter 31
Role of Biotechnology in Management of Solid Waste

Mujahid Farid, Wajiha Sarfraz, Zaki ul Zaman Asam, Mohsin Abbas, Fatima Abdul Sattar, Khadija Asif, Shabnam Abbas, Malaika Iqbal, Zahra Afgun, and Aiza Afzal

Abstract Managing solid waste has been challenging for decades. Multiple sources contribute to the production of metric tons of solid waste. Wastes generated by civilization, urbanization, and industrialization are disposed of annually into the environment. It also lacks proper disposal strategies. The disposal and management of industrial and municipal wastes have been discussed. Solid waste is usually disposed of in an open dump, which requires a large area and releases toxic, hazardous substances into the environment. For several years, significant research has been carried out to find a feasible approach to replace these unsuitable solid waste disposal methods, which can finally be replaced by biotechnological techniques such as composting, biodegradation, and bioremediation. The methodologies mentioned above are discussed briefly in this chapter. Composting is a biotechnological method of organic decomposition of solid waste in a controlled environment by microbes. Bioremediation utilizes living things to remove contaminants from soils, sludges, and groundwater. Moreover, biodegradation considers biological organisms to degrade solid waste. These are biotechnological advances in solid waste disposal.

Keywords Solid waste management · Biotechnological approaches · Bioremediation · Composting · Biodegradation

M. Farid (✉) · Z. u. Z. Asam · M. Abbas · F. A. Sattar · K. Asif · S. Abbas · M. Iqbal
Z. Afgun · A. Afzal
Department of Environmental Sciences, University of Gujrat, Hafiz Hayat Campus, Gujrat, Pakistan

W. Sarfraz
Department of Botany, Government College Women University, Sialkot, Pakistan

M. Hasanuzzaman (ed.), *Climate-Resilient Agriculture, Vol 2*,
https://doi.org/10.1007/978-3-031-37428-9_31

1 Introduction

Biotechnology has attracted considerable attention to new technologies which have emerged in the last few years. It has expressed the ability to produce enormous wealth and impact every substantial part of the economy. Biotechnology has a negative reputation because it is likely to be associated with agricultural and geriatric crops, known as agrotechnology. Nevertheless, biotechnology has many additional applications, such as environmental management and pharmaceuticals (Sufficiency et al. 2022). One of the most promising applications of biotechnology is waste management. Waste management is the management of various methods of removing waste material from the environment, like degradation, waste removal, and processing (Iqbal et al. 2013). Solid waste has been increasing over time. By 2025, solid waste generation will range from 1.2 to 1.42 kg per person per day. The cause of the increase in solid waste is a lack of waste management infrastructure. It involves the handling and disposal of different kinds of waste, including household waste, industrial waste, E-waste, and wastewater. Solid waste management must assist in the long-term sustainability of solid waste and its conversion into resources (Zhou et al. 2021; Salemdeeb et al. 2017). Nutrients are contaminated when solid waste is simplified with lignin, cellulose, amylose, and monosaccharides. The substantial influences for the expansion of environmentally friendly technologies based on lower food products to achieve the global targets for biofuel, chemicals, and biomaterials are also demanded for their rapid growth, although the rising prices of fossil fuels, the decline of natural resources, and the increase in taxes are unavoidable (Das et al. 2012). This chapter discusses biotechnology as a tool for solid waste management and provides proper management.

Essentially, biotechnology encompasses the use of organisms and cells, their parts, and their molecules to create products and services. But several tools and solutions available from environmental biotechnology are claimed to be sustainable and can be used to reduce contamination risks on sites, water, soil, and air (Abdel-Shafy and Mansour 2018). Biotechnology in solid waste management is the technology of biotic and non-biotic resources to control and remove solid waste while reducing environmental impact. This scenario suggests that biotechnology approaches to solid waste management can be effective and sustainable (Khan et al. 2017). The emergence of new technologies in this era has led to greater attention to biotechnology since landfilling and incineration are the most damaging and have not provided a viable solution to the waste crisis. Biotechnology assists in the decomposition of solid waste by using microorganisms to speed up the process, which is an efficient approach to reducing solid waste disposal in the environment. As degrading agents, biological materials can be used in these mechanisms. Environmental waste can come from a variety of sources. Despite their use, many waste management technologies produce other types of pollution (Akmal and Jamil 2021).

This chapter summarizes the current appropriate biotechnology approaches available for the management of solid waste and the potential application of new methods, such as composting, bioremediation, and biodegrading. Composting is ideal for a simple organic scale from individual to the community. Composting is a controlled process in which microbial populations oxidize organic matter (Cerda et al. 2018). The mineralization process converts organic matter into compost, a stable product similar to humus. This final product is a hygienic substance with no unpleasant physicochemical characteristics. Biotechnology can even assist in the conversion of industrial and other wastes into usable products. Indeed, the field of biotechnology appears to have no bounds (Samin et al. 2014). Bioremediation, which uses living organisms to remove, degrade, or contain contaminants in soils, sludges, sediments, surface water, and groundwater, is one of the most exciting discoveries (Anekwe and Isa 2021). In other words, bioremediation is the biological degradation of waste. This method can be used as an alternative to conventional cleanup processes, and it is both environmentally friendly and solar-energy-driven and natural abilities can assist in its effectiveness. It also suggests future research into the conversion and use of bioremediations to solid waste from the petroleum industry (Buyukgungor and Gurel 2009). Biodegradation is another biotechnology approach. Organic chemicals in nature are converted into simpler compounds, mineralized and distributed naturally through the elemental cycle (carbon, nitrogen, and sulfur). Because microorganisms are essential to the bioprocess, they cannot occur within a biological environment (Kulshreshtha 2013; Aleisa and Al-Jarallah 2018). This chapter discusses the different applications of biotechnology to solid waste management. Also, to characterize the effectiveness of various biotechnological procedures used in solid waste management, there has been an effort to develop suitable biotechnological technologies for the increasing number of solids wastes. Moreover, it describes the most effective and collaborative approach to solid waste management at various levels of management.

2 Composting

Composting is an ex situ remediation process in which microorganisms degrade waste at high temperatures ranging from 55 to 65 °C (Stentiford and Sánchez-Monedero 2016). Microbes release heat during this process, causing the temperature to rise and waste to become more soluble. This procedure is used for bulk material in solid waste, along with organic wastes (Liu et al. 2022).

2.1 Compost: A Natural and Eco-Friendly Approach

Composting is a microbiological aerobic process aided by bacteria and fungi. And, the final product has a high carbon and nitrogen content, similar to soil. Compost is a high-quality waste product, a great medium to grow plants, and suitable for agriculture. The composting process requires only the use of biodegradable waste

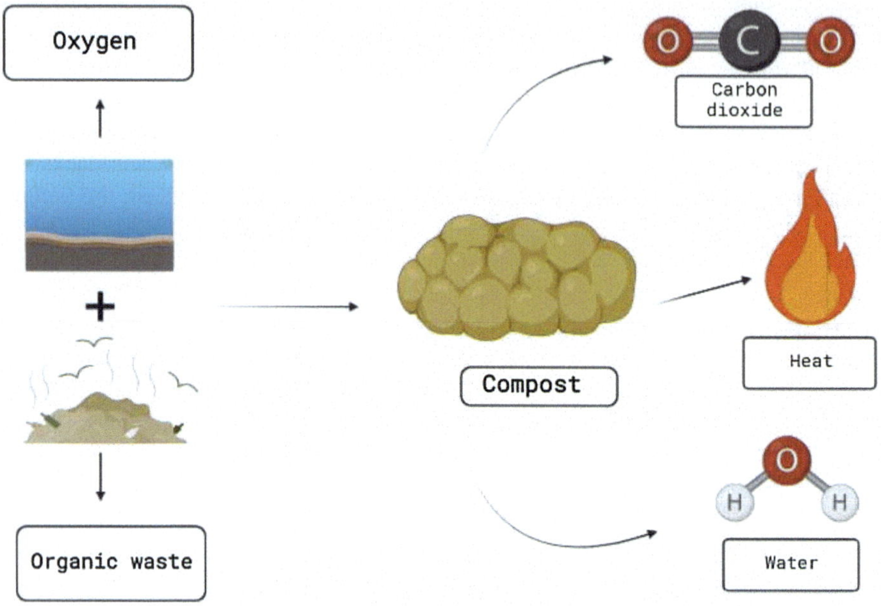

Fig. 31.1 A diagrammatic view of compost process

materials. Composting can commence as early as the raw materials are combined. Microbes quickly consume oxygen and rapidly degrade raw materials. The idea was to use the animal waste as compost and as a replacement for the pea (Inbar et al. 1993). Plant material is the most vital ingredient of any mix, whereas animal tissue or microorganisms are only a minor part but are the most nutritious. The first stage of biological activity occurs when bacteria consume readily available carbohydrates, causing a rapid temperature rise. Bacteria and actinomycetes cause cellulose breakdown in the second stage. Finally, fungi break down the tougher lignins (Waqas et al. 2018). This pathway is followed in the process of composting (Fig. 31.1).

2.2 *Biotechnological Methods of Composting and Recycled Solid Waste*

The loss of global capacity to deal with the increasing amount of organic material poses a serious threat to environmental protection. Composting and vermicomposting are widely accepted as environmentally friendly and sustainable methods of organic waste reduction (Taiwo 2011). Today, the population's demand for land has increased dramatically, as have the people's needs over the years (Lim et al. 2016). Composting is among the most prevalent and beneficial biotechnological techniques. Composting is an efficient method of converting waste into fertilizer, particularly biodegradable waste (Diaz Hepperly et al. 2009). Composting is the complex interaction between waste and microorganisms such as bacteria and fungi found in waste (Arumugam et al. 2017). Composting and recycling municipal solid waste is a comprehensive guide that identifies and explains waste disposal options. Worldwide, composting has quickly become the most effective way to manage solid waste, especially in the United States. Other biotech techniques include the utilization of incinerators, landfills, and waste disposal, but these methods are not entirely eco-friendly, even though they are damaging to the environment (Diaz Hepperly et al. 2009; Malloy et al. 1993). Therefore, this study explores the various composting processes, their benefits, and their sustainability of the composting process. Furthermore, it explains how composting is an innovative method for managing solid waste that is advantageous and affordable.

2.3 *Classification of Composting*

Composting is a natural process that generates energy from combustible biological organic matter. Composting involves vigorous waste removal, which may release high concentrations of biological aerosols into the environment. The composting system can be classified based on its structure and composting method (Bonifait et al. 2017). Some common types of compost include aerobic, anaerobic vermicomposting, windrow composting, aerated static piles, and in-vessel composting (Palaniveloo et al. 2020) (Table 31.1).

Today's world prefers organic methodologies over inorganic due to their cognizance of the chemical hazard. It can be distinguished in this way by the presence or absence of air supplied to the system (Mehta and Sirari 2018).

2.4 *Impact of Environmental Factors on Composting of Solid Waste*

Composting is controlled by some factors that impact the mechanism. The most crucial component is the balance of nutrients, water content, the size of the particles, the porosities, and the pH of the water. During the process of composting, there are several factors, such as oxygen concentration, temperature, and moisture (Keener et al. 2000). The most advanced mechanism for composting municipal solid wastes has highlighted that organic waste materials have improved by strict rules on environmental protection (Gajalakshmi and Abbasi 2008). In this circumstance, the impact of temperature, moisture content, waste particle size, and carbon to nitrogen ratio can affect the reliability of the composting. To improve the efficiency and productivity of these factors, they must be maintained regularly (Hamoda et al. 1998).

Table 31.1 Different types of composting

Types	Function	References
Aerobic composting	Aerobic composting occurs due to the presence of sufficient oxygen. Aerobic microorganisms degrade organic matter, releasing carbon dioxide (CO_2), ammonia (NH_3), water (H_2O), heat, and humus, a relatively stable organic end product	Dinel et al. (2004)
Anaerobic composting	The decomposition process occurs in the anaerobic composting process when oxygen is absent or when there is a short supply of oxygen. Under this process, the anaerobic microbes dominate the area and form intermediate compounds such as methane, organic acid, hydrogen sulfur, and other materials	Cerda et al. (2018)
Vermicomposting	The vermicomposting process breaks down organic waste safely with little odor by using worms, oxygen, and moisture. The worms do most of it, and the bacteria also help	Abu-Zahra et al. (2014)
Windrow composting	This method involves separating the organic waste into rows of large piles, known as windrows and regularly aerated by turning them manually or mechanically	Saad et al. (2014)
Aerated static pile composting	For static compost, organic material is collected and then metabolized by bacteria. As the pile dries, the nutrients and the heat return to the soil	Íñiguez-Covarrubias et al. (2018)
In-vessel composting	This method involves pumping the organic mixture into a drum, silos, a concrete trench, or other similar equipment. It is necessary to mechanically retract or manually mix the material to ensure proper aeration	Manyapu et al. (2017)

2.5 Composting as an Alternative of Solid Waste

Composting is a method used to modify various types of organic matter through the action of complex microbial communities (Mbareche et al. 2017). New livestock management, based upon intensification in farmland, produces large amounts of waste and waste with no agricultural land to be used directly as fertilizer. Composting is thus a viable option for the pyrolysis of dung as an agricultural product that is both reliable and hygienic (Kumar 2011). However, a lot of research has been performed in recent decades. Composting can change practically useless by-products into very usable materials such as humus for crop production by supplying nutrients, controlling plant disease, and so on (Keener et al. 2000).

3 Bioremediation

Bioremediation is a form of the removal process; it is a subcategory of environmental biotechnology. Its purpose is to enhance the factors in the environment (Romera et al. 2007). Bioremediation is the process by which microbes and bacteria dissolve

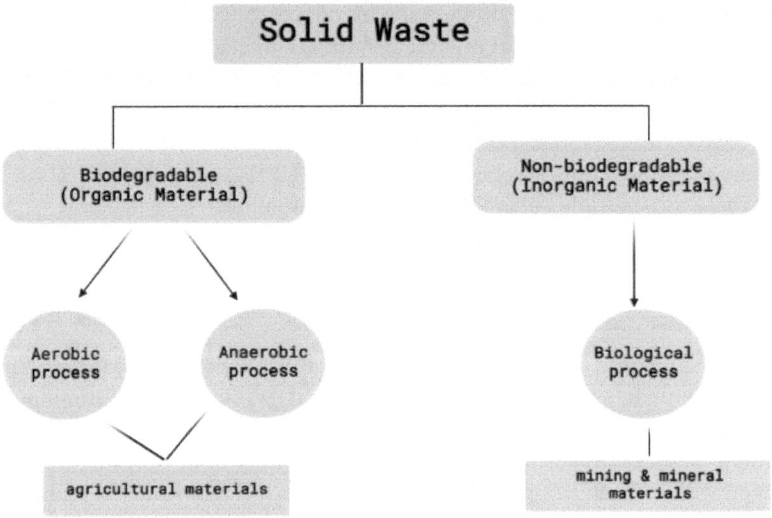

Fig. 31.2 A schematic view of the degradation of solid waste

pollutants and toxicants. It employs microbial metabolism in the availability of ideal temperature and sufficient nutrients to disintegrate toxins (Giriyan et al. 2021). Due to remediation, these biotech strategies could well gain traction in the coming decades. Bioremediations are also growing in popularity, as they are more environmentally responsible and more cost-effective than other techniques for the degradation of pollutants. It is a treatment process for soil and water that includes the emergence of naturally occurring organisms to destroy or reduce the toxicity of hazardous substances found in soil and water, according to the Environmental Protection Agency (Coelho et al. 2015; Mohapatra and Phale 2021). Rapid industrialization and urbanization have led to a rise in population and waste production. Sustained waste release from industries and urban waste have become a challenge and serious worldwide concern; hence, bioremediation is the best methodology for simply removing contaminated waste (Philp and Atlas 2005) (Fig. 31.2).

3.1 In Situ Bioremediation

In situ remediation is a type of bioremediation in which solid waste is treated on the polluting site rather than moved or transported to another location, requiring no excavation of polluting water or soil (Mangunwardoyo et al. 2013). Also, in situ bioremediation is remarkably cost-effective and easy to implement. Some reductions were seen due to inadequate site preparation, suboptimal application design, and unrealistic technology setup (Cerda et al. 2016). Microbes and bacteria converse with waste, degrading contaminants and toxins rather than transferring solid waste to another medium. In situ bioremediation has effectively removed chlorine, dye, nutrient, heavy metals, and organic residues (Gul et al. 2022). The tendency of a microbe to metabolize toxins in nontoxic or nontoxic forms depends on the presence of nutrients and electron donors. Several more engineering and scientific practices need to be utilized to achieve optimal outcomes and control the transformation of organic pollutants by microbes (Das and Dash 2014; Subashchandrabose et al. 2013). There are multiple subareas of in situ remediation.

3.1.1 Bioventing

Bioventing is the type of in situ bioremediation. It is the procedure of decomposing organic contaminants in waste by supplying air and oxygen. Venting stimulates bacterial activities by providing oxygen to the air, affecting the derivatives of hydrocarbons found in waste material (Vidali 2001). During the removal process, the air is combined with oxygen in the soil, which facilitates the removal of remaining contaminants (Gibert et al. 2005). This process eliminates pollutants with aerobic biodegradability, while providing nutrients for microorganisms to deplete contaminants while maintaining respiration rates. In addition to removing solid waste, venting can remove phenanthrene. Nitrogen and phosphorus are also transfused into the waste material during the venting process (Das and Dash 2014).

3.1.2 Biosparging

The second type of in situ bioremediation is biosparging. It biodegrades organic components using the existing microorganisms. The treatment of residual waste and petroleum waste is recognized as biosparging (Megateli et al. 2009). During biosparging, oxygen is added beneath the water table where waste exists, increasing the oxygen level in groundwater and accelerating the growth of naturally occurring bacteria; it also increases interaction between contaminated soil and groundwater. This technique is beneficial for mass removal and mass admittance into groundwater (Maitra 2018). For heavier waste, biosparging is a slow process. The convenience and low cost of setting up small-diameter air injection points enable substantial flexibility in system design and construction (Geyer et al. 2017).

3.1.3 Bioaugmentation

The process involves adding microbes with different metabolic functions to polluted areas to reduce waste emissions. Surplus contaminants such as chlorinated ethenes, tetrachloroethylene, and trichloroethylene are eliminated using bioaugmentation. Despite its effectiveness, this method is challenging to monitor (Garima and Singh 2014).

3.2 *Ex Situ Bioremediation*

Ex situ bioremediation is a bioremediation process that does not occur at the polluted site. The waste is excavated and transported to a different location for treatment (Thompson et al. 2005). Ex situ biodegradation is further classified into several categories.

3.2.1 Biopiling

Biopiling is a form of ex situ remediation in which dredged waste is mixed with compost and piled conventionally in a treatment area that contains a leachate sampling and ventilation system. It is a form of composting. A treatment bed, an aeration system, a nutrient system, and a leachate collection system are all components of the piling system. Moisture, heat, nutrients, oxygen, and pH should all be maintained in this system to ensure appropriate waste degradation. The waste in this process is covered with plastic to prevent vaporization and keep the system warm. The elimination of the contaminants from this method ended up taking 2–3 months (Khan et al. 2017).

3.2.2 Land Forming

The land formation is an ex situ bioremediation technique in which waste collected for cleanup is positioned between clean soil, clay, and concrete. Clean soil is on the bottom, and concrete is on top. Once this natural degradation practice has been initiated, it requires oxygen, nutrition, and the maximum pH near 7. This system treats sludge and municipal solid waste. It also regulates leaching (Garima and Singh 2014).

3.3 *Bioremediation of Heavy Metals*

Unless heavy metals are complicated, they contaminate our waste more. Cadmium, silver and lead, nickel, and other metals are examples. Microorganisms are pertinent in heavy metal remediation. Heavy metals (HMs) can be eliminated by biological agents such as bacteria, fungi, and algae. Metal ions were solubilized in this mechanism (Patel et al. 2022). Heavy metals are also diminished through biosorption. Heavy metals react with microbe cell membranes to create this effect. *Pseudomonas aeruginosa* and *Aspergillus* are two examples of microbes that help to remove metals (Yang et al. 2016). Many other examples of living organisms involved in bioremediation are shown in Table 31.2.

3.4 Bioremediation of Rubber Waste

Rubber makes up about 12% of solid waste. Due to its physical characteristics, rubber is not easily degraded and cannot be recycled. The tires are a great example of rubber waste and contain a black carbon that is difficult to remove (Gonçalves and Delabona 2022). When rubber waste is burned, it emits more harmful toxins into the environment. Moreover, smoke contains toxic compounds such as zinc oxide, which kills naturally occurring bacteria. Bioremediation of rubber, therefore, requires first removing the toxin, then devulcanizing and recycling the rubber (Griffiths 2020).

3.5 Bioremediation for Treatment of Agrarian Solid Waste

Among the other environmental problems, leachate production from agro solid wastes, such as contaminated soils and straws, risks surface and ground water (Gul et al. 2022). As a result, numerous cost-effective and ecologically friendly microbial initiatives for treating agrarian solid waste leachates have evolved (Ebrahimpour and Mushrifah 2008). These techniques can be classified into three types:

(i) Residual organic waste erosion using fermentation under anaerobic or anoxic environments
(ii) Solid waste removal using a combined method involving anaerobic-aerobic processes
(iii) Solid disposal or recovery using bioelectrochemical mechanisms (Yang et al. 2016)

Agriculture creates more waste; each year, livestock, humans, and crops produce 38 billion metvric tons of organic waste. It is necessary to decompose this waste. One of the fundamental steps is vermicomposting, which is the interaction of earthworms and microorganisms. Animal waste is utilized as a nutrient source. Vermicomposting makes the product vermicompost, which converts waste into manure. Crop production is improved by modifying waste material into valuable products (Akmal and Jamil 2021).

3.6 Limitation of Bioremediation

Solid waste bioremediation also entails some constraints; prevalent environmental restrictions are associated with this method. Some hazardous materials have increasing toxic effects and waste concentrations, restricting microorganisms and killing beneficial bacteria. As a result, limitations such as favorable pH, temperature, and moisture are necessary for proper microorganism upkeep (Coelho et al. 2015). Because the mechanism is functioning appropriately, the temperature is between 20

Table 31.2 Living organisms involved in bioremediation

Compounds	Species name	Life form	References
Arsenic	*Sporosarcina ginsengisoli*, *Geobacter metallireducens*, *Corrigiola telephiifolia*, *Spirogyra hyalina*		Ojuederie et al., (2017), Lee et al. (2012), Garcia-Salgado et al. (2012), Kumar and Oommen (2012)
Cadmium	*Bacillus* sp., *Ceramium virgatum*, *Alternaria alternata*, *Parthenium hysterophrous*		Rani et al. (2010), Romera et al. (2007), Bahobil et al. (2017), Naeem et al. (2021)
Polycyclic aromatic hydrocarbons	*Myceliophthora thermophila*		Mohapatra and Phale (2021)
Endosulfan	*Staphylococcus*		Liu et al. (2018)
N,N-dimethylpphenylenediamine	*Klebsiella pneumonia (RS-13)*		Giriyan et al. (2021)
Phenol	*Corynebacterium propinquum*		Gaur et al. (2018)
Lead	*Corynebacterium glutamicum*, *Asparagopsis* sp., *Agaricus bisporus*, *Aspergillus niger*, *Ricinus communis*		Choi and Yun (2004), Romera et al. (2007), Frutos et al. (2016), Nazir et al. (2023), Sarfraz et al. (2022)
Copper	*Kocuria flava*, *Desulfovibrio desulfuricans*, *Spirogyra hyaline*, *Pteris vittata*	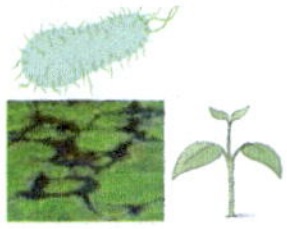	Coelho et al. (2015), Kim et al. (2015), Kumar and Oommen (2012), Wang et al. (2011)
Cadmium	*Scirpus grossus*, *Phytolacca americana*		Ebrahimpour and Mushrifah (2008), Peng et al. (2008)
Zinc	*Filipendula ulmaria*, *Typha angustifolia*		Fritioff and Greger (2003), Sricoth et al. (2018)
Mercury	*Lemna minor*		Rakhshaee et al. (2009)

(continued)

Table 31.2 (continued)

Compounds	Species name	Life form	References
Nickle	*Desulfovibrio desulfuricans*, *Lemna minor*	 	Kim et al. (2015), Al-Khafaji et al. (2018)
Cobalt	*Rhodopseudomonas palustris*, *Haumaniastrum robertii*	 	Gao et al. (2017), Marques et al. (2009)
Selenium	*Lecythis ollaria*		Marques et al. (2009)
Manganese	*Hibiscus cannabinus*, *Spirogyra* sp.	 	Shehata et al. (2019), Rajfur et al. (2010)
Azo dyes effluents	*Exiguobacterium indicum, B. cereus, E. aurantiacums*		Raquel et al. (2013)
Crude oil	*Aspergillus niger, Candida krusei*		Krab-Hüsken (2002)
Aromatic hydrocarbons	*Acinetobacter* sp., *Microbacterium* sp.		Basu et al. (2018)
Methyl parathion and chlorpyrifos	*Pseudomonas* sp.		Halak et al. (2006)
Diesel oil	*P. cepacia, B. coagulans*		Miri et al. (2022)
Vat dyes	*B. firmus, Staphylococcus aureus*		Sangkharak et al. (2020)

and 30 °C. Aside from these aspects, other factors of bioremediation include waste solubility, oxidation, reduction, and microbial interaction; limitations are also considered necessary for these factors (Basu et al. 2018). To improve this process, the researchers must find microbes that can tolerate the abovementioned attributes. Ultimately, bioremediation, part of biotech, has made the environment more eco-sustainable by reducing waste. It also has limitations that need to be addressed.

4 Biodegradation

Biodegradation is a valuable resource for biotechnology. It is a naturally occurring phenomenon in which biological organisms such as bacteria, fungi, and many other living substances break down solid waste. We can obtain products from waste and remove contaminants during this process. It can assist in minimizing environmental contaminants (Gaur et al. 2018).

4.1 Biodegradation of Plastics

Plastic is a mixture of organic and inorganic compounds, respectively synthetic and semisynthetic (Ali et al. 2014; Álvarez-Barragán et al. 2016). A combination of enzymes and nonenzymatic plastics contributes to biodegradation. Plastic biodegradation is even feasible with renewable resources. Some of these plastic items are harmful to human health (Cosgrove et al. 2007). And, findings demonstrate that it is advantageous from a variety of perspectives. Many organisms can consume starch-based polymer composites because of the activity of metabolic enzymes. The mechanism is also beneficial for the environment, as it mitigates contamination, among other issues (Geyer et al. 2017). There are long-term consequences to dumping waste materials everywhere. It directly impacts soil organisms when it leaches into the ground. Both chemical and physical elements are involved in biodegradation. Plastics can be dissolved by different types of microbes such as *Pseudomonas* sp. and *Bacillus* sp. (Ru et al. 2020).

4.2 Plastic Waste Management

Polyethylene (PE), polystyrene (PS), polypropylene (PP), polyvinyl chloride (PVC), polyurethane (PUR), and polyethylene terephthalate (PET) are examples of synthetic plastics. They have an impact on almost all of our daily activities (Jambeck et al. 2015). Globally, plastic waste production and consumption are expanding rapidly. It is estimated that up to 26 billion tons of plastic waste will be generated by 2050, with much more than half of that waste ending up in landfills and eventually entering ecospheres such as oceans and lakes, exacerbating adverse effects on the environment (Lönnstedt and Eklöv 2016). Currently, only 9–12% of plastic waste is recycled or burned, while 79% is discarded into landfills or natural habitats. This indicates that there is a critical need of research into advanced recycling methods for the removal of plastic waste. Numerous studies carried out in recent years have identified that many microbes and enzymes can break down synthetic plastics (Garcia and Robertson 2017). In the inclusive environment, a variety of plastics, including PE, PS, PP, PVC, PUR, and PET, have been sequestered, along with soil from a plastic-dumping site, film waste from leaf litter, marine water, crude oil-contaminated soil, sewage sludge, landfills, and plastic-eating worm guts. Plastic-degrading microbial screening is pivotal for identifying depolymerase and certain other essential enzymatic activities in plastic degradation (Taniguchi et al. 2019; Magnin et al. 2020).

4.3 *Impact of Solid Waste Management in the Environment*

Solid wastes are so unfriendly that if not appropriately controlled, they affect life forms and the ozone layer, damage the environment, lowering industrial activity, and so on. In other words, waste management involves managing liquid and solid waste to keep the surroundings stable and free of contamination. Agricultural waste, garbage, other waste, dead livestock, industrial waste, mine waste, mineral waste, etc., are solid waste and sewage, and industrial waste is liquid waste (Giriyan et al. 2021; Cerda et al. 2018) (Fig. 31.3).

Some other biotechnological techniques are as follows:

(i) *Activated Sludge:* This methodology employs bioweapons such as bacteria and protozoa to remove sewage and wastewater in an aerobic environment. This mechanism disposes of microorganism wastewater. An aeration tank, a settling tank, a return sludge pump, and introducing oxygen into the aeration tank are the four parts of a typical activated sludge system (Sivakumar et al. 2014).
(ii) *Biofiltration:* This is a technology for the treatment of wastewater. It is the process of controlling pollution, in which a biological reactor is filled with biomaterials to accumulate and degrade the pollutants biologically. Other implementations of biofiltration involve wastewater treatment, oxidation of contaminants in the air, and others (Das et al. 2012).
(iii) *Biosorption:* In this method, metal and nonmetallic species are absorbed by living or decayed biomass. The bacterial community thus removes the metal species from the solution (Bahobil et al. 2017).
(iv) *Bioreactor:* This apparatus wherein life forms, particularly bacteria, synthesize useful substances or degrade harmful ones like sewage (Arumugam et al. 2017).
(v) *Bioleaching:* Rather than using chemical solutions, it is the approach by which microorganisms such as bacteria solubilize their ores of metals. Nickel (Ni),

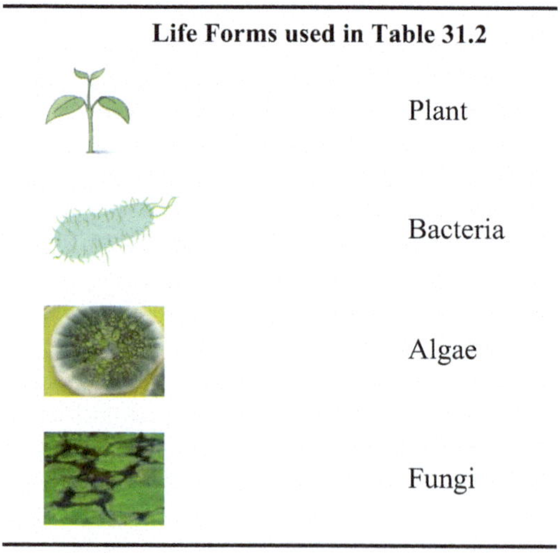

Fig. 31.3 Life forms used in Table 31.2

copper (Cu), cobalt (Co), lead (Pb), zinc (Zn), and other metals subsumed through bioleaching are examples. Bioleaching appears to work by using bacteria that can obtain the metal concentrations of their ores (Mohapatra and Phale 2021).

5 Conclusions

The approaches mentioned above are momentarily demonstrated in this section. One of the technologies and methods for the natural degradation of residual waste by microbiota in a restricted surrounding is soil treatment. These are driving solid waste disposal through the use of biotechnological methodologies. The management of solid waste has been extremely challenging for a long time. A wide range of sources generate large amounts of waste material. The disposal of modern waste and civil solid waste has been investigated. Solid waste is frequently disposed of inclusively. Long-term research has been conducted to find an efficient method to replace these ineffective expulsion strategies. Eventually, biotechnological approaches such as composting, biodegradation, and bioremediation could be used as potential substitutes. The purpose of this chapter is to recommend a multidimensional technique for organizing sustainable waste administration mechanisms, both quantitative and subjective. Organic waste collection at the curb may be the most effective way to reduce atmospheric danger and attract microbes to landfills Neverthless the combination of multi-model selection and influence appraisal cannot provide a comprehensive picture of the upcoming situation. It further can provide an initial coordinated insight into the problem and pave the way for another solid waste administration framework in the formation of urban areas, allowing for the development of a circular economy and achieving significant outcomes.

References

Abdel-Shafy HI, Mansour MSM (2018) Solid waste issue: sources, composition, disposal, recycling, and valorization. Egypt J Pet 27(4):1275–1290

Abu-Zahra TR, Ta'any RA, Arabiyyat AR (2014) Changes in compost physical and chemical properties during aerobic decomposition. Int J Curr Microbiol App Sci 3(10):479–486

Akmal T, Jamil F (2021) Testing the role of waste management and environmental quality on health indicators using structural equation modeling in Pakistan. Int J Environ Res Public Health 18(8):4193. https://doi.org/10.3390/ijerph18084193

Aleisa E, Al-Jarallah RA (2018) A triple bottom line evaluation of solid waste management strategies: a case study for an arid Gulf State, Kuwait. Int J Life Cycle Assess 23:1460–1475

Ali MI, Ahmed S, Robson G, Javed I, Ali N, Atiq N, Hameed A (2014) Isolation and molecular characterization of polyvinyl chloride (PVC) plastic degrading fungal isolates. J Basic Microbiol 54(1):18–27. https://doi.org/10.1002/jobm.201200496

Al-Khafaji MS, Al-Ani FH, Ibrahim AF (2018) Removal of some heavy metals from industrial wastewater by *Lemmna minor*. KSCE J Civ Eng 22(4):1077–1082

Álvarez-Barragán J, Domínguez-Malfavón L, Vargas-Suárez M, Gonzalez-Hernandez R, Aguilar-Osorio G, Loza-Tavera H (2016) Biodegradative activities of selected environmental fungi on a polyester polyurethane varnish and polyether polyurethane foams. Appl Environ Microbiol 82(17):5225–5235. https://doi.org/10.1128/AEM.01344-16
Anekwe IM, Isa YM (2021) Wastewater and bioventing treatment systems for acid mine drainage-contaminated soil. Soil Sediment Contam 30(5):518–531. https://doi.org/10.1080/15320383.2020.1863909
Arumugam K, Seenivasagan R, Kasimani R, Sharma N, Babalola O (2017) Enhancing the post consumer waste management through vermicomposting along with bioinoculumn. Int J Eng Trends Technol 44:179–182
Bahobil A, Bayoumi RA, Atta HM, El-Sehrawey MM (2017) Fungal biosorption for cadmium and mercury heavy metal ions isolated from some polluted localities in KSA. Int J Curr Microbiol App Sci 6(6):2138–2154
Basu S, Rabara RC, Negi S, Shukla P (2018) Engineering PGPMOs through gene editing and systems biology: a solution for phytoremediation? Trends Biotechnol 36(5):499–510
Bonifait L, Marchand G, Veillette M, M'Bareche H, Dubuis ME, Pepin C, Cloutier Y, Bernard Y, Duchaine C (2017) Workers' exposure to bioaerosols from three different types of composting facilities. J Occup Environ Hyg 14(10):815–822
Buyukgungor H, Gurel L (2009) The role of biotechnology on the treatment of waste. Afr J Biotechnol 8(25):7253–7262
Cerda A, El-Bakry M, Gea T, Sánchez A (2016) Long term enhanced solid-state fermentation: inoculation strategies for amylase production from soy and bread wastes by *Thermomyces* sp. in a sequential batch operation. J Environ Chem Eng 4(2):2394–2401
Cerda A, Artola A, Font X, Barrena R, Gea T, Sánchez A (2018) Composting of food wastes: status and challenges. Bioresour Technol 248:57–67
Choi SB, Yun YS (2004) Lead biosorption by waste biomass of *Corynebacterium glutamicum* generated from lysine fermentation process. Biotechnol Lett 26:331–336. https://doi.org/10.1023/B:BILE.0000015453.20708.fc
Coelho LM, Rezende HC, Coelho LM, de Sousa PAR, Melo DFO, Coelho NMM (2015) Bioremediation of polluted waters using microorganisms. Adv Bioremed Wastewater Pollut Soil 10:60770
Cosgrove L, McGeechan P, Robson GD, Handley PS (2007) Fungal communities associated with degradation of polyester polyurethane in soil. Appl Environ Microbiol 73(18):5817–5824. https://doi.org/10.1128/aem.01083-07
Das S, Dash HR (2014) Microbial bioremediation: a potential tool for restoration of contaminated areas. In: Microbial biodegradation and bioremediation, pp 1–21. https://doi.org/10.1016/B978-0-12-800021-2.00001-7
Das SP, Ravindran R, Ahmed S, Das D, Goyal D, Fontes CM, Goyal A (2012) Bioethanol production involving recombinant *C. thermocellum* hydrolytic hemicellulase and fermentative microbes. Appl Biochem Biotechnol 167:1475–1488
Diaz Hepperly P, Lotter D, Ulsh CZ, Seidel R, Reider C (2009) Compost, manure and synthetic fertilizer influences crop yields, soil properties, nitrate leaching and crop nutrient content. Compost Sci Util 17(2):117–126
Dinel H, Marche T, Schnitzer M, Pre T, Champagne P (2004) Co-composting of papers mill sludge and hardwood sawdust under two types of in-vessel processes. J Environ Sci Health B 39(1):139–151
Ebrahimpour M, Mushrifah I (2008) Heavy metal concentrations (Cd, Cu and Pb) in five aquatic plant species in Tasik Chini, Malaysia. Environ Geol 54:689–698
Fritioff A, Greger M (2003) Aquatic and terrestrial plant species with potential to remove heavy metals from stormwater. Int J Phytoremediation 5(3):211–224
Frutos I, Garcia-Delgado C, Garate A, Eymar E (2016) Biosorption of heavy metals by organic carbon from spent mushroom substrates and their raw materials. Int J Environ Sci Technol 13:2713–2720

Gajalakshmi S, Abbasi SA (2008) Solid waste management by composting: state of the art. Crit Rev Environ Sci Technol 38(5):311–400

Gao R, Wang Y, Zhang Y, Tong J, Dai W (2017) Cobalt (II) bioaccumulation and distribution in *Rhodopseudomonas palustris*. Biotechnol Biotechnol Equip 31(3):527–534

Garcia JM, Robertson ML (2017) The future of plastics recycling. Science 358(6365):870–872. https://doi.org/10.1126/science.aaq032

Garima T, Singh SP (2014) Application of bioremediation on solid waste management: a review. J Bioremed Biodegr 5(6). https://doi.org/10.4172/2155-6199.1000248

García-Salgado S, García-Casillas D, Quijano-Nieto MA, Bonilla-Simón MM (2012) Arsenic and heavy metal uptake and accumulation in native plant species from soils polluted by mining activities. Water Air Soil Pollut 223:559–572

Gaur N, Narasimhulu K, PydiSetty Y (2018) Recent advances in the bio-remediation of persistent organic pollutants and its effect on environment. J Clean Prod 198:1602–1631

Geyer R, Jambeck J, Law KL (2017) Production, use, and fate of all plastics ever made. Sci Adv 3(7):e1700782. https://doi.org/10.1126/sciadv.1700782

Gibert O, de Pablo J, Cortina JL, Ayora C (2005) Municipal compost-based mixture for acid mine drainage bioremediation: metal retention mechanisms. Appl Geochem 20(9):1648–1657

Giriyan AL, Berde VB, Pereira EJ, Parulekar-Berde CV (2021) Microbial bioremediation of heavy metals: a genetic and omics approach. In: Handbook of research on microbial remediation and microbial biotechnology for sustainable soil. IGI Global, Beijing, China pp 417–439

Gonçalves CR, Delabona PS (2022) Strategies for bioremediation of pesticides: challenges and perspectives of the Brazilian scenario for global application – a review. Environ Adv 8:100220. https://doi.org/10.1016/j.envadv.2022.100220

Griffiths AR (2020) Controlling secondary pollution impacts during enhanced in situ anaerobic bioremediation. In: Sustainable remediation of contaminated soil and groundwater, pp 201–220. https://doi.org/10.1016/B978-0-12-817982-6.00008-2

Gul S, Hussain S, Khan H, Khan KI, Khan S, Ullah S, Clasen B (2022) Advances in bioremediation of antibiotic pollution in the environment. In: Biological approaches to controlling pollutants, pp 49–78. https://doi.org/10.1016/B978-0-12-824316-9.00015-X

Halak S, Basta T, Bürger S, Contzen M, Stolz A (2006) Characterization of the genes encoding the 3-carboxy-cis, cis-muconatelactonizing enzymes from the 4-sulfocatechol degradative pathways of *Hydrogenophaga intermedia* S1 and *Agrobacterium radiobacter* S2. Microbiol 152(11):3207–3216

Hamoda MF, Qdais HA, Newham J (1998) Evaluation of municipal solid waste composting kinetics. Resour Conserv Recycl 23(4):209–223

Inbar Y, Hadar Y, Chen Y (1993) Recycling of cattle manure: the composting process and characterization of maturity. J Environ Qual 22(4):857–863

Íñiguez-Covarrubias G, Gómez-Rizo R, Ramírez-Meda W, Jesús Bernal-Casillas J (2018) Composting of food and yard wastes under the static aerated pile method. Adv Chem Eng Sci 8(04):271–279

Iqbal HMN, Kyazze G, Keshavarz T (2013) Advances in the valorization of lignocellulosic materials by biotechnology: an overview. Bioresources 8(2):3157–3176

Jambeck JR, Geyer R, Wilcox C, Siegler TR, Perryman M, Andrady A, Narayan R, Law KL (2015) Marine pollution. Plastic waste inputs from land into the ocean. Science 347(6223):768–771. https://doi.org/10.1126/science.1260352

Keener HM, Dick WA, Hoitink HA (2000) Composting and beneficial utilization of composted by- products materials. Land application of agricultural, industrial, and municipal by-products 1(6):315–341

Khan IU, Othman MHD, Hashim H, Matsuura T, Ismail AF, Rezaei-DashtArzhandi M, Azelee IW (2017) Biogas as a renewable energy fuel–a review of biogas upgrading, utilisation and storage. Energy Convers Manag 150:277–294

Kim IH, Choi JH, Joo JO, Kim YK, Choi JW, Oh BK (2015) Development of a microbe-zeolite carrier for the effective elimination of heavy metals from seawater. J Microb Biotechnol 25(9):1542–1546

Krab-Hüsken L (2002) Production of catechols: microbiology and technology. Ph.D. Thesis, Wageningen University and Research, Wageningen

Kulshreshtha S (2013) Genetically engineered microorganisms: a problem solving approach for bioremediation. J Bioremed Biodegr 4(4):1–12

Kumar S (2011) Composting of municipal solid waste. Crit Rev Biotechnol 3(2):112–136

Kumar JI, Oommen C (2012) Removal of heavy metals by biosorption using freshwater alga *Spirogyra hyalina*. J Environ Biol 33(1):27–31

Lee KY, Bosch J, Meckenstock RU (2012) Use of metal-reducing bacteria for bioremediation of soil contaminated with mixed organic and inorganic pollutants. Environ Geochem Health 34:135–142

Lim SL, Lee LH, Wu TY (2016) Sustainability of using composting and vermicomposting technologies for organic solid waste biotransformation: recent overview greenhouse gasses emissions and economic analysis. J Clean Prod 111:262–278

Liu Z, Shao B, Zeng G, Chen M, Li Z, Liu Y, Yan M (2018) Effects of rhamnolipids on the removal of 2, 4, 2, 4-tetrabrominated biphenyl ether (BDE-47) by *Phanerochaete chrysosporium* analyzed with a combined approach of experiments and molecular docking. Chemosphere 210:922–930

Liu T, Chen H, Zhou Y, Awasthi SK, Qin S, Liu H, Zhang Z, Pandey A, Varjani S, Awasthi MK (2022) Composting as a sustainable technology for integrated municipal solid waste management. In: Biomass, biofuels, biochemicals, pp 23–39. https://doi.org/10.1016/B978-0-323-88511-9.00002-1

Lönnstedt OM, Eklöv P (2016) Environmentally relevant concentrations of microplastic particles influence larval fish ecology. Science 352(6290):1213–1216

Magnin A, Pollet E, Phalip V, Avérous L (2020) Evaluation of biological degradation of polyurethanes. Biotechnol Adv 39:107457. https://doi.org/10.1016/j.biotechadv.2019.107457

Maitra S (2018) In situ bioremediation-an overview. Res J Life Sci Bioinfo Pharm Chem Sci 4:576–598. https://doi.org/10.26479/2018.0406.45

Malloy TA, Goldfarb TD, Surico MTJ (1993) PCDDs, PCDFs, PCBs, chlorophenols (CPs) and chlorobenzenes (CBzs) in samples from various types of composting facilities in United States. Chemosphere 27(1 3):325–334

Mangunwardoyo W, Sudjarwo T, Patria MP (2013) Bioremediation of effluent wastewater treatment plant Bojongsoang Bandung Indonesia using consortium aquatic plants and animals. Int J Res Rev Appl Sci 14(1):150–160

Manyapu V, Shukla S, Kumar S, Rajendra K (2017) In-vessel composting: a rapid technology for conversion of biowaste into compost. Int J Sci Eng 2(9):58–63

Marques AP, Rangel AO, Castro PM (2009) Remediation of heavy metal contaminated soils: phytoremediation as a potentially promising clean-up technology. Crit Rev Environ Sci Technol 39(8):622–654

Mbareche H, Veillette M, Bonifait L, Dubuis ME, Benard Y, Marchand G, Bilodeau GJ, Duchaine C (2017) A next generation sequencing approach with a suitable bioinformatics workflow to study fungal diversity in bioaerosols released from two different types of composting plants. Sci Total Environ 601:1306–1314

Megateli S, Semsari S, Couderchet M (2009) Toxicity and removal of heavy metals (cadmium, copper, and zinc) by *Lemna gibba*. Ecotoxicol Environ Saf 72(6):1774–1780

Mehta CM, Sirari K (2018) Comparative study of aerobic and anaerobic composting for better understanding of organic waste management: a mini review. Plant Arch 18(1):44–48

Miri S, Rasooli A, Brar SK, Rouissi T, Martel R (2022) Biodegradation of p-xylene—a comparison of three psychrophilic *Pseudomonas* strains through the lens of gene expression. Environ Sci Pollut Res 29:1–15

Mohapatra B, Phale PS (2021) Microbial degradation of naphthalene and substituted naphthalenes: metabolic diversity and genomic insight for bioremediation. Front Bioeng Biotechnol 9:602445

Naeem N, Khalid N, Sarfraz W, Ejaz U, Yousaf A, Rizvi ZF, Ikram S (2021) Assessment of lead and cadmium pollution in soil and wild plants at different functional areas of Sialkot. Bull Environ Contam Toxicol 107(2):336–342

Nazir A, Sarfraz W, Allah D, Khalid N, Farid M, Shafiq M, e-Bareen F, Rizvi ZF, Naeem N (2023) Synergistic impact of two autochthonous saprobic fungi (*A. niger* and *T. pseudokoningii*) on the growth, ionic contents, and metals uptake in *Brassica juncea* L. and *Vigna radiata* L. under tannery solid waste contaminated soil. Int J Phytoremediation:1–13. https://doi.org/10.1080/15226514.2023.2166

Palaniveloo K, Amran MA, Norhashim NA, Mohamad-Fauzia N, Peng-Hui F, Hui-Wen L, Kai-Lin Y, Jiale L, Chian-Y MG, Jing-Yi L, Gunasekaran B, Razak SA (2020) Food waste composting and microbial community structure profiling. Processes 8(6):723

Patel AK, Singhanis RR, Albarico FPJB, Pandey A, Chen CW, Dong CD (2022) Organic wastes bioremediation and its changing prospects. Sci Tot Environ 153889

Peng K, Luo C, You W, Lian C, Li X, Shen Z (2008) Manganese uptake and interactions with cadmium in the hyperaccumulator—Phytolacca Americana L. J Hazard Mater 154(1–3):674–681

Philp J, Atlas R (2005) Bioremediation of contaminated soil and aquifers. In: Atlas RM, Jim CP (eds) Bioremediation: applied microbial solution for real-world environmental clean up. ASM Press, Washington, DC, p 139–236.

Rajfur M, Kłos A, Wacławek M (2010) Sorption properties of algae Spirogyra sp. and their use for determination of heavy metal ions concentrations in surface water. Bioelectrochemistry 80(1):81–86

Rakhshaee R, Giahi M, Pourahmad A (2009) Studying effect of cell wall's carboxyl-carboxylate ratio change of *Lemna minor* to remove heavy metals from aqueous solution. J Hazard Mater 163(1):65–173

Rani MJ, Hemambika B, Hemapriya J, Rajeshkannan V (2010) Comparative assessment of heavy metal removal by immobilized and dead bacterial cells: a biosorption approach. Afr J Environ Sci Technol 4(2):23–30

Raquel S, Natalia G, Luis Fernando B, Maria Carmen M (2013) Biodegradation of high-molecular-weight polycyclic aromatic hydrocarbons by a wood-degrading consortium at low temperatures. FEMS Microbiol Ecol 83(2):438–449

Romera E, Gonzalez F, Ballester A, Blazquez ML, Munoz JA (2007) Comparative study of biosorption of heavy metals using different types of algae. Bioresour Technol 98(17):3344–3353

Ru J, Huo Y, Yang Y (2020) Microbial degradation and valorization of plastic wastes. Front Microbiol 11:42. https://doi.org/10.3389/fmicb.2020.00442

Saad NFBM, Baharin N, Zain SM (2014) Windrow composting of yard wastes and food waste. Aust J Basic Appl Sci 8(19):64–68

Salemdeeb R, ZuErmgassen EK, Kim MH, Balmford A, Al-Tabbaa A (2017) Environmental and health impacts of using food waste as animal feed: a comparative analysis of food waste management options. J Clean Prod 140:871–880

Samin G, Pavlova M, Arif MI, Postema CP, Damborsky J, Janssen DB (2014) A *Pseudomonas putida* strain genetically engineered for 1,2,3-trichloropropane bioremediation. Appl Environ Microbiol 80(17):5467–5476

Sangkharak K, Choonut A, Rakkan T, Prasertsan P (2020) The degradation of phenanthrene, pyrene, and fluoranthene and its conversion into medium-chain-length polyhydroxyalkanoate by novel polycyclic aromatic hydrocarbon-degrading bacteria. Curr Microbiol 77:897–909

Sarfraz W, Farid M, Ejaz U, Naeem N, Yousaf A, Kabir Z, Rizvi ZF, Hadayat N, Khalid N (2022) EDTA could enhance the bioaccumulation of Pb and Cd in *Ricinus communis* for potential decontamination of roadside soils. Fresenius Environ Bull 31(01):81–91

Shehata SM, Badawy RK, Aboulsoud YIE (2019) Phytoremediation of some heavy metals in contaminated soil. Bull Natl Res Cent 43:1–15

Sivakumar D, Kandaswamy A, Gomathi V, Rajeshwaran R, Murugan N (2014) Bioremediation studies on reduction of heavy metals toxicity. Pollut Res 33(3):553–558

Sricoth T, Meeinkuirt W, Pichtel J, Taeprayoon P, Saengwilai P (2018) Synergistic phytoremediation of wastewater by two aquatic plants (*Typha angustifolia* and *Eichhornia crassipes*) and potential as biomass fuel. Environ Sci Pollut Res 25:5344–5358

Stentiford E, Sánchez-Monedero MA (2016) Past, present and future of composting research. In: III international symposium on organic matter management and compost use in horticulture 1146, pp 1–10

Subashchandrabose SR, Ramakrishnan B, Megharaj M, Venkateswarlu K, Naidu R (2013) Mixotrophic cyanobacteria and microalgae as distinctive biological agents for organic pollutant degradation. Environ Int 51:59–72

Sufficiency E, Qamar SA, Ferreira LFR, Franco M, Iqbal HMN, Bilal M (2022) Emerging biotechnological strategies for food waste management: a green leap towards achieving high-value products and environmental abatement. Energy Nexus 6:100077

Taiwo AM (2011) Composting as a sustainable waste management technique in developing. J Environ Sci Technol 4(2):93–102

Taniguchi I, Yoshida S, Hiraga K, Miyamoto K, Kimura Y, Oda K (2019) Biodegradation of PET: current status and application aspects. ACS Catal 9(5):4089–4105

Thompson IP, van der Gast CJ, Ciric L, Singer AC (2005) Bioaugmentation for bioremediation: the challenge of strain selection. Environ Microbiol 7(7):909–915

Vidali M (2001) Bioremediation. An overview. Pure Appl Chem 73(7):1163–1172

Wang Q, Xiong D, Zhao P, Yu X, Tu B, Wang G (2011) Effect of applying an arsenic-resistant and plant growth-promoting rhizobacterium to enhance soil arsenic phytoremediation by *Populus deltoides* LH05-17. J Appl Microbiol 111(5):1065–1074

Waqas M, Nizami AS, Aburiazaiza AS, Barakat MA, Ismail IMI, Rashid MI (2018) Optimization of food waste compost with the use of biochar. J Environ Manag 216:70–81

Yang J, Tang C, Wang F, Wu Y (2016) Co-contamination of Cu and Cd in paddy fields: using periphyton to entrap heavy metals. J Hazard Mater 304:150–158

Zhou W, Zhang W, Cai Y (2021) Laccase immobilization for water purification: a comprehensive review. Chem Eng J 403:126272

Chapter 32
Bioremediation of Sites Contaminated with Heavy Metals, Techniques, and Their Application

Mujahid Farid, Muhammad Zubair, Numra, Muhammad Bilal Shakoor, Mubeen Fatima, Sheharyaar Farid, Zaiba Saher, Sehar Bashir, and Nafeesa Liaqat

Abstract Contamination in the environment is caused by a variety of inorganic elements and heavy metals (HMs) which are some of the major causes of environmental pollution. Due to their non-biodegradability, HMs are difficult to remove from the environment. Heavy metals naturally come from volcanic eruptions and rock weathering, whereas anthropogenic causes include mining, industrialization, untreated sewage, use of synthetic products, and combustion in power plants. Researchers have serious concerns about certain HMs, such as chromium, cadmium, lead, palladium, arsenic, and nickel, and they are attempting to reduce their effects on the environment and human health because of their excessive concentrations. For this, many remediation technologies have been developed to remediate harmful metals. They have been already removed using conventional techniques, but these techniques are non-economical and do not give the best outcomes. Due to the excellent activity, low cost, and easy availability, a wide variety of organisms, including bacteria, algae, yeasts, and fungi, have recently drawn attention of researchers for the remediation of HMs. Use of these biological tools has not been particularly sophisticated technically. Designing and implementing workable bioremediation programme requires a great deal of experience and practical skills. Most of the conventional techniques do not provide an effective solution for metal removal from soils. Metals can be remediated from contaminated sites using microorganisms which utilize metals as terminal electron acceptors and decrease via

M. Farid (✉) · Numra · M. Fatima · Z. Saher · S. Bashir · N. Liaqat
Department of Environmental Sciences, University of Gujrat, Hafiz Hayat Campus, Gujrat, Pakistan

M. Zubair
Department of Chemistry, University of Gujrat, Hafiz Hayat Campus, Gujrat, Pakistan

M. B. Shakoor
College of Earth and Environmental Sciences, The University of the Punjab, Lahore, Pakistan

S. Farid
School of Natural Sciences and Ryan Institute, University of Galway, Galway, Ireland

M. Hasanuzzaman (ed.), *Climate-Resilient Agriculture, Vol 2*,
https://doi.org/10.1007/978-3-031-37428-9_32

detoxification mechanism. In some instances, using plants that accumulate metal to clean up soil which are contaminated with metal is a cost-effective strategy.

Keywords Environmental pollution · Remediation techniques · Bioremediation · Heavy metals · Toxicity

1 Introduction

Due to the abundant usage of heavy metals, their concentration is increasing day by day and affecting the environment and human health, which has become a global concern (Lakherwal 2014). As pollutants, heavy metals can enter the environment through natural way or as anthropogenically (Masindi and Muedi 2018). Rapid industrialization and its processes discharging the contaminants into the environmental ecosystems has increased over the past few decades to many folds. Research community is currently looking for solutions to deal with the challenges brought on by the expanding industrialization (Sodango et al. 2018). The disposal of waste into the sites as leachate and the resulting degradation of our ecosystem is caused by controlled and uncontrolled application of waste disposal. The ecosystem's capacity to self-clean has been exceeded by the sudden arrival of pollutants. Global environmental quality has decreased as a result of the large-scale production of chemicals (Liu et al. 2015). Different problems arise from inorganic heavy metal pollutants than organic contaminants (Akram et al. 2018). Emergence of heavy metals that are produced and discarded by humans harms the environment and public health (Jaishankar et al. 2014).

The understanding of the issue, as well as the policies and technology used to address it, varies greatly among the world communities that are dealing with polluted soil (Khalid et al. 2017). However, there occurs a significant exchange of knowledge among the various nations about the management and remediation of toxic soils. Large tracts of our land have become dangerous for both the animals and human populations as the result of pollution. Most contaminated soils are result of industrial activity (Sharma 2012). Due to the low cost, the old physicochemical approaches have encouraged businesses to disregard the issue (Gunatilake 2015). Setting up the new technology that could treat toxic wastes affordably is crucial (Caliman et al. 2011). Since the public awareness has enhanced many folds over the past few years, they are now strongly demanding that clean-up methods be implemented immediately. Uncontrolled waste discharge is now strictly prohibited in law. The necessity for new technology for the treatment of these waste sites has been underlined by the recent environmental awareness (Wang et al. 2021).

Among the most dangerous contaminants damaging the environment, the heavy metals are leading (Cimboláková et al. 2019). Metals are earth-based naturally occurring substances and are persistent and non-biodegradable. Heavy metals make up the contamination identified by USEPA as trace inorganic contaminants (Gautam et al. 2016). An increase in the flux of metallic compounds in the aquatic and terrestrial environment is the result of the growing use of heavy metals (Kahlon et al. 2018). Environmental pollution has been increased by agricultural runoff, acid rain, industrial and domestic effluents, and waste (Singh and Gupta 2016). Weathering, industrial metal processing, leaching from waste dumps and landfills, livestock runoff, and the reuse of wastewater and sewage sludge are all synergetic sources of heavy metal in the environment and their abundance causes pollution (Li et al. 2019). The use of fertilizers and pesticides in agriculture is another main cause of heavy metal accumulation in the environment pollution. Problems in biology and physiology of plants are caused by the high concentration of heavy metals (Mishra et al. 2019a). Leaching, diffusion, and infiltration are the three processes that transfer these metals into the lower soil surface and ground water. They have the unusual ability to accumulate along a food chain and can be hazardous to aquatic life and humans in both short term and long term (Pandey and Madhuri 2014).

Through the ingestion of food plants cultivated in contaminated soil, heavy metals may accumulate in human body (Guerra et al. 2012). Traditional clean-up techniques such as ion exchange, chemical precipitation, and advanced oxidation/reduction processes are ineffective and expensive, making them unsuitable. For heavy metal, advanced bioremediation procedures have gained attention worldwide to remediate the problem (Gaur et al. 2014). In this process, biological activity is used to reduce toxic heavy metals pollutants. In comparison to other techniques, bioremediation is a more effective and affordable way to remove heavy metals. The best method now available for managing and reducing the polluted environment is bioremediation (Singh 2014). To remediate contaminated soil and water, it uses bacteria, yeast, or fungi. Microbes utilize the pollutants as food and energy sources during this process (Adams et al. 2015).

2 Types of Bioremediation

Bioremediation is basically divided into two types as in situ remediation of contaminated sites where these exist, and ex situ remediation is the remediation of contaminants at the specific place rather than where contaminants are present. In situ remediation is further classified into physical, chemical, and biological remediation, and ex situ remediation is classified into land forming, bio pile, composting, and bioreactor (Fig. 32.1).

2.1 In Situ Bioremediation

In situ remediation and ex situ remediation are the two main remediation techniques used to treat contaminated soils (Liu et al. 2018). The application of methods to treat pollutants in the soils and water at the location where they are found is known

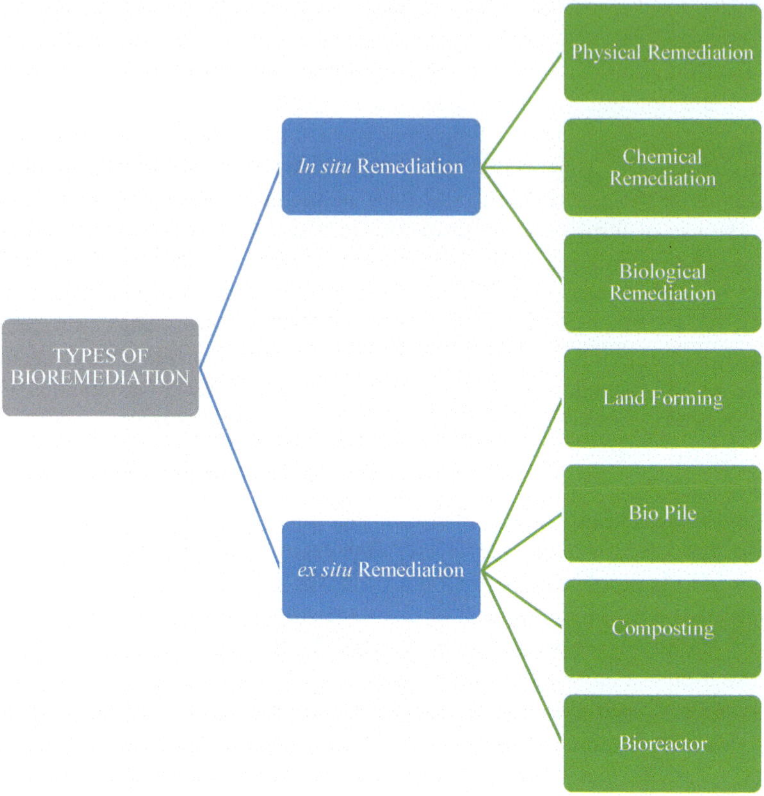

Fig. 32.1 Types of bioremediation

as in situ bioremediation. Microorganisms are employed in in situ remediation to biodegrade contaminants at the site and to encourage those microorganisms to accelerate their growth and biodegradation by microorganisms such as bacteria (Vidali 2001). The site characteristics, the types of pollutants present, their concentrations, and the intended use of the contaminated medium all play a role in determining the best remediation strategy (Azubuike et al. 2016) (Fig. 32.2).

2.1.1 Physical Remediation

This particular remediation has significant contribution in remediation and hence further sub-classified into following processes:

- *Bioventing*

In this technique, air is introduced or removed from surface or ground water to clean up polluted or contaminated areas. As vapours gradually move through biologically active soil, flow rate should be at its ideal level to accelerate the

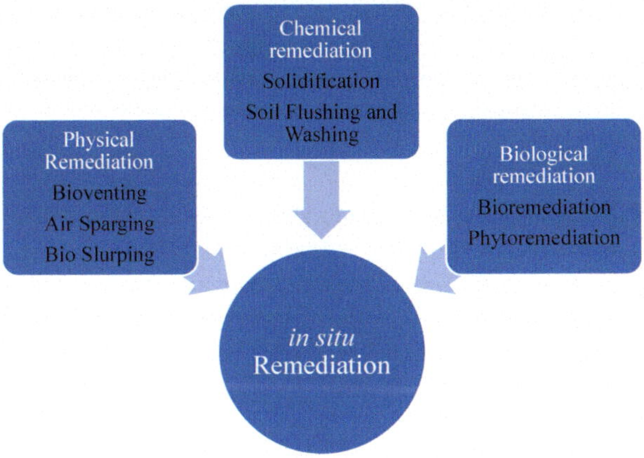

Fig. 32.2 Classification of in situ remediation

degradation process and decrease the volatilization of pollutants (Eslami and Joodat 2018). With a one-way valve and an electric blower, bioventing devices help to maintain an air flow rate that is proportional to the decline in atmospheric pressure (Rajendran et al. 2022). This research by Anekwe and Isa (2021) demonstrated the effectiveness of wastewater biostimulations and bioventing as a potentially effective technique for removing sulphate and heavy metals from contaminated soils.

- *Air sparging*

To treat contaminated locations, this approach uses high pressure oxygen injection to enhance oxygen concentration and volatilize toxins onto the soil surface and subsurface water (Hu et al. 2010). When compared to an air sparging system with a single injection well, an in situ air sparging system with multiple injection wells was performed and was more effective in collecting contaminants and remediating locations (Sudilovskiy et al. 2007).

- *Bio slurping*

The rate at which contaminants are extracted from fine-grained soils increases with multi-phase extraction, sometimes referred to as bio slurping. High vacuum pumps are employed in the bio slurping process to remove pollutants from groundwater, hydrocarbon vapour, and separate phase petroleum products from the subsurface. Fluids and vapour are subsequently re-injected into the subsurface (Paul et al. 2021). In vadose and saturated zones, bio slurping removes dissolved, vapour, residual, and non-aqueous pollutants, whereas other traditional approaches are not suitable for all contaminated sites (Naidu 2013).

2.1.2 Chemical Remediation

This particular remediation has significant contribution in remediation and hence further sub-classified into following processes:

- *Solidification/stabilization*

By introducing chemicals to the polluted areas, this technique is utilized to either remove or reduce the mobility of the contaminants. While stability happens as a result of chemical reactions by limiting the movement of pollutants, solidification is the creation of matrix by cement, asphalt, and thermoplastic. In contaminated areas, the addition of bone meal (finely ground, weakly crystalline apatite, Ca10 $(PO_4)_6OH_2)$ decreases the mobility and bioavailability of heavy metals (Yao et al. 2012).

- *Soil flushing and washing*

With this technique, pollutants are removed from the soils by adding water or another cleaning agent (Wuana et al. 2010). Water, saponin, organic acid, chelating agents, surfactants, and low-molecular-weight organic acids are some of the washing agents used to promote desorption of pollutants (Wan et al. 2015). Ethylenediaminetetraacetic acid is an efficient reagent for removing heavy metal from contaminated areas (Abumaizar and Smith 1999).

2.1.3 Biological Remediation

This particular remediation has significant contribution in remediation and hence further sub-classified into following processes:

- *Bioremediation*

This technique uses microbes, plants, or a mix of the two to treat contaminated soils. Heavy metal-contaminated locations are often remedied using bacteria and fungi; however, occasionally yeast and algae are also employed (Akhtar et al. 2013). Microorganisms such as *Pseudomonas putida*, *Bacillus subtilis*, and *Sporosarcina ginsengisoli* are utilized to treat heavy metal-contaminated locations. Compared to alternative approaches such as excavation and landfills, the bioremediation method gives the greatest savings in one acre of Pb-contaminated soil (Bai et al. 2014).

- *Phytoremediation*

Different plant species are employed in this remediation technique to treat contaminated sites and lessen their impact on the environment. This method incorporates plant species traits for treating locations, including their capacity to grow quickly, produce a lot of biomass, tolerate high metal concentrations, and have a large capacity for metal accumulating (Bhargava et al. 2012). Brassicaceae family member *Noccaea caerulescens* has a great capacity to absorb Cd without experiencing any harmful effects. Phytoremediation of contaminated sites uses phytoextraction, phytofiltration, phytostabilization, phytovolatilization, and phytodegradation (Bayçu et al. 2017).

2.2 *Ex Situ Bioremediation*

Ex situ bioremediation is a process that involves removing contaminated soil and storing it in a secure, above-ground location for treatment which involves supplying oxygen to the soil to promote the growth of naturally occurring microbes, which break down harmful organic substances (Kulshreshtha et al. 2014). Ex situ bioremediation has two phases, slurry and solid phase bioremediation. Slurry phase soil treatment process includes the slurry phase bioreactor. Land forming, soil bio piles and windrows are included in solid phase soil treatment (Nano et al. 2003).

2.2.1 Land Forming

The contaminated soil is excavated and spread out in a layer with a thickness of about 0.3 m on a treatment site that has been lined. This is the simplest method of bioremediation. Bioremediation can be accelerated by adding nutrients and changing the bed occasionally. Although it is the least expensive, this land formation method necessitates a sizable area which has little impact on the environment and takes little energy for making it unsuitable for small locations. This is because the soil layer thickness is limited to 0.3 m (Ausma et al. 2002). It also has some limitations such as decreased microbial activity because of unfavourable environmental conditions, less efficacy in the removal of inorganic pollutants, and higher costs because of excavation. As a result, land formation takes longer and is less effective than other methods (Daniel et al. 2022).

2.2.2 Bio Pile

A bio pile is a bioremediation technique in which mined toxic soils are combined with soil additives, then shaped into compost heaps, and occasionally bound for treatment and an aeration system to improve bioremediation by essentially boosting microbial activities (Rajendran et al. 2022). Aeration, irrigation, a nutrient, and leachate collecting system, and a treatment bed are the main elements of this method. The technique known as bio pile bioremediation is frequently used to reduce the concentration of petroleum components in soil (Cristorean et al. 2016). The use of this bioremediation in cold regions to clean up sites can also assist to reduce the volatilization of low-molecular-weight contaminants. For this treatment, dirt from various samples, including clay and sandy soil, is employed. High heating systems are included into bio pile designs in the bio pile technology to boost microbial activity and pollutant availability, which in turn accelerate biodegradation

(Ossai et al. 2022). Heated air is delivered to promote bioremediation by simultaneously delivering heat and air. It should be mentioned that humified bio piles have lower final total petroleum hydrocarbon concentrations (TPH) than heated and passive bio piles because they have ideal moisture content that minimizes the volatilization of contaminants that are less likely to degrade and reduce leaching. In a small amount of space, a bio pile system can be utilized to treat a lot of contaminated soil (García-Carmona et al. 2017).

2.2.3 Composting

By utilizing microorganisms in both aerobic and anaerobic environments, organic pollutants such as polycyclic aromatic hydrocarbon (PAHs) are transformed into harmless and stable by-products during the biological process of composting (Kumar et al. 2018). Temperatures between 54 and 65 °C must be maintained to successfully compost soil that has been contaminated with harmful organic contaminants. Typically natural or existing microorganisms are used in this (Myers and Williford 2000). Excavated soil is combined with bulking agents and organic amendments such as wood chips, animal and plant waste, to increase soil porosity. There are two methods for composting windrows and aerated static piles. Although it is very affordable and needs a lot of room, it could also have the highest fugitive emissions. TNT, RDX, HMX, and other dangerous explosives can be sufficiently reduced in concentration and toxicity by aerobic composting at a temperature of 50 °C (Semple et al. 2001).

2.2.4 Bioreactor

A system that uses microorganisms to digest contaminants in soil and groundwater is referred to as a bio reactor (Miller and Logan 2000). It is the option that removes organic pollutants the best. In this method, the contaminated soil is suspended in a nutrient-water solution thoroughly mixed and aerated in the presence of either exogenous or native bacteria (Balseiro-Romero et al. 2019). Wastewater is treated using bioreactors. When organic material is broken down by bacteria and forms sludge that is discarded or regenerated, contaminated water is circulated in an aeration basin. They are mostly used to clean up groundwater and soil of VOCs and fuel hydrocarbons (Ghoshdastidar et al. 2012).

3 Effects of Heavy Metals

The amount and concentration of heavy metals in the environment have significantly changed as a result of their widespread use for anthropogenic purposes such as color pigments, batteries, fertilizers, and other industrial things (Kumar and

Bharadvaja 2020). As a result, the accumulating of one or more heavy metals at a site in excess of the naturally permitted limits contaminates the air, water, and soil. Most heavy metals are toxic, cancer-causing, and mutagenic even in very small amounts (Tchounwou et al. 2012). Heavy metals can cause a number of illnesses in people and animals when they are exposed to them over an extended period of time by skin contact, inhalation, or ingesting foods containing the metals. Numerous public health organizations have released statistics showing that related diseases affect millions of people worldwide (Mahurpawar 2015).

Heavy metals are managed using physical, chemical, and biological, or a combinations of these mechanisms. However, many of these technologies are not sustainable for an environmental or economic standpoint (Mani and Kumar 2014). No single method makes the claim that it can totally reduce or reclaim heavy metals. In general, heavy metal salts are water soluble and cannot be physically separated. Physiochemical processes at the treatment site introduce secondary pollutants. Because they transform metals into a less toxic or innocuous form, microbes have been used in heavy metal clean-up for decades (Medfu Tarekegn et al. 2020). It is more effective, economical, and environment-friendly. Low-molecular-weight organic acids are a type of microbial metabolic secretion that can complex with heavy metals and soil particles containing heavy metal minerals. Microbes use biosorption, precipitation, and other processes (Gavrilescu 2004).

4 Bioremediation of Heavy Metal in Soil

Due to industrialization and the exploitation of natural resources, soil and water contamination has grown to a significant level and has become a global issue. In the contemporary era of environmental protection, there is a lot of interest in the use of microbes for heavy metal recovery from soil, sediments, and water as well as the use of plants for landfill applications (Shuaib et al. 2021). Numerous studies have demonstrated that microbes may remove heavy metals from polluted soils. According to a scientific analysis, metal can seriously harm our health by interfering with our natural processes when they are present in our bodies (Engwa et al. 2019). Arsenic, copper, iron, nickel, and other metals are advantageous to the body in trace amounts, but dangerous at excessive concentrations (Sankhla et al. 2016). The process of incorporating microorganisms to reduce toxin concentration is known as bioremediation. It is a natural interaction and its significance for biodiversity (above or below the ground) is increasingly taken into account for cleaning up metal-tainted and contaminated environment (Khan 2005). The microbial environment is significantly altered when heavy metals are present in the soil, despite the fact that their concentrations are often modest, and they are essential for the development of microorganisms (Aislabie et al. 2013).

5 Bioremediation of Heavy Metals in Water

Nowadays, natural contamination is an exceptionally enormous issue due to perilous waste that has prompted shortage of clean water and unsettling influences of soil along these lines restricting yield creation (Abbasi et al. 2014). Bioremediation utilizes natural specialists, essentially microorganisms, that is, microbes, parasites, and yeast, to tidy up contaminated water. Bioremediation is characterized as the cycle by which microorganisms are invigorated to quickly corrupt unsafe natural toxins to earth safe levels in ground water, soil, substances, materials, and residue (Muthukumaran 2022). Bioremediation is not a remedy, yet rather a characteristic cycle option in contrast to such strategies as burning, synergist annihilation, the utilization of sponges, and actual expulsion and resulting obliteration of poisons. (Sanjay and Shukla 2021).

Heavy metals present in the soil crust outer layer may get solubilized in ground water through normal cycles or by change in soil pH. In addition, groundwater can get contaminated with heavy metals from landfill leachate, sewage, leachate from mine tailings, profound well removal of fluid squanders, and leakage from modern squanders tidal ponds or from modern spills and leaks. An assortment of responses in soil climate, for example, corrosive/base, oxidation/decrease, precipitation/disintegration, sorption, or particle trade cycles, can impact the speciation and portability of metal impurities. The rate and degree of these responses will rely upon elements for example, pH, Eh, complexion with other broke down constituents, sorption and particle trade limit of the topographical materials, and natural matter substance. The harmfulness, portability, and reactivity of heavy metals rely upon its speciation, which again relies on certain circumstances, for example, pH, EC, temperature, dampness, and so forth, to decide the speciation of metals in soils; explicit extractants are utilized to solubilize various periods of metals (Engelbrecht 2020).

6 Mechanism of Heavy Metal Resistance by Microbial Cells

Significant metal contamination is an overall issue as it leaves risk to general affluence and receptiveness to metals over a particular edge level can cause health effects in all living things including all forms of life (Nassiri et al. 2021). Such challenging circumstances of a couple of creature's progress, several frameworks to use and change significant metal into a less toxic structure, achieving the improvement of profound metal remediation microorganisms. Profound metal remediation microorganisms can be used in bioremediation to remediate the degraded sites (Das and Dash 2014). Bioremediation is also used for standard regular activities, is less costly, and has high open affirmation. In this way, we see the coordinated efforts and frameworks that occur between the life forms and profound metal, including pressure responses and shield instruments, which incorporate aggregation and biofilm plan, formation of extracellular polymeric substances, and improvement of resistance characteristics and hailing pathways against significant metals (Savitha et al. 2022).

6.1 Extracellular Barrier and Sequestration

Numerous microorganisms are used for protections from metals in water, soil, and modern waste. Qualities present on chromosomes, plasmids, or transposons encode explicit protection from an assortment of metal particles (Hall et al. 2017). Some metals, such as cobalt, nickel, and copper, fill in as micronutrients and are utilized for redox processes, to balance out particles through electrostatic connections, as parts of different catalysts, and for guideline of osmotic strain (Sarode et al. 2019). Many metals are unnecessary and have no supplement concern and are possibly harmful to microorganisms. These harmful metals cooperate with fundamental cell parts through covalent and ionic holding. At undeniable levels, both fundamental and heavy metals can harm cell films, modify chemical particularity, upset cell capacities, and harm the construction level of DNA (Inyinbor Adejumoke et al. 2018). Microorganisms have adjusted to the presence of both supplement and insignificant metals by fostering an assortment of obstruction systems. Six metal opposition components show avoidance by porousness boundary, intra- and extracellular sequestration, dynamic vehicle efflux siphons, enzymatic detoxification, and decrease in the awareness of cell focuses to metal particles. The comprehension of how microorganisms oppose metals can be understood by the techniques to their detoxification or expulsion from the climate (Bruins et al. 2000).

6.2 Extracellular Sequestration

Extracellular sequestration is the accumulation of metal ions by individual cell components in the periplasm or their appearance as insoluble combinations (Sreedevi et al. 2022). In order to sequester the metal inside the periplasm, microscopic organisms can synthesize metal particles from the cytoplasm. Copper-safe periplasmic proteins, CopA, CopB, and CopC, which link copper particles and bacteria regions, are produced by *Pseudomonas syringe* strains. Zinc particles can exit the cytoplasm through the efflux framework and assemble in *Synechocystis* periplasm. Metal from the PCC 6803 strain is sequestered extracellularly. Bacterium that reduce iron is *Geobacter* spp., and that reduce sulphur is *Desulfuromonas* spp. (Guo et al. 2021).

6.3 *Intracellular Sequestration*

The organization of sub-atomic to micro-scale transporters for intracellular conveyance has colossal potential for science and medication and is used in medical particularly for in vivo treatments. The field stays restricted nonetheless by an unfortunate comprehension of how transporters get to the cell inside. In this method, we take an outline of the various types of transporters, many methods of section, putative pathways of vesicular vehicle, and destinations of endosomal escape (Ianeva 2009). We summarize this by appropriate models from the cell science of how infectious microbes and their effectors enter cells and getaway endosomal repression. We expect that experiences into the instruments of cell passage and endosomal getaway will help future exploration endeavours on compelling transporter-intervened intracellular conveyance (Mishra et al. 2019b).

Metal particles can be sequestered inside of cells by various combinations in a process known as intracellular sequestration. Metals can get concentrated inside microbial cells as a result of interaction with surface ligands and slow transport. In practice and radically in the treatment of profluent therapy, the ability of bacterial cells to gather metals intracellularly has been exploited (Borkow and Gabbay 2005). The cadmium-tolerant *P. putida* strain was capable of securing copper, cadmium, and zinc particles within its cells using cysteine-rich low atomic weight proteins. *Rhizobium leguminosarum* cells have been found to sequester cadmium particles intracellularly in a manner similar to glutathione (Corticeiro et al. 2013).

Chitin, mineral particles, lipids, nitrogen-containing polysaccharides, polyphosphates, and proteins make up the rigid cell mass of parasites (Remacle 1990). By fiery take-up of extracellular and intracellular precipitation and valence change along with a few growths that collect metals to their mycelium and spores, they can detoxify metal particles. The parasite mass on the outside of the phone functions as a ligand for identifying metal particles and facilitates the removal of inorganic metals (Sanchez et al. 2005). Amine binds to anionic metal species by electrostatic connection and cationic metal species via surface complexity, making it more dynamic in metal take-up among these practical groups (Sharma et al. 2016).

6.4 *Reduction of Metal Ions*

The harmfulness of metal particles can be reduced by microbial cells by changing them from one oxidation state to another. Metals and metalloids are used by microbes as electron donors or acceptors for energy transfer. When microorganisms breathe anaerobically, metals in the oxidized structure might act as terminal acceptors of electrons (Tsezos 2009). Enzymatic reduction of metal particles may result in the creation of less dangerous forms of mercury and chromium. Demonstration that metal hydrides, for instance, CuH^{+}, AgH^{+}, and AgH, are outlined as intermediates in the reduction of metal particles by sub-nuclear hydrogen in liquid type of action suggests that species may be related with the oxidation and lessening of metal particles by hydrogen atoms (Sun et al. 2021). For example, in the prominent reactions, related reactions with hydrogen particles and metal particles have been seen in watery plans in which hydrogen particles were made in a grouping of ways or introduced from the gas phase, but their parts are at this point questionable and the subject of broad conversation. The method involved creating metals from minerals utilizing decreased responses. The decrease of metals was initially perceived to be the responses used to get metals from their oxides by involving substances having more prominent fond for oxygen than the metal (Osgood et al. 2016).

7 Future Prospective

Bioremediation method is a better alternative as compared to other conventional methods in terms of efficiency and budget. Bioremediation has also its own limitation like other methods. Some microorganisms cannot break toxic biodegradable metals because the contaminants are resistant to their attack, for example, chlorinated macrobiotic or high aromatic hydrocarbons. The sorption of metal in water is generally faster than other processes (Singh et al. 2020). This uniqueness makes it the most useful in the treatment of wastewater to meet regulation standards before discharging it into the environment. The cost of running the treatment systems is often minimal. They may occasionally need adjustments in which case they will charge more depending on the reagents employed. The pre-treatment technique can be changed to enhance the performance of microorganisms. This method's main goal is to increase the biosorption of metal ions. Future research should concentrate on using various methods so that bioremediation can be successful in all unfavourable situations (Khoo et al. 2021).

8 Conclusions

These days, heavy metal contamination is a serious issue. As an evolving technology, bioremediation needs to establish its boundaries. It depends on several strategies that are necessary due to the complexity of contamination and problems with polluted locations. Despite the limitations, bioremediation has a bright future as research and development tool in the field accelerating heavy metal remediation (Kumar et al. 2011). The development of in situ microbial methods for metals' clean-up has accelerated significantly in recent years. Bioremediation of heavy metals would be a low-cost, environmentally accountable technique. Small-scale field tests show that the phytoextraction process is viable (Mench et al. 2010). It seems like a really capable technology for cleaning up the environment. Methods of phytoremediation are particularly suited for usage at large sites where other remediation techniques are not practical or cost-effective. In a novel cleaning technique called phytoremediation, damaged areas are cleaned or stabilized using plants. Future efforts should be made to advance ex situ clean-up. The biosorption approach of toxic heavy metal removal can be crucial in the development of new methods and technology. However, much effort needs to be done to fully comprehend the adsorption mechanism (Volesky 2007).

References

Abbasi A, Sajid A, Haq N, Rahman S, Misbah ZT, Sanober G, Kazi AG (2014) In: Ahmad P, Wani M, Azooz M, Tran LS (eds) Improvement of crops in the era of climatic changes. Springer, New York. https://doi.org/10.1007/978-1-4614-8830-9_13

Abumaizar RJ, Smith EH (1999) Heavy metal contaminants removal by soil washing. J Hazard Mater 70(1–2):71–86

Adams GO, Fufeyin PT, Okoro SE, Ehinomen I (2015) Bioremediation, biostimulation and bioaugmention: a review. Int J Environ Biorem Biodegrad 3(1):28–39

Aislabie J, Deslippe JR, Dymond J (2013) Soil microbes and their contribution to soil services. In: Dymond J (ed) Ecosystem services in New Zealand: conditions and trends. Manaaki Whenua Press, Lincoln, pp 143–161

Akhtar MS, Chali B, Azam T (2013) Bioremediation of arsenic and lead by plants and microbes from contaminated soil. Res Plant Sci 1(3):68–73

Akram R, Turan V, Hammad HM, Ahmad S, Hussain S, Hasnain A, Nasim W (2018) Fate of organic and inorganic pollutants in paddy soils. In: Environmental pollution of paddy soils, Springer, Cham 197–214. https://doi.org/10.1007/978-3-319-93671-0_13

Anekwe IM, Isa YM (2021) Wastewater and bioventing treatment systems for acid mine drainage–contaminated soil. Soil Sediment Contam 30(5):518–531. https://doi.org/10.1080/15320383.2020.1863909

Ausma S, Edwards GC, Fitzgerald-Hubble CR, Halfpenny-Mitchell L, Gillespie TJ, Mortimer WP (2002) Volatile hydrocarbon emissions from a diesel fuel-contaminated soil bioremediation facility. J Air Waste Manage Assoc 52(7):769–780

Azubuike CC, Chikere CB, Okpokwasili GC (2016) Bioremediation techniques–classification based on site of application: principles, advantages, limitations and prospects. World J Microbiol Biotechnol 32(11):1–18

Bai J, Yang X, Du R, Chen Y, Wang S, Qiu R (2014) Biosorption mechanisms involved in immobilization of soil Pb by *Bacillus subtilis* DBM in a multi-metal-contaminated soil. J Environ Sci 26(10):2056–2064

Balseiro-Romero M, Monterroso C, Kidd PS, Lu-Chau TA, Gkorezis P, Vangronsveld J, Casares JJ (2019) Modelling the ex-situ bioremediation of diesel-contaminated soil in a slurry bioreactor using a hydrocarbon-degrading inoculant. J Environ Manag 246:840–848

Bayçu G, Gevrek-Kürüm N, Moustaka J, Csatári I, Rognes SE, Moustakas M (2017) Cadmium-zinc accumulation and photosystem II responses of *Noccaea caerulescens* to Cd and Zn exposure. Environ Sci Pollut Res 24(3):2840–2850

Bhargava A, Carmona FF, Bhargava M, Srivastava S (2012) Approaches for enhanced phytoextraction of heavy metals. J Environ Manag 105:103–120

Borkow G, Gabbay J (2005) Copper as a biocidal tool. Curr Med Chem 12(18):2163–2175

Bruins MR, Kapil S, Oehme FW (2000) Microbial resistance to metals in the environment. Ecotoxicol Environ Saf 45(3):198–207

Caliman FA, Robu BM, Smaranda C, Pavel VL, Gavrilescu M (2011) Soil and groundwater cleanup: benefits and limits of emerging technologies. Clean Technol Environ Policy 13(2):241–268

Cimboláková I, Uher I, Laktičová KV, Vargová M, Kimáková T, Papajová I (2019) Heavy metals and the environment. In: Environmental factors affecting human health. IntechOpen. https://doi.org/10.5772/intechopen.86876

Corticeiro S, Freitas R, Figueira E (2013) The role of GSTs in the tolerance of *Rhizobium leguminosarum* to cadmium. Biometals 26(6):879–886

Cristorean C, Micle V, Sur IM (2016) A critical analysis of ex-situ bioremediation technologies of hydrocarbon polluted soils. ECOTERRA J Environ Res Prot 13:17–29

Daniel AJ, Enzo ER, Juliana MS, Stefanie BCG, Analia A, Claudia SB, Marta AP (2022) The current approach to soil remediation: a review of physicochemical and biological technologies, and the potential of their strategic combination. J Environ Chem Eng 10:7141

Das S, Dash HR (2014) Microbial bioremediation: a potential tool for restoration of contaminated areas. In: Microbial biodegradation bioremediation. Elsevier, Netherland, 1–21. https://doi.org/10.1016/B978-0-12-800021-2.00001-7

Engelbrecht R (2020) Environmental radioactivity monitoring. In: Handbook of radioactivity analysis, Academic Press, Elsevier, Netherland (2):1–40. https://doi.org/10.1016/B978-0-12-814395-7.00001-5

Engwa GA, Ferdinand PU, Nwalo FN, Unachukwu MN (2019) Mechanism and health effects of heavy metal toxicity in humans. In: Poisoning in the modern world-new tricks for an old dog, vol 10. IntechOpen, pp 70–90

Eslami E, Joodat SHS (2018) Bioremediation of oil and heavy metal contaminated soil in construction sites: a case study of using bioventing-biosparging and phytoextraction techniques. [online] Available at: https://arxiv.org/abs/1806.03717. Accessed 27 Mar 2022.

García-Carmona M, Romero-Freire A, Aragón MS, Garzón FM, Peinado FM (2017) Evaluation of remediation techniques in soils affected by residual contamination with heavy metals and arsenic. J Environ Manag 191:228–236

Gaur N, Flora G, Yadav M, Tiwari A (2014) A review with recent advancements on bioremediation-based abolition of heavy metals. Environ Sci Proces Impact 16(2):180–193

Gautam PK, Gautam RK, Banerjee S, Chattopadhyaya MC, Pandey JD (2016) Heavy metals in the environment: fate, transport, toxicity and remediation technologies. Nova Sci Publishers 60:101–130

Gavrilescu M (2004) Removal of heavy metals from the environment by biosorption. Eng Life Sci 4(3):219–232

Ghoshdastidar AJ, Saunders JE, Brown KH, Tong AZ (2012) Membrane bioreactor treatment of commonly used organophosphate pesticides. J Environ Sci Health Part B 47(7):742–750

Guerra F, Trevizam AR, Muraoka T, Marcante NC, Canniatti-Brazaca SG (2012) Heavy metals in vegetables and potential risk for human health. Sci Agric 69:54–60

Gunatilake SK (2015) Methods of removing heavy metals from industrial wastewater. J Multi Eng Sci Stu 1(1):14

Guo Y, Aoyagi T, Hori T (2021) Comparative insights into genome signatures of ferric iron oxide- and anode-stimulated *Desulfuromonas* spp. strains. BMC Genomics 22(1):1–17

Hall JP, Brockhurst MA, Harrison E (2017) Sampling the mobile gene pool: innovation via horizontal gene transfer in bacteria. Philos Trans R Soc B Biol Sci 372(1735):20160424

Hu L, Wu X, Liu Y, Meegoda JN, Gao S (2010) Physical modelling of air flow during air sparging remediation. Environ Sci Technol 44(10):3883–3888

Ianeva OD (2009) Mechanisms of bacteria resistance to heavy metals. Mikrobiol Z (Kiev, Ukraine: 1993) 71(6):54–65

Inyinbor Adejumoke A, Adebesin Babatunde O, Oluyori Abimbola P, Adelani Akande Tabitha A, Dada Adewumi O, Oreofe Toyin A (2018) Water pollution: effects, prevention, and climatic impact. In: Water challenges of an urbanizing world. Intechopen,Croatia 33–47

Jaishankar M, Tseten T, Anbalagan N, Mathew BB, Beeregowda KN (2014) Toxicity, mechanism and health effects of some heavy metals. Interdiscip Toxicol 7(2):60

Kahlon SK, Sharma G, Julka JM, Kumar A, Sharma S, Stadler FJ (2018) Impact of heavy metals and nanoparticles on aquatic biota. Environ Chem Lett 16(3):919–946

Khalid S, Shahid M, Niazi NK, Murtaza B, Bibi I, Dumat CA (2017) A comparison of technologies for remediation of heavy metal contaminated soils. J Geochem Explor 182:247–268

Khan AG (2005) Role of soil microbes in the rhizospheres of plants growing on trace metal contaminated soils in phytoremediation. J Trace Elem Med Biol 18(4):355–364

Khoo KS, Chia WY, Chew KW, Show PL (2021) Microalgal-bacterial consortia as future prospect in wastewater bioremediation, environmental management and bioenergy production. Indian J Microbiol 61(3):262–269

Kulshreshtha A, Agrawal R, Barar M, Saxena S (2014) A review on bioremediation of heavy metals in contaminated water. IOSR J Environ Sci Toxicol Food Technol 8(7):44–50

Kumar L, Bharadvaja N (2020) Microbial remediation of heavy metals. In: Microbial bioremediation and biodegradation. Springer, Singapore, pp 49–72

Kumar A, Bisht BS, Joshi VD, Dhewa T (2011) Review on bioremediation of polluted environment: a management tool. Int J Environ Sci 1(6):1079

Kumar V, Shahi SK, Singh S (2018) Bioremediation: an eco-sustainable approach for restoration of contaminated sites. In: Microbial bioprospecting for sustainable development. Springer, Singapore, pp 115–136

Lakherwal D (2014) Adsorption of heavy metals: a review. Int J Environ Res Develop 4(1):41–48

Li C, Zhou K, Qin W, Tian C, Qi M, Yan X, Han W (2019) A review on heavy metals contamination in soil: effects, sources, and remediation techniques. Soil Sediment Contam 28(4):380–394

Liu M, Zhang A, Liao Y, Chen B, Fan D (2015) The environment quality of heavy metals in sediments from the central Bohai Sea. Mar Pollut Bull 100(1):534–543

Liu L, Li W, Song W, Guo M (2018) Remediation techniques for heavy metal-contaminated soils: principles and applicability. Sci Total Environ 633:206–219

Mahurpawar M (2015) Effects of heavy metals on human health. Int J Res Granthaalayah 530:1–7

Mani D, Kumar C (2014) Biotechnological advances in bioremediation of heavy metals contaminated ecosystems: an overview with special reference to phytoremediation. Int J Environ Sci Technol 11(3):843–872

Masindi V, Muedi KL (2018) Environmental contamination by heavy metals. In: Heavy metals. https://doi.org/10.5772/intechopen.76082

Medfu Tarekegn M, Zewdu Salilih F, Ishetu AI (2020) Microbes used as a tool for bioremediation of heavy metal from the environment. Cogent Food Agric 6(1):1783174

Mench M, Lepp N, Bert V, Schwitzguébel JP, Gawronski SW, Schröder P, Vangronsveld J (2010) Successes and limitations of phytotechnologies at field scale: outcomes, assessment and outlook from COST Action 859. J Soil Sediment 10(6):1039–1070

Miller JP, Logan BE (2000) Sustained perchlorate degradation in an autotrophic, gas-phase, packed-bed bioreactor. Environ Sci Technol 34(14):3018–3022

Mishra S, Bharagava RN, More N, Yadav A, Zainith S, Mani S, Chowdhary P (2019a) Heavy metal contamination: an alarming threat to environment and human health. In: Environmental biotechnology: for sustainable future. Springer, Singapore, pp 103–125

Mishra V, Yadav N, Saraogi GK, Tambuwala MM, Giri N (2019b) Dendrimer based nanoarchitectures in diabetes management: an overview. Curr Pharm Des 25(23):2569–2583

Muthukumaran M (2022) Advances in bioremediation of nonaqueous phase liquid pollution in soil and water. In: Biological approaches to controlling pollutants. Woodhead Publishing, Elsevier, Netherland 191–231. https://doi.org/10.1016/B978-0-12-824316-9.00006-9

Myers TE, Williford CW (2000) Concepts and technologies for bioremediation in confined disposal facilities. https://apps.dtic.mil/sti/pdfs/ADA375964.pdf. Accessed 24 Nov 2023

Naidu R (2013) Recent advances in contaminated site remediation. Water Air Soil Pollut 224(12):1–11

Nano G, Borroni A, Rota R (2003) Combined slurry and solid-phase bioremediation of diesel contaminated soils. J Hazard Mater 100(1–3):79–94

Nassiri O, Rhoujjati A, EL Hachimi ML (2021) Contamination, sources and environmental risk assessment of heavy metals in water, sediment and soil around an abandoned Pb mine site in North East Morocco. Environ Earth Sci 80(3):1–20

Osgood H, Devaguptapu SV, Xu H, Cho J, Wu G (2016) Transition metal (Fe, Co, Ni, and Mn) oxides for oxygen reduction and evolution bifunctional catalysts in alkaline media. Nano Today 11(5):601–625

Ossai IC, Hamid FS, Hassan A (2022) Biological treatments for petroleum hydrocarbon pollutions: the eco-friendly Technol. In: Hazardous waste management. Intechopen. https://doi.org/10.5772/intechopen.102053

Pandey G, Madhuri S (2014) Heavy metals causing toxicity in animals and fishes. Res J Anim Vet Fish Sci 2(2):17–23

Paul O, Jasu A, Lahiri D, Nag M, Ray RR (2021) *In situ* and *ex situ* bioremediation of heavy metals: the present scenario. J Environ Eng Landsc Manag 29(4):454–469

Rajendran S, Priya TAK, Khoo KS, Hoang TK, Ng HS, Munawaroh HSH, Show PL (2022) A critical review on various remediation approaches for heavy metal contaminants removal from contaminated soils. Chemosphere 287:132369

Remacle J (1990) The cell wall and metal binding. In: Volesky B (ed) Biosorption of heavy metals. CRC Press, Boca Raton, pp 83–92

Sanchez C, Julián B, Belleville P, Popall M (2005) Applications of hybrid organic–inorganic nanocomposites. J Mater Chem 15(35–36):3559–3592

Sanjay SS, Shukla AK (2021) Potential therapeutic applications of nano-antioxidants. Berlin/Heidelberg, Germany: Springer. https://doi.org/10.1007/978-981-16-1143-8

Sankhla MS, Kumari M, Nandan M, Kumar R, Agrawal P (2016) Heavy metals contamination in water and their hazardous effect on human health-a review. Int J Curr Microbiol App Sci 5(10):759–766

Sarode S, Upadhyay P, Khosa MA, Mak T, Shakir A, Song S, Ullah A (2019) Overview of wastewater treatment methods with special focus on biopolymer chitin-chitosan. Int J Biol Macromol 121:1086–1100

Savitha T, Khalifa AY, Sankaranarayanan A (2022) Biofilm interceded microbial prospecting of bioremediation. In: Bioprospecting of microbial diversity, Elsevier 371–391. https://doi.org/10.1016/B978-0-323-90958-7.00028-5

Semple KT, Reid BJ, Fermor TR (2001) Impact of composting strategies on the treatment of soils contaminated with organic pollutants. Environ Pollut 112(2):269–283

Sharma S (2012) Bioremediation: features, strategies and applications. Asian J Pharm Life Sci 2231:4423

Sharma S, Rana S, Thakkar A, Baldi A, Murthy RSR, Sharma RK (2016) Physical, chemical and phytoremediation technique for removal of heavy metals. J Heavy Metal Tox Dis 1(2):1–15

Shuaib M, Azam N, Bahadur S, Romman M, Yu Q, Xuexiu C (2021) Variation and succession of microbial communities under the conditions of persistent heavy metal and their survival mechanism. Microb Pathog 150:104713

Singh R (2014) Microorganism as a tool of bioremediation technology for cleaning environment: a review. Proc Int Acad Ecol Environ Sci 4(1):1

Singh MR, Gupta A (2016) Water pollution-sources, effects and control. Centre for Biodiversity, Department of Botany, Nagaland University

Singh S, Kumar V, Datta S, Dhanjal DS, Sharma K, Samuel J, Singh J (2020) Current advancement and future prospect of biosorbents for bioremediation. Sci Total Environ 709:135895

Sodango TH, Li X, Sha J, Bao Z (2018) Review of the spatial distribution, source and extent of heavy metal pollution of soil in China: impacts and mitigation approaches. J Health Pollut 8(17):53–70

Sreedevi PR, Suresh K, Jiang G (2022) Bacterial bioremediation of heavy metals in wastewater: a review of processes and applications. J Water Proc Eng 48:102884

Sudilovskiy PS, Kagramanov GG, Trushin AM, Kolesnikov VA (2007) Use of membranes for heavy metal cationic wastewater treatment: flotation and membrane filtration. Clean Technol Environ Policy 9(3):189–198

Sun C, Teo BK, Deng C, Lin J, Luo GG, Tung CH, Sun D (2021) Hydrido-coinage-metal clusters: rational design, synthetic protocols and structural characteristics. Coord Chem Rev 427:213576

Tchounwou PB, Yedjou CG, Patlolla AK, Sutton DJ (2012) Heavy metal toxicity and the environment. In: Molecular, clinical and environmental toxicology, pp 133–164
Tsezos M (2009) Metal-microbes interactions: beyond environmental protection. Adv Mater Res 71:527–532
Vidali M (2001) Bioremediation. An overview. Pure Appl Chem 73(7):1163–1172
Volesky B (2007) Biosorption and me. Water Res 41(18):4017–4029
Wan J, Meng D, Long T, Ying R, Ye M, Zhang S, Li Q, Zhou Y, Lin Y (2015) Simultaneous removal of lindane, lead and cadmium from soils by rhamnolipids combined with citric acid. PLoS One 10(6):e0129978. https://doi.org/10.1371/journal.pone.0129978
Wang L, Rinklebe J, Tack FM, Hou D (2021) A review of green remediation strategies for heavy metal contaminated soil. Soil Use Manag 37(4):936–963
Wuana RA, Okieimen FE, Imborvungu JA (2010) Removal of heavy metals from a contaminated soil using organic chelating acids. Int J Environ Sci Technol 7(3):485–496
Yao Z, Li J, Xie H, Yu C (2012) Review on remediation technologies of soil contaminated by heavy metals. Procedia Environ Sci 16:722–729

Chapter 33
MicroRNAs (miRNAs): Crosstalk with Regulatory Networks of Abiotic Stress Tolerance in Plants

Deepu Pandita

Abstract MicroRNAs (miRNAs) are a class of small, endogenous, non-coding 20–24 nucleotide RNA, found abundantly in plants and regulate their target gene expression post-transcriptionally. Plant's miRNAs are transcribed from miRNA-encoding genes and regulate regulatory networks of processes such as development, metabolism, stress responses, and almost all aspects of plant biology. The miRNAs have regulatory roles in water deficiency, salinity, heat, cold, and other abiotic stress interactions. The expression profiling of miRNAs with high-throughput techniques reveals that miRNAs involved in the growth and development of plants show differential expression during various abiotic stress responses. The miRNA-mediated interference (miRNAi) helps in designing abiotic stress-resistant crops with controlled response to abiotic stress conditions.

Keywords Plant miRNAs · Post-transcriptional regulation · Non-coding RNA · miRNA expression profiling · Abiotic stress

1 Introduction

MicroRNAs (miRNAs) are endogenous, small (20–24 nucleotides in length), non-coding, single-stranded regulatory RNAs regulating several biological processes by gene silencing at the post-transcriptional levels (Jones-Rhoades et al. 2006; Voinnet 2009; Mittal et al. 2016; Chaudhary et al. 2021). Few miRNAs are conserved, and majority of miRNAs are species- or family-specific evolved during evolution (Axtell and Bartel 2005; Cuperus et al. 2011; Borges and Martienssen 2015; Cui et al. 2016). The miRNAs are produced by transcription, processing of precursors, and assembly of miRNA-induced silencing complex (miRISC). The miRNAs bring-up gene regulation based on sequence complementarity and pairing followed by cleavage or translation inhibition of target mRNA (Jones-Rhoades et al. 2006). The

D. Pandita (✉)
Government Department of School Education, Jammu, Jammu and Kashmir, India

M. Hasanuzzaman (ed.), *Climate-Resilient Agriculture, Vol 2*,
https://doi.org/10.1007/978-3-031-37428-9_33

miRNAs modify regulation of abiotic stress-responsive genes and have been linked to the majority of plant biology aspects (Jones-Rhoades et al. 2006; Voinnet 2009; Mittal et al. 2016; Chaudhary et al. 2021). The stress-related miRNAs reprogramme gene expression at downstream levels in various abiotic stress scenarios (e.g., water deficiency, salinity, low/high temperature) (Sunkar and Zhu 2004; Zhao et al. 2013; Gupta et al. 2014; Xie et al. 2014; Ferdous et al. 2015; Zhu 2016; Shriram et al. 2016). The crosstalk of miRNAs with various abiotic stress responsive gene networks and stress responses has been elaborated (Pandita 2019), in *Oryza sativa* (Pandita and Wani 2019), wheat (Pandita 2022a), maize (Pandita 2022b), and other plants. The stress-responsive miRNAs regulate gene networks at post-transcriptional levels to reduce growth and development in plants. The miRNAs can be engineered for abiotic stress tolerance in plant kingdom (Winter and Diederichs 2011; Iwakawa and Tomari 2013; Li and Zhang 2016). Here, in this chapter, we discuss the biogenesis, mode of action, and the roles of miRNAs in plant abiotic stress interactions.

2 Biogenesis

The biogenesis of miRNAs and the mode of action of plant miRNAs has been discussed in detail in several chapters previously (Pandita 2019, 2021, 2022c; Pandita and Wani 2019). The miRNA biogenesis initiates inside nucleus of cell. The graphical representation of biogenesis of miRNAs in plants is given in Fig. 33.1. The genes which encode for miRNAs in plants are termed microRNA genes or MIR genes. MIR genes can be localized in the intergenic regions or within introns of other genes (Jones-Rhoades et al. 2006). The transcription of MIR genes by RNA polymerase II leads to formation of a specific long primary single-stranded RNA transcript called primary miRNA (pri-miRNA) (Jones-Rhoades et al. 2006). The pri-miRNAs are modified and capped at 5′ end and polyadenylated at 3′ tail end (Pantaleo et al. 2010). The pri-miRNA fold into perfect stem-loop structure and stabilize by an RNA-binding protein known as DAWDLE (DDL). The pri-miRNA is modified by DICER-LIKE (DCL1), SERRATE (SE), and HYPONASTIC LEAVES 1 (HYL1) enzymes into hairpin-like precursor miRNA (pre-miRNA) (Jones-Rhoades et al. 2006; Khraiwesh et al. 2012). The one arm of stem-looped pre-miRNA representing mature miRNA sequence is recognized and processed by DCL1, HYL1, and SE to generate the miRNA: miRNA* duplex inside the nucleus (Bartel 2004; Jones-Rhoades et al. 2006; Pantaleo et al. 2010; Bologna and Voinnet 2014; Fukudome and Fukuhara 2017). This miRNA: miRNA* duplex is stabilized and protected by the methylation at 3′ end due to enzyme HUA ENHANCER 1 (HEN1). The miRNA: miRNA* duplex is transported into cytoplasm by HASTY (HST1) (Park et al. 2002; Yu et al. 2005; Ramachandran and Chen 2008; Sun et al. 2012; Pelaez et al. 2012; Xie et al. 2014). The mature miRNA: miRNA* duplex gets loaded into RNA-Induced Silencing Complex (RISC) forming the miR-RISC complex in cytoplasm which is further stabilized by ARGONAUTE 1 (AGO1). The AGO1 unwinds miRNA: miRNA* duplex. One strand degrades and another strand

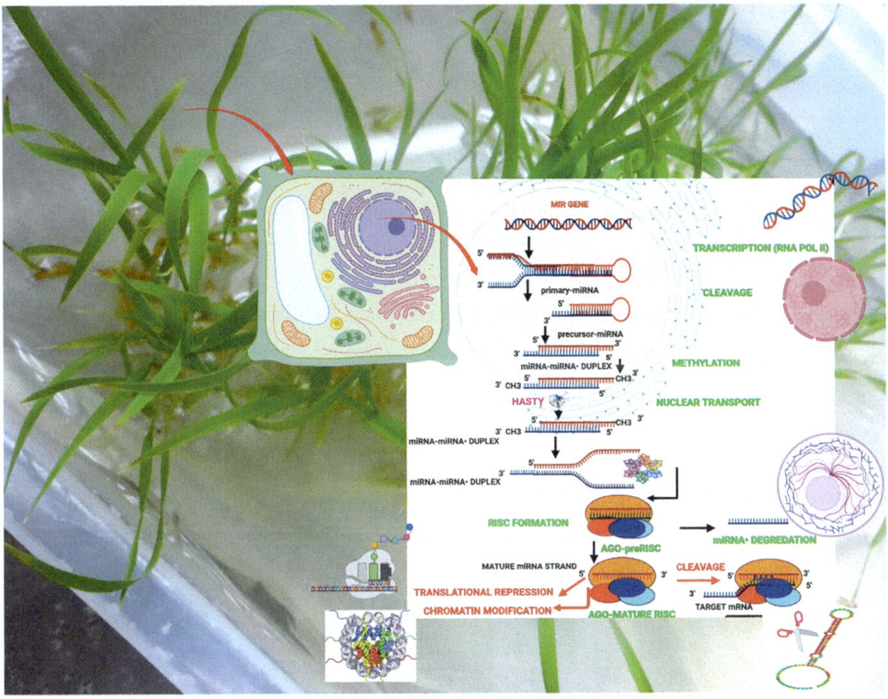

Fig. 33.1 Biogenesis and mechanism of action of microRNAs

of mature miRNA continues to attach to RISC with AGO1 (Baumberger and Baulcombe 2005; Arribas-Hernández et al. 2016; Iki 2017).

3 Mechanism of Action

The pri-miRNA encoded by MIR genes transcribed by Pol II folds into stem-loop structure to form mature miRNAs (Jones-Rhoades et al. 2006; Chen 2009; Reis et al. 2015). In plants, mature miRNA promotes the RISC loading to complementary mRNAs targets. The two modes of mechanism of miRNA action includes cleavage of target mRNA and translation inhibition. Plant mature miRNAs guides miR-RISC complex to encourage site-specific cleavage of complimentary mRNA at the poly-(A) tail with perfect pairing and destabilizes and decay it or inhibition of protein translation of the targeted mRNAs/transcripts by imperfect pairing (Fig. 33.1). Previously, only mature miRNA was supposed to inhibit mRNAs and translation. But now, miRNA* strand is also assumed to target mRNA and regulate gene expression (Jones-Rhoades et al. 2006; Chen 2009; Sun 2011; Guleria et al. 2011; Sun et al. 2012; Khraiwesh et al. 2012; Reis et al. 2015; Zhang and Unver 2018).

4 Role of miRNAs in Abiotic Stress Tolerance of Crops

The miRNAs modulate transcriptional levels in several plant crop species such as *Oryza sativa* (rice), *Zea mays* (maize), sunflower, *Triticum aestivum* (wheat), sorghum, cotton, and Arabidopsis during their life cycle under different stress scenarios. This recommends miRNA as emergent, efficient, and important candidate in controlling the abiotic stress response in plants (Sunkar et al. 2007, 2012; Khraiwesh et al. 2012; Kumar 2014; Zhang 2015; Shriram et al. 2016; Noman et al. 2017; Zhang and Unver 2018; Basso et al. 2019; Sun et al. 2019). The miRNAs and their targeted genes show differential expression under various stress cues. This permits the selection and identification of plant miRNAs and genes of several agronomic traits (Du et al. 2011; Zhang 2015; Salvador-Guirao et al. 2018). A change in miRNA expression acts as the amplifier of response by impacting plant physiology (Megraw et al. 2016; Samad et al. 2017). The pioneer scientific report providing a direct connection between miRNA levels and stress responses in plants was the miR398 which cleaves Cu/Zn superoxide dismutase coding genes (*CSD1* and *CSD2*) and miR395 targeting sulfate transporter (AST68) and miR399 which targets phosphate transporter (PHO1) (Sunkar and Zhu 2004). Table 33.1 summarizes the key findings vis-à-vis role of miRNAs in response to abiotic stress of several plants.

4.1 Drought Stress Responsive miRNAs

In maize, zma-miR159 is highly conserved. The zma-miR159a upregulates after initial drought and regulates 2 MYB gene expression. The zma-miR159 downregulated in later drought stages and targets transcription factors or DNA binding proteins. The miR319 initially up-regulates differentially and down-regulate in later stage. The zma-R319 family regulates expression and targets transcription factors of Teosinte branched1/cycloidea/proliferating cell factor1 (TCP) and enhances drought tolerance in *Zea mays* (Li et al. 2013; López-Galiano et al. 2019). The miR159 downregulates during drought in tomato. The miR160, miR164, miR166, miR169, miR529, and miR2275 respond finely in drought conditions (Upadhyay et al. 2019). The miR164 acts as an important regulator of various developmental processes in maize and down-regulates and cleaves NAC in drought conditions (Zhang et al. 2014). The miR165/166 negatively control drought response and regulate growth and development by targeting HD-ZIP III. The miR166 up-regulates during drought by targeting HD-ZIP III (Hawker and Bowman 2004). The miR165/166 downregulation enhances tolerance to water deficiency in Arabidopsis and *Oryza sativa* by HD-ZIP III–mediated increase in ABA levels (Yan et al. 2016; Zhang et al. 2018a, b). In Arabidopsis, miR167a shows downregulation and targets IAA-ALA RESISTANT 3 (Kinoshita et al. 2012). The miR169 down-regulates in maize and Arabidopsis and induces overexpression of Nuclear Transcription Factor Y Subunit

Table 33.1 The miRNAs and their targets under various abiotic stresses in plants

Crop	Stress type	microRNA	Target genes	Reference/s
Maize (*Zea mays*)	Drought	miR159	MYB, transcription factor GAMYB/serine/threonine protein phosphatase	Li et al. (2013), Aravind et al. (2017), Wei et al. (2009)
		miR156	tpa: squamosa promoter-binding (sbp domain) transcription factor/DNA binding protein	Aravind et al. (2017), Li et al. (2013)
		miR160	gdsl esterase lipase at5g45910-like/ 40S ribosomal protein S16	Aravind et al. (2017), Wei et al. (2009)
		miR168	Serine/threonine-protein phosphatase	Liu et al. (2019), Wei et al. (2009)
		miR390	Activator of 90 kda heat shock protein ATPase	Aravind et al. (2017)
		miR399	60s ribosomal protein l7a-like	Reyes and Chua (2007), Liu et al. (2019), Aravind et al. (2017)
		miR393	MYB101, MYB33	Reyes and Chua (2007)
		miR396	DEAD-box ATP-dependent RNA helicase 3	Wei et al. (2009)
		miR389a	MYB101, MYB33	Wei et al. (2009)
		miR474	PHD	Wei et al. (2009)
		miR402	MYB101, MYB33	Reyes and Chua (2007)
		miR408	Leucine-rich repeat family protein	Liu et al. (2019), Wei et al. (2009)
		miR167	PLD	Wei et al. (2009)
		miR528	POD	Wei et al. (2009)
		miR397b	MYB101, MYB33	Reyes and Chua (2007)
		miR160	ARF10, gdsl esterase lipase at5g45910-like	Ding et al. (2013), Liu et al. (2007), Upadhyay et al. (2019)
		miR164	NAC, psbp domain-containing protein chloroplastic-like	Zhang et al. (2014), Upadhyay et al. (2019)
		miR166	HD-ZIP III, tpa: homeobox lipid-binding domain family protein; Protein methyltransferase	Hawker and Bowman (2004), Upadhyay et al. (2019), Li et al. (2013)
		miR169	nfya5; Nuclear transcription factor y subunit a-3	Ding et al. (2013), Upadhyay et al. (2019)
		miR529	SPL, tpa: squamosa promoter-binding (sbp domain)	Zhou et al. (2010), Upadhyay et al. (2019)

(continued)

Table 33.1 (continued)

Crop	Stress type	microRNA	Target genes	Reference/s
	Salinity	miR156	SPLs	Ding et al. (2009)
		miR162	DCL1	Ding et al. (2009)
		miR164	Sucrose-phosphatase 1 (SPP1); NAC1	Fu et al. (2017), Ding et al. (2009)
		miR167	Translation initiation factor IF-1/ p5cs isoform 1/ glutathione peroxidase, DNA (cytosine-5-)-methyltransferase	Fu et al. (2017)
		miR168	AGO1	Ding et al. (2009)
		miR398	Cu/Zn-SODs	Li et al. (2022), Zhu et al. (2011)
		miR399	HASTY, granule-bound starch synthase	Ding et al. (2009)
	Submergence	miR159	threonine protein phosphatase, GAMYB	Zhang et al. (2008)
		miR160	Alpha/beta fold family protein	Zhang et al. (2008)
		miR162	DICER-LIKE1	Zhang et al. (2008)
		miR166	Rolledleaf1, HD-ZIP	Zhang et al. (2008)
		miR167	Auxin response factor	Zhang et al. (2008)
		miR171	Scl1, WRKY	Zhang et al. (2008)
		miR319	Peptide transporter	Zhang et al. (2008)
		miR399	Aminotransferase, Granule-bound starch synthase	Zhang et al. (2008)

Rice (*Oryza sativa*)	Drought	miR159	MYB and TCP transcription factors	Jones-Rhoades and Bartel (2004)
		miR167	Auxin response factor (ARF)	Wei et al. (2009), Liu et al. (2009)
		miR168	AGO1	Wen et al. (2016)
		miR169	CBF/DREBs transcription factors	Zhao et al. (2007)
		miR170	SCL transcription factor	Zhou et al. (2010)
		miR171	GRAS transcription factors	Liu et al. (2008)
		miR393	Auxin receptors TIR1, AFB2, AFB3	Xia et al. (2012)
		miR396	GRF	Mohsenifard et al. (2017)
		miR397	β-fructofuranosidase, Laccases	Cai et al. (2006), Zhou et al. (2010)
		miR474	Kinesin, a pentatricopeptide repeat (PPR) family protein	Zhou et al. (2010)
		miR2275		Barrera-Figueroa et al. (2012), Barakat et al. (2012)
	Salinity	miR16	Germin-like protein; Ethylene-insensitive3 (EIN3)- Like 1 protein	Sanan-Mishra et al. (2009)
		miR29	Strictosidine synthase precursor	Sanan-Mishra et al. (2009)
		miR164	CUC2 No apical meristem (NAM) protein; NAC domain-containing protein; Helicase	Sharma et al. (2015), Macovei and Tuteja (2012)
		miR169	CBF HAP2-like factor	Zhao et al. (2009)
		miR393	Phytosulfokine receptor precursor; GRF-interacting factor (GIF)	Gao et al. (2011)
		miR394	F-box protein	Barrera-Figueroa et al. (2012)
		miR408	S-receptor kinase-like; DEAD-box ATP-dependent RNA helicase	Macovei and Tuteja (2012)
		miR414	Helicase	Macovei and Tuteja (2012)
		miR1866	OsWRKY34; cytochrome P450 72A1	Barrera-Figueroa et al. (2012)
		miR1867	Putative protein; DUF1242 superfamily	Barrera-Figueroa et al. (2012)

(continued)

Table 33.1 (continued)

Crop	Stress type	microRNA	Target genes	Reference/s
	Cold	miR156k	OsP5CS Os01g22249 SPL3, SPL14, and SPL17	Cui et al. (2015), Wang et al. (2014a, b)
		miR164	NAC plant-specific transcription factors	Zhang (2015)
		miR168	Argonaute (AGO) proteins	Zhang (2015)
		miR394	F-box domain containing protein	Zhang (2015), Barrera-Figueroa et al. (2012)
		miR529	SBP-box gene family	Barrera-Figueroa et al. (2012)
		miR2871	GT family protein	Barrera-Figueroa et al. (2012)
	Oxidative stress	miR529	OsSPL2, OsSPL14	Yue et al. (2017)
	Phosphate Starvation	multiple miRNAs		Secco et al. (2013)
	Nitrogen starvation	miR399 miR530		Cai et al. (2012)
		miR156		Nischal et al. (2012)
Wheat (*Triticum aestivum*)	Drought	miR156	Squamosa-promoter binding protein (SBP)-like transcription factors	Ma et al. (2015b)
		miR159	MYB transcription factor	Ma et al. (2015b)
		miR166	Homeodomain leucine zipper (HD-Zip) transcription factor	Ma et al. (2015b)
		miR167	Auxin response factor	Ma et al. (2015b), Alptekin and Budak (2016)
		miR169	Nuclear transcription factor Y subunit	Zhou et al. (2010)
		miR393	Transport inhibitor response 1	Akdogan et al. (2016), Gupta et al. (2014), Ganie et al. (2016)
		miR395	ATP sulfurylase (Sulfate adenylyltransferase)	Zhou et al. (2010)
		miR398	Copper super oxide dismutase	Kantar et al. (2011), Liu et al. (2015)
		miR528	Superoxide dismutase [Cu-Zn], L-ascorbate oxidase	Wu et al. (2017)
		miR5156	40S ribosomal protein	Pandey et al. (2013)
		miR1117	Receptor-like protein kinase	Pandey et al. (2013)

	Salinity	miR159	phosphoinositide-specific phospholipase C, pyruvate decarboxylase	Wang et al. (2019), Gupta et al. (2014)
		miR167	Putative auxin response transcription factor (ARF6)	Lu et al. (2011a)
		miR174	Scarecrow-like transcription factor 6 (SCL6), RLK4, CRK10 (CYSTEINE-RICH RLK10), kinase	Lu et al. (2011a)
		miR855	Transcriptional activator Myb; transmembrane protein	Gupta et al. (2014), Pandey et al. (2013)
		miR1133	Fca-like protein, SnRK3.23, ATCIPK23, CIPK23 (CBLINTERACTING PROTEIN KINASE 23); kinase, pentatricopeptide (PPR) repeat-containing protein, NB-ARC domain	Pandey et al. (2013), Lu et al. (2011a)
		miR14769	Putative RSH disease resistance-related protein, T-complex protein 1 subunit alpha	Agharbaoui et al. (2015)
		miR19980	Putative membrane-associated protein	Agharbaoui et al. (2015)
		miR20602	Glutathione peroxidase, putative phosphatase phospho1	Agharbaoui et al. (2015)
	Cold	miR1122	Maf-like protein, single-strand DNA-binding protein	Lu et al. (2011a)
		miR5653	Wheat ubiquitin carrier protein	Pandey et al. (2013)

(continued)

Table 33.1 (continued)

Crop	Stress type	microRNA	Target genes	Reference/s
Barley (*Hordeum* L.)	Drought	miR169b		Ferdous et al. (2016)
		Hv-miR827		Ferdous et al. (2017)
Arabidopsis thaliana	Drought	miR165/166	HD-ZIP III	Zhang et al. (2018a, b), Yan et al. (2016)
		miR167	IAA-ALA RESISTANT 3	Kinoshita et al. (2012)
		miR169	NUCLEAR TRANSCRIPTION FACTOR Y SUBUNIT ALPHA 5 (NFYA5)	Li et al. (2008)
		miR397b	Casein kinase II; laccases	Sunkar and Zhu (2004)
		miR402	ROS1-like, a putative DNA glycosylase	Sunkar and Zhu (2004)
		miR167		Wei et al. (2009)
	Salinity	miR393	Transport Inhibitor Response protein 1 (mTIR1)	Chen et al. (2015)
		miR408	LAC	Ma et al. (2015a)
Chickpea (*Cicer arietinum*)	Salinity	miR390	protein kinases and the CZF1 TF	Kohli et al. (2014)
		miR414		Wang et al. (2019)
		miR399		Deng et al. (2015)
Potato (*Solanum tuberosum*)	Osmotic stress	mi164		Zhang et al. (2018a, b)
	Drought	miR172		Yang et al. (2013)
	Heat stress	miR159b-3p		Yang et al. (2019)

Alpha 5 (NFYA5) to increase tolerance to water deficiency. The *nfya5* mutants and miR169 overexpressed plants show hypersensitivity to water deficiency. Thus, NFYA5 leads to response of water deficiency tolerance (Ding et al. 2013). NFYA5 mediates stress-responsive transcriptional cascade (Li et al. 2008). The miR169-GmNFYA3 module in soybean plants has conserved regulatory functions in tolerance to drought (Ni et al. 2013). Plants downregulate miR169 and by other mechanisms induce NFYA5 expression during drought. NFYA5 antisense gene ENHANCING RING FINGER (NERF) in Arabidopsis produces siRNAs with sequences akin to miR169 but cannot target NFYA5. NERF siRNAs avoid miR169-mediated suppression of NFYA5. This mechanism contributes to NFYA5-mediated drought tolerance (Gao et al. 2015). The miR169i and miR169l upregulate and control expression of NFYA5 through translational activation under drought stress in Arabidopsis (Du et al. 2017). The miR529 targets Squamosa Promoter binding protein Like genes and down-regulates under abiotic stress conditions (Zhou et al. 2010). The miR2275 up-regulates during stress conditions in *Prunus persica* and *Oryza sativa* (Barrera-Figueroa et al. 2012; Barakat et al. 2012).

4.2 Heat

During high temperature conditions, Heat Shock Proteins (HSPs) defend proteins from denaturation at the level of cells in plants (Iba 2002). Plants attain tolerance to stress after being exposed to moderate stress (Mittler et al. 2012). The miRNAs crosstalk with heat stress factors. Arabidopsis miR398 plays an important role in heat stress tolerance. The heat shock transcription factors trigger expression of miR398. The expressed miR398 sequentially suppresses target genes, namely, COPPER/ZINC SUPEROXIDE DISMUTASE 1 (CSD1), CSD2, and COPPER CHAPERONE OF CSD. The downregulation of these target genes recovers tolerance to high temperature by upregulation of heat shock transcription factors and Heat Shock Proteins (HSPs). The transgenic plants on expression of miR398-resistant CSD1, CSD2, and COPPER CHAPERONE OF CSD show sensitivity to high temperature and display a sharp decline in heat shock transcription factors (Guan et al. 2013). In *Gossypium hirsutum*, miR160 down-regulation mediates auxin signaling inhibition by cleaving ARF10/16/17 and fertility of anthers turns on under temperature stress in high temperature-tolerant cotton plants. Overexpression of miR160 enhances sensitivity to heat in cotton (Ding et al. 2017). The miR156-SPLs module prims plants to high temperature. In Arabidopsis, repeated heat tempts miR156 to preserve reminiscence of high temperature stress and SPLs act as transcriptional repressors to repress the genes involved (Stief et al. 2014). In sunflower, miR396-HaWRKY6 regulatory module functions in heat defence during development. Transgenics expressing miR396-resistant HaWRKY6 show reduced tolerance to high temperature stress (Giacomelli et al. 2012).

4.3 Cold

In plants for instance sugarcane and *Oryza sativa*, cold triggers conserved miR319. The miR319 positively regulates tolerance to cold stress and downregulates PROLIFERATING CELL NUCLEAR ANTIGEN BINDING FACTOR, TEOSINTE BRANCHED1, and CYCLOIDEA genes in *Saccharum officinarum* and *Oryza sativa* (Thiebaut et al. 2012; Yang et al. 2013). The cold-inducible miR393 acts as positive regulator of tolerance to cold stress in switchgrass. The miR393 targets TIR1/AFB and miR393 overexpression or TIR1/AFB mutation increases cold tolerance in switchgrass (Liu et al. 2017). In Arabidopsis, Brachypodium and popular, miR397 and miR169 shows upregulation during cold stress (Zhang et al. 2009). The miR168 shows up-regulation in popular and Arabidopsis and down-regulation in *Oryza sativa* (Liu et al. 2008; Lv et al. 2010). The role of miRNAs under various abiotic stresses in plants is summarized in Table 33.1.

5 Genetic Engineering of miRNA for Abiotic Stress Tolerance

The plant miRNAs emerged as novel and promising tools in the arena of genetic engineering to improve crop stress tolerance. These may serve as chief candidates to design high yielding and abiotic stress tolerant transgenic crops and miRNA regulatory pathways enable understanding of the stress responses at molecular level (Zhang and Wang 2015; Zhang 2015; Djami-Tchatchou et al. 2017; Zhang and Unver 2018; Xu et al. 2019). The overexpression or knock out of both miRNAs and artificial miRNAs (amiRNAs) from diverse regulatory networks help in designing of smart crop plants tolerant to abiotic stress (Table 33.2) as well as by transgenesis or cisgenesis and intragenesis (Yang et al. 2013; Basso et al. 2019). The miR156 overexpression boosted the high temperature tolerance in transgenic Arabidopsis (Stief et al. 2014); whereas miR159 overexpression increased heat tolerance in *Oryza sativa* (Wang et al. 2012). In transgenic *Solanum lycopersicum*, miR169 overexpression causes tolerance to water deficiency by reducing stomatal aperture index, preventing opening of stomata, decline in transpiration rate and water loss (Zhang et al. 2011). The miR169 overexpression generates hypersensitivity to nitrogen starvation in Arabidopsis and enhanced tolerance to cold stress in *Oryza sativa* (Zhao et al. 2011; Yang et al. 2013). Moreover, miR169 overexpression enhanced water deficiency and salinity tolerance in *Agrotis stolonifera* (Zhou et al. 2013). Modified miR319 increases the tolerance to multiple stress scenarios in plants such as in *Oryza sativa*, over-expressed miR319 increases cold tolerance and improved leaf morphology and in bent grass, this miRNA increases tolerance to water scarcity (Zhou et al. 2013; Yang et al. 2013). The miR393 and miR396 mediate multiple stress tolerance in *Oryza sativa* (Gao et al. 2010, 2011; Zhao et al. 2018). Tolerance to salt stress in *Oryza sativa* increased by decline in expression of osa-MIR396c

Table 33.2 Genetic engineering of miRNA for abiotic stress tolerance

miRNAs	Transgenics	Effect in plant modified	Reference/s
miR156	*Panicum virgatum*	Increase of biomass	Fu et al. (2012)
miR169	*Arabidopsis thaliana*	Sensitivity to nitrogen deficiency, tolerance to UV radiation	Zhao et al. (2011), Jia et al. (2009)
	Solanum lycopersicum	Decline in rate of transpiration	Zhang et al. (2011)
miR319	*Oryza sativa*	Tolerance to cold, increase in integrity of cell membrane	Yang et al. (2013)
miR390	*Oryza sativa*	Susceptibility to cadmium	Ding et al. (2016)
miR394	*Arabidopsis thaliana*	Restricted water loss	Ni et al. (2012)
miR395	*Brassica napus*	Increased tolerance to oxidative and heavy metal stress	Zhang et al. (2013)
miR396	*Oryza sativa*	Gain in size and yield of grains	Li and Zhang (2016)
miR397	*Oryza sativa*	Promotes panicle branching	Zhang et al. (2013)
miR398	*Arabidopsis thaliana*	Rise in tolerance to oxidative stress	Sunkar et al. (2006)
	Brassica	Regulator of copper homeostasis	Marschner (1995)
miR399	*Solanum lycopersicum*	Tolerance to cold stress and phosphorus deficiency	Goa et al. (2015)
miR482	*Glycine max*	Increase in number of nodules	Li et al. (2010)
miR828	*Ipomoea batatas*	Oxidative stress	Lin et al. (2012)

(Gao et al. 2010). Moreover, transgenic *Oryza sativa* having overexpressed miR398-resistant miRNA with more Cu- or Zn superoxide dismutases enzyme showed enhanced tolerance to water and salt stresses (Lu et al. 2011b). The transgenic *Lycopersicon esculentum* holds extra water inside the cells and needs reduced amount of water from soil (Zhang et al. 2011). In sweet potato, overexpression of miR828 makes plant tolerant to oxidative stress by increasing biosynthesis of lignin and production of hydrogen peroxide (Lin et al. 2012). The miR1514 and miR1509 were edited by CRISPR/Cas 9 in *Glycine max* (Jacobs et al. 2015).

6 Conclusion

The majority of microRNAs are species- or family-specific and control biological processes at the post-transcriptional levels by cleavage or translational inhibition of target mRNAs. The stress-responsive miRNAs modify abiotic stress responsive genes and reprogramme their expression at downstream levels to reduce growth and development in plants. The miR156, miR169, miR319, miR395, miR828, miR394, miR398, miR390, miR396, miR397, miR399, miR482, etc. have been engineered for abiotic stress tolerance in plant kingdom. The miRNAs present a new ray of hope in genome editing toolkit of plants.

Acknowledgment DP acknowledges Dr. Saurabh Raghuvanshi, Department of Plant Molecular Biology, Delhi University, New Delhi, India, for acquainting her to plant microRNAs.

References

Agharbaoui Z, Leclercq M, Remita MA (2015) An integrative approach to identify hexaploid wheat miRNAome associated with development and tolerance to abiotic stress. BMC Genomics 16:339

Akdogan G, Tufekci ED, Uranbey S, Unver T (2016) miRNA-based drought regulation in wheat. Funct Integr Genomics 16:221–233

Alptekin B, Budak H (2016) Wheat miRNA ancestors: evident by transcriptome analysis of A, B, and D genome donors. Funct Integr Genomics 17(2–3):171–187. https://doi.org/10.1007/s10142-016-0487-y

Aravind J, Rinku S, Pooja B, Shikha M, Kaliyugam S, Mallikarjuna MG, Kumar A, Rao AR, Nepolean T (2017) Identification, characterization, and functional validation of drought-responsive MicroRNAs in subtropical maize inbreds. Front Plant Sci 8:941

Arribas-Hernández L, Marchais A, Poulsen C, Haase B, Hauptmann J, Benes V, Meister G, Brodersen P (2016) The slicer activity of ARGONAUTE1 is required specifically for the phasing, not production, of trans-acting short interfering RNAs in *Arabidopsis*. Plant Cell 28:1563–1580

Axtell MJ, Bartel DP (2005) Antiquity of microRNAs and their targets in land-plants. Plant Cell 17:1658–1673

Barakat A, Aditya S, Joseph P, Tetyana Z, Dorrie M, Albert A (2012) Genome wide identification of chilling responsive microRNAs in *Prunus persica*. BMC Genomics 13:481. https://doi.org/10.1186/1471-2164-13-481

Barrera-Figueroa BE, Gao L, Wu ZG, Zhou XF, Zhu JH, Jin HL, Liu R, Zhu JK (2012) High throughput sequencing reveals novel and abiotic stress-regulated microRNAs in the inflorescences of *Oryza sativa*. BMC Plant Biol 12:132. https://doi.org/10.1186/1471-2229-12-132

Bartel DP (2004) MicroRNAs: genomics, biogenesis, mechanism, and function. Cell 116:281–297

Basso MF, Ferreira PCG, Kobayashi AK, Harmon FG, Nepomuceno AL, Molinari HBC, Grossi-de-Sa MF (2019) Micro RNAs and new biotechnological tools for its modulation and improving stress tolerance in plants. Plant Biotechnol J 17(8):1482–1500

Baumberger N, Baulcombe DC (2005) Arabidopsis ARGONAUTE1 is an RNA Slicer that selectively recruits microRNAs and short interfering RNAs. Proc Natl Acad Sci U S A 102:11928–11933

Bologna NG, Voinnet O (2014) The diversity, biogenesis, and activities of endogenous silencing small RNAs in Arabidopsis. Annu Rev Plant Biol 65:473–503

Borges F, Martienssen RA (2015) The expanding world of small RNAs in plants. Nat Rev Mol Cell Biol 16:727–741

Cai X, Davis EJ, Ballif J, Liang M, Bushman E, Haroldsen V et al (2006) Mutant identification and characterization of the laccase gene family in Arabidopsis. J Exp Bot 57(11):2563–2569

Cai H, Lu Y, Xie W, Zhu T, Lian X (2012) Transcriptome response to nitrogen starvation in *Oryza sativa*. J Biosci 37:731–747

Chaudhary S, Grover A, Sharma PC (2021) MicroRNAs: potential targets for developing stress-tolerant crops. Life 11:289. https://doi.org/10.3390/life11040289

Chen X (2009) Small RNAs and their roles in plant development. Annu Rev Cell Dev Biol 25:21–44

Chen Z, Hu L, Han N, Hu J, Yang Y et al (2015) Overexpression of a miR393-resistant form of transport inhibitor response protein 1 (mTIR1) enhances salt tolerance by increased osmoregulation and Na^+ exclusion in *Arabidopsis thaliana*. Plant Cell Physiol 56:73–83

Cui N, Sun X, Sun M, Jia B, Duanmu H, Lv D, Duan X, Zhu Y (2015) Overexpression of OsmiR156k leads to reduced tolerance to cold stress in rice (*Oryza sativa*). Mol Breed 35:214
Cui J, You C, Chen X (2016) The evolution of microRNAs in plants. Curr Opin Plant Biol 35:61–67
Cuperus JT, Fahlgren N, Carrington JC (2011) Evolution and functional diversification of MIRNA genes. Plant Cell 23:431–442
Deng P, Wang L, Cui L, Feng K, Liu F, Du X, Tong W, Nie X, Ji W, Weining S (2015) Global identification of MicroRNAs and their targets in barley under salinity stress. PLoS One 10:e0137990
Ding D, Zhang L, Wang H, Liu Z, Zhang Z, Zheng Y (2009) Differential expression of miRNAs in response to salt stress in maize roots. Ann Bot 103:29–38
Ding Y, Tao Y, Zhu C (2013) Emerging roles of microRNAs in the mediation of drought stress response in plants. J Exp Bot 64:1–10. https://doi.org/10.1093/jxb/ert164
Ding Y, Ye Y, Jiang Z, Wang Y, Zhu C (2016) MicroRNA390 is involved in cadmium tolerance and accumulation in *Oryza sativa*. Front Plant Sci 7:235
Ding Y, Ma Y, Liu N, Xu J, Hu Q et al (2017) microRNAs involved in auxin signalling modulate male sterility under high-temperature stress in cotton (*Gossypium hirsutum*). Plant J 91:977–994
Djami-Tchatchou AT, Sanan-Mishra N, Ntushelo K, Dubery IA (2017) Functional roles of microRNAs in agronomically important plants—potential as targets for crop improvement and protection. Front Plant Sci 8:378
Du P, Wu J, Zhang J, Zhao S, Zheng H, Gao G, Wei L, Li Y (2011) Viral infection induces expression of novel phased MicroRNAs from conserved cellular MicroRNA precursors. PLoS Pathog 7:e1002176
Du Q, Zhao M, Gao W, Sun S, Li WX (2017) microRNA/microRNA∗ complementarity is important for the regulation pattern of NFYA5 by miR169 under dehydration shock in *Arabidopsis*. Plant J 91:22–33
Ferdous J, Hussain SS, Shi BJ (2015) Role of microRNAs in plant drought tolerance. Plant Biotechnol J 13(3):293–305. https://doi.org/10.1111/pbi.12318
Ferdous J, Sanchez-Ferrero JC, Langridge P, Milne L, Chowdhury J, Brien C, Tricker PJ (2016) Differential expression of microRNAs and potential targets under drought stress in barley. Plant Cell Environ 40:11–24
Ferdous J, Whitford R, Nguyen M, Brien C, Langridge P, Tricker PJ (2017) Drought-inducible expression of Hv-miR827 enhances drought tolerance in transgenic barley. Funct Integr Genomics 17:279–292
Fu C, Sunkar R, Zhou C et al (2012) Overexpression of miR156 in switchgrass (*Panicum virgatum* L.) results in various morphological alterations and leads to improved biomass production. Plant Biotechnol J 10:443–452
Fu R, Zhang M, Zhao Y, He X, Ding C, Wang S, Feng Y, Song X, Li P, Wang B (2017) Identification of salt tolerance-related microRNAs and their targets in maize (*Zea mays* L.) using high-throughput sequencing and degradome analysis. Front Plant Sci 8:864
Fukudome A, Fukuhara T (2017) Plant dicer-like proteins: double-stranded RNA-cleaving enzymes for small RNA biogenesis. J Plant Res 130:33–44
Ganie SA, Dey N, Mondal TK (2016) Promoter methylation regulates the abundance of Osa-miR393a in contrasting rice genotypes under salinity stress. Funct Integr Genomics 16:1–11
Gao P, Bai X, Yang L, Lv D, Li Y, Cai H, Ji W, Guo D, Zhu Y (2010) Over-expression of osa-MIR396c decreases salt and alkali stress tolerance. Planta 231:991–1001
Gao P, Bai X, Yang L, Lv D, Pan X, Li Y, Cai H, Ji W, Chen Q, Zhu Y (2011) Osa- MIR393: a salinity-and alkaline stress-related microRNA gene. Mol Biol Rep 38:237–242
Gao W, Liu W, Zhao M, Li WX (2015) NERF encodes a RING E3 ligase important for drought resistance and enhances the expression of its antisense gene *NFYA5* in Arabidopsis. Nucleic Acids Res 43:607
Giacomelli JI, Weigel D, Chan RL, Manavella PA (2012) Role of recently evolved miRNA regulation of sunflower HaWRKY6 in response to temperature damage. New Phytol 195:766–773
Guan Q, Lu X, Zeng H, Zhang Y, Zhu J (2013) Heat stress induction of miR398 triggers a regulatory loop that is critical for thermotolerance in Arabidopsis. Plant J 74:840–851

Guleria P, Mahajan M, Bhardwaj J, Yadav SK (2011) Plant small RNAs: biogenesis, mode of action and their roles in abiotic stresses. Genom Proteom Bioinform 9:183–199

Gupta OP, Meena NL, Sharma I, Sharma P (2014) Differential regulation of microRNAs in response to osmotic, salt and cold stresses in wheat. Mol Biol Rep 41:4623–4629

Hawker NP, Bowman JL (2004) Roles of class III HD-ZIP and KANAD1 genes in Arabidopsis root development. Plant Physiol 135:2261–2270. https://doi.org/10.1104/pp.104.040196

Iba K (2002) Acclimative response to temperature stress in higher plants: approaches of gene engineering for temperature tolerance. Annu Rev Plant Biol 53:225–245

Iki T (2017) Messages on small RNA duplexes in plants. J Plant Res 130:7–16

Iwakawa HO, Tomari Y (2013) Molecular insights into microRNA-mediated translational repression in plants. Mol Cell 52(4):591–601

Jacobs TB, Lafayette PR, Schmitz RJ, Parrott WA (2015) Targeted genome modifications in soybean with CRISPR/Cas9. BMC Biotechnol 15:1–10

Jia X, Ren L, Chen QJ, Li R, Tang G (2009) UV-B-responsive microRNAs in *Populus tremula*. J Plant Physiol 166(18):2046–2057

Jones-Rhoades MW, Bartel DP (2004) Computational identification of plant microRNAs and their targets, including a stress-induced miRNA. Mol Cell 14:787–799

Jones-Rhoades MW, Bartel DP, Bartel B (2006) MicroRNAs and their regulatory roles in plants. Annu Rev Plant Biol 57:19–53

Kantar M, Lucas SJ, Budak H (2011) miRNA expression patterns of *Triticum dicoccoides* in response to shock drought stress. Planta 233:471–484

Khraiwesh B, Zhu JK, Zhu J (2012) Role of miRNAs and siRNAs in biotic and abiotic stress responses of plants. Biochim Biophys Acta 1819:137–148

Kinoshita N, Wang H, Kasahara H, Liu J, MacPherson C et al (2012) IAA-Ala Resistant 3, an evolutionarily conserved target of miR167, mediates Arabidopsis root architecture changes during high osmotic stress. Plant Cell 24:3590–3602

Kohli D, Joshi G, Deokar AA, Bhardwaj AR, Agarwal M, Katiyar-Agarwal S, Srinivasan R, Jain PK (2014) Identification and characterization of wilt and salt stress-responsive MicroRNAs in chickpea through high-throughput sequencing. PLoS One 9:e108851

Kumar R (2014) Role of microRNAs in biotic and abiotic stress responses in crop plants. Appl Biochem Biotechnol 174:93–115

Li C, Zhang B (2016) MicroRNAs in control of plant development. J Cell Physiol 231(2):303–313

Li WX, Oono Y, Zhu J, He XJ, Wu JM et al (2008) The Arabidopsis NFYA5 transcription factor is regulated transcriptionally and posttranscriptionally to promote drought resistance. Plant Cell 20:2238–2251

Li H, Deng Y, Wu T, Subramanian S, Yu O (2010) Misexpression of miR482, miR1512, and miR1515 increases soybean nodulation. Plant Physiol 153(4):1759–1770

Li JS, Fu FL, An M, Zhou SF, She YH, Li WC (2013) Differential expression of microRNAs in response to drought stress in maize. J Integr Agric 12:1414–1422. https://doi.org/10.1016/S2095-3119(13)60311-1

Li J, Song Q, Zuo ZF, Liu L (2022) MicroRNA398: a master regulator of plant development and stress responses. Int J Mol Sci 23:10803. https://doi.org/10.3390/ijms231810803

Lin JS, Lin CC, Lin HH, Chen YC, Jeng ST (2012) Micro R 828 regulates lignin and H_2O_2 accumulation in sweet potato on wounding. New Phytol 196:427–440

Liu HH, Tian X, Li YJ, Wu CA, Zheng CC (2008) Microarray-based analysis of stress regulated microRNAs in *Arabidopsis thaliana*. RNA 14:836–843

Liu Q, Zhang YC, Wang CY, Luo YC, Huang QJ, Chen SY, Zhou QLH, Chen YQ (2009) Expression analysis of phytohormone-regulated microRNAs in rice, implying their regulation roles in plant hormone signaling. FEBS Lett 583:723–728

Liu H, Searle IR, Watson-Haigh NS et al (2015) Genome-wide identification of microRNAs in leaves and the developing head of four durum genotypes during water deficit stress. PLoS One 10:e0142799

Liu Y, Wang K, Li D, Yan J, Zhang W (2017) Enhanced cold tolerance and tillering in switchgrass (*Panicum virgatum* L.) by heterologous expression of Osa-miR393a. Plant Cell Physiol 58:2226–2240

Liu X, Zhang X, Sun B, Hao L, Liu C, Zhang D, Tang H, Li C, Li Y, Shi Y, Xie X, Song Y, Wang T, Li Y (2019) Genome-wide identification and comparative analysis of drought-related microRNAs in two maize inbred lines with contrasting drought tolerance by deep sequencing. PLoS One 14(7):e0219176. https://doi.org/10.1371/journal.pone.0219176

Liu PP, Montgomery T, Fahlgren N, Kasschau KD, Nonogaki H, Carrington JC, (2007) Repression of AUXIN RESPONSE FACTOR10 by microRNA160 is critical for seed germination and post-germination stages. Plant J 52:133–146. https://doi.org/10.1111/j.1365-313X.2007.03218.x

López-Galiano MJ, García-Robles I, González-Hernández AI, Camañes G, Vicedo B, Real MD, Rausell C (2019) Expression of miR159 is altered in tomato plants undergoing drought stress. Plan Theory 8:201

Lu W, Li J, Liu F et al (2011a) Expression pattern of wheat miRNAs under salinity stress and prediction of salt-inducible miRNAs targets. Front Agric China 5:413–422

Lu YZ, Feng Z, Bian LY, Xie H, Liang JS (2011b) miR398 regulation in *Oryza sativa* of the responses to abiotic and biotic stresses depends on CSD1 and CSD2 expression. Funct Plant Biol 38:44–53

Lv DK, Bai X, Li Y, Ding XD, Ge Y, Cai H, Ji W, Wu N, Zhu YM (2010) Profiling of cold-stress-responsive miRNAs in *Oryza sativa* by microarrays. Gene 459:39–47

Ma C, Burd S, Lers A (2015a) miR408 is involved in abiotic stress responses in Arabidopsis. Plant J 84:169–187

Ma X, Xin Z, Wang Z et al (2015b) Identification and comparative analysis of differentially expressed miRNAs in leaves of two wheat (*Triticum aestivum* L.) genotypes during dehydration stress. BMC Plant Biol 15:21

Macovei A, Tuteja N (2012) microRNAs targeting DEAD-box helicases are involved in salinity stress response in *Oryza sativa* L. BMC Plant Biol 12:183

Marschner H (1995) Mineral nutrition of higher plants second edition. Academic Press Edition, London

Megraw M, Cumbie J, Ivanchenko M, Filichkin S (2016) Small genetic circuits and MicroRNAs: big players in polymerase II transcriptional control in plants. Plant Cell 28:286–303

Mittal D, Sharma N, Sharma V, Sopory SK, Sanan-Mishra N (2016) Role of microRNAs in *Oryza sativa* plant under salt stress. Ann Appl Biol 168(1):2–18

Mittler R, Finka A, Goloubinoff P (2012) How do plants feel the heat? Trends Biochem Sci 37:118–125

Mohsenifard E, Ghabooli M, Mehri N, Bakhshi B (2017) Regulation of miR159 and miR396 mediated by *Piriformospora indica* confer drought tolerance in rice. J Plant Mol Breed 5:10–18

Ni Z, Hu Z, Jiang Q, Zhang H (2012) Overexpression of gma-MIR394a confers tolerance to drought in transgenic *Arabidopsis thaliana*. Biochem Biophys Res Commun 427(2):330–335

Ni Z, Hu Z, Jiang Q, Zhang H (2013) GmNFYA3, a target gene of *miR169*, is a positive regulator of plant tolerance to drought stress. Plant Mol Biol 82:113–129

Nischal L, Mohsin M, Khan I, Kardam H, Wadhwa A, Abrol YP, Iqbal M, Ahmad A (2012) Identification and comparative analysis of MicroRNAs associated with low-N tolerance in *Oryza sativa* genotypes. PLoS One 7:e50261

Noman A, Fahad S, Aqeel M, Ali U, Amanullah AS, Baloch SK, Zainab M (2017) miRNAs: major modulators for crop growth and development under abiotic stresses. Biotechnol Lett 39:685–700

Pandey B, Gupta OP, Pandey DM, Sharma I, Sharma P (2013) Identification of new stress-induced microRNA and their targets in wheat using computational approach. Plant Signal Behav 8(5):e23932

Pandita D (2019) Plant MIRnome: miRNA biogenesis and abiotic stress response. In: Hasanuzzaman M, Hakeem K, Nahar K, Alharby H (eds) Plant abiotic stress tolerance. Springer, Cham. https://doi.org/10.1007/978-3-030-06118-0_18

Pandita D (2021) Role of miRNAi technology and miRNAs in abiotic and biotic stress resilience. In: Aftab T, Roychoudhury A (eds) Plant perspectives to global climate changes. https://doi.org/10.1016/B978-0-323-85665-2.00015-7

Pandita D (2022a) How microRNAs regulate abiotic stress tolerance in wheat? A snapshot. In: Roychoudhury A, Aftab T, Acharya K (eds) Omics approach to manage abiotic stress in cereals. Springer, Singapore. https://doi.org/10.1007/978-981-19-0140-9_17
Pandita D (2022b) MicroRNAs shape the tolerance mechanisms against abiotic stress in maize. In: Roychoudhury A, Aftab T, Acharya K (eds) Omics approach to manage abiotic stress in cereals. Springer, Singapore. https://doi.org/10.1007/978-981-19-0140-9_19
Pandita D (2022c) miRNA- and RNAi-mediated metabolic engineering in plants. In: Aftab T, Hakeem KR (eds) Metabolic engineering in plants. Springer, Singapore. https://doi.org/10.1007/978-981-16-7262-0_7
Pandita D, Wani SH (2019) MicroRNA as a tool for mitigating abiotic stress in *Oryza sativa* (*Oryza sativa* L.). In: Wani SH (ed) Recent approaches in omics for plant resilience to climate change. Springer, Cham. https://doi.org/10.1007/978-3-030-21687-0_6
Pantaleo V, Szittya G, Moxon S, Miozzi L, Moulton V, Dalmay T (2010) Identification of grapevine microRNAs and their targets using high-throughput sequencing and degradome analysis. Plant J 62:960–976
Park W, Li J, Song R, Messing J, Chen X (2002) CARPEL FACTORY, a dicer homolog, and HEN1, a novel protein, act in microRNA metabolism in *Arabidopsis thaliana*. Curr Biol 12:1484–1495
Pelaez P, Trejo MS, Iñiguez LP, Estrada-Navarrete G, Covarrubias AA, Reyes JL (2012) Identification and characterization of microRNAs in *Phaseolus vulgaris* by high-throughput sequencing. BMC Genomics 13:1–8
Ramachandran V, Chen X (2008) Small RNA metabolism in Arabidopsis. Trends Plant Sci 13:368–374
Reis RS, Eamens AL, Waterhouse PM (2015) Missing pieces in the puzzle of plant MicroRNAs. Trends Plant Sci 20:721–728
Reyes JL, Chua NH (2007) ABA induction of miR159 controls transcript levels of two MYB factors during *Arabidopsis* seed germination. Plant J 49:592–606. https://doi.org/10.1111/j.1365-313X.2006.02980
Salvador-Guirao R, Baldrich P, Weigel D, Rubio-Somoza I, Segundo BS (2018) The MicroRNA miR773 is involved in the Arabidopsis immune response to fungal pathogens. Mol Plant Microbe Interact 31:249–259
Samad A, Sajad M, Nazaruddin N, Fauzi I, Murad A, Zainal Z, Ismanizan Ismail I (2017) MicroRNA and transcription factor: key players in plant regulatory network. Front Plant Sci 8:565
Sanan-Mishra N, Kumar V, Sopory SK, Mukherjee SK (2009) Cloning and validation of novel miRNA from basmati rice indicates cross talk between abiotic and biotic stresses. Mol Gen Genomics 282:463–474
Secco D, Jabnoune M, Walker H, Shou H, Wu P, Poirier Y, Whelan J (2013) Spatio-temporal transcript profiling of *Oryza sativa* roots and shoots in response to phosphate starvation and recovery. Plant Cell 25:4285–4304
Sharma N, Tripathi A, Sanan-Mishra N (2015) Profiling the expression domains of a *Oryza sativa*-specific microRNA under stress. Front Plant Sci 6:333
Shriram V, Kumar V, Devarumath RM, Khare TS, Wani SH (2016) MicroRNAs as potential targets for abiotic stress tolerance in plants. Front Plant Sci 7:817
Stief A, Altmann S, Hoffmann K, Pant BD, Scheible WR, Bäurle I (2014) Arabidopsis miR156 regulates tolerance to recurring environmental stress through SPL transcription factors. Plant Cell 26:1792–1807
Sun G (2011) MicroRNAs and their diverse functions in plants. Plant Mol Biol 80:17–36
Sun G, Stewart CN, Xiao P, Zhang B (2012) MicroRNA expression analysis in the cellulosic biofuel crop Switchgrass (*Panicum virgatum*) under abiotic stress. PLoS One 7:1–7
Sun X, Lin L, Sui N (2019) Regulation mechanism of microRNA in plant response to abiotic stress and breeding. Mol Biol Rep 46:1447–1457
Sunkar R, Zhu JK (2004) Novel and stress-regulated MicroRNAs and other small RNAs from *Arabidopsis*. Plant Cell 16:2001–2019. https://doi.org/10.1105/tpc.104.022830

Sunkar R, Kapoor A, Zhu JK (2006) Posttranscriptional induction of two cu/Zn superoxide dismutase genes in *Arabidopsis* is mediated by downregulation of miR398 and important for oxidative stress tolerance. Plant Cell 18(8):2051–2065

Sunkar R, Chinnusamy V, Zhu J, Zhu JH (2007) Small RNAs as big players in plant abiotic stress responses and nutrient deprivation. Trends Plant Sci 12:301–309

Sunkar R, Li YF, Jagadeeswaran G (2012) Functions of microRNAs in plant stress responses. Trends Plant Sci 17:196–203

Thiebaut F, Rojas CA, Almeida KL, Grativol C, Domiciano GC et al (2012) Regulation of miR319 during cold stress in sugarcane. Plant Cell Environ 35:502–512

Upadhyay U, Singh P, Verma OP (2019) Role of microRNAs in regulating drought stress tolerance in maize. J Pharmacogn Phytochem 8:328–331

Voinnet O (2009) Origin, biogenesis, and activity of plant microRNAs. Cell 136(4):669–687

Wang Y, Sun F, Cao H, Peng H, Ni Z, Sun Q, Yao Y (2012) TamiR159 directed wheat TaGAMYB cleavage and its involvement in anther development and heat response. PLoS One 7:e48445

Wang ST, Sun XL, Hoshino Y, Yu Y, Jia B, Sun ZW, Sun MZ, Duan XB, Zhu YM (2014a) MicroRNA319 positively regulates cold tolerance by targeting OsPCF6 and OsTCP21 in rice (*Oryza sativa* L.). PLoS One 9(3):e91357

Wang B, Sun YF, Song N et al (2014b) MicroRNAs involving in cold, wounding and salt stresses in *Triticum aestivum* L. Plant Physiol Biochem 80:90–96

Wang W, Liu D, Chen D, Cheng Y, Zhang X, Song L, Hu M, Dong J, Shen F (2019) MicroRNA414c affects salt tolerance of cotton by regulating reactive oxygen species metabolism under salinity stress. RNA Biol 16:362–375

Wei L, Zhang D, Xiang F, Zhang Z (2009) Differentially expressed miRNAs potentially involved in the regulation of defense mechanism to drought stress in maize seedlings. Int J Plant Sci 170:979–989. https://doi.org/10.1086/605122

Wen M, Xie M, He L, Wang Y, Shi S, Tang T (2016) Expression variations of miRNAs and mRNAs in rice (*Oryza sativa*). Genome Biol Evol 8:3529–3544. https://doi.org/10.1093/gbe/evw252

Winter J, Diederichs S (2011) Argonaute proteins regulate microRNA stability: increased microRNA abundance by Argonaute proteins is due to microRNA stabilization. RNA Biol 8(6):1149–1157

Wu J, Yang R, Yang Z, Yao S, Zhao S, Wang Y, Li P, Song X, Jin L, Zhou T et al (2017) ROS accumulation and antiviral defence control by microRNA528 in rice. Nat Plants 3:16203

Xia K, Wang R, Ou X, Fang Z, Tian C, Duan J (2012) OsTIR1 and OsAFB2 downregulation via OsmiR393 overexpression leads to more tillers, early flowering and less tolerance to salt and drought in rice. PLoS One 7:e30039

Xie M, Zhang S, Yu B (2014) microRNA biogenesis, degradation and activity in plants. Cell Mol Life Sci 72:87–99

Xu J, Hou QM, Khare T, Verma SK, Kumar V (2019) Exploring miRNAs for developing climate-resilient crops: a perspective review. Sci Total Environ 653:91–104

Yan J, Zhao C, Zhou J, Yang Y, Wang P et al (2016) The miR165/166 mediated regulatory module plays critical roles in ABA homeostasis and response in *Arabidopsis thaliana*. PLoS Genet 12:e1006416

Yang C, Li D, Mao D, Liu X, Ji C, Li X, Zhao X, Cheng Z, Chen C, Zhu L (2013) Overexpression of microRNA319 impacts leaf morphogenesis and leads to enhanced cold tolerance in *Oryza sativa* (*Oryza sativa* L.). Plant Cell Environ 36:2207–2218

Yang Z, Li W, Su X, Ge P, Zhou Y, Hao Y, Shu H, Gao C, Cheng S, Zhu G et al (2019) Early response of radish to heat stress by strand-specific transcriptome and miRNA analysis. Int J Mol Sci 20:3321

Yu B, Yang Z, Li J, Minakhina S, Yang M, Padgett RW, Steward R, Chen X (2005) Methylation as a crucial step in plant microRNA biogenesis. Science 307:932–935

Yue E, Liu Z, Li C, Li Y, Liu Q, Xu JH (2017) Overexpression of miR529a confers enhanced resistance to oxidative stress in *Oryza sativa* (*Oryza sativa* L.). Plant Cell Rep 36:1171–1182

Zhang B (2015) MicroRNA: a new target for improving plant tolerance to abiotic stress. J Exp Bot 66:1749–1761

Zhang B, Unver T (2018) A critical and speculative review on microRNA technology in crop improvement: current challenges and future directions. Plant Sci 274:193–200
Zhang B, Wang Q (2015) MicroRNA-based biotechnology for plant improvement. J Cell Physiol 230:1–15
Zhang J, Xu Y, Huan Q, Chong K (2009) Deep sequencing of Brachypodium small RNAs at the global genome level identifies microRNAs involved in cold stress response. BMC Genomics 10:449
Zhang Z, Wei L, Zou X, Tao Y, Liu Z, Zheng Y. (2008). Submergence-responsive MicroRNAs are potentially involved in the regulation of morphological and metabolic adaptations in maize root cells. Ann Bot 102(4):509–519. https://doi.org/10.1093/aob/mcn129
Zhang X, Zou Z, Gong P, Zhang J, Ziaf K, Li H, Xiao F, Ye Z (2011) Over-expression of microRNA169 confers enhanced drought tolerance to tomato. Biotechnol Lett 33:403–409
Zhang YC, Yu Y, Wang CY, Li ZY, Liu Q, Xu J et al (2013) Overexpression of microRNA OsmiR397 improves rice yield by increasing grain size and promoting panicle branching. Nat Biotechnol 31(9):848
Zhang JF, Yuan LJ, Shao Y, Du W, Yan DW, Lu YT (2014) The disturbance of small RNA pathways enhanced abscisic acid response and multiple stress responses in Arabidopsis. Plant Cell Environ 31:562–574
Zhang L, Yao L, Zhang N, Yang J, Zhu X, Tang X, Calderón-Urrea A, Si H (2018a) Lateral root development in potato is mediated by Stu-mi164 regulation of NAC transcription factor. Front Plant Sci 9:383
Zhang J, Zhang H, Srivastava AK, Pan Y, Bai J (2018b) Knock-down of *Oryza sativa* microRNA166 confers drought resistance by causing leaf rolling and altering stem xylem development. Plant Physiol 177:1691–1703
Zhao B, Liang R, Ge L, Li W, Xiao H, Lin H, Ruan K, Jin Y (2007) Identification of drought induced microRNAs in rice. Biochem Biophys Res Commun 354:585–590
Zhao B, Ge L, Liang R, Li W, Ruan K, Lin H, Jin Y (2009) Members of miR-169 family are induced by high salinity and transiently inhibit the NF-YA transcription factor. BMC Mol Biol 10:29
Zhao M, Ding H, Zhu JK, Zhang F, Li WX (2011) Involvement of miR169 in the nitrogen-starvation responses in Arabidopsis. New Phytol 190:906–915
Zhao X, Zhang H, Li L (2013) Identification and analysis of the proximal promoters of microRNA genes in Arabidopsis. Genomics 101:187–194. https://doi.org/10.1016/j.ygeno.2012.12.004
Zhao J, Yuan S, Zhou M, Yuan N, Li Z, Hu Q, Bethea FG, Liu H, Li S, Luo H (2018) Transgenic creeping bentgrass overexpressing Osa-miR393a exhibits altered plant development and improved multiple stress tolerance. Plant Biotechnol J 17:233–251
Zhou L, Liu Y, Liu Z, Kong D, Duan M, Luo L (2010) Genome-wide identification and analysis of drought-responsive microRNAs in *Oryza sativa*. J Exp Bot 61:4157–4168. https://doi.org/10.1093/jxb/erq237
Zhou M, Li D, Li Z, Hu Q, Yang C, Zhu L, Luo H (2013) Constitutive expression of a miR319 gene alters plant development and enhances salt and drought tolerance in transgenic creeping bentgrass. Plant Physiol 161:1375–1391
Zhu JK (2016) Abiotic stress signaling and responses in plants. Cell 167:313–324
Zhu C, Ding Y, Liu H (2011) MiR398 and plant stress responses. Physiol Plant 143:1–9. https://doi.org/10.1111/j.1399-3054.2011.01477.x

Chapter 34
Orchestration of Omics Technologies for Crop Improvement

Deepu Pandita

Abstract Our agriculture productivity systems face the grand challenge of global food security for the growing population which will reach 10 billion in the subsequent 30 years. We need a better characterization of the genetic diversity of entire plant germplasms or the allelic diversity in pangenomes by the novel crop improvement approaches such as next-generation sequencing (NGS) and other high-throughput omics technologies. The integration of genomics, pangenomics, transcriptomics, phenomics, proteomics, ionomics, bioinformatics, and metabolomics tools have improved our understanding of the mechanisms underlying the architecture of traits related to agricultural yields, quality and nutritive profile of crops, plant growth, senescence, and epigenetics and strategies to be adapted for crop improvement with desired characteristics. This chapter highlights the wealth of omics data available by the use of recent omics technologies for crop improvement to secure future food supplies for the snowballing global population.

Keywords Bioinformatics · Genomics · Transcriptomics · Proteomics · Metabolomics · Ionomics · Phenomics · Pangenomics

1 Introduction

Agriculture faces myriad challenges due to reduced agricultural areas and extreme environmental stressors and further threat as yield of all key crops such as rice, wheat, and maize are attaining a plateau. To cater to the future food requirements of the increasing global population, which is anticipated to rise up to 10 billion in the succeeding 30 years, global food production needs the support of traits for high and stable yielding, highly resilient crop varieties to extreme biotic and abiotic threats

D. Pandita (✉)
Government Department of School Education, Jammu, Jammu and Kashmir, India

M. Hasanuzzaman (ed.), *Climate-Resilient Agriculture, Vol 2*,
https://doi.org/10.1007/978-3-031-37428-9_34

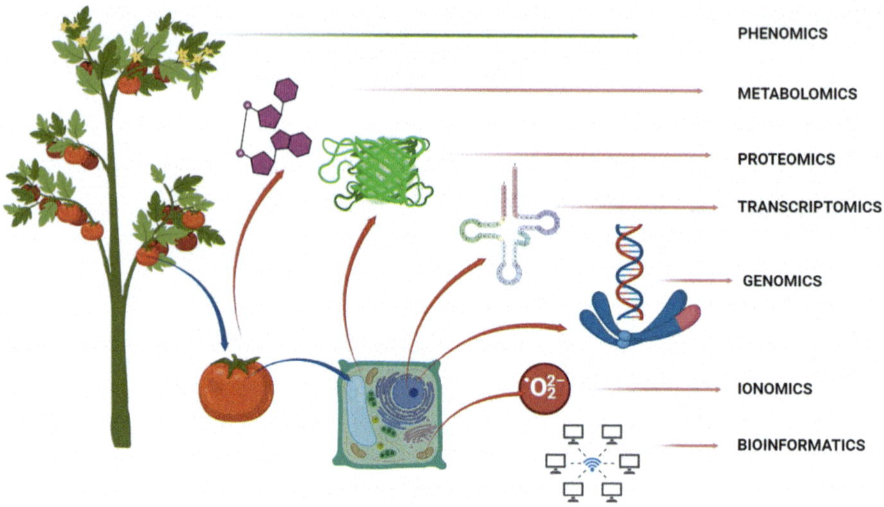

Fig. 34.1 Recent omics platforms

(Voss-Fels et al. 2019), harmless and more nutritious food crops. Approaches to accomplish these include the discovery of novel genetic variation with desired characteristics, which can be knock out or knock in in other transgenic plants, or designing new variations for specific traits in crops by genome modification and efficient transference and expression of these variations into novel elite agricultural cultivars. The omics platforms can allow significant steps for bringing a real revolution in the process of genetic refinement of crop plants by the discovery of novel gene variations for specific traits, extensive phenotyping, and elucidation of growth- and yield-related complex architecture traits, uncovering and decoding critical components of senescence, stress resilience, and metabolic traits of agricultural relevance. The current trend of next-generation sequencing (NGS) of crop plants has provided formidable achievements with a plethora of data elucidating the genes' function and traits, protein networks, and small molecule analysis epigenomic effects due to physiological and environmental stress factors, and transcriptome and proteome details. The various recent omics platforms are illustrated in Fig. 34.1.

2 Genomics

Genomics deals with the study of genes and genomes and focuses on the structure, function, evolution, mapping, epigenomic, muta-genomic, and genome editing aspects (Muthamilarasan et al. 2019) and elucidates genetic variation for subsequent genetic improvement of crops (Tettelin et al. 2005). *Arabidopsis thaliana* is the first plant in which genome sequencing was done by the traditional first-generation sequencing method of Sanger-based dideoxy sequencing in 2000

(Arabidopsis Genome 2000); however, this sequencing method has been exceeded by the NGS, single-molecule real-time (SMRT) sequencing (Kim et al. 2014; Amarasinghe et al. 2020; Dumschott et al. 2020) which have heralded new approaches in investigating plant genomes. The genomics includes multiple disciplines such as structural genomics, functional genomics, epigenomics, mutagenomics, and pangenomics.

2.1 Structural Genomics

Structural genomics depends on molecular markers. Sequence polymorphisms and chromosomal organization can create genetic and physical maps for tagging and mapping genes of interest (Tettelin et al. 2005). Structural genomics utilizes both the PCR-based and non-PCR-based molecular marker techniques. Non-PCR-based marker techniques are restriction fragment length polymorphism (RFLP) and sequence characterized amplified region (SCAR), whereas PCR-based molecular marker techniques are random amplified polymorphic DNA (RAPD), amplified fragments length polymorphism (AFLP), and single nucleotide polymorphisms (SNPs) (Williams et al. 1990; Vos et al. 1995; Rabouam et al. 1999; Agarwal et al. 2008; Appleby et al. 2009). A high-throughput technique known as diversity arrays technology (DArT) is microarray hybridization based and involves genotyping of numerous polymorphic loci (Jaccoud et al. 2001). Genomics-based NGS enables faster identification of SNPs, quantitative trait loci (QTLs), and genome-wide association studies (GWAS). These tools helped in the investigation of multiple traits in crops, genes associated with domestication, effect of abiotic stressors on content of seed oil in *Helianthus annuus* L., QTLs associated with yield (48 in maize) under heat and drought stress, SNPs in *Sorghum bicolor* linked with water deficiency and drought resistance (DR)-related loci in *Oryza sativa*, drought-responsive TFs in *Zea mays*, association of structural variants (SVs) for agronomically essential traits in *Brassica napus*, and few other crops (Lasky et al. 2015; Millet et al. 2016; Shikha et al. 2017; Mangin et al. 2017; Challa and Neelapu 2018; Guo et al. 2018; Gabur et al. 2018). Marker-assisted selection (MAS) with genotyping by sequencing (GBS) improves superiority and yield of crops. Multiparent mapping, like multiparent advanced generation inter-crosses (MAGIC) and nested association mapping (NAM), enables phenotypic diversity studies, ideal for breeding improvement (Yu et al. 2008; Kover et al. 2009; He et al. 2014).

2.2 Functional Genomics

Functional genomics delivers insights into various gene functions supplied by structural genomics and gene regulation of trait of interest for plant biologists by use of experimental methodologies (Tettelin et al. 2005). NGS in crops led to identification of genes associated with the biotic and abiotic stress resistance and yield of plant (Muthamilarasan et al. 2019). Functional genomics uses genome editing tools of mega nucleases, clustered regularly interspaced short palindromic repeats (CRISPR/Cas), zinc finger nucleases (ZFN), transcription activator-like effector nuclease (TALEN), RNA interference (RNAi), and virus-induced gene silencing (VIGS). These tools improved crop yield, pest, and insect disease resilience in genetically modified crops (GMO), e.g., mutants of wheat mildew resistance locus o (TaMlo) and SlMlo tomato, mutants linked with growth and development and stress resilience in various crop plants (Wang et al. 2014; Rinaldo and Ayliffe 2015; Talukdar and Sinjushin 2015; Nekrasov et al. 2017).

2.3 Epigenomics

Epigenomics involves the epigenetic modifications (hereditary modifications except that in DNA) of histone proteins, DNA and/or small RNA (sRNA) by methylation, acetylation, or other post-translational modifications (Tettelin et al. 2005). Epigenomics provides knowledge of how expression of genes and structure of epigenome is affected by the biotic and abiotic stress conditions. Methylation is studied by the tools of bisulfite sequencing, ChIP sequencing, or methylation-sensitive amplified polymorphism (Cokus et al. 2008). The non-coding RNA in plants growing under water deficiency and hypo- and hypermethylation states affecting the fruit yield can be studied (Gelaw and Sanan-Mishra 2021).

2.4 Mutagenomics

Mutagenomics is a recent omics approach which aids in screening of mutational events with specific sequence modifications that gives a mutant superior trait (Tettelin et al. 2005). About 3000 mutants of several plants have been developed with 776 ensuring quality of nutrition (Jain and Suprasanna 2011). The events of mutations may be characterized by use of high-throughput genomics tools along with serial analysis of gene expression (SAGE), high-resolution melt (HRM), targeted induced local lesions in genomes (TILLING), and microarray analysis (Penna and Jain 2017). It is a vital technique to allow analysis of candidate target gene functions by their transient knockdown or silencing and interruption using reverse genetic methods such as RNA interference (RNAi) and virus-induced gene silencing (VIGS) and developing a myriad of agronomic traits with desirable variations. Tools like TILLING are used as a potential alternative to transgenics by enabling high-throughput mutations in crops detection of mutations, screening of mutants with controlled seed oil composition, and improving crop breeding (McCallum et al. 2000; Henikoff et al. 2004; Dong et al. 2009; Varshney et al. 2010; Knoll et al. 2011; Kumar et al. 2013; Mba 2013; Chen et al. 2014; Talukdar and Sinjushin 2015;

Prasad et al. 2021). TILLING technology has been reported in various crop plants, for instance, rice, tomato, soybean, rapeseed, *Triticum aestivum,* and *H. annus* (Kurowska et al. 2011; Witzel et al. 2015). The analyses by microarray technology exhibited mutagenesis of plants to induce more transcriptomic modifications (Varshney et al. 2010).

2.5 *Pangenomics*

Pangenomics is a species' whole genetic makeup or sum of core genome divided into core and accessory/dispensable genes (Tettelin et al. 2005; Lei et al. 2021). Core gene sets are common in all individuals, whereas accessory/dispensable gene sets are specific to an individual or a small subset of the population. Advent of NGS has led to the discovery that accessory genes greatly contribute to crop diversity and quality as they often confer disease resistance (Wang et al. 2021a, b). Iterative assembly, "map-to-pan," and de novo are utilized to build pangenomes and has been used in rice and soybean to compare their genome to their wild relatives (Li et al. 2014). Pangenomes enhance our knowledge of breeding, domestication, polyploidization, and effect of this variable (dispensable) genome on the phenotypic variations in plants, flowering time variants or metabolic traits, and their functions associated with resilience to diseases and abiotic stressors (Gordon et al. 2017; Bayer et al. 2019; Gao et al. 2019; Song et al. 2020). Pangenomics enables screening of mutants and wild relatives to breed new crops with higher stress tolerance and crop yield. Thereby, pangenomic research might thus be utilised to select for superior genes in crop wild relatives (CWRs) for crop enhancement (Della Coletta et al. 2021). Pangenomes have been assembled for several species with agricultural importance, e.g., in Brassicaceae family, soybean and rice crop plants (Schatz et al. 2014; Li et al. 2014; Golicz et al. 2016). Pangenome studies have also been done in *Z. mays* (Hirsch et al. 2014), *O. sativa* (Wang et al. 2018), *T. aestivum* (Montenegro 2017), *Glycine max* (Liu et al. 2020), *B. napus* (Song et al. 2020), and *Lycopersicum esculentum* (Gao et al. 2019) by Illumina Hiseq technologies and a number of other crop species. In *Solanum lycopersicum*, pangenome was assembled from heterogeneous group of above 700 accessions of all stages of tomato domestication, wild red and orange-fruited species related to *S. cheesmaniae*, *S. pimpinellifolium* and *S. galapagense*, *S. lycopersicum* var. *cerasiforme*, and *S. lycopersicum* var. *lycopersicum* (Gao et al. 2019). Presence/absence variation (PAV) analyses detected promoter allele of TomLoxC, which encodes 13-lypoxygenase active in volatile biosynthesis of fruits, frequent in wild accessions (>47%), and rare in cerasiforme (8%) and heirloom varieties (1%). This rare promoter allele, in heterozygous state, confers high TomLoxC expression in orange-stage fruits due to improved release of flavor-associated volatiles of fruits and now has been re-introduced in modern elite varieties of tomato (Gao et al. 2019).

3 Transcriptomics

Transcriptomics deals with transcriptome or total RNA transcripts (mRNA, sRNA, tRNA, rRNA, and other non-coding RNA) produced by genome in cells, tissues, or organs (Raza et al. 2021; Chaturvedi et al. 2021). The profiling of transcriptome is dynamic. This is a promising technique to determine effect of any stimuli on the gene expression or differential expression of genes over some period (Duque et al. 2013; El-Metwally et al. 2014). Traditional transcriptomic profiling tools with low resolution included cDNA-AFLP, differential display-PCR (DD-PCR), and suppression subtractive hybridization (SSH) (Nataraja et al. 2017). However, robust and higher-resolution tools like NGS, microarray analysis, serial analysis of gene expression (SAGE), 3′ digital gene expression profiling (3′DGE), and RNA sequencing (RNA-seq) replaced traditional tools (Kawahara et al. 2012; Duque et al. 2013; De Cremer et al. 2013). The microarray analysis in soybean and barley (Guo et al. 2009; Le et al. 2012), soybean (Khan et al. 2017), transcription factors in *A. thaliana*, *G. max*, and *O. sativa* (Xiong et al. 2002; Wohlbach et al. 2008), Cys-2/His-2-type zinc finger (C2H2-ZF) transcription factor in *O. sativa* (Huang et al. 2009), and WRKY TFs in *T. aestivum* (Okay et al. 2014) showed differential expression of genes in response to various types of abiotic stressors. Transcriptome sequencing in sorghum revealed differential expression patterns under stress conditions (Dugas et al. 2011; Johnson et al. 2014) and OsMADS genes in growing and developing rice (Jin et al. 2013). In situ RNA-seq includes RNA sequencing of living cells or tissues (Ke et al. 2013). The spatially resolved transcriptomics tools enable detection of gene expression within cells or tissues for gaining insights into physiological processes (Burgess 2015). RNA-seq analyses stress responsive gene expression on interaction with abiotic and biotic stressors in various crops (Huang et al. 2014; Bhardwaj et al. 2015; Bonthala et al. 2016; Li et al. 2017). Comparative transcriptomic analysis unveiled multiple regulatory cross-talk pathways in *Gossypium* (cotton) and *S. tuberosum* and differential expression profiles in different crop species, e.g., in rice, wheat, and maize and switch grass under high temperature conditions and in soybean from embryonic to reproductive phases (Massa et al. 2013; Zhu et al. 2013; Ding et al. 2013; Li et al. 2013; Xu et al. 2018; Chang et al. 2019). The alternative splicing (AS) transcriptomics approach in various crops generates multiple transcripts and shows role of splicing factors in controlling abiotic stresses (Zhang et al. 2015; Laloum et al. 2018).

4 Proteomics

Proteomics is an area of protein science techniques and provides insights into the abundances and covalent modifications, post-translational modifications, large-scale profiling of whole expressed proteins in a plant and their three-dimensional (3D) structure, subcellular localisation, interactions of protein-metabolite (ligand),

and protein-protein interactions of the cell in a plant in response to varied cellular and environmental scenarios (Hu et al. 2015; Mosa et al. 2017; Aizat and Hassan 2018; Liu et al. 2019).

Proteomics has four diverse shares of sequence proteomics (amino acid sequences), structural proteomics (structure of proteins), functional proteomics (functions of proteins), and expression proteomics (for expression of proteins) (Mosa et al. 2017; Aizat and Hassan 2018). Sequences are identified by high-performance liquid chromatography (HPLC) proteomic tool (Twyman 2013). The structural proteomics involves approaches like computer-based modeling and nuclear magnetic resonance spectroscopy (NMR), crystallization, electron microscopy (EM), and X-ray diffraction crystallography of crystals of proteins (Sali et al. 2003; Aslam et al. 2017; Woolfson 2018). The functional proteomics studies involves methods like yeast one or two hybrids and protein profiling id done by microarray analysis (Lueong et al. 2014) and depict protein role in crop defense responses, e.g., in rice cultivars (Muthurajan et al. 2011; Maksup et al. 2014). Significantly, two-dimensional gel electrophoresis (2D-GE), sodium dodecyl sulphate-polyacrylamide gel electrophoresis (SDS-PAGE), matrix-assisted laser desorption/ionisation-time of flight mass spectrometry (MALDI-TOF), ESI-IT-liquid chromatography tandem mass spectrometry (LC–MS/MS), and isobaric tags for relative and absolute quantitation (iTRAQ) are some important proteomics tools (Hu et al. 2015; Ghatak et al. 2017). The iTRAQ quantitative proteomic analysis enhanced understandings of differential expression of stress-responsive proteins (Liu et al. 2015; Yang et al. 2020) and deciphering functional mechanisms in response to abiotic stressors and pathogens (Basha et al. 2010) and somatic embryogenesis (Zhu et al. 2018) in crops which can pave way for crop improvement in response to varied biotic and abiotic stress conditions and plant adaptive pathways. The differentially expressed proteins can be identified by LC-tandem MS systems (LC-MS/MS) approaches as well (Schubert et al. 2017). "Metabolite-centric" and "protein-centric" methods for characterization of protein-metabolite complexes are now deep-rooted (Veyel et al. 2018; Li and Shui 2020). Massive atlas of protein-protein interactions (PPI) has been defined in diverse species of plants in a phylogenetic framework across a variety of vascular plants, and provides unparalleled understanding of degree of PPI from above two million proteins (McWhite et al. 2020). The highly conserved assemblies with analogous function and architecture were not always composed by equal homologous subunits. Numerous conserved and formerly unknown protein complexes were identified with high significance in improving agriculture. One such example includes pathogenesis-related protein endochitinase B and osmotin-like 34 PPI interactions in response to gray mold caused by *Botrytis cinerea* which suggests role of conserved protein assemblies in improvement of crops (Zhang et al. 2020). Comparative proteomics of wild and cultivated germplasms is boon for crop improvement as is in case of wild relative of wheat known as *Dasypyrum villosum* with several abundant proteins (Wang et al. 2021a, b).

5 Metabolomics

Metabolomics denotes comprehensive investigation of metabolites involved in diverse cellular events in an organism whereas the metabolome is complete set of metabolites which are end products of gene expression or expression of proteins produced through various metabolic pathways in a plant (Weckwerth and Fiehn 2002; Fiehn 2002; Weckwerth and Morgenthal 2005; Baharum and Azizan 2018). Plant metabolomics is a mature, robust, and challenging tool of metabolite profiling or fingerprinting for elucidation of the functions of specific genes (annotation) of specific metabolic pathways (Bobik and Burch-Smith 2015; Alseekh and Fernie, 2018; Fang et al., 2019) and metabolite quantitative trait loci (mQTL) in crops, and model and medicinal plants (Tohge et al. 2017; Scossa et al. 2018; Perez de Souza et al. 2019b; Shi et al. 2020). It is challenging, as the kingdom of plants is predicted to encompass up to one million diverse metabolites which act as signalling molecules (Bobik and Burch-Smith 2015; Wang et al. 2019; Fang et al. 2019), whereas a species is expected to comprise more than 5000 metabolites (Fernie 2007). The metabolomic analytical techniques like gas/liquid-chromatography–mass spectrometry (GC/LC-MS), high-performance liquid chromatography–MS (HPLC–MS), capillary electrophoresis–MS (CE–MS), and nuclear magnetic resonance (NMR) are most frequent in plants for the quantification of various metabolites and their related pathways (Fiehn 2002; Moco et al. 2007; Kikuchi and Hirayama 2007; Weckwerth 2010; Allwood and Goodacre 2010; Kim et al. 2011; Perez de Souza et al. 2019a).

Metabolite annotation provides information about the metabolic pathways of high significance for the refinement of crops. The substrate-product pairs approaches aid in annotation of metabolites (Naake and Fernie 2019) and currently a number of metabolomics databases are available (Wang et al. 2016, 2020; Domingo-Almenara et al. 2019a, b). The pivotal role of metabolites as metabolic profile biomarkers in crops under influence of abiotic stressors or pathogen interactions has been identified (Balmerl et al. 2013), e.g., in rice against gall midge biotype 1 (GMB1) (Agarrwal et al. 2014), and *Xanthomonas oryzae* (Sana et al. 2010), in *O. sativa* and *Hordeum vulgare* against *Magnaporthe oryzae* by use of GC-MS (Parker et al. 2009), in *O. sativa* under water deficiency or salinity by GC-TOF-MS (Liu et al. 2013; Do et al. 2013; Gupta and De 2017; Ghatak et al. 2018), in *A. thaliana* under abiotic stress (Obata and Fernie 2012), and wheat under factors of disease stress (Gunnaiah et al. 2012). Plant metabolite pathway modifications improved nutritive value by accumulating β-carotene in endosperm of the genetically modified (GM) rice plants (Paine et al. 2005) and enhanced production of anthocyanin pigments in *S. lycopersicum* (tomato) (Butelli et al. 2008).

6 Ionomics

Ionomics denotes quantitative measurement or quantification of the entirety of mineral nutrient, and trace elements composition known as ionome and cellular inorganic components of a plant system (Salt et al. 2008; Satismruti et al. 2013). The ionomics analyses changes in composition of minerals activated by numerous genetic alterations, physiological stimuli, or stages of development, and various biotic and abiotic factors (Baxter 2010; Guo et al. 2017). Ionomics are accomplished in plants by diverse analytical technologies which includes inductively coupled plasma-mass/optical emission spectroscopy (ICP-MS/OES), neutron activation analysis (NAA), and X-ray crystallography (Salt et al. 2008; Kumari et al. 2015; Guo et al. 2017). The ionome of leaves identifies regulatory gene networks of iron (Fe) and phosphorus (P) homeostasis (Baxter 2015). Analysis of ionomes of leaves and grains in rice by ICP-MS generated genetic maps, QTLs, and genetic diversification of mineral elements (Norton et al. 2010; Zhang et al. 2014; Pinson et al. 2015). Analysis of ionome of maize seed showed differential gene expression and enhanced symbiotic response to mycorrhizal fungi and alteration in the phenotypes of growth during starvation of phosphate (Mascher et al. 2014). Elemental ionomics profiling elucidated QTLs associated with accumulation of minerals and grain yield in *Z. mays* (Gu et al. 2015), response of elements, mineral nutrients, and metabolites in *H. vulgare* in response to salinity (Wu et al. 2013) and nutrient balance in *Actinidia deliciosa* (kiwifruits), Citrus × sinensis (Orange), *Mangifera indica* (mango), *Malus domestica* (apple), and *Vaccinium caesariense* (blueberry) (Parent et al. 2013), suggest their importance in crop improvement. Ionomics and omics approaches help in discovery of networks for improving crop resilience under abiotic and biotic stress scenarios (Guo et al. 2017).

7 Phenomics

Phenomics is described as the characterization of phenotypes on an organism-wide scale by collecting high-dimensional phenotypic data (Houle et al. 2010). Plant phenome is phenotype en bloc and has basis on the interactions of genome (G), environment (E), and management (M) and in short, it is genotype–phenotype–envirotype (G–P–E) interactions (Gjuvsland et al. 2013; Großkinsky et al. 2018; Zhao et al. 2019). Phenomics by non-invasive, automated high-throughput color imaging approaches of biomass can be applied in gene and QTL mapping for phenotyping of the traits (phenes) and best varieties of best germplasm lines for crop genetic improvements (Kumar et al. 2013). The imaging can analyse crop phenotyping parameters such as far infrared imaging to estimate canopy, light detection and ranging (LIDAR) for the measurement of growth parameters and positron emission tomography, magnetic resonance imaging phenotyping tools for the analyses of root system architecture in crop plants (Finkel 2009; Berger et al. 2010; Furbank and

Tester 2011; Großkinsky et al. 2018; Zhao et al. 2019; McGrail et al. 2020), and high-throughput phenotyping (HTP) in breeding of crops (Yang et al. 2014; Li et al. 2021). The platforms of high-throughput plant phenotyping (HT3P), micro-image phenomics tools coupled with computer-based technologies algorithms and software facilitate phenotyping of key root traits, fluctuation in biomass, analysis of photosynthetic efficiency of field crops (laser-induced fluorescence transient (LIFT)), CO_2 assimilation from the canopy and level of leaves, plant-soil rhizosphere-microbiome interactions, and determination of agronomic traits (multi- and hyper-spectral technologies), from lab to field phenotyping data (Rascher and Pieruschka 2008; Yang et al. 2009; Pieruschka et al. 2010; Flood et al. 2016; Tardieu et al. 2017; Zhao et al. 2019; Furbank et al. 2019, 2020; Furbank et al. 2020; Demidchik et al. 2020; Li et al. 2021; Clouse and Wagner 2021; Trivedi et al. 2021). Thus, phenomics applications include various phenotypic parameters in crops, e.g., root system for *T. durum* (Bodner et al. 2017), phenotyping of shoots in response to numerous stress factors (Humplík et al. 2015), e.g., increased drought tolerance by increase in primary root length density in *T. aestivum* (Djanaguiraman et al. 2019) and decrease in lateral root branching density and increase in length for tolerance to water deficiency in *Z. mays* (Zhan et al. 2015), conductance of stomata in response to salinity in seedlings of *H. vulgare* and *T. aestivum* (Sirault et al. 2009), abiotic stress responses in tobacco, canola, and cotton (Saranga et al. 2004; Baker 2008), quantification of boron toxicity in response to abiotic stress in *H. vulgare* and *T. aestivum* (Schnurbusch et al. 2010), plant pathological system and symptoms of diseases (Mahlein 2016), and detection and quantification of disease symptoms in *H. vulgare* (Swarbrick et al. 2006; Chaerle et al. 2009). The GWAS and high-throughput rice phenotyping facility (HRPF) identified natural variation of 15 agronomic characteristics and 25 linked loci of semi-dwarf gene (*SD1*) in *O. sativa* (Yang et al. 2014). The Internet of things (IoT)-based cropsight system determines both first-class phenotyping of crops and screens the dynamics of microclimate conditions or genotype–environment interactions (GxE) (Reynolds et al. 2019).

8 Bioinformatics

Bioinformatics is an interdisciplinary computational technology for analyzing the biological data. Being interdisciplinary platform of molecular biology, statistics, mathematics, and computer science (Raza et al. 2021), computational software interpret biological queries, molecular mechanisms of biological systems, and plant abiotic stressors (Ambrosino et al. 2020; Raza et al. 2021). Bioinformatics also provides mining of data and data organization of different omics platforms (Ambrosino et al. 2020), and accessible resources for the computational modeling and simulation analysis. Bioinformatics multi-omics databases for crop sciences have been created such as gabipd, PMND, knapsack, KOMICS, planttfdb, bioleaf, easypcc, kappa-View4 KEGG, STRING, GSDS, VISTA, Gramene, Plant Reactome, GAP4, SIMCA-P 14.0, lemnalauncher, and Gromacs which provide data regarding genomics, gene structure analysis, transcriptomics, proteomics, protein-protein interactions, metabolomics, and protein, lipid simulation (Yang et al. 2021).

9 Conclusion

Omics platforms facilitate comprehensive understanding of various complex biological, physiological, and molecular process mechanisms behind the differential regulation of primary and secondary metabolites, proteins, and ions and key traits of agronomic in response to stressors. Several omics tools like pangenomics, phenomics, genomics, metabolomics, transcriptomics, proteomics, ionomics, and bioinformatics and their integration elucidated genetic and biochemical processes, identification of potential candidate genes, molecular mechanisms of genes and networks of genes, growth and development, and resilience to stress in various crop plants by utilizing various high-throughput platforms and computational techniques. This knowledge can facilitate prediction of several agronomically imperative traits for improvement of crop plants for various traits of question.

References

Agarrwal R, Bentur JS, Nair S (2014) Gas chromatography mass spectrometry based metabolic profiling reveals biomarkers involved in rice-gall midge interactions. J Integr Plant Biol 56:837–848. https://doi.org/10.1111/jipb.12244

Agarwal M, Shrivastava N, Padh H (2008) Advances in molecular marker techniques and their applications in plant sciences. Plant Cell Rep 27:617–631. https://doi.org/10.1007/s00299-008-0507-z

Aizat WM, Hassan M (2018) Proteomics in systems biology. In: Aizat W, Goh HH, Baharum S (eds) Omics applications for systems biology. Advances in experimental medicine and biology. Springer, Cham, pp 31–49. https://doi.org/10.1007/978-3-319-98758-3_3

Allwood JW, Goodacre R (2010) An introduction to liquid chromatography-mass spectrometry instrumentation applied in plant metabolomic analyses. Phytochem Anal 21:33–47. https://doi.org/10.1002/pca.1187

Alseekh S, Fernie AR (2018) Metabolomics 20 years on: what have we learned and what hurdles remain? Plant J 94(6):933–942

Amarasinghe SL, Su S, Dong X, Zappia L, Ritchie ME, Gouil Q (2020) Opportunities and challenges in long-read sequencing data analysis. Genome Biol 21(1):30

Ambrosino L, Colantuono C, Diretto G, Fiore A, Chiusano ML (2020) Bioinformatics resources for plant abiotic stress responses: state of the art and opportunities in the fast evolving-omics era. Plan Theory 9:591. https://doi.org/10.3390/plants9050591

Appleby N, Edwards D, Batley J (2009) New technologies for ultrahigh throughput genotyping in plants. In: Gustafson JP, Langridge P, Somers DJ, Totowa NJ (eds) Methods in molecular biology, plant genomics. Humana Press, New York, pp 19–39. https://doi.org/10.1007/978-1-59745-427-8_2

Aslam B, Basit M, Atif NM, Khurshid M, Rasool MH (2017) Proteomics: technologies and their applications. J Chromatogr Sci 55:182–196. https://doi.org/10.1093/chromsci/bmw167

Baharum SN, Azizan KA (2018) Metabolomics in systems biology. Adv Exp Med Biol 1102:51–68. https://doi.org/10.1007/978-3-319-98758-3_4
Baker NR (2008) Chlorophyll fluorescence: a probe of photosynthesis in vivo. Annu Rev Plant Biol 59:89–113. https://doi.org/10.1146/annurev.arplant.59.032607.092759
Balmerl D, Flors V, Glauser G, Mauch-Mani B (2013) Metabolomics of cereals under biotic stress: current knowledge and techniques. Front Plant Sci 4:82. https://doi.org/10.3389/fpls.2013.00082
Basha SM, Mazhar H, Vasanthaiah HKN (2010) Proteomics approach to identify unique xylem sap proteins in Pierce's disease-tolerant *Vitis* species. Appl Biochem Biotechnol 160:932–944. https://doi.org/10.1007/s12010-009-8620-1
Baxter I (2010) Ionomics: the functional genomics of elements. Brief Funct Genomics 9:149–156. https://doi.org/10.1093/bfgp/elp055
Baxter I (2015) Should we treat the ionome as a combination of individual elements, or should we be deriving novel combined traits? J Exp Bot 66:2127–2131. https://doi.org/10.1093/jxb/erv040
Bayer PE, Golicz AA, Tirnaz S, Chan CK, Edwards D, Batley J (2019) Variation in abundance of predicted resistance genes in the *Brassica oleracea* pangenome. Plant Biotechnol J 17(4):789–800
Berger B, Parent B, Tester M (2010) High-throughput shoot imaging to study drought responses. J Exp Bot 61:3519–3528. https://doi.org/10.1093/jxb/erq201
Bhardwaj AR, Joshi G, Kukreja B, Malik V, Arora P, Pandey R et al (2015) Global insights into high temperature and drought stress regulated genes by RNA-Seq in economically important oilseed crop Brassica juncea. BMC Plant Biol 15:9. https://doi.org/10.1186/s12870-014-0405-1
Bobik K, Burch-Smith TM (2015) Chloroplast signaling within, between and beyond cells. Front Plant Sci 6:781. https://doi.org/10.3389/fpls.2015.00781
Bodner G, Alsalem M, Nakhforoosh A, Arnold T, Leitner D (2017) RGB and spectral root imaging for plant phenotyping and physiological research: experimental setup and imaging protocols. J Vis Exp 126:56251. https://doi.org/10.3791/56251
Bonthala VS, Mayes K, Moreton J, Blythe M, Wright V, May ST et al (2016) Identification of gene modules associated with low temperatures response in Bambara groundnut by network-based analysis. PLoS One 11:e0148771. https://doi.org/10.1371/journal.pone.0148771
Burgess DJ (2015) Putting transcriptomics in its place. Nat Rev Genet 16:319. https://doi.org/10.1038/nrg3951
Butelli E, Titta L, Giorgio M, Mock HP, Matros A, Peterek S et al (2008) Enrichment of tomato fruit with health-promoting anthocyanins by expression of select transcription factors. Nat Biotechnol 26:1301–1308. https://doi.org/10.1038/nbt.1506
Chaerle L, Lenk S, Leinonen I, Jones HG, Van Der Straeten D, Buschmann C (2009) Multi-sensor plant imaging: towards the development of a stress-catalogue. Biotechnol J 4:1152–1167. https://doi.org/10.1002/biot.200800242
Challa S, Neelapu NR (2018) Genome-wide association studies (GWAS) for abiotic stress tolerance in plants. In: Wani SH (ed) Biochemical, physiological and molecular avenues for combating abiotic stress tolerance in plants. Academic Press, Cham, pp 135–150. https://doi.org/10.1016/B978-0-12-813066-7.00009-7
Chang Y, Lin H, Liu W, Yu C, Chen H, Wartini P et al (2019) Comparative transcriptomics method to infer gene coexpression networks and its applications to maize and rice leaf transcriptomes. Proc Natl Acad Sci U S A 116(8):3091–3099. https://doi.org/10.1073/pnas.1817621116
Chaturvedi P, Wiese AJ, Ghatak A, Záveská Drábková L, Weckwerth W, Honys D (2021) Heat stress response mechanisms in pollen development. New Phytol 231:571–585. https://doi.org/10.1111/nph.17380
Chen L, Hao L, Parry MAJ, Phillips AL, Hu YG (2014) Progress in TILLING as a tool for functional genomics and improvement of crops. J Integr Plant Biol 56:425–443. https://doi.org/10.1111/jipb.12192

Clouse KM, Wagner MR (2021) Plant genetics as a tool for manipulating crop microbiomes: opportunities and challenges. Front Bioeng Biotechnol 9:567548

Cokus S, Feng S, Zhang X, Chen Z, Merriman B, Haudenschild C et al (2008) Shotgun bisulphite sequencing of the Arabidopsis genome reveals DNA methylation patterning. Nature 452(7184):215–219. https://doi.org/10.1038/nature06745

De Cremer K, Mathys J, Vos C, Froenicke L, Michelmore RW, Cammue BPA et al (2013) RNAseq-based transcriptome analysis of *Lactuca sativa* infected by the fungal necrotroph Botrytis cinerea. Plant Cell Environ 36:1992–2007. https://doi.org/10.1111/pce.12106

Della Coletta R, Qiu Y, Ou S, Hufford M, Hirsch C (2021) How the pan-genome is changing crop genomics and improvement. Genome Biol 22(1):1–19. https://doi.org/10.1186/s13059-020-02224-8

Demidchik VV, Shashko AY, Bandarenka UY, Smolikova GN, Przhevalskaya DA, Charnysh MA et al (2020) Plant phenomics: fundamental bases, software and hardware platforms, and machine learning. Russ J Plant Physiol 67:397–412

Ding X, Li X, Xiong L (2013) Insight into differential responses of upland and paddy rice to drought stress by comparative expression profiling analysis. Int J Mol Sci 14:5214–5238. https://doi.org/10.3390/ijms14035214

Djanaguiraman M, Prasad PVV, Kumari J, Rengel Z (2019) Root length and root lipid composition contribute to drought tolerance of winter and spring wheat. Plant Soil 439:57–73

Do PT, Degenkolbe T, Erban A, Heyer AG, Kopka J, Kohl KI et al (2013) Dissecting rice polyamine metabolism under controlled long-term drought stress. PLoS One 8:e60325. https://doi.org/10.1371/journal.pone.0060325

Domingo-Almenara X, Guijas C, Billings E, Montenegro-Burke JR, Uritboonthai W, Aisporna AE, Chen E, Benton HP, Siuzdak G (2019a) The METLIN small molecule dataset for machine learning-based retention time prediction. Nat Commun 10(1):5811

Domingo-Almenara X, Montenegro-Burke JR, Guijas C, Majumder EL, Benton HP, Siuzdak G (2019b) Autonomous METLIN-guided in-source fragment annotation for untargeted metabolomics. Anal Chem 91(5):3246–3253

Dong C, Dalton-Morgan J, Vincent K, Sharp P (2009) A modified tilling method for wheat breeding. Plant Genome 2:39–47. https://doi.org/10.3835/plantgenome2008.10.0012

Dugas DV, Monaco MK, Olsen A, Klein RR, Kumari S, Ware D et al (2011) Functional annotation of the transcriptome of *Sorghum bicolor* in response to osmotic stress and abscisic acid. BMC Genomics 12:514. https://doi.org/10.1186/1471-2164-12-514

Dumschott K, Schmidt MH, Chawla HS, Snowdon R, Usadel B (2020) Oxford nanopore sequencing: new opportunities for plant genomics? J Exp Bot 71(18):5313–5322

Duque AS, Almeida AM, Bernardes da Silva A, Marques da Silva J, Farinha AP, Santos D et al (2013) Abiotic stress responses in plants: unraveling the complexity of genes and networks to survive. In: Vahdati K, Leslie C (eds) Abiotic stress: plant responses and applications in agriculture. INTECH Open, Rijeka, pp 49–102. https://doi.org/10.5772/45842

El-Metwally S, Ouda OM, Helmy M (2014) Next generation sequencing technologies and challenges in sequence assembly, 1st edn. Springer, New York. https://doi.org/10.1007/978-1-4939-0715-1

Fang C, Fernie AR, Luo J (2019) Exploring the diversity of plant metabolism. Trends Plant Sci 24(1):83–98

Fernie AR (2007) The future of metabolic phytochemistry: larger numbers of metabolites, higher resolution, greater understanding. Phytochemistry 68(22–24):2861–2880

Fiehn O (2002) Metabolomics — the link between genotypes and phenotypes. Plant Mol Biol 48:155–171. https://doi.org/10.1023/A:1013713905833

Finkel E (2009) With 'phenomics' plant scientists hope to shift breeding into overdrive. Science 325:380–381. https://doi.org/10.1126/science.325_380

Flood PJ, Kruijer W, Schnabel SK, van der Schoor R, Jalink H, Snel JF et al (2016) Phenomics for photosynthesis, growth and reflectance in *Arabidopsis thaliana* reveals circadian and long-term fluctuations in heritability. Plant Methods 12:14

Furbank RT, Tester M (2011) Phenomics—technologies to relieve the phenotyping bottleneck. Trends Plant Sci 16:635–644. https://doi.org/10.1016/j.tplants.2011.09.005

Furbank RT, Jimenez-Berni JA, George-Jaeggli B, Potgieter AB, Deery DM (2019) Field crop phenomics: enabling breeding for radiation use efficiency and biomass in cereal crops. New Phytol 223:1714–1727

Furbank RT, Sharwood R, Estavillo GM, Silva-Perez V, Condon AG (2020) Photons to food: genetic improvement of cereal crop photosynthesis. J Exp Bot 71:2226–2238

Gabur I, Chawla HS, Liu X, Kumar V, Faure S, von Tiedemann S et al (2018) Finding invisible quantitative trait loci with missing data. Plant Biotechnol J 16:2102–2112. https://doi.org/10.1111/pbi.12942

Gao L, Gonda I, Sun H, Ma Q, Bao K, Tieman DM et al (2019) The tomato pan-genome uncovers new genes and a rare allele regulating fruit flavor. Nat Genet 51(6):1044–1051

Gelaw T, Sanan-Mishra N (2021) Non-coding RNAs in response to drought stress. Int J Mol Sci 22(22):12519. https://doi.org/10.3390/ijms222212519

Ghatak A, Chaturvedi P, Weckwerth W (2017) Cereal crop proteomics: systemic analysis of crop drought stress responses towards marker-assisted selection breeding. Front Plant Sci 8:757. https://doi.org/10.3389/fpls.2017.00757

Ghatak A, Chaturvedi P, Weckwerth W (2018) Metabolomics in plant stress physiology. Adv Biochem Eng Biotechnol 164:187–236. https://doi.org/10.1007/10_2017_55

Gjuvsland AB, Vik JO, Beard DA, Hunter PJ, Omholt SW (2013) Bridging the genotype-phenotype gap: what does it take? J Physiol 591:2055–2066. https://doi.org/10.1113/jphysiol.2012.248864

Golicz AA, Bayer PE, Barker GC, Edger PP, Kim H, Martinez PA, Chan CK, Severn-Ellis A et al (2016) The pangenome of an agronomically important crop plant *Brassica oleracea*. Nat Commun 7:13390

Gordon SP, Contreras-Moreira B, Woods DP, Des Marais DL, Burgess D, Shu S, Stritt C et al (2017) Extensive gene content variation in the *Brachypodium distachyon* pan-genome correlates with population structure. Nat Commun 8(1):2184

Großkinsky DK, Syaifullah SJ, Roitsch T (2018) Integration of multiomics techniques and physiological phenotyping within a holistic phenomics approach to study senescence in model and crop plants. J Exp Bot 69:825–844. https://doi.org/10.1093/jxb/erx333

Gu R, Chen F, Liu B, Wang X, Liu J, Li P et al (2015) Comprehensive phenotypic analysis and quantitative trait locus identification for grain mineral concentration, content, and yield in maize (*Zea mays* L.). Theor Appl Genet 128:1777–1789. https://doi.org/10.1007/s00122-015-2546-5

Gunnaiah R, Kushalappa AC, Duggavathi R, Fox S, Somers DJ (2012) Integrated metabolo-proteomic approach to decipher the mechanisms by which wheat QTL (Fhb1) contributes to resistance against *Fusarium graminearum*. PLoS One 7:e40695. https://doi.org/10.1371/journal.pone.0040695

Guo P, Baum M, Grando S, Ceccarelli S, Bai G, Li RVK et al (2009) Differentially expressed genes between drought-tolerant and drought-sensitive barley genotypes in response to drought stress during the reproductive stage. J Exp Bot 60:3531–3544. https://doi.org/10.1093/jxb/erp194

Guo R, Shi L, Yan C, Zhong X, Gu F, Liu Q, et al. (2017) Ionomic and metabolic responses to neutral salt or alkaline salt stress in maize (Zea mays L.) seedlings. BMC Plant Biol 17:41. https://doi.org/10.1186/s12870-017-0994-6

Guo Z, Yang W, Chang Y, Ma X, Tu H, Xiong F et al (2018) Genome-wide association studies of image traits reveal genetic architecture of drought resistance in rice. Mol Plant 11(6):789–805. https://doi.org/10.1016/j.molp.2018.03.018

Gupta P, De B (2017) Metabolomics analysis of rice responses to salinity stress revealed elevation of serotonin, and gentisic acid levels in leaves of tolerant varieties. Plant Signal Behav 12:e1335845. https://doi.org/10.1080/15592324

He J, Zhao X, Laroche A, Lu ZX, Liu H, Li Z (2014) Genotyping by-sequencing (GBS), an ultimate marker-assisted selection (MAS) tool to accelerate plant breeding. Front Plant Sci 5:484. https://doi.org/10.3389/fpls.2014.00484

Henikoff S, Till BJ, Comai L (2004) TILLING. Traditional mutagenesis meets functional genomics. Plant Physiol 135:630–636. https://doi.org/10.1104/pp.104.041061

Hirsch CN, Foerster JM, Johnson JM, Sekhon RS, Muttoni G, Vaillancourt B et al (2014) Insights into the maize pan-genome and pan transcriptome. Plant Cell 26:121–135. https://doi.org/10.1105/tpc.113.119982

Houle D, Govindaraju DR, Omholt S (2010) Phenomics: the next challenge. Nat Rev Genet 11:855–866. https://doi.org/10.1038/nrg2897

Hu X, Yang Y, Gong F, Zhang D, Zhang L, Wu L, Li C, Wang W (2015) Protein sHSP26 improves chloroplast performance under heat stress by interacting with specific chloroplast proteins in maize (*Zea mays*). J Proteome 115:81–92

Huang XY, Chao DY, Gao JP, Zhu MZ, Shi M, Lin HX (2009) A previously unknown zinc finger protein, DST, regulates drought and salt tolerance in rice via stomatal aperture control. Genes Dev 23:1805–1817. https://doi.org/10.1101/gad.1812409

Huang L, Zhang F, Zhang F, Wang W, Zhou Y, Fu B et al (2014) Comparative transcriptome sequencing of tolerant rice introgression line and its parents in response to drought stress. BMC Genomics 15:1026. https://doi.org/10.1186/1471-2164-15-1026

Humplík JF, Lazár D, Husicková A, Spíchal L (2015) Automated phenotyping of plant shoots using imaging methods for analysis of plant stress responses – a review. Plant Methods 11:29. https://doi.org/10.1186/s13007-015-0072-8

Jaccoud D, Peng K, Feinstein D, Kilian A (2001) Diversity arrays: a solid-state technology for sequence information independent genotyping. Nucleic Acids Res 29:25. https://doi.org/10.1093/nar/29.4.e25

Jain SM, Suprasanna P (2011) Induced mutations for enhancing nutrition and food production. Gene Conserve 40:201–215

Jin Y, Yang H, Wei Z, Ma H, Ge X (2013) Rice male development under drought stress: phenotypic changes and stage-dependent transcriptomic reprogramming. Mol Plant 6:1630–1645. https://doi.org/10.1093/mp/sst067

Johnson SM, Lim FL, Finkler A, Fromm H, Slabas AR, Knight MR (2014) Transcriptomic analysis of *Sorghum bicolor* responding to combined heat and drought stress. BMC Genomics 15:456. https://doi.org/10.1186/1471-2164-15-456

Kawahara Y, Oono Y, Kanamori H, Matsumoto T, Itoh T, Minami E (2012) Simultaneous RNA-Seq analysis of a mixed transcriptome of rice and blast fungus interaction. PLoS One 7:e49423. https://doi.org/10.1371/journal.pone.0049423

Ke R, Marco M, Pacureanu A, Svedlund J, Botling J, Wählby C et al (2013) In situ sequencing for RNA analysis in preserved tissue and cells. Nat Methods 10:857–860. https://doi.org/10.1038/nmeth.2563

Khan F, Chai HH, Ajmera I, Hodgman C, Mayes S, Lu C (2017) A transcriptomic comparison of two bambara groundnut landraces under dehydration stress. Genes 8:1–19. https://doi.org/10.3390/genes8040121

Kikuchi J, Hirayama T (2007) Practical aspects of uniform stable isotope labeling of higher plants for heteronuclear NMR-based metabolomics. In: Weckwerth W (ed) Metabolomics, methods in molecular biology. Humana Press, New York, pp 273–286. https://doi.org/10.1007/978-1-59745-244-1_15

Kim HK, Choi YH, Verpoorte R (2011) NMR-based plant metabolomics: where do we stand, where do we go? Trends Biotechnol 29:267–275. https://doi.org/10.1016/j.tibtech.2011.02.001

Kim S, Park M, Yeom SI, Kim YM, Lee JM, Lee HA, Seo E, Choi J, Cheong K, Kim KT, Jung K et al (2014) Genome sequence of the hot pepper provides insights into the evolution of pungency in *Capsicum* species. Nat Genet 46(3):270–278

Knoll JE, Ramos ML, Zeng Y, Holbrook CC, Chow M, Chen S et al (2011) TILLING for allergen reduction and improvement of quality traits in peanut (*Arachis hypogaea* L.). BMC Plant Biol 11:81. https://doi.org/10.1186/1471-2229-11-81

Kover PX, Valdar W, Trakalo J, Scarcelli N, Ehrenreich IM, Purugganan MD et al (2009) A multiparent advanced generation inter-cross to fine-map quantitative traits in *Arabidopsis thaliana*. PLoS Genet 5:e1000551. https://doi.org/10.1371/journal.pgen.1000551

Kumar APK, Boualem A, Bhattacharya A, Parikh S, Desai N, Zambelli A et al (2013) SMART – sunflower mutant population and reverse genetic tools for crop improvement. BMC Plant Biol 13:38. https://doi.org/10.1186/1471-2229-13-38

Kumari A, Das P, Parida AK, Agarwal PK (2015) Proteomics, metabolomics, and ionomics perspectives of salinity tolerance in halophyte. Front Plan Sci 6:537. https://doi.org/10.3389/fpls.2015.00537

Kurowska M, Daszkowska-Golec A, Gruszka D, Marzec M, Szurman M, Szarejko I et al (2011) TILLING: a shortcut in functional genomics. J Appl Genet 52:371–390. https://doi.org/10.1007/s13353-011-0061-1

Lasky JR, Upadhyaya HD, Ramu P, Deshpande S, Hash CT, Bonnette J et al (2015) Genome-environment associations in sorghum landraces predict adaptive traits. Sci Adv 1:e1400218. https://doi.org/10.1126/sciadv.1400218

Laloum T, Martín G, Duque P (2018). Alternative splicing control of abiotic stress responses. Trends Plant Sci 23:140–150. https://doi.org/10.1016/j.tplants.2017.09.019

Le DT, Nishiyama R, Watanabe Y, Tanaka M, Seki M, Le H et al (2012) Differential gene expression in soybean leaf tissues at late developmental stages under drought stress revealed by genome-wide transcriptome analysis. PLoS One 7:e49522. https://doi.org/10.1371/journal.pone.0049522

Lei L, Goltsman E, Goodstein D, Wu G, Rokhsar D, Vogel J (2021) Plant pan-genomics comes of age. Annu Rev Plant Biol 72(1):411–435. https://doi.org/10.1146/annurev-arplant-080720-105454

Li S, Shui W (2020) Systematic mapping of protein–metabolite interactions with mass spectrometry-based techniques. Curr Opin Biotechnol 64:24–31

Li YF, Wang YI, Tang Y, Kakani VG, Mahalingam R (2013) Transcriptome analysis of heat stress response in switchgrass (*Panicum virgatum* L.). BMC Plant Biol 13:153. https://doi.org/10.1186/1471-2229-13-153

Li YH, Zhou G, Ma J, Jiang W, Jin LG, Zhang Z, Guo Y, Zhang J, Sui Y, Zheng L, Zhang SS et al (2014) De novo assembly of soybean wild relatives for pan-genome analysis of diversity and agronomic traits. Nat Biotechnol 32(10):1045–1052

Li Y, Zou W, Lin S, Onofua D, Yang Z, Chen H et al (2017) Transcriptome profiling and digital gene expression analysis of sweet potato for the identification of putative genes involved in the defense response against *Fusarium oxysporum* f. sp. batatas. PLoS One 12:e0187838. https://doi.org/10.1371/journal.pone.0187838

Li D, Quan C, Song Z, Li X, Yu G, Li C et al (2021) High-throughput plant phenotyping platform (ht3p) as a novel tool for estimating agronomic traits from the lab to the field. Front Bioeng Biotechnol 8:1533. https://doi.org/10.3389/fbioe.2020.623705

Liu D, Ford KL, Roessner U, Natera S, Cassin AM, Patterson JH et al (2013) Rice suspension cultured cells are evaluated as a model system to study salt responsive networks in plants using a combined proteomic and metabolomic profiling approach. Proteomics 13:2046–2206. https://doi.org/10.1002/pmic.201200425

Liu B, Zhang N, Zhao S, Chang J, Wang Z, Zhang G et al (2015) Proteomic changes during tuber dormancy release process revealed by iTRAQ quantitative proteomics in potato. Plant Physiol Biochem 86:181–190. https://doi.org/10.1016/j.plaphy.2014.12.003

Liu Y, Lu S, Liu K, Wang S, Huang L, Guo L (2019) Proteomics: a powerful tool to study plant responses to biotic stress. Plant Methods 15(1):1–20. https://doi.org/10.1186/s13007-019-0515-8

Liu Y, Du H, Li P, Shen Y, Peng H, Liu S et al (2020) Pan-genome of wild and cultivated soybeans. Cell 182:162–176. https://doi.org/10.1016/j.cell.2020.05.023

Lueong SS, Hoheisel JD, Alhamdani MSS (2014) Protein microarrays as tools for functional proteomics: achievements, promises and challenges. J Proteom Bioinform 7:1–10. https://doi.org/10.4172/jpb.S7-004

Mahlein AK (2016) Plant disease detection by imaging sensors–parallels and specific demands for precision agriculture and plant phenotyping. Plant Dis 100:241–251. https://doi.org/10.1094/PDIS-03-15-0340-FE

Maksup S, Roytrakul S, Supaibulwatana K (2014) Physiological and comparative proteomic analyses of Thai jasmine rice and two check cultivars in response to drought stress. J Plant Interact 9:43–55. https://doi.org/10.1080/17429145.2012.752042

Mangin B, Casadebaig P, Cadic E, Blanchet N, Boniface MC, Carrère S et al (2017) Genetic control of plasticity of oil yield for combined abiotic stresses using a joint approach of crop modelling and genome-wide association. Plant Cell Environ 40:2276–2291. https://doi.org/10.1111/pce.12961

Mascher M, Gerlach N, Gahrtz M, Bucher M, Scholz U, Dresselhaus T (2014) Sequence and ionomic analysis of divergent strains of maize inbred line B73 with an altered growth phenotype. PLoS One 9:e96782. https://doi.org/10.1371/journal.pone.0096782

Massa AN, Childs KL, Buell CR (2013) Abiotic and biotic stress responses in *Solanum tuberosum* group phureja DM1-3516R44 as measured through whole transcriptome sequencing. Plant Genome 6:15. https://doi.org/10.3835/plantgenome2013.05.0014

Mba C (2013) Induced mutations unleash the potentials of plant genetic resources for food and agriculture. Agronomy 3:200–231. https://doi.org/10.3390/agronomy3010200

McCallum CM, Comai L, Greene EA, Henikoff S (2000) Targeting induced local lesions IN genomes (TILLING) for plant functional genomics. Plant Physiol 123:439–442. https://doi.org/10.1104/pp.123.2.439

McGrail RK, Van Sanford DA, McNear DH Jr (2020) Trait-based root phenotyping as a necessary tool for crop selection and improvement. Agronomy 10:1328. https://doi.org/10.3390/agronomy10091328

McWhite CD, Papoulas O, Drew K, Cox RM, June V, Dong OX, Kwon T, Wan C, Salmi ML, Roux SJ, Browning KS, Chen ZJ, Ronald PC, Marcotte EM (2020) A pan-plant protein complex map reveals deep conservation and novel assemblies. Cell 181(2):460–474 e414

Millet EJ, Welcker C, Kruijer W, Negro S, Coupel-Ledru A, Nicolas SD et al (2016) Genome-wide analysis of yield in Europe: allelic effects vary with drought and heat scenarios. Plant Physiol 172:749–764. https://doi.org/10.1104/pp.16.00621

Moco S, Bino RJ, De Vos RCH, Vervoort J (2007) Metabolomics technologies and metabolite identification. Trends Anal Chem 26:855–866. https://doi.org/10.1016/j.trac.2007.08.003

Montenegro JD (2017) The pangenome of hexaploid bread wheat. Plant J 90:1007–1013. https://doi.org/10.1111/tpj.13515

Mosa KA, Ismail A, Helmy M (2017) Omics and system biology approaches in plant stress research. In: Mosa KA, Ismail A, Helmy M (eds) Plant stress tolerance: an integrated omics approach. Springer, Cham, pp 21–34. https://doi.org/10.1007/978-3-319-59379-1_2

Muthamilarasan M, Singh NK, Prasad M (2019) Multi-omics approaches for strategic improvement of stress tolerance in underutilized crop species: a climate change perspective. Adv Genet 103:1–38. https://doi.org/10.1016/bs.adgen.2019.01.001

Muthurajan R, Shobbar ZS, Jagadish S, Bruskiewich R, Ismail A, Leung H et al (2011) Physiological and proteomic responses of rice peduncles to drought stress. Mol Biotechnol 48:173–182. https://doi.org/10.1007/s12033-010-9358-2

Naake T, Fernie AR (2019) MetNet: metabolite network prediction from high-resolution mass spectrometry data in raiding metabolite annotation. Anal Chem 91(3):1768–1772

Nataraja KN, Madhura BG, Parvathi SM (2017) Omics: modern tools for precise understanding of drought adaptation in plants. In: Zargar SM, Rai V (eds) Plant OMICS and crop breeding. Apple Academic Press, Palm Bay, pp 289–320

Nekrasov V, Wang C, Win J, Lanz C, Weigel D, Kamoun S (2017) Rapid generation of a transgene-free powdery mildew resistant tomato by genome deletion. Sci Rep 7:482. https://doi.org/10.1038/s41598-017-00578-x

Norton GJ, Deacon CM, Xiong LZ, Huang SY, Meharg AA, Price AH (2010) Genetic mapping of the rice ionome in leaves and grain: identification of QTLs for 17 elements includ-

ing arsenic, cadmium, iron and selenium. Plant Soil 329:139–153. https://doi.org/10.1007/s11104-009-0141-8
Obata T, Fernie AR (2012) The use of metabolomics to dissect plant responses to abiotic stresses. Cell Mol Life Sci 69(19):3225–3243
Okay S, Derelli E, Unver T (2014) Transcriptome-wide identification of bread wheat WRKY transcription factors in response to drought stress. Mol Gen Genomics 289:765–781. https://doi.org/10.1007/s00438-014-0849-x
Paine JA, Shipton CA, Chaggar S, Howells RM, Kennedy MJ, Vernon G et al (2005) Improving the nutritional value of Golden Rice through increased pro-vitamin A content. Nat Biotechnol 23:482–487. https://doi.org/10.1038/nbt1082
Parent SE, Parent LE, Egozcue JJ, Rozane DE, Hernandes A, Lapointe L et al (2013) The plant ionome revisited by the nutrient balance concept. Front Plant Sci 4:39. https://doi.org/10.3389/fpls.2013.00039
Parker D, Beckmann M, Zubair H, Enot DP, Caracuel-Rios Z, Overy DP et al (2009) Metabolomic analysis reveals a common pattern of metabolic re-programming during invasion of three host plant species by *Magnaporthe grisea*. Plant J 59:723–737. https://doi.org/10.1111/j.1365-313X.2009.03912.x
Penna S, Jain SM (2017) Mutant resources and mutagenomics in crop plants. Emirates J Food Agric 29:651–657. https://doi.org/10.9755/ejfa.2017.v29.i9.86
Perez de Souza L, Alseekh S, Naake T, Fernie A (2019a) Mass spectrometry-based untargeted plant metabolomics. Curr Protoc Plant Biol 4(4):e20100
Perez de Souza L, Scossa F, Proost S, Bitocchi E, Papa R, Tohge T, Fernie AR (2019b) Multi-tissue integration of transcriptomic and specialized metabolite profiling provides tools for assessing the common bean (*Phaseolus vulgaris*) metabolome. Plant J 97(6):1132–1153
Pieruschka R, Klimov D, Kolber Z, Berry JA (2010) Monitoring of cold and light stress impact on photosynthesis by using the laser induced fluorescence transient (LIFT) approach. Funct Plant Biol 37:395–402. https://doi.org/10.1071/FP09266
Pinson SRM, Tarpley L, Yan W, Yeater K, Lahner B, Yakubova E et al (2015) Worldwide genetic diversity for mineral element concentrations in rice grain. Crop Sci 55:294–311. https://doi.org/10.2135/cropsci2013.10.0656
Prasad A, Hari-Gowthem G, Muthamilarasan M, Hussain Z, Yadav PK, Tripathi S, Prasad M (2021) Molecular characteriza tion of SlATG18f in response to tomato leaf curl New Delhi virus infection in tomato and development of a CAPS marker for leaf curl disease tolerance. Theor Appl Gen 134(5):1463–1474
Rabouam C, Comes AM, Bretagnolle V, Humbert JF, Periquet G, Bigot Y (1999) Features of DNA fragments obtained by random amplified polymorphic DNA (RAPD) assays. Mol Ecol 8:493–503. https://doi.org/10.1046/j.1365-294X.1999.00605.x
Rascher U, Pieruschka R (2008) Spatio-temporal variations of photosynthesis: the potential of optical remote sensing to better understand and scale light use efficiency and stresses of plant ecosystems. Precis Agric 9:355–366. https://doi.org/10.1007/s11119-008-9074-0
Raza A, Tabassum J, Kudapa H, Varshney RK (2021) Can omics deliver temperature resilient ready-to-grow crops? Crit Rev Biotechnol 41:1209. https://doi.org/10.1080/07388551.2021.1898332
Reynolds D, Ball J, Bauer A, Davey R, Griffiths S, Zhou J (2019) CropSight: a scalable and open-source information management system for distributed plant phenotyping and IoT-based crop management. Gigascience 8:1–35. https://doi.org/10.1093/gigascience/giz009
Rinaldo AR, Ayliffe M (2015) Gene targeting and editing in crop plants: a new era of precision opportunities. Mol Breed 35:40. https://doi.org/10.1007/s11032-015-0210-z
Sali A, Glaeser R, Earnest T, Baumeister W (2003) From words to literature in structural proteomics. Nature 422:216–225. https://doi.org/10.1038/nature01513
Salt DE, Baxter I, Lahner B (2008) Ionomics and the study of the plant ionome. Annu Rev Plant Biol 59:709–733. https://doi.org/10.1146/annurev.arplant.59.032607.092942

Sana TR, Fischer S, Wohlgemuth G, Katrekar A, Jung KH, Ronald PC et al (2010) Metabolomic and transcriptomic analysis of the rice response to the bacterial blight pathogen *Xanthomonas oryzae* pv. oryzae. Metabolomics 6:451–465. https://doi.org/10.1007/s11306-010-0218-7

Saranga JCX, Wright RJ, Yakir D, Paterson AH (2004) Genetic dissection of cotton physiological responses to arid conditions and their inter-relationships with productivity. Plant Cell Environ 27:263–277. https://doi.org/10.1111/j.1365-3040.2003.01134.x

Satismruti K, Senthil N, Vellaikumar S, Ranjani RV, Raveendran M (2013) Plant Ionomics: a platform for identifying novel gene regulating plant mineral nutrition. Am J Plant Sci 4:1309–1315. https://doi.org/10.4236/ajps.2013.47162

Schatz MC, Maron LG, Stein JC, Hernandez Wences A, Gurtowski J, Biggers E, Lee H et al (2014) Whole genome de novo assemblies of three divergent strains of rice, *Oryza sativa*, document novel gene space of aus and indica. Genome Biol 15(11):506

Schnurbusch T, Hayes J, Sutton T. (2010) Boron toxicity tolerance in wheat and barley: Australian perspectives. Breed Sci. 60:297–304. https://doi.org/10.1270/jsbbs.60.297

Schubert OT, Rost HL, Collins BC, Rosenberger G, Aebersold R (2017) Quantitative proteomics: challenges and opportunities in basic and applied research. Nat Protoc 12(7):1289–1294

Scossa F, Benina M, Alseekh S, Zhang Y, Fernie AR (2018) The integration of metabolomics and next-generation sequencing data to elucidate the pathways of natural product metabolism in medicinal plants. Planta Med 84(12–13):855–873

Shi T, Zhu A, Jia J, Hu X, Chen J, Liu W, Ren X, Sun D, Fernie AR, Cui F, Chen W (2020) Metabolomics analysis and metabolite-agronomic trait associations using kernels of wheat (*Triticum aestivum*) recombinant inbred lines. Plant J 103(1):279–292

Shikha M, Kanika A, Rao AR, Mallikarjuna MG, Gupta HS, Nepolean T (2017) Genomic selection for drought tolerance using genome-wide SNPs in maize. Front Plant Sci 8:550. https://doi.org/10.3389/fpls.2017.00550

Sirault XRR, James RA, Furbank RT (2009) A new screening method for osmotic component of salinity tolerance in cereals using infrared thermography. Funct Plant Biol 36:970–977. https://doi.org/10.1071/FP09182

Song JM, Guan Z, Hu J, Guo C, Yang Z, Wang S et al (2020) Eight high quality genomes reveal pan-genome architecture and ecotype differentiation of *Brassica napus*. Nat Plants 6(1):34–45. https://doi.org/10.1038/s41477-019-0577-7

Swarbrick PJ, Schulze-Lefert P, Scholes JD (2006) The metabolic consequences of susceptibility and the activation of race specific or broad-spectrum resistance pathways in barley leaves challenged with the powdery mildew fungus. Plant Cell Environ 29:1061–1076. https://doi.org/10.1111/j.1365-3040.2005.01472.x

Talukdar D, Sinjushin A (2015) Cytogenomics and mutagenomics in plant functional biology and breeding. In: Barh D, Khan M, Davies E (eds) PlantOmics: the omics of plant science. Springer, New Delhi, pp 113–156. https://doi.org/10.1007/978-81-322-2172-2_5

Tardieu F, Cabrera-Bosquet L, Pridmore T, Bennett M (2017) Plant phenomics, from sensors to knowledge. Curr Biol 27:R770–R783

Tettelin H, Masignani V, Cieslewicz MJ, Donati C, Medini D, Ward NL et al (2005) Genome analysis of multiple pathogenic isolates of Streptococcus agalactiae: implications for the microbial 'pan-genome'. Proc Natl Acad Sci U S A 102:13950–13955. https://doi.org/10.1073/pnas.0506758102

The Arabidopsis Genome Initiative (2000) Analysis of the genome sequence of the flowering plant *Arabidopsis thaliana*. Nature 408:796–815. https://doi.org/10.1038/35048692

Tohge T, de Souza LP, Fernie AR (2017) Current understanding of the pathways of flavonoid biosynthesis in model and crop plants. J Exp Bot 68(15):4013–4028

Trivedi P, Mattupalli C, Eversole K, Leach JE (2021) Enabling sustainable agriculture through understanding and enhancement of microbiomes. New Phytol 230:2129–2147

Twyman RM (2013) Principles of proteomics. Garland Science Press, Abingdon

Varshney RK, Thudi M, May GD, Jackson SA (2010) Legume genomics and breeding. Plant Breed Rev 33:257–304. https://doi.org/10.1002/9780470535486.ch6

Veyel D, Sokolowska EM, Moreno JC, Kierszniowska S, Cichon J, Wojciechowska I et al (2018) PROMIS, global analysis of PROtein-metabolite interactions using size separation in Arabidopsis thaliana. J Biol Chem 293(32):12440–12453

Vos P, Hogers R, Bleeker M, Reijans M, Lee T, Hornes M et al (1995) AFLP: a new technique for DNA fingerprinting. Nucleic Acids Res 23:4407–4414. https://doi.org/10.1093/nar/23.21.4407

Voss-Fels KP, Stahl A, Hickey LT (2019) Q&A: modern crop breeding for future food security. BMC Biol 17(1):18

Wang Y, Cheng X, Shan Q, Zhang Y, Liu J, Gao C et al (2014) Simultaneous editing of three homoeoalleles in hexaploid bread wheat confers heritable resistance to powdery mildew. Nat Biotechnol 32:947–951. https://doi.org/10.1038/nbt.2969

Wang M, Carver JJ, Phelan VV, Sanchez LM, Garg N, Peng Y, Nguyen DD, Watrous J et al (2016) Sharing and community curation of mass spectrometry data with global natural products social molecular networking. Nat Biotechnol 34(8):828–837

Wang W, Mauleon R, Hu Z, Chebotarov D, Tai S, Wu Z et al (2018) Genomic variation in 3,010 diverse accessions of Asian cultivated rice. Nature 557:43–49. https://doi.org/10.1038/s41586-018-0063-9

Wang S, Alseekh S, Fernie AR, Luo J (2019) The structure and function of major plant metabolite modifications. Mol Plant 12(7):899–919

Wang M, Jarmusch AK, Vargas F, Aksenov AA, Gauglitz JM, Weldon K, Petras D, da Silva R et al (2020) Mass spectrometry searches using MASST. Nat Biotechnol 38(1):23–26

Wang G, Li X, Dong J, Jin R, Yang L, Huo L, Chen L, Zhao W, Gao X (2021a) Proteomics unravels new candidate genes of *Dasypyrum villosum* for improving wheat quality. J Proteome 245:104292

Wang X, Wang A, Li Y, Xu Y, Wei Q, Wang J et al (2021b) A novel banana mutant "RF 1" (*Musa* spp. ABB, Pisang Awak Subgroup) for improved agronomic traits and enhanced cold tolerance and disease resistance. Front. Plant Sci 12:730718. https://doi.org/10.3389/fpls.2021.730718

Weckwerth W (2010) Metabolomics: an integral technique in systems biology. Bioanalysis 2:829–836. https://doi.org/10.4155/bio.09.192

Weckwerth W, Morgenthal K (2005). Metabolomics: from pattern recognition to biological interpretation. Drug Discov Today 10:1551–1155. https://doi.org/10.1016/S1359-6446(05)03609-3

Williams JGK, Kubelik AR, Livak KJ, Rafalski JA, Tingey SV (1990) DNA polymorphisms amplified by arbitrary primers are useful as genetic markers. Nucleic Acid Res 18:6531–6535. https://doi.org/10.1093/nar/18.22.6531

Witzel K, Neugart S, Ruppel S, Schreiner M, Wiesner M, Baldermann S (2015) Recent progress in the use of 'omics technologies in brassicaceous vegetables. Front Plant Sci 6:244. https://doi.org/10.3389/fpls.2015.00244

Wohlbach DJ, Quirino BF, Sussman MR (2008) Analysis of the Arabidopsis histidine kinase ATHK1 reveals a connection between vegetative osmotic stress sensing and seed maturation. Plant Cell 20:1101–1117. https://doi.org/10.1105/tpc.107.055871

Woolfson M (2018) The development of structural x-ray crystallography. Phys Scr 93:1–32. https://doi.org/10.1088/1402-4896/aa9c30

Wu D, Shen Q, Cai S, Chen ZH, Dai F, Zhang G (2013) Ionomic responses and correlations between elements and metabolites under salt stress in wild and cultivated barley. Plant Cell Physiol 54:1976–1988. https://doi.org/10.1093/pcp/pct134

Xiong L, Schumaker KS, Zhu JK (2002) Cell signaling during cold, drought, and salt stress. Plant Cell 14:65–S183. https://doi.org/10.1105/tpc.000596

Xu C, Xia C, Xia Z, Zhou X, Huang J, Huang Z et al (2018) Physiological and transcriptomic responses of reproductive stage soybean to drought stress. Plant Cell Rep 37(12):1611–1624. https://doi.org/10.1007/s00299-018-2332-3

Yang LN, Peng L, Zhang LM, Zhang LL, Yang SS (2009) A prediction model for population occurrence of paddy stem borer (*Scirpophaga incertulas*), based on Back propagation artificial neural network and principal components analysis. Comput Electron Agric 68:200–206. https://doi.org/10.1016/j.compag.2009.06.003

Yang W, Guo Z, Huang C, Duan L, Chen G, Jiang N et al (2014) Combining high-throughput phenotyping and genome-wide association studies to reveal natural genetic variation in rice. Nat Commun 5:5087. https://doi.org/10.1038/ncomms6087

Yang Y, Saand MA, Abdelaal WB, Zhang J, Wu YI, Li J et al (2020) TRAQ-based comparative proteomic analysis of two coconut varieties reveals aromatic coconut cold-sensitive in response to low temperature. J Proteome 220:103766. https://doi.org/10.1016/j.jprot.2020.103766

Yang Y, Saand MA, Huang L, Abdelaal WB, Zhang J, Wu Y, Li J, Sirohi MH, Wang F (2021) Applications of multi-omics technologies for crop improvement. Front Plant Sci 12:563953. https://doi.org/10.3389/fpls.2021.563953

Yu J, Holland JB, McMullen MD, Buckler ES (2008) Genetic design and statistical power of nested association mapping in maize. Genetics 178:539–551. https://doi.org/10.1534/genetics.107.074245

Zhan A, Schneider H, Lynch JP (2015) Reduced lateral root branching density improves drought tolerance in maize. Plant Physiol 168:1603–1615

Zhang M, Pinson SR, Tarpley L, Huang XY, Lahner B, Yakubova E et al (2014) Mapping and validation of quantitative trait loci associated with concentrations of 16 elements in unmilled rice grain. Theor Appl Genet 127:137–165. https://doi.org/10.1007/s00122-013-2207-5

Zhang C, Yang H, Yang H (2015) Evolutionary character of alternative splicing in plants. Bioinform. Biol. Insights 9:47–52. https://doi.org/10.4137/BBI.S33716

Zhang Y, Skirycz A, Fernie AR (2020) An abundance and interaction encyclopedia of plant protein function. Trends Plant Sci 25(7):627–630

Zhao C, Zhang Y, Du J, Guo X, Wen W, Gu S et al (2019) Crop phenomics: current status and perspectives. Front Plant Sci 10:714

Zhu YN, Shi DQ, Ruan MB, Zhang LL, Meng ZH, Liu J et al (2013) Transcriptome analysis reveals crosstalk of responsive genes to multiple abiotic stresses in cotton (*Gossypium hirsutum* L.). PLoS One 8:e80218

Zhu HG, Cheng WH, Tian WG, Li YJ, Liu F, Xue F et al (2018) iTRAQ-based comparative proteomic analysis provides insights into somatic embryogenesis in *Gossypium hirsutum* L. Plant Mol Biol 96:89–102

Chapter 35
Transgenic Approaches for Stress Tolerance in Crops

Deepu Pandita

Abstract Global food security is an alarming question in regard to upcoming scenarios of climate change. Stress (abiotic and biotic) is a major challenge in agriculture for sustainable food production, which negatively affects growth and development of flora and reduces their potential yield and productivity. The traditional breeding approaches have limitations. For the tailor-made solutions to generate climate-smart crops with a higher level of stress tolerance by use of foreign gene(s), genetic engineering allows distinct genetic changes in a faster, and more effective mode in a variety of plant crops. The plant germplasm with better adapted traits to stresses can be designed by the novel biotechnological tools and transgenic approaches. The stress induced genes and the mechanisms of stress tolerance can be improved in stress-tolerant crops without eradicating native traits. Quite a lot of stress responsive genes involved in regulatory systems of the antioxidant defence system, reactive oxygen species, sensor receptors, microRNA and other stress signaling pathways have been overexpressed or knockdown by antisense RNA, RNAi, miRNAi and induction (CRISPR activation) or repression (CRISPR interference) of CRISPR/CAS based toolbox in plants to enhance tolerance to various kind of stressors. The current chapter emphases on transgenics for amelioration of tolerance to various stressors in plants.

Keywords Abiotic and biotic stress · Stress signaling pathways · Transgenic plants · MicroRNA · Antisense RNA · RNAi · MiRNAi · CRISPR/CAS toolbox

1 Introduction

Various erratic abiotic and biotic stress fluctuations including drought, salinity, extreme temperatures, insect pests, and disease pathogens damagingly impact the growth and development of crop plants and are limiting factors for their yield and

D. Pandita (✉)
Government Department of School Education, Jammu, Jammu and Kashmir, India

M. Hasanuzzaman (ed.), *Climate-Resilient Agriculture, Vol 2*,
https://doi.org/10.1007/978-3-031-37428-9_35

productivity worldwide (Munns and Tester 2008; Pandita 2019, 2021). Climate change has unparalleled stress frequency on crop plants (Batley and Edwards 2016). Plant abiotic stresses alone account for about 70% reduction in agricultural yield (Acquaah 2007). Biotic stress agents lead to loss of 10–16% in crop harvest at global levels (Chakraborty and Newton 2011). In traditional agriculture, conventional breeding methods have limitations of resource- and time-consuming and intensive nature. Further, only plants of same or closely linked germplasm undergo cross hybridization. Such gene pool may lack resistance to stress, or resistance cannot be introgressed into new varieties. Therefore, alternative methods for improvement of crop varieties for tolerance to biotic and abiotic stresses including genetic engineering biotechnological approaches for selection and transfer of stress tolerant genes from unrelated species into crop plants hold importance (Roy et al. 2011; Varshney et al. 2011; Dhankher and Foyer 2018). Transgenic approaches and metabolic engineering of plants provide considerable potential to develop climate-smart crops suitable for hostile environmental scenarios having sustainable yield efficiency. Genetic engineering can design transgenic plants carrying genes for stress resistance and to generate climate-resilient crops (Dhankher and Foyer 2018; Pandita 2022a, b, c, d, e, f, g). In this chapter, we deliberate the transgenic approaches for abiotic and biotic stress tolerant crop plants. Transgenic plants have been designed by gene knockdown, antisense RNA and RNAi or gene overexpression and clustered regularly interspersed short palindromic repeats (CRISPR)-associated protein (CRISPR/Cas) toolbox-based induction (CRISPR activation) or repression (CRISPR interference) (Ramesh et al. 2007; Yan et al. 2007; Ali et al. 2015; Baltes et al. 2015; Zafar et al. 2020; Pandita 2022d, e, f, g). The crosstalk of stress-responsive miRNAs with different abiotic and biotic stressors in different plant species has been discussed comprehensively in various chapters (Pandita 2019, 2021, 2022a, b, c). CRISPR/Cas-mediated genome editing technologies for stress resilience in plants is a new era for tailor-made plants tolerant to varied stress factors (Pandita 2022d, e, f, g).

2 Mechanisms of Stress Response in Plants

In harsh environments, plants experiencing stress cannot run away because of being sessile but need to face stress. Plants have thus developed complex regulatory networks for avoidance or tolerance of stress by triggering numerous changes or adaptive mechanisms of plethora of morphological, physiological, cellular, biochemical and molecular levels (Munns and Tester 2008; Suprasanna et al. 2018; Ferguson 2019; Suprasanna and Ghag 2019). Maximum plants display a radical drop-in photosynthetic rate, growth, increase in reactive oxygen species (ROS) scavenging system and even face death. In stress perception, the stress signals are perceived or sensed by the receptors present on cell plasma membrane for example, heat by photoreceptor phytochrome B, cold by COLD1, salinity by glycosyl inositol phosphorylceramide sphingolipids and drought by OSCA1, which in turn activate a cascade of downstream signaling and defense pathways and at molecular levels modulate the master regulator stress-responsive genes, transcription factors, and small RNAs (Munns and Tester 2008; Zhu 2016; Patel et al. 2019; Lamers et al. 2020; Pandita 2019, 2021, 2022a, b, c). Stress-induced genes encode enzymatic proteins or proteins with structural functions, and other proteins of regulatory nature (Bhatnagar-Mathur et al. 2008).

3 Transgenic Approaches for Enhancing Stress Tolerance in Plants

Transgenic approaches allow manipulation and transfer of any desirable gene of interest from the genetic reservoir for manipulating other plant species and are widespread for improving stress tolerance or resistance in plants (Paul and Roychoudhury 2018). Recent significant developments in transgenic research to fine-tune and tailor single or multiple stress tolerance in sensitive crop plants are deliberated in subsequent sections.

3.1 Abiotic Stress

The abiotic stress responses are complex in nature with various complex, well-synchronised networks and regulatory pathways which cross-talk at the biochemical and molecular levels (Mittler and Blumwald 2010; Varshney et al. 2011; Pandita 2019, 2021, 2022a, b, c). The wild, untapped germplasm, stress-responsive regulatory genes within and across plant families, transposons, master regulators like transcriptional factors and signalling proteins, quantitative trait loci (QTL) e.g., Saltol QTL (Thomson et al. 2010) and SKC1 (Ren et al. 2005), microRNAs (miRNAs) and their expression profiling regulate abiotic stress-related signalling pathways and aid in designing of abiotic stress-tolerant crops (Sunkar et al. 2012; Bhaskara et al. 2015; Negi et al. 2016; Isayenkov 2019; Jha 2019; Patel et al. 2019; VanWallendael et al. 2019). In crop plants like rice, wheat, maize, barley, tomato, potato, and many other species, drought, salinity, extreme temperature stress-responsive miRNAs act as negative regulators (Zhou et al. 2010; Zhang and Wang 2015; Barciszewska-Pacak et al. 2015; Pandita and Wani 2019; Pandita 2022b, c). Some of the examples of transgenics generated for resistance to abiotic stress factors are summarized in Table 35.1.

Table 35.1 Transgenic crop plants resistant to abiotic stress

Transgenic plant	Gene introduced	Resistance to stress type	References
Arabidopsis thaliana	JcLEA from *Jatropha curca*	Salinity	Liang et al. (2013)
	SiLAE from *Setaria italica*	Salinity	Wang et al. (2014c)
	GhPIP2;7 from *Gossypium hirsutum*	Drought	Zhang et al. (2013)
	BvCOLD1 from *Beta vulgaris*	Cold	Porcel et al. (2018)
	Mpgsmt and Mpsdmt from *Methanohalophilus portucalensis*	Drought, salinity	Wei et al. (2017)
	NsylCBL10 from *Nicotiana sylvestris*	Salinity	Dong et al. (2015)
	TaHSP26 from *Triticum aestivum*	Heat	Chauhan et al. (2012)
	OsDREB1D from *O. sativa*	Cold, salinity	Zhang et al. (2009)
	ZmbZIP72 from *Zea Mays*	Drought, salinity	Ying et al. (2012)
	TaNAC67 from *Triticum aestivum*	Cold, salinity, drought	Mao et al. (2014)
	TaNAC47 from *T. aestivum*	Salt, cold, drought	Zhang et al. (2016)
A. thaliana, Oryza sativa	microRNA156	Drought	Kantar et al. (2011), Ren et al. (2012)
Cajanus cajan	P5CSF129A from *Vigna aconitifolia*	Salinity	Ahmad et al. (2014)
Cicer arietinum	P5CS from *V. aconitifolia*	Salinity	Ghanti et al. (2011)
Glycine max	GmPIP1;6 from *G. max*	Salinity	Zhou et al. (2014)
	AtDREB1A from *A. thaliana*	Drought	Polizel et al. (2011)
G. max, Triticum dicoccoides	microRNA 166	Drought	Kantar et al. (2011), Li et al. (2017)
Ipomoea batatas	BADH from *Spinacia oleraceu*	Salinity, cold	He et al. (2011)
Lycopersicon esculentum	BADH betaine aldehyde dehydrogenase from *Spinacia oleracea*	Heat	Luo et al. (2017)
	LeAN2 from *Lycopersicum esculentum*	Heat	Meng et al. (2015)
Medicago truncatula	MtP5CS3 from *Medicago truncatula*	Salinity	Surekha et al. (2014)
	MtWRKY76 from *Medicago truncatula*	Drought, salinity	Liu et al. (2016)
Morus indica	Hva1 from *Hordeum vulgare*	Salinity, drought, cold	Checker et al. (2012)
Nicotiana tabacum	AdLEA from *Arachis diogoi*	Salinity	Sharma et al. (2016)
	AhNAC3 from *Arachis hypogaea*	Drought	Liu et al. (2013b)
N. tabacum, Arabidopsis thaliana	GmbZIP1 from *G. max*	Drought, salinity, cold	Gao et al. (2011)

(continued)

Table 35.1 (continued)

Transgenic plant	Gene introduced	Resistance to stress type	References
O. sativa	OsPIP1 from *O. sativa*	Salinity	Liu et al. (2013a)
	P5CS from *Vigna aconitifolia*	Salinity	Kondrák et al. (2012)
	Trehalose-6-phosphate synthase (OsTSP1) from *Oryza sativa*	Salinity, drought, cold	Fan et al. (2012)
	OsCIPK23m from *O. sativa*	Drought	Yang et al. (2008)
	HSP70 from Citrus tristeza virus (CTV)	Salinity	Hoang et al. (2015)
	OsERF4a from *O. sativa*	Drought	Joo et al. (2013)
	AtDREB1A from *A. thaliana*	Drought	Ravikumar et al. (2014)
	OsEREBP1 from *O. sativa*	Drought	Jisha et al. (2015)
	microRNA 29	Salinity	Barrera-Figueroa et al. (2012)
	Osa- microRNA 319b	Cold tolerance	Wang et al. (2014a, b)
O. sativa subsp indica	SbDREB2 from *Sorghum bicolor*	Drought	Bihani et al. (2010)
O. sativa, A. thaliana	OsLEA3–2 from *O. sativa*	Salinity, Drought	Duan and Cai (2012)
P. davidiana × *P. bolleana*	PtCBL10A PtCBL10B from *Populus trichocarpa*	Salinity	Tang et al. (2014)
Poncirus trifoliata	Betaine aldehyde dehydrogenase gene (AhBADH) from *Atriplex hortensis*	Salinity	Lai et al. (2014)
Solanum tuberosum	Pyrroline-5-carboxylate synthetase (P5CS) from *A. thaliana*	Salinity	Kim and Nam (2013)
	TPS1 from yeast StDS2	Drought	Fu et al. (2011)
	Choline oxidase from *Arthrobacter globiformis*	Cold, salinity	He et al. (2010)
Zea mays	LEA Rab28 gene from *Z. mays*	Drought	Amara et al. (2013)

3.1.1 Drought

Transcriptional factors (TFs) such as MYB and MYC, dehydration-responsive-element-bindings (DREBs) abscisic acid-responsive element binding proteins (AREBs)/abscisic acid-responsive element binding factors (ABFs), nuclear factor Y-B subunits (NF-YB), and tryptophan– arginine–lysine–tyrosine (WRKY) have important roles in regulation or switching on the expression of drought responsive genes and are thus efficient targets for transgenics (Bhatnagar-Mathur et al. 2008; Lata and Prasad 2011; Wang et al. 2016).

Transgenic tomato plants resistant to drought stress were generated with DNA cassette containing nos terminator and Arabidopsis C repeat/dehydration-responsive element binding factor 1 (CBF1) (Tsai-Hung et al. 2002). The cold and drought-resistant transgenic apple lines were generated by Osmyb 4 gene (isolated from rice) over-expression (Pasquali et al. 2008). The DREB1b TF gene over-expression induced tolerance to the low temperature and water deficiency in transgenic grapevine plants (Jin et al. 2009). The transgenic banana plants showed drought tolerance by over-expression of MusaWRKY71 from Musa species (Shekhawat and Ganapathi 2013). The CmWRKY1 transcription factor (TF) isolated from *Chrysanthemum morifolium* and on over-expression in *Chrysanthemum* cv. Jinba behaves as positive stress regulator (Fan et al. 2016). Transgenic lines displayed better tolerance for drought stress in a number of other plant transgenic lines like *Agrobacterium tumefaciens*-mediated transformation of tomato plants with bacterial mannitol-1-phosphate dehydrogenase (mtlD) gene (Khare et al. 2010), potato cv. 'Superior' with choline oxidase gene (CodA) from *Arthrobacter globiformis* (Cheng et al. 2013), potato (*Solanum tuberosum* cv. Desiree) with StnsLTP1 (Gangadhar et al. 2016), mulberry with Hva1 gene from barley (Checker et al. 2012). Plant micro-RNA (miRNA) regulates drought responses. Transformation of tomato over-expressing Sly-miR169c down-regulated nuclear factor Y subunit genes (SINF-YA1/SINF-YA2/SINF-YA3) and multidrug resistance-associated protein gene (SIMRP1). This led to increased tolerance to drought traits in tomato plants (Zhang et al. 2011a, b). OsNCEB3 overexpression by CRISPRa improved accumulation of ABA and tolerance to salinity and drought in rice (Huang et al. 2018). CRISPR/dCAS9 based activation of ABA-responsive element binding proteins/ABRE binding factors (AREB/ABFs) activates gene expression and enhances drought resistance in Arabidopsis plants (Roca Paixão et al. 2019). In Arabidopsis and maize, ARGOS genes are negative regulators of ethylene signal responses and when overexpressed or by using CRISPRa approach enhance drought tolerance and grain yield in these plants (Shi et al. 2015, 2017).

3.1.2 Temperature Extremes

The high temperature (heat) stress produces hydrogen peroxide and superoxide in plant cells, and affects crop growth and productivity. The more plants have capability to scavenge ROS by enzymes like superoxide dismutase (SOD), glutathione reductase (GR) and peroxidase for normal growth and metabolism, the more is heating stress tolerance in plants (Noctor and Foyer 1998; Chaitanya et al. 2002). Transgenic apple and tomato plants developed by cytosolic ascorbate peroxidase (cAPX) gene over-expression showed enhanced heat tolerance (Wisniewski et al. 2002; Wang et al. 2006). Overexpression of Cu/Zn superoxide dismutase (Cu/Zn SOD) gene in potato enhanced tolerance to heat stress (Tang et al. 2006). The over-expression of S-adenosyl-I-methionine decarboxylase (SAMDC) from *Saccharomyces cerevisiae* in tomato plants enhanced production of polyamines namely spermidine and spermine and enhanced tolerance to high temperature stress (Cheng et al. 2009). The upregulation of heat shock protein (HSP) namely CgHSP70 gene improved peroxidase (POD) activity and proline content in chrysanthemum plants to endure heat stress (Song et al. 2014).

3.1.3 Salinity

Salinity affects various metabolic pathways and leads to osmotic and ionic/hormonal homeostasis in plant cells. Plant strategy in response to salinity is osmotic adjustment and compartmentation and exclusion of Na^+ and Cl^- ions (Munns and Tester 2008; Hasegawa 2013). Transgenic plants with over-expression of Na^+ transporter HKT1 lessen Na^+ ion toxicity in shoots and develop salt-tolerant plants (Paul and Roychoudhury 2018). GmSALT3 study between salt-tolerant and salt-sensitive soybean accessions suggested stress-related, molecular evolutionary signatures for rigorous, robust selection for alleles of salt tolerance (Guan et al. 2014; Haak et al. 2017). The approaches for the generation of transgenics include transformation with novel genes or altered expression levels of existing genes, e.g., genes of compatible organic solutes, water channel proteins, antioxidants, ion transporters, transcription factors, miRNAs etc. for salt stress tolerance. ABA-responsive element (ABRE), CBF (C-repeat binding factor)/DREBs (dehydration-responsive element binding protein, MYC (myelocytomatosis oncogene)/MYB (myeloblastosis oncogene), and NAC (NAM, ATAF and CUC) transcription factors provide tolerance to salinity (Lata et al. 2011). Overexpression of mitogen-activated protein kinase kinase (MKK) enhanced tolerance to salinity in rice plants. Overexpression of protein kinases upregulated multiple genes of osmotic-stress responses in cucumber (Oh et al. 2014). Transgenic plants of tobacco and rice overexpressing *E. coli* TPS gene (otsA) showed tolerance to salinity and increase in productivity (Garg et al. 2002; Penna 2003; Wu and Garg 2003; Jun et al. 2005; Guo et al. 2014). Choline oxidase (CodA), choline dehydrogenase (BetA), and betaine aldehyde dehydrogenases (Bet B) genes exhibited tolerance to salinity stress in *Escherichia coli*, *Arthrobacter globiformis*, and *Atriplex hortensis*. The CodA, BetA, and Bet B gene over-expression in *Nicotiana tabacum*, *A. thaliana*, *O. sativa*, *Brassica juncea*, *B. napus*, *B. oleracea*, and *B. campestris* exhibited tolerance to salt stress (Shirasawa et al. 2006; Bhattacharya et al. 2006; Ahmad et al. 2008; Duan et al. 2009; Cheng et al. 2013; Nguyen et al. 2013; Jiang et al. 2013; Lai et al. 2014). Transgenics generated with overexpression of SOD genes in rice, TaMnSOD in poplar plants, CuZnSOD and APX genes in sweet potato, DHAR gene in tobacco and potato plants exhibited enhanced salt stress tolerance (Eltayeb et al. 2006; Prashanth et al. 2008; Wang et al. 2010, 2011; Eltayeb et al. 2011), and with chloroplastic betaine aldehyde dehydrogenase (SoBADH) gene isolated from *Spinacia oleracea* and transferred to sweet potato cv. Sushu-2 (Fan et al. 2012), *Arabidopsis thaliana*-derived H+-pyrophosphatase AVP gene in bottle gourd line 'G5' (Han et al. 2015), strawberry D-galacturonic acid reductase (GalUR) gene transferred to *Solanum lycopersicum* var. cerasiforme (cherry tomato) (Lim et al. 2016) and wheat Na^+/H^+ antiporter gene (TaNHX2) in chilli pepper plants exhibited improved salinity tolerance (Bulle et al. 2016). Transgenic RNAi silencing of OsSRFP1 rice plants showed increase in salinity and cold tolerance due to increase in antioxidant potential. This can be achieved by CRISPRi as well (Fang et al. 2015, 2016). CRISPRi of transcription factor encoding *OsRR22* gene increased salinity stress tolerance in *O. sativa* L. (Zhang et al. 2019). CRISPR/Cas9 mediated downregulation and knockdown of

Auxin response factor (SlARF4) increased salinity tolerance in tomato (Bouzroud et al. 2020). CRISPRi knock-down of NADPH oxidase decreased salinity tolerance and ectopic expression of pumpkin RBOHD improved tolerance to salinity stress in *Arabidopsis thaliana* (Huang et al. 2019).

3.2 Biotic Stress

Transgenic plants resistant to insect pests, fungi, bacteria and viruses have been tailored in *Zea mays*, *O. sativa*, *Triticum aestivum*, *Glycine max*, *Solanum tuberosum*, *Solanum lycopersicum*, *Hordeum vulgare*, *Carica papaya* and *Medicago sativa*, squash, sugar beet, cotton, oilseed rape, and tobacco (James 2013). Some of the examples of transgenics generated for resistance to biotic stress factors are summarized in Table 35.2.

3.2.1 Resistance to Insect Pests

Bacillus thuringiensis produces proteinaceous crystalline (Cry) toxins which are toxic to lepidopteran insects by binding to their epithelial cell receptors of mid-gut (BANR 2000; Gahan et al. 2010). Bt genes encode Cry and these genes on transfer to various crops provide protection against insects (Gahan et al. 2010). Bt transgenic maize shows protection against *Ostrinia nubilalis* and *Sesamia nonagriodes* (with cry1Ab/cry1Ac or cry9C), *Spodoptera frugiperda* (with cry1F), and rootworms of Diabrotica (with cry3Bb, cry34Ab and cry35Ab) (James 2012). Bt cotton contains mainly cry1Ac or cry1Ac and cry1Ab fusion genes (James 2013). Bt potatoes with cry3Aa gene shows protection against *Leptinotarsa decemlineata* (Coombs et al. 2002). Bt brinjal shows protection against *Leucinodes orbonalis*. Bt crucifer vegetables show protection against *Plutella xylostella* pathogen (James 2012). Bt alfalfa with cry3a gene shows protection against *Hypera postica* (Tohidfar et al. 2013). The maize line MIR162 lines with vip3Aa20 gene showed resistance against various insects such as black cutworm, corn earworm, fall armyworm and Western bean cutworm (CERA 2010).

Lectins protect plants against Hemiptera (Powell et al. 1995). Transgenic rice plants with Allium leaf agglutinin (ASAL) gene showed resistance to hopper insect pests and with Galanthus nivalis (snow drop) agglutinin (GNA) showed tolerance against brown plant hopper (BPH) (Li et al. 2005; Saha et al. 2006).

Plant protease inhibitors (PI) are natural defense system and protect plants from predation by insects (Reeck et al. 1997). The first PI gene encoding trypsin/trypsin inhibitor CpTI (Cowpea Trypsin Inhibitor) was isolated from cowpea and transferred to a plant species which showed enhanced insect resistance (Hilder et al. 1987). Oryzacystatin 1 (OC1) cysteine PI isolated from rice seeds has been

Table 35.2 Transgenic crop plants resistant to biotic stress

Transgenic plant	Gene introduced	Resistance to stress type	References
Alfalfa	cry3a	Insect	Tohidfar et al. (2013)
Apple	NPR1 gene	Apple cedar rust caused by *V. inaequalis* and *Gymnosporangium juniperi-virginianae*; Apple Blight Disease by *Erwinia amylovora*	Thakur et al. (2018)
Apple cv. golden delicious	Knockdown of MLO-7; DIPM-1/2/4	Powdery mildew in grape cultivar Chardonnay; Fire Blight Disease	Malnoy et al. (2016)
Banana cv. 'Sukali Ndiizi' and 'Naki	Ferredoxin like protein (Pf1p) gene	Banana Xanthomonas wilt (BXW) caused by *Xanthomonas campestris* pv. musacearum	Namukwaya et al. (2012)
Chickpea	cryIAc	Insect	Sanyal et al. (2005)
Chilli Pepper	Coat Protein gene (CMVP0-CP)	CMVP0 and CMVP1 pathogen	Lee et al. (2009)
Cotton	Bean chitinase	Fungi	Tohidfar et al. (2005)
	crybaby	Cotton bollworm	Tohidfar et al. (2008)
Cucumber	Knockdown of eIF4E (eukaryotic translation initiation factor 4E) gene	CVYV, ZYMV, PRSMV-W	Chandrasekaran et al. (2016)
Eggplant	cryIIIB	Leptinotarsa decemlineata	Iannacone et al. (1997)
Grape cv. Freedom	rpfF gene (from *Xylella fastidiosa*)	Pierce's disease	Lindow et al. (2014)
Grapes Thompson Seedless grape	Stilbene synthase gene VqSTS6	Powdery mildew disease	Cheng et al. (2016)
Grapevine	knock down of susceptible (S-gene) MLO-7	Powdery Mildew	Pessina et al. (2016)
Grapevine rootstock of V. Berlandieri 9 V. Rupestris cv. Richter 110	Agrobacterium oncogene-silencing gene	Crown gall resistance	Galambos et al. (2013)
Maize	cry1H	European Corn Borer	Jansens (1997)
Papaya	Viral coat protein gene sequence and replicase (RP) gene	Papaya Ring Spot Virus (PRSV)	Gonsalves (2004), Xiangdong et al. (2007)
Potato	Snow drop lectin	Potato aphid	Gatehouse (1997)

(continued)

Table 35.2 (continued)

Transgenic plant	Gene introduced	Resistance to stress type	References
Rice	Xa21 class	Bacterial blight	Song et al. (1995)
	Chitinase	Fungi	Itoh et al. (2003)
	Cowpea serin PI	Stem borer	Duan et al. (1996)
	Barley trypsin inhibitor	Insect	Alfonso-Rubi et al. (2003)
	OsMKK6	UV radiation stress and blast infection	Wankhede et al. (2013)
	Stilbene Synthase Gene (STS) of Vst1	*Pyricularia oryzae* fungi	Coutos-Thévenot et al. (2001)
Soybean	Viral coat protein	Soybean Dwarf Virus	Tougou et al. (2006)
	Rps1-k	Phytophthora	Gao et al. (2005)
	cry1A	Insect	Macrae et al. (2005)
Strawberry	Coat Protein (CP)	Strawberry Mild Yellow Edge Potexvirus (SMYELV-CP)	Finstad and Martin (1995)
Tobacco	LpiO and Rpi-blb1 co-expression	Late blight	Vleeshouwers et al. (2008)
	Cotton ERF gene	Bacterial blight	Champion et al. (2009)
	CMV replicase derived dsRNA	Viruses	Otang Ntui et al. (2014)
Watermelon cv. Feeling, China rose, Quality	ZYMV coat protein (CP), PRSV-W (CP)	ZYMV, PRSV-W	Yu et al. (2011)

transferred to crop species like rice, wheat, oilseed rape and brinjal and these plants show protection against attacks by beetles and aphids (Duan et al. 1996; Altpeter et al. 1999; Rahbe et al. 2003; Sharma et al. 2004; Ribeiro et al. 2006). Bt-corn called Bt-Xtra has been generated with introgression of cry1Ac, bar from *Streptomyces higroscopicus* and potato proteinase inhibitor (pinII) (Oksman-Kaldentey and Barz 2002).

Alpha-amylase inhibitors control seed weevils. *Phaseolus vulgaris* amylase inhibitor gene on transference into transgenic *Pisum sativum* made their seeds resistant to larvae of bruchid beetles and pea weevil (Shade et al. 1994; Morton et al. 2000). The transgenic chickpea with this gene showed resistance to bruchid weevil (Ignacimuthu and Prakash 2006), whereas Coffea arabica seed extracts of transgenic plants inhibited amylolytic enzyme (Barbosa et al. 2010).

3.2.2 Resistance to Bacterial Pathogens

Pathogens may contain genes which encode toxin-detoxifying enzyme. The tabtoxin is produced by *Pseudomonas syringae* pv. *tabaci* which is converted to tabtoxinine-β-lactam in plants. Tabtoxinine-β-lactam stops glutamine synthase and cytotoxic ammonia accumulates in plant. *Pseudomonas syringae* pv. *tabaci* protects itself from this tabtoxin-by-tabtoxin resistance gene (ttr) expression which acetylates tabtoxin to inactive form. The ttr gene transgenic tobacco showed reduced symptoms (Batchvarova et al. 1998). Harpins (hrp) genes of phytopathogenic bacteria encode type III secretory pathways. The hrp genes elicit hypersensitive response (HR) on non-host or resistant host plants and for pathogenesis on susceptible hosts. The hrp gene secretions cause localized cell death by reactions like accumulation of reactive oxygen species (ROS). Harpin NEa (HrpNEa) (inducer of systemic acquired resistance (SAR)) is encoded by the gene hrpN of Erwinia which causes fire blight of apple. The transgenic plants with enhanced HrpNEa levels show increase in resistance to bacterial pathogens (Malnoy et al. 2005). Another approach for resistance against bacterial pathogen is by ethylene responsive transcription factors (ERF), which regulate pathogenesis-related (PR) gene expression (Grennan 2008). The ERF gene isolated from cotton and transformation in transgenic tobacco exhibits resistance to pathogen of bacterial blight (Champion et al. 2009). Transgenic Arabidopsis plants with presence of photosynthetic type Plant Ferrodoxin Like Protein (PFLP), showed enhanced resistance against bacterial pathogens (Lin et al. 2010). The other success stories include: transformation of ferredoxin like protein (Pflp) gene in banana cv. 'Sukali Ndiizi' and 'Naki' developed resistance against Banana Xanthomonas wilt (BXW) caused by *Xanthomonas campestris* pv. musacearum (Namukwaya et al. 2012), Agrobacterium oncogene-silencing gene in grapevine rootstock of V. berlandieri 9 V. rupestris cv. Richter 110 developed crown gall-resistant lines (Galambos et al. 2013), rpfF gene (isolated from *Xylella fastidiosa*) in grape cv. Freedom developed resistance against Pierce's disease (Lindow et al. 2014), stilbene synthase gene VqSTS6 from Chinese wild-type *Vitis quinquangularis* accession Danfeng-2 in *Vitis vinifera* Thompson Seedless grape developed resistance against powdery mildew disease (Cheng et al. 2016) and RNA-interference mediated knock down of susceptible (S-gene) mildew resistance locus o (MLO) in grapevine against powdery mildew (Pessina et al. 2016).

3.2.3 Resistance to Fungal Pathogens

A number of strategies have been applied for achieving fungal resistance in plants. These include protection by Chitinase and glucanase genes which encode glycolytic enzymes with cell wall degrading abilities (Jongedijk et al. 1995; Ceasar and Ignacimuthu 2012; Tohidfar et al. 2012), glycoproteins (Di et al. 2010), R genes (resistance) (Pel et al. 2009), and activation of phytoalexins in various transgenic plant species (Wankhede et al. 2013). Development of transgenic crops like grapevine cv. Neo Muscat, peanut and cotton by transformation with plant or microbial-derived genes encoding β-1,3-glucanase (degrade glucans) and chitinase which hydrolyzes chitin present in fungal cell walls after pathogen infection showed enhanced resistance to fungal diseases (Yamamoto et al. 2000; Rohini et al. 2000; Tohidfar 2012). The simultaneous chitinase plus glucanase expression in tomato, carrot, and tobacco (Neuhaus 1999; Melchers and Stuiver 2000), kiwifruit over-expressing soybean b-1,3 glucanase gene (Nakamura et al. 1999), alfalfa glucanase gene in brinjal cv. Pusa Purple Long (Singh et al. 2014), *Trichoderma harzianum* b-1,3-glucanase gene bgn13,1 in strawberry (Mercado et al. 2015), tobacco ChiC gene in carrot (Punja and Raharjo 1996), RCC2 rice gene in strawberry (Asao et al. 1997), thaumatin II gene from *Thaumatococcus danielli* (African serendipity berry) in *Fragaria* × *ananassa* (strawberry) (Schestibratov and Dolgov 2005), chitinase gene from common bean, a glucanase or a thaumatin-like protein, both from *Nicotiana tabacum* and a combination of both in strawberry cv. 'Pajaro' (Vellice et al. 2006), bacterial chitinase (chi B) gene in *Litchi chinensis* cv. Bedana (Das and Rahman 2010), CHIT42 and CHIT33 from *Trichoderma harzianum* in *Solanum tuberosum* (Lorito et al. 1998), *Malus domestica* (Bolar et al. 2001), *Brassica oleracea* var. italica (Mora and Earle 2001), and ChiC from *Streptomyces griseus* strain HUT 6037 in addition to bialaphos resistance (bar) gene into *Solanum tuberosum* (Khan et al. 2008) enhanced tolerance to fungal growth in these transformed plants.

Polygalacturonase inhibiting proteins (PGIP) of plant cell wall prevent endo-polygalacturonases activity present in fungal pathogens (Oelofse et al. 2006). The transgenic wheat exhibited improved resistance to Fusarium head blight (FHB) on transfer of L3 gene (N-terminal fragment of yeast ribosomal protein) and also improved levels of deoxynivalenol (DON) mycotoxin in kernels of wheat plants (Di et al. 2010).

To overcome the late blight in Solanaceous species, several biotechnological approaches for conferring resistance have been performed. LpiO gene effectors activate innate immunity of Solanum species and cause HR. The co-expression of LpiO (effector) and Rpi-blb1 (resistance gene) in *Nicotiana benthamiana* identified Rpi-sto1 and Rpi-pta1 as resistance genes to late blight disease (Vleeshouwers et al. 2008). The assembling of Rpi-sto1 (*Solanum stoloniferum*), Rpi-vnt1.1 (*Solanum venturii*) and Rpi-blb3 (*Solanum bulbocastanum*) in susceptible cultivar Desiree lead to HR in 4% of transgenic plants (Zhu et al. 2012). The activation of phytoalexins of defense mechanism confers disease resistance in some plant species. Rice transgenics generated by stilbene synthase gene (STS) of Vst1 showed improved resistance to *Pyricularia oryzae* (Coutos-Thévenot et al. 2001). *Hordeum vulgare* showed enhanced resistance to powdery mildew disease (Liang et al. 2000). Mitogen-activated protein kinase (MAPK) cascade regulate expression of genes of phytoalexin synthesis in rice under UV (increased) radiation stress and blast infection (Wankhede et al. 2013). OsMKK6 transgenic rice lines over-expressed phytoalexins.

3.2.4 Resistance to Viral Pathogens

Some 700 viruses cause diseases in plants and crop losses. The approach of coat protein-mediated virus resistance resists viral pathogens in crop plants. Examples include resistance to Papaya Ring Spot Virus (PRSV) was obtained in high-yielding papaya hybrid lines by transformation with viral coat protein gene sequence as transgene and mutation of replicase (RP) gene (Gonsalves 2004; Xiangdong et al.

2007), viral resistance in strawberry by expressing coat protein (cp) gene isolated from Strawberry Mild Yellow Edge Potexvirus (SMYELV-CP) (Finstad and Martin 1995), chilli pepper transformed with coat protein gene (CMVP0-CP) to generate tolerance against CMVP0 and CMVP1 pathogen (Lee et al. 2009), banana (cv. Dwarf Brazilian) tolerant to Banana Bunchy Top Virus (BBTV) (Borth et al. 2011), virus resistance in lily cv. 'Acapulco' by transformation with replicase gene of Cucumber Mosaic Virus (Azadi et al. 2011), and watermelon cv. 'Feeling', 'China rose', and 'Quality' transformed with ZYMV coat protein (CP) and PRSV W CP genes showed complete resistance to Zucchini Yellow Mosaic Virus (ZYMV) and Papaya Ring Spot Virus type W (PRSV W) (Yu et al. 2011).

Potato event HLMT15- 15 was resistant to Potato Y Virus (PYV) or potato event RBMT21-350 was resilient to PLRV (Potato Leaf Roll Virus) (James 2013). The tobacco expressing defective Cucumber Mosaic Virus (CMV) replicas derived dsRNA showed defence against viruses. The transgenic plants showed virus resistance through defective Movement Protein (MP) of virus (Hallwass et al. 2014; Otang Ntui et al. 2014; Peiró et al. 2014). Pathogen-resistant plants have been produced by engineering antibodies or rAb fragments which inactivate pathogens/proteins of pathogenesis (Cardoso et al. 2014). The pokeweed antiviral protein (PAP) depurinate extremely conserved parts of ribosome which further leads to the inhibition of viral protein translation system (Thamizhmani and Vijayachari 2014). Poinsettia plants resistant to Poinsettia Mosaic Virus have been developed by hairpin RNA gene silencing (Clarke et al. 2008). Tomato transformed with AC4 gene-RNAi construct showed suppression of activity of tomato leaf curl virus (Praveen et al. 2010). Transgenic banana resistant to BBTV expressed small interference RNA (siRNA) which targeted the viral replication initiation (Rep) gene (Shekhawat et al. 2012).

The resistance to viruses is generally attained by introduction of gene sequences from pathogenic viruses into the genome of crop plants by use of gene silencing, antisense RNA and RNAi and clustered regularly interspersed short palindromic repeats (CRISPR)-associated protein (CRISPR/Cas) toolbox (Ramesh et al. 2007; Yan et al. 2007; Ali et al. 2015; Baltes et al. 2015; Mamta and Rajam 2018). CRISPR/Cas9 technology could successfully develop resistance to Cucumber Vein Yellowing Virus (CVYV), Zucchini Yellow Mosaic Virus (ZYMV) and Papaya Ring Spot Mosaic Virus-W (PRSMV-W) resistance in cucumber plant by knockdown of eukaryotic translation initiation factor 4E (eIF4E) gene (Chandrasekaran et al. 2016), to powdery mildew in grape cv. Chardonnay by knockdown of MLO-7 and to fire blight disease by knockdown of DIPM-1/2/4 genes in apple cv golden delicious (Malnoy et al. 2016).

4 Conclusion

The dawn of transgenics has revolutionized approaches of designing stress resilient crop plants for the development of sustainable agriculture and global food security. The sensor receptors, antioxidant system, transcription factors, microRNAs and other stress signaling pathway genes as discussed above have been tailored to design overexpressed or knockdown stress tolerant-lines of crop plants with resilience to various stressors by genome editing toolboxes of antisense RNA, RNAi, miRNA and CRISPR/CAS induction (CRISPR activation) or repression (CRISPR interference).

References

Acquaah G (2007) Principles of plant genetics and breeding, 2nd edn. Blackwell Publishing, Hoboken, Oxford, p 44. ISBN 1-4051-3646-4

Ahmad R, Kim MD, Back KH, Kim HS, Lee HS, Kwon SY, Murata N, Chung WI, Kwak SS (2008) Stress-induced expression of choline oxidase in potato plant chloroplasts confers enhanced tolerance to oxidative, salt, and drought stresses. Plant Cell Rep 27:687–698

Ahmad R, Hussain J, Jamil M, Duck Kim M, Kwak S, Maroof Shah M (2014) Glycine betaine synthesizing transgenic potato plants exhibit enhanced tolerance to salt and cold stresses. http://www.pakbs.org/pjbot/PDFs/46(6)/08.pdf (accessed October 5, 2020)

Alfonso-Rubí J, Ortego F, Castañera P, Carbonero P, Díaz I (2003) Transgenic expression of trypsin inhibitor CMe from barley in indica and japonica rice, confers resistance to the rice weevil *Sitophilus oryzae*. Transgenic Res 12(1):23–31. https://doi.org/10.1023/a:1022176207180

Ali Z, Abulfaraj A, Idris A, Ali S, Tashkandi M, Mahfouz MM (2015) CRISPR/Cas9-ediated viral interference in plants. Genome Biol 16:238

Altpeter F, Diaz I, McAuslane H, Gaddour K, Carbonero P, Vasil IK (1999) Increased insect resistance in transgenic wheat stably expressing trypsin inhibitor CMe. Mol Breeding 5:53–63. https://doi.org/10.1023/A:1009659911798

Amara I, Capellades M, Ludevid MD, Pagès M, Goday A (2013) Enhanced water stress tolerance of transgenic maize plants over-expressing LEA Rab28 gene. J Plant Physiol 170:864–873. https://doi.org/10.1016/j.jplph.2013.01.004

Asao H, Nishizawa Y, Arai S, Sato T, Hirai M, Yoshida K, Shinmyo A, Hibi T (1997) Enhanced resistance against a fungal pathogen *Sphaerotheca humuli* in transgenic strawberry expressing a rice chitinase gene. Plant Biotechnol 14:145–149

Azadi P, Otang NV, Supaporn H, Khan RS, Chin DP, Nakamura I, Mii M (2011) Increased resistance to cucumber mosaic virus (CMV) in *Lilium* transformed with a defective CMV replicase gene. Biotechnol Lett 33:1249–1255

Baltes NJ, Hummel AW, Konecna E, Cegan R, Bruns AN, Bisaro DM, Voytas DF (2015) Conferring resistance to geminiviruses with the CRISPR-Cas prokaryotic immune system. Nat Plants 1(10):15145. https://doi.org/10.1038/nplants.2015.145

BANR (Board on Agriculture and Natural Resources) (2000) Genetically modified pest protected plant: science and regulation. National Academy Press, Washington

Barbosa AE, Albuquerque EV, Silva MC, Souza DS, Oliveira-Neto OB, Valencia A, Rocha TL, Grossi-de-Sa MF (2010) Alpha-amylase inhibitor-1 gene from *Phaseolus vulgaris* expressed in *Coffea arabica* plants inhibits alpha-amylases from the coffee berry borer pest. BMC Biotechnol 10:44. https://doi.org/10.1186/1472-6750-10-44

Barciszewska-Pacak M, Milanowska K, Knop K, Bielewicz D, Nuc P, Plewka P, Pacak AM, Vazquez F, Karlowski W, Jarmolowski A, Szweykowska-Kulinska Z (2015) Arabidopsis microRNA expression regulation in a wide range of abiotic stress responses. Front Plant Sci 6:410

Barrera-Figueroa BE, Gao L, Wu Z, Zhou X, Zhu J, Jin H, Liu R, Zhu JK (2012) High throughput sequencing reveals novel and abiotic stress-regulated microRNAs in the inflorescences of rice. BMC Plant Biol 12:132

Batchvarova R, Nikolaeva V, Slavov S, Bossolova S, Valkov V, Atanassova S, Guelemerov S, Atanassov A, Anzai H (1998) Transgenic tobacco cultivars resistant to *Pseudomonas syringae* pv. tabaci. Theor Appl Genet 97:986–989

Batley J, Edwards D (2016) The application of genomics and bioinformatics to accelerate crop improvement in a changing climate. Curr Opin Plant Biol 30:78–81

Bhaskara GB, Yang TH, Verslues PE (2015) Dynamic proline metabolism: importance and regulation in water limited environments. Front Plant Sci 6:484

Bhatnagar-Mathur P, Vadez V, Sharma KK (2008) Transgenic approaches for abiotic stress tolerance in plants: retrospect and prospects. Plant Cell Rep 27:411–424

Bhattacharya A, Saini U, Sharma P, Nagar PK, Ahuja PS (2006) Osmotin-regulated reserve accumulation and germination in genetically transformed tea somatic embryos: a step towards regulation of stress tolerance and seed recalcitrance. Seed Sci Res 16:203–211

Bihani P, Char B, Bhargava DS (2010) Transgenic expression of sorghum DREB2 in rice improves tolerance and yield under water limitation. J Agric Sci 149:95. https://doi.org/10.1017/S0021859610000742

Bolar JP, Norelli JL, Harman GE, Brown SK, Aldwinkle HS (2001) Synergistic activity of endochitinase and exochitinase from *Trichoderma atroviride* (*T. harzianum*) against the pathogenic fungus (*Venturia inaequalis*) in transgenic apple plants. Transgenic Res 10:533–543

Borth W, Perez E, Cheah K, Chen Y, Xie WS, Gaskill D, Khalil S, Sether D, Melzer M, Wang M, Manshardt R, Gonsalves D, Hu JS (2011) Transgenic banana plants resistant to banana bunchy top virus infection. Acta Hortic 897:449–457. https://doi.org/10.17660/ActaHortic.2011.897.61

Bouzroud S, Gasparini K, Hu G, Barbosa MAM, Rosa BL, Fahr M, Bendaou N et al (2020) Down regulation and loss of auxin response factor 4 function using CRISPR/Cas9 alters plant growth, stomatal function and improves tomato tolerance to salinity and osmotic stress. Genes 11. https://doi.org/10.3390/genes11030272

Bulle M, Yarra R, Abbagani S (2016) Enhanced salinity stress tolerance in transgenic chilli pepper (*Capsicum annuum* L.) plants overexpressing the wheat antiporter (TaNHX2) gene. Mol Breed 36:3

Cardoso FM, Ibañez LI, Van den Hoecke S, De Baets S, Smet A, Roose K, Schepens B, Descamps FJ, Fiers W, Muyldermans S, Depicker A, Saelens X (2014) Single domain antibodies targeting neuraminidase protect against an H5N1 influenza virus challenge. J Virol 88(15):8278–8296. https://doi.org/10.1128/JVI.03178-13

Ceasar SA, Ignacimuthu S (2012) Genetic engineering of crop plants for fungal resistance: role of antifungal genes. Biotechnol Lett 34:995–1002

CERA (Center for Environmental Risk Assessment) (2010) GM Crop Database, www.cera-gmc.org/index.php?action=gm_crop_database, (18.10.2010)

Chaitanya KV, Sundar D, Masilamani S, Reddy AR (2002) Variation in heat stress-induced antioxidant enzyme activities among three mulberry cultivars. Plant Growth Regul 36:175–180

Chakraborty S, Newton AC (2011) Climate change, plant diseases and food security: an overview. Plant Pathol 60(1):1–14

Champion A, Hebrard E, Parra B, Bournaud C, Marmey P, Tranchant C, Nicole M (2009) Molecular diversity and gene expression of cotton ERF transcription factors reveal that group IXa members are responsive to jasmonate, ethylene and Xanthomonas. Mol Plant Pathol 10(4):471–485. https://doi.org/10.1111/j.1364-3703.2009.00549.x

Chandrasekaran J, Brumin M, Wolf D, Leibman D, Klap C, Pearlsman M, Sherman A, Arazi T, Gal-on A (2016) Development of broad virus resistance in non-transgenic cucumber using CRISPR/Cas9 technology. Mol Plant Pathol 17(7):1140–1153

Chauhan H, Khurana N, Nijhavan A, Khurana JP, Khurana P (2012) The wheat chloroplastic small heat shock protein (sHSP26) is involved in seed maturation and germination and imparts tolerance to heat stress. Plant Cell Environ 35:1912–1931. https://doi.org/10.1111/j.1365-3040.2012.02525.x

Checker VG, Chhibbar AK, Khurana P (2012) Stress-inducible expression of barley Hva1 gene in transgenic mulberry displays enhanced tolerance against drought, salinity and cold stress. Transgenic Res 21(5):939–957. https://doi.org/10.1007/s11248-011-9577-8

Cheng L, Zou Y, Ding S, Zhang J, Yu X, Cao J, Lu G (2009) Polyamine accumulation in transgenic tomato enhances the tolerance to high temperature stress. J Integr Plant Biol 51:489–499
Cheng YJ, Deng XP, Kwak SS, Chen W, Eneji AE (2013) Enhanced tolerance of transgenic potato plants expressing choline oxidase in chloroplasts against water stress. Bot Stud 54:30
Cheng S, Xie X, Xu Y, Zhang C, Wang X, Zhang J, Wang U (2016) Genetic transformation of a fruit-specific, highly expressed stilbene synthase gene from Chinese wild *Vitis quinquangularis*. Planta 243:1041–1053
Clarke JL, Spetz C, Haugslien S, Xing S, Dees MW, Moe R, Blystad DR (2008) *Agrobacterium tumefaciens*-mediated transformation of poinsettia, *Euphorbia pulcherrima*, with virus-derived hairpin RNA constructs confers resistance to Poinsettia mosaic virus. Plant Cell Rep 27:1027–1038
Coombs JJ, Douches DS, Li W, Grafius EJ, Pett WL (2002) Combining engineered (Bt-cry3A) and natural resistance mechanisms in potato for control of Colorado potato beetle. J Am Soc Hortic Sci 127(1):62–68
Coutos-Thévenot P, Poinssot B, Bonomelli A, Yean H, Breda C, Buffard D, Esnault R, Hain R, Boulay M (2001) In vitro tolerance to *Botrytis cinerea* of grapevine 41B rootstock in transgenic plants expressing the stilbene synthase *Vst1* gene under the control of a pathogen-inducible PR 10 promoter. J Exp Bot 52(358):901–910. https://doi.org/10.1093/jexbot/52.358.901
Das DK, Rahman A (2010) Expression of a bacterial chitinase (ChiB) gene enhances antifungal potential in transgenic *Litchi chinensis* Sonn. (cv. Bedana). Curr Trends Biotechnol Pharm 41:820–833
Dhankher OP, Foyer CH (2018) Climate resilient crops for improving global food security and safety. Plant Cell Environ 41:877–884. https://doi.org/10.1111/pce.13207
Di R, Blechl A, Dill-Macky R et al (2010) Expression of a truncated form of yeast ribosomal protein L3 in transgenic wheat improves resistance to Fusarium head blight. Plant Sci 178:374–380
Dong L, Wang Q, Manik SMN, Song Y, Shi S, Su Y, Liu G, Liu H (2015) *Nicotiana sylvestris* calcineurin B-like protein NsylCBL10 enhances salt tolerance in transgenic Arabidopsis. Plant Cell Rep 34:2053–2063. https://doi.org/10.1007/s00299-015-1851-4
Duan J, Cai W (2012) OsLEA3-2, an abiotic stress induced gene of rice plays a key role in salt and drought tolerance. PLoS One 7(9):e45117. https://doi.org/10.1371/journal.pone.0045117
Duan X, Li X, Xue Q, Abo-el-Saad M, Xu D, Wu R (1996) Transgenic rice plants harboring an introduced potato proteinase inhibitor II gene are insect resistant. Nat Biotechnol 14(4):494–498. https://doi.org/10.1038/nbt0496-494
Duan X, Song Y, Yang A, Zhang J (2009) The transgene pyramiding tobacco with betaine synthesis and heterologous expression of AtNHX1 is more tolerant to salt stress than either of the tobacco lines with betaine synthesis or AtNHX1. Physiol Plant 135(3):281–295
Eltayeb AE, Kawano N, Badawi G, Kaminaka H, Sanekata T, Morishima I (2006) Enhanced tolerance to ozone and drought stresses in transgenic tobacco overexpressing dehydroascorbate reductase in cytosol. Physiol Plant 127:57–65
Eltayeb AS, Yamamoto S, Habora MEE, Yin L, Tsujimoto H, Tanak K (2011) Transgenic potato over-expressing Arabidopsis cytosolic AtDHAR1 showed higher tolerance to herbicide, drought and salt stresses. Breed Sci 611:3–10
Fan W, Zhang M, Zhang H, Zhang P (2012) Improved tolerance to various abiotic stresses in transgenic sweet potato (*Ipomoea batatas*) expressing spinach betaine aldehyde dehydrogenase. PLoS One 7:e37344. https://doi.org/10.1371/journal.pone.0037344
Fan Q, Song A, Jiang Zhang T, Sun H, Wang Y (2016) CmWRKY1 enhances the dehydration tolerance of chrysanthemum through the regulation of ABA-associated genes. PLoS One 11:e0150572
Fang H, Meng Q, Xu J, Tang H, Tang S, Zhang H, Huang J (2015) Knock-down of stress inducible OsSRFP1 encoding an E3 ubiquitin ligase with transcriptional activation activity confers abiotic stress tolerance through enhancing antioxidant protection in rice. Plant Mol Biol 87:441–458. https://doi.org/10.1007/s11103-015-0294-1
Fang H, Meng Q, Zhang H, Huang J (2016) Knock-down of a RING finger gene confers cold tolerance. Bioengineered 7:39–45. https://doi.org/10.1080/21655979.2015.1131368

Ferguson JN (2019) Climate change and abiotic stress mechanisms in plants. Emerg Topics Life Sci 3:165–181

Finstad K, Martin RR (1995) Transformation of strawberry for virus resistance. Acta Hortic 385:6–90

Fu XZ, Khan EU, Hu SS, Fan QJ, Liu JH (2011) Overexpression of the betaine aldehyde dehydrogenase gene from *Atriplex hortensis* enhances salt tolerance in the transgenic trifoliate orange (*Poncirus trifoliata* L. Raf.). Environ Exp Bot 74:106–113. https://doi.org/10.1016/j.envexpbot.2011.05.006

Gahan LJ, Pauchet Y, Vogel H, Hecke DG (2010) An ABC transporter mutation is correlated with insect resistance to *Bacillus thuringiensis* Cry1Ac toxin. PLoS Genet 6(12):1–11

Galambos A, Zok A, Kuczmog A, Olah R, Putnoky P, Ream W, Szegedi E (2013) Silencing Agrobacterium oncogenes in transgenic grapevine results in strain-specific crown gall resistance. Plant Cell Rep 32:1751–1757

Gangadhar BH, Sajeesh K, Venkatesh J, Baskar V, Abhinandan K, Yu JW, Prasad R, Mishra RK (2016) Enhanced tolerance of transgenic potato plants over-expressing non-specific lipid transfer protein-1 (StnsLTP1) against multiple abiotic stresses. Front Plant Sci 7:1228

Gao H, Narayanan NN, Ellison L, Bhattacharyya MK (2005) Two classes of highly similar coiled coilnucleotide binding-leucine rich repeat genes isolated from the Rps1-k locus encode Phytophthora resistance in soybean. Mol Plant-Microbe Interact 18:1035–1045

Gao SQ, Chen M, Xu ZS, Zhao CP, Li L, jun Xu H, miao Tang Y, Zhao X, Ma YZ (2011) The soybean GmbZIP1 transcription factor enhances multiple abiotic stress tolerances in transgenic plants. Plant Mol Biol 75:537–553. https://doi.org/10.1007/s11103-011-9738-4

Garg AK, Kim JK, Owens TG, Ranwala AP, Choi YD, Kochian LV, Wu RJ (2002) Trehalose accumulation in rice plants confers high tolerance levels to different abiotic stresses. Proc Natl Acad Sci U S A 99(25):15898–15903

Gatehouse AM, Davison GM, Newell CA et al (1997) Transgenic potato plants with enhanced resistance to the tomato moth, *Lacanobia oleracea*: growth room trials. Mol Breed 3(1):49–63. https://doi.org/10.1023/A:1009600321838

Ghanti, S. K. K., Sujata, K.G., Vijay Kumar, B.M. et al. (2011) Heterologous expression of P5CS gene in chickpea enhances salt tolerance without affecting yield. Biol Plant 55, 634. https://doi.org/10.1007/s10535-011-0161-0

Gonsalves D (2004) Transgenic papaya in Hawaii and beyond. AgBioforum 7:36–40

Grennan AK. (2008). Ethylene response factors in jasmonate signaling and defense response. Plant Physiol. 146(4):1457–1458. https://doi.org/10.1104/pp.104.900254

Guan R, Qu Y, Guo Y et al (2014) Salinity tolerance in soybean is modulated by natural variation in GmSALT3. Plant J 80:937–950

Guo BT, Wang B, Weng ML, Qiao LX, Feng YB, Wang L, Zhang PY, Wang XL, Sui JM, Liu T, Duan DL, Wang B (2014) Expression of *Porphyra yezoensis* TPS gene in transgenic rice enhanced the salt tolerance. J Plant Breed Genet 2(1):45–55

Haak DC, Fukao T, Grene R, Hua Z, Ivanov R, Perrella G, Li S (2017) Multilevel regulation of abiotic stress responses in plants. Front Plant Sci 8:1564

Hallwass M, de Oliveira AS, de Campos DE, Lohuis D, Boiteux LS, Inoue-Nagata AK, Resende RO, Kormelink R (2014) The Tomato spotted wilt virus cell-to-cell movement protein (NSM) triggers a hypersensitive response in Sw-5-containing resistant tomato lines and in *Nicotiana benthamiana* transformed with the functional Sw-5b resistance gene copy. Mol Plant Pathol 15(9):871–880

Han JS, Park LI, Jeon SM, Park S, Naing AH, Kim CK (2015) Assessments of salt tolerance in a bottle gourd line expressing the Arabidopsis H^+-pyrophosphatase AVP1 gene and in a watermelon plant grafted onto a transgenic bottle gourd rootstock. Plant Breed 134:233–238

Hasegawa PM (2013) Sodium (Na^+) homeostasis and salt tolerance of plants. Environ Exp Bot 92:19–31

He C, Yang A, Zhang W, Gao Q, Zhang J (2010) Improved salt tolerance of transgenic wheat by introducing *betA* gene for glycine betaine synthesis. Plant Cell Tissue Organ Cult 101:65–78. https://doi.org/10.1007/s11240-009-9665-0

He C, Zhang W, Gao Q, Yang A, Hu X, Zhang J (2011) Enhancement of drought resistance and biomass by increasing the amount of glycine betaine in wheat seedlings. Euphytica 177:151–167. https://doi.org/10.1007/s10681-010-0263-3

Hilder VA, Gatehouse AMR, Sheerman SE, Barker RF et al (1987) A novel mechanism of insect resistance engineered into tobacco. Nature 333:160–163

Hoang TML, Moghaddam L, Williams B, Khanna H, Dale J, Mundree SG (2015) Development of salinity tolerance in rice by constitutive overexpression of genes involved in the regulation of programmed cell death. Front Plant Sci 6:175. https://doi.org/10.3389/fpls.2015.00175

Huang Y, Guo Y, Liu Y, Zhang F, Wang Z, Wang H, Wang F, Li D, Mao D, Luan S, Liang M, Chen L (2018) 9-cis-Epoxycarotenoid dioxygenase 3 regulates plant growth and enhances multi-abiotic stress tolerance in rice. Front Plant Sci 9:162. https://doi.org/10.3389/fpls.2018.00162

Huang Y, Cao H, Yang L, Chen C, Shabala L, Xiong M, Niu M et al (2019) Tissue-specific respiratory burst oxidase homolog-dependent H_2O_2 signaling to the plasma membrane H^+-ATPase confers potassium uptake and salinity tolerance in Cucurbitaceae. J Exp Bot 70:5879–5893. https://doi.org/10.1093/jxb/erz328

Iannacone R, Grieco D, Cellini F (1997) Specific sequence modifications of a cry3B endotoxin gene result in high levels of expression and insect resistance. Plant Mol Biol 34:485–496

Ignacimuthu S, Prakash S (2006) *Agrobacterium* mediated transformation of chickpea with alpha-amylase inhibitor gene for insect resistance. J Biosci 31(3):339–345

Isayenkov SV (2019) Genetic sources for the development of salt tolerance in crops. Plant Growth Reg 89:1–17

Itoh Y, Takahashi K, Takizawa H, Nikaidou N, Tanaka H, Nishihashi H, Watanabe T, Nishizawa Y (2003) Family 19 chitinase of Streptomyces griseus HUT6037 increases plant resistance to the fungal disease. Biosci Biotechnol Biochem 67:847–855

James C (2012) Global status of commercialized biotech/GM crops. Brief No. 44. ISAAA, Ithaca

James C (2013) Global status of commercialized biotech/ GM crops. Brief No. 46. ISAAA, Ithaca

Jansens S, Vliet AV, Dickburt C, Buysse L, Piens C, Saey B, Wulf AD, Gosselé V, Paez AV, Göbel E, Peferoen M (1997) Transgenic corn expressing a Cry9C insecticidal protein from *Bacillus thuringiensis* protected from European corn borer damage. Crop Sci 37(5):1616–1624

Jha S (2019) Transgenic approaches for enhancement of salinity stress tolerance in plants. In: Singh SP et al (eds) Molecular approaches in plant biology and environmental challenges, energy, environment, and sustainability. Springer Nature, Singapore, pp 265–322

Jiang J, Li H, He G, Yin Y, Liu M, Liu B, Qiao G, Lin S, Xie L, Zhuo R (2013) Over-expression of the Cod A gene by Rd29a promoter improves salt tolerance in *Nicotiana tabacum*. Pak J Bot 45(3):821–827

Jin WM, Dong J, Hu YL, Lin ZP, Xu XF, Han ZH (2009) Improved cold-resistant performance in transgenic grape (*Vitis vinifera* L.) overexpressing cold-inducible transcription factors AtDREB1b. Hortic Sci 44:35–39

Jisha V, Dampanaboina L, Vadassery J, Mithöfer A, Kappara S, Ramanan R (2015) Overexpression of an AP2/ERF type transcription factor OsEREBP1 confers biotic and abiotic stress tolerance in rice. PLoS One 10:e0127831. https://doi.org/10.1371/journal.pone.0127831

Jongedijk E, Tigelaar H, van Roekel JSC et al (1995) Synergistic activity of chitinases and β-1,3-glucanases enhances fungal resistance in transgenic tomato plants. Euphytica 85:173–180

Joo J, Choi HJ, Lee YH, Kim YK, Song SI (2013) A transcriptional repressor of the ERF family confers drought tolerance to rice and regulates genes preferentially located on chromosome 11. Planta 238:155–170. https://doi.org/10.1007/s00425-013-1880-6

Jun SS, Yang JY, Choi HJ, Kim NR, Park MC, Hong YN (2005) Altered physiology in trehalose-producing transgenic tobacco plants: enhanced tolerance to drought and salinity stresses. J Plant Biol 48(4):456

Kantar M, Lucas S, Budak H (2011) miRNA expression patterns of *Triticum dicoccoides* in response to shock drought stress. Planta 233:471–484

Khan RS, Sjahril R, Nakamura I, Mii M (2008) Production of transgenic potato exhibiting enhanced resistance to fungal infections and herbicide applications. Plant Biotechnol Rep 2:13–20

Khare N, Goyary D, Singh NK, Shah P, Rathore M, Anandhan S (2010) Transgenic tomato cv. Pusa Uphar expressing a bacterial mannitol-1-phosphate dehydrogenase gene confers abiotic stress tolerance. Plant Cell Tissue Organ Cult 2:267–277

Kim GB, Nam YW (2013) A novel Δ1-pyrroline-5-carboxylate synthetase gene of *Medicago truncatula* plays a predominant role in stress-induced proline accumulation during symbiotic nitrogen fixation. J Plant Physiol 170:291–302. https://doi.org/10.1016/j.jplph.2012.10.004

Kondrák M, Marincs F, Antal F, Juhász Z, Bánfalvi Z (2012) Effects of yeast trehalose-6-phosphate synthase 1 on gene expression and carbohydrate contents of potato leaves under drought stress conditions. BMC Plant Biol 12:1–12. https://doi.org/10.1186/1471-2229-12-74

Lai SJ, Lai MC, Lee RJ, Chen YH, Yen HE (2014) Transgenic Arabidopsis expressing osmolyte glycine betaine synthesizing enzymes from halophilic methanogen promote tolerance to drought and salt stress. Plant Mol Biol 85(4–5):429–441

Lamers J, van der Meer T, Testerink C (2020) How plants sense and respond to stressful environments. Plant Physiol 182:1624–1635

Lata C, Prasad M (2011) Role of DREBs in regulation of abiotic stress responses in plants. J Exp Bot 62:4731–4748

Lata C, Yadav A, Prasad M (2011) Role of plant transcription factors in abiotic stress tolerance. In: Shanker A (ed) Abiotic stress response in plants – physiological, biochemical and genetic perspectives. InTech, Rijeka

Lee YH, Jung M, Shin SH, Lee JH, Choi SH, Her NH (2009) Transgenic peppers that are highly tolerant to a new CMV pathotype. Plant Cell Rep 28:223–232

Li X., Xie X., Li J., Cui Y., Hou Y., Zhai L., et al. (2017). Conservation and Diversification of the miR166 Family in Soybean and Potential Roles of Newly Identified miR166s. BMC Plant Biol. 17, 32. https://doi.org/10.1186/s12870-017-0983-9

Li G, Xu X, Xing H, Zhu H, Fan Q (2005) Insect resistance to *Nilaparvata Lugens* and *Cnaphalocrocis medinalis* in transgenic indica rice and the inheritance of gna+sbti transgenes. Pest Manag Sci 61:390–396

Li H, Dong Y, Yin H, Wang N, Yang J, Liu X, Wang Y, Wu J, Li X (2011a) Characterization of the stress associated microRNAs in *Glycine max* by deep sequencing. BMC Plant Biol 11:170

Li HW, Zang BS, Deng XW, Wang XP (2011b) Overexpression of the trehalose6-phosphate synthase gene OsTPS1 enhances abiotic stress tolerance in rice. Planta 234:1007–1018. https://doi.org/10.1007/s00425-011-1458-0

Liang H, Zheng J, Duan X et al (2000) A transgenic wheat with a stilbene synthase gene resistant to powdery mildew obtained by biolistic method. Chin Sci Bull 45:634–638

Liang J, Zhou M, Zhou X, Jin Y, Xu M, Lin J (2013) JcLEA, a novel LEA-like protein from *Jatropha curcas*, confers a high level of tolerance to dehydration and salinity in *Arabidopsis thaliana*. PLoS One 8. https://doi.org/10.1371/journal.pone.0083056

Lim MY, Jeong BR, Jung M, Harn CH (2016) Transgenic tomato plants expressing strawberry D-galacturonic acid reductase gene display enhanced tolerance to abiotic stresses. Plant Biotechnol Rep 10:105–116

Lin YH, Huang HE, Wu FS, Ger MJ, Liao PL, Chen YR, Tzeng KC, Feng TY (2010) Plant ferredoxin-like protein (PFLP) outside chloroplast in Arabidopsis enhances disease resistance against bacterial pathogens. Plant Sci 179:450–458

Lindow S, Newman K, Chatterjee S, Baccari C, Lavarone AT, Ionescu M (2014) Production of *Xylella fastidiosa* diffusible signal factor in transgenic grape causes pathogen confusion and reduction in severity of pierce's disease. Mol Plant-Microbe Interact 27:244–254

Liu C, Fukumoto T, Matsumoto T, Gena P, Frascaria D, Kaneko T, Katsuhara M, Zhong S, Sun X, Zhu Y, Iwasaki I, Ding X, Calamita G, Kitagawa Y (2013a) Aquaporin OsPIP1;1 promotes rice salt resistance and seed germination. Plant Physiol Biochem 63:151–158. https://doi.org/10.1016/j.plaphy.2012.11.018

Liu X, Liu S, Wu J, Zhang B, Li X, Yan Y, Li L (2013b) Overexpression of *Arachis hypogaea* NAC3 in tobacco enhances dehydration and drought tolerance by increasing superoxide scavenging. Plant Physiol Biochem 70:354–359. https://doi.org/10.1016/j.plaphy.2013.05.018

Liu L, Zhang Z, Dong J, Wang T (2016) Overexpression of MtWRKY76 increases both salt and drought tolerance in *Medicago truncatula*. Environ Exp Bot 123:50–58. https://doi.org/10.1016/j.envexpbot.2015.10.007

Lorito M, Woo SL, Fernandez IG, Colucci G, Harman GE, PintorToro JA, Filipone E, Muccifora S, Lawrence CB, Zoina A, Tuzun S, Scala F (1998) Genes from mycoparasitic fungi as source for improving plant resistance to fungal pathogens. Proc Natl Acad Sci U S A 95:7860–7865

Luo Q, Wei Q, Wang R, Zhang Y, Zhang F, He Y, Zhou S, Feng J, Yang G, He G (2017) BdCIPK31, a calcineurin B-like protein-interacting protein kinase, regulates plant response to drought and salt stress. Front Plant Sci 8:1184. https://doi.org/10.3389/fpls.2017.01184

MacRae TC, Baur ME, Boethel DJ, Fitzpatrick BJ, Gao AG, Gamundi JC, Harrison LA, Kabuye VT, McPherson RM, Miklos JA, Paradise MS, Toedebusch AS, Viegas A (2005) Laboratory and field evaluations of transgenic soybean exhibiting highdose expression of a synthetic *Bacillus thuringiensis* cry1A gene for control of Lepidoptera. J Econ Entomol 98:577–587

Malnoy M, Venisse JS, Chevreau E (2005) Expression of a bacterial effector hairpin N causes increased resistance to fire blight in *Pyrus communis*. Tree Genet Genomes 1:41–49

Malnoy M, Viola R, Jung MH, Koo OJ, Kim S, Kim JS, Velasco R, Nagamangala KC (2016) DNA-free genetically edited grapevine and apple protoplast using CRISPR/Cas9 ribonucleoproteins. Front Plant Sci 7:1904

Mamta B, Rajam MV (2018) RNA interference: a promising approach for crop improvement. In: Biotechnologies of crop improvement, vol 2. Springer, Cham, pp 41–65

Mao X, Chen S, Li A, Zhai C, Jing R (2014) Novel NAC transcription factor TaNAC67 confers enhanced MultiAbiotic stress tolerances in Arabidopsis. PLoS One 9:e84359. https://doi.org/10.1371/journal.pone.0084359

Melchers LS, Stuiver MH (2000) Novel genes for disease resistance breeding. Curr Opin Plant Biol 3:147–152

Meng X, Wang JR, Wang GD, Liang XQ, Li XD, Meng QW (2015) An R2R3-MYB gene, LeAN2, positively regulated the thermo-tolerance in transgenic tomato. J Plant Physiol 175:1–8. https://doi.org/10.1016/j.jplph.2014.09.018

Mercado JA, Barcelo M, Pliego C, Rey M, Caballero JL, MunozBlanco J, Ruano-Rosa D, Lopez-Herrera C, Santos B, RomeroMunoz F, Pliego-Alfaro F (2015) Expression of the b-1,3- glucanase gene bgn13,1 from *Trichoderma harzianum* in strawberry increases tolerance to crown rot diseases but interferes with plant growth. Transgenic Res 24:979–989

Mittler R, Blumwald E (2010) Genetic engineering for modern agriculture: challenges and perspectives. Annu Rev Plant Biol 61:443–462

Mora AA, Earle ED (2001) Resistance to *Alternaria brassicicola* in transgenic broccoli expressing *Trichoderma harzianum* endochitinase gene. Mol Breed 8:1–9

Morton RL, Schroeder HE, Bateman KS, Chrispeels MJ, Armstrong E, Higgins TJ (2000) Bean alpha-amylase inhibitor 1 in transgenic peas (*Pisum sativum*) provides complete protection from pea weevil (*Bruchus pisorum*) under field conditions. Proc Natl Acad Sci U S A 97:3820–3825

Munns R, Tester M (2008) Mechanisms of salinity tolerance. Annu Rev Plant Biol 59:651–681. https://doi.org/10.1146/annurev.arplant.59.032607.092911

Nakamura Y, Sawada H, Kobayashi S, Nakajima I, Yoshikawa M (1999) Expression of soybean b-1,3-glucanase cDNA and effect on disease tolerance in kiwifruit plants. Plant Cell Rep 18:527–532

Namukwaya B, Tripathi L, Tripathi JN, Arinaitwe G, Mukasa SB, Tushemereirwe WK (2012) Transgenic banana expressing *Pflp* gene confers enhanced resistance to Xanthomonas wilt disease. Transgenic Res 21:855–865

Negi P, Rai AN, Suprasanna P (2016) Moving through the stressed genome: emerging regulatory roles for transposons in plant stress response. Front Plant Sci 7:1448

Neuhaus JM (1999) Plant chitinases (PR-3, PR-4, PR-8, PR-11). In: Datta SK, Muthukrishnan S (eds) Pathogenesis-related proteins in plants. CRC Press, Boca Raton, pp 77–105

Nguyen TX, Nguyen T, Alameldin H, Goheen B, Loescher W, Sticklen M (2013) Transgene pyramiding of the HVA1 and mtlD in T3 maize (*Z. mays* L.) plants confers drought and salt tolerance, along with an increase in crop biomass. Int J Agron 10

Noctor G, Foyer CH (1998) Ascorbate and glutathione: keeping active oxygen under control. Annu Rev Plant Physiol 49:249–279

Ntui VO, Kynet K, Khan RS, Ohara M, Goto Y, Watanabe M, Fukami M, Nakamura I, Mii M (2014) Transgenic tobacco lines expressing defective CMV replicase-derived dsRNA are resistant to CMV-O and CMV-Y. Mol Biotechnol 56:50–63

Oelofse D, Dubery IA, Meyer R, Arendse MS, Gazendam I, Berger DK (2006) Apple polygalacturonase inhibition potential expressed in transgenic tobacco inhibits polygalacturonases from fungal pathogens of apple and anthracnose of lupins. Phytochemistry 67:255–263

Oh SK, Jang HA, Lee SS, Cho HS, Lee DH, Choi D, Kwon SY (2014) Cucumber Pti1-L is a cytoplasmic protein kinase involved in defense responses and salt tolerance. J Plant Physiol 171(10):817–822

Oksman-Kaldentey KM, Barz WH (2002) Plant biotechnology and transgenic plants. Marcel Dekker, Inc., New York, pp 1–20

Pandita D (2019) Plant MIRnome: miRNA biogenesis and abiotic stress response. In: Hasanuzzaman M, Hakeem K, Nahar K, Alharby H (eds) Plant abiotic stress tolerance. Springer, Cham, pp 449–474. https://doi.org/10.1007/978-3-030-06118-0_18

Pandita D (2021) Role of miRNAi technology and miRNAs in abiotic and biotic stress resilience. In: Aftab T, Roychoudhury A (eds) Plant perspectives to global climate changes. Academic Press Elsevier. https://doi.org/10.1016/B978-0-323-85665-2.00015-7

Pandita D (2022a) miRNA- and RNAi-mediated metabolic engineering in plants. In: Aftab T, Hakeem KR (eds) Metabolic engineering in plants. Springer, Singapore. https://doi.org/10.1007/978-981-16-7262-0_7

Pandita D (2022b) How microRNAs regulate abiotic stress tolerance in wheat? A snapshot. In: Roychoudhury A, Aftab T, Acharya K (eds) Omics approach to manage abiotic stress in cereals. Springer, Singapore. https://doi.org/10.1007/978-981-19-0140-9_17

Pandita D (2022c) MicroRNAs shape the tolerance mechanisms against abiotic stress in maize. In: Roychoudhury A, Aftab T, Acharya K (eds) Omics approach to manage abiotic stress in cereals. Springer, Singapore. https://doi.org/10.1007/978-981-19-0140-9_19

Pandita D (2022d) CRISPR/Cas mediated genome editing technologies in plants. In: Aftab T, Hakeem KR (eds) Plant abiotic stress physiology. Volume 1: responses and adaptations. AAP, pp 293–330. Hard ISBN: 9781774630167

Pandita D (2022e) CRISPR/Cas-mediated genome editing technologies in plants for stress resilience. In: Aftab T, Hakeem KR (eds) Antioxidant defense in plants. Springer, Singapore. https://doi.org/10.1007/978-981-16-7981-0_13

Pandita D (2022f) CRISPR/Cas-mediated genome engineering for abiotic stress resilience in plants. In: Aftab T, Hakeem KR (eds) Metabolic engineering in plants. Springer, Singapore. https://doi.org/10.1007/978-981-16-7262-0_11

Pandita D (2022g) Genetic engineering applications in inducing stress tolerance in plants through antioxidants. In: Aftab T, Hakeem KR (eds) Antioxidant Defense in Plants. Springer, Singapore. https://doi.org/10.1007/978-981-16-7981-0_5

Pandita D, Wani SH (2019) MicroRNA as a tool for mitigating abiotic stress in rice (*Oryza sativa* L.). In: Wani SH (ed) Recent approaches in omics for plant resilience to climate change. Springer, Cham. https://doi.org/10.1007/978-3-030-21687-0_6

Pasquali G, Biricolti S, Locatelli F, Baldoni E, Mattana M (2008) Osmyb4 expression improves adaptive responses to drought and cold stress in transgenic apples. Plant Cell Rep 27:1677–1686

Patel P, Yadav K, Ganapathi TR, Suprasanna P (2019) Plant miRNAome: cross talk in abiotic stressful times. In: Rajpal VR, Sehgal D, Kumar A, Raina SN (eds) Genomics-assisted breeding for crop improvement: abiotic stress tolerance. Springer, pp 25–52

Paul S, Roychoudhury A (2018) Transgenic plants for improved salinity and drought tolerance. In: Gosal SS, Wani SH (eds) Biotechnologies of crop improvement, vol 2. Transgenic approaches. Springer, Switzerland, pp 141–181. https://doi.org/10.1007/978-3-319-90650-8_7

Peiró A, Cañizares MC, Rubio L, López C, Moriones E, Aramburu J, Sánchez-Navarro J (2014) The movement protein (NSm) of Tomato spotted wilt virus is the avirulence determinant in the tomato Sw-5 gene-based resistance. Mol Plant Pathol 15(8):802–813

Pel MA, Foster SJ, Park TH, Rietman H, van Arkel G, Jones JD, Van Eck HJ, Jacobsen E, Visser RG, Van der Vossen EA (2009) Mapping and cloning of late blight resistance genes from *Solanum venturii* using an interspecific candidate gene approach. Mol Plant-Microbe Interact 22(5):601–615

Penna S (2003) Building stress tolerance through overproducing trehalose in transgenic plants. Trends Plant Sci 8(8):355–357

Pessina S, Lenzi L, Perazzolli M, Campa M, DallaCosta L, Urso S, Vale G, Salamini F, Velasco R, Malnoy M (2016) Knockdown of MLO genes reduces susceptibility to powdery mildew in grapevine. Hortic Res 3:16016

Polizel AM, Medri ME, Nakashima K, Yamanaka N, Farias JRB et al (2011) Molecular, anatomical and physiological properties of a genetically modified soybean line transformed with rd29A:AtDREB1A for the improvement of drought tolerance. Genet Mol Res 10:3641–3656. https://doi.org/10.4238/2011.October.21.4

Porcel R, Bustamante A, Ros R, Serrano R, Mulet Salort JM (2018) BvCOLD1: a novel aquaporin from sugar beet (*Beta vulgaris* L.) involved in boron homeostasis and abiotic stress. Plant Cell Environ 41:2844–2857. https://doi.org/10.1111/pce.13416

Powell KS, Gatehouse AMR, Hilder VA, Gatehouse JA (1995) Antifeedant effects of plant lectins and an enzyme on the adult stage of the rice brown plant hopper, *Nilaparvata lugens*. Entomol Exp Appl 75:51–59

Prashanth SR, Sadhasivam V, Parida A (2008) Over expression of cytosolic copper/zinc superoxide dismutase from a mangrove plant *Avicennia marina* in indica rice var Pusa Basmati-1 confers abiotic stress tolerance. Transgenic Res 17(2):281–291

Praveen S, Ramesh SV, Mishra AK, Koundal V, Palukaitis P (2010) Silencing potential of viral derived RNAi constructs in tomato leaf curl virus-AC4 gene suppression in tomato. Transgenic Res 19:45–55

Punja ZK, Raharjo V (1996) Response of transgenic cucumber and carrot plants expressing different chitinase enzymes to inoculation with pathogens. Plant Dis 80:999–1005

Rahbe Y, Deraison C, Bonadé-Bottino M, Girard C, Nardon C, Jouanin L (2003) Effects of the cysteine protease inhibitor oryzacystatin (OC-I) on different aphids and reduced performance of *Myzus persicae* on OC-I expressing transgenic oilseed rape. Plant Sci 164:441–450

Ramesh SV, Mishra AK, Praveen S (2007) Hairpin RNA-mediated strategies for silencing of tomato leaf curl virus AC1 and AC4 genes for effective resistance in plants. Oligonucleotides 17:251–257

Ravikumar G, Manimaran P, Voleti SR, Subrahmanyam D, Sundaram RM, Bansal KC, Viraktamath BC, Balachandran SM (2014) Stress-inducible expression of AtDREB1A transcription factor greatly improves drought stress tolerance in transgenic indica rice. Transgenic Res 23:421–439. https://doi.org/10.1007/s11248-013-9776-6

Reeck GR, Kramer KJ, Baker JE, Kanost MR, Fabrick JA, Behnke CA (1997) Protease inhibitors and resistance of transgenic plants to insects. In: Carozzi N, Koziel M (eds) Advances in insect control: the role of transgenic plants. Taylor and Francis, London, pp 157–183

Ren ZH, Gao JP, Li LG, Cai XL, Huang W, Chao DY, Zhu MZ, Wang ZY, Luan S, Lin HX (2005) A rice quantitative trait locus for salt tolerance encodes a sodium transporter. Nat Genet 37:1141–1146

Ren Y, Chen L, Zhang Y, Kang X, Zhang Z, Wang Y (2012) Identification of novel and conserved *Populus tomentosa* microRNA as components of a response to water stress. Funct Integr Genomics 12:327–339

Ribeiro APO, Pereira EJG, Galvan TL, Picanço MC et al (2006) Effect of eggplant transformed with oryzacystatin gene on *Myzus persicae* and *Macrosiphon euphorbiae*. J Appl Entomol 130:84–90

Roca Paixão JF, Gillet FX, Ribeiro TP, Bournaud C, Lourenço-Tessutti IT, Noriega DD, de Melo BP, de Almeida-Engler J, Grossi-de-Sa MF (2019) Improved drought stress tolerance in Arabidopsis by CRISPR/dCas9 fusion with a Histone AcetylTransferase. Sci Rep 9:1–9. https://doi.org/10.1038/s41598-019-44571-y

Rohini V, Sankara K, Rao K (2000) Transformation of peanut (*Arachis hypogaea* L.) with tobacco chitinase gene: variable response of transformants to leaf spot disease. Plant Sci 160:889–898

Roy BS, Noren K, Mandal AB, Basu AK (2011) Genetic engineering for abiotic stress tolerance in agricultural crops. Biotechnology 10:1–22

Saha P, Majumder P, Dutta I, Ray T, Roy SC, Das S (2006) Transgenic rice expressing *Allium sativum* leaf lectin with enhanced resistance against sap-sucking insect pests. Planta 23:1329–1343

Sanyal I, Singh AK, Kaushik M, Amla DV (2005) Agrobacterium-mediated transformation of chickpea (*Cicer arietinum* L.) with *Bacillus thuringiensis* cry1Ac gene for resistance against pod borer insect *Helicoverpa armigera*. Plant Sci 168(4):1135–1146

Schestibratov KA, Dolgov SV (2005) Transgenic strawberry plants expressing a thaumatin II gene demonstrate enhanced resistance to *Botrytis cinerea*. Sci Hortic 106:177–189

Shade R, Schroeder H, Pueyo J et al (1994) Transgenic pea-seeds expressing the alpha-amylase inhibitor of the common bean are resistant to bruchid beetles. Nat Biotechnol 12:793–796

Sharma HC, Sharma KK, Crouch JH (2004) Genetic transformation of crops for insect resistance: potential and limitations. Crit Rev Plant Sci 23:47–72

Sharma A, Kumar D, Kumar S, Rampuria S, Reddy AR, Kirti PB (2016) Ectopic expression of an atypical hydrophobic group 5 LEA protein from wild peanut, *Arachis diogoi* confers abiotic stress tolerance in tobacco. PLoS One 11:e0150609. https://doi.org/10.1371/journal.pone.0150609

Shekhawat UKS, Ganapathi TR (2013) MusaWRKY71 overexpression in banana plants leads to altered abiotic and biotic stress responses. PLoS One 8:e75506

Shekhawat UKS, Ganapathi TR, Hadapad AB (2012) Transgenic banana plants expressing small interfering RNAs targeted against viral replication initiation gene display high-level resistance to banana bunchy top virus infection. J Gen Virol 93:1804–1813

Shi J, Habben JE, Archibald RL, Drummond BJ, Chamberlin MA, Williams RW, Renee Lafitte H, Weers BP (2015) Overexpression of ARGOS genes modifies plant sensitivity to ethylene, leading to improved drought tolerance in both arabidopsis and maize. Plant Physiol 169:266–282. https://doi.org/10.1104/pp.15.00780

Shi J, Gao H, Wang H, Lafitte HR, Archibald RL, Yang M, Hakimi SM, Mo H, Habben JE (2017) ARGOS8 variants generated by CRISPR-Cas9 improve maize grain yield under field drought stress conditions. Plant Biotechnol J 15:207–216. https://doi.org/10.1111/pbi.12603

Shirasawa K, Takabe T, Takabe T, Kishitani S (2006) Accumulation of glycinebetaine in rice plants that overexpress choline monooxygenase from spinach and evaluation of their tolerance to abiotic stress. Ann Bot 98(3):565–571

Singh D, Ambroise A, Haicour R, Sihachakr D, Rajam MV (2014) Increased resistance to fungal wilts in transgenic eggplant expressing alfalfa glucanase gene. Physiol Mol Biol Plants 20:143–150

Song WY, Wang GL, Chen LL, Kim HS, Pi LY, Holsten T, Gardner J, Wang B, Zhai WX, Zhu LH, Fauquet C, Ronald P (1995) A receptor kinase-like protein encoded by the rice disease resistance gene, Xa21. Science 270:1804–1806

Song A, Zhu X, Chen F, Gao H, Jiang J, Chen S (2014) A chrysanthemum heat shock protein confers tolerance to abiotic stress. Int J Mol Sci 15:5063–5078

Sunkar R, Li YF, Jagadeeswaran G (2012) Functions of microRNAs in plant stress responses. Trends Plant Sci 17:196–203

Suprasanna P, Ghag SB (2019) Plant tolerance to environmental stress: translating research from lab to land. In: Roychoudhury A, Tripathi DK (eds) Molecular plant abiotic stress: biology and biotechnology. JohnWiley & Sons Ltd

Suprasanna P, Ghuge SA, Patade VY et al (2018) Genomic roadmaps for augmenting salinity stress tolerance in crop plants. In: Kumar V, Wani SH, Suprasanna P, Tran L-SP (eds) Salinity responses and tolerance in plants. Springer, Berlin, pp 189–216

Surekha C, Kumari KN, Aruna LV, Suneetha G, Arundhati A, Kavi Kishor PB (2014) Expression of the *Vigna aconitifolia* P5CSF129A gene in transgenic pigeonpea enhances proline accumulation and salt tolerance. Plant Cell Tissue Organ Cult 116:27–36. https://doi.org/10.1007/s11240-013-0378-z

Tang L, Kwon SY, Kim SH, Kim JS, Choi JS, Cho KY, Sung CK, Kwak SS, Lee HS (2006) Enhanced tolerance of transgenic potato plants expressing both superoxide dismutase and ascorbate peroxidase in chloroplasts against oxidative stress and high temperature. Plant Cell Rep 25:1380–1386

Tang RJ, Yang Y, Yang L, Liu H, Wang CT, Yu MM, Gao XS, Zhang HX (2014) Poplar calcineurin B-like proteins PtCBL10A and PtCBL10B regulate shoot salt tolerance through interaction with PtSOS2 in the vacuolar membrane. Plant Cell Environ 37:573–588. https://doi.org/10.1111/pce.12178

Thakur AK, Singh KH, Sharma D, Singh L, Parmar N, Nanjundan J, Khan YJ (2018) Transgenic development for biotic and abiotic stress management in horticultural crops. In: Genetic engineering of horticultural crops. Academic Press, London, pp 353–338

Thamizhmani R, Vijayachari P (2014) Association of dengue virus infection susceptibility with polymorphisms of 2′-5′-oligoadenylate synthetase genes: a case–control study. Braz J Infect Dis 18(5):548–550

Thomson MJ, de Ocampo M, Egdane J, Rahman MA, Sajise AG, Adorada DL (2010) Characterizing the Saltol quantitative trait locus for salinity tolerance in rice. Rice 3:148–160

Tohidfar M, Ghareyazie B, Mohammadi M, Abdmishani C (2005) Agrobacterium-mediated transformation of cotton using a chitinase gene. Plant Cell Tissue Organ Culture 83:83–96

Tohidfar M, Ghareyazie B, Mosavi M, Yazdani S, Golabchian R (2008) Agrobacterium-mediated transformation of cotton (*Gossypium hirsutum*) using a synthetic cry1Ab gene for enhanced resistance against *Heliothis armigera*. Iran J Biotechnol 6(3):164–173

Tohidfar M, Hossaini R, Shokhandan Bashir N, Tabatabaei M (2012) Enhanced resistance to *Verticillium dahliae* in transgenic cotton expressing an endochitinase gene from *Phaseolus vulgaris*. Czech J Genet Plant Breed 4:345–355

Tohidfar M, Zare N, Salhi G, Eftghari M (2013) Agrobacterium-mediated transformation of alfalfa (*Medicago sativa*) using a synthetic cry3a gene to enhance resistance against alfalfa weevil. Plant Cell Tissue Organ Culture 113:227–235

Tougou M, Furutani N, Yamagishi N, Shizukawa Y, Takahata Y, Hidaka S (2006) Development of resistant transgenic soybeans with inverted repeat-coat protein genes of soybean dwarf virus. Plant Cell Rep 25(11):1213–1218

Tsai-Hung H, Jent-turn L, Yee-yung C, Ming-Tsair C (2002) Heterology expression of the Arabidopsis C-Repeat/Dehydration Response Element Binding Factor 1 gene confers elevated tolerance to chilling and oxidative stresses in transgenic tomato. Plant Physiol 130:618–626

VanWallendael A, Soltani A, Emery NC, Peixoto MM, Olsen J, Lowry DB (2019) A molecular view of plant local adaptation: incorporating stress-response networks. Annu Rev Plant Biol 70:559–583

Varshney RK, Bansal KC, Aggarwal PK, Datta SK, Craufurd PQ (2011) Agricultural biotechnology for crop improvement in a variable climate: hope or hype? Trends Plant Sci 16:363–371

Vellice GR, Ricci JCD, Hernandez L, Castagnaro AP (2006) Enhanced resistance to Botrytis cinerea mediated by the transgenic expression of the chitinase gene ch5B in strawberry. Transgenic Res 15:57–68

Vleeshouwers VG, Rietman H, Krenek P, Champouret N, Young C, Oh SK, Wang M, Bouwmeester K, Vosman B, Visser RG, Jacobsen E, Govers F, Kamoun S, Van der Vossen EA (2008) Effector genomics accelerates discovery and functional profiling of potato disease resistance and *Phytophthora infestans* avirulence genes. PLoS One 3(8):e2875

Wang Y, Wisniewski M, Meilan R, Cu M, Fuchigami L (2006) Transgenic tomato (*Lycopersicon esculentum*) overexpressing cAPX exhibits enhanced tolerance to UV-B and heat stress. J Appl Hortic 8:87–90

Wang YC, Qu QZ, Li HY, Wu YJ, Wnag C, Liu GF, Yang CP (2010) Enhanced salt tolerance of transgenic poplar plants expressing a manganese superoxide dismutase from *Tamarix androssowii*. Mol Biol Rep 37(2):1119–1124

Wang X, Guo X, Li Q, Tang Z, Kwak S, Ma D (2011) Studies on salt tolerance of transgenic sweet potato which harbors two genes expressing CuZn super-oxide dismutase and ascorbate peroxidase with the stress-inducible SWPA2 promoter. Plant Gene Trait 3(2):6–12

Wang S, Sun X, Hoshino Y, Yu Y, Jia B, Sun Z, Zhu Y (2014a) MicroRNA319 positively regulates cold tolerance by targeting OsPCF6 and OsTCP21 in rice (*Oryza sativa* L.). PLoS One 9:e91357

Wang B, Sun YF, Song N, Wei JP, Wang XJ, Feng H, Yin ZY, Kang ZS (2014b) MicroRNAS involving in cold, wounding and salt stresses in *Triticum aestivum* L. Plant Physiol Biochem 80:90–96

Wang M, Li P, Li C, Pan Y, Jiang X, Zhu D, Zhao Q, Yu J (2014c) SiLEA14, a novel atypical LEA protein, confers abiotic stress resistance in foxtail millet. BMC Plant Biol 14:1–16. https://doi.org/10.1186/s12870-014-0290-7

Wang H, Wang H, Shao H, Tang X (2016) Recent advances in utilizing transcription factors to improve plant abiotic stress tolerance by transgenic technology. Front Plant Sci 7:67

Wankhede DP, Kumar K, Singh P, Sinha AK (2013) Involvement of mitogen activated protein kinase 6 in UV induced transcripts accumulation of genes in phytoalexin biosynthesis in rice. Rice 6(35):1–8

Wei D, Zhang W, Wang C, Meng Q, Li G, Chen THH, Yang X (2017) Genetic engineering of the biosynthesis of glycinebetaine leads to alleviate saltinduced potassium efflux and enhances salt tolerance in tomato plants. Plant Sci 257:74–83. https://doi.org/10.1016/j.plantsci.2017.01.012

Wisniewski M, Fuchigami L, Wang Y, Srinivasan C, Norilli J (2002) Overexpression of a cytosolic ascorbate peroxidase gene in apple improves resistance to heat stress. In: XXXVI International Horticultural Congress and Exhibition, p 147

Wu R, Garg A (2003) Engineering rice plants with trehalose-producing genes improves tolerance to drought, salt and low temperature, available from: http://www.seedquest.com/News/releases/2003/march/5456.htm

Xiangdong WEI, Congyul LAN, Zhijing LU, Changming YE (2007) Analysis on virus resistance and fruit quality for T4 generation of transgenic papaya. Front Biol China 2:284–290

Yamamoto T, Iketani H, Ieki H, Nishizawa Y, Notsuka K, Hibi T, Hayashi T, Matsuta N (2000) Transgenic grapevine plants expressing a rice chitinase with enhanced resistance to fungal pathogens. Plant Cell Rep 19:639–646

Yan F, Zhang WW, Xiao H, Li SF, Cheng ZM (2007) Transgenic wheat expressing virus-derived hairpin RNA is resistant to Barley yellow dwarf virus. Yi Chuan 29(1):97–102. (in Chinese). https://doi.org/10.1360/yc-007-0097

Yang W, Kong Z, Omo-Ikerodah E, Xu W, Li Q, Xue Y (2008) Calcineurin B-like interacting protein kinase OsCIPK23 functions in pollination and drought stress responses in rice (*Oryza sativa* L.). J Genet Genomics 35:531–5S2. https://doi.org/10.1016/S1673-8527(08)60073-9

Ying S, Zhang DF, Fu J, Shi YS, Song YC, Wang TY, Li Y (2012) Cloning and characterization of a maize bZIP transcription factor, ZmbZIP72, confers drought and salt tolerance in transgenic Arabidopsis. Planta 235:253–266. https://doi.org/10.1007/s00425-011-1496-7

Yu TA, Chiang CH, Wu HW, Li CM, Yang CF, Chen JH, Chen YW, Yeh SD (2011) Generation of transgenic watermelon resistant to Zucchini yellow mosaic virus and Papaya ringspot virus type W. Plant Cell Rep 30:359–371

Zafar SA, Zaidi SSEA, Gaba Y, Singla-Pareek SL, Dhankher OP, Li X, Mansoor S, Pareek A, Foyer C (2020) Engineering abiotic stress tolerance via CRISPR/Cas-mediated genome editing. J Exp Bot 71:470–479. https://doi.org/10.1093/jxb/erz476

Zhang B, Wang Q (2015) MicroRNA-based biotechnology for plant improvement. J Cell Physiol 230:1–15

Zhang G, Chen M, Li L, Xu Z, Chen X, Guo J, Ma Y (2009) Overexpression of the soybean GmERF3 gene, an AP2/ERF type transcription factor for increased tolerances to salt, drought, and diseases in transgenic tobacco. J Exp Bot 60:3781–3796. https://doi.org/10.1093/jxb/erp214

Zhang X, Zou Z, Gong P, Zhang J, Ziaf K, Li H, Xiao F, Ye Z (2011a) Over-expression of microRNA169 confers enhanced drought tolerance to tomato. Biotechnol Lett 33:403–409

Zhang Y, Li H, Shu W, Zhang C, Ye Z (2011b) RNA interference of a mitochondrial APX gene improves vitamin C accumulation in tomato fruit. Sci Hortic 129:222–226

Zhang J, Li D, Zou D, Luo F, Wang X, Zheng Y, Li X (2013) A cotton gene encoding a plasma membrane aquaporin is involved in seedling development and in response to drought stress. Acta Biochim Biophys 45:104–114. https://doi.org/10.1093/abbs/gms096

Zhang L, Zhang L, Xia C, Zhao G, Jia J, Kong X (2016) The novel wheat transcription factor TaNAC47 enhances multiple abiotic stress tolerances in transgenic plants. Front Plant Sci 6:1174. https://doi.org/10.3389/fpls.2015.01174

Zhang A, Liu Y, Wang F, Li T, Chen Z, Kong D, Bi J, Zhang F, Luo X et al (2019) Enhanced rice salinity tolerance via CRISPR/Cas9-targeted mutagenesis of the OsRR22 gene. Mol Breed 39:1–10. https://doi.org/10.1007/s11032-019-0954-y

Zhou L, Liu Y, Liu Z, Kong D, Duan M, Luo L (2010) Genome-wide identification and analysis of drought-responsive microRNAs in *Oryza sativa*. J Exp Bot 61:4157–4168

Zhou L, Wang C, Liu R, Han Q, Vandeleur RK, Du J, Tyerman S, Shou H (2014) Constitutive overexpression of soybean plasma membrane intrinsic protein GmPIP1;6 confers salt tolerance. BMC Plant Biol 14:1–13. https://doi.org/10.1186/1471-2229-14-181

Zhu JK (2016) Abiotic stress signaling and responses in plants. Cell 167:314–324

Zhu S, Li Y, Vossen JH, Visser RG, Jacobsen E (2012) Functional stacking of three resistance genes against Phytophthora infestans in potato. Transgenic Res 21:89–99

Chapter 36
Translationally Controlled Tumor Protein and Its Relationship with Responses of Plants to Abiotic Stresses

Deyvid Novaes Marques, Nicolle Louise Ferreira Barros, and Cláudia Regina Batista de Souza

Abstract Food security and plant productivity can be affected by abiotic stresses, which are associated with substantial yield losses in several plant species including crops worldwide. Thus, there is great relevance in prospecting molecular components of plants. This includes proteins that are related to endogenous defense response mechanisms, to produce tolerant plants, and to provide insights into the complexity of molecular strategies in response to these stresses. The translationally controlled tumor protein (TCTP) is a protein family highly conserved in eukaryotic organisms, which has been associated with several roles at the cellular level, such as those related to growth, development and defense. Compared to the TCTP proteins from animals, the studies focused on plant TCTPs are still incipient. However, several studies have confirmed the importance of these proteins in plant development and tolerance against abiotic stresses. In addition, studies using TCTP proteins purified from plants have shown activities, such as molecular chaperone and antioxidant activity, potentially related to abiotic stress response. In this review, we summarize current knowledge about the intracellular localization of plant TCTPs and present several evidences regarding the role of TCTP in plant response to several abiotic factors, such as drought, salinity, cold, high temperature, and heavy metal stress. Here, functional analysis of TCTPs by heterologous expression in bacteria and transgenic plants, and differential gene expression analysis, as well as putative *cis*-acting regulatory elements potentially involved in the abiotic stress found in TCTP promoter genes regulated by phytohormones are presented.

D. N. Marques (✉)
Departamento de Genética, Universidade de São Paulo (USP), Escola Superior de Agricultura "Luiz de Queiroz" (ESALQ), Piracicaba, SP, Brazil

N. L. F. Barros
Departamento de Genética, Universidade Federal do Rio Grande do Sul (UFRGS), Porto Alegre, RS, Brazil

C. R. B. de Souza
Instituto de Ciências Biológicas, Universidade Federal do Pará (UFPA), Belém, PA, Brazil

M. Hasanuzzaman (ed.), *Climate-Resilient Agriculture, Vol 2*,
https://doi.org/10.1007/978-3-031-37428-9_36

Keywords Gene expression · Plant abiotic stress tolerance · Plant breeding · Plant hormone · TCTP protein

1 Introduction

Crop productivity and yield worldwide can be impaired by environmental stresses, which include abiotic factors that affect plant metabolism, growth and physiology. Given the high socio-economic importance of plants, the efficient application of approaches for increasing abiotic stress tolerance or modulating stress-induced responses is relevant. Use of strategies for their genetic improvement focused on the tolerance against abiotic stresses is highly relevant. Within this context, providing insights associated with tolerance mechanisms-based genetic improvement and the prospection of molecules responsive to abiotic stresses constitute a potential strategy toward the production of tolerant plants in the field. However, despite the elucidation of several tolerance mechanisms, the knowledge regarding the contribution of several proteins in abiotic stress tolerance is still very limited.

Phosphatases and protein kinases, which act as signaling molecules, as well as transcription factors (such as bZIP, NAC, AP2/ERF, and MYB/MYC) and other proteins (e.g., LEA proteins, HSPs and aquaporins) encoded by stress-responsive genes are examples of proteins identified and well known as important components of regulatory networks of plant response to abiotic stress (Dos Reis et al. 2016; Marques et al. 2017a).

Among the proteins whose studies on function in plant response to abiotic stresses are still incipient, we highlight here proteins of TCTP family. Several studies have verified the role of this family in the tolerance of transgenic plants (Marques et al. 2023a). In this review, we summarize the current knowledge about the relationship of TCTP proteins with responses of plants to abiotic stresses. First of all, we introduce TCTP family. Further, we summarize the information on the role of TCTP in plant response to several abiotic factors.

2 TCTP

Translationally *c*ontrolled *t*umor *p*rotein was initially identified in mammalian tumor cells, where its regulation at translational level was confirmed (Gross et al. 1989). However, over the years, it was verified that this protein is present in normal cells and that it is transcriptionally regulated (Sanchez et al. 1997).

The TCTP constitutes a protein family highly conserved in the course of evolution and ubiquitously expressed in all eukaryotic organisms (Bommer and Thiele 2004). Two primary regions (TCTP1 and TCTP2) of high sequence homology characterize the TCTP family (Thaw et al. 2001). According to Gnanasekar and Ramaswamy (2007), the TCTP is one of the 20 proteins more expressed in eukaryotes.

Animal studies were responsible for the first identification of the TCTP function related to its response to stress conditions, as well as to physiological processes, such as those related to cell growth and development. These functions include interaction with microtubules (Gachet et al. 1999), calcium binding (Sanchez et al. 1997; Xu et al. 1999), regulation of apoptosis (Chen et al. 2007; Rho et al. 2011) and protection against heat stress (Gnanasekar et al. 2009).

In plants, the first study on TCTP gene detected its differential expression through the induction of cell division into *Pisum sativum* (Woo and Hawes 1997). Later, Sage-Ono et al. (1998) verified that the transcript levels corresponding to TCTP increased gradually in violet plants (*Pharbitis nil*) subjected to darkness, being, therefore, identified as a dark-inducible cDNA.

Some authors observed that TCTPs of some plant species can be found in phloem (Barnes et al. 2004; Hinojosa-Moya et al. 2013). Barnes et al. (2004) identified TCTP as a protein present in the phloem of *Ricinus communis*. Likewise, Toscano-Morales et al. (2014) detected that the gene products of *AtTCTP2* (mRNA or protein) are capable of long-distance movement through the phloem in tobacco plants (*Nicotiana tabacum*). Moreover, the same authors detected the *AtTCTP2* mRNA in all adventitious and primary roots of transgenic plant, providing evidence that this gene can be regulated by transcriptional or translational level (Toscano-Morales et al. 2014).

Studies have already analyzed the role of plant TCTP regarding important aspects of plant development, such as auxin homeostasis (Berkowitz et al. 2008), cell proliferation (Brioudes et al. 2010), and inhibition of apoptosis (Hoepflinger et al. 2013). However, when compared to animals, the studies related to TCTP of plants are still incipient, given the remarkable divergence among plants (Gutiérrez-Galeano et al. 2014) as well as their genomic complexity and multiple plant response mechanisms.

3 TCTP and Its Potential Role in Response to Abiotic Stresses

Differential gene expression analysis and proteomics (Table 36.1) as well as functional studies constitute an important step for prospecting plant genes, reference gene selection for RT-qPCR (Huang et al. 2014), and preliminary analysis of gene products potentially related to response to abiotic stresses. Prospecting plant genes encoding TCTPs related to abiotic stress response has been involving studies of differential expression, as well as functional analysis by means of heterologous expression, evaluation of purified protein and analysis in transgenic plants (Fig. 36.1). Knowledge and new insights regarding TCTP-related tolerance might also be improved by employing some proteomic approaches. For example, although

Table 36.1 Some examples of TCTPs in transcriptomic and proteomic studies for comprehensive understanding of plant response to abiotic stresses

Data type	Specie source	Abiotic stress	References
Proteome	*Brachypodium distachyon*	H_2O_2-induced stress	Bian et al. (2015)
Proteome	*B. distachyon*	Osmotic and cadmium stresses	Chen et al. (2018)
Transcriptome	*Morus alba*	High level of UVB irradiation with and without dark treatment	Guan et al. (2018)
Proteome	*Potentilla fruticosa* L.	Heat stress	Guo et al. (2017)
Proteome	Poplar plants	Cadmium stress	Huang et al. (2020)
Proteome	*Hordeum marinum*	Salt stress	Maršálová et al. (2016)
Proteome	*Hordeum vulgare* L.	Drought stress	Mostek et al. (2015)
Proteome	*Hordeum vulgare* L.	Low phosphorus availability	Nadira et al. (2016)
Proteome	*Arabidopsis thaliana*	Mannitol-induced stress	Nikonorova et al. (2018)
Proteome	*Oenothera glazioviana*	Copper stress	Wang et al. (2017a)

phosphoproteomics is a potential approach to provide insights into plant tolerance mechanisms and has an emerging use in experiments involving some abiotic factors-induced stresses (Marques et al. 2021), information on the phospho-regulation of TCTP during plant abiotic stress responses is still limited. Furthermore, new insights into stress-induced responses and the related organ-specific mechanisms have been provided by plant grafting (Marques et al. 2023b, c). This approach can be applied in further studies on plant TCTPs under abiotic stress conditions.

Although the transcriptional regulation of TCTP remains to be elucidated, *cis*-acting elements potentially involved in the response to abiotic stress have been detected in the promoter of TCTP genes of plants (Table 36.2). Among them, ABA-responsive elements (Kim et al. 2012), those involved in methyl jasmonate responsiveness (Zhang et al. 2013) and *cis*-elements associated with stress response, such as anaerobically induced, salt-induced, heat shock, and water stress responses (Deng et al. 2016).

Through the analysis of deduced amino acid sequences from plant TCTPs, it is possible to predict the presence of putative binding-sites to calcium, microtubules, Rab GTPase and to antiapoptotic protein MCL1, as well as phosphorylation sites, and TCTP1 and TCTP2 domains (Fig. 36.2). Phosphorylation is a post-translational modification that plays a major role in activating many proteins in response to abiotic stresses, such as calcium-dependent protein kinases (Rampitsch and Bykova 2012). These two latter domains seem to be related to protein-protein interactions of TCTPs (Ermolayev et al. 2003; Jung et al. 2004).

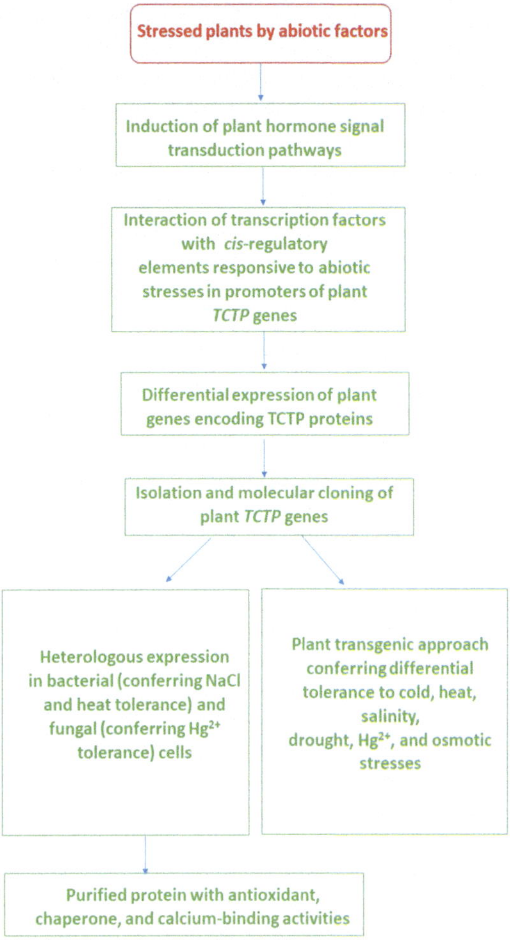

Fig. 36.1 Prospecting of plant TCTPs responsive to abiotic stresses

3.1 Differential Expression of TCTP Genes in Response to Abiotic Stresses

3.1.1 Plant Hormones

Abscisic acid (ABA) works as the central regulator of abiotic stress tolerance in plants (Sah et al. 2016). Other hormones, such as jasmonates and ethylene, also play essential roles in regulating plant response against these stresses (Kazan 2015). These hormones are important for modulating expression of transcription factors, which in turn regulate the expression of their target responsive genes (Sah et al. 2016; Kazan 2015).

Some studies concluded that the induction of plant TCTP genes by phytohormones such as abscisic acid (ABA) (Kim et al. 2012; Wang et al. 2015; de

Table 36.2 Putative *cis*-acting regulatory elements responsive to abiotic stress in promoter of plant genes encoding TCTP proteins (available on GenBank from NCBI), in sense (+) or anti-sense (−) strand, by in silico analysis search against PlantCARE (PC) (Lescot et al. 2002) and PlantPAN (PP) (Chow et al. 2016) databases

Plant TCTP gene/ GenBank accession	Element	Sequence	Database (ID)	Position upstream (−) to the codon start/strand	Relationship with abiotic stress response
AtTCTP/AF237735	ABRE	TACGTG	PC	−1066/+	Involved in the abscisic acid responsiveness
	ARE core	AGCAACGGTC	PP (0011)	−781/−	Anaerobic induction
	CRT/DRE motif	GTCGAC	PP (0322)	−2130/+− −1094/+− −565/+− −436/+− −126/+−	Preferred sequence for AP2 transcriptional activator HvCBF2 of barley; DNA binding is regulated by temperature
	DRE1 core	ACCGAGA	PP (0350)	−1427/− −1357/+ −372/+ −167/+	Found in maize *rab17* gene promoter; related to drought response
	GT-1 motif	GAAAAA	PP (0315)	−1415/− −453/− −1338/+ −1142/+ −574/+	Involved in NaCl-induced expression of promoter in soybean
	HSE	AGAAAATTCG	PC	−585/+ −205/+	Involved in heat stress responsiveness in cabbage
	HSE	CT(A/T/C/G) GAA(A/T/C/G) (A/T/C/G) TTC (A/T/C/G) AG	PP (0135)	−587/−	Involved in heat stress responsiveness in Viridiplantae family
	IDE1	ATCAAGCATGCTTCTTGC	PP (0176)	−556/−	Iron-deficiency-responsive element
	LTRE1	CCGAAA	PP (0306)	−1689/−	Low-temperature-responsive element in barley
	MBS	TAACTG	PC	−192/−	MYB binding site involved in drought-inducibility
	MYB	(A/T)AACCA	PP (0341)	−2128/+ −1799/+	Found in the promoters of the dehydration-responsive gene *rd22* and many other genes in Arabidopsis

HbTCTP/KX179468	ARE core	AGCAACGGTC	PP (0011)	−120/−	Anaerobic induction
	DRE1 core	ACCGAGA	PP (0350)	−75/+ −107/−	Found in maize *rab17* gene promoter; related to drought response
	CORE	AA(T/G)AAT(T/A)(T/C)(A/G)TA(T/A) ATAAAA(C/A) TTTTAT(T/A)TA	PP (0226)	−1096/− −816/− −814/+ −662/+ −644/+ −628/− −583/+ −570/+ −503/+ −494/− −318/−	Regulatory element conserved on the promoter regions of antioxidant defense genes in rice
	CRT/DRE motif	GTCGAC	PP (0322)	−850/+− −96/+−	Preferred sequence for AP2 transcriptional activator HvCBF2 of barley regulated by temperature
	DRE1 core	ACCGAGA	PP (0350)	−348/− −21/+	Found in maize *rab17* gene promoter; related to drought response
	GT-1 motif	GAAAAA	PP (0315)	−1049/− −549/−	Involved in NaCl-induced expression of promoter in soybean
	HSE	AAAAAATTTC	PC	−556/−	Involved in heat stress responsiveness in cabbage
	MBS	CAACTG; TAACTG	PC	−345/−; −1096/−	MYB binding site involved in drought-induced responses
	MYB	(A/T)AACCA	PP (0341)	−933/+	Found in the promoters of the dehydration-responsive gene *rd22* and many other genes in Arabidopsis
	MYBAtRD22	CTAACCA	P (0366)	−828/− −505/+	Binding site for MYB2 in dehydration-responsive gene, *rd22*, of Arabidopsis thaliana; ABA-induced responses

(continued)

Table 36.2 continued

Plant TCTP gene/ GenBank accession	Element	Sequence	Database (ID)	Position upstream (−) to the codon start/strand	Relationship with abiotic stress response
LkTCTP / KC250016	CBFHV	(A/G)(C/T)CGAC	PP (0325)	−79/+	Binding site of barley dehydration-responsive element (DRE) binding proteins
	CRT/ DRE motif	GTCGAC	PP (0322)	−79/+−	Preferred sequence for AP2 transcriptional activator HvCBF2 of barley; DNA binding is regulated by temperature
	DRE1 core	ACCGAGA	PP (0350)	−110/+ −93/−	Found in maize *rab17* gene promoter; related to drought response
	GT-1 motif	GAAAAA	PP (0315)	−48/−	Involved in NaCl-induced expression of promoter in soybean
	CRT/DRE motif	GTCGAC	PP (0322)	−97/+	Preferred sequence for AP2 transcriptional activator HvCBF2 of barley regulated by temperature
MeTCTP/JX855122	DRE1 core	ACCGAGA	PP (0350)	−77/−	Found in maize *rab17* gene promoter; related to drought response
	GT-1 motif	GAAAAA	PP (0315)	−166/− −277/+	Involved in NaCl-induced expression of promoter in soybean
	LTR Core	CCGAC	PP (0258)	−218/+	ABA responsiveness; involved in cold-induced response in Arabidopsis
	MYB	(A/T)AACCA	PP (0341)	−29/−	Found in the promoters of the dehydration-responsive gene *rd22* and many other genes in Arabidopsis
OsTCTP /KR080533	DRE1 core	ACCGAGA	PP (0350)	−107/− −75/+	Found in maize *rab17* gene promoter; related to drought response
	HSE	AAAAAATTTC	PC	−47/−	Involved in heat stress responsiveness in cabbage

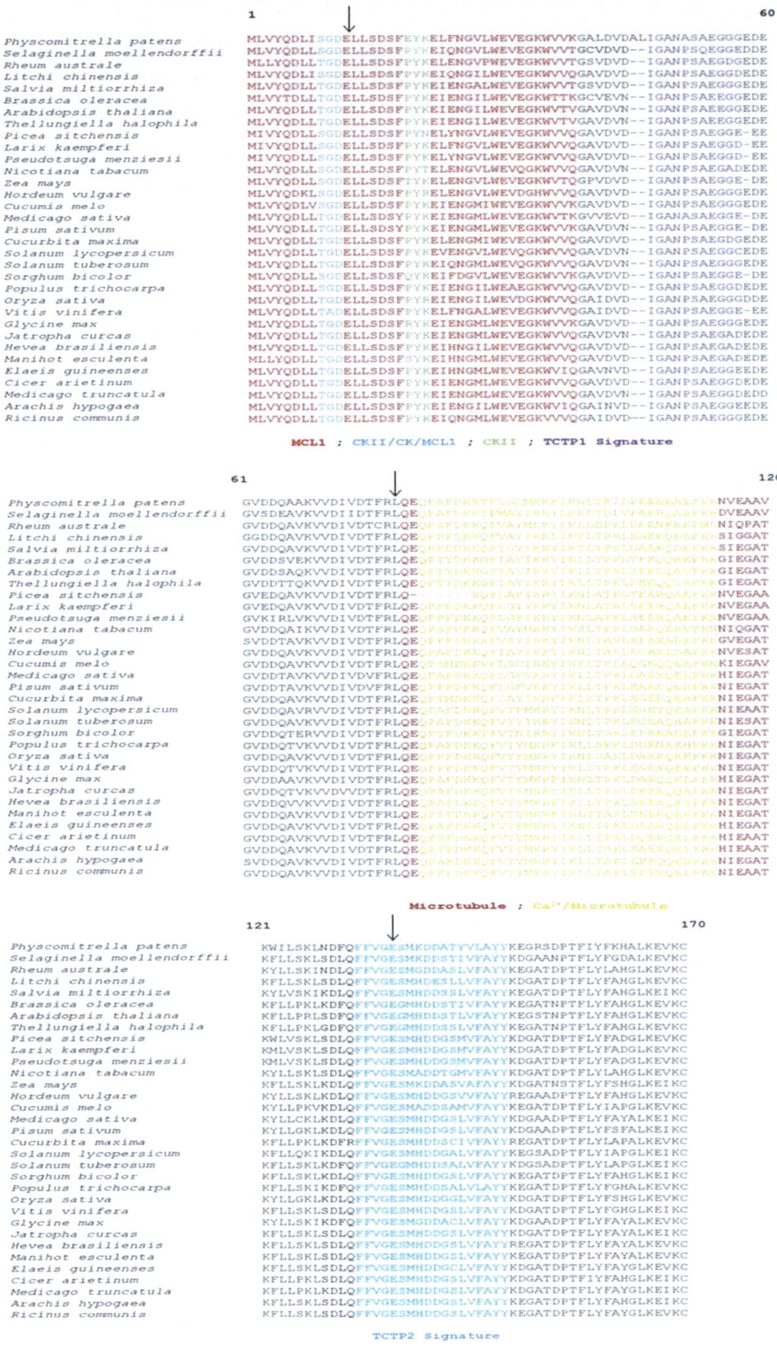

Fig. 36.2 Alignment between deduced amino acid sequence of plant TCTPs by Clustal Omega (Sievers et al. 2011). The following putative elements are highlighted: TCTP 1 and TCTP 2 signatures; Ca^{2+} and microtubule binding sites, phosphorylation sites for casein kinase II (CKII) and C kinase (CK) and MCL1 motif. Three conserved amino acids residues of binding domain to Rab GTPase are indicated by black arrows (Glu-E, Leu-L)

Carvalho et al. 2017; Meng et al. 2017), jasmonate and ethylene (Wang et al. 2012, 2015; de Carvalho et al. 2017) is a crucial step for responses to abiotic stresses. Thus, such information works as preliminary evidence on the role of TCTPs as downstream components to plant hormones in signaling pathways involved in abiotic stress response.

3.1.2 Drought, Salinity, High Temperature, and Humidity Stresses

Significant increase in plant TCTP expression under drought stress has been verified through proteomic analysis (Ghabooli et al. 2013; Zhang et al. 2014; Koh et al. 2015). Proteomic analysis of vine plants (*Vitis vinifera*) exposed to high salinity and drought allowed the identification of TCTP as being one of the proteins expressed in response to such stress (Vincent et al. 2007). Likewise, in *Jatropha curcas*, studies revealed changes in expression levels of *JcTCTP* in response to salinity (Qin et al. 2011). An elevated TCTP expression was verified in a wheat cultivar (*Triticum aestivum*) selected by its high salinity resistance (Guo et al. 2012).

Mostek et al. (2015), analyzing root proteome of salt-sensitive and tolerant barley lines, observed the TCTP upregulation in the tolerant one. Santa Brígida et al. (2014) evaluated the expression pattern of *MeTCTP* from cassava (*Manihot esculenta*) by RT-PCR and confirmed increased transcript production in response to increased NaCl concentration in cassava leaves. The differential expression of *GmTCTP* in response to high temperature and humidity stress has been verified (Wang et al. 2017b). Thus, several studies have confirmed changes in gene expression of plant TCTPs under abiotic stresses.

3.1.3 Flooding

Chen et al. (2014) analyzed the maize leaf proteome and confirmed that flooding was related to programmed cell death (PCD), and the TCTP was up-regulated in response to this stress. The TCTP protein is associated with the PCD suppression and regulated by flooding stress (Chen et al. 2014), and it represents a feature shown by other heat shock proteins (HSPs), as reported by Qi et al. (2011).

Chen et al. (2014) also concluded that under flood stress, there was a considerable increase in the accumulation of hydrogen peroxide (H_2O_2), which could be responsible for inducing TCTP expression under such stress. In accordance, Barba-Espin et al. (2010) verified that H_2O_2 pre-treatment induced the TCTP protein expression in pea seedings.

3.1.4 Heavy Metal Stress

Regarding the involvement of TCTP in response to heavy metal stress, Ermolayev et al. (2003) detected a TCTP gene differentially expressed in roots of soybean (*Glycine max*) tolerant to aluminum stress, possibly contributing to both plant development and aluminum resistance. The authors suggested that this protein could be involved in the maintenance of Ca^{2+} homeostasis in stressed plants cells, since the concentration of such ion in different cellular compartments plays an important role in easing the damage caused by aluminum.

In accordance to studies reporting a highest TCTP expression in eukaryotes under heavy metal stress (Sturzenbaum et al. 1998; Schmidt et al. 2007), Wang et al. (2012) verified that *OsTCTP* expression levels in rice (*Oryza sativa*) were substantially increased in plants subjected to mercury stress. In addition, *OsTCTP* levels were more pronounced in the cultivar tolerant to Hg^{2+} than in the wild one (Wang et al. 2012). Likewise, Li et al. (2016) verified the up-regulation of TCTP of rice roots under inorganic mercury and methyl mercury stress using metalloproteomic approaches.

3.2 *In Vitro Analysis and Heterologous Expression of Plant TCTP Genes*

Despite the wide variety of TCTP genes already identified (Table 36.3), a few studies have isolated and characterized plant TCTP genes focused on the evaluation of response to abiotic stresses so far (Table 36.4). However, the available information on functional studies confirms the potential of these proteins for plant defense against these stresses.

Li et al. (2013) verified that the supercoiled DNA protection increasing against metal-catalyzed oxidation was proportional to the increased concentration of purified HbTCTP1 of rubber tree that corresponds to the fact that *HbTCTP1* expression was regulated by the treatment with H_2O_2. Deng et al. (2016) also detected antioxidant activity of HbTCTP in metal-catalyzed oxidation system.

Studies have also highlighted the role of TCTP cell protection in non-plant organisms through its regulation by oxidative stress, even when the stress level in cells is still low (Lucibello et al. 2011). In turn, the expression induction of a TCTP from *Brugia malayi* nematode avoided cell DNA damage and *Escherichia coli* cells death caused by oxidative stress induced by H_2O_2 (Gnanasekar and Ramaswamy 2007), in accordance with the results obtained by Nagano-Ito et al. (2009), which in turn established a CHO-K1 cell line overexpressing TCTP. This line confirmed that protection against H_2O_2toxicity is an additional physiologic function of mammalian TCTP.

Structural analyses in such studies have not yet been performed. In silico analysis of the amino acid sequence of plant TCTPs do not indicate the presence of conserved domains that are typical of enzymes of antioxidant system (Fig. 36.2). According to Gnanasekar and Ramaswamy (2007), TCTP lacks typical enzymatic active site, but it might not be functioning as an antioxidant enzyme. Thus, the domains related to antioxidant activity of plant TCTPs remain to be identified, although Munirathinam and Ramaswamy (2012) have already found that small ubiquitin-like modifier is important for antioxidant function of human TCTP.

Table 36.3 Molecular weight of some plant TCTP proteins predicted by ExPASy Proteomics Server (http://ca.expasy.org/tools/pi_tool.html, http://ca.expasy.org/tools/pi_tool.html)

Plant species	Number of aa	Predicted molecular weight (kDa)	Acession (GenBank/NCBI)
Arabidopsis thaliana	168	18.8	AAM66134
Arachis hypogaea	168	19.1	ABI84255
Brassica oleracea	168	19	AAL13303
Cicer arietinum	168	19.1	XP_004495707
Cucumis melo	168	18.7	NP_001284461
Cucurbita maxima	168	19	ABC02401
Elaeis guineensis	167	19	ACF06595
Glycine max	168	18.9	NP_001237819
Hevea brasiliensis	168	19.1	ACI04518
Hordeum vulgare	168	18.8	Q9M5G3
Jatropha curcas	168	19.1	ABO25950
Larix kaempferi	167	18.7	AGW01241
Litchi chinensis	168	18.7	ADZ75463
Manihot esculenta	168	19.1	AGQ45636
Medicago sativa	167	18.9	CAA67207
Medicago truncatula	168	19.1	AES61401
Nicotiana tabacum	168	18.7	AKG62084
Oryza sativa	168	18.9	ALF37646
Physcomitrella patens	170	18.9	XP_001758666
Picea sitchensis	160	18	ABK25668
Pisum sativum	167	18.8	P50906
Populus trichocarpa	168	19.1	ERP57941
Pseudotsuga menziesii	167	18.8	CAA10048
Rheum australe	168	18.8	ACH63210
Ricinus communis	168	18.9	XP_002512437
Salvia miltiorrhiza	168	18.8	ABW16955
Selaginella moellendorffii	168	19.1	XM_002982802
Solanum lycopersicum	168	18.9	NP_001234566
Solanum tuberosum	168	18.8	NP_001275182
Sorghum bicolor	167	18.8	XP_002453140
Thellungiella halophila	168	18.8	BAJ33998
Vitis vinifera	167	18.9	XP_010649714
Zea mays	167	18.6	NP_001105104

Heterologous expression studies have also been performed to verify the plant TCTP roles. For example, Meng et al. (2017) observed positive responses of *CsTCTP1* and *CsTCTP2* from *Cucumis sativus* under salt and heat stresses. On the other hand, a negative response of these *CsTCTP* genes under drought and $HgCl_2$ stresses were observed in a prokaryotic system (Meng et al. 2017).

The overexpression of MeTCTP recombinant protein from cassava by our research group resulted in increased salinity (Santa Brígida et al. 2014) and heat

Table 36.4 Some examples of plant *TCTP* genes isolated and characterized and their relationship with abiotic stress response or tolerance

TCTP genes of plants	Source species	Relationship with response or tolerance to abiotic stress	References
AtTCTP	*Arabidopsis thaliana*	Drought tolerance with rapid ABA-induced stomatal closure	Kim et al. (2012)
BoTCTP	*Brassica oleracea*	Cold, high temperature, and salinity tolerance	Cao et al. (2010)
CsTCTP1; *CsTCTP2*	*Cucumis sativus*	In prokaryotic cells, positive responses to salt and heat stresses and a negative response to drought and $HgCl_2$ stresses	Meng et al. (2017)
HbTCTP	*Hevea brasiliensis*	Regulated by drought, cold, salinity, H_2O_2, ethylene and methyl jasmonate; antioxidant activity in metal-catalyzed oxidation system	Deng et al. (2016)
HbTCTP1	*H. brasiliensis*	Regulated by drought, cold, salinity, H_2O_2, ethylene and methyl jasmonate; antioxidant activity in metal-catalyzed oxidation system	Li et al. (2013)
JcTCTP	*Jatropha curcas*	Regulated by salinity and high temperature	Qin et al. 2011
MeTCTP	*Manihot esculenta*	Regulated by salinity; protective function against salt and heat stress in bacterial cells; chaperone activity	Santa Brígida et al. (2014), Marques et al. (2017a, b)
OsTCTP	*Oryza sativa*	Mercury tolerance; regulated by abscisic acid and H_2O_2; enhanced activity of antioxidant enzymes; protective function against Hg^{2+} stress in yeast strains	Wang et al. (2012, 2015)
SlTCTP	*Solanum lycopersicum*	Salt and osmotic tolerance	de Carvalho et al. (2017)

(Marques et al. 2017b) tolerance in *E. coli* cells. In another study, yeast strains overexpressing *OsTCTP* showed increased survival to stress. The authors also concluded that *OsTCTP* protein could form complexes with Hg^{2+} or participate in the transport regulation of such ions to the extracellular space, or even participate in the transportation of other toxic compounds related to Hg^{2+} (Wang et al. 2012).

3.3 *Plant Tolerance Against Abiotic Stress Mediated by TCTP*

After identifying genes differentially expressed under abiotic stress conditions, as well as preliminary functional analysis, the analysis in transgenic plants overexpressing or silencing a gene of interest allows a more detailed functional characterization of this gene. In this section, information obtained from studies of transgenic plants overexpressing or silencing TCTP is presented.

3.3.1 Drought

Kim et al. (2012) observed that the TCTP-related drought tolerance and stomatal closure is associated with elevated cytosolic calcium and microtubule depolymerization in *Arabidopsis thaliana*. Plant TCTP proteins are not only distributed in the cytoplasm (Table 36.5). Thus, the role of TCTP in abiotic stresses tolerance in other cellular compartments besides the cytoplasm remains to be elucidated.

de Carvalho et al. (2017) verified the function of SlTCTP from tomato as a key mediator of ABA-dependent regulation of stomatal movement, when overexpressed in tobacco, may require changes in the abundance of a subset of aquaporins, the downregulation of *BAM1* (encoding β-amylase 1), up-regulation of ASPG (a gene encoding an aspartic protease in guard cells) and a higher accumulation of unsaturated fatty acid.

Other authors have also reported the calcium-binding activity of TCTP from different plant species (Hoepflinger et al. 2013; Li et al. 2013; Nakkaew et al. 2010). Further studies may contribute to the current knowledge about the importance of calcium interaction with TCTP in their role in drought tolerance as well as other stresses, since calcium binding proteins are components of plant defense against various abiotic stresses (Tuteja and Sopory 2008). Structural studies are also important to understand the conformational modifications in plant TCTP proteins after their interaction with Ca^{2+} in the response to abiotic stresses. We have performed the prediction for calcium-binding property of MeTCTP by molecular dynamics simulations. In such investigation, the main residues related to this interaction were identified in a in silico experiment (Marques et al. 2019).

According to the results obtained by Aoki et al. (2005), the TCTP can contribute to stress resistance through interaction with other proteins located in the phloem. These authors analyzed rice and pumpkin (*Cucurbita maxima*) and detected that the TCTP interacts with proteins involved in long-distance movement through phloem (for example, RNA binding proteins, such as CmPP16) and contributes to its transport. The protein CmPP16 contributes to the increase of photosynthetic capacity during the drought, as genetic transformation assays carried out by Ramírez-Ortega et al. (2014).

Table 36.5 Some examples of cellular localization of plant TCTPs

TCTP	Localization	Reference
AtTCTP	Cytoplasm	Kim et al. (2012), Hoepflinger et al. (2013)
AtTCTP	Nucleus	Toscano-Morales et al. (2014)
OsTCTP	Cytosol and nucleus	Wang et al. (2015)
HbTCTP	Membrane, cytosol, and nucleus (rice protoplasts)	Deng et al. (2016)
SlTCTP	Cytosol and nucleus	Bruckner et al. (2017)

3.3.2 Cold, High Temperature, and Salinity

Cao et al. (2010) found that cabbage plants (*Brassica oleracea*) silenced for the *BoTCTP* gene showed a decrease in tolerance to cold, high temperature and high salinity. Proteins that act as molecular chaperones are related to the defense response of plants at the cellular level, both under stress caused by high temperature (Wang et al. 2004) and by cold (Sanghera et al. 2011).

Gnanasekar et al. (2009) observed that the TCTP of humans and of *Schistosoma mansoni* act as a heat shock protein and as molecular chaperone (by interacting with another protein and protecting it).

It commonly occurs a cross-talk among molecular chaperones and other protection mechanisms against stresses, through interaction of such proteins with apoptosis regulators, cellular cycle, differentiation and status redox (Wang et al. 2004).

As highlighted by Wang et al. (2004), the response at the cellular level occurs by the joint action of several chaperones in the performance of activities that are critical to the survival of cells subjected to stress, such as maintaining functional conformation of native proteins, preventing the aggregation of non-native proteins, refolding of denatured proteins to regain their functional conformation, and also removing polypeptides that, although non-functional, are generated from misfolding, denaturation, or aggregation.

Plant TCTPs have putative binding domain to Rab GTPase (Fig. 36.2). Thaw et al. (2001) verified that TCTP show structural similarity of TCTP to a family of guanine nucleotide-free chaperones with GDP/GTP free form Rab proteins-binding property was previously reported (Thaw et al. 2001).

Furthermore, in silico analysis indicate the presence of domains related to protein-protein interactions in plant TCTPs (Fig. 36.2). In fact, after translation, TCTP of plants have the ability to interact with other proteins. For example, AtTCTP interacts with proteins to regulate apoptosis in *Arabidopsis* (Hoepflinger et al. 2013). In tobacco, NtTCTP interacts with NTHK1 (a subfamily II ethylene receptor tobacco Histidine Kinase1) and prevents NTHK1 from proteasome-mediated protein degradation (Tao et al. 2016). In *Hevea brasiliensis*, the HbTCTP interacted with rubber elongation factor (REF), 17.5 kDa heat shock family protein, annexin, and REF-like stress related protein1 (Deng et al. 2016) proteins.

Recently, we verified the molecular chaperone activity of purified MeTCTP, in addition to ability of recombinant MeTCTP in improving tolerance of bacterial cells against heat stress (Marques et al. 2017b). It remains to be elucidated whether TCTP of other plant species also acts as a molecular chaperone, as well as whether plant TCTP can occur in cooperation with other molecular chaperones.

3.3.3 Mercury and Oxidative Stresses

Wang et al. (2015) showed that OsTCTP is related to the increased activity of several antioxidant enzymes in transgenic rice plants, reducing H_2O_2 levels induced by Hg^{2+}, by molecular mechanisms not yet elucidated. These authors also verified that the overexpression of *OsTCTP* in plants increased the tolerance of plants to Hg^{2+} stress. Thereby, the TCTP role can be related to the roots defense mechanism against the exposure to heavy metal.

Enhanced H_2O_2-mediated oxidative stress tolerance was observed in tobacco plants overexpressing *HbTCTP* or *HbTCTP1* (Deng et al. 2018). As mentioned above in this review, the activity of rubber tree TCTP was confirmed in vitro. Within this context, it is clear that TCTP has a role against oxidative stress independent of antioxidant enzymes. However, based on the results obtained by Wang et al. (2015), further studies are needed to increased knowledge on TCTP and antioxidant system enzymes. Altogether, these results confirm the potential of TCTP for the generation of tolerant plants against different abiotic stresses.

4 Conclusions and Future Perspectives

In this review, information on the high potential of TCTP proteins related to molecular mechanisms of plant endogenous response and tolerance were presented. Further studies involving molecular cloning, structural characterization, expression, protein function analyses, regulation at the cellular level, interaction with other gene products, and overexpression and/or silencing in transgenic plants will contribute to better understanding of how TCTP is functionally related to the increase and modulation of tolerance of plants to different abiotic stresses.

Acknowledgments The corresponding author (Deyvid Novaes Marques) thanks Fundação de Amparo à Pesquisa do Estado de São Paulo (FAPESP; grant numbers 2017/05544-0, 2018/20706-9, 2020/12666-7, and 2022/11018-7). Nicolle Louise Ferreira Barros thanks Conselho Nacional de Desenvolvimento Científico e Tecnológico (CNPq; grant number 140586/2021-0).

References

Aoki K, Suzui N, Fujimaki S, Dohmae N, Yonekura-Sakakibara K, Fujiwara T, Hayashi H, Yamaya T, Sakakibaraa H (2005) Destination-selective long-distance movement of phloem proteins. Plant Cell 17:1801–1814. https://doi.org/10.1105/tpc.105.031419

Barba-Espin G, Dias-Vivancos P, Clemente-Moreno MJ, Albacete A, Faize L, Faize M, Pérez-Alfocea HJA (2010) Interaction between hydrogen peroxide and plant hormones during germination and the early growth of pea seedlings. Plant Cell Environ 33:981–994. https://doi.org/10.1111/j.1365-3040.2010.02120.x

Barnes A, Bale J, Constantinidou C, Ashton P, Jones A, Pritchard J (2004) Determining protein identity from sieve element sap in *Ricinus communis* L. by quadrupole time of flight (Q-TOF) mass spectrometry. J Exp Bot 55:1473–1481. https://doi.org/10.1093/jxb/erh161

Berkowitz O, Jost R, Pollmannb S, Masle J (2008) Characterization of TCTP, the translationally controlled tumor protein, from *Arabidopsis thaliana*. Plant Cell 20:3430–3447. https://doi.org/10.1105/tpc.108.061010

Bian YW, Lv DW, Cheng ZW, Gu AQ, Cao H, Yan YM (2015) Integrative proteome analysis of *Brachypodium distachyon* roots and leaves reveals a synergetic responsive network under H_2O_2 stress. J Proteome 128:388–402. https://doi.org/10.1016/j.jprot.2015.08.020

Bommer UA, Thiele BJ (2004) The translationally controlled tumour protein (TCTP). Int J Biochem Cell Biol 36:379–385. https://doi.org/10.1016/S1357-2725(03)00213-9

Brioudes F, Thierry AM, Chambrier P, Mollereau B, Bendahmane M (2010) Translationally controlled tumor protein is a conserved mitotic growth integrator in animals and plants. Proc Natl Acad Sci U S A 107(16384–16):389. https://doi.org/10.1073/pnas.1007926107

Bruckner FP, Xavier ADS, Cascardo RS, Otoni WC, Zerbini FM, Alfenas-Zerbini P (2017) Translationally controlled tumour protein (TCTP) from tomato and *Nicotiana benthamiana* is necessary for successful infection by a potyvirus. Mol Plant Pathol 18(5):672–683. https://doi.org/10.1111/mpp.12426

Cao B, Lu Y, Chen G, Lei J (2010) Functional characterization of the translationally controlled tumor protein (TCTP) gene associated with growth and defense response in cabbage. Plant Cell Tiss Org 103:217–226. https://doi.org/10.1007/s11240-010-9769-6

Chen SH, Wu P, Chou C, Yan Y, Liu H, Weng S, Yen H (2007) A knockout mouse approach reveals that TCTP functions as an essential factor for cell proliferation and survival in a tissue- or cell type–specific manner. Mol Biol Cell 18:2525–2532. https://doi.org/10.1091/mbc.E07-02-0188

Chen Y, Chen X, Wang H, Bao Y, Zhang W (2014) Examination of the leaf proteome during flooding stress and the induction of programmed cell death in maize. Proteome Sci 12:33. https://doi.org/10.1186/1477-5956-12-33

Chen Z, Zhu D, Wu J, Cheng Z, Yan X, Deng X, Yan Y (2018) Identification of differentially accumulated proteins involved in regulating independent and combined osmosis and cadmium stress response in *Brachypodium* seedling roots. Sci Rep 8(1):7790. https://doi.org/10.1038/s41598-018-25,959-8

Chow CN, Zheng HQ, Wu NY, Chien CH, Huang HD, Lee TY, Chiang-Hsieh YF, Hou PF, Yang TY, Chang WC (2016) PlantPAN 2.0: an update of plant promoter analysis navigator for reconstructing transcriptional regulatory networks in plants. Nucleic Acids Res 44:1154–1160. https://doi.org/10.1093/nar/gkv1035

de Carvalho M, Acencio ML, Laitz AVN, de Araújo LM, de Lara Campos Arcuri M, do Nascimento LC, Maia IG (2017) Impacts of the overexpression of a tomato translationally controlled tumor protein (TCTP) in tobacco revealed by phenotypic and transcriptomic analysis. Plant Cell Rep 36:887–900. https://doi.org/10.1007/s00299-017-2117-0

Deng Z, Chen J, Leclercq J, Zhou Z, Liu C, Liu H, Yang H, Montoro P, Xia Z, Li D (2016) Expression profiles, characterization and function of *HbTCTP* in rubber tree (*Hevea brasiliensis*). Front Plant Sci 7:789. https://doi.org/10.3389/fpls.2016.00789

Deng Z, Chen J, Wei Y, Liu H, Yang H, Dai L, Li D (2018) Two translationally controlled tumor protein (TCTP) genes from *Hevea brasiliensis* play overlapping and different roles in development and stress response. Ind Crop Prod 114:137–145. https://doi.org/10.1016/j.indcrop.2018.01.061

Dos Reis SP, Marques DN, Lima AM, de Souza CRB (2016) Plant molecular adaptations and strategies under drought stress. In: Hossain M, Wani S, Bhattacharjee S, Burritt D, Tran LS (eds) Drought stress tolerance in plants. Springer, New York, pp 91–122

Ermolayev V, Weschke W, Manteuffel R (2003) Comparison of Al-induced gene expression in sensitive and tolerant soybean cultivars. J Exp Bot 54:2745–2756. https://doi.org/10.1093/jxb/erg302

Gachet Y, Tournier A, Lee M, Lazaris-Karatzas A, Poulton T, Bommer UA (1999) The growth-related, translationally controlled protein P23 has properties of a tubulin binding protein and associates transiently with microtubules during the cell cycle. J Cell Sci 112:1257–1271

Ghabooli M, Khatabi B, Ahmadi FS, Sepehri M, Mirzaei M, Amirkhani A, Jorrin-Novo JV, Salekdeh GH (2013) Proteomics study reveals the molecular mechanisms underlying water stress tolerance induced by *Piriformospora indica* in barley. J Proteome 94:289–301. https://doi.org/10.1016/j.jprot.2013.09.017

Gnanasekar M, Ramaswamy K (2007) Translationally controlled tumor protein of *Brugia malayi* functions as an antioxidant protein. Parasitol Res 101:1533–1540. https://doi.org/10.1007/s00436-007-0671-z

Gnanasekar M, Dakshinamoorthy G, Ramaswam K (2009) Translationally controlled tumor protein is a novel heat shock protein with chaperone-like activity. Biochem Biophys Res Commun 386:333–337. https://doi.org/10.1016/j.bbrc.2009.06.028

Gross B, Gaestel M, Bohm H, Bielka H (1989) cDNA sequence coding for a translationally controlled human tumor protein. Nucleic Acids Res 17:8367

Guan Q, Yu J, Zhu W, Yang B, Li Y, Zhang L, Tian J (2018) RNA-Seq transcriptomic analysis of the *Morus alba* L. leaves exposed to high-level UVB with or without dark treatment. Gene 645:60–68. https://doi.org/10.1016/j.gene.2017.12.045

Guo G, Ge P, Ma C, Li X, Lv D, Wang S, Ma W, Yan Y (2012) Comparative proteomic analysis of salt response proteins in seedling roots of two wheat varieties. J Proteome 75:1867–1885. https://doi.org/10.1016/j.jprot.2011.12.032

Guo Y, Wang Z, Guan X, Hu Z, Zhang Z, Zheng J, Lu Y (2017) Proteomic analysis of *Potentilla fruticosa* L. leaves by iTRAQ reveals responses to heat stress. PLoS One 12(8):e0182917. https://doi.org/10.1371/journal.pone.0182917

Gutiérrez-Galeano DF, Toscano-Morales R, Calderón-Pérez B, Xoconostle-Cázares B, Ruiz-Medrano R (2014) Structural divergence of plant TCTPs. Front Plant Sci 5:361. https://doi.org/10.3389/fpls.2014.00361

Hinojosa-Moya JJ, Xoconostle-Cázares B, Toscano-Morales R, Ramírez-Ortega F, Cabrera-Ponce JL, Ruiz-Medrano R (2013) Characterization of the pumpkin translationally-controlled tumor protein *CmTCTP*. Plant Signal Behav 8:e26477. https://doi.org/10.4161/psb.26477

Hoepflinger MC, Reitsamer J, Geretschlaeger AM, Mehlmer N, Tenhaken R (2013) The effect of translationally controlled tumour protein (TCTP) on programmed cell death in plants. BMC Plant Biol 13:135. https://doi.org/10.1186/1471-2229-13-135

Huang L, Yan H, Jiang X, Zhang Y, Zhang X, Ji Y, Zeng B, Xu B, Yin G, Lee S, Yan Y, Ma X, Peng Y (2014) Reference gene selection for quantitative real-time reverse-transcriptase PCR in orchardgrass subjected to various abiotic stresses. Gene 553(2):158–165. https://doi.org/10.1016/j.gene.2014.10.017

Huang J, Wu X, Tian F, Chen Q, Luo P, Zhang F, Wan X, Zhong Y, Liu Q, Lin T (2020) Changes in proteome and protein phosphorylation reveal the protective roles of exogenous nitrogen in alleviating cadmium toxicity in poplar plants. Int J Mol Sci 21(1):278. https://doi.org/10.3390/ijms21010278

Jung J, Kim M, Kim MJ, Kim J, Moon J, Lim JS, Kim M, Lee K (2004) Translationally controlled tumor protein interacts with the third cytoplasmic domain of Na, K-ATPase alpha subunit and inhibits the pump activity in HeLa cells. J Biol Chem 279(49868–49):875. https://doi.org/10.1074/jbc.M400895200

Kazan K (2015) Diverse roles of jasmonates and ethylene in abiotic stress tolerance. Trends Plant Sci 20:219–229. https://doi.org/10.1016/j.tplants.2015.02.001

Kim YM, Han YJ, Hwang OJ, Lee SS, Shin AY, Kim SY, Kim JI (2012) Overexpression of *Arabidopsis* translationally controlled tumor protein gene *AtTCTP* enhances drought tolerance with rapid ABA-induced stomatal closure. Mol Cell *33*:617–626. https://doi.org/10.1007/s10059-012-0080-8

Koh J, Chen G, Yoo M-J, Zhu N, Dufresne D, Erickson JE, Shao H, Chen S (2015) Comparative proteomic analysis of *Brassica napus* in response to drought stress. J Proteome Res 14:3068–3081. https://doi.org/10.1021/pr501323d

Lescot M, Déhais P, Thijs G, Marchal K, Moreau Y, Van de Peer Y, Rouzé P, Rombauts S (2002) PlantCARE, a database of plant cis-acting regulatory elements and a portal to tools for in silico analysis of promoter sequences. Nucleic Acids Res 30:325–327

Li D, Deng Z, Liu X, Qin B (2013) Molecular cloning, expression profiles and characterization of a novel translationally controlled tumor protein in rubber tree (*Hevea brasiliensis*). J Plant Physiol 170:497–504. https://doi.org/10.1016/j.jplph.2012.11.014

Li Y, Zhao J, Li Y, Xu X, Zhang B, Liu Y, Cui L, Li B, Gao Y, Chaia Z (2016) Comparative metalloproteomic approaches for the investigation proteins involved in the toxicity of inorganic and organic forms of mercury in rice (*Oryza sativa* L.) roots. Metallomics 8:663–671. https://doi.org/10.1039/c5mt00264h

Lucibello M, Gambacurta A, Zonfrillo M, Pierimarchi P, Serafino A, Rasi G, Rubartelli A, Garaci E (2011) TCTP is a critical survival factor that protects cancer cells from oxidative stress induced cell-death. Exp Cell Res 317:2479–2489. https://doi.org/10.1016/j.yexcr.2011.07.012

Marques DN, Reis SP, Souza CRB (2017a) Plant NAC transcription factors responsive to abiotic stresses. Plant Gene 11:170–179. https://doi.org/10.1016/j.plgene.2017.06.003

Marques DN, Reis SP, Souza CRB (2017b) Evaluation of a cassava translationally controlled tumor protein (MeTCTP) reveals its function in thermotolerance of *Escherichia coli* and in vitro chaperone-like activity. Genet Mol Res 16:gmr16039835. https://doi.org/10.4238/gmr16039835

Marques DN, Siqueira AS, Gonçalves EC, Barros NLF (2019) Homology modeling and molecular dynamics simulations of a cassava translationally controlled tumor protein (MeTCTP). Plant Gene 19:100185. https://doi.org/10.1016/j.plgene.2017.06.003

Marques DN, Stolze SC, Harzen A, Nogueira ML, Batagin-Piotto KD, Piotto FA, Mason C, Azevedo RA, Nakagami H (2021) Comparative phosphoproteomic analysis of tomato genotypes with contrasting cadmium tolerance. Plant Cell Rep 40(10):2001–2008. https://doi.org/10.1007/s00299-021-02774-6

Marques DN, Barros NLF, de Souza CRB (2023a) Does translationally controlled tumor protein (TCTP) have the potential to produce crops with increased growth and tolerance to environmental stresses? Plant Cell Rep 42(4):821–824. https://doi.org/10.1007/s00299-023-02985-z

Marques DN, Mason C, Stolze SC, Harzen A, Nakagami H, Skirycz A, Piotto FA, Azevedo RA (2023b) Grafting systems for plant cadmium research: Insights for basic plant physiology and applied mitigation. Sci Total Environ 892:164610. https://doi.org/10.1016/j.scitotenv.2023.164610

Marques DN, Nogueira ML, Gaziola SA, Batagin-Piotto KD, Freitas NC, Alcantara BK, Paiva LV, Mason C, Piotto FA, Azevedo RA (2023c) New insights into cadmium tolerance and accumulation in tomato: Dissecting root and shoot responses using cross-genotype grafting. Environ Res 216:114577. https://doi.org/10.1016/j.envres.2022.114577

Maršálová L, Vítámvás P, Hynek R, Prášil IT, Kosová K (2016) Proteomic response of *Hordeum vulgare* cv. Tadmor and *Hordeum marinum* to salinity stress: similarities and differences between a glycophyte and a halophyte. Front Plant Sci 7:1154. https://doi.org/10.3389/fpls.2016.01154

Meng XN, Chen QM, Fan HY, Song TF, Cui N, Zhao JY, Jia SM, Meng KX (2017) Molecular characterization, expression analysis and heterologous expression of two translationally controlled tumor protein genes from *Cucumis sativus*. PLoS One 12(9):e0184872. https://doi.org/10.1371/journal.pone.0184872

Mostek A, Börner A, Badowiec A, Weidner S (2015) Alterations in root proteome of salt-sensitive and tolerant barley lines under salt stress conditions. J Plant Physiol 174:166–176. https://doi.org/10.1016/j.jplph.2014.08.020

Munirathinam G, Ramaswamy K (2012) Sumoylation of human translationally controlled tumor protein is important for its nuclear transport. Biochem Res Int 831:940. https://doi.org/10.1155/2012/831940

Nadira UA, Ahmed IM, Zeng J, Wu F, Zhang G (2016) Identification of the differentially accumulated proteins associated with low phosphorus tolerance in a Tibetan wild barley accession. J Plant Physiol 198:10–22. https://doi.org/10.1016/j.jplph.2016.03.016

Nagano-Ito M, Banba A, Ichikawa S (2009) Functional cloning of genes that suppress oxidative stress-induced cell death: TCTP prevents hydrogen peroxide-induced cell death. FEBS Lett 583:1363–1367. https://doi.org/10.1016/j.febslet.2009.03.045

Nakkaew A, Chotigeat W, Phongdara A (2010) Molecular cloning and expression of *EgTCTP*, encoding a calcium binding protein, enhances the growth of callus in oil palm (*Elaeis guineensis*, Jacq). Songklanakarin J Sci Technol 32:561–569

Nikonorova N, Van den Broeck L, Zhu S, van de Cotte B, Dubois M, Gevaert K, Inzé D, De Smet I (2018) Early mannitol-triggered changes in the *Arabidopsis* leaf (phospho)proteome reveal growth regulators. J Exp Bot 69(19):4591–4607. https://doi.org/10.1093/jxb/ery261

Qi YC, Wang HJ, Zou Y, Liu C, Liu YQ, Wang Y, Zhang W (2011) Over-expression of mitochondrial heat shock protein 70 suppresses programmed cell death in rice. FEBS Lett 585:231–239. https://doi.org/10.1016/j.febslet.2010.11.051

Qin X, Gao F, Zhang J, Gao J, Lin S, Wang Y, Jiang L, Liao Y, Wang L, Jia Y, Tang L, Xu Y, Chen F (2011) Molecular cloning, characterization and expression of cDNA encoding translationally controlled tumor protein (TCTP) from *Jatropha curcas* L. Mol Biol Rep 38:3107–3112. https://doi.org/10.1007/s11033-010-9980-x

Ramírez-Ortega FA, Herrera-Pola PS, Toscano-Morales R, Xoconostle-Cázares B, Ruiz-Medrano R (2014) Overexpression of the pumpkin (*Cucurbita maxima*) 16 kDa phloem protein CmPP16 increases tolerance to water deficit. Plant Signal Behav 9:e973823. https://doi.org/10.4161/15592324.2014.973823

Rampitsch C, Bykova N (2012) The beginnings of crop phosphoproteomics: exploring early warning systems of stress. Front Plant Sci 3:144. https://doi.org/10.3389/fpls.2012.00144

Rho SB, Lee JH, Park MS, Byun HJ, Kang S, Seo S, Kim JY, Park SY (2011) Anti-apoptotic protein TCTP controls the stability of the tumor suppressor p53. FEBS Lett 585:29–35. https://doi.org/10.1016/j.febslet.2010.11.014

Sage-ono K, Ono M, Harada H, Kamada H (1998) Dark-induced accumulation of mRNA for a homolog of translationally controlled tumor protein (TCTP) in *Pharbitis*. Plant Cell Physiol 39:357–360

Sah S, Reddy KR, Li J (2016) Abscisic acid and abiotic stress tolerance in crop plants. Front Plant Sci 7:571. https://doi.org/10.3389/fpls.2016.00571

Sanchez JC, Schaller D, Ravier F, Golaz O, Jaccoud S, Belet M, Wilkins MR, James R, Deshusses J, Hochstrasser D (1997) Translationally controlled tumor protein: A protein identified in several nontumoral cells including erythrocytes. Electrophoresis 18:150–155. https://doi.org/10.1002/elps.1150180127

Sanghera GS, Wani SH, Hussain W, Singh NB (2011) Engineering cold stress tolerance in crop plants. Curr Genomics 12:30–43. https://doi.org/10.2174/138920211794520178

Santa Brígida AB, dos Reis SP, Costa CNM, Cardoso CMY, Lima AM, de Souza CRB (2014) Molecular cloning and characterization of a cassava translationally controlled tumor protein gene potentially related to salt stress response. Mol Biol Rep 41:1787–1797. https://doi.org/10.1007/s11033-014-3028-6

Schmidt I, Fahling M, Nafz B, Skalweit A, Thiele BJ (2007) Induction of translationally controlled tumor protein (TCTP) by transcriptional and post-transcriptional mechanisms. FEBS J 274:5416–5424. https://doi.org/10.1111/j.1742-4658.2007.06069.x

Sievers F, Wilm A, Dineen D, Gibson TJ, Karplus K, Li W, Lopez R, McWilliam H, Remmert M, Söding J, Thompson JD, Higgins DG (2011) Fast, scalable generation of high-quality protein multiple sequences alignments using Clustal Omega. Mol Syst Biol 7:539. https://doi.org/10.1038/msb.2011.75

Sturzenbaum SR, Kille P, Morgan AJ (1998) Identification of heavy metal induced changes in the expression patterns of the translationally controlled tumour protein (TCTP) in the earthworm *Lumbricus rubellus1*. Biochim Biophys Acta 1398:294–304. https://doi.org/10.1016/S0167-4781(98)00077-3

Tao JJ, Cao YR, Chen HW, Wei W, Li QT, Ma B, Zhang WK, Chen SY, Zhang JS (2016) Tobacco translationally controlled tumor protein interacts with ethylene receptor tobacco Histidine Kinase1 and enhances plant growth through promotion of cell proliferation. Plant Physiol 169:96–114. https://doi.org/10.1104/pp.15.00355

Thaw P, Baxter NJ, Hounslow AM, Price C, Waltho JP, Craven CJ (2001) Structure of TCTP reveals unexpected relationship with guanine nucleotide-free chaperones. Nat Struct Biol 8:701–704. https://doi.org/10.1038/90415

Toscano-Morales R, Xoconostle-Cázares B, Martínez-Navarro AC, Ruiz-Medrano R (2014) Long distance movement of an *Arabidopsis* translationally controlled tumor protein (*AtTCTP2*) mRNA and protein in tobacco. Front Plant Sci 5:705. https://doi.org/10.3389/fpls.2014.00705

Tuteja N, Sopory SK (2008) Chemical signaling under abiotic stress environment in plants. Plant Signal Behav 3:525–536

Vincent D, Ergül A, Bohlman MC, Tattersall EA, Tillett RL, Wheatley MD, Woolsey R, Quilici DR, Joets J, Schlauch K, Schooley DA, Cushman JC, Cramer GR (2007) Proteomic analysis reveals differences between *Vitis vinifera* L. cv. Chardonnay and cv. Cabernet Sauvignon and their responses to water deficit and salinity. J Exp Bot 58:1873–1892. https://doi.org/10.1093/jxb/erm012

Wang W, Vinocour B, Shoseyov O, Altman A (2004) Role of plant heat-shock proteins and molecular chaperones in the abiotic stress response. Trends Plant Sci 9:244–252. https://doi.org/10.1016/j.tplants.2004.03.006

Wang F, Shang Y, Yang L, Zhu C (2012) Comparative proteomic study and functional analysis of translationally controlled tumor protein in rice roots under Hg^{2+} stress. J Environ Sci 24:2149–2158

Wang ZQ, Li GZ, Gong QQ, Li GX, Zheng SJ (2015) *OsTCTP*, encoding a translationally controlled tumor protein, plays an important role in mercury tolerance in rice. BMC Plant Biol 15:123. https://doi.org/10.1186/s12870-015-0500-y

Wang C, Wang J, Wang X, Xia Y, Chen C, Shen Z, Chen Y (2017a) Proteomic analysis on roots of Oenothera glazioviana under copper-stress conditions. Sci Rep 7(1):10589. https://doi.org/10.1038/s41598-017-10,370-6

Wang S, Tao Y, Zhou Y, Niu J, Shu Y, Yu X, Liu S, Chen M, Gu W, Ma H (2017b) Translationally controlled tumor protein GmTCTP interacts with GmCDPKSK5 in response to high temperature and humidity stress during soybean seed development. Plant Growth Regul 82:187–200. https://doi.org/10.1007/s10725-017-0250-y

Woo HH, Hawes MC (1997) Cloning of genes whose expression is correlated with mitosis and localized in dividing cells in root caps of *Pisum sativum* L. Plant Mol Biol 35:1045–1051. https://doi.org/10.1023/A:1005930625920

Xu A, Bellamy AR, Taylor JA (1999) Expression of translationally controlled tumour protein is regulated by calcium at both the transcriptional and post-transcriptional level. Biochem J 342:683–689

Zhang L, Li W, Han S, Yang W, Qi L (2013) cDNA cloning, genomic organization and expression analysis during somatic embryogenesis of the translationally controlled tumor protein (TCTP) gene from Japanese larch (*Larix leptolepis*). Gene 529:150–158. https://doi.org/10.1016/j.gene.2013.07.076

Zhang HM, Zhang LS, Lv H, Yu ZY, Zhang DP, Zhu WN (2014) Identification of changes in *Triticum aestivum* L. leaf proteome in response to drought stress by 2D-PAGE and MALDI-TOF/TOF mass spectrometry. Acta Physiol Plant 36:1385–1398. https://doi.org/10.1007/s11738-014-1517-9

Chapter 37
Plant Tissue Culture and Crop Improvement

Amna Aslam, Ayesha Bibi, Safura Bibi, Athar Mahmood, Muhammad Mansoor Javaid, Hafeez Ur Rehman, and Muhammad Ather Nadeem

Abstract Micropropagation is used in both crop production and agriculture. Crop improvement by genetic engineering methods is possible in most cultivated crops. There is major challenge to produce more genetic diversity by vegetative method due to lack of suitable genotype, and hence the growth of crops decreases. But there are many ways to increase crop improvements by vegetative methods (tissue culture method) including micropropagation, meristem tip culture, somatic embryogenesis, somaclonal variation, ovary method, protoplast, somatic hybridization, secondary metabolites, cryopreservation, and haploid production method. Micropropagation is a vegetative method which produces the clones of cells though various techniques containing explant sources (such as root, shoot, sterilized conditions, and plant transfer to soils). For micropropagation techniques, selected genotype for vegetative growth of soft woody plants is used, for example, adventitious shoot (sugarcane nodal formation). Another meristem tip culture method has the capacity to produce virus-free plants, e.g., garlic. The protoplast method needs nutrition and culture media for plants' growth. The cryopreservation method produces storage of plants by low temperature under laboratory treatments. This chapter discusses various methods of tissue culture to increase crop growth and improvement.

Keywords Plant tissue culture · Micropropagation · Meristem culture · Somatic embryogenesis · Anther culture · Cryopreservation

A. Aslam · A. Bibi · S. Bibi · H. U. Rehman
Department of Botany, Faculty of Sciences, University of Agriculture, Faisalabad, Pakistan

A. Mahmood (✉)
Department of Agronomy, Faculty of Agriculture, University of Agriculture, Faisalabad, Pakistan
e-mail: athar.mahmood@uaf.edu.pk

M. M. Javaid (✉) · M. A. Nadeem
Department of Agronomy, College of Agriculture, University of Sargodha, Sargodha, Pakistan
e-mail: mansoor.javaid@uos.edu.pk

M. Hasanuzzaman (ed.), *Climate-Resilient Agriculture, Vol 2*,
https://doi.org/10.1007/978-3-031-37428-9_37

1 Introduction

It is expected that by 2050, the human population will be increased to one thousand million (FAO 2017). One of the biggest challenges is to feed the increasing population by the production of food at large scale. But due to short availability of fertile land and climate change, crop development seems to be stabilizing and unexpectedly reducing. It is thought that, to feed the 10 billion people of the global population, 60% rise in the production yield could be required (Springmann et al. 2018). It is known that the improvement of health of plant material and providing genetic variability are brought through a technique which is known as tissue culture technique. While most plant species have tissue culture methods, numerous crops, especially grains and trees, still require constant improvements. In vitro selection is used for the rapid identification of mutants with beneficial agronomic properties such as salt or drought resistance or disease resistance (Ali et al. 2013b; Kordrostami et al. 2022).

Tissue culture is introduced to all plant culture, including callus, cell, protoplasm, anther, meristem, embryo, and organ cultures and its benefits as a multiplication technology (Raina et al. 2018; Abbas et al. 2019). Plant cells and tissues on reproduction approaches below disease-free environment and controlled condition of light temperature and moisture are grown in this technique. It is germ-free, artificial growth of living organism tissue, and their components under specific physical and chemical conditions (Hassani et al. 2018). Vegetative propagation is one of the modern techniques to regenerate specific plant species. It has newly enhanced specific role in plant growth, disease control, and secondary metabolites production (Asif et al. 2012; Oseni et al. 2018). There are many methods involved in this technique like micropropagation, meristem culture technique, and somaclonal variations (Liu 2019). All these methods have the properties to produce the plants under germ-free conditions, and these methods can be performed in a very short duration of time (Shanee et al. 2011; Gulzar et al. 2020).

2 Methods of Tissue Culture

2.1 Micropropagation

Micropropagation is the artificial culture-based species-specific propagation of approved genetic constitution. Micropropagation is usually connected to extensive production at affordable prices (Torgbor 2020). Selected plant cells are distant, treated, and raised in a disease-free environment for cloning in this technique. Cloning isolation technique established the fact that somatic cells under suitable environment can be converted into complete plant (Gupta et al. 2020) (Fig. 37.1).

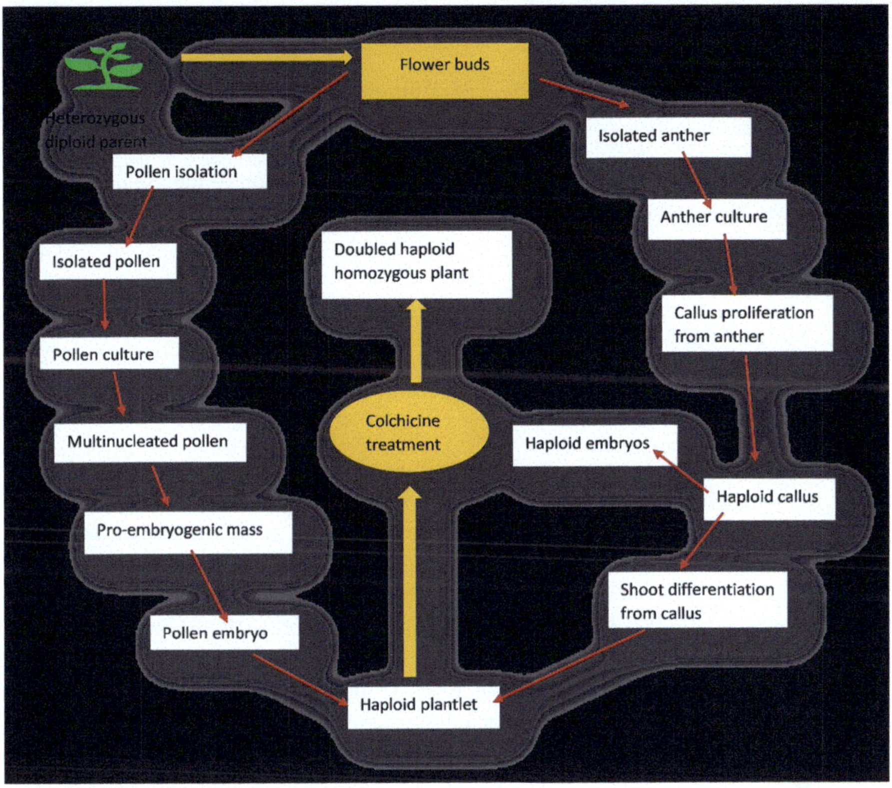

Fig. 37.1 Several plant tissue culture techniques (Mehbub et al. 2022)

2.1.1 Advantages of Micropropagation

The advantages of micropropagation (Chandana et al. 2018) are as follows:

- Fast reproduction by vegetative propagation
- Provides specific and controlled environmental conditions to plants
- Provides the availability of vegetative materials throughout the year disregarding to the regional and seasonal variation
- Helps in identification as well as development of mutant having specific properties as somaclonal variation
- Helps in development of secondary metabolites
- Provides the platform for the conservation of threatened plant species
- Provides specific range as well as succeeding to the development of specific plant-based pharmaceuticals
- Provides the tools for the preservation of genetic materials by cryopreservation method
- Produce the plants that are resistant to disease and stresses

2.1.2 Drawbacks of Micropropagation

- Expensive laboratory instruments as well as service are required (Ray and Ali 2018).
- The plants that produced are not autotrophic.
- One can face many problems such as poor acclimatization to the area.
- 'De novo' regeneration is used in micropropagation, the risk of genetic changes increases.

2.1.3 Methods of Micropropagation

2.1.3.1 Aseptic Culture

The aseptic technique is also called the method of sterilization in which plant parts are sterilized under hygienic conditions. The experimental design of sterilization is used for the identification and removal of contamination as well as for the determination of preventions during tissue culture process (Ikenganyia et al. 2017). There are many sources of contamination such as environment (airflow carries microorganisms that contaminated the culture media), culture media (that is used to produce plant tissues also can allow the growth of microorganisms such as bacteria and fungi), and plant materials (surface of plant materials can be contaminated with microorganisms). All these sources of contamination could be identified with the help of aseptic culture method (Soumare et al. 2021).

Once the contamination source is identified, the next step is to sterilize the instruments such as flask, beakers, and glass slides. For this purpose, two methods are used (chemical method and dry heat method). In chemical method, specific chemicals (ethylene oxide, chlorine dioxide, sodium hypochlorite, and mercuric chloride) are used to sterilize the tissue culture media and to identify the growth of microorganisms. While dry heat method involved the sterilization in which drying ovens are used, which allows 1 h heating process to reach the sterilization temperature (Wang et al. 2022).

2.1.3.2 Shoot Multiplication

Several micropropagation techniques use fragments such as shoots, root noodles, and buds. For micropropagation method, shoot multiplication technique is used in laboratory for many plants such as Aloe vera species, English walnut, and pineapple, etc. Many products such as lignosulphonate and other hormone like cytokinin help in shoot multiplication process (Wan Abdullah et al. 2020). This is the true proliferation stage, where the culture is removed from foregoing cycle, cut into pieces as well as disease-free re-growing on new medium in individual culture dishes. In this step, cytokinins growth regulators that speed up cell separation and germination are largely adapted (Gilbert et al. 2018). For profit of propagules, in

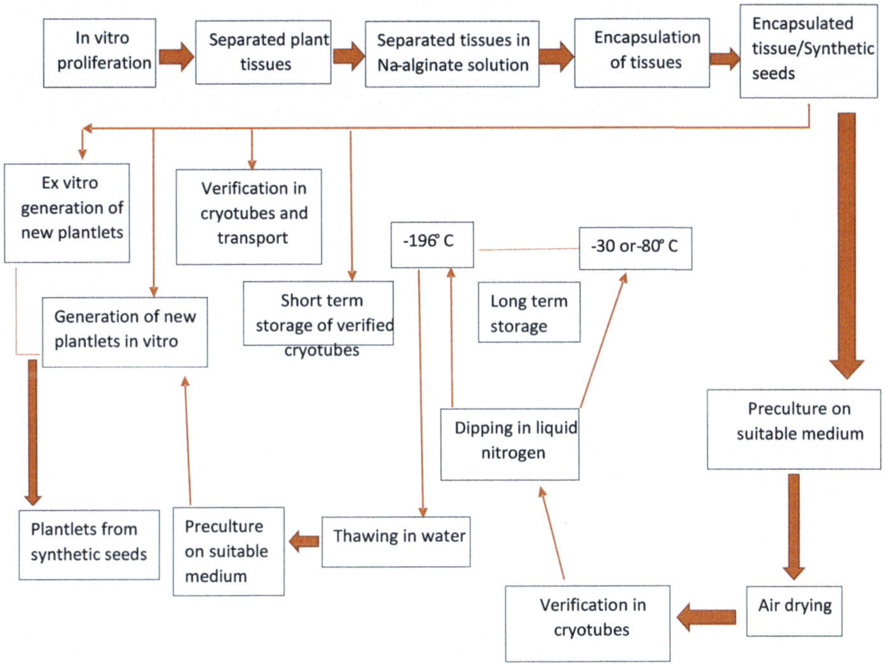

Fig. 37.2 Cytokinins effect on shoot multiplication (Nowakowska et al. 2022)

one year, 26 cycles are carried out. Because of the higher concentration of cytokinins, root growth did not take place in culture media (Fig. 37.2; Nowakowska et al. 2022). As a result, for reproduction, buds or shoots are frequently used (Gosal and Wani 2018).

Micropropagation of pineapple commercially facilities successive farming in a liquid environment for the expansion of axillary meristems and shoots (common micropropagation's focus of majority pineapple production (*Ananas comosus* L.) through a quick absorption technique is preserved). Three different steps are taken by the meaningfully short-lived falling system all over this process: shooting, shoot variation, and extension. In demand to grow many rooted shoots, it is now recognized as an active method of plant production to increase the growth of apical or axillary twigs (Sudheer et al. 2022). Sea oats, an important grass species, that dominate genetic material and cytokinin chemical can be efficiently used to *stimulate shoot* growth of oats species (Sahab et al. 2021).

2.1.3.3 Induction of Roots and Hardening

To make micropropagation methods simple and more successful at inducing roots on microshoots, rooting vessels were recognized. In demand to improve the performance of microshoots, this study highlights the advantages of osmotic potential in liquid culture over agar gel culture (Patel 2021). Through tests, the ideal root hormones and possible hardening media for *Gerbera jamesonii* micropropagation procedures were recognized. IBA (indole butyric acid) and GA3 (gibberellic acid) were used for root study. However, sand, cocopeat, vermiculite, and vermiculite mixture were used for gerbera hardening testing. Many root introduction structures were measured physically and biochemically at the hardened of the gerbera plant. Based on a diversity of variables, higher abscisic acid levels (3 and 4 ppm), lower gibberellic acid concentrations, and the best overall abscisic acid performance were recognized (van Zijl et al. 2018).

2.1.3.4 Transfer of Plantlets to Soil

Plants will die soon if placed in regular greenhouse or field conditions that are cultured on agar-based media. The ability of roots to absorb water is reduced by removed damage and poor contact with the substrate. Removing glass plants is more stimulating due to their slow growth and fast wilting. Reduced cell wall stress caused by these plants' reduced cellulose and lignin deposition increases cellular water absorption and glass turgorness in leaves and stems. While reducing damage to the root system, apply cellulose plugs to root plants in vitro and move them to soil. This can significantly increase wilting resistance (Tanveer et al. 2012; Li 2020).

2.2 *Meristem Culture*

Meristem culture is the most frequent method of get the virus-free plants from a structurally diseased plant (Shen and Hsu 2018). Various studies have established the effective production of virus-free plants of distinct species by these techniques. Meristem culture subsequently has been useful to get good plants free from virus (Javaid et al. 2012b; Adhikary et al. 2021). Instead of tissue in meristem culture, shoot tip has the maximum ability for growth into whole plant. The separation and expansion of shoot meristem is of great importance for reproducing the plants which should not be affected by pathogens. Shoot meristem culture is used over a large area for the production of different types of crops, and these crops have been grown into base of horticulture micropropagation industries (Vir and Kumar 2021). For example, shoot meristem culture is used to produce plants that are free of viruses. In the process of meristem culture, the dome and leaves present in the subapical zone are removed. If there is any damaged part of stem present outside, then shoot culture can be used rather than meristem culture (Pal et al. 2019).

Practically, the first meristem culture to find the virus-free plants of *Dahlia* from diseased individuals was cultivated through shoot tip in culture medium and some bud formation in sugarcane (Brar and Khush 2021). In the apical meristem cell, genetic stability is present, so for producing plants with same phenotype and genotype like explant source, those meristem cells have great chances in micropropagation technique. As top culture of meristem technique has ability to produce virus-free plants, pathologists are interested in it (Javaid et al. 2012a; Trivedi 2021). Top meristem culture technique is made for propagation of number of plants. It is thought as commercially workable technique currently (Johns 2019). Apical meristem culture technique is call for the eradication of shoot meristem parts, their installation on growth media in different sterilized conditions, and future arrangements of growth hormones and culture circumstances (Singh Shekhawat 2019). Meristem culture has benefits to other culture methods like disease plants can be removed. Another important method of meristem culture is the spread of haploid plant. Meristem culture of haploid plants can be exploit for their proliferation. Some drawbacks like plants are regularly affected with the germ-free plants which approved from one generation, and it can be increased if followed with the help of meristem culture disease plants (Kendurkar and Rangaswamy 2018).

2.3 Somatic Embryogenesis

Somatic embryos are very helpful tools for vegetative multiplication and for investigation of regulation of embryo development. Somatic embryogenesis is regenerating multi-step process. It starts from proembryogenic masses development, and it continues to maturation, desiccation, and regeneration of plants. Somatic embryogenesis is multifunctional process. Through somatic embryogenesis, some new structures are formed which can be produced in new whole plant. These structures are formed through cells having embryogenic capacity generated by somatic cells (Méndez-Hernández et al. 2019).

The embryo formation initiation process is still uncleared. However, it is investigated that unequal distribution of auxin is needed for embryo formation (Márquez-López et al. 2018). Development of somatic and zygotic embryo is based on classification of plants even they have morphological similarities. It was thought that zygotic embryos are nutrified through phloem tissues, while somatic embryos utilize external source of carbohydrates (Baba et al. 2009; Quinga et al. 2018). Application of plant growth regulators has vital role in culture medium especially during somatic embryogenesis induction, as they induce cell differentiation. But it depends on concentration and quality of plant growth regulator used in culture medium (Grzybkowska et al. 2018).

2.4 Somaclonal Variation

Somaclonal variation is a mutational and inherited process that arises the effect of using tissue culture techniques. Somaclonal variation depends on explant source and means of reproduction through differentiated tissues (Habib et al. 2012; Bridgen et al. 2018). It is investigated that there is more genetic diversity in vegetatively propagated plants instead of seed propagated plants. So, the presence of somaclonal variation has vital role to generate genetic diversity between plants. These plant species reproduce vegetatively. Different crops show somaclonal variations for aspects based on quality and quantity, like concentration of proteins in *Oryza sativa*, high concentration of sugar in sugarcane, early trimming of corn, changes in plants height, color of grains, tolerance of herbicides in tomato resistance of disease in maize, sugarcane, and *Solanum tuberosum,* and tolerance of salinity stress in *Oryza sativa* (Hannachi et al. 2021).

Somaclonal variation may be utilized for the selection of genotypes capable to both biotic and abiotic stresses. For example, in ornamental plant species, selected genotypes resistance to abiotic stresses as low temperature or high salinity leaf and flower morphology may be changes (Manchanda et al. 2018). Some disadvantages to manipulation of somaclonal variation as a breeding method depend upon genotype, absence of markers (DNA) that allow to recognize chosen mutants.

2.5 *Embryo/Ovary Method*

Ovary culture is a method of cultivated ovaries separate through the pollinated or un-pollinated segments. It is also called zygotic ovary/seed liberation technique which is the most primitive technique used. Major application of embryo method is for interspecies crossbreeding (Degefa 2019). Embryo culture is required for separating and producing a zygotic embryo under germ-free situations on purifying media for the purpose to acquire feasible plants. An unsuccessful embryo and ovaries of different plant species damaged by harmful chemicals can restart growth under suitable development conditions (Gupta et al. 2022; Rehman et al. 2022). Embryo culture is useful to breeders as breeding cycle becomes shorter, comparatively easy as well as demanding a simple nutrient medium.

2.6 *Protoplast Culture*

Protoplast is the spherical- or oval-shaped part of cell which is covered with cell membrane. Somatic hybridization or protoplast fusion is the process in which a hybrid is generated by the fusion of protoplast. In genetic engineering, protoplast culture is used for DNA transfer and also the transfer of chloroplast, mitochondria, and plasmid. Different methods are involved in culturing of protoplast which are given by Bridgen et al. (2018) and Javaid et al. (2022b).

2.6.1 Hanging Drop Method

In 1970, Kao proposed this technique of culturing protoplast. According to this technique, protoplast was incubated and suspended at low light intensity and at some specific position. This method is functional in the cases where inhibition in normal growth of tissues is occurred by smooth substrate through which tissues are growing (Jan et al. 2018). By reducing the evaporation rate and surface area ratio, inhibition in growth of tissues can be controlled. In order to obtain successfully cultured protoplast, their high efficiency and sufficient density are essential (Maekawa et al. 2022). The maximum density of protoplasts affects the cell division and the microcell formation. It is normally between 1.10^4 and 1.10^6 protoplast within 1 ml of media, if the densities are very high, linkage as well as joining of the cell colonies can take place (Makarem et al. 2021).

2.6.2 Agar Plating Method

Bergmann at the year of 1960 introduced this technique. In the culture plate, agar (that is melted) is mixed with the pure protoplast which is suspended in a medium (Ermert et al. 2019). At the stage of preparation, temperature is maintained at 4 °C; after that, it is placed to the contamination-free environment at temperature of 27–30 °C (Lal and Lal 2019).

2.6.2.1 Microculture Technique

In this method, very small droplets of protoplasts are placed on the microscopic glass-slide. Two cover slips are used to cover the glass slides. For protecting the protoplast suspension, the third cover slip is used. This technique is highly applied for the cultivation of tobacco crop (complete mature plant from isolated protoplasts) (Chadipiralla et al. 2020).

2.6.2.2 Factors Involved in Culture Media Formation

Major factors are involved in the formation of the media preparation, such as the sugar contents, light intensity as well as temperature (Espinosa-Leal et al. 2018). Placing liquid medium intersecting the solid medium is generally more challenging, and the absorbents within the media can be more easily controlled. For example, when the cell wall is being restored between the first division, the values of osmotic pressure should be decreased to ensure that cell division will not halt. In a fluid medium, the thickness of cells can be controlled easily and changed in the chamber of culture which is required in the process of cultivation is simple (Pantelidou et al. 2021).

Table 37.1 Antiseptics used for removal of endophytic contamination in plant tissue cultures

Applied substance	Plant species	Success of decontamination	References
Kanamycin and streptomycin sulphate at 10 µg mL^{-1} each were added to shoot multiplication medium + 2 mg L^{-1} BAP + 10 mg L^{-1} adenine sulfate	*Guadua angustifolia*	Bacterial growth was inhibited, and intensive formation of high-quality shoots was observed	Alagarsamy et al. (2018)
Antibiotic, cefotaxime at 62.5 mg L^{-1} was supplemented to Morishige and Skoog (MS) medium for establishment	*Helianthus tuberosus* L.	It recorded 0% contamination, 100% survival of stem nodes cultures	Abdalla et al. (2021)
Copper sulfate ($CuSO_4$ $5H_2O$) at 70 mg L^{-1} was supplemented to MS medium + 5 mg L^{-1} BA for shoot multiplication	*Philodendron selloum*	It eliminated the endogenous bacteria contamination to 0%, without decline in growth of in vitro shoots	Seliem et al. (2021)

2.6.2.3 External Conditions of Cultivation of Protoplast Cultures

Protoplast's development is inhibited by very high light intensities. Some species such as legumes are tolerant to light but other species are sensitive to light intensity. It depends on genotype that the protoplast culture took place at temperature from 22 to 30 °C (Verduci et al. 2020; Javaid et al. 2022a) (Table 37.1).

2.7 *Somatic Hybridization*

Somatic hybridization is the result of interspecific hybridization of the parent species (Table 37.2). In the family, Brassicaceae may control the hurdle between illustrative that may not be intersect. In fact the nucleate, mitochondrial as well as chromoplast genome of unlike and sexually opposite species can be joined into an individual genetic constitution through somatic hybridization (Pathania et al. 2021; Naqve et al. 2021). However, fertility restoration remains an issue that should be addressed in order to the better combination. The initiation of disease-resistance gene is one of the several specific aims in plant reproduction. In the genus *Brassica*, increased resistance of seedlings to infection with mycopathogens like *Alternaria*, *Phoma*, and *Plasmodiophora* is the problem as the resistance gene is only found in species that is closely associated with food crops, that is a classic crossbreeding drawback. Through protoplast fusion, this restriction may be removed to some extent (Ali et al. 2013a; Gramazio et al. 2020).

Table 37.2 Examples of somatic hybridization in different plants

No	of genus	Parent species and their chromosome number	Chromosome number of hybrid
1	**Interspecific hybrids**		
	Brassica	*B. oleracea* (2n = 18), *B. campestris* (2n = 20)	Wide variation
	Nicotiana	*N. tabacum* (2n = 48), *N. alata* (2n = 18)	66–71
2	**Intergeneric hybrids**		
	Raphanus × *Brassica*	*Raphanus sativus, B. oleracea*	Raphanobrassica
	Datura × *Atropa*	*Datura inoxia, Atropa belladonna*	Daturotropa
3	**Intertribal hybrids**		
	Arabidopsis × *Brassica*	*Arabidopsis thaliana, B. campestris*	Arabidobrassica
	Thlaspi × *Brassica*	*Thlaspi perfoliatum, B. napus*	Thlaspobrassica

2.7.1 Protoplast Fusion and Isolation

In protoplast fusion, mesophyll and hypocotyl protoplasts or mesophyll and callus protoplasts are most often used. The reasons are the optical control of fusant formation in accordance with the content of chloroplasts as well as vacuole. Plant's protoplast can be isolated mechanically or enzymatically (Klimek-Chodacka et al. 2020).

- In mechanical isolation, cutting plant part and removing protoplast from the cut surface have historically been an important technique but is rarely used due to insufficient numbers of isolated protoplasts. However, its advantage is that it eliminates the influence of unknown enzymes on the protoplast (Moon et al. 2021).
- Enzyme isolation is useful because protoplasts are obtained in large numbers, cell is not damaged, and osmotic condition can be affected (Ren et al. 2021). The cell is extracted through breaking the middle lamella as well as dissolving the tissues into single cell. Enzyme isolation can be executed into two different methods: two-step or one-step method. In the first step of a two-stage process, single cell is extracted by using commercially available enzyme preparations (e.g., macerozyme and macerase). In the second step, the free cells were processed into protoplasts with cellulase by dissolving the cell walls (Yang et al. 2020). The cells are released to the enzyme for a short time in single-step isolation. One-step isolation is much common, in which tissue loosened regularly (e.g., by cutting into strip) is placed in a mixture of enzyme (pectinase and cellulase, commercially available preparation). For each plant objects, the specific compositions of the enzyme fusion is required, and the maximum pressure of the extraction medium (Belwal et al. 2018).

2.8 In Vitro Production of Secondary Metabolites

In vivo cultivations involve the production of new cell, tissue, as well as organ derive solely from the division of the mitotic cells, resulting in cloning of cell, tissue, and persons, all with as same genetics as parental plant (Khanday and Sundaresan 2021). Plant-derived medicinal compounds (PDMC) are as similar as the secondary metabolite production from therapeutic plants (Cardoso et al. 2019). The usage of in vivo techniques for the production of secondary metabolites, particularly PDMC, has the main advantage, less impact on environment because of controlled conditions in the in vivo culture chamber, the better control is possible to the production of PDMCs through the development of protocol to quicken the accumulation of fresh biomass and rise in the production of PDMCs in tissue; seasonal self-regulating incremental manufacture from PDMC; under sterile conditions, production of PDMC and a very low risk of pollution with undesirable lethal compounds (Atif et al. 2020).

In vivo production of PDMCs consists of growing explants of target medicinal plants using plant cells and tissues culture technique. Explants are surface decontaminated and injected in vivo by using culture media that is pre-formulated and cultured in various phases by controlling the environmental conditions (Cardoso 2018). In vivo micropropagation with the help of accommodation and plant development in greenhouses or in field is an auspicious policy for the production of plants secondary metabolite, especially in infrequent or rare species, which are hard to broadcast or grow slowly, and in vulnerable species for recurring phytosanitary problems, such as in vegetative proliferation (Srivastava et al. 2022). Then, the techniques which are used to yield virus-free plants can also be useful for increasing plants biomass efficiency and PDMCs content, that is detected in *Allium sativum*. The in vivo technique enables in addition to accelerating significant propagation of particular high PDMC genotype, both native and those advanced in breeding plans (Kapoor et al. 2018).

2.9 Cryopreservation and In Vitro Germplasm Storage

Cryopreservation can be demarcated as freezing of feasible organic substantial as well as succeeding packing at lower temperature, if possible, liquid nitrogen. The growth of strategies for cryopreservation of plant's cell as well as organ follows the progress completed from mammal system. For the mammal's system, the finding of chemicals through Cryoprotective possessions is an important phase on the way to the development as well as modification of Cryopreservation Technology. The main innovation in this regard was the discovery that glycerol could protect bird sperm from freezing damage (Bustani and Baiee 2021).

Table 37.3 Shows the plant's cryopreservation bank present in different countries

Cryo-methods	Plant species	Institution (Country)	Number of shoot regrowth (%) accessions		References
DV	*Allium sativum*, A. *macrostemon*, *A. sativum Aggregatum* group, *A.* × *Proliferum*	NAC (Rep. Korea)	1158	34–100	Kim et al. (2012)
	Citrus spp., *Fortunella* spp., *Microcitrus australasica*, *Poncirus trifoliata* hybrids	USDA-ARS, NPGS (USA)	451	NS	Volk et al. (2019)
	Fragaria spp.	USDA-ARS, NPGS (USA)	NS	NS	Jenderek and Reed (2017)
	Manihot esculenta	CIAT (Colombia)	480		Machida-Hirano and Niino (2017)
		IITA (Nigeria)	9	63–100	Dumet et al. (2013)
		IITA (Nigeria)	NS	NS	Panis (2019)

Cell and tissue culture is usually kept in a well-lit area at 25 °C. Though this needs episodic transmission to new medium, which involves not just high labor as well as expenses, nevertheless also the danger of pollution in addition occasionally damage of all germplasm. In addition, cell's culture reared in batch subdivision is subject to genetic loss such as changes in ploidy level, changes, endomitosis, translocations, gene extension, etc. Therefore, the genetic solidity of nearly beneficial somaclonal cells line created in vivo cannot be guaranteed (Mat Sulaiman et al. 2020).

Additional significant feature is the preservation of Germplasm. The current plant's Germplasm bank is usually supplied from seed collections (Table 37.3). Though seed provides the efficient storing systems for genetic substantial, most of the variables and the resulting plants do not reflect the true characteristics of the parent plant (Engels and Ebert 2021). Currently, there is no dpendable method for the long-lasting preservation of germplasm, both from vegetatively propagated plants and from plants with refractory seeds. In addition, the Germplasm of occasional, leading, also threatened plants species should be preserved. To overcome this problem, in vivo method is being improved for temporary as well as long-standing storing and preservation of Germplasm. In vitro culture Cryopreservation in fluid N_2 have enabled plants retrieval of a number of species is optional for long-standing storing of germplasm, particularly vegetatively circulated cultures (Benelli 2021).

2.10 *In Vitro Production of Haploids*

Haploidic plant contains the number of gametophyte chromosomes. They are very important in mutation induction research, and they are also required in large quantities for the development of Homozygous plant (Ahmar et al. 2020). However, the traditional methods that farmers use for their production are cumbersome, and labor intensive is not very efficient. With the advent of in vitro techniques to induce androgenesis, it is clear that these methods greatly accelerate haploid production for plant breeding programs and lead to early release of cultivars (Chaikam et al. 2019).

2.10.1 Methods of Haploid Production

Several methods are available to obtain haploid and DH (double haploid), of which in vivo anther culture/isolated microspores are the utmost efficient as well as generally used.

2.10.2 Anther Culture

Anther Cultures are frequently the methods of excellent for DH productions within various plants so that the effortlessness of the method permits the establishment of large-scale Anther culture as well as implementation to a widespread variety of genotype (Richa 2022). In many genotypes, it has been observed that pre-cultivation, physical/chemical actions functional to cut bloom, entire flowerings/anthers earlier cultivation, perform as triggers for the induction of the sporophyte pathway, and thus the development of fertile pollen (gametophyte pathway) inhibits.

Early methods like refrigeration, high heat condition, high moisture, drought stress, anaerobiotic treatments, and deficiency of nitrogen, ethyl alcohol, C-irradiation, disrupting agent, electrical stimulation, average pH, and heavyweight metallic methods are very general methods in anthers and anther culture (Ali et al. 2021).

Generally, after pretreatment, flower bud is exterior pasteurized through absorption in 70% (v/v) ethanol for several records, tracked through absorption within the Na Hypochlorite solution having some droplets of Tween 20 for 10–15 min and then through washing three times for 5 min each by germ-free distilled liquid. In the final phase, the microspores are cut to the filament as well as located on a media. The exceptions were barley ears, which were sterilized only by spraying with 70% ethanol or through absorption within 70% ethyl alcohol for 5 min, then rinsing by sterile cleaned water (Ncumisa 2021).

2.10.2.1 Uses of Anther Culture

- By developing haploid plants, crops are improved through anther culture technique.
- Vegetables, cereals, and other crops such as watermelon, asparagus, and cabbage are also improved through anther culturing method (Gotame et al. 2018).
- For producing various horticulture plants, this method is highly used.
- In conservation of germplasm and cryogenic studies, anther culture is used as an alternative for developing the highly stable embryo (Bijalwan 2021).

2.10.3 Isolated Microspore Culture

This method is achieved through releasing tissue from corporal anthers, needs well apparatus as well as services. This technique provides an outstanding organization for studying microspore's instruction as well as embryogenesis also offers a stage for evolving molecular studies as well as can produce double haploid plant that is applied to quicken plant's breeding packages. In addition, isolated microspore culture has numerous benefits on anther's culture, where the occurrence of the anther's wall may lead to the development of diploid, somatic callus, and plants (Niazian and Shariatpanahi 2020).

Although the protocol for culturing isolated microspores varies from laboratory to laboratory, the basic steps for growing donor plants, collecting floral organs, isolating microspores, culturing and inducing microspores, regenerating embryos, and doubling chromosomes remain the same. Isolated microspores, given the optimal combination of culture and stress conditions, can be switched from the normal gametophyte developmental pathway to the sporophyte pathway and subsequently produce haploid or DH embryos and plants (Seguí-Simarro et al. 2021).

Microspore culture in all species has no universal procedures because of modifications between species and between genetic constitution in a species within embryogenic responses. Standard protocols are generally used in the initial screening of species for eight responses to microspore culture, named as mustard species, barleycorn, and *Nicotiana tabacum*. Although there are slight differences in technique since laboratories to laboratories, the elementary phases included within culturing microspores are constant as well as involve in developing contributor plant, collecting flowery structures, microspore cultures as well as initiation, embryo rebirth, and chromosome duplication, if needed (Shumilina et al. 2020).

3 Unpollinated Ovary Culture

Modern technique within plant tissue culture and experimental embryology has succeeded within inducing haploid plants by culturing unpollinated ovaries. This means that not only microspores but also megaspore gametophyte of angiosperm may be stimulated for sporophyte development in vitro, opening new avenues for genetic research and haploid reproduction. Meanwhile, in the finding of the first haploid in *Datura stramonium* in 1921, many attempts have been made to induce parthenogenic development of unfertilized eggs or other embryo sac cells. Many methods have been tried and can be unevenly divided into two types. One is the in vivo induction of haploid by various physical, chemical, or biological stimuli (Anju et al. 2021).

4 Embryo Rescue from Wide Cross

Embryo rescue is an in vivo technique used to save or protect the weak, immature, or hybrid embryos from its degradation. This technique is also useful for the production of seedless fruits such as grapes (Si-Hong et al. 2020). For this purpose, seedless plants are hybridized with parent seedless cultivars resulting in significant production of seedless grapes. In this technique, ovaries are divided 12–15 days after fertilization. The improved seeds are cultivated in vivo on EC3 media as well as additional rooting and germination were induced essentially. All rescued shoot-initiated plant embryos are exposed to repeated shoots also repetitive subcultures to keep plants in vivo until the following growing period. Whole plant with roots as well as shoots is toughened and brought to the ground. Embryo-saved plant hardening, rhizobium immunization, and field transmission are also carried out (Pen et al. 2018).

5 Conclusion

Plant tissue culture is widely applied for different purposes including micropropagation (most successful application at commercial level), meristem tip culture (used for the purpose of cultivation of pathogen-free plants), somaclonal variation (to produce enough genetic diversity in vegetatively propagated plant species), and some others. The most important benefit of plant tissue culture technique is that different crops that are resilient to water, salt, or cold stresses can be produced. Somaclonal differences in tissue cultures have been hired to improve many such things that are used in the group of novel diversities of crops. In vivo cultivations are the technique used to produce secondary metabolites in which novel cell, tissue as well as organ are resulting from mitotic cell partition, cloning of cell, tissue as well as individual, all through the similar inheritances as the parent plants. Cryopreservation is the freezing of biological material that are stored at very low temperature, preferably liquid nitrogen. This technique is highly used for the long-standing storing of germplasm of different plants. In short plant, tissue culture is the advance technique that helps in increment of crop production with sustainability.

References

Abbas T, Nadeem MA, Tanveer A, Matloob A, Zohaib A, Safdar ME, Ali HH, Farooq N, Javaid MM, Tabassum T (2019) Herbicide mixtures and row spacing effects on Fenoxaprop resistant Phalaris minor in wheat. Int J Agric Biol 22(4):840–840

Abdalla N, Ragab M, El-Miniawy S, Arafa N, Taha H (2021) A new aspect for in vitro propagation of Jerusalem artichoke and molecular assessment using RAPD, ISSR and SCoT marker techniques. Egypt J Bot 61(1):203–218

Adhikary D, Kulkarni M, El-Mezawy A, Mobini S, Elhiti M, Gjuric R, Ray A, Polowick P, Slaski JJ, Jones MP (2021) Medical cannabis and industrial hemp tissue culture: present status and future potential. Front Plant Sci 12:627240

Ahmar S, Gill RA, Jung K-H, Faheem A, Qasim MU, Mubeen M, Zhou W (2020) Conventional and molecular techniques from simple breeding to speed breeding in crop plants: recent advances and future outlook. Int J Mol Sci 21(7):259

Alagarsamy K, Shamala LF, Wei S (2018) Influence of media supplements on inhibition of oxidative browning and bacterial endophytes of *Camellia sinensis* var. sinensis. 3 Biotech 8(8):1–7

Ali H, Iqbal N, Shahzad AN, Sarwar N, Ahmad S, Mehmood A (2013a) Seed priming improves irrigation water use efficiency, yield, and yield components of late-sown wheat under limited water conditions. Turk J Agric For 37(5):534–544

Ali H, Tanveer A, Nadeem M, Javaid M, Kashif M, Chadhar A (2013b) Allelopathic effects of *Rhynchosia capitata* on germination and seedling growth of mungbean. Planta Daninha 31:501–509

Ali J, Nicolas KLC, Akther S, Torabi A, Ebadi AA, Marfori-Nazarea CM, Mahender A (2021) Improved anther culture media for enhanced callus formation and plant regeneration in rice (*Oryza sativa* L.). Plants 10(5):839

Anju M, Razdan M, Thomas TD (2021) In vitro production of haploids and triploids in mulberry. In: Mulberry: genetic improvement in context of climate change. CRC Press, Boca Raton, Florida, United States, pp 122–130

Asif T, Javaid M, Abdul K, Abbas R, Ahsan A (2012) Allelopathic effect of Echinochloa crus-galli on field crops. Herbologia 13(2):9–17

Atif M, Malik I, Asif M, Qamar-Uz-Zaman M, Ahmad N, Scahill S (2020) Drug safety in Pakistan. In: Drug safety in developing countries. Elsevier, London, pp 287–325

Baba BH, Tanveer A, Tahir M, Aziz A, Ahmad A-U, Nadeem MA, Javaid MM (2009) Economic Survey of Pakistan, Ministry of Finance Economic Survey of Pakistan, Ministry of Finance, 2005. Weed Biol Manag 9(2):146–151

Belwal T, Ezzat SM, Rastrelli L, Bhatt ID, Daglia M, Baldi A, Devkota HP, Orhan IE, Patra JK, Das G (2018) A critical analysis of extraction techniques used for botanicals: trends, priorities, industrial uses and optimization strategies. TrAC Trends Anal Chem 100:82–102

Benelli C (2021) Plant cryopreservation: a look at the present and the future. Plants 10. MDPI, article 2744

Bijalwan P (2021) Plant tissue culture-a new tool for vegetable improvement (Indian scenario): a review. Agric Rev 42(2):235–239

Brar D, Khush GS (2021) Cell and tissue culture for plant improvement. In: Mechanisms of plant growth and improved productivity. CRC Press, Boca Raton, Florida, United States, 229–278

Bridgen MP, Van Houtven W, Eeckhaut T (2018) Plant Tissue Culture Techniques for Breeding. In: Van Huylenbroeck J (ed) Ornamental Crops. Springer International Publishing, Cham, pp 127–144. https://doi.org/10.1007/978-3-319-90698-0_6

Bustani GS, Baiee FH (2021) Semen extenders: an evaluative overview of preservative mechanisms of semen and semen extenders. Vet World 14(5):1220

Cardoso JC (2018) Nonfood GM crops in tropical regions: a reasonable way to promote the technology for increased agricultural sustainability. Glob Chall 2(5–6):1800010

Cardoso JC, Oliveira ME, Cardoso FC (2019) Advances and challenges on the in vitro production of secondary metabolites from medicinal plants. Hortic Bras 37:124–132

Chadipiralla K, Gayathri P, Rajani V, Reddy PVB (2020) Plant tissue culture and crop improvement. In: Sustainable agriculture in the era of climate change. Springer, Cham, pp 391–412
Chaikam V, Molenaar W, Melchinger AE, Boddupalli PM (2019) Doubled haploid technology for line development in maize: technical advances and prospects. Theor Appl Genet 132(12):3227–3243
Chandana B, Kumari Nagaveni H, Lakshmana D, Shashikala SK, Heena M (2018) Role of plant tissue culture in micropropagation, secondary metabolites production and conservation of some endangered medicinal crops. J Pharmacogn Phytochem 3:246–251
Degefa I (2019) Plant breeding methods: in brief for student. Int J Agric Biol 3:156–203
Dumet D, Diebiru E, Adeyemi A, Akinyemi O, Gueye B, Franco J (2013) Cryopreservation for the 'in perpetuity' conservation of yam and cassava genetic resources. CryoLetters 34(2):107–118
Engels JM, Ebert AW (2021) A critical review of the current global ex situ conservation system for plant agrobiodiversity. II. Strengths and weaknesses of the current system and recommendations for its improvement. Plants 10(9):1904
Ermert AL, Nogué F, Stahl F, Gans T, Hughes J (2019) CRISPR/Cas9-mediated knockout of Physcomitrella patens phytochromes. In: Phytochromes. Springer, pp 237–263
Espinosa-Leal CA, Puente-Garza CA, García-Lara S (2018) In vitro plant tissue culture: means for production of biological active compounds. Planta 248(1):1–18
FAO F (2017) The future of food and agriculture–trends and challenges. Annu Rep 296:1–180
Gilbert S, Xu J, Acosta K, Poulev A, Lebeis S, Lam E (2018) Bacterial production of indole related compounds reveals their role in association between duckweeds and endophytes. Front Chem 6:265
Gosal SS, Wani SH (2018) Biotechnologies of crop improvement, volume 1: cellular approaches. Springer, Cham, pp 497
Gotame TP, Shrestha SL, Joshi BK, Karki TB (2018) Classification of crop plants based on growing season, temperature requirement and photosynthetic behavior. Working Groups of Agricultural Plant Genetic Resources (APGRs) in Nepal, 103
Gramazio P, Takayama M, Ezura H (2020) Challenges and prospects of new plant breeding techniques for GABA improvement in crops: tomato as an example. Front Plant Sci 11:577980
Grzybkowska D, Morończyk J, Wójcikowska B, Gaj MD (2018) Azacitidine (5-AzaC)-treatment and mutations in DNA methylase genes affect embryogenic response and expression of the genes that are involved in somatic embryogenesis in Arabidopsis. Plant Growth Regul. 85 (2):243–256. https://doi.org/10.1007/s10725-018-0389-1
Gulzar B, Mujib A, Malik MQ, Mamgain J, Syeed R, Zafar N (2020) Plant tissue culture: agriculture and industrial applications. In: Transgenic technology-based value addition in plant biotechnology. Elsevier, London, pp 25–49
Gupta N, Jain V, Joseph MR, Devi S (2020) A review on micropropagation culture method. Asian J Pharm Res Dev 8(1):86–93
Gupta S, Singh A, Yadav K, Pandey N, Kumar S (2022) Micropropagation for multiplication of disease-free and genetically uniform sugarcane plantlets. In: Advances in Plant Tissue Culture. Elsevier, London, pp 31–49
Habib MD, Asif M, Aziz M, Ali A, Ashraf M, Mahmood A, Javaid MM (2012) Growth performance of spring maize and soil fertility status as influenced by nutrient sources. Int J Agric Appl Sci 4(1):35–41
Hannachi S, Werbrouck S, Bahrini I, Abdelgadir A, Siddiqui HA, Van Labeke MC (2021) Obtaining salt stress-tolerant eggplant somaclonal variants from in vitro selection. Plants 10(11):2539
Hassani M, Durán P, Hacquard S (2018) Microbial interactions within the plant holobiont. Microbiome 6(1):1–17
Ikenganyia E, Anikwe M, Omeje T, Adinde J (2017) Plant tissue culture regeneration and aseptic techniques. Asian J Biotechnol Bioresour Technol 1(3):1–6
Jan N, Qazi HA, Ramzan S, John R (2018) Developing stress-tolerant plants through in vitro tissue culture: family Brassicaceae. In: Biotechnologies of crop improvement, volume 1. Springer, Cham, pp 327–372

Javaid M, Bhan M, Rathinasabapathi B, Chase CA (2012a) A non-protein amino acid from *Crotalaria juncea* foliage with allelopathic potential. In: Hortscience. vol 9. J Am Soc Hortic Sci 113 S West St, Ste 200, Alexandria, Va 22314:S332–S332

Javaid M, Tanveer A, Ahmad R, Yaseen M, Khaliq A (2012b) Optimizing activity of herbicides at reduced rate on Emex spinosa campd. with adjuvants. Planta Daninha 30:425–435

Javaid MM, Mahmood A, Alshaya D, Alkahtani M, Waheed H, Wasaya A, Khan SA, Naqve M, Haider I, Shahid MA (2022a) Influence of environmental factors on seed germination and seedling characteristics of perennial ryegrass (*Lolium perenne* L.). Sci Rep 12(1):1–11

Javaid MM, Mahmood A, Min B, Waheed H, Attia K, Aziz A, Nadeem MA, Khan N, Al-Doss AA, Fiaz S (2022b) Efficacy of metribuzin doses on physiological, growth, and yield characteristics of wheat and its associated weeds. Front Plant Sci 13:866793

Jenderek MM, Reed BM (2017) Cryopreserved storage of clonal germplasm in the USDA National Plant Germplasm System. In Vitro Cell Dev Biol 53(4):299–308

Johns AE (2019) Lessons for plant micropropagation. Educreation Publishing

Kapoor S, Raghuvanshi R, Bhardwaj P, Sood H, Saxena S, Chaurasia OP (2018) Influence of light quality on growth, secondary metabolites production and antioxidant activity in callus culture of Rhodiola imbricata Edgew. J Photochem Photobiol B Biol 183:258–265

Kendurkar SV, Rangaswamy M (2018) In vitro approaches for the improvement of eucalyptus. In: Biotechnologies of crop improvement, volume 1. Springer, Cham, pp 159–214

Khanday I, Sundaresan V (2021) Plant zygote development: recent insights and applications to clonal seeds. Curr Opin Plant Biol 59:101993

Kim H-H, Popova E, Shin D-J, Yi J-Y, Kim CH, Lee J-S, Yoon M-K, Engelmann F (2012) Cryobanking of Korean Allium germplasm collections: results from a 10 year experience. CryoLetters 33(1):45–57

Klimek-Chodacka M, Kadluczka D, Lukasiewicz A, Malec-Pala A, Baranski R, Grzebelus E (2020) Effective callus induction and plant regeneration in callus and protoplast cultures of *Nigella damascena* L. Plant Cell Tissue Organ Cult (PCTOC) 143(3):693–707

Kordrostami M, Mafakheri M, Al-Khayri JM (2022) Date palm (*Phoenix dactylifera* L.) genetic improvement via biotechnological approaches. Tree Genet Genomes 18(3):1–28

Lal R, Lal S (2019) Protoplasts in crop improvement. In: Crop improvement utilizing biotechnology. CRC Press, pp 117–194

Li H (2020) Nutrients translocation & plant growth in tissue culture. Wageningen University and Research

Liu Z (2019) In vitro propagation and cryopreservation of American chestnut (*Castanea dentata* (Marsh.) Borkh.). University of Guelph, Guelph, Ontario, Canada

Machida-Hirano R, Niino T (2017) Potato genetic resources. In: The potato genome. Springer, Cham, pp 11–30

Maekawa H, Otsubo M, Sato MP, Takahashi T, Mizoguchi K, Koyamatsu D, Inaba T, Ito-Inaba Y (2022) Establishing an efficient protoplast transient expression system for investigation of floral thermogenesis in aroids. Plant Cell Rep 41(1):263–275

Makarem MA, Moravvej Z, Rahimpour MR, Vakylabad AB (2021) Biofuel production from microalgae and process enhancement by metabolic engineering and ultrasound. In: Advances in bioenergy and microfluidic applications. Elsevier, Amsterdam, pp 209–230

Manchanda P, Kaur A, Gosal SS (2018) Somaclonal Variation for Sugarcane Improvement. In: Gosal SS, Wani SH (eds) Biotechnologies of Crop Improvement, Volume 1: Cellular Approaches. Springer International Publishing, Cham, pp 299–326. https://doi.org/10.1007/978-3-319-78283-6_9

Márquez-López RE, Pérez-Hernández C, Ku-González Á, Galaz-Ávalos RM, Loyola-Vargas VM (2018) Localization and transport of indole-3-acetic acid during somatic embryogenesis in *Coffea canephora*. Protoplasma 255(2):695–708

Mat Sulaiman NN, Rafii MY, Duangjit J, Ramlee SI, Phumichai C, Oladosu Y, Datta DR, Musa I (2020) Genetic variability of eggplant germplasm evaluated under open field and glasshouse cropping conditions. Agronomy 10(3):436

Mehbub H, Akter A, Akter MA, Mandal MSH, Hoque MA, Tuleja M, Mehraj H (2022) Tissue culture in ornamentals: cultivation factors, propagation techniques, and its application. Plants 11(23):3208

Méndez-Hernández HA, Ledezma-Rodríguez M, Avilez-Montalvo RN, Juárez-Gómez YL, Skeete A, Avilez-Montalvo J, De-la-Peña C, Loyola-Vargas VM (2019) Signaling overview of plant somatic embryogenesis. Front Plant Sci 10:77

Moon K-B, Park J-S, Park S-J, Lee H-J, Cho H-S, Min S-R, Park Y-I, Jeon J-H, Kim H-S (2021) A more accessible, time-saving, and efficient method for in vitro plant regeneration from potato protoplasts. Plants 10(4):781

Naqve M, Wang X, Shahbaz M, Mahmood A, Bibi S, Fiaz S (2021) Alpha tocopherol-induced modulations in the morphophysiological attributes of okra under saline conditions. Front Plant Sci 12:800251

Ncumisa Y (2021) Antifungal activity of endophytes from arctotis arctotoides (lf) o. hoffm against pythium and rhizoctonia root-rot diseases of maize (zea mays L.). University of KwaZulu-Natal. https://researchspace.ukzn.ac.za/handle/10413/20703

Niazian M, Shariatpanahi ME (2020) In vitro-based doubled haploid production: recent improvements. Euphytica 216(5):1–21

Nowakowska K, Pińkowska A, Siedlecka E, Pacholczak A (2022) The effect of cytokinins on shoot proliferation, biochemical changes and genetic stability of Rhododendron 'Kazimierz Odnowiciel' in the in vitro cultures. Plant Cell Tissue Organ Cult (PCTOC) 149(3):675–684

Oseni OM, Pande V, Nailwal TK (2018) A review on plant tissue culture, a technique for propagation and conservation of endangered plant species. Int J Curr Microbiol Appl Sci 7(7):3778–3786

Pal R, Meena N, Pant R, Dayamma M (2019) Cymbidium: botany, production, and uses. In: Orchids phytochemistry, biology and horticulture: fundamentals and applications. Springer, Cham, Gujrat, India, pp 1–37

Panis B Sixty years of plant cryopreservation: from freezing hardy mulberry twigs to establishing reference crop collections for future generations. In, 2019. Int Soci Hortic Sci Leuven, Belgium, pp 1–8. https://doi.org/10.17660/ActaHortic.2019.1234.1

Pantelidou D, Gerogiannis K, Goula AM, Gonas C (2021) Ultrasound-assisted osmotic dehydration as a method for supplementing potato with unused chokeberries phenolics. Food Bioprocess Technol 14(12):2231–2247

Patel SR (2021) Studies for developing shoot cultures of Leptadenia reticulata (Retz.) wight and arn and Tylophora indica (Burm. F.) Merrill with phytochemical analysis. Maharaja Sayajirao University of Baroda (MSUB) Gujrat, (India)

Pathania A, Singh L, Sharma PN (2021) Host plant resistance: an eco-friendly approach for crop disease management. In: Microbial biotechnology. Springer, Singapore, pp 395–449

Pen S, Nath UK, Song S, Goswami G, Lee J-H, Jung H-J, Kim H-T, Park J-I, Nou I-S (2018) Developmental stage and shape of embryo determine the efficacy of embryo rescue in introgressing orange/yellow color and anthocyanin genes of *Brassica* species. Plants 7(4):99

Quinga LAP, Heringer AS, de Freitas Fraga HP, do Nascimento Vieira L, Silveira V, Steinmacher DA, Guerra MP (2018) Insights into the conversion potential of *Theobroma cacao* L. somatic embryos using quantitative proteomic analysis. Sci Hortic 229:65–76

Raina RBM, Salgotra R, Kumar D (2018) Plant contributed papers. In Vitro Cell Dev Biol Anim 54(1):S28–S35

Ray SS, Ali N (2018) Biotic contamination and possible ways of sterilization: a review with reference to bamboo micropropagation. Braz Arch Biol Technol 60:1–12

Rehman S, Chattha MU, Khan I, Mahmood A, Hassan MU, Al-Huqail AA, Salem MZ, Ali HM, Hano C, El-Esawi MA (2022) Exogenously applied trehalose augments cadmium stress tolerance and yield of mung bean (*Vigna radiata* L.) grown in soil and hydroponic systems through reducing cd uptake and enhancing photosynthetic efficiency and antioxidant defense systems. Plants 11(6):822

Ren R, Gao J, Yin D, Li K, Lu C, Ahmad S, Wei Y, Jin J, Zhu G, Yang F (2021) Highly efficient leaf base protoplast isolation and transient expression systems for orchids and other important monocot crops. Front Plant Sci 12:626015

Richa U (2022) Anther culture for haploid plant production. In: Advances in plant tissue culture. Elsevier, London, pp 157–174

Sahab S, Suhani I, Srivastava V, Chauhan PS, Singh RP, Prasad V (2021) Potential risk assessment of soil salinity to agroecosystem sustainability: current status and management strategies. Sci Total Environ 764:144164

Seguí-Simarro JM, Jacquier N, Widiez T (2021) Overview of in vitro and in vivo doubled haploid technologies. In: Doubled haploid technology (DHs). Humana, New York, pp 3–22

Seliem M, El-Mahrouk M, El-Banna A, Hafez Y, Dewir Y (2021) Micropropagation of Philodendron selloum: influence of copper sulfate on endophytic bacterial contamination, antioxidant enzyme activity, electrolyte leakage, and plant survival. S Afr J Bot 139:230–240

Shanee S, Tanveer A, Javaid M, Chaudhry K, Aziz A, Khaliq A, Chaudhry M, Pervez M, Awan I (2011) Phytotoxic effects of Euphorbia dracunculoides: a weed of rainfed chickpea-chickpea cropping system. Span J Agric Res 9(2):580–588

Shen R-S, Hsu S-T (2018) Virus elimination through meristem culture and rapid clonal propagation using a temporary immersion system. In: Orchid propagation: from laboratories to greenhouses—methods and protocols. Springer, New York, pp 267–282

Shumilina D, Kornyukhin D, Domblides E, Soldatenko A, Artemyeva A (2020) Effects of genotype and culture conditions on microspore embryogenesis and plant regeneration in *Brassica Rapa* ssp. Rapa L. Plants 9(2):278

Si-Hong K, Joon-Ho K, Young-Sik P, Jae-Yun H (2020) In vitro embryo rescue for the production of hypotetraploids after cross between hypotetraploid and tetraploid grape cultivars. Not Bot Horti Agro bot Cluj-Napoca 48(1):503–508

Singh Shekhawat M (2019) Plant biotechnology: in vitro principles, techniques and applications. MJP Publisher

Soumare A, Diédhiou AG, Arora NK, Tawfeeq Al-Ani LK, Ngom M, Fall S, Hafidi M, Ouhdouch Y, Kouisni L, Sy MO (2021) Potential role and utilization of plant growth promoting microbes in plant tissue culture. Front Microbiol 12:649878

Springmann M, Clark M, Mason-D'Croz D, Wiebe K, Bodirsky BL, Lassaletta L, De Vries W, Vermeulen SJ, Herrero M, Carlson KM (2018) Options for keeping the food system within environmental limits. Nature 562(7728):519–525

Srivastava S, Nalla VK, Singh VP, Prasad R (2022) Optimization of in vitro micropropagation and root establishment through combinatorial approaches for enhanced production of secondary metabolites in the endangered species Decalepis arayalpathra KMA 05 clones. Rev Bras Bot, 45(3):869–881

Sudheer W, Thiruvengadam M, Nagella P (2022) A comprehensive review on tissue culture studies and secondary metabolite production in *Bacopa monnieri* L. Pennell: a nootropic plant. Crit Rev Biotechnol:(2):1–15

Tanveer A, Zeshan Farid M, Tahir M, Mansoor Javaid M, Khaliq A (2012) Environmental factors affecting the germination and seedling emergence of *Carthamus oxyacantha* M. Bieb. (Wild safflower). Pak J Weed Sci Res 18(2)

Torgbor MNA (2020) Transient gene editing in Potato (*Solanum tuberosum* L.) with CRISPR/Cas9 to knock-out F3H. Norwegian University of Life Sciences, Ås

Trivedi D (2021) Biotechnological Techniques to Engineer Disease Free Plants. Res Rev Biotec Biosci. 8 (1):193–214. https://doi.org/10.5281/zenodo.5166536

van Zijl PC, Lam WW, Xu J, Knutsson L, Stanisz GJ (2018) Magnetization transfer contrast and chemical exchange saturation transfer MRI. Features and analysis of the field-dependent saturation spectrum. Neuroimage 168:222–241

Verduci E, Di Profio E, Cerrato L, Nuzzi G, Riva L, Vizzari G, D'Auria E, Giannì ML, Zuccotti G, Peroni DG (2020) Use of soy-based formulas and cow's milk allergy: lights and shadows. Front Pediatr 8:591988

Vir R, Kumar V (2021) Role of plant tissue culture in improvement of horticultural crops. Ann Hortic 14(1):1–6

Volk GM, Jenderek MM, Walters C, Bonnart R, Shepherd A, Skogerboe D, Hall BD, Moreland B, Krueger R, Polek M Implementation of Citrus shoot tip cryopreservation in the USDA-ARS National Plant Germplasm System. In, 2019. Int. Soci. Hortic. Sci., Leuven, Belgium, pp 329–334. https://doi.org/10.17660/ActaHortic.2019.1234.43

Wan Abdullah WMAN, Low L-Y, Mumaiyizah SB, Chai Q-Y, Loh J-Y, Ong-Abdullah J, Lai K-S (2020) Effect of lignosulphonates on *Vanilla planifolia* shoot multiplication, regeneration and metabolism. Acta Physiol Plant 42(7):1–8

Wang D, Zhang M, Ju R, Mujumdar AS, Yu D (2022) Novel drying techniques for controlling microbial contamination in fresh food: a review. Dry Technol, 41(2):172–189

Yang H, Benatti MR, Karve RA, Fox A, Meilan R, Carpita NC, McCann MC (2020) Rhamnogalacturonan-I is a determinant of cell–cell adhesion in poplar wood. Plant Biotechnol J 18(4):1027–1040

Chapter 38
Nanotechnology for Climate-Resilient Agriculture

Somali Dhal and Harshata Pal

Abstract Current agricultural processes are no longer viable due to climate change and population expansion. The current agricultural challenges are food security, natural resource sustainability, boosting nutrient efficiency, and climate change. Therefore, we need technological interventions that can be effective as an adaptation strategy for farming issues. In an era of climate change, nanotechnology will be a viable intervention that will drive the agricultural revolution and provide us with efficient methods for dealing with climatic conditions. Because of their unique qualities at the nanoscale, nanoparticles are said to improve carbon stability and possibly soil sequestration. Nanotechnology solutions to strengthen input efficiency, reduce drought stress, and increase soil temperature are particularly relevant to poor nations. The use of nanoscale agrichemical formulations has boosted efficiency and reduced environmental losses. The purpose of this chapter was to emphasize the potential of nanomaterials for climate-resilient agriculture.

Keywords Nanotechnology · Agriculture · Nanoparticles · Climate · Resilience · Biotic and abiotic stress · Nanofertilizers

1 Introduction

Climate change is having an influence on agriculture all across the world. Still, nations such as India are particularly susceptible due to their large agricultural population, high demand for natural resources, and inadequate coping mechanisms. It is commonly assumed that the prevalence of extreme climatic events, such as

S. Dhal
Department of Biotechnology and Medical Engineering, National Institute of Technology (NIT) Rourkela, Rourkela, Odisha, India

H. Pal (✉)
Amity Institute of Biotechnology, Amity University, Newtown, Kolkata, India
e-mail: hpal@kol.amity.edu

M. Hasanuzzaman (ed.), *Climate-Resilient Agriculture, Vol 2*,
https://doi.org/10.1007/978-3-031-37428-9_38

droughts and floods, has harmed agricultural production and productivity in most sections of the nation, disproportionately harming the farming population. Due to these reasons, there is a need for such a system that can utilize existing natural resources in agricultural and animal production systems to produce long-term greater productivity and farm revenue under climate variability. Such systems can help make agriculture robust to the effects of climate change, therefore increasing agricultural production and ensuring nutritional and food security for all. Climatic-resilient agriculture (CRA) is a multifaceted method of managing the interconnected and directly affected by climate change components of agriculture and food security (Munawar et al. 2020). According to the World Bank, the CRA should achieve three major outcomes such as: (i) enhanced productivity, (ii) resilience, and (iii) carbon sequestration (Jayaraman et al. 2022). From this perspective, nanotechnology can revolutionize agriculture by offering up new opportunities, such as the development of nanosensors to safeguard agricultural produces. Nanotechnology is the synthesis and manipulation of matter at length scales of at nano-dimensions (Patel et al. 2021). This opens the possibility of engineering the material's physiochemical characteristics to boost agricultural resource efficiency. For instance, nanotechnology can help detect diseases to manage the growing frequency of their attacks due to climate change (Kumar et al. 2021). This scenario can be controlled with advanced molecular and cell biology methods linked to genetic engineering and molecular nanodevices. The utilization of nanotechnology in agriculture has commenced and will continue to expand with substantial consequences for several agricultural disciplines and the food industry. It would surely motivate scientists to create ways to help people cope with the negative impacts of climate change. Developing nations are particularly interested in nanotechnology solutions that tackle low-use efficiency of inputs (such as fertilizer, irrigation water, and pesticides), drought stress, and rising soil temperature (Fu et al. 2020). Nanoscale agrichemical formulations have improved efficiency and reduced environmental waste. Improved yields can be predicted when nutrients are distributed more effectively. Nanoporous materials capable of holding and steadily increasing water during droughts might likewise be expected to boost output. Other nano-derived systems include nanocarriers for fertilizers, macro- and micronutrients, pesticides, and genetically engineering plants (Hofmann et al. 2020). They help plants by providing resistance to biotic or abiotic stress, increasing photosynthetic ability, and enabling plants to sense and report their level of stress or nutrient needs. In this chapter, we discuss nanotechnology's prospects and how it plays a critical role in climate-resilient agriculture (Fig. 38.1).

2 Delivery of Agrochemicals Through Nanofertilizers

Several nanotechnologies, including growth regulators, molluscicides, bactericides, fungicides, and fertilizers, have been created and evaluated as possible agrochemicals. Nanofertilizers are nanometer-scale compounds that contain macro- and

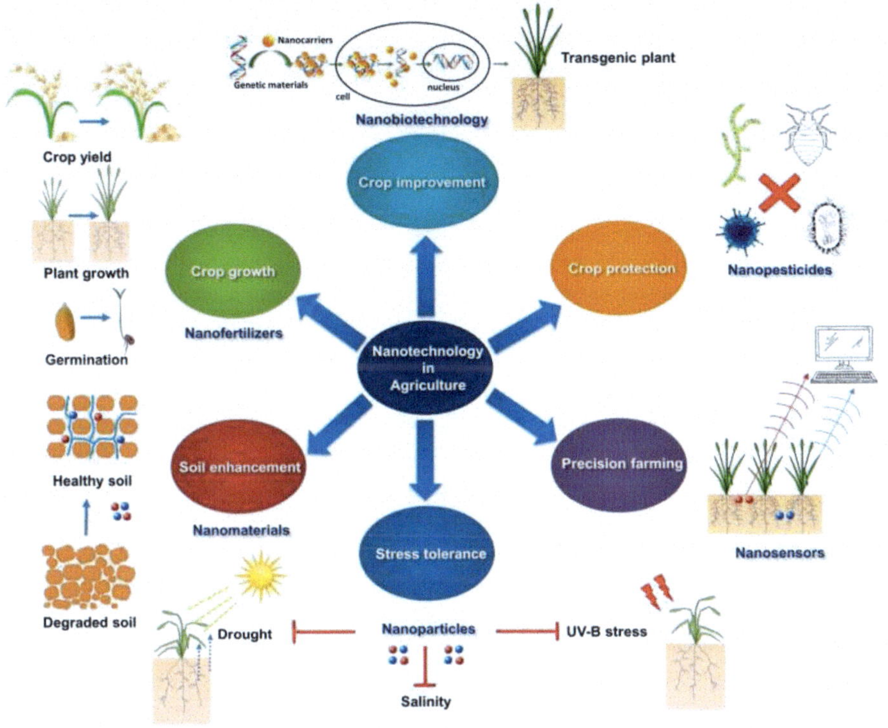

Fig. 38.1 Agriculture applications of nanotechnology. Controlled-release nanofertilizers boost crop growth, yield, and productivity. Crop enhancement is accomplished by using a nano-based target delivery technique (gene transfer). Nanopesticides can provide adequate crop protection. The use of nanosensors and digital controls considerably aids precision farming. Nanomaterials can also be employed to improve plant stress tolerance and soil quality

micronutrients and are given to crops in a regulated manner. Nanofertilizers are categorized into three types based on the kind of formulation (Tului et al. 2021):

1. Nanoscale fertilizer: It is a standard fertilizer with a smaller particle size, usually in the form of NPs.
2. Nanoscale additive fertilizer: It contains a supplemented nanomaterial.
3. Nanoscale coating fertilizer: Here, nutrients are encapsulated by nanofilms.

Different types of metal and metal-derived systems can act as nanofertilizers. For example, metal-based NPs include nanozinc, iron, silver, gold, and silica. Some quantum dot (QD) derived nanofertilizers include QDs in gold, ZnCd series/ZnS core shells, QDs in Mn/ZnSe, and QDs core-shell. The metal oxide NPs have been tested for their biological fate and toxicity in agricultural productions. Such NPs include FeO, TiO_2, CeO_2, Al_2O_3, and ZnO (Zhang et al. 2016). Liu and Lal (2015) created a nanofertilizer comprising Ca- and P-hydroxyapatite NPs that increased soybean seed yield more than traditional fertilizers. Micronutrients are nutrients that plants require in smaller amounts but are as vital for plant metabolisms, such as chloride, manganese, boron, nickel, copper, molybdenum, iron, and zinc. Nanoparticulated fertilizers are a type of different system that ate utilized to directly or indirectly to deliver important bioactive chemicals to plants. In the context of fertilizers, these nanoparticulated materials might be carbon, silica, or other organic polymer-based bases. The efficiency of chitosan NP encapsulating NPK for foliar uptake by wheat grown in sandy soil was investigated by Abdel-Aziz et al. (2016). They reported an increment in the crop yield after using chitosan NP as a fertilizer containing NPK. McGehee et al. (2017) have shown that multi-walled carbon nanotubes (MWCNT) boost fruit output in hydroponically grown tomatoes (McGehee et al. 2017). Because of its unique physicochemical properties, MWCNT has an amazing interaction with plants, considerably enhancing their development. They have the potential to collect more light energy via plant chloroplasts while increasing the pace of electron transport and, eventually, the rate of photosynthesis. MWCNT also increases plant water absorption capacity, total biomass, blooming, and yield (Joshi et al. 2018). Joshi et al. (2020) investigated the effect of oxidized MWCNTs on rice seeds and evaluated growth, germination, physiology, yield, and toxicity results. They reported that the MWCNT-treated rice plants had increased chlorophyll content in their vascular tissues and larger root length than the non-treated plants. Such morphology facilitated the absorption of nutrients from the soils, leading to a faster growth and boosting crop production (Joshi et al. 2020).

3 Plant Stress Management

Biotic (e.g., pests and diseases) and abiotic variables (e.g., drought, flood, humidity, and temperature) significantly impact crop yield. Crop improvement and output are inversely related to biotic and abiotic stressors. Farmers have relied considerably on pesticides to lessen crop losses, negatively affecting human health and environmental sustainability. Furthermore, synthetic pesticides and growth regulators utilized as a traditional technique of mitigating biotic and abiotic challenges endanger agricultural sustainability (Farooq et al. 2022). As a result, the primary objective is to promote the rapid adaptation of plants while protecting existing delicate ecosystems as they seek to survive environmental challenges. Nanomaterial engineering advances show that nanofertilizers can boost crop output in currently unfavorable settings.

Another group of researchers made an exceptional effort to investigate the applications of nanomaterials during abiotic stress conditions. They discovered that compared to traditional fertilizer-applied wheat plants, the life cycle of wheat plants with nanofertilizer was shorter for yield output from the planting date (Abdel-Aziz et al. 2016). The application of nanofertilizers to accelerate plant development and productivity demonstrates their effectiveness in agricultural operations, particularly in drought-prone locations where crop maturity is a key feature for sustainable crop production. Furthermore, NPs are useful in detoxifying or remediation of hazardous contaminants such as heavy metals. Wang et al., for example, indicate that foliar application of nano-Si considerably increases Cd stress tolerance in rice plants by controlling Cd accumulation (Wang et al. 2015).

Synthetic insecticides and growth regulators, which are often utilized to alleviate biotic challenges, endanger agricultural sustainability. In such scenarios, nanopesticides, such as nanofungicides, nanoherbicides, nanomolluscicides, and nanonematicides, can effectively mitigate biotic stresses (Hossain et al. 2020). These compounds have a similar component: a nano-based carrier or nanocarrier. For example, biosynthesized AgNPs obtained from cotton plant stem extract have strong antibacterial activity (Vanti et al. 2019). *Oncomelania hupensis* snails can be controlled with Ag nanoparticulated molluscicide, while *Biomphalaria pfeifferi* can be controlled with polylactic acid encapsulating curcuminnisin (Omobhude et al. 2017). Different plant diseases could be effectively controlled by metal oxide nanomaterials (e.g., CuO and ZnO) (Giannousi et al. 2013; Shenashen et al. 2017). As a consequence, the wise use of NPs can boost agricultural yield while protecting environmental health. Nanocomposites have been intensively investigated in crop protection in recent decades due to their great efficacy and eco-friendliness. Nanocomposite-based insecticides and fungicides are effective in crop protection. Different biopolymers, such as chitosan, alginate, and starch, have been used to formulate nanocomposites containing metal NPs (such as AgNPs) (Le et al. 2019; Sarkar et al. 2021). Graphene oxide-based nanocomposites have also been developed with acaricidal activity to get rid of mites (Wang et al. 2019).

Crop yield might be boosted more if the illness is detected early. Nanotechnology-based techniques might be effective instruments for this critical job. Nano-based biosensors, nanoimaging, nanoparticulated systems, and nanopore DNA sequencing tools have greatly improved pathogen detection accuracy, persistence, and time duration, as well as the advancement of high-throughput instruments for the qualitative detection of plant infections. It was recently shown that quantum dots with variable surface chemistry, minor photobleaching impact, and strong fluorescence could traverse plant cell wall pores and effectively detect biotic and abiotic stressors (Wu et al. 2017).

4 Role of Nanobiosenosrs in Climate-Resilient Agriculture

Nanobiosensors are non-invasive, sensitive, and particularly built sensors that use nanoscale materials and structures to detect biological molecules or cells (Abu-Salah et al. 2015). They are designed to be highly sensitive and selective and can be used for a variety of applications, including environmental monitoring and food safety testing. The real-time response signals produced by nanobiosensors are easily collected and analyzed. Nanotechnology-based sensors can transform plant security by permitting plants to interact with farmers when they are stressed and would benefit from human interference. Nanobiosensors have the potential to play a significant role in climate-resilient agriculture. Some possible uses include the following:

Soil and plant health monitoring: Nanobiosensors can be used to detect the presence of pathogens and nutrient deficiencies in soil and plants. This information can be used to optimize crop yields and reduce the use of pesticides and fertilizers.

Water quality monitoring: Nanobiosensors can be used to monitor water quality, detecting the presence of pollutants and pathogens that can harm crops.

Climate monitoring: Nanobiosensors can be used to monitor weather conditions such as temperature, humidity, and precipitation. This information can be used to optimize irrigation and other crop management practices.

Crop stress detection: Nanobiosensors can detect the presence of stress markers in plants, indicating that the plant is under stress due to environmental factors such as high temperatures, drought, or high salinity. This information can be used to take action to mitigate the stress and protect the crop. Nanobiosensors can be used to detect the presence of biotic stress, such as pests and diseases in crops, allowing for early detection and rapid response.

Overall, using nanobiosensors in agriculture can help farmers understand their crops and the environment better, allowing them to make more informed decisions and improve crop yields in a changing climate.

Although the use of nanosensors in plants is still in its infancy, intriguing studies have suggested that early detection and amplification of diseases or resource shortages using genetically encoded nanosensors might increase the efficiency of resource allocation. Below is a discussion of the most recent and significant results involving identifying abiotic stress, monitoring real-time activities, and detecting pathogen.

4.1 Abiotic Stress Detection

Important signaling molecules, including reactive oxygen species (ROS), phytohormones, inorganic elements, primary metabolites, and gaseous ROS, can all be used to reliably monitor the health status of plants under abiotic stress conditions (Mansour et al. 2019). These chemicals, especially ROS and inorganic elements, which are well-recognized secondary messengers involved in responses to a wide range of stressors, are at the forefront signaling network that elicits correct adaptation responses in plants. Hormonal signaling starts the processes that reduce stomatal conductance or permit root elongation in response to resource shortages frequently associated with ABA, IBA, and ethylene (Kim et al. 2010). The role of jasmonate signaling in plants' resistance to water stress, salt, cold, or physical injury has been documented (Tiwari et al. 2016). Resistance reactions to plant pathogens are frequently linked to ethylene, NO, and methyl salicylate. Numerous studies have also shown that carbohydrates such as glucose and sucrose help plants subjected to abiotic stress develop tolerance. By enhancing the functionality of existing monitoring equipment and facilitating the comprehension of the interaction between key molecules that influence the tolerance mechanisms, the combined use of nanomaterials and the discovered key signal molecules can certainly change our viewpoints. Genetically encoded nanosensors stand out among the several techniques used recently to create nano biosensors as a very effective strategy with a bright future.

The most recent method for creating smart biosensors uses NPs with genetically encoded nanoscale to screen important stress-responsive biomolecules. This method produces genetically encoded sensors with the capability of real-time recording of potential changes in the activity of the biomolecules (Walia et al. 2018). Physiologically linked sensitivity with temporal resolutions that may recognize dynamic essential signaling proteins and be recorded and examined on the scale of seconds are now well-understood thanks to genetically encoded sensors. This method can only be used in crop plants if there are effective DNA transformation processes for delivering DNA cassettes or plasmids, and each crop plant must be optimized gradually (Toyota et al. 2018).

4.2 *Biotic Stress Detection*

A successful and complete biotic stress management approach is attainable with the goal of modern technology for accurate intensity evaluation, forecast, and firsthand diagnosis. A wide range of molecular techniques has been developed to diagnose phytopathogens, which, while very precise, are time-consuming and expensive. Nanobiosensors, on the other hand, are considerably better than traditional approaches in terms of efficiency, non-destructiveness, cost-effectiveness, and accuracy. Some plants produce volatile organic compounds in response to insect attacks. These compounds can be detected with the help of a wireless nanosensor network for sensing insect attacks (Afsharinejad et al. 2015). These nanosensors can communicate with the gateway system attached to the plants and respond accordingly. Singh et al. (2010) revealed that Karnal bunt disease in wheat plants might be detected using a nano-gold immunosensor. Yao et al. (2009) used antibody-conjugated-fluorescent silica NPs to detect bacteria that produce leaf spots in Capsicum sp. and Solanum lycopersicum (Yao et al. 2009). Such surface-modified SiNPs are prepared through nontoxic methods. These NPs have been proven to carry antibacterial properties against pathogenic microorganisms, including *Escherichia coli* (Ananda et al. 2021) and *Alternaria longipes* (Ji et al. 2022). Modern methods have been developed to anticipate the presence of infections. For example, using nanosensors based on FRET for high temporal resolution real-time analysis of signaling molecules.

4.3 Real-Time Monitoring of Urease Activity

Nanobiosensors can be used to monitor real-time urease activity by detecting the presence of ammonia, a by-product of urease-catalyzed hydrolysis of urea. There are several different types of nanobiosensors that can be used for this purpose.

Enzyme-linked nanoparticle biosensors: In this type of sensor, nanoparticles (such as gold or silver) are coated with urease enzymes. When urea is present, the enzymes catalyze the hydrolysis of urea to produce ammonia, which binds to the nanoparticles. The change in the optical properties of the nanoparticles can then be used to detect the presence of ammonia and measure urease activity.

Carbon nanotube biosensors: Carbon nanotubes can be functionalized with urease enzymes and used as a sensing element for urease activity. The electrical properties of the carbon nanotube change when ammonia is produced, allowing for real-time monitoring of urease activity.

Microelectrode-based biosensors: Microelectrodes functionalized with urease enzymes can detect changes in pH or electrical potential resulting from the production of ammonia during urea hydrolysis.

These nanobiosensors can be used to monitor urease activity in real-time, by detecting the presence of ammonia and measuring the amount of urea hydrolysis. This can be useful in various fields such as soil nutrient management.

Works of literature have also reported using biosensors to detect various enzymatic activities in plants. The presence of urea can be determined by detecting Urease and Glutamate dehydrogenase (GLDH). Urease catalyzes the breakdown of urea to produce HCO_3^- and (NH_4^+). Ions containing NH_4^+ are usually unstable and quickly dispersed in the environment. GLDH converts NH_4^+, α-ketoglutarate (α-KG), and NAD to create NAD^+ along with glutamate right away. The immobilization of urease on an appropriate matrix is critical for constructing a urea sensor based on electrochemistry. Therefore, metal oxide NPs-chitosan-based hybrid composites have sparked the interest of researchers in developing the required biosensor. ZnO, Fe_3O_4, CeO_2, and others have been proposed as suitable matrices for biomolecule immobilization (Mandal et al. 2019). These nanomaterials have high catalytic efficiency, surface-to-volume ratio, and adsorption capacity, aiding a biosensor's stability and sensitivity.

5 Soil Remediation

Soil and water are the fundamental elements necessary for long-term agricultural production. Water pollutants may be precipitated or filtered, but soil contamination is far more difficult to remove. Nanotechnology breakthroughs provide a global opportunity to repair effectively or re-store damaged soil (Patra et al. 2016). NPs have recently been scrutinized for their ability to remove pollutants through a number of mechanisms, including adsorption, redox reactions, precipitation, and coprecipitation, which are facilitated by their huge specific surface area. Hyperaccumulators

Table 38.1 Role of different nanoparticles in the in vitro removal of heavy metals and pollutants from contaminated media

Nanoparticles (NPs) used	Contaminants remediated	Applications	Reference
Polyvinylpyrrolidone-coated Fe_3O_4 NPs	Cd and Pb	Cd and Pb-NPs were integrated with the process of bioremediation mediated by *Halomonas* species	Cao et al. (2020)
Zerovalent iron (nZVI) commercial suspension	As	Successful removal of arsenic heavy metal	Gil-Díaz et al. (2016)
Graphene oxide NPs (nGOx) and nZVI	Metals, namely, Cd, Pb, Zn, Cu, and As	NPs acted as strategies for the immobilization and stabilization	Baragaño et al. (2020)
Titanium oxide NPs-bonded-chitosan nanolayer (NTiO2-NCh)	Cd and Cu	Application of NTiO2-NCh eliminated Cu and Cd	Mahmoud et al. (2018)
Palladium (Pd), Pd NPs	Cr	The use of Pd NPs as bionanocatalyst was explored	Ha et al. (2016)
Magnetic iron oxide NPs	Cu, Ni, Pb	NPs remediated Cu^{2+}, Pb^{2+}, Ni^{2+} The study also identified this NP as an excellent biosorbent	Mahmoud et al. (2016)
ZnO NPs	Cu, Cd, Cr, and Pb	The applications of ZnO NPs showed the maximal removal of Cr, Cu, and Pb	Akhtar et al. (2021)

Adapted from open access source from Rajput et al. 2022

and indigenous soil bacteria might improve biodegradation processes with the aid of NPs, increasing the potential amount of remediation (Rajput et al. 2022). This might be referred to as nanophytoremediation or microbial-mediated nanoremediation. Plant contamination, bioavailability, and metal accumulation by plants play a role in the effectiveness of nanophytoremediation to remove heavy metals from polluted locations.

Nanobioremediation is a low-cost method of using plants and bacteria to break down toxic chemicals, thereby improving soil quality and lowering pollution. The technique may eliminate, retain, or reduce the number of pollutants present by breaking down toxins in the soil. Bioremediation effectiveness has already been examined and improved by utilizing different techniques and methods; however, nanotechnology has enhanced the process with a fresh aspect (Bharagava et al. 2020). Table 38.1 lists various NPs and their application in removing heavy metals.

6 Crop Improvement

Nanotechnology can be used to improve crop yields and resistance to environmental stressors through genetic modification, precise delivery of fertilizers, pesticides, and fungicides, and crop monitoring. This can help to increase food security in a changing climate and the sustainable use of resources. MWCNT, for instance, has a good effect on the germination of seeds from various agricultural species, including soybean, tomato, maize, peanut, corn, wheat, barley, and garlic (Srivastava and Rao 2014). Similarly, applying nanoscale SiO_2, TiO_2, and zeolite to agricultural plants favorably stimulates seed germination (Changmei et al. 2002; Manjaiah et al. 2019). Additionally, it was shown that Fe/SiO_2 nanomaterials had a great potential to enhance barley and maize seed germination (Disfani et al. 2017). The underlying processes by which nanomaterials could accelerate germination are yet unknown, despite a sizable amount of research on these valuable benefits. Studies have shown that nanomaterials can pierce the seed coat, promote water absorption and consumption, activate the enzymatic system, and eventually enhance germination and seedling development (Changmei et al. 2002; Banerjee and Kole 2016). Nanomaterials, such as ZnO, TiO_2, FeO, and hydroxyfullerenes, have been shown to improve crop quality and accelerate crop growth to germination of crop species, including mustard, onion, spinach, tomato, potato, and wheat (Dubey and Mailapalli 2016; Shojaei et al. 2019). For instance, OH-functionalized fullerenes found in carbon nanostructures called fullerols frequently benefit plant development. Varying dosages of AgNPs greatly increase the seed germination rate in maize, watermelon, and pumpkin while negatively affecting maize root elongation (Almutairi and Alharbi 2015; Szőllősi et al. 2020). The nanopriming of rice plants using a biocompatible AgNP synthesized from Kaffir lime leaf extract increases seed germination and amylase activity associated with seedling development and induces aquaporin gene upregulation (Mahakham et al. 2017). These results showed potential applications for nanomaterials in enhancing agricultural yields and product quality. These solutions provide excellent prospects to promote agricultural sustainability while improving crop yields and plant nutritional benefits.

7 Carbon Management in Soil

Nanotechnology may help prevent climate change in the modern period by storing carbon in terrestrial pools. The nanomaterials are said to improve carbon stability and potential soil sequestration because of their special nanoscale features. Contradictory reports on nanomaterials' possible effects on soil microorganisms are one of the main barriers preventing the widespread adoption of this technology for climate change mitigation. However, ongoing efforts are required to investigate the potential for using nanotechnology to sequester carbon without affecting ecosystem production to create climate-smart agriculture (Pramanik et al. 2020). This section will draw attention to the possibility of nanomaterials for better soil carbon management and the prospects for forthcoming nanotechnology research concerning soil carbon investigation.

7.1 Nanomaterials for Enhancing Soil Structure

Crop cultivation relies heavily on the soil. It is the interface between microflora and macroflora, nutrients, organic materials, humic acid, and water. Soil also retains water, minerals, and pollutants, which helps to mitigate agriculture's environmental effects. The aggregation and structure of the soil are important in the soil-atmosphere continuum's ability to regulate carbon effectively. Nano-enabled additives might improve soil quality by lowering inputs and increasing nutrient usage efficiency, for example, by using nanostructures tailored to target and release nutrients or pesticides solely in the plant rhizosphere (Frouz et al. 2009; Zheng et al. 2018; Gao et al. 2018; Okolo et al. 2020). Significant progress has been made in understanding how different management techniques impact soil carbon storage as influenced by variable soil aggregation and vice versa. Some methods include crop rotation, zero tillage, appropriate fertilization, nutrient management, and adding humic compounds to the soil (Pramanik et al. 2020). These studies have shown that soil aggregation and stability may be improved by using the right management techniques and increasing the carbon content in the soil. Research attempts that link the benefits of nanotechnology in improving soil aggregation for carbon storage are still in their infancy. Nanomaterials have been used in several attempts to enhance soil structure and related qualities. Nanomaterials can be added to soil to improve water retention by forming a network of small pores that can hold water and nutrients. This can help to reduce water loss and improve crop yields in dry or drought-prone regions. They can also be added to soil to improve aeration by creating a network of small pores that allow air to circulate. This can help to promote the growth of beneficial microorganisms and improve plant root growth. Additionally, nanomaterials can be used as a delivery system for nutrients, allowing them to be taken up more efficiently by plant roots. This can help to improve crop yields and reduce the need for excessive application of fertilizers. They stabilize soil and reduce erosion by binding soil particles together. This can help to improve the structural integrity of soil and reduce the risk of landslides and other forms of soil erosion. Most of this research has looked at how nanomaterials may improve soil structure concerning mechanical soil qualities for engineering reasons.

A few research works have only explored how nanomaterials affect soil structure in soil aggregation and carbon storage in soil (Aminiyan et al. 2015). These studies provide evidence for the beneficial impacts of nano-Fe, nano-zeolite, and nano-ZnO particles on carbon buildup in soil used for agricultural purposes. Adding nanozeolites to soil increases water-stable aggregates' mean weight diameter (MWD) and the amount of organic carbon in each aggregate size fraction. According to some research, forming cation bridges between organic matter and clay crystals in the presence of Ca^{2+} ions results in more stable aggregates that give soil microaggregates structural stability, which in turn protects organic carbon against microbial and enzymatic effects (Rowley et al. 2018). The aggregation of soil can also be improved by nano-ZnO and Fe particles. Reports also suggest that nano-ZnO and- Fe induced extracellular polysaccharide production from various microorganisms. This influences soil aggregation, organic carbon content, and moisture retention (Pramanik et al. 2020).

7.2 Nanomaterials to Sequester Carbon in Soils

Soil carbon sequestration is the method of CO_2 removal from the atmosphere and storage of CO_2 in the pool of soil carbon. Plants predominantly facilitate this method through photosynthesis, with carbon stored as soil organic carbon. This has been considered an effective technique for increasing soil functions while preserving stable CO_2 concentrations in the atmosphere. Pramanik et al. (2020) have stated different ways through which the sequestration of organic carbon in soil can be enriched through chemical, physicochemical, and biological protection to soil organic carbon (SOC). These mechanisms can be employed through practices such as adopting varied cropping systems, mulching, unified nutrient management, conservation agriculture, enhanced grazing, agroforestry, and forest management. Natural NPs, such as nanoclays, hydrous Fe oxides, or oxyhydroxides at the nanoscale, have been studied for their potential implications on soil carbon stability. The distinct electrical, magnetic, kinetic, and optical features of NPs have been linked to improved carbon stability in soil (Monreal et al. 2010). Allophane is a non-crystalline aluminosilicate and a natural nanoclay found in Andisols. Allophane nanoclay has been shown to retain a high percentage of carbon (11.8%) after intense peroxide treatment. This indicates that allophane might have a role in SOC stability and long-term sequestration. There have also been reports of high concentrations of SOC in allophonic soils (Calabi-Floody et al. 2015). Such situations might have arisen due to the protection of SOC as a result of the arrangement of SOC and minerals in the gaps of allophane aggregates (Filimonova et al. 2016).

8 Case Study: Wheat Crop and Nanotechnology

Wheat (*Triticum aestivum*) is the world's chief staple crop. Wheat is continually subjected to many biotic and abiotic stressors, resulting in massive economic fatalities. To solve these issues, breakthrough technologies that can increase wheat production while reducing the danger of different biotic and environmental pressures must be brought into modern agriculture. Among these technical developments, nanofertilizers, nanopesticides, nanoherbicides, nanosensors, and smart delivery systems have proved useful in agriculture and crop production. Carbon nanotubes, lipids, zerovalent metals, metal oxides, quantuts, nano polymers, and dendrimers with various properties such as nanofibers, nanowires, nanosheets, and so on are the most common types of nanomaterials utilized in agriculture and related industries (Kashyap et al. 2020). Thus, nanotechnology brings up new avenues for wheat production and protection. Several wheat experiments have shown that metal NPs may enter seeds without impacting germination. Wild and Jones (2009) demonstrated MWCNT breaching the cell wall of roots and reaching the cytoplasm using two-photon excitation microscopy (Wild and Jones 2009). Subsequently, the nanomaterials penetrated into epidermal cells and were involved in the circulation of the plant system from root to xylem tissues. NPs are transferred through cell wall pores by plasmodesmata and channels that link neighboring cells via an apoplastic or symplastic route.

NPs have considerable potential for managing and controlling wheat insect pests and illnesses. Savi et al. (2015) revealed that even at 100 mM concentration, ZnO NPs effectively inhibited deoxynivalenol production and *F. graminearum* infection in wheat grains (Savi et al. 2015). Aside from metal oxide NPs, the antibacterial activity of chitosan-loaded NPs has been found to suppress the *Fusarium* head blight of wheat. Chitosan NPs also substantially inhibit *F. graminearum* fungal growth, colony formation, and conidial germination (Panyuta et al. 2016). Hence, nanomaterial-derived products may provide a low-cost and dependable option for insect pest control, and further research may broaden the NP-based technologies in wheat pest management.

Nanofertilizers are a new class of synthetic fertilizers that include easily accessible nutrients on a microscopic size. Nanofertilizers are more soluble and efficient than conventional fertilizers. Using nanofertilizers enhances the solubility and dispersion of insoluble nutrients in the soil and boosts bioavailability. Furthermore, plants quickly absorb nanofertilizers and can deliver nutrients to the soil or plants for longer duration (Rameshaiah et al. 2015). Abdel-Aziz et al. (2016) studied wheat plants' foliar absorption of chitosan NPs containing nitrogen, phosphorus, and potassium. The results showed that wheat plants cultivated on sandy soil using chitosan-based nanofertilizers increased the calculated wheat yield (Abdel-Aziz et al. 2016).

Overall, it is possible to conclude that nanotechnology has the greatest potential to increase wheat growth and productivity under stressed environments. However, more testing is required to discover the optimal concentration, manner, and timing of administration and investigate the underlying physiological mechanisms responsible for increased wheat growth and production. Nanomaterials might be incorporated with existing wheat crop production and protection technology to have a greater influence on cropping systems.

9 Conclusion

Nanotechnology has profited from control over their composition, shape, and functioning by designing NPs to meet various agricultural demands. Nano-enabled technology solutions might be a critical instrument during the agri-tech insurrection required to convert agriculture into one that is more climate-resilient and sustainable. The combination of nanomaterials and other organic and inorganic substances for climate-sensitive cultivation can further broaden the use of nanomaterials to help fix current and emerging global challenges such as chemical detection, pollution emission control, decontamination, and anti-terrorism activities. Because there are certain worries regarding the toxicity of nanomaterials, future study guidelines to enhance sustainable nano-enabled agriculture research have been created. The discovery and manufacture of novel materials might improve the environment for nanotechnological applications in chemistry, material science, biology, environmental sciences, chemical engineering, and various other sectors. The discovery of nano-specific features that begin good and harmful reactions is critical to developing nanomaterial design for boosting crop productivity in the next revolution. Various laboratory and field tests will be required to create complete data to influence future choices, legislation, design, and development. While access to such knowledge remains difficult for the nano sector, progress toward more universal design guidelines for nanomaterials has been made. These discoveries and techniques can potentially translate into the creation of nanomaterials for agricultural crop production.

References

Abdel-Aziz HM, Hasaneen MN, Omer AM (2016) Nano chitosan-NPK fertilizer enhances the growth and productivity of wheat plants grown in sandy soil. Span J Agric Res 14(1):e0902–e0902

Abu-Salah KM, Zourob MM, Mouffouk F, Alrokayan SA, Alaamery MA, Ansari AA (2015) DNA-based nanobiosensors as an emerging platform for detection of disease. Sensors 15(6):14539–14568

Afsharinejad A, Davy A, Jennings B, Brennan C (2015) Performance analysis of plant monitoring nanosensor networks at THz frequencies. IEEE Internet Things J 3(1):59–69

Akhtar N, Khan S, Rehman SU, Rehman ZU, Khatoon A, Rha ES, Jamil M (2021) Synergistic effects of zinc oxide nanoparticles and bacteria reduce heavy metals toxicity in rice (Oryza sativa L.) plant. Toxics 9(5):113

Almutairi ZM, Alharbi A (2015) Effect of silver nanoparticles on seed germination of crop plants. Int J Nucl Quantum Eng 9(6):689–693

Aminiyan M, Safari Sinegani AA, Sheklabadi M (2015) Aggregation stability and organic carbon fraction in a soil amended with some plant residues, nanozeolite, and natural zeolite. Int J Recycl Org Waste Agric 4(1):11–22

Ananda PKP, Tillekaratne A, Hettiarachchi C, Lalichchandran N (2021) Sensitive detection of E. coli using bioconjugated fluorescent silica nanoparticles. Appl Surf Sci Adv 6:100159

Banerjee J, Kole C (2016) Plant nanotechnology: an overview on concepts, strategies, and tools. Plant Nano:1–14; Edition-1, Pages-XV, 383

Baragaño D, Forján R, Welte L, Gallego JLR (2020) Nanoremediation of As and metals polluted soils by means of graphene oxide nanoparticles. Sci Rep 10(1):1–10

Bharagava RN, Saxena G, Mulla SI (2020) Introduction to industrial wastes containing organic and inorganic pollutants and bioremediation approaches for environmental management. In: Bioremediation of industrial waste for environmental safety. Springer, Cham, pp 1–18

Calabi-Floody M, Rumpel C, Velásquez G, Violante A, Bol R, Condron L, Mora M (2015) Role of nanoclays in carbon stabilization in Andisols and Cambisols. J Soil Sci Plant Nutr 15(3):587–604

Cao X, Alabresm A, Chen YP, Decho AW, Lead J (2020) Improved metal remediation using a combined bacterial and nanoscience approach. Sci Total Environ 704:135378

Changmei L, Chaoying Z, Junqiang W, Guorong W, Mingxuan T (2002) Research of the effect of nanometer materials on germination and growth enhancement of *Glycine max* and its mechanism. Soyb Sci 21(3):168–171

Disfani M, Mikhak A, Kassaee MZ, Maghari A (2017) Effects of nano Fe/SiO2 fertilizers on germination and growth of barley and maize. Arch Agron Soil Sci 63(6):817–826

Dubey A, Mailapalli DR (2016) Nanofertilisers, nanopesticides, nanosensors of pest and nanotoxicity in agriculture. In: Sustainable agriculture reviews. Springer, Cham, pp 307–330

Farooq MA, Hannan F, Islam F, Ayyaz A, Zhang N, Chen W, Zhang K, Huang Q, Xu L, Zhou W (2022) The potential of nanomaterials for sustainable modern agriculture: present findings and future perspectives. Environ Sci Nano 9(6):1926–1951

Filimonova S, Kaufhold S, Wagner FE, Häusler W, Kögel-Knabner I (2016) The role of allophane nano-structure and Fe oxide speciation for hosting soil organic matter in an allophanic Andosol. Geochim Cosmochim Acta 180:284–302

Frouz J, Pižl V, Cienciala E, Kalčík J (2009) Carbon storage in post-mining forest soil, the role of tree biomass and soil bioturbation. Biogeochemistry 94(2):111–121

Fu L, Wang Z, Dhankher OP, Xing B (2020) Nanotechnology as a new sustainable approach for controlling crop diseases and increasing agricultural production. J Exp Bot 71(2):507–519

Gao X, Avellan A, Laughton S, Vaidya R, Rodrigues SM, Casman EA, Lowry GV (2018) CuO nanoparticle dissolution and toxicity to wheat (*Triticum aestivum*) in rhizosphere soil. Environ Sci Technol 52(5):2888–2897

Giannousi K, Avramidis I, Dendrinou-Samara C (2013) Synthesis, characterization and evaluation of copper based nanoparticles as agrochemicals against *Phytophthora infestans*. RSC Adv 3(44):21743–21752

Gil-Díaz M, Diez-Pascual S, González A, Alonso J, Rodríguez-Valdés E, Gallego J, Lobo MC (2016) A nanoremediation strategy for the recovery of an As-polluted soil. Chemosphere 149:137–145

Ha C, Zhu N, Shang R, Shi C, Cui J, Sohoo I, Wu P, Cao Y (2016) Biorecovery of palladium as nanoparticles by Enterococcus faecalis and its catalysis for chromate reduction. Chem Eng J 288:246–254

Hofmann T, Lowry GV, Ghoshal S, Tufenkji N, Brambilla D, Dutcher JR, Gilbertson LM, Giraldo JP, Kinsella JM, Landry MP (2020) Technology readiness and overcoming barriers to sustainably implement nanotechnology-enabled plant agriculture. Nat Food 1(7):416–425

Hossain A, Kerry RG, Farooq M, Abdullah N, Tofazzal Islam M (2020) Application of nanotechnology for sustainable crop production systems. In: Nanotechnology for food, agriculture, and environment. Springer, Cham, pp 135–159

Jayaraman S, Singh I, Shirale A, Patra A (2022) Conservation agriculture and resources conservation technologies for climate resilient rice production systems: special reference to Eastern India. Climate Resilient Technologies for Rice based Production Systems in Eastern India

Ji Y, Hu L, Xiong W, Wang Y, Yang F, Shi M, Zhang H, Shao J, Lu C, Fang D (2022) Highly sensitive time-resolved fluoroimmunoassay for the quantitative onsite detection of Alternaria longipes in tobacco. J Appl Microbiol 132(2):1250–1259

Joshi A, Kaur S, Dharamvir K, Nayyar H, Verma G (2018) Multi-walled carbon nanotubes applied through seed-priming influence early germination, root hair, growth and yield of bread wheat (*Triticum aestivum* L.). J Sci Food Agric 98(8):3148–3160

Joshi A, Sharma L, Kaur S, Dharamvir K, Nayyar H, Verma G (2020) Plant nanobionic effect of multi-walled carbon nanotubes on growth, anatomy, yield and grain composition of rice. BioNanoScience 10(2):430–445

Kashyap PL, Kumar S, Jasrotia P, Singh DP, Singh GP (2020) Nanotechnology in wheat production and protection. In: Environmental nanotechnology volume 4. Springer, Cham, pp 165–194
Kihara H (1964) Need for standardization of genetic symbols and nomenclature in rice. Paper presented at the rice genetics and cytogenetics. Proceedings of the symposium on rice genetics and cytogenetics, pp 1–12. Elsevier.
Kim T-H, Böhmer M, Hu H, Nishimura N, Schroeder JI (2010) Guard cell signal transduction network: advances in understanding abscisic acid, CO_2, and Ca^{2+} signaling. Annu Rev Plant Biol 61:561
Kumar A, Nagar S, Anand S (2021) Nanotechnology for sustainable crop production: recent development and strategies. In: Plant-microbes-engineered nano-particles (PM-ENPs) nexus in agro-ecosystems. Springer, Cham, pp 31–47
Le VT, Bach LG, Pham TT, Le NTT, Ngoc UTP, Tran D-HN, Nguyen DH (2019) Synthesis and antifungal activity of chitosan-silver nanocomposite synergize fungicide against Phytophthora capsici. J Macromol Sci 56(6):522–528
Liu R, Lal R (2015) Potentials of engineered nanoparticles as fertilizers for increasing agronomic productions. Sci Total Environ 514:131–139
Mahakham W, Sarmah AK, Maensiri S, Theerakulpisut P (2017) Nanopriming technology for enhancing germination and starch metabolism of aged rice seeds using phytosynthesized silver nanoparticles. Sci Rep 7(1):1–21
Mahmoud ME, Abdou AE, Mohamed SM, Osman MM (2016) Engineered staphylococcus aureus via immobilization on magnetic Fe_3O_4-phthalate nanoparticles for biosorption of divalent ions from aqueous solutions. J Environ Chem Eng 4(4):3810–3824
Mahmoud ME, Abou Ali SA, Elweshahy SM (2018) Microwave functionalization of titanium oxide nanoparticles with chitosan nanolayer for instantaneous microwave sorption of Cu (II) and Cd (II) from water. Int J Biol Macromol 111:393–399
Mandal N, Rakshit R, Datta SC, Singh AK (2019) Nanotechnology in the Arena of changing climate. In: Climate change and agriculture in India: impact and adaptation. Springer, Cham, pp 175–190
Manjaiah KM, Mukhopadhyay R, Paul R, Datta SC, Kumararaja P, Sarkar B (2019) Clay minerals and zeolites for environmentally sustainable agriculture. In: Modified clay and zeolite nanocomposite materials. Elsevier, Amsterdam, pp 309–329
Mansour MMF, Ali EF, Salama KHA (2019) Does seed priming play a role in regulating reactive oxygen species under saline conditions? In: Reactive oxygen, nitrogen and sulfur species in plants: production, metabolism, signaling and defense mechanisms. Wiley, Newark, pp 437–488
McGehee DL, Lahiani MH, Irin F, Green MJ, Khodakovskaya MV (2017) Multiwalled carbon nanotubes dramatically affect the fruit metabolome of exposed tomato plants. ACS Appl Mater Interfaces 9(38):32430–32435
Monreal CM, Sultan Y, Schnitzer M (2010) Soil organic matter in nanoscale structures of a cultivated Black Chernozem. Geoderma 159(1–2):237–242
Munawar S, Mustafa G, Khan MS, Joyia FA (2020) Role of biotechnology in climate resilient agriculture. In: Environment, climate, plant and vegetation growth. Springer, Cham, pp 339–365
Okolo CC, Gebresamuel G, Zenebe A, Haile M, Eze PN (2020) Accumulation of organic carbon in various soil aggregate sizes under different land use systems in a semi-arid environment. Agric Ecosyst Environ 297:106924
Omobhude ME, Morenikeji OA, Oyeyemi OT (2017) Molluscicidal activities of curcumin-nisin polylactic acid nanoparticle on Biomphalaria pfeifferi. PLoS Negl Trop Dis 11(8):e0005855
Panyuta O, Belava V, Fomaidi S, Kalinichenko O, Volkogon M, Taran N (2016) The effect of pre-sowing seed treatment with metal nanoparticles on the formation of the defensive reaction of wheat seedlings infected with the eyespot causal agent. Nanoscale Res Lett 11(1):1–5
Patel JK, Patel A, Bhatia D (2021) Introduction to nanomaterials and nanotechnology. In: Emerging technologies for nanoparticle manufacturing. Springer, Cham, pp 3–23

Patra A, Adhikari T, Bhardwaj A (2016) Enhancing crop productivity in salt-affected environments by stimulating soil biological processes and remediation using nanotechnology. In: Innovative saline agriculture. Springer, New Delhi, pp 83–103

Pramanik P, Ray P, Maity A, Das S, Ramakrishnan S, Dixit P (2020) Nanotechnology for improved carbon management in soil. In: Carbon management in tropical and sub-tropical terrestrial systems. Springer, Singapore, pp 403–415

Rajput VD, Minkina T, Upadhyay SK, Kumari A, Ranjan A, Mandzhieva S, Sushkova S, Singh RK, Verma KK (2022) Nanotechnology in the Restoration of Polluted Soil. Nanomaterials 12(5):769

Rameshaiah G, Pallavi J, Shabnam S (2015) Nano fertilizers and nano sensors–an attempt for developing smart agriculture. Int J Eng Res Gen Sci 3(1):314–320

Rowley MC, Grand S, Verrecchia ÉP (2018) Calcium-mediated stabilisation of soil organic carbon. Biogeochemistry 137(1):27–49

Sarkar S, Kundu A, Chakraborty R, Mukhopadhyay A (2021) A review on nanocomposites and their role in insecticide delivery. J Entomol Zool Stud 9(1):1985–1988

Savi GD, Piacentini KC, de Souza SR, Costa ME, Santos CM, Scussel VM (2015) Efficacy of zinc compounds in controlling Fusarium head blight and deoxynivalenol formation in wheat (Triticum aestivum L.). Int J Food Microbiol 205:98–104

Shenashen M, Derbalah A, Hamza A, Mohamed A, El Safty S (2017) Antifungal activity of fabricated mesoporous alumina nanoparticles against root rot disease of tomato caused by Fusarium oxysporium. Pest Manag Sci 73(6):1121–1126

Shojaei TR, Salleh MAM, Tabatabaei M, Mobli H, Aghbashlo M, Rashid SA, Tan T (2019) Applications of nanotechnology and carbon nanoparticles in agriculture. In: Synthesis, technology and applications of carbon nanomaterials. Elsevier, Amsterdam, pp 247–277

Singh S, Singh M, Agrawal VV, Kumar A (2010) An attempt to develop surface plasmon resonance based immunosensor for Karnal bunt (*Tilletia indica*) diagnosis based on the experience of nano-gold based lateral flow immuno-dipstick test. Thin Solid Films 519(3):1156–1159

Srivastava A, Rao D (2014) Enhancement of seed germination and plant growth of wheat, maize, peanut and garlic using multiwalled carbon nanotubes. Eur Chem Bull 3(5):502–504

Szőllősi R, Molnár Á, Kondak S, Kolbert Z (2020) Dual effect of nanomaterials on germination and seedling growth: stimulation vs. phytotoxicity. Plants 9(12):1745

Tiwari S, Lata C, Chauhan PS, Nautiyal CS (2016) Pseudomonas putida attunes morphophysiological, biochemical and molecular responses in Cicer arietinum L. during drought stress and recovery. Plant Physiol Biochem 99:108–117

Toyota M, Spencer D, Sawai-Toyota S, Jiaqi W, Zhang T, Koo AJ, Howe GA, Gilroy S (2018) Glutamate triggers long-distance, calcium-based plant defense signaling. Science 361(6407):1112–1115

Tului V, Janmohammadi M, Abbasi A, Vahdati-Khajeh S, Nouraein M (2021) Influence of iron, zinc and bimetallic Zn-Fe nanoparticles on growth and biochemical characteristics in chickpea (*Cicer arietinum*) cultivars. Poljopr Sumar 67(2):179–193

Vanti GL, Nargund VB, Vanarchi R, Kurjogi M, Mulla SI, Tubaki S, Patil RR (2019) Synthesis of Gossypium hirsutum-derived silver nanoparticles and their antibacterial efficacy against plant pathogens. App Organomet Chem 33(1):e4630

Walia A, Waadt R, Jones A (2018) Genetically encoded biosensors in plants: pathways to discovery. Annu Rev Plant Biol 69:497

Wang S, Wang F, Gao S (2015) Foliar application with nano-silicon alleviates Cd toxicity in rice seedlings. Environ Sci Pollut Res 22(4):2837–2845

Wang X, Xie H, Wang Z, He K (2019) Graphene oxide as a pesticide delivery vector for enhancing acaricidal activity against spider mites. Colloids Surf B: Biointerfaces 173:632–638

Wild E, Jones KC (2009) Novel method for the direct visualization of in vivo nanomaterials and chemical interactions in plants. Environ Sci Technol 43(14):5290–5294

Wu H, Santana I, Dansie J, Giraldo JP (2017) In vivo delivery of nanoparticles into plant leaves. Curr Protoc Chem Biol 9(4):269–284

Yao KS, Li S, Tzeng K, Cheng TC, Chang CY, Chiu C, Liao C, Hsu J, Lin Z (2009) Fluorescence silica nanoprobe as a biomarker for rapid detection of plant pathogens. In: Advanced materials research. Trans Tech Publications, Stafa-Zurich, pp 513–516

Zhang Q, Han L, Jing H, Blom DA, Lin Y, Xin HL, Wang H (2016) Facet control of gold nanorods. ACS Nano 10(2):2960–2974

Zheng H, Liu W, Zheng J, Luo Y, Li R, Wang H, Qi H (2018) Effect of long-term tillage on soil aggregates and aggregate-associated carbon in black soil of Northeast China. PLoS One 13(6):e0199523

Chapter 39
Mitigation of Plant Abiotic Stress by Plant Growth-Promoting Bacteria, Hormones, and Plant Extracts

Muhammad Saqlain, Safura Bibi, Athar Mahmood, Muhammad Anjum Zia, Muhammad Mansoor Javaid, and Javaria Nargis

Abstract Abiotic stress derives through rapid increase in the frequency and intensity of climate change causing hazardous problems. These environmental constraints induce different changes in the plants. The most common action of these environmental stresses is to generate reactive oxygen species (ROS) which alters the internal mechanism of plants. When it becomes excessive, accumulation in plant cells can lead to damage in cellular compartments and disrupt various physiological aspects in plants. To cope with these environmental problems, different strategies are introduced, such as the use of phytohormones, nanoparticles, biofertilizers, and biostimulants. Plant hormones help to stimulate plant growth under different stress conditions. Different nanoparticles applications such as ZnO, Fe_2O_3, and Fe_3O_4 also improved the plant metabolism under salt and heavy metal stress. Some PGPRs also used to mitigate abiotic stress. Biostimulant is an organic substance that promotes plant growth, nutrient uptake, and stress resistance. Plant extracts also have positive effect on the plant growth under environmental stresses such as moringa leaf extract which play crucial role in stimulating plant tolerance of abiotic stress. It enhances the photosynthetic

M. Saqlain · S. Bibi
Department of Botany, Faculty of Sciences, University of Agriculture, Faisalabad, Pakistan

A. Mahmood (✉)
Department of Biochemistry, Faculty of Sciences, University of Agriculture, Faisalabad, Pakistan
e-mail: athar.mahmood@uaf.edu.pk

M. A. Zia
Department of Biochemistry, Faculty of Agriculture, University of Agriculture, Faisalabad, Pakistan

M. M. Javaid
Department of Agronomy, College of Agriculture, University of Sargodha, Sargodha, Pakistan

J. Nargis
Department of Botany, University of Sargodha, Sargodha, Pakistan

M. Hasanuzzaman (ed.), *Climate-Resilient Agriculture, Vol 2*,
https://doi.org/10.1007/978-3-031-37428-9_39

activity and enzymatic antioxidant activity in plants. This chapter comprehensively reviews and upgrades our understandings about alleviation of abiotic stress by using different strategies, but most beneficial is use of plant extract, and their defensive mechanism against stress.

Keywords Abiotic stress · Phytohormone · Biostimulants · Moringa · Antioxidant activity

1 Introduction

Poor seedling emergence and plant growth are problems that plants face due to abiotic factors (Lamichhane et al. 2018). Numerous abiotic factors including UV radiation, heavy metals, salinity, temperature, and drought pose considerable dangers to plants and have a negative effect on agriculture and the ecosystem, causing significant losses (He et al. 2018). Plants as a sessile have adaptation to these stresses by evolving powerful strategies over the course of evolution in order to survive in these challenging circumstances. In these circumstances, plants start a variety of molecular, physiological, and cellular reactions (Fancy et al. 2017). As a result, it is important to be aware of the proper techniques for achieving effective germination, enhancing plant growth, and comprehending methods for decreasing the effect of restricting environmental factors (Dutta 2018). In plants, reactive oxygen species and reactive nitrogen species ROS and RNS (nitric oxide or nitrous oxide) play crucial signaling roles. Abiotic stress tolerance and resistance require adequate regulation of ROS/RNS equilibrium because excessive (ROS and RNS) formation may cause damaging to cell and the deactivation of signaling molecules (Ali et al. 2013; Fancy et al. 2017). Due to the increased ROS that cause for oxidative stress in plants, these environmental stresses severely damage crops by reducing productivity by up to 50% in a different variety of crops (Shafi et al. 2009). To mitigate negative impacts of the abiotic stresses and to improve the growth and production of plants, a variety of techniques have been used. These strategies include the creation of genetically engineered crops, the use of various chemicals, advanced farming methods and improving irrigation systems (Ali et al. 2015; Ali and Kim 2019). Recently, a number of biostimulants (substances or microorganisms) have been employed to boost crop output, raise tolerance of abiotic stress, and increase the effectiveness of nutrition (Luziatelli et al. 2019).

A variety of environmental stresses that plants experience, such as heavy metal, drought, heat, salinity, light, cold, and pesticide, impair their cellular and physiological functioning (Sharma et al. 2018; Jabeen et al. 2019). Environmental stress reduces plant development and growth through impacting variety of physio-chemical mechanisms, such as photosynthesis, antioxidant systems, and hormonal indicating (Asif et al. 2012; Saed-Moucheshi et al. 2019). These abiotic stresses cause crops to produce compounds that restrict growth and decrease their ability to survive in drought-stress environments. In response to ecological stresses, plants alter several molecular and cellular alterations that ultimately support their growth (Baber et al. 2009; Giri 2011). Plants produce organic compounds or osmolytes to protect the photosynthetic apparatus from various abiotic stressors. The most well-known organic solutes are polyamines, sugars such as trehalose, mannitol, and glycine betaine GB, as well as proline and sorbitol. These osmoprotectants are produced under various environmental conditions and provide resistance to cells that are barred from peering inside the plant's cellular structure (Balal et al. 2014; Ozturk et al. 2021). Accumulated sugars such as mannitol, galactinol, and trehalose under stress have reported frequently in crops, and several plant genetic factors play a significant role in the formulation of osmolytes (Ozturk et al. 2021). Plant biostimulant is a broad category of compounds that can be introduced around a plant and have beneficial impacts on growth of plant and nutrition as well as on the ability of the plant to withstand both abiotic and biotic stress. A lot of these also provide protection against environmental stresses such as water shortage, salinization of soil, and temperatures exposure that are not ideal for growth (Du Jardin 2015; Hanif et al. 2017).

2 Abiotic Stress

Extreme temperature, osmotic stress, water, and salinity are among the abiotic stresses that have impact on yield of crop. Plants can adapt to environmental stress signals by modifying metabolism and growth (Javaid and Tanveer 2014; Bhusal et al. 2019). The plant's root is an important organ which maintains crop productivity and performs a variety of activities. Because there is a direct connection between soil and roots, these are the principal sites for plants to detect changes in the environment (Javaid et al. 2016; Saini et al. 2018). Mineral nutrients, light, carbon, and water are necessary for the plant growth, nourishment, and reproductive process of plants. Extreme environmental stress can result from a variety of factors, including drought, salinity, extremely hot or low temperatures and heavy metals. The capacity of a plant to detect stimuli and react to them may be advantageous to existence (Naylor and Coleman-Derr 2018). The water stress restricts root growth, blooming, fruit production, stomatal opening, leaf size, and water potential of the leaf (Osakabe et al. 2014). Because of this, plants have developed a variety of systems for conserving water or controlling their growth until they are exposed to harmful situation (Gupta et al. 2021a, b). Understanding the metabolism, which comprise genes of regulation and their pathways of metabolism in different plant types, is essential for reducing stress in plants (Javaid et al. 2020; Manna et al. 2021).

2.1 Salinity Stress

Salinity is identified as the environmental condition that has very detrimental effects on plant development and growth, photosynthesis, and ion stability, outweighing all other environmental factors (Ahanger et al. 2014). As most agricultural plants cannot thrive in highly salinized conditions, salinity poses a danger to productivity of crop (Ikram et al. 2019; Zörb et al. 2019). Anthropogenic and other sources of salt, particularly NaCl concentrations [EC >4 dSm^{-1} or 40 mM], can make soils unfit for the use of agriculture (Bhat et al. 2020). Plant experience physiological, morphological, and changes at molecular level as a result of salinity stress, which hinders their ability to grow, develop, and survive. Furthermore, stomatal conductance and enzyme activity can also be negatively impacted by excessive salt concentrations (Kumar and Verma 2018). Salt-induced osmotic stress is the primary cause of the reduction in growth and yield. It can disrupt plant's relations with water, these modifications have an impact on the functions of numerous enzymes and plant metabolism (Munns and Tester 2008). Along with salt-induced osmotic stress, toxic salt buildup also inhibits plant growth. The most common salts are Na^+, Cl^-, SO_4^{2-}, and HCO_3, all of which are extremely hazardous. However, the crops range in how sensitive they are to various harmful ions. The common consensus is that excessive Na^+ buildup results in nutritional imbalance and specific ion toxicity. Because plants lose their ability to control Na^+ transport in salt sensitive species or at the highest levels of salt, the ionic effect of salt predominates over the osmotic effect (Munns and Tester 2008). Reactive oxygen species harm cell membranes, nucleic acids (DNA, RNA), proteins, and lipids and are increased by the presence of excessive salt and drought (Kumar et al. 2017). Programmed cell death and the buildup of too much ions (Na^+ and Cl^-) causes conditions known as hypertonic may also result from salt stress (Zhang et al. 2018; Welgama et al. 2019). One important reason to the reduction in growth of plant and output due to salinity is a crop's capacity for photosynthetic development. The efficiency with which a plant rejects or compartmentalizes the harmful ions determines how tolerant the photosynthetic system is to salinity (Yadav et al. 2011).

2.2 Drought Stress

In contrast to the water scarcity, which is the condition where water falls below the 30%, a drought situation is known as a condition where the capacity of total water is between 12% and 20% for a period of 16 days (Kränzlein et al. 2022). One of the most common factors that affect microbial activity, crop development, yield output and quality is drought (Abdul Rahman et al. 2021; Tanveer et al. 2017). Stress caused by a water shortage is a drawback on microorganisms of soil that can result in a decline in enzyme function, loss in the cycling of nutrients (such as C, N, and P), and soil fertility, which in turn can affect plant productivity, especially for crops that are susceptible to drought, and economic implications (Nguyen et al. 2018).

The relative water content, transpiration rate, and leaf water potential of plants under drought stress are all affected and all significantly decrease, and their leaf temperature also rises (Anjum et al. 2011). Wheat used water more efficiently when supplies were scarce than when it was well-watered. They linked the reduced transpiration and greater water use efficiency to stomatal closure. Plant water content and cell turgor potential are significantly lowered by drought, which often has a negative effect on rate of growth and crop output (Tanveer et al. 2013; Hu et al. 2015). Lack of water has a significant impact on how nutrients are taken by roots function and transported to the upper part of plant such as shoots. Reduced transpiration flow, disruption of the nutrient intake, unloading system, and lower inorganic nutrient absorption are all possible outcomes. Drought stress unable to absorb nutrients; nitrate and phosphate were retained in grains, whereas K^+ accumulation in stem and leaves. Under mild and severe water stress circumstances, K^+ absorption decreased by 67 and 82%, respectively (Rouphael et al. 2012).

Under drought condition, the soil state strictly relates to the development of plants. Plant physiology, biochemistry, and morphology are all directly impacted by drought (Hanaka et al. 2021). Furthermore, it slows down seed germination rate and growth of seedling. Also there was a response of plant to water scarcity at the transcriptome, metabolomic levels (You et al. 2019). The presence of microbiota to plant roots is disrupted by severe and protracted drought stress, drastically altering the microbiome composition of the plant roots and changing the structure and release of root exudates as well as the availability of beneficial nutrients (Khan et al. 2019). Drought alters the microbiome of plant roots, promoting actinobacteria and numerous other gram-positive species (Breitkreuz et al. 2021).

2.3 *Heavy Metal Stress*

Environmental contamination results from the extensive use of hazardous waste brought on by the rise of industry and agriculture (Jan et al. 2016). Each sort of pollution has an adverse effect on plants, but addition of some toxic metals to soils and streams are particularly worrisome because of their protracted environmental persistence (Briffa et al. 2020). Due to its detrimental biological effects, heavy metal-induced soil pollution in the agriculture sector has grown to be a significant ecological concern. Due to their extensive availability and their severe and long-lasting negative effects on plants growing in such polluted soils, these toxic pollutants are known as soil toxins. Each heavy metal has different negative impacts on plants (Rehman et al. 2020). Continual exposure of plants to very hazardous quantities of cadmium metal causes a reduction in the amount of water, nutrients, and photosynthesis that plants can perform. Additionally, chlorosis, growth restriction, and eventually mortality occur in plants that are grown in soil that has high levels of the element cadmium (Jin et al. 2015).

A significant amount of zinc metal is deposited in soil which cause of anthropogenic activities such as sewage sludge, fertilizers, emissions of municipal waste, and others. While cadmium is insignificant and potentially hazardous for higher plants and other living things, zinc is an essential complement for all living things. Zinc convergences in contaminated soils may result in phytotoxicity. Significant zinc concentrations in soil impair a variety of plants' metabolic processes, delay growth, and speed up plant aging. Plants that are contaminated with zinc experience altered root and shoot growth. In addition, chlorosis, which can affect other plant components, is brought on by zinc toxicity in early leaves (Mirshekali et al. 2012; Habib et al. 2016).

Chromium is a heavy element that pollutes groundwater, soil, and other environmental areas. The tanning industry is a big consumer of water, and the majority of the wastewater it produces, which has a high chromium metal content, is dumped into the environment. Chromium overexposure restricts plant growth, results in chlorosis in the leaves, unbalances nutrient intake, and damages the root tips (Ertani et al. 2017; Javaid et al. 2022a). Chromium has negative effects on plant growth, including changes to the cycle of germination and the growth of the different plant parts. As a result, excessive chromium metal exposure affects yield effectiveness and total dry matter production. Additionally, chromium harms plant physiological processes such as photosynthesis and the absorption of dietary minerals (Singh et al. 2013). The capacity of plants to produce reactive oxygen species or the ability of enzyme metabolites to be inhibited by chromium exposure both affect plant metabolism (Wakeel et al. 2020; Javaid et al. 2022b).

2.4 *Heat Stress*

Global crop productivity is seriously threatened by heat stress as a result of rising greenhouse gas emissions. It is described as a spike in temperature great enough to permanently harm plant development and growth. In general, heat shock or heat stress can be caused by a temperature increase of more than typically 10–15 °C over ambient (Liliane and Charles 2020). Leaf drop, especially in trees, is heat stress in plants. To save water, many plants will lose portion of their leaves. Many crops have production issues in hot temperatures. In hot weather, flowers fall off of plants, while under cool-season crops such as broccoli bolt (Tilley 2019; Javaid et al. 2022c).

The distribution and procreation of life on Earth are severely constrained by temperature, which is a crucial physical parameter (Richter et al. 2010). Even a small temperature increase over the ideal temperature for growth can stop the life cycle. As a result, nearly all living things had switched on pathways of signaling to detect the heat and reprogrammed cellular processes to protect themselves against heat harm. Because plants are sessile and cannot bear extremely high temperatures, a complex sense and response system had to develop to satisfy the pressing need (Takahashi and Murata 2008).

The multiple-subunit protein-pigment complex which is known as photosystem II (PSII) is made up of the reaction center, oxygen-evolving complex (OEC), light-harvesting pigment protein complex (LHCII), and other components (Shen 2015).

LHCII absorbed and transported light quantum excitation energy through pigments in the reaction center throughout the initial photochemical reaction course. The principal charge product of charge separation, [P680 + Pheo], initiates the electron transport process and utilizes the OEC to extract electrons from H_2O (Cardona et al. 2012).

There were components of the photosynthetic electron transport chain. Electron transport caused $NADP^+$ to take an electron, decrease it to NADPH, and then undergo coupled photophosphorylation to produce ATP. Calvin cycle is driven by NADPH and ATP and then CO_2 is fixed as a result. Because the entire process of photosynthesis necessitates close coordination between subunits and the supermolecular complexes in lipid bilayers, it was also susceptible to heat stress (Yamamoto et al. 2008). According to Ohira et al. (2005), the unfolding of protein or unspecific aggregation, is the most detrimental direct consequence of heat shock on PSII functions. It was made worse by the enhanced membrane fluidity brought on by the increased temperature. Additionally, the thylakoid's internal arrangement was messed up simultaneously (Horváth et al. 2012).

Only after prolonged exposure to moderately high temperatures can injuries or death occur. High temperatures can directly harm organisms by denaturing and aggregating proteins and raising the membrane lipids' fluidity. The inactivation of mitochondrial and chloroplast enzymes, the decrease in protein production, the breakdown of proteins, and the loss of membrane integrity are examples of indirect or delayed heat damage (Howarth 2005).

2.5 *Chilling Stress*

Abiotic stress, known as chilling, affects growth, yield, and quality of fruit (Zhang et al. 2021). Open-field crops are frequently affected by chilling stress; during the season of winter to early spring, the plants are grown in a greenhouse solar system (Wani et al. 2016). By interfering with the carbon-oxygen cycle, the production of photosynthetic pigments, and the CO_2 supply, the photosynthetic apparatus is either directly destroyed by cold stress or is indirectly hampered by it (Allen and Ort 2001). When vegetation is subjected to cold stress, they experience substantial changes in gene transcription, protein synthesis, and metabolism (Dou et al. 2021).

The fluidity of the plant membrane decreases, leading to a decline in enzyme activity and disturbances in the stomatal response at low temperatures (Ding et al. 2019). As a result, when plants are under cold stress, photosynthesis is frequently the first mechanism to be hindered. Because the Calvin-Benson cycle is less active under cooling stress, the overdrive of electron transport powered by light energy (Bi et al. 2019). Reactive oxygen species, which are hazardous chemicals, are created when extra electrons escape into the surrounding O_2 at the end of the electron transfer chain (Asada 2006). As a result, ROS attacking neighboring photosystems or thylakoid membranes directly, photosynthetic activity is drastically reduced, and plant growth is inhibited (Takagi et al. 2016).

3 Approaches to Mitigate Abiotic Stress

Different biological, chemical, biochemical, biostimulant, antioxidant, and plant-extracted materials are used to mitigate the plant abiotic stresses.

3.1 *Plant Growth-Promoting Bacteria*

According to a number of investigations, microbes that encourage plant development are essential for reducing climate change (Krishnamoorthy et al. 2021). Certain microbial communities, such as methanogens, which are found in wetlands, oceans, rumens, and termite guts, create methane. Methylotrophs, on the other hand, use methane as a carbon source to lower the amount of methane in the environment. The importance of plant-associated beneficial microorganisms has increased recently due to their crucial function in boosting crop output and offering resilience to stressful environmental conditions (Rani et al. 2021). Both a direct and an indirect mechanism are used by the (PGPB) to accelerate plant growth. This involves consuming minerals, producing phytohormones, fixing atmospheric nitrogen, or by creating different enzymes, antibiotics, siderophores, or improved soil texture, thus boosting the immune system's defenses against numerous infections (Prittesh et al. 2020). Microorganism, are consider as effective to mitigate the different abiotic stresses (Fig. 39.1; Rehman et al. 2021).

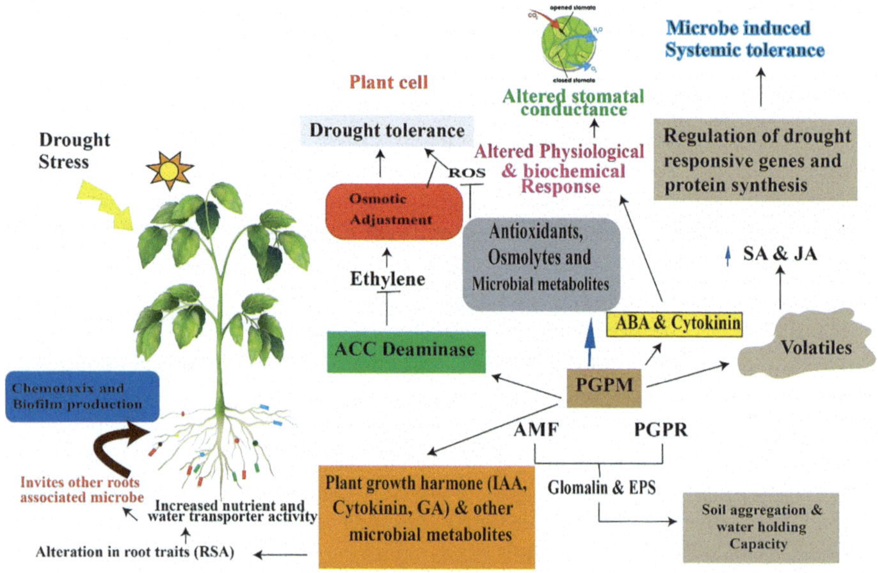

Fig. 39.1 Mechanism of microbes to mitigate abiotic stress in plants (Kumar and Verma 2018)

3.2 Plant Growth-Promoting Rhizobacteria

Plant growth-promoting rhizobacteria (PGPR) are soil microorganisms that live in the rhizosphere and work to enhance plant development and growth by secreting a range of control chemicals. They could attach to the inside of the plant, the leaves, the roots, or the phyllosphere (endosphere). Because they can communicate with the host plant much more efficiently because they reside inside the tissues of the plant, Plant growth-promoting endophytes (PGPE) are often the most successful at encouraging growth (Giauque et al. 2019). Because they are shielded from the outside world, the PGPR are also far less susceptible to the typical chemical-physical, biotic, and abiotic fluctuations in the soil. Endophytic bacteria primarily penetrate plant tissues through the growth-related natural fractures that occur in the roots. They develop from the environment of the roots' rhizosphere. Plant exudates and radical metabolites constitute a key resource for attracting and selecting the finest microorganisms (Podile and Kishore 2006). The positive role of the bacteria that reside inside plants is mostly demonstrated by directly increasing nutrient absorption through the alteration of plant hormone levels. The most significant and thoroughly studied direct process is nitrogen fixation (Fig. 39.2; Vocciante et al. 2022).

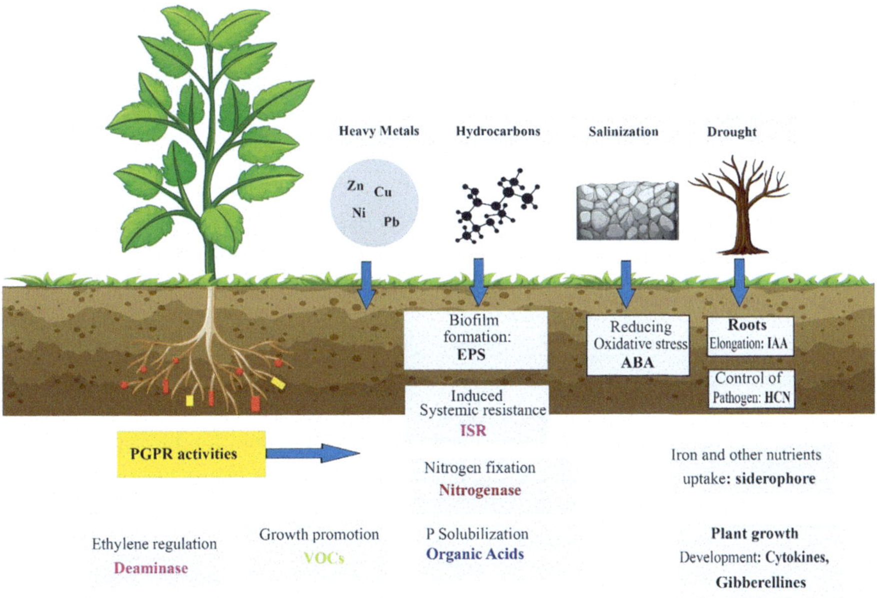

Fig. 39.2 The highly effective network of functional interactions that PGPR has built to sustain plants' health and performance in the face of environmental stressors and abiotic pressures. Exopolysaccarides (EPS), ABA (abscisic acid), volatile organic compounds (VOCs), HCN (hydrogen cyanide), IAA (indol2-3-acetic acid), and ACC (1-aminocyclopropane-1-carboxylic acid) (Vocciante et al. 2022)

3.3 Phytohormones

A phytohormone called salicylic acid governs plant growth, development, and defense against environmental threats in a number of different ways (Arif et al. 2020). Additionally, salicylic acid is involved in DNA repair and abiotic stress tolerance and is present throughout the plant's life cycle, from seed germination to fruiting (Koo et al. 2020). In fact, salicylic acid was discovered to cause oxidative stress at greater concentrations whereas lower quantities are observed to alleviate the harmful effects of abiotic stress (Miura and Tada 2014).

Polyamines, which are aliphatic polycations with low molecular weight and can be free, conjugated (attached to other small molecules, like phenolic acids), or linked, are found in all living organisms (connected with multiple macromolecules) (Pál et al. 2015). Cytokinins also regulate seed germination, leaf senescence, abiotic and biotic stress adaption, cell division, and differentiation (Wu et al. 2014).

Jasmonic acid regulates the development of plants' reproductive organs, seed germination, nitrogen storage, assimilate transfer, root growth, tuber formation, flowering, fruit ripening, senescence, and plant defense against environmental threats (Tavallali and Karimi 2019). Salicylic acid is essential for plant development improve, resistance against disease, fruit formation, and as an enzyme activator in stressed plants' antioxidant systems (El-Esawi et al. 2017).

3.4 Biostimulants

Biostimulants that include a combination of chemicals or microbes used to improve plant nutrition and lessen the effects of abiotic stress are also referred to as "plant biostimulants." The source for biostimulants is expected to grow quickly between 2017 and 2025, at a rate of 10.2% yearly. On the other hand, only around 25% of commercially available biostimulant products worldwide are based on microorganisms (Hamid et al. 2021). Crops are treated with microbial plant stimulants (MPBs) to encourage plant uptake and utilization of various nutrients, hence promoting plant growth and production. Examples of MPBs that use direct and indirect processes to encourage plant growth and development in both usual and stressful conditions include mycorrhizal and non-mycorrhizal fungi, bacterial endosymbionts, and PGPR. MPBs create and sustain sufficient activity close to plants, which regulates how they react to diverse stimuli (Ali et al. 2017b).

To improve plant growth, nutrition, and stress tolerance, the environment around a plant can be treated with a variety of compounds known as plant biostimulants, often referred to as agricultural biostimulants. Some plant stimulants also serve as environmental defense, albeit most are transported to the rhizosphere to aid with nutrient absorption. Some of the stressors include a lack of water, salinized soil, and exposure to temperatures that are unsuitable for growth (Du Jardin 2015). Instead of being nutrients in and of themselves, biostimulants help nutrients to be absorbed and either help promote development or help people resist stress (Brown and Saa 2015). Impact of a commercial biostimulant based on seaweed and specific yeast extracts on tomato *(Solanum lycopersicum* L. var. Micro-Tom) productivity, morphological characteristics, proximate composition, and nutraceutical benefits in plants raised under normal conditions. Yeast and seaweed extracts are getting more and more well-liked. They are utilised in horticulture as a plant biostimulant because of their various defensive effects against abiotic stresses (Rouphael and Colla 2018).

4 Plant Extract

Even if the growth in fertilizers has not always been accompanied by an increase in productivity, agricultural fertilizers have been misused to expand plant production systems. The majority of the mineral fertilizers that were used have been lost in the soil due to shallow roots' insufficient capacity to absorb minerals, especially in vegetable cropping systems, which has led to soil pollution and deterioration (Colla et al. 2017). The greenhouse industry is especially sensitive to this problem since it uses substantially more fertilizer per unit area than any other agricultural technique (Rouphael et al. 2018). As a result, the only way to accomplish sustainable agricultural intensification is to raise plant crops' resource efficiency (RUE) while reducing their consumption and environmental impact of chemical fertilizers (Stasio et al. 2017).

Plant biostimulants (PBs), which are commodities that promote plant nutrition processes without regard to the nutrients they contain, may be a promising and environmentally friendly strategy to improve crop quality features, abiotic stress resistance, soil and rhizosphere availability of restricted nutrients, or nutrient usage efficiency (Rouphael and Colla 2018).

The green synthesis of selenium nanoparticles (SeNPs) utilizing a range of medicinal plant extracts is more inexpensive, simple, environmentally friendly, and nontoxic when compared to alternative processes that rely on microorganisms such as bacteria and fungi. The green manufacturing of SeNPs has effectively and conveniently utilized a variety of plant extracts (Alam et al. 2019). Unexpectedly, the plant extract contains secondary metabolites, alkaloids, tannin, cinnamic acid, sesquiterpenes, phenolic acid, and monoterpenes, all of which may function as stabilizing and reducing agents in the manufacture of biocompatible SeNPs (Javed et al. 2020).

Plants can partially resist against stress naturally through chemical development or physical modification (Janská et al. 2010). In addition to activating their natural defense system, plants' capacity to endure heat stress can also be improved by exogenous delivery of several stress-relieving compounds (Zia et al. 2021). Despite significant advancements in increasing heat tolerance through the use of numerous inorganic compounds and plant growth regulators, the utilization of natural sources, particularly plant extracts, is still relatively low (Farooq et al. 2009). Plants' ability to withstand stress has been improved by moringa's phytochemical components.

Moringa leaf extract increased seed emergence in salt-stressed plants, enhanced protein synthesis, activated antioxidant activities, and eventually increased wheat grain output (Basra et al. 2011). The presence of significant numbers of phenolic compound, which are capable of reducing ROS and have high antioxidant activity (Ilyas et al. 2015).

4.1 Plant Extract Used to Mitigate Abiotic Stress in Plants

4.1.1 Moringa Plant

The family Moringaceae, which includes the drumstick, horseradish, and benzoyl trees, is home to the genus *Moringa*. Cuttings or seeds can be used to propagate these trees (Abd Rani et al. 2018). Tropical and subtropical regions of the world contain it (Asante et al. 2014). Only *Moringa oleifera* has a variety of uses among all *Moringa* species. It is indigenous to India and is resistant to a variety of environmental challenges. It is grown all over the world, including in Pacific Islands (Abd Rani et al. 2018). Because it contains large amounts of numerous essential phytochemicals, including fatty acids, terpenoids, and minerals. The roots, leaves, fruit, pods, and seeds of the *M. oleifera* plant are nutritious (Gopalakrishnan et al. 2016). As a potential substitute for proprietary compounds such as alum sulphate, *M. oleifera* seeds' active coagulant and antibacterial properties could be used to purify water (Özcan 2020).

4.1.2 Moringa Leaf Extract

Moringa leaf extract (MLE) includes essential nutrients, vitamins, flavonoids, glycosides, carbohydrates, amino acids, and phytohormones that assist the development and growth of plants, MLE acts as a powerful biostimulator or bioenhancer (Zulfiqar et al. 2020). It decreases ROS, boosts plants, and alleviates a variety abiotic stresses. It has a lot of antioxidants as well. The moringa extracts efficiently lessen environmental stressors on plants such salt by promoting the enzymatic antioxidant activity as well as total sugars and content of amino acid (Zulfiqar et al. 2020).

4.1.3 Moringa Leaf Extract Composition

For the manufacture of MLE, dried and ground leaves of moringa were extracted with ratio (4:1) of ethanol or water, then go through liquid extractions (Wright et al. 2017). Essential nutrients, including carbohydrates, lipids, proteins, fiber, minerals, and vitamins, are present in MLE. The high nutrient content of MLE has a significant impact on the physiological, therapeutic, and medicinal qualities of plants

(Abd Rani et al. 2018). Vitamins A, B, C, and E, minerals, and amino acids including arginine, lysine, phenylalanine, histidine, tryptophan, valine, and methionine are all rich in leaf extract (Yasmeen et al. 2013).

4.1.4 Effect of Moringa Leaf Extract on Plants

MLE can be given to plants through different methods, such as applied by foliar spraying, treating seeds, and rooting. Plant physio-biochemical characteristics are aided by MLE. Germination is a crucial stage in the growth of plants. MLE is necessary for seed germination, just like typical phytohormones such as GAs and IAA (Zulfiqar et al. 2020). Freshly harvested MLE contributes to improved plant growth, development, and yield attributes, which in turn enhance crop performance. Additionally, it boosts antioxidant content, plant hydration status, and membrane stability (Rehman et al. 2014).

The highly important metabolic processes of photosynthesis and gas exchange in plants control growth responses. MLE treatment increased the conductivity of stomata, rate of photosynthetic mechanism, and chlorophyll concentration (Elzaawely et al. 2017). By encouraging flowering and fruiting as well as maintaining the fruit's post-harvest quality and beauty, MLE application promotes reproductive growth (Zulfiqar et al. 2020).

Abiotic stresses that affect plants include salinity, drought, heavy metals, and temperature. Therefore, reducing environmental stress is essential for crop yield and food security. A unique biostimulant and natural growth promoter, MLE treatment increases plant tolerance to a variety of environmental signals. IAA, GAs, cytokinins, and mineral nutrients are abundant in MLE, which lessens the damaging effects of salinity (Ali et al. 2017a).

4.2 Abiotic Stress Mitigation by Moringa Leaf Extract

4.2.1 Salt Stress Alleviation by Moringa Leaf Extract

Salinity stress puts crop output and sustainable agricultural practices at risk on a global scale. By producing different ionic, oxidative, and osmotic stress, it changes many traits of plants (Arif et al. 2020). By encouraging antioxidant defense mechanisms such as the proline content, MLE improved salinity stress tolerance by lowering ROS. Thus, MLE preserved plant growth and developmental features, which reduced the damaging effect of salt (Howladar 2014). MLE treatment raised proline, total sugar, and phenol concentrations as well as CAT, APX, and SOD activity, which decreased the detrimental effects of salt (Yasmeen et al. 2013). Application of MLE to the leaves also aided yield characteristics such grain weight, biological yield, straw yield, and grain yield (Latif and Mohamed 2016).

4.2.2 Drought Stress Alleviation by Moringa Leaf Extract

Drought stress has a significant impact on plants, biomass, production, and metabolic processes. Numerous metabolic, morphological, developmental, physiological, and ecological traits, such as plant functions, are also affected by drought condition (Seleiman et al. 2021). The application of MLE improved the ability of plants to withstand drought by enhancing traits such as yield per plant, dry and fresh mass, plant height, photosynthetic efficiency, carotenoid and chlorophyll (Chl), proline content, sugar content, and phenol content (Hanafy 2017). Under drought stress, MLE treatment in *Zea mays* L. positively controlled the growth of plant, biomass, stability of membrane, cell osmoticum (Pervez et al. 2017).

4.2.3 Heavy Metals Stress Alleviation by Moringa Leaf Extract

In *Phaseolus vulgaris*, cadmium stress enhanced ROS, electrolyte leakage, peroxidation of lipid, and instability of membrane. MLE treatment reduced the impacts of Cd stress and improved the morphological traits of the plant (Howladar 2014). Catalase, peroxidases POX, polyphenol oxidase Glutathione reductase GR, and Superoxide dismutases SOD activities were all increased by MLE, which also improved growth of plants and developmental procedures (Khalofah et al. 2020).

4.2.4 Heat Stress Alleviation by Moringa Leaf Extract

Due to global warming, there has been a rise in high temperatures during the last few decades. Heat-induced oxidative stress is deleterious to plant structure, photosynthesis, and membrane stability (Zhao et al. 2020). Heat stress causes irreparable harm to a plant's photosynthetic machinery. MLE lessens the effects of heat stress by raising the activity of the photosynthetic system, total Chl content, Chl a/b ratio, and carotenoid concentration. Carotenoids prevent the oxidative damage brought on by heat stress (Batool et al. 2019). By increasing relative water content, growth, proline, total soluble sugar content, biomass, and the activity of catalase and POX, which lowered ROS, MLE injection successfully promoted heat stress tolerance in wheat. In addition to grain weight and spike length, MLE also generates other yield characteristics (Afzal et al. 2020).

5 Conclusion

It is concluded that the application of biofertilizers such as phytohormone, nanoparticles, PGPR, and other chemicals application mitigates the different environmental stresses. They also promote plant growth under stressful conditions. But there is need to increase the food security and production due to exponentially increasing population. As a result, it is important to reduce negative impacts of abiotic stress on plants. Natural biostimulants are low cost, sustainable, organic, and eco-friendly. These impart the positive result in plant growth and yield. Moringa plant extract plays a crucial role in increasing the photosynthetic mechanism. It is rich in nutrients, vitamins, sugars, flavonoids, and phytohormone. It also improves the rate of germination, growth, relative water content, water use efficiency, and activity of enzymatic antioxidant. It acts as natural bioenhancer and biostimulator, which help in increasing the morpho-physiological attributes of plants under normal and stressful condition. Moringa leaf extract application increases the tolerance and resistance against abiotic stress in plants.

References

Abd Rani NZ, Husain K, Kumolosasi E (2018) Moringa genus: a review of phytochemistry and pharmacology. Front Pharmacol 9:108

Abdul Rahman NSN, Abdul Hamid NW, Nadarajah K (2021) Effects of abiotic stress on soil microbiome. Int J Mol Sci 22(16):9036

Afzal I, Akram M, Rehman H, Rashid S, Basra S (2020) Moringa leaf and sorghum water extracts and salicylic acid to alleviate impacts of heat stress in wheat. S Afr J Bot 129:169–174

Ahanger MA, Hashem A, Abd-Allah EF, Ahmad P (2014) Arbuscular mycorrhiza in crop improvement under environmental stress. In: Emerging technologies and management of crop stress tolerance. Elsevier, San Diego, pp 69–95

Alam H, Khatoon N, Raza M, Ghosh PC, Sardar M (2019) Synthesis and characterization of nano selenium using plant biomolecules and their potential applications. BioNanoScience 9:96–104

Ali S, Kim W-C (2019) A fruitful decade using synthetic promoters in the improvement of transgenic plants. Front Plant Sci 10:1433

Ali H, Tanveer A, Nadeem M, Asghar H, Javaid M (2013) Ecologia da germinação de Rhynchosia capitata: planta daninha emergente durante o verão na Ásia. Planta Daninha 31:249–257

Ali HH, Tanveer A, Naeem M, Jamil M, Iqbal M, Javaid MM, Kashif MS (2015) Efficacy of pre-emergence herbicides in controlling Rhynchosia capitata, an emerging summer weed in Pakistan. Philipp Agric Sci 98(3):301–311

Ali A, Abbas MN, Maqbool MM, Arshad MI, Qayyum A, Lee D (2017a) Optimizing the doses of moringa (Moringa oleifera) leaf extract for salt tolerance in maize. Philipp J Crop Sci 42:1–6

Ali S, Khan AL, Ali L, Rizvi TS, Khan SA, Hussain J, Hamayun M, Al-Harrasi A (2017b) Enzyme inhibitory metabolites from endophytic Penicillium citrinum isolated from Boswellia sacra. Arch Microbiol 199:691–700

Allen DJ, Ort DR (2001) Impacts of chilling temperatures on photosynthesis in warm-climate plants. Trends Plant Sci 6(1):36–42

Anjum SA, Xie X, Wang LC, Saleem MF, Man C, Lei W (2011) Morphological, physiological and biochemical responses of plants to drought stress. Afr J Agric Res 6(9):2026–2032

Arif Y, Sami F, Siddiqui H, Bajguz A, Hayat S (2020) Salicylic acid in relation to other phytohormones in plant: a study towards physiology and signal transduction under challenging environment. Environ Exp Bot 175:104040

Asada K (2006) Production and scavenging of reactive oxygen species in chloroplasts and their functions. Plant Physiol 141(2):391–396

Asante WJ, Nasare IL, Tom-Dery D, Ochire-Boadu K, Kentil KB (2014) Nutrient composition of Moringa oleifera leaves from two agro ecological zones in Ghana. University for Development Studies, Tamale, Ghana, UDS space.

Asif T, Javaid M, Abdul K, Abbas R, Ahsan A (2012) Allelopathic effect of Echinochloa crus-galli on field crops. Herbologia 13(2):9–17

Baber B, Tanveer A, Aziz A, Javaid M, Tahir M, Sibtain M, Khaliq A, Pacanoski Z (2009) Phytotoxic influences of Asphodelus tenuifolius Cav (wild onion) on germination and seedling growth of wheat. Allelopathy J 24(2):341–350

Balal RM, Khan MM, Shahid MA, Abbas T, Aquel MA, Javaid M, Mattson N (2014) Foliar spray of phyto-extracts supplemented with silicon: an efficacious strategy to extenuate the salinity-induced deleterious effects in pea (*Pisum sativum* L.). In: 2014 ASHS annual conference, 2014

Basra S, Iftikhar M, Afzal I (2011) Potential of moringa (Moringa oleifera) leaf extract as priming agent for hybrid maize seeds. Int J Agric Biol 13(6):1006–1010

Batool A, Wahid A, Abbas G, Shah SH, Naeem M, Akhtar NP, Hassnain Z (2019) Application of Moringa oleifera plant extracts for enhancing the concentration of photosynthetic pigments leading to stable photosynthesis under heat stress in maize (Zea mays L.). Pak J Bot 51:3021–3026

Bhat MA, Kumar V, Bhat MA, Wani IA, Dar FL, Farooq I, Bhatti F, Koser R, Rahman S, Jan AT (2020) Mechanistic insights of the interaction of plant growth-promoting rhizobacteria (PGPR) with plant roots toward enhancing plant productivity by alleviating salinity stress. Front Microbiol 11:1952

Bhusal N, Han S-G, Yoon T-M (2019) Impact of drought stress on photosynthetic response, leaf water potential, and stem sap flow in two cultivars of bi-leader apple trees (Malus× domestica Borkh.). Sci Hortic 246:535–543

Bi H, Li F, Wang H, Ai X (2019) Overexpression of transketolase gene promotes chilling tolerance by increasing the activities of photosynthetic enzymes, alleviating oxidative damage and stabilizing cell structure in Cucumis sativus L. Physiol Plant 167(4):502–515

Breitkreuz C, Herzig L, Buscot F, Reitz T, Tarkka M (2021) Interactions between soil properties, agricultural management and cultivar type drive structural and functional adaptations of the wheat rhizosphere microbiome to drought. Environ Microbiol 23(10):5866–5882

Briffa J, Sinagra E, Blundell R (2020) Heavy metal pollution in the environment and their toxicological effects on humans. Heliyon 6(9):e04691

Brown P, Saa S (2015) Biostimulants in agriculture. Front Plant Sci 6:671

Cardona T, Sedoud A, Cox N, Rutherford AW (2012) Charge separation in photosystem II: a comparative and evolutionary overview. Biochim Biophys Acta (BBA)-Bioenerg 1817(1):26–43

Colla G, Cardarelli M, Bonini P, Rouphael Y (2017) Foliar applications of protein hydrolysate, plant and seaweed extracts increase yield but differentially modulate fruit quality of greenhouse tomato. HortScience 52(9):1214–1220

Ding Y, Shi Y, Yang S (2019) Advances and challenges in uncovering cold tolerance regulatory mechanisms in plants. New Phytol 222(4):1690–1704

Dou L, Sun Y, Li S, Ge C, Shen Q, Li H, Wang W, Mao J, Xiao G, Pang C (2021) Transcriptomic analyses show that 24-epibrassinolide (EBR) promotes cold tolerance in cotton seedlings. PLoS One 16(2):e0245070

Du Jardin P (2015) Plant biostimulants: definition, concept, main categories and regulation. Sci Hortic 196:3–14

Dutta P (2018) Seed priming: new vistas and contemporary perspectives. Adv Seed Priming:3–22

El-Esawi MA, Elansary HO, El-Shanhorey NA, Abdel-Hamid AM, Ali HM, Elshikh MS (2017) Salicylic acid-regulated antioxidant mechanisms and gene expression enhance rosemary performance under saline conditions. Front Physiol 8:716

Elzaawely AA, Ahmed ME, Maswada HF, Xuan TD (2017) Enhancing growth, yield, biochemical, and hormonal contents of snap bean (*Phaseolus vulgaris* L.) sprayed with moringa leaf extract. Arch Agron Soil Sci 63(5):687–699

Ertani A, Mietto A, Borin M, Nardi S (2017) Chromium in agricultural soils and crops: a review. Water Air Soil Pollut 228:1–12

Fancy NN, Bahlmann AK, Loake GJ (2017) Nitric oxide function in plant abiotic stress. Plant Cell Environ 40(4):462–472

Farooq M, Wahid A, Kobayashi N, Fujita D, Basra S (2009) Plant drought stress: effects, mechanisms and management. Sustain Agric:153–188
Giauque H, Connor EW, Hawkes CV (2019) Endophyte traits relevant to stress tolerance, resource use and habitat of origin predict effects on host plants. New Phytol 221(4):2239–2249
Giri J (2011) Glycinebetaine and abiotic stress tolerance in plants. Plant Signal Behav 6(11):1746–1751
Gopalakrishnan L, Doriya K, Kumar DS (2016) Moringa oleifera: a review on nutritive importance and its medicinal application. Food Sci Human Wellness 5(2):49–56
Gupta A, Bano A, Rai S, Dubey P, Khan F, Pathak N, Sharma S (2021a) Plant growth promoting rhizobacteria (PGPR): a sustainable agriculture to rescue the vegetation from the effect of biotic stress: a review. Lett Appl NanoBioSci 10:2459–2465
Gupta A, Bano A, Rai S, Kumar M, Ali J, Sharma S, Pathak N (2021b) ACC deaminase producing plant growth promoting rhizobacteria enhance salinity stress tolerance in Pisum sativum. 3. Biotech 11(12):514
Habib A, Mahmood A, Arshad WR, Umer M, Bilal M, Rasul F, Imran-ul-Haq I, Usman M (2016) Crop losses and varietal response of sugarcane against red rot disease in Pakistan's Punjab province. Int Sugar J 118(1411):522–526
Hamid B, Zaman M, Farooq S, Fatima S, Sayyed R, Baba ZA, Sheikh TA, Reddy MS, El Enshasy H, Gafur A (2021) Bacterial plant biostimulants: a sustainable way towards improving growth, productivity, and health of crops. Sustainability 13(5):2856
Hanafy R (2017) Using Moringa olifera leaf extract as a bio-fertilizer for drought stress mitigation of Glycine max L. plants. Egypt J Bot 57(2):281–292
Hanaka A, Ozimek E, Reszczyńska E, Jaroszuk-Ściseł J, Stolarz M (2021) Plant tolerance to drought stress in the presence of supporting bacteria and fungi: an efficient strategy in horticulture. Horticulturae 7(10):390
Hanif Z, Naeem M, Ali HH, Tanveer A, Javaid MM, Peerzada AM, Chauhan BS (2017) Effect of environmental factors on germination of Salsola foetida: potential species for rehabilitation of degraded rangelands. Rangeland Ecol Manage 70(5):638–643
He M, He C-Q, Ding N-Z (2018) Abiotic stresses: general defenses of land plants and chances for engineering multistress tolerance. Front Plant Sci 9:1771
Horváth I, Glatz A, Nakamoto H, Mishkind ML, Munnik T, Saidi Y, Goloubinoff P, Harwood JL, Vigh L (2012) Heat shock response in photosynthetic organisms: membrane and lipid connections. Prog Lipid Res 51(3):208–220
Howarth C (2005) Genetic improvements of tolerance to high temperature. In: Abiotic stresses: plant resistance through breeding and molecular approaches. Howarth Press Inc, New York, pp 277–300
Howladar SM (2014) A novel Moringa oleifera leaf extract can mitigate the stress effects of salinity and cadmium in bean (Phaseolus vulgaris L.) plants. Ecotoxicol Environ Saf 100:69–75
Hu M, Shi Z, Xu P, Li H, Zhang Z (2015) Wheat acclimate to water deficit by modifying carbohydrates metabolism, water use efficiency, and growth. Rev Bras Bot 38:505–515
Ikram RM, Tanveer A, Ali HH, Safdar ME, Javaid MM, Abbas T, Hashim S, Chauhan BS (2019) Germination Ecology of Two Troublesome Weeds of Arid Chickpea: Euphorbia dracunculoides and Astragalus Species. Planta Daninha 37:1–15. https://doi.org/10.1590/S0100-83582019370100049
Ilyas M, Arshad M, Saeed F, Iqbal M (2015) Antioxidant potential and nutritional comparison of Moringa leaf and seed powders and their tea infusions. J Anim Plant Sci 25(1):226–233
Jabeen M, Akram NA, Ashraf M, Aziz A (2019) Assessment of biochemical changes in spinach (Spinacea oleracea L.) subjected to varying water regimes. Sains Malaysiana 48(3):533–541
Jan S, Rashid B, Azooz M, Hossain MA, Ahmad P (2016) Genetic strategies for advancing phytoremediation potential in plants: a recent update. Plant Metal Interact:431–454
Janská A, Maršík P, Zelenková S, Ovesná J (2010) Cold stress and acclimation–what is important for metabolic adjustment? Plant Biol 12(3):395–405

Javaid M, Tanveer A (2014) Germination ecology of E mex spinosa and E mex australis, invasive weeds of winter crops. Weed Res 54(6):565–575
Javaid MM, Tanveer A, Ali H, Shahid M, Balal R, Aqeel M (2016) Wheat yield loss in a two species competition with Emex australis and Emex spinosa. Planta Daninha 34:35–46
Javaid MM, Abbas Z, Waheed H, Mahmood A, Abbas T, Amin MM, Asif M (2020) Studies on germination ecology and seedling characteristics of Cleome viscosa as affected by various environmental factors. Pak J Bot 52(6):1911–1919
Javaid MM, AlGwaiz HI, Waheed H, Ashraf M, Mahmood A, Li F-M, Attia KA, Nadeem MA, AlKahtani MD, Fiaz S (2022a) Ridge-furrow mulching enhances capture and utilization of rainfall for improved maize production under rain-fed conditions. Agronomy 12(5):1187
Javaid MM, Florentine SK, Ashraf M, Mahmood A, Sattar A, Wasaya A, Li FM (2022b) Photosynthetic activity and water use efficiency of Salvia verbenaca L. under elevated CO2 and water-deficit conditions. J Agron Crop Sci 208(4):536–551
Javaid MM, Mahmood A, Nadeem MA, Sarwar N, Safdar ME, Ahmad M, Hasanuzzaman M, Ahmad S (2022c) Weed management strategies in direct seeded rice. In: Modern techniques of rice crop production. Springer, Singapore, pp 327–347
Javed B, Raja NI, Nadhman A, Mashwani Z-u-R (2020) Understanding the potential of biofabricated non-oxidative silver nanoparticles to eradicate Leishmania and plant bacterial pathogens. Appl Nanosci 10:2057–2067
Jin C, Fan J, Liu R, Sun R (2015) Single and joint toxicity of sulfamonomethoxine and cadmium on three agricultural crops. Soil Sediment Contam Int J 24(4):454–470
Khalofah A, Bokhari N, Migdadi H, Alwahibi M (2020) Antioxidant responses and the role of Moringa oleifera leaf extract for mitigation of cadmium stressed Lepidium sativum L. S Afr J Bot 129:341–346
Khan MN, Zhang J, Luo T, Liu J, Ni F, Rizwan M, Fahad S, Hu L (2019) Morpho-physiological and biochemical responses of tolerant and sensitive rapeseed cultivars to drought stress during early seedling growth stage. Acta Physiol Planta 41:1–13
Koo YM, Heo AY, Choi HW (2020) Salicylic acid as a safe plant protector and growth regulator. Plant Pathol J 36(1):1
Kränzlein M, Geilfus C-M, Franzisky BL, Zhang X, Wimmer MA, Zörb C (2022) Physiological responses of contrasting maize (Zea mays L.) hybrids to repeated drought. J Plant Growth Regul 41(7):2708–2718
Krishnamoorthy R, Anandham R, Senthilkumar M, Venkatramanan V (2021) Adaptation mechanism of methylotrophic bacteria to drought condition and its strategies in mitigating plant stress caused by climate change. In: Exploring synergies and trade-offs between climate change and the sustainable development goals. Springer, Singapore, pp 145–158
Kumar A, Verma JP (2018) Does plant—microbe interaction confer stress tolerance in plants: a review? Microbiol Res 207:41–52
Kumar V, Khare T, Sharma M, Wani SH (2017) ROS-induced signaling and gene expression in crops under salinity stress. In: Reactive oxygen species and antioxidant systems in plants: role and regulation under abiotic stress. Springer, Singapore, pp 159–184
Lamichhane JR, Debaeke P, Steinberg C, You MP, Barbetti MJ, Aubertot J-N (2018) Abiotic and biotic factors affecting crop seed germination and seedling emergence: a conceptual framework. Plant Soil 432:1–28
Latif H, Mohamed H (2016) Exogenous applications of moringa leaf extract effect on retrotransposon, ultrastructural and biochemical contents of common bean plants under environmental stresses. S Afr J Bot 106:221–231
Liliane TN, Charles MS (2020) Factors affecting yield of crops. In: Agronomy-climate change & food security. IntechOpen, London, p 9
Luziatelli F, Ficca AG, Colla G, Baldassarre Švecová E, Ruzzi M (2019) Foliar application of vegetal-derived bioactive compounds stimulates the growth of beneficial bacteria and enhances microbiome biodiversity in lettuce. Front Plant Sci 10:60

Manna M, Thakur T, Chirom O, Mandlik R, Deshmukh R, Salvi P (2021) Transcription factors as key molecular target to strengthen the drought stress tolerance in plants. Physiol Plant 172 (2):847–868

Mirshekali H, Hadi H, Amirnia R, Khodaverdiloo H (2012) Effect of zinc toxicity on plant productivity, chlorophyll and Zn contents of sorghum (Sorghum bicolor) and common lambsquarter (Chenopodium album). Int J Agric Res 2(3):247–254

Miura K, Tada Y (2014) Regulation of water, salinity, and cold stress responses by salicylic acid. Front Plant Sci 5:4

Munns R, Tester M (2008) Mechanisms of salinity tolerance. Annu Rev Plant Biol 59:651–681

Naylor D, Coleman-Derr D (2018) Drought stress and root-associated bacterial communities. Front Plant Sci 8:2223

Nguyen LT, Osanai Y, Anderson IC, Bange MP, Tissue DT, Singh BK (2018) Flooding and prolonged drought have differential legacy impacts on soil nitrogen cycling, microbial communities and plant productivity. Plant Soil 431:371–387

Ohira S, Morita N, Suh H-J, Jung J, Yamamoto Y (2005) Quality control of photosystem II under light stress–turnover of aggregates of the D1 protein in vivo. Photosynth Res 84:29–33

Osakabe Y, Osakabe K, Shinozaki K, Tran L-SP (2014) Response of plants to water stress. Front Plant Sci 5:86

Özcan M (2020) Moringa spp: composition and bioactive properties. S Afr J Bot 129:25–31

Ozturk M, Turkyilmaz Unal B, García-Caparrós P, Khursheed A, Gul A, Hasanuzzaman M (2021) Osmoregulation and its actions during the drought stress in plants. Physiol Plant 172(2):1321–1335

Pál M, Szalai G, Janda T (2015) Speculation: polyamines are important in abiotic stress signaling. Plant Sci 237:16–23

Pervez K, Ullah F, Mehmood S, Khattak A (2017) Effect of Moringa oleifera Lam. leaf aqueous extract on growth attributes and cell wall bound phenolics accumulation in maize (Zea mays L.) under drought stress. Kuwait J Sci 44(4):110

Podile A, Kishore G (2006) Plant growth-promoting rhizobacteria. Plant-associated bacteria:195–230

Prittesh P, Avnika P, Kinjal P, Jinal HN, Sakthivel K, Amaresan N (2020) Amelioration effect of salt-tolerant plant growth-promoting bacteria on growth and physiological properties of rice (Oryza sativa) under salt-stressed conditions. Arch Microbiol 202:2419–2428

Rani V, Bhatia A, Kaushik R (2021) Inoculation of plant growth promoting-methane utilizing bacteria in different N-fertilizer regime influences methane emission and crop growth of flooded paddy. Sci Total Environ 775:145826

Rehman Hu, Nawaz Q, Basra SMA, Afzal I, Yasmeen A, ul-Hassan F (2014) Seed Priming Influence on Early Crop Growth, Phenological Development and Yield Performance of Linola (Linum usitatissimum L.). J Integr Agric 13 (5):990–996. https://doi.org/10.1016/S2095-3119(13)60521-3

Rehman Iu, Ishaq M, Ali L, Muhammad S, Din IU, Yaseen M, Ullah H (2020) Potentially toxic elements' occurrence and risk assessment through water and soil of Chitral urban environment, Pakistan: a case study. Environ Geochem Health 42(12):4355–4368. https://doi.org/10.1007/s10653-020-00531-4

Rehman AU, Nazir S, Irshad R, Tahir K, ur Rehman K, Islam RU, Wahab Z (2021) Toxicity of heavy metals in plants and animals and their uptake by magnetic iron oxide nanoparticles. J Mol Liq 321:114455

Richter K, Haslbeck M, Buchner J (2010) The heat shock response: life on the verge of death. Mol Cell 40(2):253–266

Rouphael Y, Colla G (2018) Synergistic biostimulatory action: designing the next generation of plant biostimulants for sustainable agriculture. Front Plant Sci 9:1655

Rouphael Y, Cardarelli M, Schwarz D, Franken P, Colla G (2012) Effects of drought on nutrient uptake and assimilation in vegetable crops. In: Plant responses to drought stress: from morphological to molecular features. Springer, Heidelberg, pp 171–195

Rouphael Y, Spíchal L, Panzarová K, Casa R, Colla G (2018) High-throughput plant phenotyping for developing novel biostimulants: from lab to field or from field to lab? Front Plant Sci 9:1197

Saed-Moucheshi A, Razi H, Dadkhodaie A, Ghodsi M, Dastfal M (2019) Association of biochemical traits with grain yield in triticale genotypes under normal irrigation and drought stress conditions. Aust J Crop Sci 13(2):272–281

Saini S, Kaur N, Pati PK (2018) Reactive oxygen species dynamics in roots of salt sensitive and salt tolerant cultivars of rice. Anal Biochem 550:99–108

Seleiman MF, Al-Suhaibani N, Ali N, Akmal M, Alotaibi M, Refay Y, Dindaroglu T, Abdul-Wajid HH, Battaglia ML (2021) Drought stress impacts on plants and different approaches to alleviate its adverse effects. Plants 10(2):259

Shafi M, Bakht J, Hassan MJ, Raziuddin M, Zhang G (2009) Effect of cadmium and salinity stresses on growth and antioxidant enzyme activities of wheat (Triticum aestivum L.). Bull Environ Contam Toxicol 82:772–776

Sharma A, Kumar V, Yuan H, Kanwar MK, Bhardwaj R, Thukral AK, Zheng B (2018) Jasmonic acid seed treatment stimulates insecticide detoxification in Brassica juncea L. Front Plant Sci 9:1609

Shen J-R (2015) The structure of photosystem II and the mechanism of water oxidation in photosynthesis. Annu Rev Plant Biol 66:23–48

Singh HP, Mahajan P, Kaur S, Batish DR, Kohli RK (2013) Chromium toxicity and tolerance in plants. Environ Chem Lett 11:229–254

Stasio ED, Rouphael Y, Colla G, Raimondi G, Giordano M, Pannico A, El-Nakhel C, Pascale SD (2017) The influence of Ecklonia maxima seaweed extract on growth, photosynthetic activity and mineral composition of Brassica rapa L. subsp. sylvestris under nutrient stress conditions. Eur J Hortic Sci 82(6):286–293

Takagi D, Takumi S, Hashiguchi M, Sejima T, Miyake C (2016) Superoxide and singlet oxygen produced within the thylakoid membranes both cause photosystem I photoinhibition. Plant Physiol 171(3):1626–1634

Takahashi S, Murata N (2008) How do environmental stresses accelerate photoinhibition? Trends Plant Sci 13(4):178–182

Tanveer A, Mumtaz K, Javaid M, Chaudhry M, Balal R, Khaliq A (2013) Effect of ecological factors on germination of horse purslane (Trianthema portulacastrum). Planta Daninha 31:587–597

Tanveer A, Javaid MM, Abbas RN, Ali HH, Nazir MQ, Balal RM, Shahid MA (2017) Ecologia de Germinação de Sementes de Silene (Silene conoidea) de Diferentes Cores. Planta Daninha 35:1–7. https://doi.org/10.1590/S0100-83582017350100030

Tavallali V, Karimi S (2019) Methyl jasmonate enhances salt tolerance of almond rootstocks by regulating endogenous phytohormones, antioxidant activity and gas-exchange. J Plant Physiol 234:98–105

Tilley MS, Heiniger RW, Crozier CR (2019) Tiller initiation and its effects on yield and yield components in winter wheat. J Agron 111(3):1323–1332

Vocciante M, Grifoni M, Fusini D, Petruzzelli G, Franchi E (2022) The role of plant growth-promoting rhizobacteria (PGPR) in mitigating plant's environmental stresses. Appl Sci 12(3):1231

Wakeel A, Xu M, Gan Y (2020) Chromium-induced reactive oxygen species accumulation by altering the enzymatic antioxidant system and associated cytotoxic, genotoxic, ultrastructural, and photosynthetic changes in plants. Int J Mol Sci 21(3):728

Wani A, Ahmad A, Hayat S, Tahir I (2016) Is foliar spray of proline sufficient for mitigation of salt stress in Brassica juncea cultivars? Environ Sci Pollut Res 23:13413–13423

Welgama A, Florentine S, Marchante H, Javaid MM, Turville C (2019) The germination success of Acacia longifolia subsp. longifolia (Fabaceae): a comparison between its native and exotic ranges. Aust J Bot 67(5):414–424

Wright RJ, Lee KS, Hyacinth HI, Hibbert JM, Reid ME, Wheatley AO, Asemota HN (2017) An investigation of the antioxidant capacity in extracts from Moringa oleifera plants grown in Jamaica. Plants 6(4):48

Wu X, He J, Chen J, Yang S, Zha D (2014) Alleviation of exogenous 6-benzyladenine on two genotypes of eggplant (Solanum melongena Mill.) growth under salt stress. Protoplasma 251:169–176

Yadav S, Irfan M, Ahmad A, Hayat S (2011) Causes of salinity and plant manifestations to salt stress: a review. J Environ Biol 32(5):667

Yamamoto Y, Aminaka R, Yoshioka M, Khatoon M, Komayama K, Takenaka D, Yamashita A, Nijo N, Inagawa K, Morita N (2008) Quality control of photosystem II: impact of light and heat stresses. Photosynth Res 98:589–608

Yasmeen A, Basra S, Farooq M, Rehman HU, Hussain N, Athar HUR (2013) Exogenous application of moringa leaf extract modulates the antioxidant enzyme system to improve wheat performance under saline conditions. Plant Growth Regul 69:225–233

You J, Zhang Y, Liu A, Li D, Wang X, Dossa K, Zhou R, Yu J, Zhang Y, Wang L (2019) Transcriptomic and metabolomic profiling of drought-tolerant and susceptible sesame genotypes in response to drought stress. BMC Plant Biol 19(1):1–16

Zhang S, Fan C, Wang Y, Xia Y, Xiao W, Cui X (2018) Salt-tolerant and plant-growth-promoting bacteria isolated from high-yield paddy soil. Can J Microbiol 64(12):968–978

Zhang L, Guo X, Zhang Z, Wang A, Zhu J (2021) Cold-regulated gene LeCOR413PM2 confers cold stress tolerance in tomato plants. Gene 764:145097

Zhao J, Lu Z, Wang L, Jin B (2020) Plant responses to heat stress: physiology, transcription, noncoding RNAs, and epigenetics. Int J Mol Sci 22(1):117

Zia R, Nawaz MS, Siddique MJ, Hakim S, Imran A (2021) Plant survival under drought stress: implications, adaptive responses, and integrated rhizosphere management strategy for stress mitigation. Microbiol Res 242:126626

Zörb C, Geilfus CM, Dietz KJ (2019) Salinity and crop yield. Plant Biol 21:31–38

Zulfiqar F, Casadesús A, Brockman H, Munné-Bosch S (2020) An overview of plant-based natural biostimulants for sustainable horticulture with a particular focus on moringa leaf extracts. Plant Sci 295:110194

Chapter 40
Bioremediation and Phytoremediation Aspects of Crop Improvement

Nadeem Iqbal, Malik Tahir Hayat, Muhammad Dawood, Waqar-Un Nisa, Shamim Akhtar, Nazneen Bangash, Abdul Nazeer, and Muhammad Ahsan Asghar

Abstract Environmental pollution due to toxic heavy metals and organic pollutants has become a global concern because of its adverse effects on plants, animals, as well as humans via food chain contamination. However, the elimination of these harmful contaminants from the polluted media requires abundant energy and complex engineering techniques. Bioremediation offers the best possible solution to reduce or remove these noxious pollutants with minimum energy costs. Microremediation, mycoremediation, and cyanoremediation have effectively eliminated or degraded toxic contaminants and resultantly promoted plant growth and development. Phytoremediation is also an environmentally friendly technique to remove pernicious pollutants using different mechanisms such as phytostabilization, phytofiltration, phytodegradation, phytoextraction, and phytostimulation.

N. Iqbal
Department of Environmental Sciences, COMSATS University Islamabad, Abbottabad Campus, Islamabad, Pakistan

Doctoral School of Environmental Sciences, University of Szeged, Szeged, Hungary

M. T. Hayat (✉) · A. Nazeer
Department of Environmental Sciences, COMSATS University Islamabad, Abbottabad Campus, Islamabad, Pakistan
e-mail: mtahir@cuiatd.edu.pk

M. Dawood
Department of Environmental Sciences, Bahauddin Zakariya University, Multan, Pakistan

W.-U. Nisa
Center for Interdisciplinary Research in Basic Sciences, International Islamic University, Islamabad, Pakistan

S. Akhtar
Department of Botany, University of Gujrat, Gujrat, Pakistan

N. Bangash
Department of Biosciences, COMSATS University Islamabad, Islamabad, Pakistan

M. A. Asghar
Department of Biological Resources, Agricultural Institute, Centre for Agricultural Research, ELKH, Martonvásár, Hungary

M. Hasanuzzaman (ed.), *Climate-Resilient Agriculture, Vol 2*,
https://doi.org/10.1007/978-3-031-37428-9_40

Further, the efficiency of phytoremediation can be enhanced using electric current, chelating agents, plant growth regulators, soil amendments, co-cropping, and bacterial inoculation resulting in more removal of pollutants due to high biomass production. Furthermore, plants deploy various stress tolerance mechanisms to mitigate the detrimental effects of harmful pollutants and maintain cellular redox homeostasis. The tolerance mechanisms exhibit the maintenance of plant physiological and biochemical processes and promote plant growth and crop improvement through various protective approaches. Phytoremediation is not only used to remove harmful pollutants but also offers the purification of soil, water, and air. In addition, phytoremediation is a green technology and economically viable. The efficacy of phytoremediation depends upon various factors including soil characteristics, type of pollutant, presence of organic matter, and properties of microflora in the rhizosphere. Additionally, disposal methods for the toxic plant waste are also of significant importance in the proper transport and dumping without any leakage of leachate from the disposal sites. However, the knowledge of the interaction between plants and microbial communities in the rhizosphere regarding the efficient removal of toxic pollutants has not been investigated fully. Therefore, a comprehensive study is still needed to investigate the involved mechanisms and more advanced technologies should be developed to purge toxic contaminants efficiently.

Keywords Bioremediation · Ecosystem services · Heavy metals · Organic pollutants · Phytoremediation · Stress tolerance

1 Introduction

Industrialization and urbanization have created many environmental problems globally. Although less production of toxic substances during different operational activities is supported worldwide, the industrial revolution and rapid urbanization have led to contamination of soil, water, and air worldwide (Azhdarpoor et al. 2014; Rostami and Azhdarpoor 2019). Soil contamination is a global concern and the toxic compounds are found in combined forms and rarely in separate forms in the environment. Further, the presence of toxic heavy metals and polycyclic aromatic hydrocarbons (PAHs) in the soil is very common (Ebrahimi et al. 2018). These toxic metals exist in ionic or colloidal forms and are commonly soil soluble but cannot be degraded like other organic compounds into less toxic ones. These metals can remain in the soil for a longer period of time with a half-life of more than 20 years (Ruiz et al. 2009). However, the toxicity of these metals depends upon their concentration ranging from 1 to 100,000 mg kg^{-1} (Karami and Shamsuddin 2010). There are various sources of these toxic metals contaminations including drilling, melting, military activities, industrial projects, and fuel production which pose adverse impacts on human health via food chain contamination (Azhdarpoor et al. 2015). The environmental degradation due to heavy metal pollution has affected more than 10 million people worldwide (Ahmadpour et al. 2012).

Similarly, PAHs are also noxious compounds with two or more benzene rings and contaminate soil through different activities such as asphalt production, petrochemical industry, aluminum production, incomplete combustion, and creosote industry. Most of the compounds are mutagens and carcinogens posing human health risks (Tsibart and Gennadiev 2013). On the other side, high-energy and more advanced engineering processes are required to purge these toxic compounds. In addition, the outcome of these processes is not satisfactory, and later the composition and biochemical properties of soil. Nevertheless, numerous methods are being employed to treat contaminated soils. Bioventing is also a method to decompose toxic compounds using microbes in the soil but the maintenance of biological conditions plays a crucial part in the process (Kalantary et al. 2014). The removal of contaminants by providing heat is another method to reclaim polluted soils and is known as thermal reclamation. During this process, generated heat will result in the evaporation of contaminants in the gaseous phase which can be easily controlled and separated (Stegmann et al. 2013). A slurry reactor is another well-known method for the reduction of contaminants in solution form in a reactor by providing optimum suitable conditions. Further, soil reclamation is considered as the best option to reduce contamination based on decomposing activities of the microbes (Pino-Herrera et al. 2017). Lastly, phytoremediation is a promising technique for the elimination of contaminants using green plants. This technique is preferred to other conventional methods due to its cost-effectiveness, soil texture conservation, and minimizing impacts on the environment (Rostami et al. 2017; Parseh et al. 2018).

2 Sources of Contaminants in the Environment

Natural, industrial, and anthropogenic activities are the potential sources of toxic pollutants and contamination in the environment. These sources such as mining, volcanic eruptions, rock weathering, and applications of agrochemicals can enhance contaminants levels such as heavy metals and organic pollutants. These pollutants result in the reduction of soil quality and consequently pose severe impacts on human health (Latif et al. 2018). There are various sources of heavy metals and dyes such as textile, pigments, paper industry, rubber, and plastics which significantly affected environmental health. The untreated wastewater from industries is directly released into nearby water bodies directly affecting the colour and quality of water leaving harmful effects on aquatic life (Sarojini et al. 2022). Metal accumulation in plants can adversely influence plant growth and resultantly yield as well. In addition, other physiological, biochemical, and metabolic processes are also perturbed badly due to the heavy metal contamination (Saravanan et al. 2020). Several heavy metals such as nickel (Ni), manganese (Mn), lead (Pb), copper (Cu), zinc (Zn), and cadmium (Cd) have been reported to have deleterious effects on living organisms. Moreover, the emissions of arsenic (As), Pb, and Cd by air have also influenced a large area (Jacob et al. 2018). Organic pollutants are also spread in the environment via various activities such as combustion of hydrocarbons, oil spills, industrial waste, shipping activities, sewage discharges, surface sediments, and lubricants (Barletta et al. 2019). In addition, other activities such as textiles, foam, plastics, carpets, television casing, furniture, petrochemicals, and leachate can cause detrimental impacts on water, soil, and air media eventually risking human health (Anim et al. 2017).

3 Toxic Effects of Contaminants on the Environment

Heavy metal contamination can perturb various metabolic processes in plants and humans as well. Such metals via food chain contamination can accumulate in the body and cause severe disorders (Sharma et al. 2018). Due to their nonbiodegradability, heavy metal pollution is a major source of contamination (Chaoua et al. 2018). Soil contaminated with heavy metals would affect living organisms through the biomagnification process and influence their biochemical pathways. Elevated concentrations of these heavy metals are taken up by plants and translocated from roots to shoots. This heavy metal pollution results in soil degradation, oxidative burst in plants, phytotoxicity, and carcinogens for humans as well (Ghori et al. 2019). In addition, metal toxicity can inhibit plant growth, reduce yield, and perturb metabolic functions (Asati et al. 2016). Heavy metal contamination can induce the generation of reactive oxygen species (ROS) which can dismantle membranes and can also affect the microbial community by destructing soil biochemical properties (Jiang et al. 2019). Organic pollutants are biodegradable and mostly spread in the environment through anthropogenic activities resulting in pollution. There are some toxic pollutants such as polychlorinated biphenyls (PCBs), PAHs, polybrominated diphenyl ethers (PBDEs), phenols, petroleum, and organochlorine pesticides (OCPs), and other pesticides (Masindi and Muedi 2018). For instance, PCBs exhibit lipophilic characteristics, and high-temperature resistance can be part of food chain contamination posing serious health risks in humans including eye and skin irritation. However, PCB exposure can significantly inhibit plant growth (Iqbal et al. 2022).

4 Translocation of Contaminants in Plants

Plants require a trace amount of metals for metabolic processes; however, higher accumulation of heavy metals pose adverse effects on plants. Cationic transporters assist in uptake of these heavy metals from soil to roots and then translocate to aerial parts (Glavac et al. 2017). In the case of hyperaccumulators, these stored heavy metals in plant vacuoles are transported to shoot system via xylem vessels. There are various steps involved in the translocation of metals from the soil to roots and thereafter to shoot system. The first step is the metal ions-binding to cell wall. Zinc and Cu ions present in the apoplast determine the number of metal ions present in root cells. Thereafter, the interface of the cell membrane and cell wall possesses higher accumulation of heavy metals (Dalcorso et al. 2013). The cationic exchange capacity of the root determines the uptake of metal ions. The cell wall is accountable for metal uptake and immobilization (Guigues et al. 2014). The next step is the movement of metals through plasma membrane transporters to the shoot system via xylem vessels or phloem vessels either binding with proteins or chelating agents (Manara 2012). During this translocation, various transporters are involved such as ZIP family, CDE, NRAMP, Cu transporter family, P-type ATPase transporter family, multidrug and toxic compound extrusion protein family, oligopeptide transporter family, and ATP-binding cassette transporter family (Lin and Aarts 2012; Kolaj-Robin et al. 2015).

5 Accumulation of Contaminants in Plants

Heavy metal transporters are implicated in the translocation of metals from soil to aerial plant parts. Some of the transporters are also present in the plasma membrane, and these metals are transported with other nutrients to plant cells (Sun et al. 2016). There are various genes responsible for the expressions of these metal transporters. The metals complex is formed in the xylem vessels with chelating agents and then translocated to shoot (Shen et al. 2017). The hyperaccumulators translocate metals from root to leaf tissues. Different factors can affect the hyperaccumulation state such as plant species, soil temperature, organic content, and pH, as well as the type of heavy metals (Ravanbakhsh et al. 2019). Accumulation of metals occurs in three steps: metal transportation from the cell membrane and then toward the shoot region via xylem loading of metals followed by cellular accumulation of metals or detoxification (Wood et al. 2016). The metal deposition rate of plants is based on uptake capacity and metals translocation in plants (Sassykova et al. 2020).

Heavy metal accumulation occurs first in plant roots. Some plants can absorb or adsorb heavy metals by roots and then transport them to aerial parts while some plants do not translocate to upper parts (Fuente et al. 2016). In addition, some plant species can accumulate heavy metals in specific compartments to prevent the toxic effects of contaminants while some store metals in plant tissues with modified oxidation states (Wei et al. 2015; Ying et al. 2019). The second accumulating site is the plant shoot system where metals are transported through xylem vessels. Different factors can influence the uptake efficacy of metals such as the plant defense system, the effectiveness of the transport system, plant physiology, and vacuoles composition (AbdElgawad et al. 2020). The metal translocation from root to shoot (leaf and stem tissues) depends upon the atmospheric conditions, metal bioavailability, soil pH, contamination level, toxicity, metal transfer, and soil texture as well (Johnsen and Aaneby 2019).

6 Bioremediation Techniques

Bioremediation refers to the contaminant removal from the polluted environment using living organisms. Heavy metals and other organic compounds are directly or indirectly released into the environment and influence human health as well as animals and plant resulting in crop productivity and yield losses. Soil degradation is also one of the detrimental consequences of these harmful contaminants. However, different bioremediation approaches such as micro-remediation, mycoremediation, cyanoremediation, and phytoremediation are extensively used to reclaim contaminated soils (Azubuike et al. 2016; Yaashikaa et al. 2022).

The main objective of bioremediation is to eliminate toxic metals and harmful organic substances from the soil in order to decrease their bioavailability. However, it is impossible to degrade heavy metals but due to their changed physical and chemical characteristics, these metals can be transformed into other less toxic forms. Nonetheless, the efficacy of bioremediation depends upon the nature of toxic chemical and plant specie, microbes, and fungi. There are numerous mechanisms for microbial remediation such as complex formation extracellularly, accumulation intracellularly, redox reactions, and precipitation of toxic pollutants (Yang et al. 2018). In parallel, microbes also utilize a biosorption mechanism for the elimination of contaminants. The immobilization of heavy metals or other organic compounds is facilitated by extracellular materials and consequently results in the binding of the cell surface with ionic functional groups (Ayangbenro and Babalola 2017; Etesami 2018). Moreover, siderophores play an essential role in microbe-assisted remediation. These siderophores are released by fungi, bacteria, and plants as well and function in binding with heavy metals because of their chelating nature resulting in the reduction of metal toxicity (Saha et al. 2016). Additionally, microbes also produce biosurfactants involved in metal complexation (Sarubbo et al. 2015). This complexity of metals causes the solubilization of metals and reduces metals in the final medium facilitating insolubility of metal salts. Microorganisms can use both enzymatic and non-enzymatic processes to eliminate heavy metals from contaminated media (soil, air, and water).

6.1 Micro-remediation

Micro-remediation is used to degrade harmful and toxic compounds into less toxic substances using different microbes, especially bacteria. This process is carried out by specific enzymes secreted by microbes, and the final products will be no toxic or less harmful compounds. Microbes utilize two pathways for defense responses against these toxic substances either by producing enzymes or through opposing the toxic metals. In addition, microbes also have the ability to eliminate toxic pollutants through different approaches including volatilization, adhesion, and transformation (Alvarez et al. 2017; Iqbal et al. 2020). Moreover, other mechanisms could also be involved such as mineralization, accumulation, biosorption, bioleaching, and microbial interactions with metals. Higher concentrations of metals or other organic compounds result in damage to the plasma membrane and induce defense responses to mitigate that damage. Further, gene editing in microbes using recombinant DNA technology has made it easier to improve contaminant removal efficacy by genetically modified microorganisms (Verma and Kuila 2019). Therefore, microbes with altered genetic makeup have shown effective results in contaminant removal from air, soil, and water. Molecular mechanisms can also efficiently increase the microbial potential to remove pollutants from different types of contaminated media. However, the limitations of utilizing microbes with genetic modifications are the mass production of effective strains and their survival in natural environments. In addition, such modified microbes could also negatively affect native microflora.

6.2 Cyanoremediation

This process is used to remove hazardous pollutants including heavy metals, pesticides, and dyes by using algae or cyanobacteria from contaminated media. Various algal species such as *Chlorella*, *Spirogyra*, Spirulina, and *Oedogonium* are being used for the amelioration of contaminated sites. However, studies also displayed that the higher levels of heavy metal accumulation in Spirulina resulted in the inhibition of its growth (Balaji et al. 2014). Despite this, other algal species have been used widely because of their availability, role in CO_2 fixation, and source of biofuel. In the case of cyanobacteria, two mechanisms have been identified for the removal of toxic compounds. Firstly, the physical adsorption of contaminants by outer cells, and the second is the accumulation inside the cells. Nevertheless, it is a time-taking process because it required time for the binding and transport of targeted proteins (Singh et al. 2019).

6.3 Mycoremediation

Various scientific reports have documented the use of certain fungal species for the elimination of toxic heavy metals as well as organic compounds. These fungal species include *Aureobasidium*, *Candida*, *Mortierella*, *Cladosporium*, *Emericella*, *Aspergillus*, *Fusarium*, *Gliocladium*, and *Ganoderma*. These fungi have the capability to release enzymes for the degradation of pollutants even in the absence of nutrient-rich environments (Mani and Kumar 2014). Interestingly, the end products of mycoremediation are nontoxic or less toxic in nature. The longer filamentous fungi assist in the quicker removal of toxic substances as compared to bacteria. Certain fungal species can also be used as a biological control in agriculture by decomposing agriculture waste such as *Trichoderma* species. Some other fungi such as *Funneliformis geosporum* can also increase soil fertility and results in higher crop production. Moreover, fungal species have also been noticed to enhance metal adsorption by roots (Khan et al. 2019). To adsorb metals, fungal species utilize two mechanisms either sequestration or avoidance. The avoidance strategy is used to reduce metal concentration via biosorption, uptake, and precipitation and results in toxicity, while chelation occurs in the processes of sequestration intracellularly (Akhtar and Mannan 2020). Hence, mycoremediation is considered an effective approach for treating heavy metals and other organic compounds. Nevertheless, less concentration of contaminants is adsorbed using these hyperaccumulators due to less biomass production, but fungi can remove or accumulate higher toxic compounds due to their diverse molecular mechanisms. There are several factors affecting mycoremediation such as pH, fungal species, growth conditions, temperature, and redox potential (Kumar et al. 2019).

6.4 Phytoremediation

In the process of phytoremediation, plants are used to remove toxic contaminants to mitigate the harmful repercussions of heavy metals, and organic compounds. To remove the organic contaminants, several mechanisms include degradation, stabilization, rhizoremediation, and volatilization but for heavy metals sequestration and extraction are utilized by plants (Patra et al. 2020; Saleem et al. 2022). As compared to other bioremediation techniques, phytoremediation is a more effective, feasible, and simpler technique. There are certain characteristics that are required for a plant to be a hyperaccumulator such as tap or fibrous roots based on pollution depth, toxic levels of the contaminant, plant growth rate, plant adaptation to specific ecological conditions, and resistant plants to diseases. The main steps involved in the removal of toxic contaminants are uptake of pollutants by roots, transportation via the xylem to upper plant parts and then the accumulation of pollutants in plant shoot (Shah and Daverey 2020). Plants that can grow in polluted sites and remove toxic pollutants are termed as phytoremediators. The addition of plant growth promoting bacteria also plays a crucial role in biomass production by enhancing plants' ability to tolerate high concentrations of pollutants. Phytoremediation is an environmentally safe approach and can be applied with other traditional approaches to eliminate pollutants from polluted sites (Khalid et al. 2017). Phytoremediation can be categorized into five groups such as phytostimulation, phytofiltration, phytotransformation, phytoextraction, and phytostabilization (Fig. 40.1; Ramanjaneyulu et al. 2017; Ashraf et al. 2019).

6.4.1 Phytostimulation

Phytostimulation refers to the breakdown of organic compounds into less toxic and simple compounds in the presence of microbial activities (Souto et al. 2020). In the rhizosphere of plants, the microbial activities can be increased in various ways such

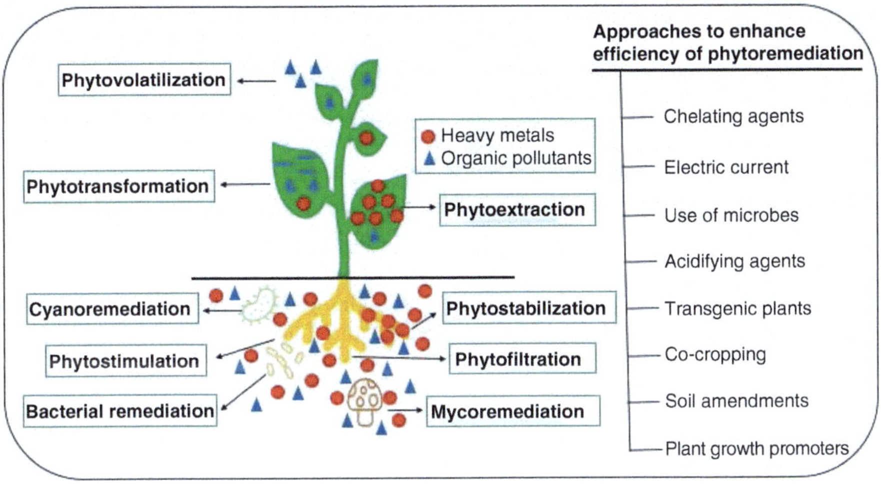

Fig. 40.1 Different types of bioremediation and phytoremediation processes along with approaches to increase the efficacy of phytoremediation of heavy metals and organic pollutants

as root exudates can supply amino acids and carbohydrates to rhizospheric microbes, and oxygen is also provided by roots to enhance the aerobic transformation of pollutants. Further, increased biomass production of roots can also elevate the availability of organic carbon and rhizospheric fungi (mycorrhizae) can degrade organic compounds which are not disintegrated by bacteria. Furthermore, plant also acts as a habitat for microbial populations (Yadav et al. 2010; Sophia and Shetty Kodialbail 2020).

6.4.2 Phytofiltration

This type of phytoremediation is used to remove harmful pollutants by using plant roots from contaminated groundwater or wastewater (Bokhari et al. 2022). In this technique, contaminated water is provided to plants for a while, then plants are shifted to polluted sites to remove toxic pollutants. In addition, plants are harvested when their roots are fully saturated. Phytofiltration can be subdivided depending upon which plant organ or part is being used for the removal of contaminants; for instance, plant roots (rhizofiltration), plant seedlings (blastofiltration), and plant shoots (caulofiltration) (da Conceição Gomes et al. 2016). This method is used to mitigate the impacts of toxic contaminants on the groundwater by absorption, adsorption, and precipitation. The precipitation of pollutants occurs when the pH of the rhizospheric zone is altered by root exudates (Javed et al. 2019). Interestingly, for the purpose of rhizofiltration, terrestrial as well as aquatic plants can be employed for the extraction of different toxic heavy metals including Cd, Pb, Ni, Cu, Zn, and chromium (Cr). However, terrestrial plants are preferably used because of their long length, and fibrous roots (Dhanwal et al. 2017).

6.4.3 Phytotransformation

This technique is the disintegration of toxic pollutants into less toxic ones through plant metabolic pathways or enzymatic catalyzed processes. However, this technique is independent of the microbial community (Kumar et al. 2018). Phytotransformation is also recognized as phytodegradation, and it is used only for the disintegration of organic pollutants. During this process, volatile compounds are also released via transpiration due to the breakdown of compounds known as phytovolatilization. Nevertheless, this technique is only confined to organic pollutants and some heavy metals including selenium (Se) and mercury (Hg). In this process, organic compounds are absorbed by plant roots and converted into less toxic substances or released into the atmosphere (Kumar and Gunasundari 2018).

6.4.4 Phytoextraction

Phytoextraction is the implementation of fast-growing plants to remove heavy metals from certain contaminated media (Pajević et al. 2016). This technique can be a natural or chemical-mediated process. The natural or continuous phytoextraction is employed to remove contaminants (heavy metals) from plant roots and then transported to upper plant parts (shoot). Following harvesting stage, plant biomass can be utilized to produce biogas and recover precious heavy metals through combustion. This technique is commonly known as biomining or phytomining (Bhargava and Singh 2017). Intriguingly, this technique offers the best suitable solution for the elimination of pollutants from contaminated soils without changing soil properties. In addition, plant parts both root and shoot result in the reduction of heavy metals which are restored from plant parts after harvesting (Bhargava and Singh 2017). Hence, continuous harvesting of such plants with the accumulation of heavy metals can significantly reduce heavy metals concentration in the field. Surprisingly, this technique is 10 times more efficient and economic than other traditional approaches (Wan et al. 2016). The plant selected for phytoextraction should exhibit fast growing, high biomass production, longer roots, and a high accumulation of heavy metals (Sytar et al. 2016). Natural hyperaccumulators are utilized in the case of continuous phytoextraction (Usman et al. 2018). Phytoextraction is comprised of different processes like the sorption of some metal fraction at the root surface and then entering roots via cellular membranes followed by immobilization of some metals in the root vacuoles. In addition, soluble metals can enter roots and move to the plant leaves from the roots through xylem vessels (Adrees et al. 2015).

6.4.5 Phytostabilization

During this process, toxic and harmful pollutants are inactivated and stabilized in plant tissues. It is also known as phytorestoration or phytoimmobilization (Ramanjaneyulu et al. 2017). This approach aims to inhibit the mobility and availability of both heavy metals and organic compounds and resultantly prevent the leachate from groundwater contamination (Khalid et al. 2017). In this technique, plants are of secondary importance as compared to soil amendments. Different toxic pollutants are physically or chemically immobilized or fixed on plant roots by applying various soil amendments (Wuana and Okieimen 2011). These soil amendments such as biosolids, organic matter, phosphate fertilizers, and clay minerals are being used for the fixation of heavy metals. However, the focus of plants is to reduce water filtration, minimize contact with pollutants, limit soil erosion, and reduce the relocation of pollutants (Akhtar et al. 2013). This strategy is not beneficial for a prolonged duration because the contaminants will remain in the soil at the end (Ashraf et al. 2019).

7 Techniques Used to Improve the Efficiency of Phytoremediation

There are various approaches that can enhance the efficacy of phytoremediation such as the use of chelating and acidic agents, microbes, electric current, transgenic plants, and plant growth regulators (Fig. 40.1).

7.1 *Use of Acidifying and Chelating Agents*

Different chelating agents are being used to improve the efficacy of phytoremediation. For instance, ethylenediaminetetraacetic acid (EDTA) helps in making a complex soil solution containing metals that are later absorbed by roots and then translocated to aerial parts of the plants (Bareen 2012). However, the ability of plants to adsorb heavy metals depends upon the accessibility of metals in the soil. Therefore, chelating or acidic agents are employed to increase metal availability in soil to be absorbed by plants. In addition to EDTA, citric acid is also used to improve phytoremediation because citric acid also has amino polycarboxylic acids. Both ETDA and citric acid have been documented to enhance metal concentrations including cobalt (Co), Zn, Pb, and Cd. Nevertheless, EDTA has more effective and efficient chelation of Cu and Zn as compared to citric acid (Lesage et al. 2005; Duarte et al. 2011). The use of chelating agents is beneficial for both increasing metal concentrations and modifying their transfer pathways to apoplastic from symplastic to facilitate the metal transfer. In addition, chelating agents have also increased plant tolerance against metal deposition (Farid et al. 2013; Ghazaryan et al. 2022). Therefore, the use of a proper and precise dose of chelating plays a critical part in the removal of metals without any negative effects.

7.2 Application of Electric Current

To improve the phytoremediation process for the elimination of heavy metals, electric current is provided in the soil to enhance the absorption of pollutants and other nutrients by plants. This method is carried out by placing two electrodes in the field for the generation of the electric field. The electrolysis mechanism is involved in the migration of contaminants from the soil (Luo et al. 2018; Gavrilescu 2022). Water molecules are oxidized into hydrogen ions and hydroxyl ions which alter the pH of the soil and assist in the enhanced heavy metal uptake by plants. This method is applied with the phytoremediation process to promote the heavy metal uptake. Nevertheless, the application of electric fields could influence the biological behaviour of plants and enhance plant growth (Bi et al. 2011).

7.3 Inoculation of Bacteria

It has been reported that several bacteria have the capability to decompose organic contaminants in the soil such as PAHs are converted into simpler and non-toxic compounds by bacterial enzymes. Nevertheless, the decomposition process is slow, but decomposition can be increased with the addition of degrading bacteria in the soil (Sharma 2021; Anerao et al. 2022). These bacteria have shown effective outcomes in the lab but would not exhibit the same decomposition rate in the field. The possible solution to this problem is the use of both plants and bacteria to promote each other simultaneously. These bacteria are also known as plant growth enhancers and are involved in the improvement of crops directly or indirectly. Subsequently, plant biomass is increased, and plant secretions are used by bacteria as nutrients for growing their population resulting in more decomposition of PAHs (Rostami et al. 2016). In addition, these bacteria are also accountable to reduce heavy metal toxicity and metal tolerance. These mechanisms include moving metal ions into outer cellular spaces, conversion of toxic metals into less or nontoxic forms, absorption or resorption of metal, and metal accumulation in the cells (Rostami and Azhdarpoor 2019).

7.4 Selection of Transgenic Plants

The phytoremediation of contaminants can be improved if the selected plants possess the characteristics of rapid growth, long and deep roots, and high biomass production. In addition, such plants should have potential ability to store high concentrations of heavy metals or organic pollutants (Rai et al. 2020). Therefore, plants with these traits can be achieved through genetic engineering processes. The use of transgenic plants for the sake phytoremediation purposes is one of the best options to enhance the efficiency of toxic contaminant removal. Hence, the transfer of desired genes capable of the accumulation of pollutants into plants will improve phytoremediation through storage, absorption, and detoxification mechanisms (Gunarathne et al. 2019). Nonetheless, transgenic plants can also be utilized for heavy metals through accumulation or tolerance mechanisms. The application of parser genes in the plant rhizosphere is also a suitable solution to increase secretions by bacteria for enhanced decomposition of pollutants. As a result, the plants do not require to absorb toxic contaminants by roots rather the contaminants are degraded by enzymatic actions of bacteria in the rhizosphere (Rostami and Azhdarpoor 2019).

7.5 Practice of Co-cropping

Numerous scientific reports illustrated the usage of only one plant species for the removal of contaminants; however, these uses of two or more different plant species (co-cropping) could enhance the phytoremediation potential for pollutants. Hence, two plants could have a more synergistic effect on the elimination of pollutants from the contaminated media (Wang et al. 2013). Nevertheless, an antagonistic interaction between two plants was also observed affecting plant growth and removal efficiency of pollutants, therefore, plants must be selected through experimental trials.

7.6 Application of Soil Amendments

The application of soil amendments is another interesting technique to remove toxic pollutants. Soils can be amended with organic matter, charcoal, sewage sludge biochar, compost, and activated carbon to increase the bioavailability of toxic pollutants (Waqas et al. 2014). The use of such amendments in polluted sites depends on their prices, availability, and toxicity level to reduce extra costs. Moreover, compost can also be a good option to increase the efficacy of phytoremediation but sometimes fails (Wang et al. 2012; Marchal et al. 2014).

7.7 Use of Plant Growth Regulators

Plant growth regulators (PGRs) are usually organic compounds that can affect the various biological processes of plants even at low concentrations. The environmental protection agency (EPA) reported that any substance that can enhance or reduce plant growth or yield by modifying the biological processes of plants is termed PGR (Shafigh et al. 2016; Chen et al. 2022). Moreover, when the plants produce these compounds then these are known as phytohormones, while these compounds are synthetically or naturally used for improving the role of hormones then called as

PGRs. Thus, the use of PGRs to increase the efficacy of phytoremediation has been widely studied and applied. The application of these PGRs to plants displayed numerous advantages such as being cost-effective, ecofriendly, and easy use (Cassina et al. 2011; Sun et al. 2020). There are different types of PGRs such as acetic acid, ethylene, auxin, gibberellins, cytokines, jasmonic acid, and salicylic acid. These compounds exhibit diverse varieties of actions in plants via different mechanisms. There are certain factors affecting their activities including environmental factors, dose concentration, and plant physiology. In addition, the use of these PGRs can enhance plant biomass as well as reduce the adverse impacts of contaminants on plants (Hadi et al. 2010; Chen et al. 2020).

8 Stress Tolerance Mechanisms in Plants for Crop Improvement

Stress tolerance mechanisms in plants consist of both physiological and molecular mechanisms. Plants use different mechanisms to purge toxic heavy metals and organic pollutants. There are various tolerance mechanisms that are adapted by plants to reduce the adverse impacts of contaminants.

8.1 Mycorrhizal Association

Recent studies reported a fungal group capable to reduce the detrimental effects of heavy metals using different mechanisms (Dhalaria et al. 2020). These mechanisms include vascular sequestration, heavy metal accumulation in cortical cells, mycelium binding, and modification in pH for metal immobilization (Garg et al. 2017). This mutualistic interaction between plants and fungi assists in plant growth and development by providing essential nutrients and protection against environmental stresses.

8.2 Role of the Cell Wall as Well as Root Exudates

The cell wall exhibits binding characteristics with heavy metals and plant roots adsorb heavy metals on their surface and confines heavy metal accumulation on the cell membrane (Bali et al. 2020). In addition, root exudates can enhance metal chelation and uptake from the soil. For instance, nickel chelators in root exudates decrease Ni uptake in plants (Chen et al. 2017).

8.3 Involvement of Plasma Membrane

Toxic heavy metals can cause damage to the plasma membrane and higher concentrations of heavy metals can affect different mechanisms such as protein inhibition, and changes in membrane composition. However, stress tolerance protects against the leakage of solutes from the membrane and maintains metal homeostasis (Morsy et al. 2012). Moreover, plants have specific heavy metal transporters for metal intake and metal tolerance for cellular homeostasis (Liu et al. 2019).

8.4 Heat Shock Proteins

Heat shock protein (HSP) plays an essential role in the regulation of appropriate growth temperature for plant growth and development. These proteins (10–200 KD) act as chaperons taking part in signal transduction during stress conditions in all living organisms (Mishra et al. 2018). Further, HSP can modulate protein repair, restoration, and prevention of misfolded proteins under stress helping in the maintenance of cellular homeostasis (Liu and Howell 2016).

8.5 Plant Chelation and Metallothioneins

Plant chelation synthase enzyme produces small chains of peptides enriched with sulphur which are further orchestrated from the reduced glutathione. However, plant chelation and metallothioneins are the two main outcomes of peptide binding ligands which are implicated in the removal of heavy metals for maintaining homeostasis (Pochodylo and Aristilde 2017). Chelation genes regulate the transport of the chelators from root to shoots for high plant tolerance against contaminants (Hasan et al. 2015). Similarly, metallothioneins are also implicated in the removal of heavy metals with their thiol group suggesting their role in cellular homeostasis, metal sequestration, and protection against oxidative stress (Liu et al. 2015).

8.6 The Presence of Amino Acids and Organic Acids

The presence of amino acids and organic acids plays a crucial role in the tolerance of heavy metals in plants. There are two main types of metal tolerance in plants such as (1) internal tolerance as well as (2) external exclusion. The former is followed by the metal chelation of organic acids in cytosol and conversion into less toxic forms while the latter the secretion of organic acids by plants facilitates metal chelation resulting in less bioavailability of heavy metals and metal deposition in plant roots

(Osmolovskaya et al. 2018). In addition, amino acids including malic acid, oxalate, citric acid, nitotianamine, and histidine act as ligands for heavy metals helping in metal tolerance and metal detoxification (Ehsan et al. 2014).

8.7 *Vacuolar Sequestration*

Plants also deploy vacuolar compartmentalization to mitigate the deleterious effects of pollutants (heavy metals and organic pollutants). Upon the entry of heavy metals into cells, two mechanisms efflux or vacuole sequestration are activated for the detoxification of pollutants. The main vacuole serves as a storage house of the deposition of heavy metals inside plants (Sharma et al. 2016).

8.8 *Trichome Structures*

Trichome is a hair-like structure in plants used for heavy metal accumulation and tolerance. These structures are also involved in the production of different secondary metabolites used for the elimination of toxic metals (Harada et al. 2010). Therefore, these trichomes play a critical part in providing protection against insects and the toxic effects of pollutants.

8.9 *Antioxidant Defense System*

The recognition of pollutants in the plants results in the oxidative burst in the form of ROS affecting plant growth and yield. Plants employ antioxidant mechanisms to detoxify ROS accumulation for the reduction of ROS-mediated toxic effects (Dixit et al. 2015). These antioxidants maintain the balance between ROS generation and ROS scavenging. Plants produced different enzymatic antioxidants including superoxide dismutase, catalase, glutathione peroxidase, and ascorbate peroxidase, and non-enzymatic antioxidants including proline, glutathione, and ascorbic acid play a key part in the protection against biotic and abiotic stresses (Gall et al. 2015).

8.10 *Using Genetic Engineering Techniques*

During this approach, genes of desired interests such as tolerance against specific pollutants are identified and used to enhance resistance in plants against specific stresses (Charfeddine et al. 2017). For this purpose, the genes responsible for uptake, translocation, and deposition of contaminants are separated and overexpressed to enhance tolerance in plants (Gangadhar et al. 2016). However, a detailed mechanistic study and interplay between plants and metal toxicity in plants require more research.

8.11 Propagation Approach

The selection of plant species matters a lot for the effective and efficient tolerance of pollutants. Commonly, plants grown in contaminated areas are identified and selected and used for treating the desired contaminated sites. However, the distribution and growth of plant species could vary geographically. To solve this issue, the tolerant genes from such plants are transferred to desired plants through the propagation technique. This approach is efficient for edible plants (Souza et al. 2016).

8.12 Application of Bioinformatic Approaches

The identification of stress-tolerant genes has always been difficult because tolerance mechanisms differ from one plant to another and are the same in the case of genes. Therefore, a segregation study is preferred to produce progeny from parental plants with distinct phenotypes. Quantitative trait loci (QTL) mapping is a promising approach to identifying tolerant genes against specific toxic pollutants. This approach compares the cDNA library of plant for the sequence resemblance (similarity) with genomic sequences (Gupta and Praveen 2019). Further, the next-generation molecular markers (SNPs) are used for the insertion or deletion of specific genes.

Moreover, the use of omics approaches has also significantly contributed to recognizing tolerant signaling at genetic levels. Omics approaches such as proteomics, transcriptomics, metabolomics, and ionomics are utilized for the identification of responsible proteins, genomic changes, metabolites, and minerals as well as trace elements, respectively. These analyses provide huge data that further required computational facilities for the storage and analysis to determine responsible genes, metabolites, minerals, and proteins involved in the reduction of pollutants-mediated toxicities (Kumar et al. 2021).

9 Enhancement of Ecosystem Services by Phytoremediation

Phytoremediation is a sustainable approach and results in no or minimum effects on the environment. It is a green technology used for the efficient removal of toxic and harmful pollutants from contaminated media such as air, water, and soil. In addition, this approach is environment friendly and also economically viable.

9.1 Sustainable and Ecofriendly

Phytoremediation is an environment friendly technique to treat contaminated sites at minimum energy costs as compared to other conventional methods. This method utilizes solar energy and the physiological mechanisms of plants for environmental clean-up. Plants have the potential to remedy pollutants from different polluted media such as air, water, and soil. Phytoremediation can increase carbon sequestration due to planting more plants eliminating toxic contaminants and resulting in reducing atmospheric CO_2 (Kafle et al. 2022). Further, phytoremediation can provide environmental, economic, and social welfare benefits when applied using site management (Burges et al. 2018). This approach can significantly improve ecosystem services such as soil fertility, nutrient recycling, carbon sequestration, water purification, and water movement (Burges et al. 2016, 2017). Moreover, phytoremediation can enhance risk mitigation in the form of phytomanagement at a specific site. The aim of phytoremediation is not only the removal of pollutants but also rehabilitate the environmental quality and resource functionality. There are various parameters to predict or determine the environmental quality such as pH, cation exchange capacity (CEC), abundance, texture, and population of microorganisms (Burges et al. 2018).

9.2 Remediation of Diverse Contaminants in Different Media

The use of phytoremediation approach has been extensively applied for treating heavy metals and organic pollutants. Approximately 400 plant species are being used to eliminate heavy metals from 50 to 500 times higher than normal plants. These plant species include Asteraceae, Caryophyllaceae, Flacountraceae, Poaceae, Violaceae, Brassicaceae, Fabaceae, Lamiaceae, and Euphorbiaceae as well as angiosperms (Kafle et al. 2022). Interestingly, the burnt ashes of these harvested toxic plants can be used to compensate for the nutrient deficiency in the fields (Koptsik 2014). Similarly, this approach can also be used for the remediation of wastewater by using different plant species such as *Azolla pinnata*, *Cyperus* spp., *Diodia virginiana*, *Eichhornia crassipes*, *Hygrophila corymbosa*, *Imperata cylindrica*, *Nymphaea* spp., *Phragmites australis*, and *Spirodela polyrhiza* (Ekperusi et al. 2019). *Azolla pinnata* exhibited a 75% removal efficiency of dairy wastewater (Goala et al. 2021). Further, poplar trees have been used for the treatment of ground water contamination due to their deep and long root structure and high evaporation rate (Barac et al. 2009). Furthermore, the phytoremediation technique can be beneficial to remove air contaminants, especially phytovolatilization mechanism. For example, trichloroethylene and perchloroethylene have been phytovolatilized using willow and poplar trees due to their higher biomass production. Additionally, *Nephrolepis obliterata* and *Chamaedorea elegans* displayed 90–100% and 65–100% removal efficacy of formaldehyde in a two-day experiment (Teiri et al. 2018).

9.3 Economic Viability

Phytoremediation also offers a cost-effective solution for treating contaminated sites. The overall costs needed for this approach are lower than other traditional methods. The costs for phytoremediation are almost 5–13-fold lesser than chemical methods. Another study reported that using phytoremediation for the treatment of dairy wastewater was more economical than using other conventional methods (Mulbry et al. 2008; Kafle et al. 2022). Therefore, the phytoremediation technique is an efficient remedy option for environmental clean-up as compared to other conventional methods.

10 Proper Disposal of Toxic Plant Waste After Phytoremediation

There are various disposal methods for plant waste after phytoremediation such as composting, pyrolysis, and incineration. The exogenous application of additives can modify heavy metal forms and reduce their bioavailability. Biochar amendment, coarse medical stone, phosphate rock, and bacterial and fungal inoculation are used in the composting to decrease the bioavailability factor of toxic heavy metals (Chen et al. 2019; Cui et al. 2021). In addition, pyrolysis can also alter the form of harmful heavy metals in biomass to a stable form. High temperature (550 °C) can change metals into stable form in acidic solutions resulting in the insolubility of metals and reducing the bioavailability of metals. An increase in volatilization of heavy metals in biochar amendments was enhanced with increasing temperature. Moreover, ferric salt can also improve the efficiency of pyrolysis due to its inhibitory characteristics of heavy metals bioavailability (Gong et al. 2018; He et al. 2020; Zhang et al. 2020). Further, incineration of toxic plant waste can also be carried out to recover precious heavy metals. Different factors can affect the pyrolysis process such as temperature, N_2/O_2 ratio, and retention time. For instance, in the case of *Pteris vittata* L., medium oxygen content and temperature around 400 °C were found optimal for incineration (Yan et al. 2008; Zhou et al. 2020). However, these disposal methods (pyrolysis, composting, and incineration) could result in secondary pollution and increase health risks for humans (Liu and Tran 2021). Therefore, more appropriate, and reasonable use of toxic waste is very important to avoid its hazardous effects on the ecosystem. Hence, more research is needed regarding disposal challenges and their possible solutions and finding good extraction strategies to leave minimum or zero impacts on the environment. Therefore, it is very important to prevent the leachate from groundwater contamination during the disposal process and construction of landfills.

11 Conclusion and Future Perspectives

The environmental pollution due to toxic heavy metals and organic pollutants poses adverse impacts on plants, animals, and humans. Anthropogenic activities and rapid population growth contribute ecosystem contamination. However, bioremediation technique is more suitable and effective than other conventional approaches. The use of micro-remediation, mycoremediation, and cyanoremediation haven been reported most productive in enhancing plant growth and development as well as promoted crop improvement. Further, phytoremediation using fast growing plants with high biomass production can be employed for the elimination of harmful pollutants from contaminated media. The application of phytoremediation approach has also significantly enhanced quality of soil and water through ecosystem services. Plant uptake these pollutants from roots and then transported to upper parts of plants through various transporters. The efficiency of phytoremediation can enhance further using the chelating agents, soil amendments, microbial inoculation, and plant growth promoters. Interestingly, plants are also equipped with stress tolerance mechanisms to avoid the toxic effects of pollutants and main cellular homeostasis. Further, the proper disposal of toxic plant waste after phytoremediation is also of significance importance. However, further research is needed to explore the interplay between soil microbial activity and plants which can affect solubility, immobilization, degradation of toxic pollutants. Further, more hyperaccumulators should be developed with characteristic of high biomass, rapid growth, and ability to remove more pollutants.

References

AbdElgawad H, Zinta G, Hamed BA, Selim S, Beemster G, Hozzein WN, Wadaan MAM, Asard H, Abuelsoud W (2020) Maize roots and shoots show distinct profiles of oxidative stress and antioxidant defense under heavy metal toxicity. Environ Pollut 258:113705

Adrees M, Ali S, Rizwan M, Zia-ur-Rehman M, Ibrahim M, Abbas F, Farid M, Qayyum MF, Irshad MK (2015) Mechanisms of silicon-mediated alleviation of heavy metal toxicity in plants: a review. Ecotoxicol Environ Saf 119:186–197

Ahmadpour P, Ahmadpour F, Mahmud TMM, Abdu A, Soleimani M, Tayefeh FH (2012) Phytoremediation of heavy metals: a green technology. Afr J Biotechnol 11:14036–14043

Akhtar N, Mannan MA (2020) Mycoremediation: expunging environmental pollutants. Biotechnol Rep 26:e00452

Akhtar MS, Chali B, Azam T (2013) Bioremediation of arsenic and lead by plants and microbes from contaminated soil. Res Plant Sci 1(3):68–73

Alvarez A, Saez JM, Costa JSD, Colin VL, Fuentes MS, Cuozzo SA, Benimeli CS, Polti MA, Amoroso MJ (2017) Actinobacter: current research and perspectives for bioremediation of pesticides and heavy metals. Chemosphere 166:41–62

Anerao P, Kaware R, Kumar Khedikar A, Kumar M, Singh L (2022) Phytoremediation of persistent organic pollutants: concept challenges and perspectives. In: Phytoremediation technology for the removal of heavy metals and other contaminants from soil and water. Elsevier, Amsterdam, pp 375–404

Anim AK, Drage DS, Goonetilleke A, Mueller JF, Ayoko GA (2017) Distribution of PBDEs, HBCDs and PCBs in the Brisbane River estuary sediment. Mar Pollut Bull 120(1–2):165–173

Asati A, Pichhode M, Nikhil K (2016) Effect of heavy metals on plants: an overview. Int J Appl Innov Eng Manag 5:2319–4847

Ashraf S, Ali Q, Zahir ZA, Ashraf S, Asghar HN (2019) Phytoremediation: environmentally sustainable way for reclamation of heavy metal polluted soils. Ecotoxicol Environ Saf 174:714–727

Ayangbenro AS, Babalola OO (2017) A new strategy for heavy metal polluted environments: a review of microbial biosorbents. Int J Environ Res Public Health 14(1):94

Azhdarpoor A, Nikmanesh R, Khademi F (2014) A study of Reactive Red 198 adsorption on iron filings from aqueous solutions. Environ Technol 35:2956–2960

Azhdarpoor A, Nikmanesh R, Samaei MR (2015) Removal of arsenic from aqueous solutions using waste iron columns inoculated with iron bacteria. Environ Technol 36:2525–2531

Azubuike CC, Chikere CB, Okpokwasili GC (2016) Bioremediation techniques classification based on site of application: principles, advantages, limitations and prospects. World J Microbiol Biotechnol 32:180

Balaji S, Kalaivani T, Rajasekaran C, Shalini M, Siva R, Singh RK, Akthar MA (2014) Arthrospira (Spirulina) species as bioadsorbents for lead, chromium, and cadmium – a comparative study. Clean-Soil Air Water 42:1790–1797

Bali AS, Sidhu GPS, Kumar V (2020) Root exudates ameliorate cadmium tolerance in plants: a review. Environ Chem Lett 18:1243–1275

Barac T, Weyens N, Oeyen L, Taghavi S, van der Lelie D, Dubin D, Spliet M, Vangronsveld J (2009) Field note: hydraulic containment of a BTEX plume using poplar trees. Int J Phytoremediation 11(5):416–424

Bareen FE (2012) Chelate assisted phytoextraction using oilseed brassicas. Environ Pollut 21:289–311

Barletta M, Lima AR, Costa MF (2019) Distribution, sources and consequences of nutrients, persistent organic pollutants, metals and microplastics in South American estuaries. Sci Total Environ 651:1199–1218

Bhargava M, Singh V (2017) Phytomining: principles and applications. In: Biotechnology. CRC Press, Boca Raton, pp 141–159

Bi R, Schlaak M, Siefert E, Lord R, Connolly H (2011) Influence of electrical fields (AC and DC) on phytoremediation of metal polluted soils with rapeseed (*Brassica napus*) and tobacco (*Nicotiana tabacum*). Chemosphere 83:318–326

Bokhari SH, Nawaz G, Azizullah A, Mahmood-Ul-Hassan M, Ali Z (2022) Heavy metals phytofiltration potential of Hydrocotyle umbellata from Nullah Lai wastewater and its environmental risk. Int J Phytoremediation 24:1–10

Burges A, Epelde L, Benito G, Artetxe U, Becerril JM, Garbisu C (2016) Enhancement of ecosystem services during endophyte-assisted aided phytostabilization of metal contaminated mine soil. Sci Total Environ 562:480–492

Burges A, Epelde L, Blanco F, Becerril JM, Garbisu C (2017) Ecosystem services and plant physiological status during endophyte-assisted phytoremediation of metal contaminated soil. Sci Total Environ 584:329–338

Burges A, Alkorta I, Epelde L, Garbisu C (2018) From phytoremediation of soil contaminants to phytomanagement of ecosystem services in metal contaminated sites. Int J Phytoremediation 20(4):384–397

Cassina L, Tassi E, Morelli E, Giorgetti L, Remorini D, Chaney RL, Barbafieri M (2011) Exogenous cytokinin treatments of an Ni hyperaccumulator, Alyssum murale, grown in a serpentine soil: implications for phytoextraction. Int J Phytoremediation 13:90–101

Chaoua S, Boussaa S, El Gharmali A, Boumezzough A (2018) Impact of irrigation with wastewater on accumulation of heavy metals in soil and crops in the region of Marrakech in Morocco. J Saudi Soc Agric Sci 18(4):429–436

Charfeddine M, Charfeddine S, Bouaziz D, Messaoud RB, Bouzid RG (2017) The effect of cadmium on transgenic potato (*Solanum tuberosum*) plants overexpressing the StDREB transcription factors. Plant Cell Tissue Organ Cult 128:521–541

Chen YT, Wang Y, Yeh KC (2017) Role of root exudates in metal acquisition and tolerance. Curr Opin Plant Biol 39:66–72

Chen YR, Chen YN, Li YP, Wu YX, Zeng ZP, Xu R, Wang S, Li H, Zhang JC (2019) Changes of heavy metal fractions during co-composting of agricultural waste and river sediment with inoculation of *Phanerochaete chrysosporium*. J Hazard Mater 378:120757

Chen L, Long C, Wang D, Yang J (2020) Phytoremediation of cadmium (Cd) and uranium (U) contaminated soils by *Brassica juncea* L. enhanced with exogenous application of plant growth regulators. Chemosphere 242:125112

Chen Z, Liu Q, Chen S, Zhang S, Wang M, Munir MAM, Feng Y, He Z, Yang X (2022) Roles of exogenous plant growth regulators on phytoextraction of Cd/Pb/Zn by Sedum alfredii Hance in contaminated soils. Environ Pollut 293:118510

Cui H, Ou Y, Wang LX, Yan BX, Li YX, Bao MW (2021) Additive grain-size: an innovative perspective to investigate the transformation among heavy metal and phosphorus fractions during aerobic composting. J Environ Manag 292:112768

da Conceição Gomes MA, Hauser-Davis RA, de Souza AN, Vitória AP (2016) Metal phytoremediation: general strategies, genetically modified plants and applications in metal nanoparticle contamination. Ecotoxicol Environ Saf 134:133–147

Dalcorso G, Manara A, Furini A (2013) An overview of heavy metal challenge in plants: from roots to shoots. Metallomics 5:1117–1132

Dhalaria R, Kumar D, Kumar H, Nepovimova E, Kuca K, Islam MT, Verma R (2020) Arbuscular mycorrhizal fungi as potential agents in ameliorating heavy metal stress in plants. Agronomy 10:815

Dhanwal P, Kumar A, Dudeja S, Chhokar V, Beniwal V (2017) Recent advances in phytoremediation technology. Adv Environ Biol 1:227–241

Dixit G, Singh AP, Kumar A, Singh PK, Kumar S, Dwivedi S, Trivedi PK, Pandey V, Norton GJ, Dhankher OP, Tripathi RD (2015) Sulfur mediated reduction of arsenic toxicity involves efficient thoil metabolism and the antioxidant defense system in rice. J Hazard Mater 298:241–251

Duarte B, Freitas J, Caçador I (2011) The role of organic acids in assisted phytoremediation processes of salt marsh sediments. Hydrobiologia 674:169–177

Ebrahimi A, Hashemi H, Eslami H, Fallahzadeh RA, Khosravi R, Askari R, Ghahramani E (2018) Kinetics of biogas production and chemical oxygen demand removal from compost leachate in an anaerobic migrating blanket reactor. J Environ Manag 206:707–714

Ehsan S, Ali S, Noureen S, Mahmood K, Farid M, Ishaque W, Shakoor MB, Rizwan M (2014) Citric acid mediated phytoremediation of cadmium by *Brassica napus* L. Ecotoxicol Environ Saf 106:164–172

Ekperusi AO, Sikoki FD, Nwachukwu EO (2019) Application of common duckweed (*Lemna minor*) in phytoremediation of chemicals in the environment: state and future perspective. Chemosphere 223:285–309

Etesami H (2018) Bacterial mediated alleviation of heavy metal stress and decreased accumulation of metals in plant tissues: mechanisms and future prospects. Ecotoxicol Environ Saf 147:175–191

Farid M, Ali S, Shakoor MB, Bharwana SA, Rizvi H, Ehsan S, Tauqeer HM, Iftikhar U, Hannan F (2013) EDTA assisted phytoremediation of cadmium, lead and zinc. Int J Agron Plant Prod 4(11):2833–2846

Fuente V, Rufo L, Ju'arez BH, Men'endez N, García-Hern'andez M, Salas-Colera E, Espinosa A (2016) Formation of biomineral iron oxides compounds in a Fe hyperaccumulator plant: *Imperata cylindrica* (L.) P. Beauv. J Struct Biol 193:23–32

Gall JE, Boyd RS, Rajakaruna N (2015) Transfer of heavy metals through terrestrial food webs: a review. Environ Monit Assess 187:201

Gangadhar BH, Sajeesh K, Venkatesh J, Basker V, Abhinandan K, Yu JW, Prasad R, Mishra RK (2016) Enhanced tolerance of transgenic potato plants over-expressing non-specific lipid transfer protein-1 (StnsLTP1) against multiple abiotic stresses. Front Plant Sci 7:1228

Garg N, Singh S, Kashyap L (2017) Arbuscular mycorrhizal fungi and heavy metal tolerance in plants: an insight into physiological and molecular mechanisms. Springer, Cham. In: Varm A, Prasad R, Tuteja N (eds) Mycorrhiza – nutrient uptake, biocontrol, ecorestoration. Springer, Cham, pp 75–97

Gavrilescu M (2022) Enhancing phytoremediation of soils polluted with heavy metals. Curr Opin Biotechnol 74:21–31

Ghazaryan KA, Movsesyan HS, Minkina TM, Nevidomskaya DG, Rajput VD (2022) Phytoremediation of copper-contaminated soil by *Artemisia absinthium*: comparative effect of chelating agents. Environ Geochem Health 44(4):1203–1215

Ghori NH, Ghori T, Hayat MQ, Imadi SR, Gul A, Altay V, Ozturk M (2019) Heavy metal stress and responses in plants. Int J Environ Sci Technol 16:1807–1828

Glavac NK, Djogo S, Razic S, Kreft S, Veber M (2017) Accumulation of heavy metals from soil in medicinal plants. Arh Hig Rada Toksikol 68:236–244

Goala M, Yadav KK, Alam J, Adelodun B, Choi KS, Cabral-Pinto MM, Hamid AA, Alhoshan M, Ali FAA, Shukla AK (2021) Phytoremediation of dairy wastewater using *Azolla pinnata*: application of image processing technique for leaflet growth simulation. J Water Process Eng 42:102152

Gong XM, Huang DL, Liu YG, Zeng GM, Wang RZ, Wei JJ, Huang C, Xu P, Wan J, Zhang C (2018) Pyrolysis and reutilization of plant residues after phytoremediation of heavy metals contaminated sediments: for heavy metals stabilization and dye adsorption. Bioresour Technol 253:64–71

Guigues S, Bravin MN, Garnier C, Masion A, Doelsch E (2014) Isolated cell walls exhibit cation binding properties distinct from those of plant roots. Plant Soil 381:367–379

Gunarathne V, Mayakaduwa S, Ashiq A, Weerakoon SR, Biswas JK, Vithanage M (2019) Transgenic plants: benefits, applications, and potential risks in phytoremediation. In: Transgenic plant technology for remediation of toxic metals and metalloids. Academic Press, London, pp 89–102

Gupta M, Praveen A (2019) Quantitative Trait Loci mapping of heavy metal accumulation and resistance in crop plants. In: Rajpal V, Sehgal D, Kumar A, Raina S (eds) Genomics assisted breeding of crops for abiotic stress tolerance, vol II. Springer, Cham

Hadi F, Bano A, Fuller MP (2010) The improved phytoextraction of lead (Pb) and the growth of maize (*Zea mays* L.): the role of plant growth regulators (GA3 and IAA) and EDTA alone and in combinations. Chemosphere 80(4):457–462

Harada E, Kim JA, Meyer AJ, Hell R, Clemens S, Choi YE (2010) Expression profiling of tobacco leaf trichomes identifies genes for biotic and abiotic stresses. Plant Cell Physiol 51:1627–1637

Hasan M, Ahammed GJ, Yin L, Shi K, Xia X, Zhou Y, Yu J, Zhou J (2015) Melatonin mitigates cadmium phytotoxicity through modulation of phytochelatins biosynthesis, vacuolar sequestration, and antioxidant potential in *Solanum lycopersium* L. Front Plant Sci 6:601

He J, Strezov V, Zhou XT, Kumar R, Kan T (2020) Pyrolysis of heavy metal contaminated biomass pre-treated with ferric salts: product characterisation and heavy metal deportment. Bioresour Technol 313:123641

Iqbal N, Ahmed S, Pervez A, Nazir R, Tang X, Irshad U (2020) The Fate of Lead (Pb) in Multitrophic Interactions Among Bacteria, Fungi, and Bacterivorous Soil Nematodes. Clean–Soil Air Water 48(11):2000307

Iqbal N, Nazir N, Numan M, Hayat MT, Mahmood Q, Zeb BS, Ma B, Abbas Z (2022) Effects of polychlorinated biphenyls on plant growth. In: Sustainable plant nutrition under contaminated environments. Springer, Cham, pp 187–208

Jacob JM, Karthik C, Saratale RG, Kumar SS, Prabakar D, Kadirvelu K, Pugazhendhi A (2018) Biological approaches to tackle heavy metal pollution: a survey of literature. J Environ Manag 217:56–70

Javed MT, Tanwir K, Akram MS, Shahid M, Niazi NK, Lindberg S (2019) Phytoremediation of cadmium-polluted water/sediment by aquatic macrophytes: role of plant-induced pH changes. In: Cadmium toxicity and tolerance in plants. Academic Press, London, pp 495–529

Jiang B, Adebayo A, Jia J, Xing Y, Deng S, Guo L, Liang Y, Zhang D (2019) Impacts of heavy metals and soil properties at a Nigerian e-waste site on soil microbial community. J Hazard Mater 362:187–195

Johnsen IV, Aaneby J (2019) Soil intake in ruminants grazing on heavy-metal contaminated shooting ranges. Sci Total Environ 687:41–49

Kafle A, Timilsina A, Gautam A, Adhikari K, Bhattarai A, Aryal N (2022) Phytoremediation: mechanisms, plant selection and enhancement by natural and synthetic agents. Environ Adv 8:100203

Kalantary RR, Mohseni-Bandpi A, Esrafili A, Nasseri S, Ashmagh FR, Jorfi S, Ja'fari M (2014) Effectiveness of biostimulation through nutrient content on the bioremediation of phenanthrene contaminated soil. J Environ Health Sci Eng 12:143

Karami A, Shamsuddin ZH (2010) Phytoremediation of heavy metals with several efficiency enhancer methods. Afr J Biotechnol 9:3689–3698

Khalid S, Shahid M, Niazi NK, Murtaza B, Bibi I, Dumat C (2017) A comparison of technologies for remediation of heavy metal contaminated soils. J Geochem Explor 182:247–268

Khan I, Ali M, Aftab M, Shakir S, Qayyum S, Haleem KS, Tauseef I (2019) Mycoremediation: a treatment for heavy metal-polluted soil using indigenous metallotolerant fungi. Environ Monit Assess 191:622

Kolaj-Robin O, Russell D, Hayes KA, Pembroke JT, Soulimane T (2015) Cation diffusion facilitator family: structure and function. FEBS Lett 589:1283–1295

Koptsik GN (2014) Problems and prospects concerning the phytoremediation of heavy metal polluted soils: a review. Eurasian Soil Sci 47(9):923–939

Kumar PS, Gunasundari E (2018) Bioremediation of heavy metals. In: Bioremediation: applications for environmental protection and management. Springer, Singapore, pp 165–195

Kumar V, Shahi K, Singh S (2018) Bioremediation: an eco-sustainable approach for restoration of contaminated sites. In: Microbial bioprospecting for sustainable development. Springer, Singapore, pp 115–136

Kumar A, Chaturvedi AK, Yadav K, Arunkumar KP, Malyan SK, Raja P, Kumar R, Khan SA, Yadav KK, Rana KL, Kour D (2019) Fungal phytoremediation of heavy metal-contaminated resources: current scenario and future prospects. In: Recent advancement in white biotechnology through fungi: Volume 3: Perspective for sustainable environments. Springer, Cham, pp 437–461

Kumar A, Chaturvedi AK, Surendran U, Shabnam AA, Singh A, Vinodakumar SN, Tamuly B, Malyan SK, Khan SA, Cabral-Pinto MMS, Raja P, and Yadav KK (2021) Mechanistic overview of metal tolerance in edible -plants: a physiological and molecular perspective. In: Handbook of bioremediation. Academic press, pp 23–47

Latif A, Bilal M, Asghar W, Azeem M, Ahmad MI, Abbas A, Ahmad MZ, Shahzad T (2018) Heavy metal accumulation in vegetables and assessment of their potential health risk. J Environ Anal Chem 5(234):2380–2391

Lesage E, Meers E, Vervaeke P, Lamsal S, Hopgood M, Tack FMG, Verloo MG (2005) Enhanced phytoextraction: II. Effect of EDTA and citric acid on heavy metal uptake by *Helianthus annuus* from a calcareous soil. Int J Phytoremediation 7:143–152

Lin YF, Aarts MG (2012) The molecular mechanism of zinc and cadmium stress response in plants. Cell Mol Life Sci 69:3187–3206

Liu JX, Howell SH (2016) Managing the protein folding demands in the endoplasmic reticulum of plants. New Phytol 211:418–428

Liu Z, Tran K (2021) A review on disposal and utilization of phytoremediation plants containing heavy metals. Ecotoxicol Environ Saf 226:112821

Liu J, Shi X, Qian M, Zheng L, Lian C, Xia Y, Shen Z (2015) Copper-induced hydrogen peroxide upregulation of a metallothionein gene, OsMT2c, from *Oryza sativa* L. confers copper tolerance in *Arabidopsis thaliana*. J Hazard Mater 294:99–108

Liu XS, Feng SJ, Zhang BQ, Wang MQ, Cao HW, Rono JK, Chen X, Yang ZM (2019) OsZIP1 functions as a metal efflux transporter limiting excess zinc, copper and cadmium accumulation in rice. BMC Plant Biol 19:283

Luo J, Cai L, Qi S, Wu J, Gu XS (2018) The interactive effects between chelator and electric fields on the leaching risk of metals and the phytoremediation efficiency of *Eucalyptus globulus*. J Clean Prod 202:830–837

Manara A (2012) Plant responses to heavy metal toxicity. In: Furini A (ed) Plants and heavy metals. Springer, New York, pp 27–53

Mani D, Kumar C (2014) Biotechnological advances in bioremediation of heavy metals contaminated ecosystems: an overview with special reference to phytoremediation. Int J Environ Sci Technol 11:843–872

Marchal G, Smith KEC, Mayer P, de Jonge LW, Karlson UG (2014) Impact of soil amendments and the plant rhizosphere on PAH behaviour in soil. Environ Pollut 188:124–131

Masindi V, Muedi KL (2018) Environmental contamination by heavy metals. In: Heavy metals. IntechOpen, Aglan, pp 115–133

Mishra D, Shekhar S, Singh D, Chakraborty S, Chakraborty N (2018) Heat shock proteins and abiotic stress tolerance in plants. In: Asea A, Kaur P (eds) Regulation of heat shock protein responses, Heat shock proteins, vol 13. Springer, Cham, pp 41–69

Morsy AA, Salama KHA, Kamel HA, Mansour MMF (2012) Effect of heavy metals on plasma membrane lipids and antioxidant enzymes of *Zygophyllum* species. EurAsian J Biosci 6:1–10

Mulbry W, Kondrad S, Pizarro C, Kebede-Westhead E (2008) Treatment of dairy manure effluent using freshwater algae: algal productivity and recovery of manure nutrients using pilot-scale algal turf scrubbers. Bioresour Technol 99(17):8137–8142

Osmolovskaya N, Dung VV, Kuchaeva L (2018) The role of organic acids in heavy metal tolerance in plants. Biol Commun 63:9–16

Pajević S, Borišev M, Nikolić N, Arsenov DD, Orlović S, Župunski M (2016) Phytoextraction of heavy metals by fast-growing trees: a review. Phytoremediation, vol 3:29–64

Parseh I, Teiri H, Hajizadeh Y, Ebrahimpour K (2018) Phytoremediation of benzene vapors from indoor air by *Schefflera arboricola* and *Spathiphyllum wallisii* plants. Atmos Pollut Res 9(6):1083–1087

Patra DK, Pradhan C, Patra HK (2020) Toxic metal decontamination by phytoremediation approach: concept, challenges, opportunities and future perspectives. Environ Technol Innov 18:100672

Pino-Herrera DO, Pechaud Y, Huguenot D, Esposito G, van Hullebusch ED, Oturan MA (2017) Removal mechanisms in aerobic slurry bioreactors for remediation of soils and sediments polluted with hydrophobic organic compounds: an overview. J Hazard Mater 339:427–449

Pochodylo AL, Aristilde L (2017) Molecular dynamics of stability and structures in phytochelatin complexes with Zn, Cu, Fe, Mg, and Ca: implications for metal detoxification. Environ Chem Lett 15:495–500

Rai PK, Kim KH, Lee SS, Lee JH (2020) Molecular mechanisms in phytoremediation of environmental contaminants and prospects of engineered transgenic plants/microbes. Sci Total Environ 705:135858

Ramanjaneyulu AV, Neelima TL, Madhavi A, Ramprakash T (2017) Phytoremediation: an overview. Appl Bot, vol 14:42–84

Ravanbakhsh M, Kowalchuk GA, Jousset A (2019) Optimization of plant hormonal balance by microorganisms prevents plant heavy metal accumulation. J Hazard Mater 379:120787

Rostami S, Azhdarpoor A (2019) The application of plant growth regulators to improve phytoremediation of contaminated soils: a review. Chemosphere 220:818–827

Rostami S, Azhdarpoor A, Rostami M, Samaei MR (2016) The effects of simultaneous application of plant growth regulators and bioaugmentation on improvement of phytoremediation of pyrene contaminated soils. Chemosphere 161:219–223
Rostami S, Azhdarpoor A, Samaei MR (2017) Removal of pyrene from soil using phytobioremediation (Sorghum Bicolor-Pseudomonas). Health Scope 6(4):e62153
Ruiz JM, Blasco B, Ríos JJ, Cervilla LM, Rosales MA, Rubio-Wilhelmi M, Sánchez-Rodríguez E, Castellano R, Romero L (2009) Distribution and efficiency of the phytoextraction of cadmium by different organic chelates. Terra Latinoam 27:295
Saha M, Sarkar S, Sarkar B, Sharma BK, Bhattacharjee S, Tribedi P (2016) Microbial siderophores and their potential applications: a review. Environ Sci Pollut Res 23(5):3984–3999
Saleem K, Asghar MA, Saleem MH, Raza A, Kocsy G, Iqbal N, Ali B, Albeshr MF, Bhat EA (2022) Chrysotile-asbestos-induced damage in *Panicum virgatum* and *Phleum pretense* species and its alleviation by organic-soil amendment. Sustainability 14(17):10824
Saravanan A, Jeevanantham S, Narayanan VA, Kumar PS, Yaashikaa PR, Muthu CM (2020) Rhizoremediation–a promising tool for the removal of soil contaminants: a review. J Environ Chem Eng 8(2):103543
Sarojini G, Babu SV, Rajamohan N, Rajasimman M (2022) Performance evaluation of polymer-marine biomass based bionanocomposite for the adsorptive removal of malachite green from synthetic wastewater. Environ Res 204:112132
Sarubbo LA, Rocha RB Jr, Luna JM, Rufino RD, Santos VA, Banat IM (2015) Some aspects of heavy metals contamination remediation and role of biosurfactants. Chem Ecol 31(8):707–723
Sassykova LR, Aubakirov YA, Akhmetkaliyeva MS, Sassykova AR, Sendilvelan S, Prabhahar M, Prakash S, Tashmukhambetova ZK, Abildin TS, Zhussupova AK (2020) Heavy metal accumulation in plants of the dry-steppe zone of the East Kazakhstan region. Mater Today 33:1187–1191
Shafigh M, Ghasemi-Fasaei R, Ronaghi A (2016) Influence of plant growth regulators and humic acid on the phytoremediation of lead by maize in a Pb-polluted calcareous soil. Arch Agron Soil Sci 62(12):1733–1740
Shah V, Daverey A (2020) Phytoremediation: a multidisciplinary approach to clean up heavy metal contaminated soil. Environ Technol Innov 18:100774
Sharma P (2021) Efficiency of bacteria and bacterial assisted phytoremediation of heavy metals: an update. Bioresour Technol 328:124835
Sharma SS, Dietz KJ, Mimuta T (2016) Vacuolar compartmentalization as indispensable component of heavy metal detoxification in plants. Plant Cell Environ 39:1112–1126
Sharma S, Tiwari S, Hasan A, Saxena V, Pandey LM (2018) Recent advances in conventional and contemporary methods for remediation of heavy metal-contaminated soils. 3 Biotech 8(4):216
Shen ZJ, Xu DC, Chen YS, Zhang Z (2017) Heavy metals translocation and accumulation from the rhizosphere soils to the edible parts of the medicinal plant Fengdan (*Paeonia ostii*) grown on a metal mining area, China. Ecotoxicol Environ Saf 143:19–27
Singh JS, Kumar A, Singh M (2019) Cyanobacteria: a sustainable and commercial bio-resource in production of bio-fertilizer and biofuel from waste waters. Environ Sustain Indic 3–4:100008
Sophia S, Shetty Kodialbail V (2020) Phytoremediation of soil for metal and organic pollutant removal. Bioprocess Eng Bioremediat 104:45–66
Souto KM, Jacques RJS, Zanella R, Machado SLDO, Balbinot A, Avila LAD (2020) Phytostimulation of lowland soil contaminated with imidazolinone herbicides. Int J Phytoremediation 22(7):774–780
Souza LTD, Cambraia J, Ribeiro C, Oliveira JAD, Silva LCD (2016) Effects of aluminium on the elongation and external morphology of root tips in two maize genotypes. Bragantia 75:19–25
Stegmann R, Brunner G, Calmano W, Matz G (eds) (2013) Treatment of contaminated soil: fundamentals, analysis, applications. Springer Science and Business Media, Berlin
Sun Z, Chen J, Wang X, Lv C (2016) Heavy metal accumulation in native plants at a metallurgy waste site in rural areas of Northern China. Ecol Eng 86:60–68

Sun S, Zhou X, Cui X, Liu C, Fan Y, McBride MB, Li Y, Li Z, Zhuang P (2020) Exogenous plant growth regulators improved phytoextraction efficiency by *Amaranths hypochondriacus* L. in cadmium contaminated soil. Plant Growth Regul 90(1):29–40

Sytar O, Brestic M, Taran N, Zivcak M (2016) Plants used for biomonitoring and phytoremediation of trace elements in soil and water. In: Plant metal interaction. Elsevier, Amsterdam, pp 361–384

Teiri H, Pourzamani H, Hajizadeh Y (2018) Phytoremediation of VOCs from indoor air by ornamental potted plants: a pilot study using a palm species under the controlled environment. Chemosphere 197:375–381

Tsibart AS, Gennadiev AN (2013) Polycyclic aromatic hydrocarbons in soils: sources, behavior, and indication significance (a review). Eurasian Soil Sci 46:728–741

Usman K, Al-Ghouti MA, Abu-Dieyeh MH (2018) Phytoremediation: halophytes as promising heavy metal hyperaccumulators. Heavy Met 27:7378

Verma S, Kuila A (2019) Bioremediation of heavy metals by microbial process. Environ Technol Innov 14:100369

Wan X, Lei M, Chen T (2016) Cost–benefit calculation of phytoremediation technology for heavy-metal-contaminated soil. Sci Total Environ 563:796–802

Wang MC, Chen YT, Chen SH, Chang Chien SW, Sunkara SV (2012) Phytoremediation of pyrene contaminated soils amended with compost and planted with ryegrass and alfalfa. Chemosphere 87:217–225

Wang K, Huang H, Zhu Z, Li T, He Z, Yang X, Alva A (2013) Phytoextraction of metals and rhizoremediation of PAHs in co-contaminated soil by co-planting of *Sedum alfredii* with ryegrass (*Lolium perenne*) or castor (*Ricinus communis*). Int J Phytoremediation 15:283–298

Waqas M, Khan S, Qing H, Reid BJ, Chao C (2014) The effects of sewage sludge and sewage sludge biochar on PAHs and potentially toxic element bioaccumulation in *Cucumis sativa* L. Chemosphere 105:53–61

Wei C, Ge Z, Chu W, Feng R (2015) Speciation of antimony and arsenic in the soils and plants in an old antimony mine. Environ Exp Bot 109:31–39

Wood JL, Tang C, Franks AE (2016) Microbial assisted plant growth and heavy metal accumulation to improve phytoextraction of contaminated soils. Soil Biol Biochem 103:131–137

Wuana RA, Okieimen FE (2011) Heavy metals in contaminated soils: a review of sources, chemistry, risks and best available strategies for remediation. Int Sch Res Notices 2011:1–20

Yaashikaa PR, Kumar PS, Jeevanantham S, Saravanan R (2022) A review on bioremediation approach for heavy metal detoxification and accumulation in plants. Environ Pollut 301:119035

Yadav R, Arora P, Kumar S, Chaudhury A (2010) Perspectives for genetic engineering of poplars for enhanced phytoremediation abilities. Ecotoxicology 19(8):1574–1588

Yan XL, Chen TB, Liao XY, Huang ZC, Pan JR, Hu TD, Nie CJ, Xie H (2008) Arsenic transformation and volatilization during incineration of the hyperaccumulator *Pteris vittata* L. Environ Sci Technol 42:1479–1484

Yang Z, Shi W, Yang W, Liang L, Yao W, Chai L, Gao S, Liao Q (2018) Combination of bioleaching by gross bacterial biosurfactants and flocculation: a potential remediation for the heavy metal contaminated soils. Chemosphere 206:83–91

Ying M, Yasuda H, Kobayashi S, Sakurai N, Kidou SI (2019) Barley cold-induced CISP proteins contribute to the accumulation of heavy metals in roots. Environ Exp Bot 165:53–58

Zhang XY, Su C, Liu XY, Liu ZG, Liang X, Zhang YM, Feng YW (2020) Effect of plant-growth-promoting rhizobacteria on phytoremediation efficiency of *Scirpus triqueter* in pyrene-Ni co-contaminated soils. Chemosphere 241:125027

Zhou C, Hu KR, Xiao XY, Guo ZH, Peng C (2020) Arsenic retention characteristics of *Pteris vittata* L. Incinerated under different atmospheric nitrogen/oxygen ratio and temperature. Min Metall Eng 40:91–94

Chapter 41
Ecofriendly Management of Insect Pests for Sustainable Agriculture

Asim Abbasi, Aqsa Asif, Zahoor Ahmad, Inzamam Ul Haq, Asad Aslam, Ramish Saleem, Aliza Saleha, and Zeenat Zafar

Abstract The sustainability in food security is the prime objective of modern world as humans are facing severe challenges of poverty and hunger at global level. Among these challenges, insect pests are considered the main competitors of humans on earth and arguably poses a significant threat to our agriculture, ecosystem and food security. Today, a number of important insect pests have occupied diverse parts of the earth and cause severe economic losses to our pre- and post-harvest agricultural commodities. An accurate assessment of anticipated risks associated with different pest control tactics is necessary to devise some viable ecofriendly pest management approaches for sustainable agricultural intensification. Hence, a detailed knowledge regarding the pest habitat and its biology is needed to identify the best suited pest control tactic. In this regard, integrated pest management techniques provide useful and viable solutions to manage notorious

A. Abbasi (✉)
Department of Environmental Sciences, Kohsar University Murree, Murree, Pakistan

A. Asif
Department of Zoology, Cholistan University of Veterinary and Animal Sciences, Bahawalpur, Pakistan

Z. Ahmad
Department of Botany, Constituent Punjab College, University of Central Punjab, Bahawalpur, Pakistan

I. U. Haq
College of Plant Protection, Gansu Agricultural University, Lanzhou, China

A. Aslam
Key Laboratory for Sustainable Forest Ecosystem Management-Ministry of Education, College of Forestry, Northeast Forestry University, Harbin, Heilongjiang, People's Republic of China

R. Saleem
Department of Zoology, Constituent Punjab College of University of Central Punjab, Bahawalpur, Pakistan

A. Saleha · Z. Zafar
Department of Zoology, The Islamia University of Bahawalpur, Bahawalpur, Pakistan

M. Hasanuzzaman (ed.), *Climate-Resilient Agriculture, Vol 2*,
https://doi.org/10.1007/978-3-031-37428-9_41

field pests and reduce crop yield losses without contaminating environment. This chapter highlights all important sustainable pest control tactics such as microbial, biological, botanicals, sterile insect technique, sex pheromones, nutrient management and biotechnological approaches in detail that provide useful insights for not only conserving beneficial biodiversity but also global environment for our forthcoming generations.

Keywords Sustainable intensification · Integrated pest management · Food security · Environment

1 Introduction

Agriculture is arguably considered as the backbone of any country's economy, which may be influenced by a number of important factors (Loizou et al. 2019). In general practices, crop production of major crops is prone to heavy insect pest infestation, resulting in the overall decline of agricultural output (Sharma et al. 2017). Statistics showed that heavy infestation of insect pests can deteriorate 70% of total crop yield (Carvalho 2017). Integrated pest management (IPM) is a key strategy for sustainable agricultural development. Although synthetic chemicals significantly suppress pest manifestation (Dangles and Casas 2019), they also negatively impact human health and their surrounding environment (Struelens and Silvie 2020). According to a report published in 2005 by the world trade of pesticides, the amount of total chemical consumed in a year was greater than 31$ billion, with 25%, 48%, and 14% share of insecticides, herbicides, and fungicides/bactericides, respectively, whereas 3% contribution is reported from all other types of synthetic chemicals (Zhang et al. 2011).

The major concerns regarding environmental degradation are the gradual accumulation of hazardous chemicals on the soil surface, the draining of these pesticides into groundwater, and their harmful impacts on non-targeted entities including humans. Residues of these chemicals contaminate the whole ecosystem by altering air, water, and soil quality as well as negatively impacting the sea life beneath the ocean, including useful flora and fauna (Saxena and Pandey 2001; Riyaz et al. 2020). Hence, in order to preserve natural habitats and environment, scientists must look for sustainable agricultural practices (Slätmo et al. 2017). Sustainable conservation of natural resources, their proper management, and use of modern and improved technologies for the benefit of humans could best define the true spirit of sustainable agricultural approaches (FAO 2014). In this regard, IPM approaches are gaining popularity owing to their efficient role in pest suppression along with their ecofriendly nature. IPM is the combination of multiple approaches, including use of naturally occurring biological agents, pheromone-trap, and infertility-causing techniques in insects, along with DNA recombinant technology. These strategies are considered environment friendly because of their target specific action thus protecting both non-target organisms and global environment (Deguine et al. 2021).

2 Integrated Pest Management

Integrated pest management is usually a combination of techniques utilized to create a sustainable ecosystem comprising of precautionary and preventive measures to turn down insect pest attack and their reproduction by using ecofriendly strategies rather than encouraging the use of hazardous chemicals. This IPM model was first introduced by the end of 1950s and is considered a successful plant protection model which promotes natural biological control programs aiming to enhance crop production without causing any deadly effects on non-target organisms including humans (FAO 2020). Globally, IPM played a major role in the development of healthy pesticide free crops and also significantly enhanced crop production. Basically, IPM is based on a number of principles which helps us to remind that this strategy is not just a combination of multiple pest control practices instead, it is created to prioritize the different crop preventive measures with minimal utilization of synthetic pesticides/insecticides (Barzman et al. 2015). Before introduction of the IPM model, use of synthetic pesticides remained the first choice of every farmer, which made it difficult to achieve the goal of sustainable agricultural development and also negatively impacted the whole agroecosystem (Naranjo and Ellsworth 2009).

3 Microbial Control Agents

There are numerous microorganisms that work as naturally occurring biological control agents (Chowdhury et al. 2015; Yoo et al. 2018; Sendi et al. 2020). These microbes are named as entomopathogens; among them, the most prominent microbes are nematodes, bacteria, viruses, and fungi which could protect various pre- and post-harvest stages of crops from insect pest infestations (Choudhary and Sindhu 2015; Mihalache et al. 2018; Yu et al. 2018). These microbes have the ability to produce antibiotics, siderophores, toxins, hydrogen cyanide, and various other metabolites, which are essential in multiple IPM programs utilized in different agroecosystems for controlling insect pests. Moreover, the chances of resistance development against these microbial agents are also low because of their dual action mechanisms (Yu et al. 2006; Sehrawat and Sindhu 2019).

3.1 Entomopathogenic Bacteria

Among different microbial agents, bacteria are the most widespread entomopathogens effective against a score of economically important insect pests (Mampallil et al. 2017). It is believed that more than 100 bacterial species have been successfully utilized in different biological control programs against notorious insect pests attacking our field crops. Most of these bacterial species belong to three main families: Pseudomonaceae, Bacillaceae, and Enterobacteriaceae (Ruiu et al. 2007). Among these bacterial entomopathogens, *Bacillus thuringiensis* is the most effective and viable microorganism targeting a wide range of important insect pests (MacGregor 2006). Upon ingestion, the bacterium secretes spores along with Bt Cry proteins which are lethal for insect pests belonging to different insect orders, including Diptera, Coleoptera, and Lepidoptera. The Bt proteins are proteolytically converted into toxic core fragments after solubilization in insect midgut and attach with receptors of the apical microvillus membrane (Bravo et al. 2013). Latterly, the toxins makes pores in the cell membrane, upset the cell osmotic balance and disturb the integrity of gut epithelium leading to septicemia (Ortiz and Sansinenea 2021). Moreover, biopesticides of *B. thuringiensis*, either natural or commercial, usually bypass the host defense system, particularly attains and cecropins which produce different antibacterial peptides in the host body to neutralize the ill effects of Bt toxins (Sellami et al. 2015).

3.2 Entomopathogenic Fungi

Similarly, entomopathogenic fungi (EPF) have the potential to be utilized as a biocontrol agent against important crop pests, offering a more ecologically friendly alternative to conventional insect pest management approaches (Khachatourians and Qazi 2008; Malik et al. 2019; Sufyan et al. 2019). EPFs are renowned for their ability to assault pests and cause infection by directly colonizing and piercing the insect body via cuticle. According to an estimate, nearly 700 fungal species fitting in 90 genera have the potential to infect insect pests. Among these, the most widely used EPF usually belongs to genus *Beauveria*, *Isaria*, *Metarhizium*, *Lecanicillium*, and *Hirsutella* (Inglis et al. 2001).

The most prevalent entomopathogenic fungi, *B. bassiana* and *M. anisopliae*, have been isolated from diseased insect cadavers found in both tropical and temperate parts of the globe (Zimmermann 2007; Sufyan et al. 2017). Various EPF species are claimed to be innocuous and environmentally sustainable since they can invade different hosts without inflicting any detrimental effects often associated with synthetic chemicals (Gao et al. 2017). It has been demonstrated that these EPFs perform significant multitrophic functions and naturally suppress pest populations by inducing epizootics. Moreover, the endophytic potential of this class of entomopathogens makes them a viable option to manage certain cryptic feeding insects, which are usually difficult to manage with synthetic chemicals (Posada and Vega 2006; Tefera and Vidal 2009).

Uptil now, multiple companies have successfully launched spore-based entomopathogen insecticides (Wraight et al. 2001). Moreover, majority of these products have fulfilled the requirements for registration and are now successfully utilized for pest management in multiple nations (de Faria and Wraight 2007; Akutse et al. 2020).

EPF also causes certain indirect deleterious impacts on certain pests through a wide range of indirect strategies, such as antixenosis and antibiosis (Shah and Pell

2003; Hartley and Gange 2009). Endophytic fungus protects their host cells against insect herbivore damage through a number of mechanisms, including pest deterrence, pest avoidance, reduction in feeding, longevity, oviposition, growth, and development rate (Lekberg and Koide 2005; Lacey and Neven 2006; Wang and Qiu 2006; Valenzuela-Soto et al. 2010). Additionally, certain insect-pathogenic fungal strains have also been produced as natural bio-pesticides and are now readily available for the farming community. Hence, EPF are currently considered as most safe alternative of synthetic chemicals to manage notorious field pests (Vega et al. 2008; Fang et al. 2014).

3.3 Entomopathogenic Nematodes

Under the wide umbrella of Nematodes lies the two most significant families, i.e., Steinernematidae and Heterorhabditidae, which comprises certain useful types of nematodes named Entomopathogenic nematodes (EPNs). These EPNs act as holoparasites and remain alive to maintain their existence; also, their lethality was observed against numerous crop pests (Bhat et al. 2020; Liu et al. 2020). So far, 1700 species of Heterorhabditis and Steinernema have reported their lethal effects on various insects (Bhat et al. 2020). Moreover, such nematodes having multiple bacteria in a symbiotic relationship are considered efficient and significant naturally occurring biological control agents around the globe (Lalramnghaki 2017). Usually, these EPNs kill the host insects. However, during the complete life cycle of a nematode, one of the specialized form, IJ3 (infective juvenile), stays outside the body of the host and transport endosymbionts, which reproduce within the intestine of IJ. The other forms of infective juveniles, IJ1 and IJ2, are completely dependent upon a host for living and do not possess any symbiotic relationship with gut bacteria. The most important types of bacteria living inside the host are Xenorhabdus and Photorhabdus, which possess almost similar life cycles to nematodes (Sicard et al. 2004).

When EPNs attack their host insects, it slowly enters the body cavity (hemocoel) to discharge endosymbionts. Soon enough, these ejected bacteria start moving inside the host and release several poisonous compounds to cause toxicity and pathogenicity (Koppenhöfer et al. 2007). These bacteria not only produce toxins but also protect the remains of host insects from predators and other pathogens by producing noxious chemicals. Similarly, the production of these guarding compounds also helps nematodes to reproduce (Gulcu et al. 2012). Eventually, after two or three successions of nematode reproduction, the corpse of host insects becomes insufficient to fulfill the nutrient requirements. Therefore, the nematode finds another host and provide us better pest control as compared to synthetic chemicals (Grewal and Georgis 1999).

3.4 Entomopathogenic Viruses

Entomopathogenic viruses served as an important class of bio-control agents. They became an integral component of different IPM programs primarily because of their target specificity, compatibility with other control tactics, and environmental safety. Besides their direct toxicity, this particular class of biological control agents also causes epizootics in field conditions, which gives better control of cryptic insect pests. More than 800 mite and insect species are vulnerable to different viral infections. However, members of the baculoviradae family were the most common viruses, which cause a number of deadly diseases in a score of important insect pests, particularly lepidopterous pests (Black et al. 1997). Granuloviruses (GV) and nucleopolyhedroviruses (NPV) are the two most important viruses of this family, infecting a number of important insect pests (Shiga et al. 1973; Sekita et al. 1984).

The viruses usually target their host after ingestion and initially infect their digestive system. However, the infection spreads to other body organs in case of severe infection, and in the end, the virus-killed insects usually resemble a virus particle. The infected insects usually feed and look like normal insects until death, as the digestive system is the last body organ to be demolished. However, their color usually changes, and their movement is restricted just before death. The virus-infected insects usually develop much slower than healthy insects (Toprak et al. 2012). The cadavers of virus-infected insects can be seen on plants on which they feed. The cadavers usually burst and disseminate the virus inoculum into the environment to infect healthy insects. The other means by which virus inoculum can spread from one place to another includes the movement of infected insects, birds, or other predators that come in contact with these insects and some abiotic factors such as rain splash or water runoff (Wang and Granados 1997; Braunagel and Summers 2007).

The target-specific action, ecofriendly nature, and potential of managing insect pests through secondary inoculum make baculoviruses an essential component of different IPM programs and a suitable replacement of synthetic chemicals (Huber 1986; Cunningham 1995; Petrik et al. 2003).

4 Biological Control Agents

Biocontrol agents, also known as "beneficial insects," are a group of insects that help to lower populations of harmful insect pests and crop damage. These insects are divided into two main groups: parasitoids, insects that inhabit their host and feed from inside, and predators, who feed directly on pests from outside (Colmenarez et al. 2018). There are two types of biological control: conservation biological control and natural biological control. The term "conservation biological control" refers to activities carried out by humans to preserve and improve the effectiveness of naturally existing beneficial insects. Similarly, insects and other harmful organisms may be kept under control by their natural predators and parasites via the process of natural biological control. This is the most economical, sustainable, and significant contribution of biological control to agriculture and mankind, and it occurs naturally in all habitats across the globe (Waage and Greathead 1988; Colmenarez et al. 2018). In another kind of biological control known as inoculative biological control, the natural enemies of the pest are gathered from an area in which they are already prevalent and then released in other locations where the pest was unintentionally introduced. The progeny of the natural enemies that were released into the new environment usually maintain their population up to a considerable number, so that it may cause a reduction in the pest populations during succeeding years. The most common use of this kind of biological pest management has been made against invasive species, which are thought to have travelled to a new location without the presence of their native predators or parasites. However, in augmentative biological control, population of a particular natural enemy is maintained in a biofactory before being released in field. This approach is usually carried out in order to get quick control of the target pests. This is done in order to maximize the performance of natural enemies in open field conditions (van Lenteren and Bueno 2003). This particular approach is considered as economically viable and ecologically sustainable alternate to the use of synthetic chemicals in certain agricultural systems (Colmenarez et al. 2018).

4.1 Predators

Predatory insects, as the name implies, feed on other insects, including their eggs, larvae, nymphs, pupae, and adults. Many insect species are predatory both as larvae and adults, albeit they may not feed exclusively on the same food substrate. Others are only predatory as larvae, with adults subsisting on floral nectar and honeydew. However, the non-predaceous adult females actively hunt out prey to lay their eggs. This is because their larvae often cannot locate the prey on their own. Beetles, lacewing larvae, and mites are just a few predators that actively seek their food on the ground, on the plants, or capture it in flight, like robber flies and dragonflies. The labial mask of aquatic dragonfly nymphs and barbed forelegs of mantids are only two examples of the specialized capturing organs used by some predators (DeBach and Rosen 1991). Predators use a wide variety of traps for capturing host insects, the well-known of which are spider webs and ant lion sand pitfalls (Beckage et al. 1993).

Both the direct introduction of predators into farms and the use of certain floral plants to entice predators to the region are viable options. Assassin bugs are generalist predators that may be observed in almost all environments and feed on eggs, caterpillars, and other species that feed on plants (Stewart et al. 2007; Grantham and Arnold 2013). Similarly, another most common predator, big-eyed bugs can survive throughout the season after successfully colonizing a new plantation. These bugs exclusively feeds on all stages of mites, whiteflies, aphids and caterpillars. Moreover, they also maintain their population by eating plants when there are no other insects around, yet they do little to no harm to the plants (Stewart et al. 2007). Similarly, the predatory stink bugs named as spined soldier bugs are sometimes misidentified for the brown stink bug, which feeds on plants. They are generalist predators, and their primary diet consists of the larvae of leaf beetles as well as soybean looper larvae. Spined soldier bugs are big enough, even fully developed caterpillars (Stewart et al. 2007; Grantham and Arnold 2013).

There are hundreds of distinct species of ladybeetles, sometimes known as ladybugs, which belong to the family Coccinellidae in the order Coleoptera. Lady beetles are among the most common beneficial insects. They are voracious predators, and both the adult and larvae prefer to feed on aphids, although they will also feed on other soft bodied insects (Stewart et al. 2007). There is a remarkable amount of diversity in the predatory behavior of these polyphagous coccinellids. Some of them are mobile and actively search for prey, while others are stationary and undergo preimaginal growth on a single leaf. Lacewings, which belong to the order Neuroptera are carnivorous predators like ladybird beetles. Aphids, whiteflies, moths, and tiny caterpillars are their main hosts (Stewart et al. 2007).

4.2 Parasitoids

Most of the parasitoids that are used in the biological management of insect pests are classified into two distinct orders, i.e., Hymenoptera and Diptera. The vast majority of parasitoid insects that feed on other insects are protelean parasites, which means that they are pathogenic only during their juvenile stages and otherwise live independently as adults. They will often devour the whole or the majority of the body of their host before going to pupate either within or outside of the host. After emerging from its cocoon as a pupa, the adult parasite begins a new cycle of reproduction by venturing out in search of living hosts where it may attach its eggs. Most adult parasitoids need food in the form of pollen, nectar or honeydew, and many feeds on the body fluids of their hosts (DeBach and Rosen 1991. On the other hand, certain adult parasitoids can also survive on free water bodies. The life cycle of a parasite may be univoltine, where it has one generation for each host generation, or multivoltine, where it can have two or more generations for each host generation (Altieri and Nicholls 1998). The parasites may either act as solitary, where only one larva can develop on a single host specimen, or cosmopolitan, where several larvae will develop simultaneously on a single host. However, this is not always the case since there are certain species can develop facultatively as solitaries on a tiny host and jovially on a bigger host. Certain adult parasitoids use a range of signals from both the target host and its surroundings to locate their hosts in the surrounding environment. Parasitoids can locate their victims via the use of visual, olfactory, and tactile cues (DeBach and Rosen 1991). Once a host has been located, the female parasitoid will deposit an egg in, on, or around the host body. Larval parasitoids feed on the host's tissues, but they might not be able to cause the death of the host until after it has grown and reached the adult stage. The parasitoids have their pupal stages either inside or close to the host, and the newly emerged adults rapidly mates and scatter in the surroundings in order to locate new hosts (Hoy 1994).

Parasitoid wasps have a long history of providing significant biological control solutions against agricultural and forest pests. There are about 5500 known species of parasitoid wasps, belonging to 48 families within 12 hymenopteran superfamilies. Many of these insects have been successfully used in different biological control programs (Chen et al. 2014). Asian corn borer is the most devastating corn pest in different parts of the globe, and *Trichogramma* species are still efficient agents for its management (Wang et al. 2015).

5 Botanical Pesticides

Naturally, plant products have vast use and application possibilities in modern agricultural production systems. It is been shown that some plants include biologically active chemicals that are very efficient against a number of important insect pests (Neeraj et al. 2017). Phytochemicals, or botanicals, are active compounds derived from plants and employed in different IPM programs with great success. Their effectiveness in lowering crop toxicity has led to widespread acceptance of their usage in modern agriculture (Neeraj et al. 2017). Phytochemicals have increased significance from planting to harvesting and storing crops because of their unique properties, such as decreased toxicity, biodegradability, diverse action mechanism, effectiveness, and wide availability of primary sources (Neeraj et al. 2017). Ecofriendly botanical pesticides have been shown to be efficient against a number of important pests, including bacteria, fungi, nematodes, and insects (Feyisa et al. 2015; Todorov et al. 2015; Muthomi et al. 2017).

Since so many different botanical pesticides are available, it is now possible for us to replace synthetic chemicals with these natural alternatives for keeping the pest population below their threshold levels. This would significantly pave the way for successful adoption of different IPM programs for environment-friendly agriculture (Das 2014). Approximately, 2500 plant species belonging to 235 plant families have been recognized to possess certain biochemical qualities and can act as toxic pesticide, a repellent to pests, or a growth regulator (Roy et al. 2016; Wanzala et al. 2016; Ahmad et al. 2017). There have also been reports of various widely used plants having pesticidal biochemical properties which includes *A. indica, T. cinerariifolium, A. sativum, C. longa, R. officinalis, Z. officinale* and *T. vulgaris* (Castillo-Sánchez et al. 2015). Since they provide effective outcomes, isolated botanical compounds like azadirachtin and pyrethrum are frequent examples of distinct phytochemicals that have been successfully marketed in different parts of the globe (Kumar et al. 2015; Laxmishree and Nandita 2017).

The physio-chemical processes, metabolic activities, behavioral patterns, and morphological characteristics of insect pests change with the application of different botanicals, especially metabolites. The blockage of glucose in chemo-sensory receptor cells of lepidopterous pests associated with triterpenoids and the chemo-sterilant activity of certain plant-based essential oils are the best examples of unique acion mechanisms of these plant-based products (Lengai et al. 2020).

6 The Sterile Insect Technique

The sterile insect technique (SIT) basically works for the reduction in the population of the specific insect through sexually sterilizing, mass-producing, and eventually enabling interaction of conspecific individuals in the wild (Dyck et al. 2021). The transformed males will compete for females by their counterparts, but this type of mating will not produce fertile progeny. The wild populations of a particular species can be eliminated by releasing ample number of sterile insects in a locality. This will ultimately diminish the ratio of fertile insects in that area leading to the local extinction of that particular species. The target-specific action of SIT makes it a viable option to use in different IPM programs. This technique can be considered for managing different insects, including certain field pests such as fruit flies, beetles, and many other vectors that carry disease-causing pathogens (Klassen et al. 2021).

The biological and ecological state of the targeted insect, the landscape or area features, and the feasibility of producing and releasing sterile insects will decide the volume of an area over which the SIT can be applied successfully (Lance and McInnis 2021). The success of SIT technique depends on its compatibility with other types of insect control methods, such as mating disruption, and is commonly used in different IPM and area-wide (AW) control programs. The integrated action of SIT and mating disruption can be synergistic even though both techniques do not necessarily kill the target pests but cause a significant reduction in the pest population, usually below their threshold level (Bloem et al. 2007). This leads to extinction due to the unavailability of locating potential mates for progeny development and ecological stochasticity (Marec and Vreysen 2019). Similarly, the integration of biological control and SIT is also synergistic in action (Rendón et al. 2006) as the former performs better when the pest population is high (Vargas et al. 2004) and cause a reduction in pest population to a level that can be efficiently controlled with SIT technique (Wong et al. 1992; Gurr and Kvedaras 2010).

7 Management of Pests Using Sex Pheromones

Sex pheromones are usually the chemicals signals which are relased by members of a specie to provoke a sexual response in opposite sex member of same specie (Smart et al. 2014; Seybold et al. 2018). The first sex pheromone was characterized in 1959 from a silkworm specie 1959 (Butenandt 1959; Karlson and Lüscher 1959). Until now, about 600 species of lepidopterous pheromones have been recognized and used in different area wide control programs of different field pests (Petkevicius et al. 2020).

Certain promising features of sex pheromones, such as target specificity, rapid degradation, and high efficiency even at low doses, make them a better option than most other pest management agents. These characteristics not only help in managing insect pests but also aid in estimating their field population and spotting alien species' entry points and progress (Reddy and Guerrero 2010; Tewari et al. 2014; Yew and Chung 2015; Larsson 2016). Recently, sex pheromones have been deployed in different IPM programs with different motives, such as monitoring pest populations, mating disruption, and mass trapping (Trematerra 2012; Tewari et al. 2014).

7.1 Monitoring

Sex pheromones have been most widely deployed in monitoring pest populations (Witzgall et al. 2010). Monitoring systems primarily work on the interaction between the trap and pest population or the damage done by the pest species. The number of trapped males is used to assess the threshold level of that particular species for taking necessary management actions. Pests with low population levels can also be detected with sex pheromones because of their target specificity. Invasive species can also be detected with this technology, allowing farmers to devise timely management approaches, thereby reducing ecological and economic costs (Smart et al. 2014).

7.2 Mass Trapping

Mass trapping is taken as a direct control plan for mitigating the population density of a specific specie by using a large number of pheromones (Jones 1998; Trematerra and Colacci 2019). Mass trapping is a more effective way of managing insect pests than mating disruption, especially when both have the same number of pheromone sources (El-Sayed et al. 2006; Byers 2012). Mating disruption only prolongs the time to find a mate, whereas mass trapping delays this activity for an indefinite time (Hegazi et al. 2009; Trematerra 2012).

7.3 Mating Disruption

This is a technique in which crop plants are permeated with synthetic sex pheromones to avoid communication between opposite genders of a similar insect species so that mating between them can be prevented. This technique has been broadly used for managing different moth insects. The ecological safety and target specificity of this particular technique make it a reliable and viable tool to be used in different area-wide pest management programs, including invasive species (Lance et al. 2016).

7.4 *Push-Pull Strategy*

The push-pull strategy is becoming reliable and can be easily used as an alternative to synthetic pesticides. In this particular strategy, attractive or repellent stimuli are used simultaneously for the diversion of pests. This strategy works on the simple principle in which the pests are directed away from a target crop by using the repellent stimulus. On the other hand, these pests are attracted towards another area by the use of attractant stimulus. The agricultural system is now using this push and pull technique to deal with the issues related to pesticide resistance and injudicious use of pesticidal applications. Certain ecological and biological parameter of insects along with relations between host plants and their natural enemies largely determine the success of implementing this ecofriendly pest management approach (Cook et al. 2007). Sex pheromones can be integrated with different push-pull techniques to initiate the timing of a stimuli and other population management actions (Dickens 2006; Reddy and Guerrero 2004).

8 Management of Insect Pests with Nutrients

Resistance of crop plants against different biotic and abiotic stresses and their yield can be significantly enhanced with proper nutrient management programs (Van Bockhaven et al. 2013; Golubkina et al. 2014; Liang et al. 2015; Sakr 2016). Adding proper mineral nutrients to crop plants is a sustainable alternative for managing cryptic insect pests that have become resistant against synthetic pesticides. Some specific nutrients can increase crop resistance against a number of important insect pests, including sap-sucking and chewing insects (Golubkina et al. 2014; Liang et al. 2015; Reynolds et al. 2016).

8.1 *Silicon*

Insect pests usually possess the capability to redesign their metabolism against management approaches that have a single action mechanism; hence, plant protection strategies need instant improvements for a sustainable future (Reynolds et al. 2009; Kvedaras et al. 2010; Dias et al. 2014; Grover et al. 2019). Application of silicon (Si) induces resistance in crop plants against a score of economically important insect pests with diverse feeding habits (He et al. 2015; Abbasi et al. 2022).

Moreover, due to its low environmental toxicity and beneficial nature, it has become a vital part of different IPM programs (Bellotti and Arias 2001; Epstein 2001; Hou and Han 2010). Silicon is taken up from soil in the form of H_4SiO_4 (Ma 2004; Epstein 2009), which ultimately concentrates in the epidermal cells of plant tissues. After deposition in plant cells, Si limits insect feeding by increasing the hardiness and abrasiveness of plant tissues making them less preferable for feeding and egg laying (Massey and Hartley 2009; Hartley et al. 2015). Similarly, the chemical defenses induced by Si deposition include the activation of plant defensive enzymes (Ye et al. 2013) including trypsin protease, polyphenol oxidase, and phenylalanine ammonia-lyase (Reynolds et al. 2016; Yang et al. 2017). These enzymes significantly extends the developmental period of insect pest making them vulnerable to biological control or other pest control tactics. Moreover, the application of Si also acts at the third trophic level and is also involved in attraction of natural enemies of target pests in that area through production of certain herbivore-induced plant volatiles (Liu et al. 2017 Abbasi et al. 2020).

8.2 *Selenium*

Selenium in low concentrations is highly effective for normal plant growth as it enhances plant resistance against salinity, water scarcity, and UV-induced stress acts as an antioxidant, and hinders the upward movement of certain toxic heavy metals (Pedrero et al. 2008). This will ultimately cause a significant improvement in the quantity and quality of crop produce (Mechora et al. 2011). Moreover, selenium (Se) sources and their concentrations in plants also affect the population density of certain insect pests primarily through direct toxicity and impeded insect growth and development (Trumble et al. 1998). Moreover, Se enriched diet significantly affects the oviposition and feeding site of *S. exigua* (Vickerman and Trumble 1999). Similarly, the preference for *M. persicae* was recorded on *Brassica juncea* plants treated with and without Se. The trial results showed that a significant number of *M. persicae* insects avoided Se-enriched plants (Hanson et al. 2004).

9 Biotechnological Approaches in Insect Pest Management

Biotechnology is a branch of science that effectively utilizes a specific portion of living organisms or biological systems to produce or modify a product, animal, microorganism, or plant for particular desirable functioning (Persley 2000). It can also be defined as the deliberate controlled manipulation of certain biological organisms to achieve effective pest control under different ecological conditions. The role of different biotechnological approaches, such as the incorporation of novel transgenic genes in crop plants and insect resistance breeding, showed promising results in different pest management programs. There is a score of important biotechnological approaches successfully used in different insect pest management programs. However, gene transformation, RNA interference, and anther culture are of significant importance and are mentioned in detail in the below section (Talakayala et al. 2020).

9.1 Gene Transformation

Genetic engineering or gene transformation of crops for resistance development usually involves incorporating genes or DNA segments into crop plants, making them resistant to insect attack. The introduced DNA segments normally encode a protein having an insecticidal action which offers protection against a number of important field pests. Insect pests from the order Diptera, Coleoptera, and Lepidoptera are successfully managed using this gene-incorporation technology.

9.1.1 Bt Proteins

Since their first introduction in 1996, genetically modified crops, including Bt-producing insecticidal proteins derived from soil-dwelling *B. thuringiensis* bacteria have been widely used all over the globe (Abbas 2018). Particle bombardment of *Agrobacterium*-mediated transformation technologies is used to successfully incorporate Bt cry genes in different crop plants (Juturu et al. 2015). The Bt genes producing parasporal crystalline protoxins are solely responsible for insecticidal activity in Bt crop plants. First, the crystalline protoxin proteins get solubilized in the alkaline midgut of target larvae and later cleaved enzymatically to become an active toxic protein (Palma et al. 2014). The active toxin binds to the midgut epithelium receptor cells after diffusing through the host peritrophic membrane, which covers its gut. The toxins make holes in midgut epithelium which paralyzes the insect gut forcing it to stop feeding and ultimately leading to death in 2–3 days (Paul and Das 2020).

9.1.2 Vegetative Insecticidal Proteins Genes

The bacterium *B. thuringiensis* possess natural pesticidal properties against a wide range of insect pests. The bacteria is found in a variety of ecological habitats and secretes numerous proteins during different growth stages of its life history. Apart from Bt cry proteins, the other best entomocidal Bt proteins are Vips having three important subfamilies, i.e., Vip1, Vip2, and Vip3. They are considered an effective entomo toxic candidate due to their unique sequence homology and receptor binding sites. The Vip proteins are produced during the crop plant's vegetative phases and protect them against target insect pests. The heterodimer toxins belonging to Vip1 and Vip2 subfamilies are effective against insect pests belonging to the order

Coleoptera and Hemiptera; however, Vip3 toxins provide protection against lepidopterous pests and are the most extensively studied Vip toxin family. The Vip proteins can be successfully used individually or in integration with cry proteins to manage different pests and are usually known as second-generation entomocidal proteins (Gupta et al. 2021). The pathogenicity of these Vip3 proteins against pests is similar to that of cry proteins, and in some cases, they showed more toxicity against those pests which cannot be controlled with cry proteins. The Vip proteins target their host by osmotic lysis, which usually results in the interruption and swelling of midgut epithelial cells. Moreover, in different transgenic crops, the Vip proteins have been pyramided with cry proteins to provide better pest control and overcome resistance to pests (Syed et al. 2020).

9.1.3 Fusion Proteins

B. thuringiensis-based biopesticides containing different Bt toxins usually account for almost 90% of the total microbial insect control products (Tabashnik 1994). The widespread adoption of Bt-based products might trigger the development of resistance in their target hosts (Ferré et al. 1995). Hence, to address the potential limitations of conventional transgenic crops, gene pyramiding or stacking and using hybrid toxins in the same crop plant is the best option to counter the resistant insects. Fusion proteins are made with combination of various insecticidal proteins. In transgenic plants, following transcription and translation, these fusion proteins make a single polypeptide unit, providing better efficacy against phytophagous insect pests. Dow Agro-Sciences created a transgenic cotton line with a hybrid fusion protein, cry1Fa + cry1Be, which showed greater resistance against *O. nubilalis* and *S. litura* as compared to single gene Bt cotton (Meade et al. 2017). Similarly, Chakraborty et al. (2016) also reported that transgenic rice having *cry2AX1*, a derivative of *cry2Ac* and *cry2Aa* gene, showed greater resistance against lepidopterous insects. Furthermore, Koerniati et al. (2020) also reported that sugarcane cultivars having both cry1Ac + cry1Ab fusion proteins exhibited enhanced resistance against shoot borers. It was also reported that phytophagous aphids were more susceptible to those transgenic *B. juncea* lines with fusion proteins derived from protease inhibitors and lectin (Rani et al. 2017). Similarly, the susceptibility of *H. axyridis* and *S. exigua* was more pronounced on those hybrid maize lines which were transformed using cry2Aj/cry1Ab fusion proteins (Chang et al. 2017).

9.1.4 Protease Inhibitors

Protease inhibitors (PIs) are a class of plant-based inhibitors that diminish the activity of digestive proteases and limit insects to digest their food properly (Macedo and Freire 2011). The PIs prevent proteolysis, which often results in increased pest mortality, prolonged developmental duration, and reduced fecundity primarily due to deficits of essential amino acids in the insect body (Haq et al. 2004; Zhu-Salzman and Zeng 2015). Cystatins and serpins are the two most studied plant PIs effective against a number of important insect pests. Serpins having an approximate molecular mass of 39–43 kDa are usually the inhibitors of serine proteases. The insecticidal activity of serine proteases has been explored against different insects belonging to order Orthoptera, Hymenoptera, Lepidoptera, Coleoptera, and Diptera (Irving et al. 2002). Similarly, cystatins having a molecular mass of 12–16 kDa usually hinder the functioning of cysteine proteases which are prime food-digesting proteases in insects belonging to the order Hemiptera and Coleoptera.

Methyl jasmonate, a key regulator of plant defensive mechanism against insect pests, inhibits gut protease in wounded plants, causing nearby unwounded plants to produce proteases inhibitors that intimate local plantations to combat insect pest attack (Gatehouse 2011; Stevens et al. 2012; Sharma 2015; Singh et al. 2016). Similarly, legume trypsin inhibitors inhibit a wide range of proteases and possess an entomocidal activity against a score of important insect pests (Macedo et al. 2004). Furthermore, rice and wheat cultivars have been amended with PI genes to enhance their resistance against borers and stored and foliage-feeding insects, respectively (Xu et al. 1996; Altpeter et al. 1999; de Pg Gomes et al. 2005). Increased insect mortality and retarded growth were recorded when protease inhibitors were offered to insects through transgenic plants or as an artificial diet (Duan et al. 2018; Schneider et al. 2017).

9.1.5 α-Amylase Inhibitors

The digestion of carbohydrates in insects is dependent on a digestive enzyme called α-amylase. However, the inhibitors of α-amylase affect the normal working of this digestive enzyme and diminish insect digestion (Ishimoto and Kitamura 1989). A score of economically important insect pests has been controlled with these α-amylase inhibitors present in different vegetative organs of crop plants and seeds (Chrispeels et al. 1998). Similarly, the biology of *C. maculatus* and *C. chinensis* was negatively affected when offered seeds of *P. vulgaris* expressing an α-amylase inhibitor (Shade et al. 1994). Similarly, Kaur et al. (2022) stated that enhanced resistance level in maize plants against *C. partellus* infestation might be attributed to the enhanced activity of these α-amylase inhibitors.

The inhibition of α-amylase enzyme using wheat α-amylase inhibitors is also associated with reduced damage potential and fecundity of a major wheat pest, *R. hyzopertha dominica* (Priya et al. 2010). Similarly, an α-amylase inhibitor coding gene named aAI-Pc1 was isolated from *P. coccineus* and incorporated into the coffee plant. The transgenic plant later showed enhanced resistance against coffer borer, *H. hampei* (de Azevedo et al. 2006). Furthermore, *C. maculates* exhibited higher larval mortality, reduced oviposition, and adult longevity when exposed to papaya seeds enriched with α-amylase inhibitor (Farias et al. 2007). The above studies depict the successful integration of α-amylase inhibitors genes in different sustainable insect pest management programs.

9.1.6 Insect Chitinase

Insect chitinase enzymes are usually hydrolytic and can degrade or inhibit insect chitin (Pan et al. 2012). Chitin is the main constituent of insect peritrophic membrane and exoskeleton and protects insects from natural enemies, mechanical disruption, and harsh ecological conditions (Su et al. 2016; Chen et al. 2018). The expression of chitinase enzyme has been expressed in diverse transgenic crops and proved to be a viable option for managing different economically important insect pests such as *N. lugens*, *C. suppressalis*, and *P. solenopsis* (Xi et al. 2015). Similarly, transgenic maize having the chitinase gene exhibited higher resistance against *S. cretica* (Osman et al. 2016). Due to their inhibitory action on different biological parameters of insect pests, nowadays, insect chitinases have gained much attention as a transgene and commercial biopesticide (Omar et al. 2019).

10 RNA Interference for Plant Resistance to Insect

RNAi is a method of inducing resistance in plants against insects and involves the suppression of specific gene expression by suppressing its specific sequences. This technique is also renowned with certain other names, such as post-transcriptional gene silencing, concealment, and suppression (Kamthan et al. 2015). It is an advanced novel technique of gene silencing mechanisms initiated by double-stranded RNA at the cell level. The undesirable genes are usually repressed when a dsRNA is inserted into the cell. The RNAi technology works when dsRNA is inserted into the host system, where it expresses throughout the insect system either through a hairpin or by some other means (Katoch et al. 2013). This technique has been deployed to manage a score of important insect pests, especially sap-feeding insects, which are usually not controlled with transgenic crops. This particular technology opens new avenues for sustainable pest management of economically important crop pests (Mamta and Rajam 2017).

11 Anther Culture

It is a technique that involves the division of immature pollen and is later allowed to develop into embryonic tissue or callus to create haploid plants. In this technique, the pollen containing anthers is separated from flowers and placed in a suitable artificial growth medium for development. Anthers are used to produce a callus, root, shoot, and even a large plant using a suitable artificial growth medium. However, all the plants which are grown using this technique are haploid in nature. It is the most efficient way of rapidly producing a large number of insect-resistant homozygous haploid plants. Certain rice anther culture lines, such as 953510, 953508, 952836, 953527, 953509, 953541, and 953511, have been produced using this technique. All these lines exhibited a moderate level of resistance against *L. oryzophilus* (N'guessan et al. 1994). Similarly, certain rice lines which showed resistance against brown planthopper, stripe virus and bacterial blight have been developed using this anther culture technology and have been successfully tested in field conditions (Park et al. 2014).

12 Conclusion

Crop insect pests pose a considerable threat to global food security and poverty mitigation program all around the globe. Integrated pest management techniques include a diverse array of methods by which pests can be kept below their threshold level, and damage can be reduced. A wide range of off- and on-farm advantages, such as better livelihood and improved ecological and public health, can be achieved with the mitigated use of synthetic pesticides. Therefore, farmers must consider all the possible IPM techniques so that pests can be managed in a sustainable way without inflicting any damage to our environment, and the needs of future generations may not be compromised. Benefits go beyond improved crop yields and may include better soil health, improved biodiversity, livestock integration, and income diversification. These, coupled with better cost-effectiveness and reduced crop losses, make IPM techniques viable for all groups of farming communities all around the globe. However, efforts are still needed in this regard to enhance the efficacy of different IMP tools to make our agroecosystem more sustainable for our forthcoming generations.

References

Abbas MST (2018) Genetically engineered (modified) crops (*Bacillus thuringiensis* Crops) and the world controversy on their safety. Egypt J Biol Pest Control 28:1–12

Abbasi A, Sufyan M, Arif MJ, Sahi ST (2020) Effect of silicon on tri-trophic interaction of cotton, *Gossypium hirsutum* (Linnaeus), *Bemisia tabaci* (Gennadius) (Homoptera: Aleyrodidae) and the predator, *Chrysoperla carnea* (Stephens) (Neuroptera: Chrysopidae). Arthropod-Plant Inter 14:717–725

Abbasi A, Sufyan M, Ashraf HJ, Zaman Q, Haq IU, Ahmad Z, Saleem R, Hashmi MR, Jaremko M, Abdelsalam NR (2022) Determination of silicon accumulation in non-bt cotton (*Gossypium hirsutum*) plants and its impact on fecundity and biology of whitefly (*Bemisia tabaci*) under controlled conditions. Sustainability 14:10996

Ahmad W, Shilpa S, Sanjay K (2017) Phytochemical screening and antimicrobial study of Euphorbia hirta extracts. J Med Plants Stud 2:183–186

Akutse K, Subramanian S, Maniania N, Dubois T, Ekesi S (2020) Biopesticide research and product development in Africa for sustainable agriculture and food security – experiences from the International Centre of Insect Physiology and Ecology (icipe). Front Sustain Food Syst 4:152. https://doi.org/10.3389/fsufs.2020.563016

Altieri MA, Nicholls CI (1998) Biological control in agro-ecosystems through management of entomophagous insects. In: Dhaliwal GS, Heinrichs EA (eds) Critical issues in insect pest management. Commonwealth Publishers, New Delhi, pp 67–86

Altpeter F, Diaz I, McAuslane H, Gaddour K, Carbonero P, Vasil IK (1999) Increased insect resistance in transgenic wheat stably expressing trypsin inhibitor CMe. Mol Breed 5:53–63

Barzman M, Bàrberi P, Birch ANE, Boonekamp P, Dachbrodt-Saaydeh S, Graf B, Hommel B, Jensen JE, Kiss J, Kudsk P, Lamichhane JR, Messéan A, Moonen AC, Ratnadass A, Ricci P, Sarah JL, Sattin M (2015) Eight principles of integrated pest management. Agron Sustain Dev 35:1199–1215

Beckage NE, Thompson SN, Frederici BA (eds) (1993) Parasites and pathogens of insects, vol. 1, Parasites. Academic Press, London

Bellotti AC, Arias B (2001) Host plant resistance to whiteflies with emphasis on cassava as a case study. Crop Prot 20:813–823

Bhat AH, Chaubey AK, Askary TH (2020) Global distribution of entomopathogenic nematodes, Steinernema and Heterorhabditis. Egypt J Biol Pest Control 30:1–15

Black BC, Brennan LA, Dierks PM, Gard IE (1997) Commercialization of baculoviral insecticides. In: The baculoviruses. Springer, Berlin, pp 341–387

Bloem S, McCluskey A, Fugger R, Arthur S, Wood S, Carpenter J (2007) Suppression of the codling moth Cydia pomonella in British Columbia, Canada using an area-wide integrated approach with an SIT components. In: Vreysen MJB, Robinson AS, Hendrichs J (eds) Area-wide control of insect pests: from research to field implementation. Springer, Dordrecht, pp 591–601

Braunagel SC, Summers MD (2007) Molecular biology of the baculovirus occlusion-derived virus envelope. Available from: https://www.ingentaconnect.com/content/ben/cdt/2007/00000008/00000010/art00006

Bravo A, Gómez I, Porta H, García-Gómez BI, Rodriguez-Almazan C, Pardo L, Soberón M (2013) Evolution of *Bacillus thuringiensis* Cry toxins insecticidal activity. Microb Biotechnol 6:17–26

Butenandt VA (1959) Uber den sexsual-lockstoff des seidenspinners *Bombyx mori*. Reindarstellung und Konstitution. Z Naturforsch B 14:283

Byers JA (2012) Modelling female mating success during mass trapping and natural competitive attraction of searching males or females. Entomol Exp Appl 145:228–237

Carvalho FP (2017) Pesticides, environment, and food safety. Food Energy Secur 6:48–60

Castillo-Sánchez LE, Jiménez-Osornio JJ, Delgado-Herrera MA, Candelaria-Martínez B, Sandoval-Gío JJ (2015) Effects of the hexanic extract of neem *Azadirachta indica* against adult whitefy *Bemisia tabaci*. J Entomol Zool Stud 5:95–99

Chakraborty M, Reddy PS, Mustafa G, Rajesh G, Narasu VML, Udayasuriyan V, Debashis R (2016) Transgenic rice expressing the *cry2AX1* gene confers resistance to multiple lepidopteran pests. Transgenic Res 25:665–678

Chang X, Lu Z, Shen Z, Peng Y, Ye G (2017) Bitrophic and tritrophic effects of transgenic cry1Ab/cry2Aj maize on the beneficial, nontarget *Harmonia axyridis* (Coleoptera: Coccinellidae). Environ Entomol 46:1171–1176

Chen XX, Liu YQ, Ren SX, Zhang F, Zhang WQ, Ge F (2014) Plant-mediated support system for natural enemies of insect pests. Chin Bull Entomol 51:1–12

Chen W, Liu C, Lu G, Cheng H, Shen Z, Wu K (2018) Effects of Vip3AcAa+ Cry1Ac cotton on midgut tissue in *Helicoverpa Armigera* (Lepidoptera: Noctuidae). J Insect Sci 18:13

Choudhary SR, Sindhu SS (2015) Suppression of *Rhizoctonia solani* root rot disease of clusterbean (*Cyamopsis tetragonoloba*) and plant growth promotion by rhizosphere bacteria. Plant Pathol J 14:48–57

Chowdhury SP, Uhl J, Grosch R, Alquéres S, Pittroff S, Dietel K, SchmittKopplin P, Borriss R, Hartmann A (2015) Cyclic lipopeptides of *Bacillus amyloliquefaciens* subsp. plantarum colonizing the lettuce rhizosphere enhance plant defense responses toward the bottom rot pathogen *Rhizoctonia solani*. Mol Plant Microbe Interact 28:984–995

Chrispeels MJ, Fatima Grossi de Sa M, Higgins TJV (1998) Genetic engineering with α-amylase inhibitors makes seeds resistant to bruchids. Seed Sci Res 8:257–264

Colmenarez YC, Corniani N, Jahnke SM, Sampaio M, Vásquez C (2018) Use of parasitoids as a biocontrol agent in the Neotropical region: challenges and potential. In: Baimey HK, Hamamouch N, Kolombia YA (eds) Horticultural crops. IntechOpen. https://doi.org/10.5772/intechopen.80720

Cook SM, Khan ZR, Pickett JA (2007) The use of push-pull strategies in integrated pest management. Annu Rev Entomol 52:375–400

Cunningham JC (1995) Baculoviruses as microbial insecticides. Available from: https://www.cfs.nrcan.gc.ca/publications?id=21623

Dangles O, Casas J (2019) Ecosystem services provided by insects for achieving sustainable development goals. Ecosyst Serv 35:109–115

Das SK (2014) Recent development and future of botanical pesticides in India. Popular Kheti 2:93–99

de Azevedo PR, Nogueira Batista JA, Mattar da Silva MC, Brilhante de Oliveira Neto O, Zangrando Figueira EL, Valencia Jiménez A (2006) An alpha-amylase inhibitor gene from Phaseolus coccineus encodes a protein with potential for control of coffee Berry borer (*Hypothenemus Hampei*). Phytochemistry 67:2009–2016

de Faria MR, Wraight SP (2007) Mycoinsecticides and mycoacaricides: a comprehensive list with worldwide coverage and international classification of formulation types. Biol Control 43:237–256

DeBach P, Rosen D (1991) Biological control by natural enemies. Cambridge University Press, Cambridge

Deguine JP, Aubertot JN, Flor RJ (2021) Integrated pest management: good intentions, hard realities. a review. Agron Sustain Dev 41:38

Dias PAS, Sampaio MV, Rodrigues MP, Korndorfer AP, Oliveira RS, Ferreira SE, Korndorfer GH (2014) Induction of resistance by silicon in wheat plants to alate and apterous morphs of *Sitobion avenae* (Hemiptera: Aphididae). Environ Entomol 43:949–956

Dickens J (2006) Plant volatiles moderate response to aggregation pheromone in Colorado potato beetle. J Appl Entomol 130:26–31

Duan X, Hou Q, Liu G, Pang X, Niu Z, Wang X (2018) Expression of *Pinellia Pedatisecta* lectin gene in transgenic wheat enhances resistance to wheat aphids. Molecules 23:748

Dyck VA, Hendrichs J, Robinson AS (2021) Sterile insect technique: principles and practice in area-wide integrated pest management, 2nd edn. CRC Press, Boca Raton

El-Sayed AM, Suckling DM, Wearing CH, Byers JA (2006) Potential of mass trapping for long-term pest management and eradication of invasive species. J Econ Entomol 99:1550–1564

Epstein E (2001) Silicon in plants: facts vs concepts. In: Datnoff LE, Snyder GN, Korndorfer GH (eds) Silicon in agriculture. Elsevier, Amsterdam, pp 1–15

Epstein E (2009) Silicon: its manifold roles in plants. Ann Appl Biol 155:155–160

Fang W, Lu HL, King GF, Leger RJS (2014) Construction of a hypervirulent and specific mycoinsecticide for locust control. Sci Rep 4:7345. https://doi.org/10.1038/srep07345

FAO (2020) NSP – integrated pest management, FAO definition. http://www.fao.org/agriculture/crops/thematic-sitemap/theme/pests/ipm/en/

Farias LR, Costa FT, Souza LA, Pelegrini PB, Grossi-de-Sá MF, Neto SM (2007) Isolation of a novel Carica papaya α-amylase inhibitor with deleterious activity toward *Callosobruchus maculatus*. Pest Biochem Physiol 87:255–260

Ferré J, Escriche B, Bel Y, Van Rie J (1995) Biochemistry and genetics of insect resistance to Bacillus thuringiensis insecticidal crystal proteins. FEMS Microbiol Lett 132:1–7

Feyisa B, Lencho A, Selvaraj T, Getaneh G (2015) Evaluation of some botanicals and *Trichoderma harzianum* for the management of tomato root-knot nematode (*Meloidogyne incognita* (Kofoid and White) chit wood). Adv Crop Sci Technol 1:1–10

Food and Agriculture Organization (FAO) (ed) (2014) State of world fisheries and aquaculture, 2014: opportunities and challenges. FAO, Rome

Gao T, Wang Z, Huang Y, Keyhani NO, Huang Z (2017) Lack of resistance development in *Bemisia tabaci* to *Isaria fumosorosea* after multiple generations of selection. Sci Rep 7:42727

Gatehouse JA (2011) Prospects for using proteinase inhibitors to protect transgenic plants against attack by herbivorous insects. Curr Protein Pept Sci 12:409–416

Golubkina N, Sheshnitsan S, Kapitalchuk M (2014) Ecological importance of insects in selenium biogenic cycling. Int J Ecol 5:1–6

Gomes ADPG, Dias SC, Bloch C Jr, Melo FR, Furtado JR, Monnerat RJ, Grossi-de-Sá MF, Franco OL (2005) Toxicity to cotton boll weevil Anthonomus grandis of a trypsin inhibitor from chick-pea seeds. Comp Biochem Physiol Part B Biochem Mol Biol 140:313–319

Grantham R, Arnold DC (2013) Beneficial insects. Okla. State Ext. 2013, Epp-7307. Available online: www.Osufacts.Okstate.Edu/docushare/dsweb/get/document-8688/

Grewal P, Georgis R (1999) Entomopathogenic nematodes. In: Hall FR, Menn JJ (eds) Biopesticides: use and delivery. Springer, Totowa, pp 271–299

Grover S, Jindal V, Banta G (2019) RNA interference mediated knockdown of juvenile hormone esterase gene in *Bemisia tabaci* (Gennadius): effects on adults and their progeny. J Asia-Pac Entomol 22:56–62

Gulcu B, Hazir S, Kaya HK (2012) Scavenger deterrent factor (SDF) from symbiotic bacteria of entomopathogenic nematodes. J Invertebr Pathol 110:326–333

Gupta M, Kumar H, Kaur S (2021) Vegetative insecticidal protein (Vip): a potential contender from Bacillus thuringiensis for efficient management of various detrimental agricultural pests. Front Microbiol 12:1139

Gurr GM, Kvedaras OL (2010) Synergizing biological control: scope for sterile insect technique, induced plant defences and cultural techniques to enhance natural enemy impact. Biol Control 52:198–207

Hanson B, Lindblom SD, Loeffler ML, Pilon-Smits EAH (2004) Selenium protects plants from phloem-feeding aphids due to both deterrence and toxicity. New Phytol 162:655–662

Haq SK, Atif SM, Khan RH (2004) Protein proteinase inhibitor genes in combat against insects, pests, and pathogens: natural and engineered phytoprotection. Arch Biochem Biophys 431:145–159

Hartley SE, Gange AC (2009) Impacts of plant symbiotic fungi on insect herbivores: mutualism in a multitrophic context. Annu Rev Entomol 54:323–342

Hartley SE, Fitt RN, McLarnon EL, Wade RN (2015) Defending the leaf surface: intra- and inter-specific differences in silicon deposition in grasses in response to damage and silicon supply. Front Plant Sci 6:35

He W, Yang M, Li Z, Qiu J, Liu F, Qu X, Qiu Y, Li R (2015) High levels of silicon provided as a nutrient in hydroponic culture enhances rice plant resistance to brown planthopper. Crop Prot 67:20–25

Hegazi E, Khafagi WE, Konstantopoulou M, Raptopoulos D, Tawfik H, El-Aziz GMA, El-Rahman SMA, Atwa A, Aggamy E, Showeil S (2009) Efficient mass-trapping method as an alternative tactic for suppressing populations of leopard moth (Lepidoptera: Cossidae). Ann Entomol Soc Am 102:809–818

Hou ML, Han YQ (2010) Silicon-mediated rice plant resistance to the Asiatic rice borer (Lepidoptera: Crambidae): effects of silicon amendment and rice varietal resistance. J Econ Entomol 103:1412–1419

Hoy MA (1994) Parasitoids and predators in management of arthropod pests. In: Metcalf RL, Luckmann WH (eds) Introduction to insect pest management. Wiley, New York, pp 129–198

Huber J (1986) Use of baculoviruses in pest management programs. http://agris.fao.org/agris-search/search.do?recordID=US19880053890

Inglis GD, Goettel MS, Butt TM, Strasser H (2001) Use of hyphomycetous fungi for managing insect pests. In: Butt TM, Jackson C, Magan N (eds) Fungi as biocontrol agents. CABI, Wallingford, pp 23–69

Irving JA, Pike RN, Dai W, Brömme D, Worrall DM, Silverman GA, Coetzer THT, Dennison C, Bottomley SP, Whisstock JC (2002) Evidence that serpin architecture intrinsically supports

papainlike cysteine protease inhibition: engineering α1-antitrypsin to inhibit cathepsin proteases. Biochemistry 41:4998–5004
Ishimoto M, Kitamura K (1989) Growth inhibitory effects of an α-amylase inhibitor from the kidney bean, *Phaseolus vulgaris* (L.) on three species of bruchids (Coleoptera: Bruchidae). Appl Entomol Zool 24:281–286
Jones OT (1998) Practical applications of pheromones and other semiochemicals. In: Howse P, Stevens I, Jones O (eds) Insect pheromones and their use in pest management. Chapman and Hall, London, pp 261–355
Juturu VN, Mekala GK, Kirti PB (2015) Current status of tissue culture and genetic transformation research in cotton (*Gossypium* spp.). Plant Cell Tissue Organ Cult 120:813–839
Kamthan A, Chaudhuri A, Kamthan M, Datta A (2015) Small RNAs in plants: recent development and application for crop improvement. Front Plant Sci 06:208
Karlson P, Lüscher M (1959) 'Pheromones': a new term for a class of biologically active substances. Nature 183:55–56
Katoch R, Sethi A, Thakur N, Murdock LL (2013) RNAi for insect control: current perspective and future challenges. Appl Biochem Biotechnol 171:847–873
Kaur S, Kaur K, Jindal J (2022) Status of phenolic metabolism and α-amylase inhibitor in maize under Chilo partellus infestation. Cereal Res Commun 4:1–9
Khachatourians GG, Qazi SS (2008) Entomopathogenic fungi: biochemistry and molecular biology. In: Brakhage AA, Zipfel PF (eds) Human and animal relationships. Springer, Berlin/Heidelberg, pp 33–61
Klassen W, Curtis CF, Hendrichs J (2021) History of the sterile insect technique. In: Dyck VA, Hendrichs J, Robinson AS (eds) Sterile insect technique: principles and practice in area wide integrated pest management. CRC Press, Boca Raton, pp 1–44
Koerniati S, Sukmadjaja D, Samudra IM (2020) C synthetic gene of CryIAb-CryIAc fusion to generate resistant sugarcane to shoot or stem borer. IOP Conf Ser Earth Environ Sci 418:012069
Koppenhöfer AM, Grewal PS, Fuzy EM (2007) Diferences in penetration routes and establishment rates of four entomopathogenic nematode species into four white grub species. J Invertebr Pathol 94:184–195
Kumar MM, Kumar S, Prasad CS, Kumar P (2015) Management of gram pod borer, Helicoverpa armigera (Hubner) in chickpea with botanical and chemical insecticide. J Exp Zool 18:741
Kvedaras OL, An M, Choi Y, Gurr G (2010) Silicon enhances natural enemy attraction and biological control through induced plant defences. Bull Entomol Res 100:367–371
Lacey LA, Neven LG (2006) The potential of the fungus, Muscodor albus, as a microbial control agent of potato tuber moth (Lepidoptera: Gelechiidae) in stored potatoes. J Invertebr Pathol 91:195–198
Lalramnghaki H (2017) Characterization of a new isolate of entomopathogenic nematode, *Steinernema sangi* (Rhabditida, Steinernematidae), and its symbiotic bacteria *Xenorhabdus vietnamensis* (γ-Proteobacteria) from Mizoram, northeastern India. J Parasit Dis 41:1123–1131
Lance DR, McInnis DO (2021) Biological basis of the sterile insect technique. In: Dyck VA, Hendrichs J, Robinson AS (eds) Sterile insect technique: principles and practice in area-wide integrated pest management. CRC Press, Boca Raton, pp 113–142
Lance DR, Leonard DS, Mastro VC, Walters ML (2016) Mating disruption as a suppression tactic in programs targeting regulated lepidopteran pests in US. J Chem Ecol 42:590–605
Larsson MC (2016) Pheromones and other semiochemicals for monitoring rare and endangered species. J Chem Ecol 42:853–868
Laxmishree C, Nandita S (2017) Botanical pesticides – a major alternative to chemical pesticides: a review. Int J Life Sci 4:722–729
Lekberg Y, Koide R (2005) Is plant performance limited by abundance of arbuscular mycorrhizal fungi? A meta-analysis of studies published between 1988 and 2003. New Phytol 168:189–204
Lengai GMW, Muthomi JW, Mbega ER (2020) Phytochemical activity and role of botanical pesticides in pest management for sustainable agricultural crop production. Sci Afr 7:00239. https://doi.org/10.1016/j.sciaf.2019.e00239

Liang Y, Nikolic M, Belanger R, Gong H, Song A (2015) Silicon in agriculture, from theory to practice. Springer, Dordrecht, pp 1–325
Liu J, Zhu J, Zhang P, Reynolds OL, Han L, Wu J, Shao Y, You M, Gurr GM (2017) Silicon supplementation alters the composition of herbivore induced plant volatiles and enhances attraction of parasitoids to infested rice plants. Front Plant Sci 8:1265
Liu W-T, Chen T-L, Hou RF, Chen C-C, Tu W-C (2020) The invasion and encapsulation of the entomopathogenic nematode, *Steinernema abbasi*, in *Aedes albopictus* (Diptera: Culicidae) Larvae. Insects 11:832
Loizou E, Karelakis C, Galanopoulos K, Mattas K (2019) The role of agriculture as a development tool for a regional economy. Agric Syst 173:482–490
Ma JF (2004) Role of silicon in enhancing the resistance of plants to biotic and abiotic stresses. Soil Sci Plant Nutr 50:11–18
Macedo MR, Freire MDG (2011) Insect digestive enzymes as a target for pest control. Invertebr Surviv J 8:190–198
Macedo ML, de Sá CM, Freire MD, Parra JR (2004) A Kunitz-type inhibitor of coleopteran proteases, isolated from Adenanthera pavonina L seeds and its effect on *Callosobruchus Maculatus*. J Agric Food Chem 52:2533–2540
MacGregor JT (2006) Genetic toxicity assessment of microbial pesticides: needs and recommended approaches. Organization for Economic Cooperation and Development, Paris
Malik MA, Ahmad SJN, Ahmad JN, Abbasi A, Sufyan M, Arif MJ (2019) Efficacy of *Bacillus thuringiensis* and *Beauveria bassiana* against red palm weevil *Rhynchophorus ferrugineus* Olivier (Coleoptera: Curculionidae). Afr Entomol 27:386–394
Mampallil LJ, Faizal MH, Anith KN (2017) Bacterial bioagents for insect pest management. J Entomol Zool Stud 5:2237–2244
Mamta B, Rajam MV (2017) RNAi technology: a new platform for crop pest control. Physiol Mol Biol Plants 23:487–501
Marec F, Vreysen MJB (2019) Advances and challenges of using the sterile insect technique for the management of pest lepidoptera. Insects 10:371
Massey FP, Hartley SE (2009) Physical defences wear you down: progressive and irreversible impacts of silica on insect herbivores. J Anim Ecol 78:281–291
Meade T, Narva K, Storer NP, Sheets JJ, Burton SL, Woosley AT (2017) U.S. Patent No. 9,556,453. U.S. Patent and Trademark Office, Washington, DC
Mechora Š, Stibilj V, Kreft I, Radešcˇek T, Gaberšcˇik A, Germ M (2011) Impact of Se (VI) fertilization on Se concentration in different parts of red cabbage plants. J Food Agric Environ 9:357–361
Mihalache G, Balaes T, Gostin I, Stefan M, Coutte F, Krier F (2018) Lipopeptides produced by Bacillus subtilis as new biocontrol products against fusariosis in ornamental plants. Environ Sci Pollut Res 25:29784–29793
Muthomi J, Fulano AM, Wagacha JM, Mwang'ombe AW (2017) Management of snap bean insect pests and diseases by use of antagonistic fungi and plant extracts. Sustain Agric Res 6:52
Naranjo SE, Ellsworth PC (2009) Fifty years of the integrated control concept: moving the model and implementation forward in Arizona. Pest Manag Sci 65:1267–1286
Neeraj GS, Kumar A, Ram S, Kumar V (2017) Evaluation of nematicidal activity of ethanolic extracts of medicinal plants to *Meloidogyne incognita* (Kofoid and White) Chitwood under lab conditions. Indian J Pure Appl Biosci 1:827–831
N'guessan FK, Quisenberry SS, Croughan TP (1994) Evaluation of rice anther culture lines for tolerance to the rice water weevil (Coleoptera: Curculionidae). Environ Entomol 23:331–336
Omar MAA, Ao Y, Li M, He K, Xu L, Tong H (2019) The functional difference of eight Chitinase genes between male and female of the cotton mealybug, *Phenacoccus solenopsis*. Insect Mol Biol 28:550–567
Ortiz A, Sansinenea E (2021) Bacillus thuringiensis based biopesticides for integrated crop management. In: Rakshit A, Meena V, Abhilash PC, Sarma BK, Singh HB, Fraceto L, Parihar

M, Kumar A (eds) Biopesticides. Volume 2. Advances in bio-inoculants. Elsevier/Woodhead Publishing, Amsterdam/Sawston, pp 1–6

Osman GH, Assem SK, Alreedy RM, El-Ghareeb DK, Basry MA, Rastogi A (2016) Erratum: Development of insect resistant maize plants expressing a chitinase gene from the cotton leaf worm, *Spodoptera littoralis*. Sci Rep 6:20449

Palma L, Muñoz D, Berry C, Murillo J, Caballero P (2014) Bacillus thuringiensis toxins: an overview of their biocidal activity. Toxins 6:3296–3325

Pan Y, Lü P, Wang Y, Yin L, Ma H, Ma G (2012) In silico identification of novel Chitinase-like proteins in the silkworm, *Bombyx mori*, genome. J Insect Sci 12:1–14

Park HS, Baek SH, Kim WJ, Jeung JU, Lee JH, Ha KY (2014) Development of multi-resistant lines to brown planthopper, bacterial blight, and rice stripe virus using anther culture in rice. Korean J Breed Sci 46:78–89

Paul S, Das S (2020) Natural insecticidal proteins, the promising bio-control compounds for future crop protection. Nucleus 64:7–20

Pedrero Z, Madrid Y, Hartikainen H, Camara C (2008) Selenium species bioaccessibility in enriched radish (*Raphanus sativus*): a potential dietary source of selenium. J Agric Food Chem 56:266–271

Persley GJ (2000) Agricultural biotechnology and the poor: promethean science. Consultative Group on International Agricultural Research (CGIAR), Washington, DC, pp 3–21

Petkevicius K, Löfstedt C, Borodina I (2020) Insect sex pheromone production in yeasts and plants. Curr Opin Biotechnol 65:259–267

Petrik DT, Iseli A, Montelone BA, Van Etten JL, Clem RJ (2003) Improving baculovirus resistance to UV inactivation: increased virulence resulting from expression of a DNA repair enzyme. J Invertebr Pathol 82:50–56

Posada F, Vega FE (2006) Inoculation and colonization of coffee seedlings (*Coffea arabica* L.) with the fungal entomopathogen *Beauveria bassiana* (Ascomycota: Hypocreales). Mycoscience 47:284–289

Priya S, Kaur N, Gupta AK (2010) Purification, characterization and inhibition studies of α-amylase of *Rhyzopertha dominica*. Pest Biochem Physiol 98:231–237

Rani S, Sharma V, Hada A, Bhattacharya RC, Koundal KR (2017) Fusion gene construct preparation with lectin and protease inhibitor genes against aphids and efficient genetic transformation of *Brassica juncea* using cotyledons explants. Acta Physiol Plant 39:1–13

Reddy GV, Guerrero A (2004) Interactions of insect pheromones and plant semiochemicals. Trends Plant Sci 9:253–261

Reddy GV, Guerrero A (2010) New pheromones and insect control strategies. Vitam Horm 83:493–519

Rendón P, Sivinski J, Holler T, Bloem K, Lopez M, Martinez A, Aluja M (2006) The effects of sterile males and two braconid parasitoids, *Fopius arisanus* (Sonan) and *Diachasmimorpha krausii* (Fullaway) (Hymenoptera), on caged populations of Mediterranean fruit flies, *Ceratitis capitata* (Wied.) (Diptera: Tephritidae) at various sites. Biol Control 36:224–231

Reynolds OL, Keeping MG, Meyer JH (2009) Silicon-augmented resistance of plants to herbivorous insects: a review. Ann Appl Biol 155:171–186

Reynolds OL, Padula MP, Zeng R, Gurr GM (2016) Silicon: potential to promote direct and indirect effects on plant defense against arthropod pests in agriculture. Front Plant Sci 7:744

Riyaz M, Aamir Iqbal WA, Sivasankaran K, Ignacimuthu S (2020) Impact on farmers' health due to the pesticide exposure in the agrarian zones of Kashmir valley: a review. Acta Sci Agric 4:01–07

Roy S, Handique G, Muraleedharan N, Dashora K, Roy SM, Mukhopadhyay A, Babu A (2016) Use of plant extracts for tea pest management in India. Appl Microbiol Biotechnol 100:4831–4844

Ruiu L, Floris I, Satta A, Fois X (2007) Susceptibilité du miel (*Apis mellifera* L.) aux toxines bactériennes entomopathogènes utilisées pour le contrôle biologique. REDIA XC:87–90

Sakr N (2016) Silicon control of bacterial and viral diseases in plants. J Plant Prot Res 56:331–336

Saxena S, Pandey AK (2001) Microbial metabolites as ecofriendly agrochemicals for the next millennium. Appl Microbiol Biotechnol 55:395–403

Schneider VK, Soares-Costa A, Chakravarthi M, Ribeiro C, Chabregas SM, Falco MC (2017) Transgenic sugarcane overexpressing CaneCPI-1 negatively affects the growth and development of the sugarcane weevil *Sphenophorus levis*. Plant Cell Rep 36:193–201

Sehrawat A, Sindhu SS (2019) Potential of biocontrol agents in plant disease control for improving food safety. Def Life Sci J 4:220–225

Sekita N, Kawashima K, Aizu H, Shirasaki S, Yamada M (1984) A short term control of *Adoxophyes orana* fasciata Walsingham (Lepidoptera: Tortricidae) by a granulosis virus in apple orchards. Appl Entomol Zool 19:498–508

Sellami S, Tounsi S, Jamoussi K (2015) *Bacillus thuringiensis*: un biopesticide environnemental. Revue Marocaine de Protection des Plantes 7:25–42

Sendi Y, Pfeiffer T, Koch E, Mhadhbi H, Mrabet M (2020) Potential of common bean (*Phaseolus vulgaris* L.) root microbiome in the biocontrol of root rot disease and traits of performance. J Plant Dis Prot 127:453–462

Seybold SJ, Bentz BJ, Fettig CJ, Lundquist JE, Progar RA, Gillette NE (2018) Management of western North American bark beetles with semiochemicals. Annu Rev Entomol 63:407–432

Shade RE, Schroeder HE, Pueyo JJ, Tabe LM, Murdock LL, Higgins TJV (1994) Transgenic pea seeds expressing the α-amylase inhibitor of the common bean are resistant to bruchid beetles. Nat Biotechnol 12:793–796

Shah P, Pell J (2003) Entomopathogenic fungi as biological control agents. Appl Microbiol Biotechnol 61:413–423

Sharma K (2015) Protease inhibitors in crop protection from insects. Int J Curr Res Acad Rev 3:55–70

Sharma S, Kooner R, Arora R (2017) Insect pests and crop losses. In: Inbreeding insect resistant crops for sustainable agriculture. Springer, Singapore, pp 45–66

Shiga M, Yamada H, Oho H, Nakazawa H, Ito Y (1973) A granulosis virus, possible biological agent for control of *Adoxophyes orana* (Lepidoptera: Tortricidae) in apple orchards. Semipersistent effect of artificial dissemination into an apple orchard. J Invertebr Pathol 21:149–157

Sicard M, Brugirard-Ricaud K, Pagès S, Lanois A, Boemare NE, Brehélin M, Givaudan A (2004) Stages of infection during the tripartite interaction between *Xenorhabdus nematophila*, its nematode vector, and insect hosts. Appl Environ Microbiol 70:6473–6480

Singh A, Singh S, Singh IK (2016) Recent insights into the molecular mechanism of jasmonate signaling during insect-plant interaction. Australas Plant Pathol 45:123–133

Slätmo E, Fischer K, Röös E (2017) The framing of sustainability in sustainability assessment frameworks for agriculture. Sociol Rur 57:378–395

Smart L, Aradottir G, Bruce T (2014) Role of semiochemicals in integrated pest management. In: Abrol DP (ed) Integrated pest management. Academic Press, Amsterdam, pp 93–109

Stevens J, Dunse K, Fox J, Evans S, Anderso M (2012) Biotechnological approaches for the control of insect pests in crop plants. In: Soundararajan RP (ed) Pesticides advances in chemical and botanical pesticides. InTech Open, pp 269–308

Stewart SD, Layton B, Catchot A (2007) Common beneficial arthropods found in field crops. Univ. Tenn. Ext. 2007, W127. Available online: www.extension.org/mediawiki/files/0/00/W127-Beneficials.pdf

Struelens Q, Silvie P (2020) Orienting insecticide research in the tropics to meet the sustainable development goals. Curr Opin Insect Sci 40:24–30

Su C, Tu G, Huang S, Yang Q, Shahzad MF, Li F (2016) Genome-wide analysis of chitinase genes and their varied functions in larval moult, pupation and eclosion in the rice striped stem borer, Chilo Suppressalis. Insect Mol Biol 25:401–412

Sufyan M, Abbasi A, Gogi MD, Arshad M, Nawaz A, Neuhoff D (2017) Efficacy of *Beauveria bassiana* for the management of economically important wireworm species (Coleoptera: Elateridae) in organic farming. Gesunde Pflanzen 69:197–202

Sufyan M, Abbasi A, Wakil W, Gogi MD, Arshad M, Nawaz A, Shabbir Z (2019) Efficacy of *Beauveria bassiana* and *Bacillus thuringiensis* against maize stem borer *Chilo partellus* (Swinhoe) (Lepidoptera: Pyralidae). Gesunde Pflanzen 71:197–204

Syed T, Askari M, Meng Z, Li Y, Abid MA, Wei Y (2020) Current insights on vegetative insecticidal proteins (Vip) as next generation pest killers. Toxins 12:522

Tabashnik BE (1994) Evolution of resistance to *Bacillus thuringiensis*. Annu Rev Entomol 39:47–79

Talakayala A, Katta S, Garladinne M (2020) Genetic engineering of crops for insect resistance: an overview. J Biosci 45:1–12

Tefera T, Vidal S (2009) Effect of inoculation method and plant growth medium on endophytic colonization of sorghum by the entomopathogenic fungus *Beauveria bassiana*. BioControl 54:663–669

Tewari S, Leskey TC, Nielsen AL, Piñero JC, Rodriguez-Saona CR (2014) Use of pheromones in insect pest management, with special attention to weevil pheromones. In: Abrol DP (ed) Integrated pest management. Academic Press, Amsterdam, pp 141–168

Todorov D, Shishkova K, Dragolova D, Hinkov A, Kapchina-Toteva V, Shishkov S (2015) Antiviral activity of medicinal plant Nepeta nuda. Biotechnol Biotechnol Equip 1:39–43

Toprak U, Harris S, Baldwin D, Theilmann D, Gillott C, Hegedus DD (2012) Role of enhancin in *Mamestra configurata* nucleopolyhedrovirus virulence: selective degradation of host peritrophic matrix proteins. J Gen Virol 93:744–753

Trematerra P (2012) Advances in the use of pheromones for stored-product protection. J Pest Sci 85:285–299

Trematerra P, Colacci M (2019) Recent advances in management by pheromones of Thaumetopoea moths in urban parks and woodland recreational areas. Insects 10:395

Trumble JT, Kund GS, White KK (1998) Influence of form and quantity of selenium on the development and survival of an insect herbivore. Environ Pollut 101:175–182

Valenzuela-Soto JH, Estrada-Hernández MG, Ibarra-Laclette E, Délano-Frier JP (2010) Inoculation of tomato plants (*Solanum lycopersicum*) with growth-promoting *Bacillus subtilis* retards whitefly *Bemisia tabaci* development. Planta 231:397–410

Van Bockhaven J, Vleesschauwer DD, Hofte M (2013) Towards establishing broad-spectrum disease resistance in plants: silicon leads the way. J Exp Bot 64:1281–1293

van Lenteren JC, Bueno VHP (2003) Augmentative biological control of arthropods in Latin America. BioControl 8:123–139

Vargas RI, Long J, Miller NW, Delate K, Jackson CG, Uchida GK, Bautista RC, Harris EJ (2004) Releases of *Psyttalia fletcheri* (Hymenoptera: Braconidae) and sterile flies to suppress melon fly (Diptera: Tephritidae) in Hawaii. J Econ Entomol 97:1531–1539

Vega FE, Posada F, Aime MC, Pava-Ripoll M, Infante F, Rehner SA (2008) Entomopathogenic fungal endophytes. Biol Control 46:72–82

Vickerman DB, Trumble JT (1999) Feeding preferences of *Spodoptera exigua* in response to form and concentration of selenium. Arch Insect Biochem Physiol 42:64–73

Waage JK, Greathead DJ (1988) Biological control: challenges and opportunities. Phil Trans Royal Soc London B Biol Sci 318:111–112

Wang P, Granados RR (1997) An intestinal mucin is the target substrate for a baculovirus enhancin. PNAS 94:6977–6982

Wang B, Qiu YL (2006) Phylogenetic distribution and evolution of mycorrhizas in land plants. Mycorrhiza 16:299–363. https://doi.org/10.1007/s00572-005-0033-6

Wang GR, Han RP, Shen Q, Yang WT, Wang GD, Lu ZX, Zheng XS (2015) Effects of modified fertilizer application on rice pests, diseases and yields. China Rice 21:94–97

Wanzala W, Wagacha JM, Dossaji SF, Gakuubi MM (2016) Bioactive properties of *Tagetes minuta* L. (Asteraceae) essential oils: a review. Am J Essent Oils Nat Prod 4:27–36

Witzgall P, Kirsch P, Cork A (2010) Sex pheromones and their impact on pest management. J Chem Ecol 36:80–100

Wong TTY, Ramadan MM, Herr JC, McInnis DO (1992) Suppression of a Mediterranean fruit fly (Diptera: Tephritidae) population with concurrent parasitoid and sterile fly releases in Kula, Maui, Hawaii. J Econ Entomol 85:1671–1681

Wraight S, Jackson M, De Kock S (2001) Production, stabilization and formulation of fungal biocontrol agents. In: Butt TM, Jackson C, Magan N (eds) Fungi as biocontrol agents: progress, problems and potential. CABI, Wallingford, pp 253–287

Xi Y, Pan PL, Ye YX, Yu B, Xu HJ, Zhang CX (2015) Chitinaselike gene family in the brown planthopper, *Nilaparvata lugens*. Insect Mol Biol 24:29–40

Xu D, Xue Q, McElroy D, Mawal Y, Hilder VA, Wu R (1996) Constitutive expression of a cowpea trypsin inhibitor gene, *CpTi*, in transgenic rice plants confers resistance to two major rice insect pests. Mol Breed 2:167–173

Yang L, Han Y, Li P, Wen L, Hou M (2017) Silicon amendment to rice plants impairs sucking behaviors and population growth in the phloem feeder *Nilaparvata lugens* (Hemiptera: Delphacidae). Sci Rep 7:1101

Ye M, Song YY, Long J, Wang RL, Baerson SR, Pan Z, Zhu-Salzman K, Xie J, Cai K, Luo S (2013) Priming of jasmonate-mediated antiherbivore defense responses in rice by silicon. Proc Natl Acad Sci U S A 38:3631–3639

Yew JY, Chung H (2015) Insect pheromones: an overview of function, form, and discovery. Prog Lipid Res 59:88–105

Yoo JY, Jang EJ, Park SY, Son HJ (2018) Production of HCN, weed control substance, by Pseudomonas koreensis and its plant growth-promoting and termiticidal activities. J Environ Sci Int 27:771–780

Yu H, Gouge DH, Baker P (2006) Parasitism of subterranean termites (Isoptera: Rhinotermitidae: Termitidae) by entomopathogenic nematodes (Rhabditida: Steinernematidae; Heterorhabditidae). J Econ Entomol 99:1112–1119

Yu JM, Wang DP, Pierson LS, Pierson EA (2018) Effect of producing different phenazines on bacterial fitness and biological control in *Pseudomonas chlororaphis*. Plant Pathol J 34:44–58

Zhang WJ, Jiang FB, Ou JF (2011) Global pesticide consumption and pollution: with China as a focus. Proc Int Acad Ecol Env Sci 1:125–144

Zhu-Salzman K, Zeng R (2015) Insect response to plant defensive protease inhibitors. Annu Rev Entomol 60:233–252

Zimmermann G (2007) Review on safety of the entomopathogenic fungi *Beauveria bassiana* and *Beauveria brongniartii*. Biocont Sci Technol 17:553–596

Chapter 42
Ecofriendly Management of Disease for Sustainable Agriculture

Dipika Mal, Gunuguntla Veera Narayana, and Malige Bhavani

Abstract Currently, the indiscriminate use of insect pesticides has made a severe threat to plants by causing diseases and polluting the environment day by day. Organic farming is now recognized as a key area on a global scale due to the rising environmental pollution issues brought by a greater understanding of the drawbacks of the indiscriminate use of agro-chemicals. With each passing day, there is a growing demand for wholesome meals. The main worries are the negative consequences of chemicals on people, animals, plants, and the environment as a whole. It is possible to stimulate organic production in some high-value fruit and vegetable crops in order to satisfy both domestic and international demand for fresh produce. Chemical inputs must be used in agriculture to manage some plant diseases and meet the raising food demand in a populous country like India. A well-designed organic production technology module for a particular crop may sensibly and strategically incorporate a number of non-chemical environmental strategies. Cultural shifts, destruction of inoculum sources by mechanical means, bio-fumigation and clean cultivation using organic additives, development of organic insecticides, encouragement of naturally occurring biological agents, use of cover and trap crops, heating, low temperatures, and solar energy, etc. These plant disease control approaches are sustainable, environmentally safe, and excellent for the environment.

Keywords Ecofriendly · Organic management · Organic manures, organic farming, sustainable agriculture · Plant disease

D. Mal (✉) · G. V. Narayana · M. Bhavani
Lovely professional University, Domain of Horticulture, School of Agriculture, Phagwara, Punjab, India
e-mail: dipika.21885@lpu.co.in

M. Hasanuzzaman (ed.), *Climate-Resilient Agriculture, Vol 2*,
https://doi.org/10.1007/978-3-031-37428-9_42

1 Introduction

Scientists and decision-makers are being compelled to carefully consider alternatives to chemical agriculture because of the rising concern over environmental degradation, the degradation of natural resources, and to fulfilling the food demand of the growing population. A viable and possible path has been identified as sustainable agriculture supported by green technologies in an integrated farming system. The widespread erosion of natural resources, including water and energy, is one of the two issues limiting agricultural production. Population growth speeds up the first, whereas agronomic deficiencies speed up the second. As we all known that insect pesticides have a severe negative impact on the ecosystem, and in many areas, the entire system that supports life is at risk due to pesticide residues in the food chain. In addition to endangering the ecosystem through nitrate poisoning, chemical fertilizers have also destroyed the beneficial soil microflora and microfauna by negatively impacting the chemical and physical composition of the soil.

Ecological agriculture will start to stop the misuse of natural resources when it successfully combines input efficiency and economic yield maximization (Manimozhi and Gayathri 2012). Its advancement is necessary for the achievement of food security and the sustainability of farming systems. Ecological agriculture will revive the worn-out green revolution and advance through the promotion of soil health, a collection of agronomic techniques that each integrate traditional and modern practices, and a strategy to value each input based on its intrinsic worth free from subsidies. It is necessary to regain regional expertise on framing systems. Higher input usage efficiency and sustainable crop yield maximization will be made possible by combining traditional knowledge and ecological prudence of tropical agriculture with contemporary technological advancements. This is an example of low-input sustainable agriculture (LISA) and sustainable agricultural research and education program (SAREP) promoted by the USA.

The most significant concerns of agriculture in recent years have been the rise in importance of sustainable agriculture. Additionally, plant diseases continue to be a significant production-minimizing factor in agriculture. Alternative pest management methods are now required because of the use of traditional insect pesticides to control plant diseases raises severe questions regarding food quality, environmental safety, and insect pesticide resistance (Dubey et al. 2010). In order to fulfil important ecological, economic, and social functions now and in the future at the local, national, and global levels without harming other ecosystems, sustainable agriculture means managing and using the agricultural ecosystem in a way that preserves its biological diversity, productivity, regeneration capacity, vitality, and ability to function (Sahoo et al. 2013). The sustainability of agriculture recently underwent its most challenging test. The population is expanding more quickly than it used to, and there is a growing need for agricultural land and resources. Another is the misuse of fossil fuels globally and the rising costs of exploiting non-renewable resources in terms of both the economy and the environment. A fourth issue is globalization.

Like never before in history, these pressing concerns present a challenge for agriculturalists to create more sustainable management practices.

Agriculture needs to expand its focus beyond the previous concentration on productivity to include and enhance public health in order to achieve the nutritional and food needs of a increasing population. For social harmony and for healthy environment, finding alternatives to conventional plant disease control methods that boost yield and enhance product quality without harming the environment is also crucial (Gupta et al. 2017).

1.1 Plant Disease Management Strategies in Eco-Friendly Management of Sustainable Agriculture

Not just the provision of nutrients like fertilizers, but any action that might increase the availability and decrease the imbalance of particular components can affect growth and the tolerance of diseases. The majority of sustainable farming techniques have been demonstrated to provide plants with a balanced diet, as well as to increase the availability of some nutrients and improve the plants' resilience to disease (Ram et al. 2018). Plant growth and disease resistance can be affected by practices like crop rotation, green manure, manure application, intercropping, and tillage. Most of these techniques have the potential to significantly increase soil organic matter, which is essential for sustainable agriculture.

2 Soil Organic Matter

The amount and quality of soil organic matter affects a variety of soil processes that are linked to soil health, including moisture retention, infiltration, release, and plant health (SOM). A cultural technique known as "field-applied organic residues" (also known as "crop residues," "cover crops," and "organic wastes") has the potential to change the way that nutrients are available as well as the health of the soil. It has been shown that suppressive soils may be made using techniques including the addition of sphagnum peat, green manures, and animal manures, which prevent the development and persistence of diseases as well as the harming of agricultural plants. Sphagnum peat has been shown to help prevent Pythium spp. disease when added to the soil. Additionally, it has been shown that adding different organic amendments may make many plants susceptible to Phytophthora root rot (Reddy 2017). Dairy manure can reduce the pathogens that cause sweet corn (*Drechslera* spp., *Phoma* spp., and *Pythium arrhenomanes*) and snap beans (causal agents *Fusarium solani* and *Pythium* spp.). Many mechanisms have been proposed to be involved in biologically and organic material-mediated disease suppression, including microbiotas, microbial colonisation of pathogen propagules, destruction of pathogen propagules, antibiosis, competition for substrate colonization, competition for root infection sites, and induced systemic resistance (or systemic acquired resistance, or SAR). The level and nature of soil organic matter can impact nutritional status of the plants. Plant nutrition can be affected by the quantity and quality of soil organic matter (SOM). SOM can affect nutrient availability in addition to the total amount of nutrients in the soil by influencing the activity of soil microorganisms. Therefore, nutrients can improve plant resistance to disease, improve plant development (which aids the plant in resisting disease), and alter the environment in which the pathogen thrives. Despite the fact that quantity and quality can have a substantial impact on soil and plant nutrient content, very few studies that focus on soil features and disease incidence examine the function of soil or tissue nutrient content in disease-suppressive effects. Fields that had previously received yearly organic amendments exhibited higher levels of microbial activity and K. Lower NO_3 level and the prevalence of corky roots were associated with positive relationships between soil NO_3 and plant tissue N and negative correlations between soil N mineralization potential, microbial activity, total soil N, and soil pH. In a different study, biosolids improved N nutrition in the modified soil, which promoted ryegrass germination, growth, and resistance to Puccinia species-caused leaf rot.

3 Levels of Pest and Disease Management

3.1 Level-A (Preventive)

Level-A systems-based strategy is the first line of defense for the management of insects and diseases. Naturally, a healthy, organic system with a good design will have fewer pest issues. This system is designed to stop the spread of illness and pests.

3.1.1 Choosing Adapted and Resistant Varieties

Pick cultivars that are compatible with the regional climate. This enables them to develop healthily and strengthens their resistance to pests and diseases. For instance, tomato varieties marketed as VFN are nematode, verticillium, and fusarium resistant.

3.1.2 Selection of Clean Seed and Planting Material

- Choose pathogen and weed free seeds for crop production.
- Choose planting material from healthy sources.

3.1.3 Choose Suitable Cropping Systems

Crop Rotation

Crops of different families are taken up in rotations. This minimizes the risk of the incidence of a family specific disease (Mishra 2013), e.g., cereal crops like wheat, barley, etc. are taken up in rotation with legumes. This breaks the disease cycle of *Fusarium* Blight.

Intercropping

Systems of intercropping have the ability to lower the prevalence of illnesses (Mishra 2013). However, with various intercropping systems, variable reactions to disease severity have been noted. An intercropping system has four strategies which can minimize disease effect, all of this entail slowing the attacking organism's population growth rate: (1) Plants of the attacked component become inferior hosts as a result of the associated crop. (2) The associated crop directly obstructs the invading pathogen. (3) The associated crop modifies the host's environment in a way that the attacking organism's natural enemies are encouraged. (4) Resistant or non-host growing close to susceptible plants can actually stop the inoculum from getting to the hosts that are vulnerable (i.e., the non-host functions as a physical barrier to the pathogen inoculum). Francis (1989) noted that intercropping reduced pests and diseases in 53% of experiments while causing them to increase in 18% of them. Some of the reasons for this increase in pests include decreased cultivation and increasing shade, which favour some pests and pathogens, allied species serving as substitute hosts, and crop remnants serving as a source of pathogen inoculums. Additionally, it was found that intercropping increased nutrients by increasing the consumption of phosphate and potassium or N from legumes.

3.1.4 Use of Balanced Nutrient Management Moderate Fertilization

A plant becomes less susceptible to infection as it grows steadily. Too much fertilization damages roots with salt, which opens the door for secondary infections (Kumari et al. 2019) For instance, too much nitrogen fertilizer makes the plant succulent and thus more prone to illness and pest infestation. The plant is resistant to many illnesses when it receives a balanced potassium supply.

3.1.5 Supply of Organic Matter

It decreases the density and activity of soil's micropathogens while increasing the density and activity of soil's microorganisms (Ahirwar et al. 2020). soil structure is stabilised Through the release of certain chemicals, it boosts the plants' defensive mechanisms and increases the aeration and water holding capacity of the soil.

3.1.6 Application of Suitable Soil Cultivation Methods

It speeds up the breakdown of diseased plant components. It controls weeds, which act as hosts for various pests and illnesses. It safeguards the advantageous soil microorganisms.

3.1.7 Use of Proper Water Management

Avoid standing water since it creates ideal conditions for various pests and diseases. Avoid getting water on the plants since fungi and water-borne diseases can both spread through droplets.

3.1.8 Conservation and Promotion of Natural Enemies

Providing an ideal environment for the growth and reproduction of natural enemies. Avoid products that may harm to the natural enemies, e.g. the growing of some plants species that will attract ladybugs or other natural predators which will help to reduce populations of plant pests.

3.1.9 Selection of Optimum Planting Time and Spacing

Most of the pests and diseases are attack the plants at a certain growth stage called the "critical stage". This vulnerable stage should not coincide with the period of high pest density, so optimal planting time should be chosen. Optimum spacing reduces the spread of a disease. Good aeration of the foliage hinders pathogen development.

3.1.10 Use of Proper Sanitation Measure

To stop the illness from spreading, remove affected plant components (leaves, fruits) from the ground. Once the plant has been harvested, throw away the leftovers. For instance, (a) Apple and pear growers cut off any branches that have fire blight (strikes). To eradicate the microorganisms that cause fire blight, the strikes are taken from the orchard and torched. (b) In order to eliminate fungus that might harm the mushroom crop, mushroom producers pasteurize the bedding material.

3.1.11 Cover Crops

Sudan grass, rapeseed, and mustard are a few cover crops that work well to control nematodes. Cover crops can boost the amount of active OM, microbial biomass, and microbial activity in the soil as well as help with suppression. Cover crops have an indirect impact on plant health by changing the composition of the soil's microbial population and rhizosphere. By enhancing the soil's ability to act as a buffer, denying the pathogen a host during the absence of compatible species, and influencing nitrification, which affects the type of N that predominates in the soil, crop rotation might reduce the severity of soilborne diseases (Reddy 2017).

3.1.12 Trap Crops

According to Reddy 2017, these are minor plantings of a crop or crop variety meant to deter a specific pest from the main crop. For instance, alfalfa planted in strips among cotton drives lygus bugs away from the cotton crop. To get rid of the pests that have been drawn to trap crops, they must be exterminated.

3.2 *Level-B (Physical and Mechanical)*

If level A techniques are insufficient to control the weed, insect, or disease problem, the second line of defence is used. Standard mechanical and physical techniques used in organic farming are typically included in Level B.

3.2.1 Mulching

Organic Mulches
Organic mulches are those items of natural origin that can naturally decompose, such as agricultural waste, and give nutrients and humus to the soil as they break down, enhancing the soil's tilth and moisture-holding capacity (Table 42.1).

Different Types of Organic Mulches
Grass clippings, straw, dry leaves, bark clippings, sawdust, compost, feathers, hay, mushroom compost, pine, wood chips, peanut hulls, manure, buckwheat hulls, and cocoa bean hulls.

Mulching can reduce disease to some extent by minimizing soil contact and maintaining soil moisture, e.g., mulching in tomatoes can reduce certain diseases.

3.2.2 Canopy Management

The airflow is improved, and disease breakout is reduced by training and pruning trees. For instance, viticulturists remove leaves to prevent Botrytis bunch rot in grapes.

Table 42.1 Different organic mulches used and their thickness (in inches)

Sl. No.	Different mulches	Thickness (inches)
(i)	Saw dust	2 inches layer
(ii)	Hay or straw	6–8 inches layer
(iii)	Pine needles	4–5 inches layer
(iv)	Grass clippings	2 inches layer
(v)	Leaves	2–3 inches layer
(vi)	Peat moss	2–3 inches layer
(vii)	Compost	2–3 inches layer
(viii)	Bark and wood chips	2–3 inches layer
(ix)	Hulls and ground concords	2–4 inches layer

3.2.3 Soil Solarization

Hydrothermal disinfection is accomplished through soil solarization. This is accomplished by exposing moist soil to direct sunlight at the hottest time of the year while covered in transparent polythene sheets.

Benefits of Soil Solarization

Fungal pathogens can be controlled by solarization, as can a number of other soil-borne pathogens. Fungi like Pythium, Phytophthora, Fusarium, Rhizoctonia, etc. are included in this.

Nematode control: Solarization can be used to reduce the population of nematodes like *Meloidogyne*, *Heterodera*, *Xiphinema*, etc.

Several regularly occurring weeds, especially annuals, can be successfully managed by solarization. *Cynodon dactylon*, *Cyperus rotundus*, and *Digitaria ciliaris* are among the monocots, whereas *Crotalaria mucronata*, *Indigofera hirsuta*, and *Nexia* sp. are among the dicots.

Plant reaction to solarization: Plants grown in solarized soil have been found to have a greater response to solarization (increase in plant height, number of leaves, better root formation, increased root nodulation in legumes and yield).

3.2.4 Mass Trapping of Insects

Light trap used to catch moths, viz., armyworms, cutworms, stem borers, and other night flying insects. It is most efficient when placed before the adult moths start laying eggs.

Yellow Sticky Trap

Yellow sticky traps can be used to control whiteflies, aphids, and leaf mining flies. The yellow color attracts many insects, even beneficial ones. So, it should be used only when needed. Motor oil or transparent car grease are used. Placed around 10 cm above the foliage, 2–5 sticky traps per 500 m^2 field area.

Pheromone Trap
It uses pheromones to lure insects. Successful in mating disruption yellow sticky trap, light trap, and pheromone trap.

Fruit Bagging
Additionally, it stops fruit flies from hatching their eggs on the fruit. The bag shields the user's body from mechanical harm (scars and scratches). Works well with fruits including star fruit, guava, mango, melon, and bitter gourd.

3.3 Level-C (Allowed Biopesticide)

If after A and B the needed level of pest control is not reached, the third line of defence is deployed. Inputs including biological and botanical insecticides are used in Level C procedures to manage pests.

3.3.1 Biological Control

The management of pest populations through the use of natural enemies is known as biological control, or biocontrol. Although it frequently includes requires an active human management role, it typically relies on predation, parasitism, herbivory, or other natural mechanisms (such as ladybird beetles, predatory gall midges, or hoverfly larvae against aphids and psyllids).

Types of Biological Pest Control
Introduction: It involves the introduction of a natural enemies to a new habitats where they do not available previously.

Augmentation It entails the additional release of naturally occurring adversaries that exist in a certain location, enhancing the naturally existing populations there. For instance, the Trichogramma spp are very important egg parasitoids for many of the agricultural pests.

Conservation Natural enemies may be easily and cheaply conserved because they are already adapted to the environment and the pest that they are meant to control. For instance, crop plants that produce nectar that are grown around the edges of rice fields supply nectar to parasitoids and predators of planthopper pests.

3.3.2 Releasing Natural Enemies

Fungi and bacteria are frequently the natural enemies that kill or reduce pests and diseases. They are referred to as antagonists, microbial insecticides, or biopesticides.

3.3.2.1 Microbial Insecticides

Bacteria*:* *Bacillus thuringiensis* (Bt)

control of beetles and caterpillars in agricultural crops, particularly in vegetables. The lepidopteran pests *Bacillus thuringiensis* var. Kurstaki and *Bacillus thuringiensis* var. Aizawai are controlled by mosquito and blackfly. Black bollworm Spiny bollworms, green loopers, tomato loopers, bean armyworms, beet armyworms, cabbage webworms, etc.).

4 Bio-pesticides

Some plant species contain substances that are poisonous to insects. Botanical pesticides are these plant extracts when they are applied to infested crops (Prabha et al. 2016). As natural pesticides, people have utilised azadirachtin (from neem), pyrethrin (from *Chrysanthemum* spp.), rotenone (from *Derris* spp.), nicotine (from tobacco), and limonene (from citrus). Contact, respiratory, or stomach toxins make up the majority of plant pesticides. They therefore target a wide variety of insects and are not particularly discriminating. Since botanical pesticides are often very biodegradable, they lose their effectiveness after a few hours or days. This once again lessens the adverse effects on beneficial organisms, and they are more environmentally safe than chemical pesticides.

4.1 Commonly Used Botanical Pesticides

4.1.1 Neem

Azadirachtin is the substance that is active. Many types of caterpillars, thrips, and whiteflies are both deterred and killed. Neem oil is more abundant in neem seeds. Approximately 8 hours after preparation and when exposed to sunshine, a neem solution loses its potency. Neem is best applied in the evening immediately following preparation in humid conditions (Table 42.2).

Neem Applications

In traditional medicine, neem has been used to cure ailments like malaria, ulcers, cardiovascular disease, and skin issues. Neem is utilized in the formulation of face masks, lotions, sunscreens, soaps, and toothpastes in the cosmetics and hygiene industry. It possesses a potent insect growth regulator (IGR) that can fertilize plants and has a wide range of additional effects on many living things, including nematodes and fungi. Neem seed cake can be used as a biofertilizer since it contains the macronutrients needed for plant growth, which boost the quality of the soil and the crops. They function as sterilant, growth regulators, anti-pedants, anti-oviposition agents, and repellents as a result of the intricacy of their composition (Table 42.3).

Table 42.2 Neem applications and commercial products available in India

Application	Product	Manufacturer in India
Fertilizer	Ozoneem cake	Ozone biotech
	Ozoneem coat	Ozone biotech
	Parker neem coat	Parker neem
	Neem urea guard	Neemex
Agrochemical	SubhdeepNeem oil	King agro food
	Ozoneem oil	Ozone biotech

Table 42.3 Neem applications and commercial products available worldwide

Application	Product	Manufacturer in world
Fertilizer	Fortune neem cake	Fortune biotech (USA)
	Fortune neem coat	Fortune biotech (USA)
	Neem cake	Uniball Corporation (Russia)
Agrochemical	Fortune Aza 3% EC	Fortune biotech (USA)
	Azamax	UPL Ltd. (Brazil)
	Safer brand 3 in 1	Wood stream Corp. (Canada)

4.1.2 Pyrethrum and Pyrethrin

Pyrethrum is a chrysanthemum that resembles a daisy. Chemicals known as pyrethrin are obtained from dried pyrethrum flowers. To generate dust, the flower heads are ground into a powder. This dust can be sprayed directly on surfaces or mixed with water. Most insects are instantly paralysed by pyrethrin. Low doses produce a "knockdown" effect rather than being lethal. Higher doses are lethal. Pyrethrin should be stored in darkness because they degrade quickly in sunlight.

4.1.3 Rotenone

Rotenone, an insecticidal substance, is found in the roots of numerous closely related tropical legumes as well as Derris species in Asia. Currently, Peruvian Lonchocarpus, sometimes known as cube root, is the primary commercial source of rotenone. In acetone or ether, rotenone is removed from cube roots. Instead of using pure rotenone, the complex resin is used to create the majority of rotenone products. To create an insecticidal dust instead, cube roots can be dried, ground, and combined directly with an inert carrier.

4.1.4 Nicotine

Nicotiana tabacum and other Nicotiana plants are the sources of nicotine, a simple alkaloid found in tobacco. 2–8% of the dried tobacco leaves are nicotine. Forty percent nicotine sulphate is a common type of nicotine found in insecticidal formulations. It works well against tiny soft-bodied insects with soft bodies, such as aphids, white flies, fruit tree borers, termites, and cabbage butterfly larvae.

4.1.5 Citrus Oil

Orange and other citrus fruit peels are used to extract the refined chemicals limonene and linalool as well as the crude citrus oils. About 90% of crude citrus oil is the terpene limonene, which is removed from the oil via steam distillation. In addition to more than 200 other herbs, flowers, fruits, and woods, citrus peel contains trace amounts of the terpene alcohol linalool. One of the main ingredients of many plant volatiles, or essential oils, is terpenes, as well as terpene alcohols. The majority of a plant's flavors and smells are produced by volatile molecules called essential oils.

5 Biofertilizers

Nitrogen-fixing organisms are sometimes provided to farmers in the form of biofertilizers, often referred to as microbial inoculants. These preparations are typically those that contain live or latent cells of efficient strains of helpful microorganisms. They are applied to seeds or soil with the goal of increasing their numbers and speeding up certain microbial processes to increase the extent of the plants that require the availability of nutrients in an assimilable form by the plants (Sahoo et al. 2013). Due to the following benefits, biofertilizers are containing biological nitrogen-fixing organisms which are extremely important in agriculture:

I. They aid in the establishment and growth of trees and crop plants.
II. They increase grain yields and biomass production by 10–20%.
III. They assist in the practice of sustainable agriculture.
IV. They are appropriate for organic farming.
V. They are very crucial to agroforestry and silvopastoral systems.

5.1 Types of Biofertilizers

5.1.1 Rhizobium

The most common type of biofertilizer, which available on the roots of some legumes to produce root nodules, which resemble tumors. These nodules serve as ammonia producing facilities. In one crop season, the Rhizobium-legume relationship can fix up to 100–300 kg of nitrogen per hectare, and in some cases, it can also leave behind a significant amount of nitrogen for the crop that follows.

5.1.2 Azotobacter

The benefits of Azotobacter biofertilizer on cereals, millets, vegetables, cotton, and sugarcane have been verified and documented in both irrigated and rainfed field conditions. Azotobacter application has been shown to boost wheat, rice, maize, pearl millet, and sorghum yields by 0–30% in comparison to controls. This organism has the ability to create siderophores, antibacterial and antifungal compounds, as well as nitrogen.

5.1.3 Azospirillum

Microbes that may utilize atmospheric nitrogen and transfer it to agricultural plants include bacteria and blue-green algae. While other nitrogen fixers enter the root zones and cooperate loosely with plants, rhizobia, for instance, is an obligatory symbiont of leguminous plants. A very important member of this group is *Azospirillum*, a bacterium from the latter category that was identified by a Brazilian scientist and garnered media attention in the middle of the 1970s. Feed crops, maize, barley, oats, sorghum, and pearl millet all react to *Azospirillum* inoculation. Applications increase grain productivity by 30% in millets, 5–20% in cereals, and over 50% in fodder.

5.1.4 Blue-Green Algae

It looks highly promising to use blue-green algae as a biofertilizer for rice. With careful use, these algae may give the nation's whole rice farmland the same amount of nitrogen as 15–17 lakh tonnes of urea. Algal biofertilizer can now be produced in large quantities, and rice farmers around the world have already begun employing these algae. Recent studies have demonstrated that algae can assist lower soil alkalinity, which creates opportunities for bio-reclamation of such hostile conditions.

5.1.5 *Azolla*

Little floating water ferns called azolla are common in shallow freshwater areas and low-lying meadows. This fern contains the blue-green algae *Anabaena azollae*. The Azolla-Anabaena association is a living, floating nitrogen factory that uses photosynthesis and about 40–60 tonnes of biomass to fix atmospheric nitrogen at a rate of 100–150 kg N/ha/year. It has reportedly been used as a biofertilizer for rice in China, Vietnam, India, Indonesia, Thailand, and other East and South Asian countries. Azolla fish made of rice has a sophisticated infrastructure in place in China.

Table 42.4 Effect of mycorrhizal (VAM) inoculation on crop yields

Crop	P-fertilizer supplementation (kg P_2O_5/ha)	% Yield increase over control
Finger millet	19	18
Soyabean	25–50	19
Chillies	37.5	55
Chickpea	40	25
Groundnut	–	10–25

5.1.6 Mycorrhizae

Mycorrhizae are symbiotic relationships between fungi and vascular plant roots. The primary benefit of mycorrhizae to the host plants is the expansion of the root fungal system's penetration zone in the soil, which makes it easier for plants to absorb more phosphorus. In the soil beyond the depletion zone, which is otherwise inaccessible to the plant roots, the interconnected network of external hyphae serves as an additional catchment and absorbing surface. Numerous horticultural species, including apple, walnut, almond, citrus, avocado, strawberry, and grape, have been proven to contain endotrophic mycorrhizae. Mycorrhizae have been demonstrated in numerous instances to significantly enhance plant development (Table 42.4).

Fruit crops like citrus, papaya, and litchi have benefited from vesicular-arbuscular mycorrhizae (VAM) in India. Different citrus rootstocks were observed to absorb more Fe, Zn, and Mn after being inoculated with the *Glomus fasciculatum* fungus.

6 Systemic Induced Resistance or Systemic Acquired Resistance

A common occurrence in plant pathology is the generation of pathogen resistance responses in plants. It was initially defined as defense against a nonvirulent pathogen attack. As a result, it is a persistent, non-specific resistance to diseases that is brought on by pathogens that produce necrosis on the diseased leaves. Systemic acquired resistance is used to describe a resistance that has spread across the entire plant (SAR). SAR can also be brought on by virulent microorganisms in addition to pharmacological agents like salicylic acid (SA), which is implicated in the signal transduction pathway leading to SAR. Wiese et al. introduced the term chemically-induced resistance (CIR) in 2003, which is used to describe the systemic resistance that develops after the application of synthetic chemicals. This resistance is linked to the lignification of structural barriers, the activation of pathogen-related proteins, and the conditioning of the plants. It has been demonstrated that foliar sprays of nutrients including phosphates, K, and N result in systemic induced resistance

(SIR). Theoretically, during SIR, an immunity signals generated or released at the inducer leaf's induction site is carried systemically to the challenged leaves, where it activates the defense systems. When exogenously administered, salicylic acid (SA), which has been suggested as a possible signal, generates the resistance and PR proteins that are frequently found with SIR. SA was found in the phloem sap of the upper leaves that were not infected, but it was not found in the phloem sap obtained from the petioles of the lower leaves that were infected with *Pseudomonas syringae*. This shows that SA might not be the primary systemic indication for SIR. *Sphaerotheca fuliginea*, which causes powdery mildew in cucumbers, can be managed by a foliar spray of one phosphate. The same outcomes were seen in maize, where foliar phosphate administration led to a systemic defense against both common (caused by Puccinia sorghi) and northern leaf blight (NLB) diseases (caused by *Exserohilum turcicum*).

7 Future Perspectives

Finding the foods or vitamin combinations that can lessen illness severity will require more investigation. Finding the finest integrated pest management strategies with disease-resistant plant varieties that can be paired with particular cultural management techniques and effectively control plant disease is also required. Additionally, additional research is needed to determine how nutrients affect plant metabolism, disease tolerance, or resistance, and how this might be used to control plant diseases.

8 Conclusion

In the last few years, the importance of sustainable agriculture has been increasing day by day. To achieve the goal of sustainable agriculture, it is very important to minimize the usage of chemical pesticides or fertilizers in order to ensure a sustainable future. Ecological management of disease for sustainable agriculture is one of the best practices to minimize chemical pesticides in today's agricultural practices. Ecological disease management for sustainable agriculture helps management researchers advance soil well-being by improving soil physical, chemical, and natural properties, bringing about improved sustenance, upgraded yield, and disease suppression by expanding the biocontrol agents in the soil.

Acknowledgements We are highly thankful to Dr. Savita, Associate Professor, Graphic Era Hill University, Dehradun, for critical reading and formatting of the manuscript draft. Special thanks to Mirza Hasanuzzaman, Professor, Department of Agronomy, Faculty of Agriculture, Sher-e-Bangla Agricultural University (SAU), for the linguistic editing of the manuscript.

References

Ahirwar NK, Singh R, Chaurasia S, Chandra R, Ramana S (2020) Effective role of beneficial microbes in achieving the sustainable agriculture and eco-friendly environment development goals: a review. Front Microbiol 5:111–123

Dubey NK, Shukla R, Kumar A, Singh P, Prakash B (2010) Prospects of botanical pesticides in sustainable agriculture. Curr Sci 98(4):479–480

Francis CA (1989) Biolgical efficiencies in multiple cropping systems. Adv Agron 42:1–42

Gupta N, Debnath S, Sharma S, Sharma P, Purohit J (2017) Role of nutrients in controlling the plant diseases in sustainable agriculture. In: Agriculturally important microbes for sustainable agriculture. Springer, Singapore, pp 217–262

Kumari B, Mallick MA, Solanki MK, Solanki AC, Hora A, Guo W (2019) Plant growth promoting rhizobacteria (PGPR): modern prospects for sustainable agriculture. In: Plant health under biotic stress. Springer, Singapore, pp 109–127

Manimozhi K, Gayathri D (2012) Eco friendly approaches for sustainable agriculture. J Environ Res Develop 7(1):166–173

Mishra M (2013) Role of eco-friendly agricultural practices in Indian agriculture development. Int J Agric Food Sci Technol 4(2):11–15

Prabha S, Yadav A, Kumar A, Yadav A, Yadav HK, Kumar S, Kumar R (2016) Biopesticides—an alternative and eco-friendly source for the control of pests in agricultural crops. Plant Arch 16(2):902–906

Ram RM, Keswani C, Bisen K, Tripathi R, Singh SP, Singh HB (2018) Biocontrol technology: eco-friendly approaches for sustainable agriculture. In: Omics technologies and bio-engineering. Academic, pp 177–190

Reddy PP (2017) Agro-ecological approaches to pest management for sustainable agriculture. Springer, Singapore, pp 29–41

Sahoo RK, Bhardwaj D, Tuteja N (2013) Biofertilizers: a sustainable, eco-friendly agricultural approach to crop improvement. In: Plant acclimation to environmental stress. Springer, New York, pp 403–432

Chapter 43
Use of Advance Composting Techniques and Areas of Improvement in Pakistan

Mujahid Farid, Nouman Munawar, Gulfam Anjum, Muhammad Abubakar, Umar Shahzad, Muhammd Sheraz Asif, Hina Batool, Sajal Mir, Sara Javed, and Hafiz Khuzama Ishaq

Abstract Waste management (WM) is the world's most serious environmental challenge. A significant portion of waste is biodegradable material (organic). Composting is a viable and cost-effective method of converting waste into a valuable product. Composting is the regulated transformation of wastes and biodegradable organic materials into stable products with the help of microbes. Although composting has been around for a while, there are significant drawbacks that have limited its widespread application and effectiveness. The drawbacks include the inability to detect pathogens, a deficiency in nutrients, a lengthy composting and mineralization process, and odor creation. Over time, these difficulties have drawn attention to the use of synthetic fertilizers created using the coagulation–flocculation process as an alternative to compost. Synthetic fertilizers enable plants to easily access nutrients, but their drawbacks exceed their benefits. Chemical fertilizers, for instance, have a negative impact on human health, ozone layer depletion, environmental pollution, greenhouse effects, and marine and soil life. Farmers have turned back to applying composts to restore soil fertility as a result of this. Composting is an essential agricultural procedure that aids in the recycling of farm waste. Due to the presence of components that take a while to breakdown, composting takes a while, especially when co-composting. This study describes the appropriate management of wastes through composting, various composting procedures, the factors effecting composting, long-duration composting, the mechanism underpinning it, current trends in composting, and potential outcomes. The ability to separate mono-fertilizers from compost, create test strips to detect heavy metals and pathogens, and use an odor-trapping approach can all significantly enhance composting procedures. The nutritional value of compost can be enhanced by adding catalysts to the basic

---

M. Farid (✉) · N. Munawar · G. Anjum · M. Abubakar · U. Shahzad · M. S. Asif · H. Batool
S. Mir · S. Javed
Department of Environmental Sciences, University of Gujrat, Hafiz Hayat Campus, Gujrat, Pakistan

H. K. Ishaq
Dipartimento di Scienze e Tecnologie Ambientali, Biologiche e Farmaceutiche, Università degli Studi della Campania Luigi Vanvitelli, Caserta, Italy

M. Hasanuzzaman (ed.), *Climate-Resilient Agriculture, Vol 2*,
https://doi.org/10.1007/978-3-031-37428-9_43

ingredients. To enhance the quality of the compost, you might also add pesticides, nematicides, fungicides, anti-bacterial, and viricides from plant or organic sources. The prolonging of the composting process will also be beneficial.

Keywords Biodegradable · Coagulation · Chemical fertilizers · Catalyst · Composting

1 Introduction

Inadequate waste disposal is hazardous to human health. In addition to being unsightly, it pollutes the air, harms waterbodies when dumped into them, and destroys the ozone layer when burned, exacerbating the effects of climate change. Using traditional approaches, waste is frequently mismanaged (Alam and Ahmade 2013; Aruna et al. 2018). Waste is burned, disposed of in seas and rivers, and discarded along roadsides (Ogwueleka 2009). These methods encourage the reproduction of insects and vermin, emit terrible aromas, are ugly, additionally drive global warming (during combustion). Organic waste can decompose either aerobically or anaerobically. Compost is generated when organic matter is converted under aerobic conditions (Lasaridi et al. 2018). Methane and effluents that can be used as biofertilizers are created during anaerobic processing (Khan et al. 2018). Waste may be safely disposed of by composting. Microorganisms break down and transform complex degradable materials into organic and inorganic trash during the aerobic process of composting (Toledo et al. 2018). The byproducts differ from natural soil, coals, and peats by having “humic-like” elements. Various biodegradable wastes may be transformed into compounds through the process of composting, which can then be used safely and effectively as bio fertilizers and soil supplements (Cai et al. 2007; Yu et al. 2019a). The composting procedure aids in preventing groundwater pollution as opposed to the practice of landfilling garbage. Which may jeopardise subsurface water. This is because composting reduces the amount of polluting chemicals and microorganisms. These are the infectious microorganisms found in trash that are harmful to humans. Pollutants like persistent organic pollutants POPs that are left in the soil after composting are absorbed by the beneficial microorganisms. A subsequent portion of this evaluation goes into greater detail. Because composted manure contains vital nutrients and organisms that promote plant growth, using it boosts agricultural production and the organic matter content of the soil (Luo et al. 2017). This makes a significant contribution to ensuring food security. In addition to being used as fertilizer, compost is beneficial for bioremediation (Ventorino et al. 2019), control plant diseases (Pane et al. 2019) and weeds (Coelho et al. 2019), prevents pollution (Uyizeye et al. 2019), controls erosion, and restoring wetland areas. Additionally, composting boosts soil biodiversity and lowers the environmental hazards of synthetic fertilizer (Pose-Juan et al. 2017). Instead of a naturally occurring and unmanaged process, composting is initiated and maintained in a controlled atmosphere (Cáceres et al. 2018). Composting and decomposition, a naturally occurring process, are distinguished by the regulating process (Hoitink and Fahy 1986). Although composting offers benefits, it also has drawbacks, including a longer completion time, an unpleasant odor, a slow mineralization process, and the potential for diseases that can at least partially withstand high temperatures, or thermotolerant pathogens, and has insufficient nutritional content. Farmers have been deterred from utilizing them as a method of sustainable agriculture by all of these. Because of this, artificial fertilizers, which are easily accessible, were chosen over composting as the organic alternative.

Composting is decomposition and mineralization of organic residue in the presence of oxygen, resulting in a finally stabilized product called compost which is just like humus and free of pathogens and phytotoxicity (Zucconi 1987). Composting is a biotic degradation of biological materials that occurs naturally in a primarily aerobic atmosphere. Bacteria, fungus, and other microbes are present during the process. Micro-arthropods, for example, compost, a stable, useful organic product, is produced by decomposing organic materials. Composting is also beneficial, indicates waste volume reduction and waste annihilation, also control on pathogenic microbes and weed seeds (Bernal et al. 2009). The cattle business generates a large quantity of decomposable waste, which need to handle according to suitable removal methods to prevent having a negative influence on the environment. Examples of such emissions include odor and gaseous ones, soil and water contamination, and so on (Burton 2003). Composting is a cost-effective and environmentally friendly solution to the problem of disposing of organic waste (Gutiérrez et al. 2017). Composting animal manure can cost significantly more than utilizing it directly. Since compost is created to a high grade and is paid to pay its production costs, for manures that need to be partially disinfected, composting is excellent. (Parkinson et al. 2004). In order to produce high-quality compost with greater agricultural value, the current study explores the parameters influencing the composting of animal manures with a focus on the nutrient content, organic matter (OM) humification, and maturity level. More information is given on the impacts of manure composting on the environment and public safety by Moral et al. (2009). Despite the numerous reported pluses of composting that have already been mentioned, there are a number of drawbacks to this waste management strategy, including its effects on global warming, the emission oxygen depletion and the emission of carbon dioxide into the environment, and the generation of unpleasant odors brought on by the release of hydrogen sulphide created by anaerobic activity. Various laws regarding the usage of the procedure have been put in place by various agencies in various nations as a result of these health effects. These drawbacks highlight the need for further development of this approach to solve the numerous issues raised. The management of oxygen flow and temperature regulation are two important areas that need improvement. These play a crucial role in how the bacteria that carry out the composting process work. Each stage's different bacteria have a certain operating temperature; therefore, they must be carefully regulated. They also require oxygen in order to operate, which reduces the anaerobes' efficiency. Hydrogen sulphide is released together with carbon dioxide during these processes, both of which have potentially harmful consequences on health, are encouraged by anaerobe activity. If these can be separated, there are several ways that this technique might contribute to proper waste management.

2 Waste Characterization

Solid waste management is as old as human civilization. Management system of integrated waste for sustainable development is the greatest challenge of whole globe. For a good and successful waste management, the first step is waste characterization. Characterization and composition of waste both are different according to their generation source or it depend site to site (De Vega et al. 2008). Waste is not by birth property of anything or item but it depends on certain situation in which item appear and it defined by owner values to items. Solid waste characterization is tough and hard task because of waste heterogeneity and its spatial as well as temporal variations (Carlsson et al. 2015). According to source of waste generation, waste is categorized into four major categories with common characteristics. Major categories of trash include municipal garbage, business and institutional garbage, scrap metal, and debris from building and destruction (Eisted and Christensen 2011).

Waste categories are further divided into subclasses known as waste type. Municipal waste is further divided into types in which organic waste include mixed municipal waste, septic tank sludge, garden and park waste, household waste, restaurant, canteen, and bars waste. Waste comes from the commercial/industrial is different according to their activity in industry. Organic waste comes from the commercial and industrial include wood pulp, paper, cardboard production and process, and food and beverage processing plants waste. Waste from agriculture, horticulture, aquaculture, forestry, and hunting is also organic waste (Gentil et al. 2011).

2.1 *Effects of Wastes*

Wastes have a negative influence on the environment, which puts human lives in grave danger. These negative effects, which can lead to disease outbreaks, decreased life expectancy, and dangerous environments, have an impact on both humans and animals. Some wastes could decay, but those that do not produce methane gas which is essential for the greenhouse effect and stench. The effects of wastes on the environment and human health will next be discussed. Wastes contaminate the soil, water, and air. Dust, smoke, and odor all contribute to air pollution. Burning solid trash releases greenhouse gases including carbon dioxide and nitrous oxide, which harm the ozone layer and have a warming effect (Bhat et al. 2018).

Additionally, emitted into the atmosphere are methane and hydrogen sulphide. These materials are harmful to human life. Water contamination is another negative environmental impact of garbage. According to reports, 1400 individuals each day pass away as a result of illnesses and difficulties associated to water (Khan et al. 2019). Wastes that end up in water bodies, such as rivers, streams, and seas, can disturb the ecosystem by reducing pH levels and making the water harmful to aquatic life and those who use it. Some of these contaminants have a low water solubility and a high lipophilicity (Varjani et al. 2017).

Untreated wastes that contain disease vectors cause human illnesses. Mosquitoes thrive in stagnant water, clogged drains, tires that catch rainfall, empty food cans, plastics, etc. Furthermore, tissue damage, lung infections, injuries from glass, razor blades, and needles, as well as parasite diseases brought on by direct contact with waste, are problems that trash workers must contend with (Alam and Ahmade 2013) ASD. To protect refuse workers from mishaps connected to trash treatment, novel automated solutions should be pushed even while staff use safety equipment such gloves and nasal masks (Langdon et al. 2019). Conducted research in Australia on the risk evaluation of organic pollutants in solid municipal garbage that has been composted. They were allowed to assign a low, medium, or high priority to each of the risk categories hazards in their study based on the possible negative consequences on health. This allows for the establishment of procedures for the effective and proper disposal of the various toxicants from these wastes. In an another research, Gangwar et al. (2019) examined how electronic trash affects people's health and reported that it releases a number of harmful metals into the air, producing air pollution. These wastes' effects on the environment and human health have been covered in other research as well (Ali et al. 2019; Yu et al. 2019b).

3 What Is Composting

As a result of microbial activity, waste decomposes and stabilizes into compost through a biological breakdown of organic substances known as composting (Kapley 2021). Composting is a sustainable method of recycling solid waste that uses microbial (bacteria, archaea, and fungus) potential to break down trash and create compost, a stable byproduct. Organic fertilizers frequently employ compost since it typically has a high percentage of organic material (Kausar 2010).

4 Composting Methods

There are different types of composting that are occurring across the globe. Here is some important technique of composting which are common in all over the world.

4.1 *Vermicomposting*

When organic wastes are turned into a humus-like substance known as compost, the process is called vermicomposting. Processing the data as quickly as is practical is the goal. These two processes are comparable such as windrow composting and vermicomposting, but they are not the same. If you want to make vermicomposting, you should always have the highest worm population density possible (Munroe 2007).

4.1.1 The Compost Worm

Earthworms are found in over 1800 different species around the world (Malley et al. 2006). *Eisenia fetida* (Sevigny), manure worm, red worm and red wiggler is also known as the compost worm. This robust and adaptable worm is native to most of the world, and it may be found on most farms where manure piles have been permitted to age for more than a few months (Gunadi and Edwards 2003).

4.1.2 Types of Worms

Anecic Worms

Greek phrase anecic, which meaning "out of the soil," is used. In order to get food down into their permanent burrows deep under the soil's mineral layers, burrowing worms come to the surface at night. A nice illustration is the Canadian Night Crawler (Beetz 1999; Mishra et al. 2022).

Endogeic Worms

They are also burrowing worms, but because they feed on organic materials already existing in the soil and often dig deeper tunnels, they seldom emerge from the surface (Knowles et al. 2016).

Epigeic Worms

These worms feed on decomposing organic waste and reside in the surface litter. They do not have burrows that are permanent. The worms employed in vermicomposting are known as decomposers (Ganin and Atopkin 2018).

4.1.3 Advantage of Vermicomposting

There are various options, and not all of them will apply to all organic farmers. In conclusion, they are the following:

- In general, vermicomposting looks to be superior to conventionally generated compost. There are several reasons why compost is advantageous.
- When it comes to inoculants, vermicomposting outperforms most composts, teas made from compost.

- Worms can be used in a variety of ways on farms, including as a source of protein.
- Organic farms can benefit from vermicomposting and vermiculture as sources of nutrients for additional income (Mundiyara and Jat 2017).

4.1.4 Disadvantages of Vermicomposting

Composting is a more sophisticated process than it used to be due to the following reasons:

- It can be faster, but doing so usually involves more effort.
- It need greater space because worms are surface feeders and will not thrive in enclosed spaces.
- It is more susceptible to environmental stresses like frigid temperatures as well as drought.
- Most importantly, it requires greater initial capital, either in cash, or in kind to acquire the grubs or in terms of time and effort (Malakar and BISWAS 2019).

4.1.5 The Five Crucial Elements

1. A comfortable living environment, also referred to as bedding.
2. A source of food.
3. Adequate moisture (water content more than 50% by weight).
4. Proper aeration.
5. Temperature extremes are guarded against (Munroe 2007).

4.2 Windrows Composting

The organic waste is broken down throughout the 22-week windrow composting process in open, slender heaps. Recyclable waste must be disturbed and cycled often for a desirable outcome. Regular stirring is done to the feedstock. A 1.5 m tall by 2 m wide windrow is where the food waste is dragged ahead and stored. The compost has to regularly be exposed to ideal processing conditions, such as air, light, and temperature. Turn the compost occasionally during stabilization to maintain an active bacterial population. The tides changing. It is crucial to keep in mind that composting frequency has a substantial influence on the potential for microbial count increase and degradation. It is recommended to turn the compost once a week or once every 3 days rather than every day (Awasthi et al. 2014). Composting with natural aeration has nonetheless generally been highly effective. Microorganisms can successfully be given oxygen by using an air blower in a forced aeration system. During productive metabolic activity, carbon is converted into carbon dioxide, which produces heat energy (Gopikumar et al. 2020).

4.2.1 Windrow for Aeration

With the addition of a supporting material, the 5-week process of static pile composting can last for 12 weeks. Aeration system that consists of a blower connected to an open or covered stack. Once the pile is formed, there is no need for agitation since both the uphill and downward segments of the biofilter aerate the water. In contrast, turning is required. Static heaps with natural aeration cost a lot less money than static piles with forced aeration. (Gopikumar et al. 2020; Larney and Olson 2006).

4.3 Composting with In-vessel

Composting occurs in the in-vessel system in a closed-circuit channel with high-pressure aeration. It is emphasized that the rate of aeration change, which determines the quality of the compost, is a carefully regulated process. The closed system regularly facilitates the microbial community's proliferation and disintegration, and the ongoing agitation in this module guarantees that the system as a whole is aerated. This method may be optimized for composting by including a mechanical turning procedure (Manyapu et al. 2018; Rihani et al. 2010).

4.4 Static Composting

This is a conventional composting technique that uses passive aeration to aerobically compost wastes. This method is being contrasted with vermicomposting, windrow, vessel, and Indian Bangalore composting, it has lower operating and capital expenditures but is still time-consuming. The only labor- and resource-intensive part of this technique is the simple construction of a mound of basic materials. Aeration is primarily dependent on the passive passage of air through the pile, which causes the organic materials to deteriorate gradually (Meng et al. 2019; Gonawala and Jardosh 2018).

4.5 Sheet Composting

Without creating a compost stack, sheet composting allows you to get the benefits of decomposed organic matter. The leaves, orchard trash, grass clippings, weeds, and vegetative food utilized in this method are thinly spread as a mulch directly onto the soil. Then, using a hoe, spade, or garden fork, you till the organic materials into the ground and leave them there to degrade rather than putting them in a mound or container. It is customary to cover the growing area with one or more layers of organic material, which are then well watered and left to decompose until it is time to plant. More organic molecules, which fully dissolve, are piled in the lower layers (Misra et al. 2003; Stratton et al. 2000). It is a quick and inexpensive procedure.

4.6 Berkley Rapid Composting

This composting process is quick. In this case, materials that range in size from 0.5 to 1.5 inches compost more quickly. Soft, succulent tissues do not require being cut into extremely tiny pieces since they degrade quickly. To improve composition, tissues must be cut into smaller pieces as they get tougher. Since it takes time for the first components to degrade, anything added would have to begin that process, lengthening the time it would take for the entire pile to decay. As a result, nothing should be added to a pile after it has started (Behera et al. 2020; Misra et al. 2003).

5 Uses of Composting

5.1 Crop Yield, Soil Amendment, and Soil Fertility Increase; Furthermore, Erosion Control

Compost-based compound fertilizers are currently a welcome idea since there has been a recent campaign to ban the usage of artificial pesticides. Majbar et al. (2018) asserting that compost improves plant productivity and soil fertility. Complementing artificial fertilizers with compost is another technique to employ it for plant growth. We encourage mixing the two in the correct amounts because studies show synthetic fertilizers may be more helpful than compost in boosting plant development (Pampuro et al. 2017). Additionally, composts include microorganisms that support plant growth, enhancing soil fertility and plant growth. The fertility of the soil is lost as a result of erosion. Significant amounts of potassium, phosphorus, and nitrogen are lost through erosion. It has been found that organic surface-applied supplements work quite well to stop erosion. Compost improves the soil's ability to store water and its aggregate stability (Gonawala and Jardosh 2018). This is because the soil contains humus, which binds to the soil and acts as a sort of "glue" to hold the various soil elements together (Epelde et al. 2018). Humus is a stable residue left behind after a significant amount of organic matter has decomposed.

Compost is end product of this process, and it can be used as fertilizer and soil ingredient because it contains a lot of organic and inorganic nutrients in it. Compost can be used in gardens. It is also used as top cover of solid waste landfills as it has the ability to reduce methane emissions from that solid waste decomposition process (Keener et al. 2000). Compost has the ability to increase the fertility of land and it reduces the need of inorganic fertilizers which in long term cause damage to soil.

6 Composting Process

Mesophilic, thermophilic, cooling, and maturation are the four phases of the composting process. There is a release of greenhouse gases such carbon dioxide (CO_2), nitrous oxide (N_2O), and methane (CH_4) as well as ammonia (NH_3), volatile organic compounds (VOCs), hydrogen supplied (H_2S), and nitrous oxide (N_2O) are released during the complicated aerobic and anaerobic digestion process that takes place during composting (Rincón 2019). The fundamental difference between composting and aerobic fermentation is the regulated circumstances. It has been demonstrated that controlling variables like bulk density, porosity, particle size, nutrient content, C/N ratio, temperature, pH, moisture, and oxygen supply is crucial for composting optimization. These conditions are then used to determine the best methods for microbial growth and OM decomposition (Agnew 2003).

7 Characteristics of the Organic Waste for Composting

Manure is less concentrated and homogenous than compost, which distributes more readily and disperses evenly in the soil while obliterating diseases and weed seeds. The compost may also be used as a soil substitute for areas lacking soil and as fertilizers for pots. Composting animal manures has the following advantages over direct application:

Moisture and volume reduction
Odor elimination and management
Simple to use, store, and transport
Manufacture of high-quality fertilizer or substrate

Animal manures and other nitrogen-rich wastes are frequently co-composted with lignocellulose from agriculture and forestry byproducts. The materials that are most frequently utilized are cereal straws (Bernal 1998). To produce a high-quality compost, the manure must be composted with sufficient management. This has led to the application of various aeration techniques. Manure composting substrate conditioning-feedstock formulae, bulking agents, and process control alternatives are explored in an effort to speed up and lower the cost of the process while improving the quality of the final products (Lau et al. 1992; Michel et al. 2004; Solano et al. 2001). All of them have high levels of organic carbon, little moisture, and high C/N ratios (an average of 50 for wheat straw and >80 for wood by-products), which can compensate for the absence of considerable levels of animal excrement.

8 Organic Matter Degradation and Nitrogen Losses

Because there are less carbon sources available as composting progresses, the rate of OM breakdown gradually decreases. During the maturation phase, synthesis activities of novel, complex, and polymerized organic molecules (humification) overwhelm mineralization. This method's stabilized end products are used as agricultural slow-release fertilizers. However, C and N losses are the key issues in manure composting since they reduce the agronomic value of compost and increase greenhouse gas emissions (Hao et al. 2004).

The nature of the OM affects how quickly the substrate changes during the composting process, depending on how easily it degrades (Dias et al. 2010). Bovine manure, chicken dung, and pig manure can all experience organic-C losses of up to 67%, 52%, and 72%, respectively, throughout the composting process. The assessment of the main emphasis of this research is on the variables influencing the composting of animal manures for the creation of high-quality compost with enhanced agricultural value. The nutritional value, humification of organic matter (OM), and maturity level are the main themes. Further data on the effects of composting manure on the environment and public safety are analyzed by Moral et al. (2009). According to the OM's degradability, the type of the material will have an impact on the process' duration and scope, rate of decomposition, gas emissions, and oxygen requirements. Simple sugars, lipids, and amino acids, which are labile organic materials, degrade quickly in the early phases of composting. Nitrogen is formed during nitrification when the temperature falls below the thermophilic threshold (40 °C), which is indicated by the production of NO_3-N. The amount of NH_4-N that the nitrifying bacteria can access determines how quickly nitrification takes place (Tiquia et al. 2002).

9 Humification Process

Since it is the portion of the soil that is least susceptible to microbial degradation, the humified fraction of the soil, also known as OM, is the one that is most crucial for the soil's organic fertility activities. Therefore, one agronomic criteria for compost quality is assessing the degree of OM humification during composting. Agriculturally speaking, compost becomes more valuable when the OM reaches a high humification level. According to the widely recognized humification theories of soil OM, the creation of humic acids with enhanced molecular weight, aromatic properties, oxygen and nitrogen concentrations, and functional groups during composting is proof that the OM has been humified (Senesi 1989; García-Gómez et al. 2005). Composting results in the formation of alkali-extractable Organic-C, also known as CEX, and Organic-C that mimic's humic acid (CHA). Organic-C (CFA) and water-extractable

Organic-C, two folic acid derivatives, degrade due to microbial activity. To determine how much a material has been humified throughout the composting process, there are a number of indicators that may be employed (Roletto et al. 1985; Xu et al. 2019).

10 Maturity Assessment for Quality Compost

According to Iannotti et al. (1993), stability is frequently correlated with the microbiological activity of the compost, whereas maturity is connected to plant growth potential or phytotoxicity. But since the microorganisms in unstable composts create phytotoxic chemicals, stability and maturity typically go hand in hand (Hachicha et al. 2009). Maturity, which refers to the degree or amount of composting completion, suggests that a product has increased attributes as a consequence of "ageing" or "curing." Maturity is simply "the degree to which a compost has developed" (Antil et al. 2014). According to the Slater and Frederickson (2001), mature compost is "compost that does not have a detrimental influence on seed germination or plant development." Maturity is described by the California Compost Quality Council as "the degree or level of completion of composting", Bernal et al. (1998). It is best to measure two or more compost properties since maturity cannot be expressed by a single attribute. The relative stability of the substance has an influence on maturity, but maturity also refers to how other chemical qualities of compost affect plant growth. Certain organic acids, free ammonia, and other water-soluble substances may be present in high concentrations in some immature composts, which may prevent root growth and seed germination (Qiao et al. 2021).

A water soluble Organic-C/Organic-N ratio of 5–6 was established by as a critical indicator of compost maturity. This ratio, however, can be difficult to detect because to the frequently very low concentration of Organic-N in the water extract of mature samples. Because of this, Bernal et al. (1998) and Hue and Liu (1995) suggested using the ratio of water-soluble organic carbon to total organic nitrogen to gauge the maturity of a compost. Dissolved organic carbon, the most energetic form of carbon, is a reliable sign of compost stability (Xu et al. 2020). It is possible to determine the maturity of a compost by looking at its microbial stability, microbial activity characteristics including microbial biomass count and metabolic activity, and the proportion of readily biodegradable components. Previously, the aerobic respiration rate was determined to be the optimal measure for assessing aerobic biological activity and, consequently, stability. (Kumar et al. 2010) Under aerobic conditions, one catabolized carbon atom bonds with two oxygen atoms to make carbon dioxide, releasing heat and other types of energy in the process. As a result, there are several methods for assessing respiration, such as self-heating, carbon dioxide evolution, and oxygen consumption, which all serve as markers of how much OM is still able to degrade and are inversely related to stability (Gao et al. 2010).

11 Conclusions and Recommendations

A safer waste management technique, like composting, can take the place of a prevalent practice called improper waste management. The global trend is toward bettering both human and environmental health. Composting, which is an organic fertilizer, can be very helpful in attaining this objective. A focus on composting will lead to a move away from chemical fertilizers and toward compost. The decrease in the amount of hazardous substances discharged into the environment as a result of this change would unavoidably improve both environmental and human health. The promise of this technology still has to be widely publicized if farmers are to fully embrace it in its current form. Some suggestions are made here on improving technologies to help them advance. It is advised that poly nutrients be removed from composts due to the high nutrient content in composts. When a pre-planting soil study is done, it happens frequently that one nutrient may be deficient. In order to avoid over-applying nutrients that are not required, mono fertilizers can be extracted from compost's compound fertilizer form. In addition, organisms that can degrade Composting operations can be avoided by providing farmers with complicated degradable materials as inoculum. To address the issue of air pollution brought on by compost production, more study should be carried out to identify the odor-trapping mechanism. In order to stop greenhouse gas emissions, composting should have CO_2 capturing systems. The anti-nematode, vermicide, bactericide, and fungicide generated by plants can be added to compost to increase its efficacy. By doing away with chemical pesticides, this will support the growth of organic farming alone. Items that degrade slowly should be composted separately from other materials so that the composting periods of other materials are not increased by the slower decaying materials. More study should be done to determine whether compounds that take longer to decompose also mineralize progressively. Biennial and perennial crops, which are long-term sources of nutrients, may benefit from minerals that slowly mineralize. It is necessary to look at this theory's viability more thoroughly. Since this will help in selecting whether or not to compost them, research should include information on the nutritional content of particular leaves that are decaying slowly. In underdeveloped countries like Nigeria, composting may be utilized to minimize the quantity of agricultural waste generated. Before being put to the ground, composts should always be checked for maturity and pathogens to minimize potential threats to the environment and other living things. In order to increase the composting process' length, further study should be done. Even if the Berkley method was previously discovered and is still the fastest, the creation of faster approaches will help to sustain the composting process.

References

Agnew JL (2003) The physical properties of compost. Compost Sci Util 11(3):238–264

Alam P, Ahmade K (2013) Impact of solid waste on health and the environment. Int J Sust Develp Green Econ 2:165–168

Ali IH, Siddeeg SM, Idris AM, Brima EI, Ibrahim KA, Ebraheem SA, Arshad M (2019) Contamination and human health risk assessment of heavy metals in soil of a municipal solid waste dump site in Khamees-Mushait. Saudi Arabia Toxin Rev 40(1):1–14

Antil, R., Raj, D., Abdalla, N., Inubushi, K. (2014). Physical, Chemical and Biological Parameters for Compost Maturity Assessment: A Review. In: Maheshwari, D. (eds) Composting for Sustainable Agriculture. Sustainable Development and Biodiversity, vol 3. Springer, Cham. https://doi.org/10.1007/978-3-319-08004-8_5

Aruna G, Kavitha B, Subashini N (2018) Indira S An observational study on practices of disposal of waste Garbages in Kamakshi Nagar at Nellore. Int J Appl Res 4:392–394

Awasthi MK, Pandey AK, Khan J, Bundela PS, Wong JW, Selvam A (2014) Evaluation of thermophilic fungal consortium for organic municipal solid waste composting. Bioresour Technol 168:214–221

Beetz A (1999) Worms for Composting (Vermicomposting)." ATTRA-National Sustainable Agriculture Information Service, Livestock Technical Note, June, 1999

Behera SK, Vasundhara L, Saisudha G, Haariharan NC (2020) AI based waste classifier with thermo-rapid composting. In: International conference on power, energy, control and transmission systems, pp 1–4

Bernal MP-M (1998) Maturity and stability parameters of composts prepared with a wide range of organic wastes. Bioresour Technol 63(1):91–99

Bernal MP, Sanchez-Monedero MA, Paredes C, Roig A (1998) Carbon mineralization from organic wastes at different composting stages during their incubation with soil. Agri Ecosys Environ 69(3):175–189

Bernal MP, Alburquerque JA, Moral R (2009) Composting of animal manures and chemical criteria for compost maturity assessment. A review. Bioresour Technol 100(22):5444–5453

Bhat RA, Dar SA, Dar DA, Dar G (2018) Municipal solid waste generation and current scenario of its management in India. Int J Adv Res Sci 7:419–431

Burton HT (2003) Manure management, second treatment strategies for sustainable agriculture. Silsoe Research Institute. Lister and Durling Printers, Flitwick, Elsevier, Bedford

Cáceres R, Malin'ska K, Marfà OJWM (2018) Nitrification within composting: a review. Front Microbiol 72:119–137

Cai QY, Mo CH, Wu QT, Zeng QY, Katsoyiannis A (2007) Concentration and speciation of heavy metals in six different sewage sludge-composts. J Hazard Mater 147:1063–1072

Carlsson M, Holmström D, Bohn I, Bisaillon M, Morgan-Sagastume F, Lagerkvist A (2015) Impact of physical pre-treatment of source-sorted organic fraction of municipal solid waste on greenhouse-gas emissions and the economy in a Swedish anaerobic digestion system. Waste Manag 38:117–125

Coelho L, Osório J, Beltrão J, Reis M (2019) Organic compost effects on Stevia rebaudiana weed control and on soil properties in the Mediterranean region. Rev Ciênc Agrár 42:109–121

De Vega CA, Benítez SO, Barreto MER (2008) Solid waste characterization and recycling potential for a university campus. Waste Manag 28:21–26

Dias BO, Silva CA, Higashikawa FS, Roig A, Sánchez-Monedero MA (2010) Use of biochar as bulking agent for the composting of poultry manure: effect on organic matter degradation and humification. Bioresour Technol 101(4):1239–1246

Eisted R, Christensen TH (2011) Characterization of household waste in Greenland. Waste Manage 31(7):1461–1466

Epelde L, Jauregi L, Urra J, Ibarretxe L, Romo J, Goikoetxea I, Garbisu C (2018) Characterization of composted organic amendments for agricultural use. Front Sust Food Syst 2:44

Food Waste to Valuable Resources Aerobic biodegradation of food wastes Elsevier 235–250

Gangwar C, Choudhari R, Chauhan A, Kumar A, Singh A, Tripathi A (2019) Assessment of air pollution caused by illegal e-waste burning to evaluate the human health risk. Environ Int 125:191–199

Ganin GN, Atopkin DM (2018) Molecular differentiation of epigeic and anceic forms of Drawida ghilarovi Gates, 1969 (Moniligastridae, Clitellata) in the Russian Far East: Sequence data of two mitochondrial genes. European J Soil Biol 86:1–7

Gao M, Liang F, Yu A, Li B, Yang L (2010) Evaluation of stability and maturity during forced-aeration composting of chicken manure and sawdust at different C/N ratios. Chemosphere 78(5):614–619

García-Gómez A, Bernal MP, Roig A (2005) Organic matter fractions involved in degradation and humification processes during composting. Compost Sci Util 13(2):127–135

Gentil EC, Gallo D, Christensen TH (2011) Environmental evaluation of municipal waste prevention. Waste Manag 31(12):2371–2379

Gonawala SS, Jardosh H (2018) Organic waste in composting: a brief review. Int J Curr Eng Technol 8:36–38

Gopikumar S, Tharanyalakshmi R, Kannah RY, Selvam A, Banu JR (2020) Aerobic biodegradation of food wastes. In Food Waste to Valuable Resources (pp. 235–250). Academic Press. Elsevier https://doi.org/10.1016/C2018-0-03868-2

Gunadi B, Edwards CA (2003) The effects of multiple applications of different organic wastes on the growth, fecundity and survival of Eisenia fetida (Savigny)(Lumbricidae). Pedobiol 47(4):321–329.

Gutiérrez MC, Siles JA, Diz J, Chica AF, Martín MA (2017) Modelling of composting process of different organic waste at pilot scale: biodegradability and odor emissions. Waste Manag 59:48–58

Hachicha R, Hachicha S, Trabelsi I, Woodward S, Mechichi T (2009) Evolution of the fatty fraction during co-composting of olive oil industry wastes with animal manure: maturity assessment of the end product. Chemosphere 75(10):1382–1386

Hao X, Chang C, Larney FJ (2004) Carbon, nitrogen balances and greenhouse gas emission during cattle feedlot manure composting. J Environ Qual 33:37–44

Hoitink HA, Fahy PC (1986) Basis for the control of soil borne plant pathogens with composts. Annu Rev Phytopathol 24:93–114

Hue NV, Liu J (1995) Predicting compost stability. Compost Sci Util 3(2):8–15

Iannotti DA, Pang T, Toth BL, Elwell DL, Keener HM, Hoitink HAJ (1993) A quantitative respirometric method for monitoring compost stability. Compost Sci Util 1(3):52–65

Kapley JA (2021) Challenges and control strategies of odor emission from composting operation. Appl Biochem Biotechnol 193:2331–2356

Kausar HS (2010) Development of compatible lignocellulolytic fungal consortium for rapid composting of rice straw. Int Biodeter Biodegrad 64(7):594–600

Keener HM, Dick WA, Hoitink HA (2000) Composting and beneficial utilization of composted by-product materials. Land Appl Agri Indust Munc Prod 6:315–341

Khan M, Chniti S, Owaid M (2018) An overview on properties and internal characteristics of anaerobic bioreactors of food waste. J Nutr Health Food Eng 8:319–322

Khan SA, Suleman M, Asad M (2019) Assessment of pollution load in marble wastewater in Khairabad, district Nowshera, Khyber Pukhtunkhwa, Pakistan. Int J Econ Environ Geol 8:35–39

Knowles ME, Ross DS, Görres JH (2016) Effect of the endogeic earthworm Aporrectodea tuberculata on aggregation and carbon redistribution in uninvaded forest soil columns. Soil Biol Biochem 100:192–200

Kumar M, Ou YL, Lin JG (2010) Co-composting of green waste and food waste at low C/N ratio. Waste Manag 30(4):602–609

Langdon KA, Chandra A, Bowles K, Symons A, Pablo F, Osborne KA (2019) Preliminary ecological and human health risk assessment for organic contaminants in composted municipal solid waste generated in New South Wales, Australia. Waste Manag 100:199–207

Larney FJ, Olson AF (2006) Windrow temperatures and chemical properties during active and passive aeration composting of beef cattle feedlot manure. Canadian J soil Sci 86(5):783–797

Lasaridi KE, Manios T, Stamatiadis S, Chroni C, Kyriacou A (2018) The evaluation of hazards to man and the environment during the composting of sewage sludge. Sustainability 10:2618

Lau J, Antman EM, Jimenez-Silva J, Kupelnick B, Mosteller F, Chalmers TC (1992) Cumulative meta-analysis of therapeutic trials for myocardial infarction. New Engl J Med 327(4):248–254

Luo X, Liu G, Xia Y, Chen L, Jiang Z, Zheng H, Wang ZJ (2017) Use of biochar-compost to improve properties and productivity of the degraded coastal soil in the Yellow River Delta, China. J Soils Sediments 17:780–789

Majbar Z, Lahlou K, Ben Abbou M, Ammar E, Triki A, Abid W, Nawdali M, Bouka H, Taleb M, El Haji M (2018) Co-composting of olive mill waste and wine-processing waste: an application of compost as soil amendment. J Chem 79:185–183

Malakar M, BISWAS PAS (2019) Rejoinder of diverse organic growing media on morphological attributes of nine Heliconia species and varieties under West Bengal condition. J Crop Weed 15:35–44

Malley C, Nair J, Ho G (2006) Impact of heavy metals on enzymatic activity of substrate and on composting worms *Eisenia fetida*. Bioresour Technol 97(13):1498–1502

Manyapu V, Mandpe A, Kumar S (2018) Synergistic effect of fly ash in in-vessel composting of biomass and kitchen waste. Bioresour Technol 251:114–120

Meng X, Liu B, Zhang H, Wu J, Yuan X, Cui Z (2019) Co-composting of the biogas residues and spent mushroom substrate: Physicochemical properties and maturity assessment. Bioresour Technol 276:281–287

Michel Jr FC, Pecchia JA, Rigot J, Keener HM (2004) Mass and nutrient losses during the composting of dairy manure amended with sawdust or straw. Compost Sci Utilizat 12(4):323–334

Misra R, Roy R, Hiraoka H (2003) On-farm composting methods. UN-FAO, Rome, pp 7–26

Mishra CSK, Samal S, Samal RR (2022) Evaluating earthworms as candidates for remediating pesticide contaminated agricultural soil: A review. Front Environ Sci 10:924480

Moral R, Paredes C, Bustamante MA, Marhuenda-Egea F, Bernal MP (2009) Utilisation of manure composts by high-value crops. Safety and environmental challenges. Bioresour Technol 100(22):5454–5460

Mundiyara R, Jat MK (2017) Importance of vermicompost. Biotech. Retrieved from: https://www.biotecharticles.com/Agriculture-Article/Importance-of-Vermicompost-4013.html

Munroe G (2007) Manual of on-farm vermicomposting and vermiculture. Organic Agriculture Centre of Canada 39–40.

Ogwueleka TC (2009) Municipal solid waste characteristics and management in Nigeria. Iran J Environ Health Sci Eng 6:173–180

Pampuro N, Bertora C, Sacco D, Dinuccio E, Grignani C, Balsari P, Cavallo E, Bernal MP (2017) Fertilizer value and greenhouse gas emissions from solid fraction pig slurry compost pellets. J Agric Sci 155:1646–1658

Pane C, Spaccini R, Piccolo A, Celano G, Zaccardelli M (2019) Disease suppressiveness of agricultural green waste composts as related to chemical and bio-based properties shaped by different on- farm composting methods. Biol Control 137:104026

Parkinson R, Gibbs P, Burchett S, Misselbrook T (2004) Effect of turning regime and seasonal weather conditions on nitrogen and phosphorus losses during aerobic composting of cattle manure. Bioresour Technol 91(2):171–178

Pose-Juan E, Igual JM, Sánchez-Martín MJ, Rodríguez-Cruz MS (2017) Influence of herbicide triasulfuron on soil microbial community in an unamended soil and a soil amended with organic residues. Front Microbiol 8:378

Qiao C, Penton CR, Liu C, Tao C, Deng X, Ou Y, Li R (2021) Patterns of fungal community succession triggered by C/N ratios during composting. J Hazard Mater 401:123344

Rihani M, Malamis D, Bihaoui B, Etahiri S, Loizidou M, Assobhei O (2010) In-vessel treatment of urban primary sludge by aerobic composting. Bioresour Technol 101(15):5988–5995

Rincón CA (2019) Chemical and odor characterization of gas emissions released during composting of solid wastes and digestates. J Environ Manag 233:39–53

Roletto E, Cerruti M, Barberis R (1985) Investigation on humic substances from decomposing spruce bark. Agric Waste 13(2):137–148

Senesi N (1989) Composted materials as organic fertilizers. Sci Total Environ 81:521–542

Slater RA, Frederickson J (2001) Composting municipal waste in the UK. Some lessons from Europe. Res Conserv Recycl 32(3–4):359–374

Solano ML, Iriarte F, Ciria P, Negro MJ (2001) SE—structure and environment: performance characteristics of three aeration systems in the composting of sheep manure and straw. J Agri Eng Res 79(3):317–329.

Stratton ML, Barker A, Ragsdale J (2000) Sheet composting overpowers weeds in restoration project. BioCycle 41(4):57–57

Tiquia SM, Wan HC, Tam NF (2002) Microbial population dynamics and enzyme activities during composting. Compost Sci Util 10(2):150–161

Toledo M, Siles J, Gutiérrez M, Martín M (2018) Monitoring of the composting process of different agro industrial waste. Influence of the operational variables on the odorous impact. Waste Manage 76:266–274

Uyizeye OC, Thiet RK, Knorr MA (2019) Effects of community-accessible biochar and compost on diesel-contaminated soil. Biorem J 23:107–117

Varjani SJ, Gnansounou E, Pandey A (2017) Comprehensive review on toxicity of persistent organic pollutants from petroleum refinery waste and their degradation by microorganisms. Chemosphere 188:280–291

Ventorino V, Pascale A, Fagnano M, Adamo P, Faraco V, Rocco C, Fiorentino N, Pepe O (2019) Soil tillage and compost amendment promote bioremediation and biofertility of polluted area. J Clean Prod 239:118087

Xu J, Jiang Z, Li M, Li Q (2019) A compost-derived thermophilic microbial consortium enhances the humification process and alters the microbial diversity during composting. J Environ Manag 243:240–249

Xu Z, Li G, Huda N, Zhang B, Wang M, Luo W (2020) Effects of moisture and carbon/nitrogen ratio on gaseous emissions and maturity during direct composting of cornstalks used for filtration of anaerobically digested manure centrate. Bioresour Technol 298:122–503

Yu H, Xie B, Khan R, Shen G (2019a) The changes in carbon, nitrogen components and humic substances during organic-inorganic aerobic co-composting. Bioresour Technol 271:228–235

Yu Y, Zhu X, Li L, Lin B, Xiang M, Zhang X, Chen X, Yu Z, Wang Z, Wan Y (2019b) Health implication of heavy metals exposure via multiple pathways for residents living near a former e-waste recycling area in China: a comparative study. Ecotoxicol Environ Safe 169:178–184

Zucconi F (1987) Compost specifications for the production and characterization of compost from municipal solid waste. In: Compost: production, quality and use. Elsevier, New York, pp 30–50

Correction to: Advances in Biotechnological Tools and Their Impact on Global Climate Change and Food Security

Zafar Iqbal, Asad Azeem, Sami Ul-Allah, Ahmad Sher, Muhammad Qadir Ahmad, Bilal Haider, and Muhammad Asghar

Correction to: Chapter 28 in: M. Hasanuzzaman (ed.), *Climate-Resilient Agriculture, Vol 2*, https://doi.org/10.1007/978-3-031-37428-9_28

The name of the chapter author Muhammad Qadir Ahmad was unfortunately published with an error as Muhammad Qadeer Ahmad. The initially published version has now been corrected.

The updated version of this chapter can be found at
https://doi.org/10.1007/978-3-031-37428-9_28

M. Hasanuzzaman (ed.), *Climate-Resilient Agriculture, Vol 2*,
https://doi.org/10.1007/978-3-031-37428-9_44

Index

A

Abiotic stress, 9, 43, 114, 164, 171, 190, 210, 247, 249, 260, 269, 284, 298, 303, 320, 370, 392, 439, 456, 469, 507, 548, 572, 610, 642, 678, 752, 778, 795, 820, 866, 882, 918, 942
Abiotic stress tolerance, 15, 17, 303, 311, 370, 375–378, 623, 624, 642–644, 646, 649, 652–657, 752, 754–763, 820, 823, 882, 890
Agriculture, 2, 35, 52, 61, 82, 108, 129, 160, 192, 211, 264, 282, 296, 323, 374, 392, 456, 500, 536, 583, 657, 668, 679, 707, 722, 735, 771, 863, 882, 909, 932, 960, 984
Anther culture, 854, 855, 944, 947–948
Antioxidant activity, 193, 249, 251, 269, 291, 306, 311, 374, 377, 378, 415, 417, 533, 559, 577, 580, 583, 586, 588, 591, 593, 594, 611, 617–619, 621, 829, 831, 892
Antioxidants, 11, 193, 249, 283, 296, 337, 374, 418, 459, 500, 528, 550, 578, 605, 642, 681, 829, 892, 918, 943
Antisense RNA, 794, 805, 806
Application method, 223, 224, 272

B

Biochar, 36, 68, 246–255, 288–289, 452–459, 465, 479–481, 915, 921
Biochemical mechanism, 195
Biodegradable, 69, 219, 309, 706, 715, 717, 745, 906, 968, 976, 986
Biodegradation, 310, 695, 715, 721, 724–727, 736, 739, 871
Biofertilizer, 76, 213, 282, 590–592, 673, 674, 705–706, 894, 968, 970–972, 976
Bioinformatics, 441, 486, 780–781, 919
Biopolymer, 261, 298, 301–302, 573, 679–681, 867
Bioremediation, 465, 477, 483–486, 529, 535, 537, 538, 690–693, 701, 715, 718–724, 727, 735–742, 745, 871, 908–913, 922, 976
Biostimulants, 264, 282, 283, 289–291, 296–303, 306–311, 364–381, 573, 579, 582–590, 592–595, 679, 680, 686, 882, 883, 888, 890–891, 893, 895
Bio stimulate enzymes, 283
Biotechnological approaches, 727, 794, 804, 943–947
Biotechnology, 159, 165, 172, 428, 441–445, 511, 669–675, 690, 696, 704, 705, 714, 715, 718, 724, 943
Biotic stress, 15, 17, 114, 118, 167, 195, 197, 303, 304, 306, 503, 522, 573, 578, 606–608, 611, 614, 615, 617, 626, 646, 669, 670, 681, 774, 776, 777, 779, 793, 794, 848, 883, 890, 918, 942

C

Catalyst, 228, 288, 743
Cereal crops, 193, 229, 327–330, 342

M. Hasanuzzaman (ed.), *Climate-Resilient Agriculture, Vol 2*,
https://doi.org/10.1007/978-3-031-37428-9

Chemical fertilizers, 36, 66–72, 76, 93, 163, 311, 399, 678, 679, 685, 705, 960, 987
Chitosan, 261–275, 282, 297, 298, 301–302, 309, 369–370, 573, 583, 584, 586, 647, 651–652, 679–681, 866, 867, 875
Climate, 2, 36, 51, 62, 83, 108, 126, 158, 190, 220, 247, 303, 320, 399, 428, 502, 524, 548, 620, 668, 685, 743, 864
Climate change, 2, 35, 62, 83, 108, 126, 158, 190, 211, 247, 274, 296, 320, 500, 522, 572, 604, 668, 678, 691, 794, 842, 863, 888, 976
Climate resilience, 15, 17, 112, 132–144, 176, 190
Climate-resilient agriculture (CRA), 52, 117–118, 864, 867–870
Climate risks, 83, 90–92
Climate-tolerant, 177, 178
Climatic changes, 3, 52, 55, 83, 88, 108, 118, 130, 169, 174, 192, 194, 229, 548, 551
Coagulation, 892
Compost, 36, 68, 75, 76, 114, 133, 135, 223, 371, 396, 674, 715–718, 721, 739, 740, 915, 965, 966, 976, 977, 979–987
Composting, 69, 93, 114, 396, 715–718, 721, 727, 735, 740, 921, 976, 977, 979–987
CRISPR/CAS toolbox, 794, 805
Crop tolerance, 298
Cryopreservation, 843, 852–854, 856
Cytokinins, 11, 39, 297, 300, 308, 433, 435, 529, 552, 554–555, 563, 581, 585, 589, 614, 649, 682, 844, 845, 890, 893

D

Drought, 9, 36, 109, 129, 158, 191, 216, 247, 269, 296, 320, 392, 428, 457, 501, 548, 572, 604, 650, 675, 678, 754, 773, 795, 828, 842, 882
Drought stress, 9, 36, 116, 193, 216, 249, 269, 283, 305, 327, 408, 444, 456, 501, 604, 647, 678, 773, 797, 828, 854, 864, 884–885

E

Eco-friendly, 115, 141, 198, 248, 379, 479, 528, 529, 535, 582, 619, 675, 686, 690, 694, 715–717, 895, 916, 920, 932, 933, 936, 939, 942, 961
Ecosystem services, 58, 919–922
Environment, 3, 36, 51, 66, 82, 116, 129, 194, 218, 249, 275, 284, 296, 321, 364, 414, 430, 510, 522, 552, 649, 668, 678, 715, 779, 842, 868, 905, 932, 961, 977
Environmental pollution, 90, 211, 219, 221, 522, 534, 537, 694, 735, 922
Environmental sustainability, 66, 75, 249, 282, 296, 690, 866
Erosion, 52, 53, 57, 63–65, 73, 76, 83, 95, 96, 112, 139, 163, 194, 215, 409, 722, 873, 913, 960, 976, 983
Ethylene, 11, 39, 253, 306, 307, 397, 416, 431, 433, 435, 552, 557, 558, 563, 578, 589, 614, 618, 645, 650, 655, 681–683, 798, 803, 823, 828, 831, 833, 844, 868, 916

F

Feedstock, 248, 452–455, 691, 699, 705, 981
Field efficacy, 284–287
Food security, 13, 14, 22, 35, 44–46, 62, 76, 83, 96, 108, 109, 115, 118, 131, 160–162, 176, 190–192, 200, 296, 321, 322, 336, 346, 364, 456, 500, 523, 668–671, 673–675, 805, 864, 871, 893, 894, 948, 960, 976

G

Gene expression, 167, 269, 320, 372, 373, 378, 417, 433, 436, 439, 527, 532, 533, 537, 538, 572, 591, 604, 606, 623, 654, 682, 702, 752–754, 774, 776, 778, 779, 798, 803, 821, 828, 947
Genetic modification (GM), 111, 112, 163, 165–168, 172, 428, 441, 442, 623, 778, 871, 909
Genomics, 16, 17, 111, 112, 116, 117, 159, 165, 170, 171, 177, 200, 343, 381, 430, 431, 441, 529, 671–672, 772–775, 781, 821, 919
Genomic technologies, 197

Green house gases (GHGs), 93, 165, 166, 174, 220, 221, 523, 691

H

Heat, 8, 44, 66, 83, 108, 127, 190, 251, 268, 296, 322, 392, 436, 452, 500, 534, 548, 575, 643, 668, 678, 715, 740, 773, 794, 821, 844, 886, 905, 981
Heat stress, 9–11, 18, 19, 85, 112, 127, 164, 173, 192, 193, 195, 198, 199, 251, 252, 268, 274, 307, 322, 327, 333, 335, 338, 343, 345, 368, 378, 392, 500–508, 510–514, 548–550, 552, 557, 561, 654–655, 760, 761, 798, 821, 822, 824–826, 830, 831, 833, 886–887, 891, 894
Heavy metals (HMs), 63, 126, 247, 268, 291, 296, 324, 392, 466, 523, 548, 644, 678, 720, 740, 828, 871, 894, 904, 943
Hormones, 70, 283, 297, 299, 304, 309, 369, 373, 430, 435, 440, 465, 477, 482, 509, 527, 548, 557, 585, 590, 594, 614, 615, 642, 649, 650, 655, 681, 823–828, 846, 847, 915
Hydro-priming, 215
Hydrothermal carbonization, 452, 453

I

Industrialization, 158, 269, 522, 528, 668, 719, 734, 741, 904
Integrated pest management (IPM), 77, 932, 933, 936, 939–941, 943, 948, 973
Ion homeostasis, 588, 591
Ionomics, 779, 781, 919

L

Land use system, 56

M

Melatonin, 306–308, 372, 525, 530, 557, 559–561, 563, 586–589, 619, 626, 646, 651
Meristem culture, 842, 846, 847
Metabolic engineering, 369, 619, 620, 623, 624, 626, 794
Metabolomics, 17, 135, 220, 373, 528, 594, 623, 681, 778, 781, 885, 919
Micropropagation, 842–846, 852, 856
MicroRNA (miRNA), 531, 751–764, 794–799, 805, 806
miRNA expression profiling, 754
Mitigating strategies, 477–485, 585
Modern agricultural system, 109, 939
Modern biotechnology, 705
Moringa, 375, 377, 578, 579, 581, 582, 891–895

N

Nanofertilizers, 69, 399, 536, 585, 864–866, 874, 875
Nanoparticles (NPs), 218–220, 266, 268, 311, 346, 371, 400, 444, 445, 480, 532, 534–536, 583–585, 866, 870, 871, 875, 891, 894
Nanotechnology, 69, 218, 261, 346, 444–445, 536–538, 864, 865, 870–876
Nitrogen, 37–39, 42, 62, 66, 67, 70–73, 75, 88, 90, 93, 97, 110, 136, 137, 145, 146, 148, 165, 211, 214–217, 219, 221, 224, 227–229, 247, 267, 285, 297, 300, 304, 306–308, 310, 334, 337, 338, 340, 344, 345, 347, 375, 379, 393, 397, 398, 479, 513, 522–524, 532, 536, 554, 559, 562, 579, 588, 590, 604, 616, 618–620, 626, 674, 682, 705, 706, 715, 717, 720, 758, 762, 763, 852, 854, 856, 875, 882, 888–890, 963, 970, 971, 983, 985, 986
Non-coding RNA, 531, 774, 776
Nutrient, 16, 36, 65, 93, 126, 193, 210, 261, 282, 296, 323, 364, 397, 433, 456, 465, 500, 530, 550, 578, 604, 648, 671, 679, 721, 739, 779, 848, 864, 885, 920, 935, 962, 977
Nutrient use efficiency (NUE), 36–39, 44, 74, 210–218, 220, 230, 364, 375, 679

O

Organic agricultural system, 88, 109
Organic farming, 71, 92–94, 98, 115, 141, 142, 965, 970, 987
Organic manures, 21, 69, 70, 75, 76, 119, 140, 284, 344
Organic pollutants, 529, 694, 720, 740, 905, 906, 911, 912, 914, 916, 918, 920, 922, 976, 979

Osmo protectant, 549, 681
Oxidative stress, 13, 255, 272, 307, 308, 310, 341, 346, 375, 378, 394, 397, 409, 434, 466, 478, 500, 502, 505, 506, 514, 523, 528–530, 532, 535, 536, 550, 560, 578, 579, 581, 582, 588, 590, 607, 611, 613, 615, 618–622, 642, 646, 649, 655, 656, 678, 758, 763, 829, 834, 882, 890, 894, 917
Ozone (O_3), 3, 96, 162, 398, 400, 522, 523, 531, 533, 537, 538, 670, 969, 976, 978

P
Pangenomics, 773, 775, 781
Pesticides and insecticides, 85–87
Phenomics, 779–781
Phosphorus, 39–42, 62, 67, 70, 95, 96, 211, 215, 220, 285, 286, 297, 299, 375, 483, 513, 534, 579, 590, 620, 720, 763, 779, 822, 875, 972, 983
Phytohormones, 11, 39, 284, 297, 306, 308–309, 367, 400, 416, 445, 469, 514, 525, 529, 548, 552, 554, 557, 559, 561–563, 572, 573, 577, 578, 583, 585, 588, 589, 606, 609, 614–615, 620, 648, 649, 823, 868, 888, 890, 892–895, 915
Phytoremediation, 21, 397, 477, 482–483, 525, 526, 528, 529, 535, 537, 538, 692, 693, 738, 745, 746, 905, 908, 910–916, 919–922
Phytotoxicity, 131–133, 216, 273, 301, 345, 464, 886, 906, 977, 986
Phytotoxic substances, 131, 273
Plant abiotic stress tolerance, 648
Plant breeding, 15–17, 82, 177, 197, 230, 343, 442, 500, 511, 512, 672, 854
Plant defense mechanism, 303–306
Plant disease, 94, 96, 287, 306, 706, 718, 867, 960, 961, 973, 976
Plant extracts, 282, 537, 573, 577–579, 585, 594–595, 891–895, 968
Plant growth promoting microorganism (PGPM), 682
Plant hormone, 283, 297, 309, 369, 370, 440, 469, 527, 532, 561, 563, 655, 681, 682, 828, 889
Plant miRNAs, 752, 754, 762
Plant stress, 246–255, 320, 343, 347, 456, 555, 559, 608, 625, 642, 649, 866–867
Plant stress tolerance, 343, 574, 625, 642, 865
Plant tissue culture, 842–856
Pollution, 2, 41, 52, 53, 57, 85, 90, 211, 218, 221, 249, 327, 364, 379, 381, 454, 466, 468, 481, 482, 522–524, 526, 529, 531, 534, 536, 538, 595, 669, 678, 685, 690, 691, 694, 699, 706, 714, 726, 734, 852, 853, 871, 876, 885, 904, 906, 976, 978, 979, 987
Post-transcriptional-regulation, 751, 752, 763, 947
Potassium, 19, 42–45, 62, 72, 96, 219, 222, 262, 336, 375, 397, 410, 412, 417, 418, 434, 435, 513, 550, 579, 581, 590, 645, 651, 653, 682, 705, 875, 963, 983
Priming, 216, 264, 345, 400, 513, 529, 588, 642
Proteomics, 302, 373, 441, 528, 776–777, 781, 821, 822, 828, 830, 919
Pyrolysis, 68, 246, 248, 253, 288, 452–455, 479, 718, 921

R
Reactive oxygen species (ROS), 13, 44, 249, 251, 255, 269, 287, 303, 304, 306, 308, 310, 311, 330, 340–342, 376, 378, 392, 395, 409, 412, 415, 417, 418, 430, 433, 435, 436, 438, 464, 484, 502, 505–509, 511, 513, 523, 527, 528, 530, 532, 533, 550, 553, 557–560, 572, 578, 580, 583, 584, 610–613, 615, 618, 619, 621, 642, 643, 646, 647, 650, 652–654, 656, 678, 683, 794, 798, 803, 868, 882, 884, 886, 887, 892–894, 906, 918
Remediation techniques, 481, 735, 738, 745
Resilience, 15, 17, 93, 109, 112, 115, 132–144, 165, 170, 173, 220, 283, 304, 342, 364, 590, 592, 772, 774, 775, 779, 781, 794, 806, 864, 961
Rice, 5, 41, 68, 111, 128, 214, 251, 264, 328, 417, 457, 464, 506, 523, 550, 580, 611, 649, 674, 771, 795, 829, 866, 946, 967
RNA interference (RNAi), 774, 794, 799, 805, 806, 944, 947
Rural-urban interaction, 54

S
Salinity, 21, 36, 52, 114, 216, 247, 271, 283, 310, 321, 369, 394, 408, 458, 528, 548, 572, 624, 643, 678, 778, 793, 828, 848, 883

Salt stress, 10, 43, 162, 271, 272, 287, 325, 331, 332, 335, 336, 338, 340, 346, 367, 392, 394–396, 399, 400, 408–419, 457, 458, 550, 554, 555, 557, 572, 578, 582–585, 587–589, 591, 648–651, 672, 678, 679, 681, 684, 685, 762, 799, 822, 884, 893
Secondary metabolism, 364, 530, 532, 593, 604–626
Seed priming, 215, 216, 345, 346, 513, 535, 578, 582, 586, 588, 619, 642–657
Signaling molecules, 296–311, 473, 514, 586, 587, 645, 820, 868, 869, 882
Soil, 3, 36, 62, 82, 110, 126, 191, 211, 246, 260, 282, 296, 320, 364, 392, 408, 434, 452, 465, 522, 550, 572, 609, 655, 673, 678, 719, 734, 763, 864, 883, 904, 932, 960, 976
Soil health, 65, 66, 70, 72–73, 93–96, 98, 112, 114, 115, 118, 191, 219, 282, 285, 287, 289, 655, 673, 948, 960, 961
Soil microbes, 254, 458
Soil salinity, 21, 52, 320, 323, 325, 336, 392–400, 408, 409, 550
Soil temperature, 19, 136, 137, 140, 220, 224, 227, 336, 864, 907
Solid waste management, 699, 700, 714, 715, 726–727, 978
Somatic embryogenesis, 442, 777, 847
Strategies, 3, 41, 76, 86, 112, 136, 192, 212, 255, 289, 297, 343, 379, 416, 428, 465, 508, 524, 590, 605, 642, 675, 686, 719, 742, 803, 820, 852, 882, 921, 932, 963
Stressors, 3, 13, 23, 84, 158, 159, 164, 165, 167, 169, 176, 195, 247, 268, 283, 289, 296, 321, 336, 339, 340, 342, 343, 345–347, 399, 417, 501, 502, 509, 510, 512, 513, 538, 583, 587, 590, 607, 617, 626, 643, 652, 657, 679, 771, 773, 775–778, 780, 781, 794, 806, 866–868, 871, 874, 883, 889, 890, 892
Stress signaling pathways, 806
Stress tolerance, 11, 12, 17, 111, 246–255, 260–275, 296, 301, 303, 370, 375, 376, 400, 408–419, 428–445, 500–514, 604–626, 680–685, 754–763, 795, 820, 834, 867, 882, 890, 917
Stress types, 560, 606, 755, 796, 801
Strigolactones, 552, 557, 561, 563
Sustainable, v, vi, 12, 17–22, 37, 41, 52, 54, 56, 68, 69, 77, 92–98, 107–119, 126, 138, 159, 161, 177, 210, 262, 264, 297, 363, 379, 381, 391–400, 408, 409, 512, 529, 537, 538, 594, 595, 657, 677, 680, 689–707, 714, 716, 724, 727, 741, 794, 805, 866, 872, 876, 895, 919, 920, 931–948, 959–973, 978, 979
Sustainable agriculture, 35, 45, 76, 92, 230, 247, 275, 325, 364, 381, 392–400, 536, 680, 805, 960, 961, 970, 973, 977
Sustainable intensification, 891
Synthetic chemicals, 265, 932, 934–937, 939, 972

T

TCTP proteins, 820, 824–826, 828, 830, 832, 834
Tolerance, 11, 111, 139, 191, 250, 268, 283, 296, 332, 369, 408, 472, 500, 522, 550, 573, 605, 643, 672, 679, 775, 794, 820, 848, 869, 882, 913, 961
Tolerance mechanism, 11, 12, 415, 416
Tolerant crop, 17, 44, 45, 794
Toxicity, 36, 87, 216, 217, 219, 220, 249, 260, 268, 272–274, 291, 321, 323, 325–327, 395, 397, 409–411, 418, 467, 469–472, 524, 528, 529, 533, 535, 551, 559, 575, 585, 586, 607, 678, 701–703, 719, 740, 780, 799, 865, 866, 876, 884, 886, 904, 906–908, 910, 914, 915, 919, 935, 936, 939, 943, 945
Transcription, 167, 169–170, 291, 305, 439, 440, 444, 500, 511, 531–533, 553, 614, 622, 623, 649, 652, 656, 657, 751, 752, 754, 755, 757–761, 774, 776, 795, 798, 799, 803, 805, 820, 823, 887, 945
Transcriptomics, 302, 373, 381, 441, 623, 681, 775, 776, 781, 822, 919
Transgenic plants, 304, 623, 761, 772, 794, 796, 799, 801–805, 820, 821, 831, 834, 913–915, 945, 946
Trends in biotechnology, 695–698

V

Vermicompost, 70, 71, 76, 299, 593, 722

W

Waste treatment, 68, 529, 574, 698
Water management, 18, 19, 82, 92, 95–98, 113, 116, 162, 465, 477, 964
Wheat, 5, 7, 38, 88, 109, 128, 190, 215, 250, 251, 266, 269, 286, 309, 327, 397, 416, 431, 456, 465, 500, 526, 549, 579, 644, 752, 774, 795, 828, 866, 885, 892, 946, 984

Y

Yield, 3, 36, 51, 66, 83, 108, 126, 192, 211, 246, 261, 284, 296, 322, 364, 392, 408, 430, 452, 464, 500, 523, 550, 572, 618, 654, 673, 681, 690, 742, 771, 793, 820, 842, 865, 883, 905, 932, 960

SPRINGER NATURE

GPSR Compliance

The European Union's (EU) General Product Safety Regulation (GPSR) is a set of rules that requires consumer products to be safe and our obligations to ensure this.

If you have any concerns about our products, you can contact us on ProductSafety@springernature.com

In case Publisher is established outside the EU, the EU authorized representative is:

Springer Nature Customer Service Center GmbH
Europaplatz 3
69115 Heidelberg, Germany

The manufacturer's authorised representative in the EU is Springer Nature Customer Service Centre GmbH, Europaplatz 3, 69115 Heidelberg, Germany. If you have any concerns regarding our products, please contact ProductSafety@springernature.com

Printed and bound by CPI Group (UK) Ltd, Croydon, CR0 4YY
01/10/2026
02225855-0017